Living in the Environment

Concepts, Connections, and Solutions

SIXTEENTH EDITION

G. Tyler Miller, Jr.

Scott E. Spoolman

BROOKS/COLE
CENGAGE Learning™

Australia • Brazil • Japan • Korea • Mexico • Singapore • Spain • United Kingdom • United States

BROOKS/COLE
CENGAGE Learning˙

Living in the Environment: Concepts, Connections, and Solutions, 16e
G. Tyler Miller, Jr. and Scott E. Spoolman

Vice President, Editor-in-Chief: Michelle Julet

Publisher: Yolanda Cossio

Development Editor: Christopher Delgado

Assistant Editor: Lauren Oliveira

Editorial Assistant: Samantha Arvin

Media Editor: Kristina Razmara

Marketing Manager: Amanda Jellerichs

Marketing Assistant: Katherine Malatesta

Marketing Communications Manager:
Linda Yip

Project Manager, Editorial Production:
Andy Marinkovich

Creative Director: Rob Hugel

Art Director: John Walker

Print Buyer: Karen Hunt

Permissions Editor: John Hill

Production Service/Compositor:
Thompson-Steele, Inc.

Text Designer: Carolyn Deacy

Photo Researcher: Abigail Reip

Copy Editor: Andrea Fincke

Illustrator: Patrick Lane, ScEYEence Studios;
Rachel Ciemma

Cover Image: Wind farm in a crop field in
Canada © Dave Reede/Getty Images. For more
information about this photo, see p. iv.

For product information and technology assistance, contact us at
Cengage Learning Customer & Sales Support, 1-800-354-9706
For permission to use material from this text or product,
submit all requests online at **www.cengage.com/permissions**
Further permissions questions can be e-mailed to
permissionrequest@cengage.com

Library of Congress Control Number: 2008933000

ISBN-13: 978-0-495-55671-8

ISBN-10: 0-495-55671-8

Brooks/Cole
10 Davis Drive
Belmont, CA 94002-3098
USA

Cengage Learning is a leading provider of customized learning solutions with office
locations around the globe, including Singapore, the United Kingdom, Australia, Mexico,
Brazil, and Japan. Locate your local office at **international.cengage.com/region**

Cengage Learning products are represented in Canada by Nelson Education, Ltd.

For your course and learning solutions, visit **academic.cengage.com**.

Purchase any of our products at your local college store or at our preferred online store
www.ichapters.com.

Printed in Canada
3 4 5 6 7 12 11 10

Brief Contents

About the Cover Photo

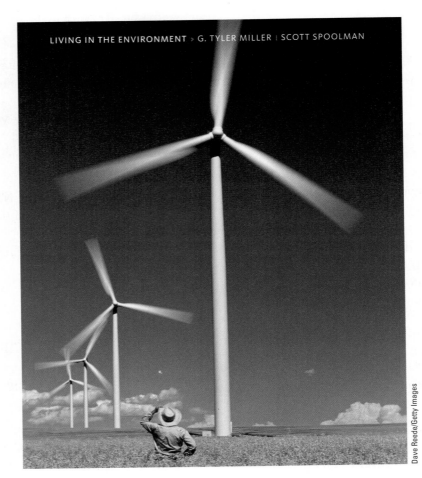

LIVING IN THE ENVIRONMENT > G. TYLER MILLER | SCOTT SPOOLMAN

Dave Reede/Getty Images

The cover photo shows a farmer viewing a *wind farm*—a group of large wind turbines—from a canola field in St. Leon, Manitoba, Canada. Wind captured by these turbines can be converted into electricity. Wind farms can be located on the land or at sea, and land lying under the turbines can still be used to grow crops (as in the photo) or to raise cattle. According to the U.S. Department of Energy, the United States is the "Saudi Arabia of wind power," capable of producing nearly three times its current output of electricity by using wind. Since 1990, wind power has been one of the world's fastest growing sources of energy, with Europe leading the way into the *age of wind energy*. For more details on wind energy see pp. 419–422 and Figures 14, 15, and 16 on pp. S68–69 in Supplement 10.

Detailed Contents

Photo 1 The endangered brown pelican was protected in the first U.S. wildlife refuge in Florida.

SuperStock

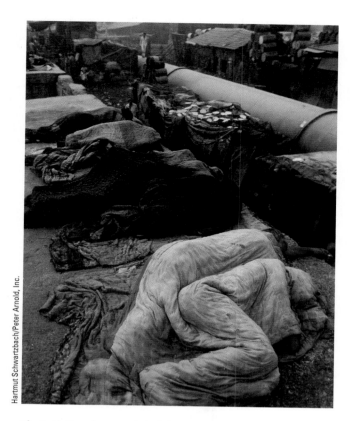

Photo 2 Homeless people in Calcutta India

Photo 3 Endangered ring-tailed lemur in Madagascar

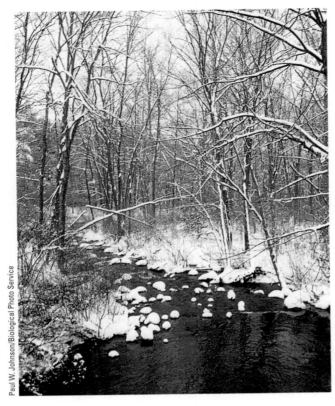

Photo 4 Temperate deciduous forest, winter, Rhode Island (USA)

Paul W. Johnson/Biological Photo Service

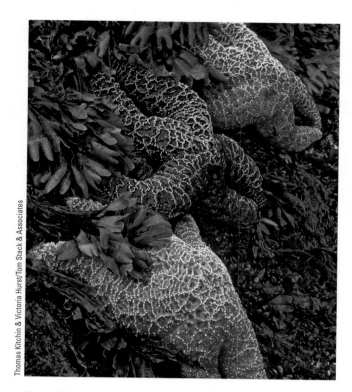

Photo 5 Sea star species helps to control mussel populations in intertidal zone communities in the U.S. Pacific northwest.

Photo 6 Flexible solar cells manufactured with use of nanotechnology

Photo 7 Cypress swamp, an inland wetland in U.S. state of Tennessee

SUSTAINING ENVIRONMENTAL QUALITY

Photo 8 Treadle pump used to supply irrigation water in parts of Bangladesh and India

Photo 9 Energy efficient *straw bale house* in Crested Butte, Colorado (USA) during construction.

Photo 10 Completed energy efficient *straw bale house* in Crested Butte, Colorado (USA)

Photo 12 Photochemical smog in Mexico City, Mexico

SUSTAINING HUMAN SOCIETIES

Photo 11 Roof garden in Wales, Machynlleth (UK)

Photo 13 Cow dung is collected and burned as a fuel for cooking and heating in India

Pierre A. Pittet/UN Food and Agriculture Organization

25 Environmental Worldviews, Ethics, and Sustainability 660

SUPPLEMENTS

For Instructors

What's New

In this edition, we build on proven strengths of past editions with the following major new features:

- New concept-centered approach
- Quantitative Data Analysis or Ecological Footprint Analysis exercise at the end of each chapter and 68 additional Data Analysis exercises in the Supplements
- New design with 254 pieces of new or upgraded art and 89 new photos
- Comprehensive review section at the end of each chapter with review questions that include all chapter key terms in boldface

This edition also introduces a new coauthor, **Scott Spoolman,** who worked as a contributing editor on this and other environmental science textbooks by Tyler Miller for more than 4 years. (See About the Authors, p. xxiii.)

New Concept–Centered Approach

Each major chapter section is built around one to three 9 **key concepts**—a major new feature of this edition. These concepts state the most important take-away messages of each chapter. They are listed at the front of each chapter (see Chapter 9, p. 184), and each chapter section begins with a key question and concepts (see Chapter 9, pp. 189, 193, and 206), which are highlighted and referenced throughout each chapter.

A logo ⤴ CONCEPT LINKS in the margin links the material in each chapter to appropriate key concepts in foregoing chapters (see pp. 124, 219, and 315).

New Design

The concepts approach is well-served by our new design, which showcases the concepts, core case studies, and other new features as well as proven strengths of this textbook. The new design (see Chapter 1, pp. 5–27), which enhances visual learning, also incorporates a thoroughly updated art program with 254 new or upgraded diagrams and 89 new photos—amounting to 56% of the book's 617 figures.

Sustainability Remains as the Integrating Theme of This Book

Sustainability, a watchword in the 21st Century for those concerned about the environment, is the overarching theme of this *concept-centered, solutions-oriented, science-based,* and *flexible* introductory environmental science textbook. In this edition, we showcase the sustainability theme even more to reflect its growing importance in understanding environmental problems and their solutions. You can see the sustainability emphasis by looking at the Brief Contents (p. iii).

Four scientific principles of sustainability play a major role in carrying out this book's sustainability theme. These principles, depicted in Figure 1-17 (p. 23 and the back cover of the student edition), are used throughout the book. Each reference is marked in the margin by 🍃. (See Chapter 12, pp. 279, 280, 292, 293, 294, 296, 300, 306, 307, 308, 310, and 311.)

Core Case Studies and the Sustainability Theme

Each chapter opens with a **Core Case Study** (see Chapter 21, p. 560), which is applied throughout the chapter. These connections to the **Core Case Study** are indicated in the book's margin by ⤴ CORE CASE STUDY. (See Chapter 21, pp. 561, 563, 564, 567, 574, 577, 582, 585, and 586.)

Each chapter ends with a *Revisiting* box (see Chapter 21, p. 586), which connects the **Core Case Study** and other material in the chapter to the four scientific principles of sustainability. **Thinking About** exercises placed throughout each chapter (see Chapter 21, pp. 564, 578, 582, and 584) challenge students to make these and other connections for themselves.

Five Subthemes Guide the Way toward Sustainability

In the previous edition of this book, we used five major subthemes, which are carried on in this new edition: *natural capital, natural capital degradation, solutions, trade-offs,* and *individuals matter* (see diagram on back cover).

- **Natural capital.** Sustainability focuses on the natural resources and natural services that support all life and economies. Examples of diagrams that illustrate this subtheme are Figures 1-3 (p. 8), 8-4 (p. 165), and 10-4 (p. 217).
- **Natural capital degradation.** We describe how human activities can degrade natural capital. Examples of diagrams that illustrate this subtheme are Figures 1-7 (p. 12), 7-20 (p. 158), and 10-15 (p. 225).
- **Solutions.** Next comes the search for *solutions* to natural capital degradation and other environmental problems. We present proposed solutions in a balanced manner and challenge students to use critical thinking to evaluate them. Some 33 figures and many chapter sections and subsections present proven and possible solutions to various environmental problems. Examples are Figures 12-32 (p. 308), 13-23 (p. 337), and 18-23 (p. 491).
- **Trade-Offs.** The search for solutions involves *trade-offs,* because any solution requires weighing advantages against disadvantages. Some 35 Trade-Offs boxes present advantages and disadvantages of various environmental technologies and solutions to environmental problems. Examples are Figures 12-20 (p. 295), 15-15 (p. 385), and 16-21 (p. 418).
- **Individuals Matter.** Throughout the book *Individuals Matter* boxes describe what various concerned citizens and scientists have done to help us work toward sustainability. (See pp. 230, 537, 572, and 622). Also, 14 *What Can You Do?* boxes describe how readers can deal with the problems we face. Examples are Figures 12-33 (p. 310), 19-16 (p. 522), and 20-21 (p. 557). Twelve especially important things individuals can do—the *sustainability dozen*—are summarized in Figure 25-8 (p. 669).

Science–Based Global Coverage

Chapters 2–8 present the basic scientific principles of ecology (see Brief Contents, p. iii) needed for a basic understanding of how the earth works and for evaluating proposed solutions to environmental problems. Important environmental science topics are explored in depth in 54 **Science Focus** boxes distributed among the chapters (see pp. 104, 309, and 333). (See items in BOLD type in the Detailed Contents, pp. v–xiv.) Science is also integrated throughout the book in various **Case Studies** (see pp. 177–178, 202–203, and 259–260) and in figures (see Figures 15-A, p. 374; 15-12, p. 383; and 15-17, p. 387). In addition, 58 **Research Frontier** boxes list key areas of cutting-edge research, with links to such research provided on the website for this book (see pp. 74, 217, and 240). **Green Career** items in the text list various green careers with further information found on the website.

This book also provides a *global perspective* on two levels. *First,* ecological principles reveal how all the world's life is connected and sustained within the biosphere (Chapter 3). *Second,* the book integrates information and images from around the world into its presentation of environmental problems and their possible solutions. This includes 32 global maps and 28 U.S. maps in the basic text and in Supplements 3, 4, and 10.

CengageNOW, an online visual learning supplement, allows students to enhance their scientific understanding by viewing animations, many of them interactive. Some *CengageNOW* animations are related to figures (see Figures 5-17, p. 117 and 8-15, p. 175) and others to text (see pp. 132 and 144).

Three Levels of Flexibility

There are hundreds of ways to organize the content of this course to fit the needs of different instructors with a wide variety of professional backgrounds and course lengths and goals. Our solution to this problem has been to design a highly flexible book that allows instructors to vary the order of chapters and sections within chapters without exposing students to terms and concepts that would hinder their learning.

We recommend that instructors start with Chapter 1 because it defines basic terms and gives an overview of sustainability, population, pollution, resources, and economic development issues that are treated throughout the book. This provides a springboard for instructors to use other chapters in almost any order.

One often-used strategy is to follow Chapter 1 with Chapters 2–8, which introduce basic science and ecological concepts. Instructors can then use the remaining chapters in any order desired. Some instructors follow Chapter 1 with any or all of Chapters 23, 24, and 25 on environmental economics, politics, and worldviews, respectively, before proceeding to the chapters on basic science and ecological concepts.

We provide a *second level of flexibility* in 13 Supplements (see p. xiv in the Detailed Contents and p. S1), which instructors can assign as desired to meet the needs of their specific courses. Examples include basic environmental data (Supplement 3), maps of biodiversity and ecological footprints (Supplement 4), environmental history (Supplement 5), basic chemistry (Supplement 6), weather basics (Supplement 8), and graphs and maps related to energy resources (Supplement 10). These supplements contain 34 graphs and 34 U.S. and global maps (each with Data Analysis questions) that make up a large database of environmental information.

There is also a third level of flexibilty. Instructors wanting to emphasize a mastery of basic environmental concepts and information can rely mostly on the detailed review questions at the end of each chapter and leave out any or all of the book's numerous critical thinking questions in Thinking About boxes, many figure captions, and at the end of each chapter.

Case Studies

In addition to the 25 Core Case Studies described above, 51 additional **Case Studies** (see pp. 93–95, 177–178, and 257–259) appear throughout the book. (See items in BOLD type in the Detailed Contents, pp. v–xiv.) The total of 76 case studies provides an in-depth look at specific environmental problems and their possible solutions.

Critical Thinking

The introduction on *Learning Skills* describes critical thinking skills (pp. 2–4). Specific critical thinking exercises are used throughout the book in several ways:

- As 159 **Thinking About** exercises. This *interactive approach to learning* reinforces textual and graphic information and concepts by asking students to analyze material immediately after it is presented throughout each chapter rather than at the end of the chapter (see pp. 87, 380, and 405).

- In all Science Focus boxes.

- In the captions of many of the book's figures (see Figures 7-12, p. 151; 20-B, p. 550; and 23-9, p. 623).

- In the nearly 70 *How Would You Vote?* exercises (see pp. 114, 426, and 514).

- As end-of-chapter questions (see pp. 120 and 609).

Visual Learning

This book's 498 diagrams—218 of them new to this edition—are designed to present complex ideas in understandable ways relating to the real world. (See Figures 1-7, p. 12; 3-18, p. 68; 4-2, p. 79; and 7-11, p. 149.) We have also carefully selected 209 photographs—89 of them new to this edition—to illustrate key ideas. (See Figures 4-10, p. 89; 10-11, p. 222; and 12-12, p. 288.) We have avoided the common practice of including numerous "filler" photographs that are not very effective or that show the obvious.

Finally, to enhance visual learning, more than 80 *CengageNOW* animations, many referenced in the text and diagrams, are available online. *CengageNOW* provides students with a more complete learning experience that takes what students read on the page and places it into a more interactive environment.

Major Changes in This Edition: A Closer Look

Major changes in this new edition include the following:

- New co-author (see p. xxiii)

- Concept-centered approach with each chapter section built around one to three **Key Concepts** that provide the most important messages of each chapter. Each chapter also links material to related key concepts from previous chapters. All of the Key Concepts, listed by chapter, can be found in Supplement 14 on page S83.

- New design serving the concept-centered approach and integration of Core Case Studies, with an upgraded art program, which includes 254 new or upgraded diagrams. (See Figures 1-3, p. 8; 5-8, p. 107; and 10-15, p. 225.)

- 89 carefully selected new photographs. (See Figures 3-1, p. 50; 3-4, p. 53; and 14-21, p. 359.)

- Expansion of the sustainability theme built around the four scientific principles of sustainability (Figure 1-17, p. 23 and the back cover of the student edition)

- 7 new chapter opening **Core Case Studies** (pp. 28, 399, and 468)

- 54 **Science Focus** boxes that provide greater depth on scientific concepts and on the work of environmental scientists (see pp. 197, 253, and 296).

- Connections to *The Habitable Planet,* a set of 13 videos produced by Annenberg Media. Each half-hour video describes research that two different scientists are doing on a particular environmental problem (see pp. 67, 217, and 254).

- *Review* section at the end of each chapter with comprehensive review questions that include all key terms in boldface. (See pp. 180 and 528.)

- A Data Analysis or Ecological Footprint Analysis exercise at the end of each chapter (see pp. 76, 98–99, and 274) and 64 additional exercises analyzing graphs or maps in the book's Supplements (see pp. S14, S24–S25, and S60).

- Student projects listed by chapter are found in Supplement 13 p. S78. Some instructors may find these useful for getting students more deeply involved in key environmental issues.

- *Active Graphing exercises* in *CengageNOW* for many chapters that involve students in the graphing and evaluation of data.

- Improved flow and content based on input from 47 new reviewers (identified by an asterisk in the List of Reviewers on pp. xx–xxii).

- More than 4,000 updates based on information and data published in 2005, 2006, 2007, and 2008.

- New *Laboratory Manual* developed by a team of teachers led by Ed Wells.

- Integration of material on the growing ecological and economic impacts of China throughout much of the book. (See Index citations for China.)

- More than 100 new or expanded topics including expanded treatment of ecological footprints (Figures 1-9, p. 14; and 1-10, p. 15 and ecological footprint calculations at the end of a number of chapters); additional maps of global economic, population, hunger, health, and waste production data

(Supplement 3, pp. S10–S19); additional energy resource and climate data maps (Supplement 10, pp. S59–S72); revisiting Easter Island (p. 31); tipping points (p. 46); tropical forest losses (p. 50); hurricanes and New Orleans (p. 177–178); tropical forest fragmentation (p. 195); vultures and rabies (p. 197); disappearing honeybees (pp. 202–203); threatened polar bears (p. 203); Jane Goodall (p. 205); effects of gray wolves on the Yellowstone ecosystem (p. 235); Blackfoot reconciliation ecology (pp. 244–245); restoring mangroves (p. 255); endangered marine turtles (pp. 259–260); crop weed management (p. 296); organic agriculture (p. 307); perennial crops (p. 309); deep aquifers (p. 324); water transfer projects in China (pp. 331–332); improving desalination (p. 333); environmental effects of gold mining (p. 344); ecological restoration (p. 359); coal consumption in China (pp. 384–385); Iceland's renewable energy economy (p. 399); plug-in hybrid cars (p. 405); better electric car batteries (p. 406); solar tower power plants (p. 414); solar cell power plants (p. 416); update on wind energy (pp. 419–422); updates on biodiesel (pp. 423–424) and ethanol (pp. 424–426) biofuels; improving energy efficiency in California (pp. 433–434); ecological medicine (p. 446); harmful chemicals in a home (p. 458); South Asia's brown cloud (p. 468); Hubbard Brook forest and acid deposition (p. 482); updates on global warming and climate change (pp. 499–503, 507); melting ice in Greenland (p. 508); climate stabilization wedges (p. 515); expanded coverage of carbon capture and storage (pp. 516–517); geoengineering and climate change (pp. 517–518); Lake Washington (pp. 531, 540); LifeStraw water purifier (p. 545); E-Waste (pp. 560, 577); integrated waste management (p. 565); recycling plastics (p. 572); urban indoor farming (p. 607); an ecocity in China (pp. 607–608); greening American campuses (p. 651); environmental literacy (p. 666); and sustainability dozen (p. 669).

In-Text Study Aids

Each chapter begins with a list of *key questions and concepts* showing how the chapter is organized and what students will be learning. When a new term is introduced and defined, it is printed in boldface type, and all such terms are summarized in the glossary at the end of the book and highlighted in review questions at the end of each chapter.

One hundred and fifty-nine *Thinking About* exercises reinforce learning by asking students to think critically about the implications of various environmental issues and solutions immediately after they are discussed in the text. The captions of many figures contain questions that involve students in thinking about and evaluating their content.

Each chapter ends with a *Review* section containing a detailed set of review questions that include all chapter key terms in boldface (p. 558), followed by a set of *Critical Thinking* (pp. 558–559) questions to encourage students to think critically and apply what they have learned to their lives.

Supplements for Students

A multitude of electronic supplements available to students take the learning experience beyond the textbook:

■ *CengageNOW* is an online learning tool that helps students access their unique study needs. Students take a pre-test and a personalized study plan provides them with specific resources for review. A post-test then identifies content that might require further study. *How Do I Prepare* tutorials, another feature of *CengageNOW,* walk students through basic math, chemistry, and study skills to help them brush up quickly and be ready to succeed in their course.

■ *WebTutor* on WebCT or Blackboard provides access to a full array of study tools, including flash cards, practice quizzes, animations, exercises, and web links.

■ *Audio Study Tools.* Students can download these useful study aids, which contain valuable information such as reviews of important concepts, key terms, questions, clarifications of common misconceptions, and study tips.

■ Access to *InfoTrac® College Edition* for teachers and students using *CengageNOW* and *WebTutor* on WebCT or Blackboard. This fully searchable online library gives users access to complete environmental articles from several hundred current periodicals and others dating back over 20 years.

The following materials for this textbook are available on the companion website at

academic.cengage.com/biology/miller

■ *Chapter Summaries* help guide student reading and study of each chapter.

■ *Flash Cards* and *Glossary* allow students to test their mastery of each chapter's Key Terms.

■ *Chapter Tests* provide multiple-choice practice quizzes.

■ Information on a variety of *Green Careers.*

■ *Readings* list major books and articles consulted in writing each chapter and include suggestions for articles, books, and websites that provide additional information.

■ *What Can You Do?* offers students resources for what they can do to effect individual change on key environmental issues.

■ *Weblinks* and *Research Frontier Links* offer an extensive list of websites with news and research related to each chapter.

Other student learning tools include:

- *Essential Study Skills for Science Students* by Daniel D. Chiras. This book includes chapters on developing good study habits, sharpening memory, getting the most out of lectures, labs, and reading assignments, improving test-taking abilities, and becoming a critical thinker. Available for students on instructor request.

- *Lab Manual.* New to this edition, this lab manual includes both hands-on and data analysis labs to help your students develop a range of skills. Create a custom version of this Lab Manual by adding labs you have written or ones from our collection with Cengage Custom Publishing. An Instructor's Manual for the labs will be available to adopters.

- *What Can You Do?* This guide presents students with a variety of ways that they can affect the environment, and shows them how to track the effect their actions have on their ecological footprint. Available for students on instructor request.

Supplements for Instructors

- *PowerLecture.* This DVD, available to adopters, allows you to create custom lectures in Microsoft® PowerPoint using lecture outlines, all of the figures and photos from the text, bonus photos, and animations from *CengageNOW*. PowerPoint's editing tools allow use of slides from other lectures, modification or removal of figure labels and leaders, insertion of your own slides, saving slides as JPEG images, and preparation of lectures for use on the Web.

- *Instructor's Manual.* Available to adopters. Updated and reorganized, the Instructor's Manual has been thoughtfully revised to make creating your lectures even easier. Some of the features new to this edition include the integration of the case studies and feature boxes into the lecture outline, a new section on teaching tips, and a revised video reference list with web resources. Also available on PowerLecture.

- *Test Bank.* Available to Adopters. The test bank contains thousands of questions and answers in a variety of formats. New to this edition are short essay questions to further challenge your students' understanding of the topics. Also available on PowerLecture.

- *Transparencies.* Featuring all the illustrations from the chapters, this set contains 250 printed Transparencies of key figures, and 250 electronic Masters. These electronic Masters will allow you to print, in color, only those additional figures you need.

- *ABC Videos for Environmental Science.* The 45 informative and short video clips cover current news stories on environmental issues from around the world. These clips are a great way to start a lecture or spark a discussion. Available on DVD with a workbook, on the PowerLecture, and in *CengageNOW* with additional internet activities.

- *ExamView.* This full-featured program helps you create and deliver customized tests (both print and online) in minutes, using its complete word processing capabilities.

Other Textbook Options

Instructors wanting a book with a different length and emphasis can use one of our three other books that we have written for various types of environmental science courses: *Environmental Science,* 12th edition (430 pages, Brooks/Cole 2008), *Sustaining the Earth: An Integrated Approach,* 9th edition (339 pages, Brooks/Cole, 2009) and *Essentials of Ecology,* 5th edition (274 pages, Brooks/Cole, 2009).

Help Us Improve This Book

Let us know how you think this book can be improved. If you find any errors, bias, or confusing explanations, please e-mail us about them at

mtg89@hotmail.com
spoolman@tds.net

Most errors can be corrected in subsequent printings of this edition, as well as in future editions.

Acknowledgments

We wish to thank the many students and teachers who have responded so favorably to the 15 previous editions of *Living in the Environment,* the 12 editions of *Environmental Science,* the 8 editions of *Sustaining the Earth,* and the 4 editions of *Essentials of Ecology,* and who have corrected errors and offered many helpful suggestions for improvement. We are also deeply indebted to the more than 295 reviewers, who pointed out errors and suggested many important improvements in the various editions of these four books. We especially want to thank the reviewers of the fifteenth edition of this book, who are identified by an asterisk in the master list of reviewers on pp. xx–xxii.

It takes a village to produce a textbook, and the members of the talented production team, listed on the copyright page, have made vital contributions as well. Our special thanks go to developmental editor Christopher Delgado, production editors Andy Marinkovich and Nicole Barone, copy editor Andrea Fincke, layout expert Bonnie Van Slyke, photo researcher Abigail Reip, artist Patrick Lane, media editor Kristina Razmara, assistant editor Lauren Oliveira, editorial assistant Samantha Arvin, and Brooks/Cole's hard-working sales staff.

We also thank Ed Wells and the dedicated team who developed the *Laboratory Manual* to accompany this book, and the people who have translated this book into eight languages for use throughout much of the world.

We also deeply appreciate having had the opportunity to work with Jack Carey, former biology publisher at Brooks/Cole, for 40 years before his recent retirement. We now are fortunate and excited to be working with Yolanda Cossio, the biology publisher at Brooks/Cole.

G. Tyler Miller, Jr.
Scott Spoolman

Guest Essayists

Guest essays by the following authors are available on *CengageNOW:* **M. Kat Anderson,** ethnoecologist with the National Plant Center of the USDA's Natural Resource Conservation Center; **Lester R. Brown,** president, Earth Policy Institute; **Alberto Ruz Buenfil,** environmental activist, writer, and performer; **Robert D. Bullard,** professor of sociology and director of the Environmental Justice Resource Center at Clark Atlanta University; **Michael Cain,** ecologist and adjunct professor at Bowdoin College; **Herman E. Daly,** senior research scholar at the School of Public Affairs, University of Maryland; **Lois Marie Gibbs,** director, Center for Health, Environment, and Justice; **Garrett Hardin,** professor emeritus (now deceased) of human ecology, University of California, Santa Barbara; **John Harte,** professor of energy and resources, University of California, Berkeley; **Paul G. Hawken,** environmental author and business leader; **Jane Heinze-Fry,** environmental educator; **Paul F. Kamitsuja,** infectious disease expert and physician; **Amory B. Lovins,** energy policy consultant and director of research, Rocky Mountain Institute; **Bobbi S. Low,** professor of resource ecology, University of Michigan; **Lester W. Milbrath,** former director of the research program in environment and society, State University of New York, Buffalo; **Peter Montague,** director, Environmental Research Foundation; **Norman Myers,** tropical ecologist and consultant in environment and development; **David W. Orr,** professor of environmental studies, Oberlin College; **Noel Perrin,** adjunct professor of environmental studies, Dartmouth College; **David Pimentel,** professor of insect ecology and agricultural sciences, Cornell University; **John Pichtel,** Ball State University; **Andrew C. Revkin,** environmental author and environmental reporter for the *New York Times;* **Vandana Shiva,** physicist, educator, environmental consultant; **Nancy Wicks,** ecopioneer and director of Round Mountain Organics; **Donald Worster,** environmental historian and professor of American history, University of Kansas.

Quantitative Exercise Contributors

Dr. Dean Goodwin and his colleagues, Berry Cobb, Deborah Stevens, Jeannette Adkins, Jim Lehner, Judy Treharne, Lonnie Miller, and Tom Mowbray, provided excellent contributions to the Data Analysis and Ecological Footprint Analysis exercises.

Cumulative Reviewers (Reviewers of the 15th edition are indicated by an asterisk.)

Barbara J. Abraham, Hampton College; Donald D. Adams, State University of New York at Plattsburgh; Larry G. Allen, California State University, Northridge; Susan Allen-Gil, Ithaca College; James R. Anderson, U.S. Geological Survey; Mark W. Anderson, University of Maine; Kenneth B. Armitage, University of Kansas; Samuel Arthur, Bowling Green State University; Gary J. Atchison, Iowa State University; *Thomas W. H. Backman, Lewis Clark State University; Marvin W. Baker, Jr., University of Oklahoma; Virgil R. Baker, Arizona State University; *Stephen W. Banks, Louisiana State University in Shreveport; Ian G. Barbour, Carleton College; Albert J. Beck, California State University, Chico; *Eugene C. Beckham, Northwood University; *Diane B. Beechinor, Northeast Lakeview College; W. Behan, Northern Arizona University; *David Belt, Johnson County Community College; Keith L. Bildstein, Winthrop College; *Andrea Bixler, Clarke College; Jeff Bland, University of Puget Sound; Roger G. Bland, Central Michigan University; Grady Blount II, Texas A&M University, Corpus Christi; *Lisa K. Bonneau, University of Missouri-Kansas City; Georg Borgstrom, Michigan State University; Arthur C. Borror, University of New Hampshire; John H. Bounds, Sam Houston State University; Leon F. Bouvier, Population Reference Bureau; Daniel J. Bovin, Universitè Laval; *Jan Boyle, University of Great Falls; *James A. Brenneman, University of Evansville; Michael F. Brewer, Resources for the Future, Inc.; Mark

M. Brinson, East Carolina University; Dale Brown, University of Hartford; Patrick E. Brunelle, Contra Costa College; Terrence J. Burgess, Saddleback College North; David Byman, Pennsylvania State University, Worthington–Scranton; Michael L. Cain, Bowdoin College, Lynton K. Caldwell, Indiana University; Faith Thompson Campbell, Natural Resources Defense Council, Inc.; *John S. Campbell, Northwest College; Ray Canterbery, Florida State University; Ted J. Case, University of San Diego; Ann Causey, Auburn University; Richard A. Cellarius, Evergreen State University; William U. Chandler, Worldwatch Institute; F. Christman, University of North Carolina, Chapel Hill; Lu Anne Clark, Lansing Community College; Preston Cloud, University of California, Santa Barbara; Bernard C. Cohen, University of Pittsburgh; Richard A. Cooley, University of California, Santa Cruz; Dennis J. Corrigan; George Cox, San Diego State University; John D. Cunningham, Keene State College; Herman E. Daly, University of Maryland; Raymond F. Dasmann, University of California, Santa Cruz; Kingsley Davis, Hoover Institution; Edward E. DeMartini, University of California, Santa Barbara; *James Demastes, University of Northern Iowa; Charles E. DePoe, Northeast Louisiana University; Thomas R. Detwyler, University of Wisconsin; *Bruce DeVantier, Southern Illinois University Carbondale; Peter H. Diage, University of California, Riverside; *Stephanie Dockstader, Monroe Community College; Lon D. Drake, University of Iowa; *Michael Draney, University of Wisconsin–Green Bay; David DuBose, Shasta College; Dietrich Earnhart, University of Kansas; *Robert East, Washington & Jefferson College; T. Edmonson, University of Washington; Thomas Eisner, Cornell University; Michael Esler, Southern Illinois University; David E. Fairbrothers, Rutgers University; Paul P. Feeny, Cornell University; Richard S. Feldman, Marist College; *Vicki Fella-Pleier, La Salle University; Nancy Field, Bellevue Community College; Allan Fitzsimmons, University of Kentucky; Andrew J. Friedland, Dartmouth College; Kenneth O. Fulgham, Humboldt State University; Lowell L. Getz, University of Illinois at Urbana–Champaign; Frederick F. Gilbert, Washington State University; Jay Glassman, Los Angeles Valley College; Harold Goetz, North Dakota State University; *Srikanth Gogineni, Axia College of University of Phoenix; Jeffery J. Gordon, Bowling Green State University; Eville Gorham, University of Minnesota; Michael Gough, Resources for the Future; Ernest M. Gould, Jr., Harvard University; Peter Green, Golden West College; Katharine B. Gregg, West Virginia Wesleyan College; Paul K. Grogger, University of Colorado at Colorado Springs; L. Guernsey, Indiana State University; Ralph Guzman, University of California, Santa Cruz; Raymond Hames, University of Nebraska, Lincoln; *Robert Hamilton IV, Kent State University, Stark Campus; Raymond E. Hampton, Central Michigan University; Ted L. Hanes, California State University, Fullerton; William S. Hardenbergh, Southern Illinois University at Carbondale; John P. Harley, Eastern Kentucky University; Neil A. Harriman, University of Wisconsin, Oshkosh; Grant A. Harris, Washington State University; Harry S. Hass, San Jose City College; Arthur N. Haupt, Population Reference Bureau; Denis A. Hayes, environmental consultant; Stephen Heard, University of Iowa; Gene Heinze-Fry, Department of Utilities, Commonwealth of Massachusetts; Jane Heinze-Fry, environmental educator; John G. Hewston, Humboldt State University; David L. Hicks, Whitworth College; Kenneth M. Hinkel, University of Cincinnati; Eric Hirst, Oak Ridge National Laboratory; Doug Hix, University of Hartford; S. Holling, University of British Columbia; Sue Holt, Cabrillo College; Donald Holtgrieve, California State University, Hayward; *Michelle Homan, Gannon University; Michael H. Horn, California State University, Fullerton; Mark A. Hornberger, Bloomsburg University; Marilyn Houck, Pennsylvania State University; Richard D. Houk, Winthrop College; Robert J. Huggett, College of William and Mary; Donald Huisingh, North Carolina State University; *Catherine Hurlbut, Florida Community College at Jacksonville; Marlene K. Hutt, IBM; David R. Inglis, University of Massachusetts; Robert Janiskee, University of South Carolina; Hugo H. John, University of Connecticut; Brian A. Johnson, University of Pennsylvania, Bloomsburg; David I. Johnson, Michigan State University; Mark Jonasson, Crafton Hills College; *Zoghlul Kabir, Rutgers/New Brunswick; Agnes Kadar, Nassau Community College; Thomas L. Keefe, Eastern Kentucky University; *David Kelley, University of St. Thomas; William E. Kelso, Louisiana State University; Nathan Keyfitz, Harvard University; David Kidd, University of New Mexico; Pamela S. Kimbrough; Jesse Klingebiel, Kent School; Edward J. Kormondy, University of Hawaii–Hilo/West Oahu College; John V. Krutilla, Resources for the Future, Inc.; Judith Kunofsky, Sierra Club; E. Kurtz; Theodore Kury, State University of New York at Buffalo; Steve Ladochy, University of Winnipeg; *Troy A. Ladine, East Texas Baptist University; *Anna J. Lang, Weber State University; Mark B. Lapping, Kansas State University; *Michael L. Larsen, Campbell University; *Linda Lee, University of Connecticut; Tom Leege, Idaho Department of Fish and Game; *Maureen Leupold, Genesee Community College; William S. Lindsay, Monterey Peninsula College; E. S. Lindstrom, Pennsylvania State University; M. Lippiman, New York University Medical Center; Valerie A. Liston, University of Minnesota; Dennis Livingston, Rensselaer Polytechnic Institute; James P. Lodge, air pollution consultant; Raymond C. Loehr, University of Texas at Austin; Ruth Logan, Santa Monica City College; Robert D. Loring, DePauw University; Paul F. Love, Angelo State University; Thomas Lovering, University of California, Santa Barbara; Amory B. Lovins, Rocky Mountain Institute; Hunter Lovins, Rocky Mountain Institute; Gene A. Lucas, Drake University; Claudia Luke; David Lynn; Timothy F. Lyon, Ball State University; Stephen Malcolm, Western Michigan University; Melvin G. Marcus, Arizona State University; Gordon E. Matzke, Oregon State University; Parker Mauldin, Rockefeller Foundation; Marie McClune, The Agnes Irwin School (Rosemont, Pennsylvania);

Theodore R. McDowell, California State University; Vincent E. McKelvey, U.S. Geological Survey; Robert T. McMaster, Smith College; John G. Merriam, Bowling Green State University; A. Steven Messenger, Northern Illinois University; John Meyers, Middlesex Community College; Raymond W. Miller, Utah State University; Arthur B. Millman, University of Massachusetts, Boston; *Sheila Miracle, Southeast Kentucky Community & Technical College; Fred Montague, University of Utah; Rolf Monteen, California Polytechnic State University; *Debbie Moore, Troy University Dothan Campus; *Michael K. Moore, Mercer University; Ralph Morris, Brock University, St. Catherine's, Ontario, Canada; Angela Morrow, Auburn University; William W. Murdoch, University of California, Santa Barbara; Norman Myers, environmental consultant; Brian C. Myres, Cypress College; A. Neale, Illinois State University; Duane Nellis, Kansas State University; Jan Newhouse, University of Hawaii, Manoa; Jim Norwine, Texas A&M University, Kingsville; John E. Oliver, Indiana State University; *Mark Olsen, University of Notre Dame; Carol Page, copyeditor; Eric Pallant, Allegheny College; *Bill Paletski, Penn State University; Charles F. Park, Stanford University; Richard J. Pedersen, U.S. Department of Agriculture, Forest Service; David Pelliam, Bureau of Land Management, U.S. Department of Interior; *Murray Paton Pendarvis, Southeastern Louisiana University; *Dave Perault, Lynchburg College; Rodney Peterson, Colorado State University; Julie Phillips, De Anza College; John Pichtel, Ball State University; William S. Pierce, Case Western Reserve University; David Pimentel, Cornell University; Peter Pizor, Northwest Community College; Mark D. Plunkett, Bellevue Community College; Grace L. Powell, University of Akron; James H. Price, Oklahoma College; Marian E. Reeve, Merritt College; Carl H. Reidel, University of Vermont; Charles C. Reith, Tulane University; Roger Revelle, California State University, San Diego; L. Reynolds, University of Central Arkansas; Ronald R. Rhein, Kutztown University of Pennsylvania; Charles Rhyne, Jackson State University; Robert A. Richardson, University of Wisconsin; Benjamin F. Richason III, St. Cloud State University; Jennifer Rivers, Northeastern University; Ronald Robberecht, University of Idaho; William Van B. Robertson, School of Medicine, Stanford University; C. Lee Rockett, Bowling Green State University; Terry D. Roelofs, Humboldt State University; *Daniel Ropek, Columbia George Community College; Christopher Rose, California Polytechnic State University; Richard G. Rose, West Valley College; Stephen T. Ross, University of Southern Mississippi; Robert E. Roth, Ohio State University; *Dorna Sakurai, Santa Monica College; Arthur N. Samel, Bowling Green State University; *Shamili Sandiford, College of DuPage; Floyd Sanford, Coe College; David Satterthwaite, I.E.E.D., London; Stephen W. Sawyer, University of Maryland; Arnold Schecter, State University of New York; Frank Schiavo, San Jose State University; William H. Schlesinger, Ecological Society of America; Stephen H. Schneider, National Center for Atmospheric Research; Clarence A. Schoenfeld, University of Wisconsin, Madison; *Madeline Schreiber, Virginia Polytechnic Institute; Henry A. Schroeder, Dartmouth Medical School; Lauren A. Schroeder, Youngstown State University; Norman B. Schwartz, University of Delaware; George Sessions, Sierra College; David J. Severn, Clement Associates; *Don Sheets, Gardner-Webb University; Paul Shepard, Pitzer College and Claremont Graduate School; Michael P. Shields, Southern Illinois University at Carbondale; Kenneth Shiovitz; F. Siewert, Ball State University; E. K. Silbergold, Environmental Defense Fund; Joseph L. Simon, University of South Florida; William E. Sloey, University of Wisconsin, Oshkosh; Robert L. Smith, West Virginia University; Val Smith, University of Kansas; Howard M. Smolkin, U.S. Environmental Protection Agency; Patricia M. Sparks, Glassboro State College; John E. Stanley, University of Virginia; Mel Stanley, California State Polytechnic University, Pomona; *Richard Stevens, Monroe Community College; Norman R. Stewart, University of Wisconsin, Milwaukee; Frank E. Studnicka, University of Wisconsin, Platteville; Chris Tarp, Contra Costa College; Roger E. Thibault, Bowling Green State University; William L. Thomas, California State University, Hayward; Shari Turney, copyeditor; John D. Usis, Youngstown State University; Tinco E. A. van Hylckama, Texas Tech University; Robert R. Van Kirk, Humboldt State University; Donald E. Van Meter, Ball State University; *Rick Van Schoik, San Diego State University; Gary Varner, Texas A&M University; John D. Vitek, Oklahoma State University; Harry A. Wagner, Victoria College; Lee B. Waian, Saddleback College; Warren C. Walker, Stephen F. Austin State University; Thomas D. Warner, South Dakota State University; Kenneth E. F. Watt, University of California, Davis; Alvin M. Weinberg, Institute of Energy Analysis, Oak Ridge Associated Universities; Brian Weiss; Margery Weitkamp, James Monroe High School (Granada Hills, California); Anthony Weston, State University of New York at Stony Brook; Raymond White, San Francisco City College; Douglas Wickum, University of Wisconsin, Stout; Charles G. Wilber, Colorado State University; Nancy Lee Wilkinson, San Francisco State University; John C. Williams, College of San Mateo; Ray Williams, Rio Hondo College; Roberta Williams, University of Nevada, Las Vegas; Samuel J. Williamson, New York University; *Dwina Willis, Freed-Hardeman University; Ted L. Willrich, Oregon State University; James Winsor, Pennsylvania State University; Fred Witzig, University of Minnesota at Duluth; *Martha Wolfe, Elizabethtown Community and Technical College; George M. Woodwell, Woods Hole Research Center; *Todd Yetter, University of the Cumberlands; Robert Yoerg, Belmont Hills Hospital; Hideo Yonenaka, San Francisco State University; *Brenda Young, Daemen College; *Anita Zvodsk, Barry University; Malcolm J. Zwolinski, University of Arizona.

About the Authors

G. Tyler Miller, Jr.

G. Tyler Miller, Jr., has written 58 textbooks for introductory courses in environmental science, basic ecology, energy, and environmental chemistry. Since 1975, Miller's books have been the most widely used textbooks for environmental science in the United States and throughout the world. They have been used by almost 3 million students and have been translated into eight languages.

Miller has a Ph.D. from the University of Virginia and has received two honorary doctorate degrees for his contributions to environmental education. He taught college for 20 years and developed an innovative interdisciplinary undergraduate science program before deciding to write environmental science textbooks full time since 1975. Currently, he is the President of Earth Education and Research, devoted to improving environmental education.

He describes his hopes for the future as follows:

If I had to pick a time to be alive, it would be the next 75 years. Why? First, there is overwhelming scientific evidence that we are in the process of seriously degrading our own life support system. In other words, we are living unsustainably. Second, within your lifetime we have the opportunity to learn how to live more sustainably by working with the rest of nature, as described in this book.

I am fortunate to have three smart, talented, and wonderful sons—Greg, David, and Bill. I am especially privileged to have Kathleen as my wife, best friend, and research associate. It is inspiring to have a brilliant, beautiful (inside and out), and strong woman who cares deeply about nature as a lifemate. She is my hero. I dedicate this book to her and to the earth.

Scott E. Spoolman

Scott Spoolman is a writer and textbook editor with over 25 years of experience in educational publishing. He has worked with Tyler Miller since 2003 as a contributing editor on earlier editions of *Living in the Environment, Environmental Science,* and *Sustaining the Earth.*

Spoolman holds a master's degree in science journalism from the University of Minnesota. He has authored numerous articles in the fields of science, environmental engineering, politics, and business. He worked as an acquisitions editor on a series of college forestry textbooks. He has also worked as a consulting editor in the development of over 70 college and high school textbooks in fields of the natural and social sciences.

In his free time, he enjoys exploring the forests and waters of his native Wisconsin along with his family—his wife, environmental educator Gail Martinelli, and his children, Will and Katie.

Spoolman has the following to say about his collaboration with Tyler Miller:

I am honored to be joining with Tyler Miller as a coauthor to continue the Miller tradition of thorough, clear, and engaging writing about the vast and complex field of environmental science. I share Tyler Miller's passion for ensuring that these textbooks and their multimedia supplements will be valuable tools for students and instructors. To that end, we strive to introduce this interdisciplinary field in ways that will be informative and sobering, but also tantalizing and motivational.

If the flip side of any problem is indeed an opportunity, then this truly is one of the most exciting times in history for students to start an environmental career. Environmental problems are numerous, serious, and daunting, but their possible solutions generate exciting new career opportunities. We place high priorities on inspiring students with these possibilities, challenging them to maintain a scientific focus, pointing them toward rewarding and fulfilling careers, and in doing so, working to help sustain life on earth.

Learning Skills

Students who can begin early in their lives to think of things as connected, even if they revise their views every year, have begun the life of learning.

MARK VAN DOREN

Why Is It Important to Study Environmental Science?

Welcome to **environmental science**—an *interdisciplinary* study of how the earth works, how we interact with the earth, and how we can deal with the environmental problems we face. Because environmental issues affect every part of your life, the concepts, information, and issues discussed in this book and the course you are taking will be useful to you now and throughout your life.

Understandably, we are biased, but *we strongly believe that environmental science is the single most important course in your education.* What could be more important than learning how the earth works, how we are affecting its life support system, and how we can reduce our environmental impact?

We live in an incredibly challenging era. We are becoming increasingly aware that during this century we need to make a new cultural transition in which we learn how to live more sustainably by sharply reducing the degradation of our life-support system. We hope this book will inspire you to become involved in this change in the way we view and treat the earth, which sustains us and our economies and all other living things.

You Can Improve Your Study and Learning Skills

Maximizing your ability to learn should be one of your most important lifetime educational goals. It involves continually trying to *improve your study and learning skills.* Here are some suggestions for doing so:

Develop a passion for learning. As the famous physicist and philosopher Albert Einstein put it, "I have no special talent. I am only passionately curious."

Get organized. Becoming more efficient at studying gives you more time for other interests.

Make daily to-do lists in writing. Put items in order of importance, focus on the most important tasks, and assign a time to work on these items. Because life is full of uncertainties, you might be lucky to accomplish half of the items on your daily list. Shift your schedule as needed to accomplish the most important items.

Set up a study routine in a distraction-free environment. Develop a written daily study schedule and stick to it. Study in a quiet, well-lighted space. Work while sitting at a desk or table—not lying down on a couch or bed. Take breaks every hour or so. During each break, take several deep breaths and move around; this will help you to stay more alert and focused.

Avoid procrastination—putting work off until another time. Do not fall behind on your reading and other assignments. Set aside a particular time for studying each day and make it a part of your daily routine.

Do not eat dessert first. Otherwise, you may never get to the main meal (studying). When you have accomplished your study goals, reward yourself with dessert (play or leisure).

Make hills out of mountains. It is psychologically difficult to climb a mountain, which is what reading an entire book, reading a chapter in a book, writing a paper, or cramming to study for a test can feel like. Instead, break these large tasks (mountains) down into a series of small tasks (hills). Each day, read a few pages of a book or chapter, write a few paragraphs of a paper, and review what you have studied and learned. As American automobile designer and builder Henry Ford put it, "Nothing is particularly hard if you divide it into small jobs."

Look at the big picture first. Get an overview of an assigned reading in this book by looking at the *Key Questions and Concepts* box at the beginning of each chapter. It lists key questions explored in the chapter sections and the corresponding key concepts, which are the critical lessons to be learned in the chapter. Use this list as a chapter roadmap. When you finish a chapter you can also use it to review.

Ask and answer questions as you read. For example, "What is the main point of a particular subsection or paragraph?" Relate your own questions to the key questions and key concepts being addressed in each major chapter section. In this way, you can flesh out a chapter outline to help you understand the chapter material. You may even want to do such an outline in writing.

Focus on key terms. Use the glossary in this textbook to look up the meanings of terms or words you do not understand. This book shows all key terms in **boldface** type and lesser, but still important, terms in *italicized* type. The review

questions at the end of each chapter also include the chapter's key terms in boldface. Flash cards for testing your mastery of key terms for each chapter are available on the website for this book, or you can make your own by putting a term on one side of an index card or piece of paper and its meaning on the other side.

Interact with what you read. We suggest that you mark key sentences and paragraphs with a highlighter or pen. Consider putting an asterisk in the margin next to material you think is important and double asterisks next to material you think is especially important. Write comments in the margins, such as *beautiful, confusing, misleading,* or *wrong.* You might fold down the top corners of pages on which you highlighted passages and the top and bottom corners of especially important pages. This way, you can flip through a chapter or book and quickly review the key ideas.

Review to reinforce learning. Before each class session, review the material you learned in the previous session and read the assigned material.

Become a good note taker. Do not try to take down everything your instructor says. Instead, write down main points and key facts using your own shorthand system. Review, fill in, and organize your notes as soon as possible after each class.

Write out answers to questions to focus and reinforce learning. Answer the critical thinking questions found in *Thinking About* boxes throughout chapters, in many figure captions, and at the end of each chapter. These questions are designed to inspire you to think critically about key ideas and connect them to other ideas and to your own life. Also answer the review questions found at the end of each chapter. The website for each chapter has an additional detailed list of review questions. Writing out your answers to the critical thinking and review questions can reinforce your learning. Save your answers for review and preparation for tests.

Use the buddy system. Study with a friend or become a member of a study group to compare notes, review material, and prepare for tests. Explaining something to someone else is a great way to focus your thoughts and reinforce your learning. Attend any review sessions offered by instructors or teaching assistants.

Learn your instructor's test style. Does your instructor emphasize multiple-choice, fill-in-the-blank, true-or-false, factual, or essay questions? How much of the test will come from the textbook and how much from lecture material? Adapt your learning and studying methods to your instructor's style. It may not exactly match your own, but the reality is that your instructor is in charge.

Become a good test taker. Avoid cramming. Eat well and get plenty of sleep before a test. Arrive on time or early. Calm yourself and increase your oxygen intake by taking several deep breaths. (Do this also about every 10–15 minutes while taking the test.) Look over the test and answer the questions you know well first.

Then work on the harder ones. Use the process of elimination to narrow down the choices for multiple-choice questions. Paring them down to two choices gives you a 50% chance of guessing the right answer. For essay questions, organize your thoughts before you start writing. If you have no idea what a question means, make an educated guess. You might get some partial credit and avoid getting a zero. Another strategy for getting some credit is to show your knowledge and reasoning by writing something like this: "If this question means so and so, then my answer is _____."

Develop an optimistic but realistic outlook. Try to be a "glass is half-full" rather than a "glass is half-empty" person. Pessimism, fear, anxiety, and excessive worrying (especially over things you cannot control) are destructive and lead to inaction. Try to keep your energizing feelings of realistic optimism slightly ahead of any immobilizing feelings of pessimism. Then you will always be moving forward.

Take time to enjoy life. Every day, take time to laugh and enjoy nature, beauty, and friendship. You can do this without falling behind in your work and living under a cloud of guilt and anxiety if you become an effective and efficient learner.

You Can Improve Your Critical Thinking Skills: Becoming a Good Baloney Detector

Critical thinking involves developing skills to analyze information and ideas, judge their validity, and make decisions. Critical thinking helps you to distinguish between facts and opinions, evaluate evidence and arguments, take and defend informed positions on issues, integrate information, see relationships, and apply your knowledge to dealing with new and different problems and to your own lifestyle choices. Here are some basic skills for learning how to think more critically.

Question everything and everybody. Be skeptical, as any good scientist is. Do not believe everything you hear and read, including the content of this textbook, without evaluating the information you receive. Seek other sources and opinions. As Albert Einstein put it, "The important thing is not to stop questioning."

Identify and evaluate your personal biases and beliefs. Each of us has biases and beliefs taught to us by our parents, teachers, friends, role models, and experience. What are your basic beliefs, values, and biases? Where did they come from? What assumptions are they based on? How sure are you that your beliefs, values, and assumptions are right and why? According to the American psychologist and philosopher William James, "A great many people think they are thinking when they are merely rearranging their prejudices."

Be open-minded and flexible. Be open to considering different points of view. Suspend judgment until you

gather more evidence, and be capable of changing your mind. Recognize that there may be a number of useful and acceptable solutions to a problem and that very few issues are black or white. There are trade-offs involved in dealing with any environmental issue, as you will learn in this book. One way to evaluate divergent views is to try to take the viewpoints of other people. How do they see the world? What are their basic assumptions and beliefs? Are their positions logically consistent with their assumptions and beliefs?

Be humble about what you know. Some people are so confident in what they know that they stop thinking and questioning. To paraphrase American writer Mark Twain, "It's not what we don't know that's so bad. It's what we know is true, but just ain't so, that hurts us."

Evaluate how the information related to an issue was obtained. Are the statements you heard or read based on firsthand knowledge and research or on hearsay? Are unnamed sources used? Is the information based on reproducible and widely accepted scientific studies (*reliable science*, p. 33) or on preliminary scientific results that may be valid but need further testing (*tentative* or *frontier science*, p. 33)? Is the information based on a few isolated stories or experiences (*anecdotal information*) or on carefully controlled studies with the results reviewed by experts in the field involved (*peer review*)? Is it based on unsubstantiated and dubious scientific information or beliefs (*unreliable science*, p. 34)?

Question the evidence and conclusions presented. What are the conclusions or claims? What evidence is presented to support them? Does the evidence support them? Is there a need to gather more evidence to test the conclusions? Are there other, more reasonable conclusions?

Try to uncover differences in basic beliefs and assumptions. On the surface most arguments or disagreements involve differences in opinions about the validity or meaning of certain facts or conclusions. Scratch a little deeper and you will find that most disagreements are usually based on different (and often hidden) basic assumptions concerning how we look at and interpret the world around us. Uncovering these basic differences can allow the parties involved to understand where each is "coming from" and to agree to disagree about their basic assumptions, beliefs, or principles.

Try to identify and assess any motives on the part of those presenting evidence and drawing conclusions. What is their expertise in this area? Do they have any unstated assumptions, beliefs, biases, or values? Do they have a personal agenda? Can they benefit financially or politically from acceptance of their evidence and conclusions? Would investigators with different basic assumptions or beliefs take the same data and come to different conclusions?

Expect and tolerate uncertainty. Recognize that science is an ever-changing adventure that provides only a degree of certainty. Scientists can disprove things but they cannot establish absolute proof or certainty. However, the widely accepted results of reliable science have a high degree of certainty.

Do the arguments used involve logical fallacies or debating tricks? Here are six of many examples. *First,* attack the presenter of an argument rather than the argument itself. *Second,* appeal to emotion rather than facts and logic. *Third,* claim that if one piece of evidence or one conclusion is false, then all other related pieces of evidence and conclusions are false. *Fourth,* say that a conclusion is false because it has not been scientifically proven. (Scientists never prove anything absolutely, but they can often establish high degrees of certainty, as discussed on pp. 33–34.) *Fifth,* inject irrelevant or misleading information to divert attention from important points. *Sixth,* present only either/or alternatives when there may be a number of options.

Do not believe everything you read on the Internet. The Internet is a wonderful and easily accessible source of information, providing alternative explanations and opinions on almost any subject or issue—much of it not available in the mainstream media and scholarly articles. Web logs, or blogs, have become a major source of information, even more important than standard news media for some people. However, because the Internet is so open, anyone can post anything they want to a blog or other website with no editorial control or review by experts. As a result, evaluating information on the Internet is one of the best ways to put into practice the principles of critical thinking discussed here. Use and enjoy the Internet, but think critically and proceed with caution.

Develop principles or rules for evaluating evidence. Develop a written list of principles to serve as guidelines for evaluating evidence and claims. Continually evaluate and modify this list on the basis of your experience.

Become a seeker of wisdom, not a vessel of information. Many people believe that the main goal of education is to learn as much as you can by gathering more and more information. We believe that the primary goal is to learn how to sift through mountains of facts and ideas to find the few *nuggets of wisdom* that are the most useful for understanding the world and for making decisions. This book is full of facts and numbers, but they are useful only to the extent that they lead to an understanding of key ideas, scientific laws, theories, concepts, and connections. The major goals of the study of environmental science are to find out how nature works and sustains itself (*environmental wisdom*) and to use *principles of environmental wisdom* to help make human societies and economies more sustainable, more just, and more beneficial and enjoyable for all. As writer Sandra Carey put it, "Never mistake knowledge for wisdom. One helps you make a living; the other helps you make a life." Or as American writer Walker Percy suggested "some individuals with a high intelligence but lacking wisdom can get all A's and flunk life."

To help you practice critical thinking, we have supplied questions throughout this book—at the end of each chapter, and throughout each chapter in brief boxes labeled *Thinking About* and in the captions of many figures. There are no right or wrong answers to many of these questions. A good way to improve your critical thinking skills is to compare your answers with those of your classmates and to discuss how you arrived at your answers.

Know Your Own Learning Style

People have different ways of learning and it can be helpful to know your own learning style. *Visual learners* learn best from reading and viewing illustrations and diagrams. They can benefit from using flash cards (available on the website for this book) to memorize key terms and ideas. This is a highly visual book with many carefully selected photographs and diagrams designed to illustrate important ideas, concepts, and processes.

Auditory learners learn best by listening and discussing. They might benefit from reading aloud while studying and using a tape recorder in lectures for study and review. *Logical learners* learn best by using concepts and logic to uncover and understand a subject rather than relying mostly on memory.

Part of what determines your learning style is how your brain works. According to the *split-brain hypothesis,* the left hemisphere of your brain is good at logic, analysis, and evaluation, and the right half of the brain is good at visualizing, synthesizing, and creating. Our goal is to provide material that stimulates both sides of your brain.

The study and critical thinking skills encouraged in this book and in most courses largely involve the left brain. However, you can improve these skills by giving your left brain a break and letting your creative side loose. You can do this by brainstorming ideas with classmates with the rule that no left-brain criticism is allowed until the session is over.

When you are trying to solve a problem, rest, meditate, take a walk, exercise, or do something to shut down your controlling left-brain activity, and allow the right side of your brain to work on the problem in a less controlled and more creative manner.

This Book Presents a Positive and Realistic Environmental Vision of the Future

There are always *trade-offs* involved in making and implementing environmental decisions. Our challenge is to give a fair and balanced presentation of different viewpoints, advantages and disadvantages of various technologies and proposed solutions to environmental problems, and good and bad news about environmental problems without injecting personal bias.

Studying a subject as important as environmental science and ending up with no conclusions, opinions, and beliefs means that both teacher and student have failed. However, any conclusions one does reach must result from a process of thinking critically to evaluate different ideas and understand the trade-offs involved. Our goal is to present a positive vision of our environmental future based on realistic optimism.

Help Us Improve This Book

Researching and writing a book that covers and connects ideas in a wide variety of disciplines is a challenging and exciting task. Almost every day, we learn about some new connection in nature.

In a book this complex, there are bound to be some errors—some typographical mistakes that slip through and some statements that you might question, based on your knowledge and research. We invite you to contact us and point out any bias, correct any errors you find, and suggest ways to improve this book. Please e-mail your suggestions to Tyler Miller at **mtg@hotmail.com** or Scott Spoolman at **spoolman@tds.net**.

Now start your journey into this fascinating and important study of how the earth works and how we can leave the planet in a condition at least as good as what we found. Have fun.

Study nature, love nature, stay close to nature. It will never fail you.
FRANK LLOYD WRIGHT

Environmental Problems, Their Causes, and Sustainability

1

Living in an Exponential Age

Two ancient kings enjoyed playing chess. The winner claimed a prize from the loser. After one match, the winning king asked the losing king to pay him by placing one grain of wheat on the first square of the chessboard, two grains on the second square, four on the third, and so on, with the number doubling on each square until all 64 squares were filled.

The losing king, thinking he was getting off easy, agreed with delight. It was the biggest mistake he ever made. He bankrupted his kingdom because the number of grains of wheat he had promised was probably more than all the wheat that has ever been harvested!

This fictional story illustrates the concept of **exponential growth,** by which a quantity increases at a *fixed percentage* per unit of time, such as 2% per year. Exponential growth is deceptive. It starts off slowly, but after only a few doublings, it grows to enormous numbers because each doubling is more than the total of all earlier growth.

Here is another example. Fold a piece of paper in half to double its thickness. If you could continue doubling the thickness of the paper 42 times, the stack would reach from the earth to the moon—386,400 kilometers (240,000 miles) away. If you could double it 50 times, the folded paper would almost reach the sun—149 million kilometers (93 million miles) away!

Because of exponential growth in the human population (Figure 1-1), in 2008 there were 6.7 billion people on the planet. Collectively, these people consume vast amounts of food, water, raw materials, and energy and in the process produce huge amounts of pollution and wastes. Unless death rates rise sharply, there will probably be 9.3 billion of us by 2050 and perhaps as many as 10 billion by the end of this century.

The exponential rate of global population growth has declined since 1963. Even so, each day we add an average of 225,000 more people to the earth's population. This is roughly equivalent to adding a new U.S. city of Los Angeles, California, every 2 months, a new France every 9 months, and a new United States—the world's third most populous country—about every 4 years.

No one knows how many people the earth can support, and at what level of resource consumption or affluence, without seriously degrading the ability of the planet to support us and other forms of life and our economies. But there are some disturbing warning signs. Biologists estimate that, by the end of this century, our exponentially increasing population and resource consumption could cause the irreversible loss of one-third to one-half of the world's known different types of plants and animals.

There is also growing evidence and concern that continued exponential growth in human activities such as burning *fossil fuels* (carbon-based fuels such as coal, natural gas, and gasoline) and clearing forests will change the earth's climate during this century. This could ruin some areas for farming, shift water supplies, eliminate many of the earth's unique forms of life, and disrupt economies in various parts of the world.

Great news: We have solutions to these problems that we could implement within a few decades, as you will learn in this book.

Figure 1-1 *Exponential growth:* the J-shaped curve of past exponential world population growth, with projections to 2100 showing possible population stabilization with the J-shaped curve of growth changing to an S-shaped curve. (This figure is not to scale.) (Data from the World Bank and United Nations; photo L. Yong/UNEP/Peter Arnold, Inc)

1-1 What is an environmentally sustainable society?

CONCEPT 1-1A Our lives and economies depend on energy from the sun (*solar capital*) and on natural resources and natural services (*natural capital*) provided by the earth.

CONCEPT 1-1B Living sustainably means living off the earth's natural income without depleting or degrading the natural capital that supplies it.

1-2 How can environmentally sustainable societies grow economically?

CONCEPT 1-2 Societies can become more environmentally sustainable through economic development dedicated to improving the quality of life for everyone without degrading the earth's life support systems.

1-3 How are our ecological footprints affecting the earth?

CONCEPT 1-3 As our ecological footprints grow, we are depleting and degrading more of the earth's natural capital.

1-4 What is pollution, and what can we do about it?

CONCEPT 1-4 Preventing pollution is more effective and less costly than cleaning up pollution.

1-5 Why do we have environmental problems?

CONCEPT 1-5A Major causes of environmental problems are population growth, wasteful and unsustainable resource use, poverty, exclusion of environmental costs of resource use from the market prices of goods and services, and attempts to manage nature with insufficient knowledge.

CONCEPT 1-5B People with different environmental worldviews often disagree about the seriousness of environmental problems and what we should do about them.

1-6 What are four scientific principles of sustainability?

CONCEPT 1-6 Nature has sustained itself for billions of years by using solar energy, biodiversity, population control, and nutrient cycling—lessons from nature that we can apply to our lifestyles and economies.

*This is a *concept-centered* book, with each major chapter section built around one to three key concepts derived from the natural or social sciences. Key questions and concepts are summarized at the beginning of each chapter. You can use this list as a preview and as a review of the key ideas in each chapter.
Note: Supplements 2 (p. S4), 3 (p. S10), 4 (p. S20), 5 (p. S31), and 6 (p. S39) can be used with this chapter.

Alone in space, alone in its life-supporting systems,
powered by inconceivable energies,
mediating them to us through the most delicate adjustments,
wayward, unlikely, unpredictable, but nourishing, enlivening, and enriching
in the largest degree—is this not a precious home for all of us?
Is it not worth our love?

BARBARA WARD AND RENÉ DUBOS

1-1 What Is an Environmentally Sustainable Society?

▶ **CONCEPT 1-1A** Our lives and economies depend on energy from the sun (*solar capital*) and on natural resources and natural services (*natural capital*) provided by the earth.

▶ **CONCEPT 1-1B** Living sustainably means living off the earth's natural income without depleting or degrading the natural capital that supplies it.

Environmental Science Is a Study of Connections in Nature

The **environment** is everything around us. It includes all of the living and the nonliving things with which we interact. And it includes a complex web of relationships that connect us with one another and with the world we live in.

Despite our many scientific and technological advances, we are utterly dependent on the environment for air, water, food, shelter, energy, and everything else we need to stay alive and healthy. As a result, we are part of, and not apart from, the rest of nature.

This textbook is an introduction to **environmental science,** an *interdisciplinary* study of how humans interact with the environment of living and nonliving

Table 1-1

Major Fields of Study Related to Environmental Science

Major Fields	Subfields
Biology: study of living things (organisms)	**Ecology:** study of how organisms interact with one another and with their nonliving environment
	Botany: study of plants
	Zoology: study of animals
Chemistry: study of chemicals and their interactions	**Biochemistry:** study of the chemistry of living things
Earth science: study of the planet as a whole and its nonliving systems	**Climatology:** study of the earth's atmosphere and climate
	Geology: study of the earth's origin, history, surface, and interior processes
	Hydrology: study of the earth's water resources
	Paleontology: study of fossils and ancient life
Social sciences: studies of human society	**Anthropology:** study of human cultures
	Demography: study of the characteristics of human populations
	Geography: study of the relationships between human populations and the earth's surface features
	Economics: study of the production, distribution, and consumption of goods and services
	Political Science: study of the principles, processes, and structure of government and political institutions
Humanities: study of the aspects of the human condition not covered by the physical and social sciences	**History:** study of information and ideas about humanity's past
	Ethics: study of moral values and concepts concerning right and wrong human behavior and responsibilities
	Philosophy: study of knowledge and wisdom about the nature of reality, values, and human conduct

things. It integrates information and ideas from the *natural sciences,* such as biology, chemistry, and geology, the *social sciences,* such as geography, economics, political science, and demography (the study of populations), and the *humanities,* including philosophy and ethics (Table 1-1 and Figure 1-2). The goals of environmental science are to learn *how nature works, how the environment affects us, how we affect the environment, and how to deal with environmental problems and live more sustainably.*

A key subfield of environmental science is **ecology,** the biological science that studies how **organisms,** or living things, interact with their environment and with each other. Every organism is a member of a certain **species:** a group of organisms with distinctive traits and, for sexually reproducing organisms, can mate and produce fertile offspring. For example, all humans are members of a species that biologists have named *Homo sapiens sapiens.* A major focus of ecology is the study of ecosystems. An **ecosystem** is a set of

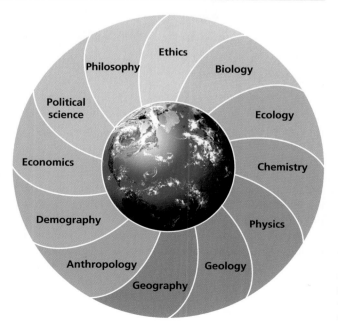

Figure 1-2
Environmental science is an interdisciplinary study of connections between the earth's life-support system and human activities.

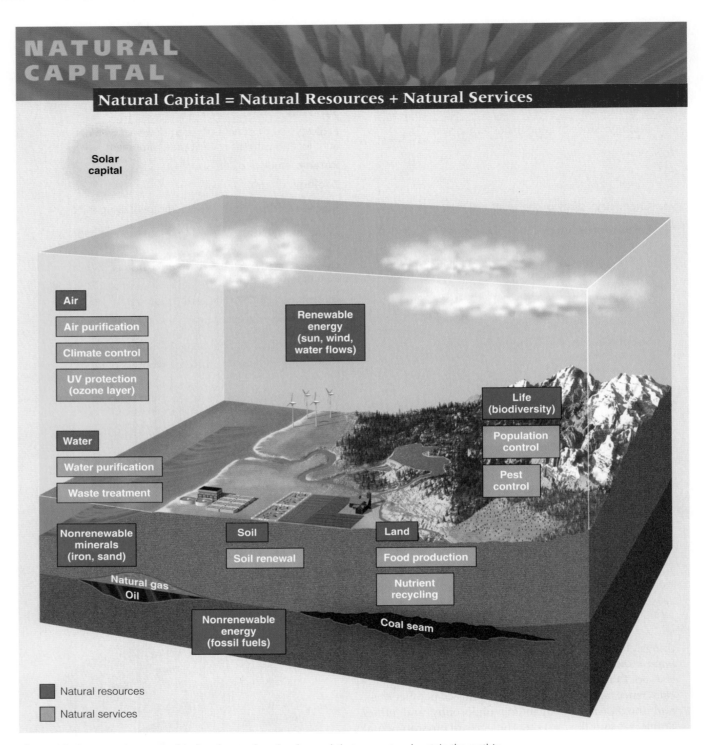

NATURAL CAPITAL

Natural Capital = Natural Resources + Natural Services

Solar capital

Air
Air purification
Climate control
UV protection (ozone layer)

Water
Water purification
Waste treatment

Nonrenewable minerals (iron, sand)

Natural gas
Oil

Soil
Soil renewal

Nonrenewable energy (fossil fuels)

Renewable energy (sun, wind, water flows)

Life (biodiversity)
Population control
Pest control

Land
Food production
Nutrient recycling

Coal seam

Natural resources
Natural services

Figure 1-3 Key *natural resources* (blue) and *natural services* (orange) that support and sustain the earth's life and economies (**Concept 1-1A**).

organisms interacting with one another and with their environment of nonliving matter and energy within a defined area or volume.

We should not confuse environmental science and ecology with **environmentalism,** a social movement dedicated to protecting the earth's life-support systems for us and all other forms of life. Environmentalism is practiced more in the political and ethical arenas than in the realm of science.

Sustainability Is the Central Theme of This Book

Sustainability is the ability of the earth's various natural systems and human cultural systems and economies to survive and adapt to changing environmental conditions indefinitely. It is the central theme of this book, and its components provide the subthemes of this book.

A critical component of sustainability is **natural capital**—the natural resources and natural services that keep us and other forms of life alive and support our economies (Figure 1-3). **Natural resources** are materials and energy in nature that are essential or useful to humans. These resources are often classified as *renewable* (such as air, water, soil, plants, and wind) or *nonrenewable* (such as copper, oil, and coal). **Natural services** are functions of nature, such as purification of air and water, which support life and human economies. Ecosystems provide us with these essential services at no cost.

One vital natural service is **nutrient cycling,** the circulation of chemicals necessary for life, from the environment (mostly from soil and water) through organisms and back to the environment (Figure 1-4). For example, *topsoil,* the upper layer of the earth's crust, provides the nutrients that support the plants, animals, and microorganisms that live on land; when they die and decay, they resupply the soil with these nutrients. Without this service, life as we know it could not exist.

Natural capital is supported by **solar capital:** energy from the sun (Figure 1-3). Take away solar energy, and all natural capital would collapse. Solar energy warms the planet and supports *photosynthesis*—a complex chemical process that plants use to provide food for themselves and for us and most other animals. This direct input of solar energy also produces indirect forms of renewable solar energy such as wind, flowing water, and biofuels made from plants and plant residues. Thus, our lives and economies depend on energy from the sun (*solar capital*) and natural resources and natural services (*natural capital*) provided by the earth (**Concept 1-1A**).

A second component of sustainability—and another sub-theme of this text—is to recognize that many human activities can *degrade natural capital* by using normally renewable resources faster than nature can renew them. For example, in parts of the world, we are clearing mature forests much faster than nature can replenish them. We are also harvesting many species of ocean fish faster than they can replenish themselves.

This leads us to a third component of sustainability. Environmental scientists search for *solutions* to problems such as the degradation of natural capital. However, their work is limited to finding the scientific solutions, while the political solutions are left to political processes. For example, scientific solutions might be to stop chopping down biologically diverse, mature forests, and to harvest fish no faster than they can replenish themselves. But implementing such solutions could require government laws and regulations.

The search for solutions often involves conflicts. When scientists argue for protecting a diverse natural forest to help prevent the premature extinction of various life forms, for example, the timber company that had planned to harvest trees in that forest might protest. Dealing with such conflicts often involves making *trade-offs,* or compromises—a fourth component of sustainability. In the case of the timber company, it might be persuaded to plant a tree farm in an area that had

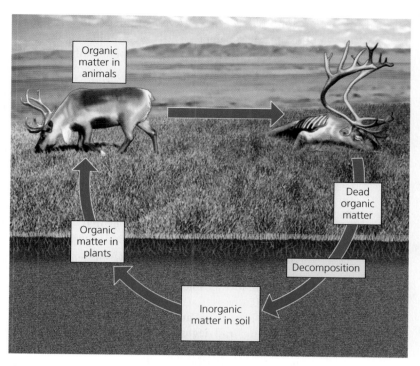

Figure 1-4 *Nutrient cycling:* an important natural service that recycles chemicals needed by organisms from the environment (mostly from soil and water) through organisms and back to the environment.

already been cleared or degraded, in exchange for preserving the natural forest.

Any shift toward environmental sustainability should be based on scientific concepts and results that are widely accepted by experts in a particular field, as discussed in more detail in Chapter 2. In making such a shift, *individuals matter*—another subtheme of this book. Some people are good at thinking of new ideas and inventing innovative technologies or solutions. Others are good at putting political pressure on government officials and business leaders, acting either alone or in groups to implement those solutions. In any case, a shift toward sustainability for a society ultimately depends on the actions of individuals within that society.

Environmentally Sustainable Societies Protect Natural Capital and Live Off Its Income

The ultimate goal is an **environmentally sustainable society**—one that meets the current and future basic resource needs of its people in a just and equitable manner without compromising the ability of future generations to meet their basic needs.

Imagine you win $1 million in a lottery. If you invest this money and earn 10% interest per year, you will have a sustainable income of $100,000 a year that you can live off of indefinitely, while allowing interest to accumulate on what is left after each withdrawal, without depleting your capital. However, if you spend

$200,000 per year, even while allowing interest to accumulate, your capital of $1 million will be gone early in the seventh year. Even if you spend only $110,000 per year and still allow the interest to accumulate, you will be bankrupt early in the eighteenth year.

The lesson here is an old one: *Protect your capital and live off the income it provides.* Deplete or waste your capital, and you will move from a sustainable to an unsustainable lifestyle.

The same lesson applies to our use of the earth's natural capital—the global trust fund that nature provides for us. *Living sustainably* means living off **natural income,** the renewable resources such as plants, animals, and soil provided by natural capital. This means preserving the earth's natural capital, which supplies this income, while providing the human population with adequate and equitable access to this natural income for the foreseeable future (**Concept 1-1B**).

The bad news is that, according to a growing body of scientific evidence, we are living unsustainably by wasting, depleting, and degrading the earth's natural capital at an exponentially accelerating rate (**Core Case Study**).* In 2005, the United Nations (U.N.) released its *Millennium Ecosystem Assessment.*

*The opening Core Case Study is used as a theme to connect and integrate much of the material in each chapter. The logo indicates these connections

According to this 4-year study by 1,360 experts from 95 countries, human activities are degrading or overusing about 62% of the earth's natural services (Figure 1-3). In its summary statement, the report warned that "human activity is putting such a strain on the natural functions of Earth that the ability of the planet's ecosystems to sustain future generations can no longer be taken for granted." The good news is that the report suggests we have the knowledge and tools to conserve the planet's natural capital, and it describes commonsense strategies for doing this.

RESEARCH FRONTIER*

A crash program to gain better and more comprehensive information about the health of the world's life-support systems. See **academic.cengage.com/biology/miller**.

HOW WOULD YOU VOTE?* ☑

Do you believe that the society you live in is on an unsustainable path? Cast your vote online at **academic .cengage.com/biology/miller**.

*Environmental science is a developing field with many exciting research frontiers that are identified throughout this book.

**To cast your vote, go the website for this book and then to the appropriate chapter (in this case, Chapter 1). In most cases, you will be able to compare how you voted with others using this book.

1-2 How Can Environmentally Sustainable Societies Grow Economically?

▶ **CONCEPT 1-2** Societies can become more environmentally sustainable through economic development dedicated to improving the quality of life for everyone without degrading the earth's life support systems.

There Is a Wide Economic Gap between Rich and Poor Countries

Economic growth is an increase in a nation's output of goods and services. It is usually measured by the percentage of change in a country's **gross domestic product (GDP):** the annual market value of all goods and services produced by all firms and organizations, foreign and domestic, operating within a country. Changes in a country's economic growth per person are measured by **per capita GDP:** the GDP divided by the total population at midyear.

The value of any country's currency changes when it is used in other countries. Because of such differences, a basic unit of currency in one country can buy more of a particular thing than the basic unit of currency of another country can buy. Consumers in the

first country are said to have more *purchasing power* than consumers in the second country have. To help compare countries, economists use a tool called *purchasing power parity (PPP)*. By combining per capita GDP and PPP, for any given country, they arrive at a **per capita GDP PPP**—a measure of the amount of goods and services that a country's average citizen could buy in the United States.

While economic growth provides people with more goods and services, **economic development** has the goal of using economic growth to improve living standards. The United Nations classifies the world's countries as economically developed or developing based primarily on their degree of industrialization and their per capita GDP PPP. The **developed countries** (with 1.2 billion people) include the United States, Canada, Japan, Australia, New Zealand, and most countries of

Europe. Most are highly industrialized and have a high per capita GDP PPP.

All other nations (with 5.5 billion people) are classified as **developing countries,** most of them in Africa, Asia, and Latin America. Some are *middle-income, moderately developed countries* such as China, India, Brazil, Turkey, Thailand, and Mexico. Others are *low-income, least developed countries* where per capita GDP PPP is steadily declining. These 49 countries with 11% of the world's population include Angola, Congo, Belarus, Nigeria, Nicaragua, and Jordan. Figure 2 on p. S10 in Supplement 3 is a map of high-, upper middle-, lower middle-, and low-income countries.

Figure 1-5 compares some key characteristics of developed and developing countries. About 97% of the projected increase in the world's population between 2008 and 2050 is expected to take place in developing countries, which are least equipped to handle such large population increases.

We live in a world of haves and have-nots. Despite a 40-fold increase in economic growth since 1900, *more than half of the people in the world live in extreme poverty and try to survive on a daily income of less than $2. And one of every six people, classified as desperately poor, struggle to survive on less than $1 a day.* (All dollar figures are in U.S. dollars.) (Figure 1-6)

Figure 1-6 *Extreme poverty:* boy searching for items to sell in an open dump in Rio de Janeiro, Brazil. Many children of poor families who live in makeshift shantytowns in or near such dumps often scavenge all day for food and other items to help their families survive. This means that they cannot go to school.

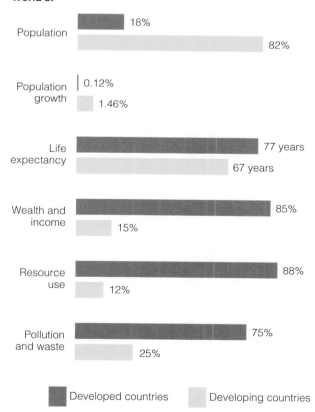

Percentage of World's:

Population	18% / 82%
Population growth	0.12% / 1.46%
Life expectancy	77 years / 67 years
Wealth and income	85% / 15%
Resource use	88% / 12%
Pollution and waste	75% / 25%

Developed countries Developing countries

Figure 1-5 *Global outlook:* comparison of developed and developing countries, 2008. (Data from the United Nations and the World Bank)

Some economists call for continuing conventional economic growth, which has helped to increase food supplies, allowed people to live longer, and stimulated mass production of an array of useful goods and services for many people. They also see such growth as a cure for poverty, maintaining that some of the resulting increase in wealth trickles down to countries and people near the bottom of the economic ladder.

Other economists call for us to put much greater emphasis on **environmentally sustainable economic development.** This involves using political and economic systems to *discourage* environmentally harmful and unsustainable forms of economic growth that degrade natural capital, and to *encourage* environmentally beneficial and sustainable forms of economic development that help sustain natural capital (**Concept 1-2**).

THINKING ABOUT
Economic Growth and Sustainability

Is exponential economic growth incompatible with environmental sustainability? What are three types of goods whose exponential growth would promote environmental sustainability?

1-3 How Are Our Ecological Footprints Affecting the Earth?

> **CONCEPT 1-3** As our ecological footprints grow, we are depleting and degrading more of the earth's natural capital.

Some Resources Are Renewable

From a human standpoint, a **resource** is anything obtained from the environment to meet our needs and wants. **Conservation** is the management of natural resources with the goal of minimizing resource waste and sustaining resource supplies for current and future generations.

Some resources, such as solar energy, fresh air, wind, fresh surface water, fertile soil, and wild edible plants, are directly available for use. Other resources such as petroleum, iron, water found underground, and cultivated crops, are not directly available. They become useful to us only with some effort and technological ingenuity. For example, petroleum was a mysterious fluid until we learned how to find, extract, and convert (refine) it into gasoline, heating oil, and other products that could be sold.

Solar energy is called a **perpetual resource** because it is renewed continuously and is expected to last at least 6 billion years as the sun completes its life cycle.

On a human time scale, a **renewable resource** can be replenished fairly quickly (from hours to hundreds of years) through natural processes as long as it is not used up faster than it is renewed. Examples include forests, grasslands, fisheries, freshwater, fresh air, and fertile soil.

The highest rate at which a renewable resource can be used *indefinitely* without reducing its available supply is called its **sustainable yield.** When we exceed a renewable resource's natural replacement rate, the available supply begins to shrink, a process known as **environmental degradation,** as shown in Figure 1-7.

We Can Overexploit Commonly Shared Renewable Resources: The Tragedy of the Commons

There are three types of property or resource rights. One is *private property* where individuals or firms own

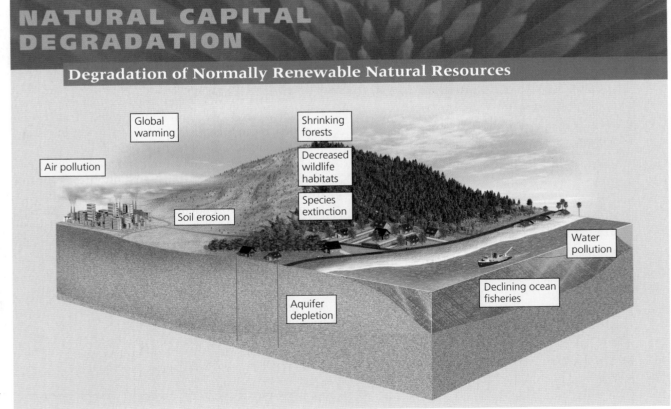

NATURAL CAPITAL DEGRADATION

Degradation of Normally Renewable Natural Resources

- Global warming
- Air pollution
- Shrinking forests
- Decreased wildlife habitats
- Species extinction
- Soil erosion
- Aquifer depletion
- Water pollution
- Declining ocean fisheries

Figure 1-7 Degradation of normally renewable natural resources and services in parts of the world, mostly as a result of rising population and resource use per person.

the rights to land, minerals, or other resources. Another is *common property* where the rights to certain resources are held by large groups of individuals. For example, roughly one-third of the land in the United States is owned jointly by all U.S. citizens and held and managed for them by the government. Another example is land that belongs to a whole village and can be used by anyone for activities such as grazing cows or sheep.

A third category consists of *open access renewable resources,* owned by no one and available for use by anyone at little or no charge. Examples of such shared renewable resources include clean air, underground water supplies, and the open ocean and its fish.

Many common property and open access renewable resources have been degraded. In 1968, biologist Garrett Hardin (1915–2003) called such degradation the *tragedy of the commons.* It occurs because each user of a shared common resource or open-access resource reasons, "If I do not use this resource, someone else will. The little bit that I use or pollute is not enough to matter, and anyway, it's a renewable resource."

When the number of users is small, this logic works. Eventually, however, the cumulative effect of many people trying to exploit a shared resource can exhaust or ruin it. Then no one can benefit from it. Such resource degradation results from the push to satisfy the short-term needs and wants of a growing number of people. It threatens our ability to ensure the long-term economic and environmental sustainability of open-access resources such as clean air or an open-ocean fishery.

One solution is to *use shared resources at rates well below their estimated sustainable yields* by reducing use of the resources, regulating access to the resources, or doing both. For example, the most common approach is for governments to establish laws and regulations limiting the annual harvests of various types of ocean fish that are being harvested at unsustainable levels in their coastal waters. Another approach is for nations to enter into agreements that regulate access to open-access renewable resources such as the fish in the open ocean.

Another solution is to *convert open-access resources to private ownership.* The reasoning is that if you own something, you are more likely to protect your investment. That sounds good, but this approach is not practical for global open-access resources—such as the atmosphere, the open ocean, and most wildlife species—that cannot be divided up and converted to private property.

THINKING ABOUT
Degradation of Commonly Shared Resources
CORE CASE STUDY

How is the degradation of shared renewable resources related to exponential growth (**Core Case Study**) of the world's population and economies? What are three examples of how most of us contribute to this environmental degradation?

Some Resources Are Not Renewable

Nonrenewable resources exist in a fixed quantity, or *stock,* in the earth's crust. On a time scale of millions to billions of years, geological processes can renew such resources. But on the much shorter human time scale of hundreds to thousands of years, these resources can be depleted much faster than they are formed. Such exhaustible resources include *energy resources* (such as coal and oil), *metallic mineral resources* (such as copper and aluminum), and *nonmetallic mineral resources* (such as salt and sand).

As such resources are depleted, human ingenuity can often find substitutes. For example, during this century, a mix of renewable energy resources such as wind, the sun, flowing water, and the heat in the earth's interior could reduce our dependence on nonrenewable fossil fuels such as oil and coal. Also, various types of plastics and composite materials can replace certain metals. But sometimes there is no acceptable or affordable substitute.

Some nonrenewable resources, such as copper and aluminum, can be recycled or reused to extend supplies. **Reuse** is using a resource over and over in the same form. For example, glass bottles can be collected, washed, and refilled many times (Figure 1-8). **Recycling** involves collecting waste materials and processing them into new materials. For example, discarded aluminum cans can be crushed and melted to make new

Mark Edwards/Peter Arnold, Inc.

Figure 1-8 *Reuse:* This child and his family in Katmandu, Nepal, collect beer bottles and sell them for cash to a brewery where they will be reused.

aluminum cans or other aluminum products. But energy resources such as oil and coal cannot be recycled. Once burned, their energy is no longer available to us.

Recycling nonrenewable metallic resources takes much less energy, water, and other resources and produces much less pollution and environmental degradation than exploiting virgin metallic resources. Reusing such resources takes even less energy and other resources and produces less pollution and environmental degradation than recycling does.

Our Ecological Footprints Are Growing

Many people in developing countries struggle to survive. Their individual use of resources and the resulting environmental impact is low and is devoted mostly to meeting their basic needs (Figure 1-9, top). By contrast, many individuals in more affluent nations consume large amounts of resources way beyond their basic needs (Figure 1-9, bottom).

Supplying people with resources and dealing with the resulting wastes and pollution can have a large environmental impact. We can think of it as an **ecological footprint**—the amount of biologically productive land and water needed to supply the people in a particular country or area with resources and to absorb and recycle the wastes and pollution produced by such resource use. The **per capita ecological footprint** is the average ecological footprint of an individual in a given country or area.

If a country's, or the world's, total ecological footprint is larger than its *biological capacity* to replenish its renewable resources and absorb the resulting waste products and pollution, it is said to have an *ecological deficit*. The World Wildlife Fund (WWF) and the Global Footprint Network estimated that in 2003 (the latest data available) humanity's global ecological footprint exceeded the earth's *biological capacity* by about 25% (Figure 1-10, right). That figure was about 88% in the world's high-income countries, with the United States having the world's largest total ecological footprint. If the current exponential growth in the use of renewable resources continues, the Global Footprint Network estimates that by 2050 humanity will be trying to use twice as many renewable resources as the planet can supply (Figure 1-10, bottom) (**Concept 1-3**). See Figure 3 on p. S24 and Figure 5 on pp. S27 in Supplement 4 for maps of the human ecological footprints for the world and the United States, and Figure 4 on p. S26 for a map of countries that are ecological debtors and those that are ecological creditors.

The per capita ecological footprint is an estimate of how much of the earth's renewable resources an individual consumes. After the oil-rich United Arab Emirates, the United States has the world's second largest per capita ecological footprint. In 2003 (the latest data available), its per capita ecological footprint was about 4.5 times the average global footprint per person, 6 times larger than China's per capita footprint, and 12 times the average per capita footprint in the world's low-income countries.

According to William Rees and Mathis Wackernagel, the developers of the ecological footprint concept, it would take the land area of about *five more planet earths* for the rest of the world to reach current U.S. levels of consumption with existing technology. Put another way, if everyone consumed as much as the average American does today, the earth's natural capital could support only about 1.3 billion people—not

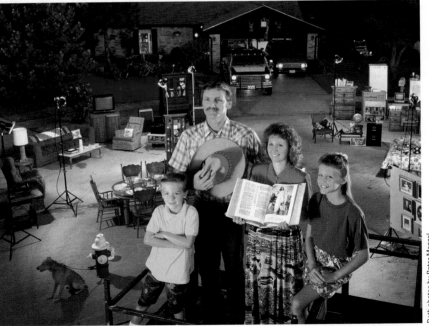

Figure 1-9 *Consumption of natural resources.* The top photo shows a family of five subsistence farmers with all their possessions. They live in the village of Shingkhey, Bhutan, in the Himalaya Mountains, which are sandwiched between China and India in South Asia. The bottom photo shows a typical U.S. family of four living in Pearland, Texas, with their possessions .

Both photos by Peter Menzel

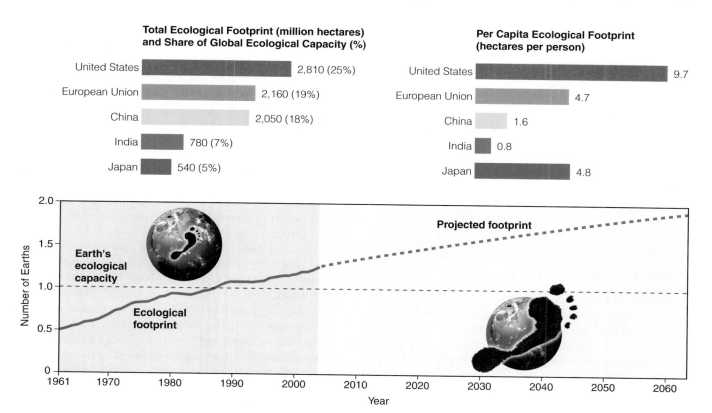

Total Ecological Footprint (million hectares) and Share of Global Ecological Capacity (%)

United States — 2,810 (25%)
European Union — 2,160 (19%)
China — 2,050 (18%)
India — 780 (7%)
Japan — 540 (5%)

Per Capita Ecological Footprint (hectares per person)

United States — 9.7
European Union — 4.7
China — 1.6
India — 0.8
Japan — 4.8

Figure 1-10 Natural capital use and degradation: total and per capita ecological footprints of selected countries (top). In 2003, humanity's total or global ecological footprint was about 25% higher than the earth's ecological capacity (bottom) and is projected to be twice the planet's ecological capacity by 2050. **Question:** If we are living beyond the earth's biological capacity, why do you think the human population and per capita resource consumption are still growing exponentially? (Data from Worldwide Fund for Nature, Global Footprint Network)

today's 6.7 billion. In other words, we are living unsustainably by depleting and degrading some of the earth's irreplaceable natural capital and the natural renewable income it provides as our ecological footprints grow and spread across the earth's surface (**Concept 1-3**). For more on this subject, see the Guest Essay by Michael Cain at CengageNOW™. See the Case Study that follows about the growing ecological footprint of China.

> **THINKING ABOUT**
> **Your Ecological Footprint**
>
> Estimate your own ecological footprint by visiting the website **www.myfootprint.org/**. What are three things you could do to reduce your ecological footprint?

■ CASE STUDY
China's New Affluent Consumers

More than a billion super-affluent consumers in developed countries are putting immense pressure on the earth's natural capital. Another billion consumers are attaining middle-class, affluent lifestyles in rapidly developing countries such as China, India, Brazil, South Korea, and Mexico. The 700 million middle-class consumers in China and India number more than twice the size of the entire U.S. population, and the number is growing rapidly. In 2006, the World Bank projected that by 2030 the number of middle-class consumers

living in today's developing nations will reach 1.2 billion—about four times the current U.S. population.

China is now the world's leading consumer of wheat, rice, meat, coal, fertilizers, steel, and cement, and it is the second largest consumer of oil after the United States. China leads the world in consumption of goods such as television sets, cell phones, refrigerators, and soon, personal computers. On the other hand, after 20 years of industrialization, two-thirds of the world's most polluted cities are in China; this pollution threatens the health of urban dwellers. By 2020, China is projected to be the world's largest producer and consumer of cars and to have the world's leading economy in terms of GDP PPP.

Suppose that China's economy continues growing exponentially at a rapid rate and its projected population size reaches 1.5 billion by 2033. Then China will need two-thirds of the world's current grain harvest, twice the world's current paper consumption, and more than the current global production of oil.

According to environmental policy expert Lester R. Brown:

> *The western economic model—the fossil fuel–based, automobile-centered, throwaway economy—is not going to work for China. Nor will it work for India, which by 2033 is projected to have a population even larger than China's, or for the other 3 billion people in developing countries who are also dreaming the "American dream."*

For more details on the growing ecological footprint of China, see the Guest Essay by Norman Myers for this chapter at CengageNOW.

┌─ THINKING ABOUT
China and Sustainability

What are three things China could do to shift toward more sustainable consumption? What are three things the United States, Japan, and the European Union could do to shift toward more sustainable consumption?

Cultural Changes Have Increased Our Ecological Footprints

Culture is the whole of a society's knowledge, beliefs, technology, and practices, and human cultural changes have had profound effects on the earth.

Evidence of organisms from the past and studies of ancient cultures suggest that the current form of our species, *Homo sapiens sapiens,* has walked the earth for perhaps 90,000–195,000 years—less than an eye-blink in the 3.56 billion years of life on the earth. Until about 12,000 years ago, we were mostly *hunter–gatherers* who obtained food by hunting wild animals or scavenging their remains and gathering wild plants. Early hunter–gathers lived in small groups and moved as needed to find enough food for survival.

Since then, three major cultural changes have occurred. *First* was the *agricultural revolution,* which began 10,000–12,000 years ago when humans learned how to grow and breed plants and animals for food, clothing, and other purposes. *Second* was the *industrial–medical revolution,* beginning about 275 years ago when people invented machines for the large-scale production of goods in factories. This involved learning how to get energy from fossil fuels, such as coal and oil, and how to grow large quantities of food in an efficient manner. *Finally,* the *information–globalization revolution* began about 50 years ago, when we developed new technologies for gaining rapid access to much more information and resources on a global scale.

Each of these cultural changes gave us more energy and new technologies with which to alter and control more of the planet to meet our basic needs and increasing wants. They also allowed expansion of the human population, mostly because of increased food supplies and longer life spans. In addition, they each resulted in greater resource use, pollution, and environmental degradation as our ecological footprints expanded (Figure 1-10) and allowed us to dominate the planet.

Many environmental scientists and other analysts call for us to bring about a new **environmental,** or **sustainability, revolution** during this century. It would involve learning how to reduce our ecological footprints and live more sustainability.

For more background and details on environmental history, see Supplement 5 (p. S31).

1-4 What Is Pollution and What Can We Do about It?

▶ **CONCEPT 1-4** Preventing pollution is more effective and less costly than cleaning up pollution.

Pollution Comes from a Number of Sources

Pollution is any in the environment that is harmful to the health, survival, or activities of humans or other organisms. Pollutants can enter the environment naturally, such as from volcanic eruptions, or through human activities, such as burning coal and gasoline and discharging chemicals into rivers and the ocean.

The pollutants we produce come from two types of sources. **Point sources** are single, identifiable sources. Examples are the smokestack of a coal-burning power or industrial plant (Figure 1-11), the drainpipe of a factory, and the exhaust pipe of an automobile. **Nonpoint sources** are dispersed and often difficult to identify. Examples are pesticides blown from the land into the air and the runoff of fertilizers and pesticides from farmlands, lawns, gardens, and golf courses into streams and lakes. It is much easier and cheaper to identify and control or prevent pollution from point sources than from widely dispersed nonpoint sources.

There are two main types of pollutants. **Biodegradable pollutants** are harmful materials that can be broken down by natural processes. Examples are human sewage and newspapers. **Nondegradable pollutants** are harmful materials that natural processes cannot break down. Examples are toxic chemical elements such as lead, mercury, and arsenic (see Supplement 6, p. S39, for an introduction to basic chemistry).

Pollutants can have three types of unwanted effects. *First,* they can disrupt or degrade life-support systems for humans and other species. *Second,* they can damage wildlife, human health, and property. *Third,* they can

create nuisances such as noise and unpleasant smells, tastes, and sights.

We Can Clean Up Pollution or Prevent It

Consider the smoke produced by a steel mill. We can try to deal with this problem by asking two entirely different questions. One question is "how can we clean up the smoke?" The other is "how can we avoid producing the smoke in the first place?"

The answers to these questions involve two different ways of dealing with pollution. One is **pollution cleanup,** or **output pollution control,** which involves cleaning up or diluting pollutants after they have been produced. The other is **pollution prevention,** or **input pollution control,** which reduces or eliminates the production of pollutants.

Environmental scientists have identified three problems with relying primarily on pollution cleanup. *First,* it is only a temporary bandage as long as population and consumption levels grow without corresponding improvements in pollution control technology. For example, adding catalytic converters to car exhaust systems has reduced some forms of air pollution. At the same time, increases in the number of cars and the total distance each car travels have reduced the effectiveness of this cleanup approach.

Second, cleanup often removes a pollutant from one part of the environment only to cause pollution in another. For example, we can collect garbage, but the garbage is then *burned* (perhaps causing air pollution and leaving toxic ash that must be put somewhere), *dumped*

Ray Pfortner/Peter Arnold, Inc.

Figure 1-11 *Point-source air pollution* from a pulp mill in New York State (USA).

on the land (perhaps causing water pollution through runoff or seepage into groundwater), or *buried* (perhaps causing soil and groundwater pollution).

Third, once pollutants become dispersed into the environment at harmful levels, it usually costs too much or is impossible to reduce them to acceptable levels.

Pollution prevention (front-of-the-pipe) and pollution cleanup (end-of-the-pipe) solutions are both needed. But environmental scientists, some economists, and some major companies urge us to put more emphasis on prevention because it works better and in the long run is cheaper than cleanup (**Concept 1-4**).

1-5 Why Do We Have Environmental Problems?

▶ **CONCEPT 1-5A** Major causes of environmental problems are population growth, wasteful and unsustainable resource use, poverty, exclusion of environmental costs of resource use from the market prices of goods and services, and attempts to manage nature with insufficient knowledge.

▶ **CONCEPT 1-5B** People with different environmental worldviews often disagree about the seriousness of environmental problems and what we should do about them.

Experts Have Identified Five Basic Causes of Environmental Problems

As we run more and more of the earth's natural resources through the global economy, in many parts of the world, forests are shrinking, deserts are expanding, soils are eroding, and agricultural lands are deteriorat-

ing. In addition, the lower atmosphere is warming, glaciers are melting, sea levels are rising, and storms are becoming more destructive. And in many areas, water tables are falling, rivers are running dry, fisheries are collapsing, coral reefs are disappearing, and various species are becoming extinct.

According to a number of environmental and social scientists, the major causes of these and other

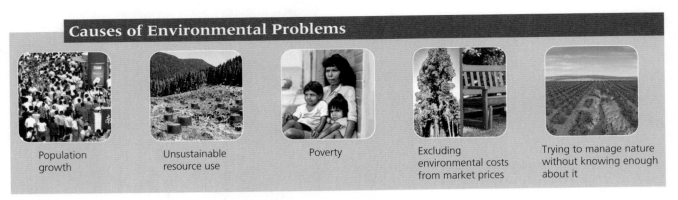

Causes of Environmental Problems

Population growth

Unsustainable resource use

Poverty

Excluding environmental costs from market prices

Trying to manage nature without knowing enough about it

Figure 1-12 Environmental and social scientists have identified five basic causes of the environmental problems we face (**Concept 1-5A**). **Question:** What are three ways in which your lifestyle contributes to these causes?

environmental problems are population growth, wasteful and unsustainable resource use, poverty, failure to include the harmful environmental costs of goods and services in their market prices, and insufficient knowledge of how nature works (Figure 1-12 and **Concept 1-5A**).

We have discussed the exponential growth of the human population (**Core Case Study**), and here we will examine other major causes of environmental problems in more detail.

Poverty Has Harmful Environmental and Health Effects

Poverty occurs when people are unable to meet their basic needs for adequate food, water, shelter, health, and education. Poverty has a number of harmful environmental and health effects (Figure 1-13). The daily lives of half of the world's people, who are trying to live on the equivalent of less than $2 a day, are focused on getting enough food, water, and cooking and heating fuel to survive. Desperate for short-term survival, some of these people deplete and degrade forests, soil, grasslands, fisheries, and wildlife, at an ever-increasing rate. They do not have the luxury of worrying about long-term environmental quality or sustainability.

Poverty affects population growth. To many poor people, having more children is a matter of survival. Their children help them gather fuel (mostly wood and animal dung), haul drinking water, and tend crops and livestock. Their children also help to care for them in their old age (which is their 40s or 50s in the poorest countries) because they do not have social security, health care, and retirement funds.

While poverty can increase some types of environmental degradation, the reverse is also true. Pollution and environmental degradation have a severe impact on the poor and can increase poverty. Consequently, many of the world's desperately poor people die prematurely from several preventable health problems.

One such problem is *malnutrition* from a lack of protein and other nutrients needed for good health

(Figure 1-14). The resulting weakened condition can increase the chances of death from normally nonfatal illnesses, such as diarrhea and measles. A second problem is limited access to adequate sanitation facilities and clean drinking water. More than 2.6 billion people (38% of the world's population) have no decent bathroom facilities. They are forced to use fields, backyards, ditches, and streams. As a result, more than 1 billion people—one of every seven—get water for drinking, washing, and cooking from sources polluted by human and animal feces. A third problem is severe respiratory disease and premature death from inhaling indoor air pollutants produced by burning wood or coal in open fires or in poorly vented stoves for heat and cooking.

According to the World Health Organization, these factors cause premature death for at least 7 million people each year. *This amounts to about 19,200 premature deaths per day, equivalent to 96 fully loaded 200-passenger airliners crashing every day with no survivors!* Two-thirds of those dying are children younger than age 5. The news media rarely cover this ongoing human tragedy.

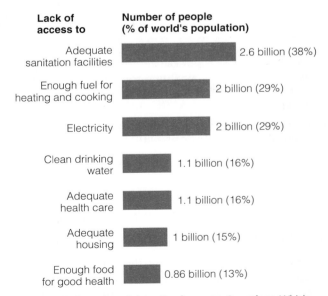

Lack of access to	Number of people (% of world's population)
Adequate sanitation facilities	2.6 billion (38%)
Enough fuel for heating and cooking	2 billion (29%)
Electricity	2 billion (29%)
Clean drinking water	1.1 billion (16%)
Adequate health care	1.1 billion (16%)
Adequate housing	1 billion (15%)
Enough food for good health	0.86 billion (13%)

Figure 1-13 Some harmful results of poverty. **Question:** Which two of these effects do you think are the most harmful? Why? (Data from United Nations, World Bank, and World Health Organization)

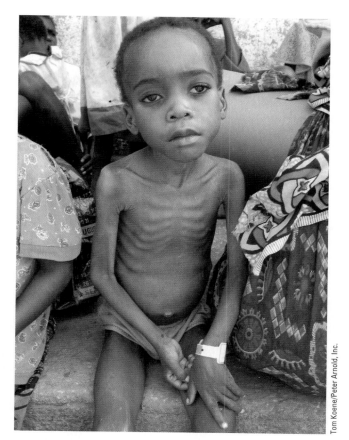

Figure 1-14 *Global Outlook:* in developing countries, one of every three children under age 5, such as this child in Lunda, Angola, suffers from severe malnutrition caused by a lack of calories and protein. According to the World Health Organization, each day at least 13,700 children under age 5 die prematurely from malnutrition and infectious diseases, most from drinking contaminated water and being weakened by malnutrition.

The *great news* is that we have the means to solve the environmental, health, and social problems resulting from poverty within 20–30 years if we can find the political and ethical will to act.

Affluence Has Harmful and Beneficial Environmental Effects

The harmful environmental effects of poverty are serious, but those of affluence are much worse (Figure 1-10, top). The lifestyles of many affluent consumers in developed countries and in rapidly developing countries such as India and China (p. 15) are built upon high levels of consumption and unnecessary waste of resources. Such affluence is based mostly on the assumption—fueled by mass advertising—that buying more and more things will bring happiness.

This type of affluence has an enormous harmful environmental impact. It takes about 27 tractor-trailer loads of resources per year to support one American, or 7.9 billion truckloads per year to support the entire U.S. population. Stretched end-to-end, each year these trucks would reach beyond the sun!

While the United States has far fewer people than India, the average American consumes about 30 times as much as the average citizen of India and 100 times as much as the average person in the world's poorest countries. As a result, the average environmental impact, or ecological footprint per person, in the United States is much larger than the average impact per person in developing countries (Figure 1-10, top).

On the other hand, affluence can lead people to become more concerned about environmental quality. It also provides money for developing technologies to reduce pollution, environmental degradation, and resource waste.

In the United States and most other affluent countries, the air is cleaner, drinking water is purer, and most rivers and lakes are cleaner than they were in the 1970s. In addition, the food supply is more abundant and safer, the incidence of life-threatening infectious diseases has been greatly reduced, lifespans are longer, and some endangered species are being rescued from premature extinction.

Affluence financed these improvements in environmental quality, based on greatly increased scientific research and technological advances. And education spurred citizens insist that businesses and elected officials improve environmental quality. Affluence and education have also helped to reduce population growth in most developed countries. However, a downside to wealth is that it allows the affluent to obtain the resources they need from almost anywhere in the world without seeing the harmful environmental impacts of their high-consumption life styles.

> **THINKING ABOUT**
> **The Poor, the Affluent, and Exponentially Increasing Population Growth**
>
> CORE CASE STUDY
>
> Some see rapid population growth of the poor in developing countries as the primary cause of our environmental problems. Others say that the much higher resource use per person in developed countries is a more important factor. Which factor do you think is more important? Why?

Prices Do Not Include the Value of Natural Capital

When companies use resources to create goods and services for consumers, they are generally not required to pay the environmental costs of such resource use. For example, fishing companies pay the costs of catching fish but do not pay for the depletion of fish stocks. Timber companies pay for clear-cutting forests but not for the resulting environmental degradation and loss of wildlife habitat. The primary goal of these companies is to maximize their profits, so they do not voluntarily pay these harmful environmental costs or even try to assess them, unless required to do so by government laws or regulations.

As a result, the prices of goods and services do not include their harmful environmental costs. Thus, consumers are generally not aware of them and have no effective way to evaluate the resulting harmful effects on the earth's life-support systems and on their own health.

Another problem is that governments give companies tax breaks and payments called *subsidies* to assist them in using resources to run their businesses. This helps to create jobs and stimulate economies, but it can also result in degradation of natural capital, again because the value of the natural capital is not included in the market prices of goods and services. We explore this problem and some possible solutions in later chapters.

People Have Different Views about Environmental Problems and Their Solutions

Differing views about the seriousness of our environmental problems and what we should do about them arise mostly out of differing environmental worldviews. Your **environmental worldview** is a set of assumptions and values reflecting how you think the world works and what you think your role in the world should be. This involves **environmental ethics,** which are our beliefs about what is right and wrong with how we treat the environment. Here are some important *ethical questions* relating to the environment:

- Why should we care about the environment?

- Are we the most important beings on the planet or are we just one of the earth's millions of different forms of life?

- Do we have an obligation to see that our activities do not cause the premature extinction of other species? Should we try to protect all species or only some? How do we decide which species to protect?

- Do we have an ethical obligation to pass on to future generations the extraordinary natural world in a condition at least as good as what we inherited?

- Should every person be entitled to equal protection from environmental hazards regardless of race, gender, age, national origin, income, social class, or any other factor?

THINKING ABOUT
Our Responsibilities

How would you answer each of the questions above? Compare your answers with those of your classmates. Record your answers and, at the end of this course, return to these questions to see if your answers have changed.

People with widely differing environmental worldviews can take the same data, be logically consistent, and arrive at quite different conclusions because they start with different assumptions and moral, ethical, or religious beliefs (**Concept 1-5B**). Environmental worldviews are discussed in detail in Chapter 25, but here is a brief introduction.

The **planetary management worldview** holds that we are separate from nature, that nature exists mainly to meet our needs and increasing wants, and that we can use our ingenuity and technology to manage the earth's life-support systems, mostly for our benefit, indefinitely.

The **stewardship worldview** holds that we can and should manage the earth for our benefit, but that we have an ethical responsibility to be caring and responsible managers, or *stewards,* of the earth. It says we should encourage environmentally beneficial forms of economic growth and development and discourage environmentally harmful forms.

The **environmental wisdom worldview** holds that we are part of, and totally dependent on, nature and that nature exists for all species, not just for us. It also calls for encouraging earth-sustaining forms of economic growth and development and discouraging earth-degrading forms. According to this view, our success depends on learning how life on earth sustains itself and integrating such *environmental wisdom* into the ways we think and act.

Many of the ideas for the stewardship and environmental wisdom worldviews are derived from the writings of Aldo Leopold (Individuals Matter, p. 22).

We Can Learn to Make Informed Environmental Decisions

The first step for dealing with an environmental problem is to carry out scientific research on the nature of the problem and to evaluate possible solutions to the problem. Once this is done, other factors involving the social sciences and the humanities (Table 1-1) must be used to evaluate each proposed solution. This involves considering various *human values.* What are its projected short-term and long-term beneficial and harmful environmental, economic, and health effects? How much will it cost? Is it ethical? Figure 1-15 shows the major steps involved in making an environmental decision.

We Can Work Together to Solve Environmental Problems

Making the shift to more sustainable societies and economies involves building what sociologists call **social capital.** This involves getting people with different views and values to talk and listen to one another, find common ground based on understanding and trust, and work together to solve environmental and other

problems. This means nurturing openness, communication, cooperation, and hope and discouraging close-mindedness, polarization, confrontation, and fear.

Solutions to environmental problems are not black and white, but rather all shades of gray because proponents of all sides of these issues have some legitimate and useful insights. In addition, any proposed solution has short- and long-term advantages and disadvantages that must be evaluated (Figure 1-15). This means that citizens who strive to build social capital also search for *trade-off solutions* to environmental problems—an im-

Figure 1-15 Steps involved in making an environmental decision.

portant theme of this book. They can also try to agree on shared visions of the future and work together to develop strategies for implementing such visions beginning at the local level, as citizens of Chattanooga, Tennessee (USA), have done.

■ CASE STUDY

The Environmental Transformation of Chattanooga, Tennessee

Local officials, business leaders, and citizens have worked together to transform Chattanooga, Tennessee (USA), from a highly polluted city to one of the most sustainable and livable cities in the United States (Figure 1-16).

During the 1960s, U.S. government officials rated Chattanooga as having the dirtiest air in the United States. Its air was so polluted by smoke from its coke ovens and steel mills that people sometimes had to turn on their vehicle headlights in the middle of the day. The Tennessee River, flowing through the city's industrial center, bubbled with toxic waste. People and industries fled the downtown area and left a wasteland of abandoned and polluting factories, boarded-up buildings, high unemployment, and crime.

In 1984, the city decided to get serious about improving its environmental quality. Civic leaders started a *Vision 2000* process with a 20-week series of community meetings in which more than 1,700 citizens from all walks of life gathered to build a consensus about what the city could be at the turn of the century. Citizens identified the city's main problems, set goals, and brainstormed thousands of ideas for solutions.

Chattanooga Area Convention and Visitors Bureau

Figure 1-16 Since 1984, citizens have worked together to make the city of Chattanooga, Tennessee, one of the most sustainable and best places to live in the United States.

INDIVIDUALS MATTER

Aldo Leopold's Environmental Ethics

According to *Aldo Leopold* (Figure 1-A), the role of the human species should be to protect nature, not conquer it.

In 1933, Leopold became a professor at the University of Wisconsin and in 1935, he was one of the founders of the U.S. Wilderness Society. Through his writings and teachings, he became one of the leaders of the *conservation* and *environmental movements* of the 20th century. In doing this, he laid important groundwork for the field of environmental ethics.

Leopold's weekends of planting, hiking, and observing nature at his farm in Wisconsin provided material for his most famous book, *A Sand County Almanac*, published after his death in 1949. Since then, more than 2 million copies of this environmental classic have been sold.

The following quotations from his writings reflect Leopold's *land ethic,* and they form the basis for many of the beliefs of the modern stewardship and environmental wisdom worldviews:

- *All ethics so far evolved rest upon a single premise: that the individual is*

Courtesy of the University of Wisconsin—Madison Archives

Figure 1-A Individuals Matter: *Aldo Leopold* (1887–1948) was a forester, writer, and conservationist. His book *A Sand County Almanac* (published after his death) is considered an environmental classic that inspired the modern environmental and conservation movement.

a member of a community of interdependent parts.

- *To keep every cog and wheel is the first precaution of intelligent tinkering.*
- *That land is a community is the basic concept of ecology, but that land is to be loved and respected is an extension of ethics.*
- *The land ethic changes the role of Homo sapiens from conqueror of the*

land-community to plain member and citizen of it.

- *We abuse land because we regard it as a commodity belonging to us. When we see land as a community to which we belong, we may begin to use it with love and respect.*
- *Anything is right when it tends to preserve the integrity, stability, and beauty of the biotic community. It is wrong when it tends otherwise.*

By 1995, Chattanooga had met most of its original goals. The city had encouraged zero-emission industries to locate there and replaced its diesel buses with a fleet of quiet, zero-emission electric buses, made by a new local firm.

The city also launched an innovative recycling program after environmentally concerned citizens blocked construction of a garbage incinerator that would have emitted harmful air pollutants. These efforts paid off. Since 1989, the levels of the seven major air pollutants in Chattanooga have been lower than those required by federal standards.

Another project involved renovating much of the city's low-income housing and building new low-income rental units. Chattanooga also built the nation's largest freshwater aquarium, which became the centerpiece for downtown renewal. The city developed a riverfront park along both banks of the Tennessee River running through downtown. The park draws more than 1 million visitors per year. As property values and living conditions have improved, people and businesses have moved back downtown.

In 1993, the community began the process again in *Revision 2000*. Goals included transforming an abandoned and blighted area in South Chattanooga into a mixed community of residences, retail stores, and zero-

emission industries where employees can live near their workplaces. Most of these goals have been implemented.

Chattanooga's environmental success story, enacted by people working together to produce a more livable and sustainable city, is a shining example of what other cities can do by building their social capital.

Individuals Matter

Chattanooga's story shows that a key to finding solutions to environmental problems is to recognize that most social change results from individual actions and individuals acting together (using *social capital*) to bring about change through *bottom-up* grassroots action. In other words, *individuals matter*—another important theme of this book. Here are two pieces of good news. First, research by social scientists suggests that it takes only 5–10% of the population of a community, a country, or the world to bring about major social change. Second, such research also shows that significant social change can occur much more quickly than most people think.

Anthropologist Margaret Mead summarized our potential for social change: "Never doubt that a small group of thoughtful, committed citizens can change the world. Indeed, it is the only thing that ever has."

1-6 What Are Four Scientific Principles of Sustainability?

▶ **CONCEPT 1-6** Nature has sustained itself for billions of years by using solar energy, biodiversity, population control, and nutrient cycling—lessons from nature that we can apply to our lifestyles and economies.

Studying Nature Reveals Four Scientific Principles of Sustainability

How can we live more sustainably? According to environmental scientists, we should study how life on the earth has survived and adapted to major changes in environmental conditions for billions of years. We could

make the transition to more sustainable societies by applying these *lessons from nature* to our lifestyles and economies, as summarized below and in Figure 1-17 (**Concept 1-6**).

- *Reliance on Solar Energy:* the sun (solar capital) warms the planet and supports photosynthesis used by plants to provide food for themselves and for us and most other animals.

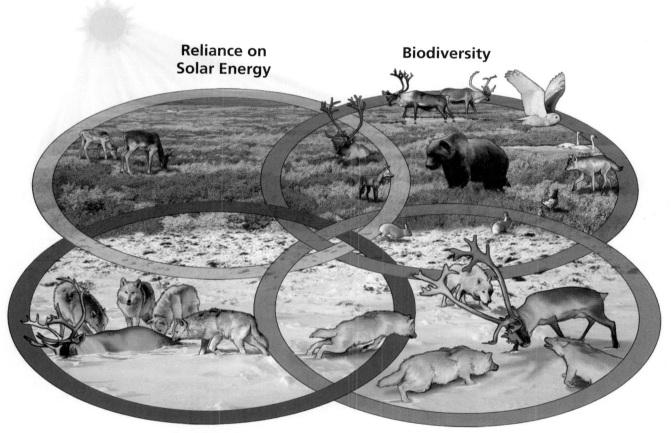

Figure 1-17 Four scientific principles of sustainability: These four interconnected principles of sustainability are derived from learning how nature has sustained a variety of life forms on the earth for about 3.56 billion years. The top left oval shows sunlight stimulating the production of vegetation in the arctic tundra during its brief summer (*solar energy*) and the top right oval shows some of the diversity of species found there during the summer (*biodiversity*). The bottom right oval shows arctic gray wolves stalking a caribou during the long cold winter (*population control*). The bottom left oval shows arctic gray wolves feeding on their kill. This, plus huge numbers of tiny decomposers that convert dead matter to soil nutrients, recycle all materials needed to support the plant growth shown in the top left and right ovals (*nutrient cycling*).

Current Emphasis	Sustainability Emphasis
Pollution cleanup	Pollution prevention
Waste disposal (bury or burn)	Waste prevention
Protecting species	Protecting habitat
Environmental degradation	Environmental restoration
Increasing resource use	Less resource waste
Population growth	Population stabilization
Depleting and degrading natural capital	Protecting natural capital

Figure 1-18 Solutions: some shifts involved in bringing about the *environmental* or *sustainability revolution*. **Question:** Which three of these shifts do you think are most important? Why?

- *Biodiversity* (short for *biological diversity*): the astounding variety of different organisms, the genes they contain, the ecosystems in which they exist, and the natural services they provide have yielded countless ways for life to adapt to changing environmental conditions throughout the earth's history.

- *Population Control:* competition for limited resources among different species places a limit on how much their populations can grow.

- *Nutrient Cycling:* natural processes recycle chemicals that plants and animals need to stay alive and reproduce (Figure 1-4). There is little or no waste in natural systems.

Using the four **scientific principles of sustainability** to guide our lifestyles and economies could help us bring about an *environmental* or *sustainability revolution* during your lifetime (see the Guest Essay by Lester R. Brown at CengageNOW). Figure 1-18 lists some of the shifts involved in bringing about this new cultural change by learning how to live more sustainably.

Scientific evidence indicates that we have perhaps 50 years and no more than 100 years to make such crucial cultural changes. If this is correct, sometime during this century we could come to a critical fork in the road, at which point we will choose a path toward sustainability or continue on our current unsustainable course. Everything you do, or do not do, will play a role in our collective choice of which path we will take. One of the goals of this book is to provide a realistic environmental vision of the future that, instead of immobilizing you with fear, gloom, and doom, will energize you by inspiring realistic hope.

REVISITING Exponential Growth and Sustainability

We face an array of serious environmental problems. This book is about *solutions* to these problems. Making the transition to more sustainable societies and economies challenges us to devise ways to slow down the harmful effects of exponential growth (**Core Case Study**) and to use the same power of exponential growth to implement more sustainable lifestyles and economies.

The key is to apply the four **scientific principles of sustainability** (Figure 1-17 and **Concept 1-6**) to the design of our economic and social systems and to our individual lifestyles. We can use such information to help slow human population growth, sharply reduce poverty, curb the unsustainable forms of resource use that are eating away at the earth's natural capital, build social capital, and create a better world for ourselves, our children, and future generations.

Exponential growth is a double-edged sword. It can cause environmental harm. But we can also use it positively to amplify beneficial changes in our lifestyles and economies by applying the four **scientific principles of sustainability**. Through our individual and collective actions or inactions, we choose which side of that sword to use.

We are rapidly altering the planet that is our only home. If we make the right choices during this century, we can create an extraordinary and sustainable future on our planetary home. If we get it wrong, we face irreversible ecological disruption that could set humanity back for centuries and wipe out as many as half of the world's species.

You have the good fortune to be a member of the 21st century *transition generation*, which will decide what path humanity takes. What a challenging and exciting time to be alive!

What's the use of a house
if you don't have a decent planet to put it on?
HENRY DAVID THOREAU

1. Review the Key Questions and Concepts for this chapter on p. 6. What is **exponential growth?** Why is living in an exponential age a cause for concern for everyone living on the planet?

2. Define **environment.** Distinguish among **environmental science, ecology,** and **environmentalism.** Distinguish between an **organism** and a **species.** What is an **ecosystem?** What is **sustainability?** Explain the terms **natural capital, natural resources, natural services, solar capital,** and **natural capital degradation.** What is **nutrient cycling** and why is it important? Describe the ultimate goal of an **environmentally sustainable society.** What is **natural income?**

3. What is the difference between **economic growth** and **economic development?** Distinguish among **gross domestic product (GDP), per capita GDP,** and **per capita GDP PPP.** Distinguish between **developed countries** and **developing countries** and describe their key characteristics. What is **environmentally sustainable economic development?**

4. What is a **resource?** What is **conservation?** Distinguish among a **renewable resource, nonrenewable resource,** and **perpetual resource** and give an example of each. What is **sustainable yield?** Define and give three examples of **environmental degradation.** What is the tragedy of the commons? Distinguish between **recycling** and **reuse** and give an example of each. What is an **ecological footprint?** What is a **per capita ecological footprint?** Compare the total and per capita ecological footprints of the United States and China.

5. What is **culture?** Describe three major cultural changes that have occurred since humans arrived on the earth.

Why has each change led to more environmental degradation? What is the **environmental** or **sustainability revolution?**

6. Define **pollution.** Distinguish between **point sources** and **nonpoint sources** of pollution. Distinguish between **biodegradable pollutants** and **nondegradable pollutants** and give an example of each. Distinguish between **pollution cleanup** and **pollution prevention** and give an example of each. Describe three problems with solutions that rely mostly on pollution cleanup.

7. Identify five basic causes of the environmental problems that we face today. What is **poverty?** In what ways do poverty and affluence affect the environment? Explain the problems we face by not including the harmful environmental costs in the prices of goods and services.

8. What is an **environmental worldview?** What is **environmental ethics?** Distinguish among the **planetary management, stewardship,** and **environmental wisdom worldviews.** Describe Aldo Leopold's environmental ethics. What major steps are involved in making an environmental decision? What is **social capital?**

9. Discuss the lessons we can learn from the environmental transformation of Chattanooga, Tennessee (USA). Explain why individuals matter in dealing with the environmental problems we face.

10. What are four scientific principles of sustainability? Explain how exponential growth (**Core Case Study**) affects them.

Note: Key Terms are in **bold** type.

1. List three ways in which you could apply **Concepts 1-5A** and **1-6** to making your lifestyle more environmentally sustainable.

2. Describe two environmentally beneficial forms of exponential growth (**Core Case Study**).

3. Explain why you agree or disagree with the following propositions:
 a. Stabilizing population is not desirable because, without more consumers, economic growth would stop.
 b. The world will never run out of resources because we can use technology to find substitutes and to help us reduce resource waste.

4. Suppose the world's population stopped growing today. What environmental problems might this help solve? What environmental problems would remain? What economic problems might population stabilization make worse?

5. When you read that at least 19,200 people die prematurely each day (13 per minute) from preventable malnutrition and infectious disease, do you **(a)** doubt that it is true, **(b)** not want to think about it, **(c)** feel hopeless, **(d)** feel sad, **(e)** feel guilty, or **(f)** want to do something about this problem?

6. What do you think when you read that **(a)** the average American consumes 30 times more resources than

the average citizen of India, and **(b)** human activities are projected to make the earth's climate warmer? Are you skeptical, indifferent, sad, helpless, guilty, concerned, or outraged? Which of these feelings help perpetuate such problems, and which can help solve them?

7. For each of the following actions, state one or more of the four **scientific principles of sustainability** (Figure 1-17) that are involved: **(a)** recycling soda cans; **(b)** using a rake instead of leaf blower; **(c)** choosing to have no more than one child; **(d)** walking to class instead of driving; **(e)** taking your own reusable bags to the grocery store to carry things home in; **(f)** volunteering to help restore a prairie ; and **(g)** lobbying elected officials to require that 20% of your country's electricity be produced by renewable wind power by 2020.

8. Explain why you agree or disagree with each of the following statements: **(a)** humans are superior to other forms of life, **(b)** humans are in charge of the earth, **(c)** all economic growth is good, **(d)** the value of other forms of life depends only on whether they are useful to us, **(e)** because all forms of life eventually become extinct we should not worry about whether our activities cause their premature extinction, **(f)** all forms of life have an inherent right to exist, **(g)** nature has an almost unlimited storehouse of resources for human use, **(h)** technology can solve our environmental problems, **(i)** I do not believe I have any obligation to future generations, and **(j)** I do not believe I have any obligation to other forms of life.

9. What are the basic beliefs of your environmental worldview (p. 20)? Record your answer. Then at the end of this course, return to your answer to see if your environmental worldview has changed. Are the beliefs included in your environmental worldview consistent with your answers to question 8? Are your environmental actions consistent with your environmental worldview?

10. List two questions that you would like to have answered as a result of reading this chapter.

Note: See Supplement 13 (p. S78) for a list of Projects related to this chapter.

ECOLOGICAL FOOTPRINT ANALYSIS

If a country's or the world's *ecological footprint per person* (Figure 1-10, p. 15) is larger than its *biological capacity per person* to replenish its renewable resources and to absorb the resulting waste products and pollution, it is said to have an *ecological* *deficit*. If the reverse is true, it has an *ecological credit* or *reserve*. Use the data below to calculate the ecological deficit or credit for various countries. (For a map of ecological creditors and debtors, see Figure 4 on p. S26 in Supplement 4.)

Place	Per Capita Ecological Footprint (hectares per person)*	Per Capita Biocapacity (hectares per person)	Ecological Credit (+) or Debit (−) (hectares per person)
World	2.2	1.8	− 0.4
United States	9.8	4.7	
China	1.6	0.8	
India	0.8	0.4	
Russia	4.4	0.9	
Japan	4.4	0.7	
Brazil	2.1	9.9	
Germany	4.5	1.7	
United Kingdom	5.6	1.6	
Mexico	2.6	1.7	
Canada	7.6	14.5	

Source: Date from WWF, *Living Planet Report 2006.*

*1 hectare = 2.47 acres

1. Which two countries have the largest ecological deficits?

2. Which two countries have an ecological credit?

3. Rank the countries in order from the largest to the smallest per capita footprint.

LEARNING ONLINE

Log on to the Student Companion Site for this book at **academic.cengage.com/biology/miller**, and choose Chapter 1 for many study aids and ideas for further reading and research. These include flash cards, practice quizzing, Weblinks, information on Green Careers, and InfoTrac® College Edition articles.

Science, Matter, Energy, and Systems

Carrying Out a Controlled Scientific Experiment

One way in which scientists learn about how nature works is to conduct a *controlled experiment*. To begin, scientists isolate *variables,* or factors that can change within a system or situation being studied. An experiment involving *single-variable analysis* is designed to isolate and study the effects of one variable at a time.

To do such an experiment, scientists set up two groups. One is the *experimental group* in which a chosen variable is changed in a known way, and the other is the *control group* in which the chosen variable is not changed. If the experiment is designed and run properly, differences between the two groups should result from the variable that was changed in the experimental group.

In 1963, botanist F. Herbert Bormann, forest ecologist Gene Likens, and their colleagues began carrying out a classic controlled experiment. The goal was to compare the loss of water and nutrients from an uncut forest ecosystem (the *control site*) with one that was stripped of its trees (the *experimental site*).

They built V-shaped concrete dams across the creeks at the bottoms of several forested valleys in the Hubbard Brook Experimental Forest in New Hampshire (Figure 2-1). The dams were anchored on impenetrable bedrock, so that all surface water

leaving each forested valley had to flow across a dam where scientists could measure its volume and dissolved nutrient content.

In the first experiment, the investigators measured the amounts of water and dissolved plant nutrients that entered and left an undisturbed forested area (the control site) (Figure 2-1, left). These measurements showed that an undisturbed mature forest is very efficient at storing water and retaining chemical nutrients in its soils.

The next experiment involved setting up an experimental forested area. One winter, the investigators cut down all trees and shrubs in one valley (the experimental site), left them where they fell, and sprayed the area with herbicides to prevent the regrowth of vegetation. Then they compared the inflow and outflow of water and nutrients in this experimental site (Figure 2-1, right) with those in the control site (Figure 2-1, left) for 3 years.

With no plants to help absorb and retain water, the amount of water flowing out of the deforested valley increased by 30–40%. As this excess water ran rapidly over the ground, it eroded soil and carried dissolved nutrients out of the deforested site. Overall, the loss of key nutrients from the experimental forest was six to eight times that in the nearby control forest.

Figure 2-1 Controlled field experiment to measure the effects of deforestation on the loss of water and soil nutrients from a forest. V–notched dams were built into the impenetrable bedrock at the bottoms of several forested valleys (left) so that all water and nutrients flowing from each valley could be collected and measured for volume and mineral content. These measurements were recorded for the forested valley (left), which acted as the control site. Then all the trees in another valley (the experimental site) were cut (right) and the flows of water and soil nutrients from this experimental valley were measured for 3 years.

2-1 What is science?

CONCEPT 2-1 Scientists collect data and develop theories, models, and laws about how nature works.

2-2 What is matter?

CONCEPT 2-2 Matter consists of elements and compounds, which are in turn made up of atoms, ions, or molecules.

2-3 How can matter change?

CONCEPT 2-3 When matter undergoes a physical or chemical change, no atoms are created or destroyed (the law of conservation of matter).

2-4 What is energy and how can it be changed?

CONCEPT 2-4A When energy is converted from one form to another in a physical or chemical change, no energy is created or destroyed (first law of thermodynamics).

CONCEPT 2-4B Whenever energy is changed from one form to another, we end up with lower-quality or less usable energy than we started with (second law of thermodynamics).

2-5 What are systems and how do they respond to change?

CONCEPT 2-5A Systems have inputs, flows, and outputs of matter and energy, and their behavior can be affected by feedback.

CONCEPT 2-5B Life, human systems, and the earth's life-support systems must conform to the law of conservation of matter and the two laws of thermodynamics.

Note: Supplements 1 (p. S2), 2 (p. S4), 5 (p. S31), and 6 (p. S39) can be used with this chapter.

Science is an adventure of the human spirit.
It is essentially an artistic enterprise, stimulated largely by curiosity,
served largely by disciplined imagination,
and based largely on faith in the reasonableness, order,
and beauty of the universe.

WARREN WEAVER

2-1 What Is Science?

▶ **CONCEPT 2-1** Scientists collect data and develop theories, models, and laws about how nature works.

Science Is a Search for Order in Nature

Have you ever seen an area in a forest where all the trees were cut down? If so, you might wonder about the effects of cutting down all those trees. You might wonder how it affected the animals and people living in that area and how it affected the land itself. That is what scientists Bormann and Likens (**Core Case Study**) thought about when they designed their experiment.

Such curiosity is what motivates scientists. **Science** is an endeavor to discover how nature works and to use that knowledge to make predictions about what is likely to happen in nature. It is based on the assumption that events in the natural world follow or-

derly cause-and-effect patterns that can be understood through careful observation, measurements, experimentation, and modeling. Figure 2-2 (p. 30) summarizes the scientific process.

There is nothing mysterious about this process. You use it all the time in making decisions. Here is an example of applying the scientific process to an everyday situation:

Observation: You try to switch on your flashlight and nothing happens.

Question: Why didn't the light come on?

Hypothesis: Maybe the batteries are dead.

Test the hypothesis: Put in new batteries and try to switch on the flashlight.

Figure 2-2 *What scientists do.* The essence of science is this process for testing ideas about how nature works. Scientists do not necessarily follow the exact order of steps shown here. For example, sometimes a scientist might start by formulating a hypothesis to answer the initial question and then run experiments to test the hypothesis.

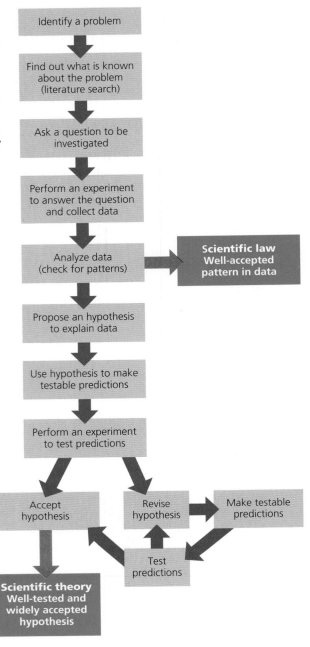

Identify a problem

↓

Find out what is known about the problem (literature search)

↓

Ask a question to be investigated

↓

Perform an experiment to answer the question and collect data

↓

Analyze data (check for patterns) → **Scientific law** Well-accepted pattern in data

↓

Propose an hypothesis to explain data

↓

Use hypothesis to make testable predictions

↓

Perform an experiment to test predictions

↓

Accept hypothesis Revise hypothesis → Make testable predictions

Test predictions

Scientific theory Well-tested and widely accepted hypothesis

Result: Flashlight still does not work.

New hypothesis: Maybe the bulb is burned out.

Experiment: Replace bulb with a new bulb.

Result: Flashlight works when switched on.

Conclusion: Second hypothesis is verified.

Here is a more formal outline of steps scientists often take in trying to understand nature, although not always in the order listed:

- *Identify a problem.* Bormann and Likens (**Core Case Study**) identified the loss of water and soil nutrients from cutover forests as a problem worth studying. CORE CASE STUDY

- *Find out what is known about the problem.* Bormann and Likens searched the scientific literature to find out what was known about retention and loss of water and soil nutrients in forests.

- *Ask a question to be investigated.* The scientists asked: "How does clearing forested land affect its ability to store water and retain soil nutrients?

- *Collect data to answer the question.* To collect **data**— information needed to answer their questions— scientists make observations of the subject area they are studying. Scientific observations involve gathering information by using human senses of sight, smell, hearing, and touch and extending those senses by using tools such as rulers, microscopes, and satellites. Often scientists conduct **experiments,** or procedures carried out under controlled conditions to gather information and test ideas. Bormann and Likens collected and analyzed data on the water and soil nutrients flowing from a patch of an undisturbed forest (Figure 2-1, left) and from a nearby patch of forest where they had cleared the trees for their experiment (Figure 2-1, right).

- *Propose a hypothesis to explain the data.* Scientists suggest a **scientific hypothesis,** a possible and testable explanation of what they observe in nature or in the results of their experiments. The data collected by Bormann and Likens show a decrease in the ability of a cleared forest to store water and retain soil nutrients such as nitrogen. They came up with the following hypothesis to explain their data: When a forest is cleared, it retains less water and loses large quantities of its soil nutrients when water from rain and melting snow flows across its exposed soil.

- *Make testable predictions.* Scientists use a hypothesis to make testable or logical predictions about what should happen if the hypothesis is valid. They often do this by making "If . . . then" predictions. Bormann and Likens predicted that *if* their original hypothesis was valid for nitrogen, *then* a cleared forest should also lose other soil nutrients such as phosphorus.

- *Test the predictions with further experiments, models, or observations.* To test their prediction, Bormann and Likens repeated their controlled experiment and measured the phosphorus content of the soil. Another way to test predictions is to develop a **model,** an approximate representation or simulation of a system being studied. Since Bormann and Likens performed their experiments, scientists have developed increasingly sophisticated mathematical and computer models of how forest systems work. Data from Bormann and Likens's research and that of other scientists can be fed into such models and

used to predict the loss of phosphorus and other types of soil nutrients. These predictions can be compared with the actual measured losses to test the validity of the models.

- *Accept or reject the hypothesis.* If their new data do not support their hypotheses, scientists come up with other testable explanations. This process continues until there is general agreement among scientists in the field being studied that a particular hypothesis is the best explanation of the data. After Bormann and Likens confirmed that the soil in a cleared forest also loses phosphorus, they measured losses of other soil nutrients, which also supported their hypothesis. A well-tested and widely accepted scientific hypothesis or a group of related hypotheses is called a **scientific theory.** Thus, Bormann and Likens and their colleagues developed a theory that trees and other plants hold soil in place and help it

to retain water and nutrients needed by the plants for their growth.

Important features of the scientific process are *curiosity, skepticism, peer review, reproducibility,* and *openness to new ideas.* Good scientists are extremely curious about how nature works. But they tend to be highly skeptical of new data, hypotheses, and models until they can be tested and verified. **Peer review** happens when scientists report details of the methods and models they used, the results of their experiments, and the reasoning behind their hypotheses for other scientists working in the same field (their peers) to examine and criticize. Ideally, other scientists repeat and analyze the work to see if the data can be reproduced and whether the proposed hypothesis is reasonable and useful (Science Focus, below).

For example, Bormann and Likens (**Core Case Study**) submitted the results of their for- CORE CASE STUDY

SCIENCE FOCUS

Easter Island: Some Revisions to a Popular Environmental Story

For years, the story of Easter Island has been used in textbooks as an example of how humans can seriously degrade their own life-support system. It concerns a civilization that once thrived and then largely disappeared from a small, isolated island in the great expanse of the South Pacific, located about 3,600 kilometers (2,200 miles) off the coast of Chile.

Scientists used anthropological evidence and scientific measurements to estimate the ages of certain artifacts found on Easter Island (also called Rapa Nui). They hypothesized that about 2,900 years ago, Polynesians used double-hulled, seagoing canoes to colonize the island. The settlers probably found a paradise with fertile soil that supported dense and diverse forests and lush grasses. According to this hypothesis, the islanders thrived, and their population increased to as many as 15,000 people.

Measurements made by scientists seemed to indicate that over time, the Polynesians began living unsustainably by using the island's forest and soil resources faster than they could be renewed. When they used up the large trees, the islanders could no longer build their traditional seagoing canoes for fishing in deeper offshore waters, and no one could escape the island by boat.

Without the once-great forests to absorb and slowly release water, springs and streams dried up, exposed soils were

eroded, crop yields plummeted, and famine struck. There was no firewood for cooking or keeping warm. According to the original hypothesis, the population and the civilization collapsed as rival clans fought one another for dwindling food supplies, and the island's population dropped sharply. By the late 1870s, only about 100 native islanders were left.

In 2006, anthropologist Terry L. Hunt, Director of the University of Hawaii Rapa Nui Archeological Field School, evaluated the accuracy of past measurements and other evidence and carried out new measurements to estimate the ages of various artifacts. He used these data to formulate an alternative hypothesis describing the human tragedy on Easter Island.

Hunt came to several new conclusions. *First,* the Polynesians arrived on the island about 800 years ago, not 2,900 years ago. *Second,* their population size probably never exceeded 3,000, contrary to the earlier estimate of up to 15,000. *Third,* the Polynesians did use the island's trees and other vegetation in an unsustainable manner, and by 1722, visitors reported that most of the island's trees were gone.

But one question not answered by the earlier hypothesis was, why did the trees never grow back? Recent evidence and Hunt's new hypothesis suggest that rats (which either came along with the original settlers as stowaways or were brought along

as a source of protein for the long voyage) played a key role in the island's permanent deforestation. Over the years, the rats multiplied rapidly into the millions and devoured the seeds that would have regenerated the forests.

Another of Hunt's conclusions was that after 1722, the population of Polynesians on the island dropped to about 100, mostly from contact with European visitors and invaders. Hunt hypothesized that these newcomers introduced fatal diseases, killed off some of the islanders, and took large numbers of them away to be sold as slaves.

This story is an excellent example of how science works. The gathering of new scientific data and reevaluation of older data led to a revised hypothesis that challenges our thinking about the decline of civilization on Easter Island. As a result, the tragedy may not be as clear an example of human-caused ecological collapse as was once thought. However, there is evidence that other earlier civilizations did suffer ecological collapse largely from unsustainable use of soil, water, and other resources, as described in Supplement 5 on p. S31.

Critical Thinking

Does the new doubt about the original Easter Island hypothesis mean that we should not be concerned about using resources unsustainably on the island in space we call Earth? Explain.

est experiments to a respected scientific journal. Before publishing this report, the journal editors had it reviewed by other soil and forest experts. Other scientists have repeated the measurements of soil content in undisturbed and cleared forests of the same type and also in different types of forests. Their results have also been subjected to peer review. In addition, computer models of forest systems have been used to evaluate this problem, with the results subjected to peer review.

Scientific knowledge advances in this way, with scientists continually questioning measurements, making new measurements, and sometimes coming up with new and better hypotheses (Science Focus, p. 31). As a result, good scientists are *open to new ideas* that have survived the rigors of the scientific process.

Scientists Use Reasoning, Imagination, and Creativity to Learn How Nature Works

Scientists arrive at conclusions, with varying degrees of certainty, by using two major types of reasoning. **Inductive reasoning** involves using specific observations and measurements to arrive at a general conclusion or hypothesis. It is a form of "bottom-up" reasoning that goes from the specific to the general. For example, suppose we observe that a variety of different objects fall to the ground when we drop them from various heights. We can then use inductive reasoning to propose that *all objects fall to the earth's surface when dropped.*

Depending on the number of observations made, there may be a high degree of certainty in this conclusion. However, what we are really saying is "All objects that we or other observers have dropped from various heights have fallen to the earth's surface." Although it is extremely unlikely, we cannot be *absolutely sure* that no one will ever drop an object that does not fall to the earth's surface.

Deductive reasoning involves using logic to arrive at a specific conclusion based on a generalization or premise. It is a form of "top-down" reasoning that goes from the general to the specific. For example,

Generalization or premise: All birds have feathers.

Example: Eagles are birds.

Deductive conclusion: Eagles have feathers.

THINKING ABOUT
The Hubbard Brook Experiment and Scientific Reasoning CORE CASE STUDY

In carrying out and interpreting their experiment, did Bormann and Likens rely primarily on inductive or deductive reasoning?

Deductive and inductive reasoning and critical thinking skills (pp. 2–3) are important scientific tools. But scientists also use intuition, imagination, and creativity

to explain some of their observations in nature. Often such ideas defy conventional logic and current scientific knowledge. According to physicist Albert Einstein, "There is no completely logical way to a new scientific idea." Intuition, imagination, and creativity are as important in science as they are in poetry, art, music, and other great adventures of the human spirit, as reflected by scientist Warren Weaver's quotation found at the opening of this chapter.

Scientific Theories and Laws Are the Most Important Results of Science

If an overwhelming body of observations and measurements supports a scientific hypothesis, it becomes a *scientific theory. Scientific theories are not to be taken lightly.* They have been tested widely, are supported by extensive evidence, and are accepted by most scientists in a particular field or related fields of study.

Nonscientists often use the word *theory* incorrectly when they actually mean *scientific hypothesis,* a tentative explanation that needs further evaluation. The statement, "Oh, that's just a theory," made in everyday conversation, implies that the theory was stated without proper investigation and careful testing—the opposite of the scientific meaning of the word.

Another important and reliable outcome of science is a **scientific law,** or **law of nature:** a well-tested and widely accepted description of what we find happening over and over again in the same way in nature. An example is the *law of gravity,* based on countless observations and measurements of objects falling from different heights. According to this law, all objects fall to the earth's surface at predictable speeds.

A scientific law is no better than the accuracy of the observations or measurements upon which it is based (see Figure 1 in Supplement 1 on p. S3). But if the data are accurate, a scientific law cannot be broken, unless and until we get contradictory new data.

Scientific theories and laws have a high probability of being valid, but they are not infallible. Occasionally, new discoveries and new ideas can overthrow a well-accepted scientific theory or law in what is called a **paradigm shift.** It occurs when the majority of scientists in a field or related fields accept a new *paradigm,* or framework for theories and laws in a particular field.

A good way to summarize the most important outcomes of science is to say that scientists collect data and develop theories, models, and laws that describe and explain how nature works (**Concept 2-1**). Scientists use reasoning and critical thinking skills. But the best scientists also use intuition, imagination, and creativity in asking important questions, developing hypotheses, and designing ways to test them.

For a superb look at how science works and what scientists do, see the Annenberg video series, *The Habitable Planet: A Systems Approach to Environmental Science* (see

the website at **www.learner.org/resources/series209 .html**). Each of the 13 videos describes how scientists working on two different problems related to a certain subject are learning about how nature works. Also see Video 2, *Thinking Like Scientists*, in another Annenberg series, *Teaching High School Science* (see the website at **www.learner.org/resources/series126.html**).

The Results of Science Can Be Tentative, Reliable, or Unreliable

A fundamental part of science is *testing*. Scientists insist on testing their hypotheses, models, methods, and results over and over again to establish the reliability of these scientific tools and the resulting conclusions.

Media news reports often focus on disputes among scientists over the validity of data, hypotheses, models, methods, or results (see Science Focus, below). This helps to reveal differences in the reliability of various

scientific tools and results. Simply put, some science is more reliable than other science, depending on how carefully it has been done and on how thoroughly the hypotheses, models, methods, and results have been tested.

Sometimes, preliminary results that capture news headlines are controversial because they have not been widely tested and accepted by peer review. They are not yet considered reliable, and can be thought of as **tentative science** or **frontier science.** Some of these results will be validated and classified as reliable and some will be discredited and classified as unreliable. At the frontier stage, it is normal for scientists to disagree about the meaning and accuracy of data and the validity of hypotheses and results. This is how scientific knowledge advances.

By contrast, **reliable science** consists of data, hypotheses, theories, and laws that are widely accepted by scientists who are considered experts in the field under study. The results of reliable science are based on

SCIENCE FOCUS

The Scientific Consensus over Global Warming

Based on measurements and models, it is clear that carbon dioxide and other gases in the atmosphere play a major role in determining the temperature of the atmosphere through a natural warming process called the *natural greenhouse effect.* Without the presence of these *greenhouse gases* in the atmosphere, the earth would be too cold for most life as we know it to exist, and you would not be reading these words. The earth's natural greenhouse effect is one of the most widely accepted theories in the atmospheric sciences and is an example of *reliable science.*

Since 1980, many climate scientists have been focusing their studies on three major questions:

- How much has the earth's atmosphere warmed during the past 50 years?

- How much of the warming is the result of human activities such as burning oil, gas, and coal and clearing forests, which add carbon dioxide and other greenhouse gases to the atmosphere?

- How much is the atmosphere likely to warm in the future and how might this affect the climate of different parts of the world?

To help clarify these issues, in 1988, the United Nations and the World Meteorological Organization established the Intergovernmental Panel on Climate Change (IPCC) to study how the climate system works, document past

climate changes, and project future climate changes. The IPCC network includes more than 2,500 climate experts from 70 nations.

Since 1990, the IPCC has published four major reports summarizing the scientific consensus among these climate experts. In its 2007 report, the IPCC came to three major conclusions:

- It is *very likely* (a 90–99% probability) that the lower atmosphere is getting warmer and has warmed by about 0.74 C° (1.3 F°) between 1906 and 2005.

- Based on analysis of past climate data and use of 19 climate models, it is *very likely* (a 90–99% probability) that human activities, led by emissions of carbon dioxide from burning fossil fuels, have been the main cause of the observed atmospheric warming during the past 50 years.

- It is *very likely* that the earth's mean surface temperature will increase by about 3 C° (5.4 F°) between 2005 and 2100, unless we make drastic cuts in greenhouse gas emissions from power plants, factories, and cars that burn fossil fuels.

This scientific consensus among most of the world's climate experts is currently considered the most *reliable science* we have on this subject.

As always, there are individual scientists who disagree with the scientific consensus

view. Typically, they question the reliability of certain data, say we don't have enough data to come to reliable conclusions, or question some of the hypotheses or models involved. However, in the case of global warming, they are in a distinct and declining minority.

Media reports are sometimes confusing or misleading because they present reliable science along with a quote from a scientist in the field who disagrees with the consensus view, or from someone who is not an expert in the field. This can cause public distrust of well-established reliable science, such as that reported by the IPCC, and may sometimes lead to a belief in ideas that are not widely accepted by the scientific community. (See the Guest Essay on environmental reporting by Andrew C. Revkin at CengageNOW.)

Critical Thinking

Find a newspaper article or other media report that presents the scientific consensus view on global warming and then attempts to balance it with a quote from a scientist who disagrees with the consensus view. Try to determine: **(a)** whether the dissenting scientist is considered an expert in climate science, **(b)** whether the scientist has published any peer reviewed papers on the subject, and **(c)** what organizations or industries are supporting the dissenting scientist.

the self-correcting process of testing, open peer review, reproducibility, and debate. New evidence and better hypotheses (Science Focus, p. 31) may discredit or alter tried and accepted views and even result in paradigm shifts. But unless that happens, those views are considered to be the results of reliable science.

Scientific hypotheses and results that are presented as reliable without having undergone the rigors of peer review, or that have been discarded as a result of peer review, are considered to be **unreliable science.** Here are some critical thinking questions you can use to uncover unreliable science:

- Was the experiment well designed? Did it involve enough testing? Did it involve a control group? (**Core Case Study**)

 ↰ CORE CASE STUDY

- Have the data supporting the proposed hypotheses been verified? Have the results been reproduced by other scientists?

- Do the conclusions and hypotheses follow logically from the data?

- Are the investigators unbiased in their interpretations of the results? Are they free of a hid-

den agenda? Were they funded by an unbiased source?

- Have the conclusions been verified by impartial peer review?

- Are the conclusions of the research widely accepted by other experts in this field?

If the answer to each of these questions is "yes," then the results can be classified as reliable science. Otherwise, the results may represent tentative science that needs further testing and evaluation, or they can be classified as unreliable science.

Environmental Science Has Some Limitations

Before continuing our study of environmental science, we need to recognize some of its limitations, as well as those of science in general. *First,* scientists can disprove things but they cannot prove anything absolutely, because there is always some degree of uncertainty in scientific measurements, observations, and models.

SCIENCE FOCUS

Statistics and Probability

Statistics consists of mathematical tools used to collect, organize, and interpret numerical data. For example, suppose we weigh each individual in a population of 15 rabbits. We can use statistics to calculate the *average* weight of the population. To do this, we add up the weights of the 15 rabbits and divide the total by 15. Similarly, Bormann and Likens (**Core Case Study**) made many measurements of nitrate levels in the water flowing from their undisturbed and cut patches of forests (Figure 2-1) and then averaged the results to get the most reliable value.

↰ CORE CASE STUDY

Scientists also use the statistical concept of probability to evaluate their results. **Probability** is the chance that something will happen. For example, if you toss a nickel, what is the probability or chance that it will come up heads? If your answer is 50%, you are correct. The chance of the nickel coming up heads is ½, which can also be expressed as 50% or 0.5. Probability is often expressed as a number between 0 and 1 written as a decimal (such as 0.5).

Now suppose you toss the coin 10 times and it comes up heads 6 times. Does this mean that the probability of it coming up

heads is 0.6 or 60%? The answer is no because the *sample size*—the number of objects or events studied—was too small to yield a statistically accurate result. If you increase your sample size to 1,000 by tossing the coin 1,000 times, you are almost certain to get heads 50% of the time and tails 50% of the time.

It is important when doing scientific research to take samples in different places, in order to get a comprehensive evaluation of the variable being studied. It is also critical to have a large enough sample size to give an accurate estimate of the overall probability of an event happening.

For example, if you wanted to study the effects of a certain air pollutant on the needles of pine trees, you would need to locate different stands of the same type of pine tree that are all exposed to the pollutant over a certain period of time. At each location, you would need to measure the levels of the pollutant in the atmosphere at different times and average the results. You would also need to make measurements of the damage (such as needle loss) to a large enough sample of trees in each location over a certain time period. Then you would average the results in

each location and compare the results from all locations.

If the average results were consistent in different locations, you could then say that there is a certain probability, say 60% (or 0.6), that this type of pine tree suffered a certain percentage loss of its needles when exposed to a specified average level of the pollutant over a given time. You would also need to run other experiments to determine that natural needle loss, extreme temperatures, insects, plant diseases, drought, or other factors did not cause the needle losses you observed. As you can see, getting reliable scientific results is not a simple process.

Critical Thinking

What does it mean when an international body of the world's climate experts says that there is a 90–99% chance (probability of 0.9–0.99) that human activities, led by emissions of carbon dioxide from burning fossil fuels, have been the main cause of the observed atmospheric warming during the past 50 years? Why would the probability never be 100%?

Instead scientists try to establish that a particular hypothesis, theory, or law has a very high *probability* (90–99%) of being true and thus is classified as reliable science. Most scientists rarely say something like, "Cigarettes cause lung cancer." Rather, they might say, "Overwhelming evidence from thousands of studies indicates that people who smoke have an increased risk of developing lung cancer."

THINKING ABOUT
Scientific Proof

Does the fact that science can never prove anything absolutely mean that its results are not valid or useful? Explain.

Second, scientists are human and cannot be expected to be totally free of bias about their results and hypotheses. However, bias can be minimized and often uncovered by the high standards of evidence required through peer review, although some scientists are bypassing traditional peer review by publishing their results online.

A *third* limitation involves use of statistical tools. There is no way to measure accurately how much soil is eroded annually worldwide, for example. Instead, scientists use statistical sampling and methods to estimate such numbers (Science Focus, at left). Such results should not be dismissed as "only estimates" because they can indicate important trends.

A *fourth* problem is that many environmental phenomena involve a huge number of interacting variables and complex interactions, which makes it too costly to test one variable at a time in controlled experiments such as the one described in the **Core Case Study** that opens this chapter. To help deal with this problem, scientists develop mathematical models that include the interactions of many variables. Running such models on computers can sometimes overcome this limitation and save both time and money. In addition, computer models can be used to simulate global experiments on phenomena like climate change, which are impossible to do in a controlled physical experiment.

Finally, the scientific process is limited to understanding the natural world. It cannot be applied to moral or ethical questions, because such questions are about matters for which we cannot collect data from the natural world. For example, we can use the scientific process to understand the effects of removing trees from an ecosystem, but this process does not tell us whether it is right or wrong to remove the trees.

Much progress has been made, but we still know too little about how the earth works, its current state of environmental health, and the environmental impacts of our activities. These knowledge gaps point to important *research frontiers*, several of which are highlighted throughout this text.

2-2 What Is Matter?

> **CONCEPT 2-2** Matter consists of elements and compounds, which are in turn made up of atoms, ions, or molecules.

Matter Consists of Elements and Compounds

To begin our study of environmental science, we start at the most basic level, looking at matter—the stuff that makes up life and its environment. **Matter** is anything that has mass and takes up space. It is made up of **elements,** each of which is a fundamental substance that has a unique set of properties and cannot be broken down into simpler substances by chemical means. For example, gold is an element; it cannot be broken down chemically into any other substance.

Some matter is composed of one element, such as gold or silver, but most matter consists of **compounds:** combinations of two or more different elements held together in fixed proportions. For example, water is a compound made of the elements hydrogen and oxygen, which have chemically combined with one another. (See Supplement 6 on p. S39 for an expanded discussion of basic chemistry.)

To simplify things, chemists represent each element by a one- or two-letter symbol. Table 2-1 (p. 36), lists the elements and their symbols that you need to know to understand the material in this book. Just four elements—oxygen, carbon, hydrogen, and nitrogen—make up about 96% of your body weight and that of most other living things.

Atoms, Ions, and Molecules Are the Building Blocks of Matter

The most basic building block of matter is an **atom:** the smallest unit of matter into which an element can be divided and still retain its chemical properties. The idea that all elements are made up of atoms is called the **atomic theory** and is the most widely accepted scientific theory in chemistry.

Table 2-1

Elements Important to the Study of Environmental Science

Element	Symbol	Element	Symbol
Hydrogen	H	Bromine	Br
Carbon	C	Sodium	Na
Oxygen	O	Calcium	Ca
Nitrogen	N	Lead	Pb
Phosphorus	P	Mercury	Hg
Sulfur	S	Arsenic	As
Chlorine	Cl	Uranium	U
Fluorine	F		

Atoms are incredibly small. In fact, more than 3 million hydrogen atoms could sit side by side on the period at the end of this sentence. If you could view them with a supermicroscope, you would find that each different type of atom contains a certain number of three different types of *subatomic particles:* positively charged **protons (p), neutrons (n)** with no electrical charge, and negatively charged **electrons (e).**

Each atom consists of an extremely small and dense center called its **nucleus**—which contains one or more protons and, in most cases, one or more neutrons— and one or more electrons moving rapidly somewhere around the nucleus in what is called an *electron probability cloud* (Figure 2-3). Each atom (except for *ions,* expained at right) has equal numbers of positively

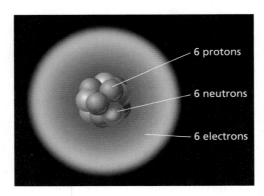

Figure 2-3 Greatly simplified model of a carbon-12 atom. It consists of a nucleus containing six positively charge protons and six neutral neutrons. There are six negatively charged electrons found outside its nucleus. We cannot determine the exact locations of the electrons. Instead, we can estimate the *probability* that they will be found at various locations outside the nucleus—sometimes called an *electron probability cloud.* This is somewhat like saying that there are six airplanes flying around inside a cloud. We don't know their exact location, but the cloud represents an area where we can probably find them.

charged protons and negatively charged electrons. Because these electrical charges cancel one another, *atoms as a whole have no net electrical charge.*

Each element has a unique **atomic number,** equal to the number of protons in the nucleus of its atom. Carbon (C), with 6 protons in its nucleus (Figure 2-3), has an atomic number of 6, whereas uranium (U), a much larger atom, has 92 protons in its nucleus and an atomic number of 92.

Because electrons have so little mass compared to protons and neutrons, *most of an atom's mass is concentrated in its nucleus.* The mass of an atom is described by its **mass number:** the total number of neutrons and protons in its nucleus. For example, a carbon atom with 6 protons and 6 neutrons in its nucleus has a mass number of 12, and a uranium atom with 92 protons and 143 neutrons in its nucleus has a mass number of 235 (92 + 143 = 235).

Each atom of a particular element has the same number of protons in its nucleus. But the nuclei of atoms of a particular element can vary in the number of neutrons they contain, and therefore, in their mass numbers. Forms of an element having the same atomic number but different mass numbers are called **isotopes** of that element. Scientists identify isotopes by attaching their mass numbers to the name or symbol of the element. For example, the three most common isotopes of carbon are carbon-12 (Figure 2-3, with six protons and six neutrons), carbon-13 (with six protons and seven neutrons), and carbon-14 (with six protons and eight neutrons). Carbon-12 makes up about 98.9% of all naturally occurring carbon.

A second building block of matter is an **ion**—an atom or groups of atoms with one or more net positive or negative electrical charges. An ion forms when an atom gains or loses one or more electrons. An atom that loses one or more of its electrons becomes an ion with one or more positive electrical charges, because the number of positively charged protons in its nucleus is now greater than the number of negatively charged electrons outside its nucleus. Similarly, when an atom gains one or more electrons, it becomes an ion with one or more negative electrical charges, because the number of negatively charged electrons is greater than the number of positively charged protons in its nucleus.

Ions containing atoms of more than one element are the basic units found in some compounds (called *ionic compounds*). For more details on how ions form see p. S39 in Supplement 6.

The number of positive or negative charges carried by an ion is shown as a superscript after the symbol for an atom or a group of atoms. Examples encountered in this book include a *positive* hydrogen ion (H^+), with one positive charge, an aluminum ion (Al^{3+}) with three positive charges, and a *negative* chloride ion (Cl^-) with one negative charge. These and other ions listed in Table 2-2 are used in other chapters in this book.

One example of the importance of ions in our study of environmental science is the nitrate ion (NO_3^-), a nutrient essential for plant growth. Figure 2-4 shows measurements of the loss of nitrate ions from the deforested area (Figure 2-1, right) in the controlled experiment run by Bormann and Likens (**Core Case Study**). Numerous chemical analyses of the water flowing through the dams of the cleared forest area showed an average 60-fold rise in the concentration of NO_3^- compared to water running off of the uncleared forest area. The stream below this valley became covered with algae whose populations soared as a result of an excess of nitrate plant nutrients. After a few years, however, vegetation began growing back on the cleared valley and nitrate levels in its runoff returned to normal levels.

Ions are also important for measuring a substance's **acidity** in a water solution, a chemical characteristic that helps determine how a substance dissolved in water will interact with and affect its environment. Scientists use **pH** as a measure of acidity, based on the amount of hydrogen ions (H^+) and hydroxide ions (OH^-) contained in a particular volume of a solution. Pure water (not tap water or rainwater) has an equal number of H^+ and OH^- ions. It is called a *neutral solution* and has a pH of 7. An *acidic solution* has more hydrogen ions than hydroxide ions and has a pH less than 7. A *basic solution* has more hydroxide ions than hydrogen ions and has a pH greater than 7. (See Figure 5 on p. S41 in Supplement 6 for more details.)

The third building block of matter is a **molecule:** a combination of two or more atoms of the same or different elements held together by forces called *chemical bonds*. Molecules are the basic units of some compounds

Table 2-2

Ions Important to the Study of Environmental Science

Positive Ion	Symbol	Negative Ion	Symbol
hydrogen ion	H^+	chloride ion	Cl^-
sodium ion	Na^+	hydroxide ion	OH^-
calcium ion	Ca^{2+}	nitrate ion	NO_3^-
aluminum ion	Al^{3+}	sulfate ion	SO_4^{2-}
ammonium ion	NH_4^+	phosphate ion	PO_4^{3-}

Table 2-3

Compounds Important to the Study of Environmental Science

Compound	Formula	Compound	Formula
sodium chloride	NaCl	methane	CH_4
carbon monoxide	CO	glucose	$C_6H_{12}O_6$
carbon dioxide	CO_2	water	H_2O
nitric oxide	NO	hydrogen sulfide	H_2S
nitrogen dioxide	NO_2	sulfur dioxide	SO_2
nitrous oxide	N_2O	sulfuric acid	H_2SO_4
nitric acid	HNO_3	ammonia	NH_3

(called *molecular compounds*). Examples are shown in Figure 4 on p. S41 in Supplement 6.

Chemists use a **chemical formula** to show the number of each type of atom or ion in a compound. This shorthand contains the symbol for each element present and uses subscripts to represent the number of atoms or ions of each element in the compound's basic structural unit. Examples of compounds and their formulas encountered in this book are sodium chloride (NaCl) and water (H_2O, read as "H-two-O"). These and other compounds important to our study of environmental science are listed in Table 2-3.

You may wish to mark the pages containing Tables 2-1 through 2-3, as they could be useful references for understanding material in other chapters.

Figure 2-4 Loss of nitrate ions (NO_3^-) from a deforested watershed in the Hubbard Brook Experimental Forest in New Hampshire (Figure 2-1, right). The average concentration of nitrate ions in runoff from the deforested experimental watershed was 60 times greater than in a nearby unlogged watershed used as a control (Figure 2-1, left). (Data from F. H. Bormann and Gene Likens)

CENGAGENOW™ Examine atoms—their parts, how they work, and how they bond together to form molecules—at CengageNOW™.

Organic Compounds Are the Chemicals of Life

Table sugar, vitamins, plastics, aspirin, penicillin, and most of the chemicals in your body are **organic compounds,** which contain at least two carbon atoms combined with atoms of one or more other elements. All other compounds are called **inorganic compounds.** One exception, methane (CH_4), has only one carbon atom but is considered an organic compound.

The millions of known organic (carbon-based) compounds include the following:

- *Hydrocarbons:* compounds of carbon and hydrogen atoms. One example is methane (CH_4), the main component of natural gas, and the simplest organic compound. Another is octane (C_8H_{18}), a major component of gasoline.

- *Chlorinated hydrocarbons:* compounds of carbon, hydrogen, and chlorine atoms. An example is the insecticide DDT ($C_{14}H_9Cl_5$).

- *Simple carbohydrates* (simple sugars): certain types of compounds of carbon, hydrogen, and oxygen atoms. An example is glucose ($C_6H_{12}O_6$), which most plants and animals break down in their cells to obtain energy. (For more details see Figure 8 on p. S42 in Supplement 6.)

Larger and more complex organic compounds, essential to life, are composed of *macromolecules.* Some of these molecules, called *polymers,* are formed when a number of simple organic molecules (*monomers*) are linked together by chemical bonds, somewhat like rail cars linked in a freight train. The three major types of organic polymers are

- *complex carbohydrates* such as cellulose and starch, which consist of two or more monomers of simple sugars such as glucose (see Figure 8 on p. S42 in Supplement 6),

- *proteins* formed by monomers called *amino acids* (see Figure 9 on p. S42 in Supplement 6), and

- *nucleic acids* (DNA and RNA) formed by monomers called *nucleotides* (see Figures 10 and 11 on p. S43 in Supplement 6).

Lipids, which include fats and waxes, are a fourth type of macromolecule essential for life (see Figure 12 on p. S43 in Supplement 6).

Matter Comes to Life through Genes, Chromosomes, and Cells

The story of matter, starting with the hydrogen atom, becomes more complex as molecules grow in complexity. This is no less true when we examine the fundamental components of life. The bridge between nonliving and living matter lies somewhere between macromole-cules and **cells**—the fundamental structural units of life, which we explore in more detail in the next chapter.

Above, we mentioned nucleotides in DNA (see Figures 10 and 11 on p. S43 in Supplement 6). Within some DNA molecules are certain sequences of nucleotides called **genes.** Each of these distinct pieces of DNA contains instructions, called *genetic information,* for making specific proteins. Each of these coded units of genetic information concerns a specific **trait,** or characteristic passed on from parents to offspring during reproduction in an animal or plant.

Thousands of genes, in turn, make up a single **chromosome,** a special DNA molecule together with a number of proteins. Genetic information coded in your chromosomal DNA is what makes you different from an oak leaf, an alligator, or a flea, and from your parents. In other words, it makes you human, but it also makes you unique. The relationships of genetic material to cells are depicted in Figure 2-5.

A human body contains trillions of cells, each with an identical set of genes.

Each human cell (except for red blood cells) contains a nucleus.

Each cell nucleus has an identical set of chromosomes, which are found in pairs.

A specific pair of chromosomes contains one chromosome from each parent.

Each chromosome contains a long DNA molecule in the form of a coiled double helix.

Genes are segments of DNA on chromosomes that contain instructions to make proteins—the building blocks of life.

Figure 2-5 Relationships among cells, nuclei, chromosomes, DNA, and genes.

Matter Occurs in Various Physical Forms

The atoms, ions, and molecules that make up matter are found in three *physical states:* solid, liquid, and gas. For example, water exists as ice, liquid water, or water vapor depending on its temperature and the surrounding air pressure. The three physical states of any sample of matter differ in the spacing and orderliness of its atoms, ions, or molecules. A solid has the most compact and orderly arrangement, and a gas the least compact and orderly arrangement. Liquids are somewhere in between.

Some Forms of Matter Are More Useful than Others

Matter quality is a measure of how useful a form of matter is to humans as a resource, based on its availability and *concentration,* or amount of it that is contained in a given area or volume. **High-quality matter** is highly concentrated, is typically found near the earth's surface, and has great potential for use as a resource. Low-quality matter is not highly concentrated, is often located deep underground or dispersed in the ocean or atmosphere, and usually has little potential for use as a resource. See Figure 2-6 for examples illustrating differences in matter quality.

Figure 2-6 Examples of differences in matter quality. *High-quality matter* (left column) is fairly easy to extract and is highly concentrated; *low-quality matter* (right column) is not highly concentrated and is more difficult to extract than high-quality matter.

High Quality **Low Quality**

Solid Gas

Salt Solution of salt in water

Coal Coal-fired power plant emissions

Gasoline Automobile emissions

Aluminum can Aluminum ore

2-3 How Can Matter Change?

▶ **CONCEPT 2-3** When matter undergoes a physical or chemical change, no atoms are created or destroyed (the law of conservation of matter).

Matter Undergoes Physical, Chemical, and Nuclear Changes

When a sample of matter undergoes a **physical change,** its *chemical composition,* or the arrangement of its atoms or ions within molecules does not change. A piece of aluminum foil cut into small pieces is still aluminum foil. When solid water (ice) melts or liquid water boils, none of the H_2O molecules are changed. The molecules are simply arranged in different spatial (physical) patterns.

> **THINKING ABOUT**
> **Controlled Experiments and Physical Changes** CORE CASE STUDY
>
> How would you set up a controlled experiment (**Core Case Study**) to verify that when water changes from one physical state to another, its chemical composition does not change?

In a **chemical change,** or **chemical reaction,** there is a change in the arrangement of atoms or ions within molecules of the substances involved. Chemists use *chemical equations* to represent what happens in a chemical reaction. For example, when coal burns completely, the solid carbon (C) in the coal combines with oxygen gas (O_2) from the atmosphere to form the gaseous compound carbon dioxide (CO_2).

In addition to physical and chemical changes, matter can undergo three types of **nuclear changes,** or changes in the nuclei of its atoms (Figure 2-7). In the first type, called **natural radioactive decay,** isotopes spontaneously emit fast-moving subatomic particles, high-energy radiation such as gamma rays, or both (Figure 2-7, top). The unstable isotopes are called **radioactive isotopes** or **radioisotopes.**

Nuclear fission is a nuclear change in which the nuclei of certain isotopes with large mass numbers (such as uranium-235) are split apart into lighter nuclei when struck by neutrons; each fission releases two or three neutrons plus energy (Figure 2-7, middle). Each of these neutrons, in turn, can trigger an additional fission reaction. Multiple fissions within a certain amount of mass produce a **chain reaction,** which releases an enormous amount of energy.

Nuclear fusion is a nuclear change in which two isotopes of light elements, such as hydrogen, are forced together at extremely high temperatures until they fuse to form a heavier nucleus (Figure 2-7, bottom). A tremendous amount of energy is released in this process. Fusion of hydrogen nuclei to form helium nuclei is the source of energy in the sun and other stars.

We Cannot Create or Destroy Matter

We can change elements and compounds from one physical, chemical, or nuclear form to another, but we can never create or destroy any of the atoms involved in any physical or chemical change. All we can do is rearrange the atoms, ions, or molecules into different spatial patterns (physical changes) or combinations (chemical changes). These statements, based on many thousands of measurements, describe a scientific law known as the **law of conservation of matter:** when a physical or chemical change occurs, no atoms are created or destroyed (**Concept 2-3**).

This law means there is no "away" as in "to throw away." *Everything we think we have thrown away remains here with us in some form.* We can reuse or recycle some materials and chemicals, but the law of conservation of matter means we will always face the problem of what to do with some quantity of the wastes and pollutants we produce.

We talk about consuming matter as if matter is being used up or destroyed, but the law of conservation of matter says that this is impossible. What is meant by *matter consumption,* is not destruction of matter, but rather conversion of matter from one form to another.

2-4 What Is Energy and How Can It Be Changed?

▶ **CONCEPT 2-4A** When energy is converted from one form to another in a physical or chemical change, no energy is created or destroyed (first law of thermodynamics).

▶ **CONCEPT 2-4B** Whenever energy is changed from one form to another, we end up with lower-quality or less usable energy than we started with (second law of thermodynamics).

Energy Comes in Many Forms

Energy is the capacity to do work or transfer heat. Work is done when something is moved. The amount of work done is the product of the force applied to an object to move it a certain distance (work = force × distance).

For example, it takes a certain amount of muscular force to lift this book to a certain height.

There are two major types of energy: *moving energy* (called *kinetic energy*) and *stored energy* (called *potential energy*). Moving matter has **kinetic energy** because it has mass and velocity. Examples are wind (a mov-

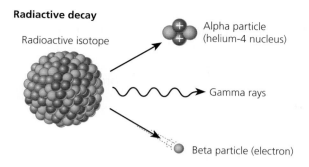

Radiactive decay

Radioactive isotope

Alpha particle
(helium-4 nucleus)

Gamma rays

Beta particle (electron)

Radioactive decay occurs when nuclei of unstable isotopes spontaneously emit fast-moving chunks of matter (alpha particles or beta particles), high-energy radiation (gamma rays), or both at a fixed rate. A particular radioactive isotope may emit any one or a combination of the three items shown in the diagram.

Nuclear fission

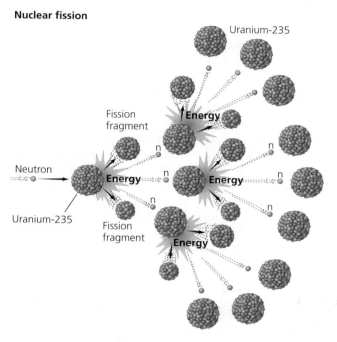

Uranium-235

Fission fragment

Energy

Neutron

Energy

Energy

Uranium-235

Fission fragment

Energy

Nuclear fission occurs when the nuclei of certain isotopes with large mass numbers (such as uranium-235) are split apart into lighter nuclei when struck by a neutron and release energy plus two or three more neutrons. Each neutron can trigger an additional fission reaction and lead to a *chain reaction*, which releases an enormous amount of energy.

Nuclear fusion

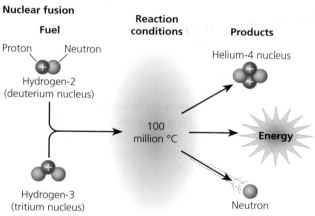

Fuel

Reaction conditions

Products

Proton Neutron

Hydrogen-2
(deuterium nucleus)

Helium-4 nucleus

100 million °C

Energy

Hydrogen-3
(tritium nucleus)

Neutron

Nuclear fusion occurs when two isotopes of light elements, such as hydrogen, are forced together at extremely high temperatures until they fuse to form a heavier nucleus and release a tremendous amount of energy.

Figure 2-7 Types of nuclear changes: natural radioactive decay (top), nuclear fission (middle), and nuclear fusion (bottom).

ing mass of air), flowing water, and electricity (flowing electrons).

Another form of kinetic energy is **heat:** the total kinetic energy of all moving atoms, ions, or molecules within a given substance. When two objects at differ-

ent temperatures contact one another, heat flows from the warmer object to the cooler object.

Heat can be transferred from one place to another by three different methods: *radiation* (the emission of electromagnetic energy), *conduction* (the transfer of

Energy emitted from sun (kcal/cm²/min)

Visible

Ultraviolet

Infrared

Wavelength (micrometers)

CENGAGENOW **Active Figure 2-8** *Solar capital:* the spectrum of electromagnetic radiation released by the sun consists mostly of visible light. *See an animation based on this figure at CengageNOW.*

kinetic energy between substances in contact with one another), and *convection* (the movement of heat within liquids and gases from warmer to cooler portions).

In **electromagnetic radiation,** another form of kinetic energy, energy travels in the form of a *wave* as a result of changes in electric and magnetic fields. There are many different forms of electromagnetic radiation, each having a different *wavelength* (distance between successive peaks or troughs in the wave) and *energy content.* Forms of electromagnetic radiation with short wavelengths, such as gamma rays, X rays, and ultraviolet (UV) radiation, have a higher energy content than do forms with longer wavelengths, such as visible light and infrared (IR) radiation (Figure 2-8). Visible light makes up most of the spectrum of electromagnetic radiation emitted by the sun (Figure 2-8).

CENGAGENOW Find out how color, wavelengths, and energy intensities of visible light are related at CengageNOW.

The other major type of energy is **potential energy,** which is stored and potentially available for use. Examples of potential energy include a rock held in your hand, an unlit match, the chemical energy stored in gasoline molecules, and the nuclear energy stored in the nuclei of atoms.

Potential energy can be changed to kinetic energy. Hold this book up, and it has potential energy; drop it on your foot, and its potential energy changes to kinetic energy. When a car engine burns gasoline, the potential energy stored in the chemical bonds of gasoline molecules changes into mechanical (kinetic) energy, which propels the car, and heat. Potential energy stored in the

molecules of carbohydrates you eat becomes kinetic energy when your body uses it to move and do other forms of work.

CENGAGENOW Witness how a Martian might use kinetic and potential energy at CengageNOW.

Some Types of Energy Are More Useful Than Others

Energy quality is a measure of an energy source's capacity to do useful work. **High-quality energy** is concentrated and has a high capacity to do useful work. Examples are very high-temperature heat, nuclear fission, concentrated sunlight, high-velocity wind, and energy released by burning natural gas, gasoline, or coal.

By contrast, **low-quality energy** is dispersed and has little capacity to do useful work. An example is heat dispersed in the moving molecules of a large amount of matter (such as the atmosphere or an ocean) so that its temperature is low. The total amount of heat stored in the Atlantic Ocean is greater than the amount of high-quality chemical energy stored in all the oil deposits of Saudi Arabia. Yet because the ocean's heat is so widely dispersed, it cannot be used to move things or to heat things to high temperatures.

Energy Changes Are Governed by Two Scientific Laws

Thermodynamics is the study of energy transformations. Scientists have observed energy being changed from one form to another in millions of physical and chemical changes. But they have never been able to detect the creation or destruction of any energy in such changes. The results of these experiments have been summarized in the **law of conservation of energy,** also known as the **first law of thermodynamics:** When energy is converted from one form to another in a physical or chemical change, no energy is created or destroyed (**Concept 2-4A**).

This scientific law tells us that when one form of energy is converted to another form in any physical or chemical change, *energy input always equals energy output.* No matter how hard we try or how clever we are, we cannot get more energy out of a system than we put in. This is one of nature's basic rules.

People talk about consuming energy but the first law says that it is impossible to use up energy. *Energy consumption,* then, means converting energy from one form to another with no energy being destroyed or created in the process.

Because the first law of thermodynamics states that energy cannot be created or destroyed, only converted from one form to another, you may be tempted to think

CENGAGENOW" **Active Figure 2-9** The second law of thermodynamics in action in living systems. Each time energy changes from one form to another, some of the initial input of high-quality energy is degraded, usually to low-quality heat that is dispersed into the environment. *See an animation based on this figure at* CengageNOW.
Question: What are three things that you did during the past hour that degraded high-quality energy?

there will always be enough energy. Yet if you fill a car's tank with gasoline and drive around or use a flashlight battery until it is dead, something has been lost. But what is it? The answer is *energy quality*, the amount of energy available that can perform useful work.

Countless experiments have shown that whenever energy changes from one form to another, we always end up with less usable energy than we started with. These results have been summarized in the **second law of thermodynamics:** When energy changes from one form to another, we always end up with lower-quality or less usable energy than we started with (**Concept 2-4B**). This lower-quality energy usually takes the form of heat given off at a low temperature to the environment. There it is dispersed by the random motion of air or water molecules and becomes even less useful as a resource.

In other words, *energy always goes from a more useful to a less useful form when it is changed from one form to another.* No one has ever found a violation of this fundamental scientific law. It is another one of nature's basic rules.

Consider three examples of the second law of thermodynamics in action. *First,* when you drive a car, only about 6% of the high-quality energy available in its gasoline fuel actually moves the car, according to energy expert Amory Lovins. (See his Guest Essay at CengageNOW.) The remaining 94% is degraded to low-quality heat that is released into the environment. Thus, 94% of the money you spend for gasoline is not used to transport you anywhere.

Second, when electrical energy in the form of moving electrons flows through filament wires in an incandescent lightbulb, about 5% of it changes into useful

light, and 95% flows into the environment as low-quality heat. In other words, the *incandescent lightbulb* is really an energy-wasting *heat bulb.*

Third, in living systems, solar energy is converted into chemical energy (food molecules) and then into mechanical energy (used for moving, thinking, and living). During each conversion, high-quality energy is degraded and flows into the environment as low-quality heat. Trace the flows and energy conversions in Figure 2-9 to see how this happens.

The second law of thermodynamics also means *we can never recycle or reuse high-quality energy to perform useful work.* Once the concentrated energy in a serving of food, a liter of gasoline, or a chunk of uranium is released, it is degraded to low-quality heat that is dispersed into the environment.

Energy efficiency, or **energy productivity,** is a measure of how much useful work is accomplished by a particular input of energy into a system. There is plenty of room for improving energy efficiency. Scientists estimate that only 16% of the energy used in the United States ends up performing useful work. The remaining 84% is either unavoidably wasted because of the second law of thermodynamics (41%) or unnecessarily wasted (43%). Thus, thermodynamics teaches us an important lesson: the cheapest and quickest way to get more energy is to stop wasting almost half the energy we use. We explore energy waste and energy efficiency in depth in Chapters 15 and 16.

CENGAGENOW" See examples of how the first and second laws of thermodynamics apply in our world at CengageNOW.

2-5 What Are Systems and How Do They Respond to Change?

> ▶ **CONCEPT 2-5A** Systems have inputs, flows, and outputs of matter and energy, and their behavior can be affected by feedback.
>
> ▶ **CONCEPT 2-5B** Life, human systems, and the earth's life-support systems must conform to the law of conservation of matter and the two laws of thermodynamics.

Systems Have Inputs, Flows, and Outputs

A **system** is a set of components that function and interact in some regular way. The human body, a river, an economy, and the earth are all systems.

Most systems have the following key components: **inputs** from the environment, **flows** or **throughputs** of matter and energy within the system at certain rates, and **outputs** to the environment (Figure 2-10) (**Concept 2-5A**). One of the most powerful tools used by environmental scientists to study how these components of systems interact is computer modeling. (Science Focus, below)

Systems Respond to Change through Feedback Loops

When people ask you for feedback, they are usually seeking your response to something they said or did. They might feed this information back into their mental processes to help them decide whether and how to change what they are saying or doing.

Figure 2-10 Inputs, throughput, and outputs of an economic system. Such systems depend on inputs of matter and energy resources and outputs of waste and heat to the environment. Such a system can become unsustainable if the throughput of matter and energy resources exceeds the ability of the earth's natural capital to provide the required resource inputs or the ability of the environment to assimilate or dilute the resulting heat, pollution, and environmental degradation.

Similarly, most systems are affected one way or another by **feedback,** any process that increases (positive feedback) or decreases (negative feedback) a change to a system (**Concept 2-5A**). Such a process, called a **feedback loop,** occurs when an output of matter, energy,

SCIENCE FOCUS

The Usefulness of Models

Scientists use *models,* or simulations, to learn how systems work. Some of our most powerful and useful technologies are mathematical and computer models.

Making a mathematical model usually requires going through three steps many times. *First,* scientists make guesses about systems they are modeling and write down equations to express these estimates. *Second,* they compute the likely behavior of a system implied by such equations. *Third,* they compare the system's projected behavior with observations of its actual behavior, also considering existing experimental data.

Mathematical models are particularly useful when there are many interacting variables, when the time frame of events being modeled is long, and when controlled experiments are impossible or too expensive to conduct.

After building and testing a mathematical model, scientists use it to predict what is *likely* to happen under a variety of conditions. In effect, they use mathematical models to answer *if–then* questions: "*If* we do such and such, *then* what is likely to happen now and in the future?" This process can give us a variety of projections or scenarios of possible futures or outcomes based on different assumptions. Mathematical models (like all other models) are no better than the assumptions on which they are built and the data fed into them.

Using data collected by Bormann and Likens in their Hubbard Brook experiment (**Core Case Study**), scientists created mathematical models to describe a forest and evaluate what happens to soil nutrients or other variables if the forest is disturbed or cut down.

Other areas of environmental science where computer modeling is becoming increasingly important include the studies of climate change, deforestation, biodiversity loss, and ocean systems.

Critical Thinking

What are two limitations of computer models? Do their limitations mean that we should not rely on such models? Explain.

or information is fed back into the system as an input and leads to changes in that system.

A **positive feedback loop** causes a system to change further in the same direction (Figure 2-11). In the Hubbard Brook experiments, for example (**Core Case Study**), researchers found that when vegetation was removed from a stream valley, flowing water from precipitation caused erosion and loss of nutrients, which caused more vegetation to die. With even less vegetation to hold soil in place, flowing water caused even more erosion and nutrient loss, which caused even more plants to die.

Such accelerating positive feedback loops are of great concern in several areas of environmental science. One of the most alarming is the melting of polar ice, which has occurred as the temperature of the atmosphere has risen during the past few decades. As that ice melts, there is less of it to reflect sunlight, and more water is exposed to sunlight. Because water is darker, it absorbs more solar energy, making the area warmer and causing the ice to melt faster, thus exposing more water. The melting of polar ice thus accelerates, causing a number of serious problems that we explore further in Chapter 19.

A **negative,** or **corrective, feedback loop** causes a system to change in the opposite direction from which is it moving. A simple example is a thermostat, a device that controls how often, and how long a heating or cooling system runs (Figure 2-12). When the furnace in a house is turned on and begins heating the house, the thermostat can be set to turn the furnace off when the temperature in the house reaches the set number. The house then stops getting warmer and starts to cool.

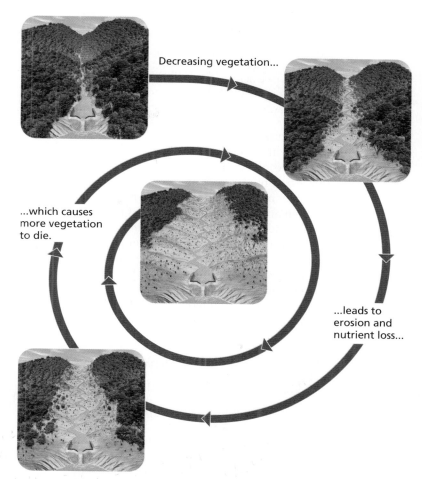

Figure 2-11 *Positive feedback loop.* Decreasing vegetation in a valley causes increasing erosion and nutrient losses, which in turn causes more vegetation to die, which allows for more erosion and nutrient losses. The system receives feedback that continues the process of deforestation.

Figure 2-12 *Negative feedback loop.* When a house being heated by a furnace gets to a certain temperature, its thermostat is set to turn off the furnace, and the house begins to cool instead of continuing to get warmer. When the house temperature drops below the set point, this information is fed back, and the furnace is turned on and runs until the desired temperature is reached. The system receives feedback that reverses the process of heating or cooling.

THINKING ABOUT
Hubbard Brook and Feedback Loops

CORE
CASE
STUDY

How might experimenters have employed a negative feedback loop to stop, or correct, the positive feedback loop that resulted in increasing erosion and nutrient losses in the Hubbard Brook experimental forest?

An important case of a negative feedback loop is the recycling and reuse of some resources such as aluminum, copper, and glass. For example, an aluminum can is one output of a mining and manufacturing system. When that output becomes an input, as the can is recycled and used in place of raw aluminum to make a new product, that much less aluminum is mined and the environmental impact of the mining-manufacturing system is lessened. Such a negative feedback loop therefore can promote sustainability and reduce the environmental impact of human activities by reducing the use of matter and energy resources and the amount of pollution and solid waste produced by use of such material.

Time Delays Can Allow a System to Reach a Tipping Point

Complex systems often show **time delays** between the input of a feedback stimulus and the response to it. For example, scientists could plant trees in a degraded area such as the Hubbard Brook experimental forest to slow erosion and nutrient losses (**Core Case Study**), but it would take years for the trees and other vegetation to grow enough to accomplish this purpose.

CORE
CASE
STUDY

Time delays can also allow an environmental problem to build slowly until it reaches a *threshold level*, or **tipping point,** causing a fundamental shift in the behavior of a system. Prolonged delays dampen the negative feedback mechanisms that might slow, prevent, or halt environmental problems. In the Hubbard Brook example, if erosion and nutrient losses reached a certain point where the land could not support vegetation, then an irreversible tipping point would have been reached, and it would be futile to plant trees to try to restore the system. Other environmental problems that can reach tipping point levels are population growth, leaks from toxic waste dumps, global climate change, and degradation of forests from prolonged exposure to air pollutants.

System Effects Can Be Amplified through Synergy

A **synergistic interaction,** or **synergy,** occurs when two or more processes interact so that the combined effect is greater than the sum of their separate effects. Scientific studies reveal such an interaction between smoking and inhaling asbestos particles. Lifetime smokers have ten times the risk that nonsmokers have of getting lung cancer. And individuals exposed to asbes-

tos particles for long periods increase their risk of getting lung cancer fivefold. But people who smoke and are exposed to asbestos have 50 times the risk that nonsmokers have of getting lung cancer.

Similar dangers can result from combinations of certain air pollutants that, when combined, are more hazardous to human health than they would be acting independently. We examine such hazards further in Chapter 17.

On the other hand, synergy can be helpful. Suppose we want to persuade an elected official to vote for a certain environmental law. You could write, e-mail, or visit the official. But you may have more success if you can get a group of potential voters to do such things. In other words, the combined or synergistic efforts of people working together can be more effective than the efforts of each person acting alone.

RESEARCH FRONTIER

Identifying environmentally harmful and beneficial synergistic interactions. See **academic.cengage.com/biology/miller**.

Human Activities Can Have Unintended Harmful Results

One of the lessons we can derive from the four **scientific principles of sustainability** (see back cover) is that *everything we do affects someone or something in the environment in some way*. In other words, any action in a complex system has multiple and often unintended, unpredictable effects. As a result, most of the environmental problems we face today are unintended results of activities designed to increase the quality of human life.

For example, clearing trees from the land to plant crops can increase food production and feed more people. But it can also lead to soil erosion, flooding, and a loss of biodiversity, as Easter Islanders and other civilizations learned the hard way (Science Focus, p. 31, and Supplement 5 on p. S31).

One factor that can lead to an environmental surprise is a *discontinuity* or abrupt change in a previously stable system when some *environmental threshold* or *tipping point* is crossed. Scientific evidence indicates that we are now reaching an increasing number of such tipping points. For example, we have depleted fish stocks in some parts of the world to the point where it is not profitable to harvest them. Other examples, such as deforested areas turning to desert, coral reefs dying, species disappearing, glaciers melting, and sea levels rising, will be discussed in later chapters.

RESEARCH FRONTIER

Tipping points for various environmental systems such as fisheries, forests coral reefs, and the earth's climate system. See **academic.cengage.com/biology/miller**.

Life, economic and other human systems, and the earth's life support systems depend on matter and energy, and therefore they must obey the law of conservation of matter and the two laws of thermodynamics (**Concept 2-5B**). Without these laws, economic growth based on using matter and energy resources to produce goods and services (Figure 2-10) could be expanded indefinitely and cause even more serious environmental problems. But these scientific laws place limits on what we can do with matter and energy resources.

A Look Ahead

In the next six chapters, we apply the three basic laws of matter and thermodynamics and the four scientific principles of sustainability (see

back cover) to living systems. Chapter 3 shows how the sustainability principles related to solar energy and nutrient cycling apply in ecosystems. Chapter 4 focuses on using the biodiversity principle to understand the relationships between species diversity and evolution. Chapter 5 examines how the biodiversity and population control principles relate to interactions among species and how such interactions regulate population size. In Chapter 6, we apply the principles of biodiversity and population control to the growth of the human population. In Chapter 7, we look more closely at terrestrial biodiversity in different types of deserts, grasslands, and forests. Chapter 8 examines aquatic biodiversity in aquatic systems such as oceans, lakes, wetlands, and rivers.

REVISITING 　 The Hubbard Brook Experimental Forest and Sustainability 　

The controlled experiment discussed in the Core Case Study that opened this chapter revealed that clearing a mature forest degrades some of its natural capital (Figure 1-7, p. 12). Specifically, the loss of trees and vegetation altered the ability of the forest to retain and recycle water and other critical plant nutrients—a crucial ecological function based on one of the four scientific principles of sustainability (see back cover). In other words, the uncleared forest was a more sustainable system than a similar area of cleared forest (Figures 2-1 and 2-4).

This loss of vegetation also violated the other three scientific principles of sustainability. For example, the cleared forest had fewer plants that could use solar energy to produce food for

animals. And the loss of plants and animals reduced the life-sustaining biodiversity of the cleared forest. This in turn reduced some of the interactions between different types of plants and animals that help control their populations.

Humans clear forests to grow food and build cities. The key question is, how far can we go in expanding our ecological footprints (Figure 1-10, p. 15) without threatening the quality of life for our own species and the other species that keep us alive and support our economies? To live sustainably, we need to find and maintain a balance between preserving undisturbed natural systems and modifying other natural systems for our use.

The second law of thermodynamics holds, I think, the supreme position among laws of nature. . . . If your theory is found to be against the second law of thermodynamics, I can give you no hope.

ARTHUR S. EDDINGTON

REVIEW

1. Review the Key Questions and Concepts for this chapter on p. 29. Describe the controlled scientific experiment carried out at the Hubbard Brook Experimental Forest. What is **science?** Describe the steps involved in the scientific process. What is **data?** What is an **experiment?** What is a **model?** Distinguish among a **scientific hypothesis, scientific theory,** and **scientific law** (**law of nature**). What is **peer review** and why is it important? Explain why scientific theories are not to be taken lightly and why people often use the term "theory" incorrectly.

2. Distinguish between **inductive reasoning** and **deductive reasoning** and give an example of each. Explain

why scientific theories and laws are the most important results of science.

3. What is a **paradigm shift?** Distinguish among **tentative science (frontier science), reliable science,** and **unreliable science.** Describe the scientific consensus concerning global warming. What is **statistics?** What is **probability** and what is its role in scientific conclusions? What are five limitations of science and environmental science?

4. What is **matter?** Distinguish between an **element** and a **compound** and give an example of each. Distinguish among **atoms, ions,** and **molecules** and give an

example of each. What is the **atomic theory?** Distinguish among **protons, neutrons,** and **electrons.** What is the **nucleus** of an atom? Distinguish between the **atomic number** and the **mass number** of an element. What is an **isotope?** What is **acidity?** What is **pH?**

5. What is a **chemical formula?** Distinguish between **organic compounds** and **inorganic compounds** and give an example of each. Distinguish among complex carbohydrates, proteins, nucleic acids, and lipids. What is a **cell?** Distinguish among **genes, traits,** and **chromosomes.** What is **matter quality?** Distinguish between **high-quality matter** and **low-quality matter** and give an example of each.

6. Distinguish between a **physical change** and a **chemical change (chemical reaction)** and give an example of each. What is a **nuclear change?** Explain the differences among **natural radioactive decay, nuclear fission,** and **nuclear fusion.** What is a **radioactive isotope (radioisotope)?** What is a **chain reaction?** What is the **law of conservation of matter** and why is it important?

7. What is **energy?** Distinguish between **kinetic energy** and **potential energy** and give an example of each. What is **heat?** Define and give two examples of **electromagnetic radiation.** What is **energy quality?** Distinguish between **high-quality energy** and **low-quality energy** and give an example of each.

8. What is the **law of conservation of energy (first law of thermodynamics)** and why is it important? What is the **second law of thermodynamics** and why is it important? Explain why this law means that we can never recycle or reuse high-quality energy. What is **energy efficiency (energy productivity)** and why is it important?

9. Define and give an example of a **system?** Distinguish among the **input, flow (throughput),** and **output** of a system. Why are scientific models useful? What is **feedback?** What is a **feedback loop?** Distinguish between a **positive feedback loop** and a **negative (corrective) feedback loop** in a system, and give an example of each. Distinguish between a **time delay** and a **synergistic interaction (synergy)** in a system and give an example of each. What is a **tipping point?**

10. Explain how human activities can have unintended harmful environmental results. Relate the four **scientific principles of sustainability** to the Hubbard Brook Experimental Forest controlled experiment (**Core Case Study**).

Note: Key Terms are in **bold** type.

CRITICAL THINKING

1. What ecological lesson can we learn from the controlled experiment on the clearing of forests described in the **Core Case Study** that opened this chapter?

2. Think of an area you have seen where some significant change has occurred to a natural system. What is a question you might ask in order to start a scientific process to evaluate the effects of this change, similar to the process described in the **Core Case Study**?

3. Describe a way in which you have applied the scientific process described in this chapter (Figure 2-2) in your own life, and state the conclusion you drew from this process. Describe a new problem that you would like to solve using this process.

4. Respond to the following statements:
 a. Scientists have not absolutely proven that anyone has ever died from smoking cigarettes.
 b. The natural greenhouse theory—that certain gases (such as water vapor and carbon dioxide) warm the lower atmosphere—is not a reliable idea because it is just a scientific theory.

5. A tree grows and increases its mass. Explain why this phenomenon is not a violation of the law of conservation of matter.

6. If there is no "away" where organisms can get rid of their wastes, why is the world not filled with waste matter?

7. Someone wants you to invest money in an automobile engine, claiming that it will produce more energy than the energy in the fuel used to run it. What is your response? Explain.

8. Use the second law of thermodynamics to explain why a barrel of oil can be used only once as a fuel, or in other words, why we cannot recycle high-quality energy.

9. a. Imagine you have the power to revoke the law of conservation of matter for one day. What are three things you would do with this power?
 b. Imagine you have the power to violate the first law of thermodynamics for one day. What are three things you would do with this power?

10. List two questions that you would like to have answered as a result of reading this chapter.

Note: See Supplement 13 (p. S78) for a list of Projects related to this chapter.

Marine scientists from the U.S. state of Maryland have produced the following two graphs as part of a report on the current health of the Chesapeake Bay. They are pleased with the recovery of the striped bass population but are concerned about the decline of the blue crab population, because blue crabs are consumed by mature striped bass. Their hypothesis is that as the population of striped bass increases, the population of blue crab decreases.

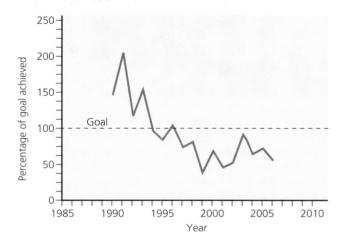

Using the data in the above graphs, answer the following questions:

1. Which years confirm their hypothesis?

2. Which years do not support their hypothesis?

3. If the crab population reaches 100% of the goal figure, what would you predict the striped bass goal figure would be?

3

Ecosystems: What Are They and How Do They Work?

Tropical Rain Forests Are Disappearing

Tropical rain forests are found near the earth's equator and contain an incredible variety of life. These lush forests are warm year round and have high humidity and heavy rainfall almost daily. Although they cover only about 2% of the earth's land surface, studies indicate that they contain up to half of the world's known terrestrial plant and animal species. For these reasons, they make an excellent natural laboratory for the study of *ecosystems*—communities of organisms interacting with one another and with the physical environment of matter and energy in which they live.

So far, at least half of these forests have been destroyed or disturbed by humans cutting down trees, growing crops, grazing cattle, and building settlements (Figure 3-1), and the degradation of these centers of life (biodiversity) is increasing. Ecologists warn that without strong conservation measures, most of these forests will probably be gone or severely degraded within your lifetime.

Scientists project that disrupting these ecosystems will have three major harmful effects. *First,* it will reduce the earth's vital biodiversity by destroying or degrading the habitats of many of their unique plant and animal species, thereby causing their premature extinction. *Second,* it will help to accelerate climate change due to global warming by eliminating large areas of trees faster than they can grow back, thereby reducing the trees' overall uptake of the greenhouse gas carbon dioxide.

Third, it will change regional weather patterns in ways that will prevent the return of diverse tropical rain forests in cleared or degraded areas. Once this tipping point is reached, tropical rain forest in such areas will become less diverse tropical grassland.

Ecosystems recycle materials and provide humans and other organisms with essential natural services (Figure 1-3, p. 8) and natural resources such as nutrients (Figure 1-4, p. 9). In this chapter, we look more closely at how ecosystems work and how human activities, such as stripping a large area of its trees, can disrupt the cycling of nutrients within ecosystems and the flow of energy through them.

Figure 3-1 Natural capital degradation: satellite image of the loss of tropical rain forest, cleared for farming, cattle grazing, and settlements, near the Bolivian city of Santa Cruz between June 1975 (left) and May 2003 (right).

3-1 What is ecology?

CONCEPT 3-1 Ecology is the study of how organisms interact with one another and with their physical environment of matter and energy.

3-2 What keeps us and other organisms alive?

CONCEPT 3-2 Life is sustained by the flow of energy from the sun through the biosphere, the cycling of nutrients within the biosphere, and gravity.

3-3 What are the major components of an ecosystem?

CONCEPT 3-3A Ecosystems contain living (biotic) and nonliving (abiotic) components.

CONCEPT 3-3B Some organisms produce the nutrients they need, others get their nutrients by consuming other organisms, and some recycle nutrients back to producers by decomposing the wastes and remains of organisms.

3-4 What happens to energy in an ecosystem?

CONCEPT 3-4A Energy flows through ecosystems in food chains and webs.

CONCEPT 3-4B As energy flows through ecosystems in food chains and webs, the amount of chemical energy available to organisms at each succeeding feeding level decreases.

3-5 What happens to matter in an ecosystem?

CONCEPT 3-5 Matter, in the form of nutrients, cycles within and among ecosystems and the biosphere, and human activities are altering these chemical cycles.

3-6 How do scientists study ecosystems?

CONCEPT 3-6 Scientists use field research, laboratory research, and mathematical and other models to learn about ecosystems.

Note: Supplements 2 (p. S4), 4 (p. S20), 6 (p. S39), 7 (p. S46), and 13 (p. S78) can be used with this chapter.

The earth's thin film of living matter is sustained by grand-scale cycles of chemical elements.

G. EVELYN HUTCHINSON

3-1 What Is Ecology?

▶ **CONCEPT 3-1 Ecology is the study of how organisms interact with one another and with their physical environment of matter and energy.**

Cells Are the Basic Units of Life

All organisms (living things) are composed of **cells:** the smallest and most fundamental structural and functional units of life. They are minute compartments covered with a thin membrane and within which the processes of life occur. The idea that all living things are composed of cells is called the **cell theory** and it is the most widely accepted scientific theory in biology. Organisms may consist of a single cell (bacteria, for instance) or huge numbers of cells, as is the case for most plants and animals.

On the basis of their cell structure, organisms can be classified as either *eukaryotic* or *prokaryotic.* A **eukaryotic cell** is surrounded by a membrane and has a distinct *nucleus* (a membrane-bounded structure containing genetic material in the form of DNA) and

several other internal parts called *organelles,* which are also surrounded by membranes (Figure 3-2a, p. 52). Most organisms consist of eukaryotic cells. A **prokaryotic cell** is also surrounded by a membrane, but it has no distinct nucleus and no other internal parts surrounded by membranes (Figure 3-2b, p. 52). All bacteria consist of a single prokaryotic cell. The relationships among cells and their genetic material were shown in Figure 2-5 (p. 38).

Species Make Up the Encyclopedia of Life

For a group of sexually reproducing organisms, a **species** is a set of individuals that can mate and produce fertile offspring. Every organism is a member of a certain species with certain traits. Scientists have developed a

Figure 3-2 Natural capital: (a) generalized structure of a *eukaryotic cell* and (b) *prokaryotic cell.* Note that a prokaryotic cell lacks a distinct nucleus and generalized structure of a *eukaryotic cell.*

(a) Eukaryotic Cell

Energy conversion

Nucleus (DNA)

Protein construction

Cell membrane

(b) Prokaryotic Cell

DNA (no nucleus)

Cell membrane

Protein construction and energy conversion occur without specialized internal structures

Biosphere Parts of the earth's air, water, and soil where life is found

Ecosystem A community of different species interacting with one another and with their nonliving environment of matter and energy

Community Populations of different species living in a particular place, and potentially interacting with each other

Population A group of individuals of the same species living in a particular place

Organism An individual living being

Cell The fundemental structural and functional unit of life

Molecule Chemical combination of two or more atoms of the same or different elements

Water

Atom Smallest unit of a chemical element that exhibits its chemical properties

Hydrogen Oxygen

distinctive system for classifying and naming each species, as discussed in Supplement 7 on p. S46.

We do not know how many species are on the earth. Estimates range from 4 million to 100 million. The best guess is that there are 10–14 million species. So far biologists have identified about 1.8 million species. These and millions of species still to be classified are the entries in the encyclopedia of life found on the earth. Up to half of the world's plant and animal species live in tropical rain forests that are being cleared rapidly (**Core Case Study**). Insects make up most of the world's known species (Science Focus, p. 54). In 2007, scientists began a $100 million, 10-year project to list and describe all 1.8 million known species in a free Internet encyclopedia (**www.eol.org**).

CORE CASE STUDY

Ecologists Study Connections in Nature

Ecology (from the Greek words *oikos,* meaning "house" or "place to live," and *logos,* meaning "study of") is the study of how organisms interact with their living (biotic) environment of other organisms and with their nonliving (abiotic) environment of soil, water, other forms of matter, and energy mostly from the sun (**Concept 3-1**). In effect, it is a study of *connections in nature.*

To enhance their understanding of nature, scientists classify matter into levels of organization from atoms to the biosphere (Figure 3-3). Ecologists focus on organisms, populations, communities, ecosystems, and the biosphere.

A **population** is a group of individuals of the same species that live in the same place at the same time. Examples include a school of glassfish in the Red Sea (Figure 3-4), the field mice living in a cornfield, monarch butterflies clustered in a tree, and people in a country.

CENGAGENOW™ **Active Figure 3-3** Some levels of organization of matter in nature. Ecology focuses on the top five of these levels. *See an animation based on this figure at CengageNOW.*

Wolfgang Poelzer/Peter Arnold, Inc.

In most natural populations, individuals vary slightly in their genetic makeup, which is why they do not all look or act alike. This variation in a population is called **genetic diversity** (Figure 3-5).

The place where a population or an individual organism normally lives is its **habitat.** It may be as large as an ocean or as small as the intestine of a termite. An organism's habitat can be thought of as its natural "address." Each habitat, such as a tropical rain forest (Figure 3-1, **Core Case Study**), a desert, or a pond, has certain resources, such as water, and environmental conditions, such as temperature and light, that its organisms need in order to survive.

A **community,** or **biological community,** consists of all the populations of different species that live in a particular place. For example, a catfish species in a pond usually shares the pond with other fish species, and with plants, insects, ducks, and many other species that make up the community. Many of the organisms in a community interact with one another in feeding and other relationships.

An **ecosystem** is a community of different species interacting with one another and with their nonliving environment of soil, water, other forms of matter, and energy, mostly from the sun. Ecosystems can range in size from a puddle of water to an ocean, or from a patch of woods to a forest. Ecosystems can be natural or artificial (human created). Examples of artificial ecosystems are crop fields, tree farms, and reservoirs.

Ecosystems do not have clear boundaries and are not isolated from one another. Matter and energy move from one ecosystem to another. For example, soil can wash from a grassland or crop field into a nearby river or lake. Water flows from forests into nearby rivers and crop fields. Birds and various other species migrate from one ecosystem to another. And winds can blow pollen from a forest into a grassland.

The **biosphere** consists of the parts of the earth's air, water, and soil where life is found. In effect, it is the global ecosystem in which all organisms exist and can interact with one another.

CENGAGENOW⁻ Learn more about how the earth's life is organized in five levels in the study of ecology at CengageNOW™.

Figure 3-5 *Genetic diversity* among individuals in a population of a species of Caribbean snail is reflected in the variations in shell color and banding patterns. Genetic diversity can also include other variations such as slight differences in chemical makeup, sensitivity to various chemicals, and behavior.

SCIENCE FOCUS

Have You Thanked the Insects Today?

Insects are an important part of the earth's natural capital, although they generally have a bad reputation. We classify many insect species as *pests* because they compete with us for food, spread human diseases such as malaria, bite or sting us, and invade our lawns, gardens, and houses. Some people fear insects and think the only good bug is a dead bug. They fail to recognize the vital roles insects play in helping to sustain life on earth.

For example, pollination is a natural service that allows plants to reproduce sexually when pollen grains are transferred from one plant to a receptive part of another plant. Some plants are pollinated by species such as humming-birds and bats, and by pollen being transmit-ted by wind or flowing water. But many of the earth's plant species depend on insects to pollinate their flowers (Figure 3-A, left).

Insects that eat other insects—such as the praying mantis (Figure 3-A, right)—help control the populations of at least half the species of insects we call pests. This free pest control service is an important part of the earth's natural capital. Some insects also play a key role in loosening and renewing the soil that supports terrestrial plant life. In 2006, scientists John Losey and Mace Vaughan estimated that the value of the ecological services provided by insects in the United States is at least $57 billion a year.

Figure 3-A *Importance of insects:* The monarch butterfly, which feeds on pollen in a flower (left), and other insects pollinate flowering plants that serve as food for many plant eaters. The praying mantis, which is eating a house cricket (right), and many other insect species help to control the populations of most of the insect species we classify as pests.

Insects have been around for at least 400 million years—about 4,000 times longer than the latest version of the human species that we belong to. They are phenomenally successful forms of life. Some reproduce at an astounding rate and can rapidly develop new genetic traits, such as resistance to pes-ticides. They also have an exceptional ability to evolve into new species when faced with new environmental conditions, and they are very resistant to extinction. This is fortunate because, according to ant specialist and biodiversity expert E. O. Wilson, if all insects disappeared, parts of the life support systems for us and other species would be greatly disrupted.

The environmental lesson: although in-sects do not need newcomer species such as us, we and most other land organisms need them.

Critical Thinking

Identify three insect species not discussed above that benefit your life.

3-2 What Keeps Us and Other Organisms Alive?

▶ **CONCEPT 3-2** Life is sustained by the flow of energy from the sun through the biosphere, the cycling of nutrients within the biosphere, and gravity.

The Earth's Life–Support System Has Four Major Components

Scientific studies reveal that the earth's life-support system consists of four main spherical systems that in-teract with one another—the atmosphere (air), the hy-drosphere (water), the geosphere (rock, soil, sediment), and the biosphere (living things) (Figure 3-6).

The **atmosphere** is a thin spherical envelope of gases surrounding the earth's surface. Its inner layer, the **troposphere,** extends only about 17 kilometers (11 miles) above sea level at the tropics and about 7 kilometers (4 miles) above the earth's north and south poles. It contains the majority of the air that we breathe, consisting mostly of nitrogen (78% of the total volume) and oxygen (21%). The remaining 1% of the air includes water vapor, carbon dioxide, and methane, all of which are called **greenhouse gases,** because they trap heat and thus warm the lower atmosphere. Almost all of the earth's weather occurs in this layer.

The next layer, stretching 17–50 kilometers (11–31 miles) above the earth's surface, is the **strato-sphere.** Its lower portion contains enough ozone (O_3) gas to filter out most of the sun's harmful ultraviolet radiation. This global sunscreen allows life to exist on land and in the surface layers of bodies of water.

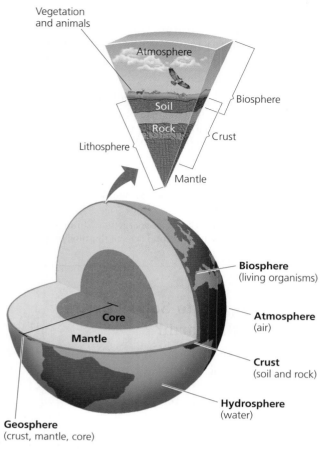

Figure 3-6 **Natural capital:** general structure of the earth showing that it consists of a land sphere, air sphere, water sphere, and life sphere.

The **hydrosphere** consists of all of the water on or near the earth's surface. It is found as *liquid water* (on the surface and underground), *ice* (polar ice, icebergs, and ice in frozen soil layers called *permafrost*), and *water*

vapor in the atmosphere. Most of this water is in the oceans, which cover about 71% of the globe.

The **geosphere** consists of the earth's intensely hot *core,* a thick *mantle* composed mostly of rock, and a thin outer *crust.* Most of the geosphere is located in the earth's interior. Its upper portion contains nonrenewable fossil fuels and minerals that we use, as well as renewable soil chemicals that organisms need in order to live, grow, and reproduce.

The *biosphere* occupies those parts of the atmosphere, hydrosphere, and geosphere where life exists. This thin layer of the earth extends from about 9 kilometers (6 miles) above the earth's surface down to the bottom of the ocean, and it includes the lower part of the atmosphere, most of the hydrosphere, and the uppermost part of the geosphere. If the earth were an apple, the biosphere would be no thicker than the apple's skin. *The goal of ecology is to understand the interactions in this thin layer of air, water, soil, and organisms.*

Life Exists on Land and in Water

Biologists have classified the terrestrial (land) portion of the biosphere into **biomes**—large regions such as forests, deserts, and grasslands, with distinct climates and certain species (especially vegetation) adapted to them. Figure 3-7 shows different major biomes along the 39th parallel spanning the United States (see Figure 5 on p. S27 in Supplement 4 for a map of the major biomes

Average annual precipitation
- 100–125 cm (40–50 in.)
- 75–100 cm (30–40 in.)
- 50–75 cm (20–30 in.)
- 25–50 cm (10–20 in.)
- below 25 cm (0–10 in.)

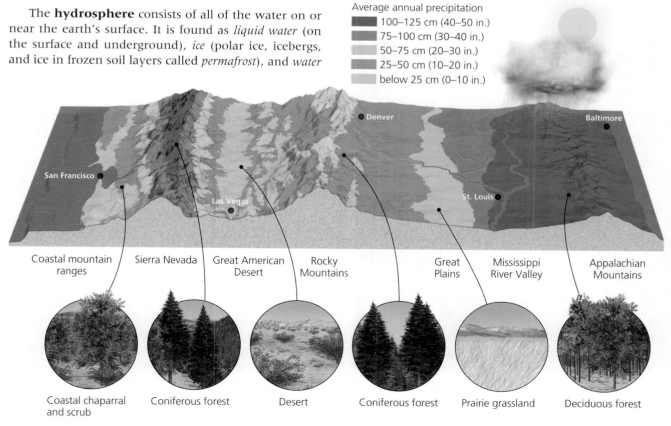

Coastal mountain ranges — Coastal chaparral and scrub

Sierra Nevada — Coniferous forest

Great American Desert — Desert

Rocky Mountains — Coniferous forest

Great Plains — Prairie grassland

Mississippi River Valley

Appalachian Mountains — Deciduous forest

Figure 3-7 Major biomes found along the 39th parallel across the United States. The differences reflect changes in climate, mainly differences in average annual precipitation and temperature.

of North America). We discuss biomes in more detail in Chapter 7.

Scientists divide the watery parts of the biosphere into **aquatic life zones,** each containing numerous ecosystems. There are *freshwater life zones* (such as lakes and streams) and *ocean* or *marine life zones* (such as coral reefs and coastal estuaries). The earth is mostly a water planet with saltwater covering about 71% of its surface and freshwater covering just 2%.

Three Factors Sustain Life on Earth

Life on the earth depends on three interconnected factors (**Concept 3-2**):

- The *one-way flow of high-quality energy* from the sun, through living things in their feeding interactions, into the environment as low-quality energy (mostly heat dispersed into air or water at a low temperature), and eventually back into space as heat. No round-trips are allowed because high-quality energy cannot be recycled. The first and second laws of thermodynamics (**Concepts 2-4A** and **2-4B**, p. 40) govern this energy flow. **CONCEPT LINKS**

- The *cycling of matter* or *nutrients* (the atoms, ions, and compounds needed for survival by living organisms) through parts of the biosphere. Because the earth is closed to significant inputs of matter from space, its essentially fixed supply of nutrients must be continually recycled to support life (Figure 1-4,

p. 9). Nutrient movements in ecosystems and in the biosphere are round-trips, which can take from seconds to centuries to complete. The law of conservation of matter (**Concept 2-3**, p. 39) governs this nutrient cycling process. **CONCEPT LINK**

- *Gravity*, which allows the planet to hold onto its atmosphere and helps to enable the movement and cycling of chemicals through the air, water, soil, and organisms.

THINKING ABOUT
Energy Flow and the First and Second Laws of Thermodynamics

Explain the relationship between energy flow through the biosphere and the first and second laws of thermodynamics (pp. 42–43).

What Happens to Solar Energy Reaching the Earth?

Millions of kilometers from the earth, in the immense nuclear fusion reactor that is the sun, nuclei of hydrogen fuse together to form larger helium nuclei (Figure 2-7, bottom, p. 41), releasing tremendous amounts of energy into space. Only a very small amount of this output of energy reaches the earth—a tiny sphere in the vastness of space. This energy reaches the earth in the form of electromagnetic waves, mostly as visible light, ultraviolet (UV) radiation, and heat (infrared radiation) (Figure 2-8, p. 42). Much of this energy is absorbed or reflected back into space by the earth's atmosphere, clouds, and surface (Figure 3-8). Ozone gas (O_3) in the lower stratosphere absorbs about 95% of the sun's harmful incoming UV radiation. Without this ozone layer, life as we know it on the land and in the upper layer of water would not exist.

The UV, visible, and infrared energy that reaches the atmosphere lights the earth during daytime, warms the air, and evaporates and cycles water through the biosphere. Approximately 1% of this incoming energy generates winds. Green plants, algae, and some types of bacteria use less than 0.1% of it to produce the nutrients they need through photosynthesis and in turn to feed animals that eat plants and flesh.

Of the total solar radiation intercepted by the earth, about 1% reaches the earth's surface, and most of it is then reflected as longer-wavelength infrared radiation. As this infrared radiation travels back up through the lower atmosphere toward space, it encounters greenhouse gases such as water vapor, carbon dioxide, methane, nitrous oxide, and ozone. It causes these gaseous molecules to vibrate and release infrared radiation with even longer wavelengths. The vibrating gaseous molecules then have higher kinetic energy, which helps to warm the lower atmosphere and the earth's surface. Without this **natural greenhouse effect,** the earth

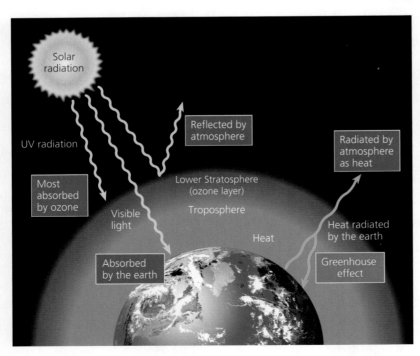

CENGAGENOW **Active Figure 3-8** *Solar capital:* flow of energy to and from the earth. *See an animation based on this figure at* CengageNOW.

would be too cold to support the forms of life we find here today. (See *The Habitable Planet,* Video 2, **www.learner.org/resources/series209.html**.)

Human activities add greenhouse gases to the atmosphere. For example, burning carbon-containing fuels releases huge amounts of carbon dioxide (CO_2) into the atmosphere. Growing crops and raising livestock release large amounts of methane (CH_4) and nitrous oxide (N_2O). Clearing CO_2-absorbing tropical rain forests (**Core Case Study**) faster than they can

CORE CASE STUDY

grow back also increases the amount of CO_2 in the atmosphere. There is considerable and growing evidence that these activities are increasing the natural greenhouse effect and warming the earth's atmosphere (Science Focus, p. 33). This in turn is changing the earth's climate as we discuss at length in Chapter 19.

CENGAGENOW™ Learn more about the flow of energy—from sun to earth and within the earth's systems—at CengageNOW.

3-3 What Are the Major Components of an Ecosystem?

▶ **CONCEPT 3-3A** Ecosystems contain living (biotic) and nonliving (abiotic) components.

▶ **CONCEPT 3-3B** Some organisms produce the nutrients they need, others get their nutrients by consuming other organisms, and some recycle nutrients back to producers by decomposing the wastes and remains of organisms.

Ecosystems Have Living and Nonliving Components

Two types of components make up the biosphere and its ecosystems: One type, called **abiotic,** consists of nonliving components such as water, air, nutrients, rocks, heat, and solar energy. The other type, called **biotic,** consists of living and once living biological components—plants, animals, and microbes (**Concept 3-3A**). Biotic factors also include dead organisms, dead parts of organisms, and the waste products of organisms. Figure 3-9 is a greatly simplified diagram of some of the biotic and abiotic components of a terrestrial ecosystem.

Different species and their populations thrive under different physical and chemical conditions. Some need bright sunlight; others flourish in shade. Some need a hot environment; others prefer a cool or cold one. Some do best under wet conditions; others thrive under dry conditions.

Each population in an ecosystem has a **range of tolerance** to variations in its physical and chemical environment, as shown in Figure 3-10 (p. 58). Individuals within a population may also have slightly different tolerance ranges for temperature or other factors because of small differences in genetic makeup, health, and age. For example, a trout population may do best within

CENGAGENOW™ **Active Figure 3-9** Major living (biotic) and nonliving (abiotic) components of an ecosystem in a field. *See an animation based on this figure at* CengageNOW.

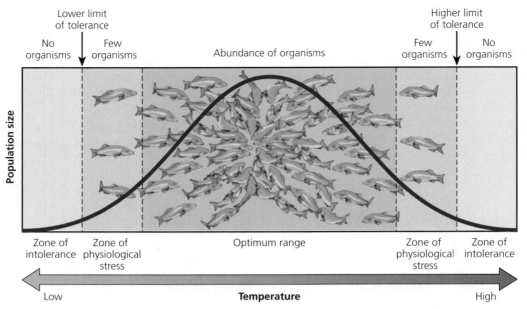

Lower limit
of tolerance

Higher limit
of tolerance

No
organisms

Few
organisms

Abundance of organisms

Few
organisms

No
organisms

Zone of
intolerance

Zone of
physiological
stress

Optimum range

Zone of
physiological
stress

Zone of
intolerance

Low

Temperature

High

Figure 3-10 Range of tolerance for a population of organisms, such as fish, to an abiotic environmental factor—in this case, temperature. These restrictions keep particular species from taking over an ecosystem by keeping their population size in check. **Question:** Which **scientific principle of sustainability** (see back cover) is related to the range of tolerance concept?

a narrow band of temperatures (*optimum level* or *range*), but a few individuals can survive above and below that band. Of course, if the water becomes much too hot or too cold, none of the trout can survive.

Several Abiotic Factors Can Limit Population Growth

A variety of abiotic factors can affect the number of organisms in a population. Sometimes one or more factors, known as **limiting factors,** are more important in regulating population growth than other factors are. This ecological principle is called the **limiting factor principle:** *Too much or too little of any abiotic factor can limit or prevent growth of a population, even if all other factors are at or near the optimal range of tolerance.* This principle describes one way in which population control—a **scientific principle of sustainability** (see back cover)—is achieved.

On land, precipitation often is the limiting abiotic factor. Lack of water in a desert limits plant growth. Soil nutrients also can act as a limiting factor on land. Suppose a farmer plants corn in phosphorus-poor soil. Even if water, nitrogen, potassium, and other nutrients are at optimal levels, the corn will stop growing when it uses up the available phosphorus. Too much of an abiotic factor can also be limiting. For example, too much water or fertilizer can kill plants. Temperature can also be a limiting factor. Both high and low temperatures can limit the survival and population sizes of various terrestrial species, especially plants.

Important limiting abiotic factors in aquatic life zones include temperature, sunlight, nutrient availability, and the low solubility of oxygen gas in water (*dissolved oxygen content*). Another such factor is *salinity*—the amounts of various inorganic minerals or salts dissolved in a given volume of water.

Producers and Consumers Are the Living Components of Ecosystems

Ecologists assign every organism in an ecosystem to a *feeding level,* or **trophic level,** depending on its source of food or nutrients. The organisms that transfer energy and nutrients from one trophic level to another in an ecosystem can be broadly classified as producers and consumers.

Producers, sometimes called **autotrophs** (self-feeders), make the nutrients they need from compounds and energy obtained from their environment. On land, most producers are green plants, which generally capture about 1% of the solar energy that falls on their leaves and convert it to chemical energy stored in organic molecules such as carbohydrates. In freshwater and marine ecosystems, algae and aquatic plants are the major producers near shorelines. In open water, the dominant producers are *phytoplankton*—mostly microscopic organisms that float or drift in the water.

Most producers capture sunlight to produce energy-rich carbohydrates (such as glucose, $C_6H_{12}O_6$) by **photosynthesis,** which is the way energy enters most ecosystems. Although hundreds of chemical changes

take place during photosynthesis, the overall reaction can be summarized as follows:

carbon dioxide + water + solar energy $\longrightarrow$ glucose + oxygen

$6\,CO_2\ \ \ + 6\,H_2O$ + solar energy $\longrightarrow C_6H_{12}O_6\ +\ 6\,O_2$

(See p. S44 in Supplement 6 for information on how to balance chemical equations such as this one and p. S44 in Supplement 6 for more details on photosynthesis.)

A few producers, mostly specialized bacteria, can convert simple inorganic compounds from their environment into more complex nutrient compounds without using sunlight, through a process called **chemosynthesis.** In 1977, scientists discovered a community of bacteria living in the extremely hot water around *hydrothermal vents* on the deep ocean floor. These bacteria serve as producers for their ecosystems without the use of sunlight. They draw energy and produce carbohydrates from hydrogen sulfide (H_2S) gas escaping through fissures in the ocean floor. Most of the earth's organisms get their energy indirectly from the sun. But chemosynthetic organisms in these dark and deep-sea habitats survive indirectly on *geothermal energy* from the earth's interior and represent an exception to the first **scientific principle of sustainability.**

All other organisms in an ecosystem are **consumers,** or **heterotrophs** ("other-feeders"), that cannot produce the nutrients they need through photosynthesis or other processes and must obtain their nutrients by feeding on other organisms (producers or other consumers) or their remains. In other words, all consumers (including humans) are directly or indirectly dependent on producers for their food or nutrients.

There are several types of consumers:

- **Primary consumers,** or **herbivores** (plant eaters), are animals such as rabbits, grasshoppers, deer, and zooplankton that eat producers, mostly by feeding on green plants.

- **Secondary consumers,** or **carnivores** (meat eaters), are animals such as spiders, hyenas, birds, frogs, and some zooplankton-eating fish, all of which feed on the flesh of herbivores.

- **Third-** and **higher-level consumers** are carnivores such as tigers, wolves, mice-eating snakes, hawks, and killer whales (orcas) that feed on the flesh of other carnivores. Some of these relationships are shown in Figure 3-9.

- **Omnivores** such as pigs, foxes, cockroaches, and humans, play dual roles by feeding on both plants and animals.

- **Decomposers,** primarily certain types of bacteria and fungi, are consumers that release nutrients from the dead bodies of plants and animals and return them to the soil, water, and air for reuse by producers. They feed by secreting enzymes that speed up the break down of bodies of dead organisms into nutrient compounds such as water, carbon dioxide, minerals, and simpler organic compounds.

- **Detritus feeders,** or **detritivores,** feed on the wastes or dead bodies of other organisms, called **detritus** ("di-TRI-tus," meaning debris). Examples include small organisms such as mites and earthworms, some insects, catfish, and larger scavenger organisms such as vultures.

Hordes of decomposers and detritus feeders can transform a fallen tree trunk into a powder and finally into simple inorganic molecules that plants can absorb as nutrients (Figure 3-11, p. 60). In summary, some organisms produce the nutrients they need, others get their nutrients by consuming other organisms, and still others recycle the nutrients in the wastes and remains of organisms so that producers can use them again (**Concept 3-3B**).

THINKING ABOUT
What You Eat

When you had your most recent meal, were you an herbivore, a carnivore, or an omnivore?

Producers, consumers, and decomposers use the chemical energy stored in glucose and other organic compounds to fuel their life processes. In most cells this energy is released by **aerobic respiration,** which uses oxygen to convert glucose (or other organic nutrient molecules) back into carbon dioxide and water. The net effect of the hundreds of steps in this complex process is represented by the following reaction:

glucose + oxygen $\longrightarrow$ carbon dioxide + water + energy

$C_6H_{12}O_6\ +\ 6\,O_2 \longrightarrow\ \ \ 6\,CO_2\ \ \ + 6\,H_2O$ + energy

Although the detailed steps differ, the net chemical change for aerobic respiration is the opposite of that for photosynthesis.

Some decomposers get the energy they need by breaking down glucose (or other organic compounds) in the *absence* of oxygen. This form of cellular respiration is called **anaerobic respiration,** or **fermentation.** Instead of carbon dioxide and water, the end products of this process are compounds such as methane gas (CH_4, the main component of natural gas), ethyl alcohol (C_2H_6O), acetic acid ($C_2H_4O_2$, the key component of vinegar), and hydrogen sulfide (H_2S, when sulfur compounds are broken down). Note that all organisms get their energy from aerobic or anaerobic respiration but only plants carry out photosynthesis.

CENGAGENOW‾ Explore the components of ecosystems, how they interact, the roles of bugs and plants, and what a fox will eat at CengageNOW.

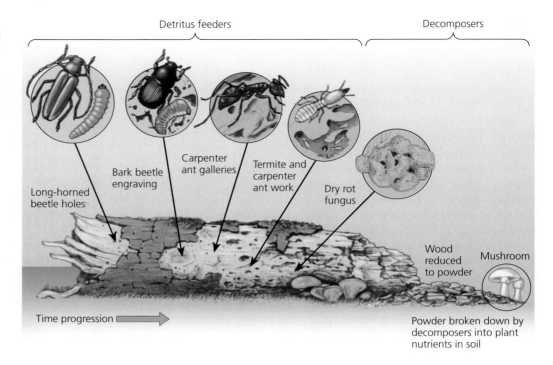

Figure 3-11 Various detritivores and decomposers (mostly fungi and bacteria) can "feed on" or digest parts of a log and eventually convert its complex organic chemicals into simpler inorganic nutrients that can be taken up by producers.

Detritus feeders

Decomposers

Long-horned beetle holes

Bark beetle engraving

Carpenter ant galleries

Termite and carpenter ant work

Dry rot fungus

Wood reduced to powder

Mushroom

Time progression

Powder broken down by decomposers into plant nutrients in soil

Energy Flow and Nutrient Cycling Sustain Ecosystems and the Biosphere

Ecosystems and the biosphere are sustained through a combination of *one-way energy flow* from the sun through these systems and *nutrient cycling* of key mate-rials within them—two important natural services that are components of the earth's natural capital. These two **scientific principles of sustainability** arise from the structure and function of natural ecosystems (Figure 3-12), the law of conservation of matter (**Concept 2-3**, p. 39), and the two laws of thermodynamics (**Concepts 2-4A** and **2-4B**, p. 40).

SUSTAINABILITY

CONCEPT LINKS

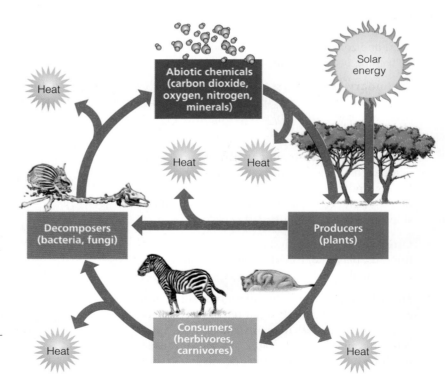

CENGAGENOW **Active Figure 3-12** **Natural capital:** the main structural components of an ecosystem (energy, chemicals, and organisms). Nutrient cycling and the flow of energy—first from the sun, then through organisms, and finally into the environment as low-quality heat—link these components. *See an animation based on this figure at* CengageNOW.

Abiotic chemicals (carbon dioxide, oxygen, nitrogen, minerals)

Solar energy

Heat

Heat

Heat

Decomposers (bacteria, fungi)

Producers (plants)

Consumers (herbivores, carnivores)

Heat

Heat

SCIENCE FOCUS

They are everywhere. Billions of them can be found inside your body, on your body, in a handful of soil, and in a cup of ocean water.

These mostly invisible rulers of the earth are *microbes,* or *microorganisms,* catchall terms for many thousands of species of bacteria, protozoa, fungi, and floating phytoplankton—most too small to be seen with the naked eye.

Microbes do not get the respect they deserve. Most of us view them primarily as threats to our health in the form of infectious bacteria or "germs," fungi that cause athlete's foot and other skin diseases, and protozoa that cause diseases such as malaria. But these harmful microbes are in the minority.

We are alive because of multitudes of microbes toiling away mostly out of sight.

Bacteria in our intestinal tracts help to break down the food we eat and microbes in our noses help to prevent harmful bacteria from reaching our lungs.

Bacteria and other microbes help to purify the water we drink by breaking down wastes. Bacteria also help to produce foods such as bread, cheese, yogurt, soy sauce, beer, and wine. Bacteria and fungi in the soil decompose organic wastes into nutrients that can be taken up by plants that we and most other animals eat. Without these tiny creatures, we would go hungry and be up to our necks in waste matter.

Microbes, particularly phytoplankton in the ocean, provide much of the planet's oxygen, and help slow global warming by removing some of the carbon dioxide produced when we burn coal, natural gas, and

gasoline. (See *The Habitable Planet,* Video 3, www.learner.org/resources/series209 .html.) Scientists are working on using microbes to develop new medicines and fuels. Genetic engineers are inserting genetic material into existing microbes to convert them to microbes that can help clean up polluted water and soils.

Some microbes help control diseases that affect plants and populations of insect species that attack our food crops. Relying more on these microbes for pest control could reduce the use of potentially harmful chemical pesticides. In other words, microbes are a vital part of the earth's natural capital.

Critical Thinking

What are three advantages that microbes have over us for thriving in the world?

Decomposers and detritus feeders, many of which are microscopic organisms, (Science Focus, above) are the key to nutrient cycling because they break down organic matter into simpler nutrients that can be reused by producers. Without decomposers and detritus feeders, there would be little, if any, nutrient cycling and the planet would be overwhelmed with plant litter, dead animal bodies, animal wastes, and garbage. In addition, most life as we know it could not exist because the nutrients stored in such wastes and dead

bodies would be locked up and unavailable for use by other organisms.

> **THINKING ABOUT**
> **Chemical Cycling and the Law**
> **of Conservation of Matter**
> CONCEPT LINK
>
> Explain the relationship between chemical cycling in ecosystems and in the biosphere and the law of conservation of matter (**Concept 2-3**, p. 39).

3-4 What Happens to Energy in an Ecosystem?

▶ **CONCEPT 3-4A** Energy flows through ecosystems in food chains and webs.

▶ **CONCEPT 3-4B** As energy flows through ecosystems in food chains and webs, the amount of chemical energy available to organisms at each succeeding feeding level decreases.

Energy Flows through Ecosystems in Food Chains and Food Webs

The chemical energy stored as nutrients in the bodies and wastes of organisms flows through ecosystems from one trophic (feeding) level to another. For example, a plant uses solar energy to store chemical energy in a leaf. A caterpillar eats the leaf, a robin eats the caterpillar, and a hawk eats the robin. Decomposers and detritus feeders consume the leaf, caterpillar, robin, and hawk after they die and return their nutrients to the soil for reuse by producers.

A sequence of organisms, each of which serves as a source of food or energy for the next, is called a **food chain.** It determines how chemical energy and nutrients move from one organism to another through the trophic levels in an ecosystem—primarily through photosynthesis, feeding, and decomposition—as shown

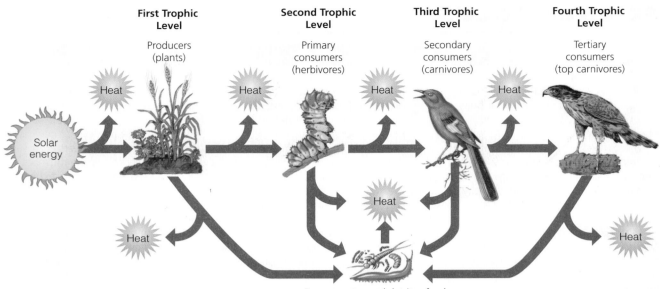

First Trophic Level	Second Trophic Level	Third Trophic Level	Fourth Trophic Level
Producers (plants)	Primary consumers (herbivores)	Secondary consumers (carnivores)	Tertiary consumers (top carnivores)

Heat

Heat

Heat

Heat

Solar energy

Heat

Heat

Heat

Decomposers and detritus feeders

CENGAGENOW™ **Active Figure 3-13** A *food chain*. The arrows show how chemical energy in nutrients flows through various *trophic levels* in energy transfers; most of the energy is degraded to heat, in accordance with the second law of thermodynamics. *See an animation based on this figure at* CengageNOW. **Question:** Think about what you ate for breakfast. At what level or levels on a food chain were you eating?

in Figure 3-13. Every use and transfer of energy by organisms involves a loss of some useful energy to the environment as heat. Thus, eventually an ecosystem and the biosphere would run out of energy if they were not powered by a continuous inflow of energy from an outside source, ultimately the sun.

In natural ecosystems, most consumers feed on more than one type of organism, and most organisms are eaten or decomposed by more than one type of consumer. Because of this, organisms in most ecosystems form a complex network of interconnected food chains called a **food web** (Figure 3-14). Trophic levels can be assigned in food webs just as in food chains. Food chains and webs show how producers, consumers, and decomposers are connected to one another as energy flows through trophic levels in an ecosystem.

> **THINKING ABOUT**
> **Energy Flow and Tropical Rain forests**
> What happens to the flow of energy through tropical rain forest ecosystems when such forests are degraded (**Core Case Study**)?
>
> CORE CASE STUDY

Usable Energy Decreases with Each Link in a Food Chain or Web

Each trophic level in a food chain or web contains a certain amount of **biomass,** the dry weight of all organic matter contained in its organisms. In a food chain or web, chemical energy stored in biomass is transferred from one trophic level to another.

Energy transfer through food chains and food webs is not very efficient because, with each transfer, some usable chemical energy is degraded and lost to the environment as low-quality heat, as a result of the second law of thermodynamics. In other words, as energy flows through ecosystems in food chains and webs, there is a decrease in the amount of chemical energy available to organisms at each succeeding feeding level (**Concept 3-4B**).

The percentage of usable chemical energy transferred as biomass from one trophic level to the next is called **ecological efficiency.** It ranges from 2% to 40% (that is, a loss of 60–98%) depending on what types of species and ecosystems are involved, but 10% is typical.

Assuming 10% ecological efficiency (90% loss of usable energy) at each trophic transfer, if green plants in an area manage to capture 10,000 units of energy from the sun, then only about 1,000 units of chemical energy will be available to support herbivores, and only about 100 units will be available to support carnivores.

The more trophic levels there are in a food chain or web, the greater is the cumulative loss of usable chemical energy as it flows through the trophic levels. The **pyramid of energy flow** in Figure 3-15 illustrates this energy loss for a simple food chain, assuming a 90% energy loss with each transfer.

> **THINKING ABOUT**
> **Energy Flow and the Second Law of Thermodynamics**
> Explain the relationship between the second law of thermodynamics (**Concept 2-4B**, p. 40) and the flow of energy through a food chain or web.
>
> CONCEPT LINK

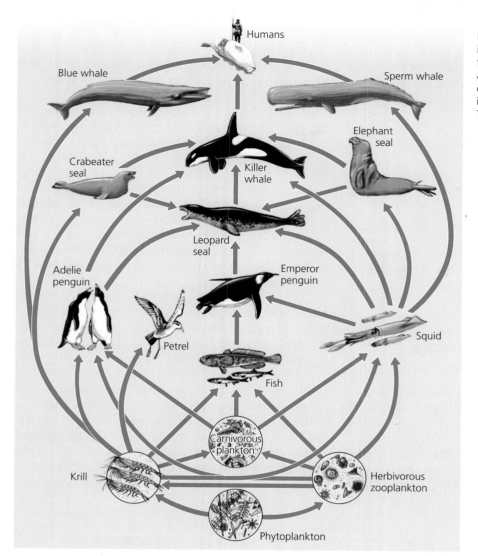

Figure 3-14 *Greatly simplified food web* in the Antarctic. Many more participants in the web, including an array of decomposer and detritus feeder organisms, are not depicted here. **Question:** Can you imagine a food web of which you are a part? Try drawing a simple diagram of it.

Energy flow pyramids explain why the earth can support more people if they eat at lower trophic levels by consuming grains, vegetables, and fruits directly, rather than passing such crops through another trophic level and eating grain eaters or herbivores such as cattle. About two-thirds of the world's people survive primarily by eating wheat, rice, and corn at the first trophic level, mostly because they cannot afford meat.

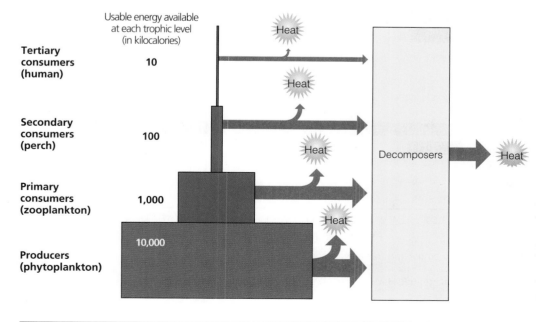

Figure 3-15 Generalized *pyramid of energy flow* showing the decrease in usable chemical energy available at each succeeding trophic level in a food chain or web. In nature, ecological efficiency varies from 2% to 40%, with 10% efficiency being common. This model assumes a 10% ecological efficiency (90% loss of usable energy to the environment, in the form of low-quality heat) with each transfer from one trophic level to another. **Question:** Why is a vegetarian diet more energy efficient than a meat-based diet?

The large loss in chemical energy between successive trophic levels also explains why food chains and webs rarely have more than four or five trophic levels. In most cases, too little chemical energy is left after four or five transfers to support organisms feeding at these high trophic levels.

┌─ THINKING ABOUT ──────────────────
│ **Food Webs, Tigers, and Insects**
│
│ Use Figure 3-15 to help explain **(a)** why there are not many
│ tigers in the world and why they are vulnerable to premature
│ extinction because of human activities, and **(b)** why there are
│ so many insects (Science Focus, p. 54) in the world.
└──────────────────────────────────

CENGAGENOW™ Examine how energy flows among organisms at different trophic levels and through food webs in tropical rain forests, prairies, and other ecosystems at CengageNOW.

Some Ecosystems Produce Plant Matter Faster Than Others Do

The amount, or mass, of living organic material (biomass) that a particular ecosystem can support is determined by the amount of energy captured and stored as chemical energy by the producers of that ecosystem and by how rapidly they can produce and store such chemical energy. **Gross primary productivity (GPP)** is the *rate* at which an ecosystem's producers (usually plants) convert solar energy into chemical energy as biomass found in their tissues. It is usually measured in terms of energy production per unit area over a given time span, such as kilocalories per square meter per year (kcal/m²/yr).

To stay alive, grow, and reproduce, producers must use some of the chemical energy stored in the biomass they make for their own respiration. **Net primary productivity (NPP)** is the *rate* at which producers use photosynthesis to produce and store chemical energy *minus* the *rate* at which they use some of this stored chemical energy through aerobic respiration. In other words, NPP = GPP − R, where R is energy used in respiration. NPP measures how fast producers can provide the chemical energy stored in their tissue that is potentially available to other organisms (consumers) in an ecosystem.

Ecosystems and life zones differ in their NPP as illustrated in Figure 3-16. On land, NPP generally decreases from the equator toward the poles because the amount of solar radiation available to terrestrial plant producers is highest at the equator and lowest at the poles.

In the ocean, the highest NPP is found in estuaries where high inputs of plant nutrients flow from nutrient-laden rivers, which also stir up nutrients in bottom sediments. Because of the lack of nutrients, the open ocean has a low NPP, except at occasional areas where an *upwelling* (water moving up from the depths toward the surface) brings nutrients in bottom sediments to the surface. Despite its low NPP, the open

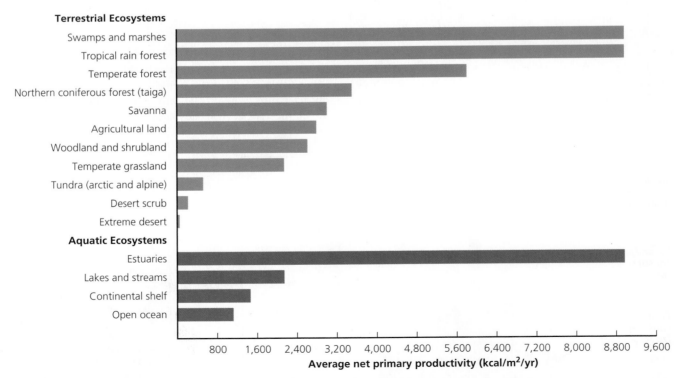

Figure 3-16 Estimated annual average *net primary productivity* in major life zones and ecosystems, expressed as kilocalories of energy produced per square meter per year (kcal/m²/yr). **Question:** What are nature's three most productive and three least productive systems? (Data from R. H. Whittaker, *Communities and Ecosystems,* 2nd ed., New York: Macmillan, 1975)

ocean produces more of the earth's biomass per year than any other ecosystem or life zone, simply because there is so much open ocean.

As we have seen, producers are the source of all nutrients or chemical energy in an ecosystem for themselves and for the animals and decomposers that feed on them. Only the biomass represented by NPP is available as nutrients for consumers, and they use only a portion of this amount. Thus, *the planet's NPP ultimately limits the number of consumers (including humans) that can survive on the earth.* This is an important lesson from nature.

Peter Vitousek, Stuart Rojstaczer, and other ecologists estimate that humans now use, waste, or destroy about 20–32% of the earth's total potential NPP. This is a remarkably high value, considering that the human population makes up less than 1% of the total biomass of all of the earth's consumers that depend on producers for their nutrients.

3-5 What Happens to Matter in an Ecosystem?

▶ **CONCEPT 3-5** Matter, in the form of nutrients, cycles within and among ecosystems and the biosphere, and human activities are altering these chemical cycles.

Nutrients Cycle in the Biosphere

The elements and compounds that make up nutrients move continually through air, water, soil, rock, and living organisms in ecosystems and in the biosphere in cycles called **biogeochemical cycles** (literally, life-earth-chemical cycles) or **nutrient cycles**—prime examples of one of the four scientific principles of sustainability (see back cover). These cycles, driven directly or indirectly by incoming solar energy and gravity, include the hydrologic (water), carbon, nitrogen, phosphorus, and sulfur cycles. These cycles are an important component of the earth's natural capital (Figure 1-3, p. 8), and human activities are altering them (**Concept 3-5**).

As nutrients move through the biogeochemical cycles, they may accumulate in one portion of the cycle and remain there for different lengths of time. These temporary storage sites such as the atmosphere, the oceans and other waters, and underground deposits are called *reservoirs.*

Nutrient cycles connect past, present, and future forms of life. Some of the carbon atoms in your skin may once have been part of an oak leaf, a dinosaur's skin, or a layer of limestone rock. Your grandmother, Attila the Hun, or a hunter–gatherer who lived 25,000 years ago may have inhaled some of the oxygen molecules you just inhaled.

Water Cycles through the Biosphere

The **hydrologic cycle,** or **water cycle,** collects, purifies, and distributes the earth's fixed supply of water, as shown in Figure 3-17 (p. 66). The water cycle is a global cycle because there is a large reservoir of water in the atmosphere as well as in the hydrosphere, especially the oceans. Water is an amazing substance (Science Focus, p. 67), which makes the water cycle critical to life on earth.

The water cycle is powered by energy from the sun and involves three major processes—evaporation, precipitation, and transpiration. Incoming solar energy causes *evaporation* of water from the oceans, lakes, rivers, and soil. Evaporation changes liquid water into water vapor in the atmosphere, and gravity draws the water back to the earth's surface as *precipitation* (rain, snow, sleet, and dew). About 84% of water vapor in the atmosphere comes from the oceans, which cover almost three-fourths of the earth's surface; the rest comes from land. Over land, about 90% of the water that reaches the atmosphere evaporates from the surfaces of plants through a process called **transpiration.**

Water returning to the earth's surface as precipitation takes various paths. Most precipitation falling on terrestrial ecosystems becomes *surface runoff.* This water flows into streams and lakes, which eventually carry water back to the oceans, from which it can evaporate to repeat the cycle. Some surface water also seeps into the upper layer of soils and some evaporates from soil, lakes, and streams back into the atmosphere.

Some precipitation is converted to ice that is stored in *glaciers,* usually for long periods of time. Some precipitation sinks through soil and permeable rock formations to underground layers of rock, sand, and gravel called *aquifers,* where it is stored as *groundwater.*

A small amount of the earth's water ends up in the living components of ecosystems. Roots of plants absorb some of this water, most of which evaporates from plant leaves back into the atmosphere. Some combines with carbon dioxide during photosynthesis to produce high-energy organic compounds such as carbohydrates. Eventually these compounds are broken down in plant cells, which release water back into the environment. Consumers get their water from their food or by drinking it.

Surface runoff replenishes streams and lakes, but also causes soil erosion, which moves soil and rock fragments from one place to another. Water is the primary sculptor of the earth's landscape. Because water

Condensation · Global warming · Ice and snow · Precipitation to land · Transpiration from plants · Evaporation from land · Condensation · Surface runoff · Increased flooding from wetland destruction · Evaporation from ocean · Precipitation to ocean · Runoff · Lakes and reservoirs · Reduced recharge of aquifers and flooding from covering land with crops and buildings · Point source pollution · Infiltration and percolation into aquifer · Surface runoff · Groundwater movement (slow) · Aquifer depletion from overpumping · Ocean

☐ Processes
☐ Processes affected by humans
☐ Reservoir
▶ Pathway affected by humans
▶ Natural pathway

CENGAGENOW™ **Active Figure 3-17 Natural capital:** simplified model of the *hydrologic cycle* with major harmful impacts of human activities shown in red. *See an animation based on this figure at* CengageNOW. **Question:** What are three ways in which your lifestyle directly or indirectly affects the hydrologic cycle?

dissolves many nutrient compounds, it is a major medium for transporting nutrients within and between ecosystems.

Throughout the hydrologic cycle, many natural processes purify water. Evaporation and subsequent precipitation act as a natural distillation process that removes impurities dissolved in water. Water flowing above ground through streams and lakes and below ground in aquifers is naturally filtered and partially purified by chemical and biological processes—mostly by the actions of decomposer bacteria—as long as these natural processes are not overloaded. Thus, *the hydrologic cycle can be viewed as a cycle of natural renewal of water quality.*

Only about 0.024% of the earth's vast water supply is available to us as liquid freshwater in accessible groundwater deposits and in lakes, rivers, and streams. The rest is too salty for us to use, is stored as ice, or is too deep underground to extract at affordable prices using current technology.

We alter the water cycle in three major ways (see red arrows and boxes in Figure 3-17). *First,* we withdraw large quantities of freshwater from streams, lakes, and underground sources, sometimes at rates faster than nature can replace it.

Second, we clear vegetation from land for agriculture, mining, road building, and other activities, and cover much of the land with buildings, concrete, and asphalt. This increases runoff, reduces infiltration that would normally recharge groundwater supplies, in-

creases the risk of flooding, and accelerates soil erosion and landslides.

Clearing vegetation can also alter weather patterns by reducing transpiration. This is especially important in dense tropical rain forests (**Core Case Study** and Figure 3-1). Because so many plants in a tropical rain forest transpire water into the atmosphere, vegetation is the primary source of local rainfall. In other words, as part of the water cycle, these plants create their own rain.

Cutting down the forest raises ground temperatures (because it reduces shade) and can reduce local rainfall so much that the forest cannot grow back. When such a tipping point is reached, these biologically diverse forests are converted into much less diverse tropical grasslands, as a 2005 study showed in parts of Brazil's huge Amazon basin. Models project that if current burning and deforestation rates continue, 20–30% of the Amazon rain forests will turn into tropical grassland in the next 50 years.

The *third* way in which we alter the water cycle is by increasing flooding. This happens when we drain wetlands for farming and other purposes. Left undisturbed, wetlands provide the natural service of flood control, acting like sponges to absorb and hold overflows of water from drenching rains or rapidly melting snow. We also cover much of the land with roads, parking lots, and buildings, eliminating the land's ability to absorb water and dramatically increasing runoff and flooding.

SCIENCE FOCUS

Water's Unique Properties

Water is a remarkable substance with a unique combination of properties:

- *Forces of attraction,* called *hydrogen bonds* (see Figure 7 on p. S42 in Supplement 6), *hold water molecules together*— the major factor determining water's distinctive properties.

- *Water exists as a liquid over a wide temperature range because of the hydrogen bonds.* Without water's high boiling point the oceans would have evaporated long ago.

- *Liquid water changes temperature slowly because it can store a large amount of heat without a large change in temperature.* This high heat storage capacity helps protect living organisms from temperature changes, moderates the earth's climate, and makes water an excellent coolant for car engines and power plants.

- *It takes a large amount of energy to evaporate water because of the hydrogen bonds.* Water absorbs large amounts of heat as it changes into water vapor and releases this heat as the vapor condenses back to liquid water. This helps to distribute heat throughout the world and to determine regional and local climates. It also makes evaporation a cooling process—explaining why you feel cooler when perspiration evaporates from your skin.

- *Liquid water can dissolve a variety of compounds* (see Figure 3, p. S40, in Supplement 6). It carries dissolved nutrients into the tissues of living organisms, flushes waste products out of those tissues, serves as an all-purpose cleanser, and helps remove and dilute the water-soluble wastes of civilization. This property also means that water-soluble wastes can easily pollute water.

- *Water filters out some of the sun's ultraviolet radiation* (Figure 2-8, p. 42) *that would harm some aquatic organisms.* However, up to a certain depth it is transparent to visible light needed for photosynthesis.

- *Hydrogen bonds allow water to adhere to a solid surface.* This enables narrow columns of water to rise through a plant from its roots to its leaves (a process called capillary action).

- *Unlike most liquids, water expands when it freezes.* This means that ice floats on water because it has a lower density (mass per unit of volume) than liquid water. Otherwise, lakes and streams in cold climates would freeze solid, losing most of their aquatic life. Because water expands upon freezing, it can break pipes, crack a car's engine block (if it doesn't contain antifreeze), break up street pavements, and fracture rocks.

Critical Thinking

Water is a bent molecule (see Figure 4 on p. S40 in Supplement 6) and this allows it to form hydrogen bonds (Figure 7, p. S42, in Supplement 6) between its molecules. What are three ways in which your life would be different if water were a linear or straight molecule?

Carbon Cycles through the Biosphere and Depends on Photosynthesis and Respiration

Carbon is the basic building block of the carbohydrates, fats, proteins, DNA, and other organic compounds necessary for life. It circulates through the biosphere, the atmosphere, and parts of the hydrosphere, in the **carbon cycle** shown in Figure 3-18 (p. 68). It depends on photosynthesis and aerobic respiration by the earth's living organisms.

The carbon cycle is based on carbon dioxide (CO_2) gas, which makes up 0.038% of the volume of the atmosphere and is also dissolved in water. Carbon dioxide is a key component of nature's thermostat. If the carbon cycle removes too much CO_2 from the atmosphere, the atmosphere will cool, and if it generates too much CO_2, the atmosphere will get warmer. Thus, even slight changes in this cycle caused by natural or human factors can affect climate and ultimately help determine the types of life that can exist in various places.

Terrestrial producers remove CO_2 from the atmosphere and aquatic producers remove it from the water. (See *The Habitable Planet,* Video 3, **www.learner.org/ resources/series209.html**.) These producers then use photosynthesis to convert CO_2 into complex carbohydrates such as glucose ($C_6H_{12}O_6$).

The cells in oxygen-consuming producers, consumers, and decomposers then carry out aerobic respiration. This process breaks down glucose and other complex organic compounds and converts the carbon back to CO_2 in the atmosphere or water for reuse by producers. This linkage between *photosynthesis* in producers and *aerobic respiration* in producers, consumers, and decomposers circulates carbon in the biosphere. Oxygen and hydrogen—the other elements in carbohydrates—cycle almost in step with carbon.

Some carbon atoms take a long time to recycle. Decomposers release the carbon stored in the bodies of dead organisms on land back into the air as CO_2. However, in water, decomposers can release carbon that is stored as insoluble carbonates in bottom sediment. Indeed, marine sediments are the earth's largest store of carbon. Over millions of years, buried deposits of dead plant matter and bacteria are compressed between layers of sediment, where high pressure and heat convert them to carbon-containing *fossil fuels* such as coal, oil, and natural gas (Figure 3-18). This carbon is not released to the atmosphere as CO_2 for recycling until these fuels are extracted and burned, or until long-term geological processes expose these deposits to air. In only a few hundred years, we have extracted and burned large quantities of fossil fuels that took millions of years to form. This is why, on a human time scale, fossil fuels are nonrenewable resources.

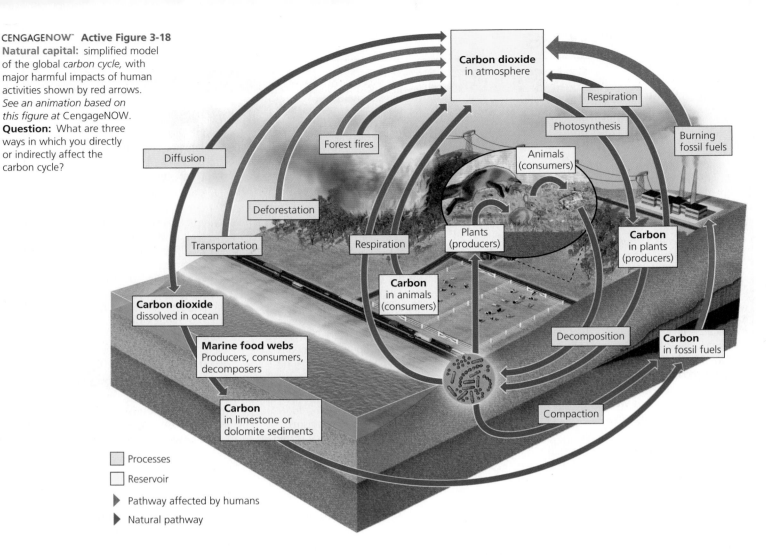

Natural capital: simplified model of the global *carbon cycle,* with major harmful impacts of human activities shown by red arrows. *See an animation based on this figure at* CengageNOW. **Question:** What are three ways in which you directly or indirectly affect the carbon cycle?

Carbon dioxide in atmosphere

Respiration

Photosynthesis

Burning fossil fuels

Forest fires

Diffusion

Animals (consumers)

Deforestation

Carbon in plants (producers)

Transportation

Respiration

Plants (producers)

Carbon dioxide dissolved in ocean

Carbon in animals (consumers)

Decomposition

Carbon in fossil fuels

Marine food webs Producers, consumers, decomposers

Compaction

Carbon in limestone or dolomite sediments

☐ Processes

☐ Reservoir

▶ Pathway affected by humans

▶ Natural pathway

Since 1800, and especially since 1950, we have been intervening in the earth's carbon cycle by adding carbon dioxide to the atmosphere in two ways (shown by red arrows in Figure 3-18). *First,* in some areas, especially in tropical forests, we clear trees and other plants, which absorb CO_2 through photosynthesis, faster than they can grow back (**Core Case Study**). *Second,* we add large amounts of CO_2 to the atmosphere by burning carbon-containing fossil fuels and wood.

> **THINKING ABOUT**
> **The Carbon Cycle, Tropical Deforestation, and Global Warming**
>
> Use Figure 3-18 and Figure 3-8 to explain why clearing tropical rain forests faster than they can grow back (**Core Case Study**) can warm the earth's atmosphere. What are two ways in which this could affect the survival of remaining tropical forests? What are two ways in which it could affect your lifestyle?

Computer models of the earth's climate systems indicate that increased concentrations of atmospheric CO_2 and other gases are very likely (90–99% probability) to enhance the planet's natural greenhouse effect, which

will cause global warming and change the earth's climate (see Science Focus, p. 33).

Nitrogen Cycles through the Biosphere: Bacteria in Action

The major reservoir for nitrogen is the atmosphere. Chemically unreactive nitrogen gas (N_2) makes up 78% of the volume of the atmosphere. Nitrogen is a crucial component of proteins, many vitamins, and nucleic acids such as DNA. However, N_2 cannot be absorbed and used directly as a nutrient by multicellular plants or animals.

Fortunately, two natural processes convert or *fix* N_2 into compounds useful as nutrients for plants and animals. One is electrical discharges, or lightning, taking place in the atmosphere. The other takes place in aquatic systems, soil, and the roots of some plants, where specialized bacteria, called *nitrogen-fixing bacteria,* complete this conversion as part of the **nitrogen cycle,** which is depicted in Figure 3-19.

The nitrogen cycle consists of several major steps. In *nitrogen fixation,* specialized bacteria in soil and blue-green algae (cyanobacteria) in aquatic environments

combine gaseous N_2 with hydrogen to make ammonia (NH_3). The bacteria use some of the ammonia they produce as a nutrient and excrete the rest to the soil or water. Some of the ammonia is converted to ammonium ions (NH_4^+) that can be used as a nutrient by plants.

Ammonia not taken up by plants may undergo *nitrification*. In this two-step process, specialized soil bacteria convert most of the NH_3 and NH_4^+ in soil to *nitrate ions* (NO_3^-), which are easily taken up by the roots of plants. The plants then use these forms of nitrogen to produce various amino acids, proteins, nucleic acids, and vitamins (see Supplement 6, p. S39). Animals that eat plants eventually consume these nitrogen-containing compounds, as do detritus feeders, or decomposers.

Plants and animals return nitrogen-rich organic compounds to the environment as wastes, cast-off particles, and through their bodies when they die and are decomposed or eaten by detritus feeders. In *ammonification*, vast armies of specialized decomposer bacteria convert this detritus into simpler nitrogen-containing inorganic compounds such as ammonia (NH_3) and water-soluble salts containing ammonium ions (NH_4^+).

In *denitrification*, specialized bacteria in waterlogged soil and in the bottom sediments of lakes, oceans, swamps, and bogs convert NH_3 and NH_4^+ back into nitrite and nitrate ions, and then into nitrogen gas (N_2) and nitrous oxide gas (N_2O). These gases are released to the atmosphere to begin the nitrogen cycle again.

We intervene in the nitrogen cycle in several ways (as shown by red arrows in Figure 3-19). *First,* we add large amounts of nitric oxide (NO) into the atmosphere when N_2 and O_2 combine as we burn any fuel at high temperatures, such as in car, truck, and jet engines. In the atmosphere, this gas can be converted to nitrogen dioxide gas (NO_2) and nitric acid vapor (HNO_3), which can return to the earth's surface as damaging *acid deposition*, commonly called *acid rain*.

Second, we add nitrous oxide (N_2O) to the atmosphere through the action of anaerobic bacteria on livestock wastes and commercial inorganic fertilizers applied to the soil. This greenhouse gas can warm the atmosphere and deplete stratospheric ozone, which keeps most of the sun's harmful ultraviolet radiation from reaching the earth's surface.

Third, we release large quantities of nitrogen stored in soils and plants as gaseous compounds into the atmosphere through destruction of forests, grasslands, and wetlands.

Fourth, we upset the nitrogen cycle in aquatic ecosystems by adding excess nitrates to bodies of water through agricultural runoff and discharges from municipal sewage systems.

Fifth, we remove nitrogen from topsoil when we harvest nitrogen-rich crops, irrigate crops (washing nitrates out of the soil), and burn or clear grasslands and forests before planting crops.

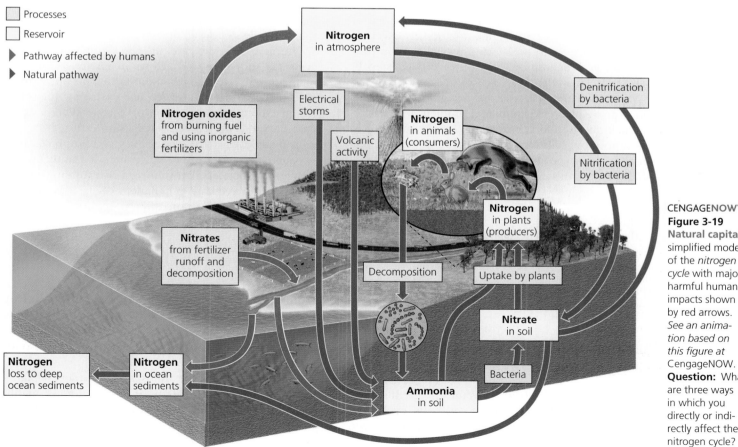

CENGAGENOW™
Figure 3-19
Natural capital: simplified model of the *nitrogen cycle* with major harmful human impacts shown by red arrows. *See an animation based on this figure at* CengageNOW.
Question: What are three ways in which you directly or indirectly affect the nitrogen cycle?

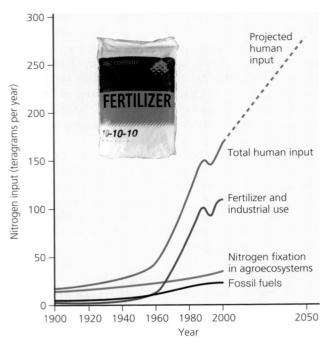

Figure 3-20 Global trends in the annual inputs of nitrogen into the environment from human activities, with projections to 2050. (Data from 2005 Millennium Ecosystem Assessment)

According to the 2005 Millennium Ecosystem Assessment, since 1950, human activities have more than doubled the annual release of nitrogen from the land into the rest of the environment. Most of this is from the greatly increased use of inorganic fertilizer to grow crops, and the amount released is projected to double again by 2050 (Figure 3-20). This excessive input of nitrogen into the air and water contributes to pollution, acid deposition, and other problems to be discussed in later chapters.

Nitrogen overload is a serious and growing local, regional, and global environmental problem that has attracted little attention. Princeton University physicist Robert Socolow calls for countries around the world to work out some type of nitrogen management agreement to help prevent this problem from reaching crisis levels.

THINKING ABOUT
The Nitrogen Cycle and Tropical Deforestation CORE CASE STUDY

What effects might the clearing and degrading of tropical rain forests (**Core Case Study**) have on the nitrogen cycle in such ecosystems and on any nearby water systems (see Figure 2-1, p. 28, and Figure 2-4, p. 37).

Phosphorus Cycles through the Biosphere

Phosphorus circulates through water, the earth's crust, and living organisms in the **phosphorus cycle,** depicted in Figure 3-21. In contrast to the cycles of water,

carbon, and nitrogen, the phosphorus cycle does not include the atmosphere. The major reservoir for phosphorous is phosphate salts containing phosphate ions (PO_4^{3-}) in terrestrial rock formations and ocean bottom sediments. The phosphorus cycle is slow compared to the water, carbon, and nitrogen cycles.

As water runs over exposed phosphorus-containing rocks, it slowly erodes away inorganic compounds that contain phosphate ions (PO_4^{3-}). The dissolved phosphate can be absorbed by the roots of plants and by other producers. Phosphorous is transferred by food webs from such producers to consumers, eventually including detritus feeders and decomposers. In both producers and consumers, phosphorous is a component of biologically important molecules such as nucleic acids (Figure 10, p. S43, in Supplement 6) and energy transfer molecules such as ADP and ATP (Figure 14, p. S44, in Supplement 6). It is also a major component of vertebrate bones and teeth.

Phosphate can be lost from the cycle for long periods when it washes from the land into streams and rivers and is carried to the ocean. There it can be deposited as marine sediment and remain trapped for millions of years. Someday, geological processes may uplift and expose these seafloor deposits, from which phosphate can be eroded to start the cycle again.

Because most soils contain little phosphate, it is often the *limiting factor* for plant growth on land unless phosphorus (as phosphate salts mined from the earth) is applied to the soil as an inorganic fertilizer. Phosphorus also limits the growth of producer populations in many freshwater streams and lakes because phosphate salts are only slightly soluble in water.

Human activities are affecting the phosphorous cycle (as shown by red arrows in Figure 3-21). This includes removing large amounts of phosphate from the earth to make fertilizer and reducing phosphorus in tropical soils by clearing forests (**Core Case Study**). Soil that is eroded from fertilized crop fields carries large quantities of phosphates into streams, lakes, and the ocean, where it stimulates the growth of producers. Phosphorous-rich runoff from the land can produce huge populations of algae, which can upset chemical cycling and other processes in lakes.

Sulfur Cycles through the Biosphere

Sulfur circulates through the biosphere in the **sulfur cycle,** shown in Figure 3-22 (p. 72). Much of the earth's sulfur is stored underground in rocks and minerals, including sulfate (SO_4^{2-}) salts buried deep under ocean sediments.

Sulfur also enters the atmosphere from several natural sources. Hydrogen sulfide (H_2S)—a colorless, highly poisonous gas with a rotten-egg smell—is released from active volcanoes and from organic matter broken down

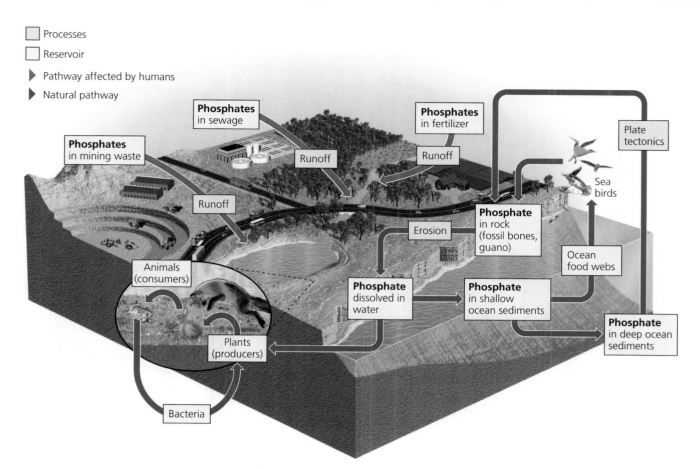

Figure 3-21 Natural capital: simplified model of the *phosphorus cycle*, with major harmful human impacts shown by red arrows. **Question:** What are three ways in which you directly or indirectly affect the phosphorus cycle?

by anaerobic decomposers in flooded swamps, bogs, and tidal flats. Sulfur dioxide (SO_2), a colorless and suffocating gas, also comes from volcanoes.

Particles of sulfate (SO_4^{2-}) salts, such as ammonium sulfate, enter the atmosphere from sea spray, dust storms, and forest fires. Plant roots absorb sulfate ions and incorporate the sulfur as an essential component of many proteins.

Certain marine algae produce large amounts of volatile dimethyl sulfide, or DMS (CH_3SCH_3). Tiny droplets of DMS serve as nuclei for the condensation of water into droplets found in clouds. In this way, changes in DMS emissions can affect cloud cover and climate.

In the atmosphere, DMS is converted to sulfur dioxide, some of which in turn is converted to sulfur trioxide gas (SO_3) and to tiny droplets of sulfuric acid (H_2SO_4). DMS also reacts with other atmospheric chemicals such as ammonia to produce tiny particles of sulfate salts. These droplets and particles fall to the earth as components of *acid deposition*, which along with other air pollutants can harm trees and aquatic life.

In the oxygen-deficient environments of flooded soils, freshwater wetlands, and tidal flats, specialized bacteria convert sulfate ions to sulfide ions (S^{2-}). The sulfide ions can then react with metal ions to form insol-

uble metallic sulfides, which are deposited as rock, and the cycle continues.

Human activities have affected the sulfur cycle primarily by releasing large amounts of sulfur dioxide (SO_2) into the atmosphere (as shown by red arrows in Figure 3-22). We add sulfur dioxide to the atmosphere in three ways. *First,* we burn sulfur-containing coal and oil to produce electric power. *Second,* we refine sulfur-containing petroleum to make gasoline, heating oil, and other useful products. *Third,* we convert sulfur-containing metallic mineral ores into free metals such as copper, lead, and zinc. Once in the atmosphere, SO_2 is converted to droplets of sulfuric acid (H_2SO_4) and particles of sulfate (SO_4^{2-}) salts, which return to the earth as acid deposition.

— **RESEARCH FRONTIER** —

The effects of human activities on the major nutrient cycles and how we can reduce these effects. See **academic .cengage.com/biology/miller.**

CENGAGENOW˘ Learn more about the water, carbon, nitrogen, phosphorus, and sulfur cycles using interactive animations at CengageNOW.

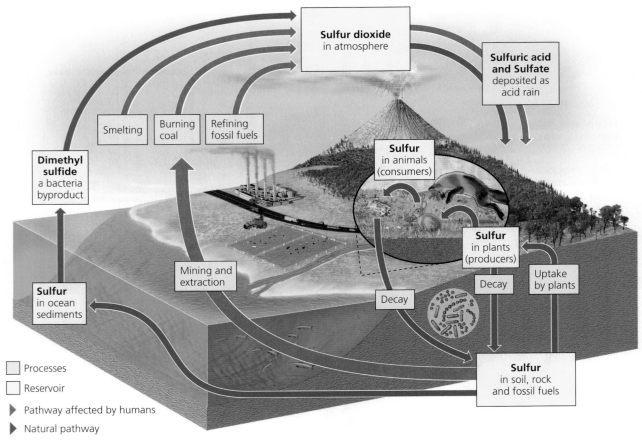

Processes

Reservoir

▶ Pathway affected by humans

▶ Natural pathway

CENGAGENOW **Active Figure 3-22** **Natural capital:** simplified model of the *sulfur cycle,* with major harmful impacts of human activities shown by red arrows. *See an animation based on this figure at* CengageNOW. **Question:** What are three ways in which your lifestyle directly or indirectly affects the sulfur cycle?

3-6 How Do Scientists Study Ecosystems?

▶ **CONCEPT 3-6** **Scientists use field research, laboratory research, and mathematical and other models to learn about ecosystems.**

Some Scientists Study Nature Directly

Scientists use field research, laboratory research, and mathematical and other models to learn about ecosystems (**Concept 3-6**). *Field research,* sometimes called "muddy-boots biology," involves observing and measuring the structure of natural ecosystems and what happens in them. Most of what we know about structure and functioning of ecosystems has come from such research. **GREEN CAREER:** Ecologist. See **academic .cengage.com/biology/miller** for details on various green careers.

Ecologists trek through forests, deserts, and grasslands and wade or boat through wetlands, lakes, streams, and oceans collecting and observing species.

Sometimes they carry out controlled experiments by isolating and changing a variable in part of an area and comparing the results with nearby unchanged areas (Chapter 2 Core Case Study, p. 28, and see *The Habitable Planet,* Videos 4 and 9, **www.learner.org/resources/ series209.html**). Tropical ecologists have erected tall construction cranes over the canopies of tropical forests from which they observe the rich diversity of species living or feeding in these treetop habitats.

Increasingly, new technologies are being used to collect ecological data. Scientists use aircraft and satellites equipped with sophisticated cameras and other *remote sensing* devices to scan and collect data on the earth's surface. Then they use *geographic information system* (GIS) software to capture, store, analyze, and display such geographically or spatially based information.

In a GIS, geographic and ecological data can be stored electronically as numbers or as images in computer databases. For example, a GIS can convert digital satellite images generated through remote sensing into global, regional, and local maps showing variations in vegetation (Figure 1, pp. S20–S21, and Figure 2, pp. S22–S23, in Supplement 4), gross primary productivity (Figure 6, p. S27, in Supplement 4), temperature patterns, air pollution emissions, and other variables.

Scientists also use GIS programs and digital satellite images to produce two- or three-dimensional maps combining information about a variable such as land use with other data. Separate layers within such maps, each showing how a certain factor varies over an area, can be combined to show a composite effect (Figure 3-23). Such composites of information can lead to a better understanding of environmental problems and to better decision making about how to deal with such problems.

In 2005, scientists launched the Global Earth Observation System of Systems (GEOSS)—a 10-year program to integrate data from sensors, gauges, buoys, and satellites that monitor the earth's surface, atmosphere, and oceans. **GREEN CAREERS:** GIS analyst; remote sensing analyst

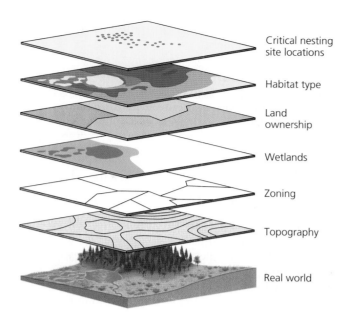

Figure 3-23 *Geographic information systems* (*GIS*) provide the computer technology for storing, organizing, and analyzing complex data collected over broad geographic areas. They enable scientists to produce maps of various geographic data sets and then to overlay and compare the layers of data (such as soils, topography, distribution of endangered populations, and land protection status).

Some Scientists Study Ecosystems in the Laboratory

During the past 50 years, ecologists have increasingly supplemented field research by using *laboratory research* to set up, observe, and make measurements of model ecosystems and populations under laboratory conditions. Such simplified systems have been created in containers such as culture tubes, bottles, aquaria tanks, and greenhouses, and in indoor and outdoor chambers where temperature, light, CO_2, humidity, and other variables can be controlled.

Such systems make it easier for scientists to carry out controlled experiments. In addition, laboratory experiments often are quicker and less costly than similar experiments in the field.

THINKING ABOUT

Greenhouse Experiments and Tropical Rain Forests

CORE CASE STUDY

How would you design an experiment, including an experimental group and a control group, that uses a greenhouse to determine the effect of clearing a patch of tropical rain forest vegetation (**Core Case Study**) on the temperature above the cleared patch?

But there is a catch. Scientists must consider how well their scientific observations and measurements in a simplified, controlled system under laboratory conditions reflect what takes place under the more complex and dynamic conditions found in nature. Thus, the re-

sults of laboratory research must be coupled with and supported by field research. (See *The Habitable Planet*, Videos, 2, 3, and 12, **www.learner.org/resources/series209.html**.)

Some Scientists Use Models to Simulate Ecosystems

Since the late 1960s, ecologists have developed mathematical and other models that simulate ecosystems. Computer simulations can help scientists understand large and very complex systems that cannot be adequately studied and modeled in field and laboratory research. Examples include rivers, lakes, oceans, forests, grasslands, cities, and the earth's climate system. Scientists are learning a lot about how the earth works by feeding data into increasingly sophisticated models of the earth's systems and running them on supercomputers.

Researchers can change values of the variables in their computer models to project possible changes in environmental conditions, to help them anticipate environmental surprises, and to analyze the effectiveness of various alternative solutions to environmental problems. **GREEN CAREER:** Ecosystem modeler

Of course, simulations and projections made with ecosystem models are no better than the data and assumptions used to develop the models. Ecologists must do careful field and laboratory research to get *baseline data,* or beginning measurements, of variables being studied. They also must determine the relationships

among key variables that they will use to develop and test ecosystem models.

┌─ RESEARCH FRONTIER ─────────────────────────┐
│ Improved computer modeling for understanding complex │
│ environmental systems. See **academic.cengage.com/** │
│ **biology/miller**. │
└──┘

We Need to Learn More about the Health of the World's Ecosystems

We need baseline data on the condition of the world's ecosystems to see how they are changing and to develop effective strategies for preventing or slowing their degradation.

By analogy, your doctor needs baseline data on your blood pressure, weight, and functioning of your organs and other systems, as revealed through basic tests. If your health declines in some way, the doctor can run new tests and compare the results with the baseline data to identify changes and come up with a treatment.

According to a 2002 ecological study published by the Heinz Foundation and the 2005 Millennium Ecosystem Assessment, scientists have less than half of the basic ecological data they need to evaluate the status of ecosystems in the United States. Even fewer data are available for most other parts of the world. Ecologists call for a massive program to develop baseline data for the world's ecosystems.

┌─ RESEARCH FRONTIER ─────────────────────────┐
│ A crash program to gather and evaluate baseline data for │
│ all of the world's major terrestrial and aquatic systems. See │
│ **academic.cengage.com/biology/miller**. │
└──┘

REVISITING — Tropical Rain Forests and Sustainability

This chapter applied two of the **scientific principles of sustainability** (see back cover and **Concept 1-6**, p. 23) by which the biosphere and the ecosystems it contains have been sustained over the long term. *First,* the biosphere and almost all of its ecosystems use *solar energy* as their energy source, and this energy flows through the biosphere. *Second,* they *recycle the chemical nutrients* that their organisms need for survival, growth, and reproduction.

These two principles arise from the structure and function of natural ecosystems (Figure 3-12), the law of conservation of matter (**Concept 2-3**, p. 39), and the two laws of thermodynamics (**Concepts 2-4A** and **2-4B**, p. 40). Nature's required adherence to these principles is enhanced by *biodiversity,* another sustainability principle, which also helps to *regulate population levels* of interacting species in the world's ecosystems—yet another of the sustainability principles.

This chapter started with a discussion of the importance of incredibly diverse tropical rain forests (**Core Case Study**), which showcase the functioning of the four **scientific principles of sustainability**. Producers within rain forests rely on solar energy to produce a vast amount of biomass through photosynthesis. Species living in the forests take part in, and depend on cycling of nutrients in the biosphere and the flow of energy through the biosphere. Tropical forests contain a huge and vital part of the earth's biodiversity, and interactions among species living in these forests help to control the populations of the species living there.

All things come from earth,
and to earth they all return.

MENANDER (342–290 B.C.)

REVIEW

1. Review the Key Questions and Concepts for this chapter on p. 51. What are three harmful effects resulting from the clearing and degradation of tropical rain forests?

2. What is a **cell?** What is the **cell theory?** Distinguish between a **eukaryotic cell** and a **prokaryotic cell.** What is a **species?** Explain the importance of insects.

Define **ecology.** What is **genetic diversity?** Distinguish among a **species, population, community (biological community), habitat, ecosystem,** and the **biosphere.**

3. Distinguish among the **atmosphere, troposphere, stratosphere, greenhouse gases, hydrosphere,** and

geosphere. Distinguish between **biomes** and **aquatic life zones** and give an example of each. What three interconnected factors sustain life on earth?

4. Describe what happens to solar energy as it flows to and from the earth. What is the **natural greenhouse effect** and why is it important for life on earth?

5. Distinguish between the **abiotic** and **biotic components** in ecosystems and give two examples of each. What is the **range of tolerance** for an abiotic factor? Define and give an example of a **limiting factor.** What is the **limiting factor principle?**

6. What is a **trophic level?** Distinguish among **producers (autotrophs), consumers (heterotrophs),** and **decomposers** and give an example of each in an ecosystem. Distinguish among **primary consumers (herbivores), secondary consumers (carnivores), high-level (third-level) consumers, omnivores, decomposers,** and **detritus feeders (detritivores),** and give an example of each.

7. Distinguish among **photosynthesis, chemosynthesis, aerobic respiration,** and **anaerobic respiration (fermentation).** What two processes sustain ecosystems and the biosphere and how are they linked? Explain the importance of microbes.

8. Explain what happens to energy as it flows through the food chains and food webs of an ecosystem. Distinguish between a **food chain** and a **food web.** What is **biomass?** What is **ecological efficiency?** What is the **pyramid of energy flow?** Discuss the difference between **gross primary productivity (GPP)** and **net primary productivity (NPP)** and explain their importance.

9. What happens to matter in an ecosystem? What is a **biogeochemical cycle (nutrient cycle)?** Describe the unique properties of water. What is **transpiration?** Describe the **hydrologic (water), carbon, nitrogen, phosphorus,** and **sulfur cycles** and describe how human activities are affecting each cycle.

10. Describe three ways in which scientists study ecosystems. Explain why we need much more basic data about the structure and condition of the world's ecosystems. How are the four scientific principles of sustainability showcased in tropical rain forests (Core Case Study)?

Note: Key Terms are in **bold** type.

CRITICAL THINKING

1. List three ways in which you could apply **Concept 3-4B** and **Concept 3-5** to making your lifestyle more environmentally sustainable.

2. How would you explain the importance of tropical rain forests (**Core Case Study**) to people who think that such forests have no connection to their lives?

3. Explain why **(a)** the flow of energy through the biosphere (**Concept 3-2**) depends on the cycling of nutrients, and **(b)** the cycling of nutrients depends on gravity.

4. Explain why microbes are so important. List two beneficial and two harmful effects of microbes on your health and lifestyle.

5. Make a list of the food you ate for lunch or dinner today. Trace each type of food back to a particular producer species.

6. Use the second law of thermodynamics (**Concept 2-5B**, p. 44) to explain why many poor people in developing countries live on a mostly vegetarian diet.

7. Why do farmers not need to apply carbon to grow their crops but often need to add fertilizer containing nitrogen and phosphorus?

8. What changes might take place in the hydrologic cycle if the earth's climate becomes **(a)** hotter or **(b)** cooler? In each case, what are two ways in which these changes might affect your lifestyle?

9. What would happen to an ecosystem if **(a)** all its decomposers and detritus feeders were eliminated, **(b)** all its producers were eliminated, or **(c)** all of its insects were eliminated? Could a balanced ecosystem exist with only producers and decomposers and no consumers such as humans and other animals? Explain.

10. List two questions that you would like to have answered as a result of reading this chapter.

Note: See Supplement 13 (p. S78) for a list of Projects related to this chapter.

Based on the following carbon dioxide emissions data and 2007 population data, answer the questions below.

Country	Total Carbon Footprint—Carbon Dioxide Emissions (in metric gigatons per year*)	Population (in billions, 2007)	Per Capita Carbon Footprint—Per Capita Carbon Dioxide Emissions Per Year
China	5.0 (5.5)	1.3	
India	1.3 (1.4)	1.1	
Japan	1.3 (1.4)	0.13	
Russia	1.5 (1.6)	1.14	
United States	6.0 (6.6)	0.30	
WORLD	29 (32)	6.6	

Souce: Data from World Resources Institute and International Energy Agency

*The prefix *giga* stands for "1 billion."

1. Calculate the per capita carbon footprint for each country and the world and complete the table.

2. It has been suggested that a sustainable average world-wide carbon footprint per person should be no more than 2.0 metric tons per person per year (2.2 tons per person per year). How many times larger is the U.S. carbon footprint per person than are **(a)** the sustainable level, and **(b)** the world average?

3. By what percentage will China, Japan, Russia, the United States, and the world each have to reduce their carbon footprints per person to achieve the estimated maximum sustainable carbon footprint per person of 2.0 metric tons (2.2 tons) per person per year?

Biodiversity and Evolution

Why Should We Care about the American Alligator?

The American alligator (Figure 4-1), North America's largest reptile, has no natural predators except for humans, and it plays a number of important roles in the ecosystems where it is found. This species outlived the dinosaurs and has been able to survive numerous dramatic changes in the earth's environmental conditions.

But starting in the 1930s, these alligators faced a new challenge. Hunters began killing them in large numbers for their exotic meat and their supple belly skin, used to make shoes, belts, and pocketbooks. Other people hunted alligators for sport or out of hatred. By the 1960s, hunters and poachers had wiped out 90% of the alligators in the U.S. state of Louisiana, and the alligator population in the Florida Everglades was also near extinction.

Those who did not care much for the American alligator were probably not aware of its important ecological role—its *niche*—in subtropical wetland communities. These alligators dig deep depressions, or gator holes, which hold freshwater during dry spells, serve as refuges for aquatic life, and supply freshwater and food for fish, insects, snakes, turtles, birds, and other animals. Large alligator nesting mounds provide nesting and feeding sites for species of herons and egrets, and red-bellied turtles use old gator nests for incubating their eggs. These alligators eat large numbers of gar, a predatory fish. This helps maintain populations of game fish such as bass and bream.

As alligators move from gator holes to nesting mounds, they help keep areas of open water free of invading vegetation. Without these free ecosystem services, freshwater ponds and coastal wetlands where these alligators live would be filled in with shrubs and trees, and dozens of species would disappear from these ecosystems. Some ecologists classify the American alligator as a *keystone species* because of its important ecological role in helping to maintain the structure, function, and sustainability of the ecosystems where it is found. And, in 2008, scientists began analyzing the blood of the American alligator to identify compounds that could kill a variety of harmful bacteria, including those that have become resistant to commonly used antibiotics.

In 1967, the U.S. government placed the American alligator on the endangered species list. Protected from hunters, the population made a strong comeback in many areas by 1975—too strong, according to those who find alligators in their backyards and swimming pools, and to duck hunters whose retriever dogs are sometimes eaten by alligators. In 1977, the U.S. Fish and Wildlife Service reclassified the American alligator as a *threatened* species in the U.S. states of Florida, Louisiana, and Texas, where 90% of the animals live. Today there are 1–2 million American alligators in Florida, and the state now allows property owners to kill alligators that stray onto their land.

To biologists, the comeback of the American alligator is an important success story in wildlife conservation. This tale illustrates how each species in a community or ecosystem fills a unique role, and it highlights how interactions between species can affect ecosystem structure and function. In this chapter, we will examine biodiversity, with an emphasis on species diversity, and the theory of how the earth's diverse species arose.

A. & J. Visage/Peter Arnold, Inc.

Figure 4-1 The American alligator plays an important ecological role in its marsh and swamp habitats in the southeastern United States. Since being classified as an endangered species in 1967, it has recovered enough to have its status changed from endangered to threatened—an outstanding success story in wildlife conservation.

4-1 What is biodiversity and why is it important?

CONCEPT 4-1 The biodiversity found in genes, species, ecosystems, and ecosystem processes is vital to sustaining life on earth.

4-2 Where do species come from?

CONCEPT 4-2A The scientific theory of evolution explains how life on earth changes over time through changes in the genes of populations.

CONCEPT 4-2B Populations evolve when genes mutate and give some individuals genetic traits that enhance their abilities to survive and to produce offspring with these traits (natural selection).

4-3 How do geological processes and climate change affect evolution?

CONCEPT 4-3 Tectonic plate movements, volcanic eruptions, earthquakes, and climate change have shifted wildlife habitats, wiped out large numbers of species, and created opportunities for the evolution of new species.

4-4 How do speciation, extinction, and human activities affect biodiversity?

CONCEPT 4-4A As environmental conditions change, the balance between formation of new species and extinction of existing species determines the earth's biodiversity.

CONCEPT 4-4B Human activities can decrease biodiversity by causing the premature extinction of species and by destroying or degrading habitats needed for the development of new species.

4-5 What is species diversity and why is it important?

CONCEPT 4-5 Species diversity is a major component of biodiversity and tends to increase the sustainability of ecosystems.

4-6 What roles do species play in ecosystems?

CONCEPT 4-6A Each species plays a specific ecological role called its niche.

CONCEPT 4-6B Any given species may play one or more of five important roles—native, nonnative, indicator, keystone, or foundation roles—in a particular ecosystem.

Note: Supplements 2 (p. S4), 4 (p. S20), 6 (p. S39), 7 (p. S46), 8 (p. S47), and 13 (p. S78) can be used with this chapter.

There is grandeur to this view of life...
that, whilst this planet has gone cycling on...
endless forms most beautiful and most wonderful
have been, and are being, evolved.

CHARLES DARWIN

4-1 What Is Biodiversity and Why Is It Important?

▶ **CONCEPT 4-1** The biodiversity found in genes, species, ecosystems, and ecosystem processes is vital to sustaining life on earth.

Biodiversity Is a Crucial Part of the Earth's Natural Capital

Biological diversity, or **biodiversity,** is the variety of the earth's species, the genes they contain, the ecosystems in which they live, and the ecosystem processes such as energy flow and nutrient cycling that sustain all life (Figure 4-2). Biodiversity is a vital renewable resource (**Concept 4-1**).

So far, scientists have identified about 1.8 million of the earth's 4 million to 100 million species, and every year, thousands of new species are identified. The identified species include almost a million species of insects, 270,000 plant species, and 45,000 vertebrate animal species. Later in this chapter, we look at the scientific theory of how such a great variety of life forms came to be, and we consider the importance of species diversity.

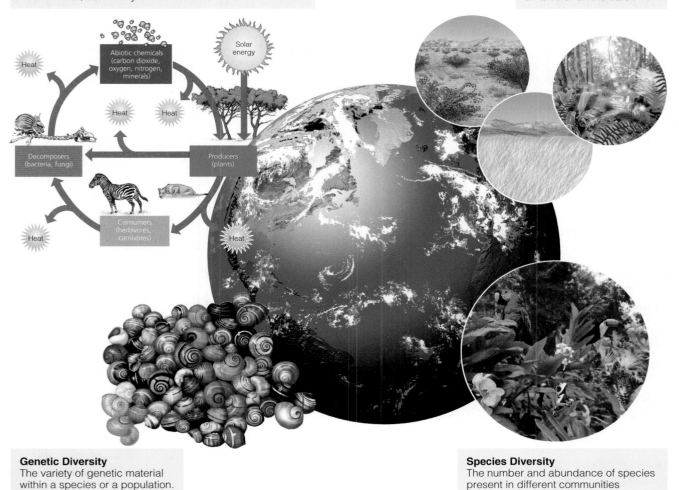

Functional Diversity
The biological and chemical processes such as energy flow and matter recycling needed for the survival of species, communities, and ecosystems.

Ecological Diversity
The variety of terrestrial and aquatic ecosystems found in an area or on the earth.

Heat

Abiotic chemicals (carbon dioxide, oxygen, nitrogen, minerals)

Solar energy

Heat

Heat

Decomposers (bacteria, fungi)

Producers (plants)

Consumers (herbivores, carnivores)

Heat

Heat

Genetic Diversity
The variety of genetic material within a species or a population.

Species Diversity
The number and abundance of species present in different communities

CENGAGENOW™ **Active Figure 4-2 Natural capital:** the major components of the earth's *biodiversity*—one of the earth's most important renewable resources. *See an animation based on this figure at* CengageNOW™.
Question: What are three examples of how people, in their daily living, intentionally or unintentionally degrade each of these types of biodiversity?

Species diversity is the most obvious, but not the only, component of biodiversity. Another important component is *genetic diversity* (Figure 3-5, p. 53). The earth's variety of species contains an even greater variety of genes. Genetic diversity enables life on the earth to adapt to and survive dramatic environmental changes. In other words, genetic diversity is vital to the sustainability of life on earth.

Ecosystem diversity—the earth's variety of deserts, grasslands, forests, mountains, oceans, lakes, rivers, and wetlands is another major component of biodiversity. Each of these ecosystems is a storehouse of genetic and species diversity.

Yet another important component of biodiversity is *functional diversity*—the variety of processes such as

matter cycling and energy flow taking place within ecosystems (Figure 3-12, p. 60) as species interact with one another in food chains and webs. Part of the importance of the American alligator (Figure 4-1) is its role in supporting these processes within its ecosystems, which help to maintain other species of animals and plants that live there.

THINKING ABOUT
Alligators and Biodiversity

 CORE CASE STUDY

What are three ways in which the American alligator (**Core Case Study**) supports one or more of the four components of biodiversity within its environment?

The earth's biodiversity is a vital part of the natural capital that keeps us alive. It supplies us with food, wood, fibers, energy, and medicines—all of which represent hundreds of billions of dollars in the world economy each year. Biodiversity also plays a role in preserving the quality of the air and water and maintaining the fertility of soils. It helps us to dispose of wastes and to control populations of pests. In carrying out these free ecological services, which are also part of the earth's natural capital (**Concept 1-1A**, p. 6), **CONCEPT LINK** biodiversity helps to sustain life on the earth.

Because biodiversity is such an important concept and so vital to sustainability, we are going to take a grand tour of biodiversity in this and the next seven chapters. This chapter focuses on the earth's variety of species, how these species evolved, and the major roles that species play in ecosystems. Chapter 5 examines how different interactions among species help to control population sizes and promote biodiversity. Chapter 6 uses principles of population dynamics developed in Chapter 5 to look at human population growth and its effects on biodiversity. Chapters 7 and 8, respectively, look at the major types of terrestrial and aquatic ecosystems that make up a key component of biodiversity. Then, the next three chapters examine major threats to species diversity (Chapter 9), terrestrial biodiversity (Chapter 10), and aquatic biodiversity (Chapter 11), and solutions for dealing with these threats.

4-2 Where Do Species Come From?

▶ **CONCEPT 4-2A** The scientific theory of evolution explains how life on earth changes over time through changes in the genes of populations.

▶ **CONCEPT 4-2B** Populations evolve when genes mutate and give some individuals genetic traits that enhance their abilities to survive and to produce offspring with these traits (natural selection).

Biological Evolution by Natural Selection Explains How Life Changes over Time

How did we end up with an amazing array of 4 million to 100 million species? The scientific answer involves **biological evolution:** the process whereby earth's life changes over time through changes in the genes of populations (**Concept 4-2A**).

The idea that organisms change over time and are descended from a single common ancestor has been around in one form or another since the early Greek philosophers. But no one had come up with a credible explanation of how this could happen until 1858 when naturalists Charles Darwin (1809–1882) and Alfred Russel Wallace (1823–1913) independently proposed the concept of *natural selection* as a mechanism for biological evolution. Although Wallace also proposed the idea of natural selection, it was Darwin, who meticulously gathered evidence for this idea and published it in 1859 in his book, *On the Origin of Species by Means of Natural Selection.*

Darwin and Wallace observed that organisms must constantly struggle to obtain enough food and other resources to survive and reproduce. They also observed that individuals in a population with a specific advantage over other individuals are more likely to survive, reproduce, and have offspring with similar survival skills. The advantage was due to a characteristic, or *trait,* possessed by these individuals but not by others.

Darwin and Wallace concluded that these survival traits would become more prevalent in future populations of the species through a process called **natural selection,** which occurs when some individuals of a population have genetically based traits that enhance their ability to survive and produce offspring with the same traits. A change in the genetic characteristics of a population from one generation to another is known as *biological evolution,* or simply *evolution.*

A huge body of field and laboratory evidence has supported this idea. As a result, biological evolution through natural selection has become an important scientific theory. According to this theory, life has evolved into six major groups of species, called *kingdoms,* as a result of natural selection. This view sees the development of life as an ever-branching tree of species diversity, sometimes called the *tree of life* (Figure 4-3).

This scientific theory generally explains how life has changed over the past 3.7 billion years and why life is so diverse today. However, there are still many unanswered questions and scientific debates about the details of evolution by natural selection. Such continual questioning and discussion is an important way in which science advances our knowledge of how the earth works.

CENGAGENOW™ Get a detailed look at early biological evolution by natural selection—the roots of the tree of life—at CengageNOW™.

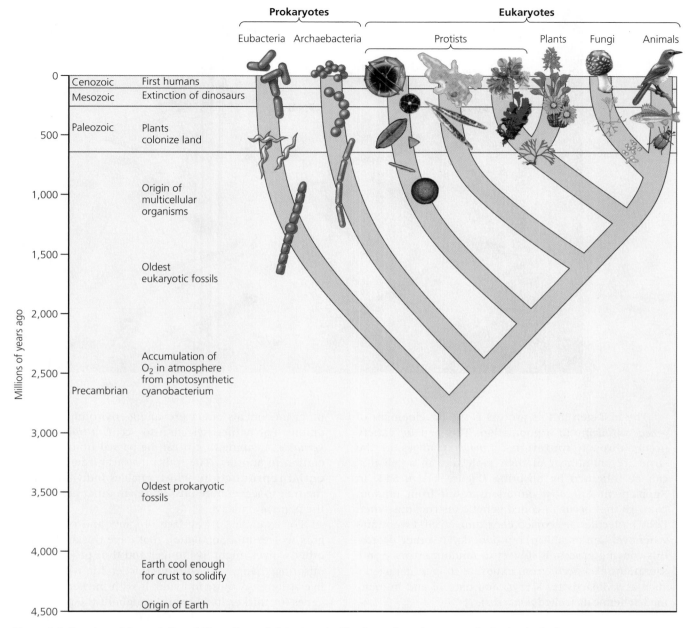

Prokaryotes

Eubacteria Archaebacteria

Eukaryotes

Protists Plants Fungi Animals

Millions of years ago

0		
	Cenozoic	First humans
	Mesozoic	Extinction of dinosaurs
	Paleozoic	Plants colonize land
500		
1,000		Origin of multicellular organisms
1,500		Oldest eukaryotic fossils
2,000		
2,500	Precambrian	Accumulation of O₂ in atmosphere from photosynthetic cyanobacterium
3,000		
3,500		Oldest prokaryotic fossils
4,000		Earth cool enough for crust to solidify
4,500		Origin of Earth

Figure 4-3 Overview of the evolution of life on the earth into six major kingdoms of species as a result of natural selection. For more details, see p. S46 in Supplement 7.

The Fossil Record Tells Much of the Story of Evolution

Most of what we know of the earth's life history comes from **fossils:** mineralized or petrified replicas of skeletons, bones, teeth, shells, leaves, and seeds, or impressions of such items found in rocks. Also, scientists drill cores from glacial ice at the earth's poles and on mountaintops and examine the kinds of life found at different layers. Fossils provide physical evidence of ancient organisms and reveal what their internal structures looked like (Figure 4-4, p. 82).

The world's cumulative body of fossils found is called the *fossil record.* This record is uneven and incomplete. Some forms of life left no fossils, and some fossils have decomposed. The fossils found so far probably represent only 1% of all species that have ever lived. Trying to reconstruct the development of life with so little evidence—a challenging scientific detective game—is the work of paleontologists. **GREEN CAREER:** Paleontologist

The Genetic Makeup of a Population Can Change

The process of biological evolution by natural selection involves changes in a population's genetic makeup through successive generations. Note that *populations— not individuals—evolve by becoming genetically different.*

Figure 4-4 Fossilized skeleton of an herbivore that lived during the Cenozoic era from 26–66 million years ago.

The first step in this process is the development of *genetic variability* in a population. This genetic variety occurs through **mutations:** *random* changes in the structure or number of DNA molecules in a cell that can be inherited by offspring (Figure 11, p. S43, in Supplement 6). Most mutations result from random changes that occur in coded genetic instructions when DNA molecules are copied each time a cell divides and whenever an organism reproduces. In other words, this copying process is subject to random errors. Some mutations also occur from exposure to external agents such as radioactivity, X rays, and natural and human-made chemicals (called *mutagens*).

Mutations can occur in any cell, but only those taking place in reproductive cells are passed on to offspring. Sometimes a mutation can result in a new genetic trait that gives an individual and its offspring better chances for survival and reproduction under existing environmental conditions or when such conditions change.

Individuals in Populations with Beneficial Genetic Traits Can Leave More Offspring

The next step in biological evolution is *natural selection,* which occurs when some individuals of a population have genetically based traits (resulting from mutations) that enhance their ability to survive and produce offspring with these traits (**Concept 4-2B**).

An **adaptation,** or **adaptive trait,** is any heritable trait that enables an individual organism to survive through natural selection and to reproduce more than other individuals under prevailing environmental conditions. For natural selection to occur, a trait must be *heritable,* meaning that it can be passed from one generation to another. The trait must also lead to **differential reproduction,** which enables individuals with the trait to leave more offspring than other members of the population leave.

For example, in the face of snow and cold, a few gray wolves in a population that have thicker fur than other wolves might live longer and thus produce more offspring than those without thicker fur who do not live as long. As those individuals with thicker fur mate, genes for thicker fur spread throughout the population and individuals with those genes increase in number and pass this helpful trait on to their offspring. Thus, the concept of natural selection explains how populations adapt to changes in environmental conditions.

Genetic resistance is the ability of one or more organisms in a population to tolerate a chemical designed to kill it. For example, an organism might have a gene that allows it to break the chemical down into harmless substances. Another important example of natural selection at work is the evolution of antibiotic resistance in disease-causing bacteria. Scientists have developed antibacterial drugs (antibiotics) to fight these bacteria, and the drugs have become a driving force of natural selection. The few bacteria that are genetically resistant to the drugs (because of some trait they possess) survive and produce more offspring than the bacteria that were killed by the drugs could have produced. Thus, the antibiotic eventually loses its effectiveness, as resistant bacteria rapidly reproduce and those that are susceptible to the drug die off (Figure 4-5).

(a)
A group of bacteria, including genetically resistant ones, are exposed to an antibiotic

(b)
Most of the normal bacteria die

(c)
The genetically resistant bacteria start multiplying

(d)
Eventually the resistant strain replaces the strain affected by the antibiotic

Normal bacterium Resistant bacterium

Figure 4-5 *Evolution by natural selection.* **(a)** A population of bacteria is exposed to an antibiotic, which **(b)** kills all but those possessing a trait that makes them resistant to the drug. **(c)** The resistant bacteria multiply and eventually **(d)** replace the nonresistant bacteria.

Note that natural selection acts on individuals, but evolution occurs in populations. In other words, populations can evolve when genes change or mutate and give some individuals genetic traits that enhance their ability to survive and to produce offspring with these traits (natural selection) (**Concept 4-2B**).

CENGAGENOW™ How many moths can you eat? Find out and learn more about adaptation at CengageNOW.

Another way to summarize the process of biological evolution by natural selection is: *Genes mutate, individuals are selected, and populations evolve that are better adapted to survive and reproduce under existing environmental conditions.*

When environmental conditions change, a population of a species faces three possible futures: *adapt* to the new conditions through natural selection, *migrate* (if possible) to an area with more favorable conditions, or *become extinct.*

A remarkable example of evolution by natural selection is human beings. We have evolved certain traits that have allowed us to take over much of the world (see Case Study below).

■ CASE STUDY

How Did Humans Become Such a Powerful Species?

Like many other species, humans have survived and thrived because we have certain traits that allow us to adapt to and modify parts of the environment to increase our survival chances.

Evolutionary biologists attribute our success to three adaptations: *strong opposable thumbs* that allow us to grip and use tools better than the few other animals that have thumbs can do, an ability to *walk upright*, and a *complex brain*. These adaptations have helped us develop

weapons, protective devices, and technologies that extend our limited senses and make up for some of our deficiencies. Thus, in just a twitch of the 3.56-billion-year history of life on earth, we have developed powerful technologies and taken over much of the earth's life-support systems and net primary productivity.

But adaptations that make a species successful during one period of time may not be enough to ensure the species' survival when environmental conditions change. This is no less true for humans, and some environmental conditions are now changing rapidly, largely due to our own actions.

The *good news* is that one of our adaptations—our powerful brain—may enable us to live more sustainably by understanding and copying the ways in which nature has sustained itself for billions of years, despite major changes in environmental conditions (**Concept 1-6**, p. 23).

CONCEPT LINK

┌─ **THINKING ABOUT**
 Human Adaptations

An important adaptation of humans is strong opposable thumbs, which allow us to grip and manipulate things with our hands. Make a list of the things you could not do without the use of your thumbs.

Adaptation through Natural Selection Has Limits

In the not-too-distant future, will adaptations to new environmental conditions through natural selection allow our skin to become more resistant to the harmful effects of ultraviolet radiation, our lungs to cope with air pollutants, and our livers to better detoxify pollutants?

According to scientists in this field, the answer is *no* because of two limits to adaptations in nature through natural selection. *First,* a change in environmental conditions can lead to such an adaptation only for genetic

traits already present in a population's gene pool or for traits resulting from mutations.

Second, even if a beneficial heritable trait is present in a population, the population's ability to adapt may be limited by its reproductive capacity. Populations of genetically diverse species that reproduce quickly—such as weeds, mosquitoes, rats, bacteria, or cockroaches—often adapt to a change in environmental conditions in a short time. In contrast, species that cannot produce large numbers of offspring rapidly—such as elephants, tigers, sharks, and humans—take a long time (typically thousands or even millions of years) to adapt through natural selection.

Three Common Myths about Evolution through Natural Selection

According to evolution experts, there are three common misconceptions about biological evolution through natural selection. One is that "survival of the fittest" means "survival of the strongest." To biologists, *fitness* is a measure of reproductive success, not strength. Thus, the fittest individuals are those that leave the most descendants.

Another misconception is that organisms develop certain traits because they need or want them. A giraffe does not have a very long neck because it needs or wants it in order to feed on vegetation high in trees. Rather, some ancestor had a gene for long necks that gave it an advantage over other members of its population in getting food, and that giraffe produced more offspring with long necks.

A third misconception is that evolution by natural selection involves some grand plan of nature in which species become more perfectly adapted. From a scientific standpoint, no plan or goal of genetic perfection has been identified in the evolutionary process. Rather, it appears to be a random, branching process that results in a great variety of species (Figure 4-3).

4-3 How Do Geological Processes and Climate Change Affect Evolution?

▶ **CONCEPT 4-3** Tectonic plate movements, volcanic eruptions, earthquakes, and climate change have shifted wildlife habitats, wiped out large numbers of species, and created opportunities for the evolution of new species.

Geological Processes Affect Natural Selection

The earth's surface has changed dramatically over its long history. Scientists have discovered that huge flows of molten rock within the earth's interior break its surface into a series of gigantic solid plates, called *tectonic plates*. For hundreds of millions of years, these plates have drifted slowly atop the planet's mantle (Figure 4-6).

This process has had two important effects on the evolution and location of life on the earth. *First*, the locations of continents and oceanic basins greatly influence the earth's climate and thus help determine where plants and animals can live.

Second, the movement of continents has allowed species to move, adapt to new environments, and form new species through natural selection. When continents join together, populations can disperse to new areas and adapt to new environmental conditions. And when continents separate, populations either evolve under the new conditions or become extinct.

Earthquakes can also affect biological evolution by causing fissures in the earth's crust that can separate and isolate populations of species. Over long periods of time, this can lead to the formation of new species as each isolated population changes genetically in response to new environmental conditions. And *volcanic eruptions* affect biological evolution by destroying habitats and reducing or wiping out populations of species (**Concept 4-3**).

Climate Change and Catastrophes Affect Natural Selection

Throughout its long history, the earth's climate has changed drastically. Sometimes it has cooled and covered much of the earth with ice. At other times it has warmed, melted ice, and drastically raised sea levels. Such alternating periods of cooling and heating have led to advances and retreats of ice sheets at high latitudes over much of the northern hemisphere, most recently, about 18,000 years ago (Figure 4-7).

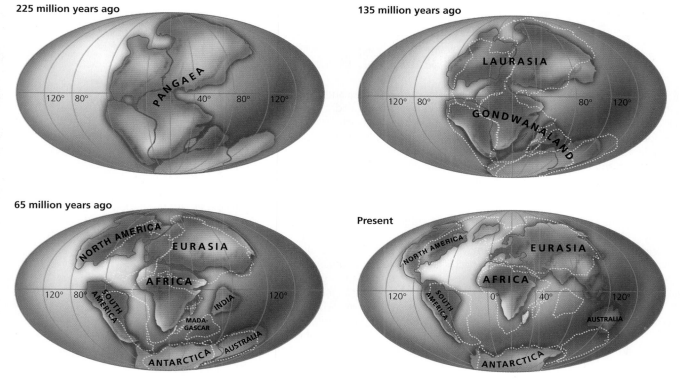

225 million years ago

135 million years ago

65 million years ago

Present

Figure 4-6 Over millions of years, the earth's continents have moved very slowly on several gigantic tectonic plates. This process plays a role in the extinction of species, as land areas split apart, and also in the rise of new species when isolated land areas combine. Rock and fossil evidence indicates that 200–250 million years ago, all of the earth's present-day continents were locked together in a supercontinent called Pangaea (top left). About 180 million years ago, Pangaea began splitting apart as the earth's tectonic plates separated, eventually resulting in today's locations of the continents (bottom right). **Question:** How might an area of land splitting apart cause the extinction of a species?

These long-term climate changes have a major effect on biological evolution by determining where different types of plants and animals can survive and thrive and by changing the locations of different types of ecosystems such as deserts, grasslands, and forests (**Concept 4-3**). Some species became extinct because the climate changed too rapidly for them to survive, and new species evolved to fill their ecological roles.

Another force affecting natural selection has been catastrophic events such as collisions between the earth and large asteroids. There have probably been many of these collisions during the earth's 4.5 billion years. Such impacts have caused widespread destruction of ecosystems and wiped out large numbers of species. But they have also caused shifts in the locations of ecosystems and created opportunities for the evolution of new species. On a long-term basis, the four **scientific principles of sustainability** (see back cover), especially biodiversity (Figure 4-2) have enabled life on earth to adapt to drastic changes in environmental conditions (Science Focus, p. 86). In other words, we live on a habitable planet. (See *The Habitable Planet*, Video 1, **www.learner.org/ resources/series209.html**.)

Figure 4-7 Changes in ice coverage in the northern hemisphere during the past 18,000 years. **Question:** What are two characteristics of an animal and two characteristics of a plant that natural selection would have favored as these ice sheets (left) advanced? (Data from the National Oceanic and Atmospheric Administration)

SCIENCE FOCUS

Earth Is Just Right for Life to Thrive

Life on the earth, as we know it, can thrive only within a certain temperature range, which depends on the liquid water that dominates the earth's surface. Most life on the earth requires average temperatures between the freezing and boiling points of water.

The earth's orbit is the right distance from the sun to provide these conditions. If the earth were much closer to the sun, it would be too hot—like Venus—for water vapor to condense and form rain. If it were much farther away, the earth's surface would be so cold—like Mars—that its water would exist only as ice. The earth also spins; if it did not, the side facing the sun would be too hot and the other side too cold for water-based life to exist.

The size of the earth is also just right for life. It has enough gravitational mass to keep its iron and nickel core molten and to keep the atmosphere—made up of light gaseous molecules required for life (such as N_2, O_2, CO_2, and H_2O)—from flying off into space.

Although life on earth has been enormously resilient and adaptive, it has benefited from a favorable temperature range. During the 3.7 billion years since life arose, the average surface temperature of the earth has remained within the narrow range of 10–20 °C (50–68 °F), even with a 30–40% increase in the sun's energy output. One reason for this is the evolution of organisms that modify levels of the temperature-regulating gas carbon dioxide in the atmosphere as a part of the carbon cycle (Figure 3-18, p. 68).

For almost 600 million years, oxygen has made up about 21% of the volume of earth's atmosphere. If this oxygen content dropped to about 15%, it would be lethal for most forms of life. If it increased to about 25%, oxygen in the atmosphere would probably ignite into a giant fireball. The current oxygen content of the atmosphere is largely the result of producer and consumer organisms interacting in the carbon cycle. Also, because of the development of photosynthesizing bacteria that have been adding oxygen to the atmosphere for more than 2 billion years, an ozone sunscreen in the stratosphere protects us and many other forms of life from an overdose of ultraviolet radiation.

In short, this remarkable planet we live on is uniquely suited for life as we know it.

Critical Thinking

Design an experiment to test the hypothesis that various forms of life can maintain the oxygen content in the atmosphere at around 21% of its volume.

4-4 How Do Speciation, Extinction, and Human Activities Affect Biodiversity?

▶ **CONCEPT 4-4A** As environmental conditions change, the balance between formation of new species and extinction of existing species determines the earth's biodiversity.

▶ **CONCEPT 4-4B** Human activities can decrease biodiversity by causing the premature extinction of species and by destroying or degrading habitats needed for the development of new species.

How Do New Species Evolve?

Under certain circumstances, natural selection can lead to an entirely new species. In this process, called **speciation,** two species arise from one. For sexually reproducing species, a new species is formed when some members of a population have evolved to the point where they no longer can breed with other members to produce fertile offspring.

The most common mechanism of speciation (especially among sexually reproducing animals) takes place in two phases: geographic isolation and reproductive isolation. **Geographic isolation** occurs when different groups of the same population of a species become physically isolated from one another for long periods. For example, part of a population may migrate in search of food and then begin living in another area

with different environmental conditions. Separation of populations can occur because of a physical barrier (such as a mountain range, stream, or road), a volcanic eruption or earthquake, or when a few individuals are carried to a new area by wind or flowing water.

In **reproductive isolation,** mutation and change by natural selection operate independently in the gene pools of geographically isolated populations. If this process continues long enough, members of the geographically and reproductively isolated populations may become so different in genetic makeup that they cannot produce live, fertile offspring if they are rejoined. Then one species has become two, and speciation has occurred (Figure 4-8).

For some rapidly reproducing organisms, this type of speciation may occur within hundreds of years. For most species, it takes from tens of thousands to millions

I notice there's excessive filler. Let me just emit the clean footer.

Arctic Fox Adapted to cold through heavier fur, short ears, short legs, and short nose. White fur matches snow for camouflage.

Early fox population → Spreads northward and southward and separates → Northern population / Southern population

Different environmental conditions lead to different selective pressures and evolution into two different species.

Gray Fox Adapted to heat through lightweight fur and long ears, legs, and nose, which give off more heat.

Figure 4-8 *Geographic isolation* can lead to reproductive isolation, divergence of gene pools, and speciation.

of years—making it difficult to observe and document the appearance of a new species.

CENGAGENOW™ Learn more about different types of speciation and ways in which they occur at CengageNOW.

Humans are playing an increasing role in the process of speciation. We have learned to shuffle genes from one species to another though artificial selection and more recently through genetic engineering (Science Focus, p. 88).

┌─ **THINKING ABOUT** ───────┐
Speciation and American Alligators CORE CASE STUDY

Imagine how a population of American alligators (**Core Case Study**) might have evolved into two species had they become separated, with one group evolving in a more northern climate. Describe some of the traits of the hypothetical northern species.
└──────────────────────────────┘

Michael P. Fogden/Bruce Coleman USA

Figure 4-9 Male golden toad in Costa Rica's high-altitude Monteverde Cloud Forest Reserve. This species has recently become extinct, primarily because changes in climate dried up its habitat.

Extinction Is Forever

Another process affecting the number and types of species on the earth is **extinction**, in which an entire species ceases to exist. Species that are found in only one area are called **endemic species** and are especially vulnerable to extinction. They exist on islands and in other unique small areas, especially in tropical rain forests where most species are highly specialized.

One example is the brilliantly colored golden toad (Figure 4-9) once found only in a small area of lush cloud rain forests in Costa Rica's mountainous region. Despite living in the country's well-protected Monteverde Cloud Forest Reserve, by 1989, the golden toad had apparently become extinct. Much of the moisture that supported its rain forest habitat came in the form of moisture-laden clouds blowing in from the Caribbean Sea. But warmer air from global climate change caused

these clouds to rise, depriving the forests of moisture, and the habitat for the golden toad and many other species dried up. The golden toad appears to be one of the first victims of climate change caused largely by global warming. A 2007 study found that global warming has also contributed to the extinction of five other toad and frog species in the jungles of Costa Rica.

Extinction Can Affect One Species or Many Species at a Time

All species eventually become extinct, but drastic changes in environmental conditions can eliminate large groups of species. Throughout most of history, species have disappeared at a low rate, called **background**

SCIENCE FOCUS

We Have Developed Two Ways to Change the Genetic Traits of Populations

We have used **artificial selection** to change the genetic characteristics of populations with similar genes. In this process, we select one or more desirable genetic traits in the population of a plant or animal, such as a type of wheat, fruit, or dog. Then we use *selective breeding* to generate populations of the species containing large numbers of individuals with the desired traits. Note that artificial selection involves crossbreeding between genetic varieties of the same species and thus is not a form of speciation. Most, of the grains, fruits, and vegetables we eat are produced by artificial selection.

Artificial selection has given us food crops with higher yields, cows that give more milk, trees that grow faster, and many different types of dogs and cats. But traditional crossbreeding is a slow process. Also, it can combine traits only from species that are close to one another genetically.

Now scientists are using genetic engineering to speed up our ability to manipulate genes. **Genetic engineering, or gene splicing,** is the alteration of an organism's genetic material, through adding, deleting, or changing segments of its DNA (Figure 11, p. S43, in Supplement 6), to produce desirable traits or eliminate undesirable ones. It enables scientists to transfer genes between different species that would not interbreed in nature. For example, genes from a fish species can be put into a tomato plant to give it certain properties.

Scientists have used gene splicing to develop modified crop plants, new drugs, pest-resistant plants, and animals that grow rapidly (Figure 4-A). They have also created genetically engineered bacteria to extract minerals such as copper from their underground ores and to clean up spills of oil and other toxic pollutants.

Application of our increasing genetic knowledge is filled with great promise, but it

Figure 4-A An example of genetic engineering. The 6-month-old mouse on the left is normal; the same-age mouse on the right has a human growth hormone gene inserted in its cells. Mice with the human growth hormone gene grow two to three times faster and twice as large as mice without the gene. **Question:** How do you think the creation of such species might change the process of evolution by natural selection?

R. L. Brinster and R. E. Hammer/School of Veterinary Medicine, University of Pennsylvania

raises some serious ethical and privacy issues. For example, some people have genes that make them more likely to develop certain genetic diseases or disorders. We now have the power to detect these genetic deficiencies, even before birth. Questions of ethics and morality arise over how this knowledge and technology will be applied, who will benefit, and who might suffer from it.

Further, what will be the environmental impacts of such applications? If genetic engineering could help all humans live in good health much longer than we do now, it might increase pollution, environmental degradation, and the strain on natural resources. More and more affluent people living longer and longer could create an enormous and ever-growing ecological footprint.

Some people dream of a day when our genetic engineering prowess could eliminate death and aging altogether. As one's cells, organs, or other parts wear out or become damaged, they could be replaced with new ones grown in genetic engineering facilities.

Assuming this is scientifically possible, is it morally acceptable to take this path? Who will decide? Who will regulate this new industry? Sometime in the not-too-distant future, will we be able to change the nature of

what it means to be human? If so, how will it change? These are some of the most important and controversial ethical questions of the 21st century.

Another concern is that most new technologies have had unintended harmful consequences. For example, pesticides have helped protect crops from insect pests and disease. But their overuse has accelerated the evolution of pesticide-resistant species and has wiped out many natural predator insects that had helped to keep pest populations under control.

For these and other reasons, a backlash developed in the 1990s against the increasing use of genetically modified food plants and animals. Some protesters argue against using this new technology, mostly for ethical reasons. Others advocate slowing down the technological rush and taking a closer look at the short- and long-term advantages and disadvantages of genetic technologies.

Critical Thinking

What might be some beneficial and harmful effects on the evolutionary process if genetic engineering is widely applied to plants and animals?

extinction. Based on the fossil record and analysis of ice cores, biologists estimate that the average annual background extinction rate is one to five species for each million species on the earth.

In contrast, **mass extinction** is a significant rise in extinction rates above the background level. In such a catastrophic, widespread (often global) event, large groups of species (perhaps 25–70%) are wiped out in a geological period lasting up to 5 million years. Fossil and geological evidence indicate that the earth's species have experienced five mass extinctions (20–60 million years apart) during the past 500 million years. For example, about 250 million years ago, as much as 95% of all existing species became extinct.

Some biologists argue that a mass extinction should be distinguished by a low speciation rate as well as by a high rate of extinction. Under this more strict definition, there have been only three mass extinctions. As this subject is debated, the definitions will be refined, and one argument or the other will be adopted as the working hypothesis. Either way, there is substantial evidence that large numbers of species have become extinct several times in the past.

A mass extinction provides an opportunity for the evolution of new species that can fill unoccupied ecological roles or newly created ones. As environmental conditions change, the balance between formation of new species (speciation) and extinction of existing species determines the earth's biodiversity (**Concept 4-4A**). The existence of millions of species today means that speciation, on average, has kept ahead of extinction.

Extinction is a natural process. But much evidence indicates that humans have become a major force in the premature extinction of a growing number of species, as discussed further in Chapter 9.

4-5 What Is Species Diversity and Why Is It Important?

▶ **CONCEPT 4-5** Species diversity is a major component of biodiversity and tends to increase the sustainability of ecosystems.

Species Diversity Includes the Variety and Abundance of Species in a Particular Place

An important characteristic of a community and the ecosystem to which it belongs is its **species diversity**: the number of different species it contains (**species richness**) combined with the relative abundance of individuals within each of those species (**species evenness**).

For example, a biologically diverse community such as a tropical rain forest or a coral reef (Figure 4-10, left) with a large number of different species (high species richness) generally has only a few members of each

Figure 4-10 *Variations in species richness and species evenness.* A coral reef (left), with a large number of different species (high species richness), generally has only a few members of each species (low species evenness). In contrast, a grove of aspen trees in Alberta, Canada, in the fall (right) has a small number of different species (low species richness), but large numbers of individuals of each species (high species evenness).

SCIENCE FOCUS

Species Richness on Islands

In the 1960s, ecologists Robert MacArthur and Edward O. Wilson began studying communities on islands to discover why large islands tend to have more species of a certain category such as insects, birds, or ferns than do small islands.

To explain these differences in species richness among islands of varying sizes, MacArthur and Wilson carried out research and used their findings to propose what is called the **species equilibrium model,** or the **theory of island biogeography.** According to this widely accepted scientific theory, the number of different species (species richness) found on an island is determined by the interactions of two factors: the rate at which new species immigrate to the island and the rate at which species become *extinct,* or cease to exist, on the island.

The model projects that, at some point, the rates of species immigration and species extinction should balance so that neither rate is increasing or decreasing sharply. This balance point is the equilibrium point that determines the island's average number of different species (species richness) over time.

(The website **CengageNOW** has a great interactive animation of this model. Go to the end of any chapter for instructions on how to use it.)

According to the model, two features of an island affect the immigration and extinction rates of its species and thus its species diversity. One is the island's *size.* Small islands tend to have fewer species than large islands do because they make smaller targets for potential colonizers flying or floating toward them. Thus, they have lower immigration rates than larger islands do. In addition, a small island should have a higher extinction rate because it usually has fewer resources and less diverse habitats for its species.

A second factor is an island's *distance from the nearest mainland.* Suppose we have two islands about equal in size, extinction rates, and other factors. According to the model, the island closer to a mainland source of immigrant species should have the higher immigration rate and thus a higher species richness. The farther a potential colonizing species has to travel, the less likely it is to reach the island.

These factors interact to influence the relative species richness of different islands. Thus, larger islands closer to a mainland tend to have the most species, while smaller islands farther away from a mainland tend to have the fewest. Since MacArthur and Wilson presented their hypothesis and did their experiments, others have conducted more scientific studies that have born out their hypothesis, making it a widely accepted scientific theory.

Scientists have used this theory to study and make predictions about wildlife in *habitat islands*—areas of natural habitat, such as national parks and mountain ecosystems, surrounded by developed and fragmented land. These studies and predictions have helped scientists to preserve these ecosystems and protect their resident wildlife.

Critical Thinking

Suppose we have two national parks surrounded by development. One is a large park and the other is much smaller. Which park is likely to have the highest species richness? Why?

species (low species evenness). Biologist Terry Erwin found an estimated 1,700 different beetle species in a single tree in a tropical forest in Panama but only a few individuals of each species. On the other hand, an aspen forest community in Canada (Figure 4-10, right) may have only a few plant species (low species richness) but large numbers of each species (high species evenness).

The species diversity of communities varies with their *geographical location.* For most terrestrial plants and animals, species diversity (primarily species richness) is highest in the tropics and declines as we move from the equator toward the poles (see Figure 2, pp. S22–S23, in Supplement 4). The most species-rich environments are tropical rain forests, coral reefs (Figure 4-10, left), the ocean bottom zone, and large tropical lakes.

Scientists have sought to learn more about species richness by studying species on islands (Science Focus, above). Islands make good study areas because they are relatively isolated, and it is easier to observe species arriving and disappearing from islands than it would be to make such a study in other less isolated ecosystems.

CENGAGENOW™ Learn about how latitude affects species diversity and about the differences between big and small islands at CengageNOW.

Species-Rich Ecosystems Tend to Be Productive and Sustainable

How does species richness affect an ecosystem? In trying to answer this question, ecologists have been conducting research to answer two related questions: Is plant productivity higher in species-rich ecosystems? And does species richness enhance the *stability,* or sustainability of an ecosystem? Research suggests that the answers to both questions may be "yes" but more research is needed before these scientific hypotheses can be accepted as scientific theories.

According to the first hypothesis, the more diverse an ecosystem is, the more productive it will be. That is, with a greater variety of producer species, an ecosystem will produce more plant biomass, which in turn will support a greater variety of consumer species.

A related hypothesis is that greater species richness and productivity will make an ecosystem more stable or sustainable. In other words, the greater the species richness and the accompanying web of feeding and biotic interactions in an ecosystem, the greater its sustainability, or ability to withstand environmental disturbances such as drought or insect infestations. According to this hypothesis, a complex ecosystem with

many different species (high species richness) and the resulting variety of feeding paths has more ways to respond to most environmental stresses because it does not have "all its eggs in one basket."

Many studies support the idea that some level of species richness and productivity can provide insurance against catastrophe. In one prominent 11-year study, David Tilman and his colleagues at the University of Minnesota found that communities with high plant species richness produced a certain amount of biomass more consistently than did communities with fewer species. The species-rich communities were also less affected by drought and more resistant to invasions by new insect species. Because of their higher level of biomass, the species-rich communities also consumed more carbon dioxide and took up more nitrogen, thus taking more robust roles in the carbon and nitrogen cycles (**Concept 3-5**, p. 65). Later laboratory studies involved setting up artificial ecosystems in growth chambers where key variables such as temperature, light, and atmospheric gas concentrations could be controlled and varied. These studies have supported Tilman's findings.

Ecologists hypothesize that in a species-rich ecosystem, each species can exploit a different portion of the resources available. For example, some plants will bloom early and others will bloom late. Some have shallow roots to absorb water and nutrients in shallow soils, and others use deeper roots to tap into deeper soils.

There is some debate among scientists about how much species richness is needed to help sustain various ecosystems. Some research suggests that the average annual net primary productivity of an ecosystem reaches a peak with 10–40 producer species. Many ecosystems contain more than 40 producer species, but do not necessarily produce more biomass or reach a higher level of stability. Scientists are still trying to determine how many producer species are needed to enhance the sustainability of particular ecosystems and which producer species are the most important in providing such stability.

Bottom line: species richness appears to increase the productivity and stability or sustainability of an ecosystem (**Concept 4-5**). While there may be some exceptions to this, most ecologists now accept it as a useful hypothesis.

RESEARCH FRONTIER

Learning more about how biodiversity is related to ecosystem stability and sustainability. See **academic.cengage.com/biology/miller**.

4-6 What Roles Do Species Play in Ecosystems?

▶ **CONCEPT 4-6A** Each species plays a specific ecological role called its niche.

▶ **CONCEPT 4-6B** Any given species may play one or more of five important roles—native, nonnative, indicator, keystone, or foundation roles—in a particular ecosystem.

Each Species Plays a Unique Role in Its Ecosystem

An important principle of ecology is that *each species has a distinct role to play in the ecosystems where it is found* (**Concept 4-6A**). Scientists describe the role that a species plays in its ecosystem as its **ecological niche,** or simply **niche** (pronounced "nitch"). It is a species' way of life in a community and includes everything that affects its survival and reproduction, such as how much water and sunlight it needs, how much space it requires, and the temperatures it can tolerate. A species' niche should not be confused with its *habitat,* which is the place where it lives. Its niche is its pattern of living.

Scientists use the niches of species to classify them broadly as *generalists* or *specialists.* **Generalist species** have broad niches (Figure 4-11, right curve). They can live in many different places, eat a variety of foods, and

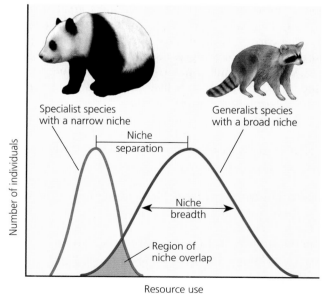

Figure 4-11 Specialist species such as the giant panda have a narrow niche (left) and generalist species such as a raccoon have a broad niche (right).

often tolerate a wide range of environmental conditions. Flies, cockroaches (see Case Study below), mice, rats, white-tailed deer, raccoons, and humans are generalist species.

In contrast, **specialist species** occupy narrow niches (Figure 4-11, left curve). They may be able to live in only one type of habitat, use one or a few types of food, or tolerate a narrow range of climatic and other environmental conditions. This makes specialists more prone to extinction when environmental conditions change.

For example, *tiger salamanders* breed only in fishless ponds where their larvae will not be eaten. China's *giant panda* (Figure 4-11, left) is highly endangered because of a combination of habitat loss, low birth rate, and its specialized diet consisting mostly of bamboo. Some shorebirds occupy specialized niches, feeding on crustaceans, insects, and other organisms on sandy beaches and their adjoining coastal wetlands (Figure 4-13).

Is it better to be a generalist or a specialist? It depends. When environmental conditions are fairly constant, as in a tropical rain forest, specialists have an advantage because they have fewer competitors. But under rapidly changing environmental conditions, the generalist usually is better off than the specialist.

THINKING ABOUT
The American Alligator's Niche

 CORE CASE STUDY

Does the American alligator (**Core Case Study**) have a specialist or a generalist niche? Explain.

■ CASE STUDY
Cockroaches: Nature's Ultimate Survivors

Cockroaches (Figure 4-12), the bugs many people love to hate, have been around for 350 million years, outliving the dinosaurs. One of evolution's great success stories, they have thrived because they are *generalists.*

The earth's 3,500 cockroach species can eat almost anything, including algae, dead insects, fingernail clippings, salts in tennis shoes, electrical cords, glue, paper, and soap. They can also live and breed almost anywhere except in polar regions.

Some cockroach species can go for a month without food, survive for a month on a drop of water from a dishrag, and withstand massive doses of radiation. One species can survive being frozen for 48 hours.

Cockroaches usually can evade their predators—and a human foot in hot pursuit—because most species have antennae that can detect minute movements of air. They also have vibration sensors in their knee joints, and they can respond faster than you can blink your eye. Some even have wings. They have compound eyes that allow them to see in almost all directions at once. Each eye has about 2,000 lenses, compared to one in each of your eyes.

And, perhaps most significantly, they have high reproductive rates. In only a year, a single Asian cockroach and its offspring can add about 10 million new cockroaches to the world. Their high reproductive rate also helps them to quickly develop genetic resistance to almost any poison we throw at them.

Most cockroaches sample food before it enters their mouths and learn to shun foul-tasting poisons. They also clean up after themselves by eating their own dead and, if food is scarce enough, their living.

About 25 species of cockroach live in homes and can carry viruses and bacteria that cause diseases. On the other hand, cockroaches play a role in nature's food webs. They make a tasty meal for birds and lizards.

Niches Can Be Occupied by Native and Nonnative Species

Niches can be classified further in terms of specific roles that certain species play within ecosystems. Ecologists describe *native, nonnative, indicator, keystone,* and *foundation species.* Any given species may play one or more of these five roles in a particular community (**Concept 4-6B**).

Native species are those species that normally live and thrive in a particular ecosystem. Other species that migrate into or are deliberately or accidentally introduced into an ecosystem are called **nonnative species,** also referred to as *invasive, alien,* or *exotic species.*

Figure 4-12 As generalists, cockroaches are among the earth's most adaptable and prolific species. This is a photo of an American cockroach.

Clemson University–USDA Cooperative Extension Slide Series

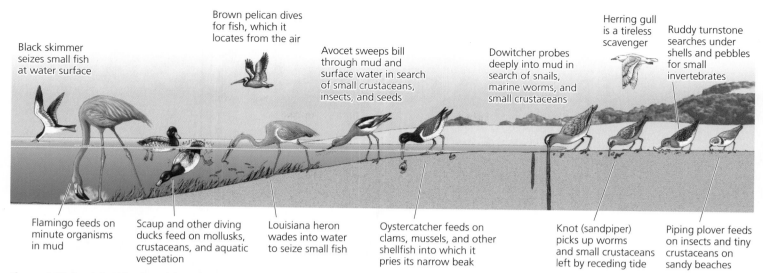

Black skimmer seizes small fish at water surface

Brown pelican dives for fish, which it locates from the air

Avocet sweeps bill through mud and surface water in search of small crustaceans, insects, and seeds

Dowitcher probes deeply into mud in search of snails, marine worms, and small crustaceans

Herring gull is a tireless scavenger

Ruddy turnstone searches under shells and pebbles for small invertebrates

Flamingo feeds on minute organisms in mud

Scaup and other diving ducks feed on mollusks, crustaceans, and aquatic vegetation

Louisiana heron wades into water to seize small fish

Oystercatcher feeds on clams, mussels, and other shellfish into which it pries its narrow beak

Knot (sandpiper) picks up worms and small crustaceans left by receding tide

Piping plover feeds on insects and tiny crustaceans on sandy beaches

Figure 4-13 Specialized feeding niches of various bird species in a coastal wetland. This specialization reduces competition and allows sharing of limited resources.

Some people tend to think of nonnative species as villains. In fact, most introduced and domesticated species of crops and animals, such as chickens, cattle, and fish from around the world, are beneficial to us. However, some nonnative species can threaten a community's native species and cause unintended and unexpected consequences. In 1957, for example, Brazil imported wild African bees to help increase honey production. Instead, the bees displaced domestic honeybees and reduced the honey supply.

Since then, these nonnative bee species—popularly known as "killer bees"—have moved northward into Central America and parts of the southwestern and southeastern United States. The wild African bees are not the fearsome killers portrayed in some horror movies, but they are aggressive and unpredictable. They have killed thousands of domesticated animals and an estimated 1,000 people in the western hemisphere, many of whom were allergic to bee stings.

Nonnative species can spread rapidly if they find a new more favorable niche. In their new niches, these species often do not face the predators and diseases they faced before, or they may be able to out-compete some native species in their new niches. We will examine this environmental threat in greater detail in Chapter 9.

Indicator Species Serve as Biological Smoke Alarms

Species that provide early warnings of damage to a community or an ecosystem are called **indicator species.** For example, the presence or absence of trout species in water at temperatures within their range of tolerance (Figure 3-10, p. 58) is an indicator of water quality because trout need clean water with high levels of dissolved oxygen.

Birds are excellent biological indicators because they are found almost everywhere and are affected quickly by environmental changes such as loss or fragmentation of their habitats and introduction of chemical pesticides. The populations of many bird species are declining. Butterflies are also good indicator species because their association with various plant species makes them vulnerable to habitat loss and fragmentation. Some amphibians are also classified as indicator species (Case Study below).

Using a living organism to monitor environmental quality is not new. Coal mining is a dangerous occupation, partly because of the underground presence of poisonous and explosive gases, many of which have no detectable odor. In the 1800s and early 1900s, coal miners took caged canaries into mines to act as early-warning sentinels. These birds sing loudly and often. If they quit singing for a long period and appeared to be distressed, miners took this as an indicator of the presence of poisonous or explosive gases and got out of the mine.

■ CASE STUDY
Why Are Amphibians Vanishing?

Amphibians (frogs, toads, and salamanders) live part of their lives in water and part on land. Populations of some amphibians, also believed to be indicator species, are declining throughout the world.

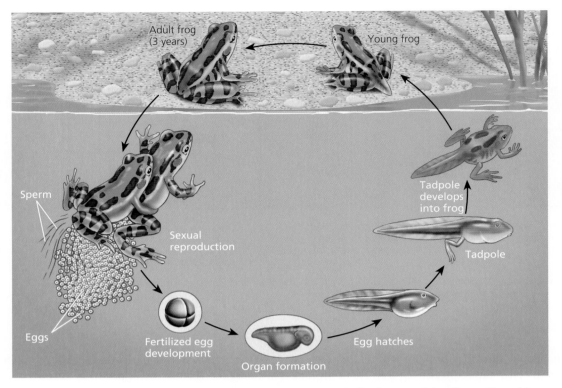

Figure 4-14 *Life cycle of a frog.* Populations of various frog species can decline because of the effects of harmful factors at different points in their life cycle. Such factors include habitat loss, drought, pollution, increased ultraviolet radiation, parasitism, disease, overhunting by humans, and nonnative predators and competitors.

Amphibians were the first vertebrates to set foot on the earth. Historically, they have been better than many other species have been at adapting to environmental changes through evolution. But many amphibian species apparently are having difficulty adapting to some of the rapid environmental changes that have taken place in the air and water and on the land during the past few decades—changes resulting mostly from human activities. Evolution takes time and some amphibians have traits that can make them vulnerable to certain changes in environmental conditions. Frogs, for example, are especially vulnerable to environmental disruption at various points in their life cycle (Figure 4-14).

As tadpoles, frogs live in water and eat plants; as adults, they live mostly on land and eat insects, which can expose them to pesticides. The eggs of frogs have no protective shells to block UV radiation or pollution. As adults, they take in water and air through their thin, permeable skins, which can readily absorb pollutants from water, air, or soil. During their life cycle, frogs and many other amphibian species also seek sunlight, which warms them and helps them to grow and develop, but which also increases their exposure to UV radiation.

Since 1980, populations of hundreds of the world's almost 6,000 amphibian species have been vanishing or declining in almost every part of the world, even in protected wildlife reserves and parks. Accord-

ing to the 2004 Global Amphibian Assessment, about 33% of all known amphibian species (and more than 80% of those in the Caribbean) are threatened with extinction, and populations of 43% of the species are declining.

No single cause has been identified to explain these amphibian declines. However, scientists have identified a number of factors that can affect frogs and other amphibians at various points in their life cycles:

- *Habitat loss and fragmentation,* especially from draining and filling of inland wetlands, deforestation, and urban development

- *Prolonged drought,* which can dry up breeding pools so that few tadpoles survive

- *Pollution,* especially exposure to pesticides, which can make frogs more vulnerable to bacterial, viral, and fungal diseases

- *Increases in UV radiation* caused by reductions in stratospheric ozone during the past few decades, caused by chemicals we have put into the air that have ended up in the stratosphere

- *Parasites* such as flatbed worms, which feed on the amphibian eggs laid in water, apparently have caused an increase in births of amphibians with missing or extra limbs

- *Viral and fungal diseases,* especially the chytrid fungus, which attacks the skin of frogs, apparently reducing their ability to take in water, which leads to death from dehydration. Such diseases spread when adults of many amphibian species congregate in large numbers to breed.

- *Climate change.* A 2005 study found an apparent correlation between global warming and the extinction of about two-thirds of the 110 known species of harlequin frog in tropical forests in Central and South America by creating favorable conditions for the spread of the deadly chytrid fungus to the frogs. But a 2008 study cast doubt on this hypothesis—another example of how science works. Climate change from global warming has also been identified as the primary cause of the extinction of the golden toad in Costa Rica (Figure 4-9).

- *Overhunting,* especially in Asia and France, where frog legs are a delicacy

- *Natural immigration of, or deliberate introduction of, nonnative predators and competitors* (such as certain fish species)

A combination of such factors probably is responsible for the decline or disappearance of most amphibian species.

RESEARCH FRONTIER

Learning more about why amphibians are disappearing and applying this knowledge to other threatened species. See **academic.cengage.com/biology/miller**.

Why should we care if some amphibian species become extinct? Scientists give three reasons. *First,* amphibians are sensitive biological indicators of changes in environmental conditions such as habitat loss and degradation, air and water pollution, exposure to ultraviolet light, and climate change. Their possible extinction suggests that environmental health is deteriorating in parts of the world.

Second, adult amphibians play important ecological roles in biological communities. For example, amphibians eat more insects (including mosquitoes) than do birds. In some habitats, extinction of certain amphibian species could lead to extinction of other species, such as reptiles, birds, aquatic insects, fish, mammals, and other amphibians that feed on them or their larvae.

Third, amphibians are a genetic storehouse of pharmaceutical products waiting to be discovered. Compounds in secretions from amphibian skin have been isolated and used as painkillers and antibiotics and as treatment for burns and heart disease.

The rapidly increasing global extinction of a variety of amphibian species is a warning about the harmful effects of an array of environmental threats to biodiversity. Like canaries in a coal mine, these indicator species are sending us urgent distress signals.

Keystone and Foundation Species Help Determine the Structure and Functions of Their Ecosystems

A keystone is the wedge-shaped stone placed at the top of a stone archway. Remove this stone and the arch collapses. In some communities and ecosystems, ecologists hypothesize that certain species play a similar role. **Keystone species** have a large effect on the types and abundances of other species in an ecosystem.

The effects that keystone species have in their ecosystems is often much larger than their numbers would suggest, and because of their relatively limited numbers, some keystone species are more vulnerable to extinction than others are. As was shown by the near extinction of the American alligator (**Core Case Study**) in the southeastern United States, eliminating a keystone species may dramatically alter the structure and function of a community.

Keystone species can play several critical roles in helping to sustain an ecosystem. One such role is *pollination* of flowering plant species by bees, butterflies (Figure 3-A, left, p. 54), hummingbirds, bats, and other species. In addition, *top predator* keystone species feed on and help regulate the populations of other species. Examples are the alligator, wolf, leopard, lion, and some shark species (see Case Study, p. 96).

Ecologist Robert Paine conducted a controlled experiment along the rocky Pacific coast of the U.S. state of Washington that demonstrated the keystone role of the top-predator sea star *Piaster orchaceus* in an intertidal zone community. Paine removed the mussel-eating *Piaster* sea stars from one rocky shoreline community but not from an adjacent community, which served as a control group. Mussels took over and crowded out most other species in the community without the *Piaster* sea stars.

The loss of a keystone species can lead to population crashes and extinctions of other species in a community that depends on it for certain services, as we saw in the **Core Case Study** that opens this chapter. This explains why it so important for scientists to identify and protect keystone species.

Another important type of species in some ecosystems is a **foundation species,** which plays a major role in shaping communities by creating and enhancing their habitats in ways that benefit other species. For example, elephants push over, break, or uproot trees, creating forest openings in the grasslands and woodlands of Africa. This promotes the growth of grasses and other

forage plants that benefit smaller grazing species such as antelope. It also accelerates nutrient cycling rates.

Beavers are another good example of a foundation species. Acting as "ecological engineers," they build dams in streams to create ponds and other wetlands used by other species. Some bat and bird foundation species help to regenerate deforested areas and spread fruit plants by depositing plant seeds in their droppings.

Keystone and foundation species play similar roles. In general, the major difference between the two types of species is that foundation species help to create habitats and ecosystems. They often do this almost literally by providing the foundation for the ecosystem (as beavers do, for example). On the other hand, keystone species can do this and more. They sometimes play this foundation role (as do American alligators, for example), but they also play an active role in maintaining the ecosystem and keeping it functioning in a way that serves the other species living there. (Recall that the American alligator helps to keep the waters in its habitat clear of invading vegetation for use by other species that need open water.)

RESEARCH FRONTIER

Identifying and protecting keystone and foundation species. See **academic.cengage.com/biology/miller**.

THINKING ABOUT

The American Alligator

 CORE CASE STUDY

What species might disappear or suffer sharp population declines if the American alligator (**Core Case Study**) became extinct in subtropical wetland ecosystems?

■ CASE STUDY
Why Should We Protect Sharks?

The world's 370 shark species vary widely in size. The smallest is the dwarf dog shark, about the size of a large goldfish. The largest, the whale shark, can grow to 15 meters (50 feet) long and weigh as much as two full-grown African elephants.

Shark species that feed at or near the tops of food webs (Figure 3-14, p. 63) remove injured and sick animals from the ocean, and thus play an important ecological role. Without the services provided by these *keystone species,* the oceans would be teeming with dead and dying fish.

In addition to their important ecological roles, sharks could save human lives. If we can learn why they almost never get cancer, we could possibly use this information to fight cancer in our own species. Scientists are also studying their highly effective immune system, which allows wounds to heal without becoming infected.

Many people—influenced by movies, popular novels, and widespread media coverage of a fairly small number of shark attacks per year—think of sharks as

people-eating monsters. In reality, the three largest species—the whale shark, basking shark, and megamouth shark—are gentle giants. They swim through the water with their mouths open, filtering out and swallowing huge quantities of plankton.

Media coverage of shark attacks greatly distorts the danger from sharks. Every year, members of a few species—mostly great white, bull, tiger, gray reef, lemon, hammerhead, shortfin mako, and blue sharks—injure 60–100 people worldwide. Since 1990, sharks have killed an average of seven people per year. Most attacks involve great white sharks, which feed on sea lions and other marine mammals and sometimes mistake divers and surfers for their usual prey. Compare the risks: poverty prematurely kills about 11 million people a year, tobacco 5 million a year, and air pollution 3 million a year.

For every shark that injures a person, we kill at least 1 million sharks. Sharks are caught mostly for their valuable fins and then thrown back alive into the water, fins removed, to bleed to death or drown because they can no longer swim. The fins are widely used in Asia as a soup ingredient and as a pharmaceutical cure-all. A top (dorsal) fin from a large whale shark can fetch up to $10,000. In high-end restaurants in China, a bowl of shark fin soup can cost $100 or more. Ironically, shark fins have been found to contain dangerously high levels of toxic mercury.

Sharks are also killed for their livers, meat, hides, and jaws, and because we fear them. Some sharks die when they are trapped in nets or lines deployed to catch swordfish, tuna, shrimp, and other species. Sharks are especially vulnerable to overfishing because they grow slowly, mature late, and have only a few offspring per generation. Today, they are among the most vulnerable and least protected animals on earth.

In 2008, The IUCN-World Conservation Union reported that the populations of many large shark species have declined by half since the 1970s. Because of the increased demand for shark fins and meat, eleven of the world's open ocean shark species are considered critically endangered or endangered, and 81 species are threatened with extinction. In response to a public outcry over depletion of some species, the United States and several other countries have banned the hunting of sharks for their fins. But such bans apply only in territorial waters and are difficult to enforce.

Scientists call for banning shark finning in international waters and establishing a network of fully protected marine reserves to help protect coastal shark and other aquatic species from overfishing. Between 1970 and 2005, overfishing of hammerhead, bull, dusky, and other large predatory sharks in the northwest Atlantic for their fins and meat cut their numbers by 99%. In 2007, scientists Charles "Pete" Peterson and Julia Baum reported that this decline may be indirectly decimating the bay scallop fishery along the eastern coast of the United States. With fewer sharks around, populations of rays and skates, which sharks normally feed

on, have exploded and are feasting on bay scallops in seagrass beds along the Atlantic coast.

Sharks have been around for more than 400 million years. Sustaining this portion of the earth's biodiversity begins with the knowledge that sharks may not need us, but that we and other species need them.

REVISITING The American Alligator and Sustainability

The **Core Case Study** of the American alligator at the beginning of this chapter illustrates the power humans have over the environment—the power both to do harm and to make amends. As most American alligators were eliminated from their natural areas in the 1950s, scientists began pointing out the ecological benefits these animals had been providing to their ecosystems (such as building water holes, nesting mounds, and feeding sites for other species). Scientific understanding of these ecological connections led to protection of this species and to its recovery.

In this chapter, we studied the importance of biodiversity, especially the numbers and varieties of species found in different parts of the world (species richness), along with the other forms of biodiversity—functional, ecosystem, and genetic diversity. We also studied the process whereby all species came to be, according to scientific theory of biological evolution through natural selection. Taken together, these two great assets, biodiversity and evolution, represent irreplaceable natural capital. Each depends upon the other and upon whether humans can respect and preserve this natural capital. Finally, we examined the variety of roles played by species in ecosystems.

Ecosystems and the variety of species they contain are functioning examples of the four **scientific principles of sustainability** (see back cover) in action. They depend on solar energy and provide functional biodiversity in the form of energy flow and the chemical cycling of nutrients. In addition, ecosystems sustain biodiversity in all its forms, and population sizes are controlled by interactions among diverse species. In the next chapter, we delve further into this natural regulation of populations and the biodiversity of ecosystems.

All we have yet discovered is but a trifle
in comparison with what lies hid
in the great treasury of nature.

ANTOINE VAN LEEUWENHOEK

REVIEW

1. Review the Key Questions and Concepts for this chapter on p. 78. Explain why we should protect the American alligator (**Core Case Study**) from being driven to extinction as a result of our activities.

2. What are the four major components of **biodiversity (biological diversity)?** What is the importance of biodiversity?

3. What is **biological evolution?** What is **natural selection?** What is a **fossil** and why are fossils important in understanding biological evolution? What is a **mutation** and what role do mutations play in evolution by natural selection? What is an **adaptation (adaptive trait)?** What is **differential reproduction?** How did we become such a powerful species?

4. What are two limits to evolution by natural selection? What are three myths about evolution through natural selection?

5. Describe how geologic processes and climate change can affect natural selection. Describe conditions on the earth that favor the development of life as we know it.

6. What is **speciation?** Distinguish between **geographic isolation** and **reproductive isolation** and explain how they can lead to the formation of a new species. Distinguish between **artificial selection** and **genetic engineering (gene splicing)** and give an example of each. What are some possible social, ethical, and environmental problems with the widespread use of genetic

engineering? What is **extinction?** What is an **endemic species** and why is it vulnerable to extinction? Distinguish between **background extinction** and **mass extinction.**

7. What is **species diversity?** Distinguish between **species richness** and **species evenness** and give an example of each. Describe the **theory of island biogeography (species equilibrium model).** Explain why species-rich ecosystems tend to be productive and sustainable.

8. What is an **ecological niche?** Distinguish between **specialist species** and **generalist species** and give an example of each.

9. Distinguish among **native, nonnative, indicator, keystone,** and **foundation** species and give an example of each type. Explain why birds are excellent indicator species. Why are amphibians vanishing and why should we protect them? Why should we protect shark species from being driven to extinction as a result of our activities? Describe the role of the beaver as a foundation species.

10. Explain how the role of the American alligator in its ecosystem (**Core Case Study**) illustrates the biodiversity **principles of sustainability?**

Note: Key Terms are in **bold** type.

CRITICAL THINKING

1. List three ways in which you could apply **Concept 4-4B** in order to live a more environmentally sustainable lifestyle.

2. Explain what could happen to the ecosystem where American alligators (**Core Case Study**) live if the alligators went extinct. Name a plant species and an animal species that would be seriously affected, and describe how each might respond to these changes in their environmental conditions.

3. What role does each of the following processes play in helping implement the four **scientific principles of sustainability** (see back cover): **(a)** natural selection, **(b)** speciation, and **(c)** extinction?

4. Describe the major differences between the ecological niches of humans and cockroaches. Are these two species in competition? If so, how do they manage to coexist?

5. How would you experimentally determine whether an organism is a keystone species?

6. Is the human species a keystone species? Explain. If humans were to become extinct, what are three species that

might also become extinct and three species whose populations would probably grow?

7. How would you respond to someone who tells you:
 a. that he or she does not believe in biological evolution because it is "just a theory"?
 b. that we should not worry about air pollution because natural selection will enable humans to develop lungs that can detoxify pollutants?

8. How would you respond to someone who says that because extinction is a natural process, we should not worry about the loss of biodiversity when species become prematurely extinct as a result of our activities?

9. Congratulations! You are in charge of the future evolution of life on the earth. What are the three most important things you would do?

10. List two questions that you would like to have answered as a result of reading this chapter.

Note: See Supplement 13 (p. S78) for a list of Projects related to this chapter.

DATA ANALYSIS

Injuries and deaths from shark attacks are highly publicized by the media. However, the risk of injury or death from a shark attack for people going into coastal waters as swimmers, surfers, or divers is extremely small (see Case Study, p. 96). For example, according to the National Safety Council, the Centers for Disease Control and Prevention, and the International Shark Attack File, the estimated lifetime risk of dying from a shark attack in the United States is about 1 in 3,750,000 compared to risks of 1 in 1,130 from drowning, 1 in 218 from a

fall, 1 in 84 from a car accident, 1 in 63 from the flu, and 1 in 38 from a hospital infection.

Between 1998 and 2007, the United States had the world's highest percentage of deaths and injuries from unprovoked shark attacks, and the U.S. state of Florida had the country's highest percentage of deaths and injuries from unprovoked shark attacks, as shown by the following data about shark attacks in the world, in the United States, and in Florida.

Year	World			United States			Florida		
	Total Attacks	Fatal Attacks	Non-fatal Attacks	Total Attacks	Fatal Attacks	Non-fatal Attacks	Total Attacks	Fatal Attacks	Non-fatal Attacks
1998	51	6	45	26	1	25	21	1	20
1999	56	4	52	38	0	38	26	0	26
2000	79	11	68	53	1	52	37	1	36
2001	68	4	64	50	3	47	34	1	33
2002	62	3	59	47	0	47	29	0	29
2003	57	4	53	40	1	39	30	0	30
2004	65	7	58	30	2	28	12	0	12
2005	61	4	57	40	1	39	20	1	19
2006	63	4	59	40	0	40	23	0	23
2007	71	1	70	50	0	50	32	0	32

Source: Data from *International Shark Attack File,* Florida Museum of Natural History, University of Florida.

1. What is the average number for each of the nine columns of data—unprovoked shark attacks, deaths, and non-fatal injuries between 1998 and 2007 for the world, the United States, and Florida?

2. What percentage of the world's average annual unprovoked shark attacks between 1998 and 2007 occurred in (a) the United States and (b) Florida?

2. What percentage of the average annual unprovoked shark attacks in the United States between 1998 and 2007 occurred in Florida?

LEARNING ONLINE

Log on to the Student Companion Site for this book at **academic.cengage.com/biology/miller**, and choose Chapter 4 for many study aids and ideas for further reading and research. These include flash cards, practice quizzing, Weblinks, information on Green Careers, and InfoTrac® College Edition articles.

Southern Sea Otters: Are They Back from the Brink of Extinction?

Southern sea otters (Figure 5-1, top left) live in giant kelp forests (Figure 5-1, right) in shallow waters along part of the Pacific coast of North America. Most remaining members of this endangered species are found between the U.S. state of California's coastal cities of Santa Cruz and Santa Barbara.

Southern sea otters are fast and agile swimmers that dive to the ocean bottom looking for shellfish and other prey. These tool-using marine mammals use stones to pry shellfish off rocks under water. When they return to the surface they break open the shells while swimming on their backs, using their bellies as a table (Figure 5-1, top left). Each day a sea otter consumes about a fourth of its weight in clams, mussels, crabs, sea urchins, abalone, and about 40 other species of bottom-dwelling organisms.

Historically, between 16,000 and 17,000 southern sea otters are believed to have populated the waters along their habitat area of the California coast before fur traders began killing them for their thick, luxurious fur. For that reason, and because the otters competed with humans for valuable abalone and other shellfish, the species was hunted almost to extinction in this region by the early 1900s.

However, between 1938 and 2007 the population of southern sea otters off California's coast increased from about 50 to almost 3,026. This partial recovery was helped when, in 1977, the U.S. Fish and Wildlife Service declared the species endangered in most of its range. But this species has a long way to go before its population increases enough to allow removing it from the endangered species list.

Why should we care about this species? One reason is that people love to look at these charismatic, cute, and cuddly animals as they play in the water. As a result, they help to generate millions of dollars a year in tourism income in coastal areas where they are found. Another reason is *ethical.* Some people believe it is wrong to cause the premature extinction of any species.

A third reason to care about otters—and a key reason in our study of environmental science—is that biologists classify them as a *keystone species* (p. 95), which play an important ecological role through its interactions with other species. The otters help to keep sea urchins and other kelp-eating species from depleting highly productive and rapidly growing kelp forests, which provide habitats for a number of species in offshore coastal waters, as discussed in more detail later in this chapter. Without southern sea otters, sea urchins would probably destroy the kelp forests and much of the rich biodiversity associated with them.

Biodiversity, an important part of the earth's natural capital, is the focus of one of the four **scientific principles of sustainability** (see back cover).

One of its components is species diversity (Figure 4-2, p. 79), which is affected by how species interact with one another and, in the process, help control each others' population sizes.

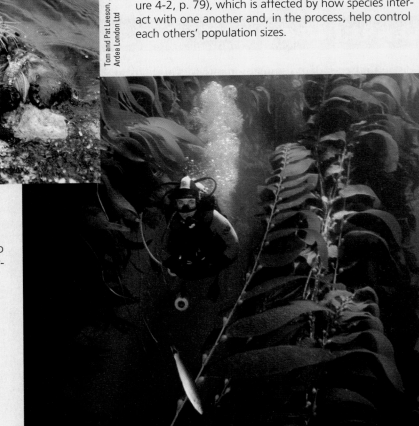

Figure 5-1 An endangered southern sea otter in Monterey Bay, California (USA), uses a stone to crack the shell of a clam (top left). It lives in a giant kelp bed near San Clemente Island, California (right). Scientific studies indicate that the otters act as a keystone species in a kelp forest system by helping to control the populations of sea urchins and other kelp-eating species.

Tom and Pat Leeson, Ardea London Ltd

Bruce Coleman USA

5-1 How do species interact?

CONCEPT 5-1 Five types of species interactions—competition, predation, parasitism, mutualism, and commensalism—affect the resource use and population sizes of the species in an ecosystem.

5-2 How can natural selection reduce competition between species?

CONCEPT 5-2 Some species develop adaptations that allow them to reduce or avoid competition with other species for resources.

5-3 What limits the growth of populations?

CONCEPT 5-3 No population can continue to grow indefinitely because of limitations on resources and because of competition among species for those resources.

5-4 How do communities and ecosystems respond to changing environmental conditions?

CONCEPT 5-4 The structure and species composition of communities and ecosystems change in response to changing environmental conditions through a process called ecological succession.

Note: Supplements 2 (p. S4), 4 (p. S20), 5 (p. S31), 6 (p. S39), and 13 (p. S78) can be used with this chapter.

In looking at nature, never forget that every single organic being around us may be said to be striving to increase its numbers.

CHARLES DARWIN, 1859

5-1 How Do Species Interact?

> **CONCEPT 5-1** Five types of species interactions—competition, predation, parasitism, mutualism, and commensalism—affect the resource use and population sizes of the species in an ecosystem.

Species Interact in Five Major Ways

Ecologists identify five basic types of interactions between species that share limited resources such as food, shelter, and space:

- **Interspecific competition** occurs when members of two or more species interact to gain access to the same limited resources such as food, light, or space.

- **Predation** occurs when a member of one species (the *predator*) feeds directly on all or part of a member of another species (the *prey*).

- **Parasitism** occurs when one organism (the *parasite*) feeds on the body of, or the energy used by, another organism (the *host*), usually by living on or in the host.

- **Mutualism** is an interaction that benefits both species by providing each with food, shelter, or some other resource.

- **Commensalism** is an interaction that benefits one species but has little, if any, effect on the other.

These interactions have significant effects on the resource use and population sizes of the species in an ecosystem (**Concept 5-1**). Interactions that help to limit population size illustrate one of the four **scientific principles of sustainability** (see back cover). These interactions also influence the abilities of the interacting species to survive and reproduce; thus the interactions serve as agents of natural selection (**Concept 4-2B**, p. 80).

Most Species Compete with One Another for Certain Resources

The most common interaction between species is *competition* for limited resources. While fighting for resources does occur, most competition involves the ability of one species to become more efficient than another species in acquiring food or other resources.

Recall that each species plays a unique role in its ecosystem called its *ecological niche* (p. 91). Some species are generalists with broad niches and some are specialists with narrow niches. When two species compete with one another for the same resources such as food, light, or space, their niches overlap (Figure 4-11, p. 91).

The greater this overlap the more intense their competition for key resources.

Although different species may share some aspects of their niches, no two species can occupy exactly the same ecological niche for very long—a concept known as the *competitive exclusion principle*. When there is intense competition between two species for the same resources, both species suffer harm by having reduced access to important resources. If one species can take over the largest share of one or more key resources, the other competing species must migrate to another area (if possible), shift its feeding habits or behavior through natural selection to reduce or alter its niche, suffer a sharp population decline, or become extinct in that area.

Humans compete with many other species for space, food, and other resources. As our ecological footprints grow and spread (Figure 1-10, p. 15) and we convert more of the earth's land, aquatic resources, and net primary productivity (Figure 3-16, p. 64) to our uses, we are taking over the habitats of many other species and depriving them of resources they need to survive.

THINKING ABOUT
Humans and the Southern Sea Otter

CORE CASE STUDY

What human activities have interfered with the ecological niche of the southern sea otter (**Core Case Study**)?

Most Consumer Species Feed on Live Organisms of Other Species

All organisms must have a source of food to survive. Recall that members of producer species, such as plants and floating phytoplankton, make their own food, mostly through photosynthesis (p. 58). Other species are consumers that interact with some species by feeding on them. Some consumers feed on live individuals of other species. They include herbivores that feed on plants, carnivores that feed on the flesh of other animals, and omnivores that feed on plants and animals. Other consumers, such as detritus feeders and decomposers, feed on the wastes or dead bodies of organisms.

In **predation,** a member of one species (the **predator**) feeds directly on all or part of a living organism of another plant or animal species (the **prey**) as part of a food web (**Concept 3-4A**, p. 61). Together, the two different species, such as lions (the predator or hunter) and zebras (the prey or hunted), form a **predator–prey relationship.** Such relationships are shown in Figures 3-13 (p. 62) and 3-14 (p. 63).

CONCEPT LINK

Herbivores, carnivores, and omnivores are predators. However, detritus feeders and decomposers, while they do feed on other organisms after they have died, are not considered predators because they do not feed on live organisms.

Sometimes predator–prey relationships can surprise us. During the summer months, the grizzly bears of the Greater Yellowstone ecosystem in the western United States eat huge amounts of army cutworm moths, which huddle in masses high on remote mountain slopes. One grizzly bear can dig out and lap up as many as 40,000 of these moths in a day. Consisting of 50–70% fat, the moths offer a nutrient that the bear can store in its fatty tissues and draw on during its winter hibernation.

In giant kelp forest ecosystems, sea urchins prey on giant kelp, a form of seaweed (**Core Case Study**, Figure 5-1, right). However, as keystone species, southern sea otters (Figure 5-1, top left) prey on the sea urchins and help to keep them from destroying the kelp forest ecosystems (Science Focus, p. 104).

CORE CASE STUDY

Predators have a variety of methods that help them capture prey. *Herbivores* can simply walk, swim, or fly up to the plants they feed on. For example, sea urchins (Science Focus, Figure 5-A, p. 104) can move along the ocean bottom to feed on the base of giant kelp plants. *Carnivores* feeding on mobile prey have two main options: *pursuit* and *ambush*. Some, such as the cheetah, catch prey by running fast; others, such as the American bald eagle, can fly and have keen eyesight; still others, such as wolves and African lions, cooperate in capturing their prey by hunting in packs.

Other predators use *camouflage* to hide in plain sight and ambush their prey. For example, praying mantises (Figure 3-A, right, p. 54) sit in flowers of a similar color and ambush visiting insects. White ermines (a type of weasel) and snowy owls hunt in snow-covered areas. People camouflage themselves to hunt wild game and use camouflaged traps to ambush wild game.

Some predators use *chemical warfare* to attack their prey. For example, spiders and poisonous snakes use venom to paralyze their prey and to deter their predators.

Prey species have evolved many ways to avoid predators, including abilities to run, swim, or fly fast, and highly developed senses of sight or smell that alert them to the presence of predators. Other avoidance adaptations include protective shells (as on armadillos and turtles), thick bark (giant sequoia), spines (porcupines), and thorns (cacti and rosebushes). Many lizards have brightly colored tails that break off when they are attacked, often giving them enough time to escape.

Other prey species use the camouflage of certain shapes or colors or the ability to change color (chameleons and cuttlefish). Some insect species have shapes that make them look like twigs (Figure 5-2a), bark, thorns, or even bird droppings on leaves. A leaf insect can be almost invisible against its background (Figure 5-2b), as can an arctic hare in its white winter fur.

Chemical warfare is another common strategy. Some prey species discourage predators with chemicals that are *poisonous* (oleander plants), *irritating* (stinging nettles and bombardier beetles, Figure 5-2c), *foul smelling* (skunks, skunk cabbages, and stinkbugs), or *bad tast-*

ing (buttercups and monarch butterflies, Figure 5-2d). When attacked, some species of squid and octopus emit clouds of black ink, allowing them to escape by confusing their predators.

Many bad-tasting, bad-smelling, toxic, or stinging prey species have evolved *warning coloration*, brightly colored advertising that enables experienced predators to recognize and avoid them. They flash a warning: "Eating me is risky." Examples are brilliantly colored poisonous frogs (Figure 5-2e); and foul-tasting monarch butterflies (Figure 5-2d). For example, when a bird such as a blue jay eats a monarch butterfly it usually vomits and learns to avoid them.

Biologist Edward O. Wilson gives us two rules, based on coloration, for evaluating possible danger from an unknown animal species we encounter in nature. *First,* if it is small and strikingly beautiful, it is probably poisonous. *Second,* if it is strikingly beautiful and easy to catch, it is probably deadly.

Some butterfly species, such as the nonpoisonous viceroy (Figure 5-2f), gain protection by looking and acting like the monarch, a protective device known as *mimicry.* Other prey species use *behavioral strategies* to avoid predation. Some attempt to scare off predators by puffing up (blowfish), spreading their wings (peacocks), or mimicking a predator (Figure 5-2h). Some moths have wings that look like the eyes of much larger animals (Figure 5-2g). Other prey species gain some protection by living in large groups such as schools of fish and herds of antelope.

THINKING ABOUT
Predation and the Southern Sea Otter

CORE CASE STUDY

Describe a trait possessed by the southern sea otter (**Core Case Study**) that helps it **(a)** catch prey and **(b)** avoid being preyed upon?

At the individual level, members of the predator species benefit and members of the prey species are harmed. At the population level, predation plays a role in evolution by natural selection (**Concept 4-2B**, p. 80). Animal predators, for example, tend to kill the sick, weak, aged, and least fit members of a population because they are the easiest to catch. This leaves behind individuals with better defenses against predation. Such individuals tend to survive longer and leave more offspring with adaptations that can help them avoid predation. Thus, predation can help increase biodiversity by promoting natural selection in which species evolve with the ability to share limited resources by reducing their niche overlap.

CONCEPT LINK

Some people tend to view certain animal predators with contempt. When a hawk tries to capture and feed on a rabbit, some root for the rabbit. Yet the hawk, like all predators, is merely trying to get enough food for itself and its young. In doing so, it plays an important ecological role in controlling rabbit populations.

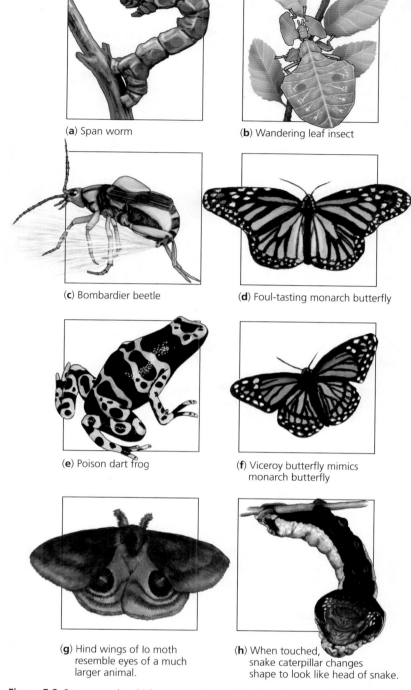

(a) Span worm

(b) Wandering leaf insect

(c) Bombardier beetle

(d) Foul-tasting monarch butterfly

(e) Poison dart frog

(f) Viceroy butterfly mimics monarch butterfly

(g) Hind wings of Io moth resemble eyes of a much larger animal.

(h) When touched, snake caterpillar changes shape to look like head of snake.

Figure 5-2 Some ways in which prey species avoid their predators: **(a, b)** *camouflage,* **(c–e)** *chemical warfare,* **(d, e)** *warning coloration,* **(f)** *mimicry,* **(g)** *deceptive looks,* and **(h)** *deceptive behavior.*

Predator and Prey Species Can Drive Each Other's Evolution

To survive, predators must eat and prey must avoid becoming a meal. As a result, predator and prey populations exert intense natural selection pressures on one another. Over time, as prey develop traits that make them more difficult to catch, predators face selection

SCIENCE FOCUS

Why Should We Care about Kelp Forests?

A kelp forest is a forest of seaweed called giant kelp whose large blades grow straight to the surface (Figure 5-1, right) (**Core Case Study**). The dependence of these plants on photosynthesis restricts their growth to cold, nutrient-rich, and fairly shallow coastal waters, which are found in various areas of the world, such as off the coast of northern California (USA).

CORE CASE STUDY

Giant kelp is one of the world's fastest growing plants. Under good conditions, its blades can grow 0.6 meter (2 feet) a day. Each blade is held up by a gas-filled bladder at its base. The blades are very flexible and can survive all but the most violent storms and waves.

Kelp forests are one of the most biologically diverse ecosystems found in marine waters, supporting large numbers of marine plants and animals. These forests help reduce shore erosion by blunting the force of incoming waves and helping to trap outgoing sand. People harvest kelp as a renewable resource, extracting a substance called algin from its blades. We use this substance in toothpaste, cosmetics, ice cream, and hundreds of other products.

Sea urchins and pollution are major threats to kelp forests. Large populations

Figure 5-A Purple sea urchin in coastal waters of the U.S. state of California.

Deborah Meeks/SuperStock

of sea urchins (Figure 5-A) can rapidly devastate a kelp forest because they eat the base of young kelp. Male southern sea otters, a keystone species, help to control populations of sea urchins. An adult male southern sea otter (Figure 5-1, top left) can eat up to 50 sea urchins a day—equivalent to a 68-kilogram (150-pound) person eating 160 quarter-pound hamburgers a day. Without southern sea otters, giant kelp for-

est ecosystems would collapse and reduce aquatic biodiversity.

A second threat to kelp forests is polluted water running off of the land into the coastal waters where kelp forests grow. The pollutants in this runoff include pesticides and herbicides, which can kill kelp plants and other kelp forest species and upset the food webs in these forests. Another runoff pollutant is fertilizer whose plant nutrients (mostly nitrates) can cause excessive growth of algae and other plants. These growths block some of the sunlight needed to support the growth of giant kelp, and thus upset these aquatic ecosystems.

A third looming threat is global warming. Giant kelp forests require fairly cool water, where the temperature stays between 28–36 °C (50–65 °F). If coastal waters warm up as projected during this century, many—perhaps most—of the world's giant kelp forests will disappear and along with them the southern sea otter and many other species.

Critical Thinking

What are three ways to protect giant kelp forests and southern sea otters?

pressures that favor traits that increase their ability to catch prey. Then prey must get better at eluding the more effective predators.

When populations of two different species interact in this way over a such long period of time, changes in the gene pool of one species can lead to changes in the gene pool of the other species. Such changes can help both sides to become more competitive or can help to avoid or reduce competition. Biologists call this process **coevolution.**

Consider the species interaction between bats (the predator) and certain species of moths (the prey). Bats like to eat moths, and they hunt at night (Figure 5-3) and use echolocation to navigate and to locate their prey, emitting pulses of extremely high-frequency and high-intensity sound. They capture and analyze the returning echoes and create a sonic "image" of their prey. (We have copied this natural technology by using sonar to detect submarines, whales, and schools of fish.)

As a countermeasure to this effective prey-detection system, certain moth species have evolved ears that are especially sensitive to the sound frequencies that bats use to find them. When the moths hear the bat fre-

quencies, they try to escape by dropping to the ground or flying evasively.

Some bat species evolved ways to counter this defense by changing the frequency of their sound pulses. In turn, some moths have evolved their own high-frequency clicks to jam the bats' echolocation systems. Some bat species then adapted by turning off their echolocation systems and using the moths' clicks to locate their prey.

Coevolution is like an arms race between interacting populations of different species. Sometimes the predators surge ahead; at other times the prey get the upper hand. Coevolution is one of nature's ways of maintaining long-term sustainability through population control (see back cover and **Concept 1-6**, p. 23), and it can promote biodiversity by increasing species diversity.

CONCEPT LINK

However, we should not think of coevolution as a process with which species can design strategies to increase their survival chances. Instead, it is a prime example of populations responding to changes in environmental conditions as part of the process of evolution by natural selection. And, unlike a human arms

race, each step in this process takes hundreds to thousands of years.

Some Species Feed off Other Species by Living on or in Them

Parasitism occurs when one species (the *parasite*) feeds on the body of, or the energy used by, another organism (the *host*), usually by living on or in the host. In this relationship, the parasite benefits and the host is harmed but not immediately killed.

Unlike the typical predator, a parasite usually is much smaller than its host and rarely kills its host. Also, most parasites remain closely associated with their hosts, draw nourishment from them, and may gradually weaken them over time.

Some parasites, such as tapeworms and some disease-causing microorganisms (pathogens), live *inside* their hosts. Other parasites attach themselves to the *outsides* of their hosts. Examples of the latter include mosquitoes, mistletoe plants (Figure 5-4, left), and sea lampreys, which use their sucker-like mouths to attach themselves to fish and feed on their blood (Figure 5-4, right). Some parasites move from one host to another, as fleas and ticks do; others, such as tapeworms, spend their adult lives with a single host.

Some parasites have little contact with their hosts. For example, North American cowbirds take over the nests of other birds by laying their eggs in them and then letting the host birds raise their young.

From the host's point of view, parasites are harmful. But at the population level, parasites can promote biodiversity by increasing species richness, and they help to keep their hosts' populations in check.

Like predator–prey interactions, parasite–host interactions can lead to coevolutionary change. For example, malaria is caused by a parasite spread by the bites of a certain mosquito species. The parasite in-

Figure 5-3 *Coevolution.* A Langohrfledermaus bat hunting a moth. Long-term interactions between bats and their prey such as moths and butterflies can lead to coevolution, as the bats evolve traits that increase their chances of getting a meal and the moths evolve traits that help them avoid being eaten.

vades red blood cells, which are destroyed every few days when they are swept into the spleen. However, through coevolution, the malaria parasite developed an adaptation that keeps it from being swept into the spleen. The parasite produces a sticky protein nodule that attaches the cell it has infected to the wall of a blood vessel.

However, the body's immune system detects the foreign protein on the blood vessel wall and sends antibodies to attack it. Through coevolution, the malaria parasite has in turn developed a defense against this attack. It produces thousands of different versions of the sticky protein that keep it attached to the blood vessel wall. By the time the immune system recognizes and attacks one type of the protein, the parasite has switched to another type.

Figure 5-4 *Parasitism:* **(a)** Healthy tree on the left and an unhealthy one on the right, which is infested with parasitic mistletoe. **(b)** Blood-sucking parasitic sea lampreys attached to an adult lake trout from the Great Lakes (USA).

Figure 5-5 Examples of *mutualism.* **(a)** Oxpeckers (or tickbirds) feed on parasitic ticks that infest large, thick-skinned animals such as the endangered black rhinoceros. **(b)** A clownfish gains protection and food by living among deadly stinging sea anemones and helps protect the anemones from some of their predators. (Oxpeckers and black rhinoceros: Joe McDonald/Tom Stack & Associates; clownfish an sea anemone: Fred Bavendam/ Peter Arnold, Inc.)

(a) Oxpeckers and black rhinoceros **(b)** Clownfish and sea anemone

In Some Interactions, Both Species Benefit

In **mutualism,** two species behave in ways that benefit both by providing each with food, shelter, or some other resource. For example, honeybees, caterpillars, butterflies, and other insects feed on a male flower's nectar, picking up pollen in the process, and then pollinating female flowers when they feed on them.

Figure 5-5 shows two examples of mutualistic relationships that combine *nutrition* and *protection.* One involves birds that ride on the backs of large animals like African buffalo, elephants, and rhinoceroses (Figure 5-5a). The birds remove and eat parasites and pests (such as ticks and flies) from the animal's body and often make noises warning the larger animals when predators approach.

A second example involves the clownfish species (Figure 5-5b), which live within sea anemones, whose tentacles sting and paralyze most fish that touch them. The clownfish, which are not harmed by the tentacles, gain protection from predators and feed on the detritus left from the anemones' meals. The sea anemones benefit because the clownfish protect them from some of their predators.

In *gut inhabitant mutualism,* vast armies of bacteria in the digestive systems of animals help to break down (digest) their hosts' food. In turn, the bacteria receive a sheltered habitat and food from their host. Hundreds of millions of bacteria in your gut secrete enzymes that help digest the food you eat. Cows and termites are able to digest the cellulose in plant tissues they eat because of the large number of microorganisms, mostly bacteria, that live in their guts.

It is tempting to think of mutualism as an example of cooperation between species. In reality, each species benefits by unintentionally exploiting the other as a result of traits they obtained through natural selection.

In Some Interactions, One Species Benefits and the Other Is Not Harmed

Commensalism is an interaction that benefits one species but has little, if any, effect on the other. For example, in tropical forests certain kinds of silverfish insects move along with columns of army ants to share the food obtained by the ants in their raids. The army ants receive no apparent harm or benefit from the silverfish.

Another example involves plants called *epiphytes* (such as certain types of orchids and bromeliads), which attach themselves to the trunks or branches of large trees in tropical and subtropical forests (Figure 5-6). These *air plants* benefit by having a solid base on which to grow. They also live in an elevated spot that gives them better access to sunlight, water from the humid air and rain, and nutrients falling from the tree's upper leaves and limbs. Their presence apparently does not harm the tree.

Figure 5-6 In an example of *commensalism,* this bromeliad—an epiphyte, or air plant, in Brazil's Atlantic tropical rain forest—roots on the trunk of a tree, rather than in soil, without penetrating or harming the tree. In this interaction, the epiphyte gains access to water, other nutrient debris, and sunlight; the tree apparently remains unharmed.

Luiz C. Marigo/Peter Arnold, Inc.

CENGAGENOW™ Review the way in which species can interact and see the results of an experiment on species interaction at CengageNOW™.

5-2 How Can Natural Selection Reduce Competition between Species?

▶ **CONCEPT 5-2** Some species develop adaptations that allow them to reduce or avoid competition with other species for resources.

Some Species Evolve Ways to Share Resources

Over a time scale long enough for natural selection to occur, populations of some species competing for the same resources develop adaptations through natural selection that allow them to reduce or avoid such competition (**Concept 5-2**). In other words, some species evolve to reduce niche overlap. One way this happens is through **resource partitioning.** It occurs when species competing for similar scarce resources evolve specialized traits that allow them to use shared resources at different times, in different ways, or in different places. For example, through natural selection, the fairly broad and overlapping niches of two competing species (Figure 5-7, top) can reduce their niche overlap by becoming more specialized (Figure 5-7, bottom).

Figure 5-8 shows resource partitioning by some insect-eating bird species. In this case, their adaptations allow them to reduce competition by feeding in different portions of the same tree species and by feeding on different insect species. Figure 4-13 (p. 93) shows how bird species in a coastal wetland have

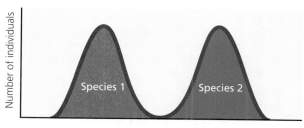

Figure 5-7 Competing species can evolve to reduce niche overlap. The top diagram shows the overlapping niches of two competing species. The bottom diagram shows that through natural selection, the niches of the two species become separated and more specialized (narrower) as the species develop adaptations that allow them to avoid or reduce competition for the same resources.

Blackburnian Warbler Black-throated Green Warbler Cape May Warbler Bay-breasted Warbler Yellow-rumped Warbler

Figure 5-8 *Sharing the wealth: resource partitioning* of five species of insect-eating warblers in the spruce forests of the U.S. state of Maine. Each species minimizes competition for food with the others by spending at least half its feeding time in a distinct portion (shaded areas) of the spruce trees, and by consuming different insect species. (After R. H. MacArthur, "Population Ecology of Some Warblers in Northeastern Coniferous Forests," *Ecology* 36 (1958): 533–536.)

evolved specialized feeding niches, thereby reducing their competition for resources.

Another example of resource partitioning through natural selection involves birds called honeycreepers that live in the U. S. state of Hawaii. Long ago these birds started from a single ancestor species. But because of evolution by natural selection, there are now numerous honeycreeper species. Each has a different type of beak specialized to feed on certain food sources, such as specific types of insects, nectar from particular types of flowers, and certain types of seeds and fruit (Figure 5-9). This is an example of a process called *evolutionary divergence*.

Figure 5-9 *Specialist species of honeycreepers.* Evolutionary divergence of honeycreepers into species with specialized ecological niches has reduced competition between these species. Each species has evolved a beak specialized to take advantage of certain types of food resources.

5-3 What Limits the Growth of Populations?

▶ **CONCEPT 5-3** No population can continue to grow indefinitely because of limitations on resources and because of competition among species for those resources.

Populations Have Certain Characteristics

Populations differ in factors such as their *distribution, numbers, age structure* (proportions of individuals in different age groups), and *density* (number of individuals in a certain space). **Population dynamics** is a study of how these characteristics of populations change in response to changes in environmental conditions. Examples of such conditions are temperature, presence of disease organisms or harmful chemicals, resource availability, and arrival or disappearance of competing species.

Studying the population dynamics of southern sea otter populations (**Core Case Study**) and their interactions with other species has helped us understand the ecological importance of this keystone species. Let's look at some of the characteristics of populations in more detail.

Most Populations Live Together in Clumps or Patches

Let's begin our study of population dynamics with how individuals in populations are distributed or dispersed within a particular area or volume. Three general patterns of *population distribution* or *dispersion* in a habitat are *clumping, uniform dispersion,* and *random dispersion* (Figure 5-10).

In most populations, individuals of a species live together in clumps or patches (Figure 5-10a). Examples are patches of desert vegetation around springs, cottonwood trees clustered along streams, wolf packs, flocks of birds, and schools of fish. The locations and sizes of these clumps and patches vary with the availability of resources. Southern sea otters (Figure 5-1, top left), for example, are usually found in groups known as rafts or pods ranging in size from a few to several hundred animals.

Why clumping? Several reasons: *First,* the resources a species needs vary greatly in availability from place to place, so the species tends to cluster where the resources are available. *Second,* individuals moving in groups have a better chance of encountering patches or clumps of resources, such as water and vegetation, than they would searching for the resources on their own. *Third,* living in groups protects some animals from predators. *Fourth,* hunting in packs gives some predators a better chance of finding and catching prey. *Fifth,* some species form temporary groups for mating and caring for young.

(a) Clumped (elephants)

(b) Uniform (creosote bush)

(c) Random (dandelions)

Figure 5-10 Generalized *dispersion patterns* for individuals in a population throughout their habitat. The most common pattern is *clumps* of members of a population scattered throughout their habitat, mostly because resources are usually found in patches. **Question:** Why do you think the creosote bushes are uniformly spaced while the dandelions are not?

> **THINKING ABOUT**
> **Population Distribution**
>
> Why do you think living in packs would help some predators to find and kill their prey?

Some species maintain a fairly constant distance between individuals. Because of its sparse distribution pattern, creosote bushes in a desert (Figure 5-10b) have better access to scarce water resources. Organisms with a random distribution (Figure 5-10c) are fairly rare. The living world is mostly clumpy and patchy.

Populations Can Grow, Shrink, or Remain Stable

Over time, the number of individuals in a population may increase, decrease, remain about the same, or go up and down in cycles in response to changes in environmental conditions. Four variables—*births, deaths, immigration,* and *emigration*—govern changes in population size. A population increases by birth and immigration (arrival of individuals from outside the population) and decreases by death and emigration (departure of individuals from the population):

$$\text{Population change} = (\text{Births} + \text{Immigration}) - (\text{Deaths} + \text{Emigration})$$

In natural systems, species that are able to move can leave or *emigrate* from an area where their habitat has been degraded or destroyed and *immigrate* to another area where resources are more plentiful.

A population's **age structure**—the proportions of individuals at various ages—can have a strong effect on how rapidly it increases or decreases in size. Age structures are usually described in terms of organisms not mature enough to reproduce (the *pre-reproductive age*), those capable of reproduction (the *reproductive stage*), and those too old to reproduce (the *post-reproductive stage*).

The size of a population will likely increase if it is made up mostly of individuals in their reproductive stage or soon to enter this stage. In contrast, a population dominated by individuals past their reproductive stage will tend to decrease over time. Excluding emigration and immigration, the size of a population with a fairly even distribution among these three age groups tends to remain stable because reproduction by younger individuals will be roughly balanced by the deaths of older individuals.

No Population Can Grow Indefinitely: J-Curves and S-Curves

Species vary in their **biotic potential** or capacity for population growth under ideal conditions. Generally, populations of species with large individuals, such as elephants and blue whales, have a low biotic potential while those of small individuals, such as bacteria and insects, have a high biotic potential.

The **intrinsic rate of increase** (*r*) is the rate at which the population of a species would grow if it had unlimited resources. Individuals in populations with a high intrinsic rate of growth typically *reproduce early in life, have short generation times* (the time between successive generations), *can reproduce many times,* and *have many offspring each time they reproduce.*

Some species have an astounding biotic potential. With no controls on population growth, a species of bacteria that can reproduce every 20 minutes would generate enough offspring to form a layer 0.3 meter (1 foot) deep over the entire earth's surface in only 36 hours!

Fortunately, this is not a realistic scenario. Research reveals that no population can grow indefinitely because of limitations on resources and competition with populations of other species for those resources (**Concept 5-3**). In the real world, a rapidly growing population reaches some size limit imposed by one or more *limiting factors,* such as light, water, space, or nutrients, or by exposure to too many competitors, predators, or infectious diseases. *There are always limits to population growth in nature.* This is one of nature's four **scientific principles of sustainability** (see back cover and **Concept 1-6**, p. 23). Sea otters, for example, face extinction because of low biotic potential and other factors (Science Focus, p. 110).

SCIENCE FOCUS

Why Are Protected Sea Otters Making a Slow Comeback?

The southern sea otter (**Core Case Study**) does not have a high biotic potential for several reasons. Female southern sea otters reach sexual maturity between 2 and 5 years of age, can reproduce until age 15, and typically each produce only one pup a year.

The population size of southern sea otters has fluctuated in response to changes in environmental conditions. One such change has been a rise in populations of orcas (killer whales) that feed on them. Scientists hypothesize that orcas feed more on southern sea otters when populations of their normal prey, sea lions and seals, decline.

Another factor may be deaths from parasites known to breed in cats. Scientists hypothesize that some sea otters may be dying because California cat owners flush used cat litter containing these parasites down their toilets or dump it in storm drains that empty into coastal waters.

Thorny-headed worms from seabirds are also known to be killing sea otters, as are toxic algae blooms triggered by urea, a key ingredient in fertilizer that washes into coastal waters. Toxins such as PCBs and other toxic chemicals released by human activities accumulate in the tissues of the shellfish on which otters feed and prove fatal to otters. The facts that sea otters feed at high trophic levels and live close to the shore makes them vulnerable to these and other pollutants in coastal waters. In other words, sea otters are *indicator species* that warn us of the condition of coastal waters in their habitat.

Some southern sea otters also die when they encounter oil spilled from ships. The entire California southern sea otter population could be wiped out by a large oil spill from a single tanker off the state's central coast. These factors plus a fairly low reproductive rate have hindered the ability of the endangered southern sea otter to rebuild its population (Figure 5-B).

Critical Thinking

How would you design a controlled experiment to test the hypothesis that parasites contained in cat litter that is flushed down toilets may be killing sea otters?

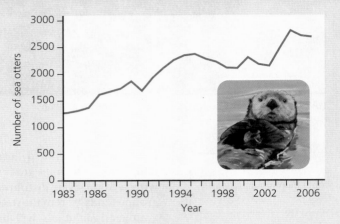

Figure 5-B Population size of southern sea otters off the coast of the U.S. state of California, 1983–2007. According to the U.S. Fish and Wildlife Service, the sea otter population would have to reach about 8,400 animals before it can be removed from the endangered species list. (Data from U.S. Geological Survey)

Environmental resistance is the combination of all factors that act to limit the growth of a population. Together, biotic potential and environmental resistance determine the **carrying capacity (K):** the maximum population of a given species that a particular habitat can sustain indefinitely without being degraded. The growth rate of a population decreases as its size nears the carrying capacity of its environment because resources such as food, water, and space begin to dwindle.

A population with few, if any, limitations on its resource supplies can grow exponentially at a fixed rate such as 1% or 2% per year. *Exponential* or *geometric growth* (Figure 1-1, p. 5) starts slowly but then accelerates as the population increases, because the base size of the population is increasing. Plotting the number of individuals against time yields a J-shaped growth curve (Figure 5-11, left half of curve).

Logistic growth involves rapid exponential population growth followed by a steady decrease in population growth until the population size levels off (Figure 5-11, right half of curve). This slowdown occurs as the population encounters environmental resistance from declining resources and other environmental factors and approaches the carrying capacity of its environment. After leveling off, a population with this type of growth typically fluctuates slightly above and below the carrying capacity. The size of such a population may also increase or decrease as the carrying capacity changes because of short- or long-term changes in environmental conditions.

A plot of the number of individuals against time yields a sigmoid, or S-shaped, logistic growth curve (the whole curve in Figure 5-11). Figure 5-12 depicts such a curve for sheep on the island of Tasmania, south of Australia, in the 19th century.

CENGAGENOW™ Learn how to estimate a population of butterflies and see a mouse population growing exponentially at CengageNOW.

Sometimes a species whose population has been kept in check mostly by natural predators is deliberately or accidentally transferred to a different ecosystem where it has few if any predators. For example, the *brown tree snake* is native to the Solomon Islands,

CENGAGENOW **Active Figure 5-11** No population can continue to increase in size indefinitely. *Exponential growth* (left half of the curve) occurs when resources are not limiting and a population can grow at its *intrinsic rate of increase* (*r*) or *biotic potential*. Such exponential growth is converted to *logistic growth*, in which the growth rate decreases as the population becomes larger and faces environmental resistance. Over time, the population size stabilizes at or near the *carrying capacity* (*K*) of its environment, which results in a sigmoid (S-shaped) population growth curve. Depending on resource availability, the size of a population often fluctuates around its carrying capacity, although a population may temporarily exceed its carrying capacity and then suffer a sharp decline or crash in its numbers. *See an animation based on this figure at* CengageNOW. **Question:** What is an example of environmental resistance that humans have not been able to overcome?

New Guinea, and Australia. After World War II, a few of these snakes stowed away on military planes going to the island of Guam. With no enemies or rivals in Guam, they have multiplied exponentially for several decades and have wiped out 8 of Guam's 11 native forest bird species. Their venomous bites have also sent large numbers of people to emergency rooms. Sooner

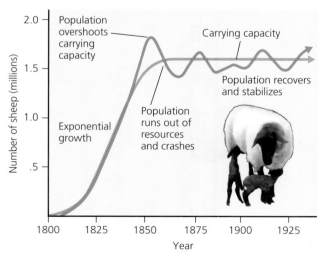

Figure 5-12 *Logistic growth* of a sheep population on the island of Tasmania between 1800 and 1925. After sheep were introduced in 1800, their population grew exponentially, thanks to an ample food supply. By 1855, they had overshot the land's carrying capacity. Their numbers then stabilized and fluctuated around a carrying capacity of about 1.6 million sheep.

or later the brown tree snake will use up most of its food supply in Guam and will then decline in numbers, but meanwhile they are causing serious ecological and economic damage. They may also end up on islands such as those in Hawaii, where they could devastate bird populations.

Changes in the population sizes of keystone species such as the southern sea otter (**Core Case Study**) and the American alligator (Chapter 4 Core Case Study, p. 77) can alter the species composition and biodiversity of an ecosystem. For example, a decline in the population of the southern sea otter caused a decline in the populations of species dependent on them, including the giant kelp. This reduced species diversity of the kelp forest and altered its functional biodiversity by upsetting its food webs and reducing energy flows and nutrient cycling within the forest.

THINKING ABOUT
Southern Sea Otters

CORE CASE STUDY

Name two species whose populations will likely decline if the population of southern sea otters in kelp beds declines sharply. Name a species whose population would increase if this happened.

When a Population Exceeds Its Habitat's Carrying Capacity, Its Population Can Crash

Some species do not make a smooth transition from exponential growth to logistic growth. Such populations use up their resource supplies and temporarily *overshoot*, or exceed, the carrying capacity of their environment. This occurs because of a *reproductive time lag*—the period needed for the birth rate to fall and the death rate to rise in response to resource overconsumption.

In such cases, the population suffers a *dieback*, or *crash*, unless the excess individuals can switch to new resources or move to an area with more resources. Such a crash occurred when reindeer were introduced onto a small island in the Bering Sea (Figure 5-13, p. 112).

The carrying capacity of an area or volume is not fixed. The carrying capacity of some areas can increase or decrease seasonally and from year to year because of variations in weather and other factors, including an abundance of predators and competitors. For example, a drought can decrease the amount of vegetation growing in an area supporting deer and other herbivores, and this would decrease the normal carrying capacity for those species.

Sometimes when a population exceeds the carrying capacity of an area, it causes damage that reduces the area's carrying capacity. For example, overgrazing by cattle on dry western lands in the United States has reduced grass cover in some areas. This has allowed sagebrush—which cattle cannot eat—to move in, thrive, and replace grasses, reducing the land's carrying capacity for cattle.

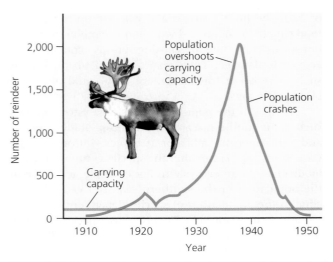

Figure 5-13 Exponential growth, overshoot, and population crash of reindeer introduced to the small Bering Sea island of St. Paul. When 26 reindeer (24 of them female) were introduced in 1910, lichens, mosses, and other food sources were plentiful. By 1935, the herd size had soared to 2,000, overshooting the island's carrying capacity. This led to a population crash, when the herd size plummeted to only 8 reindeer by 1950. **Question:** Why do you think this population grew fast and crashed, unlike the sheep in Figure 5-12?

┌───┐
THINKING ABOUT
Population Overshoot and
Human Ecological Footprints

Humanity's ecological footprint is about 25% larger than the earth's ecological capacity (Figure 1-10, bottom, p. 15) and is growing rapidly. If this keeps up, is the curve for future human population growth more likely to resemble Figure 5-12 or Figure 5-13? Explain.
└───┘

Species Have Different Reproductive Patterns

Species have different reproductive patterns that can help enhance their survival. Species with a capacity for a high rate of population increase (r) are called **r-selected species** (Figure 5-14). These species have many, usually small, offspring and give them little or no parental care or protection. They overcome typically massive losses of offspring by producing so many offspring that a few will likely survive to reproduce many more offspring to begin this reproductive pattern again. Examples include algae, bacteria, rodents, frogs, turtles, annual plants (such as dandelions), and most insects.

Such species tend to be *opportunists*. They reproduce and disperse rapidly when conditions are favorable or when a disturbance opens up a new habitat or niche for invasion. Environmental changes caused by disturbances, such as fires, clear-cutting, and volcanic eruptions, can allow opportunist species to gain a foothold. However, once established, their populations may crash because of unfavorable changes in environmental conditions or invasion by more competitive species. This helps to explain why most opportunist species go

through irregular and unstable boom-and-bust cycles in their population sizes.

At the other extreme are *competitor* or **K-selected species** (Figure 5-14). They tend to reproduce later in life and have a small number of offspring with fairly long life spans. Typically, for K-selected mammals, the offspring develop inside their mothers (where they are safe), are born fairly large, mature slowly, and are cared for and protected by one or both parents, and in some cases by living in herds or groups, until they reach reproductive age. This reproductive pattern results in a few big and strong individuals that can compete for resources and reproduce a few young to begin the cycle again.

Such species are called K-selected species because they tend to do well in competitive conditions when their population size is near the carrying capacity (K) of their environment. Their populations typically follow a logistic growth curve (Figure 5-12).

Most large mammals (such as elephants, whales, and humans), birds of prey, and large and long-lived plants (such as the saguaro cactus, and most tropical rain forest trees) are K-selected species. Ocean fish such as orange roughy and swordfish, which are now being depleted by overfishing, are also K-selected. Many of these species—especially those with long times between generations and low reproductive rates like elephants, rhinoceroses, and sharks—are prone to extinction.

Most organisms have reproductive patterns between the extremes of r-selected and K-selected species. In agriculture we raise both r-selected species (crops) and K-selected species (livestock). Individuals of species with different reproductive strategies tend to have different *life expectancies*, or expected lengths of life.

┌───┐
THINKING ABOUT
r-Selected and K-selected Species

If the earth experiences significant warming during this century as projected, is the resulting climate change likely to favor r-selected or K-selected species? Explain.
└───┘

Figure 5-14 Positions of *r-selected* and *K-selected* species on the sigmoid (S-shaped) population growth curve.

Genetic Diversity Can Affect the Size of Small Populations

In most large populations, genetic diversity is fairly constant and the loss or addition of some individuals has little effect on the total gene pool. However, several genetic factors can play a role in the loss of genetic diversity and the survival of small, isolated populations.

One such factor, called the *founder effect*, can occur when a few individuals in a population colonize a new habitat that is geographically isolated from other members of the population (Figure 4-8, p. 87). In such cases, limited genetic diversity or variability may threaten the survival of the colonizing population.

Another factor is a *demographic bottleneck*. It occurs when only a few individuals in a population survive a catastrophe such as a fire or hurricane, as if they had passed through the narrow neck of a bottle. Lack of genetic diversity may limit the ability of these individuals to rebuild the population. Even if the population is able to increase in size, its decreased genetic diversity may lead to an increase in the frequency of harmful genetic diseases.

A third factor is *genetic drift*. It involves random changes in the gene frequencies in a population that can lead to unequal reproductive success. For example, some individuals may breed more than others do and their genes may eventually dominate the gene pool of the population. This change in gene frequency could help or hinder the survival of the population. The founder effect is one cause of genetic drift.

A fourth factor is *inbreeding*. It occurs when individuals in a small population mate with one another. This can occur when a population passes through a demographic bottleneck. This can increase the frequency of defective genes within a population and affect its long-term survival.

Conservation biologists use the concepts of founder effects, demographic bottleneck, genetic drift, inbreeding, and island biogeography (Science Focus, p. 90) to estimate the *minimum viable population size* of rare and endangered species: the number of individuals such populations need for long-term survival.

Under Some Circumstances Population Density Affects Population Size

Population density is the number of individuals in a population found in a particular area or volume. Some factors that limit population growth have a greater effect as a population's density increases. Examples of such *density-dependent population controls* include predation, parasitism, infectious disease, and competition for resources.

Higher population density may help sexually reproducing individuals find mates, but it can also lead to increased competition for mates, food, living space, water, sunlight, and other resources. High population density can help to shield some members from predators, but it can also make large groups such as schools of fish vulnerable to human harvesting methods. In addition, close contact among individuals in dense populations can increase the transmission of parasites and infectious diseases. When population density decreases, the opposite effects occur. Density-dependent factors tend to regulate a population at a fairly constant size, often near the carrying capacity of its environment.

Some factors—mostly abiotic—that can kill members of a population are *density independent*. In other words, their effect is not dependent on the density of the population. For example, a severe freeze in late spring can kill many individuals in a plant population or a population of monarch butterflies (Figure 3-A, left, p. 54), regardless of their density. Other such factors include floods, hurricanes, fire, pollution, and habitat destruction, such as clearing a forest of its trees or filling in a wetland.

Several Different Types of Population Change Occur in Nature

In nature, we find four general patterns of variation in population size: *stable, irruptive, cyclic,* and *irregular*. A species whose population size fluctuates slightly above and below its carrying capacity is said to have a fairly stable population size (Figure 5-12). Such stability is characteristic of many species found in undisturbed tropical rain forests, where average temperature and rainfall vary little from year to year.

For some species, population growth may occasionally explode, or *irrupt*, to a high peak and then crash to a more stable lower level or in some cases to a very low level. Many short-lived, rapidly reproducing species such as algae and many insects have irruptive population cycles that are linked to seasonal changes in weather or nutrient availability. For example, in temperate climates, insect populations grow rapidly during the spring and summer and then crash during the hard frosts of winter.

A third type of fluctuation consists of regular *cyclic fluctuations*, or *boom-and-bust cycles*, of population size over a time period. Examples are lemmings, whose populations rise and fall every 3–4 years, and lynx and snowshoe hare, whose populations generally rise and fall in a 10-year cycle. Ecologists distinguish between *top-down population regulation*, through predation, and *bottom-up population regulation*, in which the size of predator and prey populations is controlled by the scarcity of one or more resources (Figure 5-15, p. 114).

Finally, some populations appear to have *irregular* changes in population size, with no recurring pattern. Some scientists attribute this irregularity to chaos in

Figure 5-15 Population cycles for the snow-shoe hare and Canada lynx. At one time, scientists believed these curves provided circumstantial evidence that these predator and prey populations regulated one another. More recent research suggests that the periodic swings in the hare population are caused by a combination of *top-down population control*—through predation by lynx and other predators—and *bottom-up population control,* in which changes in the availability of the food supply for hares help determine hare population size, which in turn helps determine the lynx population size. (Data from D. A. MacLulich)

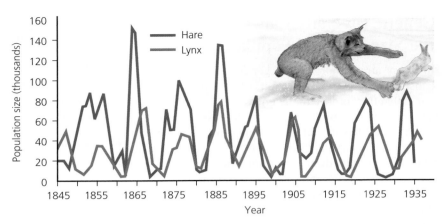

such systems. Others scientists contend that it may represent fluctuations in response to periodic catastrophic population crashes due to severe winter weather.

Humans Are Not Exempt from Nature's Population Controls

Humans are not exempt from population overshoot and dieback. Ireland experienced a population crash after a fungus destroyed the potato crop in 1845. About 1 million people died from hunger or diseases related to malnutrition, and 3 million people migrated to other countries, mostly the United States.

During the 14th century the *bubonic plague* spread through densely populated European cities and killed at least 25 million people. The bacterium causing this disease normally lives in rodents. It was transferred to humans by fleas that fed on infected rodents and then bit humans. The disease spread like wildfire through crowded cities, where sanitary conditions were poor and rats were abundant.

Currently, the world is experiencing a global epidemic of eventually fatal AIDS, caused by infection with the human immunodeficiency virus (HIV). Between 1981 and 2007, AIDS killed more than 25 million people (584,000 in the United states) and claims another 2.1 million lives each year—an average of four deaths per minute.

So far, technological, social, and other cultural changes have extended the earth's carrying capacity for the human species. We have increased food production and used large amounts of energy and matter resources to occupy normally uninhabitable areas. As humans spread into larger areas, they interact with and attempt to control the populations of other species such as alligators (Chapter 4 Core Case Study, p. 77) and white-tailed deer in the United States (Case Study, at right), as these two case studies reveal.

Some say we can keep expanding our ecological footprint indefinitely, mostly because of our technological ingenuity. Others say that sooner or later, we will reach the limits that nature always imposes on populations.

THINKING ABOUT
The Human Species

If the human species were to suffer a sharp population decline, name three types of species that might move in to occupy part of our ecological niche.

■ CASE STUDY

Exploding White-Tailed Deer Populations in the United States

By 1900, habitat destruction and uncontrolled hunting had reduced the white-tailed deer population in the United States to about 500,000 animals. In the 1920s and 1930s, laws were passed to protect the remaining deer. Hunting was restricted and predators such as wolves and mountain lions that preyed on the deer were nearly eliminated.

These protections worked, and some suburbanites and farmers say perhaps they worked too well. Today there are 25–30 million white-tailed deer in the United States. During the last 50 years, large numbers of Americans have moved into the wooded habitat of deer and provided them with flowers, garden crops, and other plants they like to eat.

Deer like to live in the woods for security and go to nearby fields, orchards, lawns, and gardens for food. Border areas between two ecosystems such as forests and fields are called *edge* habitat, and suburbanization has created an all-you-can-eat edge paradise for deer. Their populations in such areas have soared. In some forests, they are consuming native ground cover vegetation and allowing nonnative weed species to take over. Deer also spread Lyme disease (carried by deer ticks) to humans.

In addition, in deer–vehicle collisions, deer accidentally kill and injure more people each year in the United

States than do any other wild animals. Each year, these 1.5 million accidents injure at least 14,000 people and kill at least 200 (up from 101 deaths in 1993).

There are no easy answers to the deer population problem in the suburbs. Changing hunting regulations to allow killing of more female deer cuts down the overall deer population. But these actions have little effect on deer in suburban areas because it is too dangerous to allow widespread hunting with guns in such populated communities. Some areas have hired experienced and licensed archers who use bows and arrows to help reduce deer numbers. To protect nearby residents the archers hunt from elevated tree stands and shoot their arrows only downward. However, animal activists strongly oppose killing deer on ethical grounds, arguing that this is cruel and inhumane treatment.

Some communities spray the scent of deer predators or rotting deer meat in edge areas to scare off deer. Others use electronic equipment that emits high-frequency sounds, which humans cannot hear, for the same purpose. Some homeowners surround their gardens and yards with a high black plastic mesh fencing that is invisible from a distance. Such deterrents may protect one area, but cause the deer to seek food in someone else's yard or garden.

Deer can also be trapped and moved from one area to another, but this is expensive and must be repeated whenever deer move back into an area. Also, there are questions concerning where to move the deer and how to pay for such programs.

Should we put deer on birth control? Darts loaded with a contraceptive could be shot into female deer to hold down their birth rates. But this is expensive and must be repeated every year. One possibility is an experimental single-shot contraceptive vaccine that causes females to stop producing eggs for several years. Another approach is to trap dominant males and use chemical injections to sterilize them. Both these approaches will require years of testing.

Meanwhile, suburbanites can expect deer to chow down on their shrubs, flowers, and garden plants unless they can protect their properties with high, deer-proof fences or other methods. Deer have to eat every day just as we do. Suburban dwellers might consider not planting their yards with plants that deer like to eat.

THINKING ABOUT
White-Tailed Deer

Some blame the white-tailed deer for invading farms and suburban yards and gardens to eat food that humans have made easily available to them. Others say humans are mostly to blame because they have invaded deer territory, eliminated most of the predators that kept their populations under control, and provided the deer with plenty to eat in their lawns and gardens. Which view do you hold? Do you see a solution to this problem?

5-4 How Do Communities and Ecosystems Respond to Changing Environmental Conditions?

▶ **CONCEPT 5-4** The structure and species composition of communities and ecosystems change in response to changing environmental conditions through a process called ecological succession.

Communities and Ecosystems Change over Time: Ecological Succession

We have seen how changes in environmental conditions (such as loss of habitat) can reduce access to key resources and how invasions by competing species can lead to increases or decreases in species population sizes. Next, we look at how the types and numbers of species in biological communities change in response to changing environmental conditions such as a fires, climate change, and the clearing of forests to plant crops.

Mature forests and other ecosystems do not spring up from bare rock or bare soil. Instead, they undergo changes in their species composition over long periods of time. The gradual change in species composition in a given area is called **ecological succession,** during which, some species colonize an area and their populations become more numerous, while populations of other species decline and may even disappear. In this process, *colonizing* or *pioneer species* arrive first. As environmental conditions change, they are replaced by other species, and later these species may be replaced by another set of species.

Ecologists recognize two main types of ecological succession, depending on the conditions present at the beginning of the process. **Primary succession** involves the gradual establishment of biotic communities in lifeless areas where there is no soil in a terrestrial ecosystem or no bottom sediment in an aquatic ecosystem. The other more common type of ecological succession is called **secondary succession,** in which a series of communities or ecosystems with different species develop

in places containing soil or bottom sediment. As part of the earth's natural capital, both types of succession are examples of *natural ecological restoration,* in which various forms life adapt to changes in environmental conditions, resulting in changes to the species composition, population size, and biodiversity in a given area.

Some Ecosystems Start from Scratch: Primary Succession

Primary succession begins with an essentially lifeless area where there is no soil in a terrestrial system (Figure 5-16) or bottom sediment in an aquatic system. Examples include bare rock exposed by a retreating glacier or severe soil erosion, newly cooled lava from a volcanic eruption, an abandoned highway or parking lot, and a newly created shallow pond or reservoir.

Primary succession usually takes a long time because there is no fertile soil to provide the nutrients needed to establish a plant community. Over time, bare rock *weathers* by crumbling into particles and releasing nutrients. Physical weathering occurs when a rock is fragmented, as water in its cracks freezes and expands. Rocks also undergo chemical weathering, reacting with substances in the atmosphere or with precipitation, which can break down the rock's surface material.

The slow process of soil formation begins when *pioneer* or *early successional species* arrive and attach themselves to inhospitable patches of the weathered rock. Examples are lichens and mosses whose seeds or spores

are distributed by the wind and carried by animals. A lichen consists of an alga and a fungus interacting in a mutualistic relationship. The fungi in the lichens provide protection and support for the algae, which, through photosynthesis, provide sugar nutrients for both of the interacting species.

These tough *early successional plant species* start the long process of soil formation by trapping wind-blown soil particles and tiny pieces of detritus, and adding their own wastes and dead bodies. They also secrete mild acids that further fragment and break down the rock. As the lichens spread over the rock, drought-resistant and sun-loving mosses start growing in cracks. As the mosses spread, they form a mat that traps moisture, much like a sponge. When the lichens and mosses die, their decomposing remains add to the growing thin layer of nutrients.

After hundreds to thousands of years, the soil may be deep and fertile enough to store the moisture and nutrients needed to support the growth of *midsuccessional plant species* such as herbs, grasses, and low shrubs. As the shrubs grow and create shade, the lichens and mosses die and decay from lack of sunlight. Next, trees that need lots of sunlight and are adapted to the area's climate and soil usually replace the grasses and shrubs.

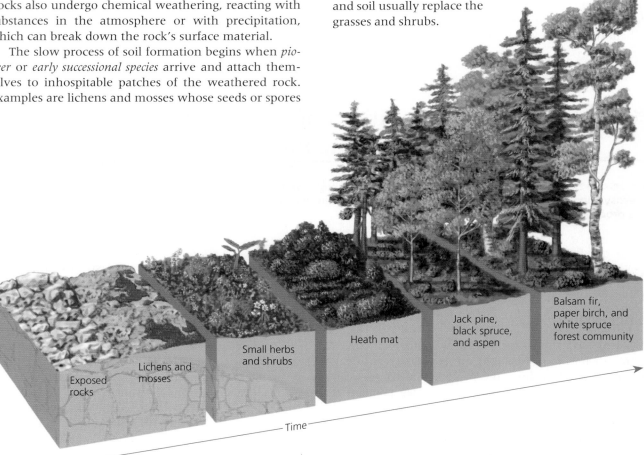

Exposed rocks

Lichens and mosses

Small herbs and shrubs

Heath mat

Jack pine, black spruce, and aspen

Balsam fir, paper birch, and white spruce forest community

Time

Figure 5-16 *Primary ecological succession.* Over almost a thousand years, plant communities developed, starting on bare rock exposed by a retreating glacier on Isle Royal, Michigan (USA) in northern Lake Superior. The details of this process vary from one site to another. **Question:** What are two ways in which lichens, mosses, and plants might get started growing on bare rock?

As these tree species grow and create shade, they are replaced by *late successional plant species* (mostly trees) that can tolerate shade. Unless fire, flooding, severe erosion, tree cutting, climate change, or other natural or human processes disturb the area, what was once bare rock becomes a complex forest community or ecosystem (Figure 5-16).

Primary succession can also take place in a newly created small pond, starting with an influx of sediments and nutrients in runoff from the surrounding land. This sediment can support seeds or spores of plants carried to the pond by winds, birds, or other animals. Over time, this process can transform the pond first into a marsh and eventually to dry land.

Some Ecosystems Do Not Have to Start from Scratch: Secondary Succession

Secondary succession begins in an area where an ecosystem has been disturbed, removed, or destroyed, but some soil or bottom sediment remains. Candidates for secondary succession include abandoned farmland,

burned or cut forests, heavily polluted streams, and land that has been flooded. In the soil that remains on disturbed land systems, new vegetation can germinate, usually within a few weeks, from seeds already in the soil and from those imported by wind, birds, and other animals.

In the central, or Piedmont, region of the U.S. state of North Carolina, European settlers cleared many of the mature native oak and hickory forests and planted the land with crops. Later, they abandoned some of this farmland because of erosion and loss of soil nutrients. Figure 5-17 shows one way in which such abandoned farmland has undergone secondary succession over 150–200 years.

CENGAGENOW™ Explore the difference between primary and secondary succession at CengageNOW.

Descriptions of ecological succession usually focus on changes in vegetation. But these changes in turn affect food and shelter for various types of animals. As a consequence, the numbers and types of animals and

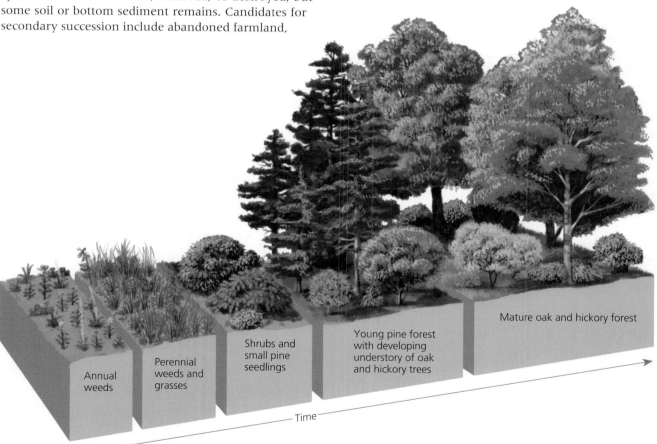

Annual weeds

Perennial weeds and grasses

Shrubs and small pine seedlings

Young pine forest with developing understory of oak and hickory trees

Mature oak and hickory forest

Time

CENGAGENOW™ **Active Figure 5-17** *Natural ecological restoration of disturbed land.* Secondary ecological succession of plant communities on an abandoned farm field in the U.S. state of North Carolina. It took 150–200 years after the farmland was abandoned for the area to become covered with a mature oak and hickory forest. A new disturbance, such as deforestation or fire, would create conditions favoring pioneer species such as annual weeds. In the absence of new disturbances, secondary succession would recur over time, but not necessarily in the same sequence shown here. **Questions:** Do you think the annual weeds (left) would continue to thrive in the mature forest (right)? Why or why not? *See an animation based on this figure at* CengageNOW.

How Do Species Replace One Another in Ecological Succession?

Ecologists have identified three factors that affect how and at what rate succession occurs. One is *facilitation*, in which one set of species makes an area suitable for species with different niche requirements, but less suitable for itself. For example, as lichens and mosses gradually build up soil on a rock in primary succession, herbs and grasses can colonize the site and crowd out the pioneer community of lichens and mosses.

A second factor is *inhibition*, in which some early species hinder the establishment and growth of other species. Inhibition often occurs when plants release toxic chemicals that reduce competition from other plants. For example, pine needles dropped by some species of pines can make the underlying soil acidic and inhospitable to other plant species. Succession then can proceed only when a fire, bulldozer, or other human or natural disturbance removes most of the inhibiting species.

A third factor is *tolerance*, in which late successional plants are largely unaffected by plants at earlier stages of succession because they are not in direct competition with them for key resources. For example, shade tolerant trees and other plants can thrive in the understory of a mature forest (Figure 5-17) because they do not need to compete with the taller species for access to sunlight.

Critical Thinking

Explain how tolerance can increase biodiversity by increasing species diversity and functional diversity (energy flow and chemical cycling) in an ecosystem.

decomposers also change. A key point is that primary succession (Figure 5-16) and secondary succession (Figure 5-17) tend to increase biodiversity and thus the sustainability of communities and ecosystems by increasing species richness and interactions among species. Such interactions in turn enhance sustainability by promoting population control and increasing the complexity of food webs for the energy flow and nutrient cycling that make up the functional component of biodiversity (Figure 4-2, p. 79). Ecologists have been conducting research to find out more about the factors involved in ecological succession (Science Focus, above)

During primary or secondary succession, environmental disturbances such as fires, hurricanes, clearcutting of forests, plowing of grasslands, and invasions by nonnative species can interrupt a particular stage of succession, setting it back to an earlier stage. For example, the American alligator (Chapter 4 Core Case Study, p. 77) lives in ponds that would normally become filled in through natural selection. The alligator's movements keep bottom vegetation from growing, thus preventing such succession.

Succession Doesn't Follow a Predictable Path

According to traditional view, succession proceeds in an orderly sequence along an expected path until a certain stable type of *climax community* occupies an area. Such a community is dominated by a few long-lived plant species and is in balance with its environment. This equilibrium model of succession is what ecologists once meant when they talked about the *balance of nature*.

Over the last several decades, many ecologists have changed their views about balance and equilibrium in nature. Under the balance-of-nature view, a large terrestrial community or ecosystem undergoing succession eventually became covered with an expected type of climax vegetation such as a mature forest (Figures 5-16 and 5-17). There is a general tendency for succession to lead to more complex, diverse, and presumably stable ecosystems. But a close look at almost any terrestrial community or ecosystem reveals that it consists of an ever-changing mosaic of patches of vegetation at different stages of succession.

The current view is that we cannot predict the course of a given succession or view it as preordained progress toward an ideally adapted climax plant community or ecosystem. Rather, succession reflects the ongoing struggle by different species for enough light, nutrients, food, and space. Most ecologists now recognize that mature late-successional ecosystems are not in a state of permanent equilibrium, but rather a state of continual disturbance and change.

Living Systems Are Sustained through Constant Change

All living systems from a cell to the biosphere are dynamic systems that are constantly changing in response to changing environmental conditions. Continents move, the climate changes, and disturbances and succession change the composition of communities and ecosystems.

Living systems contain complex networks of positive and negative feedback loops (Figure 2-11, p. 45, and Figure 2-12, p. 45) that interact to provide some degree of stability, or sustainability, over each system's expected life span. This stability, or capacity to withstand external stress and disturbance, is maintained only by constant change in response to changing environmental conditions. For example, in a mature tropical rain forest, some trees die and others take their places. However, unless the forest is cut, burned, or otherwise destroyed, you would still recognize it as a tropical rain forest 50 or 100 years from now.

It is useful to distinguish among two aspects of stability in living systems. One is **inertia,** or **persistence:** the ability of a living system, such as a grassland or a forest, to survive moderate disturbances. A second factor is **resilience:** the ability of a living system to be restored through secondary succession after a moderate disturbance.

Evidence suggests that some ecosystems have one of these properties but not the other. For example, tropical rain forests have high species diversity and high inertia and thus are resistant to significant alteration or destruction. But once a large tract of tropical rain forest is severely damaged, the resilience of the resulting degraded ecosystem may be so low that the forest may not be restored by secondary ecological succession. One reason for this is that most of the nutrients in a tropical rain forest are stored in its vegetation, not in the soil as in most other terrestrial ecosystems. Once the nutrient-rich vegetation is gone, daily rains can remove most of the other nutrients left in the soil and thus prevent a tropical rain forest from regrowing on a large cleared area.

Another reason for why the rain forest cannot recover is that large-scale deforestation can change an area's climate by decreasing the input of water vapor from its trees into the atmosphere. Without such water vapor, rain decreases and the local climate gets warmer. Over many decades, this can allow for the development of a tropical grassland in the cleared area but not for the reestablishment of a tropical rain forest.

By contrast, grasslands are much less diverse than most forests, and consequently they have low inertia and can burn easily. However, because most of their plant matter is stored in underground roots, these ecosystems have high resilience and can recover quickly after a fire as their root systems produce new grasses. Grassland can be destroyed only if its roots are plowed up and something else is planted in its place, or if it is severely overgrazed by livestock or other herbivores.

Variations among species in resilience and inertia are yet another example of biodiversity—one aspect of natural capital that has allowed life on earth to sustain itself for billions of years.

However, there are limits to the stresses that ecosystems and global systems such as climate can take. As a result, such systems can reach a **tipping point,** where any additional stress can cause the system to change in an abrupt and usually irreversible way that often involves collapse. For example, once a certain number of trees have been eliminated from a stable tropical rain forest, it can crash and become a grassland. And continuing to warm the atmosphere by burning fossil fuels that emit CO_2 and cutting down tropical forests that help remove CO_2 could eventually change the global climate system in ways that could last for thousands of years.

Exceeding a tipping point is like falling off a cliff. There is no way back. One of the most urgent scientific research priorities is to identify these and other tipping points and to develop strategies to prevent natural systems from reaching their tipping points.

REVISITING Southern Sea Otters and Sustainability

Before the arrival of European settlers on the North American west coast, the sea otter population was part of a complex ecosystem made up of bottom-dwelling creatures, kelp, otters, whales, and other species depending on one another for survival. Giant kelp forests served as food and shelter for sea urchins. Otters ate the sea urchins and other kelp eaters. Some species of whales and sharks ate the otters. And detritus from all these species helped to maintain the giant kelp forests. Each of these interacting populations was kept in check by, and helped to sustain, all others.

When humans arrived and began hunting the otters for their pelts, they probably did not know much about the intricate web of life beneath the ocean surface. But with the effects of overhunting, people realized they had done more than simply take otters. They had torn the web, disrupted an entire ecosystem, and triggered a loss of valuable natural resources and services, including biodiversity.

Populations of most plants and animals depend directly or indirectly on solar energy and each population plays a role in the cycling of nutrients in the ecosystems where they live. In addition, the biodiversity found in the variety of species in different terrestrial and aquatic ecosystems provides alternative paths for energy flow and nutrient cycling and better opportunities for natural selection as environmental conditions change. Disrupt these paths and we can decrease the various components of the biodiversity of ecosystems and the sizes of their populations.

In this chapter, we looked more closely at two **principles of sustainability**: *biodiversity promotes sustainability* and *there are always limits to population growth in nature.* Chapter 6 applies the concepts of biodiversity and population dynamics discussed in this chapter to the growth of the human population and its environmental impacts. Chapters 7 and 8 look more closely at biodiversity in the variety of terrestrial ecosystems (such as deserts, grasslands, and forests) and aquatic ecosystems (such as oceans, lakes, rivers, and wetlands) found on the earth.

We cannot command nature
except by obeying her.
SIR FRANCIS BACON

1. Review the Key Questions and Concepts for this chapter on p. 101. Explain how southern sea otters act as a keystone species in kelp beds. Explain why we should care about protecting this species from extinction. Explain why we should help to preserve kelp forests.

2. Define **interspecific competition, predation, parasitism, mutalism,** and **commensalism** and give an example of each. Explain how each of these species interactions can affect the population sizes of species in ecosystems. Distinguish between a **predator** and a **prey** and give an example of each. What is a **predator–prey relationship?** Describe four ways in which prey species can avoid their predators and four ways in which predators can capture these prey.

3. Define and give an example of **coevolution.**

4. Describe and give an example of **resource partitioning** and explain how it can increase species diversity.

5. What is **population dynamics?** Why do most populations live in clumps?

6. Describe four variables that govern changes in population size and write an equation showing how they interact. What is a population's **age structure** and what are three major age group categories? Distinguish among the **biotic potential, intrinsic rate of increase, exponential growth, environmental resistance, carrying capacity,** and **logistic growth** of a population, and use these concepts to explain why there are always limits to population growth in nature. Why are southern sea otters making a slow comeback and what factors can threaten this recovery? Define and give an example of a **population crash.** Explain why humans are not exempt from nature's population controls.

7. Distinguish between **r-selected species** and **K-selected species** and give an example of each type. Define **population density** and explain how it can affect the size of some but not all populations.

8. Describe the exploding white-tailed deer population problem in the United States and discuss options for dealing with it.

9. What is **ecological succession?** Distinguish between **primary ecological succession** and **secondary ecological succession** and give an example of each. Explain why succession does not follow a predictable path. In terms of stability, distinguish between **inertia (persistence)** and **resilience.** Explain how living systems achieve some degree of stability or sustainability by undergoing constant change in response to changing environmental conditions.

10. Explain how the role of the southern sea otter in its ecosystem (**Core Case Study**) illustrates the population control **principle of sustainability.**

Note: Key Terms are in **bold** type.

1. What difference would it make if the southern sea otter (**Core Case Study**) became prematurely extinct because of human activities? What are three things we could do to help prevent the premature extinction of this species?

2. Use the second law of thermodynamics (p. 43) to explain why predators are generally less abundant than their prey.

3. Explain why most species with a high capacity for population growth (high biotic potential) tend to have small individuals (such as bacteria and flies) while those with a low capacity for population growth tend to have large individuals (such as humans, elephants, and whales).

4. List three factors that have limited human population growth in the past that we have overcome. Describe how we overcame each of these factors. List two factors that may limit human population growth in the future.

5. Why are pest species likely to be extreme examples of r-selected species? Why are many endangered species likely to be extreme examples of K-selected species?

6. Given current environmental conditions, if you had a choice, would you rather be an r-strategist or a K-strategist? Explain your answer.

7. Is the southern sea otter an r-strategist or a K-strategist species? Explain. How does this affect our efforts to protect this species from premature extinction?

8. How would you reply to someone who argues that we should not worry about our effects on natural systems because natural succession will heal the wounds of human activities and restore the balance of nature?

9. In your own words, restate this chapter's closing quotation by Sir Francis Bacon. Do you agree with this notion? Why or why not?

10. List two questions that you would like to have answered as a result of reading this chapter.

Note: See Supplement 13 (p. S78) for a list of Projects related to this chapter.

The graph below shows the changes in size of an Emperor Penguin population in terms of breeding pairs on Terre Adelie in the Antarctic. Use the graph to answer the questions below.

Source: Data from *Nature,* May 10, 2001.

1. What was the approximate carrying capacity of the penguin population on the island from 1960 to 1975? (*Hint:* See Figure 5-11, p. 111.) What was the approximate carrying capacity of the penguin population on the island from 1980 to 1999?

2. What is the percentage decline in the penguin population from 1975 to 1999?

Log on to the Student Companion Site for this book at **academic.cengage.com/biology/miller**, and choose Chapter 5 for many study aids and ideas for further reading and research. These include flash cards, practice quizzing, Weblinks, information on Green Careers, and InfoTrac® College Edition articles.

6 The Human Population and Its Impact

Are There Too Many of Us?

Each week, about 1.6 million people are added to the world's population. As a result, the number of people on the earth is projected to increase from 6.7 to 9.3 billion or more between 2008 and 2050, with most of this growth occurring in the world's developing countries (Figure 6-1). This raises an important question: *Can the world provide an adequate standard of living for a projected 2.6 billion more people by 2050 without causing widespread environmental damage?* There is disagreement over the answer to this question.

According to one view, the planet already has too many people collectively degrading the earth's natural capital. To some analysts, the problem is the sheer number of people in developing countries with 82% of the world's population. To others, it is high per capita resource consumption rates in developed countries—and to an increasing extent in rapidly developing countries such as China and India—that magnify the environmental impact, or ecological footprint, of each person (Figure 1-10, p. 15).

Many argue that both population growth and resource consumption per person are important causes of the environmental problems we face (**Concept 1-5A**, p. 17).

Another view is that technological advances have allowed us to overcome the environmental resistance that all populations face (Figure 5-11, p. 111) and to increase the earth's carrying capacity for our species. Some analysts argue there is no reason we cannot continue doing so, and they believe that the planet can support billions more people. They also see a growing population as our most valuable resource for solving environmental and other problems and for stimulating economic growth by increasing the number of consumers. As a result, they see no need to control the world's population growth.

Some people view any form of population regulation as a violation of their religious or moral beliefs. Others see it as an intrusion into their privacy and their freedom to have as many children as they want. These people also would argue against any form of population control.

Proponents of slowing and eventually stopping population growth have a different view. They point out that we are not providing the basic necessities for about one of every five people—a total of some 1.4 billion. They ask how we will be able to do so for the projected 2.6 billion more people by 2050.

They also warn of two serious consequences we will face if we do not sharply lower birth rates. *First,* death rates may increase because of declining health and environmental conditions in some areas, as is already happening in parts of Africa. *Second,* resource use and environmental degradation may intensify as more consumers increase their already large ecological footprints in developed countries and in rapidly developing countries, such as China and India (Figure 1-10, p. 15). This could increase environmental stresses such as infectious disease, biodiversity losses, water shortages, traffic congestion, pollution of the seas, and climate change.

This debate over interactions among population growth, economic growth, politics, and moral beliefs is one of the most important and controversial issues in environmental science.

Figure 6-1 Crowded street in China. Together, China and India have 36% of the world's population and the resource use per person in these countries is projected to grow rapidly as they become more modernized (Case Study, p. 15).

6-1 How many people can the earth support?

CONCEPT 6-1 We do not know how long we can continue increasing the earth's carrying capacity for humans without seriously degrading the life-support system for humans and many other species.

6-2 What factors influence the size of the human population?

CONCEPT 6-2A Population size increases because of births and immigration and decreases through deaths and emigration.

CONCEPT 6-2B The average number of children born to women in a population (*total fertility rate*) is the key factor that determines population size.

6-3 How does a population's age structure affect its growth or decline?

CONCEPT 6-3 The numbers of males and females in young, middle, and older age groups determine how fast a population grows or declines.

6-4 How can we slow human population growth?

CONCEPT 6-4 Experience indicates that the most effective ways to slow human population growth are to encourage family planning, to reduce poverty, and to elevate the status of women.

Note: Supplements 2 (p. S4), 3 (p. S10), 4 (p. S20), and 13 (p. S78) can be used with this chapter.

The problems to be faced are vast and complex, but come down to this:
6.7 billion people are breeding exponentially.
The process of fulfilling their wants and needs is stripping earth
of its biotic capacity to support life;
a climactic burst of consumption by a single species
is overwhelming the skies, earth, waters, and fauna.

PAUL HAWKEN

6-1 How Many People Can the Earth Support?

▶ **CONCEPT 6-1** We do not know how long we can continue increasing the earth's carrying capacity for humans without seriously degrading the life-support system for humans and many other species.

Human Population Growth Continues but It Is Unevenly Distributed

For most of history, the human population grew slowly (Figure 1-1, p. 5, left part of curve). But for the past 200 years, the human population has experienced rapid exponential growth reflected in the characteristic J-curve (Figure 1-1, right part of curve).

Three major factors account for this population increase. *First,* humans developed the ability to expand into diverse new habitats and different climate zones. *Second,* the emergence of early and modern agriculture allowed more people to be fed for each unit of land area farmed. *Third,* the development of sanitation systems, antibiotics, and vaccines helped control infectious disease agents. As a result, death rates dropped sharply below birth rates and population size grew rapidly.

About 10,000 years ago when agriculture began, there were about 5 million humans on the planet; now there are 6.7 billion of us. It took from the time we arrived until about 1927 to add the first 2 billion people to the planet; less than 50 years to add the next 2 billion (by 1974); and just 25 years to add the next 2 billion (by 1999)—an illustration of the awesome power of exponential growth (Chapter 1 Core Case Study, p. 5). By 2012 we will be trying to support 7 billion people and perhaps 9.3 billion by 2050. Such growth raises the question of whether the earth is overpopulated (**Core Case Study**). (See Figure 4, p. S12, in Supplement 3 for a timeline of key events related to human population growth.)

The rate of population growth has slowed, but the world's population is still growing exponentially at a rate of 1.22% a year. This means that 82 million people were added to the world's population during 2008—an average of nearly 225,000 more people each day, or 2.4 more people every time your heart beats. (See *The Habitable Planet,* Video 5, at **www.learner.org/resources/series209.html** for a discussion of how demographers measure population size and growth.)

Geographically, this growth is unevenly distributed. About 1.2 million of these people were added to the

SCIENCE FOCUS

How Long Can the Human Population Keep Growing?

To survive and provide resources for growing numbers of people, humans have modified, cultivated, built on, or degraded a large and increasing portion of the earth's natural systems. Our activities have directly affected, to some degree, about 83% of the earth's land surface, excluding Antarctica (Figure 3, pp. S24–25, in Supplement 4), as our ecological footprints have spread across the globe (**Concept 1-3**, p. 12, and Figure 1-10, p. 15). In other words, human activities have degraded the various components of earth's biodiversity (Figure 4-2, p. 79) and such threats are expected to increase.

We have used technology to alter much of the rest of nature to meet our growing needs and wants in eight major ways (Figure 6-A).

Scientific studies of populations of other species tell us that *no population can continue growing indefinitely* (**Concept 5-3**, p.108), which is one of the four **scientific principles of sustainability** (see back cover). How long can we continue increasing the earth's carrying capacity for our species by sidestepping many of the factors that sooner or later limit the growth of any population?

The debate over this important question has been going on since 1798 when Thomas Malthus, a British economist, hypothesized that the human population tends to increase exponentially, while food supplies tend to increase more slowly at a linear rate. So far, Malthus has been proven wrong. Food pro-duction has grown at an exponential rate instead of at a linear rate because of genetic and technological advances in industrialized food production.

No one knows how close we are to the environmental limits that sooner or later will control the size of the human population, but mounting evidence indicates that we are steadily degrading the natural capital, which keeps us and other species alive and supports our economies (**Concept 6-1**).

Critical Thinking

How close do you think we are to the environmental limits of human population growth?

NATURAL CAPITAL DEGRADATION

Altering Nature to Meet Our Needs

Reduction of biodiversity

Increasing use of the earth's net primary productivity

Increasing genetic resistance of pest species and disease-causing bacteria

Elimination of many natural predators

Introduction of potentially harmful species into communities

Using some renewable resources faster than they can be replenished

Interfering with the earth's chemical cycling and energy flow processes

Relying mostly on polluting and climate-changing fossil fuels

CENGAGENOW Active Figure 6-A
Major ways in which humans have altered the rest of nature to meet our growing population's resource needs and wants. *See an animation based on this figure at* CengageNOW. **Questions:** Which three of these items do you believe have been the most harmful? Explain. How does your lifestyle contribute directly or indirectly to each of these three items?

world's developed countries, growing at 0.1% a year. About 80.8 million were added to developing countries, growing 15 times faster at 1.5% a year. In other words, most of the world's population growth takes place in already heavily populated parts of world most of which are the least equipped to deal with the pressures of such rapid growth. In our demographically divided world, roughly 1 billion people live in countries with essentially a stable population size while another billion or so live in countries whose populations are projected to at least double between 2008 and 2050.

How many of us are likely to be here in 2050? Answer: 7.8–10.7 billion people, depending mostly on projections about the average number of babies women are likely to have. The medium projection is 9.3 billion people (Figure 6-2). About 97% of this growth is projected to take place in developing countries, where acute poverty is a way of life for about 1.4 billion people.

The prospects for stabilization of the human population in the near future are nil. However, during this century, the human population may level off as it moves from a J-shaped curve of exponential growth to an S-shaped curve of logistic growth because of various factors that can limit human population growth (Figure 5-11, p. 111).

(Figure 5-11, p. 111).

HOW WOULD YOU VOTE? ☑

Should the population of the country where you live be stabilized as soon as possible? Cast your vote online at **academic.cengage.com/biology/miller**.

This raises the question posed in the Core Case Study at the beginning of this chapter: How many people can the earth support indefinitely? Some say about 2 billion. Others say as many as 30 billion. This issue has long been a topic of scientific debate (Science Focus, at left).

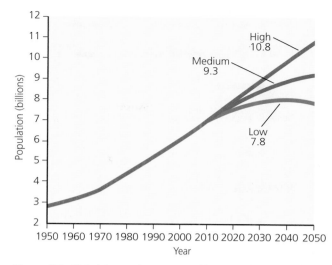

Figure 6-2 *Global connections:* UN world population projections, assuming that by 2050 women will have an average of 2.5 children (high), 2.0 children (medium), or 1.5 children (low). The most likely projection is the medium one—9.3 billion by 2050. (Data from United Nations).

Some analysts believe this is the wrong question. Instead, they say, we should ask what the *optimum sustainable population* of the earth might be, based on the planet's **cultural carrying capacity.** This would be an optimum level that would allow most people to live in reasonable comfort and freedom without impairing the ability of the planet to sustain future generations. (See the Guest Essay by Garrett Hardin at CengageNOW™.)

RESEARCH FRONTIER

Determining the optimum sustainable population size for the earth and for various regions. See **academic.cengage.com/biology/miller**.

6-2 What Factors Influence the Size of the Human Population?

▶ **CONCEPT 6-2A** Population size increases because of births and immigration and decreases through deaths and emigration.

▶ **CONCEPT 6-2B** The average number of children born to women in a population *(total fertility rate)* is the key factor that determines population size.

The Human Population Can Grow, Decline, or Remain Fairly Stable

On a global basis, if there are more births than deaths during a given period of time, the earth's population increases, and when the reverse is true, it decreases.

When births equal deaths during a particular time period population size does not change.

Human populations of countries and cities grow or decline through the interplay of three factors: *births (fertility), deaths (mortality),* and *migration.* We can calculate **population change** of an area by subtracting the

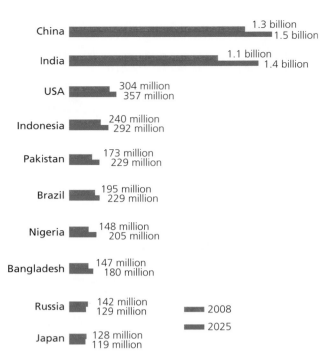

	2008	2025
China	1.3 billion	1.5 billion
India	1.1 billion	1.4 billion
USA	304 million	357 million
Indonesia	240 million	292 million
Pakistan	173 million	229 million
Brazil	195 million	229 million
Nigeria	148 million	205 million
Bangladesh	147 million	180 million
Russia	142 million	129 million
Japan	128 million	119 million

Figure 6-3 *Global connections:* the world's 10 most populous countries in 2008, with projections of their population sizes in 2025. (Data from World Bank and Population Reference Bureau)

number of people leaving a population (through death and emigration) from the number entering it (through birth and immigration) during a specified period of time (usually one year) (**Concept 6-2A**). See Figures 5 and 6, p. S13, in Supplement 3.

$$\text{Population change} = (\text{Births} + \text{Immigration}) - (\text{Deaths} + \text{Emigration})$$

When births plus immigration exceed deaths plus emigration, population increases; when the reverse is true, population declines.

Instead of using the total numbers of births and deaths per year, population experts (demographers) use the **birth rate,** or **crude birth rate** (the number of live births per 1,000 people in a population in a given year), and the **death rate,** or **crude death rate** (the number of deaths per 1,000 people in a population in a given year).

What five countries had the largest numbers of people in 2008? Number 1 was China with 1.3 billion people, or one of every five people in the world (Figures 6-1 and 6-3). Number 2 was India with 1.1 billion people, or one of every six people. Together, China and India have 36% of the world's population. The United States, with 304 million people in 2008—had the world's third largest population but only 4.5% of the world's people. Can you guess the next two most populous countries? What three countries are expected

to have the most people in 2025? Look at Figure 6-3 to see if your answers are correct.

Women Are Having Fewer Babies but Not Few Enough to Stabilize the World's Population

Another measurement used in population studies is **fertility rate,** the number of children born to a woman during her lifetime. Two types of fertility rates affect a country's population size and growth rate. The first type, called the **replacement-level fertility rate,** is the average number of children that couples in a population must bear to replace themselves. It is slightly higher than two children per couple (2.1 in developed countries and as high as 2.5 in some developing countries), mostly because some children die before reaching their reproductive years.

Does reaching replacement-level fertility bring an immediate halt to population growth? No, because so many *future* parents are alive. If each of today's couples had an average of 2.1 children, they would not be contributing to population growth. But if all of today's girl children also have 2.1 children, the world's population will continue to grow for 50 years or more (assuming death rates do not rise).

The second type of fertility rate, the **total fertility rate (TFR),** is the average number of children born to women in a population during their reproductive years. This factor plays a key role in determining population size (**Concept 6-2B**). The average fertility rate has been declining. In 2008, the average global TFR was 2.6 children per woman: 1.6 in developed countries (down from 2.5 in 1950) and 2.8 in developing countries (down from 6.5 in 1950). Although the decline in TFR in developing countries is impressive, the TFR remains far above the replacement level of 2.1, not low enough to stabilize the world's population in the near future. See Figures 7 and 8, p. S14, in Supplement 3.

> **THINKING ABOUT**
> **The Slower Rate of Population Growth**
>
> Why has the world's exponential rate of population growth slowed down in the last few decades? What would have to happen for the world's population to stop growing?

■ CASE STUDY

The U.S. Population Is Growing Rapidly

The population of the United States grew from 76 million in 1900 to 304 million in 2008, despite oscillations in the country's TFR (Figure 6-4) and birth rates (Figure 6-5). It took the country 139 years to add its

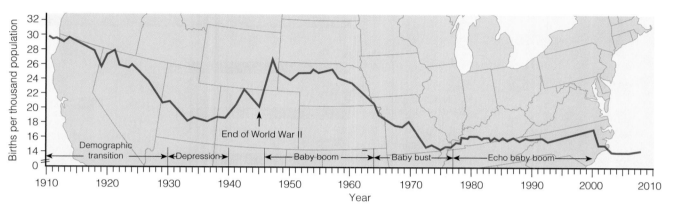

Figure 6-4 Total fertility rates for the United States between 1917 and 2008. **Question:** The U.S. fertility rate has declined and remained at or below replacement levels since 1972, so why is the population of the United States still increasing? (Data from Population Reference Bureau and U.S. Census Bureau)

Figure 6-5 Birth rates in the United States, 1910–2008. Use this figure to trace changes in crude birth rates during your lifetime. (Data from U.S. Bureau of Census and U.S. Commerce Department)

first 100 million people, 52 years to add another 100 million by 1967, and only 39 years to add the third 100 million in 2006. The period of high birth rates between 1946 and 1964 is known as the *baby boom,* when 79 million people were added to the U.S. population. In 1957, the peak of the baby boom, the TFR reached 3.7 children per woman. Since then, it has generally declined, remaining at or below replacement level since 1972. In 2008, the TFR was 2.1 children per woman, compared to 1.6 in China.

The drop in the TFR has slowed the rate of population growth in the United States. But the country's population is still growing faster than that of any other developed country, and of China, and is not close to leveling off. About 2.9 million people (one person every 11 seconds) were added to the U.S. population in 2008. About 66% (1.9 million) of this growth occurred because births outnumbered deaths and 34% (1 million) came from legal and illegal immigration (with someone migrating to the U.S. every 32 seconds).

In addition to the almost fourfold increase in population growth since 1900, some amazing changes in lifestyles took place in the United States during the 20th century (Figure 6-6, p. 128), which led to dramatic increases in per capita resource use and a much larger U.S. ecological footprint (**Concept 1-3**, p. 12, and Figure 1-10, top, p. 15). **CONCEPT LINK**

Here are a few more changes that occurred during the last century. In 1907, the three leading causes of death in the United States were pneumonia, tuberculosis, and diarrhea; 90% of U.S. doctors had no college education; one out of five adults could not read or write; only 6% of Americans graduated from high school; the average U.S. worker earned $200–400 per year and the average daily wage was 22 cents per hour; there were only 9,000 cars in the U.S., and only 232 kilometers (144 miles) of paved roads; a 3-minute phone call from Denver, Colorado, to New York city cost $11; only 30 people lived in Las Vegas, Nevada; most women washed their hair only once a month; marijuana, heroin, and morphine were available over the counter at local drugstores; and there were only 230 reported murders in the entire country.

According to U.S. Census Bureau, the U.S. population is likely to increase from 304 million in 2008 to 438 million by 2050 and then to 571 million by 2100. In contrast, population growth has slowed in other major developed countries since 1950, most of which are expected to have declining populations after 2010. Because of a high per capita rate of resource use and the resulting waste and pollution, each addition to the U.S. population has an enormous environmental impact (Figure 1-9, bottom, p. 14, Figure 1-10, p. 15, and Figure 7 on pp. S28–S29 in Supplement 4).

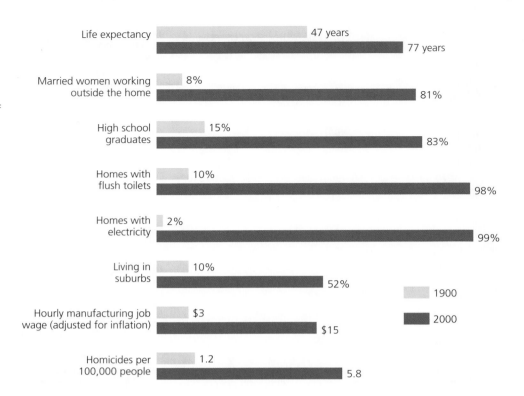

Figure 6-6 Some major changes that took place in the United States between 1900 and 2000. **Question:** Which two of these changes do you think were the most important? (Data from U.S. Census Bureau and Department of Commerce)

Life expectancy — 47 years / 77 years

Married women working outside the home — 8% / 81%

High school graduates — 15% / 83%

Homes with flush toilets — 10% / 98%

Homes with electricity — 2% / 99%

Living in suburbs — 10% / 52%

Hourly manufacturing job wage (adjusted for inflation) — $3 / $15

Homicides per 100,000 people — 1.2 / 5.8

1900 / 2000

Several Factors Affect Birth Rates and Fertility Rates

Many factors affect a country's average birth rate and TFR. One is the *importance of children as a part of the labor force.* Proportions of children working tend to be higher in developing countries.

Another economic factor is the *cost of raising and educating children.* Birth and fertility rates tend to be lower in developed countries, where raising children is much more costly because they do not enter the labor force until they are in their late teens or twenties. In the United States, it costs about $290,000 to raise a middle-class child from birth to age 18. By contrast, many children in poor countries have to work to help their families survive.

The *availability of private and public pension systems* can influence the decision for some couples on how many children to have, especially the poor in developing countries. Pensions reduce a couple's need to have many children to help support them in old age.

Urbanization plays a role. People living in urban areas usually have better access to family planning services and tend to have fewer children than do those living in rural areas (especially in developing countries) where children are often needed to help raise crops and carry daily water and fuelwood supplies.

Another important factor is the *educational and employment opportunities available for women.* TFRs tend to be low when women have access to education and paid employment outside the home. In developing countries, a woman with no education typically has two more children than does a woman with a high school edu-

cation. In nearly all societies, better-educated women tend to marry later and have fewer children.

Another factor is the **infant mortality rate**—the number of children per 1,000 live births who die before one year of age. In areas with low infant mortality rates, people tend to have fewer children because fewer children die at an early age.

Average age at marriage (or, more precisely, the average age at which a woman has her first child) also plays a role. Women normally have fewer children when their average age at marriage is 25 or older.

Birth rates and TFRs are also affected by the *availability of legal abortions.* Each year about 190 million women become pregnant. The United Nations and the World Bank estimate that 46 million of these women get abortions—26 million of them legal and 20 million illegal (and often unsafe). Also, the *availability of reliable birth control methods* allows women to control the number and spacing of the children they have.

Religious beliefs, traditions, and cultural norms also play a role. In some countries, these factors favor large families and strongly oppose abortion and some forms of birth control.

Several Factors Affect Death Rates

The rapid growth of the world's population over the past 100 years is not primarily the result of a rise in the crude birth rate. Instead, it has been caused largely by a decline in crude death rates, especially in developing countries.

More people started living longer and fewer infants died because of increased food supplies and distribu-

tion, better nutrition, medical advances such as immunizations and antibiotics, improved sanitation, and safer water supplies (which curtailed the spread of many infectious diseases).

Two useful indicators of the overall health of people in a country or region are **life expectancy** (the average number of years a newborn infant can expect to live) and the **infant mortality rate** (the number of babies out of every 1,000 born who die before their first birthday). Between 1955 and 2008, the global life expectancy increased from 48 years to 68 years (77 years in developed countries and 67 years in developing countries) and is projected to reach 74 by 2050. Between 1900 and 2008, life expectancy in the United States increased from 47 to 78 years and, by 2050, is projected to reach 82 years. In the world's poorest countries, however, life expectancy is 49 years or less and may fall further in some countries because of more deaths from AIDS.

Even though more is spent on health care per person in the United States than in any other country, people in 41 other countries including Canada, Japan, Singapore, and a number of European countries have longer life expectancies than do Americans. Analysts cite two major reasons for this. First, 45 million Americans lack health care insurance, while Canada and many European countries have universal health care insurance. Second, adults in the United States have one of the world's highest obesity rates.

Infant mortality is viewed as one of the best measures of a society's quality of life because it reflects a country's general level of nutrition and health care. A high infant mortality rate usually indicates insufficient food (undernutrition), poor nutrition (malnutrition), and a high incidence of infectious disease (usually from drinking contaminated water and having weakened disease resistance due to undernutrition and malnutrition).

Between 1965 and 2008, the world's infant mortality rate dropped from 20 to 6.3 in developed countries and from 118 to 59 in developing countries (see Figures 9 and 10, p. S15, in Supplement 3). This is good news, but annually, more than 4 million infants (most in developing countries) die of preventable causes during their first year of life—an average of 11,000 mostly unnecessary infant deaths per day. This is equivalent to 55 jet airliners, each loaded with 200 infants younger than age 1, crashing *each day* with no survivors!

The U.S. infant mortality rate declined from 165 in 1900 to 6.6 in 2008. This sharp decline was a major factor in the marked increase in U.S. average life expectancy during this period. Still, some 40 countries, including Taiwan, Cuba, and most of Europe had lower infant mortality rates than the United States had in 2008. Three factors helped keep the U.S. infant mortality rate high: *inadequate health care for poor women during pregnancy and for their babies after birth, drug addiction among pregnant women*, and *a high birth rate among teenagers* (although this rate dropped by almost half between 1991 and 2006).

Migration Affects an Area's Population Size

The third factor in population change is **migration:** the movement of people into (*immigration*) and out of (*emigration*) specific geographic areas.

Most people migrating from one area or country to another seek jobs and economic improvement. But some are driven by religious persecution, ethnic conflicts, political oppression, wars, and environmental degradations such as water and food shortages and soil erosion. According to a U.N. study, there were about 25 million *environmental refugees* in 2005 and the number could reach 50 million by 2010. Environmental scientist Norman Myers warns that, in a warmer world, the number of such refugees could soar to 250 million or more before the end of this century. (See more on this in the Guest Essay by Norman Myers at CengageNOW.)

■ CASE STUDY

The United States: A Nation of Immigrants

Since 1820, the United States has admitted almost twice as many immigrants and refugees as all other countries combined. The number of legal immigrants (including refugees) has varied during different periods because of changes in immigration laws and rates of economic growth (Figure 6-7). Currently, legal and illegal immigration account for about 40% of the country's annual population growth.

Figure 6-7 Legal immigration to the United States, 1820–2003 (the last year for which data are available). The large increase in immigration since 1989 resulted mostly from the Immigration Reform and Control Act of 1986, which granted legal status to illegal immigrants who could show they had been living in the country for several years. (Data from U.S. Immigration and Naturalization Service and the Pew Hispanic Center)

Between 1820 and 1960, most legal immigrants to the United States came from Europe. Since 1960, most have come from Latin America (53%) and Asia (25%), followed by Europe (14%). In 2007, Latinos (67% of them from Mexico) made up 15% of the U.S. population, and by 2050, are projected to make up 25% of the population. According to the Pew Hispanic Center, 53% of the 100 million Americans that were added to the population between 1967 and 2007 were either immigrants or their children.

There is controversy over whether to reduce legal immigration to the United States. Some analysts would accept new entrants only if they can support themselves, arguing that providing legal immigrants with public services makes the United States a magnet for the world's poor. Proponents of reducing legal immigration argue that it would allow the United States to stabilize its population sooner and help reduce the country's enormous environmental impact from its huge ecological footprint (Figure 1-10, p. 15; and **Concept 1-3**, p. 12).

Polls show that almost 60% of the U.S. public strongly supports reducing legal immigration. There is also intense political controversy over what to do about illegal immigration. In 2007, there were an estimated 11.3 million illegal immigrants in the United States, with about 58% of them from Mexico and 22% from other Latin American countries.

Those opposed to reducing current levels of legal immigration argue that it would diminish the histori-cal role of the United States as a place of opportunity for the world's poor and oppressed and as a source of cultural diversity that has been a hallmark of American culture since its beginnings. In addition, according to several studies, including a 2006 study by the Pew Hispanic Center, immigrants and their descendants pay taxes, take many menial and low-paying jobs that most other Americans shun, start new businesses, create jobs, add cultural vitality, and help the United States succeed in the global economy. Also, according to the U.S. Census Bureau, after 2020, higher immigration levels will be needed to supply enough workers as baby boomers retire.

According to a recent study by the U.N. Population Division, if the United States wants to maintain its current ratio of workers to retirees, it will need to absorb an average of 10.8 million immigrants each year—more than 13 times the current immigration level—through 2050. At that point, the U.S. population would total 1.1 billion people, 73% of them fairly recent immigrants or their descendants. Housing this influx of almost 11 million immigrants per year would require the equivalent of building another New York City every 10 months.

— **HOW WOULD YOU VOTE?** —

Should legal immigration into the United States be reduced? Cast your vote online at **academic.cengage .com/biology/miller**.

6-3 How Does a Population's Age Structure Affect Its Growth or Decline?

▶ **CONCEPT 6-3** The numbers of males and females in young, middle, and older age groups determine how fast a population grows or declines.

Populations Made Up Mostly of Young People Can Grow Rapidly

As mentioned earlier, even if the replacement-level fertility rate of 2.1 children per woman were magically achieved globally tomorrow, the world's population would keep growing for at least another 50 years (assuming no large increase in the death rate). This results mostly from the **age structure:** the distribution of males and females among age groups in a population—in this case, the world population (**Concept 6-3**).

Population experts construct a population *age-structure diagram* by plotting the percentages or numbers of males and females in the total population in each of three age categories: *prereproductive* (ages 0–14), *reproductive* (ages 15–44), and *postreproductive* (ages 45 and older). Figure 6-8 presents generalized age-structure diagrams for countries with rapid, slow, zero, and negative population growth rates.

Any country with many people younger than age 15 (represented by a wide base in Figure 6-8, far left) has a powerful built-in momentum to increase its population size unless death rates rise sharply. The number of births will rise even if women have only one or two children, because a large number of girls will soon be moving into their reproductive years.

What is one of the world's most important population statistics? *Nearly 28% of the people on the planet were under 15 years old in 2008.* These 1.9 billion young people are poised to move into their prime reproductive

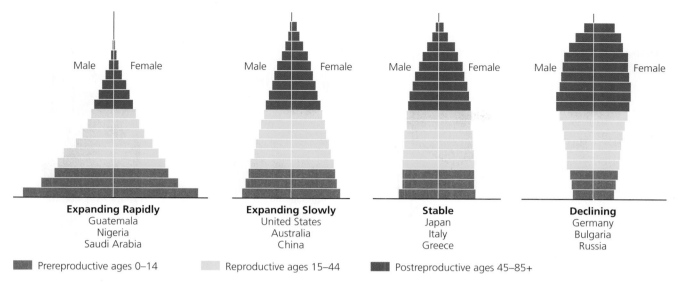

Expanding Rapidly	Expanding Slowly	Stable	Declining
Guatemala	United States	Japan	Germany
Nigeria	Australia	Italy	Bulgaria
Saudi Arabia	China	Greece	Russia

■ Prereproductive ages 0–14 ■ Reproductive ages 15–44 ■ Postreproductive ages 45–85+

CENGAGENOW™ **Active Figure 6-8** Generalized population age structure diagrams for countries with rapid (1.5–3%), slow (0.3–1.4%), zero (0–0.2%), and negative (declining) population growth rates. A population with a large proportion of its people in the prereproductive age group (far left) has a large potential for rapid population growth. *See an animation based on this figure at* CengageNOW. **Question:** Which of these figures best represents the country where you live? (Data from Population Reference Bureau)

years. In developing countries, the percentage is even higher: 30% on average (41% in Africa) compared with 17% in developed countries (20% in the United States and 16% in Europe). These differences in population age structure between developed and developing countries are dramatic, as Figure 6-9 reveals. This figure also shows why almost all of future human population growth will be in developing countries.

We Can Use Age–Structure Information to Make Population and Economic Projections

Changes in the distribution of a country's age groups have long-lasting economic and social impacts. Between 1946 and 1964, the United States had a *baby boom*, which added 79 million people to its population. Over time, this group looks like a bulge moving up through the country's age structure, as shown in Figure 6-10, p. 132.

Baby boomers now make up almost half of all adult Americans. As a result, they dominate the population's demand for goods and services and play increasingly important roles in deciding who gets elected and what laws are passed. Baby boomers who created the youth market in their teens and twenties are now creating the 50-something market and will soon move on to create a 60-something market. After 2011, when the first baby boomers will turn 65, the number of Americans older than age 65 will grow sharply through 2029 in what

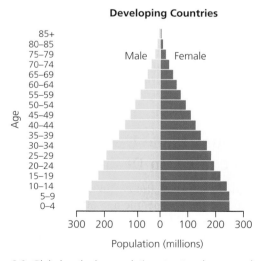

Figure 6-9 *Global outlook:* population structure by age and sex in developing countries and developed countries, 2006. **Question:** If all girls under 15 had only one child during their lifetimes, how do you think these structures would change over time? (Data from United Nations Population Division and Population Reference Bureau)

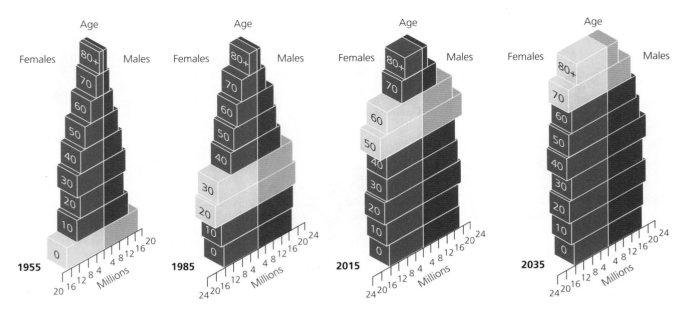

CENGAGENOW™ **Active Figure 6-10** Tracking the baby-boom generation in the United States. U.S. population by age and sex, 1955, 1985, 2015, and 2035 (projected). *See an animation based on this figure at* CengageNOW. (Data from U.S. Census Bureau)

has been called the *graying of America*. In 2008, about 13% of Americans were 65 or older, but that number is projected to increase to about 25% by 2043.

According to some analysts, the retirement of baby boomers is likely to create a shortage of workers in the United States unless immigrant workers or various forms of automation replace some of them. Retired baby boomers may use their political clout to have the smaller number of people in the baby-bust generation that followed them pay higher income, health-care, and social security taxes. However, the rapidly increasing number of immigrants and their descendants may dilute their political power. Their power may also be weakened by the rise of members of the echo baby boom generation (Figure 6-5).

CENGAGENOW™ Examine how the baby boom affects the U.S. age structure over several decades at CengageNOW.

Populations Made Up Mostly of Older People Can Decline Rapidly

As the age structure of the world's population changes and the percentage of people age 60 or older increases, more countries will begin experiencing population declines. If population decline is gradual, its harmful effects usually can be managed.

Japan has the world's highest proportion of elderly people and the lowest proportion of young people. Its population, at 128 million in 2008, is projected to decline to about 96 million by 2050.

Rapid population decline can lead to severe economic and social problems. A country that experiences a fairly rapid "baby bust" or a "birth dearth" when its TFR falls below 1.5 children per couple for a prolonged period sees a sharp rise in the proportion of older people. This puts severe strains on government budgets because these individuals consume an increasingly larger share of medical care, social security funds, and other costly public services, which are funded by a decreasing number of working taxpayers. Such countries can also face labor shortages unless they rely more heavily on automation or massive immigration of foreign workers. In the next two to three decades, countries such as the United States and many European nations with rapidly aging populations will face shortages of health workers. For example, the United Nations will need half a million more nurses by 2020 and twice as many doctors specializing in health care for the elderly (geriatrics).

Figure 6-11 lists some of the problems associated with rapid population decline. Countries faced with a rapidly declining population in the future include Japan, Russia, Germany, Bulgaria, the Czech Republic, Hungary, Poland, Ukraine, Greece, Italy, and Spain.

Populations Can Decline from a Rising Death Rate: The AIDS Tragedy

A large number of deaths from AIDS can disrupt a country's social and economic structure by removing significant numbers of young adults from its age

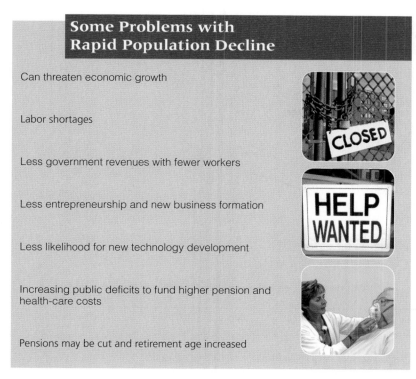

Some Problems with Rapid Population Decline

Can threaten economic growth

Labor shortages

Less government revenues with fewer workers

Less entrepreneurship and new business formation

Less likelihood for new technology development

Increasing public deficits to fund higher pension and health-care costs

Pensions may be cut and retirement age increased

Figure 6-11 Some problems with rapid population decline. **Question:** Which three of these problems do you think are the most important?

This change in the young-adult age structure of a country has a number of harmful effects. One is a sharp drop in average life expectancy. In 8 African countries, where 16–39% of the adult population is infected with HIV, life expectancy could drop to 34–40 years.

Another effect is a loss of a country's most productive young adult workers and trained personnel such as scientists, farmers, engineers, teachers, and government, business, and health-care workers. This causes a sharp drop in the number of productive adults available to support the young and the elderly and to grow food and provide essential services. Within a decade, countries such as Zimbabwe and Botswana in sub-Saharan Africa could lose more than a fifth of their adult populations.

Analysts call for the international community—especially developed countries—to create and fund a massive program to help countries ravaged by AIDS in Africa and elsewhere. This program would have two major goals. *First,* reduce the spread of HIV through a combination of improved education and health care. *Second,* provide financial assistance for education and health care as well as volunteer teachers and health-care and social workers to help compensate for the missing young-adult generation.

structure. According to the World Health Organization, AIDS had killed 25 million people by 2008. Unlike hunger and malnutrition, which kill mostly infants and children, AIDS kills many young adults.

6-4 How Can We Slow Human Population Growth?

▶ **CONCEPT 6-4** Experience indicates that the most effective ways to slow human population growth are to encourage family planning, to reduce poverty, and to elevate the status of women.

As Countries Develop, Their Populations Tend to Grow More Slowly

Demographers examining birth and death rates of western European countries that became industrialized during the 19th century developed a hypothesis of population change known as the **demographic transition:** as countries become industrialized, first their death rates and then their birth rates decline. According to the hypothesis, based on such data, this transition takes place in four distinct stages (Figure 6-12, p. 134).

Some analysts believe that most of the world's developing countries will make a demographic transi-

tion over the next few decades mostly because modern technology can bring economic development and family planning to such countries. Others fear that the still-rapid population growth in some developing countries might outstrip economic growth and overwhelm some local life-support systems. As a consequence, some of these countries could become caught in a *demographic trap* at stage 2. This is now happening as death rates rise in a number of developing countries, especially in Africa. Indeed, countries in Africa being ravaged by the HIV/AIDS epidemic are falling back to stage 1.

Other factors that could hinder the demographic transition in some developing countries are shortages of scientists and engineers (94% of them work in the industrialized world), shortages of skilled workers,

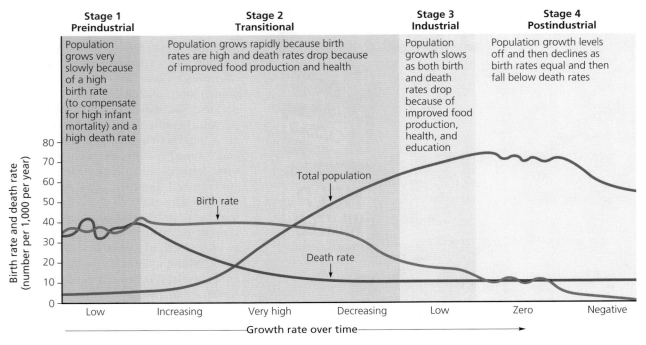

Stage 1 Preindustrial	Stage 2 Transitional	Stage 3 Industrial	Stage 4 Postindustrial
Population grows very slowly because of a high birth rate (to compensate for high infant mortality) and a high death rate	Population grows rapidly because birth rates are high and death rates drop because of improved food production and health	Population growth slows as both birth and death rates drop because of improved food production, health, and education	Population growth levels off and then declines as birth rates equal and then fall below death rates

CENGAGENOW™ **Active Figure 6-12** Four stages of the *demographic transition,* which the population of a country can experience when it becomes industrialized. There is uncertainty about whether this model will apply to some of today's developing countries. *See an animation based on this figure at* CengageNOW. **Question:** At what stage is the country where you live?

insufficient financial capital, large debts to developed countries, and a drop in economic assistance from developed countries since 1985.

CENGAGENOW™ Explore the effects of economic development on birth and death rates and population growth at CengageNOW.

Planning for Babies Works

Family planning provides educational and clinical services that help couples choose how many children to have and when to have them. Such programs vary from culture to culture, but most provide information on birth spacing, birth control, and health care for pregnant women and infants.

Family planning has been a major factor in reducing the number of births throughout most of the world, mostly because of increased knowledge and availability of contraceptives. According to the U.N. Population Division, 58% of married women ages 15–45 in developed countries and 54% in developing countries used modern contraception in 2008. Family planning has also reduced the number of legal and illegal abortions performed each year and decreased the number of deaths of mothers and fetuses during pregnancy.

Studies by the U.N. Population Division and other population agencies indicate that family planning is responsible for at least 55% of the drop in total fertility

rates (TFRs) in developing countries, from 6.0 in 1960 to 3.0 in 2008. Between 1971 and 2008, for example, Thailand used family planning to cut its annual population growth rate from 3.2% to 0.5% and its TFR from 6.4 to 1.6 children per family. Another family planning success involves Iran, which between 1989 and 2000, cut its population growth rate from 2.5% to 1.4%.

Despite such successes, two problems remain. *First,* according to the U.N. Population Fund, 42% of all pregnancies in developing countries are unplanned and 26% end with abortion. *Second,* an estimated 201 million couples in developing countries want to limit the number and determine the spacing of their children, but they lack access to family planning services. According to a recent study by the U.N. Population Fund and the Alan Guttmacher Institute, meeting women's current unmet needs for family planning and contraception could *each year* prevent 52 million unwanted pregnancies, 22 million induced abortions, 1.4 million infant deaths, and 142,000 pregnancy-related deaths.

Some analysts call for expanding family planning programs to include teenagers and sexually active unmarried women, who are excluded from many existing programs. Another suggestion is to develop programs that educate men about the importance of having fewer children and taking more responsibility for raising them. Proponents also call for greatly increased research on developing more effective and more acceptable birth control methods for men.

In 1994, the United Nations held its third Conference on Population and Development in Cairo, Egypt.

One of the conference's goals was to encourage action to stabilize the world's population at 7.8 billion by 2050 instead of the projected 9.2 billion.

The experiences of countries such as Japan, Thailand, South Korea, Taiwan, Iran, and China show that a country can achieve or come close to replacement-level fertility within a decade or two. Such experiences also suggest that the best ways to slow and stabilize population growth are through *investing in family planning, reducing poverty,* and *elevating the social and economic status of women* (**Concept 6-4**).

Empowering Women Can Slow Population Growth

Studies show that women tend to have fewer children if they are educated, hold a paying job outside the home, and live in societies where their human rights are not suppressed. Although women make up roughly half of the world's population, in most societies they do not have the same rights and educational and economic opportunities as men do.

Women do almost all of the world's domestic work and child care for little or no pay and provide more unpaid health care than all of the world's organized health services combined. They also do 60–80% of the work associated with growing food, gathering and hauling wood (Figure 6-13) and animal dung for use as fuel, and hauling water in rural areas of Africa, Latin America, and Asia. As one Brazilian woman put it, "For poor women the only holiday is when you are asleep."

Globally, women account for two-thirds of all hours worked but receive only 10% of the world's income,

and they own less than 2% of the world's land. Also, about 70% of the world's poor and 64% of all 800 million illiterate adults are women.

Because sons are more valued than daughters in many societies, girls are often kept at home to work instead of being sent to school. Globally, some 900 million girls—three times the entire U.S. population—do not attend elementary school. Teaching women to read has a major impact on fertility rates and population growth. Poor women who cannot read often have five to seven children, compared to two or fewer in societies where almost all women can read.

According to Thorya Obaid, executive director of the U.N. Population Fund, "Many women in the developing world are trapped in poverty by illiteracy, poor health, and unwanted high fertility. All of these contribute to environmental degradation and tighten the grip of poverty."

An increasing number of women in developing countries are taking charge of their lives and reproductive behavior. As it expands, such bottom-up change by individual women will play an important role in stabilizing population and reducing environmental degradation.

■ CASE STUDY
Slowing Population Growth in China: The One–Child Policy

China has made impressive efforts to feed its people, bring its population growth under control, and encourage economic growth. Between 1972 and 2008, the country cut its crude birth rate in half and trimmed its

Figure 6-13 Women from a village in the West African country of Burkina Faso returning with fuelwood. Typically they spend 2 hours a day two or three times a week searching for and hauling fuelwood.

Mark Edwards/Peter Arnold, Inc.

TFR from 5.7 to 1.6 children per woman, compared to 2.1 in the United States. Despite such drops China is the world's most populous country and in 2008 added about 6.8 million people to its population (compared to 2.9 million in the United States and 18 million in India). If current trends continue, China's population is expected to peak by about 2033 at around 1.46 billion and then to begin a slow decline.

Since 1980, China has moved 350 million people (an amount greater than the entire U.S. population) from extreme poverty to its consumer middle class and is likely to double that number by 2010. However, about 47% of its people were struggling to live on less than $2 (U.S.) a day in 2006. China also has a literacy rate of 91% and has boosted life expectancy to 73 years. By 2020, some economists project that China could become the world's leading economic power.

In the 1960s, Chinese government officials concluded that the only alternative to mass starvation was strict population control. To achieve a sharp drop in fertility, China established the most extensive, intrusive, and strict family planning and population control program in the world. It discourages premarital sex and urges people to delay marriage and limit their families to one child each. Married couples who pledge to have no more than one child receive more food, larger pensions, better housing, free health care, salary bonuses, free school tuition for their child, and preferential employment opportunities for their child. Couples who break their pledge lose such benefits.

The government also provides married couples with free sterilization, contraceptives, and abortion. However, reports of forced abortions and other coercive actions have brought condemnation from the United States and other national governments.

In China, there is a strong preference for male children, because unlike sons, daughters are likely to marry and leave their parents. A folk saying goes, "Rear a son, and protect yourself in old age." Some pregnant Chinese women use ultrasound to determine the gender of their fetuses, and some get an abortion if it is female. The result: a rapidly growing *gender imbalance* or "bride shortage" in China's population, with a projected 30–40 million surplus of men expected by 2020. Because of this skewed sex ratio, teen-age girls in some parts of rural China are being kidnapped and sold as brides for single men in other parts of the country.

With fewer children, the average age of China's population is increasing rapidly. By 2020, 31% of the Chinese population will be over 60 years old, compared to 8% in 2008. This graying of the Chinese population could lead to a declining work force, higher wages for younger workers, a shortage of funding for continuing economic development, and fewer children and grandchildren to care for the growing number of elderly people. These concerns and other factors may slow economic growth and lead to some relaxation of China's one-child population control policy in the future.

China also faces serious resource and environmental problems that could limit its economic growth. It has 19% of the world's population, but only 7% of the world's freshwater and cropland, 4% of its forests, and 2% of its oil. In 2002, only 15% of China's land area was protected on paper (compared to 23% in the United States, 51% in Japan, and 63% in Venezuela) and only 29% of its rural population had access to adequate sanitation.

In 2005, China's deputy minister of the environment summarized the country's environmental problems: "Our raw materials are scarce, we don't have enough land, and our population is constantly growing. Half of the water in our seven largest rivers is completely useless. One-third of the urban population is breathing polluted air."

China's economy is growing at one of the world's highest rates as the country undergoes rapid industrialization. More middle class Chinese (Case Study, p. 15) will consume more resources per person, increasing China's ecological footprint (Figure 1-10, p. 15) within its own borders and in other parts of the world that provide it with resources (**Concept 1-3**, p. 12). This will put a strain on the earth's natural capital unless China steers a course toward more sustainable economic development.

CONCEPT
LINK

■ CASE STUDY
Slowing Population Growth in India

For more than 5 decades, India has tried to control its population growth with only modest success. The world's first national family planning program began in India in 1952, when its population was nearly 400 million. By 2008, after 56 years of population control efforts, India had 1.1 billion people.

In 1952, India added 5 million people to its population. In 2008, it added 18 million—more than any other country. By 2015, India is projected to be the world's most populous country, with its population projected to reach 1.76 billion by 2050.

India faces a number of serious poverty, malnutrition, and environmental problems that could worsen as its population continues to grow rapidly. India has a thriving and rapidly growing middle class of more than 300 million people—roughly equal to the entire U.S. population—many of them highly skilled software developers and entrepreneurs.

By global standards, however, one of every four people in India is poor, despite the fact that since 2004 it has had the world's second fastest growing economy, and by 2007, was the world's fourth largest economy. Such prosperity and progress have not touched many of the nearly 650,000 villages where more than two-thirds of India's population lives. In 2002, only 18% of its rural population had access to adequate sanitation. In 2006, nearly half of the country's labor force was unemployed or underemployed and 80% of its people

were struggling to live on less than $2 (U.S.) day (see Photo 2 in the Detailed Contents).

Although India currently is self-sufficient in food grain production, about 40% of its population and more than half of its children suffer from malnutrition, mostly because of poverty. In 2002, only 5% of the country's land was protected on paper.

The Indian government has provided information about the advantages of small families for years and has also made family planning available throughout the country. Even so, Indian women have an average of 2.8 children. Most poor couples still believe they need many children to work and care for them in old age. As in China, the strong cultural preference for male children also means some couples keep having children until they produce one or more boys. The result: even though 90% of Indian couples know of at least one modern birth control method, only 48% actually use one.

Like China, India also faces critical resource and environmental problems. With 17% of the world's people, India has just 2.3% of the world's land resources and 2% of the forests. About half the country's cropland is degraded as a result of soil erosion and over-

grazing. In addition, more than two-thirds of its water is seriously polluted, sanitation services often are inadequate, and many of its major cities suffer from serious air pollution.

India is undergoing rapid economic growth, which is expected to accelerate. As members of its huge and growing middle class increase their resource use per person, India's ecological footprint (**Concept 1-3**) (Figure 1-10, p. 15) will expand and increase the pressure on the country's and the earth's natural capital.

On the other hand, economic growth may help to slow population growth by accelerating India's demographic transition. By 2050, India—the largest democracy the world has ever seen—could become the world's leading economic power.

THINKING ABOUT

China, India, the United States, and Overpopulation　　　CORE CASE STUDY

Based on population size and resource use per person (Figure 1-10, p. 15) is the United States more overpopulated than China? Explain. Answer the same question for the United States versus India.

REVISITING

Population Growth and Sustainability

This chapter began with a discussion of whether the world is overpopulated (**Core Case Study**). As we have noted, some experts say this is the wrong question to be asking. Instead, they believe we ought to ask, "What is the optimal level of human population that the planet can support *sustainably?*" In other words, "What is the maximum number of people that can live comfortably without seriously degrading the earth's biodiversity and other forms of natural capital and jeopardizing the earth's ability to provide the same comforts for future generations?"

In the first six chapters of this book, you have learned how ecosystems and species have been sustained throughout history in

keeping with four **scientific principles of sustainability**—relying on solar energy, biodiversity, population control, and nutrient recycling (see back cover and **Concept 1-6**, p. 23). In this chapter, you may have gained a sense of the need for humans to apply these sustainability principles to their lifestyles and economies, especially with regard to human population growth, globally and in particular countries.

In the next five chapters, you will learn how various principles of ecology and these four **scientific principles of sustainability** can be applied to help preserve the earth's biodiversity.

Our numbers expand but Earth's natural systems do not.

LESTER R. BROWN

REVIEW

1. Review the Key Questions and Concepts in this chapter on p. 123. Do you think the world is overpopulated? Explain.

2. List three factors that account for the rapid growth of the world's human population over the past 200 years. Describe eight ways in which we have used technology to alter nature to meet our growing needs and wants. How many of us are likely to be here in 2050?

3. What is the **cultural carrying capacity** of a population? How do some analysts apply this concept in considering the question of whether the earth is overpopulated?

4. List four variables that affect the **population change** of an area and write an equation showing how they are related. Distinguish between **crude birth rate** and **crude death rate.** What five countries had the largest numbers of people in 2008?

5. What is **fertility rate?** Distinguish between **replacement-level fertility rate** and **total fertility rate (TFR)**. Explain why reaching the replacement-level fertility rate will not stop global population growth until about 50 years have passed (assuming that death rates do not rise).

6. Describe population growth in the United States and explain why it is high compared to those of most other developed countries and China. Is the United States over-populated? Explain.

7. List ten factors that can affect the birth rate and fertility rate of a country. Distinguish between **life expectancy** and **infant mortality rate** and explain how they affect the population size of a country. Why does the United States have a lower life expectancy and higher infant mortality rate than a number of other countries? What is **migration?** Describe immigration into the United States and the issues it raises.

8. What is the **age structure** of a population. Explain how it affects population growth and economic growth. What

are some problems related to rapid population decline from an aging population?

9. What is the **demographic transition** and what are its four stages? What factors could hinder some developing countries from making this transition? What is **family planning?** Describe the roles of family planning, reducing poverty, and elevating the status of women in slowing population growth. Describe China's and India's efforts to control their population growth.

10. How has human population growth (**Core Case Study**) interfered with natural processes related to three of the **scientific principles of sustainability**? Name the three principles, and for each one, describe the effects of rapid human population growth.

Note: Key Terms are in **bold** type.

CRITICAL THINKING

1. List three ways in which you could apply some of what you learned in this chapter to making your lifestyle more environmentally sustainable.

2. Which of the three major environmental worldviews summarized on p. 20 do you believe underlie the two major positions on whether the world is overpopulated (**Core Case Study**)?

3. Identify a major local, national, or global environmental problem, and describe the role of population growth in this problem.

4. Is it rational for a poor couple in a developing country such as India to have four or five children? Explain.

5. Do you believe that the population is too high in **(a)** the world (**Core Case Study**), **(b)** your own country, and **(c)** the area where you live? Explain.

6. Should everyone have the right to have as many children as they want? Explain. Is your belief on this issue consistent with your environmental worldview?

7. Some people have proposed that the earth could solve its population problem by shipping people off to space colonies, each containing about 10,000 people. Assuming we could build such large-scale, self-sustaining space stations (a big assumption), how many of them would we have to build to provide living spaces for the 82 million people added to the earth's population this year? If space shuttles could each carry 100 passengers, how many shuttles would have to be launched each day for a year to

offset the 82 million people added to the population this year? According to your calculations, determine whether this proposal is a logical solution to the earth's population problem. What effect might the daily launching of these shuttles have on global warming? Explain.

8. Some people believe our most important goal should be to sharply reduce the rate of population growth in developing countries where 97% of the world's population growth is expected to take place. Others argue that the most serious environmental problems stem from high levels of resource consumption per person in developed countries, which use 88% of the world's resources and have much larger ecological footprints per person (Figure 1-10, p. 15) than do developing countries. What is your view on this issue? Explain.

9. Congratulations! You are in charge of the world. List the three most important features of your population policy.

10. List two questions that you would like to have answered as a result of reading this chapter.

Note: See Supplement 13 (p. S78) for a list of Projects related to this chapter.

The chart below shows selected population data for two different countries A and B.

	Country A	Country B
Population (millions)	144	82
Crude birth rate	43	8
Crude death rate	18	10
Infant mortality rate	100	3.8
Total fertility rate	5.9	1.3
Percentage of population under 15 years old	45	14
Percentage of population older than 65 years	3.0	19
Average life expectancy at birth	47	79
Percentage urban	44	75

Source: Data from Population Reference Bureau 2007, *World Population Data Sheet.*

1. Calculate the rates of natural increase for the populations of Country A and Country B. From the rates of natural increase and the data in the table, suggest whether A and B are developed or developing countries, and explain the reasons for your answers.

2. Describe where each of the two countries may be in terms of their stage in the demographic transition (Figure 6-12, p. 134). Discuss factors that could hinder Country A from progressing to later stages in the demographic transition.

3. Explain how the percentage of people under 15 years of age in Country A and in Country B could affect the per capita and total ecological footprints of each country.

Climate and Terrestrial Biodiversity

Blowing in the Wind: Connections between Wind, Climate, and Biomes

Terrestrial biomes such as de serts, grasslands, and forests make up one of the components of the earth's biodiversity (Figure 4-2, p. 79). Why is one area of the earth's land surface a desert, another a grassland, and another a forest? The general answer lies in differences in *climate,* resulting mostly from long-term differences in average temperature and precipitation caused by global air circulation, which we discuss in this chapter.

Wind—an indirect form of solar energy—is an important factor in the earth's climate. It is part of the planet's circulatory system for heat, moisture, plant nutrients, soil particles, and long-lived air pollutants. Without wind, the tropics would be unbearably hot and most of the rest of the planet would freeze.

Winds transport nutrients from one place to another. For example, winds carry dust that is rich in phosphates and iron across the Atlantic Ocean from the Sahara Desert in West Africa (Figure 7-1). These deposits help to build agricultural soils in the Bahamas and to supply nutrients for plants in the upper canopies of rain forests in Brazil. A 2007 study indicated that in 2006, such dust might have helped to reduce hurricane frequency in the southwestern North Atlantic and Caribbean by blocking some energizing sunlight. Dust blown from China's Gobi Desert deposits iron into the Pacific Ocean between Hawaii and Alaska. The iron stimulates the growth of phytoplankton, the minute producers that support ocean food webs. Wind is also a rapidly growing source of renewable energy, as discussed in Chapter 16.

Winds also have a downside. They transport harmful substances. Particles of reddish-brown soil and pesticides banned in the United States are blown from Africa's deserts and eroding farmlands into the sky over the U.S. state of Florida. Some types of fungi in this dust may play a role in degrading or killing coral reefs in the Florida Keys and in the Caribbean.

Particles of iron-rich dust from Africa also enhance the productivity of algae, and have been linked to outbreaks of toxic algal blooms—referred to as *red tides*—in Florida's coastal waters. People who eat shellfish contaminated by a toxin produced in red tides can become paralyzed or can even die. These red tides can also cause fish kills.

Dust, soot, and other long-lived air pollutants from rapidly industrializing China and central Asia are blown across the Pacific Ocean and degrade air quality over parts of the western United States. Asian pollution makes up as much as 10% of West Coast smog—a problem that is expected to get worse as China continues to industrialize. A 2007 study, led by Renyl Zhang, linked such Asian air pollution to intensified storms over the North Pacific Ocean and to increased warming in the polar regions.

The ecological lesson: *Everything we do affects some other part of the biosphere* because *everything is connected.* In this chapter, we examine the key role that climate, including winds, plays in the formation and location of the deserts, grasslands, and forests that make up an important part of the earth's terrestrial biodiversity.

Seawifs Project/Nasa/GSFC. Nasa

Figure 7-1 Some of the dust blown from West Africa, shown here, can end up as soil nutrients in Amazonian rain forests and toxic air pollutants in the U.S. state of Florida and in the Caribbean. It may also help to suppress hurricanes in the western Atlantic.

7-1 What factors influence climate?

CONCEPT 7-1 An area's climate is determined mostly by solar radiation, the earth's rotation, global patterns of air and water movement, gases in the atmosphere, and the earth's surface features.

7-2 How does climate affect the nature and locations of biomes?

CONCEPT 7-2 Differences in average annual precipitation and temperature lead to the formation of tropical, temperate, and cold deserts, grasslands, and forests, and largely determine their locations.

7-3 How have we affected the world's terrestrial ecosystems?

CONCEPT 7-3 In many areas, human activities are impairing ecological and economic services provided by the earth's deserts, grasslands, forests, and mountains.

Note: Supplements 2 (p. S4), 4 (p. S20), 5 (p. S31), 8 (p. 47), 10 (p. S59), and 13 (p. S78) can be used with this chapter.

To do science is to search for repeated patterns,
not simply to accumulate facts,
and to do the science of geographical ecology
is to search for patterns of plant and animal life
that can be put on a map.

ROBERT H. MACARTHUR

7-1 What Factors Influence Climate?

▶ **CONCEPT 7-1** An area's climate is determined mostly by solar radiation, the earth's rotation, global patterns of air and water movement, gases in the atmosphere, and the earth's surface features.

The Earth Has Many Different Climates

Weather is a local area's short-term temperature, precipitation, humidity, wind speed, cloud cover, and other physical conditions of the lower atmosphere as measured over hours or days. (Supplement 8, p. S47, introduces you to weather basics.) **Climate** is an area's general pattern of atmospheric or weather conditions measured over long periods of time ranging from decades to thousands of years. As American writer and humorist Mark Twain once said, "Climate is what we expect, weather is what we get." Figure 7-2 (p. 142), depicts the earth's major climate zones, an important component of the earth's natural capital (Figure 1-3, p. 8).

Climate varies in different parts of the earth mostly because patterns of global air circulation and ocean currents distribute heat and precipitation unevenly (Figure 7-3, p. 142). Three major factors determine how air circulates in the lower atmosphere, which helps to distribute heat and moisture from the tropics to other parts of the world:

- *Uneven heating of the earth's surface by the sun.* Air is heated much more at the equator, where the sun's rays strike directly, than at the poles, where sunlight strikes at a slanted angle and spreads out over a much greater area (Figure 7-3, right). These differences in the distribution of incoming solar energy help to explain why tropical regions near the equator are hot, why polar regions are cold, and why temperate regions in between generally have intermediate average temperatures.

- *Rotation of the earth on its axis.* As the earth rotates around its axis, its equator spins faster than its polar regions. As a result, heated air masses rising above the equator and moving north and south to cooler areas are deflected to the west or east over different parts of the planet's surface (Figure 7-3). The atmosphere over these different areas is divided into huge regions called *cells,* distinguished by direction of air movement. And the differing directions of air movement are called *prevailing winds*—major surface winds that blow almost continuously

| Polar (ice) | Subarctic (snow) | Cool temperate | Highland | ⟵ Warm ocean current | ⟿ River |
| Warm temperate | Dry | Tropical | Major upwelling zones | ⟵ Cold ocean current | |

CENGAGENOW™ **Active Figure 7-2 Natural capital:** generalized map of the earth's current climate zones, showing the major contributing ocean currents and drifts and upwelling areas (where currents bring nutrients from the ocean bottom to the surface). Winds play an important role in distributing heat and moisture in the atmosphere, which leads to such climate zones. Winds also cause currents that help distribute heat throughout the world's oceans. *See an animation based on this figure at* CengageNOW™. **Question:** Based on this map what is the general type of climate where you live?

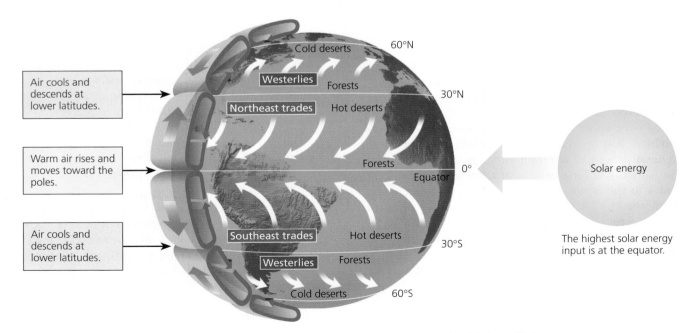

Figure 7-3 *Global air circulation.* The largest input of solar energy occurs at the equator. As this air is heated it rises and moves toward the poles. However, the earth's rotation deflects the movement of the air over different parts of the earth. This creates global patterns of prevailing winds that help distribute heat and moisture in the atmosphere.

and help distribute air, heat, moisture, and dust over the earth's surface (**Core Case Study**).

- *Properties of air, water, and land.* Heat from the sun evaporates ocean water and transfers heat from the oceans to the atmosphere, especially near the hot equator. This evaporation of water creates giant cyclical convection cells that circulate air, heat, and moisture both vertically and from place to place in the atmosphere, as shown in Figure 7-4.

Prevailing winds (Figure 7-3) blowing over the oceans produce mass movements of surface water called **currents.** Driven by prevailing winds and the earth's rotation, the earth's major ocean currents (Figure 7-2) redistribute heat from the sun from place to place, thereby influencing climate and vegetation, especially near coastal areas.

The oceans absorb heat from the earth's air circulation patterns; most of this heat is absorbed in tropical waters, which receive most of the sun's heat. This heat and differences in water *density* (mass per unit volume) create warm and cold ocean currents. Prevailing winds and irregularly shaped continents interrupt these cur-

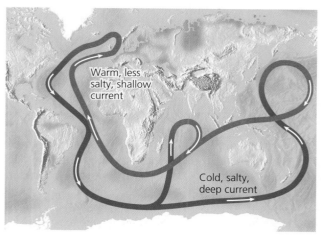

Figure 7-5 *Connected deep and shallow ocean currents.* A connected loop of shallow and deep ocean currents transports warm and cool water to various parts of the earth. This loop, which rises in some areas and falls in others, results when ocean water in the North Atlantic near Iceland is dense enough (because of its salt content and cold temperature) to sink to the ocean bottom, flow southward, and then move eastward to well up in the warmer Pacific. A shallower return current aided by winds then brings warmer, less salty—and thus less dense—water to the Atlantic. This water can cool and sink to begin this extremely slow cycle again. **Question:** How do you think this loop affects the climates of the coastal areas around it?

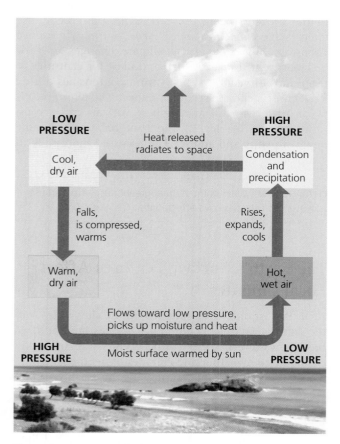

Figure 7-4 Energy transfer by convection in the atmosphere. *Convection* occurs when hot and wet warm air rises, cools, and releases heat and moisture as precipitation (right side). Then the denser cool, dry air sinks, gets warmer, and picks up moisture as it flows across the earth's surface to begin the cycle again.

rents and cause them to flow in roughly circular patterns between the continents, clockwise in the northern hemisphere and counterclockwise in the southern hemisphere.

Heat is also distributed to the different parts of the ocean and the world when ocean water mixes vertically in shallow and deep ocean currents, mostly as a result of differences in the density of seawater. Because it has a higher density, colder seawater sinks and flows beneath warmer and less dense seawater. This creates a connected loop of deep and shallow ocean currents, which act like a giant conveyer belt that moves heat to and from the deep sea and transfers warm and cold water between the tropics and the poles (Figure 7-5).

The ocean and the atmosphere are strongly linked in two ways: ocean currents are affected by winds in the atmosphere (**Core Case Study**), and heat from the ocean affects atmospheric circulation (Figure 7-4). One example of the interactions between the ocean and the atmosphere is the *El Niño–Southern Oscillation,* or *ENSO,* as discussed on pp. S48–S49 in Supplement 8 and in *The Habitable Planet,* Video 3, at **www.learner.org/resources/series209.html**. This large-scale weather phenomenon occurs every few years when prevailing winds in the tropical Pacific Ocean weaken and change direction. The resulting above-average warming of Pacific waters can affect populations of marine species by changing the distribution of plant nutrients. It also alters the weather of at least two-thirds of the earth for one or two years (see Figure 5, p. S49, in Supplement 8).

CENGAGENOW˜ Learn more about how oceans affect air movements where you live and all over the world at CengageNOW™.

The earth's air circulation patterns, prevailing winds, and configuration of continents and oceans result in six giant convection cells (like the one shown in Figure 7-4) in which warm, moist air rises and cools, and cool, dry air sinks. Three of these cells are found north of the equator and three are south of the equator. These cells lead to an irregular distribution of climates and deserts, grasslands, and forests, as shown in Figure 7-6 (**Concept 7-1**).

CENGAGENOW˜ Watch the formation of six giant convection cells and learn more about how they affect climates at CengageNOW.

THINKING ABOUT
Winds and Biomes

CORE
CASE
STUDY

How might the distribution of the world's forests, grasslands, and deserts shown in Figure 7-6 differ if the prevailing winds shown in Figure 7-3 did not exist?

Greenhouse Gases Warm the Lower Atmosphere

Figure 3-8 (p. 56) shows how energy flows to and from the earth. Small amounts of certain gases, including water vapor (H_2O), carbon dioxide (CO_2), methane

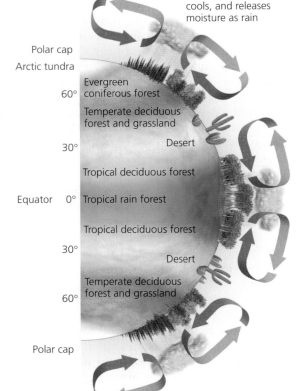

Figure 7-6 *Global air circulation, ocean currents, and biomes.* Heat and moisture are distributed over the earth's surface via six giant convection cells (like the one in Figure 7-4) at different latitudes. The resulting uneven distribution of heat and moisture over the planet's surface leads to the forests, grasslands, and deserts that make up the earth's terrestrial biomes.

Moist air rises, cools, and releases moisture as rain

Polar cap
Arctic tundra
Evergreen coniferous forest
60°
Temperate deciduous forest and grassland
30°
Desert
Tropical deciduous forest
Equator 0° Tropical rain forest
Tropical deciduous forest
30°
Desert
Temperate deciduous forest and grassland
60°
Polar cap

(CH_4), and nitrous oxide (N_2O), in the atmosphere play a role in determining the earth's average temperatures and its climates. These **greenhouse gases** allow mostly visible light and some infrared radiation and ultraviolet (UV) radiation from the sun to pass through the atmosphere. The earth's surface absorbs much of this solar energy and transforms it to longer-wavelength infrared radiation (heat), which then rises into the lower atmosphere.

Some of this heat escapes into space, but some is absorbed by molecules of greenhouse gases and emitted into the lower atmosphere as even longer-wavelength infrared radiation. Some of this released energy radiates into space, and some warms the lower atmosphere and the earth's surface. This natural warming effect of the troposphere is called the **greenhouse effect** (see Figure 3-8, p. 56, and *The Habitable Planet,* Video 2, at **www.learner.org/resources/series209.html**). Without the warming caused by these greenhouse gases, the earth would be a cold and mostly lifeless planet.

Human activities such as burning fossil fuels, clearing forests, and growing crops release carbon dioxide, methane, and nitrous oxide into the atmosphere. Considerable evidence and climate models indicate that there is a 90–99% chance that the large inputs of greenhouse gases into the atmosphere from human activities are enhancing the earth's natural greenhouse effect. This *human-enhanced global warming* (Science Focus, p. 33) could cause climate changes in various places on the earth that could last for centuries to thousands of years. As this warming intensifies during this century, climate scientists expect it to alter precipitation patterns, shift areas where we can grow crops, raise average sea levels, and shift habitats for some types of plants and animals, as discussed more fully in Chapter 19.

CENGAGENOW˜ Witness the natural greenhouse effect and see how human activities have affected it at CengageNOW.

The Earth's Surface Features Affect Local Climates

Heat is absorbed and released more slowly by water than by land. This difference creates land and sea breezes. As a result, the world's oceans and large lakes moderate the weather and climates of nearby lands.

Various topographic features of the earth's surface create local and regional weather and climatic conditions that differ from the general climate of a region. For example, mountains interrupt the flow of prevailing surface winds and the movement of storms. When moist air blowing inland from an ocean reaches a mountain range, it is forced upward. As it rises, it cools and expands and then loses most of its moisture as rain and snow on the windward slope of the mountain (the side from which the wind is blowing).

Prevailing winds pick up moisture from an ocean.

On the windward side of a mountain range, air rises, cools, and releases moisture.

On the leeward side of the mountain range, air descends, warms, and releases little moisture.

Figure 7-7 The *rain shadow effect* is a reduction of rainfall and loss of moisture from the landscape on the side of a mountain facing away from prevailing surface winds. Warm, moist air in onshore winds loses most of its moisture as rain and snow on the windward slopes of a mountain range. This leads to semiarid and arid conditions on the leeward side of the mountain range and the land beyond. The Mojave Desert in the U.S. state of California and Asia's Gobi Desert are both produced by this effect.

As the drier air mass passes over the mountaintops it flows down the leeward (away from the wind) slopes, warms up (which increases its ability to hold moisture), and sucks up moisture from the plants and soil below. The loss of moisture from the landscape and the resulting semiarid or arid conditions on the leeward side of high mountains create the **rain shadow effect** (Figure 7-7). Sometimes this leads to the formation of deserts such as Death Valley in the United States, which is in the rain shadow of Mount Whitney, the highest mountain in the Sierra Nevadas. In this way, winds (**Core Case Study**) play a key role in forming some of the earth's deserts.

Cities also create distinct microclimates. Bricks, concrete, asphalt, and other building materials absorb and hold heat, and buildings block wind flow. Motor vehi-

cles and the climate control systems of buildings release large quantities of heat and pollutants. As a result, cities tend to have more haze and smog, higher temperatures, and lower wind speeds than the surrounding countryside.

THINKING ABOUT

Winds and Your Life

CORE CASE STUDY

What are three changes in your lifestyle that would take place if there were no winds where you live?

RESEARCH FRONTIER

Modeling and other research to learn more about how human activities affect climate. See **academic.cengage.com/biology/miller**.

7-2 How Does Climate Affect the Nature and Locations of Biomes?

▶ **CONCEPT 7-2** Differences in average annual precipitation and temperature lead to the formation of tropical, temperate, and cold deserts, grasslands, and forests, and largely determine their locations.

Climate Affects Where Organisms Can Live

Different climates (Figure 7-2) explain why one area of the earth's land surface is a desert, another a grassland, and another a forest (Figure 7-6) and why global air circulation (Figure 7-3) accounts for different types of deserts, grasslands, and forests (**Concept 7-2**).

Figure 7-8 (p. 146) shows how scientists have divided the world into several major **biomes**—large ter-

restrial regions characterized by similar climate, soil, plants, and animals, regardless of where they are found in the world. The variety of terrestrial biomes and aquatic systems is one of the four components of the earth's biodiversity (Figure 4-2, p. 79, and **Concept 4-1A**, p. 78)—a vital part of the earth's natural capital.

CONCEPT LINK

By comparing Figure 7-8 with Figure 7-2 and Figure 1 on pp. S20–S21 in Supplement 4, you can see how the world's major biomes vary with climate.

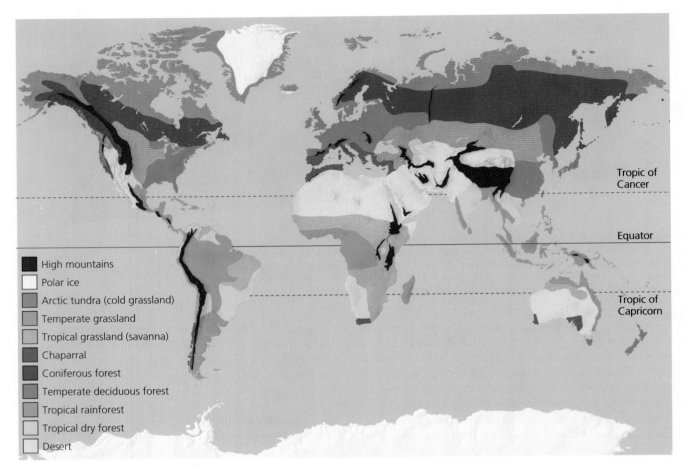

High mountains
Polar ice
Arctic tundra (cold grassland)
Temperate grassland
Tropical grassland (savanna)
Chaparral
Coniferous forest
Temperate deciduous forest
Tropical rainforest
Tropical dry forest
Desert

Tropic of Cancer

Equator

Tropic of Capricorn

CENGAGENOW **Active Figure 7-8 Natural capital:** the earth's major *biomes*—the main types of natural vegetation in various undisturbed land areas—result primarily from differences in climate. Each biome contains many ecosystems whose communities have adapted to differences in climate, soil, and other environmental factors. Figure 5 on p. S27 in Supplement 4 shows the major biomes of North America. Human activities have removed or altered much of the natural vegetation in some areas for farming, livestock grazing, lumber and fuelwood, mining, and construction of towns and cities (see Figure 3, pp. S24–S25, and Figure 7, pp. S28–S29, in Supplement 4). *See an animation based on this figure at* CengageNOW. **Question:** If you factor out human influences such as farming and urban areas, what kind of biome do you live in?

Figure 3-7 (p. 55) shows how major biomes in the United States (Figure 5, p. S27, in Supplement 4) are related to its different climates.

On maps such as the one in Figure 7-8, biomes are shown with sharp boundaries, each being covered with one general type of vegetation. In reality, *biomes are not uniform.* They consist of a *mosaic of patches*, each with somewhat different biological communities but with similarities typical of the biome. These patches occur mostly because the resources that plants and animals need are not uniformly distributed and because human activities remove and alter the natural vegetation in many areas.

Figure 7-9 shows how climate and vegetation vary with *latitude* and *elevation*. If you climb a tall mountain from its base to its summit, you can observe changes in plant life similar to those you would encounter in traveling from the equator to one of the earth's poles. For example, if you hike up a tall Andes mountain in

Ecuador, your trek can begin in tropical rain forest and end up on a glacier at the summit.

THINKING ABOUT
Biomes, Climate, and Human Activities

Use Figure 7-2 to determine the general type of climate where you live and Figure 7-8 to determine the general type of biome that should exist where you live. Then use Figure 3, pp. S24–S25, and Figure 7, pp. S28–S29, in Supplement 4 to determine how human ecological footprints have affected the general type of biome where you live.

Differences in climate, mostly from average annual precipitation and temperature, lead to the formation of tropical (hot), temperate (moderate), and polar (cold) deserts, grasslands, and forests (Figure 7-10)—another important component of the earth's natural capital (**Concept 7-2**).

Figure 7-9 Generalized effects of elevation (left) and latitude (right) on climate and biomes. Parallel changes in vegetation type occur when we travel from the equator to the poles or from lowlands to mountaintops. **Question:** How might the components of the left diagram change as the earth warms during this century? Explain.

Figure 7-10 **Natural capital:** average precipitation and average temperature, acting together as limiting factors over a long time, help to determine the type of desert, grassland, or forest biome in a particular area. Although each actual situation is much more complex, this simplified diagram explains how climate helps to determine the types and amounts of natural vegetation found in an area left undisturbed by human activities. (Used by permission of Macmillan Publishing Company, from Derek Elsom, *The Earth,* New York: Macmillan, 1992. Copyright ©1992 by Marshall Editions Developments Limited).

SCIENCE FOCUS

Staying Alive in the Desert

Adaptations for survival in the desert have two themes: *beat the heat*, and *every drop of water counts*.

Desert plants have evolved a number of strategies for doing this. During long hot and dry spells, plants such as mesquite and creosote drop their leaves to survive in a dormant state. *Succulent (fleshy) plants*, such as the saguaro ("sah-WAH-ro") cactus (Figure 7-11, middle photo), have three adaptations: they have no leaves, which can lose water by evapotranspiration; they store water and synthesize food in their expandable, fleshy tissue; and they reduce water loss by opening their pores to take up carbon dioxide (CO_2) only at night. The spines of these and many other desert plants guard them from being eaten by herbivores seeking the precious water they hold.

Some desert plants use deep roots to tap into groundwater. Others such as prickly pear and saguaro cacti use widely spread, shallow roots to collect water after brief showers and store it in their spongy tissue.

Evergreen plants conserve water by having wax-coated leaves that reduce water loss. Others, such as annual wildflowers and grasses, store much of their biomass in seeds that remain inactive, sometimes for years, until they receive enough water to germinate. Shortly after a rain, these seeds germinate, grow, and carpet some deserts with dazzling arrays of colorful flowers that last for a few weeks.

Most desert animals are small. Some beat the heat by hiding in cool burrows or rocky crevices by day and coming out at night or in the early morning. Others become dormant during periods of extreme heat or drought. Some larger animals such as camels can drink massive quantities of water when it is available and store it in their fat for use as needed. The camel is also covered with dense hair and does not sweat, which reduces heat gain and water loss through evaporation. Kangaroo rats never drink water. They get the water they need by breaking down fats in seeds that they consume.

Insects and reptiles (such as rattlesnakes and Gila monsters) have thick outer coverings to minimize water loss through evaporation, and their wastes are dry feces and a dried concentrate of urine. Many spiders and insects get their water from dew or from the food they eat.

Critical Thinking

What are three things you would do to survive in the open desert?

There Are Three Major Types of Deserts

In a **desert**, annual precipitation is low and often scattered unevenly throughout the year. During the day, the baking sun warms the ground and causes evaporation of moisture from plant leaves and soil. But at night, most of the heat stored in the ground radiates quickly into the atmosphere. Desert soils have little vegetation and moisture to help store the heat, and the skies above deserts are usually clear. This explains why, in a desert, you may roast during the day but shiver at night.

A combination of low rainfall and different average temperatures creates tropical, temperate, and cold deserts (Figures 7-10 and 7-11).

Tropical deserts (Figure 7-11, top photo), such as the Sahara and Namib of Africa, are hot and dry most of the year (Figure 7-11, top graph). They have few plants and a hard, windblown surface strewn with rocks and some sand. They are the deserts we often see in the movies. Wind-blown dust storms (**Core Case Study**) in the Sahara Desert have increased tenfold since 1950 mostly because of overgrazing and drought due to climate change and human population growth. Another reason is the *SUV connection*. Increasing numbers of four-wheel vehicles speeding over the sand (Figure 7-11, top photo) break the desert's surface crust. Wind storms can then blow the dusty material into the atmosphere.

In *temperate deserts* (Figure 7-11, center photo), such as the Mojave in the southern part of the U. S. state of California, daytime temperatures are high in summer and low in winter and there is more precipitation than in tropical deserts (Figure 7-11, center graph). The sparse vegetation consists mostly of widely dispersed, drought-resistant shrubs and cacti or other succulents adapted to the lack of water and temperature variations. (Figure 1, p. S53, in Supplement 9 shows some components and food web interactions in a temperate desert ecosystem.)

In *cold deserts*, such as the Gobi Desert in Mongolia, vegetation is sparse (Figure 7-11, bottom photo). Winters are cold, summers are warm or hot, and precipitation is low (Figure 7-11, bottom graph). Desert plants and animals have adaptations that help them to stay cool and to get enough water to survive (Science Focus, above).

Desert ecosystems are fragile. Their soils take decades to hundreds of years to recover from disturbances such as off-road vehicles. This is because of their slow plant growth, low species diversity, slow nutrient cycling (due to low bacterial activity in the soils), and lack of water.

> **THINKING ABOUT**
> **Winds and Deserts**
>
> What roles do winds (**Core Case Study**) play in creating and sustaining deserts?

CORE CASE STUDY

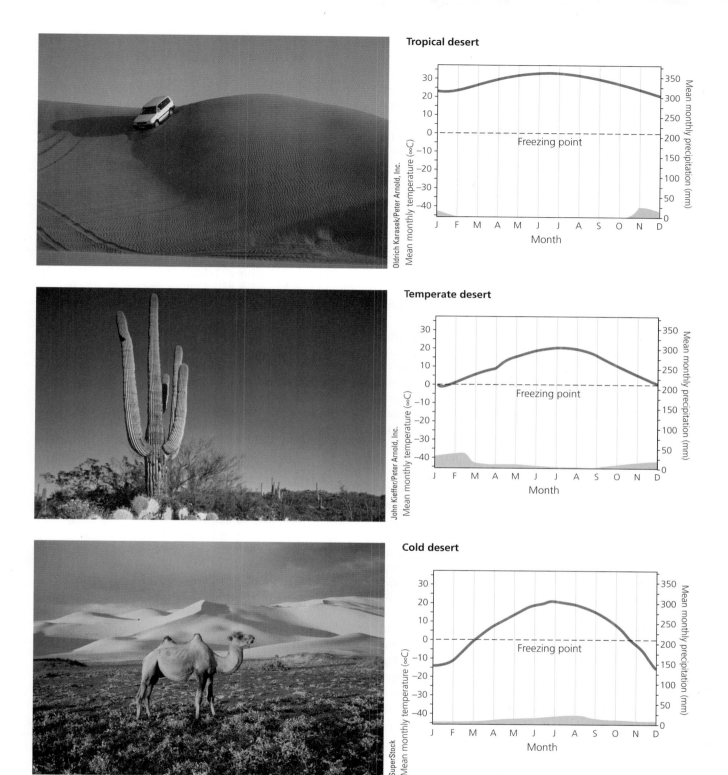

Figure 7-11 Climate graphs showing typical variations in annual temperature (red) and precipitation (blue) in tropical, temperate, and cold deserts. Top photo: a popular (but destructive) SUV rodeo in United Arab Emirates (tropical desert). Center photo: saguaro cactus in the U.S. state of Arizona (temperate desert). Bottom photo: a Bactrian camel in Mongolia's Gobi Desert (cold desert). **Question:** What month of the year has the highest temperature and the lowest rainfall for each of the three types of deserts?

There Are Three Major Types of Grasslands

Grasslands occur mostly in the interiors of continents in areas too moist for deserts and too dry for forests (Figure 7-8). Grasslands persist because of a combination of seasonal drought, grazing by large herbivores, and occasional fires—all of which keep large numbers of shrubs and trees from growing.

The three main types of grassland—tropical, temperate, and cold (arctic tundra)—result from combinations of low average precipitation and various average temperatures (Figures 7-10 and 7-12).

One type of tropical grassland, called a *savanna*, contains widely scattered clumps of trees such as acacia, (Figure 7-12, top photo), which are covered with thorns that help to keep herbivores away. This biome usually has warm temperatures year-round and alternating dry and wet seasons (Figure 7-12, top graph).

Tropical savannas in East Africa have herds of *grazing* (grass- and herb-eating) and *browsing* (twig- and leaf-nibbling) hoofed animals, including wildebeests (Figure 7-12, top photo), gazelles, zebras, giraffes, and antelopes and their predators such as lions, hyenas, and humans. Herds of these grazing and browsing animals migrate to find water and food in response to seasonal and year-to-year variations in rainfall (Figure 7-12, blue region in top graph) and food availability.

In their niches, these and other large herbivores have evolved specialized eating habits that minimize competition among species for the vegetation found on the savanna. For example, giraffes eat leaves and shoots from the tops of trees, elephants eat leaves and branches farther down, wildebeests prefer short grasses, and zebras graze on longer grasses and stems. Savanna plants, like desert plants, are adapted to survive drought and extreme heat. Many have deep roots that can tap into groundwater.

In a *temperate grassland,* winters are bitterly cold, summers are hot and dry, and annual precipitation is fairly sparse and falls unevenly through the year (Figure 7-12, center graph). Because the aboveground parts of most of the grasses die and decompose each year, organic matter accumulates to produce a deep, fertile soil. This soil is held in place by a thick network of intertwined roots of drought-tolerant grasses (unless the topsoil is plowed up, which exposes it to be blown away by high winds found in these biomes). The natural grasses are also adapted to fires, which burn the plant parts above the ground but do not harm the roots, from which new grass can grow.

Two types of temperate grasslands are *tall-grass prairies* and *short-grass prairies* (Figure 7-12, center photo), such as those of the Midwestern and western United States and Canada. Short-grass prairies typically get about 25 centimeters (10 inches) of rain a year, and the grasses have short roots. Tall-grass prairies can get up to 88 centimeters (35 inches) of rain per year, and the grasses have deep roots. Mixed or middle-grass prairies get annual rainfall between these two extremes.

In all prairies, winds blow almost continuously and evaporation is rapid, often leading to fires in the summer and fall. This combination of winds and fires helps to maintain such grasslands by hindering tree growth. (Figure 2, p. S54, in Supplement 9 shows some components and food-web interactions in a temperate tall-grass prairie ecosystem in North America.)

Many of the world's natural temperate grasslands have disappeared because their fertile soils are useful for growing crops (Figure 7-13, p. 152) and grazing cattle.

Cold grasslands, or *arctic tundra* (Russian for "marshy plain"), lie south of the arctic polar ice cap (Figure 7-8). During most of the year, these treeless plains are bitterly cold (Figure 7-12, bottom graph), swept by frigid winds, and covered by ice and snow. Winters are long and dark, and scant precipitation falls mostly as snow.

Under the snow, this biome is carpeted with a thick, spongy mat of low-growing plants, primarily grasses, mosses, lichens, and dwarf shrubs (Figure 7-12, bottom photo, and Figure 1-17, p. 23). Trees and tall plants cannot survive in the cold and windy tundra because they would lose too much of their heat. Most of the annual growth of the tundra's plants occurs during the 7- to 8-week summer, when the sun shines almost around the clock. (Figure 3, p. S55, in Supplement 9 shows some components and food-web interactions in an arctic tundra ecosystem.)

One outcome of the extreme cold is the formation of **permafrost,** underground soil in which captured water stays frozen for more than 2 consecutive years. During the brief summer, the permafrost layer keeps melted snow and ice from soaking into the ground. As a consequence, many shallow lakes, marshes, bogs, ponds, and other seasonal wetlands form when snow and frozen surface soil melt on the waterlogged tundra (Figure 7-12, bottom photo). Hordes of mosquitoes, black flies, and other insects thrive in these shallow surface pools. They serve as food for large colonies of migratory birds (especially waterfowl) that return from the south to nest and breed in the bogs and ponds.

Animals in this biome survive the intense winter cold through adaptations such as thick coats of fur (arctic wolf, arctic fox, and musk oxen) and feathers (snowy owl) and living underground (arctic lemming). In the summer, caribou migrate to the tundra to graze on its vegetation (Figure 1-17, p. 23).

Global warming is causing some of the permafrost in parts of Canada, Alaska, China, Russia, and Mongolia to melt. This disrupts these ecosystems and releases methane (CH_4) and carbon dioxide (CO_2) from the soil into the atmosphere. These two greenhouse gases can accelerate global warming and cause more permafrost to melt, which can lead to further warming and climate change. The melting permafrost causes the soil to sink

Tropical grassland (savanna)

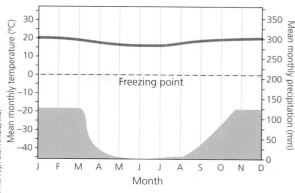

Martin Harvey/Peter Arnold, Inc.

Temperate grassland

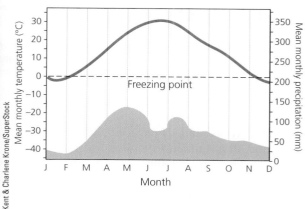

Kent & Charlene Krone/SuperStock

Cold grassland (arctic tundra)

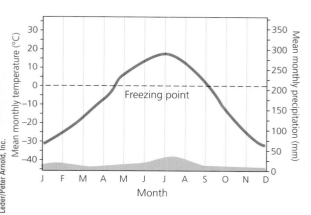

M. Leder/Peter Arnold, Inc.

Figure 7-12 Climate graphs showing typical variations in annual temperature (red) and precipitation (blue) in tropical, temperate, and cold (arctic tundra) grassland. Top photo: wildebeests grazing on a savanna in Maasai Mara National Park in Kenya, Africa (tropical grassland). Center photo: wildflowers in bloom on a prairie near East Glacier Park in the U. S. state of Montana (temperate grassland). Bottom photo: arctic tundra (cold grassland) in autumn in front of the Alaska Range, Alaska (USA). **Question:** What month of the year has the highest temperature and the lowest rainfall for each of the three types of grassland?

Figure 7-13 Natural capital degradation: replacement of a biologically diverse temperate grassland with a monoculture crop in the U.S. state of California. When humans remove the tangled root network of natural grasses, the fertile topsoil becomes subject to severe wind erosion unless it is covered with some type of vegetation.

During the brief summer, alpine tundra can be covered with an array of beautiful wildflowers.

THINKING ABOUT
Winds and Grassland
What roles do winds play in creating and sustaining grasslands?

CORE CASE STUDY

Temperate Shrubland: Nice Climate, Risky Place to Live

In many coastal regions that border on deserts we find fairly small patches of a biome known as *temperate shrubland* or *chaparral*. Closeness to the sea provides a slightly longer winter rainy season than nearby temperate deserts have, and fogs during the spring and fall reduce evaporation. These biomes are found along coastal areas of southern California in the United States, the Mediterranean Sea, central Chile, southern Australia, and southwestern South Africa (Figure 7-8).

Chaparral consists mostly of dense growths of low-growing evergreen shrubs and occasional small trees with leathery leaves that reduce evaporation (Figure 7-14). The soil is thin and not very fertile. Animal species of the chaparral include mule deer, chipmunks, jackrabbits, lizards, and a variety of birds.

During the long, warm, and dry summers, chaparral vegetation becomes very dry and highly flammable. In the late summer and fall, fires started by lightning or human activities spread with incredible swiftness. Research reveals that chaparral is adapted to and maintained by fires. Many of the shrubs store food reserves in their fire-resistant roots and produce seeds that sprout only after a hot fire. With the first rain, annual grasses and wildflowers spring up and use nutrients released by the fire. New shrubs grow quickly and crowd out the grasses.

People like living in this biome because of its moderate, sunny climate with mild, wet winters and warm, dry summers. As a result, humans have moved in and

(subside), which can damage buildings, roads, power lines, and other human structures.

Tundra is a fragile biome. Most tundra soils formed about 17,000 years ago when glaciers began retreating after the last Ice Age (Figure 4-7, p. 85). These soils usually are nutrient poor and have little detritus. Because of the short growing season, tundra soil and vegetation recover very slowly from damage or disturbance. Human activities in the arctic tundra—mostly around oil drilling sites, pipelines, mines, and military bases—leave scars that persist for centuries.

Another type of tundra, called *alpine tundra*, occurs above the limit of tree growth but below the permanent snow line on high mountains (Figure 7-9, left). The vegetation is similar to that found in arctic tundra, but it receives more sunlight than arctic vegetation gets.

Figure 7-14
Chaparral vegetation in the U.S. state of Utah and a typical climate graph.

Chaparral

modified this biome considerably (Figure 3, pp. S24–S25, in Supplement 4). The downside of its favorable climate is that people living in chaparral assume the high risk of losing their homes and possibly their lives to frequent fires during the dry season followed by mud slides during rainy seasons.

There Are Three Major Types of Forests

Forest systems are lands dominated by trees. The three main types of forest—*tropical, temperate,* and *cold* (northern coniferous and boreal)—result from combinations of the precipitation level and various average temperatures (Figures 7-10 and 7-15, p. 154).

Tropical rain forests (Figure 7-15, top photo) are found near the equator (Figure 7-8), where hot, moisture-laden air rises and dumps its moisture. These lush forests have year-round, uniformly warm temperatures, high humidity, and heavy rainfall almost daily (Figure 7-15, top graph). This fairly constant warm and wet climate is ideal for a wide variety of plants and animals.

Figure 7-16 (p. 155) shows some of the components and food web interactions in these extremely diverse ecosystems. Tropical rain forests are dominated by *broadleaf evergreen plants,* which keep most of their leaves year-round. The tops of the trees form a dense canopy, which blocks most light from reaching the forest floor, illuminating it with a dim greenish light.

The ground level in such a forest has little vegetation, except near stream banks or where a fallen tree has opened up the canopy and let in sunlight. Many of the plants that do live at the ground level have enormous leaves to capture what little sunlight filters through to the dimly lit forest floor.

Some trees are draped with vines (called lianas) that reach for the treetops to gain access to sunlight. Once in the canopy, the vines grow from one tree to another, providing walkways for many species living there. When a large tree is cut down, its lianas can pull down other trees.

Tropical rain forests have a very high net primary productivity (Figure 3-16, p. 64); they are teeming with life and boast incredible biological diversity. Although tropical rain forests cover only about 2% of the earth's land surface, ecologists estimate that they contain at least half of the earth's known terrestrial plant and animal species. For example, a single tree in a rain forest may support several thousand different insect species. Plants from tropical rain forests are a source of chemicals used as blueprints for making most of the world's prescription drugs. Thus, the plant biodiversity found in this biome saves many human lives.

Tropical rain forest life forms occupy a variety of specialized niches in distinct layers. For example, vegetation layers are structured mostly according to the plants' needs for sunlight, as shown in Figure 7-17 (p. 156). Stratification of specialized plant and animal niches in a tropical rain forest enables the coexistence of a great variety of species (high species richness; see Photo 3 in the Detailed Contents). Much of the animal life, particularly insects, bats, and birds, lives in the sunny *canopy* layer, with its abundant shelter and supplies of leaves, flowers, and fruits. To study life in the canopy, ecologists climb trees, use tall construction cranes, and build platforms and boardwalks in the upper canopy. See *The Habitable Planet,* Videos 4 and 9, at **www.learner.org/resources/series209.html** for information on how scientists gather information about tropical rain forests and the effects of human activities on such forests.

CENGAGENOW Learn more about how plants and animals in a rain forest are connected in a food web at CengageNOW.

Because of the dense vegetation, there is little wind in these forests to spread seeds and pollen. Consequently, most rain forest plant species depend on bats, butterflies, birds, bees, and other species to pollinate their flowers and to spread seeds in their droppings.

Dropped leaves, fallen trees, and dead animals decompose quickly because of the warm, moist conditions and the hordes of decomposers. This rapid recycling of scarce soil nutrients explains why there is so little plant litter on the ground. Instead of being stored in the soil, about 90% of plant nutrients released by decomposition are quickly taken up and stored by trees, vines, and other plants. This is in sharp contrast to temperate forests, where most plant nutrients are found in the soil.

Because of these ecological processes and the almost daily rainfall, which leaches nutrients from the soil, the soils in most rain tropical forests contain few plant nutrients. This helps explain why rain forests are not good places to clear and grow crops or graze cattle on a sustainable basis. Despite this ecological limitation, many of these forests are being cleared or degraded for logging, growing crops, grazing cattle, and mineral extraction (Chapter 3 Core Case Study, p. 50).

So far, at least half of these forests have been destroyed or disturbed by human activities and the pace of destruction and degradation of these centers of terrestrial biodiversity is increasing (Figure 3-1, p. 50). Ecologists warn that without strong conservation measures, most of these forests will probably be gone within your lifetime, and with them perhaps a quarter of the world's species. This will reduce the earth's biodiversity and help to accelerate global warming and the resulting climate change by eliminating large areas of trees that remove carbon dioxide from the atmosphere (see *The Habitable Planet,* Video 9, at **www.learner .org/resources/series209.html**).

Temperate deciduous forests (Figure 7-15, center photo) grow in areas with moderate average temperatures that change significantly with the season. These areas have long, warm summers, cold but not too severe winters,

Figure 7-15 Climate graphs showing typical variations in annual temperature (red) and precipitation (blue) in tropical, temperate, and cold (northern coniferous and boreal) forests. Top photo: the closed canopy of a tropical rain forest in the western Congo Basin of Gabon, Africa. Middle photo: a temperate deciduous forest in the U.S. state of Rhode Island during the fall. (Photo 4 in the Detailed Contents shows this same area of forest during winter.) Bottom photo: a northern coniferous forest in the Malheur National Forest and Strawberry Mountain Wilderness in the U.S. state of Oregon. **Question:** What month of the year has the highest temperature and the lowest rainfall for each of the three types of forest?

Producer to primary consumer

Primary to secondary consumer

Secondary to higher-level consumer

All producers and consumers to decomposers

and abundant precipitation, often spread fairly evenly throughout the year (Figure 7-15, center graph).

This biome is dominated by a few species of *broadleaf deciduous trees* such as oak, hickory, maple, poplar, and beech. They survive cold winters by dropping their leaves in the fall and becoming dormant through the winter (see Photo 4 in the Detailed Contents). Each spring, they grow new leaves whose colors change in the fall into an array of reds and golds before the leaves drop.

Because of a slow rate of decomposition, these forests accumulate a thick layer of slowly decaying leaf litter, which is a storehouse of nutrients. (Figure 4, p. S56, in Supplement 9, shows some components and food web interactions in a temperate deciduous forest ecosystem.) On a global basis, this biome has been disturbed by human activity more than any other terrestrial biome. Many forests have been cleared for growing crops or developing urban areas. However, within

100–200 years, abandoned cropland can return to a deciduous forest through secondary ecological succession (Figure 5-17, p. 117).

The temperate deciduous forests of the eastern United States were once home to such large predators as bears, wolves, foxes, wildcats, and mountain lions (pumas). Today, most of the predators have been killed or displaced, and the dominant mammal species often is the white-tailed deer (Case Study, p. 114), along with smaller mammals such as squirrels, rabbits, opossums, raccoons, and mice.

Warblers, robins, and other bird species migrate to these forests during the summer to feed and breed. Many of these species are declining in numbers because of loss or fragmentation of their summer and winter habitats.

Evergreen coniferous forests (Figure 7-15, bottom photo) are also called *boreal forests* and *taigas* ("TIE-guhs"). These cold forests are found just south of the

Height (meters)

45
40
35
30
25
20
15
10
5
0

Harpy eagle

Toco toucan

Wooly opossum

Brazilian tapir

Black-crowned antpitta

Emergent layer

Canopy

Under story

Shrub layer

Ground layer

Figure 7-17 Stratification of specialized plant and animal niches in a *tropical rain forest*. Filling such specialized niches enables species to avoid or minimize competition for resources and results in the coexistence of a great variety of species.

arctic tundra in northern regions across North America, Asia, and Europe (Figure 7-8) and above certain altitudes in the High Sierra and Rocky Mountains of the United States. In this subarctic climate, winters are long, dry, and extremely cold; in the northernmost taigas, winter sunlight is available only 6–8 hours per day. Summers are short, with cool to warm temperatures (Figure 7-15, bottom graph), and the sun shines up to 19 hours a day.

Most boreal forests are dominated by a few species of *coniferous* (cone-bearing) *evergreen trees* such as spruce, fir, cedar, hemlock, and pine that keep most of their narrow-pointed leaves (needles) year-round (Figure 7-15, bottom photo). The small, needle-shaped, waxy-coated leaves of these trees can withstand the intense cold and drought of winter, when snow blankets the ground. Such trees are ready to take advantage of the brief summers in these areas without taking time to grow new needles. Plant diversity is low because few species can survive the winters when soil moisture is frozen.

Beneath the stands of these trees is a deep layer of partially decomposed conifer needles. Decomposition is slow because of the low temperatures, waxy coating on conifer needles, and high soil acidity. The decomposing needles make the thin, nutrient-poor soil acidic, which prevents most other plants (except certain shrubs) from growing on the forest floor.

This biome contains a variety of wildlife. Year-round residents include bears, wolves, moose, lynx, and many burrowing rodent species. Caribou spend the winter in taiga and the summer in arctic tundra. During the brief summer, warblers and other insect-eating birds feed on hordes of flies, mosquitoes, and caterpillars. (Figure 5, p. S57, in Supplement 9 shows some components and food web interactions in an evergreen coniferous forest ecosystem.)

Coastal coniferous forests or *temperate rain forests* (Figure 7-18) are found in scattered coastal temperate areas that have ample rainfall or moisture from dense ocean fogs. Dense stands of large conifers such as Sitka spruce, Douglas fir, and redwoods once dominated undisturbed areas of this biome along the coast of North America, from Canada to northern California in the United States.

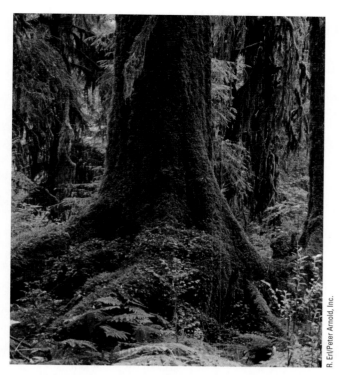

Figure 7-18 Temperate rain forest in Olympic National Park in the U.S. state of Washington.

R. Erl/Peter Arnold, Inc.

┌─ **THINKING ABOUT**
 Winds and Forests CORE
 What roles do winds play in creating temperate and CASE
 coniferous forests? STUDY

Mountains Play Important Ecological Roles

Some of the world's most spectacular environments are high on *mountains* (Figure 7-19), steep or high lands which cover about one-fourth of the earth's land surface (Figure 7-8). Mountains are places where dramatic changes in altitude, slope, climate, soil, and vegetation take place over a very short distance (Figure 7-9, left).

About 1.2 billion people (18% of the world's population) live in mountain ranges or on their edges and 4 billion people (59% of the world's population) depend on mountain systems for all or some of their water. Because of the steep slopes, mountain soils are easily eroded when the vegetation holding them in place is removed by natural disturbances, such as landslides and avalanches, or human activities, such as timber cutting and agriculture. Many freestanding mountains are *islands of biodiversity* surrounded by a sea of lower-elevation landscapes transformed by human activities.

Mountains play important ecological roles. They contain the majority of the world's forests, which are habitats for much of the planet's terrestrial biodiversity. They often provide habitats for endemic species found

nowhere else on earth. They also serve as sanctuaries for animal species driven to migrate from lowland areas to higher altitudes.

Mountains also help to regulate the earth's climate. Mountaintops covered with ice and snow affect climate by reflecting solar radiation back into space. This helps to cool the earth and offset global warming. However, many of the world's mountain glaciers are melting, mostly because of global warming. While glaciers reflect solar energy, the darker rocks exposed by melting glaciers absorb that energy. This helps to increase global warming, which melts more glaciers and warms the atmosphere more—an example of a runaway positive feedback loop.

Mountains can affect sea levels by storing and releasing water in glacial ice. As the earth gets warmer, mountaintop glaciers and other land-based glaciers can melt, adding water to the oceans and helping to raise sea levels.

Finally, mountains play a critical role in the hydrologic cycle by serving as major storehouses of water. In the warmer weather of spring and summer, much of their snow and ice melts and is released to streams for use by wildlife and by humans for drinking and irrigating crops. As the earth warms, mountaintop snowpacks and glaciers melt earlier each year. This can lower food production if water needed to irrigate crops during the summer has already been released.

Despite their ecological, economic, and cultural importance, the fate of mountain ecosystems has not been a high priority for governments or for many environmental organizations.

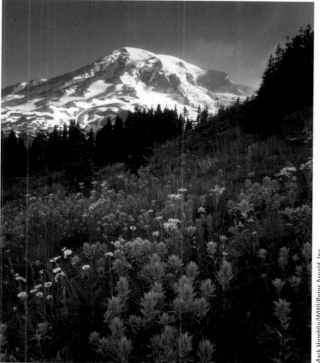

Mark Hamblin/WWI/Peter Arnold, Inc.

Figure 7-19 Mountains such as these in Mount Rainier National Park in the U.S. state of Washington play important ecological roles.

7-3 How Have We Affected the World's Terrestrial Ecosystems?

▶ **CONCEPT 7-3** In many areas, human activities are impairing ecological and economic services provided by the earth's deserts, grasslands, forests, and mountains.

Humans Have Disturbed Most of the Earth's Land

The human species dominates most of the planet. In many areas, human activities are impairing some of the ecological and economic services provided by the world's deserts, grasslands, forests, and mountains (**Concept 7-3**).

According to the 2005 Millennium Ecosystem Assessment, about 62% of the world's major terrestrial ecosystems are being degraded or used unsustainably (see Figure 3, pp. S24–S25, and Figure 7, pp. S28–S29, in Supplement 4), as the human ecological footprint intensifies and spreads across the globe (Figure 1-10, p. 15, and **Concept 1-3**, p. 12). This environ- ◀ CONCEPT LINK mental destruction and degradation is increas-

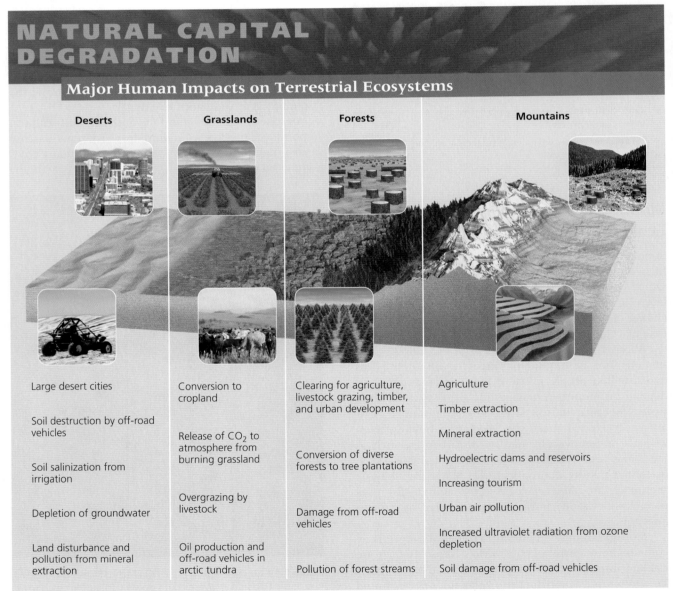

NATURAL CAPITAL DEGRADATION

Major Human Impacts on Terrestrial Ecosystems

Deserts	Grasslands	Forests	Mountains
Large desert cities	Conversion to cropland	Clearing for agriculture, livestock grazing, timber, and urban development	Agriculture
Soil destruction by off-road vehicles	Release of CO_2 to atmosphere from burning grassland	Conversion of diverse forests to tree plantations	Timber extraction
Soil salinization from irrigation			Mineral extraction
Depletion of groundwater	Overgrazing by livestock	Damage from off-road vehicles	Hydroelectric dams and reservoirs
Land disturbance and pollution from mineral extraction	Oil production and off-road vehicles in arctic tundra	Pollution of forest streams	Increasing tourism
			Urban air pollution
			Increased ultraviolet radiation from ozone depletion
			Soil damage from off-road vehicles

Figure 7-20 Major human impacts on the world's deserts, grasslands, forests, and mountains. **Question:** Which two of the impacts on each of these biomes do you think are the most harmful?

ing in many parts of the world. Figure 7-20 summarizes some of the human impacts on the world's deserts, grasslands, forests, and mountains.

How long can we keep eating away at these terrestrial forms of natural capital without threatening our economies and the long-term survival of our own and other species? No one knows. But there are increasing signs that we need to come to grips with this vital issue.

This will require protecting the world's remaining wild areas from development. In addition, many of the land areas we have degraded need to be restored. However, such efforts to achieve a balance between exploitation and conservation are highly controversial because of timber, mineral, fossil fuel, and other re-

sources found in or under many of the earth's biomes. These issues are discussed in Chapter 10.

┌─ RESEARCH FRONTIER ──────────────────
│ Better understanding of the effects of human activities on
│ terrestrial biomes and how we can reduce these impacts. See
│ **academic.cengage.com/biology/miller**.
└──────────────────────────────────────

┌─ THINKING ABOUT ─────────────────────
│ **Sustainability**
│
│ Develop four guidelines for preserving the earth's
│ terrestrial biodiversity based on the four **scientific**
│ **principles of sustainability** (see back cover).
└──────────────────────────────────────

REVISITING Winds and Sustainability

This chapter's opening **Core Case Study** described how winds connect parts of the planet to one another. Next time you feel or hear the wind blowing, think about these global connections. As part of the global climate system, winds play important roles in creating and sustaining the world's deserts, grasslands, and forests through the four **scientific principles of sustainability** (see back cover). Winds promote sustainability by helping to distribute solar energy and to recycle the earth's nutrients. In turn, this helps support biodiversity, which in turn affects species interactions that help control population sizes.

Scientists have made a good start in understanding the ecology of the world's terrestrial biomes. One of the major lessons from their research is: in nature, *everything is connected*. According to these scientists, we urgently need more research on the workings of the world's terrestrial biomes and on how they are interconnected. With such information, we will have a clearer picture of how our activities affect the earth's natural capital and of what we can do to help sustain it.

When we try to pick out anything by itself,
we find it hitched to everything else
in the universe.

JOHN MUIR

REVIEW

1. Review the Key Questions and Concepts for this chapter on p. 141. Describe the environmentally beneficial and harmful effects of the earth's winds.

2. Distinguish between **weather** and **climate.** Describe three major factors that determine how air circulates in the lower atmosphere. Describe how the properties of air, water, and land affect global air circulation. How is heat distributed to different parts of the ocean? Explain how global air circulation and ocean currents lead to the forests, grasslands, and deserts that make up the earth's terrestrial biomes.

3. Define and give four examples of a **greenhouse gas.** What is the **greenhouse effect** and why is it important to the earth's life and climate? What is the **rain shadow effect** and how can it lead to the formation of inland deserts? Why do cities tend to have more haze and smog,

higher temperatures, and lower wind speeds than the surrounding countryside?

4. What is a **biome?** Explain why there are three major types of each of the major biomes (deserts, grasslands, and forests). Describe how climate and vegetation vary with latitude and elevation.

5. Describe how the three major types of deserts differ in their climate and vegetation. How do desert plants and animals survive?

6. Describe how the three major types of grasslands differ in their climate and vegetation. What is a savanna? Why have many of the world's temperate grasslands disappeared? What is **permafrost?** Distinguish between arctic tundra and alpine tundra.

7. What is temperate shrubland or chaparral? Why is this biome a desirable place to live? Why is it a risky place to live?

8. What is a **forest system** and what are the three major types of forests? Describe how these three types differ in their climate and vegetation. Why is biodiversity so high in tropical rain forests? Why do most soils in tropical rain forests have few plant nutrients? Describe what happens in temperate deciduous forests in the winter and fall. What are coastal coniferous or temperate rain forests? What important ecological roles do mountains play?

9. Describe how human activities have affected the world's deserts, grasslands, forests, and mountains.

10. Describe the connections between the earth's winds, climates, and biomes (**Core Case Study**) and the four **scientific principles of sustainability** (see back cover).

Note: Key Terms are in **bold** type.

CRITICAL THINKING

1. What would happen to **(a)** the earth's species and **(b)** your lifestyle if the winds stopped blowing (**Core Case Study**)?

2. List a limiting factor for each of the following ecosystems: **(a)** a desert, **(b)** arctic tundra, **(c)** temperate grassland, **(d)** the floor of a tropical rain forest, and **(e)** a temperate deciduous forest.

3. Why do deserts and arctic tundra support a much smaller biomass of animals than do tropical forests?

4. Why do most animals in a tropical rain forest live in its trees?

5. Why do most species living at high latitudes or high altitudes tend to have generalist ecological niches while those living in the tropics tend to have specialist ecological niches (Figure 4-11, p. 91)?

6. Which biomes are best suited for **(a)** raising crops and **(b)** grazing livestock? Use the four **scientific principles of sustainability** (see back cover) to come

up with four guidelines for growing food and grazing livestock in these biomes on a more sustainable basis.

7. What type of biome do you live in? List three ways in which your lifestyle is harming this biome?

8. You are a defense attorney arguing in court for sparing a tropical rain forest. Give your three most important arguments for the defense of this ecosystem.

9. Congratulations! You are in charge of the world. What are the three most important features of your plan to help sustain the earth's terrestrial biodiversity?

10. List two questions that you would like to have answered as a result of reading this chapter.

Note: See Supplement 13 (p. S78) for a list of Projects related to this chapter.

DATA ANALYSIS

In this chapter, you learned how long-term variations in the average temperature and average precipitation play a major role in determining the types of deserts, forests, and grasslands

found in different parts of the world. Below are typical annual climate graphs for a tropical grassland (savanna) in Africa and a temperate grassland in the midwestern United States.

Tropical grassland (savanna)

Temperate grassland

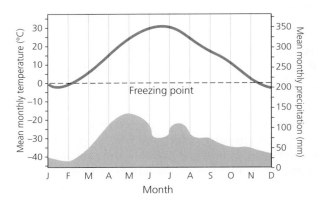

1. In what month (or months) does the most precipitation fall in each of these areas?

2. What are the driest months in each of these areas?

3. What is the coldest month in the tropical grassland?

4. What is the warmest month in the temperate grassland?

LEARNING ONLINE

Log on to the Student Companion Site for this book at **academic.cengage.com/biology/miller**, and choose Chapter 7 for many study aids and ideas for further reading and research. These include flash cards, practice quizzing, Weblinks, information on Green Careers, and InfoTrac® College Edition articles.

Aquatic Biodiversity

Why Should We Care about Coral Reefs?

Coral reefs form in clear, warm coastal waters of the tropics and subtropics (Figure 8-1, left). These stunningly beautiful natural wonders are among the world's oldest, most diverse, and most productive ecosystems. In terms of biodiversity, they are the marine equivalents of tropical rain forests.

Coral reefs are formed by massive colonies of tiny animals called *polyps* (close relatives of jellyfish). They slowly build reefs by secreting a protective crust of limestone (calcium carbonate) around their soft bodies. When the polyps die, their empty crusts remain behind as a platform for more reef growth. The resulting elaborate network of crevices, ledges, and holes serves as calcium carbonate "condominiums" for a variety of marine animals.

Coral reefs are the result of a mutually beneficial relationship between the polyps and tiny single-celled algae called *zooxanthellae* ("zoh-ZAN-thel-ee") that live in the tissues of the polyps. In this example of mutualism (p. 106), the algae provide the polyps with food and oxygen through photosynthesis, and help to produce calcium carbonate, which forms the coral skeleton. Algae also give the reefs their stunning coloration. The polyps, in turn, provide the algae with a well-protected home and some of their nutrients.

Although coral reefs occupy only about 0.2% of the ocean floor, they provide important ecological and economic services. They help moderate atmospheric temperatures by removing CO_2 from the atmosphere, and they act as natural barriers that protect 15% of the world's coastlines from erosion caused by battering waves and storms. And they provide habitats for one-quarter of all marine organisms. Economically, coral reefs produce about one-tenth of the global fish catch—one-fourth of the catch in developing countries—and they provide jobs and building materials for some of the world's poorest countries. Coral reefs also support important fishing and tourism industries.

Finally, these biological treasures give us an underwater world to study and enjoy. Each year, more than 1 million scuba divers and snorkelers visit coral reefs to experience these wonders of aquatic biodiversity.

According to a 2005 report by the World Conservation Union, 15% of the world's coral reefs have been destroyed and another 20% have been damaged by coastal development, pollution, overfishing, warmer ocean temperatures, increasing ocean acidity, and other stresses. And another 25–33% of these centers of aquatic biodiversity could be lost within 20–40 years. One problem is *coral bleaching* (Figure 8-1, upper right). It occurs when stresses such as increased temperature cause the algae, upon which corals depend for food, to die off, leaving behind a white skeleton of calcium carbonate. Another threat is the increasing acidity of ocean water as it absorbs some of the CO_2 produced by the burning of carbon-containing fossils fuels. The CO_2 reacts with ocean water to form a weak acid, which can slowly dissolve the calcium carbonate that makes up the corals.

The degradation and decline of these colorful oceanic sentinels should serve as a warning about threats to the health of the oceans, which provide us with crucial ecological and economic services.

Sergio Hanquet/Peter Arnold, Inc

Figure 8-1 *A healthy coral reef* in the Red Sea covered by colorful algae (left) and a bleached coral reef that has lost most of its algae (right) because of changes in the environment (such as cloudy water or high water temperatures). With the colorful algae gone, the white limestone of the coral skeleton becomes visible. If the environmental stress is not removed and no other alga species fill the abandoned niche, the corals die. These diverse and productive ecosystems are being damaged and destroyed at an alarming rate.

8-1 What is the general nature of aquatic systems?

CONCEPT 8-1A Saltwater and freshwater aquatic life zones cover almost three-fourths of the earth's surface with oceans dominating the planet.

CONCEPT 8-1B The key factors determining biodiversity in aquatic systems are temperature, dissolved oxygen content, availability of food, and availability of light and nutrients necessary for photosynthesis.

8-2 Why are marine aquatic systems important?

CONCEPT 8-2 Saltwater ecosystems are irreplaceable reservoirs of biodiversity and provide major ecological and economic services.

8-3 How have human activities affected marine ecosystems?

CONCEPT 8-3 Human activities threaten aquatic biodiversity and disrupt ecological and economic services provided by saltwater systems.

8-4 Why are freshwater ecosystems important?

CONCEPT 8-4 Freshwater ecosystems provide major ecological and economic services and are irreplaceable reservoirs of biodiversity.

8-5 How have human activities affected freshwater ecosystems?

CONCEPT 8-5 Human activities threaten biodiversity and disrupt ecological and economic services provided by freshwater lakes, rivers, and wetlands.

Note: Supplements 2 (p. S4), 4 (p. S20), 5 (p. S31), 7 (p. S46), 9 (p. S53), and 13 (p. S78) can be used with this chapter.

If there is magic on this planet,
it is contained in water.
LOREN EISLEY

8-1 What Is the General Nature of Aquatic Systems?

▶ **CONCEPT 8-1A** Saltwater and freshwater aquatic life zones cover almost three-fourths of the earth's surface with oceans dominating the planet.

▶ **CONCEPT 8-1B** The key factors determining biodiversity in aquatic systems are temperature, dissolved oxygen content, availability of food, and availability of light and nutrients necessary for photosynthesis.

Most of the Earth Is Covered with Water

When viewed from a certain point in outer space, the earth appears to be almost completely covered with water (Figure 8-2). Saltwater covers about 71% of the earth's surface, and freshwater occupies roughly another 2.2%. Yet, in proportion to the entire planet, it all amounts to a thin and precious film of water.

Although the *global ocean* is a single and continuous body of water, geographers divide it into four large areas—the Atlantic, Pacific, Arctic, and Indian Oceans—separated by the continents. The largest ocean is the Pacific, which contains more than half of the earth's water and covers one-third of the earth's surface.

The aquatic equivalents of biomes are called **aquatic life zones.** The distribution of many aquatic organisms is determined largely by the water's *salinity*—the amounts of various salts such as sodium chloride (NaCl) dissolved in a given volume of water. As a

Ocean hemisphere Land–ocean hemisphere

Figure 8-2 *The ocean planet.* The salty oceans cover 71% of the earth's surface. Almost all of the earth's water is in the interconnected oceans, which cover 90% of the planet's mostly ocean hemisphere (left) and half of its land–ocean hemisphere (right). Freshwater systems cover less than 2.2% of the earth's surface (**Concept 8-1A**).

result, aquatic life zones are classified into two major types: **saltwater** or **marine** (oceans and their accompanying estuaries, coastal wetlands, shorelines, coral reefs, and mangrove forests) and **freshwater** (lakes,

Figure 8-3 Natural capital: distribution of the world's major saltwater oceans, coral reefs, mangroves, and freshwater lakes and rivers. **Question:** Why do you think most coral reefs lie in the southern hemisphere?

Lakes
Rivers
Coral reefs
Mangroves

rivers, streams, and inland wetlands). Although some systems such as estuaries are a mix of saltwater and freshwater, we classify them as marine systems for purposes of discussion.

Figure 8-3 shows the distribution of the world's major oceans, coral reefs, mangroves, lakes, and rivers. These aquatic systems play vital roles in the earth's biological productivity, climate, biogeochemical cycles, and biodiversity, and they provide us with fish, shellfish, minerals, recreation, transportation routes, and many other economically important goods and services.

Most Aquatic Species Live in Top, Middle, or Bottom Layers of Water

Saltwater and freshwater life zones contain several major types of organisms. One such type consists of weakly swimming, free-floating **plankton,** which can be divided into three groups, the first of which is *phytoplankton* ("FY-toe-plank-ton," Greek for "drifting plants"; see bottom of Figure 3-14, p. 63), which includes many types of algae. They and various rooted plants near shorelines are primary producers that support most

aquatic food webs. See *The Habitable Planet,* Videos 2 and 3, at **www.learner.org/resources/series209.html**.

The second group is *zooplankton* ("ZOH-uh-plankton," Greek for "drifting animals"; see bottom of Figure 3-14, p. 63). They consist of primary consumers (herbivores) that feed on phytoplankton and secondary consumers that feed on other zooplankton. They range from single-celled protozoa to large invertebrates such as jellyfish.

A third group consists of huge populations of much smaller plankton called *ultraplankton*. These extremely small photosynthetic bacteria may be responsible for 70% of the primary productivity near the ocean surface.

A second major type of organisms is **nekton,** strongly swimming consumers such as fish, turtles, and whales. The third type, **benthos,** consists of bottom dwellers such as oysters, which anchor themselves to one spot; clams and worms, which burrow into the sand or mud; and lobsters and crabs, which walk about on the sea floor. A fourth major type is **decomposers** (mostly bacteria), which break down organic compounds in the dead bodies and wastes of aquatic organisms into nutrients that can be used by aquatic primary producers.

Most forms of aquatic life are found in the *surface, middle,* and *bottom* layers of saltwater and freshwater systems, which we explore later in this chapter. In most aquatic systems, the key factors determining the types and numbers of organisms found in these layers are *temperature, dissolved oxygen content, availability of food,* and *availability of light and nutrients required for photosynthesis,* such as carbon (as dissolved CO_2 gas), nitrogen (as NO_3^-), and phosphorus (mostly as PO_4^{3-}) (**Concept 8-1B**).

In deep aquatic systems, photosynthesis is largely confined to the upper layer—the *euphotic* or *photic* zone, through which sunlight can penetrate. The depth of the euphotic zone in oceans and deep lakes can be reduced when the water is clouded by excessive algal growth (algal blooms) resulting from nutrient overloads. This cloudiness, called **turbidity,** can occur naturally, such as from algal growth, or can result from disturbances such as clearing of land, which causes silt to flow into bodies of water. This is one of the problems plaguing coral reefs (**Core Case Study**), as excessive turbidity due to silt runoff prevents photosynthesis and causes the corals to die.

In shallow systems such as small open streams, lake edges, and ocean shorelines, ample supplies of nutrients for primary producers are usually available. By contrast, in most areas of the open ocean, nitrates, phosphates, iron, and other nutrients are often in short supply, and this limits net primary productivity (NPP) (Figure 3-16, p. 64).

8-2 Why Are Marine Aquatic Systems Important?

▶ **CONCEPT 8-2** Saltwater ecosystems are irreplaceable reservoirs of biodiversity and provide major ecological and economic services.

Oceans Provide Important Ecological and Economic Resources

Oceans provide enormously valuable ecological and economic services (Figure 8-4). One estimate of the combined value of these goods and services from all marine coastal ecosystems is over $12 trillion per year, nearly equal to the annual U.S. gross domestic product.

As land dwellers, we have a distorted and limited view of the blue aquatic wilderness that covers most of the earth's surface. We know more about the surface of the moon than about the oceans. According to aquatic scientists, the scientific investigation of poorly understood marine and freshwater aquatic systems could yield immense ecological and economic benefits.

Marine aquatic systems are huge reservoirs of biodiversity. They include many different ecosystems, which host a great variety of species, genes, and biological and chemical processes, thus sustaining four major components of the earth's biodiversity (Figure 4-2, p. 79). Marine life is found in three major *life zones:* the coastal zone, open sea, and ocean bottom (Figure 8-5, p. 166).

— RESEARCH FRONTIER —

Discovering, cataloging, and studying the huge number of unknown aquatic species and their interactions. See **academic .cengage.com/biology/miller**.

The **coastal zone** is the warm, nutrient-rich, shallow water that extends from the high-tide mark on

NATURAL CAPITAL

Marine Ecosystems

Ecological Services	Economic Services
Climate moderation	Food
CO_2 absorption	Animal and pet feed
Nutrient cycling	Pharmaceuticals
Waste treatment	Harbors and transportation routes
Reduced storm impact (mangroves, barrier islands, coastal wetlands)	Coastal habitats for humans
	Recreation
Habitats and nursery areas	Employment
Genetic resources and biodiversity	Oil and natural gas
	Minerals
Scientific information	Building materials

Figure 8-4 Major ecological and economic services provided by marine systems (**Concept 8-2**). **Question:** Which two ecological services and which two economic services do you think are the most important? Why?

Figure 8-5
Natural capital: major life zones and vertical zones (not drawn to scale) in an ocean. Actual depths of zones may vary. Available light determines the euphotic, bathyal and abyssal zones. Temperature zones also vary with depth, shown here by the red curve.
Question: How is an ocean like a rain forest? (*Hint:* see Figure 7-17, p. 156.)

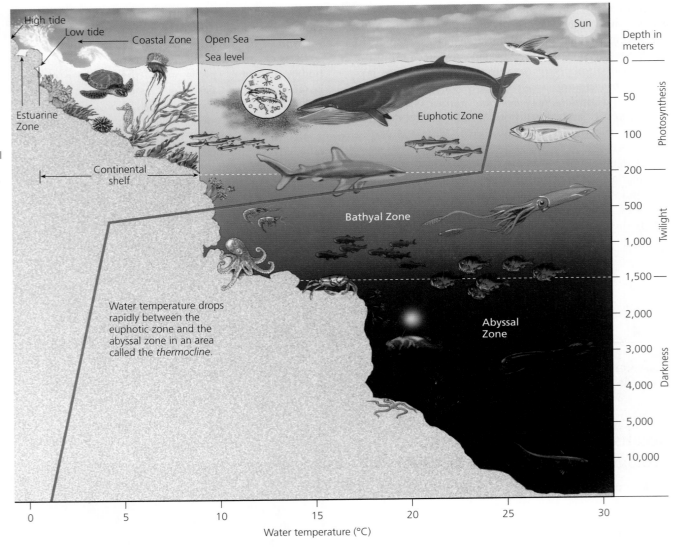

land to the gently sloping, shallow edge of the *continental shelf*. It makes up less than 10% of the world's ocean area but contains 90% of all marine species and is the site of most large commercial marine fisheries. Most coastal zone aquatic systems, such as estuaries, coastal wetlands, mangrove forests, and coral reefs, have a high NPP per unit of area (Figure 3-16, p. 64). This is the result of the zone's ample supplies of sunlight and plant nutrients that flow from land and are distributed by wind and ocean currents. Here, we look at some of these systems in more detail.

Estuaries and Coastal Wetlands Are Highly Productive

Estuaries are where rivers meet the sea (Figure 8-6). They are partially enclosed bodies of water where seawater mixes with freshwater as well as nutrients and pollutants from streams, rivers, and runoff from the land.

Estuaries and their associated **coastal wetlands**—coastal land areas covered with water all or part of the year—include river mouths, inlets, bays, sounds, and salt marshes in temperate zones (Figure 8-7), and mangrove forests in tropical zones (Figure 8-3). They are some of the earth's most productive ecosystems because of high nutrient inputs from rivers and nearby land, rapid circulation of nutrients by tidal flows, and ample sunlight penetrating the shallow waters.

Seagrass beds are another component of coastal marine biodiversity. They consist of at least 60 species of plants that grow underwater in shallow marine and estuarine areas along most continental coastlines. These highly productive and physically complex systems support a variety of marine species. They also help stabilize shorelines and reduce wave impacts.

Life in these coastal ecosystems is harsh. It must adapt to significant daily and seasonal changes in tidal and river flows, water temperatures and salinity, and runoff of eroded soil sediment and other pollutants from the land. Because of these stresses, despite their

Figure 8-6 View of an *estuary* from space. The photo shows the sediment plume (turbidity caused by runoff) at the mouth of Madagascar's Betsiboka River as it flows through the estuary and into the Mozambique Channel. Because of its topography, heavy rainfall, and the clearing of forests for agriculture, Madagascar is the world's most eroded country.

NASA

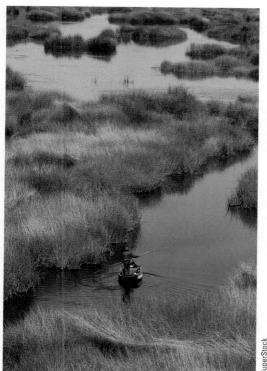

Figure 8-7 Some components and interactions in a *salt marsh ecosystem* in a temperate area such as the United States. When these organisms die, decomposers break down their organic matter into minerals used by plants. Colored arrows indicate transfers of matter and energy between consumers (herbivores), secondary or higher-level consumers (carnivores), and decomposers. Organisms are not drawn to scale. The photo shows a salt marsh in Peru.

SuperStock

 Producer to primary consumer

 Primary to secondary consumer

 Secondary to higher-level consumer

Herring gulls

Peregrine falcon

Snowy egret

Cordgrass

Short-billed dowitcher

Phytoplankton

Marsh periwinkle

Smelt

Zooplankton and small crustaceans

Soft-shelled clam

Bacteria

Clamworm

All consumers and producers to decomposers

Figure 8-8 Mangrove forest in Daintree National Park in Queensland, Australia. The tangled roots and dense vegetation in these coastal forests act like shock absorbers to reduce damage from storms and tsunamis. They also provide a highly complex habitat for a diversity of invertebrates and fishes.

Theo Allofs/Visuals Unlimited

productivity, some coastal ecosystems have low plant diversity, composed of a few species that can withstand the daily and seasonal variations.

Mangrove forests are the tropical equivalent of salt marshes. They are found along some 70% of gently sloping sandy and silty coastlines in tropical and subtropical regions, especially Southeast Asia (Figure 8-3). The dominant organisms in these nutrient-rich coastal forests are mangroves—69 different tree species that can grow in salt water. They have extensive root systems that often extend above the water, where they can obtain oxygen and support the trees during periods of changing water levels (Figure 8-8).

These coastal aquatic systems provide important ecological and economic services. They help to maintain water quality in tropical coastal zones by filtering toxic pollutants, excess plant nutrients, and sediments, and by absorbing other pollutants. They provide food, habitats, and nursery sites for a variety of aquatic and terrestrial species. They also reduce storm damage and coastal erosion by absorbing waves and storing excess water produced by storms and tsunamis. Historically, they have sustainably supplied timber and fuelwood to coastal communities.

Loss of mangroves can lead to polluted drinking water, caused by inland intrusion of saltwater into aquifers that are used to supply drinking water. Despite their ecological and economic importance, in 2008, the U.N. Food and Agriculture Organization estimated that between 1980 and 2005 at least one-fifth of the world's mangrove forests were lost mostly because of human coastal development.

Rocky and Sandy Shores Host Different Types of Organisms

The gravitational pull of the moon and sun causes *tides* to rise and fall about every 6 hours in most coastal areas. The area of shoreline between low and high tides is called the **intertidal zone.** Organisms living in this zone must be able to avoid being swept away or crushed by waves, and must deal with being immersed during high tides and left high and dry (and much hotter) at low tides. They must also survive changing levels of salinity when heavy rains dilute saltwater. To deal with such stresses, most intertidal organisms hold on to something, dig in, or hide in protective shells.

On some coasts, steep *rocky shores* are pounded by waves. The numerous pools and other habitats in their intertidal zones contain a great variety of species that occupy different niches in response to daily and seasonal changes in environmental conditions such as temperature, water flows, and salinity (Figure 8-9, top).

Other coasts have gently sloping *barrier beaches,* or *sandy shores,* that support other types of marine organisms (Figure 8-9, bottom). Most of them keep hidden from view and survive by burrowing, digging, and tunneling in the sand. These sandy beaches and their adjoining coastal wetlands are also home to a variety of shorebirds that feed in specialized niches on crustaceans, insects, and other organisms (Figure 4-13, p. 93). Many of these species also live on *barrier islands*—low, narrow, sandy islands that form offshore, parallel to some coastlines.

Rocky Shore Beach

Sea star
Hermit crab
Shore crab
High tide
Periwinkle
Sea urchin
Anemone
Mussel
Sculpin
Kelp
Sea lettuce
Barnacles
Low tide
Nudibranch
Monterey flatworm

Barrier Beach

Peanut worm
Beach flea
Blue crab
Dwarf olive
Clam
Tiger beetle
High tide
Silversides
Sandpiper
Low tide
Mole shrimp
Ghost shrimp
White sand macoma
Sand dollar
Moon snail

Figure 8-9 *Living between the tides.* Some organisms with specialized niches found in various zones on rocky shore beaches (top) and barrier or sandy beaches (bottom). Organisms are not drawn to scale.

Undisturbed barrier beaches generally have one or more rows of natural sand dunes in which the sand is held in place by plant roots (Figure 8-10, p. 170). These dunes are the first line of defense against the ravages of the sea. Such real estate is so scarce and valuable that coastal developers frequently remove the protec-tive dunes or build behind the first set of dunes and cover them with buildings and roads. Large storms can then flood and even sweep away seaside buildings and severely erode the sandy beaches. Some people incor-rectly call these human-influenced events "natural disasters."

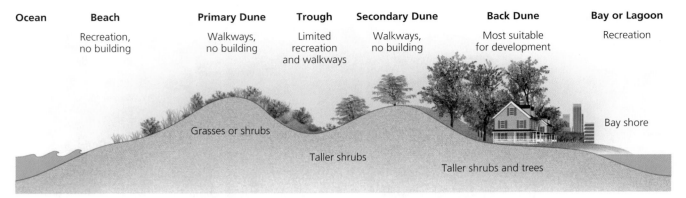

Ocean	Beach	Primary Dune	Trough	Secondary Dune	Back Dune	Bay or Lagoon
	Recreation, no building	Walkways, no building	Limited recreation and walkways	Walkways, no building	Most suitable for development	Recreation

Grasses or shrubs

Taller shrubs

Taller shrubs and trees

Bay shore

Figure 8-10 *Primary and secondary dunes* on gently sloping sandy barrier beaches help protect land from erosion by the sea. The roots of grasses that colonize the dunes hold the sand in place. Ideally, construction is allowed only behind the second strip of dunes, and walkways to the ocean beach are built so as not to damage the dunes. This helps to preserve barrier beaches and to protect buildings from damage by wind, high tides, beach erosion, and flooding from storm surges. Such protection is rare in some coastal areas because the short-term economic value of oceanfront land is considered much higher than its long-term ecological value. Rising sea levels from global warming may put many barrier beaches under water by the end of this century. **Question:** Do you think that the long- and short-term ecological values of oceanfront dunes outweigh the short-term economic value of removing them for coastal development? Explain.

Coral Reefs Are Amazing Centers of Biodiversity

As we noted in the **Core Case Study**, coral reefs are among the world's oldest, most diverse, and productive ecosystems (Figure 8-1 and Figure 4-10, left, p. 89). These amazing centers of aquatic biodiversity are the marine equivalents of tropical rain forests, with complex interactions among their diverse populations of species (Figure 8-11). Coral reefs provide homes for one-fourth of all marine species.

The Open Sea and Ocean Floor Host a Variety of Species

The sharp increase in water depth at the edge of the continental shelf separates the coastal zone from the vast volume of the ocean called the **open sea.** Primarily on the basis of the penetration of sunlight, this deep blue sea is divided into three *vertical zones* (see Figure 8-5). But temperatures also change with depth and can be used to define zones that help to determine species diversity in these layers (Figure 8-5, red curve).

The *euphotic zone* is the brightly lit upper zone where drifting phytoplankton carry out about 40% of the world's photosynthetic activity (see *The Habitable Planet,* Video 3, at **www.learner.org/resources/ series209.html**). Nutrient levels are low (except around upwellings, Figure 7-2, p. 142), and levels of dissolved oxygen are high. Large, fast-swimming predatory fishes such as swordfish, sharks, and bluefin tuna populate this zone.

The *bathyal zone* is the dimly lit middle zone, which, because it gets little sunlight, does not contain photosynthesizing producers. Zooplankton and smaller fishes,

many of which migrate to feed on the surface at night, populate this zone.

The deepest zone, called the *abyssal zone,* is dark and very cold; it has little dissolved oxygen. Nevertheless, the deep ocean floor is teeming with life—enough to be considered a major life zone—because it contains enough nutrients to support a large number of species, even though there is no sunlight to support photosynthesis.

Most organisms of the deep waters and ocean floor get their food from showers of dead and decaying organisms—called *marine snow*—drifting down from upper lighted levels of the ocean. Some of these organisms, including many types of worms, are *deposit feeders,* which take mud into their guts and extract nutrients from it. Others such as oysters, clams, and sponges are *filter feeders,* which pass water through or over their bodies and extract nutrients from it.

Average primary productivity and NPP per unit of area are quite low in the open sea. However, because open sea covers so much of the earth's surface, it makes the largest contribution to the earth's overall NPP. Also, NPP is much higher in some open sea areas where winds, ocean currents, and other factors cause water to rise from the depths to the surface. These *upwellings* bring nutrients from the ocean bottom to the surface for use by producers (Figure 7-2, p. 142).

In 2007, a team of scientists led by J. Craig Venter released a report that dramatically challenged scientists' assumptions about biodiversity in the open sea. After sailing around the world and spending 2 years collecting data, they found that the open sea contains many more bacteria, viruses, and other microbes than scientists had previously assumed.

CENGAGENOW Learn about ocean provinces where all ocean life exists at CengageNOW™.

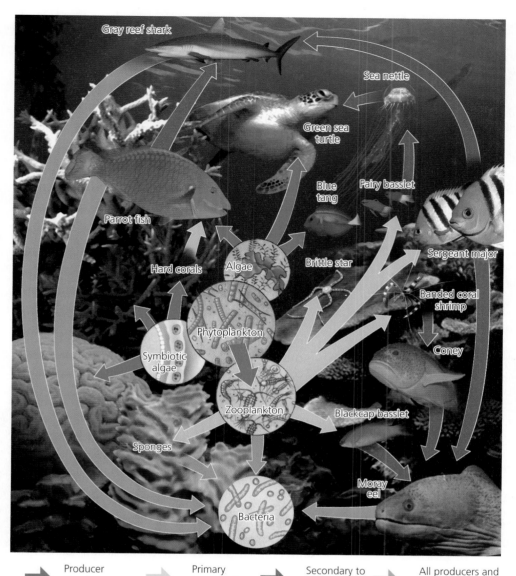

Figure 8-11 Natural capital: some components and interactions in a *coral reef ecosystem*. When these organisms die, decomposers break down their organic matter into minerals used by plants. Colored arrows indicate transfers of matter and energy between producers, primary consumers (herbivores), secondary or higher-level consumers (carnivores), and decomposers. Organisms are not drawn to scale.

Producer to primary consumer

Primary to secondary consumer

Secondary to higher-level consumer

All producers and consumers to decomposers

8-3 How Have Human Activities Affected Marine Ecosystems?

▶ **CONCEPT 8-3** Human activities threaten aquatic biodiversity and disrupt ecological and economic services provided by saltwater systems.

Human Activities Are Disrupting and Degrading Marine Systems

Human activities are disrupting and degrading some ecological and economic services provided by marine aquatic systems, especially coastal wetlands, shorelines, mangrove forests, and coral reefs (**Concept 8-3**). (See *The Hab-*

itable Planet, Video 9, at **www.learner.org/resources/series209.html**.) Thus, a single largely land-based species—humans—is increasingly threatening the biological diversity and ecosystem services provided by the oceans that cover about 71% of the earth's surface.

In 2008, the U.S. National Center for Ecological Analysis and Synthesis (NCEAS) used computer models

NATURAL CAPITAL DEGRADATION

Major Human Impacts on Marine Ecosystems and Coral Reefs

Marine Ecosystems

Half of coastal wetlands lost to agriculture and urban development

Over one-fifth of mangrove forests lost to agriculture, development, and shrimp farms since 1980

Beaches eroding because of coastal development and rising sea levels

Ocean bottom habitats degraded by dredging and trawler fishing

At least 20% of coral reefs severely damaged and 25–33% more threatened

Coral Reefs

Ocean warming

Soil erosion

Algae growth from fertilizer runoff

Bleaching

Rising sea levels

Increased UV exposure

Damage from anchors

Damage from fishing and diving

Figure 8-12 Major threats to marine ecosystems (left) and particularly coral reefs (right) resulting from human activities (**Concept 8-3**). **Questions:** Which two of the threats to marine ecosystems do you think are the most serious? Why? Which two of the threats to coral reefs do you think are the most serious? Why?

to analyze and provide the first-ever comprehensive map of the effects of 17 different types of human activities on the world's oceans. In this four-year study, the international team of scientists found that human activity has heavily affected 41% of the world's ocean area, with no area left completely untouched.

In their desire to live near the coast, people are destroying or degrading the natural resources and services (Figure 8-4) that make coastal areas so enjoyable and valuable. In 2006, about 45% of the world's population (including more than half of the U.S. population) lived along or near coasts. By 2040, up to 80% of the world's people are projected to be living in or near coastal zones.

Major threats to marine systems from human activities include:

- Coastal development, which destroys and pollutes coastal habitats (see *The Habitable Planet,* Video 5, at **www.learner.org/resources/series209.html**)

- Overfishing, which depletes populations of commercial fish species

- Runoff of nonpoint source pollution such as fertilizers, pesticides, and livestock wastes from the land (see *The Habitable Planet,* Videos 7 and 8, at **www.learner.org/resources/series209.html**)

- Point source pollution such as sewage from passenger cruise ships and spills from oil tankers

- Habitat destruction from coastal development and trawler fishing boats that drag weighted nets across the ocean bottom

- Invasive species, introduced by humans, that can deplete populations of native aquatic species and cause economic damage

- Climate change, enhanced by human activities, that could cause a rise in sea levels, which could destroy coral reefs and flood coastal marshes and coastal cities (see *The Habitable Planet,* Videos 7 and 8, at **www.learner.org/resources/series209.html**)

- Climate change from burning fossil fuels, which is also threatening marine ecosystems by warming the oceans and making them more acidic

- Pollution and degradation of coastal wetlands and estuaries (Case Study, below)

Figure 8-12 shows some of the effects of such human impacts on marine systems (left) and coral reefs (right). According to a 2007 study by O. Hoegh-Guldberg and 16 other scientists, unless we take action soon to significantly reduce carbon dioxide emissions, the oceans may become too acidic and too warm for most of the world's coral reefs to survive this century, and the important ecological and economic services they provide will be lost. We will examine some of these impacts more closely in Chapter 11.

RESEARCH FRONTIER

Learning more about harmful human impacts on marine ecosystems and how to reduce these impacts. See **academic.cengage.com/biology/miller**.

■ CASE STUDY

The Chesapeake Bay—an Estuary in Trouble

Since 1960, the Chesapeake Bay (Figure 8-13)—the largest estuary in the United States—has been in serious trouble from water pollution, mostly because of human activities. One problem is population growth. Between 1940 and 2007, the number of people living

Drainage basin

NEW YORK

Cooperstown

PENNSYLVANIA

Harrisburg

ATLANTIC
OCEAN

NEW
JERSEY

MARYLAND
Baltimore

WEST
VIRGINIA

Washington

DELAWARE

Rappahannock R.

No oxygen

Low concentrations
of oxygen

Richmond

York R.

James R.

Chesapeake Bay

VIRGINIA

Norfolk

Figure 8-13 *Chesapeake Bay*, the largest estuary in the United States, is severely degraded as a result of water pollution from point and nonpoint sources in six states and from the atmospheric deposition of air pollutants.

in the Chesapeake Bay area grew from 3.7 million to 16.6 million. And the rate of population increase in the bay area has increased since 1990. With more than 450 people moving into the watershed each day, the population may soon reach 17 million.

The estuary receives wastes from point and nonpoint sources scattered throughout a huge drainage basin that includes 9 large rivers and 141 smaller streams and creeks in parts of six states (Figure 8-13). The bay has become a huge pollution sink because only 1% of the waste entering it is flushed into the Atlantic Ocean. It is also so shallow that people can wade through much of it.

Phosphate and nitrate levels have risen sharply in many parts of the bay, causing algal blooms and oxygen depletion. Commercial harvests of its once-abundant oysters, crabs, and several important fishes have fallen sharply since 1960 because of a combination of pollution, overfishing, and disease.

Point sources, primarily sewage treatment plants and industrial plants (often in violation of their discharge permits), account for 60% by weight of the phosphates. Nonpoint sources—mostly runoff of fertilizer and animal wastes from urban, suburban, and agricultural land and deposition from the atmosphere—account for 60% by weight of the nitrates. According to a 2004 study by the Chesapeake Bay Foundation, animal manure is the largest source of nitrates and phosphates from agricultural pollution.

In 1983, the United States implemented the Chesapeake Bay Program. In this ambitious attempt at *integrated coastal management*, citizens' groups, communities, state legislatures, and the federal government are working together to reduce pollution inputs into the bay. Strategies include establishing land-use regulations in the bay's six watershed states to reduce agricultural and urban runoff, banning phosphate detergents, upgrading sewage treatment plants, and monitoring

industrial discharges more closely. In addition, wetlands are being restored and large areas of the bay are being replanted with sea grasses to help filter out nutrients and other pollutants.

A century ago, oysters were so abundant that they filtered and cleaned the Chesapeake's entire volume of water every 3 days. This important form of natural capital helped to remove excess nutrients and reduce algal blooms that decreased dissolved oxygen levels. Now the oyster population has been reduced to the point where this filtration process takes a year.

Officials in the states of Maryland and Virginia are evaluating whether to rebuild the Chesapeake's oyster population by introducing an Asian oyster that appears resistant to two parasites that have killed off many of the bay's native oysters. The Asian oysters grow bigger and faster and taste as good as native oysters. But introducing the nonnative Asian oyster is unpredictable and irreversible, and some researchers warn that this nonnative species may not help to clean the water, because it requires clean water in order to flourish.

The hard work on improving the water quality of the Chesapeake Bay has paid off. Between 1985 and 2000, phosphorus levels declined 27%, nitrogen levels dropped 16%, and grasses growing on the bay's floor have made a comeback. This is a significant achievement, given the increasing population in the watershed and the fact that nearly 40% of the nitrogen inputs come from the atmosphere.

There is still a long way to go, and a sharp drop in state and federal funding has slowed progress. During the summer of 2005, more than 40% of the bay had too little dissolved oxygen to support many kinds of aquatic life. And according to a 2006 report by the Chesapeake Bay Foundation, "the bay's health remains dangerously out of balance." Yet despite some setbacks, the Chesapeake Bay Program shows what can be done when diverse groups work together to achieve goals that benefit both wildlife and people.

THINKING ABOUT
The Chesapeake Bay

What are three ways in which Chesapeake Bay area residents could apply the four **scientific principles of sustainability** (see back cover) to try to improve the environmental quality of the bay?

8-4 Why Are Freshwater Ecosystems Important?

> ► **CONCEPT 8-4** Freshwater ecosystems provide major ecological and economic services and are irreplaceable reservoirs of biodiversity.

Water Stands in Some Freshwater Systems and Flows in Others

Freshwater life zones include *standing* (lentic) bodies of freshwater, such as lakes, ponds, and inland wetlands, and *flowing* (lotic) systems, such as streams and rivers. Although these freshwater systems cover less than 2.2% of the earth's surface, they provide a number of important ecological and economic services (Figure 8-14).

Lakes are large natural bodies of standing freshwater formed when precipitation, runoff, or groundwater seepage fills depressions in the earth's surface. Causes of such depressions include glaciations (Lake Louise, Alberta, Canada), crustal displacement (Lake Nyasa in East Africa), and volcanic activity (Crater Lake in the

U.S. state of Oregon). Lakes are supplied with water from rainfall, melting snow, and streams that drain their surrounding watershed.

Freshwater lakes vary tremendously in size, depth, and nutrient content. Deep lakes normally consist of four distinct zones that are defined by their depth and distance from shore (Figure 8-15). The top layer, called the *littoral* ("LIT-tore-el") *zone,* is near the shore and consists of the shallow sunlit waters to the depth at which rooted plants stop growing. It has a high biological diversity because of ample sunlight and inputs of nutrients from the surrounding land. Species living in the littoral zone include many rooted plants and animals such as turtles, frogs, crayfish, and many fishes such bass, perch, and carp.

Next is the *limnetic* ("lim-NET-ic") *zone:* the open, sunlit surface layer away from the shore that extends to the depth penetrated by sunlight. The main photosynthetic body of the lake, this zone produces the food and oxygen that support most of the lake's consumers. Its most abundant organisms are microscopic phytoplankton and zooplankton. Some large fishes spend most of their time in this zone, with occasional visits to the littoral zone to feed and reproduce.

Next comes the *profundal* ("pro-FUN-dahl") *zone:* the deep, open water where it is too dark for photosynthesis to occur. Without sunlight and plants, oxygen levels are often low here. Fishes adapted to the lake's cooler and darker water are found in this zone.

The bottom of the lake contains the *benthic* ("BEN-thic") *zone,* inhabited mostly by decomposers, detritus feeders, and some fishes. The benthic zone is nourished mainly by dead matter that falls from the littoral and limnetic zones and by sediment washing into the lake.

Some Lakes Have More Nutrients Than Others

Ecologists classify lakes according to their nutrient content and primary productivity. Lakes that have a small supply of plant nutrients are called **oligotrophic** (poorly nourished) **lakes** (Figure 8-16, left). Often, this type of lake is deep and has steep banks.

Glaciers and mountain streams supply water to many such lakes, bringing little in the way of sediment or microscopic life to cloud the water. These lakes usually have crystal-clear water and small populations of phytoplankton and fishes such as smallmouth bass and trout. Because of their low levels of nutrients, these lakes have a low net primary productivity.

NATURAL CAPITAL

Freshwater Systems

Ecological Services	Economic Services
Climate moderation	Food
Nutrient cycling	Drinking water
Waste treatment	Irrigation water
Flood control	Hydroelectricity
Groundwater recharge	Transportation corridors
Habitats for many species	Recreation
Genetic resources and biodiversity	
Scientific information	Employment

Figure 8-14 Major ecological and economic services provided by freshwater systems (**Concept 8-4**). **Question:** Which two ecological services and which two economic services do you think are the most important? Why?

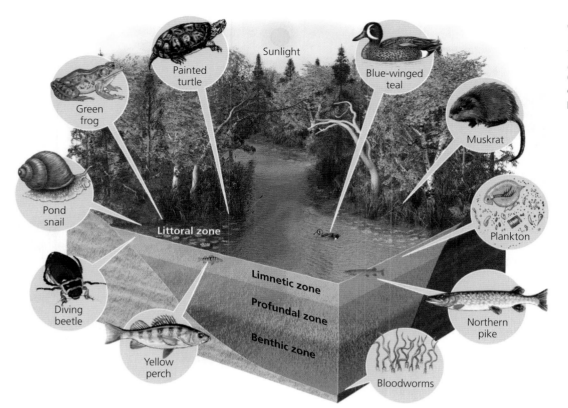

Sunlight

Painted turtle

Green frog

Blue-winged teal

Muskrat

Pond snail

Littoral zone

Plankton

Limnetic zone

Profundal zone

Benthic zone

Diving beetle

Yellow perch

Bloodworms

Northern pike

CENGAGENOW" **Active Figure 8-15**
Distinct zones of life in a fairly deep temperate zone lake. *See an animation based on this figure at* CengageNOW.
Question: How are deep lakes like tropical rain forests? (*Hint:* See Figure 7-17, p. 156)

Over time, sediment, organic material, and inorganic nutrients wash into most oligotrophic lakes, and plants grow and decompose to form bottom sediments. A lake with a large supply of nutrients needed by producers is called a **eutrophic** (well-nourished) **lake** (Figure 8-16, right). Such lakes typically are shallow and have murky brown or green water with high turbidity. Because of their high levels of nutrients, these lakes have a high net primary productivity.

Human inputs of nutrients from the atmosphere and from nearby urban and agricultural areas can accelerate the eutrophication of lakes, a process called **cultural eutrophication.** This process often puts excessive nutrients into lakes, which are then described

Figure 8-16 The effect of nutrient enrichment on a lake. Crater Lake in the U.S. state of Oregon (left) is an example of an *oligotrophic lake* that is low in nutrients. Because of the low density of plankton, its water is quite clear. The lake on the right, found in western New York State, is a *eutrophic lake.* Because of an excess of plant nutrients, its surface is covered with mats of algae and cyanobacteria.

Lake Glacier
Rain and snow Rapids
Waterfall
Tributary
Flood plain
Oxbow lake
Salt marsh
Delta
Deposited sediment
Ocean
Source Zone
Transition Zone
Floodplain Zone
Water
Sediment

Figure 8-17 Three zones in the downhill flow of water: *source zone* containing mountain (headwater) streams; *transition zone* containing wider, lower-elevation streams; and *floodplain zone* containing rivers, which empty into the ocean.

as **hypereutrophic.** Many lakes fall somewhere between the two extremes of nutrient enrichment. They are called **mesotrophic lakes.**

Freshwater Streams and Rivers Carry Water from the Mountains to the Oceans

Precipitation that does not sink into the ground or evaporate becomes **surface water.** It becomes **runoff** when it flows into streams. A **watershed,** or **drainage basin,** is the land area that delivers runoff, sediment, and dissolved substances to a stream. Small streams join to form rivers, and rivers flow downhill to the ocean (Figure 8-17).

In many areas, streams begin in mountainous or hilly areas that collect and release water falling to the earth's surface as rain or snow that melts during warm seasons. The downward flow of surface water and groundwater from mountain highlands to the sea typically takes place in three aquatic life zones characterized by different environmental conditions: the *source zone,* the *transition zone,* and the *floodplain zone.* Rivers and streams can differ somewhat from this generalized model.

In the first, narrow *source zone* (Figure 8-17, top), headwaters, or mountain highland streams are usually shallow, cold, clear, and swiftly flowing. As this

turbulent water flows and tumbles downward over waterfalls and rapids, it dissolves large amounts of oxygen from the air. Most of these streams are not very productive because of a lack of nutrients and primary producers. Their nutrients come primarily from organic matter (mostly leaves, branches, and the bodies of living and dead insects) that falls into the stream from nearby land.

The source zone is populated by cold-water fishes (such as trout in some areas), which need lots of dissolved oxygen. Many fishes and other animals in fast-flowing headwater streams have compact and flattened bodies that allow them to live under stones. Others have streamlined and muscular bodies that allow them to swim in the rapid and strong currents. Most plants are algae and mosses attached to rocks and other surfaces under water.

In the *transition zone* (Figure 8-17, middle), headwater streams merge to form wider, deeper, and warmer streams that flow down gentler slopes with fewer obstacles. They can be more turbid (from suspended sediment), slower flowing, and have less dissolved oxygen than headwater streams have. The warmer water and other conditions in this zone support more producers and cool-water and warm-water fish species (such as black bass) with slightly lower oxygen requirements.

As streams flow downhill, they shape the land through which they pass. Over millions of years, the friction of moving water may level mountains and cut deep canyons, and rock and soil removed by the water

are deposited as sediment in low-lying areas. In these *floodplain zones* (Figure 8-17, bottom), streams join into wider and deeper rivers that flow across broad, flat valleys. Water in this zone usually has higher temperatures and less dissolved oxygen than water in the two higher zones. These slow-moving rivers sometimes support fairly large populations of producers such as algae and cyanobacteria and rooted aquatic plants along the shores.

Because of increased erosion and runoff over a larger area, water in this zone often is muddy and contains high concentrations of suspended particulate matter (silt). The main channels of these slow-moving, wide, and murky rivers support distinctive varieties of fishes (such as carp and catfish), whereas their backwaters support species similar to those present in lakes. At its mouth, a river may divide into many channels as it flows through deltas, built up by deposited sediment, and coastal wetlands and estuaries, where the river water mixes with ocean water (Figure 8-6).

Coastal deltas and wetlands and inland wetlands and floodplains are important parts of the earth's natural capital. They absorb and slow the velocity of floodwaters from coastal storms, hurricanes (see Case Study below), and tsunamis. Deposits of sediments and nutrients at the mouths of rivers build up protective coastal deltas.

Streams receive many of their nutrients from bordering land ecosystems. Such nutrient inputs come from falling leaves, animal feces, insects, and other forms of biomass washed into streams during heavy rainstorms or by melting snow. Thus, the levels and types of nutrients in a stream depend on what is happening in the stream's watershed.

■ CASE STUDY

Dams, Deltas, Wetlands, Hurricanes, and New Orleans

Coastal deltas, mangrove forests, and coastal wetlands provide considerable natural protection against flood damage from coastal storms, hurricanes, typhoons, and tsunamis.

When we remove or degrade these natural speed bumps and sponges, any damage from a natural disaster such as a hurricane or typhoon is intensified. As a result, flooding in places like New Orleans, Louisiana (USA), the U.S. Gulf Coast, and Venice, Italy, are largely self-inflicted unnatural disasters. For example, the U.S. state of Louisiana, which contains about 40% of all coastal wetlands in the lower 48 states, has lost more than a fifth of such wetlands since 1950 to oil and gas wells and other forms of coastal development.

Dams and levees have been built along most of the world's rivers to control water flows and provide electricity (from hydroelectric power plants). This helps to reduce flooding along rivers, but it also traps sediments

that normally are deposited in deltas, which are continually rebuilt by such sediments. As a result, most of the world's river deltas are sinking rather than rising, and their protective coastal wetlands are being flooded. Thus, they no longer provide natural flood protection for coastal communities.

For example, the Mississippi River once delivered huge amounts of sediments to its delta each year. But the multiple dams, levees, and canals in this river system funnel much of this load through the wetlands and out into the Gulf of Mexico. Instead of building up delta lands, this causes them to subside. Other human processes that are increasing such subsidence include extraction of groundwater and oil and natural gas. As freshwater wetlands are lost, saltwater from the Gulf has intruded and killed many plants that depended on river water, further degrading this coastal aquatic system.

This helps to explain why the U.S. city of New Orleans, Louisiana, which was flooded by Hurricane Katrina in 2005 (Figure 8-18), is 3 meters (10 feet)

Figure 8-18 Much of the U.S. city of New Orleans, Louisiana, was flooded by the storm surge that accompanied Hurricane Katrina, which made landfall just east of the city on August 29, 2005. When the surging water rushed through the Mississippi River Gulf Outlet, a dredged waterway on the edge of the city, it breached a floodwall, and parts of New Orleans were flooded with 2 meters (6.5 feet) of water within a few minutes. Within a day, floodwaters reached a depth of 6 meters (nearly 20 feet) in some places; 80% of the city was under water at one point. The hurricane killed more than 1,800 people, and caused more than $100 billion in damages, making it the costliest and deadliest hurricane in the U.S. history. In addition, a variety of toxic chemicals from flooded industrial and hazardous waste sites, as well as oil and gasoline from more than 350,000 ruined cars and other vehicles, were released into the stagnant floodwaters. After the water receded, parts of New Orleans were covered with a thick oily sludge.

Figure 8-19 Projection of how a 1-meter (3.3-foot) rise in sea level from global warming by the end of this century would put New Orleans and much of Louisiana's current coast under water. (Used by permission from Jonathan Overpeck and Jeremy Weiss, University of Arizona)

below sea level and, in the not-too-distant future, will probably be 6 meters (20 feet) below sea level. Add to this the reduction of the protective effects of coastal and inland wetlands and barrier islands and you have a recipe for a major *unnatural* disaster.

To make matters worse, global sea levels have risen almost 0.3 meters (1 foot) since 1900 and are projected to rise 0.3–0.9 meter (1–3 feet) by the end of this century. Figure 8-19 shows a projection of how such a rise in sea level would put New Orleans and other current areas of the Louisiana coast under water. Most of this projected rise is due to the expansion of water and melting ice caused by global warming—another unnatural disaster helped along mostly by our burning of fossil fuels and clearing of large areas of the world's tropical forests.

Governments can spend hundreds of billions of dollars building or rebuilding higher levees around cities such as New Orleans. But sooner or later increasingly stronger hurricanes and rising sea levels will overwhelm these defenses and cause even greater damage and loss of life.

For example, much of New Orleans is a 3-meter (10-foot)-deep bathtub or bowl, with parts of the city 0.9–3 meters (3–9 feet) below sea level. According to engineers, even if we build levees high enough to make it a 6-meter (20-foot)-deep bathtub, a Category 5 hurricane and rising sea levels will eventually overwhelm such defenses and lead to a much more serious unnatural disaster.

The good news is that we now understand some of the connections between dams, deltas, wetlands, barrier islands, sea level rise, and hurricanes. The question is whether we will use such ecological and geological wisdom to change our ways or suffer the increasingly severe consequences of our own actions.

THINKING ABOUT
New Orleans

Do you think that a sinking city such as New Orleans, Louisiana (USA), should be rebuilt and protected with higher levees, or should the lower parts of the city be allowed to revert to wetlands that would help to protect nearby coastal areas? Explain.

Freshwater Inland Wetlands Are Vital Sponges

Inland wetlands are lands covered with freshwater all or part of the time (excluding lakes, reservoirs, and streams) and located away from coastal areas. They include *marshes* (dominated by grasses and reeds with few trees), *swamps* (dominated by trees and shrubs), and *prairie potholes* (depressions carved out by ancient glaciers). Other examples are *floodplains*, which receive excess water during heavy rains and floods, and the wet *arctic tundra* in summer (Figure 7-12, bottom photo, p. 151). Some wetlands are huge; others are small.

Some wetlands are covered with water year-round. Others, called *seasonal wetlands*, remain under water or are soggy for only a short time each year. The latter include prairie potholes, floodplain wetlands, and bottomland hardwood swamps. Some stay dry for years before water covers them again. In such cases, scientists must use the composition of the soil or the presence of certain plants (such as cattails, bulrushes, or red maples) to determine that a particular area is a wetland. Wetland plants are highly productive because of an abundance of nutrients. Many of these wetlands are important habitats for game fishes, muskrats, otters, beavers, migratory waterfowl, and many other bird species.

Inland wetlands provide a number of other free ecological and economic services, which include:

- Filtering and degrading toxic wastes and pollutants

- Reducing flooding and erosion by absorbing storm water and releasing it slowly and by absorbing overflows from streams and lakes

- Helping to replenish stream flows during dry periods

- Helping to recharge groundwater aquifers

- Helping to maintain biodiversity by providing habitats for a variety of species

- Supplying valuable products such as fishes and shellfish, blueberries, cranberries, wild rice, and timber

- Providing recreation for birdwatchers, nature photographers, boaters, anglers, and waterfowl hunters

8-5 How Have Human Activities Affected Freshwater Ecosystems?

▶ **CONCEPT 8-5** Human activities threaten biodiversity and disrupt ecological and economic services provided by freshwater lakes, rivers, and wetlands.

Human Activities Are Disrupting and Degrading Freshwater Systems

Human activities are disrupting and degrading many of the ecological and economic services provided by freshwater rivers, lakes, and wetlands (**Concept 8-5**). Such activities affect freshwater systems in four major ways. *First*, dams, and canals fragment about 40% of the world's 237 large rivers. They alter and destroy terrestrial and aquatic wildlife habitats along rivers and in coastal deltas and estuaries by reducing water flow and increasing damage from coastal storms (Case Study, p. 177). *Second*, flood control levees and dikes built along rivers disconnect the rivers from their floodplains, destroy aquatic habitats, and alter or reduce the functions of nearby wetlands.

For example, the 2,341-mile Missouri River in the west central United States has been harnessed by levees and a series of six dams, which have disrupted the seasonal variations in the flow of the river and changed water temperatures on some stretches. This hinders the growth of insect populations and interferes with spawning cycles of fishes and the feeding habits of shorebirds, and thus severely disrupts entire food webs. The dams and levees also have interrupted sediment flow and distribution, degrading shoreline habitats maintained by sediments. As a result of these disturbances, dozens of native species have declined, and the biodiversity of the ecosystem is threatened.

A *third* major human impact on freshwater systems comes from cities and farms, which add pollutants and excess plant nutrients to nearby streams, rivers, and lakes. For example, runoff of nutrients into a lake (cultural eutrophication, Figure 8-16, right) causes explosions in the populations of algae and cyanobacteria, which deplete the lake's dissolved oxygen. When these organisms die and sink to the lake bottom, decomposers go to work and further deplete the oxygen in deeper waters. Fishes and other species can then die off, causing a major loss in biodiversity.

Fourth, many inland wetlands have been drained or filled to grow crops or have been covered with concrete, asphalt, and buildings. See the following Case Study.

■ CASE STUDY

Inland Wetland Losses in the United States

About 95% of the wetlands in the United States contain freshwater and are found inland. The remaining 5% are saltwater or coastal wetlands. Alaska has more of the nation's inland wetlands than the other 49 states put together.

More than half of the inland wetlands estimated to have existed in the continental United States during the 1600s no longer exist. About 80% of lost wetlands were destroyed to grow crops. The rest were lost to mining, forestry, oil and gas extraction, highways, and urban development. The heavily farmed U.S. state of Iowa has lost about 99% of its original inland wetlands.

This loss of natural capital has been an important factor in increased flood and drought damage in the United States—more examples of unnatural disasters. Many other countries have suffered similar losses. For example, 80% of all wetlands in Germany and France have been destroyed.

We look further into human impacts on aquatic systems in Chapter 11. There, we also explore possible solutions to environmental problems that result from these impacts, as well as ways to sustain aquatic biodiversity.

This chapter's opening case study pointed out the ecological and economic importance of the world's incredibly diverse coral reefs. They are living examples of the four **scientific principles of sustainability** in action. They thrive on solar energy, participate in the cycling of carbon and other chemicals, are a prime example of aquatic biodiversity, and have a network of interactions among species that helps to maintain sustainable population sizes.

In this chapter, we have seen that coral reefs and other aquatic systems are being severely stressed by a variety of human activities. Research shows when such harmful human activities are reduced, coral reefs and other stressed aquatic systems can recover fairly quickly.

In other words, from a scientific standpoint, we know what to do. Whether or not we act is primarily a political and ethical problem. This requires educating leaders and citizens about the ecological and economic importance of the earth's aquatic ecosystems and about the need to include the economic values of such ecosystem services in the prices of goods and services generated with the use of these resources. Solving these problems also requires individual citizens to put pressure on elected officials and business leaders to change the ways we treat these important repositories of natural capital.

> *. . . the sea, once it casts its spell, holds one in its net of wonders forever.*
>
> JACQUES-YVES COUSTEAU

REVIEW

1. Review the Key Questions and Concepts for this chapter on p. 163. What is a **coral reef** and why should we care about coral reefs? What is coral bleaching?

2. What percentage of the earth's surface is covered with water? What is an **aquatic life zone?** Distinguish between a **saltwater (marine)** life zone and a **freshwater** life zone. What major types of organisms live in the top, middle, and bottom layers of aquatic life zones? Define **plankton** and describe three types of plankton. Distinguish among **nekton, benthos,** and **decomposers** and give an example of each. What five factors determine the types and numbers of organisms found in the three layers of aquatic life zones? What is **turbidity,** and how does it occur? Describe one of its harmful impacts.

3. What major ecological and economic services are provided by marine systems? What are the three major life zones in an ocean? Distinguish between the **coastal zone** and the **open sea.** Distinguish between an **estuary** and a **coastal wetland** and explain why they have high net primary productivities. What is a **mangrove forest** and what is its ecological and economic importance? What is the **intertidal zone?** Distinguish between rocky and sandy shores. Why does the open sea have a low net primary productivity?

4. What human activities pose major threats to marine systems and to coral reefs?

5. Explain why the Chesapeake Bay is an estuary in trouble. What is being done about some of its problems?

6. What major ecological and economic services do freshwater systems provide? What is a **lake?** What four zones are found in most lakes? Distinguish among **oligotrophic, eutrophic, hypereutrophic,** and **mesotrophic lakes.** What is **cultural eutrophication?**

7. Define **surface water, runoff,** and **watershed (drainage basin).** Describe the three zones that a stream passes through as it flows from mountains to the sea. Describe the relationships between dams, deltas, wetlands, hurricanes, and flooding in New Orleans, Louisiana (USA).

8. Give three examples of **inland wetlands** and explain the ecological importance of such wetlands.

9. What are four ways in which human activities are disrupting and degrading freshwater systems? Describe inland wetlands in the United States in terms of the area of wetlands lost and the resulting loss of ecological and economic services.

10. How is the degradation of many of the earth's coral reefs (**Core Case Study**) a reflection of our failure to follow the four **scientific principles of sustainability?** Describe this connection for each principle.

*Note: Key Terms are in **bold** type.*

1. List three ways in which you could apply **Concepts 8-3** and **8-5** to make your lifestyle more environmentally sustainable.

2. What are three steps governments and industries could take to protect remaining coral reefs (**Core Case Study**)? What are three ways in which individuals can help to protect those reefs?

3. Why do aquatic plants such as phytoplankton tend to be very small, whereas most terrestrial plants such as trees tend to be larger and have more specialized structures such as stems and leaves for growth?

4. Why are some aquatic animals, especially marine mammals such as whales, extremely large compared with terrestrial animals?

5. How would you respond to someone who proposes that we use the deep portions of the world's oceans to deposit our radioactive and other hazardous wastes because the deep oceans are vast and are located far away from human habitats? Give reasons for your response.

6. Suppose a developer builds a housing complex overlooking a coastal salt marsh and the result is pollution and degradation of the marsh. Describe the effects of such a development on the wildlife in the marsh, assuming at least one species is eliminated as a result. (See Figure 8-7.)

7. How does a levee built on a river affect species such as deer and hawks living in a forest overlooking the river?

8. Suppose you have a friend who owns property that includes a freshwater wetland, and the friend tells you he is planning to fill the wetland to make more room for his lawn and garden. What would you say to this friend?

9. Congratulations! You are in charge of the world. What are the three most important features of your plan to help sustain the earth's aquatic biodiversity?

10. List two questions that you would like to have answered as a result of reading this chapter.

Note: See Supplement 13 (p. S78) for a list of Projects related to this chapter.

CORE CASE STUDY

DATA ANALYSIS

At least 25% of the world's coral reefs have been severely damaged. A number of factors have played a role in this serious loss of aquatic biodiversity (Figure 8-12, right, p. 172), including ocean warming, sediment from coastal soil erosion, excessive algae growth from fertilizer runoff, coral bleaching, rising sea levels, overfishing, and damage from hurricanes.

In 2005, scientists Nadia Bood, Melanie McField, and Rich Aronson conducted research to evaluate the recovery of coral reefs in Belize from the combined effects of mass bleaching and Hurricane Mitch in 1998. Some of these reefs are in protected waters where no fishing is allowed. The researchers speculated that reefs in waters where no fishing is allowed should recover faster than reefs in water where fishing is allowed. The graphs below show some of the data they collected from three highly protected (no fishing) sites and three unprotected (fishing) sites to evaluate their hypothesis.

Effects of restricting fishing on the recovery of unfished and fished coral reefs damaged by the combined effects of mass bleaching and Hurricane Mitch in 1998. (Data from Melanie McField, et al., *Status of Caribbean Coral Reefs After Bleaching and Hurricanes in 2005,* NOAA, 2008. (Report available at **www.coris.noaa.gov/activities/caribbean_rpt/**)

1. By about what percentage did the mean coral cover drop in the protected (unfished) reefs between 1997 and 1999?

2. By about what percentage did the mean coral cover drop in the protected (unfished) reefs between 1997 and 2005?

3. By about what percentage did the coral cover drop in the unprotected (fished) reefs between 1997 and 1999?

4. By about what percentage did the coral cover change in the unprotected (fished) reefs between 1997 and 2005?

5. Do these data support the hypothesis that coral reef recovery should occur faster in areas where fishing is prohibited? Explain.

LEARNING ONLINE

Log on to the Student Companion Site for this book at **academic.cengage.com/biology/miller**, and choose Chapter 8 for many study aids and ideas for further reading and research. These include flash cards, practice quizzing, Weblinks, information on Green Careers, and InfoTrac® College Edition articles.

The Passenger Pigeon: Gone Forever

In 1813, bird expert John James Audubon saw a single huge flock of passenger pigeons that took three days to fly past him and was so dense that it darkened the skies.

By 1900, North America's passenger pigeon (Figure 9-1), once the most numerous bird species on earth, had disappeared from the wild because of a combination of uncontrolled commercial hunting and habitat loss as forests were cleared to make room for farms and cities. These birds were good to eat, their feathers made good pillows, and their bones were widely used for fertilizer. They were easy to kill because they flew in gigantic flocks and nested in long, narrow, densely packed colonies.

Commercial hunters would capture one pigeon alive, sew its eyes shut, and tie it to a perch called a stool. Soon a curious flock would land beside this "stool pigeon"—a term we now use to describe someone who turns in another person for breaking the law. Then the birds would be shot or ensnared by nets that could trap more than 1,000 of them at once.

Beginning in 1858, passenger pigeon hunting became a big business. Shotguns, traps, artillery, and even dynamite were used. People burned grass or sulfur below their roosts to suffocate the birds. Shooting galleries used live birds as targets. In 1878, one professional pigeon trapper made $60,000 by killing 3 million birds at their nesting grounds near Petoskey, Michigan.

By the early 1880s, only a few thousand birds remained. At that point, recovery of the species was doomed because the females laid only one egg per nest each year. On March 24, 1900, a young boy in the U.S. state of Ohio shot the last known wild passenger pigeon.

Eventually all species become extinct or evolve into new species. The archeological record reveals five *mass extinctions* since life on the earth began—each a massive impoverishment of life on the earth. These mass extinctions were caused by natural phenomena, such as major climate change or large asteroids hitting the earth, which drastically altered the earth's environmental conditions.

There is considerable evidence that we are now in the early stage of a sixth great extinction. Evidence indicates that we humans are causing this mass extinction as our population grows and as we consume more resources, disturb more land and aquatic systems, use more of the earth's net primary productivity, and cause changes to the earth's climate.

Scientists project that during this century, human activities, especially those that cause habitat destruction and climate change, will lead to the premature extinction of one-fourth to one-half of the world's plant and animal species—an incredibly rapid rate of extinction. And there will be no way to restore what we have lost, because species extinction is forever. If we keep impoverishing the earth's biodiversity, eventually, our species will also become impoverished.

Michael Sewell/Peter Arnold, Inc.

Figure 9-1 Lost natural capital: passenger pigeons have been extinct in the wild since 1900 because of human activities. The last known passenger pigeon died in the U.S. state of Ohio's Cincinnati Zoo in 1914.

9-1 What role do humans play in the premature extinction of species?

CONCEPT 9-1A We are degrading and destroying biodiversity in many parts of the world, and these threats are increasing.

CONCEPT 9-1B Species are becoming extinct 100 to 1,000 times faster than they were before modern humans arrived on the earth (the *background rate*), and by the end of this century, the extinction rate is expected to be 10,000 times the background rate.

9-2 Why should we care about preventing premature species extinction?

CONCEPT 9-2 We should prevent the premature extinction of wild species because of the economic and ecological services they provide and because they have a right to exist regardless of their usefulness to us.

9-3 How do humans accelerate species extinction?

CONCEPT 9-3 The greatest threats to any species are (in order) loss or degradation of its habitat, harmful invasive species, human population growth, pollution, climate change, and overexploitation.

9-4 How can we protect wild species from extinction resulting from our activities?

CONCEPT 9-4A We can use existing environmental laws and treaties and work to enact new laws designed to prevent premature species extinction and protect overall biodiversity.

CONCEPT 9-4B We can help to prevent premature species extinction by creating and maintaining wildlife refuges, gene banks, botanical gardens, zoos, and aquariums.

CONCEPT 9-4C According to the *precautionary principle*, we should take measures to prevent or reduce harm to the environment and to human health, even if some of the cause-and-effect relationships have not been fully established, scientifically.

Note: Supplements 2 (p. S4), 4 (p. S20), 9 (p. S53), and 13 (p. S78) can be used with this chapter.

The last word in ignorance is the person who says of an animal or plant:
"What good is it?" . . . If the land mechanism as a whole is good,
then every part of it is good, whether we understand it or not. . . .
Harmony with land is like harmony with a friend;
you cannot cherish his right hand and chop off his left.

ALDO LEOPOLD

9-1 What Role Do Humans Play in the Premature Extinction of Species?

▶ **CONCEPT 9-1A** We are degrading and destroying biodiversity in many parts of the world, and these threats are increasing.

▶ **CONCEPT 9-1B** Species are becoming extinct 100 to 1,000 times faster than they were before modern humans arrived on the earth (the *background rate*), and by the end of this century, the extinction rate is expected to be 10,000 times the background rate.

Human Activities Are Destroying and Degrading Biodiversity

We have depleted and degraded some of the earth's biodiversity, and these threats are expected to increase (**Concept 9-1A**). According to biodiversity expert Edward O. Wilson, "The natural world is everywhere disappearing before our eyes—cut to pieces, mowed down, plowed under, gobbled up, replaced by human artifacts."

According to the 2005 Millennium Ecosystem Assessment and other studies, humans have disturbed, to some extent, at least half and probably about 83% of the earth's land surface (excluding Antarctica and Greenland; see Figure 3, pp. S24–S25, in Supplement 4). Most of this disturbance involves filling in wetlands or converting grasslands and forests to crop fields and urban areas. Such disturbances eliminate large numbers of species by destroying or degrading their habitats, as discussed in more detail in Chapter 10.

Human activities are also degrading the earth's *aquatic biodiversity,* as discussed in Chapter 11.

Extinctions Are Natural but Sometimes They Increase Sharply

In due time, all species become extinct. During most of the 3.56 billion years that life has existed on the earth, there has been a continuous, low level of extinction of species known as **background extinction.**

An **extinction rate** is expressed as a percentage or number of species that go extinct within a certain time period such as a year. For example, one extinction per million species per year would be $1/1,000,000 = 0.000001$ species per year. Expressed as a percentage, this is 0.000001×100, or 0.0001%—the estimated *background extinction rate* existing before humans came on the scene.

The balance between formation of new species and extinction of existing species determines the earth's biodiversity (**Concept 4-4A**, p. 86). Overall, the earth's biodiversity has increased for several hundred million years, except during a few periods. The extinction of many species in a relatively short period of geologic time is called a **mass extinction.** Geological and other records indicate that the earth has experienced five mass extinctions when 50–95% of the world's species appear to have become extinct. After each mass extinction, biodiversity eventually returned to equal or higher levels, but each recovery required millions of years.

The causes of past mass extinctions are poorly understood but probably involved global changes in environmental conditions. Examples are major climate change or a large-scale catastrophe such as a collision between the earth and a comet or large asteroid. The last of these mass extinctions took place about 65 million years ago. One hypothesis is that the last mass extinction taking place about 65 million years ago occurred after a large asteroid hit the planet and spewed huge amounts of dust and debris into the atmosphere.

This could have reduced the input of solar energy and cooled the planet enough to wipe out the dinosaurs and many other forms of earth's life at that time.

Biologists distinguish among three levels of species extinction. *Local extinction* occurs when a species is no longer found in an area it once inhabited but is still found elsewhere in the world. Most local extinctions involve losses of one or more populations of species. *Ecological extinction* occurs when so few members of a species are left that it can no longer play its ecological roles in the biological communities where it is found.

In *biological extinction,* a species, such as the passenger pigeon (**Core Case Study**, Figure 9-1) is no longer found anywhere on the earth. Biological extinction is forever and represents an irreversible loss of natural capital.

CORE CASE STUDY

Some Human Activities Cause Premature Extinctions, and the Pace Is Speeding Up

Although extinction is a natural biological process, it has accelerated as human populations have spread over the globe, consuming large quantities of resources, and creating large ecological footprints (Figure 1-10, p. 15). As a result, human activities are destroying the earth's biodiversity at an unprecedented and accelerating rate. Figure 9-2 shows a few of the many species that have become prematurely extinct mostly because of human activities.

Using the methods described in the Science Focus box (p. 188), scientists from around the world, who published the 2005 Millennium Ecosystem Assessment, estimated that the current annual rate of species extinction is at least 100 to 1,000 times the background rate of about 0.0001%, which existed before modern humans appeared some 150,000 years ago. This amounts to an extinction rate of 0.01% to 0.1% a year. Conservation biologists project that during this century the extinction rate caused by habitat loss, climate change mostly due to global warming, and other

Passenger pigeon Great auk Dodo Golden toad Aepyornis (Madagascar)

Figure 9-2 Lost natural capital: some animal species that have become prematurely extinct largely because of human activities, mostly habitat destruction and overhunting. **Question:** Why do you think birds top the list of extinct species?

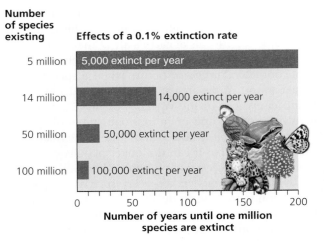

Number of species existing

Effects of a 0.1% extinction rate

5 million	5,000 extinct per year
14 million	14,000 extinct per year
50 million	50,000 extinct per year
100 million	100,000 extinct per year

0 50 100 150 200

Number of years until one million species are extinct

Figure 9-3 Effects of a 0.1% extinction rate.

human activities will increase to 10,000 times the background rate (**Concept 9-1B**). This will amount to an annual extinction rate of 1% per year.

How many species are we losing prematurely each year? The answer depends on how many species are on the earth and the rate of species extinction. Assuming that the extinction rate is 0.1%, each year we lose 5,000 species if there are 5 million species on earth and 14,000 species if there are 14 million species—the current best scientific estimate. See Figure 9-3 for more examples.

According to researchers Edward O. Wilson and Stuart Pimm, at a 1% extinction rate, at least one-fourth of the world's current animal and plant species could be gone by 2050 and half could vanish by the end of this century. In the words of biodiversity expert Norman Myers, "Within just a few human generations, we shall—in the absence of greatly expanded conservation efforts—impoverish the biosphere to an extent that will persist for at least 200,000 human generations or twenty times longer than the period since humans emerged as a species."

THINKING ABOUT

Extinction

How might your lifestyle change if human activities cause the premature extinction of up to half of the world's species in your lifetime? List three aspects of your lifestyle that contribute to this threat to the earth's natural capital.

Most extinction experts consider extinction rates of 0.01–1% to be conservative estimates (**Concept 9-1B**) for several reasons. *First*, both the rate of species loss and the extent of biodiversity loss are likely to increase during the next 50–100 years because of the projected growth of the world's human population and resource use per person (Figure 1-10, p. 15, and Figure 3, pp. S24–S25, in Supplement 4) and climate change caused mostly by global warming.

Second, current and projected extinction rates are much higher than the global average in parts of the world that are highly endangered centers of biodiver-

sity. Conservation biologists urge us to focus our efforts on slowing the much higher rates of extinction in such *hotspots* as the best and quickest way to protect much of the earth's biodiversity from being lost prematurely. (We discuss this further in Chapter 10.)

Third, we are eliminating, degrading, fragmenting, and simplifying many biologically diverse environments—such as tropical forests, tropical coral reefs, wetlands, and estuaries—that serve as potential colonization sites for the emergence of new species (**Concept 4-4B**, p. 86). Thus, in addition to increasing the rate of extinction, we may be limiting the long-term recovery of biodiversity by reducing the rate of speciation for some species. In other words, we are creating a *speciation crisis*. (See the Guest Essay by Normal Myers on this topic at CengageNOW™.) CONCEPT LINK

Philip Levin, Donald Levin, and other biologists also argue that the increasing fragmentation and disturbance of habitats throughout the world may increase the speciation rate for rapidly reproducing opportunist species such as weeds, rodents, and cockroaches and other insects. Thus, the real threat to biodiversity from current human activities may be long-term erosion in the earth's variety of species and habitats. Such a loss of biodiversity would reduce the ability of life to adapt to changing conditions by creating new species.

Endangered and Threatened Species Are Ecological Smoke Alarms

Biologists classify species heading toward biological extinction as either *endangered* or *threatened*. An **endangered species** has so few individual survivors that the species could soon become extinct over all or most of its natural range (the area in which it is normally found). Like the passenger pigeon (**Core Case Study**, Figure 9-1) and several other bird species (Figure 9-2), they may soon disappear from the earth. A **threatened species** (also known as a vulnerable species) is still abundant in its natural range but, because of declining numbers, is likely to become endangered in the near future. CORE CASE STUDY

The International Union for the Conservation of Nature and Natural Resources (IUCN)—also known as the World Conservation Union—is a coalition of the world's leading conservation groups. Since the 1960s, it has published annual *Red Lists*, which have become the world standard for listing the world's threatened species. In 2007, the list included 16,306 plants and animals that are in danger of extinction—60% higher than the number listed in 1995. Those compiling the list say it greatly underestimates the true number of threatened species because only a tiny fraction of 1.8 million known species have been assessed, and of the estimated total of 4–100 million additional species that have not been catalogued or studied. You can examine the Red Lists database online at **www.iucnredlist.org**. Figure 9-4 shows a few of the roughly 1,300 species

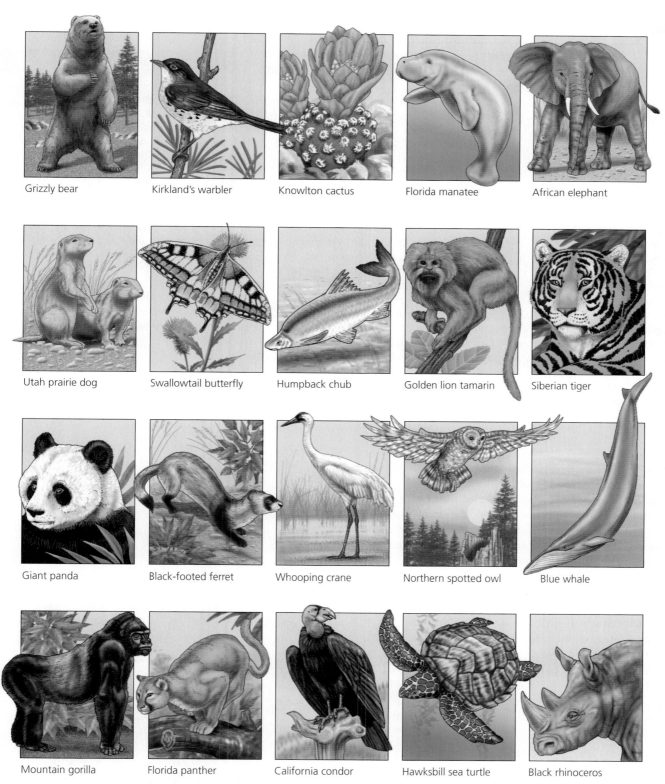

Figure 9-4 Endangered natural capital: some species that are endangered or threatened with premature extinction largely because of human activities. Almost 30,000 of the world's species and roughly 1,300 of those in the United States are officially listed as being in danger of becoming extinct. Most biologists believe the actual number of species at risk is much larger.

Grizzly bear

Kirkland's warbler

Knowlton cactus

Florida manatee

African elephant

Utah prairie dog

Swallowtail butterfly

Humpback chub

Golden lion tamarin

Siberian tiger

Giant panda

Black-footed ferret

Whooping crane

Northern spotted owl

Blue whale

Mountain gorilla

Florida panther

California condor

Hawksbill sea turtle

Black rhinoceros

SCIENCE FOCUS

Estimating Extinction Rates Is Not Easy

Conservation biologists who are trying to catalog extinctions, estimate past extinction rates, and project future rates have three problems. *First,* the extinction of a species typically takes such a long time that it is not easy to document. *Second,* we have identified only about 1.8 million of the world's estimated 4 million to 100 million species. *Third,* scientists know little about the nature and ecological roles of most of the species that have been identified.

One approach to estimating future extinction rates is to study records documenting the rates at which mammals and birds (which have been the easiest to observe) have become extinct since humans arrived and to compare this with fossil records of extinctions prior to the arrival of humans. Determining the rates at which minor DNA copying mistakes occur can help scientists to track how long various species typically last before becoming extinct. Such evidence indicates that

under normal circumstances, species survive for 1 million to 10 million years before going extinct.

Another approach is to observe how the number of species present increases with the size of an area. This *species–area relationship* suggests that, on average, a 90% loss of habitat causes the extinction of 50% of the species living in that habitat. This is based on the *theory of island biogeography* (Science Focus, p. 90). Scientists use this model to estimate the number of current and future extinctions in patches or "islands" of shrinking habitat surrounded by degraded habitats or by rapidly growing human developments.

Scientists also use mathematical models to estimate the risk of a particular species becoming endangered or extinct within a certain period of time. These models include factors such as trends in population size, changes in habitat availability, interactions with other species, and genetic factors. Re-

searchers know that their estimates of extinction rates are based on insufficient data and sampling and incomplete models. They are continually striving to get more data and to improve the models used to estimate extinction rates.

At the same time, they point to clear evidence that human activities have accelerated the rate of species extinction and that this rate is still increasing. According to these biologists, arguing over the numbers and waiting to get better data and models should not be used as excuses for inaction. They agree with the advice of Aldo Leopold (Individuals Matter, p. 22) in his thoughts about preventing premature extinction: "To keep every cog and wheel is the first precaution of intelligent tinkering."

Critical Thinking

How would you improve the estimation of extinction rates?

officially listed as endangered and protected under the U.S. Endangered Species Act.

Some species have characteristics that make them especially vulnerable to ecological and biological extinction (Figure 9-5). As biodiversity expert Edward O. Wilson puts it, "The first animal species to go are the big, the slow, the tasty, and those with valuable parts such as tusks and skins."

Some species also have *behavioral characteristics* that make them prone to extinction. The passenger pigeon (**Core Case Study**, Figure 9-1) and the Carolina parakeet nested in large flocks that made them easy to kill. Key deer, which live only in the U.S. Florida Keys, are "nicotine addicts" that get killed by cars because they forage for cigarette butts along highways. Some types of species are more threatened with premature extinction from human activities than others are (Figure 9-6).

CORE
CASE
STUDY

> — RESEARCH FRONTIER —
>
> Identifying and cataloguing the millions of unknown species and improving models for estimating extinction rates. See **academic.cengage.com/biology/miller**.

Figure 9-5 Characteristics of species that are prone to ecological and biological extinction. **Question:** Which of these characteristics helped lead to the premature extinction of the passenger pigeon within a single human lifetime?

Characteristic		Examples
Low reproductive rate (K-strategist)		Blue whale, giant panda, rhinoceros
Specialized niche		Blue whale, giant panda, Everglades kite
Narrow distribution		Elephant seal, desert pupfish
Feeds at high trophic level		Bengal tiger, bald eagle, grizzly bear
Fixed migratory patterns		Blue whale, whooping crane, sea turtle
Rare		African violet, some orchids
Commercially valuable		Snow leopard, tiger, elephant, rhinoceros, rare plants and birds
Large territories		California condor, grizzly bear, Florida panther

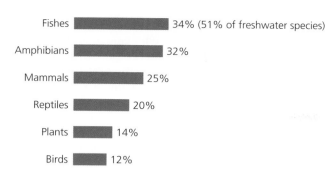

Fishes ▬▬▬▬▬▬▬ 34% (51% of freshwater species)

Amphibians ▬▬▬▬▬▬▬ 32%

Mammals ▬▬▬▬▬ 25%

Reptiles ▬▬▬▬ 20%

Plants ▬▬▬ 14%

Birds ▬▬ 12%

Figure 9-6 Endangered natural capital: percentage of various types of species threatened with premature extinction because of human activities (**Concept 9-1A**). **Question:** Why do you think fishes top this list? (Data from World Conservation Union, Conservation International, World Wide Fund for Nature, 2005 Millennium Ecosystem Assessment, and the Intergovernmental Panel on Climate Change)

9-2 Why Should We Care about Preventing Premature Species Extinction?

▶ **CONCEPT 9-2** We should prevent the premature extinction of wild species because of the economic and ecological services they provide and because they have a right to exist regardless of their usefulness to us.

Species Are a Vital Part of the Earth's Natural Capital

So what is all the fuss about? If all species eventually become extinct, why should we worry about premature extinctions? Does it matter that the passenger pigeon (**Core Case Study**) became prematurely extinct because of human activities, or that the remaining orangutans (Figure 9-7) or some unknown plant or insect in a tropical forest might suffer the same fate?

CORE CASE STUDY

New species eventually evolve to take the places of those lost through mass extinctions. So why should we care if we speed up the extinction rate over the next 50–100 years? The answer: because it will take 5–10 million years for natural speciation to rebuild the biodiversity we are likely to destroy during your lifetime.

Biodiversity researchers say we should act now to prevent premature extinction of species partly for their **instrumental value**—their usefulness to us in providing many of the ecological and economic services that

Figure 9-7 Natural capital degradation: endangered orangutans in a tropical forest. In 1900, there were over 315,000 wild orangutans. Now there are less than 20,000 and they are disappearing at a rate of over 2,000 per year because of illegal smuggling and clearing of their forest habitat in Indonesia and Malaysia to make way for oil palm plantations. An illegally smuggled orangutan typically sells for a street price of $10,000. According to 2007 study by the World Wildlife Fund (WWF), projected climate change will further devastate remaining orangutan populations in Indonesia and Malaysia. **Question:** How would you go about trying to set a price on the ecological value of an orangutan?

age fotostock/SuperStock

make up the earth's natural capital (Figure 1-3, p. 8) (**Concept 9-2**).

Instrumental values take two forms. One is *use values,* which benefit us in the form of economic goods and services, ecological services, recreation, scientific information, and the continuation of such uses for future generations. Each year, Americans, as a whole, spend more than three times as many hours watching wildlife—doing nature photography and bird watching, for example—as they spend watching movies or professional sporting events. A diversity of plant species provides economic value in the form of food crops, fuelwood, lumber, paper, drugs, and medicines (Figure 9-8). Bioprospectors search tropical forests and other ecosystems for plants and animals that have chemicals that can be converted into useful medicinal drugs. A 2005 United Nations University report concluded that 62% of all cancer drugs were derived from the discoveries of bioprospectors. **GREEN CAREER:** Bioprospecting

Species diversity also provides economic benefits from wildlife tourism, or *ecotourism,* which generates between $950,000 and $1.8 million per minute in tourist expenditures worldwide. Conservation biologist Michael Soulé estimates that one male lion living to age 7 generates $515,000 in tourist dollars in Kenya, but only $1,000 if killed for its skin. Similarly, over a lifetime of 60 years, a Kenyan elephant is worth about $1 million in ecotourist revenue—many times more than its tusks are worth when they are sold illegally for their ivory (Science Focus, at right). Ecotourism should

not cause ecological damage, but some of it does. The website for this chapter lists some guidelines for evaluating eco-tours. **GREEN CAREER:** Ecotourism guide

Another instrumental value is the *genetic information* that allows species to adapt to changing environmental conditions through evolution. Genetic engineers use this information to produce genetically modified crops and foods. Scientists warn of the alarming loss of genetic diversity resulting from our increasing reliance on a small number of crop plants for feeding the world. Such a loss also results from the premature extinction of wild plants whose genes could be used by genetic engineers to develop improved crop varieties.

One of the tragedies of the current extinction crisis is that we do not know what we are losing, because no one has ever seen or named many of the species that are becoming extinct. Consequently, we know nothing about their genetic makeup, their roles in sustaining ecosystems, or how they might be used to improve human welfare. Carelessly eliminating many of the species that make up the world's vast genetic library is like burning books that we have never read.

The other major form of instrumental value is *non-use values,* of which there are several types. For example, there is *existence value*—the satisfaction of knowing that a redwood forest, a wilderness, orangutans (Figure 9-7), and wolf packs exist, even if we will never see them or get direct use from them. For many people, biodiversity holds *aesthetic value.* For example, we can appreciate a tree, an orangutan, or a tropical bird (Figure 9-9) for its beauty. A third type is *bequest value,*

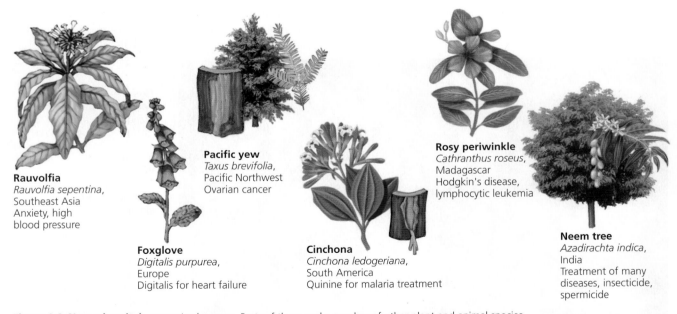

Rauvolfia
Rauvolfia sepentina,
Southeast Asia
Anxiety, high
blood pressure

Foxglove
Digitalis purpurea,
Europe
Digitalis for heart failure

Pacific yew
Taxus brevifolia,
Pacific Northwest
Ovarian cancer

Cinchona
Cinchona ledogeriana,
South America
Quinine for malaria treatment

Rosy periwinkle
Cathranthus roseus,
Madagascar
Hodgkin's disease,
lymphocytic leukemia

Neem tree
Azadirachta indica,
India
Treatment of many
diseases, insecticide,
spermicide

Figure 9-8 Natural capital: *nature's pharmacy.* Parts of these and a number of other plant and animal species (many of them found in tropical forests) are used to treat a variety of human ailments and diseases. Nine of the ten leading prescription drugs originally came from wild organisms. About 2,100 of the 3,000 plants identified by the National Cancer Institute as sources of cancer-fighting chemicals come from tropical forests. Despite their economic and health potential, fewer than 1% of the estimated 125,000 flowering plant species in tropical forests (and a mere 1,100 of the world's 260,000 known plant species) have been examined for their medicinal properties. Once the active ingredients in the plants have been identified, they can usually be produced synthetically. Many of these tropical plant species are likely to become extinct before we can study them.

SCIENCE FOCUS

Using DNA to Reduce Illegal Killing of Elephants for Their Ivory

There are about 400,000 elephants remaining in the wild, most of them in Africa and the rest in Asia. Elephants have long been valued for the ivory in their tusks, but in 1989, an international treaty instituted a ban on the trading of such ivory.

Before the 1989 treaty, poachers slaughtered about 87,000 elephants a year for their ivory. Although this treaty reduced the poaching, it did not stop it. In recent years, the illegal slaughter of elephants for ivory has increased again to about 25,000 a year, most of it in the African countries of Cameroon, Nigeria, and Democratic Republic of the Congo. This poaching has risen sharply since 2004, fueled by sharply rising prices of high-quality ivory, mostly in China. Because of the money to be made, organized crime has become more heavily involved in this illegal trade.

In 2007, scientists began developing a DNA-based map of elephant populations that allows them to use DNA analysis of seized illegal ivory to determine where the elephants were killed. They hope to use such data to identify poaching hot spots and help international law enforcement authorities to focus their anti-poaching efforts.

On the other hand, elephant populations have exploded in some areas, such as parts of South Africa, and are destroying vegetation and affecting the populations of other species.

A single adult elephant devours up to 300 kilograms (660 pounds) of grass, leaves, and twigs a day. Elephants can also uproot trees and disturb the soil. The South African government is considering killing (culling) enough elephants each year to keep the population down in selected areas. The culled ivory would be sold in the international marketplace with the proceeds used to benefit the populations of local villagers and to help pay for conservation efforts. DNA analysis could be used to distinguish such culled ivory from poached ivory.

Supporters say that increasing the amount of ivory legally available in the marketplace could help to reduce poaching by lowering market prices. Some animal rights groups oppose elephant culling on ethical grounds.

Critical Thinking

Do you favor culling elephants in areas where large populations are degrading vegetation? Explain.

based on the fact that people will pay to protect some forms of natural capital for use by future generations.

Finally, species diversity holds *ecological value*, because it is a vital component of the key ecosystem functions of energy flow, nutrient cycling (Figure 3-12, p. 60), and population control, in keeping with three of the four scientific principles of sustainability (see back cover). In other words, the species in ecosystems provide essential *ecosystem* or *natural services*, an important component of the natural capital (Figure 1-3, p. 8, orange items) that supports and sustains the earth's life and economies. Thus, in protecting species from premature extinction and in protecting their vital habitats from environmental degradation (as we discuss in the next chapter), we are helping to sustain our own health and well-being.

Are We Ethically Obligated to Prevent Premature Extinction?

Some scientists and philosophers believe that each wild species has **intrinsic** or **existence value** based on its inherent right to exist and play its ecological roles, regardless of its usefulness to us (**Concept 9-2**). According to this view, we have an ethical responsibility to protect species from becoming prematurely extinct as a result of human activities and to prevent the degradation of the world's ecosystems and its overall biodiversity.

Each species in the encyclopedia of life is a masterpiece of evolution that possesses a unique combination

Figure 9-9 Many species of wildlife, such as this endangered scarlet macaw in Brazil's Amazon rain forest, are a source of beauty and pleasure. These and other colorful species of parrots can become endangered when they are removed from the wild and sold (sometimes illegally) as pets.

Why Should We Care about Bats?

Worldwide there are 950 known species of bats—the only mammals that can fly. But bats have two traits that make them vulnerable to extinction. *First,* they reproduce slowly. *Second,* many bat species live in huge colonies in caves and abandoned mines, which people sometimes close up. This prevents them from leaving to get food, or it can interrupt their hibernation if they have to leave their shelter to escape being trapped.

Bats play important ecological roles. About 70% of all bat species feed on crop-damaging nocturnal insects (Figure 5-3, p. 105) and other insect pest species such as mosquitoes. This makes them the major nighttime SWAT team for such insects.

In some tropical forests and on many tropical islands, *pollen-eating bats* pollinate flowers, and *fruit-eating bats* distribute plants throughout these forests by excreting undigested seeds. As keystone species, such bats are vital for maintaining plant biodiversity and for regenerating large areas of tropical forest that has been cleared by humans. If you enjoy bananas, cashews, dates, figs, avocados, or mangos, you can thank bats.

Many people mistakenly view bats as fearsome, filthy, aggressive, rabies-carrying bloodsuckers. But most bat species are harmless to people, livestock, and crops. In the United States, only 10 people have died of bat-transmitted disease in more than 4 decades of record keeping; more Americans die each year from being hit by falling coconuts.

Because of unwarranted fears of bats and lack of knowledge about their vital ecological roles, several bat species have been driven to extinction. Currently, about one-fourth of the world's bat species are listed as endangered or threatened. And thousands of bats are dying from an unknown illness in the northeastern United States. Because of the important ecological and economic roles they play, conservation biologists urge us to view bats as valuable allies, not as enemies to kill.

Critical Thinking

Has reading this material changed your view of bats? Can you think of two things that could be done to help protect bat species from premature extinction?

of genetic traits. These traits allow a species to become adapted to its natural environment and to changing environmental conditions through natural selection. Many analysts believe that we have no right to prematurely erase these unique genetic packages. On this basis, we have an ethical obligation to control our resource consumption to help protect all species, which make up a key component of the earth's biodiversity (Figure 4-2, p. 79), and in the process implement one of the four **scientific principles of sustainability** (see back cover).

Biologist Edward O. Wilson contends that because of the billions of years of biological connections leading to the evolution of the human species, we have an inherent genetic kinship with the natural world. He calls this phenomenon *biophilia* (love of life).

Evidence of this natural and emotional affinity for life is seen in the preference most people have for almost any natural scene over one from an urban environment. Given a choice, most people prefer to live in an area where they can see water and natural landscapes, such as a grassland or a forest. They also have an affinity for parks, wildlife, and pets and enjoy birdwatching, hiking, camping, fishing, and other outdoor activities. More people visit zoos and aquariums than attend all professional sporting events combined.

Some have the opposite feeling—a fear of many forms of wildlife—called *biophobia.* For example, some movies, books, and TV programs condition us to fear or be repelled by certain species such as alligators (Chapter 4 Core Case Study, p. 77), cockroaches (Case Study, p. 92), sharks (Case Study, p. 96), bats (Science Focus above), and bacteria (Science Focus, p. 61). Many people have lived so long in artificial urban settings that they are largely disconnected from wildlife and from outdoor experiences in nature.

Some people distinguish between the survival rights of plants and those of animals, mostly for practical reasons. Poet Alan Watts once said he was a vegetarian "because cows scream louder than carrots."

Other people distinguish among various types of species. For example, they might think little about getting rid of the world's mosquitoes, cockroaches, rats, or disease-causing bacteria, but feel protective of panda bears, elephants, and whales.

THINKING ABOUT

The Passenger Pigeon

In earlier times, many people viewed huge flocks of passenger pigeons (**Core Case Study**) as pests that devoured grain and left massive piles of their waste. Do you think this justified the passenger pigeon's premature extinction? Explain. If you believe that premature extinction of an undesirable species is justified, what would be your three top candidates? What might be some harmful ecological effects of such extinctions?

Some biologists caution us not to focus primarily on protecting relatively large organisms—the plants and animals we can see that are familiar to us. They remind us that the true foundation of the earth's ecosystems and ecological processes is made up of invisible bacteria and the algae, fungi, and other *microorganisms* that decompose the bodies of larger organisms and recycle the nutrients needed by all life.

▶ **CONCEPT 9-3** The greatest threats to any species are (in order) loss or degradation of its habitat, harmful invasive species, human population growth, pollution, climate change, and overexploitation.

Loss of Habitat Is the Single Greatest Threat to Species: Remember HIPPCO

Figure 9-10 shows the underlying and direct causes of the endangerment and premature extinction of wild species. Conservation biologists summarize the most important causes of premature extinction using the acronym **HIPPCO: H**abitat destruction, degradation, and fragmentation; **I**nvasive (nonnative) species; **P**opulation and resource use growth (too many people consuming too many resources); **P**ollution; **C**limate change; and **O**verexploitation (**Concept 9-3**).

According to biodiversity researchers, the greatest threat to wild species is habitat loss (Figure 9-11, p. 194), degradation, and fragmentation. The passenger pigeon (**Core Case Study**, Figure 9-1) is only one of many species whose extinction was hastened by loss of habitat from forest clearing.

CORE CASE STUDY

Deforestation in tropical areas (Figure 3-1, p. 50) is the greatest eliminator of species, followed by the destruction and degradation of coral reefs and wetlands, plowing of grasslands, and pollution of streams, lakes, and oceans. Globally, temperate biomes have been affected more by habitat loss and degradation than have tropical biomes because of widespread economic development in temperate countries over the past 200 years. Such development is now shifting to many tropical biomes.

Island species—many of them *endemic species* found nowhere else on earth—are especially vulnerable to extinction when their habitats are destroyed, degraded, or fragmented. This is why the collection of islands that make up the U.S. state of Hawaii are America's "extinction capital" —with 63% of its species at risk.

Any habitat surrounded by a different one can be viewed as a *habitat island* for most of the species that live there. Most national parks and other nature reserves are habitat islands, many of them encircled by potentially damaging logging, mining, energy extraction, and industrial activities. Freshwater lakes are also habitat islands that are especially vulnerable to the introduction of nonnative species and pollution.

Habitat fragmentation—by roads, logging, agriculture, and urban development—occurs when a large,

NATURAL CAPITAL DEGRADATION

Causes of Depletion and Premature Extinction of Wild Species

Underlying Causes

- Population growth
- Rising resource use
- Undervaluing natural capital
- Poverty

Direct Causes

- Habitat loss
- Habitat degradation and fragmentation
- Introduction of nonnative species
- Pollution
- Climate change
- Overfishing
- Commercial hunting and poaching
- Sale of exotic pets and decorative plants
- Predator and pest control

Figure 9-10 Underlying and direct causes of depletion and premature extinction of wild species (**Concept 9-3**). The major direct causes of wildlife depletion and premature extinction are habitat loss, degradation, and fragmentation. This is followed by the deliberate or accidental introduction of harmful invasive (nonnative) species into ecosystems. **Question:** What are two direct causes that are related to each of the underlying causes?

Indian Tiger

Range 100 years ago

Range today

Black Rhino

Range in 1700

Range today

African Elephant

Probable range 1600

Range today

Asian or Indian Elephant

Former range

Range today

CENGAGENOW™ **Active Figure 9-11 Natural capital degradation:** reductions in the ranges of four wildlife species, mostly as the result of habitat loss and hunting. What will happen to these and millions of other species when the world's human population doubles and per capita resource consumption rises sharply in the next few decades? *See an animation based on this figure at* CengageNOW. **Question:** Would you support expanding these ranges even though this would reduce the land available for people to grow food and live on? Explain. (Data from International Union for the Conservation of Nature and World Wildlife Fund)

contiguous area of habitat is reduced in area and divided into smaller, more scattered, and isolated patches, or habitat islands. This process can decrease tree populations in forests (Science Focus, at right), block migration routes, and divide populations of a species into smaller and more isolated groups that are more vulnerable to predators, competitor species, disease, and catastrophic events such as storms and fires. Also, it creates barriers that limit the abilities of some species to disperse and colonize new areas, to get enough to eat, and to find mates. Migrating species also face dangers from fences, farms, paved areas, skyscrapers, and cell phone towers.

Certain types of species are especially vulnerable to local and regional extinction because of habitat fragmentation. They include species that are rare, species that need to roam unhindered over large areas, and species that cannot rebuild their populations because of a low reproductive capacity. Species with specialized niches and species that are sought by people for furs, food, medicines, or other uses are also especially threatened by habitat fragmentation.

Scientists use the theory of island biogeography (Science Focus, p. 90) to help them understand the effects of fragmentation on species extinction and to develop ways to help prevent such extinction.

SCIENCE FOCUS

Studying the Effects of Forest Fragmentation on Old-Growth Trees

Tropical rain forests typically consist of large numbers of different tree species with only a few members of each species in an area. Thus, most of the old-growth tree species in an area are rare and vulnerable to local or regional extinction when the forest is disturbed.

Tropical biologist Bill Laurance and his colleagues have been studying the nature of topical rain forests and how they are affected by human activities for over 25 years. One of his research interests is the effect of increasing fragmentation of tropical rain forests as people establish more roads, crop plantations, settlements, and cattle grazing areas.

The edges of forest fragments are often invaded by sun-loving species such as vines, which can gradually take over and cause the loss of a fragment's rare old-growth trees. Laurance is trying to determine how large the undisturbed inner core of a fragment must be to afford protection to its rare old-growth tree species.

His research team studies this by looking at the trees and other plant species in designated plots near the edge of a forest fragment and in the fragment's interior. In this example of muddy-boots ecology, they identify the tree species present in each plot. Then they measure the height and diameter of each tree to calculate its biomass and thus the total biomass of the trees in each plot. Such measurements are repeated every two years for two decades or more to determine changes in the species composition of the study plots.

Such painstaking research reveals that within 100 meters (330 feet) of the edge of a forest fragment, typically up to 36% of the biomass of old-growth trees is lost within 10-17 years after fragmentation. Plots in a fragment's interior show little loss of their tree biomass. Scientists can use such data to estimate how large a fragment must be in order to prevent the loss of rare trees within its protected core habitat. For more details on this research see *The Habitable Planet*, Video 9, at **www.learner.org/resources/series209.html**.

Critical Thinking

What are two ways to reduce the fragmentation of tropical rain forests?

CENGAGENOW™ See how serious the habitat fragmentation problem is for elephants, tigers, and rhinos at CengageNOW.

■ CASE STUDY

A Disturbing Message from the Birds

Approximately 70% of the world's nearly 10,000 known bird species are declining in numbers, and roughly one of every eight (12%) of these bird species is threatened with extinction, mostly because of habitat loss, degradation, and fragmentation. About three-fourths of the threatened bird species live in forests, many of which are being cleared at a rapid rate, especially in the tropical areas in Asia and Latin America (Figure 9-12).

Some 40% of Indonesia's moist, tropical forests, particularly in Borneo and Sumatra, has been cleared for lumber and palm plantations. The harvested palm oil is used as biofuel, mostly in European nations. As a result, 75% of the bird species in Sumatra's lowland forests are on the verge of extinction.

In Brazil, 115 bird species are threatened, mostly because of the burning and clearing of Amazon forests for farms and ranches. Other threats to Brazil's bird species are the loss of 93% of Brazil's Atlantic coastal rain forest and, most recently, the clearing of the country's savannah-like *cerrado* area to establish soybean plantations.

A 2007 joint study by the National Audubon Society and the American Bird Conservancy found that 30% of

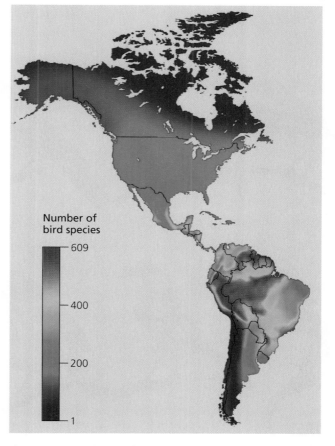

Figure 9-12 Distribution of bird species in North America and Latin America. **Question:** Why do you think more bird species are found in Latin America than in North America? (Data from The Nature Conservancy, Conservation International, World Wildlife Fund, and Environment Canada).

Number of bird species

- 609
- 400
- 200
- 1

all North American bird species (25% of those living in the United States) and 70% of those living in grasslands are declining in numbers or are at risk of disappearing. A 2007 study by the American Bird Conservancy found that Hawaii's forests are the most threatened bird habitat. The government lists 30 of 71 remaining bird species in the Hawaiian Islands as endangered or threatened. Three other seriously threatened U.S. bird habitats are stream watersheds in the Southwest, tallgrass prairies in the Midwest, and beaches and marsh areas along the country's coastlines.

The numbers and distribution of North American bird species that can prosper around humans, such as robins, blackbirds, and starlings, have increased over the last 35 years. But populations of many forest songbirds have declined sharply. The greatest decline has occurred among long-distance migrant species such as tanagers, orioles, thrushes, vireos, and warblers that nest deep in North American woods in the summer and spend their winters in Central or South America or the Caribbean Islands. Figure 9-13 shows the 10 most threatened U.S. songbird species.

The primary culprit for these declines appears to be habitat loss and fragmentation of the birds' breeding habitats. In North America, woodlands are being cleared and broken up by roads and developments. In Central and South America, tropical forest habitats, mangroves, and wetland forests are suffering the same fate.

After habitat loss, the intentional or accidental introduction of nonnative species such as bird-eating cats, rats, snakes, and mongooses is the second greatest danger, affecting about 28% of the world's threatened birds. Fifty-two of the world's 388 parrot species (Figure 9-9) are threatened by a combination of habitat loss and capture for the pet trade (often illegal), especially in Europe and the United States.

At least 23 species of seabirds face extinction. Many seabirds drown after becoming hooked on one of the many baited lines put out by fishing boats. And populations of 40% of the world's water birds are in decline because of the global loss of wetlands.

Millions of migrating birds are killed each year when they collide with power lines, communications towers, and skyscrapers that have been erected in the middle of their migration routes. While U.S. hunters kill about 121 million birds a year, as many as 1 billion birds in the United States die each year when they fly into glass windows, especially those in tall city buildings that are lit up at night—the number one cause of U.S. bird mortality. Other threats to birds are oil spills, exposure to pesticides, herbicides that destroy their habi-

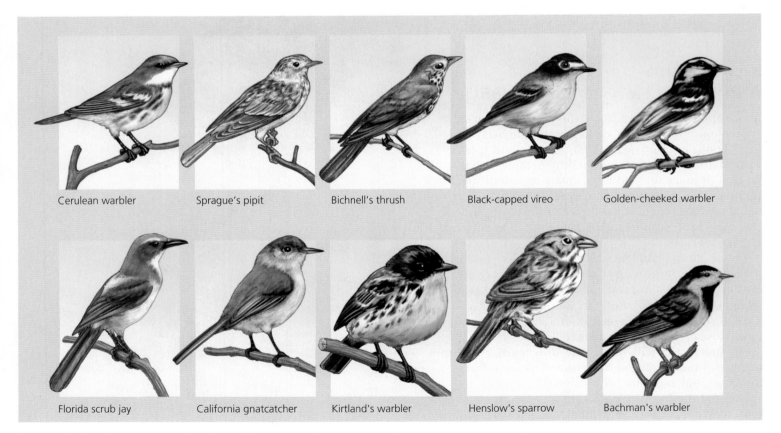

Figure 9-13 The 10 most threatened species of U.S. songbirds. Most of these species are vulnerable because of habitat loss and fragmentation from human activities. An estimated 12% of the world's known bird species may face premature extinction due mostly to human activities during this century. (Data from National Audubon Society)

SCIENCE FOCUS

Vultures, Wild Dogs, and Rabies: Some Unexpected Scientific Connections

In 2004, the World Conservation Union placed three species of vultures found in India and South Asia on the critically endangered list. During the early 1990s, there were more than 40 million of these carcass-eating vultures. But within a few years their populations had fallen by more than 97%.

This is an interesting scientific mystery, but should anyone care if various vulture species disappear? The answer is yes.

Scientists were puzzled, but they eventually discovered that the vultures were being poisoned by *diclofenac*. This anti-inflammatory drug reduces pain in cows and in humans and is used to increase milk production in cows.

But it causes kidney failure in vultures that feed on the carcasses of these cows.

As the vultures died off, huge numbers of cow carcasses, normally a source of food for the vultures, were consumed by wild dogs and rats whose populations the vultures helped control by reducing their food supply. As wild dog populations exploded due to a greatly increased food supply, the number of dogs with rabies also increased. This increased the risks to people bitten by rabid dogs. In 1997 alone, more than 30,000 people in India died of rabies—more than half the world's total number of rabies deaths that year.

Thus, protecting these vulture species from extinction can end up protecting millions of people from a life-threatening disease. Unraveling often unexpected ecological connections in nature is not only fascinating but also vital to our own lives and health.

Some who argue against protecting species and ecosystems from harmful human activities frame the issue as a choice between protecting people or wildlife. Conservation biologists reject this as a misleading conclusion. To them, the goal is to protect both wildlife and people because their fates and well-being are interconnected.

Critical Thinking

What would happen to your life and lifestyle if most of the world's vultures disappeared?

tats, and ingestion of toxic lead shotgun pellets, which fall into wetlands, and lead sinkers left by anglers.

The greatest new threat to birds is climate change. A 2006 review, done for the World Wildlife Fund (WWF), of more than 200 scientific articles found that climate change is causing declines of bird populations in every part of the world. And climate change is expected to increase sharply during this century.

Migratory, mountain, island, wetland, Antarctic, Arctic, and sea birds are especially at risk from climate change. The researchers have warned that protecting many current areas with high bird diversity will not help, because climate change will force many bird species to shift to unprotected zones. Island and mountain birds may simply have nowhere to go.

Conservation biologists view this decline of bird species with alarm. One reason is that birds are excellent *environmental indicators* because they live in every climate and biome, respond quickly to environmental changes in their habitats, and are relatively easy to track and count.

Furthermore, birds perform a number of important economic and ecological services in ecosystems throughout the world. They help control populations of rodents and insects (which decimate many tree species), remove dead animal carcasses (a food source for some birds), and spread plants throughout their habitats by helping with pollination and by consuming and excreting plant seeds.

Extinctions of birds that play key and specialized roles in pollination and seed dispersal, especially in tropical areas, may lead to extinctions of plants dependent on these ecological services. Then some specialized animals that feed on these plants may also become extinct. This cascade of extinctions, in turn, can affect our own food supplies and well-being. Protecting birds and their habitats is not only a conservation issue. It is an important issue for human health as well (Science Focus, above).

Biodiversity scientists urge us to listen more carefully to what birds are telling us about the state of the environment, for the birds' sake, as well as for ours.

THINKING ABOUT
Bird Extinctions

How does your lifestyle directly or indirectly contribute to the premature extinction of some bird species? What are three things that you think should be done to reduce the premature extinction of birds?

RESEARCH FRONTIER

Learning more about why birds are declining, what it implies for the biosphere, and what can be done about it. See **academic.cengage.com/biology/miller**.

Some Deliberately Introduced Species Can Disrupt Ecosystems

After habitat loss and degradation, the biggest cause of premature animal and plant extinctions is the deliberate or accidental introduction of harmful invasive species into ecosystems (**Concept 9-3**).

Most species introductions are beneficial to us, although they often displace native species. We depend heavily on introduced species for ecosystem services, food, shelter, medicine, and aesthetic enjoyment.

According to a 2000 study by ecologist David Pimentel, introduced species such as corn, wheat, rice, and other food crops, and cattle, poultry, and other livestock provide more than 98% of the U.S. food supply. Similarly, nonnative tree species are grown in about 85% of the world's tree plantations. Some deliberately introduced species have also helped to control pests.

The problem is that some introduced species have no natural predators, competitors, parasites, or pathogens to help control their numbers in their new habitats. Such nonnative species can reduce or wipe out populations of many native species, trigger ecological disruptions, cause human health problems, and lead to economic losses.

In 1988, for example, a giant African land snail was imported into Brazil as a cheap substitute for conventional escargot (snails) used as a source of food. It grows to the size of a human fist and weighs 1 kilogram (2.2 pounds) or more. When export prices for escargot fell, breeders dumped the snails into the wilds. Now it has spread to 23 of Brazil's states and devours everything from lettuce to mouse droppings. It also can carry rat lungworm, a parasite that burrows into the human brain and causes meningitis, and another parasite that can rupture the intestines. Authorities eventually banned the snail, but it was too late. So far, the snail has been unstoppable.

Figure 9-14 shows some of the estimated 7,100 harmful invasive species that, after being deliberately or accidentally introduced into the United States, have caused ecological and economic harm. Nonnative species threaten almost half of the roughly 1,300 endangered and threatened species in the United States and 95% of those in the state of Hawaii, according to the U.S. Fish and Wildlife Service. According to biologist Thomas Lovejoy, harmful invader species cost the U.S. public an average of $261,000 per minute! The situation in China is much worse. Biologist David Pimentel estimates that, globally, damage to watersheds, soils, and wildlife by bioinvaders may be costing as much $44,400 per second. And the damages are rising rapidly.

Some deliberately introduced species are plants such as kudzu (Case Study, at right). Deliberately introduced animal species have also caused ecological and economic damage. Consider the estimated 1 million European wild (feral) boars (Figure 9-14) found in parts of Florida and other U.S. states. They compete for food with endangered animals, root up farm fields, and cause traffic accidents. Game and wildlife officials have failed to control their numbers through hunting and trapping and say there is no way to stop them. Another example is the estimated 30 million feral cats and 41 million outdoor pet cats found in the United States. Most were introduced to the environment when they were abandoned by their owners and left to breed in the wild; they kill about 568 million birds per year. Because of pet overpopulation and abandonment, pet shelters in the United States are forced to kill over 14 million cats and dogs a year. Shelter officials urge owners to spay or neuter their cats and keep them indoors.

■ CASE STUDY

The Kudzu Vine

An example of a deliberately introduced plant species is the *kudzu* ("CUD-zoo") *vine*, which, in the 1930s, was imported from Japan and planted in the southeastern United States in an attempt to control soil erosion. Kudzu does control erosion. But it is so prolific and difficult to kill that it engulfs hillsides, gardens, trees, abandoned houses and cars, stream banks, patches of forest, and anything else in its path (Figure 9-15, p. 200).

This plant, which is sometimes called "the vine that ate the South," has spread throughout much of the southeastern United States. It could spread as far north as the Great Lakes by 2040 if climate change caused by global warming occurs as projected.

Kudzu is considered a menace in the United States but Asians use a powdered kudzu starch in beverages, gourmet confections, and herbal remedies for a range of diseases. A Japanese firm has built a large kudzu farm and processing plant in the U.S. state of Alabama and ships the extracted starch to Japan. And almost every part of the kudzu plant is edible. Its deep-fried leaves are delicious and contain high levels of vitamins A and C. Stuffed kudzu leaves, anyone?

Although kudzu can engulf and kill trees, it might eventually save some trees from loggers. Researchers at the Georgia Institute of Technology indicate that it could be used in place of trees as a source of fiber for making paper. And a preliminary 2005 study indicated that kudzu powder could be used to reduce alcoholism and binge drinking. Ingesting small amounts of the powder can lessen one's desire for alcohol.

Some Accidentally Introduced Species Can Also Disrupt Ecosystems

Welcome to one of the downsides of global trade, travel, and tourism. Many unwanted nonnative invaders arrive from other continents as stowaways on aircraft, in the ballast water of tankers and cargo ships, and as hitchhikers on imported products such as wooden packing crates. Cars and trucks can also spread the seeds of nonnative plant species embedded in their tire treads. Many tourists return home with living plants that can multiply and become invasive. These plants might also harbor insects that can escape, multiply rapidly, and threaten crops.

In the 1930s, the extremely aggressive Argentina fire ant (Figures 9-14 and 9-16, p. 200) was introduced accidentally into the United States in Mobile, Alabama.

Deliberately Introduced Species

Purple loosestrife

European starling

African honeybee ("Killer bee")

Nutria

Salt cedar (Tamarisk)

Marine toad (Giant toad)

Water hyacinth

Japanese beetle

Hydrilla

European wild boar (Feral pig)

Accidentally Introduced Species

Sea lamprey (attached to lake trout)

Argentina fire ant

Brown tree snake

Eurasian ruffe

Common pigeon (Rock dove)

Formosan termite

Zebra mussel

Asian long-horned beetle

Asian tiger mosquito

Gypsy moth larvae

Figure 9-14 Some of the more than 7,100 harmful invasive (nonnative) species that have been deliberately or accidentally introduced into the United States.

Figure 9-15 Kudzu taking over an abandoned house in the U.S. state of Mississippi. This vine, which can grow 5 centimeters (2 inches) per hour, was deliberately introduced into the United States for erosion control. Digging it up or burning it cannot stop it. Grazing by goats and repeated doses of herbicides can destroy it, but goats and herbicides also destroy other plants, and herbicides can contaminate water supplies. Scientists have found a common fungus that can kill kudzu within a few hours, apparently without harming other plants. Stay tuned.

The ants may have arrived on shiploads of lumber or coffee imported from South America. Without natural predators, fire ants have spread rapidly by land and water (they can float) throughout the South, from Texas to Florida and as far north as Tennessee and Virginia.

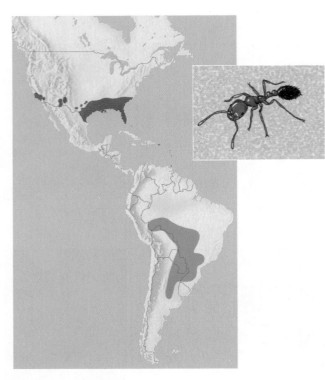

Figure 9-16 The *Argentina fire ant*, introduced accidentally into Mobile, Alabama, in the 1930s from South America (green area), has spread over much of the southern United States (red area). This invader is also found in Puerto Rico, New Mexico, and California. **Question:** How might this accidental introduction of fire ants have been prevented? (Data from S. D. Porter, Agricultural Research Service, U.S. Department of Agriculture)

When these ants invade an area, they can wipe out as much as 90% of native ant populations. Mounds containing fire ant colonies cover many farm fields and invade people's yards. Walk on one of these mounds, and as many as 100,000 ants may swarm out of their nest to attack you with painful and burning stings. They have killed deer fawns, birds, livestock, pets, and at least 80 people who were allergic to their venom. In the United States, they also do an estimated $68,000 of economic damage per hour to crops and phone and power lines.

Widespread pesticide spraying in the 1950s and 1960s temporarily reduced fire ant populations. But this chemical warfare actually hastened the advance of the rapidly multiplying fire ants by reducing populations of many native ant species. Even worse, it promoted development of genetic resistance to pesticides in the fire ants through natural selection (**Concept 4-2B**, p. 80). In other words, we helped wipe out their competitors and made them more genetically resistant to pesticides.

In the Everglades in the U.S. state of Florida, the population of the huge *Burmese python* snake is increasing. This native of Southeast Asia was imported as a pet, and many ended up being dumped in the Everglades by people who learned that, when they get larger, pythons do not make good pets. They can live 25 years, reach 6 meters (20 feet) in length, weigh more than 90 kilograms (200 pounds), and have the girth of a telephone pole. They have razor-sharp teeth and can catch, squeeze to death, and swallow whole practically anything that moves and is warm-blooded, including raccoons, a variety of birds, and full-grown deer. They are also known to have survived alligator attacks. They are slowly spreading to other areas and, by 2100, could be found in most of the southern half of the continental United States.

Prevention Is the Best Way to Reduce Threats from Invasive Species

Once a harmful nonnative species becomes established in an ecosystem, its removal is almost impossible—somewhat like trying to get smoke back into a chimney. Clearly, the best way to limit the harmful impacts of nonnative species is to prevent them from being introduced and becoming established.

Scientists suggest several ways to do this:

- Fund a massive research program to identify the major characteristics of successful invader species and the types of ecosystems that are vulnerable to invaders (Figure 9-17).

- Greatly increase ground surveys and satellite observations to detect and monitor species invasions and develop better models for predicting how they will spread.

- Step up inspection of imported goods and goods carried by travelers that are likely to contain invader species.

- Identify major harmful invader species and establish international treaties banning their transfer from one country to another, as is now done for endangered species.

- Require cargo ships to discharge their ballast water and replace it with saltwater at sea before entering ports, or require them to sterilize such water or pump nitrogen into the water to displace dissolved oxygen and kill most invader organisms.

- Increase research to find and introduce natural predators, parasites, bacteria, and viruses to control populations of established invaders.

Characteristics of Successful Invader Species

- High reproductive rate, short generation time (r-selected species)
- Pioneer species
- Long lived
- High dispersal rate
- Generalists
- High genetic variability

Characteristics of Ecosystems Vulnerable to Invader Species

- Climate similar to habitat of invader
- Absence of predators on invading species
- Early successional systems
- Low diversity of native species
- Absence of fire
- Disturbed by human activities

Figure 9-17 Some general characteristics of successful invader species and ecosystems vulnerable to invading species. **Question:** Which, if any, of the characteristics on the right-hand side could humans influence?

WHAT CAN YOU DO?

Controlling Invasive Species

- Do not capture or buy wild plants and animals
- Do not remove wild plants from their natural areas
- Do not dump the contents of an aquarium into waterways, wetlands, or storm drains
- When camping, use wood found near your camp site instead of bringing firewood from somewhere else
- Do not dump unused bait into any waterways
- After dogs visit woods or the water, brush them before taking them home
- After each use, clean your mountain bike, canoe, boat, hiking boots, and other gear before heading for home

Figure 9-18 Individuals Matter: ways to prevent or slow the spread of harmful invasive species. **Questions:** Which two of these actions do you think are the most important? Why? Which of these actions do you plan to take?

RESEARCH FRONTIER

Learning more about invasive species, why they thrive, and how to control them. See **academic.cengage.com/biology/ miller**.

Figure 9-18 shows some of the things you can do to help prevent or slow the spread of these harmful invaders.

Population Growth, Overconsumption, Pollution, and Climate Change Can Cause Species Extinctions

Human population growth and excessive and wasteful consumption of resources have greatly expanded the human ecological footprint (Figure 1-10, p. 15; and Figure 3, pp. S24–S25, in Supplement 4), which has eliminated vast areas of wildlife habitat (Figure 9-11). Acting together, these factors have caused premature extinction of many species (**Concept 9-3**). (See *The Habitable Planet*, Video 13, at **www.learner.org/resources/series209 .html**).

Pollution also threatens some species with extinction (**Concept 9-3**), as has been shown by the unintended effects of certain pesticides. According to the U.S. Fish and Wildlife Service, each year pesticides kill about one-fifth of the beneficial honeybee colonies in United States (Case Study, p. 202), more than 67 million birds, and 6–14 million fish. They also threaten about one-fifth of the country's endangered and threatened species.

Figure 9-19 *Bioaccumulation* and *biomagnification*. DDT is a fat-soluble chemical that can accumulate in the fatty tissues of animals. In a food chain or web, the accumulated DDT can be biologically magnified in the bodies of animals at each higher trophic level. The concentration of DDT in the fatty tissues of organisms was biomagnified about 10 million times in this food chain in an estuary near Long Island Sound in the U.S. state of New York. If each phytoplankton organism takes up from the water and retains one unit of DDT, a small fish eating thousands of zooplankton (which feed on the phytoplankton) will store thousands of units of DDT in its fatty tissue. Each large fish that eats 10 of the smaller fish will ingest and store tens of thousands of units, and each bird (or human) that eats several large fish will ingest hundreds of thousands of units. Dots represent DDT. **Question:** How does this story demonstrate the value of pollution prevention?

During the 1950s and 1960s, populations of fish-eating birds such as ospreys, brown pelicans (see Photo 1 in the Detailed Contents), and bald eagles plummeted. A chemical derived from the pesticide DDT, when biologically magnified in food webs (Figure 9-19), made the birds' eggshells so fragile that they could not reproduce successfully. Also hard hit were such predatory birds as the prairie falcon, sparrow hawk, and peregrine falcon, which help to control rabbits, ground squirrels, and other crop eaters.

Since the U.S. ban on DDT in 1972, most of these species have made a comeback. For example, after eliminating DDT and after crackdowns on hunting and habitat destruction, the American bald eagle has rebounded from only 417 breeding pairs in the lower 48 states in 1963 to almost 10,000 breeding pairs in 2007. This was enough to have it removed from the endangered species list. The comeback of this species from the brink of extinction is one of the greatest wildlife protection successes in U.S. history.

A 2004 study by Conservation International predicted that climate change caused mostly by global warming (Science Focus, p. 33) could drive more than a quarter of all land animals and plants to extinction by the end of this century. Some scientific studies in-

dicate that polar bears (Case Study, p. 203) and 10 of the world's 17 penguin species are already threatened because of higher temperatures and melting sea ice in their polar habitats.

■ **CASE STUDY**

Where Have All the Honeybees Gone?

Three-quarters of all flowering plants in North America—including most fruit and vegetable crops—rely on pollinators such as bees, bats (Science Focus, p. 192), butterflies, and hummingbirds for fertilization.

According to a 2006 report by the U.S. National Academy of Sciences (NAS), populations of such vital pollinators are declining across North America, and honeybee populations have been in increasing trouble for over 2 decades. The report warns that continued decreases in wild populations of such pollinators could disrupt food production and ecosystems.

The report includes a specific warning on the decline of the honeybee, which pollinates more than 110 commercially grown crops that are vital to U.S. agriculture, including up to one-third of U.S. fruit and vegetable crops. Globally, about one-third of the human diet comes from insect-pollinated plants, and the honeybee is responsible for 80% of that pollination, according to the U.S. Department of Agriculture. Adult honeybees live on honey that they make from nectar they collect from flowering plants, and they feed their young with protein-rich pollen.

Honeybees are big business. In the United States, honeybee colonies are managed by beekeepers who rent the bees out for their pollination services, especially for major crops such as almonds, apples, and blueberries. By 1994, such colonies had replaced an estimated 98% of the wild, free-range honeybees in the United States.

According to the NAS report, there has been a 30% drop in U.S. honeybee populations since the 1980s. Causes include pesticide exposure (the wax in beehives absorbs these and other airborne toxins), attacks by parasitic mites that can wipe out a colony in hours, and invasion by African honeybees (killer bees, p. 93).

Since 2006, a growing number of bee colonies in 27 states have suffered what researchers call "bee colony collapse disorder," or what French scientists call "mad bee disease," in which the worker bees in a colony vanish without a trace. When beekeepers inspect what were once healthy and strong colonies, they find all of the adult worker bees gone and an abandoned queen bee.

More than a quarter of the country's 2.4 million honeybee colonies (each with 30,000 to 100,000 individual bees) have been lost. Nobody knows where the missing bees went or what is causing them to leave the hives. Possible causes include parasites, fungi, bacteria,

and pesticides. Scientists also suspect a virus that paralyzes bees, which may have come from Israel or from bees imported from Australia to help replace declining U.S. honeybee populations. Another problem is poor nutrition and stress caused when colonies of bees are fed an artificial diet while being trucked around the country and rented out for pollination.

Scientists are also finding sharp declines in some species of bumblebees found in the United States. These bees are responsible for pollinating an estimated 15% of the crops grown in the United States, especially those that are raised in greenhouses, such as tomatoes, peppers, and strawberries. Bumblebees collect pollen and nectar to feed their young, but make very little honey.

Declining bee populations have also been reported in Brazil, Taiwan, Guatemala, and parts of Europe. China, where some argue that pesticides are overused, gives us a glimpse of a future without honeybees. Apple orchards in China are now largely hand-fertilized by humans in the absence of bees. Some scientists warn that if it continues and grows, *bee colony collapse disorder* could lead to *agricultural collapse disorder* in parts of the world.

THINKING ABOUT
Bees

What difference would it make to your lifestyle if most of the honeybees or bumblebees disappeared? What two things would you do to help reduce the loss of honeybees?

■ CASE STUDY
Polar Bears and Global Warming

The world's 20,000–25,000 polar bears are found in 19 populations distributed across the frozen Arctic. About 60% of them are in Canada, and the rest are found in arctic areas in Denmark, Norway, Russia, and the U.S. state of Alaska.

Throughout the winter, the bears hunt for seals on floating sea ice (Figure 9-20), which expands southward each winter and contracts as the temperature rises each summer. Normally the bears swim from one patch of sea ice to another to hunt and eat seals during winter as their body fat accumulates. In the summer and fall as sea ice breaks up, the animals fast and live off their body fat for several months until hunting resumes when the ice again expands.

Evidence shows that the Arctic is warming twice as fast as the rest of the world and that the average annual area of floating summer sea ice in the Arctic is declining and is breaking up earlier each year. Scientists project that summer sea ice could be gone by 2030 and perhaps as soon as 2012.

This means that polar bears have less time to feed and to store the fat they need in order to survive their summer and fall months of fasting. As a result, they must fast longer, which weakens them. As females become weaker, their ability to reproduce and keep their young cubs alive declines. And as bears grow hungrier, they are more likely to go to human settlements looking for food. The resulting increase in bear sightings gives people the false impression that their populations are increasing.

Polar bears are strong swimmers, but ice shrinkage has forced them to swim longer distances to find enough food and to spend more time during winter hunting on land where prey is nearly impossible to find. Several studies link global warming and diminished sea ice to polar bears drowning or starving while in search of prey and, in some cases, to cannibalism among the bears.

According to a 2006 study by the IUCN–World Conservation Union, the world's total polar bear population is likely to decline by 30–35% by 2050, and by the end of this century, the bears may be found only in zoos.

Another threat to the bears in some areas is the buildup of toxic PCBs, DDT (Figure 9-19), and other pesticides in their fatty tissue, which can adversely affect their development, behavior, and reproduction. And in 2007, the U.S. Fish and Wildlife Service estimated that Russian poachers are killing 100–250 polar bears a year.

In 2007, the IUCN listed polar bears as threatened in their annual red list of endangered species, and in 2008, the U.S. government listed the polar bear as threatened under the Endangered Species Act.

Figure 9-20 Polar bear with seal prey on floating ice in Svalbard, Norway. Polar bears in the Arctic are likely to become extinct sometime during this century because global warming is melting the floating sea ice on which they hunt seals.

THINKING ABOUT
Polar Bears

What difference would it make if all of the world's polar bears disappeared? List two things you would do to help protect the world's remaining polar bears from premature extinction.

Illegal Killing, Capturing, and Selling of Wild Species Threatens Biodiversity

Some protected species are illegally killed for their valuable parts or are sold live to collectors (**Concept 9-3**). Such *poaching* endangers many larger animals and some rare plants. Globally, this illegal trade in wildlife earns smugglers at least $10 billion a year—an average of $19,000 a minute. Organized crime has moved into illegal wildlife smuggling because of the huge profits involved—surpassed only by the illegal international trade in drugs and weapons. Rapidly growing wildlife smuggling is a high-profit, low-risk business because few of the smugglers are caught or punished. At least two-thirds of all live animals smuggled around the world die in transit.

To poachers, a live mountain gorilla is worth $150,000, a giant panda pelt $100,000, a chimpanzee $50,000, an Imperial Amazon macaw $30,000, and a Komodo dragon reptile from Indonesia $30,000. A poached rhinoceros horn (Figure 9-21) can be worth as much as $55,500 per kilogram ($25,000 per pound). It is used to make dagger handles in the Middle East and as a fever reducer and alleged aphrodisiac in China and other parts of Asia.

According to a 2005 study by the International Fund for Animal Welfare, the Internet has become a key market for the illegal global trade in thousands of live threatened and endangered animals and products made from such animals. For example, U.S. websites offered live chimpanzees dressed as dolls for $60,000–65,000 each and a 2-year-old highly endangered Siberian tiger for $70,000.

An important way to combat the illegal trade in these species is through research and education. Some people are dedicating their time and energy to this problem. For example, scientist Jane Goodall has devoted her life to understanding and protecting chimpanzees (Individuals Matter, at right).

In 1900, an estimated 100,000 tigers roamed free in the world. Despite international protection, only about 3,500 tigers remain in the wild, on an ever shrinking range (Figure 9-11, top left), according to a 2006 study by the World Conservation Union. Between 1900 and 2007, the estimated number of tigers in India plunged from 40,000 to about 1,400. In 2007, Eric Dinerstein and his colleagues estimated that tigers have 41% less habitat than they had in 1997, mostly because of deforestation and forest fragmentation, and now live in

Figure 9-21 White rhinoceros killed by a poacher for its horn in South Africa. **Question:** What would you say if you could talk to the poacher of this animal?

Martin Harvey/Peter Arnold, Inc

just 7% of their historic range around the world. Today all five tiger subspecies are endangered in the wild, although at least 11,000 captive tigers of mixed ancestry exist behind bars.

Tigers are also threatened because they are killed for their coats and body parts. The Bengal or Indian tiger is at risk because a coat made from its fur can sell for as much as $100,000 in Tokyo. Wealthy collectors have paid $10,000 to $20,000 for a Bengal tiger rug. With the body parts of a single tiger worth as much as $25,000, and because few of the poachers are caught or punished, it is not surprising that illegal hunting has skyrocketed. According to a 2006 study by tiger experts, without emergency action to curtail poaching and preserve their habitat, few if any tigers may be left in the wild within 20 years.

THINKING ABOUT
Tigers

What difference would it make if all the world's tigers disappeared? What two things would you do to help protect the world's remaining tigers from premature extinction?

The global legal and illegal trade in wild species for use as pets is also a huge and very profitable business. Many owners of wild pets do not know that, for every live animal captured and sold in the pet market, an es-

timated 50 others are killed or die in transit. Most are also unaware that some imported exotic animals can carry dangerous infectious diseases.

About 25 million U.S. households have exotic birds as pets, 85% of them imported. More than 60 bird species, mostly parrots, (Figure 9-9), are endangered or threatened because of this wild bird trade. Ironically, keeping birds as pets can also be dangerous for people. A 1992 study suggested that keeping a pet bird indoors for more than 10 years doubles a person's chances of getting lung cancer from inhaling tiny particles of bird dander.

Other wild species whose populations are depleted because of the pet trade include amphibians, reptiles, mammals, and tropical fishes (taken mostly from the coral reefs of Indonesia and the Philippines). Divers catch tropical fish by using plastic squeeze bottles of poisonous cyanide to stun them. For each fish caught alive, many more die. In addition, the cyanide solution kills the coral animals that create the reef.

Some exotic plants, especially orchids and cacti, are endangered because they are gathered (often illegally) and sold to collectors to decorate houses, offices, and landscapes. A collector may pay $5,000 for a single rare orchid. A mature crested saguaro cactus can earn cactus rustlers as much as $15,000.

THINKING ABOUT
Collecting Wild Species

Some people believe it is unethical to collect wild animals and plants for display and personal pleasure. They believe we should leave most exotic wild species in the wild. Explain why you agree or disagree with this view.

As commercially valuable species become endangered, their black market demand soars. This positive feedback loop increases their chances of premature extinction from poaching. Most poachers are not caught and the money they can make far outweighs the small risk of being caught, fined, or imprisoned.

On the other hand, species also hold great value by surviving in the wild. According to the U.S. Fish and Wildlife Service, collectors of exotic birds may pay $10,000 for a threatened hyacinth macaw smuggled out of Brazil. But during its lifetime, a single macaw left in the wild might yield as much as $165,000 in tourist revenues.

In Thailand, biologist Pilai Poonswad decided to do something about poachers taking Great Indian hornbills—large, beautiful, and rare birds—from a rain forest. She visited the poachers in their villages and showed them why the birds are worth more alive than dead. Today, some former poachers earn money by taking ecotourists into the forest to see these magnificent birds. Because of their vested financial interest in preserving the hornbills, they now help to protect the birds from poachers. Individuals matter.

INDIVIDUALS MATTER

Jane Goodall

Primatologist and anthropologist Jane Goodall (Figure 9-A) spent 45 years studying chimpanzee social and family life in Gombe National Park in the African country of Tanzania. One of her major scientific contributions was the discovery that chimpanzees have tool-making skills. She observed that some chimpanzees modified twigs or blades of grass and then poked them into termite mounds. When the termites latched onto these primitive tools, the chimpanzees would pull them out and eat the termites.

M. Gunther/BIOS/Peter Arnold, Inc.

Figure 9-A Jane Goodall with a young chimpanzee living in Tanzania's Gombe National Park.

In 1977, she established the Jane Goodall Institute, which supports the research at Gombe National Park and, with 19 offices around the world, works to protect chimpanzees and their habitats. Dr. Goodall spends nearly 300 days a year traveling and educating people throughout the world about chimpanzees and the need to protect the environment.

Goodall is also president of Advocates for Animals, an animal rights organization based in Edinburgh, Scotland, that campaigns against the use of animals for medical research, zoos, farming, and sport hunting. She has received many awards and prizes for her scientific and conservation contributions. She has also written 23 books for adults and children and has produced 14 films about the lives and importance of chimpanzees.

Rising Demand for Bush Meat Threatens Some African Species

Indigenous people in much of West and Central Africa have sustainably hunted wildlife for *bush meat*, a source of food, for centuries. But in the last two decades bush meat hunting in some areas has skyrocketed as local people try to provide food for rapidly growing populations or seek to make a living by supplying restaurants with exotic meat (Figure 9-22, p. 206). Logging roads have enabled miners, ranchers, and settlers to move into once inaccessible forests, which has made it easier to hunt animals for bush meat. And a 2004 study showed that people living in coastal areas of West Africa have increased bush meat hunting because local fish harvests have declined due to overfishing by heavily subsidized European Union fishing fleets.

So what is the big deal? After all, people have to eat. For most of our existence, humans have survived by hunting and gathering wild species.

Figure 9-22 *Bush meat,* such as this severed head of a lowland gorilla in the Congo, is consumed as a source of protein by local people in parts of West Africa and sold in the national and international marketplace. You can find bush meat on the menu in Cameroon and the Congo in West Africa as well as in Paris, London, Toronto, New York, and Washington, D.C. It is often supplied by poaching. Wealthy patrons of some restaurants regard gorilla meat as a source of status and power. **Question:** How, if at all, is this different from killing a cow for food?

One problem today is that bush meat hunting has led to the local extinction of many wild animals in parts of West Africa and has driven one species—Miss Waldron's red colobus monkey—to complete extinction. It is also a factor in reducing gorilla, orangutan (Figure 9-7), chimpanzee, elephant, and hippopotamus populations. This practice also threatens forest carnivores, such as crowned eagles and leopards, by depleting their main prey species.

Some conservationists fear that within 1 or 2 decades, the Congo basin's rain forest—the world's second largest remaining tropical forest—will contain few large mammals, and most of Africa's great apes will be extinct. Another problem is that butchering and eating some forms of bush meat has helped to spread fatal diseases such as HIV/AIDS and the Ebola virus to humans.

The U.S. Agency for International Development is trying to reduce unsustainable hunting for bush meat in some areas by introducing alternative sources of food, such as fish farms. They are also showing villagers how to breed large rodents such as cane rats as a source of food.

THINKING ABOUT
The Passenger Pigeon and Humans

CORE CASE STUDY

Humans exterminated the passenger pigeon (Figure 9-1) within a single human lifetime because it was considered a pest and because of its economic value. Suppose a species superior to us arrived and began taking over the earth with the goal of using the planet more sustainably. The first thing they might do is to exterminate us. Do you think such an action would be justified? Explain.

9-4 How Can We Protect Wild Species from Extinction Resulting from Our Activities?

▶ **CONCEPT 9-4A** We can use existing environmental laws and treaties and work to enact new laws designed to prevent premature species extinction and protect overall biodiversity.

▶ **CONCEPT 9-4B** We can help to prevent premature species extinction by creating and maintaining wildlife refuges, gene banks, botanical gardens, zoos, and aquariums.

▶ **CONCEPT 9-4C** According to the *precautionary principle*, we should take measures to prevent or reduce harm to the environment and to human health, even if some of the cause-and-effect relationships have not been fully established, scientifically.

International Treaties Help to Protect Species

Several international treaties and conventions help to protect endangered and threatened wild species (**Concept 9-4A**). One of the most far reaching is the 1975 *Convention on International Trade in Endangered Species* (CITES). This treaty, now signed by 172 countries, bans hunting, capturing, and selling of threatened or endangered species. It lists some 900 species that cannot be commercially traded as live specimens or wildlife products because they are in danger of extinction. It also restricts international trade of roughly 5,000 species of animals and 28,000 plants species that are at risk of becoming threatened.

CITES has helped reduce international trade in many threatened animals, including elephants, crocodiles, cheetahs, and chimpanzees. But the effects of

this treaty are limited because enforcement varies from country to country, and convicted violators often pay only small fines. Also, member countries can exempt themselves from protecting any listed species, and much of the highly profitable illegal trade in wildlife and wildlife products goes on in countries that have not signed the treaty.

The *Convention on Biological Diversity* (CBD), ratified by 190 countries (but not by the United States), legally commits participating governments to reversing the global decline of biodiversity and to equitably sharing the benefits from use of the world's genetic resources. This includes efforts to prevent or control the spread of ecologically harmful invasive species.

This convention is a landmark in international law because it focuses on ecosystems rather than on individual species and it links biodiversity protection to issues such as the traditional rights of indigenous peoples. However, because some key countries including the United States have not ratified it, implementation has been slow. Also, the law contains no severe penalties or other enforcement mechanisms.

■ CASE STUDY
The U.S. Endangered Species Act

The *Endangered Species Act of 1973* (ESA; amended in 1982, 1985, and 1988) was designed to identify and protect endangered species in the United States and abroad (**Concept 9-4A**). This act is probably the most far-reaching environmental law ever adopted by any nation, which has made it controversial. Canada and a number of other countries have similar laws.

Under the ESA, the National Marine Fisheries Service (NMFS) is responsible for identifying and listing endangered and threatened ocean species, while the U.S. Fish and Wildlife Services (USFWS) is to identify and list all other endangered and threatened species. Any decision by either agency to add a species to, or remove one from, the list must be based on biological factors alone, without consideration of economic or political factors. However, economic factors can be used in deciding whether and how to protect endangered habitat and in developing recovery plans for listed species.

The ESA also forbids federal agencies (except the Defense Department) to carry out, fund, or authorize projects that would jeopardize an endangered or threatened species or destroy or modify its critical habitat. For offenses committed on private lands, fines as high as $100,000 and 1 year in prison can be imposed to ensure protection of the habitats of endangered species. This part of the act has been controversial because at least 90% of the listed species live totally or partially on private land. The ESA also makes it illegal for Americans to sell or buy any product made from an endangered or threatened species or to hunt, kill, collect, or injure such species in the United States.

Between 1973 and 2007, the number of U.S. species on the official endangered and threatened lists increased from 92 to about 1,350 species—55% of them plants and 45% animals. According to a 2000 study by The Nature Conservancy, one-third of the country's species are at risk of extinction, and 15% of all species are at high risk—far more than the roughly 1,350 species on the country's endangered species list. The study also found that many of the country's rarest and most imperiled species are concentrated in a few areas, called *hotspots*. To conservation biologists, protecting such areas should be a top priority.

For each species listed, the USFWS or the NMFS is supposed to prepare a recovery plan that includes designation and protection of critical habitat. Examples of successful recovery plans include those for the American alligator (Chapter 4 Core Case Study, p. 77), the gray wolf, the peregrine falcon, and the bald eagle.

The ESA also requires that all commercial shipments of wildlife and wildlife products enter or leave the country through one of nine designated ports. Only 120 full-time USFWS inspectors examine shipments of wild animals that enter the United States through these ports, airports, and border crossings. They can inspect only a small fraction of the more than 200 million wild animals brought legally into the United States each year. Also, tens of millions of such animals are brought in illegally, but few illegal shipments of endangered or threatened animals or plants are confiscated (Figure 9-23, p. 208). Even when they are caught, many violators are not prosecuted, and convicted violators often pay only a small fine.

In addition, people who smuggle or buy imported exotic animals are rarely aware that many of them carry dangerous infectious diseases, such as hantavirus, Ebola virus, Asian bird flu, herpes B virus (carried by most adult macaques), and salmonella (from pets such as hamsters, turtles, and iguanas), which can spread from pets to humans. The country's small number of wildlife inspectors does not have the capability or budget to detect such diseases.

Since 1982, the ESA has been amended to give private landowners economic incentives to help save endangered species living on their lands. The goal is to strike a compromise between the interests of private landowners and those of endangered and threatened species.

With one of these approaches, called a *habitat conservation plan* (HCP), a landowner, developer, or logger is allowed to destroy some critical habitat in exchange for taking steps to protect members of the species. Such measures might include setting aside a part of the species' habitat as a protected area, paying to relocate the species to another suitable habitat, or contributing money to have the government buy suitable habitat elsewhere. Once the plan is approved, it cannot be changed, even if new data show that the plan is inadequate to protect a species and help it recover. The ESA has also been used to protect endangered and threatened marine reptiles,

Figure 9-23 Confiscated products made from endangered species. Because of a scarcity of funds and inspectors, probably no more than one-tenth of the illegal wildlife trade in the United States is discovered. The situation is even worse in most other countries.

Steve Hillebrand/U.S. Fish and Wildlife Service.

such as turtles, and mammals, especially whales, seals, and sea lions, as discussed in Chapter 11.

Some believe that the Endangered Species Act should be weakened or repealed, and others believe it should be strengthened and modified to focus on protecting ecosystems. Opponents of the act contend that it puts the rights and welfare of endangered plants and animals above those of people. They argue that it has not been effective in protecting endangered species and has caused severe economic losses by hindering development on private lands. Since 1995, efforts to weaken the ESA have included the following suggested changes:

- Make protection of endangered species on private land voluntary.

- Have the government compensate landowners if they are forced to stop using part of their land to protect endangered species.

- Make it harder and more expensive to list newly endangered species by requiring government wildlife officials to navigate through a series of hearings and peer-review panels and requiring hard data instead of computer-based models.

- Eliminate the need to designate critical habitats.

- Allow the secretary of the interior to permit a listed species to become extinct without trying to save it.

- Allow the secretary of the interior to give any state, county, or landowner a permanent exemption from the law, with no requirement for public notification or comment.

Other critics would go further and do away with this act entirely. But because this step is politically unpopular with the American public, most efforts are designed to weaken the act and reduce its meager funding. In 2007, the USFWS issued a new interpretation of the ESA that would allow it to protect plants and animals only in areas where they are struggling to survive, rather than listing a species over its entire range. If this new policy survives court tests, it would remove about 80% of the roughly 1,350 U.S. species listed as endangered or threatened. Many wildlife biologists see this proposed policy as a deceptive way to gut the ESA by weakening existing protections and making it more difficult to list new species.

Most conservation biologists and wildlife scientists agree that the ESA needs to be simplified and streamlined. But they contend that it has not been a failure (Science Focus, at right).

We Can Establish Wildlife Refuges and Other Protected Areas

In 1903, President Theodore Roosevelt established the first U.S. federal wildlife refuge at Pelican Island, Florida, to help protect birds such as the brown pelican (see Photo 1 in the Detailed Contents) from extinction. Since then, the National Wildlife Refuge System has grown to include 547 refuges. Each year, more than 35 million Americans visit these refuges to hunt, fish, hike, and watch birds and other wildlife.

More than three-fourths of the refuges serve as wetland sanctuaries vital for protecting migratory waterfowl. One-fifth of U.S. endangered and threatened species have habitats in the refuge system, and some refuges have been set aside for specific endangered species (**Concept 9-5B**). These areas have helped Florida's key deer, the brown pelican, and the trumpeter swan to recover. According to a General Accounting Office study, however, activities considered harmful to wildlife occur in nearly 60% of the nation's wildlife refuges.

Conservation biologists call for setting aside more refuges for endangered plants. They also urge Congress and state legislatures to allow abandoned military lands that contain significant wildlife habitat to become national or state wildlife refuges.

Gene Banks, Botanical Gardens, and Wildlife Farms Can Help Protect Species

Gene or *seed banks* preserve genetic information and endangered plant species by storing their seeds in refrigerated, low-humidity environments (**Concept 9-4B**). More than 100 seed banks around the world collectively hold about 3 million samples.

SCIENCE FOCUS

Accomplishments of the Endangered Species Act

Critics of the ESA call it an expensive failure because only 37 species have been removed from the endangered list. Most biologists insist that it has not been a failure, for four reasons.

First, species are listed only when they face serious danger of extinction. This is like setting up a poorly funded hospital emergency room that takes only the most desperate cases, often with little hope for recovery, and saying it should be shut down because it has not saved enough patients.

Second, it takes decades for most species to become endangered or threatened. Not surprisingly, it also takes decades to bring a species in critical condition back to the point where it can be removed from the critical list. Expecting the ESA—which has been in existence only since 1973—to quickly repair the biological depletion of many decades is unrealistic.

Third, according to federal data, the conditions of more than half of the listed species are stable or improving, and 99% of the protected species are still surviving. A hospital emergency room taking only the most desperate cases and then stabilizing or improving the conditions of more than half of its patients and keeping 99% of them alive would be considered an astounding success.

Fourth, the ESA budget included only $58 million in 2005—about what the Department of Defense spends in a little more than an hour—or 20¢ per year per U.S. citizen. To its supporters, it is amazing that the ESA, on such a small budget, has managed to stabilize or improve the conditions of more than half of the listed species.

Its supporters would agree that the act can be improved and that federal regulators have sometimes been too heavy handed in enforcing it. But instead of gutting or doing away with the ESA, biologists call for it to be strengthened and modified to help protect ecosystems and the nation's overall biodiversity.

A study by the U.S. National Academy of Sciences recommended three major changes to make the ESA more scientifically sound and effective:

- Greatly increase the meager funding for implementing the act.

- Develop recovery plans more quickly. A 2006 study by the Government Accountability Office, found that species with recovery plans have a better chance of getting off the endangered list, and it recommended that any efforts to reform the law should continue to require recovery plans.

- When a species is first listed, establish a core of its survival habitat as critical, as a temporary emergency measure that could support the species for 25–50 years.

Most biologists and wildlife conservationists believe that the United States needs a new law that emphasizes protecting and sustaining biological diversity and ecosystem functioning (**Concept 9-4A**) rather than focusing mostly on saving individual species. We discuss this idea further in Chapter 10.

Critical Thinking

Should the U.S. Endangered Species Act be modified to more effectively protect and sustain the nation's overall biodiversity? Explain.

Scientists urge the establishment of many more such banks, especially in developing countries. But some species cannot be preserved in gene banks. The banks are also expensive to operate and can be destroyed by fires and other mishaps.

The world's 1,600 *botanical gardens* and *arboreta* contain living plants representing almost one-third of the world's known plant species. However, they contain only about 3% of the world's rare and threatened plant species and have too little space and funding to preserve most of those species.

We can take pressure off some endangered or threatened species by raising individuals on *farms* for commercial sale. Farms in Florida raise alligators for their meat and hides. Butterfly farms flourish in Papua New Guinea, where many butterfly species are threatened by development activities.

Zoos and Aquariums Can Protect Some Species

Zoos, aquariums, game parks, and animal research centers are being used to preserve some individuals of critically endangered animal species, with the long-term goal of reintroducing the species into protected wild habitats (**Concept 9-4B**).

Two techniques for preserving endangered terrestrial species are egg pulling and captive breeding. *Egg pulling* involves collecting wild eggs laid by critically endangered birds and then hatching them in zoos or research centers. In *captive breeding,* some or all of the wild individuals of a critically endangered species are captured for breeding in captivity, with the aim of reintroducing the offspring into the wild. Captive breeding has been used to save the peregrine falcon and the California condor (Case Study, p. 210).

Other techniques for increasing the populations of captive species include artificial insemination, embryo transfer (surgical implantation of eggs of one species into a surrogate mother of another species), use of incubators, and cross-fostering (in which the young of a rare species are raised by parents of a similar species). Scientists also use computer databases, which hold information on family lineages of species in zoos, and DNA analysis to match individuals for mating—a computer dating service for zoo animals—and to prevent genetic erosion through inbreeding.

The ultimate goal of captive breeding programs is to build up populations to a level where they can be reintroduced into the wild. But after more than two

decades of captive breeding efforts, only a handful of endangered species have been returned to the wild. Examples shown in Figure 9-3 include the black-footed ferret, the California condor (Case Study, right), and the golden lion tamarin. Most reintroductions fail because of lack of suitable habitat, inability of individuals bred in captivity to survive in the wild, renewed overhunting, or poaching of some of the returned individuals.

Lack of space and money limits efforts to maintain breeding populations of endangered animal species in zoos and research centers. The captive population of each species must number 100–500 individuals to avoid extinction through accident, disease, or loss of genetic diversity through inbreeding. Recent genetic research indicates that 10,000 or more individuals are needed for an endangered species to maintain its capacity for biological evolution.

Public aquariums that exhibit unusual and attractive fish and some marine animals such as seals and dolphins help to educate the public about the need to protect such species. But mostly because of limited funds, public aquariums have not served as effective gene banks for endangered marine species, especially marine mammals that need large volumes of water.

Instead of seeing zoos and aquariums as sanctuaries, some critics claim that most of them imprison once-wild animals. They also contend that zoos and aquariums can foster the false notion that we do not need to preserve large numbers of wild species in their natural habitats. Proponents counter that these facilities play an important role in educating the public about wildlife and the need to protect biodiversity.

Regardless of their benefits and drawbacks, zoos, aquariums, and botanical gardens are not biologically or economically feasible solutions for the growing problem of premature extinction of species. Figure 9-24 lists some things you can do to deal with this problem.

■ CASE STUDY
Trying to Save the California Condor

At one time the California condor (Figure 9-4), North America's largest bird, was nearly extinct with only 22 birds remaining in the wild. To save the species, one approach was to capture the remaining birds and breed them in captivity at zoos.

The captured birds were isolated from human contact as much as possible, and to reduce genetic defects, closely related individuals were prevented from breeding. As of 2007, 135 condors had been released back into the wild throughout the southwestern United States.

A major threat to these birds is lead poisoning resulting when they ingest lead pellets from ammunition in animal carcasses or gut piles left behind by hunters. A lead-poisoned condor quickly becomes weak and mentally impaired and dies of starvation, or is killed by predators.

A coalition of conservationist and health organizations is lobbying state game commissions and legislatures to ban the use of lead in ammunition and to require use of less harmful substitutes. They also urge people who hunt in condor ranges to remove all killed animals or to hide carcasses and gut piles by burying them, covering them with brush or rocks, or putting them in inaccessible areas.

WHAT CAN YOU DO?

Protecting Species

- Do not buy furs, ivory products, or other items made from endangered or threatened animal species
- Do not buy wood or paper products produced by cutting old-growth forests in the tropics
- Do not buy birds, snakes, turtles, tropical fish, and other animals that are taken from the wild
- Do not buy orchids, cacti, or other plants that are taken from the wild
- Spread the word. Talk to your friends and relatives about this problem and what they can do about it

Figure 9-24 Individuals matter: ways to help prevent premature extinction of species. **Question:** Which two of these suggestions do you believe are the most important? Why?

THINKING ABOUT

The California Condor's Comeback

CORE CASE STUDY

What are some differences between the stories of the condor and the passenger pigeon (**Core Case Study**) that might give the condor a better chance of avoiding premature extinction than the passenger pigeon had?

The Precautionary Principle

Some might argue that, because we have identified fewer than 2 million of the estimated 5–100 million species on the earth, it makes little sense to take drastic measures to preserve them.

Conservation biologists disagree. They remind us that the earth's species are the primary components of its biodiversity, which should not be degraded because of the economic and ecological services it provides. They call for us to use great caution in making potentially harmful changes to communities and ecosystems and to take precautionary action to help *prevent* poten-

tially serious environmental problems, including premature extinctions.

This approach is based on the **precautionary principle:** When substantial preliminary evidence indicates that an activity can harm human health or the environment, we should take precautionary measures to prevent or reduce such harm, even if some of the cause-and-effect relationships have not been fully established, scientifically (**Concept 9-4C**). It is based on the commonsense idea behind many adages such as "Better safe than sorry" and "Look before you leap."

Scientists use the precautionary principle to argue for preservation of species, and also for preserving entire ecosystems, which is the focus of the next two chapters.

*The great challenge of the twenty-first century
is to raise people everywhere to a decent standard of living
while preserving as much of the rest of life as possible.*

EDWARD O. WILSON

REVIEW

1. Review the Key Questions and Concepts for this chapter on p. 184. What factors led to the premature extinction of the passenger pigeon in the United States?

2. Distinguish between **background extinction** and **mass extinction.** What is the **extinction rate** of a species? Describe how scientists estimate extinction rates. Give four reasons why many extinction experts believe that human activities are now causing a sixth mass extinction. Distinguish between **endangered species** and **threatened species.** List some characteristics that make some species especially vulnerable to extinction.

3. What are two reasons for trying to prevent the premature extinction of wild species? What is the **instrumental value** of a species? List six types of instrumental values provided by wild species. How are scientists using DNA analysis to reduce the illegal killing of elephants? What is the **intrinsic (existence) value** of a species?

4. What is biophilia? Why should we care about bats?

5. What is **HIPPCO?** In order, what are the six largest causes of premature extinction of species resulting from human activities? Why are island species especially vulnerable to extinction? What is habitat fragmentation, and how does it threaten many species?

6. Describe the threats to bird species in the world and in the United States. List three reasons why we should be alarmed by the decline of bird species.

7. Give two examples of the harmful effects of nonnative species that have been introduced **(a)** deliberately and **(b)** accidentally. List ways to limit the harmful impacts of nonnative species. Describe the roles of population growth, overconsumption, pollution, and climate change in the premature extinction of wild species. Describe what is happening to many of the honeybees in the United States and what economic and ecological roles they play. Explain how pesticides such as DDT can be biomagnified in food chains and webs. Explain how global warming is threatening polar bears.

8. Describe the poaching of wild species and give three examples of species that are threatened by this illegal activity. Describe the work of Jane Goodall in protecting wild primates. Why are tigers likely to disappear from the wild by the end of this century? Describe the threat to some forms of wildlife from increased hunting for bush meat.

9. Describe two international treaties that are used to help protect species. Describe the U.S. Endangered Species Act, how successful it has been, and the controversy over this act. Describe the roles of wildlife refuges, gene banks, botanical gardens, wildlife farms, zoos, and aquariums in protecting some species.

10. Describe how protecting wild species from premature extinction (**Core Case Study**) is in keeping with the four **scientific principles of sustainability**.

Note: Key Terms are in **bold** type.

CRITICAL THINKING

1. List three ways in which you could apply **Concept 9-3** to make your lifestyle more environmentally sustainable.

2. What are three aspects of your lifestyle that directly or indirectly contribute to the premature extinction of some bird species (Case Study, p. 195)? What are three things that you think should be done to reduce the premature extinction of birds?

3. Discuss your gut-level reaction to the following statement: "Eventually, all species become extinct. Thus, it does not really matter that the passenger pigeon (**Core Case Study**) is extinct, and that the whooping crane and the world's remaining tiger species are endangered mostly because of human activities." Be honest about your reaction, and give arguments for your position.

4. Do you accept the ethical position that each species has the inherent right to survive without human interference, regardless of whether it serves any useful purpose for humans? Explain. Would you extend this right to the *Anopheles* mosquito, which transmits malaria, and to infectious bacteria? Explain.

5. Wildlife ecologist and environmental philosopher Aldo Leopold wrote, "To keep every cog and wheel is the first precaution of intelligent tinkering." Explain how this statement relates to the material in this chapter. Explain how protecting wild species and their habitats is an important way to protect the health and well-being of people.

6. What would you do if **(a)** your yard and house were invaded by fire ants, **(b)** you found bats flying around your yard at night, and **(c)** deer invaded your yard and ate your shrubs, flowers, and vegetables?

7. Which of the following statements best describes your feelings toward wildlife?
 a. As long as it stays in its space, wildlife is okay.
 b. As long as I do not need its space, wildlife is okay.
 c. I have the right to use wildlife habitat to meet my own needs.
 d. When you have seen one redwood tree, elephant, or some other form of wildlife, you have seen them all, so lock up a few of each species in a zoo or wildlife park and do not worry about protecting the rest.
 e. Wildlife should be protected.

8. Environmental groups in a heavily forested state want to restrict logging in some areas to save the habitat of an endangered squirrel. Timber company officials argue that the well-being of one type of squirrel is not as important as the well-being of the many families who will be affected if the restriction causes the company to lay off hundreds of workers. If you had the power to decide this issue, what would you do and why? Can you come up with a compromise?

9. Congratulations! You are in charge of preventing the premature extinction, caused by human activities, of the world's existing species. What are three things you would do to accomplish this goal?

10. List two questions that you would like to have answered as a result of reading this chapter.

Note: See Supplement 13 (p. S78) for a list of Projects related to this chapter.

Examine these data released by the World Resources Institute and answer the following questions.

Country	Total Land Area in Square Kilometers (square miles)	Protected Area as Percent of Total Land Area (2003)	Total Number of Known Breeding Bird Species (1992–2002)	Number of Threatened Breeding Bird Species (2002)	Threatened Breeding Bird Species as Percent of Total Number of Known Breeding Bird Species
Afghanistan	647,668 (250,000)	0.3	181	11	
Cambodia	181,088 (69,900)	23.7	183	19	
China	9,599,445 (3,705,386)	7.8	218	74	
Costa Rica	50,110 (19,730)	23.4	279	13	
Haiti	27,756 (10,714)	0.3	62	14	
India	3.287,570 (1,269,388)	5.2	458	72	
Rwanda	26,344 (10,169)	7.7	200	9	
United States	9,633,915 (3,781,691)	15.8	508	55	

Source: Data from **earthtrends.wri.org/country_profiles/index.php?theme=7**.
World Resources Institute, *Earth Trends, Biodiversity and Protected Areas, Country Profiles*

1. For each of the eight countries, complete the table by filling in the last column.

2. Arrange the countries from largest to smallest according to Total Land Area. Does there appear to be any correlation between the size of country and the percentage of Threatened Breeding Bird Species? Explain your reasoning.

10 Sustaining Terrestrial Biodiversity: The Ecosystem Approach

Reintroducing Gray Wolves to Yellowstone

Around 1800 at least 350,000 gray wolves (Figure 10-1), roamed over the lower 48 states, especially in the West, and survived mostly by preying on bison, elk, caribou, and mule deer. But between 1850 and 1900, most of them were shot, trapped, and poisoned by ranchers, hunters, and government employees. When Congress passed the U.S. Endangered Species Act in 1973, only a few hundred gray wolves remained outside of Alaska, primarily in Minnesota and Michigan. In 1974,

the gray wolf was listed as an endangered species in the lower 48 states.

Ecologists recognize the important role this keystone predator species once played in parts of the West , especially in the northern Rocky Mountain states of Montana, Wyoming, and Idaho where Yellowstone National Park is located. The wolves culled herds of bison, elk, caribou, and mule deer, and kept down coyote populations. They also provided uneaten meat for scavengers such as ravens, bald eagles, ermines, grizzly bears, and foxes. When wolves declined, herds of plant-browsing elk, moose, and mule deer expanded and devastated vegetation such as willow and aspen trees often found growing near streams and rivers. This increased soil erosion and threatened habitats of other wildlife species such as beavers, which, as foundation species (p. 96), helped to maintain wetlands.

In 1987, the U.S. Fish and Wildlife Service (USFWS) proposed reintroducing gray wolves into the Yellowstone National Park ecosystem to help restore and sustain its biodiversity. The proposal brought angry protests, some from area ranchers who feared the wolves would leave the park and attack their cattle and sheep. Other objections came from hunters who feared the wolves would kill too many big-game animals, and from mining and logging companies fearing that the government would halt their operations on wolf-populated federal lands.

In 1995 and 1996, federal wildlife officials caught gray wolves in Canada and relocated 41 of them in Yellowstone National Park. Scientists estimate that the long-term carrying capacity of the park is 110 to 150 gray wolves. In 2007, the park had 171 gray wolves. Overall, this experiment in ecosystem restoration has helped to re-establish and sustain some of the biodiversity that the Yellowstone ecosystem once had, as discussed later in this chapter.

In 2008, the USFWS decided to remove the gray wolf from protection under the Endangered Species Act in the states of Montana, Wyoming, and Utah. Several conservation groups filed suits to have the courts overturn this decision. The wolves in the park will remain protected. But 6 of the park's 11 wolf packs travel outside of the park boundaries during part of every year. If the courts allow removing the wolves from the endangered species list, it will be legal to kill any of these packs' individuals found outside the park.

Biologists warn that human population growth, economic development, and poverty are exerting increasing pressure on ecosystems and the services they provide to sustain biodiversity. This chapter is devoted to helping us understand and sustain the earth's forests, grasslands, and other storehouses of terrestrial biodiversity.

Figure 10-1 Natural capital restoration: the *gray wolf*. After becoming almost extinct in much of the western United States, in 1974 the gray wolf was listed and protected as an endangered species. Despite intense opposition by ranchers, hunters, miners, and loggers 41 members of this keystone species were reintroduced to their former habitat in the Yellowstone National Park and central Idaho in 1995 and 1996. By 2007, there were about 171 gray wolves in the park.

Tom Kitchin/Tom Stack & Associates

10-1 What are the major threats to forest ecosystems?

CONCEPT 10-1A Forest ecosystems provide ecological services far greater in value than the value of raw materials obtained from forests.

CONCEPT 10-1B Unsustainable cutting and burning of forests, along with diseases and insects, made worse by global warming, are the chief threats to forest ecosystems.

CONCEPT 10-1C Tropical deforestation is a potentially catastrophic problem because of the vital ecological services at risk, the high rate of tropical deforestation, and its growing contribution to global warming.

10-2 How should we manage and sustain forests?

CONCEPT 10-2 We can sustain forests by emphasizing the economic value of their ecological services, protecting old-growth forests, harvesting trees no faster than they are replenished, and using sustainable substitute resources.

10-3 How should we manage and sustain grasslands?

CONCEPT 10-3 We can sustain the productivity of grasslands by controlling the number and distribution of grazing livestock and by restoring degraded grasslands.

10-4 How should we manage and sustain parks and nature reserves?

CONCEPT 10-4 Sustaining biodiversity will require protecting much more of the earth's remaining undisturbed land area as parks and nature reserves.

10-5 What is the ecosystem approach to sustaining biodiversity?

CONCEPT 10-5A We can help to sustain biodiversity by identifying severely threatened areas and protecting those with high plant diversity (biodiversity hotspots) and those where ecosystem services are being impaired.

CONCEPT 10-5B Sustaining biodiversity will require a global effort to rehabilitate and restore damaged ecosystems.

CONCEPT 10-5C Humans dominate most of the earth's land, and preserving biodiversity will require sharing as much of it as possible with other species.

Note: Supplements 2 (p. S4), 4 (p. S20), 5 (p. S31), 9 (p. S53), and 13 (p. S78) can be used with this chapter.

*There is no solution, I assure you,
to save Earth's biodiversity other than preservation
of natural environments in reserves large enough
to maintain wild populations sustainably.*

EDWARD O. WILSON

10-1 What Are the Major Threats to Forest Ecosystems

▶ **CONCEPT 10-1A** Forest ecosystems provide ecological services far greater in value than the value of raw materials obtained from forests.

▶ **CONCEPT 10-1B** Unsustainable cutting and burning of forests, along with diseases and insects, made worse by global warming, are the chief threats to forest ecosystems.

▶ **CONCEPT 10-1C** Tropical deforestation is a potentially catastrophic problem because of the vital ecological services at risk, the high rate of tropical deforestation, and its growing contribution to global warming.

Forests Vary in Their Make-Up, Age, and Origins

Natural and planted forests occupy about 30% of the earth's land surface (excluding Greenland and Antarctica). Figure 7-8 (p. 146) shows the distribution of the world's boreal, temperate, and tropical forests. Tropical forests (Figure 7-15, top, p. 154) account for more than half of the world's forest area, and boreal (northern coniferous) forests (Figure 7-15, bottom) account for one quarter.

Forest managers and ecologists classify natural forests into two major types based on their age and structure. The first type is an **old-growth forest:** an uncut or regenerated primary forest that has not been seriously disturbed by human activities or natural disasters for

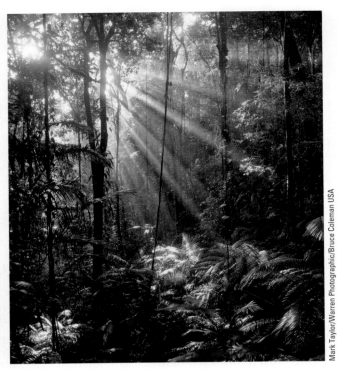

Figure 10-2 **Natural capital:** an old-growth forest in the U.S. state of Washington's Olympic National Forest (left) and an old-growth tropical rain forest in Queensland, Australia (right).

200 years or more (Figure 10-2). Old-growth or primary forests are reservoirs of biodiversity because they provide ecological niches for a multitude of wildlife species (Figure 7-16, p. 155, and Figure 7-17, p. 156).

The second type is a **second-growth forest:** a stand of trees resulting from secondary ecological succession (Figure 5-17, p. 117, and Figure 7-15, center photo, p. 154). These forests develop after the trees in an area have been removed by human activities, such as clear-cutting for timber or cropland, or by natural forces, such as fire, hurricanes, or volcanic eruption.

A **tree plantation,** also called a **tree farm** or **commercial forest** (Figure 10-3), is a managed tract with uniformly aged trees of one or two genetically uniform species that usually are harvested by clear-cutting as soon as they become commercially valuable. The land

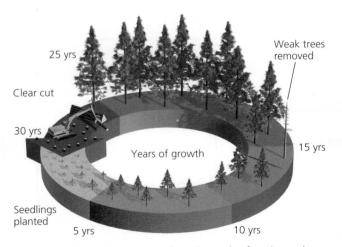

Clear cut

25 yrs

Weak trees removed

30 yrs

Years of growth

15 yrs

Seedlings planted

5 yrs

10 yrs

Figure 10-3 Short (25- to 30-year) rotation cycle of cutting and regrowth of a monoculture tree plantation used in modern industrial forestry. In tropical countries, where trees can grow more rapidly year-round, the rotation cycle can be 6–10 years. Old-growth or second-growth forests are clear-cut to provide land for growing most tree plantations (see photo, right). **Question:** What are two ways in which this process can degrade an ecosystem?

is then replanted and clear-cut again in a regular cycle. When managed carefully, such plantations can produce wood at a fast rate and thus increase their owners' profits. Most of this wood goes to paper mills and to mills that produce composites used as a substitute for natural wood.

But tree plantations with only one or two tree species are much less biologically diverse and probably less sustainable than old-growth and second-growth forests because they violate nature's biodiversity **principle of sustainability** (see back cover). And repeated cycles of cutting and replanting can eventually deplete the soil of nutrients and lead to an irreversible ecological tipping point that can hinder the growth of any type of forest on the land. There is also controversy over the increased use of genetically engineered tree species whose seeds could spread to other areas and threaten the diversity of second- and old-growth forests.

According to 2007 estimates by the FAO, about 60% of the world's forests are second-growth forests, 36% are old-growth or primary forests, and 4% are tree plantations (6% in the United States). In order, five countries—Russia, Canada, Brazil, Indonesia, and Papua New Guinea—have more than three-fourths of the world's remaining old-growth forests. In order, China (which has little original forest left), India, the United States, Russia, Canada, and Sweden account for about 60% of the world's tree plantations. Some conservation biologists urge establishing tree plantations only on land that has already been cleared or degraded instead of putting them in place of existing old-growth or secondary forests. One day, tree plantations may supply most of the world's demand for industrial wood, and this will help to protect the world's remaining forests.

Forests Provide Important Economic and Ecological Services

Forests provide highly valuable ecological and economic services (Figure 10-4 and **Concept 10-1A**). For example, through photosynthesis, forests remove CO_2 from the atmosphere and store it in organic compounds (biomass). By performing this ecological service, forests help to stabilize the earth's temperature and to slow global warming as a part of the global carbon cycle (Figure 3-18, p. 68). Scientists have attempted to estimate the economic value of the ecological services provided by the world's forests and other ecosystems (Science Focus, p. 218).

RESEARCH FRONTIER

Refining estimates of the economic values of ecological services provided by forests and other major ecosystems. See **academic.cengage.com/biology/miller**.

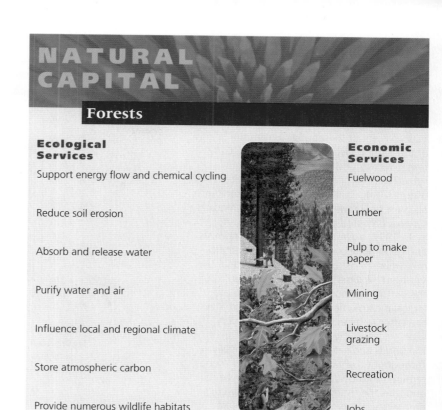

Figure 10-4 Major ecological and economic services provided by forests (**Concept 10-1A**). **Question:** Which two ecological services and which two economic services do you think are the most important?

Most biologists believe that the clearing and degrading of the world's remaining old-growth forests is a serious global environmental threat because of the important ecological and economic services they provide (**Concept 10-1A**). For example, traditional medicines, used by 80% of the world's people, are derived mostly from natural plants in forests, and chemicals found in tropical forest plants are used as blueprints for making most of the world's prescription drugs (Figure 9-8, p. 190). Forests are also habitats for about two-thirds of the earth's terrestrial species. In addition, they are home to more than 300 million people, and one of every four people depend on forests for their livelihoods.

Unsustainable Logging Is a Major Threat to Forest Ecosystems

Along with highly valuable ecological services, forests provide us with raw materials, especially wood.

The first step in harvesting trees is to build roads for access and timber removal. Even carefully designed logging roads have a number of harmful effects (Figure 10-5, p. 218)—namely, increased erosion and sediment runoff into waterways, habitat fragmentation (see Science Focus, p. 195, and *The Habitable Planet*,

SCIENCE FOCUS

Putting a Price Tag on Nature's Ecological Services

The long-term health of an economy cannot be separated from the health of the natural systems that support it. Currently, forests and other ecosystems are valued mostly for their economic services (Figure 10-4, right). But suppose we took into account the monetary value of the ecological services provided by forests (Figure 10-4, left).

In 1997, a team of ecologists, economists, and geographers—led by ecological economist Robert Costanza of the University of Vermont—estimated the monetary worth of the earth's ecological services and the biological income they provide. They estimated the latter to be at least $33.2 trillion per year—close to the economic value of all of the goods and services produced annually throughout the world. The amount of money required to provide such interest income, and thus the estimated value of the world's natural capital, would have to be at least $500 trillion—an average of about $73,500 for each person on earth!

According to this study, the world's forests provide us with ecological services worth at least $4.7 trillion per year—hundreds of times more than their economic value. And these are very conservative estimates.

Costanza's team examined many studies and a variety of methods used to estimate the values of ecosystems. For example, some researchers estimated people's *willingness to pay* for ecosystem services that are not marketed, such as natural flood control and carbon storage. These estimates were added to the known values of marketed goods like timber to arrive at a total value for an ecosystem.

The researchers estimated total global areas of 16 major categories of ecosystems, including forests, grasslands, and other ter-

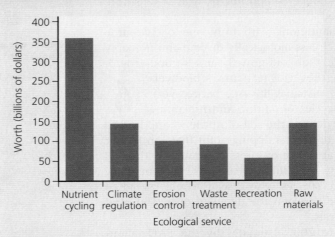

Figure 10-A Estimated annual global economic values of some ecological services provided by forests compared to the raw materials they produce (in billions of dollars).

restrial and aquatic systems. They multiplied those areas by the values per hectare of various ecosystem services to get the estimated economic values of these forms of natural capital. Some of the results for forests are shown in Figure 10-A. Note that the collective value of these ecosystem services is much greater than the value of timber and other raw materials extracted from forests (**Concept 10-1A**).

These researchers hope their estimates will alert people to three important facts: the earth's ecosystem services are essential for all humans and their economies; their economic value is huge; and they are an ongoing source of ecological income as long as they are used sustainably.

However, unless such estimates are included in the market prices of goods and services—through market tools such as

regulations and taxes that discourage biodiversity degradation and through subsidies that protect biodiversity—the world's forests and other ecosystems will continue to be degraded. For example, the governments of countries such as Brazil and Indonesia provide subsidies that encourage the clearing or burning of tropical forests to plant vast soybean and oil palm plantations. This sends the powerful message that it makes more economic sense to destroy or degrade these centers of biodiversity than it does to leave them intact.

Critical Thinking

Some analysts believe that we should not try to put economic values on the world's irreplaceable ecological services because their value is infinite. Do you agree with this view? Explain. What is the alternative?

Figure 10-5 Natural capital degradation: Building roads into previously inaccessible forests paves the way to fragmentation, destruction, and degradation.

(a) Selective cutting

Clear stream

(b) Clear-cutting

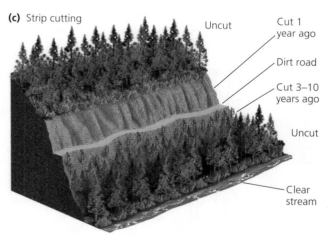

Muddy stream

(c) Strip cutting

Uncut

Cut 1 year ago

Dirt road

Cut 3–10 years ago

Uncut

Clear stream

Figure 10-6 Major tree harvesting methods. **Question:** If you were cutting trees in a forest you owned, which method would you choose and why?

Video 9, at **www.learner.org/resources/series209. html**), and loss of biodiversity. Logging roads also expose forests to invasion by nonnative pests, diseases, and wildlife species. And they open once-inaccessible forests to farmers, miners, ranchers, hunters, and off-road vehicle users.

Once loggers reach a forest area, they use a variety of methods to harvest the trees (Figure 10-6). With *selective cutting*, intermediate-aged or mature trees in an uneven-aged forest are cut singly or in small groups

(Figure 10-6a). But often, loggers remove all the trees from an area in what is called a *clear-cut* (Figures 10-6b and 10-7). Clear-cutting is the most efficient way for a logging operation to harvest trees, but it can do considerable harm to an ecosystem.

For example, scientists found that removing all the tree cover from a watershed greatly increases water runoff and loss of soil nutrients (Chapter 2 Core Case Study, p. 28). This increases soil erosion, which in turn causes more vegetation to die, leaving barren ground that can be eroded further, an example of a harmful positive feedback loop (Figure 2-11, p. 45, and **Concept 2-5A**, p. 44). More erosion also means more pollution of streams in the watershed. CONCEPT LINK And loss of vegetation destroys habitat and degrades biodiversity. Figure 10-8 (p. 220) summarizes some advantages and disadvantages of clear-cutting.

CENGAGENOW™ Learn more about how deforestation can affect the drainage of a watershed and disturb its ecosystem at CengageNOW™.

A variation of clear-cutting that allows a more sustainable timber yield without widespread destruction is *strip cutting* (Figure 10-6c). It involves clear-cutting a strip of trees along the contour of the land within a corridor narrow enough to allow natural regeneration within a few years. After regeneration, loggers cut another strip next to the first, and so on.

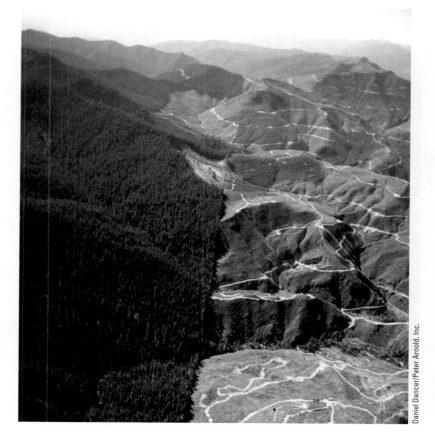

Daniel Dancer/Peter Arnold, Inc.

Figure 10-7 Clear-cut logging in the U.S. state of Washington.

TRADE-OFFS

Clear-Cutting Forests

Advantages	Disadvantages
Higher timber yields	Reduces biodiversity
Maximum profits in shortest time	Destroys and fragments wildlife habitats
Can reforest with fast-growing trees	Increases water pollution, flooding, and erosion on steep slopes
Good for tree species needing full or moderate sunlight	Eliminates most recreational value

Figure 10-8 Advantages and disadvantages of clear-cutting forests. **Question:** Which single advantage and which single disadvantage do you think are the most important? Why?

Biodiversity experts are alarmed at the growing practice of illegal, uncontrolled, and unsustainable logging taking place in 70 countries, especially in Africa and Southeast Asia (**Concept 10-1B**). Such logging has ravaged 37 of the 41 national parks in the African country of Kenya and now makes up 73–80% of all logging in Indonesia.

Complicating this issue is global trade in timber and wood products. For example, China, which has cut most of its own natural forests, imports more tropical rain forest timber than any other nation. Much of this timber is harvested illegally and unsustainably and is used to make furniture, plywood, flooring, and other products that are sold in the global marketplace.

Fire, Insects, and Climate Change Can Threaten Forest Ecosystems

Two types of fires can affect forest ecosystems. *Surface fires* (Figure 10-9, left) usually burn only undergrowth and leaf litter on the forest floor. They may kill seedlings and small trees but spare most mature trees and allow most wild animals to escape.

Occasional surface fires have a number of ecological benefits. They burn away flammable ground material and help to prevent more destructive fires. They also free valuable mineral nutrients tied up in slowly decomposing litter and undergrowth, release seeds from the cones of lodgepole pines, stimulate the germination of certain tree seeds such as those of the giant sequoia and jack pine, and help control tree diseases and insects. Wildlife species such as deer, moose, muskrat, and quail depend on occasional surface fires to maintain their habitats and provide food in the form of vegetation that sprouts after fires.

Another type of fire, called a *crown fire* (Figure 10-9, right), is an extremely hot fire that leaps from treetop

David J. Moorhead/The University of Georgia

© age footstock/SuperStock

Figure 10-9 Surface fires (left) usually burn undergrowth and leaf litter on a forest floor. They can help to prevent more destructive crown fires (right) by removing flammable ground material. In fact, carefully controlled surface fires are deliberately set sometimes to prevent buildup of flammable ground material in forests. They also recycle nutrients and thus help to maintain the productivity of a variety of forest ecosystems. **Question:** What is another way in which a surface fire might benefit a forest?

White pine blister rust

Pine shoot beetle

Beech bark disease

Sudden oak death

Hemlock woolly adelgid

Figure 10-10 Natural capital degradation: some of the nonnative insect species and disease organisms that have invaded U.S. forests and are causing billions of dollars in damages and tree loss. The light green and orange colors in the map show areas where green or red overlap with yellow. (Data from U.S. Forest Service)

to treetop, burning whole trees. Crown fires usually occur in forests that have not experienced surface fires for several decades, a situation that allows dead wood, leaves, and other flammable ground litter to accumulate. These rapidly burning fires can destroy most vegetation, kill wildlife, increase soil erosion, and burn or damage human structures in their paths.

As part of a natural cycle, forest fires are not a major threat to forest ecosystems. But they are serious threats in parts of the world where people intentionally burn forests to clear the land, mostly to make way for crop plantations (**Concept 10-1B**). This can result in dramatic habitat losses, air pollution, and increases in atmospheric CO_2.

Accidental or deliberate introductions of foreign diseases and insects are a major threat to forests in the United States and elsewhere. Figure 10-10 shows some nonnative species of pests and disease organisms that are causing serious damage to certain tree species in parts of the United States.

There are several ways to reduce the harmful impacts of tree diseases and insect pests on forests. One is to ban imported timber that might introduce harmful new diseases or insects; another is to remove or clearcut infected and infested trees. We can also develop tree species that are genetically resistant to common tree diseases. Another approach is to control insect pests by applying conventional pesticides. Scientists also use

biological control (bugs that eat harmful bugs) combined with very small amounts of conventional pesticides.

On top of these threats, projected climate change from global warming could harm many forests. For example, sugar maples are sensitive to heat, and in the U.S. region of New England, rising temperatures could kill these trees and, consequently, a productive maple syrup industry. Rising temperatures would also make many forest areas more suitable for insect pests and increase the size of pest populations. The resulting combination of drier forests and more dead trees could increase the incidence and intensity of forest fires (**Concept 10-1B**). This would add more of the greenhouse gas CO_2 to the atmosphere, which would further increase atmospheric temperatures and cause even more forest fires in a runaway positive feedback loop (Figure 2-11, p. 45).

We Have Cut Down Almost Half of the World's Forests

Deforestation is the temporary or permanent removal of large expanses of forest for agriculture, settlements, or other uses. Surveys by the World Resources Institute (WRI) indicate that over the past 8,000 years, human activities have reduced the earth's original forest cover by about 46%, with most of this loss occurring in the last 60 years.

Deforestation continues at a rapid rate in many parts of the world. The U.N. Food and Agricultural Organization (FAO) and World Resources Institute (WRI) surveys indicate that the global rate of forest cover loss between 1990 and 2005 was between 0.2%

and 0.5% per year, and that at least another 0.1–0.3% of the world's forests were degraded every year, mostly to grow crops and graze cattle. If these estimates are correct, the world's forests are being cleared or degraded exponentially at a rate of 0.3–0.8% per year, with much higher rates in some areas.

These losses are concentrated in developing countries, especially those in the tropical areas of Latin America, Indonesia, and Africa (Figure 3, pp. S24–S25, in Supplement 4). In its 2007 *State of the World's Forests* report, U.N. Food and Agricultural Organization estimated that about 130,000 square kilometers (50,000 square miles) of tropical forests are cleared each year (Figure 10-11)—equivalent to the total area of Greece or the U.S. state of Mississippi. We examine tropical forest losses further in the next subsection.

In addition to losses of tropical forests, scientists are concerned about the increased clearing of the northern boreal forests of Alaska, Canada, Scandinavia, and Russia, which together make up about one-fourth of the world's forested area. These vast coniferous forests (Figure 7-15, bottom photo, p. 154) are the world's greatest terrestrial storehouse of organic carbon and play a major role in the carbon cycle (Figure 3-18, p. 68) and in climate regulation for the entire planet. They also contain more than 70,000 plant and animal species. Surveys indicate that the total area of boreal forests lost every year is about twice the total area of Brazil's vast rain forests. In 2007, a group of 1,500 scientists from around the world signed a letter calling for the Canadian government to protect half of Canada's threatened boreal forests (of which only 10% are protected now) from logging, mining, and oil and gas extraction.

Figure 10-11 Natural capital degradation: extreme tropical deforestation in Chiang Mai, Thailand. The clearing of trees that absorb carbon dioxide increases global warming. It also dehydrates the soil by exposing it to sunlight. The dry topsoil blows away, which prevents the reestablishment of a forest in this area.

According to the WRI, if current deforestation rates continue, about 40% of the world's remaining intact forests will have been logged or converted to other uses within 2 decades, if not sooner. Clearing large areas of forests, especially old-growth forests, has important short-term economic benefits (Figure 10-4, right), but it also has a number of harmful environmental effects (Figure 10-12).

HOW WOULD YOU VOTE? ☑

Should there be a global effort to sharply reduce the cutting of old-growth forests? Cast your vote online at **academic .cengage.com/biology/miller**.

In some countries, there is encouraging news about forest use. In 2007, the FAO reported that the net total forest cover in several countries, including the United States (see Case Study below), changed very little or increased between 2000 and 2005. Some of the increases resulted from natural reforestation by secondary ecological succession on cleared forest areas and abandoned croplands. But such increases were also due to the spread of commercial tree plantations.

■ CASE STUDY

Many Cleared Forests in the United States Have Grown Back

Forests that cover about 30% of the U.S. land area provide habitats for more than 80% of the country's wildlife species and supply about two-thirds of the nation's surface water. Old-growth forests once covered more than half of the nation's land area. But between 1620 when Europeans first arrived and 1920, the old-growth forests of the eastern United States were decimated.

Today, forests (including tree plantations) cover more area in the United States than they did in 1920. Many of the old-growth forests that were cleared or partially cleared between 1620 and 1920 have grown back naturally through secondary ecological succession (Figure 5-17, p. 117). There are fairly diverse second-growth (and in some cases third-growth) forests in every region of the United States, except much of the West. In 1995, environmental writer Bill McKibben cited forest regrowth in the United States—especially in the East—as "the great environmental story of the United States, and in some ways, the whole world."

Every year, more wood is grown in the United States than is cut and the total area planted with trees increases. Protected forests make up about 40% of the country's total forest area, mostly in the *National Forest System*, which consists of 155 national forests managed by the U.S. Forest Service (USFS).

On the other hand, since the mid-1960s, an increasing area of the nation's remaining old-growth and fairly diverse second-growth forests has been cut down and replaced with biologically simplified tree planta-

NATURAL CAPITAL DEGRADATION

Deforestation

- Decreased soil fertility from erosion
- Runoff of eroded soil into aquatic systems
- Premature extinction of species with specialized niches
- Loss of habitat for native species and migratory species such as birds and butterflies
- Regional climate change from extensive clearing
- Release of CO_2 into atmosphere
- Acceleration of flooding

Figure 10-12 Harmful environmental effects of deforestation, which can reduce biodiversity and the ecological services provided by forests (Figure 10-4, left). **Question:** What are three products you have used recently that might have come from old-growth forests?

tions. According to biodiversity researchers, this reduces overall forest biodiversity and disrupts ecosystem processes such as energy flow and chemical cycling. And if such plantations are harvested too frequently, it could also deplete forest soils of key nutrients. Many biodiversity researchers favor establishing tree plantations only on land that has already been degraded instead of cutting old-growth and second-growth forests in order to replace them with tree plantations.

Tropical Forests Are Disappearing Rapidly

Tropical forests (Figure 7-15, top photo, p. 154) cover about 6% of the earth's land area—roughly the area of the lower 48 U.S. states. Climatic and biological data suggest that mature tropical forests once covered at least twice as much area as they do today; the majority of tropical forest loss has taken place since 1950 (Chapter 3 Core Case Study, p. 50).

Satellite scans and ground-level surveys indicate that large areas of tropical rain forests and tropical dry forests are being cut rapidly in parts of Africa, Southeast Asia (Figure 10-11), and South America (Figure 3-1, p. 50, and Figure 10-13, p. 224). A 2006 study by the U.S. National Academy of Sciences found that between 1990 and 2005, Brazil and Indonesia led the world in tropical forest loss. Illegal tree felling in 37 of 41 of Indonesia's supposedly protected parks account for three-quarters of the country's logging. According to the United Nations, Indonesia, which currently has the world's most diverse combination of plants, animals, and marine life, has already lost an estimated

Figure 10-13
Satellite images
of Amazon
deforestation
in the state
of Rondônia,
Brazil, between
1975 and 2001.

UNEP/GRID-Sioux Falls

UNEP/GRID-Sioux Falls

19 Jun 1975

19 Sep 2001

72% of its original intact forest, and 98% of its remaining forests will be gone by 2022.

Studies indicate that at least half of the world's known species of terrestrial plants and animals live in tropical rain forests (Figure 10-14). Because of their specialized niches (Figure 7-17, p. 156, and **Concept 4-6A**, p. 91) these species are highly vul-

CONCEPT LINK

nerable to extinction when their forest habitats are destroyed or degraded. Tropical deforestation is the main reason that more than 8,000 tree species—10% of the world's total—are threatened with extinction.

Brazil has more than 30% of the world's remaining tropical rain forest and an estimated 30% of the world's terrestrial plant and animal species (Figure 10-14, left)

Compost/Peter Arnold, Inc

Compost/Peter Arnold, Inc

Figure 10-14 *Species diversity:* two species found in tropical forests are part of the earth's biodiversity. On the left is an endangered *white ukari* in a Brazilian tropical forest. On the right is the world's largest flower, the *flesh flower* (Rafflesia) growing in a tropical rain forest of West Sumatra, Indonesia. The flower of this leafless plant can be as large 1 meter (3.3 feet) in diameter and weigh 7 kilograms (15 pounds). The plant gives off a smell like rotting meat, presumably to attract flies and beetles that pollinate the flower. After blossoming once a year for a few weeks, the blood red flower dissolves into a slimy black mass.

in its vast Amazon basin, which covers an area larger than India.

According to Brazil's government and forest experts, the percentage of its Amazon basin that had been deforested or degraded increased from 1% in 1970 to 16–20% in 2005 (Figure 10-13). Between 2005 and 2007, this rate increased sharply, with people cutting forests mostly to make way for cattle ranching and large plantations of crops such as soybeans used for cattle feed. In 2004, researchers at the Smithsonian Tropical Research Institute estimated that loggers, ranchers, and farmers in Brazil were clearing and burning an area equivalent to a loss of 11 football fields a minute! (**Concept 10-1C**)

Because of difficulties in estimating tropical forest loss, yearly estimates of global tropical deforestation vary widely from 50,000 square kilometers (19,300 square miles)—roughly the size of Costa Rica or the U.S. state of West Virginia—to 170,000 square kilometers (65,600 square miles)—about the size of the South American country of Uruguay or the U.S. state

of Florida. At such rates, half of the world's remaining tropical forests will be gone in 35–117 years, resulting in a dramatic loss and degradation of biodiversity and the ecosystem services it provides.

> **RESEARCH FRONTIER**
>
> Improving estimates of rates of tropical deforestation. See **academic.cengage.com/biology/miller**.

Causes of Tropical Deforestation Are Varied and Complex

Tropical deforestation results from a number of interconnected basic and secondary causes (Figure 10-15). Population growth and poverty combine to drive subsistence farmers and the landless poor to tropical forests, where they try to grow enough food to survive. Government subsidies can accelerate deforestation by reducing the costs of timber harvesting, cattle grazing,

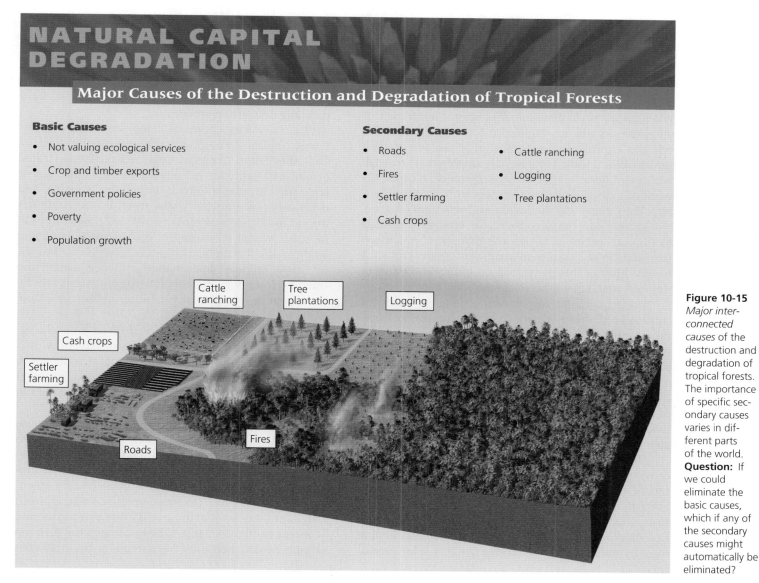

NATURAL CAPITAL DEGRADATION

Major Causes of the Destruction and Degradation of Tropical Forests

Basic Causes

- Not valuing ecological services
- Crop and timber exports
- Government policies
- Poverty
- Population growth

Secondary Causes

- Roads
- Fires
- Settler farming
- Cash crops
- Cattle ranching
- Logging
- Tree plantations

Cattle ranching · Tree plantations · Logging · Cash crops · Settler farming · Roads · Fires

Figure 10-15 *Major interconnected causes* of the destruction and degradation of tropical forests. The importance of specific secondary causes varies in different parts of the world. **Question:** If we could eliminate the basic causes, which if any of the secondary causes might automatically be eliminated?

Figure 10-16 Natural capital degradation: large areas of tropical forest in Brazil's Amazon basin are burned each year to make way for cattle ranches, small-scale farms, and plantation crops such as soybeans. **Questions:** What are three ways in which your lifestyle probably contributes to this process? How, in turn, might this process affect your life?

Herbert Girardet/Peter Arnold, Inc.

and establishing vast plantations of crops such as soybeans and oil palm.

The degradation of a tropical forest usually begins when a road is cut deep into the forest interior for logging and settlement (Figure 10-5). Loggers then use selective cutting (Figure 10-6, top) to remove the best timber. When these big trees fall, many other trees fall with them because of their shallow roots and the network of vines connecting the trees in the forest's canopy. Thus, removing the largest and best trees by selective cutting can cause considerable ecological damage in tropical forests, although the damage is much less than that from burning or clear-cutting areas of such forests. Most of this timber is used locally and much of it is exported, but a great deal is also left to rot.

Foreign corporations operating under government concession contracts do much of the logging in tropical countries. Once a country's forests are gone, the companies move on to another country, leaving ecological devastation behind. For example, the Philippines and Nigeria have lost most of their once-abundant tropical hardwood forests and now are net importers of forest products. Several other tropical countries are following this ecologically and economically unsustainable path.

After the best timber has been removed, timber companies or the government often sell the land to ranchers. Within a few years, the cattle typically overgraze the land and the ranchers move their operations to another forest area. Then they sell the degraded land to settlers who have migrated to tropical forests hoping to grow enough food to survive. After a few years of crop growing and erosion from rain, the nutrient-poor tropical soil is depleted of nutrients. Then the settlers move on to newly cleared land to repeat this environmentally destructive process.

The secondary causes of deforestation vary in different tropical areas. Tropical forests in the Amazon and other South American countries are being cleared or burned (Figure 10-16) mostly for cattle grazing and large soybean plantations. In Indonesia, Malaysia, and other areas of Southeast Asia, tropical forests are being cut or burned and replaced with vast plantations of oil palm, whose oil is used in cooking, cosmetics, and biodiesel fuel for motor vehicles (especially in Europe). In Africa, tropical deforestation and degradation are caused primarily by individuals struggling to survive by clearing plots for small-scale farming and by harvesting wood for fuel.

Burning is widely used to clear forest areas for agriculture, settlement, and other purposes. Healthy rain forests do not burn naturally. But roads, settlements, and farming, grazing, and logging operations fragment them (Science Focus, p. 195). The resulting patches of forest dry out and readily ignite.

The burning of tropical forests is a major component of human-enhanced global warming, which is projected to change the global climate at an increasing rate. Scientists estimate that globally, these fires account for at least 20% of all human-created greenhouse gas emissions. They also produce twice as much CO_2, annually, as all of the world's cars and trucks emit (**Concept 10-1C**). The large-scale burning of the Amazon (Figure 10-16) accounts for three-fourths of Brazil's

greenhouse gas emissions, making Brazil the world's fourth largest emitter of such gases.

A 2005 study by forest scientists found that widespread fires in the Amazon are changing weather patterns by raising temperatures and reducing rainfall. Resulting droughts dry out the forests and make them more likely to burn—another example of a runaway positive feedback loop **Concept 2-5A**, p. 44, and Figure 2-11, p. 45). This process is converting deforested areas of tropical forests to tropical grassland (savanna)—an example of reaching an irreversible ecological *tipping point*. When the forests disappear, rainfall declines and yields of the crops planted on the land drop sharply. Models project that if current burning and deforestation rates continue, 20–30% of the Amazon will be turned into savanna in the next 50 years, and perhaps all of it could be so converted by 2080.

┌─ THINKING ABOUT ─────────────────────
 Tropical Forests

 Why should you care whether most of the world's remaining tropical forests are burned or cleared and converted to savanna within your lifetime? What are three things you could do to help slow the rate of this depletion and degradation of the earth's natural capital?
└──────────────────────────────────────

10-2 How Should We Manage and Sustain Forests?

▶ **CONCEPT 10-2** We can sustain forests by emphasizing the economic value of their ecological services, protecting old-growth forests, harvesting trees no faster than they are replenished, and using sustainable substitute resources.

We Can Manage Forests More Sustainably

Biodiversity researchers and a growing number of foresters have called for more sustainable forest management. Figure 10-17 lists ways to achieve this goal (**Concept 10-2**). Certification of sustainably grown timber and of sustainably produced forest products can help people consume forest products more sustainably (Science Focus, p. 228). **GREEN CAREER:** Sustainable forestry

SOLUTIONS
Sustainable Forestry

- Identify and protect forest areas high in biodiversity

- Rely more on selective cutting and strip cutting

- No clear-cutting on steep slopes

- No logging of old-growth forests

- Sharply reduce road building into uncut forest areas

- Leave most standing dead trees and fallen timber for wildlife habitat and nutrient recycling

- Plant tree plantations primarily on deforested and degraded land

- Certify timber grown by sustainable methods

- Include ecological services of forests in estimating their economic value

Figure 10-17 Ways to grow and harvest trees more sustainably (**Concept 10-2**). **Question:** Which three of these solutions do you think are the most important? Why?

We Can Improve the Management of Forest Fires

In the United States, the Smokey Bear educational campaign undertaken by the Forest Service and the National Advertising Council has prevented countless forest fires. It has also saved many lives and prevented billions of dollars in losses of trees, wildlife, and human structures.

At the same time, this educational program has convinced much of the public that all forest fires are bad and should be prevented or put out. Ecologists warn that trying to prevent all forest fires increases the likelihood of destructive crown fires (Figure 10-9, right) by allowing accumulation of highly flammable underbrush and smaller trees in some forests.

According to the U.S. Forest Service, severe fires could threaten 40% of all federal forest lands, mainly because of fuel buildup resulting from past rigorous fire protection programs (the Smokey Bear era), increased logging in the 1980s that left behind highly flammable logging debris (called *slash*), and greater public use of federal forest lands.

Ecologists and forest fire experts have proposed several strategies for reducing fire-related harm to forests and people. One approach is to set small, contained surface fires to remove flammable small trees

Certifying Sustainably Grown Timber

Collins Pine owns and manages a large area of productive timberland in the northeastern part of the U.S. state of California. Since 1940, the company has used selective cutting to help maintain the ecological and economic sustainability of its timberland.

Since 1993, Scientific Certification Systems (SCS) has evaluated the company's timber production. SCS, which is part of the non-profit Forest Stewardship Council (FSC), was formed to develop a list of environmentally sound practices for use in certifying timber and products made from such timber.

Each year, SCS evaluates Collins Pine's landholdings and has consistently found that: their cutting of trees has not exceeded long-term forest regeneration; roads and harvesting systems have not caused unreasonable ecological damage; soils are not damaged; and downed wood (boles) and standing dead trees (snags) are left to provide wildlife habitat. As a result, SCS judges the company to be a good employer and a good steward of its land and water resources.

According to the FSC, between 1995 and 2007, the area of the world's forests in 76 countries that meets its international certification standards grew more than 16-fold. The countries with the largest areas of FSC-certified forests are, in order, Canada, Russia, Sweden, the United States, Poland, and Brazil. Despite this progress, by 2007, less than 10% of the world's forested area was certified. FSC also certifies 5,400 manufacturers and distributors of wood products.

Critical Thinking

Should governments provide tax breaks for sustainably grown timber to encourage this practice? Explain.

and underbrush in the highest-risk forest areas. Such *prescribed fires* require careful planning and monitoring to try to keep them from getting out of control. As an alternative to prescribed burns, local officials in populated parts of fire-prone California use herds of goats (kept in moveable pens) to eat away underbrush.

A second strategy is to allow many fires on public lands to burn, thereby removing flammable underbrush and smaller trees, as long as the fires do not threaten human structures and life. A third approach is to protect houses and other buildings in fire-prone areas by thinning a zone of about 60 meters (200 feet) around them and eliminating the use of flammable materials such as wooden roofs.

A fourth approach is to thin forest areas vulnerable to fire by clearing away small fire-prone trees and underbrush under careful environmental controls. Many forest fire scientists warn that such thinning should not involve removing economically valuable medium-size and large trees for two reasons. First, these are the most fire-resistant trees. Second, their removal encourages dense growth of more flammable young trees and underbrush and leaves behind highly flammable slash. Many of the worst fires in U.S. history—including some of those during the 1990s—burned through cleared forest areas containing slash. A 2006 study by U.S. Forest Service researchers found that thinning forests without using prescribed burning to remove accumulated brush and deadwood can greatly increase rather than decrease fire damage.

Despite such warnings from forest scientists, the U.S. Congress under lobbying pressure from timber companies passed the 2003 *Healthy Forests Restoration Act*. It allows timber companies to cut down economically valuable medium-size and large trees in 71% of the country's national forests in return for clearing away smaller, more fire-prone trees and underbrush.

However, the companies are not required to conduct prescribed burns after completing the thinning process.

This law also exempts most thinning projects from environmental reviews, which are currently required by forest protection laws in the national forests. According to many biologists and forest fire scientists, this law is likely to *increase* the chances of severe forest fires because it ignores the four strategies scientists have suggested for better management of forest fires. Critics of the Healthy Forests Restoration Act of 2003 say that healthier forests could be maintained at a much lower cost to taxpayers by giving communities in fire prone areas grants to implement these recommendations.

HOW WOULD YOU VOTE?

Do you support repealing or modifying the Healthy Forests Restoration Act of 2003? Cast your vote online at **academic .cengage.com/biology/miller**.

We Can Reduce the Demand for Harvested Trees

One way to reduce the pressure on forest ecosystems is to improve the efficiency of wood use. According to the Worldwatch Institute and forestry analysts, *up to 60% of the wood consumed in the United States is wasted unnecessarily*. This results from inefficient use of construction materials, excess packaging, overuse of junk mail, inadequate paper recycling, and failure to reuse wooden shipping containers.

One reason for cutting trees is to provide pulp for making paper, but paper can be made out of fiber that does not come from trees. China uses rice straw and other agricultural residues to make much of its paper.

Figure 10-18 Solutions: pressure to cut trees to make paper could be greatly reduced by planting and harvesting a fast-growing plant known as kenaf. According to the USDA, kenaf is "the best option for tree-free papermaking in the United States" and could replace wood-based paper within 20–30 years. **Question:** Would you invest in a kenaf plantation? Explain.

Most of the small amount of tree-free paper produced in the United States is made from the fibers of a rapidly growing woody annual plant called *kenaf* (pronounced "kuh-NAHF"; Figure 10-18). Kenaf and other nontree fibers such as hemp yield more paper pulp per hectare than tree farms and require fewer pesticides and herbicides. It is estimated, that within 2 to 3 decades we could essentially eliminate the need to use trees to make paper. However, while timber companies successfully lobby for government subsidies to grow and harvest trees to make paper, there are no major lobbying efforts or subsidies for producing paper from kenaf or kudzu (Figure 9-15, p. 200).

■ **CASE STUDY**

Deforestation and the Fuelwood Crisis

Another major strain on forests, especially in tropical areas, is the practice of cutting of trees for fuelwood. About half of the wood harvested each year and three-fourths of that in developing countries is used for fuel.

Fuelwood and charcoal made from wood are used for heating and cooking by more than 2 billion people in developing countries (Figure 6-13, p. 135). As the demand for fuelwood in urban areas exceeds the sustainable yield of nearby forests, expanding rings of deforested land encircle such cities. By 2050, the demand for fuelwood could easily be 50% greater than the amount that can be sustainably supplied.

Haiti, a country with 9 million people, was once a tropical paradise covered largely with forests. Now it is an ecological disaster. Largely because its trees were cut for fuelwood, only about 2% of its land is forested. With the trees gone, soils have eroded away, making it much more difficult to grow crops. This unsustainable use of natural capital has led to a downward spiral of environmental degradation, poverty, disease, social injustice, crime, and violence. As a result, Haiti is classified as one of the world's leading *failing states* (Figure 17, p. S19, Supplement 3).

One way to reduce the severity of the fuelwood crisis in developing countries is to establish small plantations of fast-growing fuelwood trees and shrubs around farms and in community woodlots. Another approach to this problem is to burn wood more efficiently by providing villagers with cheap, more fuel-efficient, and less-polluting wood stoves, household biogas units that run on methane produced from crop and animal wastes, solar ovens, and electric hotplates powered by solar- or wind-generated electricity. This will also greatly reduce premature deaths from indoor air pollution caused by open fires and poorly designed stoves.

In addition, villagers can switch to burning the renewable sun-dried roots of various gourds and squash plants. Scientists are also looking for ways to produce charcoal for heating and cooking without cutting down trees. For example, Professor Amy Smith, of MIT in Cambridge, Massachusetts (USA), is developing a way to make charcoal from the fibers in a waste product called bagasse, which is left over from sugar cane processing in Haiti. Because sugarcane charcoal burns cleaner than wood charcoal, using it could help Haitians reduce indoor air pollution.

Countries such as South Korea, China, Nepal, and Senegal, have used such methods to reduce fuelwood shortages, sustain biodiversity through reforestation, and reduce soil erosion. Indeed, the mountainous country of South Korea is a global model for its successful reforestation following severe deforestation during the war between North and South Korea, which ended in 1953. Today, forests cover almost two-thirds of the country, and tree plantations near villages supply fuelwood on a sustainable basis. However, most countries suffering from fuelwood shortages are cutting trees for fuelwood and forest products 10–20 times faster than new trees are being planted. Shifting government subsidies from the building of logging roads to the planting of trees would help to increase forest cover worldwide.

Wangari Maathai and Kenya's Green Belt Movement

In the mid-1970s, Wangari Maathai (Figure 10-B) took stock of environmental conditions in her native Kenya. Tree-lined streams she had known as a child had dried up. Farms and plantations that were draining the watersheds and degrading the soil had replaced vast areas of forest. The Sahara Desert was encroaching from the north.

Something inside her told Maathai she had to do something about this degradation. Starting with a small tree nursery in her backyard, she founded the Green Belt Movement in 1977. The main goal of this highly regarded women's self-help group is to organize poor women in rural Kenya to plant and protect millions of trees in order to combat deforestation and provide fuelwood. By 2004, the 50,000 members of this grassroots group had established 6,000 village nurseries and planted and protected more than 30 million trees.

The women are paid a small amount for each seedling they plant that survives. This gives them an income to help break the cycle of poverty. It also improves the environment because trees reduce soil erosion and provide fruits, fuel, building materials, fodder for livestock, shade, and beauty. Having more trees also reduces the distances women and children have to walk to get fuelwood for cooking and heating. The success of this project has sparked the creation of similar programs in more than 30 other African countries.

Charlotte Thege/Peter Arnold, Inc.

In 2004, Maathai became the first African woman and the first environmentalist to be awarded the Nobel Peace Prize for her life-long efforts. Within an hour of learning that she had won the prize, Maathai planted a tree, telling onlookers it was "the best way to celebrate." In her speech accepting the award, she said the purpose of the Green Belt program was to help people "make the connections between their own personal actions and the problems they witness in their environment and society." She urged everyone in the world to plant a tree as a symbol of commitment and hope.

Figure 10-B Wangari Maathai was the first Kenyan woman to earn a Ph.D. and to head an academic department at the University of Nairobi. In 1977, she organized the internationally acclaimed Green Belt Movement. For her work in protecting the environment, she has received many honors, including the Goldman Prize, the Right Livelihood Award, the U.N. Africa Prize for Leadership, and the 2004 Nobel Peace Prize. After years of being harassed, beaten, and jailed for opposing government policies, she was elected to Kenya's parliament as a member of the Green Party in 2002. In 2003, she was appointed Assistant Minister for Environment, Natural Resources, and Wildlife.

In 2006, she launched a project to plant a billion trees worldwide in 2007 to help fight poverty and climate change. The project greatly exceeded expectations with the planting of 2 billlion trees in 55 countries. In 2008, the UNEP set a goal of planting an additional 5 billion trees.

Wangari tells her story in her book *The Green Belt Movement: Sharing the Approach and the Experience,* published by Lantern Books in 2003.

This in turn would help to slow global warming, as more trees would remove more of the carbon dioxide that we are adding to the atmosphere.

Governments and Individuals Can Act to Reduce Tropical Deforestation

In addition to reducing fuelwood demand, analysts have suggested other ways to protect tropical forests and use them more sustainably. One way is to help new settlers in tropical forests to learn how to practice small-scale sustainable agriculture and forestry. Another is to harvest some of the renewable resources such as fruits and nuts in rain forests on a sustainable basis. And strip cutting (Figure 10-6c) can be used to harvest tropical trees for lumber.

In Africa's northern Congo Republic, some nomadic forest-dwelling pygmies go into the forests carrying hand-held satellite tracking devices in addition to their traditional spears and bows. They use these Global Positioning System (GPS) devices to identify their hunting grounds, burial grounds, water holes, sacred areas, and areas rich in medicinal plants. They then download such information on computers to provide a map of areas that need to be protected from logging, mining, and other destructive activities.

Debt-for-nature swaps can make it financially attractive for countries to protect their tropical forests. In such swaps, participating countries act as custodians

of protected forest reserves in return for foreign aid or debt relief. In a similar strategy called *conservation concessions,* governments or private conservation organizations pay nations for concessions to preserve their natural resources.

Loggers can also use tropical forests more sustainably by using gentler methods for harvesting trees. For example, cutting canopy vines (lianas) before felling a tree and using the least obstructed paths to remove the logs can sharply reduce damage to neighboring trees. In addition, governments and individuals can mount efforts to reforest and rehabilitate degraded tropical forests and watersheds (see Individuals Matter, at left) and clamp down on illegal logging.

Finally, each of us as consumers can reduce the demand that fuels illegal and unsustainable logging in tropical forests. For building projects, we can use substitutes for wood such as bamboo and recycled plastic building materials (**Concept 10-2**). Recycled waste lumber is another alternative, now marketed by companies such as TerraMai and EcoTimber.

We can also buy only lumber and wood products that are certified as sustainably produced (Science Focus, p. 228). Growing awareness of tropical deforestation and the resulting consumer pressure caused the giant retail company Home Depot to take action. It reported in 2007 that 80% of the wood it carries meets such certification standards.

These and other ways to protect tropical forests are summarized in (Figure 10-19).

SOLUTIONS

Sustaining Tropical Forests

Prevention

Protect the most diverse and endangered areas

Educate settlers about sustainable agriculture and forestry

Subsidize only sustainable forest use

Protect forests with debt-for-nature swaps and conservation concessions

Certify sustainably grown timber

Reduce poverty

Slow population growth

Restoration

Encourage regrowth through secondary succession

Rehabilitate degraded areas

Concentrate farming and ranching in already-cleared areas

Figure 10-19 Ways to protect tropical forests and use them more sustainably (**Concept 10-2**). **Question:** Which three of these solutions do you think are the most important? Why?

10-3 How Should We Manage and Sustain Grasslands?

▶ **CONCEPT 10-3** We can sustain the productivity of grasslands by controlling the number and distribution of grazing livestock and by restoring degraded grasslands.

Some Rangelands Are Overgrazed

Grasslands provide many important ecological services, including soil formation, erosion control, nutrient cycling, storage of atmospheric carbon dioxide in biomass, and maintenance of biodiversity.

After forests, the ecosystems most widely used and altered by human activities are grasslands. **Rangelands** are unfenced grasslands in temperate and tropical climates that supply *forage,* or vegetation, for grazing (grass-eating) and browsing (shrub-eating) animals. Cattle, sheep, and goats graze on about 42% of the world's grassland. The 2005 Millennium Ecosystem Assessment estimated that continuing on our present course will increase that percentage to 70% by 2050. Livestock also graze in **pastures**—managed grasslands or enclosed meadows usually planted with domesticated grasses or other forage.

Blades of rangeland grass grow from the base, not at the tip. So as long as only the upper half of the blade is eaten and its lower half remains, rangeland grass is a renewable resource that can be grazed again and again.

Moderate levels of grazing are healthy for grasslands, because removal of mature vegetation stimulates rapid regrowth and encourages greater plant diversity. The key is to prevent both overgrazing and undergrazing by domesticated livestock and wild herbivores. **Overgrazing** occurs when too many animals graze for too long and exceed the carrying capacity of a rangeland area (Figure 10-20, left, p. 232). It reduces grass cover, exposes the soil to erosion by water and wind, and compacts the soil (which diminishes its capacity

Figure 10-20 Natural capital degradation: overgrazed (left) and lightly grazed (right) rangeland.

USDA, Natural Resources Conservation Service

to hold water). Overgrazing also enhances invasion by species such as sagebrush, mesquite, cactus, and cheatgrass, which cattle will not eat.

Scientists have also learned that, before settlers made them into rangeland, natural grassland ecosystems were maintained partially by periodic wildfires sparked by lightning. Fires were important because they burned away mesquite and other invasive shrubs, keeping the land open for grasses. Ecologists have studied grasslands of the Malpai Borderlands—an area on the border between the southwestern U.S. states of Arizona and New Mexico—where ranchers, with the help of the federal government, not only allowed overgrazing for more than a century, but also fought back fires and kept the grasslands from burning. Consequently, trees and shrubs replaced grasses, the soil was badly eroded, and the area lost most of its value for grazing.

Since 1993, ranchers, scientists, environmentalists, and government agencies have joined forces to restore the native grasses and animal species to the Malpai Borderlands. Land managers conduct periodic controlled burns on the grasslands, and the ecosystem has now been largely reestablished. What was once a classic example of unsustainable resource management became a valuable scientific learning experience and a management success story.

About 200 years ago, grass may have covered nearly half the land in the southwestern United States. Today, it covers only about 20%, mostly because of a combination of prolonged droughts and overgrazing, which created footholds for invader species that now cover many former grasslands.

Limited data from FAO surveys in various countries indicate that overgrazing by livestock has caused as much as a fifth of the world's rangeland to lose productivity. Some grasslands suffer from **undergrazing,** where absence of grazing for long periods (at least 5 years) can reduce the net primary productivity of grassland vegetation and grass cover.

We Can Manage Rangelands More Sustainably

The most widely used method for more sustainable management of rangeland is to control the number of grazing animals and the duration of their grazing in a given area so that the carrying capacity of the area is not exceeded (**Concept 10-3**). One way of doing this is *rotational grazing* in which cattle are confined by portable fencing to one area for a short time (often only 1–2 days) and then moved to a new location.

Livestock tend to aggregate around natural water sources, especially thin strips of lush vegetation along streams or rivers known as *riparian zones,* and around ponds established to provide water for livestock. Overgrazing by cattle can destroy the vegetation in such areas (Figure 10-21, left). Protecting overgrazed land from further grazing by moving livestock around and by fencing off these areas can eventually lead to its natural ecological restoration (Figure 10-21, right). Ranchers can also move cattle around by providing supplemental feed at selected sites and by strategically locating water holes and tanks and salt blocks.

A more expensive and less widely used method of rangeland management is to suppress the growth of unwanted invader plants by use of herbicides, mechanical

Figure 10-21 Natural capital restoration: in the mid-1980s, cattle had degraded the vegetation and soil on this stream bank along the San Pedro River in the U.S. state of Arizona (left). Within 10 years, the area was restored through natural regeneration after the banning of grazing and off-road vehicles (right) (**Concept 10-3**).

removal, or controlled burning. A cheaper way to discourage unwanted vegetation in some areas is through controlled, short-term trampling by large numbers of livestock.

Replanting barren areas with native grass seeds and applying fertilizer can increase growth of desirable vegetation and reduce soil erosion. But this is an expensive way to restore severely degraded rangeland. The better option is to prevent degradation by using the methods described above and in the following case study.

■ CASE STUDY

Grazing and Urban Development in the American West—Cows or Condos?

The landscape is changing in ranch country. Since 1980, millions of people have moved to parts of the southwestern United States, and a growing number of ranchers have sold their land to developers. Housing developments, condos, and small "ranchettes" are creeping out from the edges of many southwestern cities and towns. Most people moving to the southwestern states value the landscape for its scenery and recreational opportunities, but uncontrolled urban development can degrade these very qualities.

For decades some environmental scientists and environmentalists have sought to reduce overgrazing on these lands and, in particular, to reduce or eliminate livestock grazing permits on public lands. They have not had the support of ranchers or of the government. They have also pushed for decreased timber cutting and increased recreational opportunities in the national forests and grasslands. These efforts have made private tracts of land, especially near protected public lands, more desirable and valuable to people who enjoy outdoor activities and can afford to live in scenic areas.

Now, because of this population surge, ranchers, ecologists, and environmentalists are joining together to help preserve cattle ranches as the best hope for sustaining the key remaining grasslands and the habitats they provide for native species. They are working together to identify areas that are best for sustainable grazing, areas best for sustainable urban development, and areas that should be neither grazed nor developed. One strategy involves land trust groups, which pay ranchers for *conservation easements*—deed restrictions that bar future owners from developing the land. These groups are also pressuring local governments to zone the land in order to prevent large-scale development in ecologically fragile rangeland areas.

Some ranchers are also reducing the harmful environmental impacts of their herds. They rotate their cattle away from riparian areas (Figure 10-21), use far less fertilizer and pesticides, and consult with range and wildlife scientists about ways to make their ranch operations more economically and ecologically sustainable.

10-4 How Should We Manage and Sustain Parks and Nature Reserves?

▶ **CONCEPT 10-4** Sustaining biodiversity will require protecting much more of the earth's remaining undisturbed land area as parks and nature reserves.

National Parks Face Many Environmental Threats

Today, more than 1,100 major national parks are located in more than 120 countries (see Figure 7-12, top, p. 151; Figure 7-18, p. 157; Figure 7-19, p. 157; and Figure 8-8, p. 168). However, most of these national parks are too small to sustain a lot of large animal species. And many parks suffer from invasions by nonnative species that compete with and reduce the populations of native species and worsen ecological disruption.

Parks in developing countries possess the greatest biodiversity of all parks, but only about 1% of these parklands are protected. Local people in many of these countries enter the parks illegally in search of wood, cropland, game animals, and other natural products for their daily survival. Loggers and miners operate illegally in many of these parks, as do wildlife poachers who kill animals to obtain and sell items such as rhino horns, elephant tusks, and furs. Park services in most developing countries have too little money and too few personnel to fight these invasions, either by force or through education.

■ CASE STUDY
Stresses on U.S. Public Parks

The U.S. national park system, established in 1912, includes 58 major national parks, sometimes called the country's crown jewels. States, counties, and cities also operate public parks.

Popularity is one of the biggest problems for many parks. Between 1960 and 2007, the number of visitors to U.S. national parks more than tripled, reaching 273 million. The Great Smoky Mountains National Park in the states of Tennessee and North Carolina, the country's most frequently visited national park, hosts about 9 million visitors each year. Many state parks are located near urban areas and receive about twice as many visitors per year as do the national parks. Visitors often expect parks to have grocery stores, laundries, bars, and other such conveniences.

During the summer, users entering the most popular parks face long backups and experience noise, congestion, eroded trails, and stress instead of peaceful solitude. In some parks and other public lands, noisy and polluting dirt bikes, dune buggies, jet skis, snowmobiles, and other off-road vehicles degrade the aesthetic experience for many visitors, destroy or damage fragile vegetation (Figure 10-22), and disturb wildlife. There is controversy over whether these machines should be allowed in national parks.

THINKING ABOUT
National Parks and Off-Road Vehicles

Do you support allowing off-road vehicles in national parks? Explain. If you do, what restrictions, if any, would you put on their use?

Parks also suffer damage from the migration or deliberate introduction of nonnative species. European wild boars (imported to the state of North Carolina in 1912 for hunting) threaten vegetation in parts of the Great Smoky Mountains National Park. Nonnative mountain goats in Washington State's Olympic National Park trample native vegetation and accelerate soil erosion. Nonnative species of plants, insects, and worms entering the parks on vehicle tires and hikers' gear also degrade the biodiversity of parklands.

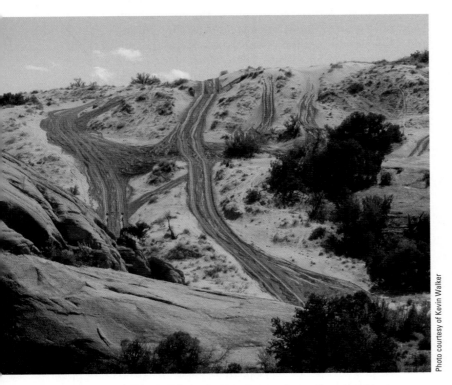

Figure 10-22 Natural capital degradation: damage from off-road vehicles in a proposed wilderness area near the U.S. city of Moab, Utah. Such vehicles pollute the air, damage soils and vegetation, threaten wildlife, and degrade wetlands and streams.

Photo courtesy of Kevin Walker

SCIENCE FOCUS

Effects of Reintroducing the Gray Wolf to Yellowstone National Park

For over a decade, wildlife ecologist Robert Crabtree and a number of other scientists have been studying the effects of reintroducing the gray wolf into the Yellowstone National Park (**Core Case Study**). They have put radio-collars on most of the wolves to gather data and track their movements. They have also studied changes in vegetation and the populations of various plant and animal species. Results of this research have suggested that the return of the gray wolf, a keystone predator species, has sent ecological ripples through the park's ecosystem.

Elk, the main herbivores in the Yellowstone system, are the primary food source for the wolves, but wolves also kill some moose, mule deer, and bison. Not surprisingly, elk populations have declined with the return of wolves. However, drought, grizzly bears (which kill elk calves), and a severe winter in 1997 have contributed to this decline. Leftovers of elk killed by wolves provide an important food source for grizzly bears

and other scavengers such as bald eagles and ravens.

Before the wolves returned, elk had been browsing on willow shoots and other vegetation near the banks of streams and rivers. With the return of wolves, the elk retreated to higher ground. This has spurred the regrowth of aspen, cottonwoods, and willow trees in these riparian areas and increased populations of riparian songbirds.

This regrowth of trees has in turn helped to stabilize and shade stream banks, which lowered the water temperature and made it better habitat for trout. Beavers seeking willow and aspen for food and dam construction have returned. The beaver dams established wetlands and created more favorable habitat for aspens.

The wolves have also cut in half the population of coyotes—the top predators in the absence of wolves. This has resulted in fewer coyote attacks on cattle and sheep on surrounding ranches and has increased populations of red fox and smaller animals such as

ground squirrels, mice, and gophers hunted by coyotes, eagles, and hawks.

Elk in the park are hunted in limited numbers, but wolves, as a protected species, are not hunted. However, wolves kill one another in clashes between packs, and a few have been killed by cars. The wolves also face threats from dogs that visitors bring to the park. The dogs carry parvovirus, which can kill wolf pups.

Wolves are an important factor in the Yellowstone ecosystem. But there are many other interacting factors involved in the structure and functioning of this complex ecosystem. Decades of research will be needed to unravel and better understand these interactions. For more information, see *The Habitable Planet*, Video 4, at **www.learner.org/resources/series209.html**.

Critical Thinking

Do you approve or disapprove of the reintroduction of the gray wolf into the Yellowstone National Park system? Explain.

At the same time, native species—some of them threatened or endangered—are killed or removed illegally in almost half of all U.S. national parks. This is what happened with the gray wolf until it was successfully reintroduced into Yellowstone National Park after a half century's absence (Science Focus, above). Not all park visitors understand the rules that protect species, and rangers have to spend an increasing amount of their time on law enforcement and crowd control instead of on conservation management and education.

Many U.S. national parks have become threatened islands of biodiversity surrounded by a sea of commercial development. Nearby human activities that threaten wildlife and recreational values in many national parks include mining, logging, livestock grazing, use of coal-burning power plants, oil drilling, water diversion, and urban development.

Polluted air, drifting hundreds of kilometers from cities, kills ancient trees in California's Sequoia National Park and often degrades the awesome views at Arizona's Grand Canyon. The Great Smoky Mountains, named for the natural haze emitted by their lush vegetation, ironically have air quality similar to that of Los Angeles, California, and vegetation on their highest peaks has been damaged by acid rain. According to the National Park Service, air pollution, mostly from coal-fired power plants and dense vehicle traffic, degrades scenic views in U.S. national parks more than 90% of the time.

Another problem, reported by the U.S. General Accounting Office, is that the national parks need at least $6 billion for long overdue repairs of trails, buildings, and other infrastructure. Some analysts say more of these funds could come from private concessionaires who provide campgrounds, restaurants, hotels, and other services for park visitors. They pay franchise fees averaging only about 6–7% of their gross receipts, and many large concessionaires with long-term contracts pay as little as 0.75%. Analysts say these percentages could reasonably be increased to around 20%.

Figure 10-23 (p. 236) lists other suggestions made by various analysts for sustaining and expanding the national park system in the United States.

Nature Reserves Occupy Only a Small Part of the Earth's Land

Most ecologists and conservation biologists believe the best way to preserve biodiversity is to create a worldwide network of protected areas. (See the chapter opening quote on p. 215.) Currently, only 12% of the earth's land area is protected strictly or partially in nature reserves, parks, wildlife refuges, wilderness, and other areas. This 12% figure is misleading because no more than 5% of the earth's land is strictly protected from potentially harmful human activities. In other words, *we have*

National Parks

- Integrate plans for managing parks and nearby federal lands

- Add new parkland near threatened parks

- Buy private land inside parks

- Locate visitor parking outside parks and provide shuttle buses for people touring heavily used parks

- Increase federal funds for park maintenance and repairs

- Raise entry fees for visitors and use resulting funds for park management and maintenance

- Seek private donations for park maintenance and repairs

- Limit the number of visitors in crowded park areas

- Increase the number of park rangers and their pay

- Encourage volunteers to give visitor lectures and tours

Figure 10-23 Suggestions for sustaining and expanding the national park system in the United States. **Question:** Which two of these solutions do you think are the most important? Why? (Data from Wilderness Society and National Parks and Conservation Association).

reserved 95% of the earth's land for human use, and most of the remaining area consists of ice, tundra, or desert—places where most people do not want to live.

Conservation biologists call for full protection of at least 20% of the earth's land area in a global system of biodiversity reserves that would include multiple examples of all the earth's biomes (**Concept 10-4**). But powerful economic and political interests oppose this idea.

Protecting more of the earth's land from unsustainable use will require action and funding by national governments and private groups, bottom-up political pressure by concerned individuals, and cooperative ventures involving governments, businesses, and private conservation organizations. Such groups play an important role in establishing wildlife refuges and other reserves to protect biological diversity.

For example, since its founding by a group of professional ecologists in 1951, The Nature Conservancy—with more than 1 million members worldwide—has created the world's largest system of private natural areas and wildlife sanctuaries in 30 countries. In the United States, efforts by The Nature Conservancy and private landowners have protected land, waterways, and wetlands in local and state trusts totaling roughly the area of the U.S. state of Georgia.

Eco-philanthropists are using some of their wealth to buy up wilderness areas in South America, and they are donating the preserved land to the governments of various countries. For example, Douglas and Kris Tompkins have created 11 wilderness parks in Latin America. In 2005, they donated two new national parks to Chile and Argentina.

In the United States, private, nonprofit *land trust groups* have protected large areas of land. Members pool their financial resources and accept tax-deductible donations to buy and protect farmlands, grasslands, woodlands, and urban green spaces.

Some governments are also making progress. By 2007, the Brazilian government had officially protected 23% of the Amazon—an area the size of France—from development. However, many of these areas are protected only on paper and are not always secure from illegal resource removal and degradation.

Most developers and resource extractors oppose protecting even the current 12% of the earth's remaining undisturbed ecosystems. They contend that these areas might contain valuable resources that would add to economic growth. Ecologists and conservation biologists disagree. They view protected areas as islands of biodiversity and natural capital that help to sustain all life and economies and serve as centers of future evolution. See Norman Myer's Guest Essay on this topic at CengageNOW.

— **HOW WOULD YOU VOTE?**

Should at least 20% of the earth's land area be strictly protected from economic development? Cast your vote online at **academic.cengage.com/biology/miller**.

Designing and Connecting Nature Reserves

Large reserves sustain more species and provide greater habitat diversity than do small reserves. They also minimize exposure to natural disturbances (such as fires and hurricanes), invading species, and human disturbances from nearby developed areas.

In 2007, scientists reported on the world's largest and longest running study of forest fragmentation, which took place in the Amazon. They found that conservation of large reserves in the Amazon was even more important than was previously thought. Because the Amazon rain forest is so diverse, a large expanse of it may contain dozens of ecosystem types, each of which is different enough from the others to support unique species. Therefore, developing just a part of such a large area could result in the elimination of many types of habitats and species.

However, research indicates that in other locales, several well-placed, medium-sized reserves may better protect a wider variety of habitats and preserve more biodiversity than would a single large reserve of the same total area. When deciding on whether to recommend large- or medium-sized reserves in a particular area, conservation biologists must carefully consider its various ecosystems.

Whenever possible, conservation biologists call for using the *buffer zone concept* to design and manage nature

Biosphere Reserve

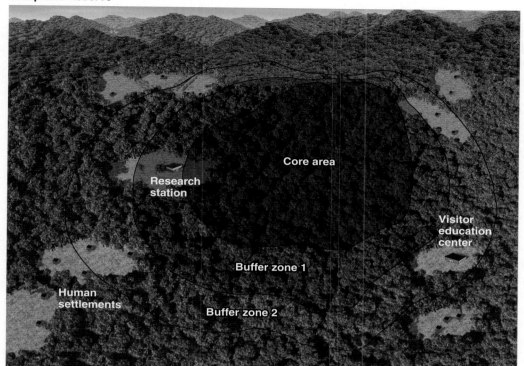

Research
station

Core area

Visitor
education
center

Buffer zone 1

Human
settlements

Buffer zone 2

Figure 10-24 **Solutions:** a model *biosphere reserve.* Each reserve contains a protected inner core surrounded by two buffer zones that local and indigenous people can use for sustainable logging, growing limited crops, grazing cattle, hunting, fishing, and ecotourism. **Question:** Do you think some of these reserves should be free of all human activity, including ecotourism? Why or why not?

reserves. This means protecting an inner core of a reserve by usually establishing two buffer zones in which local people can extract resources sustainably without harming the inner core. Instead of shutting people out of the protected areas and likely creating enemies, this approach enlists local people as partners in protecting a reserve from unsustainable uses such as illegal logging and poaching.

The United Nations has used this principle in creating its global network of 529 biosphere reserves in 105 countries (Figure 10-24). According to Craig Leisher, an economist for The Nature Conservancy, "Local people are often the best people in developing countries to manage these conservation areas, because they want them to survive in the long term as well."

So far, most biosphere reserves fall short of these ideals and receive too little funding for their protection and management. An international fund to help make up the shortfall would cost about $100 million per year—about the amount spent every 90 minutes on weapons by the world's nations.

Establishing protected *habitat corridors* between isolated reserves helps to support more species and allows migration by vertebrates that need large ranges. Corridors also permit migration of individuals and populations when environmental conditions in a reserve deteriorate, forcing animals to move to a new location, and they support animals that must make seasonal migrations to obtain food. Corridors may also enable some species to shift their ranges if global climate change makes their current ranges uninhabitable.

On the other hand, corridors can threaten isolated populations by allowing movement of pest species, disease, fire, and invasive species between reserves. They also increase exposure of migrating species to natural predators, human hunters, and pollution. In addition, corridors can be costly to acquire, protect, and manage. Nevertheless, an extensive study, reported in 2006, showed that areas connected by corridors host a greater variety of birds, insects, small mammals, and plant species. And in that study, nonnative species did not invade the connected areas.

The creation of large reserves connected by corridors on an eco-regional scale is the grand goal of many conservation biologists. This idea is being put into practice in places such as Costa Rica (see the following Case Study).

RESEARCH FRONTIER

Learning how to design, locate, connect, and manage networks of effective nature preserves. See **academic.cengage .com/biology/miller**.

■ **CASE STUDY**

Costa Rica—A Global Conservation Leader

Tropical forests once completely covered Central America's Costa Rica, which is smaller in area than the U.S. state of West Virginia and about one-tenth the size of France. Between 1963 and 1983, politically powerful

ranching families cleared much of the country's forests to graze cattle.

Despite such widespread forest loss, tiny Costa Rica is a superpower of biodiversity, with an estimated 500,000 plant and animal species. A single park in Costa Rica is home to more bird species than are found in all of North America.

In the mid-1970s, Costa Rica established a system of nature reserves and national parks that, by 2006, included about a quarter of its land—6% of it reserved for indigenous peoples. Costa Rica now devotes a larger proportion of its land to biodiversity conservation than does any other country.

The country's parks and reserves are consolidated into eight zoned *megareserves* (Figure 10-25). Each reserve contains a protected inner core surrounded by two buffer zones that local and indigenous people can use for sustainable logging, crop farming, cattle grazing, hunting, fishing, and ecotourism.

Costa Rica's biodiversity conservation strategy has paid off. Today, the country's largest source of income is its $1-billion-a-year tourism business, almost two-thirds of which involves ecotourism.

To reduce deforestation, the government has eliminated subsidies for converting forest to rangeland. It also pays landowners to maintain or restore tree coverage. The goal is to make it profitable to sustain forests. Between 2007 and 2008, the government planted nearly 14 million trees, which helps to preserve the country's biodiversity. As they grow, the trees also remove carbon dioxide from the air and help the country to meet its goal of reducing net CO_2 emissions to zero by 2021.

The strategy has worked: Costa Rica has gone from having one of the world's highest deforestation rates to having one of the lowest. Between 1986 and 2006, the country's forest cover grew from 26% to 51%.

Protecting Wilderness Is an Important Way to Preserve Biodiversity

One way to protect undeveloped lands from human exploitation is by legally setting them aside as large areas of undeveloped land called **wilderness** (Concept 10-4). Theodore Roosevelt (Figure 4, p. S34, in Supplement 5), the first U.S. president to set aside protected areas, summarized what we should do with wilderness: "Leave it as it is. You cannot improve it."

Wilderness protection is not without controversy (see the following Case Study). Some critics oppose protecting large areas for their scenic and recreational value for a relatively small number of people. They believe this is an outmoded ideal that keeps some areas of the planet from being economically useful to people here today. But to most biologists, the most important reasons for protecting wilderness and other areas from exploitation and degradation involve long-term needs. One such need is to *preserve biodiversity* as a vital part of the earth's natural capital. Another is to protect wilderness areas *as centers for evolution* in response to mostly unpredictable changes in environmental conditions. In other words, wilderness serves as a biodiversity bank and an eco-insurance policy.

■ CASE STUDY

Controversy over Wilderness Protection in the United States

In the United States, conservationists have been trying to save wild areas from development since 1900. Overall, they have fought a losing battle. Not until 1964 did Congress pass the Wilderness Act (Figure 6, p. S35, in Supplement 5). It allowed the government to protect undeveloped tracts of public land from development as part of the National Wilderness Preservation System.

The area of protected wilderness in the United States increased tenfold between 1970 and 2007. Even so, only about 4.6% of U.S. land is protected as wilderness—almost three-fourths of it in Alaska. Only 1.8% of the land area of the lower 48 states is protected, most of it in the West. In other words, Americans have reserved at least 98% of the continental United States to be used as they see fit and have protected less than 2% as wilderness. According to a 1999 study by the World Conservation Union, the United States ranks 42nd among nations in terms of terrestrial area protected as wilderness, and Canada is in 36th place.

In addition, only 4 of the 413 wilderness areas in the lower 48 states are large enough to sustain the species they contain. The system includes only 81 of the country's 233 distinct ecosystems. Most wilderness ar-

Figure 10-25 Solutions: Costa Rica has consolidated its parks and reserves into eight zoned *megareserves* designed to sustain about 80% of the country's rich biodiversity. Green areas are protected reserves and yellow areas are nearby buffer zones, which can be used for sustainable forms of forestry, agriculture, hydropower, hunting, and other human activities.

eas in the lower 48 states are threatened habitat islands in a sea of development.

Scattered blocks of public lands with a total area roughly equal to that of the U.S. state of Montana could qualify for designation as wilderness. About 60% of such land in the national forests. For more than 20 years, these areas were temporarily protected under the Roadless Rule—a federal regulation that put undeveloped areas of national forests off-limits to road building and logging while they were evaluated for wilderness protection.

For decades, politically powerful oil, gas, mining, and timber industries have sought entry to these areas—which are owned jointly by all citizens of the United States—to develop resources there. Their efforts paid off in 2005 when the secretary of the interior ended protection of roadless areas within the national forest system that were being considered for classification as wilderness. The secretary also began allowing states to classify old cow paths and off-road vehicle trails as roads (Figure 10-22), which would disqualify their surrounding areas from protection as wilderness.

THINKING ABOUT
Protecting Wolves and Wild Lands

CORE CASE STUDY

How do you think protecting wolves, in part by reintroducing them to areas such as Yellowstone National Park (**Core Case Study**), helps to protect the forest areas where they live?

10-5 What Is the Ecosystem Approach to Sustaining Biodiversity?

▶ **CONCEPT 10-5A** We can help to sustain biodiversity by identifying severely threatened areas and protecting those with high plant diversity (biodiversity hotspots) and those where ecosystem services are being impaired.

▶ **CONCEPT 10-5B** Sustaining biodiversity will require a global effort to rehabilitate and restore damaged ecosystems.

▶ **CONCEPT 10-5C** Humans dominate most of the earth's land, and preserving biodiversity will require sharing as much of it as possible with other species.

We Can Use a Four-Point Strategy to Protect Ecosystems

Most biologists and wildlife conservationists believe that we must focus more on protecting and sustaining ecosystems, and the biodiversity contained within them, than on saving individual species. Their goals certainly include preventing premature extinction of species, but they argue the best way to do that is to protect threatened habitats and ecosystem services. This *ecosystems approach* generally would employ the following four-point plan:

- Map global ecosystems and create an inventory of the species contained in each of them and the ecosystem services they provide.

- Locate and protect the most endangered ecosystems and species, with emphasis on protecting plant biodiversity and ecosystem services.

- Seek to restore as many degraded ecosystems as possible.

- Make development *biodiversity-friendly* by providing significant financial incentives (such as tax breaks and write-offs) and technical help to private landowners who agree to help protect endangered ecosystems.

Some scientists have argued that we need new laws to embody this strategy. In the United States, for example, there is support for amending the Endangered Species Act, or possibly even replacing it with a new law focused on protection of ecosystems and biodiversity.

Protecting Global Biodiversity Hotspots Is an Urgent Priority

In reality, few countries are physically, politically, or financially able to set aside and protect large biodiversity reserves. To protect as much of the earth's remaining biodiversity as possible, some conservation biologists urge adoption of an *emergency action* strategy to identify and quickly protect **biodiversity hotspots** (**Concept 10-5A**)—an idea first proposed in 1988 by environmental scientist Norman Myers. (See his Guest Essay on this topic at CengageNOW.) These "ecological arks" are areas especially rich in plant species that are found nowhere else and are in great danger of extinction. These areas suffer serious ecological disruption, mostly because of rapid human population growth and the resulting pressure on natural resources. (See Case Study p. 240.) Myers and his colleagues at Conservation International relied primarily on the diversity of plant

species to identify biodiversity hotspot areas because data on plant diversity was more readily available and was also thought to be an indicator of animal diversity.

Figure 10-26 shows 34 global terrestrial biodiversity hotspots identified by conservation biologists and Figure 10-27 shows major biodiversity hotspots in the United States. In the 34 global areas, a total of 86% of the habitat has been destroyed. They cover only a little more than 2% of the earth's land surface, but they contain an estimated 50% of the world's flowering plant species and 42% of all terrestrial vertebrates (mammals, birds, reptiles, and amphibians). They are also home for a large majority of the world's endangered or critically endangered species. Says Norman Myers, "I can think of no other biodiversity initiative that could achieve so much at a comparatively small cost, as the hotspots strategy."

One drawback of the biodiversity hotspots approach is that some areas rich in plant diversity are not necessarily rich in animal diversity. And when hotspots are protected, local people can be displaced and lose access to important resources. However, the goal of this approach—to protect the unique biodiversity in areas under great stress from human activities—remains urgent. Despite its importance, this approach has not succeeded in capturing sufficient public support and funding.

■ CASE STUDY
A Biodiversity Hotspot in East Africa

The forests covering the flanks of the Eastern Arc Mountains of the African nation of Tanzania contain the highest concentration of endangered animals on earth.

Plants and animals that exist nowhere else (species *endemic* to this area) live in these mountainside forests in considerable numbers. They include 96 species of vertebrates—10 mammal, 19 bird, 29 reptile, and 38 amphibian—43 species of butterflies, and at least 800 endemic species of plants, including most species of African violets.

An international network of scientists, who had extensively surveyed these mountain forests, reported these findings in 2007. They also reported newly discovered species in these forests, including a tree-dwelling monkey called the Kipunji and some surprisingly large reptiles and amphibians.

This area is a major biodiversity hotspot because humans now threaten to do what the ice ages could not do—kill off its forests. Farmers and loggers have cleared 70% of the ancient forests. This loss of habitat, along with hunting, has killed off many species, including elephants and buffalo, and now 71 of the 96 endemic species are threatened, 8 of them critically, with biological extinction.

These species are now forced to survive within 13 patches of forest that total an area about the size of the U.S. state of Rhode Island. Most of these forests are contained within 150 government reserves. New settlements are not allowed, but people still forage in these reserves for fuelwood and building materials, severely degrading some of the forests. Fire is also a threat, because the shrinking, degraded patches of forest are drying out, and this is likely to get worse as global warming takes hold.

CENGAGENOW™ Learn more about biodiversity hotspots around the world, what is at stake there, and how they are threatened at CengageNOW.

--- RESEARCH FRONTIER ---
Identifying and preserving all of the world's terrestrial and aquatic biodiversity hotspots. See **academic.cengage.com/ biology/miller**.

Protecting Ecosystem Services Is Also an Urgent Priority

Another way to help sustain the earth's biodiversity and its people is to identify and protect areas where vital *ecosystem services* (orange items in Figure 1-3, p. 8) are being impaired enough to reduce biodiversity or harm local residents. This approach has gotten more attention since the release in 2005 of the U.N. *Millennium Ecosystem Assessment*—a 4-year study by 1,360 experts from 95 countries. It identified key ecosystem services that provide numerous ecological and economic benefits. (Those provided by forests are summarized in Figure 10-4.) The study pointed out that human activities are degrading or overusing about 62% of the earth's natural services in various ecosystems around the world, and it outlined ways to help sustain these vital ecosystem services for human and nonhuman life.

This approach recognizes that most of the world's ecosystems are already dominated or influenced by human activities and that such pressures are increasing as population, urbanization, and resource use increase and the human ecological footprint increases (Figure 1-10, p. 15, and Figure 3, p. S24–S25, in Supplement 4). Proponents of this approach recognize that it is vital to set aside and protect reserves and wilderness areas and to protect highly endangered biodiversity hotspots (Figure 10-26). But they contend that such efforts by themselves will not significantly slow the steady erosion of the earth's biodiversity and ecosystem services.

These analysts argue that we must also identify highly stressed *life raft ecosystems*. In such areas, people live in severe poverty, and a large part of the economy depends on various ecosystem services that are being degraded severely enough to threaten the well-being of people and other forms of life. In these areas, residents, public officials, and conservation scientists are urged to work together to develop strategies to protect both biodiversity and human communities. Instead of emphasizing

CENGAGENOW™ **Active Figure 10-26 Endangered natural capital:** 34 biodiversity hotspots identified by ecologists as important and endangered centers of terrestrial biodiversity that contain a large number of species found nowhere else. Identifying and saving these critical habitats requires a vital emergency response (**Concept 10-5A**). Compare these areas with those on the map of the human ecological footprint in the world as shown in Figure 3, pp. S24–S25, in Supplement 4. According to the IUCN, the average proportion of biodiversity hotspot areas truly protected with funding and enforcement is only 5%. *See an animation based on this figure at* CengageNOW. **Questions:** Are any of these hotspots near where you live? Is there a smaller, localized hotspot in the area where you live? (Data from Center for Applied Biodiversity Science at Conservation International).

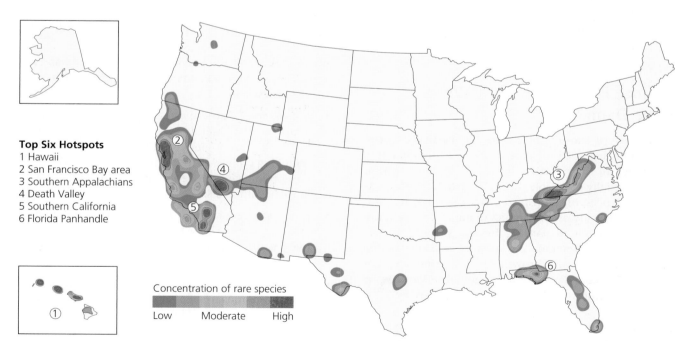

Top Six Hotspots
1 Hawaii
2 San Francisco Bay area
3 Southern Appalachians
4 Death Valley
5 Southern California
6 Florida Panhandle

Concentration of rare species

Low Moderate High

Figure 10-27 Endangered natural capital: biodiversity hotspots in the United States that need emergency protection. The shaded areas contain the largest concentrations of rare and potentially endangered species. Compare these areas with those on the map of the human ecological footprint in North America shown in Figure 7, pp. S28–S29, in Supplement 4. **Question:** Do you think that hotspots near urban areas would be harder to protect than those in rural areas? Explain. (Data from State Natural Heritage Programs, The Nature Conservancy, and Association for Biodiversity Information)

Ecological Restoration of a Tropical Dry Forest in Costa Rica

Costa Rica is the site of one of the world's largest *ecological restoration* projects. In the lowlands of its Guanacaste National Park (Figure 10-25), a small tropical dry forest was burned, degraded, and fragmented by large-scale conversion to cattle ranches and farms. Now it is being restored and relinked to the rain forest on adjacent mountain slopes. The goal is to eliminate damaging nonnative grasses and reestablish a tropical dry forest ecosystem over the next 100–300 years.

Daniel Janzen, professor of biology at the University of Pennsylvania and a leader in the field of restoration ecology, helped galvanize international support for this restoration project. He used his own MacArthur grant money to purchase this Costa Rican land to be set aside as a national park. He also raised more than $10 million for restoring the park.

Janzen realized that the original forests had been maintained partly by large native animals that ate the fruit of the Guanacaste tree and spread its seeds in their droppings.

But these animals disappeared about 10,000 years ago. About 500 years ago, horses and cattle introduced by Europeans also spread the seeds, but farming and ranching took their toll on the forest's trees. Janzen decided to speed up restoration of this tropical dry forest by incorporating limited numbers of horses and cattle as seed dispersers in his recovery plan.

Janzen recognizes that ecological restoration and protection of the park will fail unless the people in the surrounding area believe they will benefit from such efforts. His vision is to see the nearly 40,000 people who live near the park play an essential role in the restoration of the degraded forest, a concept he calls *biocultural restoration.*

By actively participating in the project, local residents reap educational, economic, and environmental benefits. Local farmers make money by sowing large areas with tree seeds and planting seedlings started in Janzen's lab. Local grade school, high school, and university students and citizens' groups study the park's ecology during field trips. The park's location near the Pan American Highway makes it an ideal area for ecotourism, which stimulates the local economy.

The project also serves as a training ground in tropical forest restoration for scientists from all over the world. Research scientists working on the project give guest classroom lectures and lead field trips.

In a few decades, today's children will be running the park and the local political system. If they understand the ecological importance of their local environment, they will be more likely to protect and sustain its biological resources. Janzen believes that education, awareness, and involvement—not guards and fences—are the best ways to restore degraded ecosystems and to protect largely intact ecosystems from unsustainable use.

Critical Thinking

Would such an ecological restoration project be possible in the area where you live? Explain.

nature-versus-people, this approach focuses on finding win–win ways to protect both people and the ecosystem services that support all life and economies.

We Can Rehabilitate and Restore Ecosystems That We Have Damaged

Almost every natural place on the earth has been affected or degraded to some degree by human activities. Much of the harm we have inflicted on nature is at least partially reversible through **ecological restoration:** the process of repairing damage caused by humans to the biodiversity and dynamics of natural ecosystems. Examples include replanting forests, restoring grasslands, restoring wetlands and stream banks, reclaiming urban industrial areas (brownfields), reintroducing native species (**Core Case Study**), removing invasive species, and freeing river flows by removing dams.

CORE
CASE
STUDY

Evidence indicates that in order to sustain biodiversity, we must make a global effort to rehabilitate and restore ecosystems we have damaged (**Concept 10-5B**). An important strategy is to mimic nature and natural processes and let nature do most of the work, usually through secondary ecological succession (Figure 5-17, p. 117).

By studying how natural ecosystems recover, scientists are learning how to speed up repair operations using a variety of approaches. They include the following measures:

- *Restoration:* returning a particular degraded habitat or ecosystem to a condition as similar as possible to its natural state.

- *Rehabilitation:* turning a degraded ecosystem into a functional or useful ecosystem without trying to restore it to its original condition. Examples include removing pollutants and replanting to reduce soil erosion in abandoned mining sites and landfills and in clear-cut forests.

- *Replacement:* replacing a degraded ecosystem with another type of ecosystem. For example, a productive pasture or tree plantation may replace a degraded forest.

- *Creating artificial ecosystems:* for example, creating artificial wetlands to help reduce flooding or to treat sewage.

Researchers have suggested a science-based four-point strategy for carrying out most forms of ecological restoration and rehabilitation.

- Identify what caused the degradation (such as pollution, farming, overgrazing, mining, or invasive species).

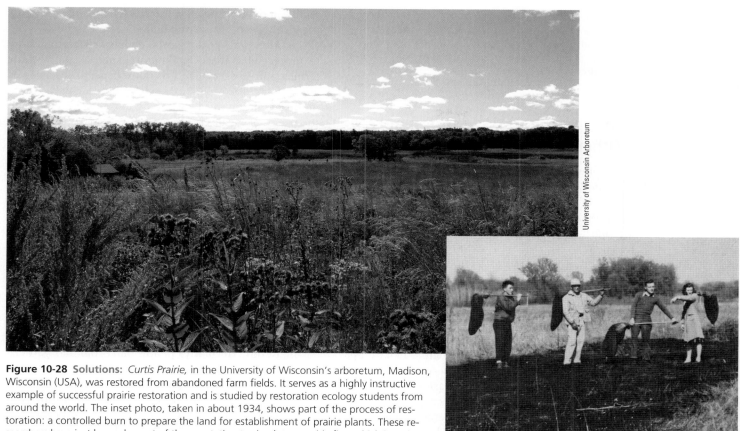

Figure 10-28 Solutions: *Curtis Prairie,* in the University of Wisconsin's arboretum, Madison, Wisconsin (USA), was restored from abandoned farm fields. It serves as a highly instructive example of successful prairie restoration and is studied by restoration ecology students from around the world. The inset photo, taken in about 1934, shows part of the process of restoration: a controlled burn to prepare the land for establishment of prairie plants. These researchers have just burned a part of the vegetation to simulate a prairie fire, which is an important natural event in the prairie ecosystem. The second person from the left in this photo is the pioneering conservation biologist Aldo Leopold (Individuals Matter, p. 22).

- Stop the abuse by eliminating or sharply reducing these factors. This would include removing toxic soil pollutants, adding nutrients to depleted soil, adding new topsoil, preventing fires, and controlling or eliminating disruptive nonnative species (Science Focus, left).

- If necessary, reintroduce species—especially pioneer, keystone, and foundation species—to help restore natural ecological processes, as was done with wolves in the Yellowstone ecosystem (**Core Case Study**). CORE CASE STUDY

- Protect the area from further degradation (Figure 10-21, right).

Most of the tall-grass prairies in the United States have been plowed up and converted to crop fields. However, these prairies are ideal subjects for ecological restoration for three reasons. *First,* many residual or transplanted native plant species can be established within a few years. *Second,* the technology involved is similar to that of gardening and agriculture. *Third,* the process is well suited for volunteer labor needed to plant native species and weed out invading species until the natural species can take over. There are a number of prairie restoration projects in the United States, a prime example of which is Curtis Prairie in the U. S. state of Wisconsin (Figure 10-28).

┌─ RESEARCH FRONTIER ─────────────────────────┐
Exploring ways to improve ecological restoration efforts. See **academic.cengage.com/biology/miller**.
└──┘

Will Restoration Encourage Further Destruction?

Some analysts worry that ecological restoration could encourage continuing environmental destruction and degradation by suggesting that any ecological harm we do can be undone. Restoration ecologists disagree with that suggestion. They point out that preventing ecosystem damage in the first place is cheaper and more effective than any form of ecological restoration. But they agree that restoration should not be used as an excuse for environmental destruction.

Restoration ecologists note that so far, we have been able to protect or preserve only about 5% of the earth's land from the effects of human activities, so ecological restoration is badly needed for many of the world's ecosystems. Even if a restored ecosystem differs from the original system, they argue, the result is better than no restoration at all. In time, further experience with ecological restoration will improve its effectiveness. Chapter 12 describes examples of the ecological restoration of aquatic systems such as wetlands and rivers.

We Can Share Areas We Dominate with Other Species

In 2003, ecologist Michael L. Rosenzweig wrote a book entitled *Win–Win Ecology: How Earth's Species Can Survive in the Midst of Human Enterprise*. Rosenzweig strongly supports proposals to help sustain the earth's biodiversity through species protection strategies such as the U.S. Endangered Species Act (Case Study, p. 207).

But Rosenzweig contends that, in the long run, these approaches will fail for two reasons. *First,* fully protected reserves currently are devoted to saving only about 5% of the world's terrestrial area, excluding polar and other uninhabitable areas. To Rosenzweig, the real challenge is to sustain wild species in more of the human-dominated portion of nature that makes up 95% of the planet's terrestrial area (**Concept 10-5C**).

Second, Rosenzweig says, setting aside funds and refuges and passing laws to protect endangered and threatened species are essentially desperate attempts to save species that are in deep trouble. These emergency efforts can help a few species, but it is equally important to learn how to keep more species away from the brink of extinction. This is a prevention approach.

Rosenzweig suggests that we develop a new form of conservation biology, called **reconciliation** or **applied ecology.** This science focuses on inventing, establishing, and maintaining new habitats to conserve species diversity in places where people live, work, or play. In other words, we need to learn how to share with other species some of the spaces we dominate.

Implementing reconciliation ecology will involve the growing practice of *community-based conservation*, in which conservation biologists work with people to help them protect biodiversity in their local communities. With this approach, scientists, citizens, and sometimes national and international conservation organizations seek ways to preserve local biodiversity while allowing people who live in or near protected areas to make sustainable use of some of the resources there (Case Study, right).

For example, people learn how protecting local wildlife and ecosystems can provide economic resources for their communities by encouraging sustainable forms of ecotourism. In the Central American country of Belize, conservation biologist Robert Horwich has helped to establish a local sanctuary for the black howler monkey. He convinced local farmers to set aside strips of forest to serve as habitats and corridors through which these monkeys can travel. The reserve, run by a local women's cooperative, has attracted ecotourists and biologists. The community has built a black howler museum,

and local residents receive income by housing and guiding visiting ecotourists and biological researchers.

In other parts of the world, people are learning how to protect vital insect pollinators, such as native butterflies and bees, which are vulnerable to insecticides and habitat loss. Neighborhoods and municipal governments are doing this by agreeing to reduce or eliminate the use of pesticides on their lawns, fields, golf courses, and parks. Neighbors also work together in planting gardens of flowering plants as a source of food for pollinating insect species. And neighborhoods and farmers build devices using wood and plastic straws, which serve as hives for increasingly threatened pollinating bees.

People have also worked together to help protect bluebirds within human-dominated habitats where most of the bluebirds' nesting trees have been cut and the bluebird populations have declined. Special boxes were designed to accommodate nesting bluebirds, and the North American Bluebird Society has encouraged Canadians and Americans to use these boxes on their properties and to keep house cats away from nesting bluebirds. Now bluebird numbers are growing again.

In Berlin, Germany, people have planted gardens on many large rooftops. These gardens support a variety of wild species by containing varying depths and types of soil and exposures to sunlight. Such roofs also save energy by providing insulation and absorbing less heat than conventional rooftops do, thereby helping to keep cities cooler. They also conserve water by reducing evapotranspiration. Some reconciliation ecology proponents call for a global campaign to use the roofs of the world to help sustain biodiversity. **GREEN CAREER:** Rooftop garden designer

In the U.S. state of California, San Francisco's Golden Gate Park is a large oasis of gardens and trees in the midst of a major city. It is a good example of reconciliation ecology, because it was designed and planted by people who transformed it from a system of sand dunes. There are many other examples of individuals and groups working together on projects to restore grasslands, wetlands, streams, and other degraded areas rain forest Case Study below). **GREEN CAREER:** Reconciliation ecology specialist

■ CASE STUDY

The Blackfoot Challenge— Reconciliation Ecology in Action

The Blackfoot River flows among beautiful mountain ranges in the west central part of the U.S. state of Montana. This large watershed is home to more than 600 species of plants, 21 species of waterfowl, bald eagles, peregrine falcons, grizzly bears, and rare species of trout. Some species, such as the Howell's gumweed and the bull trout, are threatened with extinction.

The Blackfoot River Valley is also home to people who live in seven communities and 2,500 rural households. A book and movie, both entitled *A River Runs*

Through It, tell of how residents of the valley cherish their lifestyles.

In the 1970s, many of these people recognized that their beloved valley was threatened by poor mining, logging, and grazing practices, water and air pollution, and unsustainable commercial and residential development. They also understood that their way of life depended on wildlife and wild ecosystems located on private and public lands. They began meeting informally over kitchen tables to discuss how to maintain their way of life while sustaining the other species living in the valley. These small gatherings spawned community meetings attended by individual and corporate landowners, state and federal land managers, scientists, and local government officials.

Out of these meetings came action. Teams of residents organized weed-pulling parties, built nesting structures for waterfowl, and developed sustainable grazing systems. Landowners agreed to create perpetual conservation easements, setting land aside for only conservation and sustainable uses such as hunting and fishing. They also created corridors between large tracts of undeveloped land. In 1993, these efforts were organized under a charter called the Blackfoot Challenge.

The results were dramatic. Blackfoot Challenge members have restored and enhanced large areas of wetlands, streams, and native grasslands. They have reserved large tracts of private land under perpetual conservation easements.

These pioneers might not have known it, but they were initiating what has become a classic example of *reconciliation ecology.* They worked together, respected each other's views, accepted compromises, and found ways to share their land with the area's plants and animals. They understood that all sustainability is local.

THINKING ABOUT
Wolves and Reconciliation Ecology — CORE CASE STUDY

What are some ways in which the wolf restoration project in Yellowstone National Park (**Core Case Study**) is similar to some reconciliation ecology examples described above?

RESEARCH FRONTIER

Determining where and how reconciliation ecology can work best. See **academic.cengage.com/biology/miller.**

Figure 10-29 lists some ways in which you can help sustain the earth's terrestrial biodiversity.

WHAT CAN YOU DO?

Sustaining Terrestrial Biodiversity

- Adopt a forest
- Plant trees and take care of them
- Recycle paper and buy recycled paper products
- Buy sustainably produced wood and wood products
- Choose wood substitutes such as bamboo furniture and recycled plastic outdoor furniture, decking, and fencing
- Help to restore a nearby degraded forest or grassland
- Landscape your yard with a diversity of plants natural to the area

Figure 10-29 Individuals Matter: ways to help sustain terrestrial biodiversity. **Questions:** Which two of these actions do you think are the most important? Why? Which of these things do you already do?

REVISITING — Yellowstone Wolves and Sustainability

CORE CASE STUDY

In this chapter, we looked at how terrestrial biodiversity is being destroyed or degraded. We also saw how we can reduce this destruction and degradation by using forests and grasslands more sustainably, protecting species and ecosystems in parks, wilderness, and other nature reserves, and protecting ecosystem services that support all life and economies. We learned the importance of preserving what remains of richly biodiverse and highly endangered ecosystems (biodiversity hotspots) and identifying and protecting areas where deteriorating ecosystem services threaten people and other forms of life.

We also learned about the value of restoring or rehabilitating some of the ecosystems we have degraded. Reintroducing keystone species such as the gray wolf into ecosystems they once inhabited (**Core Case Study**) is a form of ecological

restoration that can result in the reestablishment of certain ecological functions and species interactions in such systems, thereby helping to preserve biodiversity. Finally, we explored ways in which people can share with other species some of the land they occupy (95% of all the earth's land) in order to help sustain biodiversity.

Preserving terrestrial biodiversity involves applying the four **scientific principles of sustainability** (see back cover). First, it means respecting biodiversity by trying to sustain it. If we are successful, we will also be restoring and preserving the flows of energy from the sun through food webs, the cycling of nutrients in ecosystems, and the species interactions in food webs that help prevent excessive population growth of any species, including our own.

We abuse land because we regard it as a commodity belonging to us.
When we see land as a community to which we belong,
we may begin to use it with love and respect.
ALDO LEOPOLD

1. Review the Key Questions and Concepts for this chapter on p. 215. Describe the beneficial effects of reintroducing the keystone gray wolf species (Figure 10-1) into Yellowstone National Park in the United States (**Core Case Study**). **CORE CASE STUDY**

2. Distinguish among an **old-growth forest,** a **second-growth forest,** and a **tree plantation (tree farm or commercial forest).** What major ecological and economic benefits do forests provide? Describe the efforts of scientists and economists to put a price tag on the major ecological services provided by forests and other ecosystems.

3. What harm is caused by building roads into previously inaccessible forests? Distinguish among *selective cutting, clear-cutting,* and *strip cutting* in the harvesting of trees. What are the major advantages and disadvantages of clear-cutting forests?

4. What are two types of forest fires? What are some ecological benefits of occasional surface fires? What are four ways to reduce the harmful impacts of diseases and insects on forests? What effects might projected global warming have on forests?

5. What parts of the world are experiencing the greatest forest losses? Define **deforestation** and list some of its major harmful environmental effects. Describe the encouraging news about deforestation in the United States. What are the major basic and secondary causes of tropical deforestation?

6. Describe four ways to manage forests more sustainably. What is certified timber? What are four ways to reduce the harms to forests and to people from forest fires? What are three ways to reduce the need to harvest trees? What is the fuelwood crisis and what are three ways to reduce its severity? Describe the Green Belt Movement. What are five ways to protect tropical forests and use them more sustainably?

7. Distinguish between **rangelands** and **pastures.** Distinguish between the **overgrazing** and **undergrazing** of rangelands. What are three ways to reduce overgrazing and use rangelands more sustainably? Describe the conflict between ranching and urban development in the American West.

8. What major environmental threats affect national parks? How could national parks in the United States be used more sustainably? Describe some of the ecological effects of reintroducing the gray wolf to Yellowstone National Park in the United States (**Core Case Study**). What percentage of the world's land has been set aside and protected as nature reserves, and what percentage do conservation biologists believe should be protected? **CORE CASE STUDY**

9. How should nature reserves be designed and connected? Describe what Costa Rica has done to establish nature reserves. What is **wilderness** and why is it important? Describe the controversy over protecting wilderness in the United States. What is a **biological hotspot** and why is it important to protect such areas? Why is it also important to protect areas where deteriorating ecosystem services threaten people and other forms of life?

10. What is **ecological restoration?** What are the four parts of a prominent strategy for carrying out ecological restoration and rehabilitation? Describe the ecological restoration of a tropical dry forest in Costa Rica. Define and give three examples of **reconciliation (applied) ecology.** Describe the relationship between reestablishing wolves in Yellowstone National park (**Core Case Study**) and the four **scientific principles of sustainability.** **CORE CASE STUDY** **SUSTAINABILITY**

Note: Key Terms are in **bold** type.

1. List three ways in which you could apply **Concept 10-5C** to help sustain terrestrial ecosystems and biodiversity.

2. Do you support the reintroduction of the gray wolf into the Yellowstone ecosystem in the United States (**Core Case Study**)? Explain. Do you think **CORE CASE STUDY** the reintroduction of wolves should be expanded to areas outside of the park? Explain.

3. Some argue that growing oil palm trees in plantations in order to produce biodiesel fuel will help us to lessen our dependence on oil and will cut vehicle CO_2 emis-

sions. Do you think these benefits are important enough to justify burning and clearing some tropical rain forests? Why or why not? Can you think of ways to produce biofuels, other than cutting trees? What are they?

4. In the early 1990s, Miguel Sanchez, a subsistence farmer in Costa Rica, was offered $600,000 by a hotel developer for a piece of land that he and his family had been using sustainably for many years. The land contained an old-growth rain forest and a black sand beach in an area under rapid development. Sanchez refused the offer. What would you have done if you were in Miguel Sanchez's position? Explain your decision.

5. There is controversy over whether Yellowstone National Park in the United States should be accessible by snowmobile during winter. Conservationists and backpackers, who use cross-country skis or snowshoes for excursions in the park during winter, say no. They contend that snowmobiles are noisy, pollute the air, and can destroy vegetation and disrupt some of the park's wildlife. Proponents say that snowmobiles should be allowed so that snowmobilers can enjoy the park during winter when cars are mostly banned. They point out that new snowmobiles are made to cut pollution and noise. A proposed compromise plan would allow no more than 950 of these new machines into the park per day, only on roads, and primarily on guided tours. What is your view on this issue? Explain.

6. In 2007, Lester R. Brown estimated that reforesting the earth and restoring the earth's degraded rangelands would cost about $15 billion a year. Suppose the United States, the world's most affluent country, agreed to put up half this money, at an average annual cost of $25 per American. Would you support doing this? Explain. What other part or parts of the federal budget would you decrease to come up with these funds?

7. Should developed countries provide most of the money needed to help preserve remaining tropical forests in developing countries? Explain.

8. Are you in favor of establishing more wilderness areas in the United States, especially in the lower 48 states (or in the country where you live)? Explain. What might be some drawbacks of doing this?

9. Congratulations! You are in charge of the world. List the three most important features of your policies for using and managing (a) forests, (b) grasslands, (c) nature reserves such as parks and wildlife refuges, (d) biological hotspots, and (e) areas with deteriorating ecosystem services.

10. List two questions that you would like to have answered as a result of reading this chapter.

Note: See Supplement 13 (p. S78) for a list of Projects related to this chapter.

ECOLOGICAL FOOTPRINT ANALYSIS

Study the data below on deforestation in five countries, and answer the questions that follow.

Country	Area of Tropical Rain Forest (square kilometers)	Area of Deforestation per Year (square kilometers)	Annual Rate of Tropical Forest Loss
A	1,800,000	50,000	
B	55,000	3,000	
C	22,000	6,000	
D	530,000	12,000	
E	80,000	700	

1. What is the annual rate of tropical rain forest loss, as a percentage of total forest area, in each of the five countries? Answer by filling in the blank column in the table.

2. What is the annual rate of tropical deforestation collectively in all of the countries represented in the table?

3. According to the table, and assuming the rates of deforestation remain constant, which country's tropical rain forest will be completely destroyed first?

4. Assuming the rate of deforestation in Country C remains constant, how many years will it take for all of its tropical rain forests to be destroyed?

5. Assuming that a hectare (1.0 hectare = 0.01 square kilometer) of tropical rain forest absorbs 0.85 metric tons of carbon dioxide per year, what would be the total annual growth in the carbon footprint (carbon emitted but not absorbed by vegetation because of deforestation) in metric tons of carbon dioxide per year from deforestation for each of the five countries in the table?

LEARNING ONLINE

Log on to the Student Companion Site for this book at **academic.cengage.com/biology/miller**, and choose Chapter 10 for many study aids and ideas for further reading and research. These include flash cards, practice quizzing, Weblinks, information on Green Careers, and InfoTrac® College Edition articles.

Sustaining Aquatic Biodiversity

A Biological Roller Coaster Ride in Lake Victoria

Lake Victoria, a large, shallow lake in East Africa (Figure 11-1), has been in ecological trouble for more than 2 decades.

Until the early 1980s, the lake had 500 species of fish found nowhere else. About 80% of them were small fish known as cichlids (pronounced "SIK-lids"), which feed mostly on detritus, algae, and zooplankton. Since 1980, some 200 of the cichlid species have become extinct, and some of those that remain are in trouble.

Several factors caused this dramatic loss of aquatic biodiversity. First, there was a large increase in the population of the Nile perch (Figure 11-2). This large predatory fish was deliberately introduced into the lake during the 1950s and 1960s to stimulate exports of the fish to several European countries, despite warnings by biologists that this huge fish with a big appetite would reduce or eliminate many defenseless native fish species. The population of this large and prolific fish exploded, devoured the cichlids and by 1986 had wiped out over 200 cichlid species.

Introducing the perch had other social and ecological effects. The new mechanized fishing industry increased poverty and malnutrition by putting most small-scale fishers and fish vendors out of business. And because the oily flesh of the perch are preserved by use of a wood smoker, local forests were depleted for firewood.

Another factor in loss of biodiversity in Lake Victoria was frequent algal blooms. These blooms became more common in the 1980s, due to nutrient runoff from surrounding farms and deforested land, spills of untreated sewage, and declines in the populations of algae-eating cichlids.

Also, the Nile perch population is decreasing because it severely reduced its own food supply of smaller fishes—an example of one of the four **scientific principles of sustainability** (see back cover) in action—and it also shows signs of being overfished. This may allow a gradual increase in the populations of some of the remaining cichlids.

This ecological story about the dynamics of large aquatic systems illustrates that there are unintended consequences when we intrude into a poorly understood ecosystem.

This chapter is devoted to helping us to understand the threats to aquatic biodiversity and what we can do to help sustain this vital part of the earth's natural capital.

Figure 11-1 Lake Victoria is a large lake in East Africa.

Courtesy of the African Angler

Figure 11-2 Natural capital degradation: the Nile perch is a fine food fish that can weigh more than 91 kilograms (200 pounds). However, this deliberately introduced fish has played a key role in a major loss of biodiversity in East Africa's Lake Victoria (left).

11-1 What are the major threats to aquatic biodiversity?

CONCEPT 11-1 Aquatic species are threatened by habitat loss, invasive species, pollution, climate change, and over-exploitation, all made worse by the growth of the human population.

11-2 How can we protect and sustain marine biodiversity?

CONCEPT 11-2 We can help to sustain marine biodiversity by using laws and economic incentives to protect species, setting aside marine reserves to protect ecosystems, and using community-based integrated coastal management.

11-3 How should we manage and sustain marine fisheries?

CONCEPT 11-3 Sustaining marine fisheries will require improved monitoring of fish populations, cooperative fisheries management among communities and nations, reduction of fishing subsidies, and careful consumer choices in seafood markets.

11-4 How can we protect and sustain wetlands?

CONCEPT 11-4 To maintain the ecological and economic services of wetlands, we must maximize preservation of remaining wetlands and restoration of degraded and destroyed wetlands.

11-5 How can we protect and sustain freshwater lakes, rivers, and fisheries?

CONCEPT 11-5 Freshwater ecosystems are strongly affected by human activities on adjacent lands, and protecting these ecosystems must include protection of their watersheds.

11-6 What should be our priorities for sustaining biodiversity and ecosystem services?

CONCEPT 11-6 Sustaining the world's biodiversity and ecosystem services will require mapping terrestrial and aquatic biodiversity, maximizing protection of undeveloped terrestrial and aquatic areas, and carrying out ecological restoration projects worldwide.

Note: Supplements 2 (p. S4), 8 (p. S47), and 13 (p. S78) can be used with this chapter.

The coastal zone may be the single most important portion of our planet.
The loss of its biodiversity may have repercussions
far beyond our worst fears.

G. CARLETON RAY

11-1 What Are the Major Threats to Aquatic Biodiversity?

▶ **CONCEPT 11-1** Aquatic species are threatened by habitat loss, invasive species, pollution, climate change, and overexploitation, all made worse by the growth of the human population.

We Have Much to Learn about Aquatic Biodiversity

Although we live on a watery planet, we have explored only about 5% of the earth's global ocean (Figure 8-2, p. 163) and know relatively little about its biodiversity and how it works. We also have limited knowledge about freshwater biodiversity.

However, scientists have observed three general patterns of marine biodiversity. *First,* the greatest marine biodiversity occurs in coral reefs (Chapter 8 Core Case Study, p. 162), estuaries, and the deep-ocean floor. *Second,* biodiversity is higher near coasts than in the open sea because of the greater variety of producers and habitats in coastal areas (Figure 8-5, p. 166). *Third,* biodiversity is higher in the bottom region of the ocean than in the surface region because of the greater variety of habitats and food sources on the ocean bottom.

The world's marine systems provide important ecological and economic services (Figure 8-4, p. 165). Thus, scientific investigation of poorly understood marine aquatic systems is a *research frontier* that could lead to

immense ecological and economic benefits. Freshwater systems, which occupy only 1% of the earth's surface, also provide important ecological and economic services (Figure 8-14, p. 174).

─ RESEARCH FRONTIERS ─

Exploring marine and freshwater ecosystems, their species, and species interactions. See **academic.cengage.com/ biology/miller**.

Human Activities Are Destroying and Degrading Aquatic Habitats

As with terrestrial biodiversity, the greatest threats to the biodiversity of the world's marine and freshwater ecosystems (**Concept 11-1**) can be remembered with the aid of the acronym HIPPCO, with H standing for *habitat loss and degradation.* Some 90% of fish living in the ocean spawn in coral reefs (Figure 4-10, left, p. 89, and Figure 8-1, p. 162), mangrove forests (Figure 8-8, p. 168), coastal wetlands (Figure 8-7, p. 167), or rivers (Figure 8-17, p. 176). And these areas are under intense pressure from human activities (Figure 8-12, p. 172). Scientists reported in 2006 that these coastal habitats are disappearing at rates 2–10 times higher than the rate of tropical forest loss.

A major threat is loss and degradation of many sea-bottom habitats by dredging operations and trawler fishing boats. Trawlers drag huge nets weighted down with heavy chains and steel plates like giant submerged bulldozers over ocean bottoms to harvest a few species of bottom fish and shellfish (Figure 11-3). Trawling nets reduce coral reef habitats to rubble and kill a variety of creatures on the bottom by crushing them, burying them in sediment, and exposing them to predators. Each year, thousands of trawlers scrape and disturb an area of ocean floor about 150 times larger than the area of forests that are clear-cut annually.

In 2004, some 1,134 scientists signed a statement urging the United Nations to declare a moratorium on bottom trawling on the high seas by 2006 and to eliminate it globally by 2010. Fishing nations led by Iceland, Russia, China, and South Korea blocked such a ban. But in 2007, those countries (except for Iceland) and 18 others agreed to voluntary restrictions on bottom trawling in the South Pacific. The agreement will partially protect about one-quarter of the world's ocean bottom but monitoring and enforcement will be difficult.

Habitat disruption is also a problem in freshwater aquatic zones. Dams and excessive water withdrawal from rivers and lakes (mostly for agriculture) destroy aquatic habitats and water flows and disrupt freshwater biodiversity. As a result of these and other human activities, 51% of freshwater fish species—more than any other major type of species—are threatened with premature extinction (Figure 9-6, p. 189).

Figure 11-3 Natural capital degradation: area of ocean bottom before (left) and after (right) a trawler net scraped it like a gigantic plow. These ocean floor communities could take decades or centuries to recover. According to marine scientist Elliot Norse, "Bottom trawling is probably the largest human-caused disturbance to the biosphere." Trawler fishers claim that ocean bottom life recovers after trawling and that they are helping to satisfy the increasing consumer demand for fish. **Question:** What land activities are comparable to this?

Peter J. Auster/National Undersea Research Center

Invasive Species Are Degrading Aquatic Biodiversity

Another problem is the deliberate or accidental introduction of hundreds of harmful invasive species—the I in HIPPCO—(Figure 9-14, p. 199) into coastal waters, wetlands, and lakes throughout the world (**Concept 11-1**). These bioinvaders can displace or cause the extinction of native species and disrupt ecosystem services. For example, since the late 1980s, Lake Victoria (**Core Case Study**) has been invaded by the water hyacinth (Figure 11-4). This rapidly growing plant has carpeted large areas of the lake, blocked sunlight, deprived fish and plankton of oxygen, and reduced aquatic plant diversity.

CORE CASE STUDY

THINKING ABOUT

The Nile Perch and Lake Victoria CORE CASE STUDY

Would most of the now extinct cichlid fish species in Lake Victoria (**Core Case Study**) still exist today if the Nile perch had not been introduced? Or might other factors come into play? Explain.

According to a 2008 study by The Nature Conservancy, 84% of the world's coastal waters are being colonized by invasive species. Bioinvaders are blamed for about two-thirds of fish extinctions in the United States between 1900 and 2000. They cost the country an average of about $14 million *per hour*. Many of these invaders arrive in the ballast water stored in tanks in large cargo ships to keep them stable. These ships take in ballast water—along with whatever microorganisms and tiny species it contains—in one harbor and dump it in another.

Consumers also introduce invasive species. For example, the *Asian swamp eel* has invaded the waterways of south Florida (USA), probably from the dumping of a home aquarium. This rapidly reproducing eel eats almost anything—including many prized fish species—by sucking them in like a vacuum cleaner. It can elude cold weather, drought, and predators by burrowing into mud banks. It is also resistant to waterborne poisons because it can breathe air, and it can wriggle across dry land to invade new waterways, ditches, canals, and marshes. Eventually, this eel could take over much of the waterways of the southeastern United States.

Another example is the *purple loosestrife*, a perennial plant that grows in wetlands in parts of Europe. Since the 1880s, it has been imported and used in gardens as an ornamental plant in many parts of the world. A single plant can produce more than 2.5 million seeds a year, which are spread by flowing water and by becoming attached to wildlife, livestock, hikers, and vehicle tire treads. It reduces wetland biodiversity by displacing native vegetation and reducing habitat for some forms of wetland wildlife.

Some U.S. states have recently introduced two natural predators of loosestrife from Europe: a weevil species and a leaf-eating beetle. It will take some time to determine the effectiveness of this biological control approach and to be sure the introduced predators themselves do not become pests.

While threatening native species, invasive species can also disrupt and degrade whole ecosystems. This is the focus of study for a growing number of researchers (Science Focus, at right).

Population Growth and Pollution Can Reduce Aquatic Biodiversity

The U.N. Environment Programme (UNEP) projects that, by 2020, 80% of the world's people will be living along or near the coasts, mostly in gigantic coastal cities. This coastal population growth—the first P in HIPPCO—will add to the already intense pressure on the world's coastal zones, primarily by destroying more aquatic habitat and increasing pollution (**Concept 11-1**).

A 2008 study by Benjamin S. Halpern and other scientists found that only 4% of the world's oceans are not affected by pollution—the second P in HIPPCO—and 40% are strongly affected. In 2004, the UNEP estimated that 80% of all ocean pollution comes from land-based coastal activities. Humans have doubled the flow of nitrogen, mostly from nitrate fertilizers, into the oceans since 1860, and the 2005 Millennium Ecosystem Assessment estimated that this flow will increase by another two-thirds by 2050. These inputs of nitrogen (and similar inputs of phosphorus) result in eutrophication

Courtesy of Patrick Agaba and Clean Lakes, Inc.

Figure 11-4 *Invasive water hyacinths*, supported by nutrient runoff, clogged a ferry terminal on the Kenyan part of Lake Victoria in 1997. By blocking sunlight and consuming oxygen, this invasion has reduced biodiversity in the lake. Scientists reduced the problem at strategic locations by mechanical removal and by introducing two weevils for biological control of the hyacinth.

SCIENCE FOCUS

How Carp Have Muddied Some Waters

Lake Wingra lies within the city of Madison, Wisconsin (USA), surrounded mostly by a forest preserve. While almost all of its shoreline is undeveloped, the lake receives excessive nutrient inputs from runoff, containing fertilizers from area farms and lawns, and storm water flowing in from city streets and parking lots. Its waters are green and murky throughout the warmer months of the year.

Lake Wingra also contains a number of invasive plant and fish species, including purple loosestrife and common carp. The carp, which were introduced in the late 1800s, now make up about half of the fish biomass in the lake. They devour algae called chara, which would normally cover the lake bottom and stabilize sediments. Consequently, fish movements and currents stir these sediments, which accounts for much of the water's excessive turbidity, or cloudiness.

Knowing this, Dr. Richard Lathrup, a limnologist (lake scientist) who works with Wisconsin's Department of Natural Resources, hypothesized that removing the carp would help to restore the natural ecosystem of Lake Wingra. Lathrop speculated that with the carp gone, the bottom sediments would settle and become stabilized, allowing the water to clear. Clearer water would in turn allow native plants to receive more sunlight and become reestablished on the lake bottom, replacing purple loosestrife and other invasive plants that now dominate the shallow shoreline waters.

Lathrop and his colleagues built a *fish exclosure* by installing a thick, heavy vinyl curtain around a 1-hectare (2.5-acre), square-

Figure 11-A *Lake Wingra* in Madison, Wisconsin (USA) has become eutrophic largely because of the introductions of invasive species, including the common carp, which now represents half of the fish biomass in the lake. Removal of carp in the experimental area shown here resulted in a dramatic improvement in the clarity of the water and subsequent regrowth of native plant species in shallow water.

Mike Kakuska

shaped perimeter extending out from the shore (Figure 11-A). This barrier hangs from buoys on the surface to the bottom of the lake, isolating the volume of water within it. The researchers then removed all of the carp from this study area and began observing results. Within one month, the waters within the exclosure were noticeably clearer, and within a year, the difference in clarity was dramatic, as Figure 11-A shows.

In 2008, the scientists began removing carp from the rest of the lake. Lathrop notes

that keeping carp out of Lake Wingra will be a daunting task, but his controlled scientific experiment clearly shows the effects that an invasive species can have on an aquatic ecosystem. And it reminds us that preventing the introduction of invasive species in the first place is the best way to avoid such effects.

Critical Thinking

What are two other results of this controlled experiment that you might expect? (*Hint:* think food webs.)

of marine and freshwater systems, which can lead to algal blooms (Figure 8-16, right, p. 175), fish die-offs, and degradation of ecosystem services.

In Lake Victoria, such eutrophication was a key development in the takeover by invasive Nile perch and the loss of cichlid populations, as described in the **Core Case Study**. With increased runoff generated by the growing human population in nearby towns and farms came algal blooms. Because cichlids feed on algae, their populations rose dramatically. Before Nile perch were introduced, such population explosions would have ended in die-offs of the cichlids. But the Nile perch suddenly had a bigger food supply, and thus their population grew and led to changes in the lake's ecosystem and all the resulting problems.

Similar pressures are growing in freshwater systems, as more people seek homes and places for recreation near lakes and streams. The result is massive inputs of sediment and other wastes from land into these aquatic systems.

Toxic pollutants from industrial and urban areas can kill some forms of aquatic life by poisoning them. And each year, plastic items dumped from ships and left as litter on beaches kill up to 1 million seabirds and 100,000 mammals and sea turtles. Such pollutants and debris threaten the lives of millions of marine mammals (Figure 11-5, p. 254) and countless fish that ingest, become entangled in, or are poisoned by them. These forms of pollution lead to an overall reduction in aquatic biodiversity and degradation of ecosystem services.

Figure 11-5 This Hawaiian monk seal was slowly starving to death before this discarded piece of plastic was removed from its snout. Each year, plastic items dumped from ships and left as litter on beaches threaten the lives of millions of marine mammals, turtles, and seabirds that ingest, become entangled in, or are poisoned by such debris.

Climate Change Is a Growing Threat

Climate change—the C in HIPPCO—threatens aquatic biodiversity (**Concept 11-1**) and ecosystem services partly by causing sea levels to rise. During the past 100 years, average sea levels have risen by 10–20 centimeters (4–8 inches), and scientists estimate they will rise another 18–59 centimeters (0.6–1.9 feet) and perhaps as high as 1–1.6 meters (3.2–5.2 feet) between 2050 and 2100 mostly, because of projected global warming. This would destroy more coral reefs, swamp some low-lying islands, drown many highly productive coastal wetlands, and put much of the U.S. state of Louisiana's coast, including New Orleans, under water (Figure 8-18, p. 177). And some Pacific island nations could lose more than half of their protective coastal mangrove forests by 2100, according to a 2006 study by UNEP (Science Focus, at right). See *The Habitable Planet,* Video 5, at **www.learner.org/resources/series209.html** for projected sea level changes in densely populated coastal areas such as Vietnam and New York City.

Overfishing and Extinction: Gone Fishing, Fish Gone

Overfishing—the O in HIPPCO—is not new. Archaeological evidence indicates that for thousands of years, humans living in some coastal areas have overharvested fishes, shellfish, seals, turtles, whales, and other marine mammals (**Concept 11-1**). But today's industrialized fishing fleets can overfish much more of the oceans and deplete marine life at a much faster rate. Today, fish are hunted throughout the world's oceans by a global fleet of millions of fishing boats—some of them longer than a football field. Modern industrial

fishing can cause 80% depletion of a target fish species in only 10–15 years (Case Study, p. 256).

The human demand for seafood is outgrowing the sustainable yield of most ocean fisheries. To keep consuming seafood at the current rate, we will need 2.5 times the area of the earth's oceans, according to the *Fishprint of Nations 2006,* a study based on the concept of the human ecological footprint (**Concept 1-3**, p. 12, and Figure 1-10, p. 15). The **fishprint** is defined as the area of ocean needed to sustain the consumption of an average person, a nation, or the world. The study found that all nations together are overfishing the world's global oceans by an unsustainable 157%.

In most cases, overfishing leads to *commercial extinction,* which occurs when it is no longer profitable to continue fishing the affected species. Overfishing usually results in only a temporary depletion of fish stocks, as long as depleted areas and fisheries are allowed to recover. But as industrialized fishing fleets vacuum up more and more of the world's available fish and shellfish, recovery times for severely depleted populations are increasing and can take 2 decades or more. In 1992, for example, Canada's 500-year-old Atlantic cod fishery off the coast of Newfoundland collapsed and was closed. This put at least 20,000 fishers and fish processors out of work and severely damaged Newfoundland's economy. As Figure 11-6 shows,

Figure 11-6 Natural capital degradation: this graph illustrates the collapse of the cod fishery in the northwest Atlantic off the Canadian coast. Beginning in the late 1950s, fishers used bottom trawlers to capture more of the stock, reflected in the sharp rise in this graph. This resulted in extreme overexploitation of the fishery, which began a steady fall throughout the 1970s, followed by a slight recovery in the 1980s and total collapse by 1992 when the site was closed to fishing. Canadian attempts to regulate fishing through a quota system had failed to stop the sharp decline. The fishery was reopened on a limited basis in 1998 but then closed indefinitely in 2003. (Data from Millennium Ecosystem Assessment)

Sustaining Ecosystem Services by Protecting and Restoring Mangroves

Mangroves are not among the world's biodiversity hotspots and do not contain a large number of species, as do endangered tropical rain forests. However, they too require protection because of the important ecosystem services they provide for coastal dwellers.

For example, protecting mangroves and restoring them in areas where they have been destroyed are important ways to reduce the impacts of rising sea levels and more intense storm surges. These ecosystem services will become more important in the face of tropical storms, possibly becoming more intense as a result of global warming, and of tsunamis, caused mostly by earthquakes on ocean seafloors. Protecting and restoring these natural coastal barriers is also

much cheaper than building concrete sea walls or moving threatened coastal towns and cities inland.

Indonesia, a sprawling nation of about 17,000 islands, is especially vulnerable to rising sea levels and storm surges. But decades of rampant development along its island coastlines have destroyed or degraded about 70% of its mangrove forests. Even so, Indonesia still has the world's largest area of mangroves, amounting to about one-fourth of the world's remaining mangrove forests. In the 1990s, it started a program to protect more of these areas and to restore large areas of degraded mangrove forests.

Expanding mangrove protection and restoration in Indonesia and in other nations will require educating citizens,

government officials, and business leaders about the huge economic value of the natural ecosystem services they provide. These economic benefits should be considered during any decision-making process concerning development of these fragile coastal areas.

Critical Thinking

Do you agree that the estimated economic value of ecosystem services provided by mangroves should be considered in making coastal development decisions? If you agree, how would you accomplish this politically?

this cod population has not recovered, despite the fishing ban.

Such a collapse can create a domino effect, leading to collapses of other species. After the cod were fished out in the North Atlantic, fishers turned to sharks, which provide important ecosystem services and help to control the populations of other species (Case Study, p. 96). Since then, overfishing of big sharks has cut Atlantic stocks by 99%, according to a 2007 Canadian fisheries study. Scientists reported that with the large sharks essentially gone, the northwest Atlantic populations of rays and skates, which the sharks once fed on, have exploded and have wiped out most of the bay scallops.

One result of the increasingly efficient global hunt for fish is that big individuals in many populations of commercially valuable predatory species—including cod, salmon, mackerel, herring, and dogfish—are becoming scarce. And according to a 2003 study by conservation biologist Boris Worm and his colleagues, 90% or more of the large, predatory, open-ocean fishes such as tuna, swordfish, and marlin have disappeared since 1950 (see *The Habitable Planet*, Video 9, at **www.learner.org/resources/series209. html**). The large bluefin tuna, with a typical weight of 340 kilograms (750 pounds) and a length of 2 meters (6.5 feet), is the premier choice for sushi and sashimi, and, as the world's most desirable fish, can sell for as much as $880 per kilogram ($400 per pound). As a result, it is probably the most endangered of all large fish species.

The fishing industry has begun working its way down marine food webs by shifting from large species to smaller ones. This practice reduces the breeding stock needed for recovery of depleted species, which unravels marine food webs and disrupts marine ecosystems and their ecosystem services.

Most fishing boats hunt and capture one or a small number of commercially valuable species. However, their gigantic nets and incredibly long lines of hooks also catch nontarget species, called *bycatch*. Almost one-third of the world's annual fish catch, by weight, consists of these nontarget species, which are thrown overboard dead or dying. This can deplete the populations of bycatch species that play important roles in marine food webs. Marine mammals such as seals and dolphins can also become part of bycatch.

Fish species are also threatened with *biological extinction*, mostly from overfishing, water pollution, wetlands destruction, and excessive removal of water from rivers and lakes. According to the IUCN, 34% of the world's known marine fish species and 71% of the world's freshwater fish species face extinction within your lifetime. Indeed, *marine and freshwater fishes are threatened with extinction by human activities more than any other group of species.*

RESEARCH FRONTIER

Learning more about how aquatic systems work and how human activities affect aquatic biodiversity and aquatic ecosystem services. See **academic.cengage.com/biology/ miller**.

■ CASE STUDY
Industrial Fish Harvesting Methods

Industrial fishing fleets dominate the world's marine fishing industry. They use global satellite positioning equipment, sonar, huge nets and long fishing lines, spotter planes, and gigantic refrigerated factory ships that can process and freeze their catches. These fleets help meet the growing demand for seafood. But critics say that these highly efficient fleets are *vacuuming the seas,* decreasing marine biodiversity, and degrading important marine ecosystem services. Today 75% of the world's commercial fisheries are being fished at or beyond their estimated sustainable yields, according to the U.N. Food and Agricultural Organization.

Figure 11-7 shows the major methods used for the commercial harvesting of various marine fishes and shellfish. Until the mid-1980s, fishing fleets from de-veloped countries dominated the ocean catch. Today, most of these fleets come from developing countries, especially in Asia.

Let us look at a few of these methods. *Trawler fishing* is used to catch fishes and shellfish—especially shrimp, cod, flounder, and scallops—that live on or near the ocean floor. It involves dragging a funnel-shaped net held open at the neck along the ocean bottom. It is weighted down with chains or metal plates and scrapes up almost everything that lies on the ocean floor and often destroys bottom habitats—somewhat like clear-cutting the ocean floor (Figure 11-3). Newer trawling nets are large enough to swallow 12 jumbo jet planes and even larger ones are on the way.

Another method, *purse-seine fishing,* is used to catch surface-dwelling species such as tuna, mackerel, ancho-vies, and herring, which tend to feed in schools near the surface or in shallow areas. After a spotter plane lo-

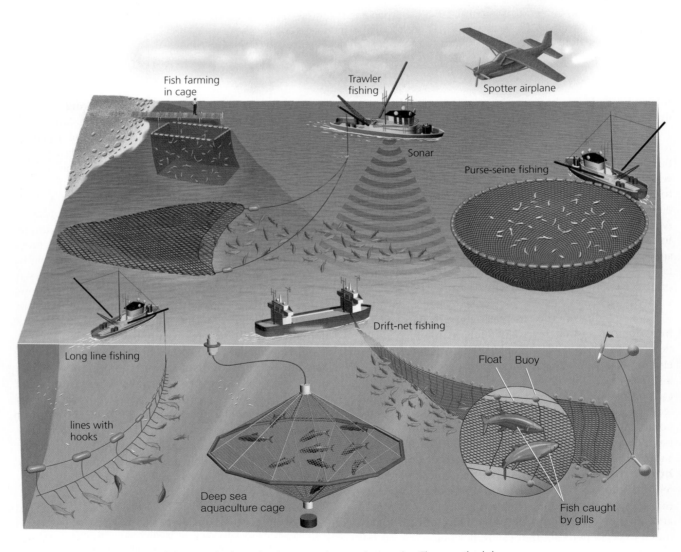

Figure 11-7 Major commercial fishing methods used to harvest various marine species. These methods have become so effective that many fish species have become commercially extinct.

cates a school, the fishing vessel encloses it with a large net called a purse seine. Nets used to capture yellow fin tuna in the eastern tropical Pacific Ocean have killed large numbers of dolphins that swim on the surface above schools of tuna.

Fishing vessels also use *longlining,* which involves putting out lines up to 130 kilometers (80 miles) long, hung with thousands of baited hooks. The depth of the lines can be adjusted to catch open-ocean fish species such as swordfish, tuna, and sharks or bottom fishes such as halibut and cod. Longlines also hook and kill large numbers endangered sea turtles, dolphins, and seabirds each year. Making simple modifications to fishing gear and practices can decrease seabird deaths.

With *drift-net fishing,* fish are caught by huge drifting nets that can hang as deep as 15 meters (50 feet) below the surface and extend to 64 kilometers (40 miles) long. This method can lead to overfishing of the desired species and may trap and kill large quantities of unwanted fish, marine mammals, sea turtles, and seabirds.

Since 1992, a U.N. ban on the use of drift nets longer than 2.5 kilometers (1.6 miles) in international waters has sharply reduced use of this technique. But longer nets continue to be used because compliance is voluntary and it is difficult to monitor fishing fleets over vast ocean areas. Also, the decrease in drift net use has led to increased use of longlines, which often have similar harmful effects on marine wildlife.

11-2 How Can We Protect and Sustain Marine Biodiversity?

▶ **CONCEPT 11-2** We can help to sustain marine biodiversity by using laws and economic incentives to protect species, setting aside marine reserves to protect ecosystems, and using community-based integrated coastal management.

Laws and Treaties Have Protected Some Endangered and Threatened Marine Species

Protecting marine biodiversity is difficult for several reasons. First, the human ecological footprint (Figure 1-10, p. 15) and fishprint are expanding so rapidly into aquatic areas that it is difficult to monitor the impacts (**Concept 1-3**, p. 12). Second, much of the damage to the oceans and other bodies of water is not visible to most people. Third, many people incorrectly view the seas as an inexhaustible resource that can absorb an almost infinite amount of waste and pollution and still produce all the seafood we want. Finally, most of the world's ocean area lies outside the legal jurisdiction of any country. Thus, it is an open-access resource, subject to overexploitation.

Nevertheless, there are ways to protect and sustain marine biodiversity, one of which is the regulatory approach (**Concept 11-2**). National and international laws and treaties to help protect marine species include the 1975 Convention on International Trade in Endangered Species (CITES), the 1979 Global Treaty on Migratory Species, the U.S. Marine Mammal Protection Act of 1972, the U.S. Endangered Species Act of 1973, the U.S. Whale Conservation and Protection Act of 1976, and the 1995 International Convention on Biological Diversity.

The U.S. Endangered Species Act (Case Study, p. 207) and international agreements have been used to identify and protect endangered and threatened marine species such as whales (see the following Case Study), seals, sea lions, and sea turtles.

■ CASE STUDY
Protecting Whales: A Success Story . . . So Far

Cetaceans are an order of mostly marine mammals ranging in size from the 0.9-meter (3-foot) porpoise to the giant 15- to 30-meter (50- to 100-foot) blue whale. They are divided into two major groups: *toothed whales* and *baleen whales* (Figure 11-8, p. 258).

Toothed whales, such as the porpoise, sperm whale, and killer whale (orca), bite and chew their food and feed mostly on squid, octopus, and other marine animals. *Baleen whales,* such as the blue, gray, humpback, minke, and fin, are filter feeders. Attached to their upper jaws are plates made of baleen, or whalebone, which they use to filter plankton, especially tiny shrimp-like krill (Figure 3-14, p. 63), from the seawater.

Whales are fairly easy to kill because of their large size and their need to come to the surface to breathe. Whale hunters became efficient at hunting and killing whales using radar, spotters in airplanes, fast ships,

CONCEPT LINK

and harpoon guns. Whale harvesting, mostly in international waters, has followed the classic pattern of a tragedy of the commons, with whalers killing an estimated 1.5 million whales between 1925 and 1975. This overharvesting drove 8 of the 11 major species to commercial extinction.

Overharvesting also drove some commercially prized species such as the giant blue whale (Figure 11-8) to the brink of biological extinction. The endangered blue whale is the world's largest animal. Fully grown, it is longer than three train boxcars and weighs more than 25 adult elephants. The adult has a heart the size of a Volkswagen Beetle, and some of its arteries are big enough for a child to swim through.

Blue whales spend about 8 months a year in Antarctic waters. During the winter, they migrate to warmer waters where their young are born. Before commercial whaling began, an estimated 250,000 blue whales roamed the Antarctic Ocean. Today, the species has been hunted to near biological extinction for its oil, meat, and bone. There are probably fewer than 5,000 blue whales left. They take 25 years to mature sexually and have only one offspring every 2–5 years. This low reproductive rate will make it difficult for the species to recover.

Blue whales have not been hunted commercially since 1964 and have been classified as an endangered species since 1975. Despite this protection, some marine biologists fear that too few blue whales remain for the species to recover and avoid premature extinction. Others believe that with continued protection they will make a slow comeback.

In 1946, the International Convention for the Regulation of Whaling established the International Whaling Commission (IWC). Its mission was to regulate the whaling industry by setting annual quotas to prevent overharvesting and commercial extinction. But IWC quotas often were based on inadequate data or were ignored by whaling countries. Without powers of enforcement, the IWC was not able to stop the decline of most commercially hunted whale species.

In 1970, the United States stopped all commercial whaling and banned all imports of whale products. Under pressure from conservationists, the U.S. government, and governments of many nonwhaling countries, the IWC imposed a moratorium on commercial whaling starting in 1986. It worked. The estimated number of whales killed commercially worldwide dropped from 42,480 in 1970 to about 1,300 in 2007. However, despite the ban, more than 26,000 whales were hunted and killed between 1986 and 2007.

Japan hunts and kills at least 1,000 minke and fin whales each year for what it calls "scientific purposes." Critics see this annual whale hunt as poorly disguised commercial whaling because the whale meat is sold to restaurants. Each whale is worth up to $30,000 wholesale. Norway openly defies the international ban on commercial whaling and hunts and kills up to 1,000 minke whales a year (Figure 11-9).

Japan, Norway, Iceland, Russia, and a growing number of small tropical island countries—which Japan brought into the IWC to support its position—hope to overthrow the IWC ban on commercial whaling and reverse the international ban on buying and selling whale products. They argue that commercial whaling should be allowed because it has been a traditional part of their economies and cultures.

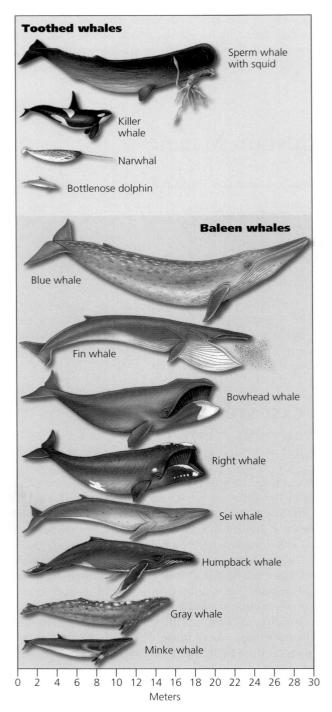

Figure 11-8 *Examples of cetaceans,* which can be classified as either toothed whales or baleen whales.

Figure 11-9 Norwegian whalers harpooning a sperm whale. Norway and Japan kill up to 2,000 whales a year. They also believe that increased but sustainable commercial whaling should be allowed for sperm, minke, and pilot whales whose stocks have built back to large numbers.

Proponents of whaling also contend that the ban is emotionally motivated and not supported by current scientific estimates of whale populations. The moratorium on commercial whaling has led to a sharp rebound in these estimates for sperm, pilot, and minke whales.

Most conservationists disagree. Some argue that whales are peaceful, intelligent, sensitive, and highly social mammals that should be protected for ethical reasons. Others question IWC estimates of the allegedly recovered whale populations, noting the inaccuracy of such estimates in the past. And many conservationists fear that opening the door to any commercial whaling may eventually weaken current international disapproval and legal sanctions against commercial whaling and lead to widespread harvests of most whale species.

> **HOW WOULD YOU VOTE?** ☑
>
> Should controlled commercial whaling be resumed for species with populations judged to be stable? Cast your vote online at **academic.cengage.com/biology/miller**.

Some coastal communities have an interest in maintaining the ban on whaling because they can provide jobs and income through increasingly popular whale watching. For example, The Nature Conservancy promoted whale watching in the town of Samaná in the Dominican Republic and has trained fishermen to work as whale-watching guides. The once run-down town has become a tourist hotspot, with spruced up houses, hotels, and inns, largely because of the popularity of whale watching. Local residents now have an economic interest in protecting the whales.

Economic Incentives Can Be Used to Sustain Aquatic Biodiversity

Other ways to protect endangered and threatened aquatic species involve using economic incentives (**Concept 11-2**). For example, according to a 2004 World Wildlife Fund study, sea turtles are worth more to local communities alive than dead. The report estimates that sea turtle tourism brings in almost three times more money than does the sale of turtle products such as meat, leather, and eggs.

The problem is that individuals seeking to make a quick gain take the turtles before their surrounding communities can realize the longer-term economic benefits by protecting them. Educating citizens about this issue could inspire communities to protect the turtles (Case Study, below).

Some individuals find economic rewards in restoring and sustaining aquatic systems. One example is an application of *reconciliation ecology* by a restaurant owner (Individuals Matter, p. 261).

■ CASE STUDY

Holding Out Hope for Marine Turtles

Of the seven species of marine turtles, six are either critically endangered or endangered. Among the latter is the leatherback sea turtle (Figure 11-10, p. 260), a species that has survived 100 million years, but now faces possible extinction. While their population is stable in the Atlantic Ocean, their numbers have declined by 95% in the Pacific.

Naturalist Carl Safina wrote about his studies of the leatherback in his book *Voyage of the Turtle*. He describes the leatherback as "the last living dinosaur." The largest of all sea turtles, and the only warm-blooded one, an adult turtle can weigh as much as 91 kilograms (200 pounds). It swims great distances, migrating across the Atlantic and Pacific Oceans, and it can dive as deep as 1,200 meters (3,900 feet). It is named for its leathery shell.

The leatherback female lays her eggs on sandy ocean beaches in the dark of night and then returns to the sea. The babies hatch simultaneously in large numbers and immediately scamper across the sand to try to survive to adulthood in ocean waters. As Safina describes it, "They start the size of a cookie, and come back the size of a dinosaur."

While leatherbacks survived the impact of the giant asteroid that probably wiped out the dinosaurs, they may not survive the growing human impact on their environment. Bottom trawlers are destroying the coral

Figure 11-10 An endangered *leatherback sea turtle* is entangled in a fishing net and lines and could have starved to death had it not been rescued.

long-line swordfish fishing off the Pacific coast to help save dwindling sea turtle populations.

On Costa Rica's northwest coast in the community of Playa Junquillal, an important leatherback nesting area, residents learned that tourism can bring in almost three times as much money as selling turtle products can earn. Biologists working with the World Wildlife Fund there directed a community-based program to educate people about the importance of protecting leatherbacks and to create revenue sources for local residents based on tourism instead of on harvesting turtle eggs. Volunteers were enlisted to find and rescue nests before they could be poached and to build hatcheries to protect the eggs.

For the leatherback turtles, this program was a success. In 2004, on the local beaches, all known nests had been poached. The following year, all known nests were protected and none were poached. The leatherback had become an important economic resource for all, not for just a few, of the residents of Playa Junquillal.

THINKING ABOUT
The Leatherback Sea Turtle

Why should we care if the leatherback sea turtle becomes extinct? What are three things you would do to help protect this species from premature extinction?

gardens that serve as their feeding grounds. The turtles are hunted for meat and leather, and their eggs are taken for food. They often drown after becoming entangled in fishing nets and lines (Figure 11-10) and lobster and crab traps. A 2004 study by R. I. Lewison and his colleagues estimated that in 2000 alone, longline fishing operations killed an estimated 50,000 leatherback and 200,000 loggerhead sea turtles. Shrimp trawling fisheries also kill large numbers of leatherback and other sea turtle species.

Pollution is another threat. Sea turtles can mistake discarded plastic bags for jellyfish and choke to death on them. Beachgoers sometimes trample their nests. And artificial lights can disorient hatchlings as they try to find their way to the ocean; going in the wrong direction increases their chances of ending up as food for predators.

Add to this the threat of climate change. Global warming will raise sea levels, which will flood nesting and feeding habitats, and change ocean currents, which could disrupt the turtles' migration routes.

Many people are working to protect the leatherbacks. On some Florida beaches, lights are turned off or blacked out during hatching season. Nesting areas are roped off, and people respect the turtles, according to Safina. Since 1991, the U.S. government has required offshore shrimp trawlers to use turtle excluder devices (TEDs), which help to keep sea turtles out of their nets and to allow caught turtles to escape. TEDs have been adopted in 15 countries that export shrimp to the United States. And, in 2004, the United States banned

Marine Sanctuaries Protect Ecosystems and Species

By international law, a country's offshore fishing zone extends to 370 kilometers (200 statute miles) from its shores. Foreign fishing vessels can take certain quotas of fish within such zones, called *exclusive economic zones,* but only with a government's permission. Ocean areas beyond the legal jurisdiction of any country are known as the *high seas,* and laws and treaties pertaining to them are difficult to monitor and enforce.

Through the Law of the Sea Treaty, the world's coastal nations have jurisdiction over 36% of the ocean surface and 90% of the world's fish stocks. Instead of using this law to protect their fishing grounds, many governments have promoted overfishing, subsidized new fishing fleets, and failed to establish and enforce stricter regulation of fish catches in their coastal waters.

Some countries are attempting to protect marine biodiversity and sustain fisheries by establishing marine sanctuaries. Since 1986, the IUCN has helped to establish a global system of *marine protected areas* (MPAs)—areas of ocean partially protected from human activities. There are more than 4,000 MPAs worldwide, 200 of them in U.S. waters. Despite their name, most MPAs are only partially protected. Nearly all allow dredging, trawler fishing, and other ecologically harmful resource extraction activities. However, the U.S. state

Renatura Congo. www.renatura.osso.eu.org

INDIVIDUALS MATTER

Creating an Artificial Coral Reef in Israel

Near the city of Eliat, Israel, at the northern tip of the Red Sea, is a magnificent coral reef, which is a major tourist attraction. To help protect the reef from excessive development and destructive tourism, Israel set aside part of the reef as a nature reserve. But tourism, industrial pollution, and inadequate sewage treatment have destroyed most of the unprotected part of the reef.

Enter Reuven Yosef, a pioneer in coral reef restoration and reconciliation ecology, who has developed an underwater restaurant called the Red Sea Star Restaurant. Patrons take an elevator down two floors and walk into a restaurant surrounded with windows looking out on a beautiful coral reef.

This reef was created from pieces of broken coral. Typically, when coral breaks, the pieces become infected and die. But researchers have learned how to treat the coral fragments with antibiotics and to store them while they are healing in large tanks of fresh seawater. Yosef has such a facility, and when divers find broken pieces of coral in the reserve near Yosef's restaurant, they bring them to his coral hospital. After several months of healing, the fragments are taken to the underwater area outside the Red Sea Star Restaurant's windows where they are wired to panels of iron mesh cloth. The corals grow and cover the iron matrix. Then fish and other creatures show up.

Similarly, other damaged coral reefs are being restored. In 20 different countries, scientists have increased the revival and growth rate of coral on submerged metal structures by exposing them to low-voltage electricity.

Using his creativity and working with nature, Yosef has helped to create a marine ecosystem that people can view and enjoy while they dine at his restaurant. At the same time, he has helped to restore and preserve aquatic biodiversity.

of California in 2007 began establishing the nation's most extensive network of MPAs where fishing will be banned or strictly limited. Conservation biologists say this could be a model for other MPAs.

Establishing a Global Network of Marine Reserves: An Ecosystem Approach to Marine Sustainability

Many scientists and policymakers call for adopting an entirely new approach to managing and sustaining marine biodiversity and the important ecological and economic services provided by the seas. The primary objective of this *ecosystem approach* is to protect and sustain whole marine ecosystems for current and future generations instead of focusing primarily on protecting individual species.

The cornerstone of this ecological approach is to establish a global network of fully protected *marines reserves,* some of which already exist. These areas are put off-limits to destructive human activities in order to enable their ecosystems to recover and flourish. This global network would include large reserves on the high seas, especially near highly productive nutrient upwelling areas (Figure 7-2, p. 142), and a mixture of smaller reserves in coastal zones that are adjacent to well-managed, sustainable commercial fishing areas. This would encourage local fishers and coastal communities to support them and participate in determining their locations. Some reserves could be made temporary or moveable to protect migrating species such as turtles. Governments could use satellite technologies to update fishing fleets about the locations of designated reserves.

Such reserves would be closed to extractive activities such as commercial fishing, dredging, and mining, as well as to waste disposal. Most reserves in the proposed global network would permit less harmful activities such as recreational boating, shipping, and in some cases, certain levels of small-scale, nondestructive fishing. However, most reserves would contain core zones where no human activity is allowed. Outside the reserves, commercial fisheries would be managed more sustainably by use of an ecosystem approach instead of the current approach, which focuses on individual species without considering their roles in the marine ecosystems where they live.

Marine reserves work and they work fast (see *The Habitable Planet,* Video 9, at **www.learner.org/ resources/series209.html**). Scientific studies show that within fully protected marine reserves, fish populations double, fish size grows by almost a third, reproduction triples, and species diversity increases by almost one-fourth. Furthermore, this improvement occurs within 2–4 years after strict protection begins, and it lasts for decades (**Concept 11-2**). Research also shows that reserves benefit nearby fisheries, because fish move into and out of the reserves, and currents carry fish larvae produced inside reserves to adjacent fishing grounds, thereby bolstering the populations there.

In 2008, the Pacific island nation of Kiribati created the world's largest protected marine reserve. This California-sized area is found about halfway between the Pacific islands of Fiji and Hawaii. In 2006, the United States created the world's second largest protected reserve northwest of the U.S. state of Hawaii. The area is about the size of the U.S. state of Montana and supports more than 7,000 marine species, including the endangered Hawaiian monk seal (Figure 11-5) and the

endangered green sea turtle (see photo on the title page of this book).

Still, less than 1% of the world's oceans are closed to fishing and other harmful human activities in marine reserves and only 0.1% is fully protected—compared to 5% of the world's land. Thus, we have reserved essentially 99.9% of the world's oceans to use as we see fit. Furthermore, many current marine reserves are too small to protect most of the species within them and do not provide adequate protection from illegal fishing or from pollution that flows from the land into coastal waters.

In 2006, a statement signed by 161 leading marine scientists called for urgent action to create a global network of fully protected marine reserves. Many marine scientists call for fully protecting at least 30% of the world's oceans as marine reserves, and some call for protecting up to 50%. They also urge connecting the global network of marine reserves, especially those in coastal waters, with protected corridors. This would also help species to move to different habitats in the process of adapting to the effects of ocean warming, acidification, and many forms of ocean pollution.

Establishing and managing a global network of marine reserves would cost an estimated $12–14 billion a year and create more than 1 million jobs, according to a 2004 study by the World Wildlife Fund International and Great Britain's Royal Society for Protection of Birds. This investment in protecting aquatic biodiversity and regenerating fisheries is roughly equal to the amount currently spent by governments on subsidies for the fishing industry, which conservationists say encourage overfishing.

── RESEARCH FRONTIER ──────────

Determining characteristics and locations of fully protected marine reserves that will maximize their effectiveness. See **academic.cengage.com/biology/miller**.

── THINKING ABOUT ──────────
Marine Reserves

Do you support setting aside at least 30% of the world's oceans as fully protected marine reserves? Explain. How would this affect your life?

Protecting Marine Biodiversity Requires Commitments from Individuals and Communities

There is hope for significant progress in sustaining marine biodiversity, but it will require that we change our ways—and soon. For example, IUCN and The Nature Conservancy scientists reported in 2006 that the world's coral reefs and mangrove forests could survive currently projected global warming if we relieve other stressors such as overfishing and pollution. And while some coral species may be able to adapt to warmer temperatures, they may not have enough time to do this unless we act now to slow down the projected rate of global warming.

Increasing ocean acidity could have a major impact on corals and other marine organisms that build shells and skeletal structures out of calcium carbonate, which can dissolve at certain acidity levels. Increasing ocean acidity is likely to have serious impacts on the biodiversity and functioning of coral reefs. A 2005 report by the United Kingdom's Royal Society concluded that there was no way to reverse the widespread chemical and biological affects of increasing ocean acidification except by sharply reducing human inputs of CO_2 into the atmosphere, without delay.

To deal with these problems, communities must closely monitor and regulate fishing and coastal land development and prevent pollution from land-based activities. More important, each of us can make careful choices in purchasing only sustainably harvested seafood. Coastal residents must also think carefully about the chemicals they put on their lawns, and the kinds of waste they generate and where it ends up. And individuals can reduce their carbon footprints to slow climate change and its numerous harmful effects on marine and other ecosystems, as discussed in more detail in Chapter 19.

One strategy emerging in some coastal communities is *integrated coastal management*—a community-based ef-

Figure 11-11 An atoll of Australia's Great Barrier Reef.

D. Parer & E. Parer-Cook/Ardea

fort to develop and use coastal resources more sustainably (**Concept 11-2**). Australia manages its huge Great Barrier Reef Marine Park this way, and more than 100 integrated coastal management programs are being developed throughout the world. Figure 11-11 shows an atoll of Australia's Great Barrier Reef, which employs integrated coastal management programs.

The overall aim of such programs is for fishers, business owners, developers, scientists, citizens, and politicians to identify shared problems and goals in their use of marine resources. The idea is to develop workable, cost-effective, and adaptable solutions that help to preserve biodiversity and environmental quality while also meeting various economic and social goals.

This requires all participants to seek reasonable short-term trade-offs that can lead to long-term ecological and economic benefits. For example, fishers might have to give up fishing in certain areas until stocks recover enough to restore biodiversity in those areas, which might then provide fishers with a more sustainable future for their businesses.

11-3 How Should We Manage and Sustain Marine Fisheries?

▶ **CONCEPT 11-3** Sustaining marine fisheries will require improved monitoring of fish populations, cooperative fisheries management among communities and nations, reduction of fishing subsidies, and careful consumer choices in seafood markets.

Estimating and Monitoring Fishery Populations Is the First Step

The first step in protecting and sustaining the world's marine fisheries is to make the best possible estimates of their fish populations (**Concept 11-3**). The traditional approach has used a *maximum sustained yield* (MSY) model to project the maximum number of fish that can be harvested annually from a fish stock without causing a population drop. However, the MSY concept has not worked very well because of the difficulty in estimating the populations and growth rates of fish stocks. Also, harvesting a particular species at its estimated maximum sustainable level can affect the populations of other target and nontarget fish species and other marine organisms.

In recent years, some fishery biologists and managers have begun using the *optimum sustained yield* (OSY) concept. It attempts to take into account interactions among species and to provide more room for error. Similarly, another approach is *multispecies management* of a number of interacting species, which takes into account their competitive and predator–prey interactions. An even more ambitious approach is to develop complex computer models for managing multispecies fisheries in *large marine systems*. However, it is a political challenge to get groups of nations to cooperate in planning and managing such large systems.

There are uncertainties built into any of these approaches because there is much to learn about the biology of fishes and because of changing ocean conditions.

As a result, many fishery and environmental scientists are increasingly interested in using the *precautionary principle* (**Concept 9-4C**, p. 210) for managing fisheries and large marine systems. This means sharply reducing fish harvests and closing some overfished areas until they recover and until we have more information about what levels of fishing can be sustained. ◀ CONCEPT LINK

RESEARCH FRONTIER

Studying fish and their habitats to make better estimates of optimum sustained yields for fisheries. See **academic .cengage.com/biology/miller**.

Some Communities Cooperate to Regulate Fish Harvests

An obvious step to take in protecting marine biodiversity—and therefore fisheries—is to regulate fishing. Traditionally, many coastal fishing communities have developed allotment and enforcement systems that have sustained their fisheries, jobs, and communities for hundreds and sometimes thousands of years. An example is Norway's Lofoten fishery, one of the world's largest cod fisheries. For 100 years, it has been self-regulated, with no participation by the Norwegian government. Cooperation can work (**Concept 11-3**).

However, the influx of large modern fishing boats and international fishing fleets has weakened the ability of many coastal communities to regulate and sustain

local fisheries. Community management systems have often been replaced by *comanagement*, in which coastal communities and the government work together to manage fisheries.

In comanagement, a central government typically sets quotas for various species and divides the quotas among communities. The government may also limit fishing seasons and regulate the types of fishing gear that can be used to harvest a particular species. Each community then allocates and enforces its quota among its members based on its own rules. Often communities focus on managing inshore fisheries, and the central government manages the offshore fisheries. When it works, community-based comanagement proves that overfishing is not inevitable.

Government Subsidies Can Encourage Overfishing

A 2006 study by fishery experts U. R. Sumaila and Daniel Pauly estimated that governments around the world give a total of about $30–34 billion per year to fishers to help them keep their businesses running. That represents about a third of all revenues earned through commercial fishing. Of that amount, about $20 billion helps fishers to buy ships, fuel, and fishing equipment; the remaining money pays for research and management of fisheries.

Some marine scientists argue that, each year, $10–14 billion of these subsidies are spent to encourage overfishing and expansion of the fishing industry. At a 2007 meeting of the World Trade Organization, the United States proposed a ban on such subsidies. Actions to slash fishing subsidies were supported by a group of 125 marine scientists from 27 countries.

Many marine scientists also call for stronger global efforts to reduce illegal fishing on the high seas and in coastal waters. Actions could include closing ports and markets to such fishers, checking on the authenticity of ship flags, and prosecuting companies that carry out illegal fishing.

THINKING ABOUT
Fishing Subsidies

What are three possible harmful effects of eliminating government fishing subsidies? Do you think they outweigh the benefits of such an action? Explain.

Some Countries Use the Marketplace to Control Overfishing

Some countries use a market-based system called *individual transfer rights* (ITRs) to control access to fisheries. In such a system, the government gives each fishing

vessel owner a specified percentage of the total allowable catch (TAC) for a fishery in a given year. Owners are permitted to buy, sell, or lease their fishing rights as private property.

The ITR market-based system was introduced in New Zealand in 1986 and in Iceland in 1990. In these countries, there has been some reduction in overfishing and in the sizes of their fishing fleets, and the governments have ended fishing subsidies that encourage overfishing. But enforcement has been difficult, some fishers illegally exceed their quotas, and the wasteful bycatch has not been reduced.

In 1995, the United States introduced tradable quotas to regulate Alaska's halibut fishery, which had declined so much that the fishing season had been cut to only 2 days per year. Some fishers sold their quotas and retired, and the number of fishers declined. Halibut prices and fisher income rose, and with less pressure from the fishing industry, the halibut population recovered. By 2005, the season was 258 days long.

Critics have identified three problems with the ITR approach and have made suggestions for its improvement. *First,* in effect, it transfers ownership of fisheries in publicly owned waters to private commercial fishers but still makes the public responsible for the costs of enforcing and managing the system. Critics suggest collecting fees of up to 5% of the value of any catch from quota holders to pay for these costs.

Second, an ITR system can squeeze out small fishing companies that do not have the capital to buy ITRs from others, and it can promote illegal fishing by companies squeezed out of the market. For example, 20 years after the ITR system was implemented in New Zealand, five companies controlled 85% of the ITRs. Critics suggest limiting the number of rights that any one company can obtain.

Third, TACs are often set too high to prevent overfishing. Scientists argue the limit should be set at 50–90% of the estimated *optimal* sustainable yield. Most fishing industry interests oppose setting stricter rules for ITR systems.

THINKING ABOUT
Individual Transfer Rights

Do you support or oppose widespread use of ITR systems to help control access to fisheries? Explain.

Consumer Choices Can Help to Sustain Fisheries and Aquatic Biodiversity

An important component of sustaining aquatic biodiversity and ecosystem services is bottom-up pressure from consumers demanding *sustainable seafood,* which will encourage more responsible fishing practices. In

choosing seafood in markets and restaurants, consumers can make choices that will further help to sustain fisheries (**Concept 11-3**).

One way to enable this is through labeling of fresh and frozen seafood to inform consumers about how and where the fish and shellfish were caught. In the United Kingdom, the Waitrose supermarket food chain provides such information for all of the seafood sold at its fresh fish counters. See information on more sustainable seafood choices and download a convenient pocket guide at **www.seafoodwatch.org**.

Another important component is certification of sustainably caught seafood. The London-based Marine Stewardship Council (MSC) was created in 1997 to support sustainable fishing and to certify sustainably produced seafood. It operates in more than 20 nations. Only certified fisheries are allowed to use the MSC's "Fish Forever" eco-label. This certification shows that the fish were caught using environmentally sound and socially responsible practices. By 2007, some 21 wild capture fisheries worldwide were MSC-certified, 18 more were being assessed, and more than 600 seafood products were available with the MSC eco-label. Even so, by 2007 only about 6% of the world's wild capture fisheries were certified.

In 2006, Wal-Mart, the world's largest food retailer, pledged to sell only "MSC-certified" wild-caught fresh and frozen fish in North America within 3–5 years. If implemented, this will have a significant impact on the sustainability of the fresh and frozen seafood market.

Figure 11-12 summarizes actions that individuals, organizations, and governments can take to manage global fisheries more sustainably and to protect marine biodiversity and ecosystem services.

SOLUTIONS

Managing Fisheries

Fishery Regulations

Set catch limits well below the maximum sustainable yield

Improve monitoring and enforcement of regulations

Economic Approaches

Sharply reduce or eliminate fishing subsidies

Charge fees for harvesting fish and shellfish from publicly owned offshore waters

Certify sustainable fisheries

Protect Areas

Establish no-fishing areas

Establish more marine protected areas

Rely more on integrated coastal management

Consumer Information

Label sustainably harvested fish

Publicize overfished and threatened species

Bycatch

Use wide-meshed nets to allow escape of smaller fish

Use net escape devices for seabirds and sea turtles

Ban throwing edible and marketable fish back into the sea

Aquaculture

Restrict coastal locations for fish farms

Control pollution more strictly

Depend more on herbivorous fish species

Nonnative Invasions

Kill organisms in ship ballast water

Filter organisms from ship ballast water

Dump ballast water far at sea and replace with deep-sea water

Figure 11-12 Ways to manage fisheries more sustainably and protect marine biodiversity and ecosystem services. **Question:** Which four of these solutions do you think are the most important? Why?

11-4 How Should We Protect and Sustain Wetlands?

▶ **CONCEPT 11-4** To maintain the ecological and economic services of wetlands, we must maximize preservation of remaining wetlands and restoration of degraded and destroyed wetlands.

Coastal and Inland Wetlands Are Disappearing around the World

Coastal and inland wetlands are important reservoirs of aquatic biodiversity that provide vital ecological and economic services. Despite their ecological value, the United States has lost more than half of its coastal and inland wetlands since 1900, and other countries have lost even more. New Zealand, for example, has lost

92% of its original coastal wetlands, and Italy has lost 95%.

People have drained, filled in, or covered over swamps, marshes, and other wetlands for centuries to create rice fields and to make land available for growing crops, expanding cities, and building roads. Wetlands have also been destroyed in the process of extracting minerals, oil, and natural gas, and in order to reduce diseases such as malaria by eliminating breeding grounds for disease-causing insects.

Wetlands serve as natural filters. Those around Lake Victoria (**Core Case Study**) have historically captured human and animal wastes and kept the lake water clean enough to be used as drinking water for millions of Africans. In 2006, the director of Uganda's wetlands program reported that extensive draining and building on Lake Victoria's coastal wetlands had led to serious water pollution that was killing fish and contaminating drinking water supplies for several countries. He noted that as the waste flow increases, still more wetlands are being destroyed. The Ugandan government is now working to protect its remaining wetlands.

To make matters worse, coastal wetlands in many parts of the world will probably be under water during your lifetime because of rising sea levels caused by global warming. This could seriously degrade aquatic biodiversity supported by coastal wetlands, including commercially important fishes and shellfish and millions of migratory ducks and other birds. It will also diminish the many other ecological and economic services provided by these wetlands.

We Can Preserve and Restore Wetlands

Scientists, land managers, landowners, and environmental groups are involved in intensive efforts to preserve existing wetlands and restore degraded ones (**Concept 11-4**). Laws have been passed to protect existing wetlands. Zoning laws, for example, can be used to steer development away from wetlands.

A U.S. law requires a federal permit to fill in or to deposit dredged material into wetlands occupying more than 1.2 hectares (3 acres). According to the U.S. Fish and Wildlife Service, this law helped cut the average annual wetland loss by 80% since 1969. However, there are continuing attempts by land developers to weaken such wetlands protection. Only about 6% of remaining U.S. inland wetlands are under federal protection, and state and local wetland protection is inconsistent and generally weak.

The stated goal of current U.S. federal policy is zero net loss in the function and value of coastal and inland wetlands. A policy known as *mitigation banking* allows destruction of existing wetlands as long as an equal area of the same type of wetland is created or restored. However, a 2001 study by the National Academy of Sciences found that at least half of the attempts to create new wetlands failed to replace lost ones, and most of the created wetlands did not provide the ecological functions of natural wetlands. The study also found that wetland creation projects often fail to meet the standards set for them and are not adequately monitored.

Creating and restoring wetlands can be profitable. Private investment bankers make money by buying wetland areas and restoring or upgrading them, working with the U.S. Army Corps of Engineers and the EPA. They thus create wetland banks or credits that they can sell to developers. Currently, there are more than 400 wetland banks in the United States with a total of more than $3 billion a year in sales.

It is difficult to restore or create wetlands. Thus, most U.S. wetland banking systems require replacing each hectare of destroyed wetland with 2–3 or more hectares of restored or created wetlands as a built-in ecological insurance policy. **GREEN CAREER:** Wetlands restoration expert

Ecologists argue that mitigation banking should be used only as a last resort. They also call for making sure that new replacement wetlands are created and evaluated *before* existing wetlands are to be destroyed. This example of applying the precautionary principle is often the reverse of what is actually done.

INDIVIDUALS MATTER

Restoring a Wetland

As we learn more about the ecological and economic importance of coastal and inland wetlands, some people have begun to question common practices that damage or destroy these ecosystems. Can we turn back the clock to restore or rehabilitate lost wetlands?

California rancher Jim Callender decided to try. In 1982, he bought 20 hectares (50 acres) of a Sacramento Valley rice field that had been a marsh until the early 1970s. To grow rice, the previous owner had destroyed the marsh by bulldozing, draining, and leveling it, uprooting the native plants, and spraying with chemicals to kill snails.

Callender and his friends set out to restore the marsh. They hollowed out low areas, built up islands, replanted bulrushes and other plants that once were there, reintroduced smartweed and other plants used by migrating and marsh-dwelling birds, and planted fast-growing Peking willows. After years of care, hand planting, and annual seeding with a mixture of watergrass, smartweed, and rice, the land is once again a marsh used by migratory waterfowl.

Jim Callender and others have shown that at least some of the continent's degraded or destroyed wetlands can be reclaimed with scientific knowledge and hard work. Such restoration is useful, but to most ecologists, the real challenge is to protect remaining wetlands from harm in the first place.

Figure 11-13 **Natural capital restoration:** wetland restoration at Fort Erie, Ontario, Canada before (right) and after (left).

<div style="border:1px solid #000; padding:8px;">
THINKING ABOUT
Wetlands Mitigation

Should a new wetland be created and evaluated before anyone is allowed to destroy the wetland it is supposed to replace? Explain.
</div>

Wetlands restoration is becoming a big business. While some wetlands restoration projects have failed, others have been very successful (Figure 11-13 and Individuals Matter, at left).

RESEARCH FRONTIER

Evaluating ecological services provided by wetlands, human impacts on wetlands, and how to preserve and restore wetlands. See **academic.cengage.com/biology/miller**.

A good example of an attempt to restore a once vast wetland is that of the Everglades in the U.S. state of Florida, as described in the following case study.

■ CASE STUDY
Can We Restore the Florida Everglades?

South Florida's Everglades (USA) was once a 100-kilometer-wide (60-mile-wide), knee-deep sheet of water flowing slowly south from Lake Okeechobee to Florida Bay (Figure 11-14, p. 268). As this shallow body of water—known as the "River of Grass"—trickled south it created a vast network of wetlands with a variety of wildlife habitats.

Since 1948, a massive water control project has provided south Florida's rapidly growing population with a reliable water supply and flood protection. But is has also contributed to widespread degradation of the original Everglades ecosystem.

Much of the original Everglades has been drained, diverted, paved over, ravaged by nutrient pollution from agriculture, and invaded by a number of plant species. As a result, the Everglades is now less than half its original size. Much of it has also dried out, leaving large areas vulnerable to summer wildfires. And much of its biodiversity has been lost because of reduced water flows, invasive species, and habitat loss and fragmentation from urbanization.

Between 1962 and 1971, the U.S. Army Corps of Engineers transformed the wandering 166-kilometer-long (103-mile-long) Kissimmee River (Figure 11-14, p. 268) into a straight 84-kilometer (56-mile) canal flowing into Lake Okeechobee. The canal provided flood control by speeding the flow of water but it drained large wetlands north of Lake Okeechobee, which farmers then turned into cow pastures.

To help preserve the wilderness in the lower end of the Everglades system, in 1947, the U.S. government established Everglades National Park, which contains about a fifth of the remaining Everglades. But this protection effort did not work—as conservationists had predicted—because the massive water distribution and land development project to the north cut off much of the water flow needed to sustain the park's wildlife.

As a result, 90% of the park's wading birds have vanished, and populations of other vertebrates, from deer to turtles, are down 75–95%. Florida Bay, south of the Everglades is a shallow estuary with many tiny islands, or *keys*. Large volumes of freshwater that once

Lake Kissimmee

Kissimmee River

Lake Istokpoga

FLORIDA

Caloosahatchee River

Lake Okeechobee

Atlantic Ocean

West Palm Beach

Fort Meyers

Gulf of Mexico

Naples

Fort Lauderdale

Miami

- Agricultural area
- Water conservation area
— Canal
--- Channelized river bed

Everglades National Park

Key Largo

FLORIDA

Area of detail

Florida Bay

| 0 | 20 | 40 | 60 miles |
| 0 | 20 | 40 | 60 kilometers |

Figure 11-14 The world's largest ecological restoration project is an attempt to undo and redo an engineering project that has been destroying Florida's Everglades (USA) and threatening water supplies for south Florida's rapidly growing population.

flowed through the park into Florida Bay have been diverted for crops and cities, causing the bay to become saltier and warmer. This, along with increased nutrient input from crop fields and cities, has stimulated the growth of large algal blooms that sometimes cover 40% of the bay. This has threatened the coral reefs and the diving, fishing, and tourism industries of the bay and the Florida Keys—another example of harmful unintended consequences.

By the 1970s, state and federal officials recognized that this huge plumbing project was reducing wildlife populations—a major source of tourism income for Florida—and cutting the water supply for the 6 million residents of south Florida. After more than 20 years of political haggling, in 1990, Florida's state government and the federal government agreed on the world's largest ecological restoration project, known as the Comprehensive Everglades Restoration Plan (CERP). The U.S. Army Corps of Engineers is supposed to carry out this joint federal and state plan to partially restore the Everglades.

The project has several ambitious goals. *First,* restore the curving flow of more than half of the Kissimmee

River. *Second,* remove 400 kilometers (250 miles) of canals and levees blocking water flow south of Lake Okeechobee. *Third,* buy 240 square kilometers (93 square miles) of farmland and allow it to be flooded to create artificial marshes that will filter agricultural runoff before it reaches Everglades National Park. *Fourth,* create 18 large reservoirs and underground water storage areas to ensure an adequate water supply for south Florida's current and projected population and for the lower Everglades. *Fifth,* build new canals, reservoirs, and huge pumping systems to capture 80% of the water currently flowing out to sea and return it to the Everglades.

Will this huge ecological restoration project work? It depends not only on the abilities of scientists and engineers but also on prolonged political and economic support from citizens, the state's powerful sugarcane and agricultural industries, and elected state and federal officials.

The carefully negotiated plan has begun to unravel. In 2003, sugarcane growers persuaded the Florida legislature to increase the amount of phosphorus they could discharge and to extend the deadline for reducing such discharges from 2006 to 2016. The project had originally been estimated to cost $7.8 billion and to take 30 years. By 2007, the price tag had risen to $10.5 billion and was expected to go much higher, mostly because of an almost tenfold increase in land prices in South Florida between 2000 and 2007. Overall, funding for the project, especially federal funding, has fallen short of the projected needs, and federal and state agencies are far behind on almost every component of the project. Now the project could take 50 years to complete, or it could be abandoned because of a lack of funding.

According to critics, the main goal of the Everglades restoration plan is to provide water for urban and agricultural development with ecological restoration as a secondary goal. Also, the plan does not specify how much of the water rerouted toward south and central Florida will go to the parched park instead of to increased industrial, agricultural, and urban development. And a National Academy of Sciences panel has found that the plan would probably not clear up Florida Bay's nutrient enrichment problems.

The need to make expensive and politically controversial efforts to undo some of the ecological damage done to the Everglades, caused by 120 years of agricultural and urban development, is another example of failure to heed two fundamental lessons from nature: prevention is the cheapest and best way to go; and when we intervene in nature, unintended and often harmful consequences always occur.

┌─ **THINKING ABOUT**
Everglades Restoration

Do you support carrying out the proposed plan for partially restoring the Florida Everglades, including having the federal government provide half of the funding? Explain.

11-5 How Can We Protect and Sustain Freshwater Lakes, Rivers, and Fisheries?

> **CONCEPT 11-5** Freshwater ecosystems are strongly affected by human activities on adjacent lands, and protecting these ecosystems must include protection of their watersheds.

Freshwater Ecosystems Are under Major Threats

The ecological and economic services provided by freshwater lakes, rivers, and fisheries (Figure 8-14, p. 174) are severely threatened by human activities (**Concept 8-5**).

Again, we can use the acronym HIPPCO to summarize these threats. As 40% of the world's rivers have been dammed or otherwise engineered, and as vast portions of the world's freshwater wetlands have been destroyed, aquatic species have been crowded out of at least half of their habitat areas, worldwide. Invasive species, pollution, and climate change threaten the ecosystems of lakes (Case Study, below), rivers, and wetlands. Freshwater fish stocks are overharvested. And increasing human population pressures and global warming make these threats worse.

Sustaining and restoring the biodiversity and ecological services provided by freshwater lakes and rivers is a complex and challenging task, as shown by the story of Lake Victoria (**Core Case Study**) as well as by the following Case Study.

■ CASE STUDY

Can the Great Lakes Survive Repeated Invasions by Alien Species?

Invasions by nonnative species is a major threat to the biodiversity and ecological functioning of lakes, as illustrated by what has happened to the five Great Lakes, located between the United States and Canada.

Collectively, the Great Lakes are the world's largest body of fresh water. Since the 1920s, they have been invaded by at least 162 nonnative species, and the number keeps rising. Many of the alien invaders arrive on the hulls or in bilge water discharges of oceangoing ships that have been entering the Great Lakes through the St. Lawrence Seaway for almost 50 years.

One of the biggest threats, the *sea lamprey*, reached the western lakes through the Welland Canal in Canada as early as 1920. This parasite attaches itself to almost any kind of fish and kills the victim by sucking out its blood (Figure 5-4b, p. 105). Over the years it has depleted populations of many important sport fish species such as lake trout. The United States and Canada keep the lamprey population down by applying a chemical that kills lamprey larvae in their spawning streams—at a cost of about $15 million a year.

In 1986, larvae of the *zebra mussel* (Figure 9-14, p. 199) arrived in ballast water discharged from a European ship near Detroit, Michigan (USA). This thumbnail-sized mollusk reproduces rapidly and has no known natural enemies in the Great Lakes. As a result, it has displaced other mussel species and depleted the food supply for some other Great Lakes species. The mussels have also clogged irrigation pipes, shut down water intake pipes for power plants and city water supplies, and fouled beaches. They have jammed ship rudders and grown in huge masses on boat hulls, piers, pipes, rocks, and almost any exposed aquatic surface (Figure 11-15). This mussel has also spread to freshwater communities in parts of southern Canada and 18 U.S. states. Currently, the mussels cost the two

Figure 11-15 *Zebra mussels* attached to a water current meter in Lake Michigan. This invader entered the Great Lakes through ballast water dumped from a European ship. It has become a major nuisance and a threat to commerce as well as to biodiversity in the Great Lakes.

countries about $140 million a year—an average of $16,000 per hour.

Sometimes, nature aids us in controlling an invasive alien species. For example, populations of zebra mussels are declining in some parts of the Great Lakes because a native sponge growing on their shells is preventing them from opening up their shells to breathe. However, it is not clear whether the sponges will be effective in controlling the invasive mussels in the long run.

Zebra mussels may not be good for some fish species or for us, but they can benefit a number of aquatic plants. By consuming algae and other microorganisms, the mussels increase water clarity, which permits deeper penetration of sunlight and more photosynthesis. This allows some native plants to thrive and could return the plant composition of Lake Erie (and presumably other lakes) closer to what it was 100 years ago. Because the plants provide food and increase dissolved oxygen, their comeback may benefit certain aquatic animals.

In 1989, a larger and potentially more destructive species, the *quagga mussel,* invaded the Great Lakes, probably discharged in the ballast water of a Russian freighter. It can survive at greater depths and tolerate more extreme temperatures than the zebra mussel can. There is concern that it may spread by river transport and eventually colonize eastern U.S. ecosystems such as Chesapeake Bay and waterways in parts of Florida. In 2007, it was found to have crossed the United States, probably hitching a ride on a boat or trailer being hauled cross-country. It now resides in the Colorado River and reservoir system.

The *Asian carp* may be the next invader. These highly prolific fish, which can quickly grow as long as 1.2 meters (4 feet) and weigh up to 50 kilograms (110 pounds), have no natural predators in the Great Lakes. In less than a decade, this hearty fish with a voracious appetite has dominated sections of the Mississippi River and its tributaries and is spreading toward the Great Lakes. The only barriers are a few kilometers of waterway and a little-tested underwater electric barrier spanning a canal near Chicago, Illinois.

THINKING ABOUT

Invasive Species in Lakes

CORE CASE STUDY

What role did invasive species play in the degradation of Lake Victoria (**Core Case Study**)? What are three ways in which people could avoid introducing more harmful invasive species into lakes?

Managing River Basins Is Complex and Controversial

Rivers and streams provide important ecological and economic services (Figure 11-16). But overfishing, pollution, dams, and water withdrawal for irrigation disrupt these services.

An example of such disruption—one that especially illustrates biodiversity loss—is what happened in the

NATURAL CAPITAL

Ecological Services of Rivers

- Deliver nutrients to sea to help sustain coastal fisheries

- Deposit silt that maintains deltas

- Purify water

- Renew and renourish wetlands

- Provide habitats for wildlife

Figure 11-16 Important ecological services provided by rivers. Currently, these services are given little or no monetary value when the costs and benefits of dam and reservoir projects are assessed. According to environmental economists, attaching even crudely estimated monetary values to these ecosystem services would help to sustain them. **Questions:** Which two of these services do you believe are the most important? Why? Which two of these services do you think we are most likely to decline? Why?

Columbia River, which runs through parts of southwestern Canada and the northwestern United States. It has 119 dams, 19 of which are major generators of inexpensive hydroelectric power. It also supplies water for several major urban areas and for irrigating large areas of agricultural land.

The Columbia River dam system has benefited many people, but it has sharply reduced populations of wild salmon. These migratory fish hatch in the upper reaches of streams and rivers, migrate to the ocean where they spend most of their adult lives, and then swim upstream to return to the places where they were hatched to spawn and die. Dams interrupt their life cycle.

Since the dams were built, the Columbia River's wild Pacific salmon population has dropped by 94% and nine Pacific Northwest salmon species are listed as endangered or threatened. Since 1980, the U.S. federal government has spent more than $3 billion in efforts to save the salmon, but none have been effective.

In another such case—on the lower Snake River in the U.S. state of Washington—conservationists, Native American tribes, and commercial salmon fishers want the government to remove four small hydroelectric dams to restore salmon spawning habitat. Farmers, barge operators, and aluminum workers argue that removing the dams would hurt local economies by reducing irrigation water, eliminating shipping in the affected areas, and reducing the supply of cheap electricity for industries and consumers.

HOW WOULD YOU VOTE?

Should U.S. government efforts to rebuild wild salmon populations in the Columbia River Basin be abandoned? Cast your vote online at **academic.cengage.com/biology/ miller.**

We Can Protect Freshwater Ecosystems by Protecting Watersheds

Sustaining freshwater aquatic systems begins with our realizing that whatever each of us does on land and in the water has some effect on those systems (Concept 11-5).

In other words, land and water are always connected in some way. For example, lakes and streams receive many of their nutrients from the ecosystems of bordering land. Such nutrient inputs come from falling leaves, animal feces, and pollutants generated by people, all of which are washed into bodies of water by rainstorms and melting snow. Therefore, to protect a stream or lake from excessive inputs of nutrients and pollutants, we must protect its watershed.

As with marine systems, freshwater ecosystems can be protected through laws, economic incentives, and restoration efforts. For example, restoring and sustaining the ecological and economic services of rivers will probably require taking down some dams and restoring river flows, as may be the case with the Snake River, as mentioned above. And some scientists and politicians have argued for protecting all remaining free-flowing rivers.

With that in mind, in 1968, the U.S. Congress passed the National Wild and Scenic Rivers Act to establish protection of rivers with outstanding scenic, recreational, geological, wildlife, historical, or cultural values. The law classified *wild rivers* as those that are relatively inaccessible (except by trail), and *scenic rivers* as rivers of great scenic value that are free of dams, mostly undeveloped, and accessible in only a few places by roads. These rivers are now protected from widen-ing, straightening, dredging, filling, and damming. But the Wild and Scenic Rivers System keeps only 2% of U.S. rivers free-flowing and protects only 0.2% of the country's total river length.

Sustainable management of freshwater fishes involves supporting populations of commercial and sport fish species, preventing such species from being overfished, and reducing or eliminating populations of harmful invasive species. The traditional way of managing freshwater fish species is to regulate the time and length of fishing seasons and the number and size of fish that can be taken.

Other techniques include building reservoirs and farm ponds and stocking them with fish, fertilizing nutrient-poor lakes and ponds, and protecting and creating fish spawning sites. In addition, fishery managers can protect fish habitats from sediment buildup and other forms of pollution and from excessive growth of aquatic plants due to large inputs of plant nutrients.

Some fishery managers seek to control predators, parasites, and diseases by improving habitats, breeding genetically resistant fish varieties, and using antibiotics and disinfectants. Hatcheries can be used to restock ponds, lakes, and streams with prized species such as trout, and entire river basins can be managed to protect valued species such as salmon. However, all of these practices should be based on on-going studies of their effects on aquatic ecosystems and biodiversity. **GREEN CAREERS:** limnology, fishery management, and wildlife biology

RESEARCH FRONTIER

Studying the effects of resource management techniques on aquatic ecosystems. See **academic.cengage.com/biology/ miller**.

11-6 What Should Be Our Priorities for Sustaining Biodiversity and Ecosystem Services?

▶ **CONCEPT 11-6** Sustaining the world's biodiversity and ecosystem services will require mapping terrestrial and aquatic biodiversity, maximizing protection of undeveloped terrestrial and aquatic areas, and carrying out ecological restoration projects worldwide.

We Need to Establish Priorities for Protecting Biodiversity and Ecosystem Services

In 2002, Edward O. Wilson, considered to be one of the world's foremost experts on biodiversity, proposed the following priorities for protecting most of the world's remaining ecosystems and species (Concept 11-6):

- Complete the mapping of the world's terrestrial and aquatic biodiversity so we know what we have and therefore can make conservation efforts more precise and cost-effective.

- Keep intact the world's remaining old-growth forests and cease all logging of such forests.

- Identify and preserve the world's terrestrial and aquatic biodiversity hotspots and areas where

deteriorating ecosystem services threaten people and many other forms of life.

- Protect and restore the world's lakes and river systems, which are the most threatened ecosystems of all.

- Carry out ecological restoration projects worldwide to heal some of the damage we have done and to increase the share of the earth's land and water allotted to the rest of nature.

- Find ways to make conservation financially rewarding for people who live in or near terrestrial and aquatic reserves so they can become partners in the protection and sustainable use of the reserves.

There is growing evidence that the current harmful effects of human activities on the earth's terrestrial and aquatic biodiversity and ecosystem services could be reversed over the next 2 decades. Doing this will require implementing an ecosystem approach to protecting and sustaining terrestrial and aquatic ecosystems. According to biologist Edward O. Wilson, such a conservation strategy would cost about $30 billion per year—an amount that could be provided by a tax of one penny per cup of coffee consumed in the world each year.

This strategy for protecting the earth's precious biodiversity will not be implemented without bottom-up political pressure on elected officials from individual citizens and groups. People will also have to vote with their wallets by not buying products and services that destroy or degrade biodiversity. Finally, implementing this strategy will require concerted efforts and cooperation among scientists, engineers, and key people in government and the private sector.

REVISITING Lake Victoria and Sustainability

This chapter began with a look at how human activities have upset the ecological processes of Africa's Lake Victoria (**Core Case Study**).

Lake Victoria and other cases examined in this chapter illustrate the significant human impacts that have contributed to habitat loss, the spread of invasive species, pollution, climate change, and depletion of commercially valuable fish populations, as well as degradation of aquatic biodiversity in general. We have seen that these threats are growing and are even greater than threats to terrestrial biodiversity.

We also explored ways to manage the world's oceans, fisheries, wetlands, lakes, and rivers more sustainably by applying the four **scientific principles of sustainability**. This means reducing inputs of sediments and excess nutrients, which cloud water, lessen the input of solar energy, and upset the natural cycling of nutrients in aquatic systems. It means placing a high priority on preserving the biodiversity and ecological functioning of aquatic systems and on maintaining natural species interactions that help to prevent excessive population growth of any one species, as happened in Lake Victoria.

*By treating the oceans with more respect
and by using them more wisely,
we can obtain more from these life-supporting waters
while also maintaining healthy
and diverse marine ecosystems.*

BRIAN HALWEIL

REVIEW

1. Review the Key Questions and Concepts for this chapter on p. 250. Describe how human activities have upset ecological processes in East Africa's Lake Victoria (**Core Case Study**).

2. What are three general patterns of marine biodiversity? Why is marine biodiversity higher **(a)** near coasts than in the open sea and **(b)** on the ocean's bottom than at its surface? Describe the threat to marine biodiversity from bottom trawling. Give two examples of threats to aquatic systems from invasive species. Describe the eco-logical experiment involving carp removal in Wisconsin's Lake Wingra. How does climate change threaten aquatic biodiversity?

3. What is a **fishprint?** Describe the collapse of the cod fishery in the northwest Atlantic and some of its side effects. Describe the effects of trawler fishing, purse-seine fishing, longlining, and drift-net fishing.

4. How have laws and treaties been used to help sustain aquatic species? Describe international efforts to protect

whales from overfishing and premature extinction. Describe threats to sea turtles and efforts to protect them.

5. Describe the use of marine protected areas and marine reserves to help sustain aquatic biodiversity and ecosystem services. What percentage of the world's oceans is fully protected from harmful human activities in marine reserves? Describe the roles of fishing communities and individual consumers in regulating fishing and coastal development. What is integrated coastal management?

6. Describe and discuss the limitations of three ways to estimate the sizes of fish populations. How can the precautionary principle help in managing fisheries and large marine systems? Describe the efforts of local fishing communities in helping to sustain fisheries. How can government subsidies encourage overfishing? Describe the advantages and disadvantages of using individual transfer rights to help manage fisheries.

7. Describe how consumers can help to sustain fisheries, aquatic biodiversity, and ecosystem services by making careful choices in purchasing seafood.

8. What percentage of the U.S. coastal and inland wetlands has been destroyed since 1900? What are three major ecological services provided by wetlands? How does the United States attempt to reduce wetland losses? Describe efforts to restore the Florida Everglades.

9. Describe the major threats to the world's rivers and other freshwater systems. What major ecological services do rivers provide? Describe invasions of the U.S. Great Lakes by nonnative species. Describe ways to help sustain rivers.

10. What are six priorities for protecting terrestrial and aquatic biodiversity? Relate the ecological problems of Lake Victoria (**Core Case Study**) to the four **scientific principles of sustainability**.

Note: Key Terms are in **bold** type.

CRITICAL THINKING

1. Explain how introducing the Nile perch into Lake Victoria (**Core Case Study**) violated all four **scientific principles of sustainability** (see back cover).

2. What difference does it make that the introduction of the Nile perch into Lake Victoria (**Core Case Study**) caused the extinction of more than 200 cichlid fish species? Explain.

3. What do you think are the three greatest threats to aquatic biodiversity and ecosystem services? Why? Why are aquatic species overall more vulnerable to premature extinction resulting from human activities than terrestrial species are? Why is it more difficult to identify and protect endangered marine species than to protect endangered species on land?

4. Why do you think no-fishing marine reserves recover their biodiversity faster and more surely than do areas where fishing is allowed but restricted?

5. Should fishers who harvest fish from a country's publicly owned waters be required to pay the government (taxpayers) fees for the fish they catch? Explain. If your livelihood depended on commercial fishing, would you be for or against such fees?

6. Why do you think that about half of all attempts to create new wetlands fail to replace lost wetlands? Give three reasons why a constructed wetland might not provide the same level of ecological services as a natural wetland. Do you agree with some ecologists' argument that mitiga-

tion wetland banking should be used only as a last resort? Explain.

7. Do you think the plan for restoring Florida's Everglades will succeed? Give three reasons why or why not?

8. Dams on some rivers provide inexpensive hydroelectric power, but they also disrupt aquatic ecosystems. For example, production of hydroelectric power on the Columbia River has resulted in the degradation of the river's Pacific salmon population. Do you think the benefits of these dams justify the ecological damage they cause? Explain. If you see this as a problem, describe a possible solution.

9. Congratulations! You are in charge of protecting the world's aquatic biodiversity and ecosystem services. List the three most important points of your policy to accomplish this goal.

10. List two questions that you would like to have answered as a result of reading this chapter.

Note: See Supplement 13 (p. S78) for a list of Projects related to this chapter.

ECOLOGICAL FOOTPRINT ANALYSIS

A *fishprint* provides a measure of a country's fish harvest in terms of area. The unit of area used in fishprint analysis is the global hectare (gha), a unit weighted to reflect the relative ecological productivity of the area fished. When compared with the fishing area's sustainable *biocapacity*, its ability to provide a stable supply of fish year after year in terms of area, its fishprint indicates whether the country's fishing intensity is sustainable. The fishprint and biocapacity are calculated using the following formulae:

$$\text{Fishprint (in gha)} = \frac{\text{metric tons of fish harvested per year}}{\text{productivity in metric tons per hectare} \times \text{weighting factor}}$$

$$\text{Biocapacity (in gha)} = \frac{\text{sustained yield of fish in metric tons per year}}{\text{productivity in metric tons per hectare} \times \text{weighting factor}}$$

The following graph shows the earth's total fishprint and biocapacity. Study it and answer the following questions.

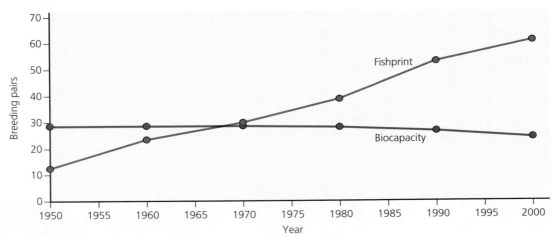

1. Based on the graph,
 a. What is the current status of the global fisheries with respect to sustainability?
 b. In what year did the global fishprint begin exceeding the biological capacity of the world's oceans?
 c. By how much did the global fishprint exceed the biological capacity of the world's oceans in 2000?

2. Assume a country harvests 18 million metric tons of fish annually from an ocean area with an average productivity of 1.3 metric tons per hectare and a weighting factor of 2.68. What is the annual fishprint of that country?

3. If biologists determine that this country's sustained yield of fish is 17 million metric tons per year,
 a. What is the country's sustainable biological capacity?
 b. Is the county's fishing intensity sustainable?
 c. To what extent, as a percentage, is the country under- or overshooting its biological capacity?

LEARNING ONLINE

Log on to the Student Companion Site for this book at **academic.cengage.com/biology/miller**, and choose Chapter 11 for many study aids and ideas for further reading and research. These include flash cards, practice quizzing, Weblinks, information on Green Careers, and InfoTrac® College Edition articles.

Food, Soil, and Pest Management

Golden Rice: Grains of Hope or an Illusion?

Many of the world's poor do not have enough land or money to obtain foods that give them enough protein and essential vitamins and minerals to prevent malnutrition. For example, according to the World Health Organization (WHO), 120–140 million children in developing countries, mostly in Africa and Southeast Asia, do not get enough vitamin A. This makes them more susceptible to common childhood infectious diseases. Some 250,000 to 500,000 children younger than age 6 go blind each year from a lack of vitamin A, and within a year, more than half of them die.

Preschool children can get their daily requirement of vitamin A by eating a small mango or small amounts of vegetables, yellow sweet potatoes, or coriander. But these foods are too expensive for most poor people to grow or buy. Children can also be given two vitamin A capsules a year, but so far, there has been no large-scale effort to provide such supplements to the world's poor.

In 1999, scientists Ingo Potrykus and Peter Beyer decided to tackle this problem by genetically engineering a form of rice that contained beta-carotene, a substance that the body can convert to vitamin A. They transferred snippets of DNA from genes of the common daffodil (Figure 12-1, right) and from a soil bacterium into conventional rice strains to produce a strain of rice containing beta-carotene. It also contains more iron than do conventional rice strains.

Potrykus and Beyer estimate that eating 200–300 grams (about 1 to 1.5 cups) of their latest golden rice strain per day should provide enough vitamin A to prevent blindness and susceptibility to common childhood infections. They have worked out agreements for poor subsistence farmers in several developing countries to get the new strain free of charge. The International Rice Research Institute is conducting field trials of golden rice and commercial stains of this rice may be released by 2011. However, there is controversy over the new rice strain, as we discuss later in this chapter.

We face important challenges to increase food production without causing serious environmental harm. Each day, there are about 225,000 more mouths to feed. Between 2008 and 2050, the world's population is projected to increase by 2.6 billion people. To provide enough food for these individuals, we will have to sharply reduce poverty, grow and distribute more food than has been produced since agriculture began about 10,000 years ago, and reduce the harmful environmental impacts of food production. In this chapter, we consider these challenges.

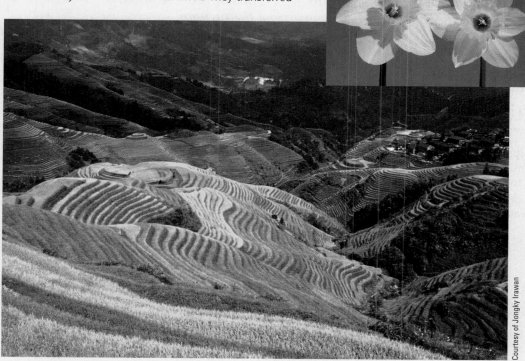

Figure 12-1 *Golden rice* is a new genetically engineered strain of rice containing beta-carotene (giving it its yellow color), which the body can convert to vitamin A. By 2011, some farmers in countries such as India, Bangladesh, and China may begin replacing some conventional strains of rice (left) with this new golden rice strain.

age fotostock/SuperStock

Courtesy of Jongky Irawan

12-1 What is food security and why is it difficult to attain?

CONCEPT 12-1A Many of the poor suffer health problems from chronic lack of food and poor nutrition, while many people in developed countries have health problems from eating too much food.

CONCEPT 12-1B The greatest obstacles to providing enough food for everyone are poverty, political upheaval, corruption, war, and the harmful environmental effects of food production.

12-2 How is food produced?

CONCEPT 12-2A We have sharply increased crop production using a mix of industrialized and traditional agriculture.

CONCEPT 12-2B We have used industrialized and traditional methods to greatly increase supplies of meat, fish, and shellfish.

12-3 What environmental problems arise from food production?

CONCEPT 12-3 Food production in the future may be limited by its serious environmental impacts, including soil erosion and degradation, desertification, water and air pollution, greenhouse gas emissions, and degradation and destruction of biodiversity.

12-4 How can we protect crops from pests more sustainably?

CONCEPT 12-4 We can sharply cut pesticide use without decreasing crop yields by using a mix of cultivation techniques,

biological pest controls, and small amounts of selected chemical pesticides as a last resort (integrated pest management).

12-5 How can we improve food security?

CONCEPT 12-5 We can improve food security by creating programs to reduce poverty and chronic malnutrition, relying more on locally grown food, and cutting food waste.

12-6 How can we produce food more sustainably?

CONCEPT 12-6A Sustainable food production will require reducing topsoil erosion, sharply reducing overgrazing and overfishing, irrigating more efficiently, using integrated pest management, promoting agrobiodiversity, and providing government subsidies for more sustainable farming, fishing, and aquaculture.

CONCEPT 12-6B Producing enough food to feed the rapidly growing human population will require growing crops in a mix of monocultures and polycultures and decreasing the enormous environmental impacts of industrialized food production.

Note: Supplements 2 (p. S4), 3 (p. S10), 4 (p. S20), 5 (p. S31), and 13 (p. 78) can be used with this chapter.

There are two spiritual dangers in not owning a farm.
One is the danger of supposing that breakfast comes from the grocery,
and the other that heat comes from the furnace.

ALDO LEOPOLD

12-1 What Is Food Security and Why Is It Difficult to Attain?

▶ **CONCEPT 12-1A** Many of the poor suffer health problems from chronic lack of food and poor nutrition, while many people in developed countries have health problems from eating too much food.

▶ **CONCEPT 12-1B** The greatest obstacles to providing enough food for everyone are poverty, political upheaval, corruption, war, and the harmful environmental effects of food production.

Many of the Poor Have Health Problems Because They Do Not Get Enough to Eat

Having **food security** means that every person in a given area has daily access to enough nutritious food

to have an active and healthy life. Today we produce more than enough food to meet the basic nutritional needs of every person on the earth. But even with this surplus of food, one of every six people in developing countries is not getting enough to eat. These people face **food insecurity,** living with chronic hunger and

poor nutrition, which threatens their ability to lead healthy and productive lives (**Concept 12-1A**).

Most agricultural experts agree that *the root cause of food insecurity is poverty*, which prevents poor people from growing or buying enough food (**Concept 12-1B**). For example, since 1990, India has produced enough grain to feed its entire population. But more than 200 million Indians—about one-fifth of the country's population—are hungry because they cannot afford to buy or grow enough food.

Other obstacles to food security are political upheaval, corruption, and war (**Concept 12-1B**). These problems interfere with food distribution and transportation systems and can result in people going hungry while stored foods spoil or are distributed unevenly in a country or region. Achieving food security on regional and global levels for both poor and affluent people also depends on greatly reducing the harmful environmental effects of agriculture, which we explore further in this chapter.

Many People Suffer from Chronic Hunger and Malnutrition

To maintain good health and resist disease, individuals need fairly large amounts of *macronutrients* (such as carbohydrates, proteins, and fats, see Table 12-1 and Figures 8, 9, and 12, pp. S42–S43, in Supplement 6), and smaller amounts of *micronutrients*—vitamins, such as A, C, and E, and minerals, such as iron, iodine, and calcium.

People who cannot grow or buy enough food to meet their basic energy needs suffer from **chronic undernutrition,** or **hunger** (**Concept 12-1A**). Most of the world's chronically undernourished children live in developing countries. They face the possibilities of suffering from mental retardation and stunted growth and dying from infectious diseases such as measles and diarrhea, which rarely kill children in developed countries.

Many of the world's poor can afford only to live on a low-protein, high-carbohydrate, vegetarian diet consisting mainly of grains such as wheat, rice, or corn. They often suffer from **chronic malnutrition**—deficiencies of protein and other key nutrients—which weakens them, makes them more susceptible to disease, and hinders the normal physical and mental development of children.

According to the U.N. Food and Agriculture Organization (FAO), the estimated number of chronically undernourished or malnourished people fell from 918 million in 1970 to 862 million in 2006. (See Figure 13, p. S17, in Supplement 3 for a map of the countries with the most undernourished people.) This is a good start, but it is still high—nearly three times the population of the United States. A 2007 study by University of Minnesota economists Ford Runge and Benjamin Senauer estimated that increased food prices

Table 12-1

Key Nutrients for a Healthy Human Life

Nutrient	Food source	Function
Proteins	Animals and some plants	Help to build and repair body tissues
Carbohydrates	Wheat, corn and rice	Provide short-term energy
Lipids (oils and fats)	Animal fats, nuts, oils	Help to build membrane tissues and create hormones

from the massive diversion of corn to produce ethanol for fueling cars could increase the number of hungry and malnourished people to 1.2 billion by 2025 instead of decreasing it to 625 million as projected by the FAO.

Despite some progress, one of every six people in developing countries (including about one of every three children younger than age 5) is chronically undernourished or malnourished (Figure 1-14, p. 19). About 75% of these people live in rural areas of developing countries. In 2005, the FAO estimated that each year, nearly 6 million children die prematurely from chronic undernutrition and malnutrition and increased susceptibility to normally nonfatal infectious diseases (such as measles and diarrhea) because of their weakened condition (**Concept 12-1A**). This means that each day, an average of 16,400 children die prematurely from these mostly poverty-related causes. How many people died from such causes during your lunch hour?

Many People Do Not Get Enough Vitamins and Minerals

According to the World Health Organization (WHO), one of every three people suffers from a deficiency of one or more vitamins and minerals, most often in developing countries and involving *iron, vitamin A,* and *iodine*.

Having too little *iron*—a component of the hemoglobin that transports oxygen in the blood—causes *anemia*, which results in a general lack of vitality. According to the WHO, one of every five people in the world—mostly women and children in tropical developing countries—suffers from iron deficiency. It causes fatigue, makes infection more likely, and increases a woman's chances of dying from hemorrhage in childbirth.

New strains of golden rice (**Core Case Study**) contain more iron than conventional strains and could help to reduce the severity of iron and vitamin A deficiencies. However, some critics view these claims as mostly a public relations ploy financed by the seed industry to soften up widespread consumer opposition to genetically engineered crops. They contend that golden rice is drawing funding and attention away from a possibly quicker and cheaper option of giving

John Paul Kay/Peter Arnold, Inc.

Figure 12-2 Woman with goiter in Bangladesh. A diet with insufficient iodine can cause this enlargement of the thyroid gland. Adding traces of iodine to salt has largely eliminated this problem in developed countries.

two vitamin A capsules per year to each of the millions of children suffering from vitamin A deficiency.

In addition, scientists want more evidence on how much of the beta-carotene in the golden rice will actually be converted to vitamin A in the body. And they want evidence that golden rice strains that perform well in the laboratory will perform as well in nature where many more factors come into play. Current field trials of golden rice in the Philippines may provide such information.

Elemental *iodine* is essential for proper functioning of the thyroid gland, which produces hormones that control the body's rate of metabolism. Iodine is found in seafood and in crops grown in iodine-rich soils. Chronic lack of iodine can cause stunted growth, mental retardation, and goiter—a swollen thyroid gland that can lead to deafness (Figure 12-2). According to the United Nations, some 600 million people—mostly in rural areas of south and southeast Asia—suffer from goiter, and 26 million children suffer brain damage each year from lack of iodine.

Acute Food Shortages Can Lead to Famines

A **famine** occurs when there is a severe shortage of food in an area accompanied by mass starvation, many deaths, economic chaos, and social disruption. Faced with starvation, desperate people eat the seed grain they have stored to grow crops in future years and slaughter their breeding livestock. Famines often result in mass migrations of starving people to other areas or to refugee camps in a search for food, water, and medical help. Many die during these journeys or in these camps. Famines are usually caused by crop failures from drought, flooding, war (Figure 12-3), and other catastrophic events (**Concept 12-1B**).

┌─ **RESEARCH FRONTIER** ────────────────

Learning more about connections between food supplies, poverty, and environmental problems. See **academic.cengage .com/biology/miller**.

Many People Have Health Problems from Eating Too Much

Overnutrition occurs when food energy intake exceeds energy use and causes excess body fat. Too many calories, too little exercise, or both can cause overnutrition. People who are underfed and underweight and those who are overfed and overweight face similar health problems: *lower life expectancy, greater susceptibility to disease and illness,* and *lower productivity and life quality* (**Concept 12-1A**).

We live in a world where 1 billion people have health problems because they do not get enough to eat and another 1.6 billion face health problems from eating too much. According to 2004 study by the International Obesity Task Force, one of every four people in the world is overweight (with a body mass index of 25 or more) and one of every 20 is obese (with a body mass index of 30 or more).

Harmut Schwartzbach/Peter Arnold, Inc.

Figure 12-3 *War and the environment:* starving children collecting ants to eat in famine-stricken Sudan, Africa, where a civil war has been going on since 1983.

In the United States, a 2005 study at Boston University found that about 66% of American adults are overweight and 33% are obese (up from 15% in 1980), the highest overnutrition rate in any developed country. The more than $50 billion that Americans spend each year trying to lose weight (according to the MarketData research firm) is more than two times the $24 billion per year that the United Nations estimates is needed to eliminate undernutrition and malnutrition in the world.

12-2 How Is Food Produced?

▶ **CONCEPT 12-2A** We have sharply increased crop production using a mix of industrialized and traditional agriculture.

▶ **CONCEPT 12-2B** We have used industrialized and traditional methods to greatly increase supplies of meat, fish, and shellfish.

Food Production Has Increased Dramatically

Currently, three systems supply most of our food. *Croplands* produce mostly grains and provide about 77% of the world's food using 11% of its land area. *Rangelands, pastures,* and *feedlots* produce meat and supply about 16% of the world's food using about 29% the world's land area. *Oceanic fisheries,* and more recently *aquaculture,* supply about 7% of the world's food.

These three systems depend on a small number of plant and animal species. Of the estimated 50,000 plant species that people can eat, only 14 of them supply an estimated 90% of the world's food calories. Just three types of grain crops—*wheat, rice,* and *corn*—provide about 47% of the calories and 42% of the protein people consume. Two-thirds of the world's people survive primarily on these three grains. A small number of species also provide most of the world's meat and seafood.

Such food specialization puts us in a vulnerable position should the small number of crop strains, livestock breeds, and fish and shellfish species we depend on fail as a result of factors such as disease, environmental degradation, and climate change. This violates the biodiversity **principle of sustainability** (see back cover), which calls for depending on a variety of food sources as an ecological insurance policy for dealing with environmental change.

Despite such vulnerability, since 1960, there has been a staggering increase in global food production from all three of the major food production systems (**Concepts 12-2A** and **12-2B**). This occurred because of technological advances such as increased use of tractors and farm machinery and high-tech fishing equipment (Figure 11-7, p. 256). Other technological developments include inorganic chemical fertilizers, irrigation, pesticides, high-yield grain varieties, and raising large numbers of livestock, poultry, and fish in factory-like conditions. These developments have led to increasing reliance on industrialized production of a fairly small number of crops.

Industrialized Crop Production Relies on High-Input Monocultures

Agriculture used to grow crops can be divided roughly into two types: industrialized agriculture and subsistence agriculture. **Industrialized agriculture,** or **high-input agriculture,** uses heavy equipment and large amounts of financial capital, fossil fuel, water, commercial fertilizers, and pesticides to produce single crops, or *monocultures,* (Figure 7-13, p. 152). The major goal of industrialized agriculture for any crop has been to steadily increase its *yield*—the amount of food produced per unit of land. Industrialized agriculture is practiced on one-fourth of all cropland, mostly in developed countries, but has spread since the mid-1960s to some developing countries and now produces about 80% of the world's food (**Concept 12-2A**).

Plantation agriculture is a form of industrialized agriculture used primarily in tropical developing countries. It involves growing *cash crops,* such as bananas, soybeans (mostly to feed livestock), sugarcane (to produce sugar and ethanol fuel), coffee, palm oil (used as a cooking oil and to produce biodiesel fuel), and vegetables on large monoculture plantations, mostly for export to developed countries. Producing such monoculture crops in the tropics increases yields but decreases biodiversity when tropical forests are cleared or burned (Figure 10-16, p. 226) for crop plantations.

A new form of industrialized agriculture involves widespread use of greenhouses to raise crops. Figure 12-4 (p. 280) shows the spread of this type of agriculture along Spain's south coast. It is intended as a way to make arid lands, such as those pictured (Figure 12-4, left) productive. But it requires large inputs of water and energy to move water from one part of the country to another.

On a small scale, however, greenhouse production of crops can be water-efficient. *Hydroponics* is a method whereby plants are grown with their roots in troughs of water inside a greenhouse. Water is pumped from the troughs and sprayed on the plants as artificial rain.

Figure 12-4 Satellite images of extensive and rapid development of greenhouse production of food crops in the Almeria province along Spain's southern coast between 1974 and 2000. Greenhouse-dominated land appears as whitish gray patches in the photo on the right. To provide the water needed to grow these crops, Spain built 118 dams and 22 water transfer projects to move water to this arid region from water-rich parts of the country.

Because the water is recycled, such an operation uses one-fifth to one-tenth less water per crop than conventional farming uses. In water-poor parts of Africa, farmers are growing tomatoes and other vegetables hydroponically in small greenhouses made of inexpensive materials.

Traditional Agriculture Often Relies on Low–Input Polycultures

Traditional agriculture consists of two main types, which together are practiced by 2.7 billion people (42% of the world's people) in developing countries, providing about one-fifth of the world's food crops on about three-fourths of its cultivated land.

Traditional subsistence agriculture uses mostly human labor and draft animals to produce only enough crops for a farm family's survival, with little left over to sell or store as a reserve in hard times. In **traditional intensive agriculture,** farmers increase their inputs of human and draft-animal labor, fertilizer, and water to obtain higher crop yields. If the weather cooperates, they produce enough food to feed their families while selling some for income.

Some traditional farmers focus on cultivating a single crop, but many grow several crops on the same plot simultaneously, a practice known as **polyculture.** Such crop diversity—an example of implementing the biodiversity **principle of sustainability** (see back cover)—reduces the chance of losing most

or all of the year's food supply to pests, bad weather, and other misfortunes.

One type of polyculture is known as **slash-and-burn agriculture.** This type of subsistence agriculture involves burning and clearing small plots in tropical forests, growing a variety of crops for a few years until the soil is depleted of nutrients, and then shifting to other plots. Early users of this method learned that each abandoned patch normally had to be left fallow (unplanted) for 10–30 years before the soil became fertile enough to grow crops again. While patches were regenerating, growers used them for tree crops, medicines, fuelwood, and other purposes. In this manner, most early growers practiced *sustainable cultivation.* However, when too many people use this approach or don't understand its limitations, it can become unsustainable and lead to depletion and degradation of tropical forests, as is now taking place in parts of Africa.

In parts of South America and Africa, some traditional slash-and-burn farmers grow as many as 20 different crops together on small cleared plots in tropical forests. The crops mature at different times, provide food throughout the year, and keep the soil covered to reduce erosion from wind and water. This lessens the need for fertilizer and water, because root systems at different depths in the soil capture nutrients and moisture efficiently, and ashes from the burning provide some fertilization. Insecticides and herbicides are rarely needed because multiple habitats are created for natural predators of crop-eating insects, and weeds have trouble competing with the multitude of crop plants.

SCIENCE FOCUS

Soil Is the Base of Life on Land

Soil is a complex mixture of eroded rock, mineral nutrients, decaying organic matter, water, air, and billions of living organisms, most of them microscopic decomposers. Soil formation begins when bedrock is slowly broken down into fragments and particles by physical, chemical, and biological processes called *weathering*. Figure 12-A shows a profile of different-aged soils.

Soil, the base of life on land, is a key component of the earth's natural capital. It supplies most of the nutrients needed for plant growth (Figure 1-4, p. 9), purifies and stores water, and helps to control the earth's climate by removing carbon dioxide from the atmosphere and storing it as carbon compounds.

Most mature soils—soils that have developed over a long period of time—contain at least three horizontal layers, or *horizons*, (Figure 12-A), each with a distinct texture and composition that varies with different types of soils. Think of them as floors in the geological building of life underneath your feet.

The roots of most plants and the majority of a soil's organic matter are concentrated in a soil's two upper layers, the O horizon of leaf litter and the A horizon of topsoil. In most mature soils, these two layers teem with bacteria, fungi, earthworms, and small insects all interacting in complex ways. Bacteria and other decomposer microorganisms found by the billions in every handful of topsoil break down some of its complex organic compounds into a porous mixture of the partially decomposed bodies of dead plants and animals, called *humus*. Top soil also usually includes inorganic materials such as clay, silt, and sand. Soil moisture carrying these dissolved nutrients is drawn up by the roots of plants and transported through stems and into leaves as part of the earth's chemical cycling processes.

The *B horizon* (*subsoil*) and the *C horizon* (*parent material*) contain most of a soil's inorganic matter, mostly broken-down rock consisting of varying mixtures of sand, silt, clay, and gravel. Much of it is transported by water from the A horizon (Figure 12-A). The C horizon lies on a base of parent material, which is often *bedrock*.

The spaces, or *pores*, between the solid organic and inorganic particles in the upper and lower soil layers contain varying amounts of air (mostly nitrogen and oxygen gas) and water. Plant roots use the oxygen for cellular respiration. As long as the O and A horizons are anchored by vegetation, the soil layers as a whole act as a sponge, storing water and releasing it in a nourishing trickle.

Although topsoil is a renewable resource, it is renewed very slowly, which means it can be depleted. Just 1 centimeter (0.4 inch) of topsoil can take hundreds of years to form, but it can be washed or blown away in a matter of weeks or months when we plow grassland or clear a forest and leave its topsoil unprotected.

Since the beginning of agriculture, human activities have accelerated natural soil erosion. We discuss erosion and ways to prevent or control it later in this chapter.

Critical Thinking

How does soil contribute to each of the four components of biodiversity described in Figure 4-2, p. 79?

Oak tree
Wood sorrel
Earthworm
Fern
Millipede
Honey fungus
Mole
Grasses and small shrubs
Organic debris builds up
Moss and lichen
Rock fragments

O horizon Leaf litter
A horizon Topsoil
B horizon Subsoil
C horizon Parent material

Root system
Red earth mite
Bacteria
Mite
Nematode
Fungus

Mature soil
Young soil
Immature soil
Bedrock

CENGAGENOW™ **Active Figure 12-A** Soil formation and generalized soil profile. Horizons, or layers, vary in number, composition, and thickness, depending on the type of soil. *See an animation based on this figure at CengageNOW™.* **Questions:** What role do you think the tree in this figure plays in soil formation? How might the picture change if the tree were removed? (Used by permission of Macmillan Publishing Company from Derek Elsom, *Earth*, New York: Macmillan, 1992. Copyright © 1992 by Marshall Editions Developments Limited)

Recent research shows that, on average, low-input polyculture produces higher yields than does high-input monoculture. For example, a 2001 study by ecologists Peter Reich and David Tilman found that carefully controlled polyculture plots with 16 different species of plants consistently outproduced plots with 9, 4, or only 1 type of plant species. Therefore, some analysts argue for increasing use of polyculture, along with monocultures, to produce food more sustainably in the future. Both industrialized and conventional production of crops depend on having fertile soil (Science Focus, above).

─── RESEARCH FRONTIER ───
Investigating the design and benefits of polyculture. See
academic.cengage.com/biology/miller.

CENGAGENOW™ Compare soil profiles from grassland, desert, and three types of forests at CengageNOW.

A Closer Look at Industrialized Crop Production

Farmers can produce more food by farming more land or by getting higher yields from existing cropland. Since 1950, about 88% of the increase in global food production has come from using high-input industrialized agriculture to increase yields in a process called the **green revolution.**

The green revolution involves three steps. *First,* develop and plant monocultures of selectively bred or genetically engineered high-yield varieties of key crops such as rice, wheat, and corn. *Second,* produce high yields by using large inputs of fertilizers, pesticides, and water. *Third,* increase the number of crops grown per year on a plot of land through *multiple cropping,* or *multicropping.* Between 1950 and 1970, this high-input approach dramatically increased crop yields in most developed countries, especially the United States (Case Study, right) in what was called the *first green revolution.*

A *second green revolution* has been taking place since 1967. Fast-growing dwarf varieties of rice and wheat, specially bred for tropical and subtropical climates, have been introduced into India and China and several developing countries in Central and South America. Producing more food on less land has the benefit of protecting biodiversity by saving large areas of forests, grasslands, wetlands, and easily eroded mountain terrain from being used to grow food crops.

Between 1950 and 1996, mostly because of the two green revolutions, world grain production tripled (Figure 12-5, left). Per capita food production increased by 31% between 1961 and 1985, but since then it has generally declined (Figure 12-5, right).

■ CASE STUDY
Industrialized Food Production in the United States

In the United States, industrialized farming has evolved into *agribusiness,* as a small number of giant multinational corporations increasingly control the growing, processing, distribution, and sale of food in the United States and in the global marketplace.

In total annual sales, agriculture is bigger than the country's automotive, steel, and housing industries combined. It generates almost one-fifth of the nation's gross domestic product. Although agriculture employs more people than any other industry, U.S. farms use industrialized agriculture to produce about 17% of the world's grain with only 0.3% of the world's farm labor force.

Since 1950, U.S. industrialized agriculture has more than doubled the yields of key crops such as wheat, corn, and soybeans without cultivating more land. Such yield increases have kept large areas of forests, grasslands, and wetlands from being converted to farmland.

U.S. food production is very efficient. While the U.S. output of crops, meat, and dairy products has been increasing steadily since 1975, the major inputs of labor and resources—with the exception of pesticides—to produce each unit of that output have fallen steadily since 1950.

U.S. consumers now spend about 2% of their disposable income on food, compared to about 11% in 1948. People in developing countries typically spend up to 40% of their income on food. And the 1.2 billion poorest people, struggling to live on less than $1 a day, typically spend about 70% of this meager income on food. Thus, to the poor any rise in the prices of wheat, rice, or corn can be disastrous.

Total World Grain Production

World Grain Production per Capita

Figure 12-5 *Global outlook:* total worldwide grain production of wheat, corn, and rice (left), and per capita grain production (right), 1961–2006. In order, the world's three largest grain-producing countries are China, the United States, and India. (Data from U.S. Department of Agriculture, Worldwatch Institute, U.N. Food and Agriculture Organization, and Earth Policy Institute) **Question:** Why do you think grain production per capita has grown less consistently than total grain production?

However, the actual food costs paid by consumers in the United States and other developed countries are much higher than the prices paid in grocery stores. In addition to the direct market prices, consumers pay taxes to give subsidies to food producers and distributors and to help deal with the massive pollution and environmental degradation caused by agriculture (discussed later in this chapter). They also face higher health and insurance bills related to the harmful environmental and health effects of agriculture. Including these harmful costs in the market prices for food would bring about a shift to more sustainable and less environmentally harmful agriculture.

Crossbreeding and Genetic Engineering Can Produce New Crop Varieties

For centuries, farmers and scientists have used *crossbreeding* through *artificial selection* to develop genetically improved varieties of crops and livestock animals. Such selective breeding in this first *gene revolution* has yielded amazing results. Ancient ears of corn were about the size of your little finger and wild tomatoes were once the size of grapes.

Traditional crossbreeding is a slow process, typically taking 15 years or more to produce a commercially valuable new crop variety, and it can combine traits only from species that are genetically similar. Resulting varieties remain useful for only 5–10 years before pests and diseases reduce their effectiveness. But important advances are still being made with this method.

Today, scientists are creating a second *gene revolution* by using *genetic engineering* to develop genetically improved strains of crops and livestock animals. It involves altering an organism's genetic material through adding, deleting, or changing segments of its DNA (Figure 10, p. S43, in Supplement 6), to produce desirable traits or to eliminate undesirable ones. It enables scientists to transfer genes between different species that would not interbreed in nature. The resulting organisms are called *genetically modified organisms* (GMOs) or *transgenic organisms*.

Figure 12-6 outlines the steps involved in developing a genetically modified plant. For example, genetic engineers used genes from ordinary daffodils (Figure 12-1, right) and a soil bacterium to produce golden rice (**Core Case Study**). Compared to traditional crossbreeding, gene splicing takes about half as long to develop a new crop variety, usually costs less, and allows for the insertion of genes from almost any other organism into crop cells.

Ready or not, much of the world is entering the *age of genetic engineering*. Bioengineers are developing, or planning to develop, new varieties of crops that are resistant to heat, cold, herbicides, insect pests, parasites, viral diseases, drought, and salty or acidic soil. They

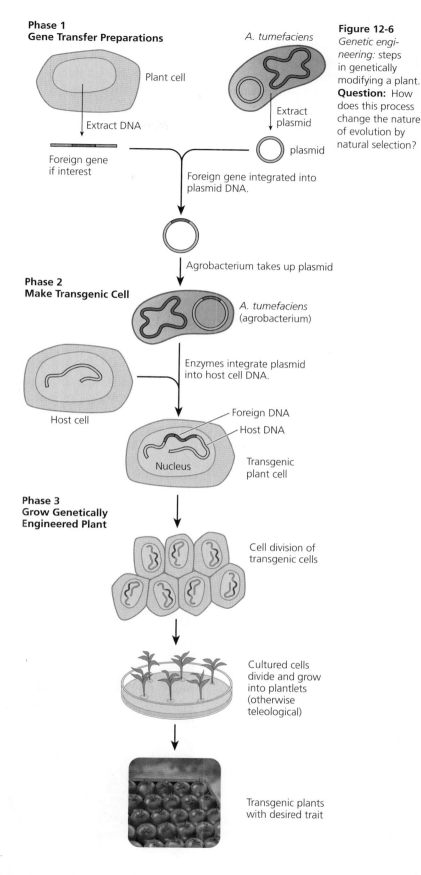

Phase 1
Gene Transfer Preparations

Plant cell

A. tumefaciens

Extract DNA

Extract plasmid

Foreign gene if interest

plasmid

Foreign gene integrated into plasmid DNA.

Agrobacterium takes up plasmid

Phase 2
Make Transgenic Cell

A. tumefaciens (agrobacterium)

Host cell

Enzymes integrate plasmid into host cell DNA.

Foreign DNA
Host DNA

Nucleus

Transgenic plant cell

Phase 3
Grow Genetically Engineered Plant

Cell division of transgenic cells

Cultured cells divide and grow into plantlets (otherwise teleological)

Transgenic plants with desired trait

Figure 12-6
Genetic engineering: steps in genetically modifying a plant. **Question:** How does this process change the nature of evolution by natural selection?

also hope to develop crop plants that can grow faster and survive with little or no irrigation and with less fertilizer and pesticides.

Bioengineers have altered citrus trees, which normally take 6 years to produce fruit, to yield fruit in only 1 year. They also hope to use *advanced tissue culture* techniques to mass-produce only orange juice sacs, which would eliminate the need for citrus orchards.

─── RESEARCH FRONTIER ───

Genetic engineering and using cell cultures for food factory systems. See **academic.cengage.com/biology/miller**.

Many scientists believe that such innovations hold great promise for helping to improve global food security. Others warn that genetic engineering is not free of drawbacks, which we examine later in this chapter.

Meat Production and Consumption Have Grown Steadily

About half of the world's meat comes from livestock grazing on grass in unfenced rangelands and enclosed pastures. The other half is produced through an industrialized system in which animals are raised in densely packed *feedlots* and *confined animal feeding operations* where they are fed grain or meal produced from fish. For example, large numbers of cattle are brought to feedlots where they are fattened up for about 4 months before slaughter (Figure 12-7). Most pigs and chickens in developed countries spend their lives crowded in pens and cages, often in huge buildings where they eat mostly grain grown on cropland.

Between 1961 and 2007 world meat production—mostly pork, poultry, and beef—increased fourfold, and is likely to more than double again by 2050 as affluence rises and middle-income people begin consuming more meat and meat products in developing countries. During this same period, average meat consumption per person more than doubled.

Industrialized meat production may face certain limits in the future. For example, if we include land used to grow grain fed to livestock, the FAO estimates that 30% of the earth's ice-free land is already directly or indirectly involved in livestock production. And this percentage is likely to increase.

Furthermore, as incomes grow in a country, more of its people tend to eat more meat, with much of it produced by feeding grain to livestock. This increases the demand for grain and can lead to increased reliance on grain imports. As industrialization increases, there is also a loss of cropland to urban development. Within a few decades, a rapidly industrializing country can go from producing all the grain it needs to importing most of its grain. As a result of this process, Japan, Taiwan, and South Korea now import 70% of the grain they consume. China and India, which currently are the world's leading producers of wheat, are likely to follow this trend as they become more industrialized.

If China were to import just one-fifth of the grain it needs, this would equal the amount of grain the United States typically exports each year—roughly half the world's grain exports. The Earth Policy Institute and the U.S. Central Intelligence Agency warn that if such a scenario comes to pass, no country or combination of countries will have the ability to supply even a small fraction of China's potential food supply deficit.

Figure 12-7 *Industrialized meat production.* A cattle feedlot in Imperial Valley, California (USA), where 40,000 cattle are being fattened up on grain before being slaughtered. With only about 4.5% of the world's population, the United States grows and kills nearly 10 billion animals a year as a source of beef, pork, chicken, and other forms of meat.

Matt Meadows/Peter Arnold, Inc.

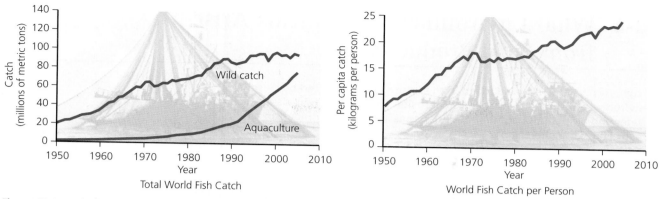

Figure 12-8 World fish catch, including both wild catch and aquaculture, 1950–2004 (**Concept 12-2B**). **Question:** What are two trends that you can see in these data? (Data from U.N. Food and Agriculture Organization, U.S. Census Bureau, and Worldwatch Institute)

This does not take into account huge grain deficits that are projected in other parts of the world by 2025, especially in Africa and India. It also does not include the possibility that climate change from global warming could decrease food production in China and in countries the world relies on for grain exports.

Fish and Shellfish Production Have Increased Dramatically

The world's third major food-producing system consists of fisheries and aquaculture. A **fishery** is a concentration of particular aquatic species suitable for commercial harvesting in a given ocean area or inland body of water. Industrial fishing fleets catch most of the world's marine fish (Case Study and Figure 11-7, p. 256). In 2006, 43% of the fish and shellfish consumed were produced through **aquaculture**—raising marine and freshwater fish in ponds and underwater cages—and this percentage has been growing steadily.

Figure 12-8 shows the effects of the global efforts to boost the seafood harvest through fishing and aquaculture (**Concept 12-2B**). Since 1950, the world fish catch (marine and freshwater harvests, excluding aquaculture) has increased almost sevenfold. Aquacultural production in the same period increased over 40-fold.

In Chapter 11, we examined the global fishing industry and its effects on aquatic biodiversity (pp. 254–257 and Case Study, p. 256). A 2007 report by the FAO found that 25% of the world's ocean fisheries are virtually depleted and 52% have been fully exploited. In other words, 77% of the world's commercially valuable fish stocks are overexploited. According to a 2006 FAO study and research by a number of fishery scientists, unless we reduce overfishing and ocean pollution and slow climate change, ocean fisheries worldwide could essentially collapse by 2050, thereby sharply decreasing food security for up to 2.6 billion people who depend on fish for protein.

Using large ships to find, catch, and freeze ocean fish consumes huge amounts of energy. According to a 2005 study by ecological economist Peter Tyedmeers and his colleagues, the energy used by the world's fishing fleets to catch and deliver the fish is 12.5 times the food energy gained by the people who eat the fish. In the process, the fishing fleets also add large amounts of greenhouse gases to the atmosphere. According to ecological economists, without government-provided fuel and other subsidies, these fleets could not afford to exist.

The other major mode of fish production is aquaculture, which involves raising fish and shellfish for food instead of hunting and gathering them. This industrialized production of fish and shellfish is the world's fastest-growing type of food production. Some call aquaculture the *blue revolution.*

Aquaculture involves cultivating fish in freshwater ponds, lakes, reservoirs, and rice paddies, or in underwater cages in coastal saltwater lagoons, estuaries, and the deep ocean (Figure 11-7, p. 256). The fish are harvested when they reach the desired size. China raises 70% of the world's farmed fish, mostly in inland ponds and rice fields.

World aquaculture is dominated by operations that raise herbivorous species—mainly carp in China and India, catfish in the United States, tilapia in several countries, and shellfish in several coastal countries. Most Chinese farmers integrate crop growing and aquaculture by using rice straw, pig and duck manure, and other agricultural wastes to fertilize farm ponds and rice paddies in order to produce phytoplankton eaten by various herbivorous species of carp—an example of *polyaquaculture.*

The continuing expansion of aquaculture faces economic and environmental challenges. For example, much of it depends on using grain and fishmeal as feed, and growing competition for these resources between food and biofuel producers could put limits on aquacultural production. The environmental impacts of aquaculture, which we examine in the next section, could also put limits on its growth.

12-3 What Environmental Problems Arise from Food Production?

▶ **CONCEPT 12-3** Food production in the future may be limited by its serious environmental impacts, including soil erosion and degradation, desertification, water and air pollution, greenhouse gas emissions, and degradation and destruction of biodiversity.

Producing Food Has Major Environmental Impacts

According to many analysts, agriculture has a greater harmful environmental impact than any human activity and these environmental effects may limit future food production. Crop yields in some areas may decline because of environmental factors such as erosion and degradation of soil, depletion and pollution of underground and surface water supplies used for irrigation, emission of greenhouse gases that contribute to climate change, and loss of croplands (**Concept 12-3**). According to the U.S. Environmental Protection Agency (EPA), agriculture is responsible for three quarters of the water quality problems in U.S. rivers and streams.

Figure 12-9 summarizes harmful effects of agriculture on air, soil, water, and biodiversity. We do not know how close we are to environmental limits on food production, but some scientists say alarm bells are going off now. Here, we explore such effects in greater depth, starting with the problems of erosion and degradation of soils.

NATURAL CAPITAL DEGRADATION

Food Production

Biodiversity Loss	Soil	Water	Air Pollution	Human Health
Loss and degradation of grasslands, forests, and wetlands	Erosion	Water waste	Greenhouse gas emissions (CO_2) from fossil fuel use	Nitrates in drinking water (blue baby)
Fish kills from pesticide runoff	Loss of fertility	Aquifer depletion	Greenhouse gas emissions (N_2O) from use of inorganic fertilizers	Pesticide residues in drinking water, food, and air
Killing wild predators to protect livestock	Salinization	Increased runoff, sediment pollution, and flooding from cleared land		Contamination of drinking and swimming water from livestock wastes
Loss of genetic diversity of wild crop strains replaced by monoculture strains	Waterlogging	Pollution from pesticides and fertilizers	Greenhouse gas emissions of methane (CH_4) by cattle (mostly belching)	
	Desertification	Algal blooms and fish kills in lakes and rivers caused by runoff of fertilizers and agricultural wastes	Other air pollutants from fossil fuel use and pesticide sprays	Bacterial contamination of meat

Figure 12-9 Major harmful environmental effects of food production (**Concept 12-3**). According to a 2002 study by the United Nations, nearly 30% of the world's cropland has been degraded to some degree by soil erosion, salt buildup, and chemical pollution, and 17% has been seriously degraded. **Question:** Which item in each of these categories do you believe is the most harmful?

Topsoil Erosion Is a Serious Problem in Parts of the World

Soil erosion is the movement of soil components, especially surface litter and topsoil (Figure 12-A), from one place to another by the actions of wind and water. Some soil erosion is natural, and some is caused by human activities. In undisturbed, vegetated ecosystems, the roots of plants help to anchor the soil, the soil stores water and releases it in a nourishing trickle, and soil usually is not lost faster than it forms. Since the beginning of agriculture, human activities have accelerated natural soil erosion. Entire civilizations have collapsed because they mismanaged the topsoil that supported their populations (see p. S31 in Supplement 5).

Flowing water, the largest cause of erosion, carries away particles of topsoil that have been loosened by rainfall; severe erosion leads to the formation of gullies (Figure 12-10). Wind loosens and blows topsoil particles away, especially in areas with a dry climate and relatively flat and exposed land (Chapter 7 Core Case Study, Figure 7-1, p. 140). We lose natural capital in the form of fertile topsoil when we destroy soil-holding grasses through activities such as farming (Figure 7-13, p. 152), clear-cut logging (Figure 10-7, p. 219), overgrazing (Figure 10-21, left, p. 233), and off-road vehicle use (Figure 10-22, p. 234).

Accurate numbers are hard to come by, but some analysts estimate that perhaps a third or more of the world's cropland is losing soil faster than new soil is forming. In the Saharan region of Africa, dust storms that were once rare are now commonplace and have increased 10-fold since 1950. Countries in this region that face serious topsoil loss from wind erosion are Chad, Burkina Faso, Mauritania, and northern Nigeria. Many

Ron Giling/Peter Arnold, Inc.

Figure 12-10 Natural capital degradation: Severe gully erosion on cropland in Bolivia.

of the world's 862 million malnourished people try to survive on land suffering from severe soil erosion.

Soil erosion has two major harmful effects. One is *loss of soil fertility* through depletion of plant nutrients in topsoil. The other is *water pollution* in nearby surface waters, where eroded soil ends up as sediment. This can kill fish and shellfish and clog irrigation ditches, boat channels, reservoirs, and lakes. Additional water pollution occurs when the eroded sediment contains residues of fertilizers and pesticides (**Concept 12-3**). A joint survey by the U.N. Environment Programme (UNEP) and the World Resources Institute estimated that topsoil is eroding faster than it forms on about 38% of the world's cropland (Figure 12-11).

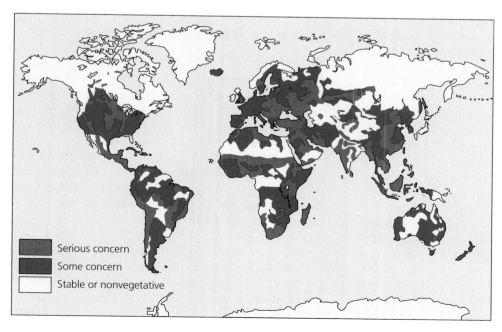

Serious concern

Some concern

Stable or nonvegetative

Figure 12-11 Natural capital degradation: global soil erosion. **Question:** Can you see any geographical pattern associated with this problem? (Data from U.N. Environment Programme and the World Resources Institute).

Figure 12-12 *Severe desertification.* Sand dunes threaten to engulf an oasis in the Sahel region of West Africa. Such severe desertification is the result of prolonged drought from natural climate change and destruction of natural vegetation from human activities such as farming and overgrazing.

Voltchev-UNEP/Peter Arnold, Inc.

Some analysts contend that erosion estimates are overstated because they do not account for the abilities of some local farmers to restore degraded land. The FAO also points out that eroded topsoil does not always move very far and is sometimes deposited on the same slope, valley, or plain from which it came. Consequently, in some places, the loss in crop yields in one area can be offset by increased yields elsewhere.

Drought and Human Activities Are Degrading Drylands

In arid and semiarid parts of the world, the contribution to the world's food supply from livestock and crops is being threatened by **desertification.** It occurs when the productive potential of soil, especially on arid or semiarid land, falls by 10% or more because of a combination of prolonged drought and human activities that reduce or degrade topsoil. The process can be *moderate* (a 10–25% drop in productivity), *severe* (a 25–50% drop), or *very severe* (a drop of more than 50%, usually creating huge gullies and sand dunes). Only in extreme cases does desertification lead to what we call desert. But when severe desertification occurs, it can expand existing desert area (Figure 12-12) or create new desert in areas that once were fertile land.

Over thousands of years, the earth's deserts have expanded and contracted, mostly because of natural climate change. However, human activities have accelerated desertification in some parts of the world (**Concept 12-3**) (Figure 12-13).

In its 2007 report on the *Status of the World's Forests,* the FAO estimated that some 70% of world's drylands used for agriculture are degraded and threatened by desertification. Most of these lands are in Africa and Asia, with countries including Nigeria, Iran, Afghanistan, and China suffering from serious desertification. Increasing desertification is also a threat in dryland areas of Brazil and Mexico.

According to a 2007 study by the Intergovernmental Panel on Climate Change (IPCC), projected climate change from global warming during this century (mostly the result of human activities) is expected to greatly increase severe and prolonged drought and, consequently, desertification in many arid and semiarid parts of the world. This could result in sharp drops in food production, water shortages for 1 billion–3 billion people, and huge numbers of environmental refugees.

Excessive Irrigation Has Serious Consequences

Between 1950 and 2007, the world's area of irrigated cropland tripled, with most of the growth occurring from 1950 to 1978, and little growth occurring since 1990. The 20% of the world's cropland that is irrigated produces almost 40% of the world's food.

But irrigation has a downside. Most irrigation water is a dilute solution of various salts, picked up as the water flows over or through soil and rocks. Irrigation water that has not been absorbed into the soil evaporates, leaving behind a thin crust of dissolved salts in the topsoil.

Repeated annual applications of irrigation water in dry climates lead to the gradual accumulation of salts in the upper soil layers—a soil degradation process called **salinization.** It stunts crop growth, lowers crop yields, and can eventually kill plants and ruin the land (**Con-**

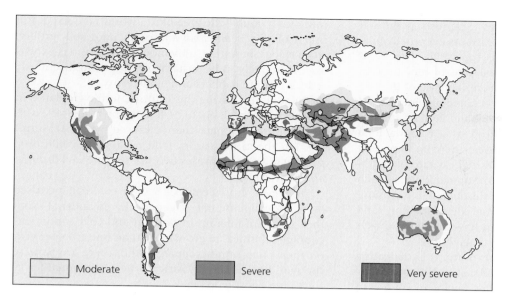

Figure 12-13 Natural capital degradation: desertification of arid and semiarid lands. It is caused by a combination of prolonged drought and human activities that expose soil to erosion. **Question:** Can you see any geographical pattern associated with this problem? (Data from U.N. Environment Programme and Harold E. Drengue)

Moderate Severe Very severe

cept 12-3). The United Nations estimates that severe salinization has reduced yields on at least one-tenth of the world's irrigated cropland, and the problem is getting worse. The most severe salinization occurs in Asia, especially in China, India, Egypt, Pakistan, and Iraq. Salinization affects almost one-fourth of irrigated cropland in the United States, especially in the western states (Figure 12-14).

Another problem with irrigation is **waterlogging,** in which water accumulates underground and gradually raises the water table. Farmers often apply large amounts of irrigation water to leach salts deeper into the soil. Without adequate drainage, waterlogging occurs and saline water then envelops the deep roots of plants, lowering their productivity and killing them

after prolonged exposure. At least one-tenth of the world's irrigated land suffers from waterlogging, and the problem is getting worse (**Concept 12-3**).

There May Be Limits to Expanding the Green Revolutions

Analysts point to several factors that have limited the success of the green revolutions to date and that may limit them in the future. Without huge inputs of fertilizer, pesticides, and water, most green revolution crop varieties produce yields that are no higher (and are sometimes lower) than those from traditional strains. These high inputs cost too much for most subsistence

U.S. Natural Resources Conservation Service

Figure 12-14 Natural capital degradation: Because of high evaporation, poor drainage, and severe salinization, white alkaline salts have displaced crops that once grew on this heavily irrigated land in the U.S. state of Colorado.

farmers in developing countries. Scientists point out that continuing to increase these inputs eventually produces no additional increase in crop yields. This helps to explain the slowdown in the rate of global grain yields per hectare from an average increase of 2.1% a year between 1950 and 1990 to an average increase of only 1.2% annually between 1990 and 2007.

Since 1970, the sharpest drop in per capita food production has occurred in Africa, the continent that for decades has had the world's highest rate of population growth (Figure 5, p. S13, Supplement 3). Such growth, plus poor soils, water shortages, soil erosion, desertification, limited economic development, and warfare in many parts of Africa, have hindered development of the two green revolutions there.

Can we expand the green revolutions by irrigating more cropland? In 2006, the International Water Management Institute projected that between 2005 and 2050, water use for agriculture will have to increase by 80% in order for farmers to provide food for 2.6 billion more people.

However since 1978, the amount of irrigated land per person has been declining, and it is projected to fall much more between 2008 and 2050. One reason for this is that since 1978, the world's population has grown faster than has use of irrigation. Other factors are depletion of underground water supplies (aquifers), wasteful use of irrigation water, soil salinization, and climate change, which is melting mountain glaciers that provide countries such as China and India, the world's two largest producers of wheat and rice, with irrigation water. As the glaciers melt, water flowing in major rivers during the dry season, when irrigation water needs are greatest, will decrease sharply. In addition, most of the world's farmers do not have enough money to irrigate their crops.

However, we can get more crop per drop by using known methods and technologies to greatly improve the efficiency of irrigation, which accounts for 70% of the world's water use. We discuss this more fully in Chapter 13.

Is cultivating more land the answer? Theoretically, clearing tropical forests and irrigating arid land could more than double the area of the world's cropland. But much of this is *marginal land* with poor soil fertility, steep slopes, or both. Cultivation of such land is expensive, is unlikely to be sustainable, and would seriously reduce wildlife habitats and biodiversity.

In addition, during this century, fertile croplands in coastal areas are likely to be flooded by rising sea levels resulting from climate change caused by global warming. For the same reason, food production could drop sharply in some countries because of increased drought and heat waves.

However, crop yields per hectare could be increased by using conventional or genetically engineered crops that are more tolerant of drought and cold and by encouraging farmers to grow more than one crop per year on their plots (multicropping). And, there are parts of the world, especially in Africa, where additional fertilizer could boost crop yields.

Industrialized Food Production Requires Huge Inputs of Energy

The industrialization of agriculture has been made possible by the availability of cheap energy (mostly from oil) used to run farm machinery, irrigate crops, process food, and produce commercial inorganic fertilizers (mostly by burning natural gas) and pesticides. Putting food on the table consumes about 17% of all commercial energy used in the United States each year (Figure 12-15).

The input of energy needed to produce a unit of food has fallen considerably, so that most crops in the United States provide an amount of energy in the form of food that is greater than the amount of energy used to grow them. However, when we consider the energy used to grow, store, process, package, transport, refrigerate, and cook all plant and animal food, *about 10 units of nonrenewable fossil fuel energy are needed to put 1 unit of food energy on the table.* In other words, industrialized food production and consumption has a large *net energy loss*. By comparison, every unit of energy from human labor in traditional farming provides 1 to 10 units of food energy.

| 4% | 2% | 6% | 5% |

| Crops | Livestock | Food processing | Food distribution and preparation |

Food production

Figure 12-15 Industrialized agriculture uses about 17% of all commercial energy used in the United States, and food travels an average 2,400 kilometers (1,300 miles) from farm to plate. The resulting pollution degrades the air and water and contributes to global warming. (Data from David Pimentel and Worldwatch Institute)

THINKING ABOUT
Food and Oil

What might happen to food production and to your lifestyle if oil prices keep rising sharply in the next 2 decades, as many analysts predict? How would you reduce this risk?

There Is Controversy over Genetically Engineered Foods

Despite its promise, controversy has arisen over the use of *genetically modified (GM) food* (Figure 12-6) and other forms of genetic engineering. Its producers and investors see this kind of food as a potentially sustainable way to solve world hunger problems and improve human health. Some critics consider it potentially dangerous "Frankenfood." Figure 12-16 summarizes projected advantages and disadvantages of this new technology.

Critics recognize the potential benefits of GM crops. But they warn that we know too little about the long-term potential harm to human health and ecosystems from the widespread use of such crops. They point out that genetic engineering mixes genes from widely differing species, which has never occurred in nature or even in selective breeding. They warn that if GM organisms released into the environment cause some unintended harmful genetic and ecological effects, as some scientists expect, they cannot be recalled.

For example, there is the potential for GM plant pollen to spread among nonengineered species, threatening crop biodiversity. In a 2006 study by the Union of Concerned Scientists, half of the nonengineered corn and soybean varieties tested by one laboratory contained DNA from GM varieties. Once experimental GM pollens are in the environment, they are impossible to retrieve.

Most scientists and economists who have studied the genetic engineering of crops believe that its potential benefits outweigh its risks. They also contend that some of the potential problems associated with GM crops can be eliminated by genetically engineering plants without inserting a gene from another species. This new technique, called *chimeraplasty,* involves inserting instead a chemical instruction that attaches to a gene, altering it to give desired genetic traits.

Critics call for more controlled field experiments, long-term testing to better understand the risks, and stricter regulation of this rapidly growing technology. A 2004 study by the Ecological Society of America recommended more caution in releasing genetically engineered organisms into the environment without adequate testing.

THINKING ABOUT
Golden Rice

Do you think that the advantages of genetically engineered golden rice (**Core Case Study**) outweigh its disadvantages? Explain.

CORE
CASE
STUDY

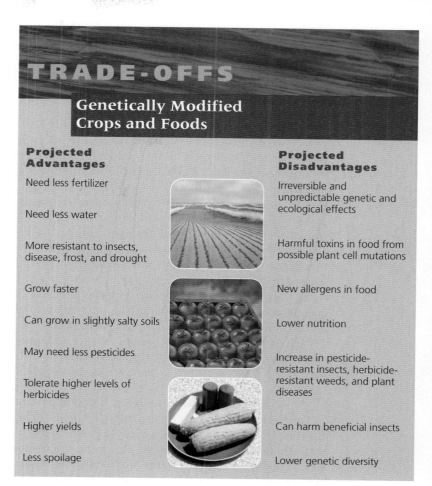

Figure 12-16 Projected advantages and disadvantages of genetically modified crops and foods. **Question:** Which two advantages and which two disadvantages do you think are the most important? Why?

Food and Biofuel Production Systems Have Caused Major Losses of Biodiversity

Biodiversity and some ecological services (Figure 10-4, left, p. 217) are threatened when forests are cleared and grasslands are plowed up and replaced with croplands (**Concept 12-3**). For example, one of the fastest-growing threats to the world's biodiversity is the clearing or burning of large areas of tropical forest in Brazil's Amazon basin (Figure 10-13, p. 224) and in its *cerrado,* a huge savanna-like region south of the Amazon basin. This land is being burned or cleared for large plantations of soybeans, grown for use as cattle feed, and sugarcane used for making ethanol fuel for cars. Other tropical forests are burned to make way for plantations of oil palm trees increasingly used to produce biodiesel fuel. These activities threaten biodiversity and contribute to climate change by releasing carbon dioxide into the atmosphere.

A closely related problem is the increasing loss of *agrobiodiversity*—the world's genetic variety of animals

and plants used to provide food. But we are replacing nature's resilient genetic diversity developed through millions of years of natural selection with human-engineered monocultures. Scientists estimate that since 1900, we have lost three-fourths of the genetic diversity of agricultural crops. For example, India once planted 30,000 varieties of rice. Now more than 75% of its rice production comes from only 10 varieties and soon, almost all of its production may come from just one variety. Rice varieties around the world may drop even more if there is a shift toward genetically engineered golden rice (Core Case Study) and other genetically engineered crops. In the United States, about 97% of the food plant varieties available to farmers in the 1940s no longer exist, except perhaps in small amounts in seed banks and in the backyards of a few gardeners.

In other words, we are rapidly shrinking the world's genetic "library," which is critical for increasing food yields. In fact, we might need it more than ever to develop new plant and livestock varieties that can adapt to climate change. This failure to preserve agrobiodiversity is a violation of the biodiversity **principle of sustainability**.

Wild and endangered varieties of crops important to the world's food supply are stored in about 1,400 refrigerated gene or seed banks, agricultural research centers, and botanical gardens scattered around the world. But increasingly, power failures, fires, storms, and war are causing irreversible losses of these stored plants and seeds. However, a new ice vault built underground in the Norwegian Arctic will eventually contain duplicates of much of the world's stock of seeds and will not be vulnerable to such hazards.

Shortages of storage space and money severely limit the number of species that can be preserved, and the seeds of many plants (such as potatoes) cannot be stored successfully in gene banks. Because stored seeds do not remain alive indefinitely, periodically they must be planted (germinated) and new seeds collected for storage. Unless this is done, seed banks become *seed morgues.*

Because of such limitations on seed banks, ecologists and plant scientists warn that the only effective way to preserve the genetic diversity of most plant and animal species is to protect representative ecosystems throughout the world from agriculture and other forms of development.

Industrialized Meat Production Has Harmful Environmental Consequences

Industrialized meat production has advantages and disadvantages (Figure 12-17). Environmental scientists point out that industrialized systems, such as feedlots and other confined animal production facilities, use large amounts of energy and water and produce huge amounts of animal waste that sometimes pollute surface water and groundwater and saturate the air with their odors.

For example, much of the fertilizer used in the midwestern United States to produce corn for animal feed and conversion to ethanol fuel runs off into the Mississippi and Ohio Rivers and ends up overfertilizing coastal waters in the Gulf of Mexico. Each year this creates a "dead zone" about the size of the U.S. state of New Jersey and threatens one-fifth of the nation's seafood yield. In other words, growing corn in the Midwest to produce ethanol and protein-rich meat can decrease the production of protein-rich seafood in the Gulf of Mexico—another example of unintended consequences.

Energy (mostly from oil) is an essential ingredient in industrialized meat production. Using this energy pollutes the air and water and contributes to global warming, with livestock production generating about 20% of the world's greenhouse gases—more than transportation generates—according to the FAO. Cattle and dairy cows release methane, the second most powerful greenhouse gas after carbon dioxide, mostly through belching. This accounts for 16% of the global annual emissions of methane. In 2007, scientists at the University of Wales in the United Kingdom found that adding garlic to the diet of cows can cut their daily methane emissions by up to 50%. They are now investigating whether adding this to cow chow will taint the taste of meat and milk. And a 2003 Swedish study found that raising beef cattle on grass emits 40% less greenhouse gases and used 85% less energy than raising the cattle on grain.

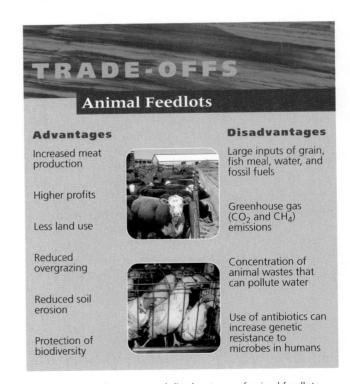

Figure 12-17 Advantages and disadvantages of animal feedlots. **Question:** Which single advantage and which single disadvantage do you think are the most important? Why?

Livestock in the United States produce 20 times more waste than is produced by the country's human population. Globally, only about half of all manure is returned to the land as nutrient-rich fertilizer—a violation of the recycling **principle of sustainability**. Much of the other half of this waste ends up polluting the air, water, and soil and producing foul odors.

Experts expect industrialized meat production to expand rapidly, especially as countries such as China and India become more affluent. This will increase pressure on the world's grain supply. Producing more meat will also increase pressure on the world's fish supply, because about 37% of the marine fish catch is converted to fish meal and fish oil, which are used to feed livestock and carnivorous fish raised by aquaculture.

Producing Fish through Aquaculture Can Harm Aquatic Ecosystems

Figure 12-18 lists the major advantages and disadvantages of aquaculture. Some analysts project that freshwater and saltwater aquaculture production could provide at least half of the world's seafood by 2025. Others warn that the harmful environmental effects of aquaculture could limit future production (**Concept 12-3**).

One problem is that fish raised on fish meal or fish oil can be contaminated with toxins such as PCBs found on ocean bottoms. In 2003, samples from various U.S. grocery stores revealed that farmed salmon had 7 times more PCBs than wild salmon and 4 times more than feedlot beef. A 2004 study found that farmed salmon also had levels of toxic dioxin 11 times higher than wild-caught salmon. Another problem is that wastes, pesticides, and antibiotics from aquaculture operations can pollute aquatic ecosystems.

TRADE-OFFS

Aquaculture

Advantages	Disadvantages
High efficiency	Needs large inputs of land, feed, and water
High yield in small volume of water	Large waste output
Can reduce overharvesting of fisheries	Can destroy mangrove forests and estuaries
Low fuel use	Uses grain to feed some species
High profits	Dense populations vulnerable to disease

Figure 12-18 Advantages and disadvantages of *aquaculture*. **Question:** Which single advantage and which single disadvantage do you think are the most important? Why?

HOW WOULD YOU VOTE? ☑

Do the advantages of aquaculture outweigh its disadvantages? Cast your vote online at **academic.cengage.com/biology/miller**.

Later in this chapter (Section 12-5), we consider some possible solutions to the environmental problems resulting from food production and ways to produce food more sustainably. But first, let us consider a special set of environmental problems and solutions related to protecting food supply systems from pests.

12-4 How Can We Protect Crops from Pests More Sustainably?

▶ **CONCEPT 12-4** We can sharply cut pesticide use without decreasing crop yields by using a mix of cultivation techniques, biological pest controls, and small amounts of selected chemical pesticides as a last resort (integrated pest management).

Nature Controls the Populations of Most Pests

A **pest** is any species that interferes with human welfare by competing with us for food, invading lawns and gardens, destroying building materials, spreading disease, invading ecosystems, or simply being a nuisance. Worldwide, only about 100 species of plants ("weeds"), animals (mostly insects), fungi, and microbes cause most of the damage to the crops we grow.

In natural ecosystems and many polyculture agroecosystems, *natural enemies* (predators, parasites, and

Figure 12-19 Natural capital: Spiders are important insect predators. Most spiders, including this ferocious-looking wolf spider, do not harm humans.

disease organisms) control the populations of most potential pest species in keeping with one of the earth's **scientific principles of sustainability** (see back cover). For example, the world's 30,000 known species of spiders, including the wolf spider (Figure 12-19), kill far more insects every year than humans do by using chemicals.

When we clear forests and grasslands, plant monoculture crops, and douse fields with chemicals that kill pests, we upset many of these natural population checks and balances, which implement the population control principle of sustainability. Then we must devise ways to protect our monoculture crops, tree plantations, lawns, and golf courses from insects and other pests that nature once controlled at no charge.

We Use Pesticides to Try to Control Pest Populations

To try to control pest organisms, we have developed a variety of **pesticides**—chemicals used to kill or control populations of organisms that humans consider undesirable. Common types of pesticides include *insecticides* (insect killers), *herbicides* (weed killers), *fungicides* (fungus killers), and *rodenticides* (rat and mouse killers).

We did not invent the use of chemicals to repel or kill other species. For nearly 225 million years, plants have been producing chemicals to ward off, deceive, or poison herbivores that feed on them. This battle produces a never-ending, ever-changing coevolutionary process: herbivores overcome various plant defenses through natural selection (**Concept 4-2B**, p. 80); then new plant defenses are favored by natural selection in this ongoing cycle of evolutionary punch and counterpunch.

In the 1600s, farmers used nicotine sulfate, extracted from tobacco leaves, as an insecticide. Eventually, other

first-generation pesticides—mainly natural chemicals borrowed from plants—were developed. Farmers were copying nature.

A major pest control revolution began in 1939, when entomologist Paul Müller discovered that DDT (dichlorodiphenyltrichloroethane), a chemical known since 1874, was a potent insecticide. DDT was the first of the so-called *second-generation pesticides* produced in the laboratory. It soon became the world's most used pesticide, and Müller received the Nobel Prize in 1948 for his discovery. Since then, chemists have made hundreds of other pesticides by making slight modifications in the molecules in various classes of chemicals.

Second-generation pesticides, however, turned out to be hazardous as well as helpful. In 1962, biologist Rachel Carson (Individuals Matter, at right) sounded a warning that eventually led to strict controls on use of DDT and several other widely used pesticides. Since 1970, chemists have returned to natural repellents and poisons produced by plants, again copying nature, to improve first-generation botanical pesticides.

Since 1950, pesticide use has increased more than 50-fold, and most of today's pesticides are 10–100 times more toxic than those used in the 1950s. About three-fourths of these chemicals are used in developed countries, but their use in developing countries is soaring.

Some pesticides, called *broad-spectrum agents*, are toxic to many pest and nonpest species. Examples are chlorinated hydrocarbon compounds such as DDT and organophosphate compounds such as malathion and parathion. Others, called *selective*, or *narrow-spectrum, agents*, are effective against a narrowly defined group of organisms.

Pesticides vary in their *persistence*, the length of time they remain deadly in the environment. Some, such as DDT and related compounds, remain in the environment for years and can be biologically magnified in food chains and webs (Figure 9-19, p. 202). Others, such as organophosphates, are active for days or weeks and are not biologically magnified but can be highly toxic to humans. About one-fourth of pesticide use in the United States is devoted to trying to rid houses, gardens, lawns, parks, playing fields, swimming pools, and golf courses of pests. According to the EPA, the average lawn in the United States is doused with ten times more synthetic pesticides per unit of land area than are put on U.S. cropland.

Modern Synthetic Pesticides Have Several Advantages

Conventional chemical pesticides have advantages and disadvantages. Proponents contend that their benefits (Figure 12-20, left) outweigh their harmful effects (Figure 12-20, right). They point to the following benefits:

- *They save human lives.* Since 1945, DDT and other insecticides probably have prevented the prema-

INDIVIDUALS MATTER

Rachel Carson

Rachel Carson (Figure 12-B) began her professional career as a biologist for the Bureau of U.S. Fisheries (now called the U.S. Fish and Wildlife Service). In that capacity, she carried out research in oceanography and marine biology and wrote articles and books about the oceans and topics related to the environment.

Throughout the 1950s, DDT use was increasing, and in 1958, DDT was sprayed to control mosquitoes near the home and private bird sanctuary of one of Carson's friends. After the spraying, her friend witnessed the agonizing deaths of several birds. She begged Carson to find someone to investigate the effects of pesticides on birds and other wildlife.

Carson decided to look into the issue herself and found that independent research on the environmental effects of pesticides was almost nonexistent. As a well-trained scientist, she surveyed the scientific literature, became convinced that pesticides could harm wildlife and humans, and methodically gathered information about the harmful effects of widespread use of pesticides.

In 1962, she published her findings in popular form in *Silent Spring*, a book whose title alluded to the silencing of "robins, catbirds, doves, jays, wrens, and scores of other bird voices" because of their exposure to pesticides. Many scientists, politicians, and policy makers read *Silent Spring*, and the public embraced it.

Chemical manufacturers viewed the book as a serious threat to their booming pesticide business and mounted a campaign to discredit Carson. A parade of critical reviewers and industry scientists claimed her book was full of inaccuracies, made selective use of research findings, and failed to give a balanced account of the benefits of pesticides.

Some critics even claimed that, as a woman, Carson was incapable of understanding such a highly scientific and technical subject. Others charged that she was an hysterical woman and a radical nature lover trying to scare the public in an effort to sell books.

During these intense attacks, Carson was a single mother, the sole caretaker of an aged parent, and was suffering from terminal breast cancer. Yet she strongly defended her research and countered her critics. She died in 1964—about 18 months after the publication of *Silent Spring*—without knowing that many historians consider her work an important contribution to the modern environmental movement then emerging in the United States.

It has been correctly noted that Carson made some errors in writing *Silent Spring*. But critics concede that the threat to birds and ecosystems—one of Carson's main messages—was real and that her errors can

Figure 12-B Biologist Rachel Carson (1907–1964) greatly increased our understanding of the importance of nature and the harmful effects of widespread use of pesticides.

be attributed to the primitive state of research on the topics she covered in her day. And her critics cannot dispute the fact that her wake-up call got the public and the scientific community focused on the potential threats from uncontrolled use of pesticides, which led to many of them being banned for use in the United States and many other countries.

ture deaths of at least 7 million people (some say as many as 500 million) from insect-transmitted diseases such as malaria (carried by the Anopheles mosquito), bubonic plague (carried by rat fleas), and typhus (carried by body lice and fleas).

- *They increase food supplies.* According to the FAO, 55% of the world's potential human food supply is lost to pests. Without pesticides, these losses would be worse and food prices would rise.

- *They increase profits for farmers.* Officials of pesticide companies estimate that every $1 spent on pesticides leads to an increase in U.S. crop yields worth approximately $4. (However, studies have shown that this benefit drops to about $2 if the harmful effects of pesticides are included.)

- *They work fast.* Pesticides control most pests quickly, have a long shelf life, and are easily shipped and applied. When genetic resistance occurs, farmers can use stronger doses or switch to other pesticides.

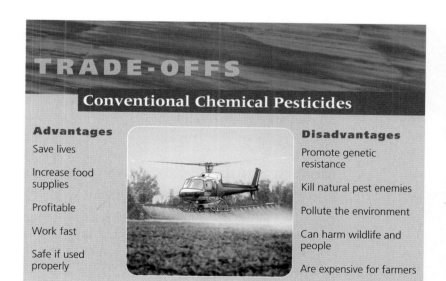

TRADE-OFFS

Conventional Chemical Pesticides

Advantages	Disadvantages
Save lives	Promote genetic resistance
Increase food supplies	Kill natural pest enemies
Profitable	Pollute the environment
Work fast	Can harm wildlife and people
Safe if used properly	Are expensive for farmers

Figure 12-20 Advantages and disadvantages of conventional chemical pesticides. **Question:** Which single advantage and which single disadvantage do you think are the most important? Why?

SCIENCE FOCUS

The Glyphosate-Resistant Crop Weed Management System: A Dilemma for Farmers and Scientists

Glyphosate is the world's best-selling herbicide, considered by farmers to be safe to use and highly effective in killing hundreds of species of weeds. It does not appear to harm animals or insects. And it does not leach into aquatic systems, but instead binds to soil particles and degrades into harmless substances within weeks.

Monsanto started selling glyphosate in 1974 under the trade name Roundup. Within 20 years, seed companies had engineered a variety of crops that are not affected by the herbicide. The idea was to create a system though which farmers could manage weeds easily and increase their yields, while herbicide and seed companies could substantially increase their sales.

The idea worked. The combination of glyphosate and glyphosate-resistant (GR) crops is a system now widely used in 22 countries. Of all the world's genetically modified crops planted in 2006, about 80% were glyphosate-resistant crops.

Another benefit was discovered. Farmers do not have to plow before planting GR crops. Instead they plant crop seeds with little or no tilling and spray emerging crops with glyphosate. This has led to a dramatic reduction in topsoil erosion.

Researchers once doubted that plants could ever evolve to be glyphosate-resistant, because they thought that the complex mutations required for such resistance were unlikely. But those researchers now have learned a lesson in unintended consequences, because GR weeds either evolved quickly or had existed already. By 2007, there were 12 known varieties of GR weeds, found in various parts of the world.

The problem is that glyphosate now dominates the herbicide market, and much of today's agriculture relies on this one chemical—a clear violation of the biodiversity **principle of sustainability** (see back cover).

If we end up with an epidemic of GR weeds, as some researchers fear, it could cost farmers and seed companies billions of dollars in lost productivity. More topsoil could be lost as farmers return to plowing, and farmers could be forced to use more harmful herbicides. Fossil fuel use would increase with more plowing, as would air pollution and greenhouse gas emissions.

The question is, are the benefits worth the costs of continuing the effort to engineer species interactions with use of herbicides and GM crops, or will that effort only result in an ever-escalating, ever more costly evolutionary battle between genetic engineers and weeds?

Critical Thinking

Do you think the benefits of using glyphosate, and new generations of similar herbicides to follow, outweigh their harmful effects on economies and the environment? Explain.

- *When used properly, their health risks are very low relative to their benefits.* Pesticide industry scientists argue that when pesticides are used in the approved regulatory manner, they pose no major risk to farm workers and consumers.

- *Newer pest control methods are safer and more effective than many older ones.* Greater use is being made of chemicals derived originally from plants. They are safer to use and less damaging to the environment than are many older pesticides. Genetic engineering is also being used to develop pest-resistant crop strains and genetically altered crops that produce natural pesticides.

Modern Synthetic Pesticides Have Several Disadvantages

Opponents of widespread pesticide use believe that the harmful effects of these chemicals (Figure 12-20, right) outweigh their benefits (Figure 12-20, left). They cite several serious problems with the use of conventional pesticides.

- *They accelerate the development of genetic resistance to pesticides by pest organisms.* Insects breed rapidly, and within 5 to 10 years (much sooner in tropical areas) they can develop immunity to widely used pesticides through natural selection and then come back stronger than before. Since 1945, about 1,000 species of insects and rodents (mostly rats) and 550 types of weeds and plant diseases have developed genetic resistance to one or more pesticides (Science Focus, above).

- *They can put farmers on a financial treadmill.* Because of genetic resistance, farmers can pay more and more for a pest control program that often becomes less and less effective.

- *Some insecticides kill natural predators and parasites that help control the pest populations.* Wiping out natural predators, such as spiders, can unleash new pests whose populations their predators had previously held in check and can cause other unexpected effects (Case Study, at right). Of the 300 most destructive insect pests in the United States, 100 were once minor pests that became major pests after widespread use of insecticides.

- *Pesticides do not stay put and can pollute the environment.* According to the U.S. Department of Agriculture (USDA), only 0.1–2% of the insecticides applied to crops by aerial spraying or ground spraying reaches the target pests, and less than 5% of herbicides applied to crops reaches the target weeds.

In other words, 98–99.9% of the insecticides and more than 95% of the herbicides we apply end up in the air, surface water, groundwater, bottom sediments, food, and nontarget organisms, including humans and wildlife.

- *Some pesticides harm wildlife.* According to the USDA and the U.S. Fish and Wildlife Service, each year, pesticides applied to cropland in the United States wipe out about 20% of U.S. honeybee colonies and damage another 15%. Pesticides also kill more than 67 million birds and 6–14 million fish. According to a 2004 study by the Center for Biological Diversity, pesticides also menace one of every three endangered and threatened species in the United States.

- *Some pesticides threaten human health.* The WHO and UNEP estimate that, each year, pesticides seriously poison at least 3 million agricultural workers in developing countries and at least 300,000 people in the United States. They also cause 20,000–40,000 deaths per year, worldwide. Each year, more than 250,000 people in the United States become ill because of household pesticide use. Such pesticides are a major source of accidental poisonings and deaths of young children.

According to studies by the National Academy of Sciences, exposure to legally allowed pesticide residues in food causes 4,000–20,000 cases of cancer per year in the United States. Some scientists are concerned about possible genetic mutations, birth defects, nervous system and behavioral disorders, and effects on the immune and endocrine systems from long-term exposure to low levels of various pesticides. (See more on this topic in Chapter 17 and *The Habitable Planet,* Video 7, at **www.learner.org/resources/series209.html**.) The pesticide industry disputes these claims, arguing that the exposures are not high enough to cause serious harm.

Children are much more susceptible to low levels of pesticides and other toxic chemicals, because on an amount-per-weight basis, they eat more food, drink more water, and breathe more air. They also put their fingers in their mouths more often and spend more time playing on grass, carpeting, and soil where pesticides can accumulate.

Figure 12-21 lists some ways in which you can reduce your exposure to pesticides.

Pesticide use has not reduced U.S. crop losses to pests, mostly because of genetic resistance and reduction of natural predators. When David Pimentel, an expert on insect ecology, evaluated data from more than 300 agricultural scientists and economists, he reached three major conclusions.

- *First,* although the use of synthetic pesticides has increased 33-fold since 1942, 37% of the U.S. food supply is lost to pests today compared to 31% in the 1940s. Since 1942, losses attributed to insects almost doubled from 7% to 13%, despite a tenfold increase in the use of synthetic insecticides.

WHAT CAN YOU DO?
Reducing Exposure to Pesticides

- Grow some of your food using organic methods
- Buy organic food
- Wash and scrub all fresh fruits, vegetables, and wild foods you pick
- Eat less meat or no meat
- Trim the fat from meat

Figure 12-21 Individuals matter: ways to reduce your exposure to pesticides. **Question:** Which three of these measures do you think are the most important? Why?

- *Second,* estimated environmental, health, and social costs of pesticide use in the United States, according to the International Food Policy Research Institute, are $5–10 in damages for every dollar spent on pesticides.

- *Third,* alternative pest management practices could halve the use of chemical pesticides on 40 major U.S. crops without reducing crop yields (**Concept 12-4**).

The pesticide industry disputes these findings. Trying to balance the economic and environmental advantages and disadvantages of pesticides is not easy and presents farmers and scientists with serious dilemmas (Science Focus, at left). However, numerous studies and experience show that pesticide use can be reduced sharply without reducing yields. Sweden has cut pesticide use in half with almost no decrease in crop yields. Campbell Soup uses no pesticides on tomatoes it grows in Mexico, and yields have not dropped. After a two-thirds cut in pesticide use on rice in Indonesia, yields increased by 15%.

■ CASE STUDY
Ecological Surprises

Malaria once infected nine of every ten people in North Borneo, now known as the eastern Malaysian state of Sabah. In 1955, the WHO began spraying the island with dieldrin (a relative of DDT) to kill malaria-carrying mosquitoes. The program was so successful that the dreaded disease was nearly eliminated.

Then unexpected things began to happen. The dieldrin also killed other insects, including flies and cockroaches living in houses. The islanders applauded. Next, small insect-eating lizards that also lived in the houses died after gorging themselves on dieldrin-contaminated insects.

Cats began dying after feeding on the lizards. In the absence of cats, rats flourished and overran the villages. When the people became threatened by sylvatic plague carried by rat fleas, the WHO parachuted healthy cats onto the island to help control the rats. Operation Cat Drop worked.

But then the villagers' roofs began to fall in. The dieldrin had killed wasps and other insects that fed on a type of caterpillar that either avoided or was not affected by the insecticide. With most of its predators eliminated, the caterpillar population exploded, munching its way through its favorite food: the leaves used to thatch roofs.

Ultimately, this episode ended well: both malaria and the unexpected effects of the spraying program were brought under control. Nevertheless, this chain of unintended and unforeseen events emphasizes the unpredictability of using insecticides. It reminds us that when we intervene in nature, we need to ask, "Now what will happen?"

HOW WOULD YOU VOTE?

Do the advantages of using synthetic chemical pesticides outweigh their disadvantages? Cast your vote online at **academic.cengage.com/biology/miller**.

Laws and Treaties Can Help to Protect Us from the Harmful Effects of Pesticides

Each year in the United States, about 2.4 million metric tons (2.6 million tons) of pesticides are used—an average of 4.6 metric tons (4.9 tons) per minute. They consist of 600 active (pest-killing) chemicals mixed with 1,200 solvents, preservatives, and other supposedly inactive ingredients in about 25,000 commercial pesticide products. Three U.S. federal agencies, the EPA, the USDA, and the Food and Drug Administration (FDA), regulate the sale and use of pesticides under the Federal Insecticide, Fungicide, and Rodenticide Act (FIFRA), first passed in 1947 and amended in 1972.

Under FIFRA, the EPA was supposed to assess the health risks of the active ingredients in pesticide products already in use. But after more than 30 years, less than 10% of the active ingredients in pesticide products have been evaluated using tests for chronic health effects. Serious evaluation of the health effects of the 1,200 inactive ingredients used in pesticide products began only recently. The EPA says that it has not had the funds to carry out this complex and lengthy evaluation process.

In 1996, Congress passed the Food Quality Protection Act, mostly because of growing scientific evidence and citizen pressure concerning the effects of small amounts of pesticides on children. This act re-

quires the EPA to reduce the allowed levels of pesticide residues in food by a factor of 10 when there is inadequate information on the potentially harmful effects on children.

There is controversy over how well U.S. citizens are protected from the harmful effects of pesticides. Between 1972 and 2007, the EPA used FIFRA to ban or severely restrict the use of 64 active pesticide ingredients, including DDT and most other chlorinated hydrocarbon insecticides. However, according to studies by the National Academy of Sciences, federal laws regulating pesticide use in the United States are inadequate and poorly enforced by the three agencies. One study by the National Academy of Sciences found that as much as 98% of the potential risk of developing cancer from pesticide residues on food grown in the United States would be eliminated if EPA standards were as strict for pre-1972 pesticides as they are for newer ones.

Although laws within countries protect citizens to some extent, banned or unregistered pesticides may be manufactured in one country and exported to other countries. For example, U.S. pesticide companies make and export to other countries pesticides that have been banned or severely restricted—or never even evaluated—in the United States. Other industrial countries also export banned or unapproved pesticides.

But what goes around can come around. In what environmental scientists call a *circle of poison* or the *boomerang effect*, residues of some banned or unapproved chemicals exported to other countries can return to the exporting countries on imported food. The wind can also carry persistent pesticides such as DDT from one country to another (Chapter 7 Core Case Study, p. 140).

Environmental scientists and environmentalists have urged the U.S. Congress—without success—to ban such exports. Supporters of the exports argue that such sales increase economic growth and provide jobs, and that banned pesticides are exported only with the consent of the importing countries. They also contend that if the United States did not export pesticides, other countries would.

In 1998, more than 50 countries developed an international treaty that requires exporting countries to have informed consent from importing countries for exports of 22 pesticides and 5 industrial chemicals. In 2000, more than 100 countries developed an international agreement to ban or phase out the use of 12 especially hazardous persistent organic pollutants (POPs)—9 of them persistent hydrocarbon pesticides such as DDT and other chemically similar pesticides. The United States has not signed this agreement.

THINKING ABOUT
Exporting Pesticides

Should companies be allowed to export pesticides that have been banned, severely restricted, or not approved for use in their home countries? Explain.

There Are Alternatives to Using Pesticides

Many scientists believe we should greatly increase the use of biological, ecological, and other alternative methods for controlling pests and diseases that affect crops and human health (**Concept 12-4**). Here are some of these alternatives:

- *Fool the pest.* A variety of *cultivation practices* can be used to fake out pest species. For example, rotating the types of crops planted in a field each year will keep specialists that eat one type of crop searching from year to year for their food. Other methods include adjusting planting times, so major insect pests either starve or get eaten by their natural predators, and growing crops in areas where their major pests do not exist.

- *Provide homes for pest enemies.* Farmers can increase the use of polyculture, which uses plant diversity to reduce losses to pests. Homeowners can reduce weed invasions by cutting grass no lower than 8 centimeters (3 inches) high. This height provides a dense enough cover to keep out crabgrass and many other undesirable weeds.

- *Implant genetic resistance.* Use genetic engineering to speed up the development of pest- and disease-resistant crop strains (Figure 12-22). But controversy persists over whether the projected advantages of using genetically modified plants and foods such as golden rice (**Core Case Study**) outweigh their projected disadvantages (Figure 12-16, right).

 🔙 CORE CASE STUDY

- *Bring in natural enemies.* Use *biological control* by importing natural predators (Figures 12-19 and 12-23), parasites, and disease-causing bacteria and viruses to help regulate pest populations. This approach is nontoxic to other species, minimizes genetic resistance, and can save large amounts of money—about $25 for every $1 invested in controlling 70 pest species in the United States. However, biological control agents cannot always be mass-produced, are often slower acting and more difficult to apply than conventional pesticides, can sometimes multiply and become pests themselves, and must be protected from pesticides sprayed in nearby fields.

- *Use insect perfumes.* Sex attractants (called *pheromones*) can lure pests into traps or attract their natural predators into crop fields (usually the more effective approach). These chemicals attract only one species, work in trace amounts, have little chance of causing genetic resistance, and are not harmful to nontarget species. However, it is costly and time-consuming to identify, isolate, and produce the specific sex attractant for each pest or predator.

Monsanto

Figure 12-22 Solutions: the results of one example of using *genetic engineering* to reduce pest damage. Both tomato plants were exposed to destructive caterpillars. The normal plant's leaves are almost gone (left), whereas the genetically altered plant shows little damage (right). **Question:** Would you have any concerns about eating the genetically engineered tomato? Why or why not?

- *Bring in the hormones. Hormones* are chemicals produced by animals to control developmental processes at different stages of life. Scientists have learned how to identify and use hormones that disrupt an insect's normal life cycle, thereby preventing it from reaching maturity and reproducing. Insect hormones have the same advantages as sex

Scott Bauer/USDA Agricultural Research Service

Figure 12-23 Natural capital: biological pest control. Wasp parasitizing a gypsy moth caterpillar.

attractants. But they take weeks to kill an insect, often are ineffective with large infestations of insects, and sometimes break down before they can act. In addition, they must be applied at exactly the right time in the target insect's life cycle, can sometimes affect the target's predators and other beneficial species, and are difficult and costly to produce.

- *Scald them.* Some farmers have controlled certain insect pests by *spraying them with hot water.* This approach has worked well on cotton, alfalfa, and potato fields and in citrus groves in the U.S. state of Florida, and its cost is roughly equal to that of using chemical pesticides. However, it requires large amounts of water and energy to heat it.

Integrated Pest Management Is a Component of Sustainable Agriculture

Many pest control experts and farmers believe the best way to control crop pests is a carefully designed **integrated pest management (IPM)** program. In this more sustainable approach, each crop and its pests are evaluated as parts of an ecological system. Then farmers develop a control program that uses a combination of cultivation, biological controls, and chemical tools and techniques, applied in a carefully coordinated way (**Concept 12-4**).

The overall aim of IPM is to reduce crop damage to an economically tolerable level. Each year, crops are moved from field to field to disrupt pest infestations, and fields are monitored carefully. When an economically damaging level of pests is reached, farmers first use biological methods (natural predators, parasites, and disease organisms) and cultivation controls. They also use large machines to vacuum up harmful bugs. They apply small amounts of insecticides—mostly those naturally produced by plants—only as a last resort and in the smallest amounts possible. Broad-spectrum, long-lived pesticides are not used, and different chemicals are used alternately to slow the development of genetic resistance and to reduce the killing predators of pest species.

In 1986, the Indonesian government banned 57 of the 66 pesticides used on rice and phased out pesticide subsidies over a 2-year period. It also launched a nationwide education program to help farmers switch to IPM. The results were dramatic: between 1987 and 1992, pesticide use dropped by 65%, rice production rose by 15%, and more than 250,000 farmers were trained in IPM techniques. For more information and animations, see *The Habitable Planet*, Video 7, at **www .learner.org/resources/series209.html**. Sweden and Denmark have used IPM to cut their pesticide use by more than half. Cuba, which uses organic farming to grow its crops, makes extensive use of IPM. In Brazil, IPM has reduced pesticide use on soybeans by as much as 90%.

According to a 2003 study by the U.S. National Academy of Sciences, these and other experiences show that a well-designed IPM program can reduce pesticide use and pest control costs by 50–65% without reducing crop yields and food quality. IPM can also reduce inputs of fertilizer and irrigation water, and slow the development of genetic resistance, because pests are assaulted less often and with lower doses of pesticides. IPM is an important form of *pollution prevention* that reduces risks to wildlife and human health and applies the population control **principle of sustainability** (see back cover).

Despite its promise, IPM—like any other form of pest control—has some disadvantages. It requires expert knowledge about each pest situation and takes more time than does using conventional pesticides. Methods developed for a crop in one area might not apply to areas with even slightly different growing conditions. Initial costs may be higher, although long-term costs typically are lower than those of using conventional pesticides. Widespread use of IPM is hindered by government subsidies for using conventional chemical pesticides, opposition by pesticide manufacturers, and a shortage of IPM experts. **GREEN CAREER:** Integrated pest management

A 1996 study by the National Academy of Sciences recommended that the United States shift from chemical-based approaches to ecological-based approaches for managing pests. According to this study, within 5–10 years, such a shift could cut U.S. pesticide use in half, as it has in several other countries.

A growing number of scientists urge the USDA to use three strategies to promote IPM in the United States:

- Add a 2% sales tax on pesticides and use the revenue to fund IPM research and education.

- Set up a federally supported IPM demonstration project on at least one farm in every U. S. county.

- Train USDA field personnel and county farm agents in IPM so they can help farmers use IPM.

The pesticide industry has successfully opposed such measures.

HOW WOULD YOU VOTE? ☑

Should governments heavily subsidize a switch to integrated pest management? Cast your vote online at **academic .cengage.com/biology/miller**.

Several U.N. agencies and the World Bank have joined together to establish an IPM facility. Its goal is to promote the use of IPM by disseminating information and establishing networks among researchers, farmers, and agricultural extension agents involved in IPM.

12-5 How Can We Improve Food Security?

▶ **CONCEPT 12-5** We can improve food security by creating programs to reduce poverty and chronic malnutrition, relying more on locally grown food, and cutting food waste.

Use Government Policies to Improve Food Production and Security

Agriculture is a financially risky business. Whether farmers have a good or bad year depends on factors over which they have little control: weather, crop prices, crop pests and diseases, interest rates, and global markets. Because of the need for reliable food supplies despite fluctuations in these factors, most governments provide various forms of assistance to farmers, fishers, and consumers.

Governments use three main approaches to influence food production:

- *Control prices.* Use price controls to keep food prices artificially low. Consumers are happy, but farmers may not be able to make a living.

- *Provide subsidies.* Give farmers price supports, tax breaks, and other subsidies to keep them in business and to encourage them to increase food production. According to the United Nations, government subsidies account for about 31% of global farm income. Globally, farm subsidies in the developed countries average about $280 billion a year—an average of $530,000 a minute! If government subsidies are too generous and the weather is good, farmers and livestock producers may produce more food than can be sold. The resulting surplus depresses food prices, which reduces the financial incentive for farmers in developing countries to increase domestic food production.

 Some analysts call for phasing out such subsidies. For example, in 1984, New Zealand ended farm subsidies. After the shock wore off, innovation took over, and production of some foods such as milk quadrupled. Other analysts call for replacing traditional farming subsidies with subsidies that promote more sustainable farming practices.

 Similarly, government subsidies to fishing fleets (p. 264) can promote overfishing and reduction of aquatic biodiversity. For example, governments give the highly destructive bottom trawling industry (Figure 11-3, p. 251) about $152 million in subsidies a year, which is the main reason they can stay in business. Many analysts call for replacing those subsidies with payments that promote more sustainable fishing and aquaculture.

- *Let the marketplace decide.* Another approach is to eliminate most or all price controls and subsidies and let farmers and fishers respond to market de-

mand without government interference. Some analysts urge that any phase-out of farm and fishery subsidies should be coupled with increased aid for the poor and the lower middle class, who would suffer the most from any increase in food prices.

HOW WOULD YOU VOTE?

Should governments phase out all or most subsidies for conventional agriculture and phase in subsidies for more sustainable agriculture? Cast your vote online at **academic .cengage.com/biology/miller**.

Government programs that reduce poverty by helping the poor to help themselves can improve food security (**Concept 12-1B**). An essential part of this would be to reduce the population growth rate. Such programs should thus promote family planning, education and jobs (especially for women), and small loans to poor people to help them start businesses or buy land for growing their own food.

Some analysts urge special programs focused on saving children from poverty. Studies by the United Nations Children's Fund (UNICEF) indicate that one-half to two-thirds of nutrition-related childhood deaths could be prevented at an average annual cost of $5–$10 per child with the following measures:

- Immunizing children against childhood diseases such as measles

- Encouraging breast-feeding (except for mothers with AIDS)

- Preventing dehydration from diarrhea by giving infants a mixture of sugar and salt in a glass of water

- Preventing blindness by giving children a vitamin A capsule twice a year at a cost of about 75¢ per child. Other options are fortifying common foods with vitamin A and other micronutrients at a cost of about 10¢ per child annually and widespread planting of golden rice in developing countries (**Core Case Study**). 🔁 CORE CASE STUDY

- Providing family planning services to help mothers space births at least 2 years apart

- Increasing education for women, with emphasis on nutrition, drinking water sterilization, family planning, and childcare

Analysts note that *preventing* health problems in children is an important step toward global food security.

12-6 How Can We Produce Food More Sustainably?

▶ **CONCEPT 12-6A** Sustainable food production will require reducing topsoil erosion, sharply reducing overgrazing and overfishing, irrigating more efficiently, using integrated pest management, promoting agrobiodiversity, and providing government subsidies for more sustainable farming, fishing, and aquaculture.

▶ **CONCEPT 12-6B** Producing enough food to feed the rapidly growing human population will require growing crops in a mix of monocultures and polycultures and decreasing the enormous environmental impacts of industrialized food production.

Reduce Soil Erosion

Soil conservation involves using a variety of ways to reduce soil erosion and restore soil fertility, mostly by keeping the soil covered with vegetation.

Figure 12-24 shows some of the methods farmers have used to reduce soil erosion (**Concept 12-6A**). For example, *terracing* is a way to grow food on steep slopes without depleting topsoil. It is done by converting steeply sloped land into a series of broad, nearly level terraces that run across the land's contours (Figure 12-24a and Figure 12-1). This retains water for crops at each level and reduces soil erosion by controlling runoff.

When the ground has a significant slope, a technique known a *contour planting* (Figures 12-24b and 12-25) can be used to reduce soil erosion. It involves plowing and planting crops in rows across the slope of the land rather than up and down. Each row acts as a small dam to help hold topsoil and to slow water runoff.

Strip cropping (Figure 12-24b) involves planting alternating strips of a row crop (such as corn or cotton) and another crop that completely covers the soil, called a *cover crop* (such as alfalfa, clover, rye, or a grass–

(a) Terracing

(b) Contour planting and strip cropping

(c) Alley cropping

(d) Windbreaks

Figure 12-24 Soil conservation methods include **(a)** terracing, **(b)** contour planting and strip cropping, **(c)** alley cropping, and **(d)** windbreaks (**Concept 12-6A**).

Figure 12-25 Solutions: mixture of monoculture crops planted in strips on a farm in central Wisconsin (USA). This farm practices sound soil conservation by planting crops with the contours of the land (*contour planting*) and alternating crops in strips (*strip cropping*) to help reduce soil erosion and depletion of soil nutrients.

Ron Nichols/USDA Natural Resources Conservation Service

legume mixture). The cover crop traps topsoil that erodes from the row crop and catches and reduces water runoff (Figure 12-25). When one crop is harvested the other strip is left to catch and reduce water runoff. Other ways to reduce erosion are to leave crop residues on the land after the crops are harvested or to plant cover crops immediately after harvest to help protect and hold the topsoil.

Alley cropping, or *agroforestry* (Figure 12-24c), is yet another way to slow erosion. One or more crops are planted together in strips or alleys between trees and shrubs, which provide shade. This reduces water loss by evaporation and helps retain and slowly release soil moisture—an insurance policy during prolonged drought. The trees also can provide fruit, fuelwood, and trimmings that can be used as mulch for the crops (green manure) and as feed for livestock.

Some farmers establish *windbreaks,* or *shelterbelts,* of trees around crop fields to reduce wind erosion (Figure 12-24d). These trees also retain soil moisture, supply wood for fuel, increase crop productivity by 5–10%, and provide habitats for birds, pest-eating and pollinating insects, and other animals.

However, many farmers do not practice these known ways to reduce soil erosion, because they are in a desperate struggle to survive, or they are more interested in increasing short-term income, even if it leads to long-term environmental degradation.

Eliminating the plowing and tilling of soil greatly reduces soil erosion. Many farmers in the United States and several other countries practice *conservation-tillage farming* by using special tillers and planting machines that drill seeds directly through crop residues into the undisturbed soil. The only soil disturbance is a narrow slit, and weeds are controlled with herbicides. Such

no-till and *minimum-tillage* farming also increases crop yields, reduces the threat of global warming by storing more carbon in the soil, and lowers use of water and tractor fuel.

In 2007, farmers used conservation tillage on about 41% of U.S. cropland, helped by the use of herbicides such as glyphosate (Science Focus, p. 296). The USDA estimates that using conservation tillage on 80% of U.S. cropland would reduce soil erosion by at least half. No-till cultivation is now spreading rapidly into countries such as Brazil, Argentina, Canada, and Australia. It also has great potential to reduce soil erosion and raise crop yields in dry regions in Africa and the Middle East. Conservation tillage, however, is not a cure-all. It requires costly machinery and increases use of herbicides, works better in some areas than in others, and is more useful for some crops than for others.

An additional way to conserve the earth's topsoil is to retire the estimated one-tenth of the world's cropland that is highly erodible and accounts for the majority of the world's soil erosion. The goal would be to identify these *erosion hotspots,* withdraw them from cultivation, and plant them with grasses or trees.

■ CASE STUDY
Soil Erosion in the United States— Learning from the Past

In the United States, a third of the country's original topsoil is gone and much of the rest is degraded. In the state of Iowa, which has the world's highest concentration of prime farmland, half of the topsoil is gone after a century of farming. According to the Natural Resources

Figure 12-26 **Natural capital degradation:** dust storm, driven by wind blowing across eroded soil, approaching Stratford, Texas (USA), in 1935.

NOAA George E. Marsh Album

Conservation Service, 90% of American farmland is, on average, losing topsoil 17 times faster than new topsoil is being formed.

In the 1930s, Americans learned a harsh environmental lesson when much of the topsoil in several dry and windy midwestern states was lost because of a combination of poor cultivation practices and prolonged drought. This threatened to turn much of the U.S. Great Plains into a vast desert. Before settlers began grazing livestock and planting crops there in the 1870s, the deep and tangled root systems of native prairie grasses anchored the fertile topsoil firmly in place. But plowing the prairie tore up these roots, and the crops that settlers planted annually in their place had less extensive root systems. After each harvest, the land was plowed and left bare for several months, exposing it to high winds. Overgrazing by livestock in some areas also destroyed large expanses of grass, denuding the ground.

The stage was set for severe wind erosion and crop failures; all that was needed was a long drought. It came between 1926 and 1937 when the annual precipitation dropped by almost two-thirds. In the 1930s, dust clouds created by hot, dry windstorms blowing across the barren exposed soil darkened the sky at midday in some areas (Figure 12-26). Rabbits and birds choked to death on the dust.

During May 1934, a cloud of topsoil blown off the Great Plains traveled some 2,400 kilometers (1,500 miles) and blanketed most of the eastern United States with dust. Laundry hung out to dry by women in the state of Georgia quickly became covered with dust blown in from the Midwest. Journalists gave the most

eroded part of the Great Plains a new name: the *Dust Bowl* (Figure 12-27).

This calamity triggered one of the largest internal migrations in U.S. history. Thousands of farm families from the states of Oklahoma, Texas, Kansas, and Colorado abandoned their dust-choked farms and dead livestock and migrated to California or to the industrial cities of the Midwest and East. Most found no jobs because the country was in the midst of the Great Depression.

In May 1934, Hugh Bennett of the USDA went before a congressional hearing in Washington, D.C., to plead for new programs to protect the country's topsoil. Lawmakers took action when Great Plains dust began seeping into the hearing room. As Bennett put it, "This nation and civilization is founded upon nine inches of topsoil. And when that is gone there will no longer be any nation or any civilization."

In 1935, the United States passed the *Soil Erosion Act,* which established the Soil Conservation Service (SCS) as part of the USDA. Soil conservation districts were formed throughout the country, and farmers and ranchers were given technical assistance to set up soil conservation programs. (The SCS is now called the Natural Resources Conservation Service.)

Of the world's major food-producing nations, only the United States is sharply reducing some of its soil losses through a combination of conservation-tillage farming and government-sponsored soil conservation programs (**Concept 12-6A**). Under the 1985 Food Security Act (Farm Act), farmers participating in the Conservation Reserve Program receive a subsidy for taking highly erodible land out of production and replanting it with topsoil-saving grass or trees for 10–15 years. Since 1985, these efforts have cut soil losses on U.S. cropland by 40%.

However, effective soil conservation is practiced today on only half of all U.S. agricultural land. Also, in recent years, some farmers have been taking erodible

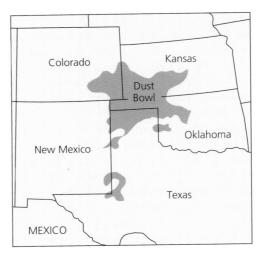

Figure 12-27 **Natural capital degradation:** the *Dust Bowl* of the Great Plains, where a combination of extreme drought and poor soil conservation practices led to severe wind erosion of topsoil in the 1930s.

land out of the conservation reserve in order to receive government subsidies for planting corn to make ethanol for use as a motor vehicle fuel.

Restore Soil Fertility

The best way to maintain soil fertility is through soil conservation. The next best option is to restore some of the plant nutrients that have been washed, blown, or leached out of the soil, or that have been removed by repeated crop harvesting. To do this, farmers can use **organic fertilizer** made from plant and animal wastes or **commercial inorganic fertilizer** produced from various minerals.

There are several types of *organic fertilizers.* One is **animal manure:** the dung and urine of cattle, horses, poultry, and other farm animals. It improves soil structure, adds organic nitrogen, and stimulates beneficial soil bacteria and fungi. Another type, called **green manure** consists of freshly cut or growing green vegetation that is plowed into the topsoil to increase the organic matter and humus available to the next crop. A third type is **compost,** produced when microorganisms in soil break down organic matter such as leaves, crop residues, food wastes, paper, and wood in the presence of oxygen. (See the website for this chapter for more details on composting.)

Crops such as corn and cotton can deplete nutrients in the topsoil (especially nitrogen) if they are planted on the same land several years in a row. *Crop rotation* provides one way to reduce these losses. Farmers plant areas or strips with nutrient-depleting crops one year. The next year, they plant the same areas with legumes whose root nodules add nitrogen to the soil. This not only helps to restore soil nutrients but also reduces erosion by keeping the topsoil covered with vegetation.

Many farmers (especially in developed countries) rely on *commercial inorganic fertilizers.* The active ingredients typically are inorganic compounds that contain *nitrogen, phosphorus,* and *potassium.* Other plant nutrients may be present in low or trace amounts. Inorganic fertilizer use has grown more than elevenfold since 1950 and now accounts for about one-fourth of the world's crop yield. Without careful control, these fertilizers can run off the land and pollute nearby bodies of water and coastal estuaries where rivers empty into the sea.

These fertilizers can replace depleted inorganic nutrients, but they do not replace organic matter. To completely restore nutrients to soil, both inorganic and organic fertilizers should be used.

Reduce Soil Salinization and Desertification

We know how to prevent and deal with soil salinization, as summarized in Figure 12-28.

SOLUTIONS

Soil Salinization

Prevention

Reduce irrigation

Switch to salt-tolerant crops (such as barley, cotton, and sugar beet)

Cleanup

Flush soil (expensive and wastes water)

Stop growing crops for 2–5 years

Install underground drainage systems (expensive)

Figure 12-28 Methods for preventing and cleaning up soil salinization (**Concept 12-6A**). **Question:** Which two of these solutions do you think are the most important? Why?

Desertification is not an easy problem to deal with. We cannot control the timing and location of prolonged droughts caused by natural factors. But we can reduce population growth, overgrazing, deforestation, and destructive forms of planting, irrigation, and mining, which have left much land vulnerable to soil erosion and thus desertification.

We can also work to decrease the human contribution to global warming, which is expected to increase severe and prolonged droughts in larger areas of the world during this century. It is possible to restore land suffering from desertification by planting trees and grasses that anchor topsoil and hold water, by establishing windbreaks (Figure 12-24d), and by growing trees and crops together (Figure 12-24c).

Practice More Sustainable Aquaculture

Figure 12-29 (p. 306) lists some ways to make aquaculture more sustainable and to reduce its harmful environmental effects. Open-ocean aquaculture, which the United States is planning to develop, is one such alternative. It involves raising large carnivorous fish in underwater pens located up to 300 kilometers (190 miles) offshore (Figure 11-7, p. 256). The fish are fattened with fish meal supplied by automated buoys. Using another approach, scientists are eliminating damage to coastal areas in Florida by raising shrimp far inland in zero-discharge freshwater ponds. **GREEN CAREER:** Aquaculture

Figure 12-29 Ways to make aquaculture more sustainable and to reduce its harmful effects. **Question:** Which two of these solutions do you think are the most important? Why?

However, making aquaculture more sustainable will require some fundamental changes, one of which is for consumers to choose more often from species that are lower on the food chain—those that feed on plants rather than on other fish. For example, raising carnivorous fishes such as salmon, trout, tuna, grouper, and cod contributes to overfishing of species used to feed these carnivores. Raising plant-eating fishes such as carp and tilapia (called the chicken of fish farming) does not add to this problem.

On the production end, fish farmers could employ *polyaquaculture,* which in fact has been part of aquaculture for centuries, especially in Southeast Asia. Polyaquaculture operations raise fish or shrimp along with algae, seaweeds, and shellfish in coastal lagoons, ponds, and tanks. The wastes of the fish or shrimp feed the other species, and in the best of these operations, there are just enough wastes from the first group to feed the second group. This applies the nutrient recycling and biodiversity **principles of sustainability.**

Produce Meat More Efficiently and Humanely

Ultimately, consumers need to evaluate their use of meat in their diets relative to the harmful environmental impacts caused by its increasingly industrialized production. Currently, about 38% of the world's grain harvest and 37% of the world's fish catch are used to produce animal protein. If everyone in the world today had the average U.S. meat-based diet, the world's current annual grain harvest could feed only about 2.5 billion people. People in many European countries consume about half as much grain per person as people in the United States consume. If everyone followed this diet, the current grain harvest would support about 5 billion people.

A more sustainable form of meat production and consumption would involve shifting from less grain-efficient forms of animal protein, such as beef and pork, to more grain-efficient forms, such as poultry and herbivorous farmed fish (Figure 12-30). Such a shift is under way. Since 1996, poultry has taken the lead over beef in the marketplace, and within a decade or so, production of farmed herbivorous fish may exceed beef production.

We can find alternatives to growing grain to feed livestock. In 1997, India became the world's largest producer of milk and other dairy products. The entire industry is based mostly on feeding dairy cows roughage such as rice straw, wheat straw, corn stalks, and grass gathered from roadsides. Another approach is to develop meat substitutes. Researchers at Vrjie University in the Netherlands are using peas and other protein-rich legumes to develop alternatives to meats that are affordable, highly nutritious, easy to prepare, and tasty.

Some individuals and groups consider it unethical and inhumane to raise livestock in crowded feedlots and pens under factory-like conditions. In 2004, Whole Foods Market, a natural and organic foods supermarket chain in the United States, committed $500,000 to establish a foundation to look for more humane ways to raise livestock in feedlots. And McDonald's, Wendy's, and Burger King have hired specialists to develop new standards for improving animal welfare. In 2005, the 165 member countries of the World Organization for Animal Health adopted voluntary standards for the more humane transportation and slaughter of livestock animals.

Move Down the Food Chain and Slow Population Growth

Food production, especially producing meat, has a huge environmental impact (Figure 12-9). Indeed, eating meat, driving fuel-inefficient motor vehicles, and liv-

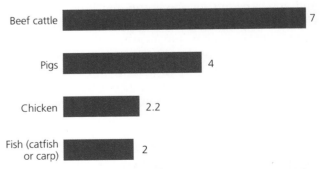

Figure 12-30 Efficiency of converting grain into animal protein. Data in kilograms of grain per kilogram of body weight added. **Question:** If you eat meat, what changes could you make in your meat-eating habits to reduce your environmental impact? (Data from U.S. Department of Agriculture)

ing and working in energy-inefficient buildings make the three largest contributions to the ecological and carbon footprints of most individuals in affluent nations. According to a 2006 study by the U.N. Food and Agriculture Organization, *Livestock's Long Shadow*, meat production and delivery cause 40% more greenhouse gas emissions than all of the world's planes, cars, trucks, and other forms of transport combined.

As a result, some people are eating less meat and switching from energy-inefficient beef, pork, and carnivorous fish produced by aquaculture to more energy-efficient chicken and herbivorous fish produced by aquaculture (Figure 12-29). In addition to reducing their ecological and carbon footprints, this switch can improve their health and increase their life expectancy. Research indicates that people who live on a Mediterranean-type diet, which includes moderate amounts of meat, cheese, and seafood, are healthier and live longer than those who live on a diet that includes high levels of meat consumption.

Some people in affluent societies are going further and eliminating most or all meat from their diet. They are replacing a meat diet with a vegetarian diet that includes a healthy combination of fruits, vegetables, protein-rich foods such as legumes, and in some cases, moderate amounts of fish.

THINKING ABOUT

Meat Consumption CONCEPT LINK

Would you be willing to live lower on the food chain (**Concept 3-4B**, p. 61) by eating less meat or no meat? Explain.

Slowing population growth (pp. 133–137) reduces the harmful environmental impacts of agriculture by reducing the number of people consuming meat and other types of food. It also applies the population control **principle of sustainability**.

Shift to More Sustainable Agriculture

Sustainability experts agree that sustainable world food production will require us to develop and phase in more sustainable, low-input agricultural systems over the next few decades. One component of this is increased use of **organic agriculture** in which crops are grown with little or no use of synthetic pesticides, synthetic fertilizers, or genetically engineered seeds. Livestock are raised without use of genetic engineering, synthetic growth regulators, or feed additives. Fields must be free of chemicals for 3 years before crops grown there can be certified as organic. Since 1990, organic farming has been the fastest growing sector of the agricultural economy.

SOLUTIONS

Sustainable Organic Agriculture

More	Less
High-yield polyculture	Soil erosion
Organic fertilizers	Aquifer depletion
Biological pest control	Overgrazing
Integrated pest management	Overfishing
Efficient irrigation	Loss of biodiversity
Perennial crops	Food waste
Crop rotation	Subsidies for unsustainable farming and fishing
Water-efficient crops	
Soil conservation	Soil salinization
Subsidies for sustainable farming and fishing	Population growth
	Poverty

Figure 12-31 Major components of more sustainable, low-throughput agriculture based mostly on mimicking and working with nature (**Concept 12-6B**). **Question:** Which two solutions do you think are the most important? Why?

Figure 12-31 lists the major components of more sustainable agriculture. Compared to high-input farming, low-input agriculture produces similar yields with lower carbon dioxide emissions, uses less energy per unit of yield, improves soil fertility, reduces soil erosion, and can often be more profitable for farmers (**Concept 12-6B**).

Since 1969, Cubans have grown most of their food using low-input sustainable organic agriculture. The government has established centers where organisms used for biological pest control are produced. The government also encourages people to grow organic food in urban gardens.

In 2002, agricultural scientists Paul Mader and David Dubois reported the results of a 22-year study comparing organic and conventional farming at the Rodale Institute in Kutztown, Pennsylvania (USA). Figure 12-32 (p. 308) summarizes their conclusions, along with those from a 2005 evaluation of the study by David Pimentel and other researchers. They concluded that yields of organic crops can be as much as 20% lower than yields of conventionally raised crops. But organic farmers can often make up for this difference by not having to use or pay for expensive pesticides, herbicides, and synthetic fertilizers, and usually,

Organic Farming

- Improves soil fertility

- Reduces soil erosion

- Retains more water in soil during drought years

- Uses about 30% less energy per unit of yield

- Lowers CO_2 emissions

- Reduces water pollution by recycling livestock wastes

- Eliminates pollution from pesticides

- Increases biodiversity above and below ground

- Benefits wildlife such as birds and bats

Figure 12-32 Environmental benefits of organic farming over conventional farming, based on 22 years of research comparing these two systems at the Rodale Institute in Kutztown, Pennsylvania USA. (Data from Paul Mader, David Dubois, and David Pimentel).

by getting higher prices for their crops. As a result, the net economic return per unit of land from organic crop production is often equal to or higher than that from conventional crop production. However, there is a need for more research to evaluate the benefits of organic farming (Science Focus, below).

Currently, certified organic farming is used on less than 1% of the world's cropland—on only 0.3% in the United States, but on 6–10% of the cropland in many European countries. Tiny Liechtenstein devotes 18% of its land to certified organic agriculture.

> **RESEARCH FRONTIER:**
> Organic agriculture. See **academic.cengage.com/biology/ miller**.

Some farmers have shown that they can use energy from the sun, wind, and flowing water, and natural gas produced from farm wastes in biogas digesters, to produce most or all of the energy they need for food production—an application of the solar energy **principle of sustainability**. What is more, they can make money by selling their excess electricity to power companies.

Most proponents of more sustainable agriculture are not opposed to high-yield agriculture. Instead, they see it as vital for protecting the earth's biodiversity by reducing the need to cultivate new and often marginal land. They call for using more environmentally sustainable forms of both high-yield polyculture and high-yield monoculture, with increasing emphasis on using organic methods for growing crops (Figure 12-31, left)

SCIENCE FOCUS

Scientists Are Studying the Benefits and Costs of Organic Farming

Are organically produced foods really safer to consume and better for the environment? Not all scientists agree on the answer, and much research remains to be done.

In one study reported in 2006, researchers John B. Reganold and Sasha B. Kramer used controlled experiments (Chapter 2 Core Case Study, p. 28) to compare some benefits of growing apples organically and growing them conventionally. In particular, they wanted to compare the amount of nitrogen leaching into soils from orchards treated in different ways. Nitrogen in the form of nitrates can contaminate underground drinking water supplies, which can cause illnesses in people who drink that water, especially small children.

The researchers worked in an apple orchard in the Yakima Valley in the central part of the U.S. state of Washington. They studied three groups of apple trees—one group treated with conventional synthetic fertilizers, including calcium nitrate; one treated organically and fertilized with composted chicken manure and alfalfa residues; and a third group treated through an *integrated approach,* (combining organic and conventional techniques) using equal amounts of chicken manure and calcium nitrate. Each of the fertilizers used—organic and synthetic—contained nitrogen. All three study plots were fertilized once in May and once in October, and each tree received the same amount of nitrogen each time, regardless of the source.

The researchers placed a plastic resin bag about 1 meter (40 inches) below each tree to collect water for measuring nitrate leaching. They found that water from under the conventionally grown trees contained nitrate concentrations 4.4 to 5.6 times higher than those in water from under the organically treated trees. Water from the third group contained concentrations of nitrates at levels between those of the other two groups. This led Reganold to conclude that the study supports organic farming techniques but also gives credibility to the integrated approach.

Critical Thinking

Can you think of any factors, other than the fertilizers applied, that might have affected the results of this experiment? Explain.

SCIENCE FOCUS

Sustainable Polycultures of Perennial Crops

Some plant scientists believe we should rely more on *polycultures of perennial crops*, which can live for many years without having to be replanted and are better adapted to some regional soil and climate conditions than most annual crops.

When you think about farms in the U.S. state of Kansas, you probably picture seemingly endless fields of wheat or corn plowed up and planted each year. By 2040, the picture might change, thanks to over 3 decades of pioneering research at the nonprofit *Land Institute* near Salina, Kansas.

The institute, founded by plant geneticist Wes Jackson, is experimenting with an ecological approach to agriculture on the midwestern prairie. The goal is to grow a diverse mixture (polyculture) of edible perennial plants to supplement traditional annual monoculture crops and to help reduce the latter's harmful environmental effects. Examples in this polyculture mix include perennial grasses, legumes (a source of nitrogen fertilizer), sunflowers, grain crops, and plants that provide natural insecticides. Some of these plants could also be used as a source of renewable biofuel for motor vehicles.

This approach, called *natural systems agriculture*, copies nature by growing a di-

Figure 12-C Comparison of the roots of an annual wheat crop plant (left) with the roots of a tallgrass prairie perennial plant, big bluestem (right). The perennial plant is in the ground year-round and is much better at using water and nutrients and making and maintaining healthy soil. It also needs less fertilizer.

versity of perennial crops. It has a number of environmental benefits. Because these plants are perennials, there is no need to till the soil and replant seeds each year. This reduces soil erosion and water pollution from eroded sediment, because the unplowed soil is not exposed to wind and rain. And it reduces the need for irrigation because the deep roots of such perennials retain more water than do the shorter roots of annuals (Figure 12-C). Also, there is little or no need for chemical fertilizers and pesticides, and thus little or no pollution from these sources.

Critical Thinking

Why do you think large seed companies generally oppose this form of more sustainable agriculture?

(**Concept 12-6B**). Some scientists also call for greater reliance on perennial crops as a component of more sustainable agriculture (Science Focus, above).

Analysts suggest four major strategies to help farmers make the transition to more sustainable agriculture (**Concept 12-6B**). *First,* greatly increase research on sustainable organic agriculture (Science Focus, at left) and on improving human nutrition. *Second,* set up demonstration projects so that farmers can see how organic agricultural systems work. *Third,* provide subsidies and foreign aid to encourage its use. *Fourth,* establish training programs in sustainable agriculture for farmers and government agricultural officials, and encourage the creation of college curricula in sustainable agriculture.

Buy Locally Grown Food

Consumers can help farmers to make a transition to more sustainable farming by buying food from local producers in farmers' markets or other outlets. This supports local economies and reduces the environmental impact of food production. Locally grown food does not have to be transported very far from producer to consumer. A typical meal using food imported from other parts of a country or from other countries can easily account for 5 to 17 times more transportation-related greenhouse gas emissions than the same meal using locally produced food would represent. So the option with the least environmental impact is to grow your own organic food or buy organic food that is produced locally. More food can also be grown locally in the world's urban areas in vacant lots, in yards, and on rooftops, as discussed in more detail in Chapter 23.

There are more than 4,300 farmers' markets in the United States, and people are flocking to them. Another growing trend is community-supported agriculture (CSA). Supporters of CSA buy shares in a farm's produce before it is harvested, and these shares are delivered to central points in nearby cities, where shareholders collect them at appointed times each week.

Figure 12-33 Individuals matter: ways to promote more sustainable agriculture (**Concepts 12-6A** and **12-6B**). According to Rajendra Pachauri, head of the Intergovernmental Panel on Climate Change, we need to start eating as if the earth's climate mattered. **Question:** Which three of these measures do you think are the most important? Why?

(Local Harvest has a searchable online directory to CSA farms in the United States at **localharvest.org/csa**.)

Figure 12-33 lists these and other ways in which you can promote more sustainable agriculture.

WHAT CAN YOU DO?

Sustainable Organic Agriculture

- Waste less food
- Eat less meat or no meat
- Use organic farming to grow some of your food
- Buy organic food
- Eat locally grown food
- Compost food wastes

REVISITING | Golden Rice and Sustainability

This chapter began with a look at how we might use genetically engineered golden rice (**Core Case Study**) to help prevent blindness in children and increase their resistance to infectious diseases. Carefully evaluated and monitored genetic engineering is only one of many tools discussed in this chapter for developing more sustainable agriculture.

Making this transition involves applying the four **scientific principles of sustainability** (see back cover). All of these principles are violated by modern industrial agriculture because it depends heavily on nonrenewable fossil fuels, includes too little recycling of crop and animal wastes, accelerates soil erosion, does too little to preserve agrobiodiversity, and can destroy or degrade wildlife habitats and disrupt natural species interactions that help to control pest population sizes.

Making the transition to more sustainable agriculture means relying more on solar energy and less on oil, and sustaining nutrient cycling by soil conservation and by returning crop residues and animal wastes to the soil. It also means helping to sustain natural and agricultural biodiversity by relying on a greater variety of crop and animal strains and by controlling pest populations through broader use of polyculture and integrated pest management. All such efforts will be enhanced if we can control the growth of the human population and our use of resources.

The goal is to feed the world's people while sustaining and restoring the earth's natural capital and living off the natural income it provides. This will not be easy, but it can be done if we heed these ecological lessons from nature.

*The sector of the economy that seems likely to unravel first is food.
Eroding soils, deteriorating rangelands, collapsing fisheries, falling water tables,
and rising temperatures are converging to make it difficult
to expand food production fast enough to keep up with the demand.*

LESTER R. BROWN

REVIEW

1. Review the Key Questions and Concepts for this chapter on p. 276. Describe the use of genetically engineered golden rice (**Core Case Study**) as a way to decrease vitamin A deficiency in children.

2. Define **food security** and **food insecurity.** What is the root cause of food insecurity? Distinguish between **chronic undernutrition (hunger)** and **chronic mal-**

nutrition and describe their harmful effects. Describe the effects of diet deficiencies in vitamin A, iron, and iodine. What is a **famine?** What is **overnutrition,** and what are its harmful effects?

3. What three systems supply most of the world's food? Distinguish among **industrialized agriculture (high-input agriculture), plantation agriculture, traditional**

subsistence agriculture, traditional intensive agriculture, polyculture, and **slash-and-burn agriculture.** Define **soil** and describe its formation and the major layers in mature soils. What is a **green revolution?** Describe industrialized food production in the United States.

4. Distinguish between *crossbreeding* and *genetic engineering.* Describe industrialized meat production. What is a **fishery?** What is **aquaculture?**

5. What are the major harmful environmental impacts of agriculture? What is **soil erosion** and what are its two major harmful environmental effects? What is **desertification** and what are its harmful environmental effects? Distinguish between **salinization** and **waterlogging** of soil and describe their harmful environmental effects.

6. What factors can limit green revolutions? Describe the use of energy in industrialized agriculture. Describe the advantages and disadvantages of genetically engineered foods. Explain how most food production systems reduce biodiversity. Describe the advantages and disadvantages of industrialized meat production. Describe the advantages and disadvantages of aquaculture.

7. What is a **pest?** Define and give two examples of a **pesticide.** Describe Rachel Carson's contribution to environmental science. Describe the advantages and disadvantages of modern pesticides. Describe the dilemma over widespread use of glyphosate as an herbicide. Describe

the use of laws and treaties to help protect us from the harmful effects of pesticides. Describe seven alternatives to conventional pesticides. Define **integrated pest management (IPM)** and discuss its advantages.

8. Describe three ways in which governments influence food production. What is **soil conservation?** Describe seven ways to reduce soil erosion. Describe soil erosion and soil conservation in the United States. Distinguish among the use of **organic fertilizer, commercial inorganic fertilizer, animal manure, green manure,** and **compost** as ways to help restore soil fertility. Describe ways to prevent and clean up soil salinization.

9. Describe ways to produce meat more efficiently, humanely, and sustainably. How can we make aquaculture more sustainable? Define **organic agriculture** and describe its advantages over conventional agriculture. Describe the advantages of relying more on polycultures of perennial crops. What can individuals do to promote more sustainable agriculture?

10. Describe the relationships among golden rice (**Core Case Study**), sustainable agriculture, and the four **scientific principles of sustainability.**

Note: Key Terms are in **bold** type.

CRITICAL THINKING

1. List three ways in which you could apply material from this chapter to make your lifestyle more environmentally sustainable.

2. What are two safeguards that you would want in place before large areas of the world were planted with golden rice (**Core Case Study**)?

3. What are the three most important actions you would take to reduce chronic hunger and malnutrition **(a)** in the country where you live and **(b)** in the world?

4. Explain why you support or oppose greatly increased use of **(a)** genetically modified food and **(b)** polyculture.

5. Suppose you live near a coastal area and a company wants to use a fairly large area of coastal marshland for an aquaculture operation. If you were an elected local official, would you support or oppose such a project? Explain. What safeguards or regulations would you impose on the operation?

6. Explain how widespread use of a pesticide can **(a)** increase the damage done by a particular pest and **(b)** create new pest organisms. If increased mosquito populations threatened you with malaria or West Nile virus,

would you want to spray DDT in your yard and inside your home to reduce the risk? Explain. What are the alternatives?

7. According to physicist and philosopher Albert Einstein, "Nothing will benefit human health and increase chances of survival of life on Earth as much as the evolution to a vegetarian diet." Are you willing to eat less meat or no meat? Explain.

8. Imagine that everyone in the world began eating only locally produced foods tomorrow. Describe a benefit and a harm that would arise from such a change to **(a)** yourself, **(b)** your community, and **(c)** the world.

9. Congratulations! You are in charge of the world. List the three most important features of your **(a)** agricultural policy, **(b)** policy to reduce soil erosion, **(c)** policy for more sustainable harvesting and farming of fish and shellfish, and **(d)** global pest management strategy.

10. List two questions that you would like to have answered as a result of reading this chapter.

Note: See Supplement 13 (p. S78) for a list of Projects related to this chapter.

The following table gives the world's fish harvest and population data.

Years	World Fish Harvest			World Population (billions)	Per Capita Fish Catch (kilograms)
	Fish Catch (million tons)	Aquaculture (million tons)	Total (million tons)		
1990	84.8	13.1	97.9	5.270	
1991	83.7	13.7	97.4	5.36	
1992	85.2	15.4	100.6	5.44	
1993	86.6	17.8	104.4	5.52	
1994	92.1	20.8	112.9	5.60	
1995	92.4	24.4	116.8	5.68	
1996	93.8	26.6	120.4	5.76	
1997	94.3	28.6	122.9	5.84	
1998	87.6	30.5	118.1	5.92	
1999	93.7	33.4	127.1	5.995	
2000	95.5	35.5	131.0	6.07	
2001	92.8	37.8	130.6	6.15	
2002	93.0	40.0	133.0	6.22	
2003	90.2	42.3	132.5	6.20	

1. Use the world fish harvest and population data in the table to calculate the per capita fish consumption from 1990–2003 in kilograms/person. (*Hints:* 1 million metric tons equals 1 billion kilograms; the human population data is measured in billions; and per capita consumption can be calculated directly by dividing.)

2. Has per capita fish consumption generally increased or generally decreased between 1990 and 2003?

3. In what years has per capita fish consumption decreased?

Log on to the Student Companion Site for this book at **academic.cengage.com/biology/miller**, and choose Chapter 12 for many study aids and ideas for further reading and research. These include flash cards, practice quizzing, Weblinks, information on Green Careers, and InfoTrac® College Edition articles.

Water Resources

Water Conflicts in the Middle East: A Preview of the Future?

Many countries in the Middle East face water shortages and rising tensions over water sources they must share. Most water in this dry region comes from three river basins: the Nile, the Jordan, and the Tigris–Euphrates (Figure 13-1).

Africa's Nile River flows through seven countries, each of which draws on it for irrigation and drinking water and flushes mostly untreated sewage and other effluents into it. Three countries—Ethiopia, Sudan, and Egypt—use most of the water that flows in the Nile. Egypt, where it rarely rains, gets more than 97% of its freshwater from the river and is last in line to tap this precious source. To meet the water and food needs of their rapidly growing populations, Ethiopia and Sudan plan to divert more water from the river. Such upstream diversions would reduce the amount of water available to Egypt, which cannot exist without irrigation water from the Nile.

Egypt has several choices. It could go to war with Sudan and Ethiopia for more water, cut its rapid population growth, or reduce irrigation water waste. Other options are to import more grain to reduce the need for irrigation water, work out water-sharing agreements with other countries, or suffer the harsh human and economic consequences of *hydrological poverty.*

The Jordan basin is by far the most water-short region, with fierce competition for its water among Jordan, Syria, Palestine (Gaza and the West Bank), and Israel. Syria, which is projected to increase its population by 70% between 2008 and 2050, plans to build dams and withdraw more water from the Jordan River, decreasing the downstream water supply for Jordan and Israel. If Syria goes through with its plans, Israel warns that it may destroy the largest dam. In contrast, Israel has cooperated with Jordan and Palestine over their shared water resources.

Turkey, located at the headwaters of the Tigris and Euphrates Rivers (Figure 13-1), controls water flowing downstream through Syria and Iraq and into the Persian Gulf. Turkey is building 24 dams along the upper Tigris and Euphrates to generate electricity and irrigate a large area of land.

If completed, these dams will reduce the flow of water downstream to Syria and Iraq by as much as 35% in normal years and by much more in dry years. Syria also plans to build a large dam on the Euphrates to divert water arriving from Turkey. This will leave little water for Iraq and could lead to a water war between Iraq and Syria.

Resolving these water distribution problems will require negotiating agreements to share water supplies, slowing population growth, cutting water waste, raising water prices to help improve irrigation efficiency, and increasing grain imports to reduce irrigation water needs.

Two or more countries share some 263 of the world's water basins, but countries in only 158 of the basins have water-sharing agreements. This explains why conflicts among nations over shared water resources are likely to increase as populations grow, demand for water increases, and supplies decrease in many parts of the world.

To many analysts, emerging water shortages in many parts of the world—along with the related problems of biodiversity loss and climate change—are the three most serious environmental problems the world faces during this century.

This chapter discusses the world's water problems and ways to use this irreplaceable resource more sustainably. About two-thirds of the water used by the world's people is wasted. A key solution to the world's water supply problems is to sharply reduce such waste.

Figure 13-1 Many countries in the Middle East, which has one of the world's highest population growth rates, face water shortages and conflicts over access to water because they share water from three major river basins.

13-1 Will we have enough usable water?

CONCEPT 13-1A We are using available freshwater unsustainably by wasting it, polluting it, and charging too little for this irreplaceable natural resource.

CONCEPT 13-1B One of every six people does not have sufficient access to clean water, and this situation will almost certainly get worse.

13-2 Is extracting groundwater the answer?

CONCEPT 13-2 Groundwater that is used to supply cities and grow food is being pumped from aquifers in some areas faster than it is renewed by precipitation.

13-3 Is building more dams the answer?

CONCEPT 13-3 Building dam and reservoir systems has greatly increased water supplies in some areas, but it has disrupted ecosystems and displaced people.

13-4 Is transferring water from one place to another the answer?

CONCEPT 13-4 Transferring water from one place to another has greatly increased water supplies in some areas, but it has also disrupted ecosystems.

13-5 Is converting salty seawater to freshwater the answer?

CONCEPT 13-5 We can convert salty ocean water to freshwater, but the cost is high, and the resulting salty brine must be disposed of without harming aquatic or terrestrial ecosystems.

13-6 How can we use water more sustainably?

CONCEPT 13-6 We can use water more sustainably by cutting water waste, raising water prices, slowing population growth, and protecting aquifers, forests, and other ecosystems that store and release water.

13-7 How can we reduce the threat of flooding?

CONCEPT 13-7 We can lessen the threat of flooding by protecting more wetlands and natural vegetation in watersheds and by not building in areas subject to frequent flooding.

Note: Supplements 2 (p. S4), 3 (p. S10), 4 (p. S20), 5 (p. S31) and 13 (p. S78) can be used with this chapter.

Our liquid planet glows like a soft blue sapphire
in the hard-edged darkness of space.
There is nothing else like it in the solar system.
It is because of water.

JOHN TODD

13-1 Will We Have Enough Usable Water?

▶ **CONCEPT 13-1A** We are using available freshwater unsustainably by wasting it, polluting it, and charging too little for this irreplaceable natural resource.

▶ **CONCEPT 13-1B** One of every six people does not have sufficient access to clean water, and this situation will almost certainly get worse.

Freshwater Is an Irreplaceable Resource That We Are Managing Poorly

We live on a water planet, with a precious layer of water—most of it saltwater—covering about 71% of the earth's surface (Figure 8-2, p. 163). Look in the mirror. What you see is about 60% water, most of it inside your cells.

Water is an amazing substance with unique properties that affect life on earth (Science Focus, p. 67). You could survive for several weeks without food but for only a few days without water. And it takes huge amounts of water to supply you with food, provide

you with shelter, and meet your other daily needs and wants. Water also plays a key role in sculpting the earth's surface, moderating climate, and removing and diluting wastes and pollutants.

Despite its importance, water is one of our most poorly managed resources. We waste it and pollute it. We also charge too little for making it available. This encourages still greater waste and pollution of this resource, for which we have no substitute (**Concept 13-1A**).

Access to water is a *global health issue* because lack of water that is safe for drinking and sanitation is the world's single largest cause of illness. In 2007, the World Health Organization (WHO) estimated that each year more than 1.6 million people—90% of them children under age five—die from largely preventable waterborne diseases such as diarrhea, typhoid fever, and hepatitis.

Water is an *economic issue* because it is vital for reducing poverty and producing food and energy. It is a *women's and children's issue* in developing countries because poor women and girls often are responsible for finding and carrying daily supplies of water (Figure 13-2). And water is a *national and global security issue* because of increasing tensions within and between nations over access to limited but shared water resources in the Middle East (**Core Case Study**) and other areas of the world.

Water is an *environmental issue* because excessive withdrawal of water from rivers and aquifers and pollution of water result in lower water tables, lower river flows, shrinking lakes, losses of wetlands, declining water quality, declining fish populations, species extinctions, and degradation of ecosystem services provided by aquatic systems. Virtually all of these environmental indicators related to water availability and quality are worsening in some regions.

Most of the Earth's Freshwater Is Not Available to Us

Only a tiny fraction of the planet's abundant water supply—*about 0.024%*—is readily available to us as liquid freshwater in accessible groundwater deposits and in lakes, rivers, and streams. The rest is in the salty oceans, frozen in polar ice caps and glaciers, or deep underground and inaccessible.

Fortunately, the world's freshwater supply is continually collected, purified, recycled, and distributed in the earth's *hydrologic cycle*—the movement of water in the seas, in the air, and on land, which is driven by solar energy and gravity (Figure 3-17, p. 66). This irreplaceable water recycling and purification system works well, unless we overload it with slowly degradable and nondegradable wastes or withdraw water from underground and surface water supplies faster than it is replenished. We also interfere with this cycle

when we destroy wetlands and cut down forests that store and slowly release water, or alter the cycle's rate and distribution patterns as a result of climate change caused by global warming. In parts of the world, we are doing all of these things (**Concept 8-3**, p. 171, and **Concept 8-5**, p. 179), mostly because we have placed little or no value on the earth's natural ecological services (Figure 1-3, p. 8, and Science Focus, p. 218).

On a global basis we have plenty of freshwater. But differences in average annual precipitation and economic resources divide the world's continents, countries, and people into water *haves* and *have-nots*. For example, Canada, with only 0.5% of the world's population, has 20% of the world's liquid freshwater, while China, with 20% of the world's people, has only 7% of the supply. Asia with 60% of the world's population has only 30% of the supply.

We Get Freshwater from Groundwater and Surface Water

Some precipitation infiltrates the ground and percolates downward through spaces in soil, gravel, and rock until

Figure 13-2 Girl carrying water from a well across dried out and cracked earth during a severe drought in India.

Figure 13-3 **Natural capital**: groundwater system. An *unconfined aquifer* is an aquifer with a permeable water table. A *confined aquifer* is bounded above and below by less permeable beds of rock, and its water is confined under pressure. Some aquifers are replenished by precipitation; others are not.

an impenetrable layer of rock stops it (Figure 13-3). The water in these spaces is called **groundwater**—one of our most important sources of freshwater and a key component of the earth's natural capital.

The spaces in soil and rock close to the earth's surface hold little moisture. Below a certain depth, in the **zone of saturation,** these spaces are completely filled with water. The top of this groundwater zone is the **water table.** It falls in dry weather, or when we remove groundwater faster than nature can replenish it, and it rises in wet weather.

Deeper down are geological layers called **aquifers:** underground caverns and porous layers of sand, gravel, or bedrock through which groundwater flows. Groundwater normally moves from points of high elevation and pressure to points of lower elevation and pressure. Some caverns have rivers of groundwater flowing through them. But the porous layers of sand, gravel, or bedrock in most aquifers are like large elongated sponges through which groundwater seeps—typically moving only a meter or so (about 3 feet) per year and rarely more than 0.3 meter (1 foot) per day. Watertight layers of rock or clay below such aquifers keep the water from escaping deeper into the earth.

Most aquifers are replenished naturally by precipitation that percolates downward through soil and rock, a process called *natural recharge.* Others are recharged from the side by *lateral recharge* from nearby rivers and streams. Most aquifers recharge extremely slowly. Because so much of the urban landscape has been built on or paved over, water can no longer penetrate the ground to recharge aquifers in some urban areas.

Nonrenewable aquifers get very little, if any, recharge. They are found deep underground and were formed tens of thousands of years ago. Withdrawing water from these aquifers amounts to *mining* a nonrenewable resource. Such withdrawals will deplete these ancient deposits of liquid natural capital, and they represent an unsustainable use of this resource (**Concept 1-1B**, p. 6).

CONCEPT LINK

One of our most important resources is **surface water,** the freshwater from precipitation and snowmelt that flows across the earth's land surface and into rivers, streams, lakes, wetlands, estuaries, and ultimately to the oceans. Precipitation that does not infiltrate the ground or return to the atmosphere by evaporation is called **surface runoff.** Surface water replenished by runoff is classified as a renewable but finite

resource. The land from which surface water drains into a particular river, lake, wetland, or other body of water is called its **watershed** or **drainage basin.**

There is a *hydrological connection* between surface water and groundwater because eventually most groundwater flows into rivers, lakes, estuaries, and wetlands. Thus, if we disrupt the hydrologic cycle by removing groundwater faster than it is replenished, nearby streams, lakes, and wetlands could dry up, resulting in degradation of aquatic biodiversity and various ecological services (see *The Habitable Planet,* Video 8, at **www.learner.org/resources/series209.html**).

We Use a Large and Growing Portion of the World's Reliable Runoff

According to hydrologists (scientists who study water supplies), two-thirds of the annual surface runoff in rivers and streams is lost by seasonal floods and is not available for human use. The remaining one-third is **reliable surface runoff:** the amount of surface runoff that we can generally count on as a source of freshwater from year to year.

During the last century, the human population tripled, global water withdrawal increased sevenfold, and per capita withdrawal quadrupled. As a result, we now withdraw about 34% of the world's reliable runoff. Because of increased population growth alone, global withdrawal rates of surface water could reach more than 70% of the reliable runoff by 2025, and 90% if per capita withdrawal of water continues increasing at the current rate. This is a global average. Withdrawal rates already severely strain the reliable runoff in some areas. For example, in the arid American Southwest, up to 70% of the reliable runoff is withdrawn for human purposes. In other words, many areas are using their water resources unsustainably as our ecological footprints spread across the earth (Figure 1-10, p. 15; **Concept 1-3**, p. 12; and Figure 3 in Supplement 4, pp. S24–S25). And because of climate change, caused mostly by the burning of fossil fuels and clearing of forests, once reliable runoff in many areas may become unreliable.

Worldwide, we use 70% of the water we withdraw each year from rivers, lakes, and aquifers to irrigate cropland. Industry uses another 20% of the water withdrawn each year, and cities and residences use the remaining 10%.

Affluent lifestyles require large amounts of water, and much of it is used inefficiently. According to environmental consultant Fred Pearce, author of *When the Rivers Run Dry,* it takes about 450,000 liters (120,000 gallons) or 2,400 bathtubs full of water to produce a small car. It takes 140 liters (37 gallons) to produce a cup of coffee, and 25 bathtubs full of water to produce a typical T-shirt.

■ CASE STUDY

Freshwater Resources in the United States

The United States has more than enough renewable freshwater; but it is unevenly distributed, and much of it is contaminated by agricultural and industrial practices. The eastern states usually have ample precipitation, whereas many western and southwestern states have little (Figure 13-4, top). Many people think of Seattle, Washington, as the rainiest city in the United States. But, in order, the three rainiest cities in the lower 48 states over the past 30 years were Mobile, Alabama; Pensacola, Florida; and New Orleans, Louisiana.

In the East, most water is used for energy production, power plant cooling, and manufacturing. In various parts of the eastern United States, the most serious water problems are flooding and occasional urban

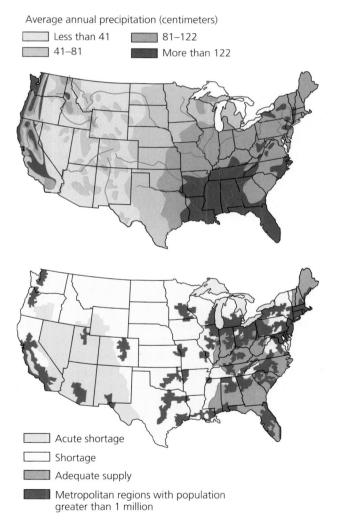

Average annual precipitation (centimeters)

☐ Less than 41		☐ 81–122	
☐ 41–81		■ More than 122	

☐ Acute shortage

☐ Shortage

▨ Adequate supply

■ Metropolitan regions with population greater than 1 million

Figure 13-4 Average annual precipitation and major rivers (top) and water-deficit regions in the continental United States and their proximity to metropolitan areas having populations greater than 1 million (bottom). **Question:** Why do you think some areas with moderate precipitation still suffer from water shortages? (Data from U.S. Water Resources Council and U.S. Geological Survey)

shortages as a result of pollution. Another occasional problem is **drought:** a prolonged period in which precipitation is at least 70% lower and evaporation is higher than normal in an area that is normally not dry.

In the arid and semiarid areas of the western half of the United States (Figure 13-4, bottom), irrigation counts for 85% of water use, much of it unnecessarily wasted. The major water problem is a shortage of run-off caused by low precipitation (Figure 13-4, top), high evaporation, and recurring severe drought.

Almost half the water used in the United States comes from groundwater sources, and the rest comes from rivers, lakes, and reservoirs. Water tables in many water-short areas, especially in the arid and semiarid western half of the lower 48 states, are dropping quickly as farmers and rapidly growing urban areas (Figure 13-4, bottom) deplete many aquifers faster than they can be recharged. Excessive withdrawal of groundwater from an aquifer near a river or stream can decrease or deplete these sources of surface water. When the water table drops below the level of the bottom of a stream, its water drains into the aquifer (see *The Habitable Planet,* Video 8, at **www.learner.org/resources/series209.html**).

In 2007, the U.S. Geological Survey projected that at least 36 U.S. states are likely to face water shortages by 2013 because of a combination of drought, rising temperatures, population growth, urban sprawl, and excessive use and waste of water. In 2003, the U.S. Department of the Interior mapped out *water hotspots* in 17 western states (Figure 13-5). In these areas, competition for scarce water to support growing urban areas, irrigation, recreation, and wildlife could trigger intense political and legal conflicts between states and between rural and urban areas within states during the next 20 years.

This is already happening, in fact. In 2007, the U.S. Supreme Court heard a case in which the state of Montana accused the state of Wyoming of taking more than its fair share of water from two tributaries of the Yellowstone River that supply water for wells and farms in both states. Predicted long-term drought and global warming will increase the number and intensity of such disputes in the United States and other parts of the world in coming years.

Water Shortages Will Grow

The main factors causing water scarcity are a dry climate, drought, too many people using a normally reliable supply of water, and wasteful use of water.

Figure 13-6 shows the current degree of stress faced by the world's major river systems, based on a comparison of the amount of available surface water with the amount used per person. More than 30 countries—most of them in the Middle East (**Core Case Study**) and Africa—now face water scarcity. By 2050, some 60 countries, many of them in Asia, are likely to

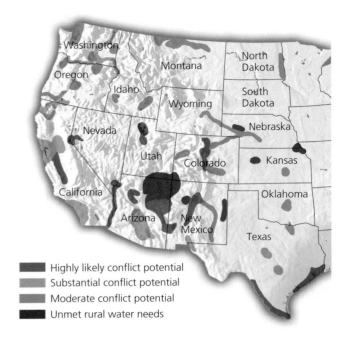

- ■ Highly likely conflict potential
- ▨ Substantial conflict potential
- ▦ Moderate conflict potential
- ■ Unmet rural water needs

Figure 13-5 Water *hotspots* in 17 western states that, by 2025, could face intense conflicts over scarce water needed for urban growth, irrigation, recreation, and wildlife. Some analysts suggest that this is a map of places not to live during the next 25 years. **Question:** If you live in one of these hotspot areas, have you noticed any signs of conflict over water supplies? (Data from U.S. Department of the Interior)

be suffering from water stress. In 2006, the Chinese government reported that two-thirds of Chinese cities faced water shortages. Rapid urbanization and economic growth are expected to put further strains on China's already short water resources (**Concept 13-1B**).

THINKING ABOUT
Water and the Middle East

How might scarcity of water in the Middle East (**Core Case Study**) affect nations that are dependent on oil from the Middle East? How could this impact your lifestyle?

CORE CASE STUDY

Because only 5% of Egypt is inhabitable, almost all of the country's 75 million people live along the Nile River (Figure 13-1, **Core Case Study**) or the Mediterranean Sea in cities such as Cairo—one of the world's most densely populated cities. Egypt's already severe water supply problem will likely get worse as the country's population is expected to increase to 118 million by 2050.

Poor people bear the brunt of water shortages. In 2005, the United Nations reported that 1.1 billion people—one of every six—lacked regular access to enough clean water for drinking, cooking, and washing, and 2.6 billion people did not have access to even basic sanitation. This already serious situation will almost certainly get worse. In 2007, the United Nations estimated that by 2025, *at least 3 billion of the world's projected 7.9 billion people will lack access to safe water* (**Concept 13-1B**). Most of these individuals suffering from hydrological poverty will be located in arid and semi-

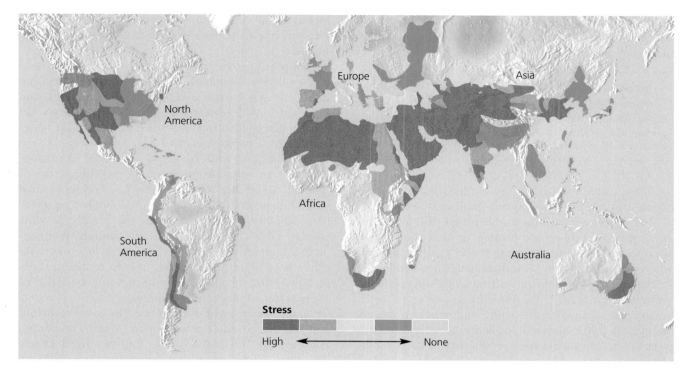

Figure 13-6 Natural capital degradation: stress on the world's major river basins, based on a comparison of the amount of water available with the amount used by humans (**Concept 13-1B**). **Questions:** If you live in a water-stressed area, what signs of stress have you noticed? In what ways, if any, has it affected your life? (Data from World Commission on Water Use in the 21st century)

arid regions with rapidly-growing populations. Many of these countries are on the growing list of failing states (Figure 17, p. S19, Supplement 3). The likely results: a flood of refugees from arid and semiarid regions searching for water, land, and food, and intense conflicts within and between countries—especially in the Middle East (**Core Case Study**) and Asia—over dwindling shared water resources.

CORE CASE STUDY

Long-Term Severe Drought Is Increasing

Drought is usually caused by an extended period of below-normal rainfall. But it can also result from diminished groundwater due to falling water tables or other factors, such as climate change, that reduce the normally available water supply.

Severe drought has a number of harmful environmental effects. It dries out soils (Figure 13-2), reduces stream flows, decreases tree growth and biomass, lowers net primary productivity (Figure 3-16, p. 64), reduces crop yields, and causes a shift in some biomes toward relatively dry conditions such as those found in savannas and deserts.

Between 1979 and 2008, the total area of the earth experiencing severe or extreme drought more than tripled in what some observers call "The Big Dry." Currently, about 30% of the earth's land area—roughly equal to the size of Asia—experiences severe drought. In 2007, climate researcher David Rind and his colleagues

at the NASA Goddard Institute for Space Studies, projected that by 2059, as much as 45% of the earth's land surface could experience extreme drought, mostly as a result of climate change caused by global warming.

Since 1999, except for a couple of wet years, much of the United States, especially the arid and already water-short Southwest (Figure 13-5), has experienced severe drought. The prolonged drought, coupled with greatly increased population and development in the U.S. Southwest, has led to drops in crop productivity and increasing squabbles over water supplies between farmers and rapidly growing populations of urban dwellers.

This drought is likely to continue. After looking at 19 global climate models used in a 2007 report from the U.N. Intergovernmental Panel on Climate Change (IPCC), climate researcher Richard Seager and his colleagues at the Lamont-Doherty Earth Observatory of Columbia University predicted that the southwestern United States and parts of northern Mexico will have long periods of extreme drought throughout the rest of this century.

In Water–Short Areas Farmers and Cities Compete for Water Resources

In water-short areas, tempers are flaring as farmers and cities compete for available water from rivers and aquifers. Farmers are losing this battle, and thus have

decreased access to water resources in many parts of the world at a time when they are trying to feed 82 million more people each year.

THINKING ABOUT
Access to Water

Between farmers and city dwellers, which group should have greater access to water in water-short areas? Explain. Can you think of a compromise for resolving this dilemma?

Some water-short countries are reducing their irrigation water needs by importing grain to free up more of their water for industrial and urban development. Thus, water shortages are leading to increasing competition among countries in the world's grain markets, where financially strong countries have an advantage.

Adding to the intensity of this global competition for water and grain is the growing use of crops such as soybeans, sugarcane, oil palms, and corn to produce biofuels. A 2007 National Academy of Science study warned that increased corn production in the United States to produce ethanol will decrease water supplies and encourage aquifer depletion in some areas and increase pollution of streams and aquifers from use of pesticides and fertilizers. Bottom line: *farmers, cities, and car owners are increasingly competing for access to the world's grain and water supplies, which in turn can degrade some of the natural capital that provides these resources.*

Who Should Own and Manage Freshwater Resources?

Most people believe that everyone should have access to clean water. But who will pay for making this water available?

Most water resources are owned by governments and managed as publicly owned resources for their citizens. An increasing number of governments, while retaining ownership of these public resources, are hiring private companies to manage them. But some are selling their water resources to private companies.

Two large French companies—Veolia and Suez—realize that water scarcity is rapidly becoming one of world's most urgent environmental problems and that there is a lot of money to be made in owning and managing water resources. Their long-range strategy is to buy or manage as much of the world's privately and publicly owned water suppliers as possible, especially in Europe, North America, and China. For example, Veolia with 70,000 employees provides water for 108 million people in 57 countries, including 17 Chinese cities.

Currently, 85% of Americans get their water from publicly owned utilities. This may change. Within 10 years, these two European-based water companies aim to control 70% of the water supply in the United States by buying up American water companies and entering into agreements with most cities to manage their water supplies.

Some argue that private companies have the money and expertise to manage these resources better and more efficiently than government bureaucracies can. Experience with this public–private partnership approach has yielded mixed results. Some private water management companies have done a good job.

For example, in 2002 Veolia was hired to manage the water and wastewater system in the Chinese city of Pudong. Since then, the company has laid more than 1,400 kilometers (900 miles) of pipe, hooked up more than 300,000 new structures to the rapidly growing system, built water-treatment and sewage treatment plants, and hired 7,000 local workers. It also established a 24-hour customer call-in help center (a rarity in China) and a control room to monitor the status of every water connection in Pudong.

Some government officials want to go even further and sell public water resources to private companies. For example, in Cochabamba, Bolivia's third largest city, 60% of the water was being lost through leaky pipes. With no money to fix the pipes, the Bolivian government sold the city's water system to a subsidiary of Bechtel Corporation, an American owned global engineering and construction company. Within 6 months, the company sharply increased water rates and began seizing and selling the houses of people who did not pay their water bills. A general strike ensued in this city of about 500,000 people, and violent street clashes between protesters and government troops led to 10,000 people being injured and 7 deaths. Bechtel executives fled the city and the Bolivian government tore up the contract with Bechtel.

Today, the city's water management system is in shambles and most of the leaks persist because of insufficient funds. Bechtel filed a $50 million suit against Bolivia in the World Bank's trade court for not honoring the contract, but decided to drop the suit in 2005 after being subjected to 4 years of international protests by citizen groups.

A similar problem occurred when a subsidiary of Bechtel signed a 30-year contract in 2000 to run the water and sanitation services for Guayaquil, Ecuador's largest city. In 2007, residents of the city were demanding damages from the company for contract violations, including broken pipelines, exorbitant water rates, malfunctioning sewage systems, unsafe tap water, repeated water cutoffs, failure to extend services to low-income neighborhoods, and public health problems from a lack of wastewater treatment. Citizens are calling for a return to local, public control of water resources and sanitation.

Many people oppose full privatization of water resources because they believe that water is a public resource too important to be left solely in private hands. Also, once a city's water systems have been taken over

by a foreign-based corporation, efforts to return the systems to public control can be quite costly for taxpayers and can lead to severe economic penalties under the rules of the World Trade Organization (WTO).

Some analysts point to two other potential problems in a fully privatized water system. *First,* because private companies make money by delivering water, they have an incentive to sell as much water as they can rather than to conserve it. *Second,* because they have too little money to pay water bills, the poor will continue to be left out. There are no easy answers for managing the water that everyone needs.

The most common ways to increase the supply of freshwater in a particular area are to withdraw groundwater, to build dams and reservoirs to store runoff in rivers for release as needed, to transport surface water from one area to another, and to use water more efficiently. The rest of this chapter evaluates these options.

13-2 Is Extracting Groundwater the Answer?

▶ **CONCEPT 13-2** Groundwater that is used to supply cities and grow food is being pumped from aquifers in some areas faster than it is renewed by precipitation.

Water Tables Fall When Groundwater Is Withdrawn Faster Than It Is Replenished

Most aquifers are renewable sources unless their water becomes contaminated or is removed faster than it is replenished by rainfall, as is occurring in many parts of the world. Aquifers provide drinking water for nearly half of the world's people. In the United States, aquifers supply almost all of the drinking water in rural areas, one-fifth of that in urban areas, and 37% of irrigation water. Relying more on groundwater has advantages and disadvantages (Figure 13-7).

Water tables are falling in many areas of the world because the rate of pumping water from aquifers (mostly to irrigate crops) exceeds the rate of natural recharge from rainfall and snowmelt (**Concept 13-2**). The world's three largest grain producers—India, China, and the United States—and several other countries such as Saudi Arabia, Iran, Yemen, Israel, Mexico, and Pakistan are overpumping many of their aquifers.

Currently, more than half a billion people are being fed by grain produced through the unsustainable use of groundwater. This number is expected to grow until the water runs out or until governments put caps on aquifer withdrawal rates, stop providing subsidies that encourage overpumping and water waste, and raise water prices.

The widespread drilling of inexpensive tubewells by small farmers, especially in India and China, has accelerated aquifer overpumping. As water in aquifers is removed faster than it is renewed, water tables fall. Then farmers drill deeper wells, buy larger pumps, and use more energy to pump water to the surface, eventually depleting aquifers or making it too costly

to pump the remaining water. In China and India, the increasing demand for electricity to run the pumps is met mostly by building more coal-fired power plants, which add greenhouse gases and other pollutants to the atmosphere.

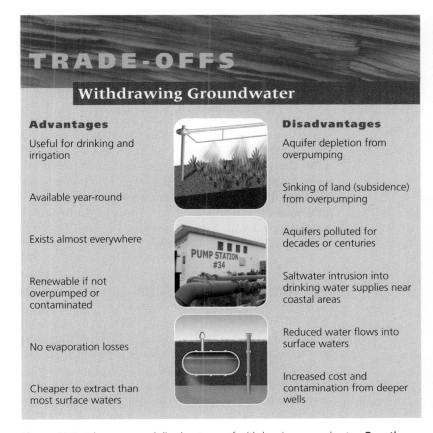

TRADE-OFFS

Withdrawing Groundwater

Advantages	Disadvantages
Useful for drinking and irrigation	Aquifer depletion from overpumping
Available year-round	Sinking of land (subsidence) from overpumping
Exists almost everywhere	Aquifers polluted for decades or centuries
Renewable if not overpumped or contaminated	Saltwater intrusion into drinking water supplies near coastal areas
No evaporation losses	Reduced water flows into surface waters
Cheaper to extract than most surface waters	Increased cost and contamination from deeper wells

PUMP STATION #34

Figure 13-7 Advantages and disadvantages of withdrawing groundwater. **Question:** Which two advantages and which two disadvantages do you think are the most important? Why?

1986

2004

UNEP/GRID-Sioux Falls

UNEP/GRID-Sioux Falls

Figure 13-8 Natural capital degradation: development of irrigation by pumping groundwater from an ancient and nonrenewable aquifer in a vast desert region of Saudi Arabia between 1986 (left) and 2004 (right). Irrigated areas appear as green dots (each representing a circular spray system), and brown dots show areas where wells have gone dry and the land has returned to desert. Hydrologists estimate that because of aquifer depletion, most irrigated agriculture in Saudi Arabia may disappear within 10–20 years.

Arid Saudi Arabia is as water-poor as it is oil-rich. It gets about 70% of its drinking water at a high cost from the world's largest system for removing salt from seawater, located on its eastern coast. The rest of the country's water is pumped from deep aquifers—most as nonrenewable as the country's oil—and used to grow irrigated crops on desert land (Figure 13-8) and to fill large numbers of fountains and swimming pools from which precious water evaporates into the hot, dry desert air. Hydrologists estimate that because of aquifer depletion, most irrigated agriculture in Saudi Arabia may disappear within 1 to 2 decades.

■ CASE STUDY
Aquifer Depletion in the United States

In the United States, groundwater is being withdrawn, on average, four times faster than it is replenished, according to the U.S. Geological Survey (**Concept 13-2**). Figure 13-9 shows the areas of greatest depletion. One of the most serious overdrafts is in the lower half of the Ogallala, the world's largest known aquifer, which lies under eight Midwestern states from South Dakota to Texas (Figure 13-10). Serious groundwater depletion is also taking place in California's Central Valley, which supplies half of the country's fruit and vegeta-

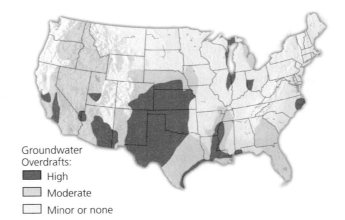

Groundwater Overdrafts:
■ High
▢ Moderate
▢ Minor or none

CENGAGENOW™ **Active Figure 13-9 Natural capital degradation:** areas of greatest aquifer depletion from groundwater overdraft in the continental United States, including the vast, central Ogallala aquifer. Aquifer depletion is also high in Hawaii and Puerto Rico (not shown on map). *See an animation based on this figure at* CengageNOW™. **Question:** If you live in the United States, how is your lifestyle affected directly or indirectly by water withdrawn from the essentially nonrenewable Ogallala aquifer? (Data from U.S. Water Resources Council and U.S. Geological Survey)

bles (the long red area in the California portion of Figure 13-9).

The gigantic Ogallala aquifer supplies about one-third of all the groundwater used in the United States and has helped to turn the Great Plains into one of

Saturated thickness of Ogallala Aquifer

☐	Less than 61 meters (200 ft.)
▨	61–183 meters (200–600 ft.)
■	More than 183 meters (600 ft.) (as much as 370 meters or 1,200 ft. in places)

Figure 13-10 Natural capital degradation: The Ogallala is the world's largest known aquifer. If the water in this aquifer were above ground, it could cover all of the lower 48 states with 0.5 meter (1.5 feet) of water. Water withdrawn from this aquifer is used to grow crops, raise cattle, and provide cities and industries with water. As a result, this aquifer, which is renewed very slowly, is being depleted, especially at its thin southern end in parts of Texas, New Mexico, Oklahoma, and Kansas. (Data from U.S. Geological Survey)

world's most productive irrigated agricultural regions. The problem is that the Ogallala is essentially a one-time deposit of liquid natural capital with a very slow rate of recharge.

The northernmost states (Wyoming, Nebraska, South Dakota, and parts of Colorado) still have ample water supplies from the aquifer. However, in parts of the southern states, where the aquifer is thinner, the water is being depleted rapidly—especially in the Texas–Oklahoma High Plains. In some of these areas water is being pumped out at a rate that is 10–40 times higher than the natural recharge rate. This has lowered the water table more than 30 meters (100 feet). The resulting higher pumping costs make it too expensive to irrigate crops in some areas. As a result, the amount of irrigated farmland in Texas has decreased by about 11% in recent years, as farmers abandon agriculture or return to lower yield dryland farming.

Government subsidies designed to increase crop production and to encourage farmers to grow water-thirsty crops have only accelerated depletion of the Ogallala. They include crop-disaster payments and tax breaks in the form of groundwater depletion allowances (with larger breaks for heavier groundwater use).

Depletion of the Ogallala threatens biodiversity. In some places, the Ogallala flows onto the land and creates wetlands, which are vital habitats for many species, especially birds. When the water tables fall, many of these aquatic oases of biodiversity will disappear, along with the ecosystem services they provide.

THINKING ABOUT
The Ogallala

What are three things you would do to promote more sustainable use of the Ogallala aquifer?

Groundwater Overpumping Has Other Harmful Effects

Overpumping of aquifers not only limits future food production, but also increases the gap between the rich and poor in some areas. As water tables drop, farmers must drill deeper wells, buy larger pumps, and use more electricity to run the pumps. Poor farmers cannot afford to do this and end up losing their land and working for richer farmers or migrating to cities already crowded with poor people struggling to survive.

Withdrawing large amounts of water sometimes causes the sand and rock in aquifers to collapse, which causes the land above the aquifer to *subside* or sink, a phenomenon known as *land subsidence*. Once an aquifer becomes compressed from subsidence, recharge is impossible. Land subsidence can damage roadways, water and sewer lines, and building foundations. Areas in the United States suffering from this problem include California's San Joaquin Valley, Baton Rouge in Louisiana, the Phoenix area in Arizona, and the Houston–Galveston area in Texas.

Mexico City, built on an old lakebed, has one of the world's worse subsidence problems because of increasing groundwater overdrafts due to rapid population growth and urbanization. Some parts of the city have sunk as much as 10 meters (33 feet). Parts of Beijing in China and Bangkok in Thailand are also sinking.

Sinkholes, which can appear suddenly and unexpectedly, are another kind of land subsidence. They are large craters that form when the roof of an underground cavern collapses after being drained of the groundwater that supports it. Sinkholes can appear suddenly and can swallow homes, businesses, roads, cars, and trees.

Groundwater overdrafts near coastal areas, where many of the world's largest cities and industrial areas are found, can cause contamination of groundwater supplies by pulling saltwater into freshwater aquifers

SCIENCE FOCUS

Are Deep Aquifers the Answer?

With global water shortages looming, scientists are evaluating *deep aquifers* as future water sources. Preliminary results suggest that some of these aquifers hold enough water to support billions of people for centuries. And the water quality in these aquifers may be much higher than that of most rivers and lakes.

The first problem is to locate such aquifers and determine whether they contain freshwater or saline water. One way to do this is to drill a borehole and measure the electrical resistance of layers of geological material at different depths. An aquifer containing freshwater has a higher electrical resistance than more saline aquifers. Measurements of the natural radioactive emissions of gamma rays (Figure 2-7, top, p. 41) emitted by rocks at different levels can also help locate aquifers. In addition, scientists send vibrations into the earth and use a seismograph to evaluate the water content of various geological layers.

A series of boreholes are used to estimate properties such as the size, depth, and flow rate of an aquifer. Such data can be mapped and evaluated by using mathematical models. The radioisotope carbon-14 can be used to determine the age of the water.

Once an aquifer is located, deep wells must be drilled to extract water samples, which are analyzed for salt content, oxygen levels, and presence of other chemicals. Because most of these aquifers are huge, samples must be withdrawn from different areas of such aquifers.

There are two major concerns about tapping these nonrenewable deposits of water. *First,* little is known about the geological and ecological impacts of pumping water from deep aquifers. *Second,* some deep aquifers flow beneath several different countries, and no international water treaties govern rights to such water. Thus, conflicts could arise over who has the right to tap these aquifers.

Critical Thinking

What are some other possible problems that could arise from tapping deep aquifers?

SOLUTIONS

Groundwater Depletion

Prevention		Control
Waste less water		Raise price of water to discourage waste
Subsidize water conservation		Tax water pumped from wells near surface waters
Limit number of wells		Set and enforce minimum stream flow levels
Do not grow water-intensive crops in dry areas		Divert surface water in wet years to recharge aquifers

thus making the groundwater undrinkable and unusable for irrigation. This problem is especially serious in coastal areas of the U.S. states of Florida, California, South Carolina, Georgia, New Jersey, and Texas, as well as in coastal areas of Turkey, Manila in the Philippines, and Bangkok in Thailand. Rising sea levels from global warming will increase saltwater intrusion and decrease the amount of groundwater available in heavily populated coastal areas.

Figure 13-11 lists ways to prevent or to slow the problem of groundwater depletion by using this largely renewable resource more sustainably. **GREEN CAREER:** Hydrogeologist

Scientists are also evaluating the use of deep aquifers as a source of freshwater (Science Focus, above).

Figure 13-11 Ways to prevent or slow groundwater depletion by using water more sustainably. **Question:** Which two of these solutions do you think are the most important? Why?

13-3 Is Building More Dams the Answer?

▶ **CONCEPT 13-3** Building dam and reservoir systems has greatly increased water supplies in some areas, but it has disrupted ecosystems and displaced people.

Large Dams and Reservoirs Have Advantages and Disadvantages

A **dam** is a structure built across a river to control the river's water flow. After a river is dammed, the river's flow creates an artificial lake, or **reservoir,** behind the dam. The main goals of a dam and reservoir system are to capture and store runoff and release it as needed to control floods, generate electricity (hydroelectricity), and supply water for irrigation and for towns and cit-

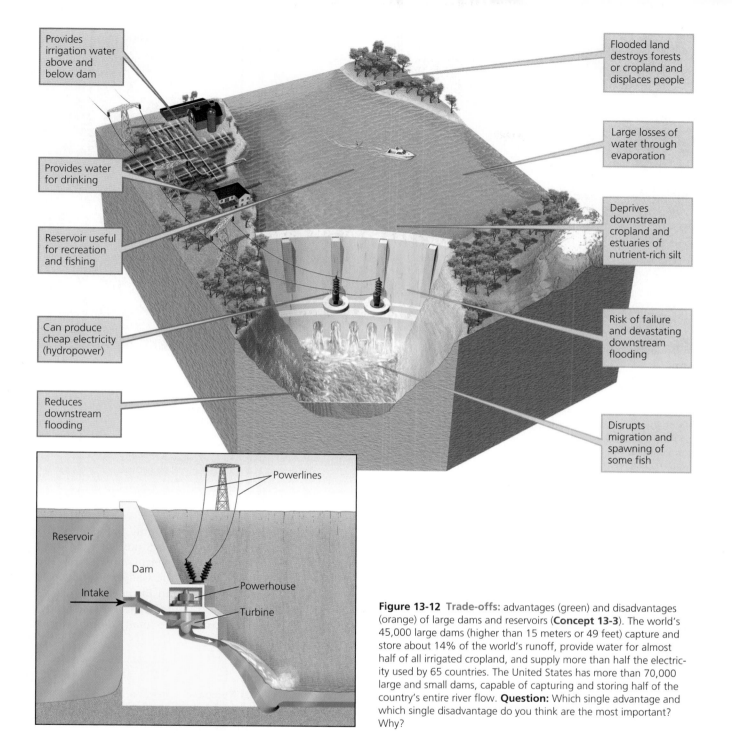

Provides irrigation water above and below dam

Provides water for drinking

Reservoir useful for recreation and fishing

Can produce cheap electricity (hydropower)

Reduces downstream flooding

Flooded land destroys forests or cropland and displaces people

Large losses of water through evaporation

Deprives downstream cropland and estuaries of nutrient-rich silt

Risk of failure and devastating downstream flooding

Disrupts migration and spawning of some fish

Powerlines

Reservoir

Dam

Intake

Powerhouse

Turbine

Figure 13-12 Trade-offs: advantages (green) and disadvantages (orange) of large dams and reservoirs (**Concept 13-3**). The world's 45,000 large dams (higher than 15 meters or 49 feet) capture and store about 14% of the world's runoff, provide water for almost half of all irrigated cropland, and supply more than half the electricity used by 65 countries. The United States has more than 70,000 large and small dams, capable of capturing and storing half of the country's entire river flow. **Question:** Which single advantage and which single disadvantage do you think are the most important? Why?

ies. Reservoirs also provide recreational activities such as swimming, fishing, and boating. Today, there are about 800,000 dams worldwide, 45,000 of them large dams—at least 15 meters (49 feet) high. Large dams and reservoirs have both benefits and drawbacks (Figure 13-12).

Large dams (22,000 of them in China) have increased the annual reliable runoff available for human use by nearly one-third. As a result, reservoirs now hold 3 to 6 times more water than flows in the world's natural rivers. These dams and reservoirs have helped to reduce flooding and grow crops in arid areas

(Figure 13-13, p. 326). One-fourth of the world's dams produce about 20% of the world's electricity.

Large dams and reservoirs also have disadvantages. Worldwide, this engineering approach to river management has displaced 40–80 million people from their homes. It has flooded an area of mostly productive land roughly equal to the area of the U.S. state of California, and it has impaired some of the important ecological services rivers provide (Figure 11-16, p. 270, and **Concept 13-3**). And according to the 2007 WWF study, about one-fifth of the world's freshwater fish and plant species are either extinct or endangered primarily

1976
1999

Figure 13-13 In 1976, a rural area of eastern Turkey was too dry for growing crops, as illustrated by the brown areas and the small number of crop fields shown in the satellite photo (left). In 1990, the Atatürk Dam was constructed on the Euphrates River (Figure 13-1). It is the largest of 22 dams built on the Euphrates and Tigris rivers in the 1980s and 1990s to provide irrigation water and hydroelectricity to arid southeastern Turkey. The satellite photo on the right shows the same area in 1999. The large areas of green show irrigated crop fields, especially around the town of Harran at the bottom right. The project has increased food production and provided electricity. However, Syria and Iraq are concerned that water diverted by this and other proposed dams on the Euphrates will leave too little water to meet each country's needs.

because dams and water withdrawals have destroyed many free-flowing rivers.

Because of evaporation and seepage into porous rock beds, the reservoirs behind some dams lose large amounts of water. Reservoirs also eventually fill up with sediment, mud, and silt, usually within 50 years, which makes them useless for storing water or producing electricity. About 85% of dam and reservoir systems in the United States will be 50 years old by 2020.

Some Rivers Are Running Dry and Some Lakes Are Shrinking

A series of dams on a river and withdrawals of river water for agricultural and urban uses, especially in arid areas, can disrupt the hydrologic cycle by reducing downstream flow to a trickle and preventing river water from reaching the sea. According to a 2007 study by the World Wildlife Fund (WWF), only 21 of the planet's 177 longest rivers run freely from their sources to the sea (Figure 8-17, p. 176) because of dams, excessive water withdrawals, and in some areas, prolonged severe drought. Projected climate change will worsen this situation in many areas because the earth's climate and hydrological cycles are intertwined.

Examples of major rivers running dry part of the year include the Colorado (Case Study, at right) and the Rio Grande in the southwestern United States, the Yangtze and Yellow in China, the Ganges in India, the Indus in Pakistan, the Danube in Europe, and Africa's Nile River–Lake Victoria (Chapter 11 Core Case Study, p. 249). Many smaller rivers have totally disappeared.

THINKING ABOUT
Dams and Egypt

CORE CASE STUDY

Upstream dams and diversions of water from the Nile River by Ethiopia and Sudan will reduce the water available to Egypt, which cannot exist without such water. Which one or more of the options discussed in the Core Case Study do you think Egypt should pursue? Explain.

Some of the world's lakes and inland seas are shrinking and, in some cases, disappearing because of excessive diversion of water from rivers that feed them and overpumping of adjacent aquifers. An example is Lake Chad in central Africa. Because of greatly increased use of river water for irrigation, the lake has shrunk by 96% since the 1960s and may soon disappear completely.

Large numbers of lakes have disappeared in China. One reason is that since 1985, China's Yellow River,

which flows through five provinces before emptying into the Yellow Sea, has often failed to reach the sea. As a result, China's Qinhai province has lost more than 2,000 of its 4,077 lakes since 1985. The following two case studies illustrate the benefits and harmful effects of damming rivers (**Concept 13-3**).

■ CASE STUDY

The Colorado River Basin—An Overtapped Resource

The Colorado River, the major river of the arid southwestern United States, flows 2,300 kilometers (1,400 miles) through seven states and eventually to the Gulf of California (Figure 13-14). During the past 50 years, this once free-flowing river, which gets its water mostly from snowmelt from the Rocky Mountains, has been tamed by a gigantic plumbing system consisting of 14 major dams and reservoirs (Figure 13-15) and canals that supply water to farmers, ranchers, and cities.

This system lies in a mostly desert area within the rain shadow (Figure 7-7, p. 145) of California's mountain ranges. It provides water and electricity for more than 25 million people in seven states. The river's water is used to produce about 15% of the nation's crops and livestock. It also supports a multibillion-dollar recreation industry of whitewater rafting, boating, fishing, camping, and hiking.

The river supplies water to some of the nation's driest and hottest cities. Take away this tamed river and Las Vegas, Nevada, would be a mostly uninhabited desert area; San Diego and Los Angeles, California, could not support their present populations; and California's Imperial Valley, which grows a major portion of the nation's vegetables, would consist mostly of cactus and mesquite plants.

Lake Mead behind Hoover Dam (Figure 13-14) supplies about 90% of water for the desert city of Las Vegas, Nevada—one of the country's fastest growing urban areas. In 2008, U.S. researchers Tim Barnett and David Pierce estimated that there is a 50% chance that by 2021, Lake Mead will run dry because of greater water demands and decreased flow of the Colorado River due to increased drought and evaporation of water.

There are four major problems associated with use of this river's water. *First,* the Colorado River basin includes some of the driest lands in the United States and Mexico (Figure 13-4). *Second,* for its size, the river has only a modest flow of water. *Third,* legal pacts signed in 1922 and 1944 allocated more water for human use in the United States and Mexico than the river can

Figure 13-14 The *Colorado River basin*. The area drained by this basin is equal to more than one-twelfth of the land area of the lower 48 states. Two large reservoirs—Lake Mead behind the Hoover Dam and Lake Powell behind the Glen Canyon Dam—store about 80% of the water in this basin.

Jim Wark/Peter Arnold, Inc.

Figure 13-15 Aerial view of Glen Canyon dam built across the Colorado River in 1963 and its reservoir called Lake Powell, the second largest reservoir in the United States.

Figure 13-16 The flow of the Colorado River measured at its mouth has dropped sharply since 1905 as a result of multiple dams, water withdrawals for agriculture and urban areas, and prolonged drought (Data from U.S. Geological Survey).

supply—even in rare years without a drought. The pacts allocated no water for environmental purposes.

Fourth, since 1905, the amount of water flowing to the mouth of the Colorado River has dropped dramatically (Figure 13-16). Since 1960, the river has rarely made it to the Gulf of California because of the many dams, the increased water withdrawals, and a prolonged drought in the American Southwest, which is projected to last throughout this century. This threatens the survival of species that spawn in the river and species that live in its estuary near the coast.

The water available from the Colorado is likely to decrease even more because of climate change caused by global warming. According to projections, as the climate continues to warm, mountain snows that feed the river will melt faster and earlier and evaporate in greater amounts.

If some of the Southwest's largest reservoirs empty out during this century, the region could experience an economic and ecological catastrophe with political and legal battles over who will get how much of the region's greatly diminished water supply. Agricultural production would drop sharply and many people in the region's booming desert cities, such as Phoenix, Arizona, and Las Vegas, Nevada, likely would have to migrate to other areas. Withdrawing more groundwater is not a solution because water tables beneath much of the area served by the Colorado River have been dropping, sometimes drastically, due to overpumping (Figure 13-9).

Traditionally, about 80% of the water withdrawn from the Colorado has been used to irrigate crops and raise cattle. The U.S. government paid for the dams and reservoirs and has supplied many farmers and ranchers with water at low prices. These government subsidies have led to inefficient use of irrigation water on thirsty crops such as rice, cotton, and alfalfa in this water-short area.

In addition, as the flow of the Colorado River slows in large reservoirs behind dams it drops much of its load of suspended silt. This deprives the river's coastal delta of much needed sediment and causes flooding and loss of ecologically important coastal wetlands. The amount of silt being deposited on the bottoms of the Lake Powell and Lake Mead reservoirs is roughly 20,000 dump truck loads every day. Sometime during this century, these reservoirs will probably be too full of silt to store enough water for generating hydroelectric power or controlling floods.

Some researchers argue that the seven states using the Colorado River should enact and enforce strict water conservation measures and sharply increase the price of water from the river. They also suggest measures to slow population growth and urban development, stop subsidizing agriculture in this region, shift water-thirsty crops to less arid areas, and stop building golf courses and growing green lawns in desert areas. People living in or moving to areas served by the river would have to realize that they are living in a desert and must adapt to their environment by drastically reducing their water usage.

These problems illustrate the challenges faced by governments and people living in arid and semiarid regions with shared river systems, as population growth and economic development place increasing demands on limited or decreasing supplies of surface water. Without major changes in water policies and management to drastically increase the efficiency of water use, the future for large numbers of people in such areas may dry up.

THINKING ABOUT
The Colorado River

What are three measures you would take to deal with the problems of the Colorado River system? How would you implement them politically?

■ CASE STUDY
China's Three Gorges Dam

China's Three Gorges Dam, built across the Yangtze River, is the world's largest hydroelectric dam and reservoir. Two kilometers (1.2 miles) long, the dam was built at a cost of at least $25 billion. Like all dams, it has created benefits and harmful effects (Figure 13-12).

When its giant reservoir fills, the dam will be able to produce an amount of electricity equal to that of 22 large coal-burning or nuclear power plants, enough electricity to provide power for a city 10 times as large as the U.S. city of Los Angeles, California. This will help to reduce China's dependence on coal and its emissions of the greenhouse gas CO_2.

The dam will also help to hold back the Yangtze River's floodwaters, which have killed more than 500,000 people during the past 100 years—including 4,000 people in 1998. The estimated cost of damage

caused by the 1998 flood alone equals the cost of the entire dam project. In addition, the dam will enable large cargo-carrying ships to travel deep into China's interior, greatly reducing transportation costs, and increasing trade in a relatively poor region of China.

When filled, the 600-kilometer-long (370-mile-long) reservoir behind the dam will cover an area almost equal in length to the distance between the U.S. cities of San Francisco and Los Angeles, California. This enormous reservoir will flood one of China's most beautiful areas, including 1,350 cities and villages and thousands of archeological and cultural sites. It will displace about 5.4 million people.

Some scientists contend that the slower flow of water in the reservoir will continually release huge amounts of sediment that will shorten the projected life of the reservoir and limit the flood-prevention capacity of the dam. In addition, productive farming regions below the dam will no longer receive annual deposits of nutrient-rich sediment.

Because the dam is built over a seismic fault, some geologists worry that it might collapse and cause a major flood that would kill millions of people. Engineers claim that the dam can withstand the maximum projected earthquake. Yet, some 80 small cracks have already been discovered in the dam.

Another problem is that plant and animal matter rotting underwater in the gigantic reservoir will release methane gas—which is a much more potent greenhouse gas than CO_2—into the atmosphere. The hydroelectric power generated by the dam will allow the Chinese to avoid some additional CO_2 emissions by not having to build as many new coal-burning power plants. But the reservoir's methane emissions will offset these savings in CO_2 emissions.

Opponents of the dam also claim that it will convert the Yangtze River to the world's largest sewer, because most cities and factories along the river dump their untreated sewage and other wastes directly into it. Opponents estimate that the hidden long-term harmful environmental and social costs of the dam project will be close to $75 billion—three times the direct cost of the project. Critics claim that it would have been cheaper, less disruptive, and safer to build a series of smaller dams.

THINKING ABOUT
The Three Gorges Dam

Do you think that the benefits of the Three Gorges Dam will outweigh its harmful effects? Explain.

13-4 Is Transferring Water from One Place to Another the Answer?

▶ **CONCEPT 13-4** Transferring water from one place to another has greatly increased water supplies in some areas, but it has also disrupted ecosystems.

California Transfers Massive Amounts of Water from Water-Rich Areas to Water-Poor Areas

Tunnels, aqueducts, and underground pipes can transfer stream runoff collected by dams and reservoirs from water-rich areas to water-poor areas, but they also create environmental problems (**Concept 13-4**).

One of the world's largest water transfer projects is the *California Water Project* (Figure 13-17, p. 330). It uses a maze of giant dams, pumps, and aqueducts to transport water from water-rich northern California to water-poor southern California's heavily populated agricultural regions and cities. This project supplies massive amounts of water to areas that, without such water transfers, would be mostly desert.

For decades, northern and southern Californians have feuded over how the state's water should be allocated under this project. Southern Californians want more water from the north to grow more crops and to support Los Angeles, San Diego, and other growing urban areas. Agriculture consumes three-fourths of the water withdrawn in California, much of it used inefficiently for water-thirsty crops such as rice and alfalfa growing in desert-like conditions.

Northern Californians counter that sending more water south degrades the Sacramento River, threatens fisheries, and reduces the river's power to flush pollutants out of San Francisco Bay. They also argue that much of the water sent south is wasted. They point to studies showing that making irrigation just 10% more efficient would provide enough water for domestic and industrial uses in southern California. But low water prices, mostly because government subsidies make it uneconomical for farmers to invest in improving irrigation efficiency.

According to a 2002 study by a group of scientists and engineers, projected global warming will sharply reduce water availability in California (especially

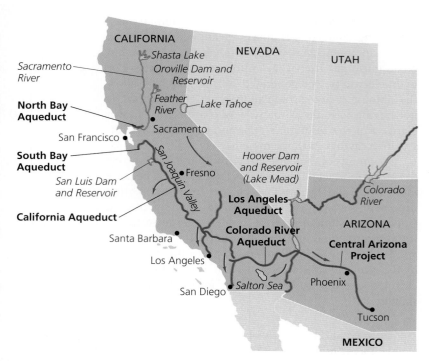

Figure 13-17 The California Water Project and the Central Arizona Project. These projects involve large-scale water transfers from one watershed to another. Arrows show the general direction of water flow.

southern California) and other water-short states in the western United States, even in the best-case scenario. Some analysts project that sometime during this century, many people living in arid southern California cities, as well as farmers in this area, may have to move elsewhere because of a shortage of water.

Pumping more groundwater is not the answer, because groundwater is already being withdrawn faster than it is replenished in much of central and southern California (Figure 13-9). According to many analysts, it would be quicker and cheaper to reduce water waste by improving irrigation efficiency, not growing water-thirsty crops in arid areas, and raising the historically low price of water to encourage water conservation.

■ CASE STUDY
The Aral Sea Disaster

The shrinking of the Aral Sea (Figure 13-18) is the result of a large-scale water transfer project in an area of the former Soviet Union with the driest climate in central Asia. Since 1960, enormous amounts of irrigation water have been diverted from the inland Aral Sea and its two feeder rivers to create one of the world's largest irrigated areas, mostly for raising cotton and rice. The irrigation canal, the world's longest, stretches more than 1,300 kilometers (800 miles).

This large-scale water diversion project, coupled with droughts and high evaporation rates due to the area's hot and dry climate, has caused a regional ecological and economic disaster. Since 1961, the sea's salinity has risen sevenfold and the average level of its

water has dropped by 22 meters (72 feet). It has lost 89% of its volume of water and has split into two major parts (Figure 13-18, right). Water withdrawal for agriculture has reduced the two rivers feeding the sea to mere trickles.

About 85% of the area's wetlands have been eliminated and roughly half the local bird and mammal species have disappeared. In addition, a huge area of former lake bottom has been converted to a human-made desert covered with glistening white salt. The sea's increased salt concentration—three times saltier than ocean water—caused the presumed extinction of 26 of the area's 32 native fish species. This has devastated the area's fishing industry, which once provided work for more than 60,000 people. Fishing villages and boats once located on the sea's coastline now sit abandoned in the middle of a salt desert (Figure 13-19).

Winds pick up the sand and salty dust and blow it onto fields as far as 500 kilometers (310 miles) away. As the salt spreads, it pollutes water and kills wildlife, crops, and other vegetation. Aral Sea dust settling on glaciers in the Himalayas is causing them to melt at a faster than normal rate—a prime example of unexpected connections and unintended consequences.

Shrinkage of the Aral Sea has also altered the area's climate. The once-huge sea acted as a thermal buffer that moderated the heat of summer and the extreme cold of winter. Now there is less rain, summers are hotter and drier, winters are colder, and the growing season is shorter. The combination of such climate change and severe salinization has reduced crop yields by 20–50% on almost one-third of the area's cropland.

To raise yields, farmers have used more herbicides, insecticides, and fertilizers, which have percolated downward and accumulated to dangerous levels in the groundwater—the source of most of the region's drinking water. Many of the 45 million people living in the Aral Sea's watershed have experienced increasing health problems—including anemia, respiratory illnesses, liver and kidney disease, eye problems, and various cancers—from a combination of toxic dust, salt, and contaminated water.

To make matters worse, Soviet scientists who were engaged in top-secret biological warfare studies buried hundreds of metric tons of deadly anthrax bacterial spores and other deadly toxins on an island in the Aral Sea. As the lakeshores receded, this island grew and joined the mainland in 2001. There is concern that any surviving disease organisms could reach people via fleas or infected rodents or by their removal and use by terrorists.

Since 1999, the United Nations and the World Bank have spent about $600 million to purify the area's drinking water and upgrade irrigation and drainage systems. This has improved irrigation efficiency and flushed salts from croplands. A dike completed in 2005 has raised the average level of the small Aral by 3 meters (10 feet) and decreased its salinity, allowing substantial fishing of several species. Some artificial wet-

1976

2006

Figure 13-18 Natural capital degradation: the *Aral Sea* was once the world's fourth largest freshwater lake. Since 1960, it has been shrinking and getting saltier because most of the water from the rivers that replenish it has been diverted to grow cotton and food crops (**Concept 13-4**). These satellite photos show the sea in 1976 and in 2006. It has split into two major parts, little Aral on the left and big Aral on the right. As the lake shrank, it left behind a salty desert, economic ruin, increasing health problems, and severe ecological disruption. **Question:** What do you think should be done to help prevent further shrinkage of the Aral Sea?

lands and lakes have been constructed to help restore aquatic vegetation, wildlife, and fisheries.

The five countries surrounding the lake and its two feeder rivers have worked to improve irrigation efficiency and to partially replace water-thirsty crops with others requiring less irrigation water. As a result, the total annual volume of water in the Aral Sea basin has been stabilized. Nevertheless, experts expect the largest portion of the Aral Sea to continue shrinking.

China Plans a Massive Transfer of Water

The Chinese government has begun a massive engineering project to transfer water from three of its southern river basins to its populous and parched northern provinces. This $62.5 billion South–North Water Transfer Project will use a series of canals, dams, reservoirs, and pumping stations to supply water to the north for irrigation and drinking.

One route, which is more than 1,000 kilometers (620 miles) long, will transfer about 2% of the flow of the Yangtze River to the north; it began supplying water to Beijing in 2008. Work on a second, equally long route is proceeding, while construction of a third,

Figure 13-19 Ship stranded in desert formed by shrinkage of the Aral Sea.

more complex route linking watersheds in western China and Tibet is not expected to start for at least another decade. If everything goes as planned, by 2050, the system will transfer ten times more water than is transferred from northern to southern California (Figure 13-17). It will provide water for almost half a billion people in northern China.

Critics of this huge project are worried that high levels of pollution in the three rivers from which water will be extracted and transported will contaminate northern water supplies. They are also concerned that the large-scale removal of water could damage southern ecosystems, especially the fragile watersheds in Tibet.

But many Chinese scientists believe that the potential economic, health, and environmental benefits of the project outweigh its risks. They point out that increasing desertification (Figure 12-13, p. 289) in north-ern China is eroding biodiversity and reducing supplies of drinking water for at least 96 million Chinese, because pollutants are becoming concentrated in small bodies of water. In addition, northern China needs much more water to support its large population, intensive agriculture, and rapid industrialization. The additional water will also help to reduce overpumping of aquifers. The central government has required northern cities receiving the water to implement water conservation measures to keep the additional water from being squandered.

However, Chinese and U.S. climate models predict that northern China will become increasingly wet as average temperatures rise from projected global warming. This shift could begin within a decade and render the water transfer project, especially the third route, unnecessary.

13-5 Is Converting Salty Seawater to Freshwater the Answer?

▶ **CONCEPT 13-5** We can convert salty ocean water to freshwater, but the cost is high, and the resulting salty brine must be disposed of without harming aquatic or terrestrial ecosystems.

Removing Salt from Seawater Seems Promising but Is Costly

Desalination involves removing dissolved salts from ocean water or from brackish (slightly salty) water in aquifers or lakes for domestic use. It is another way to increase supplies of freshwater (**Concept 13-5**).

One method for desalinating water is *distillation*—heating saltwater until it evaporates (leaving behind salts in solid form) and condenses as freshwater. Another method is *reverse osmosis* (or *microfiltration*), which uses high pressure to force saltwater through a membrane filter with pores small enough to remove the salt.

According to a 2004 report by the U.S. National Academy of Sciences, about 15,000 desalination plants operate in more than 125 countries, especially in the arid nations of the Middle East, North Africa, the Caribbean, and the Mediterranean. They meet less than 0.3% of the world's demand for freshwater.

Saudi Arabia has the world's largest number of desalination plants, and the United States has the world's second-largest desalination capacity. By 2008, water-short Israel was getting almost half of its water from desalination. And China plans to build enough desalination plants to provide 16–25% of the water used in its coastal areas by 2010.

There are three major problems with the widespread use of desalination. One is the high cost and energy footprint. It takes a lot of energy to desalinate water, with distillation requiring about ten times as much energy as reverse osmosis requires. A second problem is that pumping large volumes of seawater through pipes and using chemicals to sterilize the water and keep down algal growth kills many marine organisms.

A third problem is that desalination produces large quantities of briny wastewater that contain lots of salt and other minerals. Dumping this concentrated brine into nearby coastal ocean waters increases the salinity of the ocean water, which threatens food resources and aquatic life in the vicinity. Disposing of it on land could contaminate groundwater and surface water.

Using desalinated water produced by reverse osmosis to irrigate crops could reduce soil salinization (Figure 12-14, p. 289). However, Israeli scientist U. Yermiyahu and his colleagues reported in 2007 that desalinization removes ions such as calcium (Ca^{2+}), magnesium (Mg^{2+}), and sulfate (SO_4^{2-}) that are essential to plant growth. Thus, these ions must be added to desalinated irrigation water, which raises its already high cost, or the desalinated water must be blended with conventional irrigation water.

Bottom line: Currently, significant desalination is practical only for water-short, wealthy countries and

SCIENCE FOCUS

The Search for Improved Desalination Technology

Scientists are working to develop new membranes for reverse osmosis that can separate water from salt more efficiently and under less pressure and thus use less energy. Such technologies might bring down the cost of using desalination to produce drinking water but not to the point where they are cheap enough to irrigate conventional crops.

This may change if scientists can figure out how to use solar energy or other means to desalinate seawater cheaply and how to safely dispose of the resulting salty wastewater. Another possibility is the development of molecular size nanofilters to desalinate water at an affordable cost.

A consortium of companies is proposing to build ships that would carry desalination equipment. The ships would operate offshore, transferring the water to shore through seabed pipelines or in food grade shuttle tankers. The ships would have lower fuel costs than do land-based desalination

plants. They would also not be affected by onshore power failures, because they would generate their own power by using clean-burning biodiesel oil in gas turbine engines. In addition, they would be able to move away from the paths of storms and draw water from depths below where most marine organisms are found. And the resulting brine could be returned to the ocean and diluted far away from coastal waters. California officials are evaluating this approach.

Two Australian companies, Energetech and H2AU, have joined forces to build an experimental desalination plant that uses the power generated by ocean waves—an indirect form of solar energy—to drive reverse-osmosis desalination. This approach eliminates air pollution created from burning fossil fuels to produce electricity, and it conserves energy. Thus, costs should be much lower than for other desalination methods. The plant will be located offshore where the resulting brine can be mixed with ocean

water without affecting near-shore ecosystems.

In 2005, General Electric, one of the world's largest companies, began focusing on finding better ways to desalinate water. Because about 40% of the world is going to be short of freshwater in the next 10–20 years, GE sees desalination as one of the world's major growth businesses. **GREEN CAREER:** Desalination engineer

Before building a desalination plant, scientists urge reducing water needs by enacting strict water conservation regulations, fixing leaks in urban water supply systems, educating customers about water conservation, and cleaning up and reusing urban waste water. All of these solutions are much cheaper than desalination.

Critical Thinking

Use the second law of thermodynamics (p. 43) to explain why desalinated water is so costly.

cities that can afford its high cost (**Concept 13-5**). But scientists and engineers are working to develop better and more affordable desalination technologies (Science Focus, above).

RESEARCH FRONTIER

Developing better and more affordable desalination technologies. See **academic.cengage.com/biology/miller**.

13-6 How Can We Use Water More Sustainably?

▶ **CONCEPT 13-6** We can use water more sustainably by cutting water waste, raising water prices, slowing population growth, and protecting aquifers, forests, and other ecosystems that store and release water.

Reducing Water Waste Has Many Benefits

Mohamed El-Ashry of the World Resources Institute estimates that *65–70% of the water people use throughout the world is wasted through evaporation, leaks, and other losses.* The United States, the world's largest user of water, does slightly better but still loses about half of the water it withdraws. El-Ashry believes it is economically and technically feasible to reduce such water losses to 15%, thereby meeting most of the world's water needs for the foreseeable future (**Concept 13-6**).

Improving water use efficiency would decrease the burden on wastewater plants and reduce the need for

expensive dams and water transfer projects that destroy wildlife habitats and displace people. It would also slow depletion of groundwater aquifers and save both energy and money.

According to water resource experts, the main cause of water waste is *its low cost to users. Underpricing* is mostly the result of government subsidies that provide irrigation water, electricity, and diesel fuel used by farmers to pump water from rivers and aquifers at below-market prices.

Because these subsidies keep water prices low, users have little or no financial incentive to invest in water-saving technologies. According to water resource expert Sandra Postel, "By heavily subsidizing water,

governments give out the false message that it is abundant and can afford to be wasted—even as rivers are drying up, aquifers are being depleted, fisheries are collapsing, and species are going extinct."

But farmers, industries, and others benefiting from government water subsidies argue that the subsidies promote settlement and farming of arid, unproductive land; stimulate local economies; and help keep the prices of food, manufactured goods, and electricity low.

> **THINKING ABOUT**
> **Government Water Subsidies**
>
> Should governments provide subsidies to farmers and cities to help keep the price of water artificially low? Explain.

Most water resource experts believe that when water scarcity afflicts many areas in this century, governments will have to make the unpopular decision to raise water prices. China did so in 2002 because it faced water shortages in most of its major cities, with rivers running dry, lakes disappearing, and water tables falling in key agricultural areas.

Higher water prices encourage water conservation but make it difficult for low-income farmers and city dwellers to buy enough water to meet their needs. When South Africa raised water prices, it dealt with this problem by establishing *lifeline* rates that give each household a set amount of free or low-priced water to meet basic needs. When users exceed this amount, they pay higher prices as their water use increases—*a user-pays approach*.

The second major cause of water waste is *a lack of government subsidies for improving the efficiency of water use*. A basic rule of economics is that you get more of what you reward. Withdrawing subsidies that encourage water waste and providing subsidies for efficient water use would sharply reduce water waste and help reduce water shortages. Two goals of such subsidies should be to greatly improve the efficiency of irrigation, which accounts for 70% of the world's water use, and to use inexpensive means to collect rainwater and pipe it to where it is needed.

> **HOW WOULD YOU VOTE?** ☑
>
> Should water prices be raised sharply to help reduce water waste? Cast your vote online at **academic.cengage.com/biology/miller**.

We Can Cut Water Waste in Irrigation

About 60% of the irrigation water applied throughout the world does not reach the targeted crops. Most irrigation systems obtain water from a groundwater well or a surface water source. The water then flows by gravity through unlined ditches in crop fields so the crops can absorb it (Figure 13-20, left). This *flood irrigation* method delivers far more water than is needed for crop growth and typically loses 40% of the water through evaporation, seepage, and runoff. This wasteful method is used on 97% of China's irrigated land.

More efficient and environmentally sound irrigation technologies can greatly reduce water demands and water waste on farms by delivering water more precisely to crops—a more crop per drop strategy. For example, the *center-pivot, low-pressure sprinkler* (Figure 13-20, right) uses pumps to spray water on a crop. This results in a series of circular irrigated areas as shown by the green dots in Figure 13-8, right. Typically, it allows 80% of the water to reach crops. *Low-energy, precision application (LEPA) sprinklers*, another form of center-pivot irrigation, put 90–95% of the water where crops need it.

Drip or *trickle irrigation*, also called *microirrigation* (Figure 13-20, center), is the most efficient way to deliver small amounts of water precisely to crops. It consists of a network of perforated plastic tubing installed at or below the ground level. Small pinholes in the tubing deliver drops of water at a slow and steady rate, close to the roots of individual plants.

Current drip irrigation systems are costly but they drastically reduce water waste; 90–95% of the water input reaches the crops, and crop yields are 20–90% more than those of conventional gravity flow systems. By using less water, they also reduce the amount of salt that irrigation water leaves in the soil.

Drip irrigation is used on just over 1% of the world's irrigated crop fields and 4% of those in the United States. This percentage rises to 90% in Cyprus, 66% in Israel, and 13% in California. If water were priced closer to the value of the ecological services it provides and if government subsidies that encourage water waste were reduced or eliminated, drip irrigation would quickly be used to irrigate most of the world's crops.

Good news. The capital cost of a new type of drip irrigation system developed by the nonprofit International Development Enterprises (IDE) is one-tenth as much per hectare as that of conventional drip systems. Increased use of this inexpensive system designed for poor farmers will raise crop yields in water-short areas and help to lift poor families out of poverty. According to the United Nations, reducing current global withdrawal of water for irrigation by just 10% would save enough water to grow crops and meet the estimated additional water demands of cities and industries through 2025.

> **RESEARCH FRONTIER**
>
> Developing more efficient and affordable irrigation systems. See **academic.cengage.com/biology/miller**.

Gravity flow
(efficiency 60% and 80% with surge valves)

Water usually comes from an
aqueduct system or a nearby river.

Drip irrigation
(efficiency 90–95%)

Above- or below-ground pipes
or tubes deliver water to
individual plant roots.

Center pivot
(efficiency 80% with low-pressure
sprinkler and 90–95% with LEPA sprinkler)

Water usually pumped from
underground and sprayed from
mobile boom with sprinklers.

Figure 13-20 Major *irrigation systems*. Because of high initial costs, center-pivot irrigation and drip irrigation are not widely used. The development of new, low-cost, drip-irrigation systems may change this situation.

Figure 13-21 lists other ways to reduce water waste in crop irrigation. Since 1950, Israel has used many of these techniques to slash irrigation water waste by 84% while irrigating 44% more land. Israel now treats and reuses 30% of its municipal sewage water for crop production and plans to increase this to 80% by 2025. The government also gradually eliminated most water subsidies to raise Israel's price of irrigation water to one of the highest in the world. Israelis also import most of their wheat and meat and concentrate on growing fruits, vegetables, and flowers that need less water. For decades, wastewater from sewage treatment plants in many parts of the world has been used to water parks and golf courses and to irrigate flowers and other nonfood crops.

SOLUTIONS

Reducing Irrigation Water Waste

■ Line canals bringing water to irrigation ditches

■ Irrigate at night to reduce evaporation

■ Monitor soil moisture to add water only when necessary

■ Grow several crops on each plot of land (polyculture)

■ Encourage organic farming

■ Avoid growing water-thirsty crops in dry areas

■ Irrigate with treated urban wastewater

■ Import water-intensive crops and meat

Figure 13-21 Methods for reducing water waste in irrigation. **Question:** Which two of these solutions do you think are the most important? Why?

Developing Countries Use Low-Tech Methods for Irrigation

Many of the world's poor farmers use small-scale and low-cost traditional irrigation technologies. For example, millions of farmers in countries such as Bangladesh use human-powered treadle pumps to pump groundwater close to the earth's surface through irrigation ditches (see Photo 6 in the Detailed Contents). These wooden devices are cheap (about $25), easy to build from local materials, and operate on leg power. Other farmers in some developing countries use buckets, small tanks with holes, or cheap plastic tubing systems for drip irrigation.

Rainwater harvesting is another simple and inexpensive way to provide water for drinking and for growing crops throughout most of the world. It involves running pipes from rooftops and digging channels to catch rainwater that would otherwise run off the land. It is stored in shallow aquifers, ponds, and water tanks for use during dry spells. This is especially useful in developing countries, such as India, where much of the rain comes in a short monsoon season. In south Australia, over 40% of households use rainwater stored in tanks as their main source of drinking water. And half a million households and buildings in Germany harvest rainwater.

SOLUTIONS

Reducing Water Waste

- Redesign manufacturing processes to use less water

- Recycle water in industry

- Landscape yards with plants that require little water

- Use drip irrigation

- Fix water leaks

- Use water meters

- Raise water prices

- Use waterless composting toilets

- Require water conservation in water-short cities

- Use water-saving toilets, showerheads, and front-loading clothes washers

- Collect and reuse household water to irrigate lawns and nonedible plants

- Purify and reuse water for houses, apartments, and office buildings

Figure 13-22 Methods of reducing water waste in industries, homes, and businesses (**Concept 13-6**). **Question:** Which three of these solutions do you think are the most important? Why?

Other strategies used by poor farmers to increase the amount of crop per drop of rainfall include polyculture farming and agroforestry to create canopy cover over crops and to reduce water losses by evaporation. Farmers also plant deep-rooted perennial crop varieties (Science Focus, Figure 12-C, p. 309), control weeds, and mulch fields to retain more moisture.

In foggy mountainous areas, fog-catcher nets, developed by scientists in Chile, are used to harvest water. These upright volleyball-style plastic nets are placed on hilltops to intercept and remove water droplets from passing fog and clouds. The water is channeled into a storage tank or into gutters and then to a network of pipes that transfer it to wherever it is used.

We Can Cut Water Waste in Industry and Homes

Producers of chemicals, paper, oil, coal, primary metals, and processed food use almost 90% of the water used by industry in the United States. Some of these industries recapture, purify, and recycle water to reduce their water use and water treatment costs. For example, more than 95% of the water used to make steel is recycled. However, most industrial processes could be redesigned to use much less water. Raising water prices would spur such research. Figure 13-22 lists ways to use water more efficiently in industries, homes, and businesses (**Concept 13-6**).

Flushing toilets with water (most of it clean enough to drink) is the single largest use of domestic water in the United States. Since 1992, U.S. government standards have required that new toilets use no more than 6.1 liters (1.6 gallons) of water per flush. Architect and designer William McDonough has designed a toilet with a bowl so smooth that nothing sticks to it, including bacteria. Only a light mist is needed to flush it. Low-flow showerheads can cut shower water flow in half and save about 19,000 liters (5,000 gallons) per person each year.

According to U.N. studies, 40–60% of the water supplied in nearly all of the world's major cities in developing countries is lost mostly through leakage of water mains, pipes, pumps, and valves. Water experts say that fixing these leaks should be a high government priority that would cost less than building dams or importing water. Even in advanced industrialized countries such as the United States these losses average 10–30%. However, leakage losses have been reduced to about 3% in Copenhagen, Denmark, and to 5% in Fukuoka, Japan.

It does not require a government program to detect and fix leaks. To detect a silent toilet water leak, add a few drops of food coloring to the toilet tank and wait 5 minutes. If the color shows up in the bowl, you have a leak. Be sure to fix leaking faucets. A faucet dripping

once per second wastes up to 8,200 liters (3,000 gallons) of water a year.

Many homeowners and businesses in water-short areas are using drip irrigation and copying nature by replacing green lawns with plants that need little water. This water-thrifty landscaping, called *xeriscaping* (pronounced "ZEER-i-scaping"), reduces water use by 30–85% and sharply reduces needs for labor, fertilizer, and fuel. It is also an example of reconciliation ecology (p. 244), so it helps preserve biodiversity, and it reduces polluted runoff, air pollution, and yard wastes.

About 50–75% of the slightly dirtied water from bathtubs, showers, sinks, dishwashers, and clothes washers in a typical house could be stored in a holding tank and then reused as *gray water* to irrigate lawns and nonedible plants, to flush toilets, and to wash cars. Israel reuses 70% of its wastewater (sewage water) to irrigate nonfood crops. In Singapore, all sewage water is treated at reclamation plants for reuse by industry. Such measures mimic the way nature purifies water by recycling it, and thus they follow one of the four **scientific principles of sustainability** (see back cover).

Underpricing is also a major cause of excessive water use and waste in homes and industries (**Concept 13-6**). Many water utility and irrigation authorities charge a flat fee for water use and some charge less for the largest users of water. About one-fifth of all U.S. public water systems do not have water meters and charge a single low rate for almost unlimited use of high-quality water. Also, many apartment dwellers have little incentive to conserve water because water use charges are included in their rent. When the U.S. city of Boulder, Colorado, introduced water meters, water use per person dropped by 40%.

In Brazil, an electronic device called a water manager allows customers to obtain water on a pay-as-you-go basis. Water users buy a smart card (like a long-distance phone card) that contains a certain number of water credits. When they punch in the card's code on their water manager device, the water company automatically supplies them with a specified amount of water. Brazilian officials say this approach saves water and electrical power and typically reduces household water bills by 40%. **GREEN CAREER:** Water conservation specialist

We Can Use Less Water to Remove Wastes

Currently, we use large amounts of freshwater good enough to drink to flush away industrial, animal, and household wastes. According to the FAO, if current trends continue, within 40 years we will need the world's entire reliable flow of river water just to dilute and transport the wastes we produce. We could save much of this water by using systems that mimic the way nature deals with wastes.

For example, sewage treatment plants remove valuable plant nutrients and dump most of them into rivers, lakes, and oceans. This overloads aquatic systems with plant nutrients that could be recycled to the soil. We could mimic nature and return the nutrient-rich sludge produced by conventional waste treatment plants to the soil as a fertilizer, instead of using freshwater to transport it. Banning the discharge of industrial toxic chemicals into sewage treatment plants would help to make this feasible.

Another way to recycle waste is to rely more on waterless composting toilets that convert human fecal matter to a small amount of dry and odorless soil-like humus material that can be removed from a composting chamber every year or so and returned to the soil as fertilizer. One of the authors (Miller) successfully used a composting toilet for 15 years.

We Need to Use Water More Sustainably

More sustainable use of water is based on the commonsense principle stated in an old Inca proverb: "The frog does not drink up the pond in which it lives." Figure 13-23 lists strategies that scientists have suggested for using water more sustainably (**Concept 13-6**).

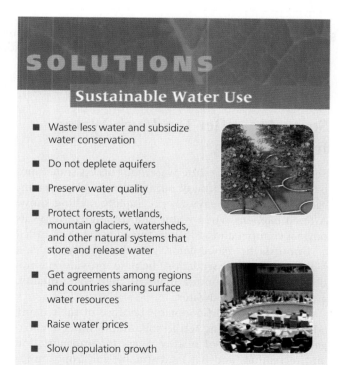

SOLUTIONS

Sustainable Water Use

- Waste less water and subsidize water conservation
- Do not deplete aquifers
- Preserve water quality
- Protect forests, wetlands, mountain glaciers, watersheds, and other natural systems that store and release water
- Get agreements among regions and countries sharing surface water resources
- Raise water prices
- Slow population growth

Figure 13-23 Methods for achieving more sustainable use of the earth's water resources (**Concept 13-6**). **Question:** Which two of these solutions do you think are the most important? Why?

Each of us can help to bring about such a *blue revolution* by using less water and cutting our water waste to reduce our water footprints (Figure 13-24). As with other problems, the solution starts with thinking globally and acting locally.

Figure 13-24 Individuals matter: ways in which you can reduce your use and waste of water. Visit www.h2ouse.org for an array of tips from the EPA and the California Urban Water Conservation Council on how to save water, no matter where you live. **Questions:** Which four of these actions do you think are the most important? Why? Which of these things do you plan to do?

WHAT CAN YOU DO?
Water Use and Waste

- Use water-saving toilets, showerheads, and faucet aerators
- Shower instead of taking baths, and take short showers
- Repair water leaks
- Turn off sink faucets while brushing teeth, shaving, or washing
- Wash only full loads of clothes or use the lowest possible water-level setting for smaller loads
- Use recycled (gray) water for watering lawns and houseplants and for washing cars
- Wash a car from a bucket of soapy water, and use the hose for rinsing only
- If you use a commercial car wash, try to find one that recycles its water
- Replace your lawn with native plants that need little if any watering
- Water lawns and yards only in the early morning or evening
- Use drip irrigation and mulch for gardens and flowerbeds

13-7 How Can We Reduce the Threat of Flooding?

▶ **CONCEPT 13-7** We can lessen the threat of flooding by protecting more wetlands and natural vegetation in watersheds and by not building in areas subject to frequent flooding.

Some Areas Get Too Much Water from Flooding

Some areas have too little water, but others sometimes have too much because of natural flooding by streams, caused mostly by heavy rain or rapidly melting snow. A flood happens when water in a stream overflows its normal channel and spills into the adjacent area, called a **floodplain.** Floodplains, which usually include highly productive wetlands, help to provide natural flood and erosion control, maintain high water quality, and recharge groundwater.

People settle on floodplains because of their many advantages, including fertile soil, ample water for irrigation, availability of nearby rivers for transportation and recreation, and flat land suitable for crops, buildings, highways, and railroads. To reduce the threat of flooding for people who live on floodplains, rivers have been narrowed and straightened (channelized), lined with protective levees and walls, and dammed to create reservoirs that store and release water as needed (Fig-

ures 13-12 and 13-15). However, in the long run, such measures can lead to greatly increased flood damage when prolonged rains overwhelm them.

Floods provide several benefits. They have created the world's most productive farmland by depositing nutrient-rich silt on floodplains. They also recharge groundwater and help refill wetlands, thereby supporting biodiversity and ecological services.

But floods kill thousands of people each year and cause tens of billions of dollars in property damage. Indeed, floods annually affect more people than the combined numbers affected by drought, tropical cyclones, famine, earthquakes, tsunamis, and volcanic eruptions. Floods usually are considered natural disasters, but since the 1960s, human activities have contributed to a sharp rise in flood deaths and damages.

One such human activity is *removal of water-absorbing vegetation*, especially on hillsides (Figure 13-25). Farm fields, pastures, pavement, or buildings usually replace such vegetation and absorb far less rainwater, or none at all. In 1998, for example, severe flooding in

China's Yangtze River watershed, home to 400 million people, killed at least 15 million people and caused massive economic losses. Scientists identified the causes as heavy rainfall, rapid snowmelt, and deforestation that had removed 85% of the watershed's tree cover. They also estimated the value of flood control, which the forested hills would have provided, to be three times the economic value of the lumber obtained by cutting the trees down. Chinese officials banned tree cutting in the watershed and accelerated tree replanting with the long-term goal of restoring some of the area's natural flood-control ecological services.

Removing protective mangrove forests (Figure 8-8, p. 168) in coastal areas can also allow for increased flooding from storms and tsunamis. Many such forests have been cleared to make way for shrimp farms.

Draining and building on wetlands, which naturally absorb floodwaters, is the second major human activity that increases the severity of flooding. When Hurricane Katrina struck the Gulf Coast of the United States in August 2005 and flooded the city of New Orleans (Figure 8-18, p. 177) and surrounding areas, the damage was intensified because of the degradation or removal of coastal wetlands that had historically helped to buffer the land from storm surges. Bottom line: Hurricane Katrina was a natural tragedy but its destructive effects were intensified because human activities had removed and degraded an important form of protective natural capital (Case Study, p. 177). Some scientists argue that, in order to help prevent future devastation of the Gulf

Coast, a major priority should be to restore its coastal wetlands.

Obviously, people who choose to live on floodplains increase their own risk of damage from flooding. Choosing to live elsewhere would thus lower this threat, but many of the poor are forced to live in such risky areas (Case Study, below).

Another human-related factor that will increase flooding is global warming caused mostly by people burning fossil fuels and clearing forests. Reports in 2007 by the Organization for Economic Cooperation and Development (OECD) and the IPCC projected that, by the 2070s, as many as 150 million people living in many of the world's largest coastal cities—an amount equal to half of the current U.S. population—are likely to be at risk from flooding as a result of rising sea levels caused by climate change.

■ CASE STUDY

Living Dangerously on Floodplains in Bangladesh

Bangladesh is one of the world's most densely populated countries. In 2008, its 147 million people were packed into an area roughly the size of the U.S. state of Wisconsin. And the country's population is projected to increase to 215 million by 2050. Bangladesh is a very flat country, only slightly above sea level, and it is one of the world's poorest countries.

Forested Hillside

After Deforestation

CENGAGENOW™ **Active Figure 13-25 Natural capital degradation:** hillside before and after deforestation. Once a hillside has been deforested for timber, fuelwood, livestock grazing, or unsustainable farming, water from precipitation rushes down the denuded slopes, erodes precious topsoil, and can increase flooding and pollution in local streams. Such deforestation can also increase landslides and mudflows. A 3,000-year-old Chinese proverb says, "To protect your rivers, protect your mountains." *See an animation based on this figure at* CengageNOW. **Question:** How might a drought in this area make these effects even worse?

The people of Bangladesh depend on moderate annual flooding during the summer monsoon season, which helps them to grow rice and maintain soil fertility in the delta basin. The annual floods deposit eroded Himalayan soil on the country's crop fields. People have adapted to this moderate flooding. Most of their houses have flat thatch roofs on which families can take refuge with their belongings in case of flooding. The roofs can also be detached from the walls, if necessary, and floated like rafts. After the waters have subsided, a roof can be reattached to the walls of its house. But great floods can overwhelm such cultural adaptations.

In the past, great floods occurred every 50 years or so. But since the 1970s, they have come roughly every 4 years. Bangladesh's flooding problems begin in the Himalayan watershed, where rapid population growth, deforestation, overgrazing, and unsustainable farming on steep and easily erodible slopes have increased flows of water during monsoon season. Monsoon rains now run more quickly off the denuded Himalayan foothills, carrying vital topsoil with them (Figure 13-25).

This increased runoff of soil, combined with heavier-than-normal monsoon rains, has increased the severity of flooding along Himalayan rivers and downstream in Bangladesh. In 1998, a disastrous flood covered two-thirds of Bangladesh's land area for 9 months, drowned at least 2,000 people, and left 30 million people homeless. It also destroyed more than one-fourth of the country's crops, which caused thousands of people to die of starvation. In 2002, another flood left 5 million people homeless and flooded large areas of rice fields. Another major flood occurred in 2004.

Living on Bangladesh's coastal floodplain at sea level means coping with storm surges, cyclones, and tsunamis. In 1970, as many as 1 million people drowned as a result of one tropical cyclone. Another cyclone in 2003 killed more than a million people and left tens of millions homeless. And another in 2007 killed at least 3,200 people, left millions homeless, destroyed crop lands, and killed at least 1.25 million livestock.

In their struggle to survive, the poor in Bangladesh have cleared many of the country's coastal mangrove forests for fuelwood, farming, and aquaculture ponds for raising shrimp. The result: more severe flooding, because these coastal wetlands that had sheltered Bangladesh's low-lying coastal areas from storm surges, cyclones, and tsunamis were now gone. Damages and deaths from cyclones in areas of Bangladesh still protected by mangrove forests have been much lower than in areas where the forests have been cleared.

A rise in sea level and an increase in storm intensity, mostly because of climate change caused by global warming, will be a major threat in coming years to millions of Bangladeshis who live along its largely flat delta, bound by the Bay of Bengal. In 2007, the IPCC projected that rises in the world's sea level during this century, mostly from the expansion of water as it warms, could permanently flood as much as 11% of Bangladesh's land area. This would create millions of refugees with no place to go in this already densely populated country.

THINKING ABOUT
Bangladesh

What are three important measures that could be taken to help reduce the threat of flooding in Bangladesh?

We Can Reduce Flood Risks

Figure 13-26 lists some ways to reduce flooding risks (**Concept 11-7**). To improve flood control, we can rely less on engineering devices such as dams and levees and more on nature's systems, such as wetlands and natural vegetation in watersheds.

Channelizing streams reduces upstream flooding. But it also eliminates aquatic habitats, reduces groundwater discharge, and results in a faster flow, which can increase downstream flooding and sediment deposition. And channelization encourages human settlement in floodplains, which increases the risk of damages and deaths from major floods.

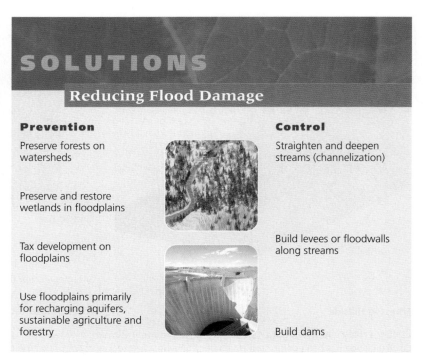

SOLUTIONS
Reducing Flood Damage

Prevention	Control
Preserve forests on watersheds	Straighten and deepen streams (channelization)
Preserve and restore wetlands in floodplains	
Tax development on floodplains	Build levees or floodwalls along streams
Use floodplains primarily for recharging aquifers, sustainable agriculture and forestry	Build dams

Figure 13-26 Methods for reducing the harmful effects of flooding (**Concept 13-7**). **Question:** Which two of these solutions do you think are the most important? Why?

Similarly, levees or floodwalls along the sides of streams contain and speed up stream flow, but they increase the water's capacity for doing damage downstream. And they do not protect against unusually high and powerful floodwaters, such as those that occurred in 1993 when two-thirds of the levees built along the Mississippi River in the United States were damaged or destroyed. Dams can reduce the threat of flooding by storing water in a reservoir and releasing it gradually, but they also have a number of disadvantages (Figure 13-12).

Perhaps one of the most important ways to reduce flooding is to *preserve existing wetlands* and *restore degraded wetlands* to take advantage of the natural flood control they provide in floodplains. This helps to protect biodiversity and to restore degraded ecological services provided by these aquatic ecosystems.

On a personal level, we can use the precautionary approach and *think carefully about where we choose to live.* Many poor people live in flood-prone areas because they have nowhere else to go. Most people, however, can choose not to live in areas especially subject to flooding or to water shortages.

REVISITING

Water Conflicts in the Middle East and Sustainability

The **Core Case Study** that opens this chapter discussed the problems and tensions of Middle Eastern countries trying to share limited water resources. Israel has become the world's most water-efficient nation and has led the world in developing technologies and a water pricing system that help the country to use its limited water supplies more sustainably. Other water-short nations would benefit from following its example in saving water.

Generally the water resource strategies of the 20th century worked against natural ecological cycles and processes. Large dams, river diversions, levees, and other big engineering schemes have helped to provide much of the world with electricity, food, drinking water, and flood control. But they have also degraded the aquatic natural capital needed for long-term economic and ecological sustainability by seriously disrupting rivers, streams, wetlands, aquifers, and other aquatic systems.

The four **scientific principles of sustainability** (see back cover) can guide us in using water more sustainably during this century. Scientists hope to use solar energy to desalinate water and increase supplies, and solar energy is already used to purify water for drinking in some areas. Recycling more water will help us to reduce water waste. Preserving biodiversity by avoiding disruption of aquatic systems and their bordering terrestrial systems is a key factor in maintaining water supplies and water quality. And controlling human population growth is fundamental to using water resources more sustainably while maintaining water quality.

This *blue revolution,* built mostly around cutting water waste, will provide innumerable economic and ecological benefits. There is no time to lose in implementing it.

The benefits of working with nature's water cycle, rather than further disrupting it, are too compelling to ignore.

SANDRA POSTEL

REVIEW

1. Review the Key Questions and Concepts for this Chapter on p. 314. Describe water conflicts in the Middle East and possible solutions to these problems.

2. What percentage of the earth's freshwater is available to us? Define **groundwater, zone of saturation, water table,** and **aquifer.** Define **surface water, surface runoff,** and **watershed (drainage basin).** Distinguish between surface runoff and **reliable surface runoff.** What percentage of the world's reliable runoff are we using and what percentage are we likely to be using by 2025?

3. How is most of the world's water used? Describe the availability and use of freshwater resources in the United States. How many people in the world lack regular access to safe drinking water, and how many do not have access to basic sanitation? What is **drought** and what are its causes and harmful effects? Discuss the question of who should own and manage freshwater resources.

4. What are the advantages and disadvantages of withdrawing groundwater? Describe the problem of groundwater depletion in the world and in the United States, especially

over the Ogallala aquifer. Describe ways to prevent or slow groundwater depletion.

5. What is a **dam?** What is a **reservoir?** What are the advantages and disadvantages of large dams and reservoirs? What ecological services do rivers provide? Describe some problems associated with the use of the Colorado River basin. What are the advantages and disadvantages of China's Three Gorges Dam?

6. Describe the California Water Project and the controversy over this water transfer project. Describe the Aral Sea disaster. Describe China's South–North Water Transfer Project.

7. Define **desalination** and distinguish between distillation and reverse osmosis as methods for desalinating water. What are the limitations of desalination and how might they be overcome?

8. What percentage of the world's water is unnecessarily wasted and what are two causes of such waste? Describe

four irrigation methods and describe ways to reduce water waste in irrigation in developed and developing countries. List ways to reduce water waste in industry and homes. List ways to use water more sustainably. Describe ways in which you can reduce your use and waste of water.

9. What is a **floodplain** and why do people like to live on floodplains? What are the benefits and drawbacks of floods? List three human activities that increase the risk of flooding. Describe the increased risk that many people in Bangladesh face. How can we reduce the risks of flooding?

10. Describe relationships between water conflicts in the Middle East (**Core Case Study**) and the four **scientific principles of sustainability**.

Note: Key Terms are in **bold** type.

CRITICAL THINKING

1. List three ways in which you could apply **Concept 13-6** to make your lifestyle more environmentally sustainable.

2. What do you believe are the three most important priorities for dealing with water shortages in parts of the Middle East as discussed in the **Core Case Study** that opens this chapter? Explain your choices.

3. List three ways in which human activities are affecting the water cycle. How might these effects impact your lifestyle? How might your lifestyle be contributing to these effects?

4. What role does population growth play in water supply problems? Relate this to water supply problems in the Middle East (**Core Case Study**) and in the U.S. city of Las Vegas, Nevada.

5. How do human activities increase the harmful effects of prolonged drought? Which of these activities are historical, and which are ongoing? Suggest ways to alter some of these activities in order to make them less harmful.

6. Explain why you are for or against **(a)** raising the price of water while providing lower rates for the poor and lower middle class, and **(b)** providing government subsidies to farmers for improving irrigation efficiency.

7. Calculate how many liters and gallons of water are wasted in 1 month by a toilet that leaks 2 drops of water per second. (1 liter of water equals about 3,500 drops and 1 liter equals 0.265 gallon.)

8. How do human activities increase the harmful effects of flooding? Describe an example from the news of how people are trying to change any of these activities to reduce their harmful effects. What are some ways in which other activities can be similarly altered?

9. Congratulations! You are in charge of the world. What are three actions you would take to provide sustainable supplies of water for the world's population?

10. List two questions that you would like to have answered as a result of reading this chapter.

Note: See Supplement 13 (p. S78) for a list of Projects related to this chapter.

In 2005, the population of the U.S. state of Florida consumed 24.5 billion liters (6.5 billion gallons) of fresh water daily. It is projected that in 2025, the daily consumption will increase to 32.1 billion liters (8.5 billion gallons) per day. Between 2005 and 2025 the population of Florida is projected to increase from 17.5 million to 25.9 million.

1. Based on total fresh water use,
 a. Calculate the per capita consumption of water per *day* in Florida in 2005 and the projected per capita consumption per day for 2025.
 b. Calculate the per capita consumption of water per *year* in Florida in 2005 and the projected per capita consumption per year for 2025.

2. In 2005, how did the Florida's *average water footprint* (consumption per person per year), based only on water used within the state, compare with the average U.S. water footprint of approximately 249,000 liters (66,000 gallons) per person per year and the global average water footprint of 123,770 liters (32,800 gallons) per person per year?

Environmental Effects of Gold Mining

Mineral resources extracted from the earth's crust are processed into an incredible variety of products that can make life easier and provide economic benefits and jobs. But extracting minerals from the ground and converting them to such products have a number of harmful environmental effects.

For example, gold miners typically remove enough rock to equal the weight of 50 automobiles to extract an amount of gold that would fit inside your clenched fist. Most newlyweds would be surprised to know that about 5.5 metric tons (6 tons) of mining waste was created to make their two gold wedding rings. This waste is left piled near the mine sites and can pollute the air and nearby surface water.

Between 1980 and 2007, global gold production more than doubled. In 2007, South Africa, Australia, the United States, and Canada were, in order, the world's top producers of gold.

In Australia and North America, a mining technology called *cyanide heap leaching* is used by mining companies to level entire mountains of rock containing only small concentrations of gold. To extract the gold, miners spray a solution of highly toxic cyanide salts (which react with gold) onto huge open-air piles of crushed rock. The solution then drains into storage ponds (Figure 14-1). After the solution is circulated a number of times, the gold is removed from the ponds.

This cyanide is extremely toxic to birds and mammals drawn to these ponds in search of water. The ponds can also leak or overflow, posing threats to underground drinking water supplies and fish and other forms of life in nearby lakes and streams. Special liners in the collection ponds can prevent leaks, but some have failed. According to the U.S. Environmental Protection Agency, all such liners will eventually leak.

In 2000, snow and heavy rains washed out an earthen dam on one end of a cyanide leach pond at a gold mine in Romania. The dam's collapse released large amounts of water laced with cyanide and toxic metals into the Tisza and Danube Rivers flowing through parts of Romania, Hungary, and Yugoslavia.

Several hundred thousand people living along these rivers were told not to fish or to drink or withdraw water from affected rivers or from wells along the rivers. Food industries and paper mills were shut down. Thousands of fish and other aquatic plants and animals were killed. This accident and another one that occurred in January 2001 could have been prevented if the mining company had installed a stronger containment dam and a backup collection pond to prevent leakage into nearby surface water.

In this chapter, we will look at the earth's dynamic geologic processes, the minerals such as gold that some of these processes produce, environmental effects of using such resources, their potential supplies, and how we might use them more sustainably.

Figure 14-1 Gold mine with cyanide leach piles and ponds in the Black Hills of the U.S. state of South Dakota.

Creatas/SuperStock

14-1 What are the earth's major geological processes and hazards?

CONCEPT 14-1A Gigantic plates in the earth's crust move very slowly atop the planet's mantle, and wind and water move matter from place to place across the earth's surface.

CONCEPT 14-1B Natural geological hazards such as earthquakes, tsunamis, volcanoes, and landslides can cause considerable damage.

14-2 How are the earth's rocks recycled?

CONCEPT 14-2 The three major types of rocks found in the earth's crust—sedimentary, igneous, and metamorphic—are recycled very slowly by the processes of erosion, melting, and metamorphism.

14-3 What are mineral resources and what are the environmental effects of using them?

CONCEPT 14-3A Some naturally occurring materials in the earth's crust can be extracted and made into useful products in processes that provide economic benefits and jobs.

CONCEPT 14-3B Extracting and using mineral resources can disturb the land, erode soils, produce large amounts of solid waste, and pollute the air, water, and soil.

14-4 How long will supplies of nonrenewable mineral resources last?

CONCEPT 14-4A All nonrenewable mineral resources exist in finite amounts, and as we get closer to depleting any mineral resource, the environmental impacts of extracting it generally become more harmful.

CONCEPT 14-4B An increase in the price of a scarce mineral resource can lead to increased supplies and more efficient use of the mineral, but there are limits to this effect.

14-5 How can we use mineral resources more sustainably?

CONCEPT 14-5 We can try to find substitutes for scarce resources, reduce resource waste, and recycle and reuse minerals.

Note: Supplements 2 (p. S4), 6 (p. S39), and 13 (p. S78) can be used with this chapter.

*Civilization exists by geological consent,
subject to change without notice.*

WILL DURANT

14-1 What Are the Earth's Major Geological Processes and Hazards?

▶ **CONCEPT 14-1A** Gigantic plates in the earth's crust move very slowly atop the planet's mantle, and wind and water move matter from place to place across the earth's surface.

▶ **CONCEPT 14-1B** Natural geological hazards such as earthquakes, tsunamis, volcanoes, and landslides can cause considerable damage.

The Earth Is a Dynamic Planet

Geology, one of the subjects of this chapter, is the science devoted to the study of dynamic processes occurring on the earth's surface and in its interior. As the primitive earth cooled over eons, its interior separated into three major concentric zones: the *core,* the *mantle,* and the *crust* (Figure 3-6, p. 55).

The **core** is the earth's innermost zone. It is extremely hot and has a solid inner part, surrounded by a liquid core of molten or semisolid material. Surrounding the core is a thick zone called the **mantle.** Most of the mantle is solid rock, but under its rigid outermost part is the *asthenosphere*—a zone of hot, partly melted rock that flows and can be deformed like soft plastic.

Links: CORE CASE STUDY refers to the Core Case Study. SUSTAINABILITY refers to the book's sustainability theme. CONCEPT LINK indicates links to key concepts in earlier chapters.

345

Figure 14-2
Major features of the earth's crust and upper mantle. The *lithosphere,* composed of the crust and outermost mantle, is rigid and brittle. The *asthenosphere,* a zone in the mantle, can be deformed by heat and pressure.

Abyssal hills

Abyssal floor

Oceanic ridge

Abyssal floor

Trench

Volcanoes

Folded mountain belt

Craton

Abyssal plain

Oceanic crust (lithosphere)

Abyssal plain

Mantle (lithosphere)

Continental shelf

Continental slope

Continental rise

Continental crust (lithosphere)

Mantle (lithosphere)

Mantle (asthenosphere)

Active Figure 14-3
The earth's crust is made up of a mosaic of huge rigid plates, called *tectonic plates,* which move very slowly across the asthenosphere in response to forces in the mantle. *See an animation based on this figure at CengageNOW™.*

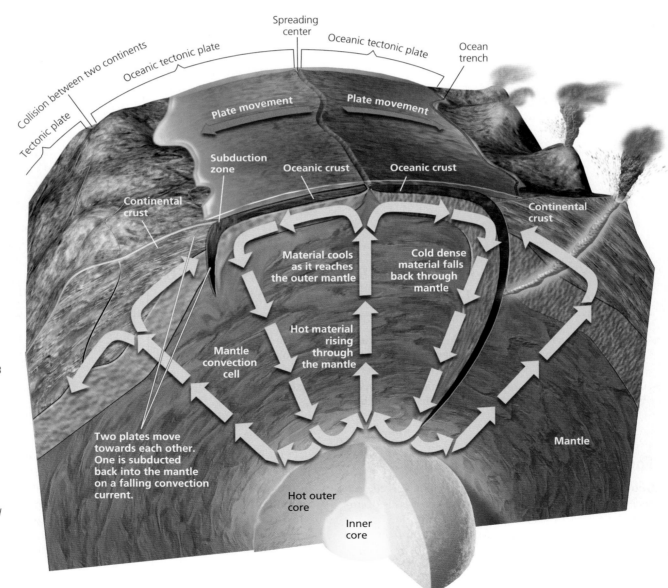

Collision between two continents

Oceanic tectonic plate

Spreading center

Oceanic tectonic plate

Ocean trench

Tectonic plate

Plate movement

Plate movement

Subduction zone

Oceanic crust

Oceanic crust

Continental crust

Continental crust

Material cools as it reaches the outer mantle

Cold dense material falls back through mantle

Hot material rising through the mantle

Mantle convection cell

Two plates move towards each other. One is subducted back into the mantle on a falling convection current.

Mantle

Hot outer core

Inner core

The outermost and thinnest zone of the earth is the **crust.** It consists of the *continental crust,* which underlies the continents (including the continental shelves extending into the oceans), and the *oceanic crust,* which underlies the ocean basins and makes up 71% of the earth's crust (Figure 14-2).

The Earth Beneath Your Feet Is Moving

We tend to think of the earth's crust, mantle, and core as fairly static. In reality, *convection cells* or *currents* move large volumes of rock and heat in loops within the mantle like gigantic conveyer belts (Figure 14-3). The flows of energy and heated material in the mantle's convection cells cause a dozen or so huge rigid plates, called **tectonic plates,** to move extremely slowly atop the denser mantle on hot soft rock in the underlying asthenosphere (Figures 14-3 and 14-4) (**Concept 14-1A**). These thick plates are composed of the continental and oceanic crust and the rigid, outermost part of the mantle (above the asthenosphere), a combination called the **lithosphere.**

These gigantic plates are somewhat like the world's largest and slowest-moving surfboards. Their typical speed is about the rate at which fingernails grow. You ride on one of these plates throughout your entire life without noticing. And throughout the earth's history, continents have split apart and joined as tectonic plates drifted thousands of kilometers back and forth atop the earth's mantle (Figure 4-6, p. 85).

Much of the geologic activity at the earth's surface takes place at the boundaries between tectonic plates as they separate, collide, or slide past one another. The tremendous forces produced at these plate boundaries can cause mountains to form, earthquakes to shake parts of the crust, and volcanoes to erupt.

When oceanic plates move apart from one another (*divergent plates*) molten rock, or *magma,* flows up through the resulting cracks. This creates *oceanic ridges* (Figure 14-2), some of which have higher peaks and deeper canyons than earth's continents have. When an oceanic plate collides with a continental plate (*convergent plates*), the continental plate usually rides up over the denser oceanic plate and pushes it down into the mantle (Figure 14-3) in a process called *subduction.* The area where this collision and subduction takes place is

／ Divergent plate boundaries ▲—▲—▲ Convergent plate boundaries —— Transform faults

CENGAGENOW **Active Figure 14-4** The earth's major tectonic plates. *See an animation based on this figure at* CengageNOW. **Question:** What plate are you riding on?

Figure 14-5 The San Andreas Fault as it crosses part of the Carrizo plain between San Francisco and Los Angeles, California (USA). This fault, which runs along almost the full length of California, is responsible for earthquakes of various magnitudes. **Question:** Is there a transform fault near where you live or go to school? (See Figure 14-4.)

Kevin Schafer/Peter Arnold, Inc.

called a *subduction zone*. Over time, the subducted plate melts and then rises again to the earth's surface as magma. A *trench* (Figure 14-2) ordinarily forms at the boundary between the two converging plates. When two continental plates collide, they push up mountain ranges, such as the Himalayas, along the collision boundary.

The third type of boundary is a *transform fault*, where plates slide and grind past one another along a fracture (fault) in the lithosphere. Most transform faults are located on the ocean floor but a few are found on land. For example, the North American Plate and the Pacific Plate slide past each other along California's San Andreas fault (Figure 14-5).

Some Parts of the Earth's Surface Build Up and Some Wear Down

Internal geologic processes, generated by heat from the earth's interior, typically build up the earth's surface in the form of continental and oceanic crust, including mountains and volcanoes (Figures 14-2 and 14-3).

By contrast, *external geologic processes*, driven directly or indirectly by energy from the sun (mostly in the form of flowing water and wind) and influenced by gravity, tend to wear down the earth's surface and move matter from one place to another (**Concept 14-1A**).

One major external geologic process is **weathering**, the physical, chemical, and biological processes that break down rocks into smaller particles that help build soil (Figure 14-6). These weathering processes play a key role in soil formation (Science, Focus, Figure 12-A, p. 281), a vital part of the earth's natural capital (**Concept 1-1A**, p. 6). ↰ CONCEPT LINK

Another major external process is *erosion,* discussed in Chapter 12 (pp. 287–288). In this process, material is dissolved, loosened, or worn away from one part of the earth's surface (Figure 12-10, p. 287) and deposited elsewhere. Flowing streams and rain cause most erosion. Wind also blows particles of soil from one area to another. Human activities—particularly those that destroy vegetation that holds soil in place—accelerate the process (Figure 7-13, p. 152, and Figure 10-11, p. 222).

Slowly flowing bodies of ice called *glaciers* also cause erosion. Under the influence of gravity, glaciers move slowly down a mountainside or over a wide area. During this movement, rock frozen to the glacial ice is pulled along or plucked out of the land surface. During the last ice age, which ended about 10,000 years ago, ice sheets called *continental glaciers* covered vast areas of North America, Europe, and Asia. The Great Lakes, the world's largest mass of freshwater, formed during this period as retreating glaciers gouged out huge basins. As the climate warmed and the glaciers melted, water filled these basins.

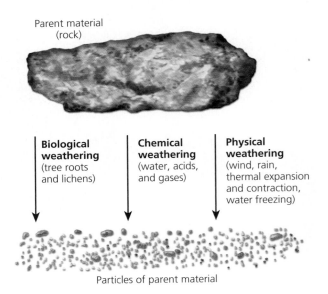

Parent material (rock)

| Biological weathering (tree roots and lichens) | Chemical weathering (water, acids, and gases) | Physical weathering (wind, rain, thermal expansion and contraction, water freezing) |

Particles of parent material

Figure 14-6 *Weathering:* Biological, chemical, and physical processes weather or convert rock into smaller fragments and particles. It is the first step in soil formation.

Volcanoes Release Molten Rock from the Earth's Interior

An active **volcano** occurs where magma reaches the earth's surface through a central vent or a long crack, called a *fissure* (Figure 14-7). Many volcanoes form when one tectonic plate slides under or moves away from another plate. Magma that reaches the earth's surface is called *lava*. Volcanic activity can release debris ranging from large chunks of lava rock to glowing hot ash, liquid lava, and gases (such as water vapor, carbon dioxide, and sulfur dioxide) into the environment (**Concept 14-1B**). Much of the world's volcanic activity is concentrated along the boundaries of the earth's tectonic plates (Figure 14-4).

The largest volcanic eruption during the 20th century occurred in 1991 when Mount Pinatubo exploded. It killed several hundred people, despite the evacuation of more than 200,000 people. Buildings covered with wet ash collapsed and the volcano ejected enough material into the atmosphere to reduce incoming solar energy and cool the earth's average temperature for 15 months.

The worst volcanic disaster in U.S. history occurred when Mount St. Helens in the state of Washington erupted on May 18, 1980. Fifty-seven people and large numbers of wild animals were killed, and large areas of forests were obliterated. However, through ecological succession, some of the vegetation has returned.

We tend to think of volcanic activity as an undesirable event, but it does provide some benefits. For example, it creates outstanding scenery in the form of majestic mountains, some lakes (such as Crater Lake in Oregon; Figure 8-16, left, p. 175), and other landforms. Perhaps the most important benefit of volcanism is the highly fertile soils produced by the weathering of lava.

We can reduce the loss of human life and some of the property damage caused by volcanic eruptions in several ways. For example, we use historical records and geologic measurements to identify high-risk areas, so that people can avoid living in those areas. We also use monitoring devices that warn us when volcanoes are likely to erupt, and we have developed evacuation plans for areas prone to volcanic activity.

Earthquakes Are Geological Rock-and-Roll Events

Forces inside the earth's mantle and along its surface push, deform, and stress rocks. At some point, the stress can cause the rocks to suddenly shift or break and produce a transform fault or fracture in earth's crust (Figure 14-5). When a fault forms, or when there is abrupt movement on an existing fault, energy that has accumulated over time is released in the form of vibrations, called *seismic waves,* which move in all directions

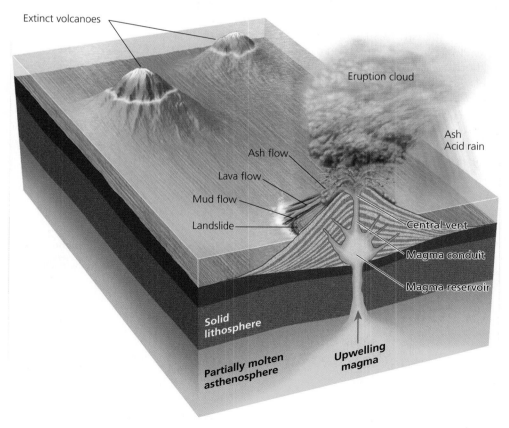

Extinct volcanoes

Eruption cloud

Ash
Acid rain

Ash flow
Lava flow
Mud flow
Landslide

Central vent
Magma conduit
Magma reservoir

Solid lithosphere

Partially molten asthenosphere

Upwelling magma

Figure 14-7 A *volcano* is created when magma in the partially molten asthenosphere rises in a plume through the lithosphere to erupt on the surface as lava, which builds a cone. Sometimes, internal pressure is high enough to cause lava, ash, and gases to be ejected into the atmosphere or to flow over land, causing considerable damage (**Concept 14-1B**). Chains of islands have been created by volcanoes that erupted and then became inactive.

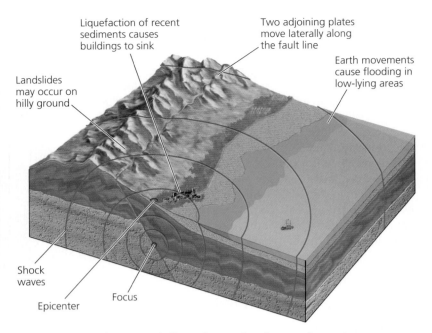

Liquefaction of recent
sediments causes
buildings to sink

Two adjoining plates
move laterally along
the fault line

Earth movements
cause flooding in
low-lying areas

Landslides
may occur on
hilly ground

Shock
waves

Epicenter

Focus

Figure 14-8 Major features and effects of an *earthquake*, one of nature's most powerful events.

through the surrounding rock. This internal geological process is called an **earthquake** (Figure 14-8 and **Concept 14-1B**). Most earthquakes occur at the boundaries of tectonic plates (Figure 14-4) when colliding plates create tremendous pressures in the earth's crust or when plates slide past one another at transform faults (Figure 14-5).

The place where an earthquake begins, often far below the earth's surface is called the *focus* (Figure 14-8). The earthquake's *epicenter* is found on the earth's surface directly above the focus. Relief of the earth's internal stress releases energy as shock (seismic) waves, which move outward from the earthquake's focus like ripples in a pool of water.

Scientists measure the severity of an earthquake by the *magnitude* of its seismic waves. The magnitude is a measure of ground motion (shaking) caused by the earthquake, as indicated by the *amplitude*, or size of the

seismic waves when they reach a recording instrument, called a *seismograph*.

Each year, scientists record the magnitudes of more than 1 million earthquakes, most of which are too small to feel. Scientists use the *Richter scale*, on which each unit has amplitude 10 times greater than the next smaller unit. Thus, a magnitude 5.0 earthquake would result in ten times more ground shaking than a magnitude 4.0 earthquake. And the amount of ground movement from a magnitude 7.0 quake is 100 times greater than that of a magnitude 5.0 quake. Seismologists rate earthquakes as *insignificant* (less than 4.0 on the Richter scale), *minor* (4.0–4.9), *damaging* (5.0–5.9), *destructive* (6.0–6.9), *major* (7.0–7.9), and *great* (over 8.0). The largest recorded earthquake occurred in Chile on May 22, 1960 and measured 9.5 on the Richter scale.

Earthquakes often release *aftershocks* that gradually decrease in frequency over periods of as long as several months. Some are preceded by *foreshocks*, which occur from seconds to weeks before the main shock.

The *primary effects of earthquakes* include shaking and sometimes a permanent vertical or horizontal displacement of the ground. These effects may have serious consequences for people and for buildings, bridges, freeway overpasses, dams, and pipelines. A major earthquake is a very large rock-and-roll geological event.

One way to reduce the loss of life and property damage from earthquakes is to examine historical records and make geologic measurements to locate active fault zones. We can then map high-risk areas and establish building codes that regulate the placement and design of buildings in such areas. Then people can evaluate the risk and factor it into their decisions about where to live. Figure 14-9 shows the areas of greatest earthquake risk in the United States and Figure 14-10 shows areas of such risk throughout the world.

We can also increase research geared toward predicting when and where earthquakes will occur. And engineers know how to make homes, large buildings, bridges, and freeways more earthquake resistant. But this can be expensive, especially the reinforcement of existing structures.

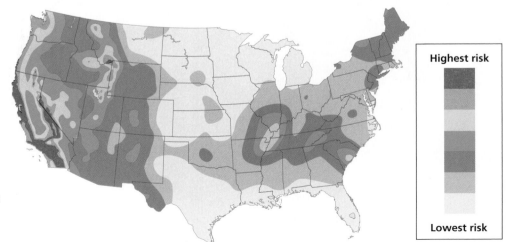

Figure 14-9 Areas of greatest earthquake (seismic) risk in the United States. **Question:** What is the degree of risk where you live or go to school? (Data from U.S. Geological Survey)

Highest risk

Lowest risk

Figure 14-10 Areas of greatest earthquake (seismic) risk in the world. **Question:** How are these areas related to the boundaries of the earth's major tectonic plates as shown in Figure 14-4? (Data from U.S. Geological Survey)

Earthquakes on the Ocean Floor Can Cause Huge Waves Called Tsunamis

A **tsunami** is a series of large waves generated when part of the ocean floor suddenly rises or drops (Figure 14-11, p. 352). Most large tsunamis are caused when thrust faults in the ocean floor move up or down as a result of a large underwater earthquake, a landslide caused by such an earthquake, or in some cases by a volcanic eruption. Such earthquakes often occur offshore in subduction zones where a tectonic plate slips under a continental plate (Figure 14-3).

Tsunamis are often called tidal waves, although they have nothing to do with tides. They can travel far across the ocean at speeds as high as those of jet planes. In deep water the waves are very far apart— sometimes hundreds of kilometers—and their crests are not very high. As a tsunami approaches a coast, it slows down, its wave crests squeeze closer together, and their heights grow rapidly. It can hit a coast as a series of towering walls of water that can level buildings (**Concept 14-1B**).

Tsunamis can be detected through a network of ocean buoys to provide some degree of early warning. Another method makes use of a pressure recorder on the ocean floor, which measures changes in water pressure as the waves of a tsunami pass over it. These data are relayed to a weather buoy, which then transmits the data via satellite to tsunami emergency warning centers.

Between 1900 and late 2007, tsunamis killed an estimated 278,000 people in some Pacific Ocean regions. The largest loss of life occurred in December 2004 when a great underwater earthquake in the Indian Ocean with a magnitude of 9.15 caused a huge tsunami that generated waves that eventually were as high as 31 meters (100 feet). It killed 228,000 people and devastated many coastal areas of Indonesia (Figure 14-12, p. 352), Thailand, Sri Lanka, South India, and eastern Africa. Tragically, no buoys or gauges were in place in the Indian Ocean to provide an early warning of this tsunami.

Satellite observations and ground studies in February 2005 by the U.N. Environment Programme pointed to the role that healthy coral reefs (Figure 8-1, p. 162) and mangrove forests (Figure 8-8, p. 168) played in reducing the death toll and destruction from the 2004 tsunami. For example, intact mangrove forests in parts of Thailand helped to protect buildings and people from the force of the huge waves. In contrast, the extensive damage and high death toll in India's Tamus state has been attributed in part to the clearing of a third of its coastal mangrove forests in recent decades. In

Earthquake in seafloor swiftly pushes water upwards, and starts a series of waves

Waves move rapidly in deep ocean reaching speeds of up to 890 kilometers per hour.

As the waves near land they slow to about 45 kilometers per hour but are squeezed upwards and increased in height.

Waves head inland causing damage in their path.

Undersea thrust fault

Upward wave

Earthquake

India
Bangladesh
Burma
Thailand
Sri Lanka
Malaysia
Sumatra
Indonesia

December 26, 2004, tsunami

Figure 14-11 Formation of a tsunami and map of area affected by a large tsunami in December 2004.

Figure 14-12 In December 2004, a great earthquake with a magnitude of 9.15 on the seafloor of the Pacific Ocean created a large tsunami that killed 168,000 people in Indonesia. These photos show the Banda Aceh Shore near Gleebruk in Indonesia on June 23, 2004 before the tsunami (left) and on December 28, 2004 after it was stuck by the tsunami (right) (**Concept 14-1B**).

Sri Lanka, some of the greatest damage occurred where illegal coral mining and reef damage had caused severe beach erosion.

Gravity and Earthquakes Can Cause Landslides

The downward pull of gravity or the forces released by an earthquake can cause detached or loose rock, soil, and mud to slide down steep slopes.

Such movement, called *mass wasting*, is most common on the steep sides of mountains above valleys and on cliffs or steep slopes near the shores of oceans or lakes. Sometimes the movement is slow but some rockslides, avalanches, and mudslides occur quickly. In 1970, an earthquake in Peru caused a massive landslide that buried the town of Yungay and killed 17,000 people.

Mass wasting can cause considerable property damage when houses and other structures slide down hillsides or when avalanches, rockslides, or mudflows hit homes and properties (**Concept 14-1B**). Human activities such as forest clearing, road building, crop growing, and building of houses on steep and unstable slopes increase the frequency of, and the damage caused by, such geological events.

14-2 How Are the Earth's Rocks Recycled?

▶ **CONCEPT 14-2** The three major types of rocks found in the earth's crust—sedimentary, igneous, and metamorphic—are recycled very slowly by the processes of erosion, melting, and metamorphism.

There Are Three Major Types of Rocks

The earth's crust consists mostly of minerals and rocks. A **mineral** is an element or inorganic compound that occurs naturally in the earth's crust as a solid with a regular internal crystalline structure. A few minerals consist of a single element, such as gold, silver, and diamonds (carbon). But most of the more than 2,000 identified minerals occur as inorganic compounds formed by various combinations of elements. Examples include salt (sodium chloride or NaCl, Figure 2 in Supplement 6, p. S40) and quartzite (silicon dioxide or SiO_2).

Rock is a solid combination of one or more minerals found in the earth's crust. Some kinds of rock, such as limestone (calcium carbonate, or $CaCO_3$) and quartzite, contain only one mineral. Most rocks consist of two or more minerals. For example, granite is a mixture of mica, feldspar, and quartz crystals.

Based on the way it forms, rock is placed in three broad classes: sedimentary, igneous, or metamorphic (**Concept 14-2**). **Sedimentary rock** is made of *sediments*—dead plant and animal remains and existing rocks that are weathered (Figure 14-6) and eroded into tiny particles. These sediments are transported by water, wind, or gravity to downstream, downwind, downhill, or underwater sites. There they are deposited in layers that accumulate over time and increase the weight and pressure on underlying layers. Examples include *sandstone* and *shale* (formed from pressure created by deposited layers of mostly sand), *dolomite* and *limestone* (formed from the compacted shells, skeletons, and other remains of dead organisms), and *lignite* and *bituminous coal* (derived from compacted plant remains).

Igneous rock forms below or on the earth's surface when magma wells up from the earth's upper mantle or deep crust and then cools and hardens. Examples include *granite* (formed underground) and *lava rock* (formed aboveground). Although often covered by sedimentary rocks or soil, igneous rocks form the bulk of the earth's crust.

Metamorphic rock forms when a preexisting rock is subjected to high temperatures (which may cause it to melt partially), high pressures, chemically active fluids, or a combination of these agents. These forces transform a rock by reshaping its internal crystalline structure and its physical properties and appearance. Examples include *anthracite* (a form of coal), *slate* (formed when shale and mudstone are heated), and *marble* (produced when limestone is exposed to heat and pressure).

The Earth's Rocks Are Recycled Very Slowly

The interaction of physical and chemical processes that change rocks from one type to another is called the **rock cycle** (Figure 14-13, p. 354). This important form of natural capital recycles the earth's three types of rocks over millions of years and is the slowest of the earth's cyclic processes (**Concept 14-2**). In

Figure 14-13 Natural capital: the *rock cycle* is the slowest of the earth's cyclic processes. Rocks are recycled over millions of years by three processes: *erosion, melting,* and *metamorphism,* which produce *sedimentary, igneous,* and *metamorphic* rocks. Rock from any of these classes can be converted to rock of either of the other two classes, or can be recycled within its own class (**Concept 14-2**). **Question:** What are three ways in which the rock cycle benefits your lifestyle?

this process, rocks are broken down, eroded, crushed, heated, melted, fused together into new forms by heat and pressure, cooled, and/or recrystallized within the earth's mantle and in the earth's crust.

The rock cycle also concentrates the planet's nonrenewable mineral resources on which our life processes depend. Without the earth's incredibly slow rock cycle, you would not exist.

14-3 What Are Mineral Resources and What Are the Environmental Effects of Using Them?

▶ **CONCEPT 14-3A** Some naturally occurring materials in the earth's crust can be extracted and made into useful products in processes that provide economic benefits and jobs.

▶ **CONCEPT 14-3B** Extracting and using mineral resources can disturb the land, erode soils, produce large amounts of solid waste, and pollute the air, water, and soil.

We Use a Variety of Nonrenewable Mineral Resources

A **mineral resource** is a concentration of naturally occurring material from the earth's crust that can be extracted and processed into useful products and raw materials at an affordable cost (**Concept 14-3A**). We know how to find and extract more than 100 minerals from

the earth's crust. Examples are *fossil fuels* (such as coal), *metallic minerals* (such as aluminum, iron, and copper), and *nonmetallic minerals* (such as sand, gravel, and limestone). Because they take so long to form, these components of the earth's natural capital are classified as *nonrenewable mineral resources.*

An **ore** is rock that contains a large enough concentration of a particular mineral—often a metal—to make

it profitable for mining and processing. A **high-grade ore** contains a fairly large amount of the desired nonrenewable mineral resource, whereas a **low-grade ore** contains a smaller amount.

Nonrenewable metal and nonmetal mineral resources are important parts of our lives. *Aluminum* (Al) is used for packaging and beverage cans and as a structural material in motor vehicles, aircraft, and buildings. *Steel,* an essential material used in buildings and motor vehicles, is a mixture (alloy) of iron (Fe) and other elements that are added to give it certain properties. *Manganese* (Mn), *cobalt* (Co), *molybdenum* (Mo) and *chromium* (Cr) are widely used in important steel alloys used in products such as lightbulbs, computers, automobiles, airplanes, oil drilling bits, pipelines, and turbines in power plants. *Copper* (Cu), a good conductor of electricity, is used for electrical and communications wiring. *Platinum* (Pt) is used in electrical equipment, as a catalyst in industry, and in automobile pollution control converters. *Gold* (Au) (**Core Case Study**) is used in electrical equipment, jewelry, coins, medical implants, and as a catalyst to speed up certain chemical reactions.

The most widely used nonmetallic minerals are sand and gravel. *Sand,* which is mostly silicon dioxide (SiO_2), is used to make glass, bricks, and concrete for construction of roads and buildings. *Gravel* is used for roadbeds and to make concrete. Another common nonmetallic mineral is *limestone* (mostly calcium carbonate, or $CaCO_3$), which is crushed to make road rock, concrete, and cement. *Phosphate salts* are mined and used in inorganic fertilizers and in some detergents.

Most published estimates of the supply of a given mineral resource refer to its **reserves**—identified resources from which the mineral can be extracted profitably at current prices. Reserves increase when new profitable deposits are found and when higher prices or improved mining technology make it profitable to extract deposits that previously were considered too expensive to extract.

Mineral Use Has Advantages and Disadvantages

Figure 14-14 depicts the typical life cycle of a metal resource. The processes of mining and converting minerals into useful products has a number of important benefits (**Concept 14-3A**). It generates significant income and provides local, state, and national revenues from taxes, fees, and royalties. It also provides employment in a variety of jobs directly and indirectly related to mineral extraction, processing, and use.

On the other hand, mining, processing, and using mineral resources take enormous amounts of energy and can disturb the land, erode soil, produce solid waste, and pollute the air, water, and soil (Figure 14-15, p. 356, and **Concept 14-3B**). Some environmental scientists and resource experts warn that the greatest danger from continually increasing our consumption of nonrenewable mineral resources may be the environmental damage caused by their extraction, processing, and conversion to products.

The environmental impacts from mining an ore are affected by its percentage of metal content, or *grade*. The more accessible and higher-grade ores are usually exploited first. As they are depleted, mining lower-grade ores takes more money, energy, water, and other materials and increases land disruption, mining waste, and pollution.

THINKING ABOUT
Low-Grade Ores

CONCEPT LINK

Use the second law of thermodynamics (**Concept 2-4B**, p. 40) to explain why mining lower-grade ores requires more energy and materials and increases land disruption, mining waste, and pollution.

There Are Several Ways to Remove Mineral Deposits

After suitable mineral deposits are located, several different mining techniques are used to remove them, depending on their location and type. Shallow deposits are removed by **surface mining,** and deep deposits are removed by **subsurface mining.**

In surface mining, gigantic mechanized equipment strips away the **overburden,** the soil and rock overlying a useful mineral deposit. It is usually discarded as waste material called **spoils.** When ore deposits that contain metals such as gold (**Core Case Study**) are dredged from streams, the unused materials or *tailings* are usually left on the land. If forests are present, they are also removed. Surface mining

Surface mining → Metal ore → Separation of ore from gangue → Smelting → Melting metal → Conversion to product → Discarding of product

Recycling

Figure 14-14 *Life cycle of a metal resource.* Each step in this process uses large amounts of energy and produces some pollution and waste.

NATURAL CAPITAL DEGRADATION

Extracting, Processing, and Using Nonrenewable Mineral and Energy Resources

Steps		Environmental Effects

Mining

Exploration, extraction

Disturbed land; mining accidents; health hazards; mine waste dumping; oil spills and blowouts; noise; ugliness; heat

Processing

Transportation, purification, manufacturing

Solid wastes; radioactive material; air, water, and soil pollution; noise; safety and health hazards; ugliness; heat

Use

Transportation or transmission to individual user, eventual use, and discarding

Noise; ugliness; thermal water pollution; pollution of air, water, and soil; solid and radioactive wastes; safety and health hazards; heat

Figure 14-15 Some harmful environmental effects of extracting, processing, and using nonrenewable mineral and energy resources (**Concept 14-3B**). Providing the energy required to carry out each step causes additional pollution and environmental degradation. **Question:** What are three mineral resources that you used today? Which of these harmful environmental effects might have resulted from obtaining and using these resources?

Andre Maslennikov/Peter Arnold, Inc.

Figure 14-16 Natural capital degradation: This *open-pit* mine, located near the city of Kalgoolie in the outback of western Australia, is the world's largest gold mine. **Question:** Should governments require mining companies to fill in and restore such sites once their ore is depleted? Explain.

is used to extract about 90% of the nonfuel mineral and rock resources and 60% of the coal used in the United States.

The type of surface mining used depends on two factors: the resource being sought and the local topography. In **open-pit mining** (Figure 14-16), machines dig holes and remove ores (of metals such as iron, copper, and gold), sand, gravel, and stone (such as limestone and marble).

Strip mining is useful and economical for extracting mineral deposits that lie close to the earth's surface in large horizontal beds. In **area strip mining,** used where the terrain is fairly flat, gigantic earthmovers strip away the overburden, and power shovels—some as tall as a 20-story building—remove the mineral deposit. The resulting trench is filled with overburden, and a new cut is made parallel to the previous one. This process is repeated over the entire site.

Contour strip mining (Figure 14-17) is used mostly to mine coal on hilly or mountainous terrain. A huge power shovel cuts a series of terraces into the side of a hill. An earthmover removes the overburden, a power shovel extracts the coal, and the overburden from each new terrace is dumped onto the one below. Unless the land is restored, a wall of dirt is left in front of a highly erodible bank of soil and rock called a *highwall.*

Another surface mining method is **mountaintop removal.** In the Appalachian Mountain area of the United States, where this form of mining is prominent, explosives, large power shovels, and huge machines called draglines are used to remove the top of a mountain and expose seams of coal, which are then removed (Figure 14-19, p. 358).

Subsurface mining is used to remove coal and metal ores that are too deep to be extracted by surface mining. Miners dig a deep vertical shaft, blast open subsurface tunnels and chambers to reach the deposit, and use machinery to remove the ore or coal and transport it to the surface.

Mining Has Harmful Environmental Effects

Mining can do long-term harm to the environment in a number of ways. One type of damage is *scarring and disruption of the land surface* (**Core Case Study**, Figure 14-1, and Figures 14-16 through 14-20). CORE CASE STUDY

For example, area strip mining often leaves a series of hills, like waves of rubble, called *spoils banks* (Figure 14-18). Spoils and tailings are very susceptible to chemical weathering and erosion by water and wind. Regrowth of vegetation on these banks is quite slow because they have no topsoil and thus have to follow the long path of primary ecological succession (Figure 5-16, p. 116, and **Concept 5-4**, p. 115). CONCEPT LINK

Figure 14-17 Natural capital degradation: *contour strip mining* of coal used in hilly or mountainous terrain.

In mountaintop removal (Figure 14-19, p. 358), colossal machines are used to plow great volumes of waste rock and dirt into valleys below the mountaintops, destroying forests, burying mountain streams, and increasing flood hazards. Toxic wastewater, produced when the coal is processed, is often stored in these

Figure 14-18 Natural capital degradation: banks of waste or spoils created by *area strip mining* of coal on an unrestored, mostly flat area near Mulla, Colorado (USA). Government laws require at least partial restoration of newly strip-mined areas in the United States. Nevertheless, many previously mined sites have not been restored and restoration is not possible in some arid areas. **Question:** Should the government require mining companies to restore such sites as fully as possible? Explain.

Figure 14-19 Natural capital degradation: mountaintop coal mining operation in the U.S. state of West Virginia. The large amount of resulting debris is deposited in the valleys and streams below. Mountaintop removal for coal is also occurring in the U.S. states of Virginia, Tennessee, Kentucky, and Pennsylvania. See other mining sites at www.appvoices.org/. **Question:** Are you for or against mountaintop coal mining? Explain.

Figure 14-20 Illegal gold mine in the Brazilian Amazon. Such mines degrade terrestrial biodiversity when numerous areas of tropical rain forest are cleared for mining sites.

valleys behind coal waste sludge dams, which can overflow or collapse, releasing toxic substances such as selenium, arsenic, and mercury. Nearby communities can be damaged economically, as well as environmentally.

According to the EPA, some 1,900 kilometers (1,200 miles) of Appalachia's rivers and streams have been buried, and 470 of its largest mountains have disappeared, leaving behind barren land and gigantic pits, some as large as Manhattan Island. You can tour some of the obliterated landscapes of Appalachia by visiting the website Appalachian Voices at **www.appvoices.org/**. In 2007, the U.S. Department of the Interior issued a rule that allows this type of mining to continue, with expanded dumping of mine waste into streams and valleys. Within a few years, this could result in the devastation of an area the size of the U.S. state of Delaware.

Since 1980, millions of miners have streamed into tropical forests and other tropical areas in search of gold (**Core Case Study**). These small-scale miners use destructive techniques to dig large pits by hand and dredge sediments from rivers. They also use hydraulic mining—a technique, outlawed in the United States—in which water cannons wash entire hillsides into collection boxes for removal of gold (Figure 14-20). Surface mining in tropical areas destroys or degrades biodiversity when forests are cleared and mining wastes pollute nearby streams or rivers.

Surface mining sites can be cleaned up and restored (Figure 14-21), but it is costly. The U.S. Department of the Interior estimates that at least 500,000 surface-mined sites dot the U.S. landscape, mostly in the West. Cleaning up these sites would cost taxpayers as much as

an estimated $70 billion. And worldwide, cleaning up abandoned mining sites would cost trillions of dollars. Most of these sites probably will never be cleaned up.

Subsurface mining disturbs less than one-tenth as much land as surface mining disturbs, and it usually produces less waste material. However, it leaves much of the resource in the ground and is more dangerous and expensive than surface mining is. Hazards include cave-ins, explosions, fires, and diseases such as black lung, caused by prolonged inhalation of mining dust.

Another problem is *subsidence*—the collapse of land above some underground mines. It can tilt and damage houses, crack sewer lines, break gas mains, and disrupt groundwater systems.

Mining operations also produce large amounts of solid waste—three-fourths of all U.S. solid waste. One example is the huge amounts of solid waste produced from mining gold (**Core Case Study**).

Finally, mining causes major pollution of water and air. This is because wind and water erosion cause toxin-laced mining wastes to be deposited in areas other than the mining site. For example, *acid mine drainage* occurs when rainwater seeping through a mine or mine waste pile carries sulfuric acid (H_2SO_4, produced when aerobic bacteria act on iron sulfide minerals in spoils) to nearby streams and groundwater (see *The Habitable Planet,* Video 6, at **www.learner.org/resources/series209.html**).

In addition, huge amounts of water used to process ore often contain pollutants such as sulfuric acid, mercury, and arsenic. This contaminates water supplies and fish used for food, and it can destroy some forms of aquatic life. According to the EPA, mining has polluted about 40% of western watersheds in the United States.

Mining operations also emit toxic chemicals into the atmosphere. In the United States, the mining industry produces more toxic emissions than any other industry—typically accounting for almost half of such emissions.

Removing Metals from Ores Has Harmful Environmental Effects

Ore extracted by mining typically has two components: the *ore mineral,* containing the desired metal, and waste material called *gangue* (pronounced "gang"). Removing the gangue from ores produces tailings. Particles of toxic metals blown by the wind or leached from tailings by rainfall can contaminate surface water and groundwater.

After removal of the gangue, heat or chemical solvents are used to extract metals from the ores. Heating ores to release metals is called **smelting** (Figure 14-14). Without effective pollution control equipment, smelters emit enormous quantities of air pollutants, including sulfur dioxide and suspended particles, which dam-

Before

After

Figure 14-21 Ecological restoration of a mining site in New Jersey (USA) by Princeton Hydro.

age vegetation and acidify soils in the surrounding area. They also cause water pollution and produce liquid and solid hazardous wastes that require safe disposal.

An example of using chemicals to remove metals from their ores is the use of solutions of highly toxic cyanide salts to extract gold from its ore (**Core Case Study**). After extracting the gold, some mining companies have declared bankruptcy and walked away from their mining operations, leaving behind

Figure 14-22 Natural capital degradation: the Summitville gold mining site near Alamosa, Colorado (USA), became a toxic waste site after the Canadian company that owned it declared bankruptcy and abandoned it rather than cleaning up the acids and toxic metals that leaked from this site into the nearby Alamosa River. Cleanup by the EPA will cost U.S. taxpayers about $120 million.

U.S. Geological Survey

large amounts of cyanide-laden water in leaking holding ponds. A glaring example is the Summitville gold mine site near Alamosa, Colorado (USA) (Figure 14-22). A Canadian company used an 1872 U.S. mining law to buy the land from the federal government for a pittance, spent $1 million developing the site, and removed $98 million worth of gold. The company then declared bankruptcy and abandoned the site without doing any clean-up or restoration.

14-4 How Long Will Supplies of Nonrenewable Mineral Resources Last?

▶ **CONCEPT 14-4A** All nonrenewable mineral resources exist in finite amounts, and as we get closer to depleting any mineral resource, the environmental impacts of extracting it generally become more harmful.

▶ **CONCEPT 14-4B** An increase in the price of a scarce mineral resource can lead to increased supplies and more efficient use of the mineral, but there are limits to this effect.

Mineral Resources Are Distributed Unevenly

The earth's crust contains fairly abundant deposits of nonrenewable mineral resources such as iron and aluminum. But deposits of important mineral resources such as manganese, chromium, cobalt, and platinum are relatively scarce. The earth's geologic processes have not distributed deposits of nonrenewable mineral resources evenly. Some countries have rich mineral deposits and others have few or none.

Massive exports can deplete the supply of a country's nonrenewable minerals. During the 1950s, for example, South Korea exported large amounts of its iron and copper. Since the 1960s, the country has not had

enough domestic iron and copper to support its rapid economic growth and now must import these metals to meet its domestic needs.

Five nations—the United States, Canada, Russia, South Africa, and Australia—supply most of the nonrenewable mineral resources used by modern societies. South Africa, for example, is nearly self-sufficient in the world's key mineral resources and is the world's largest producer of gold, chromium, and platinum. Three countries—the United States, Germany, and Russia—with only 8% of the world's population consume about 75% of the most widely used metals, but China is rapidly increasing its use of key metals.

Since 1900, and especially since 1950, there has been a sharp rise in the total and per capita use of non-

renewable mineral resources in the United States. As a result, the United States has depleted some of its once-rich deposits of metal mineral resources such as lead, aluminum, and iron. Currently, it depends on imports of 50% or more of 24 of its most important nonrenewable mineral resources. Some of these minerals are imported because they are used faster than they can be produced from domestic supplies; others are imported because foreign mineral deposits are of a higher grade and cheaper to extract than remaining U.S. reserves.

Most U.S. imports of nonrenewable metal resources come from reliable and politically stable countries. But experts are concerned about the availability of four *strategic metal resources*—manganese, cobalt, chromium, and platinum—that are essential for the country's economy and military strength. The United States has little or no reserves of these metals. Some analysts believe that nanomaterials (Science Focus, p. 362) may eventually be substituted for some of these metals.

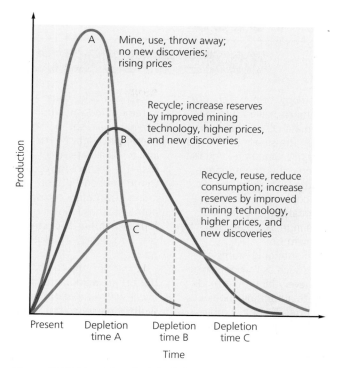

Figure 14-23 Natural capital depletion: *depletion curves* for a nonrenewable resource (such as aluminum or copper) using three sets of assumptions. Dashed vertical lines represent times when 80% depletion occurs.

Supplies of Nonrenewable Mineral Resources Can Be Economically Depleted

The future supply of nonrenewable minerals depends on two factors: the actual or potential supply of the mineral and the rate at which we use it. We never completely run out of any mineral, but a mineral becomes *economically depleted* when it costs more than it is worth to find, extract, transport, and process the remaining deposits (**Concept 14-4A**). At that point, there are five choices: *recycle or reuse existing supplies, waste less, use less, find a substitute,* or *do without.*

Depletion time is the time it takes to use up a certain proportion—usually 80%—of the reserves of a mineral at a given rate of use. When experts disagree about depletion times, it is often because they are using different assumptions about supplies and rates of use (Figure 14-23).

The shortest depletion time estimate assumes no recycling or reuse and no increase in reserves (curve A, Figure 14-23). A longer depletion time estimate assumes that recycling will stretch existing reserves and that better mining technology, higher prices, or new discoveries will increase reserves (curve B). An even longer depletion time assumes that new discoveries will further expand reserves and that recycling, reuse, and reduced consumption will extend supplies (curve C). Finding a substitute for a resource leads to a new set of depletion curves for the new resource.

According to a 2006 study by Thomas Graedel of Yale University, if all nations extract metal resources from the earth's crust at the same rate as developed nations do today, there may not be enough metal resources to meet the demand, even with extensive recycling. However, the successful development of nano-

technology, assuming we can minimize its potentially harmful environmental and health effects (Science Focus, p. 362), may help to reduce such shortages.

Market Prices Affect Supplies of Nonrenewable Minerals

Geologic processes determine the quantity and location of a mineral resource in the earth's crust. Economics determines what part of the known supply is extracted and used. An increase in the price of a scarce mineral resource can lead to increased supplies and can encourage more efficient use, but there are limits to this process (**Concept 14-4B**).

According to standard economic theory, in a competitive market system, a plentiful mineral resource is cheap when its supply exceeds demand. When a resource becomes scarce, its price rises. This can encourage exploration for new deposits, stimulate development of better mining technology, and make it profitable to mine lower-grade ores. It can also encourage a search for substitutes and promote resource conservation.

According to some economists, this price effect may no longer apply very well in most developed countries. Industry and government in such countries often use subsidies, taxes, regulations, and import tariffs to

SCIENCE FOCUS

The Nanotechnology Revolution

Nanotechnology, or *tiny tech*, uses science and engineering to manipulate and create materials out of atoms and molecules at the ultra-small scale of less than 100 nanometers. A nanometer equals one billionth of a meter. It is one hundred-thousandth the width of a human hair. At the nanoscale level, conventional materials have unconventional and unexpected properties (discussed in more detail on page S45 in Supplement 6).

Scientists envision arranging atoms of abundant substances such as carbon, oxygen, and silicon to create everything from medicines and solar cells to automobile bodies. Nanomaterials are currently used in more than 400 consumer products and the number is growing rapidly. Such products include stain-resistant and wrinkle-free coatings on clothes, odor-eating socks, self-cleaning coatings on windows and windshields, cosmetics, sunscreens with nanomolecules that block ultraviolet light, and food containers that release nanosilver ions to kill bacteria, molds, and fungi. **GREEN CAREER:** Environmental nanotechnology

Nanotechnologists envision a supercomputer the size of a sugar cube that could store all the information now found in the U.S. Library of Congress; biocomposite materials smaller than a human cell that would make our bones and tendons super strong; nanovessels that could be filled with medicines and delivered to cells anywhere in the body; and designer nanomolecules that could seek out and kill cancer cells.

Nanoparticles could also be used to remove industrial pollutants in contaminated air, soil, and groundwater, and nanofilters might be used to purify water and to desalinate water at an affordable cost. The technology could also be used to turn garbage into breakfast by mimicking how nature turns wastes into plant nutrients, thus following one of the four **principles of sustainability** (see back cover). The list could go on.

So what is the catch? Ideally, this bottom-up manufacturing process would occur with little environmental harm, without depleting nonrenewable resources, and with many potential environmental benefits. But there are concerns over some potential unintended harmful consequences.

So far, we know little about possible harmful effects of nanoparticles for workers and consumers, but a few studies have raised red flags (see Supplement 6, p. S45). As particles get smaller, they become more reactive and potentially more toxic to humans and other animals. Animal studies show that nanoparticles can move across the placenta from mother to fetus and from nasal passage to the brain. They could also penetrate deeply into the lungs, be absorbed into the bloodstream, and penetrate cell membranes.

Many analysts say we need to take two steps before unleashing nanotechnology more broadly. *First,* carefully investigate its potential ecological, economic, health, and societal risks. *Second,* develop guidelines and regulations for controlling its growing applications until we know more about the potentially harmful effects of this new technology.

So far, governments have done little to evaluate and regulate the potentially harmful effects of this rapidly emerging technology. A 2008 study by Friends of the Earth found that untested and unapproved nanotechnology materials are being used in more than 100 food products and in food packaging. However, in 2007, DuPont and the environmental group Environmental Defense, jointly developed suggested guidelines for evaluating the safety and environmental risks of nanotechnology products.

If nanotechnology lives up to its potential, the mining and processing of most mineral resources may become obsolete businesses. This would eliminate the harmful environmental effects of mining and processing of mineral resources, and it would increase profits for nanomaterial companies. However, it would also eliminate businesses, including those that export mineral resources from many developing countries. And nanotech development could cause severe economic and social stress as jobs and entire industries disappear.

Critical Thinking

How might the development of nanotechnology affect the mining of gold (**Core Case Study**)?

CORE CASE STUDY

control the supplies, demands, and prices of minerals to such an extent that a truly competitive market does not exist.

Most mineral prices are kept artificially low because governments subsidize development of their domestic mineral resources to help promote economic growth and national security. In the United States, for instance, mining companies get subsidies in the form of depletion allowances amounting to 5–22% of their gross income from mineral extraction and processing (depending on the mineral). They can also reduce their taxes by deducting much of their costs for finding and developing mineral deposits.

Most consumers are unaware that the real costs of consumer products made from mineral resources are higher than their market prices, because consumers are also paying taxes to provide government subsidies and tax breaks for mining companies and to help control the harmful environmental effects of mineral extraction, processing, and use. If these hidden extra costs were included in the market prices of such goods, the harmful environmental effects of mineral extraction and processing (Figure 14-15) would be sharply reduced. Recycling and reuse would increase dramatically, and many of these minerals would be replaced with less environmentally harmful substitutes.

Between 1982 and 2007, U.S. mining companies received more than $6 billion in government subsidies (Case Study, at right). Critics argue that eliminating or sharply reducing such environmentally harmful subsidies would promote more efficient resource use, promote waste reduction and pollution prevention, and encourage recycling and reuse of mineral resources.

Mining company representatives insist that they need taxpayer subsidies and low taxes to keep the prices of minerals low for consumers (Case Study, below). They also claim that the subsidies encourage the companies not to move their mining operations to other countries where they will not face such taxes or mining and pollution control regulations.

THINKING ABOUT
Minerals and Nanotechnology

How might arguments for and against subsidies and low taxes for mineral resource extraction be affected by a nanotechnology revolution (Science Focus, at left) during the next 20 years?

Increasingly scarce investment capital and high financial risk can also hinder the development of new supplies of mineral resources. Typically, if geologists identify 10,000 possible deposits of a given resource, only 1,000 sites are worth exploring; only 100 justify drilling, trenching, or tunneling; and only 1 becomes a producing mine or well.

■ CASE STUDY
The U.S. General Mining Law of 1872

Some people have gotten rich by using the little-known U.S. General Mining Law of 1872. It was designed to encourage mineral exploration and the mining of *hard-rock minerals* (such as gold, silver, copper, zinc, nickel, and uranium) on U.S. public lands and to help develop the then-sparsely populated West.

Under this law, a person or corporation can file a mining claim or assume legal ownership of parcels of land on essentially all U.S. public lands except national parks and wilderness. To file a claim, you say you believe the land contains valuable hard-rock minerals and you promise to spend $500 to improve it for mineral development. You must then pay $120 per year for each 8-hectare (20-acre) parcel of land used to maintain the claim, whether or not a mine is in operation.

Until 1995 when a freeze on such land transfers was declared by Congress, one could pay the federal government $6–12 per hectare ($2.50–5.00 an acre) for land owned jointly by all U.S. citizens. One could then lease the land, build on it, sell it, or use it for essentially any purpose. People have constructed golf courses, hunting lodges, hotels, and housing subdivisions on public land that they bought from taxpayers at 1872 prices. According to a 2004 study by the Environmental Working Group, public lands containing an estimated $285 billion worth of publicly owned mineral resources have been transferred to private companies—one-fifth of them foreign companies—at 1872 giveaway prices.

In 2004, the mining company Phelps Dodge bought 63 hectares (155 acres) of U.S. Forest Service land atop Colorado's Mt. Emmons near the ski resort town of Crested Butte for $875. The land could be worth as much as $155 million. The company was able to buy the land because it had claimed the option to buy before the 1995 freeze.

According to the Bureau of Land Management, mining companies remove at least $4 billion worth of hard-rock minerals per year from U.S. public land. These companies pay taxpayers royalties amounting to only 2.3% of the value of the minerals, compared to royalties of 13.2% paid for oil, natural gas, and coal, and 14% for grazing rights on public lands. After removing valuable minerals such as gold, some mining companies have abandoned severely degraded and polluted mining sites and left the expensive cleanup to taxpayers (Figure 14-22).

In 1992, the 1872 law was modified to require mining companies to post bonds to cover 100% of the estimated cleanup costs in case they go bankrupt—a requirement that mining companies are lobbying Congress to overturn or greatly weaken. Because such bonds were not required in the past, the U.S. Department of the Interior estimates that cleaning up degraded land and streams on more than 500,000 abandoned hard-rock mining sites will cost U.S. taxpayers $32–72 billion. According to former Arkansas Senator Dale Bumpers, the 1872 mining law is a "license to steal" from U.S. citizens who jointly own all public lands.

Mining companies point out that they must invest large sums (often $100 million or more) to locate and develop an ore site before they make any profits from mining hard-rock minerals. They argue that government subsidized land costs allow them to provide high-paying jobs to miners, supply vital resources for industry, stimulate the national and local economies, reduce trade deficits, and keep mineral products affordable. But critics argue that the money taxpayers give up as subsidies to mining companies offsets the lower prices they pay for these products.

Critics of this ancient law call for permanently banning such sales of public lands, but some support 20-year leases of designated public land for hard-rock mining. Critics also call for much stricter environmental controls and cleanup restrictions on hard-rock mining. They want the government to set up a fund paid for by royalties from hard-rock mining companies to clean up abandoned mining sites. They would require mining companies to pay a royalty of 8–12% on the *gross income* based on the wholesale value of all minerals removed from public land—similar to the rates paid by oil, natural gas, and coal companies. They opposed basing the royalties on the net income (income after taxes and other expenses have been subtracted), which is favored by hard-rock mining interests. Mining companies have been pressuring the U.S. Congress to lift the ban on sales of public lands

and have successfully opposed efforts to impose 8–12% royalties.

Canada, Australia, South Africa, and several other countries require higher royalty payments and have laws that make mining companies fully responsible for environmental damage.

Is Mining Lower-Grade Ores the Answer?

Some analysts contend that all we need do to increase supplies of a mineral is to extract lower grades of ore. They point to the development of new earth-moving equipment, improved techniques for removing impurities from ores, and other technological advances in mineral extraction and processing.

In 1900, the average copper ore mined in the United States was about 5% copper by weight. Today that ratio is 0.5%, and copper costs less (adjusted for inflation). New methods of mineral extraction may allow even lower-grade ores of some metals to be used.

However, several factors can limit the mining of lower-grade ores (**Concept 14-4B**). One is the increased cost of mining and processing larger volumes of ore. Another is the limited availability of freshwater needed to mine and process some minerals—especially in arid and semiarid areas. A third limiting factor is the environmental impacts of the increased land disruption, waste material, and pollution produced during mining and processing (Figure 14-15) (**Concept 14-4A**).

One way to improve mining technology is to use microorganisms that can extract minerals in a process called *in-place*, or *in situ*, (pronounced "in SY-too") *mining*. If naturally occurring bacteria cannot be found to extract a particular metal, genetic engineering techniques could be used to produce such bacteria. This biological approach, called *biomining*, removes desired metals from ores while leaving the surrounding environment undisturbed. It also reduces the air pollution associated with the smelting of metal ores and the water pollution associated with using hazardous chemicals such as cyanides and mercury to extract gold.

On the down side, microbiological ore processing is slow. It can take decades to remove the same amount of material that conventional methods can remove within months or years. However, genetic engineers are looking for ways to modify bacteria that could speed up the process. So far, biological mining methods are economically feasible only with low-grade ores for which conventional techniques are too expensive.

— **RESEARCH FRONTIER** —
Developing biomining and other new methods for extracting more minerals from ores. See **academic.cengage.com/biology/miller**.

Can We Extend Supplies by Getting More Minerals from the Ocean?

Some ocean mineral resources are dissolved in seawater. However, most of the chemical elements found in seawater occur in such low concentrations that recovering them takes more energy and money than they are worth. At current prices and with existing technology, only magnesium, bromine, and sodium chloride are abundant enough to be extracted profitably. On the other hand, deposits of minerals, mostly in sediments along the shallow continental shelf and near shorelines, are significant sources of sand, gravel, phosphates, sulfur, tin, copper, iron, tungsten, silver, titanium, platinum, and diamonds.

— **THINKING ABOUT**
Extracting Minerals from Seawater CONCEPT LINK

Use the second law of thermodynamics (**Concept 2-4B**, p. 40) to explain why it costs too much to extract most dissolved minerals from seawater.

Another potential source is hydrothermal ore deposits that form when mineral-rich superheated water shoots out of vents in solidified magma on the ocean floor. After mixing with cold seawater, particles of metal compounds (such as sulfides, silver, zinc, and copper) precipitate out and build up as mineral deposits around the vents. Currently, it costs too much to extract these minerals, even though some deposits contain large concentrations of important metals. There are also disputes over ownership of such resources located in international waters.

Another potential source of metals from the ocean floor is potato-size *manganese nodules* that cover about 25–50% of the Pacific Ocean floor. They could be sucked up from the ocean floor by giant vacuum pipes or scooped up by buckets on a continuous cable operated by a mining ship. However, marine scientists are concerned about the effects of such mining on aquatic life.

So far these nodules and resource-rich mineral beds in international waters have not been developed. As with hydrothermal ore deposits, this is because of high costs and squabbles over who owns them and how any profits made from extracting them should be distributed among the world's nations.

Some environmental scientists believe seabed mining probably would cause less environmental harm than mining on land. However, they are concerned that removing seabed mineral deposits and dumping back unwanted material will stir up ocean sediments, destroy seafloor organisms, and have potentially harmful effects on poorly understood ocean food webs and marine biodiversity. They call for more research to help evaluate such possible effects.

14-5 How Can We Use Mineral Resources More Sustainably?

▶ **CONCEPT 14-5** We can try to find substitutes for scarce resources, reduce resource waste, and recycle and reuse minerals.

We Can Find Substitutes for Some Scarce Mineral Resources

Some analysts believe that even if supplies of key minerals become too expensive or scarce due to unsustainable use, human ingenuity will find substitutes (**Concept 14-5**). They point to the current *materials revolution* in which silicon and new materials, particularly ceramics and plastics, are being used as replacements for metals. And nanotechnology (Science Focus, p. 362) may also lead to the development of materials that can serve as substitutes for various minerals.

In 2005, for example, builders began constructing houses made of Styrofoam sprayed with a ceramic spray called Grancrete. This ceramic is affordable, is twice as strong as structural concrete, and will not leak or crack. It reduces the cost of house frame construction to about one-fifteenth of current costs. It also reduces the need for timber (thereby sparing many trees) and nonrenewable mineral resources used to construct houses. Lightweight Styrofoam blocks are also being used to pave bridges.

Plastic has replaced copper, steel, and lead in much piping. Fiber-optic glass cables that transmit pulses of light are replacing copper and aluminum wires in telephone cables. And in the future, nanowires may replace the optic glass cables.

High-strength plastics and composite materials strengthened by lightweight carbon and glass fibers are beginning to transform the automobile and aerospace industries. They cost less to produce than metals because manufacturing them requires less energy, they do not need painting (which reduces pollution and costs), and they can be molded into any shape. New plastics and gels are also being developed to provide superinsulation that will not take up much space.

RESEARCH FRONTIER

Materials science and engineering. See **academic.cengage .com/biology/miller**.

Use of plastics has drawbacks, chief of which is that making them by current methods requires the use of oil and other fossil fuels. These energy resources are nonrenewable and they have their own environmental impacts, discussed in the next chapter. However, chemists are learning how to make some plastics from plant materials.

Substitution is not a cure-all. For example, currently, platinum is unrivaled as an industrial catalyst, and chromium is an essential ingredient of stainless steel. We can try to find substitutes for scare resources but this may not always be possible.

We Can Recycle and Reuse Valuable Metals

After a pure metal is produced by smelting or chemical extraction, it is usually melted and molded into desired shapes to make products that are then used and discarded or recycled (Figure 14-14). A more sustainable way to use nonrenewable mineral resources (especially valuable or scarce metals such as gold, silver, iron, copper, steel, aluminum, and platinum) is to recycle or reuse them.

Recycling also has a much lower environmental impact than that of mining and processing metals from ores. For example, recycling aluminum beverage cans and scrap aluminum produces 95% less air pollution and 97% less water pollution and uses 95% less energy than mining and processing aluminum ore.

THINKING ABOUT
Metal Recycling and Nanotechnology

How might the development of a nanotechnology revolution (Science Focus, p. 362) over the next 20 years affect the recycling of metal mineral resources?

There Are Many Ways to Use Mineral Resources More Sustainably

Some analysts say we have been asking the wrong question. Instead of asking how we can increase supplies of nonrenewable minerals, we should be asking how we can decrease our use and waste of such resources. Answering that second question could provide important ways to use mineral resources more sustainably (**Concept 14-5**). Figure 14-24 (p. 366) and the Case Study below (p. 366) describe some of these strategies.

In 1975, the U.S. based Minnesota Mining and Manufacturing Company (3M), which makes 60,000

SOLUTIONS

Sustainable Use of Nonrenewable Minerals

- Do not waste mineral resources.

- Recycle and reuse 60–80% of mineral resources.

- Include the harmful environmental costs of mining and processing minerals in the prices of items (full-cost pricing).

- Reduce mining subsidies.

- Increase subsidies for recycling, reuse, and finding substitutes.

- Redesign manufacturing processes to use less mineral resources and to produce less pollution and waste (cleaner production).

- Use mineral resource wastes of one manufacturing process as raw materials for other processes.

- Slow population growth.

Figure 14-24 Ways to achieve more sustainable use of nonrenewable mineral resources (**Concept 14-5**). **Question:** Which two of these solutions do you think are the most important? Why?

different products in 100 manufacturing plants, began a Pollution Prevention Pays (3P) program. It redesigned its equipment and processes, used fewer hazardous raw materials, identified toxic chemical outputs (and recycled or sold them as raw materials to other companies), and began making more nonpolluting products. By 1998, 3M's overall waste production was down by one-third, its air pollutant emissions per unit of production were 70% lower, and the company had saved more than $750 million in waste disposal and material costs. This is an excellent example of why pollution prevention pays (**Concept 1-4**, p. 16).

Since 1990, a growing number of companies have adopted similar pollution and waste prevention programs that lead to *cleaner production*. See the Guest Essay by Peter Montague on cleaner production on the website for this chapter.

■ **CASE STUDY**

Industrial Ecosystems: Copying Nature

An important goal for a sustainable society is to make its industrial manufacturing processes cleaner and more sustainable by redesigning them to mimic how nature deals with wastes. According to one **scientific principle of sustainability**, in nature, the waste outputs of one organism become the nutrient inputs of another organism, so that all of the earth's nutrients are endlessly recycled.

One way for industries to mimic nature is to recycle and reuse most minerals and chemicals instead of dumping them into the environment. Another is for industries to interact through *resource exchange webs* in which the wastes of one manufacturer become raw materials for another—similar to food webs in natural ecosystems (Figure 3-14, p. 63).

This is happening in Kalundborg, Denmark, where an electric power plant and nearby industries, farms, and homes are collaborating to save money and reduce their outputs of waste and pollution. They exchange waste outputs and convert them into resources, as shown in Figure 14-25. This cuts pollution and waste and reduces the flow of nonrenewable mineral and energy resources through their economy.

Today, about 20 ecoindustrial parks similar to the one in Kalundborg operate in various places in the world, including the U.S. city of Chattanooga, Tennessee (Case Study, p. 21). And more are being built or planned—some of them on abandoned industrial sites, called *brownfields*. Widespread use of the rapidly growing field of industrial ecology to develop a global network of industrial ecosystems over the next few decades could lead to an important *ecoindustrial revolution*. **GREEN CAREER:** Industrial ecology

These and other industrial forms of *biomimicry* provide many economic benefits for businesses. By encouraging recycling and pollution prevention, they reduce the costs of managing solid wastes, controlling pollution, and complying with pollution regulations. They also reduce a company's chances of being sued because of harms caused by their actions. In addition, companies improve the health and safety of workers by reducing workers' exposure to toxic and hazardous materials, thereby reducing company health-care insurance costs.

Biomimicry also stimulates companies to come up with new, environmentally beneficial and less resource-intensive chemicals, processes, and products that can be sold worldwide. Another benefit: such companies have a better image among consumers based on results rather than public relations campaigns.

THINKING ABOUT
Gold Mining

How would you apply the solutions in Figure 14-24 to decreasing the need to mine gold (**Core Case Study**) and to reducing the harmful environmental effects of gold mining?

RESEARCH FRONTIER

Developing biomimicry and other ecoindustrial tools. See **academic.cengage.com/biology/miller**.

Figure 14-25
Solutions: an *industrial ecosystem* in Kalundborg, Denmark, reduces waste production by mimicking a food web in natural ecosystems. The wastes of one business become the raw materials for another. **Question:** Is there an industrial ecosystem near where you live or go to school? If not, think about where and how such a system could be set up.

Greenhouses

Pharmaceutical plant

Local farmers

Sludge

Sludge

Waste heat

Waste heat

Waste heat

Fish farming

Oil refinery

Waste heat

Surplus natural gas

Electric power plant

Fly ash

Cement manufacturer

Surplus sulfur

Surplus natural gas

Waste calcium sulfate

Waste heat

Sulfuric acid producer

Wallboard factory

Area homes

REVISITING Gold Mining and Sustainability

In this chapter, we began with a discussion of the harmful effects of gold mining (**Core Case Study**). We also have discussed a number of exciting possibilities for extracting and using gold and other nonrenewable mineral resources in less harmful, more sustainable ways.

If developed safely, nanotechnology (Science Focus, p. 362) could provide more sustainable ways to make use of mineral resources. For example, it could be used to make cheap solar cells. This application of the first **principle of sustainability** could enable the use of solar energy to produce electricity for extracting and processing mineral resources and for a host of other environmentally beneficial processes that require electricity. Another promising technology is biomining—the use of microbes to extract mineral resources.

We can also use mineral resources more sustainably by recycling and reusing them and by reducing unnecessary resource use and waste. Industries can mimic nature by converting wastes to resources and exchanging them through a network resembling a food web (Figure 14-25). These solutions apply the recycling **principle of sustainability**.

Implementing these strategies will reduce both the need for mining and its resulting harmful environmental effects (Figure 14-15). This, in turn, will decrease the destruction and degradation of biodiversity and vital species interactions that help to keep species populations in check, in keeping with the two other **principles of sustainability**.

*Mineral resources are the building blocks
on which modern society depends.
Knowledge of their physical nature and origins,
the web they weave
between all aspects of human society and the physical earth,
can lay the foundations for a sustainable society.*

ANN DORR

1. Review the Key Questions and Concepts for this chapter on p. 345. Describe the environmental effects of gold mining.

2. Define **geology, core, mantle, crust, tectonic plate,** and **lithosphere.** What is a transform fault? What is **weathering** and why is it important? Define **volcano** and describe the nature and effects of a volcanic eruption. Define and describe the nature and effects of an **earthquake.** What is a **tsunami** and what are its effects?

3. Define **mineral, rock, sedimentary rock, igneous rock,** and **metamorphic rock** and give an example of each. Describe the nature and importance of the **rock cycle.**

4. Define **mineral resource** and list three types of such resources. Define **ore** and distinguish between a **high-grade ore** and a **low-grade ore.** What are **reserves?** Describe the life cycle of a metal resource. Describe the major harmful environmental effects of extracting, processing, and using nonrenewable mineral resources.

5. Distinguish between **surface mining** and **subsurface mining.** Define **overburden, spoils,** and **open-pit mining.** Define **strip mining** and distinguish among **area strip mining, contour strip mining,** and **mountaintop removal.** Describe the harmful environmental effects of mining. What is **smelting** and what are its major harmful environmental effects? What five nations supply most of the world's nonrenewable mineral resources? How dependent is the United States on other countries for important nonrenewable mineral resources?

6. Describe the advantages and disadvantages of the nanotechnology revolution. What are five possible options when a mineral becomes economically depleted? Define **depletion time** and describe three types of depletion curves for a mineral resource. Describe the conventional view of the relationship between the supply of a mineral resource and its market price. What factors can influence this market interaction? Describe the benefits and possible drawbacks of nanotechnology. Discuss the pros and cons of the U.S. General Mining Law of 1872.

7. Describe the opportunities and limitations of increasing mineral supplies by mining lower-grade ores. What are the advantages and disadvantages of biomining?

8. Describe the opportunities and limitations of getting more minerals from the ocean.

9. Describe the opportunities and limitations of finding substitutes for scarce mineral resources and recycling and reusing valuable metals. Describe ways of using nonrenewable mineral resources more sustainably. Describe the Pollution Prevention Pays program of the Minnesota Mining and Manufacturing Company. What is an industrial ecosystem? Describe the industrial ecosystem operating in Kalundborg, Denmark.

10. Describe the relationships between gold mining (**Core Case Study**) and the four **scientific principles of sustainability**.

Note: Key Terms are in **bold** type.

1. List three ways in which you could apply **Concept 14-5** to making your lifestyle more environmentally sustainable.

2. List three ways in which decreasing the need to mine gold and reducing its harmful environmental effects (**Core Case Study**) could benefit you.

3. What do you think would happen if **(a)** plate tectonics stopped and **(b)** erosion and weathering stopped? Explain.

4. You are an igneous rock. Write a report on what you experience as you move through the rock cycle (Figure 14-13). Repeat this exercise, assuming you are a sedimentary rock and then a metamorphic rock.

5. Use the second law of thermodynamics (**Concept 2-4B**, p. 40) to analyze the scientific and economic feasibility of each of the following processes:
 a. Extracting most minerals dissolved in seawater
 b. Mining increasingly lower-grade deposits of minerals

 c. Using inexhaustible solar energy to mine minerals
 d. Continuing to mine, use, and recycle minerals at increasing rates

6. List three ways in which a nanotechnology revolution (Science Focus, p. 362) could benefit you and three ways in which it could harm you.

7. Describe the strategy you would use to promote **(a)** the nanotechnology revolution (Science Focus, p. 362) and **(b)** the spread of ecoindustrial ecosystems (Case Study, p. 366)? As part of your promotion strategy for each of these projects, describe three benefits to your community.

8. Explain why you support or oppose each of the following proposals concerning extraction of hard-rock minerals on public land in the United States: **(a)** halting the practice of granting title to public land for actual or claimed hard-rock mineral deposits, **(b)** requiring mining companies to pay a royalty of 8–12% on the *gross* income they earn

from hard-rock minerals that they extract from public lands, and **(c)** making hard-rock mining companies legally responsible for restoring the land and cleaning up environmental damage caused by their activities.

9. Congratulations! You are in charge of the world. What are the three most important features of your policy for developing and sustaining the world's nonrenewable mineral resources?

10. List two questions that you would like to have answered as a result of reading this chapter.

Note: See Supplement 13 (p. S78) for a list of Projects related to this chapter.

DATA ANALYSIS

Uranium (U), which is used as a fuel in the reactors of nuclear power plants, is found in various rocks at different concentrations. A high-grade ore has 2% U, and a low-grade ore has 0.1% U. The estimated recoverable resources of uranium for the world add up to 4,743,000 metric tons. The United States has about 7% of the world's uranium resources, which amounts to about 332,000 metric tons.

1. Given that current worldwide usage of uranium is about 66,500 metric tons U per year, how long will the world's present recoverable resources last? (*Note:* 1 metric ton = 1,000 kg = 1.1 tons)

2. Assume U.S. usage is about 25% of world usage. If the United States were to rely only on its domestic uranium resources, how long would they last, assuming a 100% recovery rate (the amount of the resource that can be used)?

3. Assume that most U.S. ore bodies contain high-grade ore (2% U) and that recovery rates of minerals from the ore (accounting for losses in mining, extraction, and refining), average 65%, how many metric tons of ore will have to be mined to meet U.S. needs?

LEARNING ONLINE

Log on to the Student Companion Site for this book at **academic.cengage.com/biology/miller**, and choose Chapter 14 for many study aids and ideas for further reading and research. These include flash cards, practice quizzing, Weblinks, information on Green Careers, and InfoTrac® College Edition articles.

How Long Will Supplies of Conventional Oil Last?

Oil, which supplies about one-third of the world's energy, is the lifeblood of most of the world's economies and modern lifestyles. We use oil to grow most of our food, transport people and goods, and make most of the things we use every day—from plastics to asphalt on roads.

Stretched end to end, the number of barrels of oil the world used in 2007 would circle the equator 650 times! And projected oil use in 2020 would raise that number to 870. To meet this rapidly growing demand, oil companies have drilled wells on the land and at sea (Figure 15-1). They extract this oil and refine it to make gasoline, heating oil, asphalt, and other products.

Geologists project that known and projected global reserves of conventional oil will be 80% depleted sometime between 2050 and 2100, depending on consumption rates. If this estimate is correct, conventional oil should be reaching its sunset years sometime during this century. (See Figure 8, p. S64, Supplement 10 for a brief history of the Age of Oil.)

We have three options: look for more oil, reduce oil use and waste, or use other energy resources. Many analysts think we should vigorously pursue all three options. Some contend that higher prices will stimulate the search for new oil to meet global oil needs. Others doubt that oil reserves can be increased enough to meet the rapidly growing future demand for oil, despite greatly increased oil exploration. Yet, because oil companies and many governments are secretive about oil reserves, no one really knows how much oil might be available. Some geologists fear that Saudi Arabia and other OPEC countries have deliberately overestimated the size of their oil reserves to discourage a switch to other alternatives.

Others argue that even if much more conventional oil is somehow found, we are ignoring the consequences of the high exponential growth (Chapter 1 Core Case Study, p. 5) in global oil consumption. If we continue to use oil reserves at the current rate of about 2.8% per year with the unlikely assumption that the rate will not increase, then

- Saudi Arabia, with the world's largest known crude oil reserves, could supply the world's entire oil needs for about 10 years.

- The estimated reserves under Alaska's North Slope—the largest ever found in North America—would meet current world demand for only 6 months or U.S. demand alone for 3 years.

- The estimated reserves in Alaska's Arctic National Wildlife Refuge (ANWR) would meet current world oil demand for only 1–5 months and U.S. demand for 7–24 months.

Contrary to popular belief, the world is not about to run out of oil in the near future. The well-publicized debate over *peak oil production* is not about the world's ultimate supply of oil. Instead, it is about *flow rate* based on the ability of existing oil supplies to meet the annual demands for oil. An adequate annual flow of oil cannot be maintained indefinitely unless new and affordable supplies are found to replace the oil being depleted from existing reserves. And between 2000 and 2007, the world consumed nine times more oil than the amount the oil industry discovered according to British Petroleum and the International Energy Agency. *To keep using conventional oil at the projected rate of increase, we must discover global oil reserves equivalent to a new Saudi Arabian supply every 5 years.* Most oil geologists say this is highly unlikely.

The exciting and urgent challenge for this century is to sharply reduce the waste of oil and other energy resources and to find an array of substitutes for oil and other fossil fuels to slow emissions of carbon dioxide, which are warming the atmosphere and triggering global climate change. There are no easy solutions, because all energy options have advantages and disadvantages. We discuss those of nonrenewable energy sources in this chapter and those of renewable energy sources in Chapter 16.

Figure 15-1 Thunder Horse offshore floating oil production platform, located in the Gulf of Mexico.

Brad Martin/Stockphoto.com

15-1 What major sources of energy do we use?

CONCEPT 15-1A About three-quarters of the world's commercial energy comes from nonrenewable fossil fuels and the rest comes from nonrenewable nuclear fuel and renewable sources.

CONCEPT 15-1B Net energy is the amount of high-quality usable energy available from a resource after the amount of energy needed to make it available is subtracted.

15-2 What are the advantages and disadvantages of oil?

CONCEPT 15-2A Conventional oil is currently abundant, has a high net energy yield, and is relatively inexpensive, but using it causes air and water pollution and releases greenhouse gases to the atmosphere.

CONCEPT 15-2B Heavy oils from oil sand and oil shale exist in potentially large supplies but have low net energy yields and higher environmental impacts than conventional oil has.

15-3 What are the advantages and disadvantages of natural gas?

CONCEPT 15-3 Conventional natural gas is more plentiful than oil, has a high net energy yield and a fairly low cost, and has the lowest environmental impact of all fossil fuels.

15-4 What are the advantages and disadvantages of coal?

CONCEPT 15-4A Conventional coal is very plentiful and has a high net energy yield and low cost, but it has a very high environmental impact.

CONCEPT 15-4B Gaseous and liquid fuels produced from coal could be plentiful, but they have lower net energy yields and higher environmental impacts than conventional coal has.

15-5 What are the advantages and disadvantages of nuclear energy?

CONCEPT 15-5 Nuclear power has a low environmental impact and a very low accident risk, but high costs, a low net energy yield, long-lived radioactive wastes, vulnerability to sabotage, and the potential for spreading nuclear weapons technology have limited its use.

Note: Supplements 2 (p. S4), 9 (p. S53), 10 (p. S59), and 13 (p. S78) can be used with this chapter.

*Typical citizens of advanced industrialized nations
each consume as much energy in 6 months
as typical citizens in developing countries consume
during their entire life.*

MAURICE STRONG

15-1 What Major Sources of Energy Do We Use?

▶ **CONCEPT 15-1A** About three-quarters of the world's commercial energy comes from nonrenewable fossil fuels and the rest comes from nonrenewable nuclear fuel and renewable sources.

▶ **CONCEPT 15-1B** Net energy is the amount of high-quality usable energy available from a resource after the amount of energy needed to make it available is subtracted.

Fossil Fuels Supply Most of Our Commercial Energy

Almost all of the energy that heats the earth and our buildings comes from the sun at no cost to us—one of the four **scientific principles of sustainability** (see back cover). Without this essentially inexhaustible solar energy (*solar capital*, **Concept 1-1A**, p. 6), the earth's aver-

age temperature would be −240 °C (−400 °F), and life as we know it would not exist. This direct input of solar energy produces several other forms of renewable energy that can be thought of as indirect solar energy: *wind* (moving air masses heated by the sun), *hydropower* (flowing water kept fluid by heat from the sun), and *biomass* (solar energy converted to chemical energy and stored in trees and other plants); we examine these energy sources in Chapter 16.

The Case Study at right gives a brief history of human energy use. Currently, most *commercial energy*—energy sold in the marketplace—comes from extracting and burning *nonrenewable energy resources* obtained from the earth's crust, primarily carbon-containing fossil fuels—oil, natural gas, and coal (Figure 15-2).

About 82% of the commercial energy consumed in the world comes from *nonrenewable* energy resources—76% from fossil fuels (oil, natural gas, and coal) and 6% from nuclear power (Figure 15-3, left). The remaining 18% of the commercial energy we use comes from *renewable* energy resources—biomass, hydropower, geothermal, wind, and solar energy (**Concept 15-1A**).

Supplement 10 (pp. S59-S73) has several graphs showing trends in energy consumption in the world and in the United States. These graphs include world oil consumption from 1950 to 2006 (Figure 4, p. S61), global coal and natural gas consumption between 1950 and 2005 (Figure 5, p. S61), total and per capita energy consumption in the United States (Figure 2, p. S60), and energy consumption by fuel in the United States between 1980 and 2006 (Figure 3, p. S60).

In order, the three largest users of fossil fuels are the United States, China, and the European Union, together accounting for more than half of all fossil fuel consumption. Energy use per person varies through-out the world (see Figure 1, p. S59, Supplement 10). Despite its rapidly growing total energy consumption, China's per-capita energy consumption is far below that of other industrial countries.

CENGAGENOW™ Examine and compare energy sources used in developing and developed countries at CengageNOW™.

■ CASE STUDY
A Brief History of Human Energy Use

Early humans were scavengers and hunter–gatherers whose main source of energy was muscle power. A human living at this basic survival level needed about 2,000 kilocalories of energy per day, most of it in the form of food.

In a modern industrial society such as the United States, the average person uses 2,000 kilocalories of energy per day for basic energy needs, plus about 600,000 kilocalories of energy per day used by machines and systems that maintain an individual's complex lifestyle. This 300-fold increase over the minimum survival level of energy use gives individuals in the United States and other industrialized countries immense power to alter

Figure 15-2 Natural capital: important nonrenewable energy resources that can be removed from the earth's crust are coal, oil, natural gas, and some forms of geothermal energy (**Concept 15-1A**). Nonrenewable uranium ore is also extracted from the earth's crust and processed to increase its concentration of uranium-235, which serves as a fuel in nuclear reactors that produce electricity. **Question:** During a typical day, which of these resources do you use directly or indirectly?

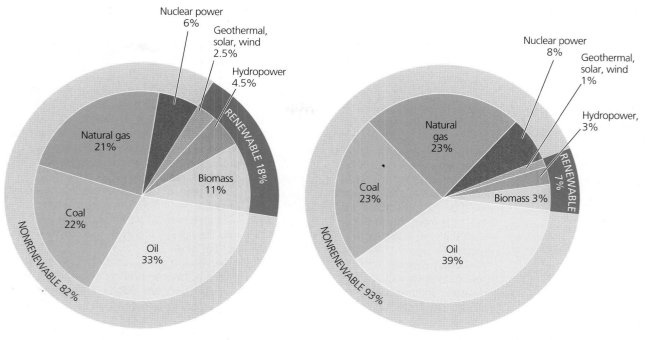

World

United States

natural systems through their huge total and per capita energy footprints.

Humanity's first step along this energy path began with the discovery of fire, which hunters–gatherers used to cook food and to light and heat their dwellings. Later they learned to use fire to burn grasslands in order to stampede animals they hunted over cliffs.

After humans began settling down as farmers about 12,000 years ago, they learned how to domesticate wild animals, using muscle power to plow fields, carry loads, and transport people from place to place. Later, they learned to tap into energy from the wind in order to pump up underground water and to transport people and goods in sailing ships. They also used the power of flowing water to move goods and people on boats, to power mills for grinding grain, and eventually to produce electricity.

About 275 years ago, we began inventing machines such as the steam engine used to power ships, tractors, locomotives, and factory machinery. Renewable firewood provided about 91% of the energy used for heating and for running steam engines. But in 1850, this began changing as many forests were depleted. In other words we used a potentially renewable energy resource—wood—unsustainably by harvesting it faster than nature replaced it.

We survived this early energy crisis by learning how to burn coal for heating and for running factories and trains. By 1900, wood provided only about 18% of our energy, and coal provided 73%. In 1859, we learned how to pump oil out of the ground and later invented ways to convert it to fuels such as gasoline and heating oil.

In 1885, Carl Benz invented the internal combustion engine to power cars and other vehicles that could run on gasoline. By 1900, we got 40% of our energy from oil, 38% from coal, and 18% from natural gas—all nonrenewable resources.

In the 1950s, we learned how to get enormous amounts of energy by splitting the nuclei of certain types of uranium atoms (Figure 2-7, top, p. 41) and to use this energy to produce electricity. Today, we continue to live in a *fossil fuel era* with 82% of our energy coming from nonrenewable oil, natural gas, and coal resources (Figure 15-3).

Now we face a new energy crisis because of the air and water pollution, greenhouse gas emissions, and environmental degradation caused by our excessive use of fossil fuels and our failure to get serious about reducing unnecessary energy waste. An urgent question is, can we greatly improve energy efficiency and shift to a variety of renewable energy resources, before we do even more serious harm to our own life-support system and to many of the world's other species.

THINKING ABOUT
The Future of Energy Use

Do you think the total use of energy by all humans, regardless of where it comes from, must keep growing as it has in the past? Explain.

How Should We Evaluate Energy Resources?

According to scientists, all energy resources should be evaluated on the basis of their supplies, the environmental impact of our using them, and how much useful energy they actually provide (Science Focus, p. 374).

SCIENCE FOCUS

Net Energy Is the Only Energy That Really Counts

It takes energy to get energy. For example, before oil becomes useful to us, it must be found, pumped up from beneath the ground or ocean floor, transferred to a refinery and converted to useful fuels, transported to users, and burned in furnaces and cars. Each of these steps uses high-quality energy. The second law of thermodynamics tells us that some of the high-quality energy used in each step is automatically wasted and degraded to lower-quality energy (**Concept 2-4B**, p. 40).

The usable amount of *high-quality energy* available from a given quantity of an energy resource is its **net energy**. It is the total amount of useful energy available from an energy resource minus the energy needed to find, extract, process, and get that energy to consumers (**Concept 15-1B**). It is calculated by estimating the total amount of energy available from the resource over its lifetime and then subtracting the amount of energy *used, automatically wasted* because of the second law of thermodynamics, and *unnecessarily wasted* in finding, processing, and transporting the useful energy to users.

Net energy is like the net profit in a business after expenses. If a business has $1 million in sales and $800,000 in expenses, its net profit is $200,000. Similarly, suppose that it takes 8 units of energy to produce 10 units of energy from a coal mine. Then the net useful energy yield is only 2 units of energy.

We can express net energy as the ratio of energy produced to the energy used to produce it. In this example, the *net energy ratio* would be 10/8, or approximately 1.25. The higher the ratio, the greater the net energy. When the ratio is less than 1, there is a net energy loss. Figure 15-A shows estimated net energy ratios for various types of space heating, high-temperature heat for industrial processes, and transportation.

Currently, conventional oil has a high net energy ratio because much of it comes from large, accessible, and cheap-to-extract land deposits such as those in the Middle East or those in shallow water. As these sources become depleted, sources that are more difficult to find and reach and therefore expensive to extract are tapped at deeper levels in the ground or under the sea bottom (Figure 15-1). As this occurs, the net energy ratio of oil declines and its price rises sharply.

Electricity produced by the *nuclear power fuel cycle* has a low net energy ratio because large amounts of energy are needed for each step in the cycle: to extract and process uranium ore, convert it into nuclear fuel, build and operate nuclear power plants, store the highly radioactive wastes they produce for thousands of years, dismantle the highly radioactive plants after their 15–60 years of useful life, and store the radioactive parts. Some analysts estimate that ultimately, we will have to put more energy into the nuclear fuel cycle than we will ever get out of it.

An honest energy accounting system would be built around net energy analysis. Otherwise, we will spend huge amounts of money and make important energy policy decisions without crucial information. We should not delude ourselves into thinking that we can somehow avoid the inevitable consequences of the first and second laws of thermodynamics (pp. 42–43).

Critical Thinking

Should governments require that all energy resources be evaluated in terms of their estimated net energy? Why do you think this is not being done?

Space Heating

Passive solar	5.8
Natural gas	4.9
Oil	4.5
Active solar	1.9
Coal gasification	1.5
Electric heating (coal-fired plant)	0.4
Electric heating (natural-gas-fired plant)	0.4
Electric heating (nuclear plant)	0.3

High-Temperature Industrial Heat

Surface-mined coal	28.2
Underground-mined coal	25.8
Natural gas	4.9
Oil	4.7
Coal gasification	1.5
Direct solar (concentrated)	0.9

Transportation

Ethanol from sugarcane residue	8.0
Ethanol from switchgrass	5.4
Natural gas	4.9
Gasoline (refined crude oil)	4.1
Coal liquefaction	1.4
Oil shale	1.2
Ethanol from corn	1.1 (but can reach 1.5)

Figure 15-A Science: *Net energy ratios* for various energy systems over their estimated lifetimes: the higher the net energy ratio, the greater the net energy available (**Concept 15-1B**). A useful rule of thumb is that any energy resource with a low net energy will need government (taxpayer) subsidies to compete in the marketplace with high net energy resources. In other words, subsidies and tax breaks must be used to keep its price artificially low. **Question:** Based on these data, which two resources in each category should we be using? Compare this with the major resources we are actually using as shown in Figure 15-3. (Data from U.S. Department of Energy, U.S. Department of Agriculture, Colorado Energy Research Institute, *Net Energy Analysis*, 1976; and Howard T. Odum and Elisabeth C. Odum, *Energy Basis for Man and Nature*, 3rd ed., New York: McGraw-Hill, 1981)

15-2 What Are the Advantages and Disadvantages of Oil?

▶ **CONCEPT 15-2A** Conventional oil is currently abundant, has a high net energy yield, and is relatively inexpensive, but using it causes air and water pollution and releases greenhouse gases to the atmosphere.

▶ **CONCEPT 15-2B** Heavy oils from oil sand and oil shale exist in potentially large supplies but have low net energy yields and higher environmental impacts than conventional oil has.

We Depend Heavily on Oil

Petroleum, or **crude oil** (oil as it comes out of the ground), is a thick and gooey liquid consisting of hundreds of different combustible hydrocarbons along with small amounts of sulfur, oxygen, and nitrogen impurities. It is also known as *conventional* or *light oil.* Crude oil and natural gas are called fossil fuels because they were formed from the decaying remains (fossils) of organisms that lived 100–500 million years ago.

Deposits of crude oil and natural gas often are trapped together under a dome deep within the earth's crust on land or under the seafloor (Figure 15-2). The crude oil is dispersed in pores and cracks in underground rock formations, somewhat like water saturating a sponge. It is extracted by means of a well drilled into the deposit. High-tech equipment can drill oil and natural gas wells on land and at sea (**Core Case Study,** Figure 15-1) to a depth of 11 kilometers (7 miles). Then oil, drawn by gravity out of the rock pores flows into the bottom of the well and is pumped to the surface.

At first oil almost squirts from many wells. But after years of pumping, usually a decade or so, the pressure drops and production starts declining at a point referred to the *peak production* of a well. For global oil production to expand, the oil output from newly found reserves must stay ahead of the declining output from wells that have passed their peak production.

After it is extracted, crude oil is transported to a *refinery* by pipeline, truck, or ship (oil tanker). There it is heated and distilled to separate it into components with different boiling points (Figure 15-4) in a process called *refining*—a technological marvel based on complex chemistry and engineering. However, refining oil decreases its net energy yield.

Some of the products of oil distillation, called **petrochemicals,** are used as raw materials in industrial organic chemicals, cleaning fluids, pesticides, plastics, synthetic fibers, paints, medicines, and many other products. Producing a desktop computer, for example, requires ten times its weight in fossil fuels, mostly oil.

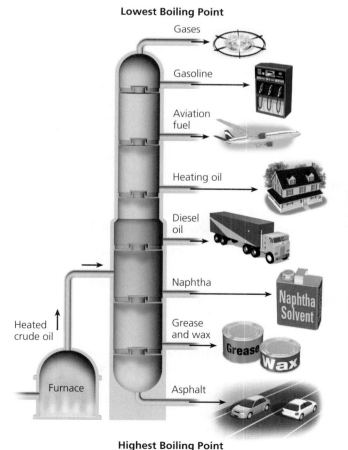

Lowest Boiling Point

Gases

Gasoline

Aviation fuel

Heating oil

Diesel oil

Naphtha

Heated crude oil

Grease and wax

Furnace

Asphalt

Highest Boiling Point

Figure 15-4 **Science:** *refining crude oil.* Components of petroleum are removed at various levels, depending on their boiling points, in a giant distillation column. The most volatile components with the lowest boiling points are removed at the top of the column. The photo shows an oil refinery in the U.S. state of Texas.

World oil consumption has been growing since 1950 (Figure 4, p. S61, Supplement 10), and oil is now the single largest source of commercial energy in the world and in the United States (Figure 15-3). In order, the world's three largest oil users in 2007 were the United States (using 24% of all oil produced), China (using 8%), and Japan (7%). A vital question is, how long can the world's oil reserves meet the growing global oil consumption (**Core Case Study**)?

↰ CORE CASE STUDY

OPEC Controls Most of the World's Oil Supplies

Oil *reserves* are identified deposits from which conventional oil can be extracted profitably at current prices with current technology. Since the world currently is greatly dependent on oil, the oil industry is the world's largest business. Thus control of oil reserves is the single greatest source of global economic and political power.

The 13 countries that make up the Organization of Petroleum Exporting Countries (OPEC) have at least 60% of the world's crude oil reserves and, in 2006, produced 43% of the world's oil. OPEC is expected to have long-term control over the supplies and prices of the world's conventional oil. Today, OPEC's members are Algeria, Angola, Indonesia, Iran, Iraq, Kuwait, Libya, Nigeria, Qatar, Saudi Arabia, the United Arab Emirates, and Venezuela. In 2007, OPEC oil revenue averaged about $1.3 million per minute.

Saudi Arabia has by far the largest portion of the world's crude oil reserves (25%). It is followed by Canada (15%), whose huge supply of oil sand was recently classified as a conventional source of oil. In order, other countries with large proven reserves are Iran, Iraq, Kuwait, the United Arab Emirates, Venezuela, and Russia. However, as mentioned earlier, because of secrecy by OPEC and other government-controlled oil companies, no one knows the real size of the world's proven oil reserves. U.N. and other agencies that project future oil production and demand have to rely mostly on figures given to them by oil-producing countries and private companies.

About 75% of the world's oil reserves, according to analysts at California's Stanford University, are in the hands of government-owned companies, not private oil companies. Private companies such as Exxon Mobil, Chevron, and BP control only a small percentage of the world's oil reserves and thus have a declining influence over oil supplies and prices.

According to oil-producing countries, the world's proven oil reserves increased by about 15% between 1996 and 2006, with the largest increases in the Middle East. The problem is that oil production from existing reserves has exceeded new oil discoveries since 1984. Global oil production has leveled off since 2005. This helps to explain why, between 2005 and 2008, oil prices climbed from $50 to more than $140 a barrel.

Global oil production peaks when the demand for oil exceeds the amount that can be produced. When the annual global demand for oil exceeds the rate at which it is produced, the *flow rate* of oil to consumers decreases, and oil prices start rising. Oil production peaked in the United States in 1970 (just as U.S. geologist M. King Hubbert predicted in 1956 that it would). It also peaked in Venezuela in 1970, the United Kingdom in 1999, Norway in 2000, and Mexico in 2004; and it may soon peak in China and even in Saudi Arabia.

Some believe that steeply rising oil prices will lead to long-overdue crash programs to reduce energy waste and shift to non-carbon energy resources such as wind, solar energy, geothermal energy, biomass, and hydrogen (all of which are discussed in the next chapter).

Other analysts have a different and more pessimistic view. As oil prices rise sharply because annual production cannot meet the demand, they project that:

- Prices of food produced by oil-dependent industrialized agriculture (Figure 12-15, p. 290) and products such as plastics, pesticides, asphalt, and other widely used materials produced from petrochemicals will rise sharply.

- Food production may become more localized—reversing the current trend toward global food production and distribution built on cheap oil.

- Airfares will go up, and air travel and air freight shipments will likely level off and perhaps decline.

- The oil-intensive automobile industry will see a steep decline in the demand for cars and trucks that run on conventional gasoline or diesel fuel and for motor vehicles with fuel economies lower than 17 kilometers per liter (40 miles per gallon).

- In countries such as the United States, which have long neglected public transportation systems, there could be a mass exodus from car-dependent suburbs as property values plummet. This migration could leave behind closed shopping malls, discount stores, and other businesses that have sprung up in far-flung suburbs where people currently must drive to get most of what they need.

The United States Uses Much More Oil Than It Produces

The United States gets about 85% of its energy from fossil fuels, with 39% coming from oil (Figure 15-3, right). About 25% of U.S. domestic oil production and 20% of domestic natural gas comes from offshore drilling, mostly off the coasts of Texas and Louisiana in the Gulf of Mexico (see Figure 15-1 and Figure 6, p. S62, Supplement 10). This area is subject to hurricanes, which on average are increasing in intensity. Another 17% of domestic oil comes from Alaska's North Slope via oil tankers and the Trans-Alaska Pipeline.

The United States produces about 9% of the world's oil but uses 24% of global oil production and has only 2.4% of the world's oil reserves. Oil use in the United States has exceeded new domestic discoveries since 1984.

The United States produces most of its dwindling domestic supply of oil at a high cost, about $7.50–$10 per barrel on dry land and four or more times this figure for tapping deep-water sources, compared to about $2 per barrel in Saudi Arabia. This helps to explain why the United States imports about 60% of its oil.

According to a 2005 report by the Institute for the Analysis of Global Security, almost one-fourth of the world's conventional oil is controlled by states that sponsor or condone terrorism. This means that, in buying oil from those countries, the United States, Great Britain, Japan, and other countries concerned with fighting terrorism are funding the enemy. According to a 2006 poll of 100 foreign policy experts in *Foreign Policy* magazine, *the highest priority in fighting terrorism must be to sharply reduce America's dependence on foreign oil.*

The U.S. Department of Energy estimates that if current trends continue, the United States will import 70% of its oil by 2025. At the same time, it will be facing stiff competition for oil imports from rapidly industrializing countries such as China, which in 2007 was the world's second largest oil user and imported nearly half of its oil.

According to the U.S. Geological Survey, potentially vast domestic oil and natural gas reserves remain to be discovered in the United States, much of it beneath federal lands and coastal waters (see Figure 6, p. S62, Supplement 10). In 2006, for example, three oil companies announced that they had found a massive oil field deep below the ocean floor in the Gulf of Mexico (**Core Case Study**, Figure 15-1) that could eventually boost U.S. oil reserves by as much as 50%. However, it will take many years and billions of dollars to bring this oil to market at a very high cost. This field will not significantly reduce the country's dependence on foreign oil, and it will not help to lower prices at the gasoline pump anytime soon. Even if fully developed, the estimated oil in this field would meet current U.S. oil needs for only about 5 years, and much less if oil consumption increases as projected. Exponential growth (Chapter 1 Core Case Study, p. 5) is a powerful force.

Searching for such oil fields is extremely difficult, expensive, and financially risky. Typically, oil companies extract only about one barrel of oil for every three they find. Two barrels are left behind either because the remaining oil is too thick to pump out or because it would cost too much to do so. New techniques to recover some of this oil include using seismic imaging to track where it is, injecting saline water into wells to increase recovery, reinjecting natural gas associated with oil fields to maintain reservoir pressures, and lowering microwave generators down boreholes to heat heavy oil so that it will flow enough to be pumped out.

Scientists are also attempting to develop genetically engineered bacteria that will react with the oil and increase its flow so that it can be pumped out or converted to natural gas. The trick is to develop organisms that can do this in 10 years, instead of the 10 million years required for oil to form naturally. This and other heavy oil recovery methods are expensive compared to getting oil from many of the world's most productive oil fields, and they decrease the already fairly low net energy yield of heavy oil.

Many geologists doubt that the United States will find enough new oil or extract enough heavy oil from older wells to come close to meeting U.S. demand. According to one analyst, if we think of U.S. conventional oil reserves as a six-pack of oil, four of the cans are empty. It has been estimated that if the country opens up virtually all of its public lands and coastal regions to oil exploration, it may find at best about half a can of new oil, which would be developed only at a very high production cost with a lower net energy yield and seriously harmful environmental effects. In other words, according to these energy analysts, *the United States cannot feed its oil addiction by trying to increase domestic oil supplies.*

■ CASE STUDY

Oil and the U.S. Arctic National Wildlife Refuge

The Arctic National Wildlife Refuge (ANWR) on Alaska's North Slope (see Figure 6, p. S62, Supplement 10) contains more than one-fifth of all land in the U.S. National Wildlife Refuge System. The refuge's coastal plain is the only stretch of Alaska's arctic coastline that is not open to oil and gas development.

This tundra biome (Figure 7-12, bottom photo, p. 151) is home to a diverse community of species (see Figure 3, p. S55, Supplement 9), including polar bears, arctic foxes, musk oxen, and peregrine falcons. During the brief arctic summer, it serves as a nesting ground for millions of tundra swans, snow geese, and other migratory birds, and as a grazing area and breeding ground for one of North America's last great herds of caribou. Partly because of its harsh climate, this is an extremely fragile ecosystem.

Since 1980, oil companies have been lobbying Congress for permission to carry out exploratory drilling in the coastal plain because they believe it might contain oil and natural gas deposits. Advocates say that the United States must use all of the oil resources it has, including ANWR, to help decrease dependence on imported oil and to help the economy. They believe that economic security should have a higher priority than environmental concerns. Advocates contend that oil production in the nearby Prudhoe Bay will soon decline and that the production facilities already in place could be used to help produce and transport oil from the nearby refuge.

Alaska's elected representatives in Congress strongly support such drilling because the state uses revenue from oil production to finance most of its budget and to provide annual dividends to its citizens. Oil company officials also say that they can now extract oil with less damage than was done in taking oil from nearby Prudhoe Bay.

Finding oil in the ANWR will increase oil company profits. But many experts say it will do little to increase domestic oil supplies or reduce U.S. dependence on oil imports. Geologists estimate that there is a moderate chance of finding enough oil in the ANWR to meet U.S. demand for only 7–24 months. If the projected supply of oil is found, it will be a tiny drop in the nation's oil bucket compared to projected future U.S. oil consumption (Figure 15-5).

Opponents say getting relatively little oil from the ANWR's coastal plain is not worth degrading this irreplaceable and fragile ecosystem. They point out that improving motor vehicle fuel efficiency is a much faster, cheaper, cleaner, and more secure way to increase future oil supplies. For example, improving fuel efficiency by just 0.4 kilometer per liter (1 mile per gallon) for new cars, SUVs, and light trucks in the United States would save more oil than is ever likely to be produced from the ANWR.

Opponents also point to the severe environmental damage that has occurred in nearby Prudhoe Bay. The Alaska pipeline and a large complex of roads and production facilities have destroyed and degraded natural habitats and exposed parts of the fragile tundra ecosystem and its wildlife to oil spills and toxic chemicals. To opponents, the large oil spill from a BP Alaska pipeline in 2006, which was caused by inadequate maintenance, cast major doubts on claims by oil companies that they can develop oil in the ANWR without serious and long-lasting environmental damage. They believe that the ANWR is a place so rare and special that it should be permanently protected from oil drilling and other forms of development.

HOW WOULD YOU VOTE?

Should the U.S. Congress allow oil and gas development in the coastal plain of Alaska's Arctic National Wildlife Refuge (ANWR) or permanently protect it from oil and gas exploration and development? Cast your vote online at **academic .cengage.com//biology/miller**.

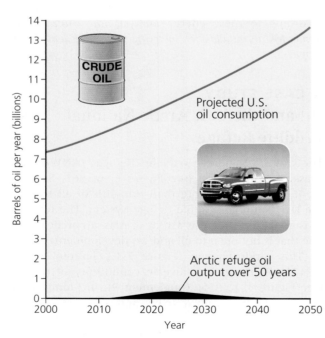

Figure 15-5 The amount of oil that might be found in the Arctic National Wildlife Refuge, if developed and extracted over 50 years, is only a tiny fraction of projected U.S. oil consumption. (Data from U.S. Department of Energy, U.S. Geological Survey, and Natural Resources Defense Council)

Conventional Oil Has Advantages and Disadvantages

Figure 15-6 lists the advantages and disadvantages of using conventional crude oil as an energy resource (**Concept 15-2A**). The extraction, processing, and burning of nonrenewable oil and other fossil fuels have a severe environmental impact (Figure 14-15, p. 356), including land disruption, air pollution, water pollution, and losses and degradation of wildlife (Figure 15-7).

A critical and growing problem is the fact that burning oil or any carbon-containing fossil fuel releases CO_2 into the atmosphere and helps to promote climate change caused by global warming. Currently, burning oil, mostly as gasoline and diesel fuel for transportation, accounts for 43% of global CO_2 emissions. Another problem with relying on oil is that its once high net energy yield is declining as oil producers are forced to turn to oil that is buried far offshore (Figure 15-1) and deep underground. In addition, much of the world's oil must be imported from unfriendly and politically unstable producer countries.

HOW WOULD YOU VOTE?

Do the advantages of relying on conventional oil as the world's major energy resource outweigh its disadvantages? Cast your vote online at **academic.cengage.com/biology/ miller**.

TRADE-OFFS

Conventional Oil

Advantages	Disadvantages
Ample supply for 42–93 years	Need to find substitutes within 50 years
Low cost	Large government subsidies
High net energy yield	Environmental costs not included in market price
Easily transported within and between countries	Artificially low price encourages waste and discourages search for alternatives
Low land use	Pollutes air when produced and burned
Technology is well developed	Releases CO_2 when burned
Efficient distribution system	Can cause water pollution

Figure 15-6 Advantages and disadvantages of using conventional crude oil as an energy resource (**Concept 15-2A**). **Question:** Which single advantage and which single disadvantage do you think are the most important? Why?

Will Heavy Oils from Oil Sand Be a Viable Option?

Oil sand, or **tar sand,** is a mixture of clay, sand, water, and a combustible organic material called *bitumen*—a thick and sticky, heavy oil with a high sulfur content that makes up about 10% of the gooey mixture.

Northeastern Alberta in Canada has three-fourths of the world's oil sand resources in sandy soil under a remote boreal forest (Figure 7-15, bottom photo, p. 154). Other deposits are in Venezuela, Colombia, Russia, and the U.S. state of Utah. Together the oil sands of Canada and Venezuela contain more oil than is found in Saudi Arabia—nearly as much as the total conventional oil reserves in the Middle East.

In 2003, the oil industry began counting Canada's oil sands as reserves of conventional oil. As a consequence, Canada has 15% of the world's oil reserves, second only to Saudi Arabia.

About 20% of Alberta's oil sand is close enough to the surface to be strip-mined, but removing it creates a serious environmental impact. Before the mining takes place, the boreal forest is clear-cut, its wetlands are drained, and its rivers and streams are diverted. Next the overburden of soil, rocks, and clay is removed to expose oil sand deposits. Then gigantic electric shovels dig up the oil sand, and load it into house-sized trucks, which carry it to energy-intensive upgrading plants. There the oil sand is mixed with hot water and steam to extract the bitumen, which is heated by natural gas in huge cookers and converted into a low-sulfur, synthetic, crude oil suitable for refining.

About 4 metric tons of overburden are removed to produce 1 metric ton of bitumen. This has a severe impact on the land as machines create open pits large enough to be seen in satellite images. The huge volumes of toxic mine tailings and other wastes are stored as slurries in ponds, also large enough to be seen from space, and the wastes are extremely toxic to aquatic life and migratory birds.

The entire process results in huge amounts of toxic sludge, as well as much more water pollution and air pollution than are created by the extraction and processing of conventional crude oil. It releases at least three times more CO_2 per barrel of oil than is released in the production of a barrel of conventional oil. Since 2003, Alberta has been Canada's industrial air pollution capital.

The process also uses large amounts of water, drawn from the Athabasca River. Each barrel of mined bitumen requires four to five barrels of water. And the Canadian government allows the tar sands industry to continue withdrawing water regardless of how low the river flow becomes.

Because of its huge environmental impact, the United Nations Environment Programme has listed Alberta's oil sand strip mines as one of the world's 100 key hotspots

Figure 15-7 Bird covered with oil from an oil spill in Brazilian waters. Unless volunteers remove the oil, it will destroy this bird's natural buoyancy and heat insulation, causing it to drown or die from exposure because of a loss of body heat.

of environmental degradation. In 2008, Environmental Defence called Canada's oil sands industry "the most destructive project on earth."

In 2006, energy economist Peter Tertzakian estimated that it takes the energy equivalent (mostly in the form of natural gas) of 0.7 barrels of oil to extract, upgrade, and produce 1 barrel of oil from oil sands. In other words, the net energy yield for producing oil from oil sands is low. To make matters worse, in Canada, there is a looming shortage of natural gas. Having to import natural gas to produce this energy resource will raise its cost and further decrease its already low net energy yield. It has been projected that by 2015, heavy oil from oil sand will meet only 4% of the world's estimated oil consumption (**Concept 15-2B**).

Will Oil Shales Be a Useable Resource?

Oily rocks are another potential supply of heavy oil. Such rocks, called *oil shales* (Figure 15-8, left), contain a solid combustible mixture of hydrocarbons called *kerogen*. It can be extracted from crushed oil shales by heating them in a large container, a process that yields a distillate called **shale oil** (Figure 15-8, right). Before the thick shale oil can be sent by pipeline to a refinery, it must be heated to increase its flow rate and processed to remove sulfur, nitrogen, and other impurities.

About 72% of the world's estimated oil shale reserves are buried deep in rock formations in the western United States beneath an area called the Green River Formation—a barren stretch of arid land covering portions of Colorado, Wyoming, and Utah. The federal government (American citizens) own about 80% of this land. The U.S. Bureau of Land Management estimates that these deposits contain an amount of potentially recoverable heavy oil equal to almost four times

Figure 15-8 Oil shale rock (left) and the shale oil (right) extracted from it. Producing shale oil has a low net energy yield and a very high environmental impact. It also requires considerable amounts of water and money (**Concept 15-2B**).

U.S. Department of Energy

TRADE-OFFS

Heavy Oils from Oil Shale and Oil Sand

Advantages	Disadvantages
Moderate cost (oil sand)	High cost (oil shale)
Large potential supplies, especially oil sands in Canada	Low net energy yield
Easily transported within and between countries	Environmental costs not included in market price
Efficient distribution system in place	Large amounts of water needed for processing
Technology well-developed (oil sand)	Severe land disruption
	Severe water pollution
	Air pollution and CO_2 emissions when produced and burned

Figure 15-9 Advantages and disadvantages of using heavy oils from oil sand and oil shale as energy resources (**Concept 15-2B**). **Question:** Which single advantage and which single disadvantage do you think are the most important? Why?

the size of Saudi Arabia's oil reserves and eleven times the size of Alberta's oil sand reserves—enough to meet the current U.S. oil demand for 110 years.

So what is the catch? One problem is that most of these deposits are locked up in rock and ore of such low grade that it would take considerable energy and money to mine and convert the kerogen to crude oil. In other words, its net energy is low, even lower than that of oil from oil sands. It also takes a lot of water to produce shale oil. The massive U.S. deposits are mostly in arid areas of the West, where water is in short supply (Figure 13-5, p. 318) and likely to become even scarcer because of intense and prolonged drought projected for this area throughout most of this century. Furthermore, producing and using shale oil has a much higher environmental impact than exploiting conventional oil. It includes digging up and processing 0.8 metric tons (1 ton) of rock to produce 1 barrel of oil.

Figure 15-9 lists the advantages and disadvantages of using heavy oil from oil sand and oil shale as energy resources (**Concept 15-2B**).

┌─ **THINKING ABOUT**
 Heavy Oils

Do the advantages of relying on heavy oils from oil sand and oil shale outweigh their disadvantages? Explain.

15-3 What Are the Advantages and Disadvantages of Natural Gas?

▶ **CONCEPT 15-3** Conventional natural gas is more plentiful than oil, has a high net energy yield and a fairly low cost, and has the lowest environmental impact of all fossil fuels.

Natural Gas Is a Useful and Clean–Burning Fossil Fuel

Natural gas is a mixture of gases of which 50–90% is methane (CH_4). It also contains smaller amounts of heavier gaseous hydrocarbons such as ethane (C_2H_6), propane (C_3H_8), and butane (C_4H_{10}), and small amounts of highly toxic hydrogen sulfide (H_2S).

Conventional natural gas lies above most reservoirs of crude oil (Figure 15-2). However, unless a natural gas pipeline has been built, these deposits cannot be used. Indeed, the natural gas found above oil reservoirs in deep-sea and remote land areas is often viewed as an unwanted by-product and is burned off. This practice wastes a valuable energy resource and releases climate-changing carbon dioxide into the atmosphere.

When a natural gas field is tapped, propane and butane gases are liquefied and removed as **liquefied petroleum gas (LPG)**. LPG is stored in pressurized tanks for use mostly in rural areas not served by natural gas pipelines. The rest of the gas (mostly methane) is dried to remove water vapor, cleansed of poisonous hydrogen sulfide and other impurities, and pumped into pressurized pipelines for distribution across land areas.

Russia—the Saudi Arabia of natural gas—has about 27% of the world's proven natural gas reserves, followed by Iran (15%) and Qatar (14%). The United States has only 3% of the world's proven natural gas reserves (see Figure 6, p. S62, Supplement 10) but uses about 27% of the world's annual production.

Natural gas is a versatile fuel that can be burned to heat space and water or produce electricity and to propel vehicles with fairly inexpensive engine modifications. In the United States, a pipeline grid delivers natural gas from domestic wells to towns and cities and directly to 60 million American homes. A homeowner can also obtain a small compressor for using natural gas to fuel cars with slightly modified engines.

Natural gas is also used to run medium-sized turbines that produce electricity. These clean-burning turbines have almost twice the energy efficiency (50–60%) of coal-burning and nuclear power plants (24–35%). They are also cheaper to build, require less time to install, and are easier and cheaper to maintain than large-scale coal and nuclear power plants.

As with any fossil fuel, burning natural gas releases carbon dioxide into the atmosphere. However, it re-leases much less CO_2 per unit of energy gained than does producing and burning coal, conventional oil, or oil from oil sand and oil shale.

So that it can be transported across oceans, natural gas is converted to **liquefied natural gas (LNG)** at a very low temperature and high pressure. This highly flammable liquid is then put aboard refrigerated tanker ships. After arriving at its destination, it is heated and converted back to the gaseous state at regasification plants before it is distributed by pipeline.

Japan imports large amounts of LNG from Russia. By 2025, the United States plans to become the world's largest importer of LNG by greatly increasing LNG port and regasification facilities in at least 40 locations. Some analysts warn that this could make the United States too dependent on countries that have not been consistently stable and friendly, such as Russia and Iran, for supplies of LNG.

In addition, LNG has a low net energy yield. The equivalent of more than a third of its energy content is required to compress, decompress, refrigerate, and transport it long distances. Like oil sands and oil shale, LNG has met limits imposed by the first and second laws of thermodynamics (pp. 42–43). This explains why some analysts do not view LNG as an economically viable alternative to conventional natural gas unless its price is kept artificially low by government (taxpayer) subsidies.

Unconventional natural gas is also found in underground sources. *Coal bed methane gas* is found in coal beds near the earth's surface across parts of the United States and Canada (most yellow areas in Figure 6, p. S62, Supplement 10). But the environmental impacts of producing it—scarring of land and pollution of air and water—are causing a public backlash against using this energy source in parts of the western United States.

Another unconventional source of natural gas is *methane hydrate*—methane trapped in icy, cage-like structures of water molecules. They are buried in some areas of tundra under arctic permafrost, in places such as Alaska and Siberia, and deep beneath the ocean bottom (see Figure 7, p. S63, Supplement 10). So far, it costs too much to get natural gas from methane hydrates, and the release of methane (a potent greenhouse gas) into the atmosphere during removal and processing will speed up global warming and the resulting climate change. In other words, this energy

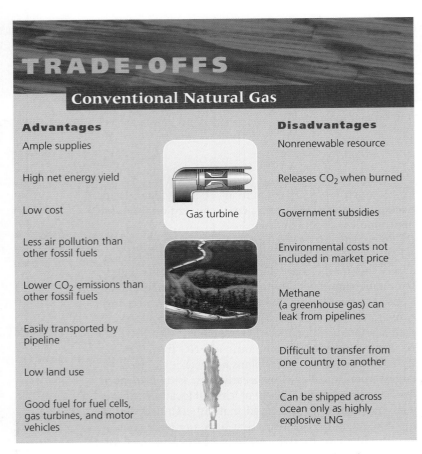

TRADE-OFFS

Conventional Natural Gas

Advantages	Disadvantages
Ample supplies	Nonrenewable resource
High net energy yield	Releases CO_2 when burned
Low cost	Government subsidies
Less air pollution than other fossil fuels	Environmental costs not included in market price
Lower CO_2 emissions than other fossil fuels	Methane (a greenhouse gas) can leak from pipelines
Easily transported by pipeline	Difficult to transfer from one country to another
Low land use	
Good fuel for fuel cells, gas turbines, and motor vehicles	Can be shipped across ocean only as highly explosive LNG

Gas turbine

Figure 15-10 Advantages and disadvantages of using conventional natural gas as an energy resource (**Concept 15-3**). **Question:** Which single advantage and which single disadvantage do you think are the most important? Why?

alternative, like several others, has a low net energy yield and a high environmental impact.

——— RESEARCH FRONTIER ———

Finding affordable and environmentally acceptable ways to tap methane hydrates. See **academic.cengage.com/biology/miller**.

Natural Gas Has More Advantages Than Disadvantages

The long-term global outlook for conventional natural gas supplies is better than that for conventional oil. At the current consumption rate, known reserves of conventional natural gas should last the world for 62–125 years depending on how rapidly they are used. In 2007, natural gas producer Robert A. Heffner III estimated that reserves of conventional natural gas in the United States should last 70–100 years at today's rate of consumption.

Figure 15-10 lists the advantages and disadvantages of using conventional natural gas as an energy resource (**Concept 15-3**). Because of its advantages over oil, coal, and nuclear energy, some analysts see natural gas (but not LNG or unconventional sources of natural gas) as a bridge fuel to help make the transition to a more sustainable energy future based on improved energy efficiency and greater reliance on a mix of renewable energy resources, as discussed in Chapter 16.

15-4 What Are the Advantages and Disadvantages of Coal?

▶ **CONCEPT 15-4A** Conventional coal is very plentiful and has a high net energy yield and low cost, but it has a very high environmental impact.

▶ **CONCEPT 15-4B** Gaseous and liquid fuels produced from coal could be plentiful, but they have lower net energy yields and higher environmental impacts than conventional coal has.

Coal Comes in Several Forms and Is Burned Mostly to Produce Electricity

Coal is a solid fossil fuel that was formed in several stages out of the remains of land plants that were buried 300–400 million years ago and subjected to intense heat and pressure over many millions of years (Figure 15-11).

Coal is burned in about 2,100 power plants (Figure 15-12) to generate about 40% of the world's elec-

tricity. Using a coal-burning power plant is essentially a complex and inefficient way to boil water and produce steam, which is used to spin turbines and produce electricity. Coal is also burned in various industrial plants. For example, bituminous coal is converted to coke, which is burned in blast furnaces to make iron.

In order, the three largest coal-burning countries are China (Case Study, p. 384), the United States, and India. By 2025, China is expected to burn twice as much coal as the United States burns, and between 2006 and 2031, India's use of coal is projected to quadruple. In the United States, coal produces 49% of the electricity,

Increasing moisture content ← → Increasing heat and carbon content

Peat (not a coal)		Lignite (brown coal)		Bituminous (soft coal)		Anthracite (hard coal)

Heat → Pressure

Heat → Pressure

Heat → Pressure

Partially decayed plant matter in swamps and bogs; low heat content

Low heat content; low sulfur content; limited supplies in most areas

Extensively used as a fuel because of its high heat content and large supplies; normally has a high sulfur content

Highly desirable fuel because of its high heat content and low sulfur content; supplies are limited in most areas

Figure 15-11 Stages in coal formation over millions of years. Peat is a soil material made of moist, partially decomposed organic matter and is not classified as a coal, although it too is used as a fuel. The different major types of coal vary in the amounts of heat, carbon dioxide, and sulfur dioxide released per unit of mass when they are burned.

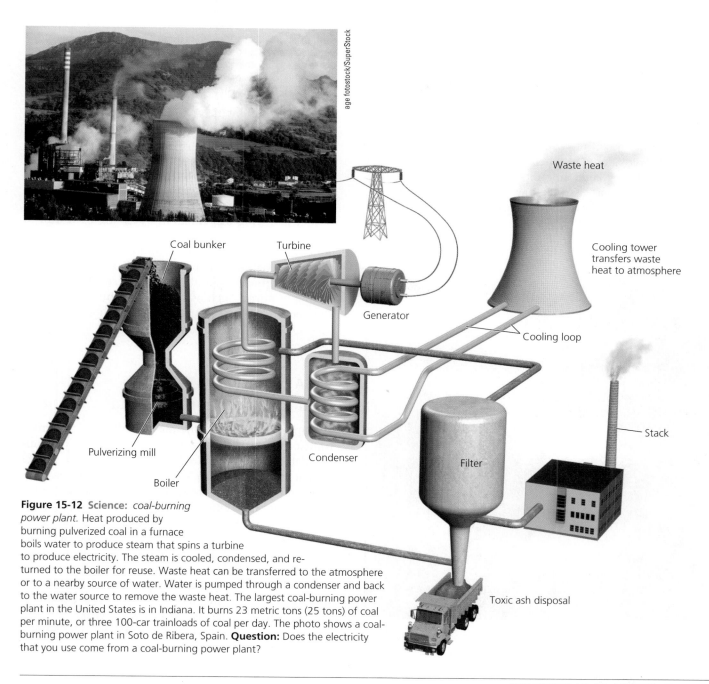

Waste heat

Cooling tower transfers waste heat to atmosphere

Coal bunker

Turbine

Generator

Cooling loop

Stack

Pulverizing mill

Condenser

Filter

Boiler

Figure 15-12 Science: *coal-burning power plant.* Heat produced by burning pulverized coal in a furnace boils water to produce steam that spins a turbine to produce electricity. The steam is cooled, condensed, and returned to the boiler for reuse. Waste heat can be transferred to the atmosphere or to a nearby source of water. Water is pumped through a condenser and back to the water source to remove the waste heat. The largest coal-burning power plant in the United States is in Indiana. It burns 23 metric tons (25 tons) of coal per minute, or three 100-car trainloads of coal per day. The photo shows a coal-burning power plant in Soto de Ribera, Spain. **Question:** Does the electricity that you use come from a coal-burning power plant?

Toxic ash disposal

followed by natural gas (21%), nuclear power (19%), renewable energy (9%, with 7% from hydroelectric power plants), and oil (2%).

Coal Is a Plentiful but Dirty Fuel

Coal is the world's most abundant fossil fuel. According to the U.S. Geological Survey, identified and unidentified global supplies of coal could last for 214–1,125 years, depending on how rapidly they are used. The United States—the Saudi Arabia of coal—has 25% of the world's proven coal reserves (see Figure 6, p. S62, Supplement 10). Russia has 15%, followed by India with 13%, China with 13%, Australia with 8%, and South Africa with 7%.

The U.S. Geological Survey estimates that identified U.S. coal reserves should last about 250 years at the current consumption rate. But a 2007 study by the U.S. National Academy of Sciences estimated that U.S. coal supplies were overestimated and would last for 100 years at current consumption rates, and for only a few decades if coal consumption continues to increase.

Without sophisticated and expensive pollution control devices, burning coal severely pollutes the air (Figure 15-13). Coal is mostly carbon but contains small amounts of sulfur, which are released into the air as sulfur dioxide (SO_2) when the coal burns. Burning coal also releases large amounts of particulates (soot), the greenhouse gas CO_2 (Figure 15-14), and trace amounts of toxic mercury and radioactive materials. According

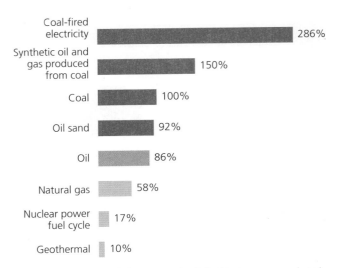

Figure 15-14 CO_2 emissions per unit of electrical energy produced for various energy resources, expressed as percentages of emissions released by burning coal directly. These emissions can enhance the earth's natural greenhouse effect (Figure 3-8, p. 56) and promote climate change (**Concept 15-4A**). **Question:** Which produces more CO_2 emissions: burning coal to heat a house, or heating with electricity generated by coal? (Data from U.S. Department of Energy)

to a 2007 report by the Center for Global Development, coal-burning power plants account for 25% of the world's emissions of CO_2 from human activities and 40% of such emissions in the United States.

Another problem is that the harmful environmental costs of using coal are not included in the price of coal-generated electricity. Environmental economists call for changing this situation by taxing each unit of carbon dioxide produced, as Norway and Sweden have done since 1991. This would promote the development of cleaner coal-burning plants as well as improvements in energy efficiency and increased use of renewable energy resources such as wind, solar, hydroelectricity, and geothermal energy. We discuss such economic issues more fully in Chapter 23.

■ CASE STUDY
Coal Consumption in China

To support its rapid economic growth, China burns a third of the world's coal to provide 70% of its commercial energy, compared to less than 25% in the United States and Japan. China gets 80% of its electricity from burning coal (compared to 49% in the United States), and is adding the equivalent of three large coal-burning power plants per week. As a result, China burns more coal than the United States, Europe, and Japan combined. A 2007 study by German scientists, led by Werner Zittel, estimated that at its current consumption rate, China has about 37 years of proven coal reserves left, and only 10–15 years if its coal consumption continues to increase by 10–15% a year.

China and other parts of the world as well are paying a heavy environmental price for its dependence

Figure 15-13 Air pollution from a coal-burning industrial plant in India with inadequate air pollution controls.

on coal. Pollution controls on older, inefficient plants in China are almost nonexistent. And even the newest coal-burning plants are inefficient and have inadequate air pollution control systems.

Since 2005, China has been the world's leading source of sulfur dioxide, which can cause respiratory and cardiovascular diseases. And sulfur dioxide and nitrogen oxides spewed by China's coal-burning power plants interact in the atmosphere to form harmful acidic compounds that fall as acid precipitation in parts of China and other countries. This pollution contributes to the air quality problems in cities such as Seoul, South Korea and Tokyo, Japan. It takes about 5 to 10 days for long-lived pollution from coal-burning plants in China to make its way to the west coast of United States. There it shows up as higher levels of ozone and other forms of air pollution in major California cities such as Los Angeles and San Francisco. Toxic mercury from China's emissions has been found in fish caught in Oregon's Willamette River. In 2008, China became the world's leading emitter of carbon dioxide, mostly from burning coal.

Major Chinese cities are in an almost perpetual haze from particulates and other pollutants released by burning coal. According to a World Bank report, China has 20 of the world's 30 most polluted cities. In 2007, another World Bank study estimated that outdoor and indoor air pollution, mostly from coal burning, were causing 650,000 to 700,000 premature deaths a year in China.

This already serious environmental and health problem is likely to get much worse if China continues to rely on coal to fuel its rapid economic growth, unless it spends the money to improve the efficiency and air pollution control systems of its new coal-fired power and industrial plants and to retrofit older plants with such equipment. Researchers across China are working on the next generation of cleaner plants, including plants that burn gases produced from coal.

THINKING ABOUT
China's Use of Coal

If you were in charge of China's energy policy, what would be your strategy for long-term coal use? How might this affect the country's rapid economic growth?

Coal Has Advantages and Disadvantages

Coal is the single biggest air polluter in coal-burning nations and accounts for at least one-fourth of the world's annual CO_2 emissions. To a growing number of scientists and economists, the burning of coal is one of the most serious environmental problems of this century.

Many are calling for finding substitutes and for finding other ways to avoid burning coal. For example,

TRADE-OFFS

Coal

Advantages	Disadvantages
Ample supplies (225–900 years)	Severe land disturbance, air pollution, and water pollution
High net energy yield	Severe threat to human health when burned
Low cost	Environmental costs not included in market price
	Large government subsidies
Well-developed technology	High CO_2 emissions when produced and burned
Air pollution can be reduced with improved technology	Radioactive particle and toxic mercury emissions

Figure 15-15 Advantages and disadvantages of using coal as an energy resource (**Concept 15-4A**). **Question:** Which single advantage and which single disadvantage do you think are the most important? Why?

German scientists spurred by strict government air pollution standards have learned how to make steel without burning coal and are selling this technology in the global marketplace.

Figure 15-15 lists the advantages and disadvantages of using coal as an energy resource (**Concept 15-4A**). *Bottom line:* Coal is cheap (as long as its harmful environmental costs are not included in its market price) and plentiful and is distributed over much of the planet. But mining and burning coal has a severe impact on the earth's air (Figure 15-13), water, land (Figures 14-17 and 14-18, p. 357, and Figure 14-19, p. 358), and climate, as well as on human health.

Because of coal's huge environmental impact and its effect on global climate, there is growing opposition by the public and many state governments to building more coal-fired power plants in the United States. In a 2007 national poll by the Opinion Research Group, only 3% of those polled chose coal as their preferred source of electricity.

In 2007, the U.S. Department of Energy listed 151 coal-fired power plants in the planning stage. By 2008, 59 of these proposed plants had either been abandoned or refused licenses by state governments because the companies planning them were not evaluating less harmful alternative ways to meet electricity demand, such as improving energy efficiency and using wind

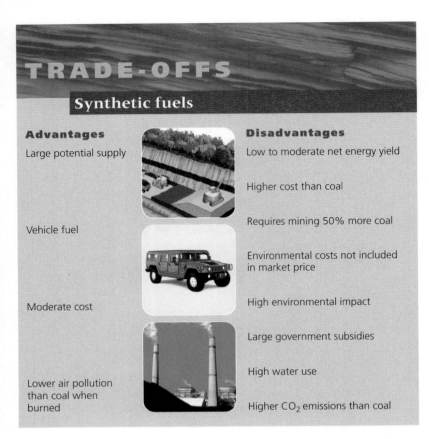

TRADE-OFFS

Synthetic fuels

Advantages	Disadvantages
Large potential supply	Low to moderate net energy yield
	Higher cost than coal
Vehicle fuel	Requires mining 50% more coal
	Environmental costs not included in market price
Moderate cost	High environmental impact
	Large government subsidies
	High water use
Lower air pollution than coal when burned	Higher CO_2 emissions than coal

Figure 15-16 Advantages and disadvantages of using synthetic natural gas (SNG) and liquid synfuels produced from coal (**Concept 15-4B**). **Question:** Which single advantage and which single disadvantage do you think are the most important? Why?

turbines and solar cells. An additional 50 of the proposed plants are being contested in the courts, as will most of the other proposed plants. Major U.S. Wall Street investment companies are also turning their backs on new coal-fired power plants. They see such plants as a bad investment because of their enormous carbon footprint, the likelihood of costly emission caps on their CO_2 emissions in the near future, and the availability of less harmful options. Coal will still be burned in many U.S. power plants, but any significant expansion of use of this resource seems unlikely.

— HOW WOULD YOU VOTE? —

Should using coal to produce electricity be phased out over the next 20–30 years? Cast your vote online at **academic.cengage.com/biology/miller**.

We Can Convert Coal into Gaseous and Liquid Fuels

Solid coal can be converted into **synthetic natural gas (SNG)** by a process called *coal gasification* and into a liquid fuel such as methanol or synthetic gasoline by *coal liquefaction*. Compared to conventional uses of coal, producing these gaseous and liquid fuels (called synfuels) requires mining 50% more coal. Producing and burning them could add 50% more carbon dioxide to the atmosphere (Figure 15-14). As a result, these synfuels have a low net energy yield and cost more to produce per unit of energy than conventional coal costs.

Figure 15-16 lists the advantages and disadvantages of using liquid and gaseous synfuels produced from coal (**Concept 15-4B**). Like energy from tar sands, oil shales, and LNG, synfuels from coal are running up against environmental limits and low net energy yields imposed by the first and second laws of thermodynamics (pp. 42–43).

— RESEARCH FRONTIER —

Improving coal gasification and coal liquefaction technologies. See **academic.cengage.com/biology/miller**.

15-5 What Are the Advantages and Disadvantages of Nuclear Energy?

▶ **CONCEPT 15-5** Nuclear power has a low environmental impact and a very low accident risk, but high costs, a low net energy yield, long-lived radioactive wastes, vulnerability to sabotage, and the potential for spreading nuclear weapons technology have limited its use.

How Does a Nuclear Fission Reactor Work?

To evaluate the advantages and disadvantages of nuclear power, we must know how a nuclear power plant and its accompanying nuclear fuel cycle work. A nuclear power plant is a highly complex and costly system designed to perform a relatively simple task: to boil water to produce steam that spins a turbine and generates electricity.

What makes it complex is the use of a controlled nuclear fission reaction (Figure 2-7, center, p. 41) to

Figure 15-17 Science: light-water–moderated and –cooled nuclear power plant with a pressurized water reactor. Some nuclear plants withdraw water for cooling from a nearby source of water and return the heated water to such a source, as shown here. Other nuclear plants that do not have access to a source of cooling water transfer the waste heat to the atmosphere by using one or more gigantic cooling towers, as shown in the insert photo of the Three Mile Island nuclear power plant near Harrisburg, Pennsylvania (USA). **Question:** How does this plant differ from the coal-burning plant in Figure 15-12?

provide the heat. The reaction takes place in a *reactor.* The most common reactors, called *light-water reactors* (LWRs, see Figure 15-17), produce 85% of the world's nuclear-generated electricity (100% in the United States). They are highly inefficient, losing about 83% of the energy available in their nuclear fuel as waste heat to the environment. About 75% of this loss occurs at the plant itself and another 9% of the energy content of the fuel is lost when it is mined, upgraded, and transported to the plant.

The fuel for a reactor is made from uranium ore mined from the earth's crust. Mined uranium ore must be enriched to increase the concentration of its fissionable uranium-235 from the normal 0.7% to about 3%. Enriched uranium-235 is processed into small pellets

of uranium dioxide. Each pellet, about the size of an eraser on a pencil, contains the energy equivalent of about a ton of coal. Large numbers of the pellets are packed into closed pipes called *fuel rods,* which are then grouped together in *fuel assemblies,* to be placed in the core of a reactor.

To control the reaction, devices called *control rods* are moved in and out of the reactor core to absorb neutrons, thereby regulating the rate of fission and amount of power produced. A *coolant,* usually water, circulates through the reactor's core to remove heat, which keeps fuel rods and other materials from melting and releasing massive amounts of radioactivity into the environment. An LWR includes an emergency core cooling system as a backup to help prevent such meltdowns.

Figure 15-18 Science: after 3 or 4 years in a reactor, spent fuel rods are removed and stored in a deep pool of water contained in a steel-lined concrete basin (left). After they have cooled considerably, some fuel rods are stored upright on concrete pads (right) in dry-storage containers made of steel or concrete. **Question:** Do you think these are safe storage methods? Why or why not?

A *containment shell* with thick, steel-reinforced, concrete walls surrounds the reactor core. It is designed to keep radioactive materials from escaping into the environment, in case there is an internal explosion or a melting of the core within the reactor. It also protects the core from some external threats such as tornadoes and impacts from airplane crashes.

When reactors are shut down and refueled about once a year, intensely hot and radioactive spent fuel rod assemblies are removed and stored outside of the nuclear reactor building in *water-filled pools* (Figure 15-18, left) or in *dry casks* (Figure 15-18, right). Spent-fuel pools or casks are not nearly as well protected as the reactor core and thus are much more vulnerable to sabotage. The long-term goal is to transport spent fuel rods and other long-lived radioactive wastes to an underground facility for long-term storage ranging from 10,000 to 240,000 years, depending on what radioactive isotopes are present. But after almost 60 years of using nuclear power, no country has developed such a facility. Meanwhile, spent fuel rods are stored at nuclear power plant sites, mostly in deep pools of water.

The overlapping and multiple safety features of a modern nuclear reactor greatly reduce the chance of a serious nuclear accident. They also make nuclear power plants very expensive to build and maintain.

What Is the Nuclear Fuel Cycle?

Nuclear power plants, each with one or more reactors, are only one part of the *nuclear fuel cycle* (Figure 15-19). This cycle includes the mining of uranium, processing and enriching the uranium to make fuel, using it in a reactor, and safely storing the resulting highly radioactive wastes until their radioactivity falls to safe levels.

The final step in the cycle occurs when, after 15–60 years, a reactor comes to the end of its useful life and must be retired, or *decommissioned*. It cannot simply be shut down and abandoned, because its structure contains large quantities of intensely radioactive materials that must be kept out of the environment for many thousands of years. Each step in the nuclear fuel cycle adds to the cost of nuclear power and reduces its net energy yield (**Concept 15-1B**). Overall, the current nuclear fuel cycle is extremely inefficient, using or wasting an amount of energy equivalent to about 92% of the energy content of its nuclear fuel.

In evaluating the safety, economic feasibility, and overall environmental impact of nuclear power, energy experts and economists caution us to look at the entire fuel cycle, not just the nuclear plant.

What Happened to Nuclear Power?

In the 1950s, researchers predicted that by the year 2000, at least 1,800 nuclear power plants would supply 21% of the world's commercial energy (25% in the United States) and most of the world's electricity.

After almost 60 years of development, enormous government subsidies, and an investment of $2 trillion, these goals have not been met. Instead, in 2007, 439 commercial nuclear reactors in 30 countries produced only 6% of the world's commercial energy and 16% of its electricity. However, France gets 77% of its electricity from nuclear power and has an excellent safety record. And Japan and South Korea each get 39% of their electricity from nuclear power.

But nuclear power is now the world's slowest-growing energy source with only 34 plants under construction and 93 ordered or planned. The International

Fuel assemblies

Decommissioning
of reactor

Enrichment
of UF$_6$

Reactor

Fuel fabrication

(conversion of enriched
UF$_6$ to UO$_2$ and fabrication
of fuel assemblies)

Temporary storage
of spent fuel assemblies
underwater or in dry casks

Conversion
of U$_3$O$_8$
to UF$_6$

Uranium-235 as UF$_6$
Plutonium-239 as PuO$_2$

Spent fuel
reprocessing

Low-level radiation
with long half-life

Geologic disposal
of moderate-
and high-level
radioactive wastes

⟶ Open fuel cycle today
⟵ - - Recycling of nuclear fuel

Figure 15-19 Science: *the nuclear fuel cycle* (**Concept 15-5**). As long as a plant is operating safely, this fuel cycle has a fairly low environmental impact and a very low risk of an accident. But costs are high, radioactive wastes must be stored safely for thousands of years, and facilities are vulnerable to terrorist attack. Also, the technology can be used to produce material for use in nuclear weapons, and an amount equal to about 92% of the energy content of the nuclear fuel is wasted in producing nuclear power. **Questions:** Do you think the market price of nuclear-generated electricity should include all the costs of the fuel cycle? Explain. If so, how would this affect the use of nuclear power to produce electricity?

Atomic Energy Agency and the U.S. Department of Energy predict that the percentage of the world's electricity produced by nuclear power will decline gradually to about 12% by 2025 because the retirement of aging plants is expected to exceed the construction of new ones.

Figure 9, p. S65, in Supplement 10 shows the trend in electricity production by nuclear power plants in the United States between 1960 and 2006. In the United States, all of the 120 plants ordered between 1973 and 2007 have been canceled. In 2007, there were 104 licensed commercial nuclear power reactors in 31 states—most of them located in the eastern half of the country (Figure 10, p. S65, Supplement 10). These reactors generate about 20% of the country's electricity. This percentage is expected to decline over the next 2–3 decades as existing reactors wear out and are retired faster than new ones are built.

Without huge government (taxpayer) subsidies, tax breaks, loan guarantees, and accident insurance guarantees—amounting to about $9 billion a year—the

nuclear power industry would not exist in the United States. In 2007, the U.S. Congress offered $10.1 billion in additional government tax breaks and loan guarantees that could lead to the construction of several new reactors, perhaps by 2015 or 2020.

According to energy analysts and economists, several reasons explain the failure of nuclear power to grow as projected. They include multibillion-dollar construction cost overruns, high operating costs, more malfunctions than expected, poor management, and the low net energy yield of the nuclear fuel cycle. There have been two other obstacles—public concerns about safety and stricter government safety regulations—especially after an accident in 1979 at the Three Mile Island nuclear plant in the U.S. state of Pennsylvania (see Case Study, p. 390), and another in 1986 at the Chernobyl nuclear plant in Ukraine (see Case Study, p. 390).

Another problem is investor concerns about the economic feasibility of nuclear power. Even with massive government subsidies and loan guarantees, the highly energy-inefficient nuclear fuel cycle (Figure 15-19)

costs more than using coal, natural gas, or wind power to produce electricity. According to economic analyses, without huge government subsidies, no existing nuclear power plant anywhere in the world can compete in the open marketplace with most other methods of producing electricity, if the entire nuclear fuel cycle is taken into account.

In 2008, analysts warned that prolonged drought projected for much of the United States during this century could lead to seasonal or permanent closure of nuclear power plants that must get huge amounts of cooling water from nearby lakes or rivers whose supplies might dry up. When water levels in these bodies of water drop below a certain level, the plant must be shut down.

Another problem that can limit the global expansion of nuclear power plants is that many of the world's shipping companies and ports are putting much tighter restrictions on the shipping of radioactive uranium fuel around the globe because of concerns over safety, terrorist attacks, and rising insurance costs. But governments say they have the power to override such restrictions. Other problems that have limited the growth of nuclear power and that could limit future growth are discussed below.

■ CASE STUDY
Three Mile Island: America's Worst Commercial Nuclear Power Plant Accident

On March 29, 1979, one of the two reactors at the Three Mile Island (TMI) nuclear plant near Harrisburg, Pennsylvania (photo in Figure 15-17), lost its coolant water because of a series of mechanical failures and human operator errors. This led to the most serious commercial nuclear power plant accident in U.S. history.

With the loss of coolant, the reactor's intensely radioactive core became partially uncovered and about half of it melted and fell to the bottom of the reactor. Had there been a complete core meltdown, large amounts of dangerous radioactivity would have been released into the surrounding countryside. Fortunately, the containment building kept most of the radioactivity released from the partially exposed core from escaping, and there were no immediate human casualties.

However, unknown amounts of radioactivity had escaped into the atmosphere. About 50,000 people were evacuated, and another 50,000 fled the area on their own. Various studies have shown no increase in cancer rates from radiation released by the accident, but there is controversy over this issue because of insufficient data.

Partial cleanup of the damaged TMI reactor, along with lawsuits and payment of damage claims, have cost $1.2 billion—almost twice the reactor's $700 million construction cost. Without significant government subsidies, loan guarantees, and accident insurance guarantees, banks and other lending institutions have shown little interest in financing new U.S. nuclear power plants, because the TMI accident showed that utility companies could lose more than $1 billion in equipment and cleanup costs, even without any established harmful effects on public health.

In raising public fears about the safety of nuclear power, the TMI accident led to improved safety regulations for U.S. nuclear plants and improved emergency and evacuation plans. Nuclear power proponents point out that there have been no notable U.S. accidents since TMI. And since 1991, the U.S. reactor fleet has operated at about 90% capacity, up from about 60% in the early 1980s.

■ CASE STUDY
Chernobyl: The World's Worst Nuclear Power Plant Accident

Chernobyl is known around the globe as the site of the world's most serious nuclear power plant accident. On April 26, 1986, a series of explosions in one of the reactors in a nuclear power plant in Ukraine (then part of the Soviet Union) blew the massive roof off a reactor building. The reactor partially melted down (Figure 15-20) and its graphite moderator caught fire and burned for 10 days, releasing more than 100 times the amount of radiation generated by the atomic bombs dropped by the United States on the Japanese cities of Hiroshima and Nagasaki at the end of World War II. The initial explosion and the prolonged fires released a huge radioactive cloud that spread over much of Belarus, Russia, Ukraine, and Europe and eventually encircled the planet. In 2008, after 22 years, areas of the Ukraine and northern Europe are still dangerously contaminated with radioactive materials as a result of the accident.

According to U.N. studies, the Chernobyl disaster was caused by poor reactor design (not used in the United States or in most other parts of the world) and by human error, and it had serious consequences. By 2005, 56 people had died prematurely from exposure to radiation released by the accident. The World Health Organization projects that eventually, this number will grow to 9,000. But the Russian Academy of Medical Sciences estimated the eventual death toll at 212,000. Because of secrecy and sparse reliable data, we will never know the real death toll.

After Chernobyl, some 350,000 people had to abandon their homes because of contamination by radioactive fallout. In addition to fear about long-term health effects such as cancers, many of these victims continue to suffer from stress and depression. In parts of Ukraine, people still cannot drink the water or eat locally produced fruits, vegetables, fish, meat, or milk. In contaminated areas, the frequency of birth defects and mental retardation in newborns has increased. There are also higher incidences of thyroid cancer, leukemia,

Figure 15-20 Remains of a nuclear reactor at the Chernobyl nuclear power plant in Ukraine 3 days after it blew up and released massive amounts of dangerous radiation into the environment. Workers and volunteers put out the fires and hastily built a concrete tomb around the reactor to contain its high-level radiation, and many of these workers died from radiation exposure.

H. Win deck-UNEP/Peter Arnold, Inc.

and immune system abnormalities in children exposed to radioactive fallout. Thyroid cancers are so common that the resulting surgical scars at the base of the neck are known as the "Chernobyl necklace."

Chernobyl taught us a hard lesson: A major nuclear accident anywhere has effects that reverberate throughout much of the world. One more major nuclear power accident anywhere in the world could have a devastating impact on the future of nuclear power.

Japan, an earthquake-prone country that gets 39% of its electricity from nuclear power, has come close to having such an accident. The country has suffered a string of nuclear accidents and cover-ups of such accidents. In 2007, a powerful earthquake in northern Japan caused severe damage to the world's largest nuclear plant (largest in terms of power output) and caused it to be shut down for at least a year. Despite the risks, Japan plans to replace 20 of its 55 aging nuclear reactors between 2010 and 2030.

CENGAGENOW™ Watch how winds carried radioactive fallout around the world after the Chernobyl meltdown at CengageNOW.

Nuclear Power Has Advantages and Disadvantages

Figure 15-21 lists the major advantages and disadvantages of the nuclear power fuel cycle (**Concept 15-5**). In particular, using nuclear power to produce electricity has some important advantages over using coal-burning power plants (Figure 15-22, p. 392).

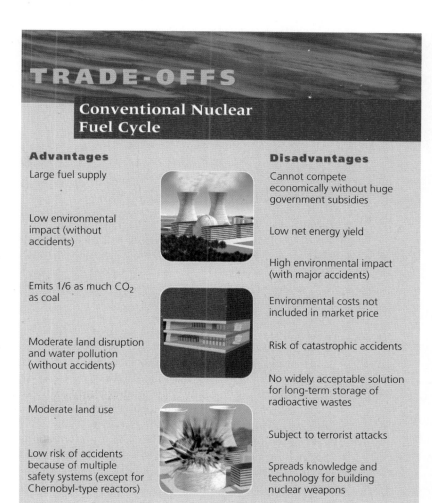

TRADE-OFFS

Conventional Nuclear Fuel Cycle

Advantages	Disadvantages
Large fuel supply	Cannot compete economically without huge government subsidies
Low environmental impact (without accidents)	Low net energy yield
Emits 1/6 as much CO_2 as coal	High environmental impact (with major accidents)
	Environmental costs not included in market price
Moderate land disruption and water pollution (without accidents)	Risk of catastrophic accidents
	No widely acceptable solution for long-term storage of radioactive wastes
Moderate land use	Subject to terrorist attacks
Low risk of accidents because of multiple safety systems (except for Chernobyl-type reactors)	Spreads knowledge and technology for building nuclear weapons

Figure 15-21 Advantages and disadvantages of using the nuclear power fuel cycle (Figure 15-19) to produce electricity (**Concept 15-5**). **Question:** Which single advantage and which single disadvantage do you think are the most important? Why?

TRADE-OFFS

Coal vs. Nuclear

Coal	Nuclear
Ample supply	Ample supply of uranium
High net energy yield	Low net energy yield
Very high air pollution	Low air pollution
High CO_2 emissions	Low CO_2 emissions
High land disruption from surface mining	Much lower land disruption from surface mining
High land use	Moderate land use
Low cost (with huge subsidies)	High cost (even with huge subsidies)

Figure 15-22 Comparison of the risks of using the nuclear power fuel cycle and coal-burning plants to produce electricity. A 1,000-megawatt nuclear plant is refueled once a year, whereas a coal plant of the same size requires 80 rail cars of coal a day. **Question:** If you had to choose, would you rather live next door to a coal-fired power plant or a nuclear power plant? Explain.

Let us examine some of these advantages and disadvantages more closely.

Nuclear Power Plants Are Vulnerable to Terrorist Acts

Because of the built-in safety features, the risk of exposure to radioactivity from nuclear power plants in the United States and most other developed countries is extremely low. However, a partial or complete meltdown or explosion is possible, as the accidents at Chernobyl and Three Mile Island taught us.

A 2005 study by the U.S. National Academy of Sciences warned that pools and casks used to store spent fuel rods at 68 nuclear power plants in 31 U.S. states are especially vulnerable to sabotage or terrorist attack. A spent-fuel pool (Figure 15-18, left) typically holds 5–10 times more long-lived radioactivity than the radioactive core inside a plant's reactor. A 2002 study by the Institute for Resource and Security Studies and the Federation of American Scientists found that about 161 million people—53% of the U.S. population—live within 121 kilometers (75 miles) of an aboveground spent-fuel storage site. For some time, critics have been calling for the immediate construction of much more secure structures to protect spent-fuel storage pools and casks, but this would add to the already high cost of the nuclear fuel cycle and has not been done.

Currently, 60 countries—1 of every 3 in the world—have nuclear weapons or the knowledge and ability to build them. The United States and 14 other countries have been selling nuclear reactors and uranium enrichment technology in the international marketplace for decades. Much of this information and equipment can be used to produce weapons-grade material. Some see this as the single most important reason for not expanding the use of nuclear power, especially when there are cheaper, quicker, and safer ways to produce electricity.

Dealing with Radioactive Wastes Produced by Nuclear Power Is a Difficult Problem

Each part of the nuclear power fuel cycle produces radioactive wastes. *High-level radioactive wastes,* which consist mainly of spent fuel rods and assemblies from commercial nuclear power plants and assorted wastes from the production of nuclear weapons, must be stored safely for 10,000–240,000 years depending on the radioactive isotopes present.

For example, wastes containing highly toxic and fissionable plutonium-239 (which can also be used to make nuclear weapons) must be stored for about 240,000 years before decaying to safe levels. And according to a Nevada state agency report, 10 years after being removed from a reactor, an unshielded spent-fuel assembly would still emit enough radiation to kill a person standing 1 meter (39 inches) away in less than 3 minutes.

Most scientists and engineers agree in principle that deep burial is the safest and cheapest way to store high-level radioactive waste. However, after almost 60 years of research and evaluation, no country has built such a repository. And some scientists contend that it is not possible to show that any method will work for 10,000–240,000 years (Case Study, at right).

U.S. scientists are working on a process involving the recycling of some wastes, which might reduce the amount of radioactive waste produced by conventional reactors by 40%. Critics say that the recycled fuel would contain as much as 90% plutonium (compared to 1% in conventional spent fuel), which would make it attractive to terrorists for making nuclear weapons. However, advocates say the recycling process would make it difficult to extract the plutonium for use in a bomb.

For decades, researchers have been looking—without success—for ways to change harmful radioactive isotopes into less harmful isotopes. Even if a method

were developed, costs would probably be extremely high, and the resulting toxic materials and low-level (but very long-lived) radioactive wastes would still require a safe disposal method.

After almost 60 years of effort, no country has come up with a scientifically and politically acceptable way to store high-level radioactive wastes safely for tens of thousands of years. An important and often ignored fact about using nuclear power to produce electricity is that, even if all the nuclear power plants in the world were shut down tomorrow, we would still have to deal with all the intensely radioactive wastes they have produced, some of which will have to be isolated safely for 240,000 years. No other existing or abandoned technology has subjected the world to such long-term health risks.

┌─ **RESEARCH FRONTIER** ─────────────────────

Safe, affordable nuclear waste storage. See **academic
.cengage.com/biology/miller**.

■ CASE STUDY

Experts Disagree about What to Do with Radioactive Wastes in the United States

In 1985, the DOE announced plans to build a repository for underground storage of high-level radioactive wastes from commercial nuclear reactors. The proposed site is on federal land in the Yucca Mountain desert region, 160 kilometers (100 miles) northwest of Las Vegas, Nevada (Figure 10, p. S65, Supplement 10).

The projected cost of this facility (financed jointly by nuclear power companies and taxpayers) is at least $58 billion and may reach $100 billion. The projected opening date is 2017 but it will probably be 2020 or later because of scientific problems with the site, a number of legal battles, and insufficient federal funding.

The idea is to encapsulate the radioactive material in a synthetic material called zircon, seal it in steel canisters, and store the canisters in underground tunnels that are supposed to be unaffected by earthquakes or a rising water table for at least 10,000 years. However, the site is located in the third most seismically active region in the United States. And a 2007 study by Ian Farman and other scientists indicated that the zircon coatings may degrade faster than originally projected.

Critics charge that the selection of the Yucca Mountain site has been based more on political convenience than on scientific suitability. Some scientists argue that the site should never be allowed to open, mostly because rock fractures and tiny cracks may allow water to leak into the site and eventually corrode the waste storage casks.

According to a 2004 review panel, any rain that percolates into the mountain could carry radioactive

wastes leaking from corroded containers into groundwater, irrigation systems, and drinking-water wells and contaminate them for thousand of years. In 1998, Jerry Szymanski, formerly the DOE's top geologist at Yucca Mountain and now an outspoken opponent of the site, said that if water flooded the site it could cause an explosion so large that "Chernobyl would be small potatoes."

In 2002, the U.S. National Academy of Sciences, in collaboration with Harvard University and University of Tokyo scientists, urged the U.S. government to slow down and rethink its nuclear waste storage process. These scientists contend that storing spent fuel rods in dry-storage casks (Figure 15-18, right) in well-protected buildings at nuclear plant sites, or at several other larger interim storage sites, is an adequate solution for at least 100 years, in terms of safety and national security. This would buy time to carry out more research on this complex problem and to evaluate other sites and storage methods.

Opponents also contend that the Yucca Mountain waste site should not be opened because it could decrease national security. The plan calls for wastes to be shipped by truck or rail cars to the Nevada site. This would require about 19,600 shipments of wastes from nuclear power plants across much of the country (Figure 10, p. S65, Supplement 10) for an estimated 38 years before the site is filled. At the end of this period, the amount of newly collected radioactive waste stored at nuclear power plant sites would be about enough to fill another such repository. Critics contend that it would be much more difficult to protect such a large number of shipments from terrorist attacks than to provide more secure ways to store such wastes at nuclear power plant sites or other centralized sites.

The U.S. government is over 10 years behind in providing a repository for radioactive wastes from commercial power plants. Because of contracts it signed with owners of nuclear reactors in the 1980s, taxpayers must now reimburse plant owners for the costs of storing spent fuel rods at 122 plant sites in 39 states. If the Yucca Mountain site opens by 2017, these little-known government expenses will cost taxpayers about $7 billion and about $11 billion if the opening of the repository is delayed until 2020. Such subsidies plus the estimated $58 billion that the government has spent on developing the Yucca Mountain site add to the already high cost of the nuclear fuel cycle.

Despite rising costs and serious objections from scientists and citizens, in 2002, the U.S. Congress approved Yucca Mountain as the official site for storing the country's commercial nuclear wastes.

┌─ **HOW WOULD YOU VOTE?** ─────────

Should highly radioactive spent fuel be stored in casks at high-security sites or near nuclear power plants instead of being shipped to a single site for burial? Cast your vote online at **academic.cengage.com/biology/miller**.

What Do We Do with Worn-out Nuclear Power Plants?

A nuclear power plant eventually comes to the end of its useful life, mostly because of corrosion and radiation damage to its metal parts. Because it contains intensely radioactive materials, it cannot simply be abandoned. Instead, it must be *decommissioned,* or retired—the last step in the nuclear power fuel cycle (Figure 15-19). Scientists have proposed three ways to do this.

One strategy is to dismantle the plant after it closes and store its large volume of highly radioactive materials in a high-level nuclear waste storage facility, which no country has built so far. A second approach is to install a physical barrier around the plant and set up full-time security for 30–100 years, until the plant can be dismantled after its radioactivity has reached safer but still quite dangerous levels.

A third option is to enclose the entire plant in a tomb that must last and be monitored for several thousand years. Such a tomb was built around the Chernobyl reactor that exploded (Figure 15-20), but after a few years, it began crumbling and leaking radioactive wastes. It is being rebuilt at great cost.

Regardless of the method chosen, decommissioning adds to the total costs of nuclear power and reduces its already low net energy yield. Dismantling a plant and storing the resulting radioactive wastes costs 2–10 times more than building the plant in the first place.

At least 228 of the world's 439 large commercial reactors (20 in the United States) are scheduled for retirement by 2012. However, under political pressure from the nuclear industry, by 2006, the Nuclear Regulatory Commission (NRC) had extended the operating licenses from 40 years to 60 years for many of the 104 U.S. nuclear reactors. Opponents contend this could increase the risk of nuclear accidents in aging reactors. At the same time, the NRC is cutting back on the frequency of safety tests from four times a year to once a year.

The nuclear industry is hoping to replace aging reactors with new second-generation reactors (see Science Focus, at right).

Can Nuclear Power Lessen Dependence on Imported Oil and Help Reduce Global Warming?

Some proponents of nuclear power in the United States claim it will reduce the country's dependence on imported oil. Other analysts disagree, pointing out that only 2–3% of the electricity in the United States (and in most other countries) is generated by burning oil.

Nuclear power advocates also contend that increased use of nuclear power will reduce the threat of global warming by greatly reducing or eliminating emissions of CO_2. Scientists point out that this argument is only partially correct. Nuclear plants themselves do not emit CO_2, but the nuclear fuel cycle does (Figure 15-19)—a fact rarely reported in media stories about nuclear power.

Such emissions are presumably much less than those produced by burning coal or natural gas to generate the same amount of electricity (Figure 15-14) and about the same as those emitted by the entire process of producing and operating solar cells and offshore wind farms. However, according to a 2004 study by German scientists, considering the entire nuclear fuel cycle, CO_2 emissions per kilowatt-hour of electricity are much higher than the numbers in Figure 15-14 indicate.

In a 2003 study "The Future of Nuclear Power," MIT researchers concluded that some 1,000 to 1,500 new reactors (compared to the 439 that exist today) would have to be built worldwide by 2025 in order to put a serious dent in projected global warming. Those plants would require a new large repository every few years to store the resulting amount of highly radioactive nuclear waste. Building and operating this many new plants would also hasten the depletion of high-grade uranium ores, and mining, processing, and transporting these ores releases CO_2. Shifting to low-grade ores to meet increased fuel demands would increase the carbon footprint of the nuclear fuel cycle.

In 2007, a leading think tank, the Oxford Research Group, said that in order to play an effective role in slowing global warming, a new nuclear reactor would have to be built somewhere in the world every week for the next 70 years—an impossibility for logistical and economic reasons. And physicist Brice C. Smith estimates that even if this were possible, because of the retirement of old reactors, the proportion of electricity coming from nuclear power would increase only slightly from its current 16% to 20%. Analysts contend that cutting energy waste and increasing the use of renewable energy resources to produce electricity are much better and faster ways to reduce CO_2 emissions.

Will Nuclear Fusion Save Us?

Nuclear fusion is a nuclear change in which two isotopes of light elements, such as hydrogen, are forced together at extremely high temperatures until they fuse to form a heavier nucleus, releasing energy in the process. Scientists hope that controlled nuclear fusion will provide an almost limitless source of high-temperature heat and electricity. Research has focused on the D–T nuclear fusion reaction, in which two isotopes of hydrogen—deuterium (D) and tritium (T)—fuse at about 100 million degrees (Figure 2-7, bottom, p. 41).

With nuclear fusion, there would be no risk of meltdown or release of large amounts of radioactive materials from a terrorist attack, and little risk of additional proliferation of nuclear weapons, because bomb-grade materials are not required for fusion energy. Fusion power might also be used to destroy toxic wastes, supply electricity for ordinary use, and decompose water

SCIENCE FOCUS

Are New Second-Generation Nuclear Reactors the Answer?

Partly to address economic and safety concerns, the U.S. nuclear industry has persuaded Congress to provide government subsidies and loan guarantees to help them build hundreds of smaller, second-generation plants using standardized designs. The industry claims these plants are safer and can be built quickly (in 3–6 years). However, experts estimate that even if everything goes well, a new round of plants in the United States will not come online before 2020 or later and will cost an estimated $4 billion each.

These *advanced light-water reactors* (ALWRs) have built-in *passive safety features* designed to make meltdowns, explosions, and the release of radioactive emissions almost impossible. Most are *high-temperature, gas-cooled reactors* (HTGC), which avoid some of the problems associated with water-cooled reactors. Proponents also contend that gas-cooled reactors could be used to decompose water to produce hydrogen fuel for powering motor vehicles and planes.

However, according to *Nucleonics Week*, an important nuclear industry publication, "Experts are flatly unconvinced that safety has been achieved—or even substantially increased—by the new designs." In addition, these new designs do not eliminate the expense and hazards of long-term radioactive waste storage, threats of terrorist attack, power plant decommissioning, and the spread of knowledge and materials for the production of nuclear weapons.

One proposed new design is called a *pebble bed modular reactor* (PBMR), an idea first proposed by physical chemist Farrington Daniels just after World War II. One advantage of the pebble design is that it does not need to be shut down for refueling. Instead, pebbles are taken from the bottom of the reactor every day and weighed to see if they still contain usable nuclear fuel; those that still have fuel are returned to the top of the pile and the others are removed. Unreactive helium gas flowing through the spaces between the spheres removes heat from the fissioning fuel. The hot gas can then be used for a number of purposes, including generation of electricity.

Proponents contend that this and other inherently safe designs eliminate the need for an expensive containment shell and emergency core cooling system. Such reactors are now in the planning stages in China and South Africa. China plans to perfect and eventually mass-produce these reactors, ship them anywhere in the world, and assemble them like Legos.

Edwin Lyman and several other nuclear physicists oppose the pebble bed reactor. They contend that

- A crack in the reactor could expose graphite protective coatings to air. At high temperatures, the graphite could burn and release massive amounts of radioactivity—similar to what happened at Chernobyl.

- A lack of a containment shell for any reactor would make it easier for terrorists to enter such facilities and steal nuclear fuel material or blow them up to release large amounts of radioactivity.

- This technology would create about ten times the volume of high-level radioactive waste per unit of electricity as a conventional nuclear reactor produces. This would greatly increase the expense and hazards of long-term radioactive waste storage and power plant decommissioning.

According to the U.S. Energy Information Administration, the costs of building next generation nuclear power plants are likely to be 2–4 times greater than the costs of building natural gas or coal plants or wind farms with equivalent output.

To be acceptable, new generation nuclear reactors must satisfy five criteria: (1) They must be so inherently safe that a runaway chain reaction is impossible; (2) their fuel must be of the sort that cannot be used to make nuclear weapons; (3) their spent fuel must be easy to dispose of without burdening future generations with radioactive waste piles; (4) taking the entire fuel cycle into account, they must generate higher net energy yields than other energy alternatives do, and compete in the open marketplace without huge government subsidies; and (5) the entire nuclear fuel cycle must generate far fewer greenhouse gas emissions than other energy alternatives do. So far, no existing or proposed reactor even comes close to meeting these requirements.

Some nuclear power proponents urge the development and widespread use of *breeder nuclear fission reactors,* which generate more nuclear fuel than they consume by converting nonfissionable uranium-238 into fissionable plutonium-239. However, a failure of the breeder reactor's safety system could result in a runaway fission chain reaction and perhaps a nuclear explosion powerful enough to blast open the containment building. This would release a cloud of highly radioactive gases and particles into the atmosphere.

In December 1986, France opened a commercial-size breeder reactor. It was so expensive to build and operate that, in 1998 after spending $13 billion, the government spent another $2.75 billion to shut it down permanently. Because of this experience, most countries have abandoned their plans to build full-size commercial breeder reactors. However, India, China, Japan, South Korea, and Russia have built or are planning to build small-scale breeder reactors to continue evaluating this technology.

Critical Thinking

Do you think we should invest in second-generation nuclear reactors? Explain.

to produce hydrogen fuel, which holds promise as an energy source.

This sounds great. So what is holding up fusion energy? In the United States, after more than 50 years of research and a $25 billion investment of mostly government funds, controlled nuclear fusion is still in the laboratory stage. None of the approaches tested so far has produced more energy than it uses.

In 2006, the United States, China, Russia, Japan, South Korea, and the European Union agreed to spend at least $12.8 billion in a joint effort to build a large-scale experimental nuclear fusion reactor by 2040 and to see if it can produce a net energy yield. If everything goes well, after 34 years, the plant is supposed to produce enough electricity to run the air conditioners in a small city for a few minutes. This helps to explain why many energy experts do not expect nuclear fusion to be a significant energy source until 2100, if then. Indeed, some skeptics joke that "nuclear fusion is the power of the future and always will be."

Experts Disagree about the Future of Nuclear Power

Proponents of nuclear power argue that governments should continue funding research, development, and pilot-plant testing of potentially safer and cheaper conventional fission reactor designs, along with breeder fission and nuclear fusion. They say we need to keep these potentially useful nuclear options available for use in the future, in case energy efficiency and renewable energy options fail to keep up with electricity demands while reducing CO_2 emissions to acceptable levels.

Others would support expansion of nuclear power only when the five criteria listed in the Science Focus on p. 395 are met. Some analysts call for phasing out all or most government subsidies, tax breaks, and loan guarantees for nuclear power. They argue that nuclear power is a complex, expensive, inefficient, and inflexible way to produce electricity. They see it as an unacceptable risk, because it is too vulnerable to terrorist attack and threatens global security by spreading knowledge and materials that can be used to build nuclear weapons.

According to many investors and World Bank economic analysts, conventional nuclear power cannot compete in today's increasingly open, decentralized, and unregulated energy market, unless it is shielded from competition by large government subsidies (as is the case in every country that has nuclear power plants). Both the U.S. Congressional Budget Office and the private investment firm Standard and Poors have concluded that investing in loans to build nuclear power plants is an unwise financial risk unless governments (taxpayers) are willing to guarantee the loans.

Opponents of nuclear power say it makes more sense to invest government funds in spurring the rapid development of energy efficiency and renewable energy resources that are much safer and can be developed more quickly. We explore these options in Chapter 16.

┌───┐
│ **— HOW WOULD YOU VOTE?** ☑ ———— │
│ Should nuclear power be phased out in the country where │
│ you live over the next 20–30 years? Cast your vote online at │
│ **academic.cengage.com/biology/miller**. │
└───┘

REVISITING Oil Supplies and Sustainability

In this chapter, we have seen that oil—the lifeblood of today's economies—may become unaffordable sometime during this century (**Core Case Study** and Figure 8, p. S64, Supplement 10). If this happens, we will need to find substitutes for oil and begin phasing them in during your lifetime, starting now. This urgent challenge is controversial, partly because of its complexity. It involves multiple problems, the solutions for which will require applications of science, economics, politics, and ethics.

A serious long-term problem is that, in using nonrenewable fossil fuels, we violate the four **scientific principles of sustainability** (see back cover). We depend not on solar energy but on

nonrenewable resources such as oil. The technologies we use to obtain energy disrupt the earth's chemical cycles by diverting huge amounts of water, disrupting land and aquatic systems, and emitting large quantities of pollutants and greenhouse gases. Using these technologies also destroys and degrades biodiversity and ecosystem services that help to control species populations.

In the next chapter, we will look at the advantages and disadvantages of reducing energy waste and relying more on renewable energy resources as ways to apply the four **principles of sustainability**.

Civilization as we know it will not survive
unless we can find a way to live without fossil fuels.

DAVID GOLDSTEIB

REVIEW

1. Review the Key Questions and Concepts for this chapter on p. 371. Summarize the issue of whether or not and when we are likely to run out of affordable oil.

2. What major energy resources do the world and the United States rely on? Give a brief history of human energy use. What is **net energy** and why is it important in evaluating

energy resources? Why does the nuclear power fuel cycle have a low net energy yield?

3. What is **crude oil (petroleum)** and how is it extracted from the earth and refined? What is a **petrochemical** and why are such chemicals important? Who controls most of the world's oil supply? What percentage of the

world's proven oil reserves does the United States have? How much of the world's annual oil production does the United States use and what percentage of the oil it uses is imported? Describe the relationship between importing oil and fighting terrorism. Explain why the United States cannot even come close to meeting its oil needs by increasing domestic oil supplies. Discuss the pros and cons of drilling for oil in Alaska's Arctic National Wildlife Refuge. What are the major advantages and disadvantages of using conventional oil as an energy resource?

4. What is **oil sand,** or **tar sand,** and how is it extracted and converted to heavy oil? What is **shale oil** and how is it produced? What are the major advantages and disadvantages of using heavy oils produced from oil sand and oil shales as energy resources?

5. Define **natural gas, liquefied petroleum gas (LPG),** and **liquefied natural gas (LNG)?** What are the major advantages and disadvantages of using natural gas as an energy resource? What are some problems involved with increasing our use of LNG?

6. What is **coal** and how is it formed? Compare the use of coal in the United States and China. What are the major advantages and disadvantages of using coal as an energy resource?

7. What is **synthetic natural gas (SNG)?** What is coal liquefaction and how can liquid fuels be produced from coal? What are the major advantages and disadvan-

tages of using liquid and gaseous synfuels produced from coal?

8. How does a nuclear fission reactor work and what are its major safety features? Describe the *nuclear fuel cycle.* What factors have hindered the development of nuclear power? Describe the nuclear power plant accidents at Three Mile Island and Chernobyl. What are the major advantages and disadvantages of relying on nuclear power as a way to produce electricity?

9. How can we deal with the highly radioactive wastes produced by nuclear power plants? Describe the controversy over this issue in the United States. What are our options for safely retiring worn out nuclear power plants? Discuss the degree to which nuclear power can reduce dependence on imported oil. Discuss the question of whether using nuclear power can help to significantly slow projected global warming. Discuss the pros and cons of building safer nuclear reactors. List the problems encountered in using breeder reactors. What is **nuclear fusion** and what is its potential as an energy resource? Summarize the arguments for and against relying more on nuclear power.

10. Discuss the relationship between relying on oil as our major source of energy (**Core Case Study**) and the four **scientific principles of sustainability**.

Note: Key Terms are in **bold** type.

CRITICAL THINKING

1. To continue using oil at the current rate (not the projected higher exponential rates), we must discover and add to global oil reserves the equivalent of two new Saudi Arabian supplies *every 10 years,* as discussed in the **Core Case Study** that opened this chapter. Do you think this is possible? If not, what effects might the failure to find such supplies have on your life and on the lives of your children or grandchildren?

2. Do you think that the United States (or the country where you live) should make every effort possible to find new domestic supplies of conventional oil (**Core Case Study**) as a way to sharply reduce its dependence on imported oil? Explain. What are other possible options?

3. List three actions you can take to reduce your dependence on oil and gasoline in order to save you money and help to slow depletion of the world's oil (**Core Case Study**). Which of these things do you already do or plan to do?

4. Explain why you are for or against increasing oil imports to the United States or in the country where you live. If you favor reducing dependence on oil imports, what do you think are the three best ways to do this?

5. Some people in China point out that the United States and European nations fueled their economic growth dur-

ing the industrial revolution by burning coal, with little effort to control the resulting air pollution, and then sought cleaner energy sources later when they became more affluent. China says it is being asked to clean up before it becomes affluent enough to do this, without greatly slowing its economic growth. How would you deal with this contradiction? Since China's air pollution has implications for the entire world, what role, if any, should the developed nations play in helping it to reduce its dependence on coal and to burn coal more cleanly and efficiently?

6. Explain why you agree or disagree with the following proposals made by various energy analysts as ways to solve U.S. energy problems: **(a)** find and develop more domestic supplies of oil; **(b)** place a heavy federal tax on gasoline and imported oil to help reduce the waste of oil resources and to encourage use of other alternatives; **(c)** increase dependence on coal; **(d)** increase dependence on nuclear power; **(e)** phase out all nuclear power plants by 2025.

7. Would you favor having high-level nuclear waste from nuclear power plants transported by truck or train through the area where you live to a centralized storage site? Explain. What are the alternatives?

8. Explain why you agree or disagree with each of the following proposals made by the U.S. nuclear power

industry: **(a)** provide up to $350 billion in government subsidies to build a large number of better-designed nuclear fission power plants, in order to reduce dependence on imported oil and slow global warming, **(b)** prevent the public from participating in hearings on licensing of new nuclear power plants and on safety issues at the nation's nuclear reactors, **(c)** restore government subsidies to develop a breeder nuclear fission reactor program, and **(d)** greatly increase federal subsidies for developing nuclear fusion.

9. Congratulations! You are in charge of the world. List the three most important features of your policy to continue relying on *nonrenewable* energy resources during the next 50 years.

10. List two questions that you would like to have answered as a result of reading this chapter.

Note: See Supplement 13 (p. S78) for a list of Projects related to this chapter.

ECOLOGICAL FOOTPRINT ANALYSIS

In 2008, the average fleet-wide fuel economy of new cars, light trucks, and SUVs in the United States was 11.4 kilometers per liter (kpl) or 26.6 miles per gallon (mpg), and the average motor vehicle in the United States was driven 19,300 kilometers (12,000 miles). There were about 250 million motor vehicles in the United States in 2008. Use these data to calculate the *gasoline consumption* and *carbon dioxide* (CO_2) *footprints* of individual motor vehicles with different fuel efficiencies and for all of the motor vehicles in the United States by answering the following questions.

1. Suppose a car has an average fuel efficiency of 8.5 kpl (20 mpg) and is driven 19,300 kilometers (12,000 miles) a year. **(a)** How many liters (and gallons) of gasoline does this vehicle consume in a year? **(b)** If gasoline costs $1.05 per liter ($4.00 per gallon), how much will the owner spend on fuel in a year? **(c)** How many liters (and gallons) of gasoline would be consumed by a U.S. fleet of 250 million such vehicles in a year? (*Note:* 1 liter = 0.265 gallons and 1 kilometer = 0.621 miles.)

2. Recalculate the values in Question 1, assuming that a car has an average fuel efficiency of 19.6 kpl (46 mpg).

3. The U.S. Environmental Protection Agency estimates that 2.4 kilograms of CO_2 are released when 1 liter of gasoline is burned (20 pounds of CO_2 are released when 1 gallon is burned). Use this information to determine the number of metric tons of CO_2 emitted annually by

(a) the car described in Question 1 with a low fuel efficiency, **(b)** a fleet of 250 million vehicles with this same fuel efficiency, **(c)** the car described in Question 2 with a high fuel efficiency, and **(d)** a fleet of 250 million vehicles with this same high fuel efficiency. These calculations provide a rough estimate of the CO_2 *footprints* for individual cars and for the entire U.S. fleet with low and high efficiency cars. (*Note:* 1 kilogram = 2.20 pounds; 1 metric ton = 1,000 kilograms = 2,200 pounds = 1.1 tons; 1 ton = 2,000 pounds).

4. If the average fuel efficiency of the U.S. fleet increased from 8.5 kpl (20 mpg) to 19.6 kpl (46 mpg), by what percentage would this reduce the CO_2 emissions from the entire fleet per year? You can think of this as the percentage reduction in the carbon footprint of the U.S. motor vehicle fleet.

LEARNING ONLINE

Log on to the Student Companion Site for this book at **academic.cengage.com/biology/miller**, and choose Chapter 15 for many study aids and ideas for further reading and research. These include flash cards, practice quizzing, Weblinks, information on Green Careers, and InfoTrac® College Edition articles.

Iceland's Vision of a Renewable-Energy Economy

Iceland, a tiny northern European island nation roughly the size of Guatemala or the U.S. state of Kentucky, is located just 320 kilometers (200 miles) south of the North Pole. Because about three-fourths of this island consists of volcanoes, glaciers, and hot springs, most of Iceland's 319,000 people live in its capital city of Reykjavik.

The island, which sits on the boundary of the Eurasian and North American tectonic plates (Figure 14-4, p. 347), has 20 active volcanoes. They are fed by molten rock (magma) that rises close to the surface (Figure 14-3, p. 346), under Iceland's abundant glaciers, lakes, and hot springs.

Iceland gets almost three-fourths of its overall energy and almost all of its electricity from two renewable sources. One is *geothermal energy* from superheated groundwater and steam found close to its surface (Figure 16-1). This energy supplies electricity and provides heat for 80% of Iceland's houses and for producing vegetables in greenhouses. The other renewable energy source is *hydroelectric power.*

Because it has no fossil fuel deposits, Iceland imports oil to run its cars, some of its factories, and the fleet of fishing boats that help to supply 60% of its income. This may change. By 2050–2060, Iceland has plans to eliminate its dependence on nonrenewable oil and to become the world's first country to run its economy entirely on renewable energy.

Bragi Arnason, a University of Iceland professor often called "Dr. Hydrogen," suggested this idea in the 1970s. He proposed that the country could eliminate its fossil fuel imports and strengthen its economy by using electricity produced by its ample geothermal, hydroelectric, and wind power resources to decompose water into hydrogen and oxygen gases. The clean-burning hydrogen could then used to power the country's transportation system, factories, and fishing boats. When hydrogen is combined with oxygen in a fuel cell to produce electricity, the only byproduct is water vapor. Thus burning hydrogen fuel avoids emissions of CO_2 and other air pollutants that are produced when fossil fuels are burned. Iceland could also boost its economy by exporting excess hydrogen to other countries.

In 1999, Daimler, Royal Dutch Shell, Norsk Hydro, and Icelandic New Energy announced plans to work together to turn Bragi Arnason's dream into reality by 2050–2060. In 2003, the world's first commercial hydrogen filling station, built and run by Royal Dutch Shell, opened in Reykjavik. Norsk Hydro is providing the electricity and technology to produce the hydrogen fuel.

Between 2003 and 2007, the station fueled three prototype fuel-cell buses provided by Daimler. The use of a new set of such buses is planned for the future. In 2008, the station began fueling a fleet of 10 Toyota Prius test vehicles converted to burn hydrogen in fuel cells; three of the cars are available for rental to tourists. Since 2008, a whale-watching boat (an important part

of the country's rapidly growing ecotourism business) has been partially powered by a hydrogen fuel cell.

These early steps in Iceland's transformation are designed to provide the experience needed to eventually run all of the country's cars, factories, and ships on hydrogen. Although they face considerable technological challenges, Icelanders hope to create a model for a renewable-energy economy within your lifetime.

In this chapter, we look at how we can cut energy waste, and we consider the potential of major renewable energy resources.

Figure 16-1 The Krafla *geothermal power* station in northern Iceland has been in operation since 1977. Beneath the station, 20 deep geothermal wells provide steam for generating electricity. There are no coal-burning power plants to emit CO_2 and other air pollutants into Iceland's skies and no nuclear power plants that produce long-lived radioactive wastes.

Birkir Fanndal

16-1 Why is energy efficiency an important energy source?

CONCEPT 16-1 We could save as much as 43% of all the energy we use by improving energy efficiency.

16-2 How can we cut energy waste?

CONCEPT 16-2 We have a variety of technologies for sharply increasing the energy efficiency of industrial operations, motor vehicles, and buildings.

16-3 What are the advantages and disadvantages of solar energy?

CONCEPT 16-3 Passive and active solar heating systems can heat water and buildings effectively, and the costs of using direct sunlight to produce high-temperature heat and electricity are coming down.

16-4 What are the advantages and disadvantages of producing electricity from the water cycle?

CONCEPT 16-4 Water flowing over dams, tidal flows, and ocean waves can be used to generate electricity, but environmental concerns and limited availability of suitable sites may limit the use of these energy resources.

16-5 What are the advantages and disadvantages of producing electricity from wind?

CONCEPT 16-5 When environmental costs of energy resources are included in market prices, wind energy is the least expensive and least polluting way to produce electricity.

16-6 What are the advantages and disadvantages of biomass as an energy source?

CONCEPT 16-6A Solid biomass is a renewable resource, but burning it faster than it is replenished produces a net gain in

atmospheric greenhouse gases, and creating biomass plantations can degrade soil and biodiversity.

CONCEPT 16-6B Liquid biofuels derived from biomass can be used in place of gasoline and diesel fuels, but creating biofuel plantations could degrade soil and biodiversity and increase food prices and greenhouse gas emissions.

16-7 What are the advantages and disadvantages of geothermal energy?

CONCEPT 16-7 Geothermal energy has great potential for supplying many areas with heat and electricity and generally has a low environmental impact, but locations where it can be exploited economically are limited.

16-8 What are the advantages and disadvantages of hydrogen as an energy source?

CONCEPT 16-8 Hydrogen fuel holds great promise for powering cars and generating electricity, but to be environmentally beneficial, it would have to be produced without the use of fossil fuels.

16-9 How can we make a transition to a more sustainable energy future?

CONCEPT 16-9 We can make a transition to a more sustainable energy future if we greatly improve energy efficiency, use a mix of renewable energy resources, and include environmental costs in the market prices of all energy resources.

Note: Supplements 2 (p. S4), 6 (p. S39), 10 (p. S59), and 13 (p. S78) can be used with this chapter.

Just as the 19th century belonged to coal and the 20th century to oil,
the 21st century will belong to the sun, the wind, and energy from within the earth.

LESTER R. BROWN

16-1 Why Is Energy Efficiency an Important Energy Resource?

▶ **CONCEPT 16-1** We could save as much as 43% of all the energy we use by improving energy efficiency.

We Waste Huge Amounts of Energy

The world will rely increasingly on a mix of renewable energy resources. In addition, however, many analysts urge us to make much greater use of a strategy

not usually thought of as a source of energy—that is, **energy conservation,** a decrease in energy use based primarily on reducing unnecessary waste of energy. The best way to conserve energy is to improve **energy efficiency**—the measure of how much work we can

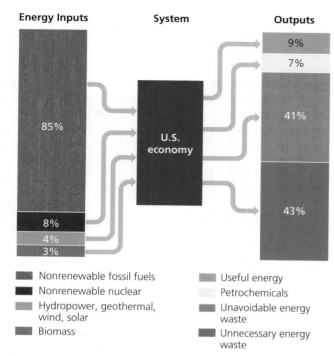

Figure 16-2 Flow of commercial energy through the U.S. economy. Only 16% of all commercial energy used in the United States ends up performing useful tasks or being converted to petrochemicals; the rest is unavoidably wasted because of the second law of thermodynamics (41%) or is wasted unnecessarily (43%). **Questions:** What is an example of unavoidable energy waste? What is an example of unnecessary energy waste? (Data from U.S. Department of Energy)

Energy Inputs | **System** | **Outputs**

85%
8%
4%
3%

U.S. economy

9%
7%
41%
43%

- Nonrenewable fossil fuels
- Nonrenewable nuclear
- Hydropower, geothermal, wind, solar
- Biomass
- Useful energy
- Petrochemicals
- Unavoidable energy waste
- Unnecessary energy waste

get from each unit of energy we use. For example, people who drive energy-efficient cars use less fuel per kilometer (or per mile) than do those who drive less efficient vehicles. You may be surprised to learn that 84% of all commercial energy used in the United States is wasted (Figure 16-2). About 41% of this energy is wasted unavoidably because of the degradation of energy quality imposed by the second law of thermodynamics (**Concept 2-4B**, p. 40).

The other 43% is wasted unnecessarily, mostly due to the inefficiency of incandescent lights, furnaces, industrial motors, coal and nuclear power plants, most motor vehicles, and other devices. Another reason is that many people live and work in leaky, poorly insulated, badly designed buildings. Unnecessary energy waste costs the United States an average of about $570,000 per minute. (See the Guest Essay by Amory Lovins at CengageNOW™.)

For many years, Americans have been buying gas-guzzling sport utility vehicles (SUVs), trucks, and minivans and moving from cities to larger and often energy-inefficient houses in far-flung suburbs, where they must depend on cars for getting around. Now, three of every four Americans commute to work, mostly in gas-guzzling vehicles, and only 5% rely on more energy-efficient mass transit. As a result, two-thirds of the oil consumed in the United States is used for transportation, and 60% of the country's oil is imported.

Canada is even less energy efficient than the United States is, and most developing countries are three times less efficient. On the other hand, Japan, Germany, and France are two to three times *more* energy efficient than the United States is.

Reducing energy waste has numerous economic and environmental advantages (Figure 16-3). To most energy analysts, *reducing energy waste is the quickest, cleanest, and usually the cheapest, way to provide more energy, reduce pollution and environmental degradation, slow global warming, and increase economic and national security* (**Concept 16-1**). The cheapest and cleanest power plant in the world—one that emits no greenhouse gases and produces no radioactive wastes—is the one we do not have to build because of reductions in energy waste. According to energy-efficiency expert Amory Lovins, improving energy efficiency is equivalent to buying oil—which now sells for $100–150 a barrel—for $15 a barrel. (See Lovins' Guest Essay on this topic at CengageNOW.)

Four widely used devices waste large amounts of energy unnecessarily:

- An *incandescent lightbulb* uses only 5–10% of the electricity it draws to produce light, while the other 90–95% is wasted as heat. It is really a *heat bulb.*

SOLUTIONS

Reducing Energy Waste

- Prolongs fossil fuel supplies
- Reduces oil imports and improves energy security
- Very high net energy yield
- Low cost
- Reduces pollution and environmental degradation
- Buys time to phase in renewable energy
- Creates local jobs

Figure 16-3 Advantages of reducing unnecessary energy waste and thereby improving energy efficiency. Global improvements in energy efficiency could save the world an average of about $1.9 million per minute! **Question:** Which two of these advantages do you think are the most important? Why?

- A motor vehicle with an *internal combustion engine* wastes about 94% of the energy in its fuel.

- A *nuclear power plant* (Figure 15-17, p. 387), producing electricity for space heating or water heating, wastes about 83% of the energy in its nuclear fuel and probably 92% when we include the additional energy used in the nuclear fuel cycle (Figure 15-19, p. 389) to dig up and to process its uranium fuel, store its radioactive wastes for thousands of years, and to retire the highly radioactive plant at the end of its useful life.

- A *coal-fired power plant* (Figure 15-12, p. 383) wastes 66% of the energy released by burning coal to produce electricity and probably 75–80% if we include the energy needed to dig up the coal and transport it to the plant.

Energy-efficiency experts say we can no longer afford to build our societies around such energy-wasting and environmentally harmful devices, which some call the *outdated four*. They call for us to replace these energy-wasting dinosaurs with more energy-efficient and less polluting and climate-changing alternatives over the next few decades.

Such alternatives include compact fluorescent and light emitting diode (LED) lights, fuel cells to power motor vehicles and provide heat and electricity for buildings, and wind and solar cell farms to produce electricity. These replacements could be coupled with a mix of existing energy-saving measures such as using more insulation; plugging air leaks; using energy-efficient windows, appliances, and building design; and recycling and reusing most materials.

In 2008, Lester Brown estimated that shifting to more energy-efficient lighting and appliances alone over the next two decades could save the world enough electricity to avoid building more than 1,400 large coal-burning power plants, thereby also avoiding great amounts of climate-changing CO_2 emissions.

Net Energy Efficiency—Honest Energy Accounting

Recall that the only energy that really counts is *net energy* (Science Focus, p. 374). The *net energy efficiency* of a system used to heat a house, for example, is determined by combining the efficiencies of all steps in the energy conversion for the entire system.

Figure 16-4 shows the net energy efficiency for heating two well-insulated homes. One is heated with electricity produced at a nuclear power plant, transported by wire to the home, and converted to heat (*electric resistance heating*). The other is heated passively: direct solar energy enters through high-efficiency windows facing the sun and strikes heat-absorbing materials that store the heat for slow release.

Electricity from Nuclear Power Plant

Passive Solar

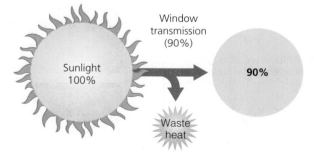

Figure 16-4 Comparison of the net energy efficiency for two types of space heating. The cumulative net efficiency is obtained by multiplying the percentage of energy available from the source (shown inside the square before each step) by the energy efficiency for that step (shown in parentheses). So 100 × 0.95 = 95%; 95 × 0.57 = 54%; and so on. About 83% of the energy used to provide space heating by electricity produced at a nuclear power plant is wasted. If the additional energy needed to deal with nuclear wastes and to retire the highly radioactive nuclear plant after its useful life is included, the net energy yield for a nuclear plant is only about 8% (or 92% waste). Burning coal or any fossil fuel to generate electricity at a power plant and transmitting electricity long distances to heat water or space is also inefficient. By contrast, with passive solar heating, only about 10% of incoming solar energy is wasted.

16-2 How Can We Cut Energy Waste?

> ▶ **CONCEPT 16-2** We have a variety of technologies for sharply increasing the energy efficiency of industrial operations, motor vehicles, and buildings.

We Can Save Energy and Money in Industry

Industry accounts for about 30% of the world's and 38% of U.S. energy consumption, mostly from production of metals, chemicals, cement, and paper. The largest consumer of energy in the manufacturing section is the petrochemical industry, which produces products such as fertilizers, plastics, and detergents. It is followed by the global steel and cement industries. There are many ways for industries to cut energy waste (**Concept 16-2**).

Some companies save energy and money by using **cogeneration,** or **combined heat and power (CHP),** systems. In such a system, two useful forms of energy (such as steam and electricity) are produced from the same fuel source. For example, the steam produced in generating electricity in a CHP system can be used to heat the plant or other nearby buildings, rather than released into the environment and wasted. The energy efficiency of these systems is as high as 80% (compared to 30–40% for coal-fired boilers and nuclear power plants), and they emit one-third as much CO_2 per unit of energy produced as do conventional coal-fired boilers. Cogeneration has been widely used in Europe for years and its use in the United States and in China is growing.

Another way to save energy and money in industry is to *replace energy-wasting electric motors,* which consume one-fourth of the electricity produced in the United States. Most of these motors are inefficient because they run only at full speed with their output throttled to match the task—somewhat like driving a car with the engine racing and keeping your foot on the brake pedal to control the speed. Each year, a heavily used electric motor consumes about ten times its purchase price in electricity—equivalent to using $200,000 worth of gasoline each year to fuel a $20,000 car. The energy savings from replacing all such motors would equal the output of 150 large coal-fired or nuclear power plants.

Recycling materials such as steel and other metals is a third way for industry to save energy and money. Producing steel from recycled scrap iron in an electric arc furnace uses 75% less energy than producing steel from virgin iron ore. Switching three-fourths of the world's steel production to such furnaces would cut energy use in the global steel industry by almost 40% and sharply reduce its CO_2 emissions. Similarly, if all of the world's energy-intensive cement producers used today's most energy-efficient dry kiln process, the global cement industry could cut its energy use by 42% and greatly reduce its CO_2 emissions.

A fourth way for industry to save energy is to *switch from low-efficiency incandescent lighting to higher-efficiency fluorescent and LED lighting.* A compact fluorescent bulb uses one-fourth as much electricity as an incandescent bulb, lasts ten times as long, and saves at least $30 in replacement costs during its lifetime. Even better, light-emitting diodes (LEDs) will be increasingly available for industrial lighting; an LED uses about one-seventh of the electricity used by an incandescent bulb and can last about 100 times as long. New York City is saving $6 million a year in maintenance and electricity costs by replacing the traditional bulbs in most of its traffic lights with LED bulbs.

The world's grid systems that transmit electricity from power plants to industries and cities waste large amounts of energy, also making them a candidate for the energy dinosaur list. Converting the outdated U.S. electrical grid system into a more responsive and energy-efficient, digitally controlled network could save the nation $100 billion a year, according to the Electric Power Research Institute. China plans to build an efficient and reliable ultra-high voltage (UHV) electricity network by 2020 and to become the global leader in ultra-high-voltage technology, equipment manufacturing, and sales.

Utility companies also waste large amounts of energy by encouraging electricity use instead of reducing electricity waste. State utility commissions in the United States have rewarded utilities for selling more kilowatt-hours by building more power plants. In the 1980s, some energy analysts proposed that state commissions instead provide financial rewards to utilities for the kilowatt-hours they save customers by helping them to improve their energy efficiency. The U.S. state of California has had great success using such a save-a-watt approach.

One outstanding case of industrial energy savings concerns the Dow Chemical Company, which operates 165 manufacturing plants in 37 countries. Dow set out to reduce its total energy consumption by 25% between 2007 and 2015. These savings will be added to the company's cuts of 22% since 1996. If it reaches its goal, the company will have cut its energy consumption in half within 20 years.

Dow's annual energy savings, due to a mix of large and small improvements in efficiency, now equal slightly more than the total amount of energy consumed each year by the nation of New Zealand. The company's CEO estimates that energy efficiency improvements cost Dow about $1 billion between 1996 and 2006, but resulted in savings of about $5 billion. And there are more savings to come.

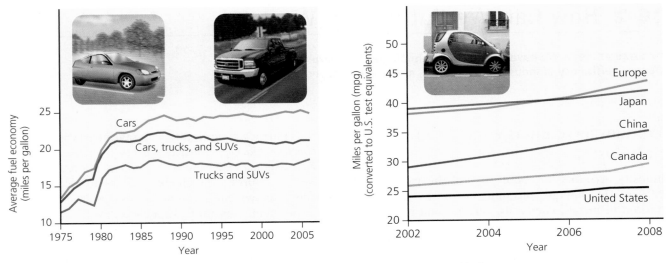

Figure 16-5 Average fuel economy of new vehicles sold in the United States, 1975–2008 (left) and fuel economy standards in other countries, 2002–2008 (right). According to scientist Joseph Romm, no gasoline-powered car assembled in North America would meet China's current fuel-efficiency standard, which is a minimum allowable standard unlike the less stringent fleet-average standard used in the United States. (U.S. Environmental Protection Agency and National Highway Traffic Safety Administration, International Council on Clean Transportation)

We Can Save Energy and Money in Transportation

Transportation accounts for two-thirds of U.S. oil consumption and is a major source of the country's urban air pollution and inputs of CO_2 into the atmosphere.

Between 1973 and 1985, average fuel efficiency for new vehicles sold in the United States rose sharply because of government-mandated *corporate average fuel economy* (*CAFE*) standards. However, since 1985, the average fuel efficiency for new vehicles sold in the United States (Figure 16-5, left) decreased to about 9 kilometers per liter (kpl) (21 miles per gallon (mpg)). This was mostly because there was no increase in the CAFE standards until 2008, and because mileage standards for popular trucks and SUVs are not as high as those for cars.

Fuel economy standards in Europe, Japan, China, and Canada are much higher than those in the United States (Figure 16-5, right). A 2008 law raised CAFE standards in the United States to 15 kpl (35 mpg) to be attained by 2020. This will still put U.S. fuel efficiency standards much lower than those of the other countries shown in Figure 16-5.

The good news is that fuel-efficient cars are available (**Concept 16-2**). Car companies like to tout their fuel-efficient vehicles. But no vehicle should be called fuel-efficient unless it gets at least 15 kpl (35 mpg), according to fuel efficiency experts. An example of such a gas-sipping vehicle is the Toyota Prius hybrid-electric car. It has a combined city and highway average fuel efficiency of 20 kpl (46 mpg)—more than twice the fuel efficiency of the average new car sold in the United States.

The bad news is that gas-sipping vehicles account for less than 1% of all car and truck sales in the United States. One reason is that the inflation-adjusted price of gasoline in the United States, despite recent increases, is still fairly low—costing less per liter than bottled water. Gasoline prices are much higher in Japan and most European nations, because their governments have set higher fuel-efficiency standards and imposed high gasoline taxes to encourage greatly improved fuel efficiency (Figure 16-5, right).

Another reason is that most U.S. consumers do not realize that gasoline costs them much more than the price they pay at the pump. According to a 2005 study by the International Center for Technology Assessment, the hidden costs of gasoline for U.S. consumers, having to do with environmental and health costs and security issues, were about $3.18 per liter ($12 per gallon). Adding this to an $1.06 per liter ($4 per gallon) pump price of gasoline for U.S. consumers would yield a *true cost* of about $4.24 per liter ($16 per gallon). In fact, in 2008, hidden costs and pump prices were rising above these levels. Filling a 20-gallon (75-liter) tank would cost about $320, and consumers would demand much more energy-efficient cars and efficient and widely available mass transportation options.

These hidden costs include government subsidies and tax breaks for oil companies, car manufacturers and road builders; costs of pollution control and cleanup; costs of military protection of oil supplies in the Middle East (not including the two Iraq wars); time wasted in traffic jams; and increased illnesses and deaths from air and water pollution, which have increased medical bills and health insurance premiums. Consumers pay for these costs, but not at the gas pump. Including all

or most of the harmful environmental and health costs of gasoline and other goods and services in their market prices would give consumers more accurate information about the environmental impacts of the items and services they buy.

One way to include more of the real costs of gasoline in its market price is through gasoline taxes, which are politically unpopular in the United States. But analysts call for reducing payroll and income taxes to balance increases in gas taxes, thereby relieving consumers of any additional financial burden.

> **THINKING ABOUT**
> **The Real Cost of Gasoline**
>
> Do you think that the estimated hidden costs of gasoline should be included in its price at the pump? Explain. Would you favor much higher gasoline taxes if payroll taxes were eliminated or sharply reduced? Explain.

A *third* reason for low fuel efficiency is that in 2008, more than half of all U.S. consumers, compared to 5% in 1990, owned SUVs, pickup trucks, minivans, and other large, inefficient vehicles. And *fourth*, the government has not given buyers large enough tax breaks to encourage them to buy more fuel-efficient vehicles. For example, in the United States, small businesses that paid $50,000 for a gas-guzzling Hummer used exclusively for business purposes in 2005 got a tax deduction of up to $25,000. People buying or leasing a $22,000, gas-sipping, hybrid car got a much smaller deduction of up to $3,100.

Energy expert Amory Lovins has proposed a *feebate* program in which fuel-inefficient vehicles would be taxed heavily and the resulting revenue would be given to buyers of efficient vehicles as rebates (not tax deductions). For example, the tax on a $50,000 Hummer H2 that averages about 5 kpl (12 mpg) might be $10,000. And the same amount could go as a rebate to the buyer of a $22,000 hybrid car that averages 20 kpl (46 mpg). Within a short time, such a program—endorsed by the U.S. National Academy of Sciences—would greatly increase sales of gas-sipping vehicles. It would also focus carmakers on producing and making their profits from such vehicles, and it would cost the government (taxpayers) nothing.

> **THINKING ABOUT**
> **Feebates**
>
> Do you support implementation of a feebate program? Explain. Why do you think this hasn't been done?

More Energy-Efficient Vehicles Are on the Way

There is growing interest in developing *superefficient and ultralight cars* that could eventually get 34–128 kpl (80–300 mpg) (**Concept 16-2**). (See Amory Lovins' Guest Essay on this topic at CengageNOW.)

One of these vehicles is the energy-efficient, gasoline–electric *hybrid car* (Figure 16-6, left), invented by Ferdinand Porsche in 1900 and improved with modern technology by Japanese automobile companies such as Toyota and Honda. It has a small traditional gasoline-powered motor and an electric motor used to provide the energy needed for acceleration and hill climbing. The most efficient models of these cars get up to 20 kpl (46 mpg) on the highway and emit about 65% less CO_2 per kilometer driven than a comparable conventional car emits.

The next step in the evolution of more energy-efficient motor vehicles will probably be the *plug-in hybrid electric vehicle*—a hybrid with a second and more powerful battery that can be plugged into a standard outlet and recharged (Figure 16-6, right). By running primarily on electricity, they could easily get the equivalent of at least 43 kpl (100 mpg) for ordinary driving and up to 430 kpl (1,000 mpg), if used only for trips of less than 64 kilometers (40 miles). Manufacturers hope to have a variety of plug-in hybrids available by

Conventional hybrid

Plug-in hybrid

Figure 16-6 Solutions: general features of a car powered by a *hybrid gasoline–electric engine* (left). A *plug-in hybrid* vehicle (right) has a smaller internal combustion engine and a second and more powerful battery that can be plugged into a standard 110-volt outlet and recharged. This allows it to run farther on electricity alone.

The Search for Better Batteries

The major obstacle standing in the way of mass-market, plug-in, hybrid-electric vehicles is the difficulty with developing a battery that can store enough energy to power a vehicle over long distances without overheating or even catching fire.

One promising type of battery is a *lithium-ion battery,* commonly used in laptop computers and cell phones. It can pack a lot of energy into a small space. It can also weigh one-sixth as much as, and take up less than half the space of, the nickel-metal hydride batteries used to power today's conventional hybrid-electric vehicles. However, current lithium-ion batteries have an occasional tendency to overheat, release

oxygen, and in rare cases burst into flames. Scientists are working to improve these batteries. General Motors plans to produce a hybrid-electric vehicle with a powerful lithium-ion battery between 2010 and 2017.

Scientists are also looking for other types of batteries that do not contain flammable components. One battery manufacturer is using nanotechnology (Science Focus, p. 362) to make electrodes out of a nanophosphate material that will not heat up and release flammable oxygen.

In the quest for lightweight, inexpensive batteries, scientist Angela Belcher at the Massachusetts Institute of Technology, is working on an entirely new type of battery.

She has genetically engineered a virus that can latch onto and coat itself with conductive materials to form a miniscule nanowire. If perfected, such viral batteries could essentially grow themselves, using water as a solvent, and yield none of the often-toxic wastes produced by the manufacture and disposal of conventional batteries.

Another approach is to power a car with an *ultracapacitor,* a battery-like device that stores and releases energy very quickly. If it works, this device would make batteries obsolete.

Critical Thinking

Would you buy a plug-in hybrid vehicle with a lithium-ion battery? Why or why not?

2010, and some analysts project that they could dominate the motor vehicle market by 2020. The key is to develop a battery that will have enough range and be strong, safe, reliable, and affordable enough to use in a mass auto market (see Science Focus, above).

Replacing the current U.S. vehicle fleet with highly efficient plug-in hybrid vehicles over 2 decades, would cut U.S. oil consumption by 70–90%, eliminate the need for oil imports, save consumers money, and reduce CO_2 emissions by 27%, according to a 2006 Department of Energy study. If the batteries in this national car fleet were recharged mostly with electricity generated by wind, solar energy, hydropower, and geothermal energy instead of by coal-burning power plants, U.S. emissions of CO_2 would drop by 80–90%, which would greatly help to slow global warming and projected climate change.

Another option is an *energy-efficient diesel car,* which accounts for 45% of new passenger car sales in Europe. Modern diesel engines are quiet and are 30% more fuel efficient, emitting 20% less CO_2 than conventional gasoline-powered engines. European car companies have also greatly reduced air pollution from diesel engines, although the engines emit more nitrogen oxides and particulates than comparable conventional and hybrid vehicles do. Diesel fuel can be made from coal, plant material, or cooking oil. Running these vehicles on a fuel called *biodiesel,* discussed later in this chapter, would reduce their air pollution emissions and increase their energy efficiency. Also, hybrid-electric diesel cars would be more fuel efficient than conventional hybrid-electric vehicles.

The next stage in the development of superefficient cars could be an electric vehicle that uses a *fuel cell.* Fuel cells are at least twice as efficient as internal combustion engines, have no moving parts, require little maintenance, and use hydrogen gas as fuel to produce electricity without emitting CO_2 and other air pollutants. Most major automobile companies have developed prototype fuel-cell cars, which if successful, could replace many energy-wasting internal combustion engines. Some fuel-cell cars could be in showrooms by 2012, but some analysts believe that it will be 2020 before these cars will be mass produced. The basic challenge is to bring the costs down. (We discuss fuel cells in more detail later in this chapter.) **GREEN CAREER:** Fuel-cell technology

The fuel efficiency for all types of cars could nearly double if car bodies were to be made of *ultralight* and *ultrastrong* composite materials such as fiberglass and the carbon-fiber composites used in bicycle helmets and in some racing cars.

RESEARCH FRONTIER

Developing better and more affordable hybrid and fuel-cell vehicles. See **academic.cengage.com/biology/miller**.

We Can Design Buildings That Save Energy and Money

We can realize huge savings in energy by designing and building for energy efficiency (**Concept 16-2**) and retrofitting existing buildings to make them more energy efficient. In fact, a 2007 U.N. study concluded that better architecture and energy savings in buildings could save 30– 40% of the energy used globally. For example, orienting a building so it can get more of its heat from the sun can save up to 20% of heating costs and as much as 75% when the building is well insulated and airtight—a simple application of the solar energy **principle of sustainability** (see back cover).

The 13-story Georgia Power Company building in the U.S. city of Atlanta, Georgia, uses 60% less energy than conventional office buildings of the same size use. The largest surface of the building faces south to capture solar energy. Each floor extends out over the one below it. This blocks out the higher summer sun to reduce air conditioning costs but allows the lower winter sun to help light and heat each floor during the day. In the building's offices, energy-efficient compact fluorescent lights focus on work areas instead of illuminating entire rooms. Such green buildings have been used widely in Europe for almost 2 decades, especially in Germany and the Netherlands, and are beginning to catch on in the United States.

Green architecture, based on energy-efficient and money-saving designs, makes use of natural lighting, passive solar heating, geothermal heat pumps for heating and cooling, cogeneration, solar hot water heaters, solar cells, fuel cells, natural ventilation, recycled building materials, energy-efficient appliances and lighting, motion sensors for lighting, rainwater collection, recycled waste water, waterless urinals, composting toilets, and nontoxic paints, glues, and building materials. Some green designs also include *living roofs,* or *green roofs,* covered with soil and vegetation (Figure 16-7). They have been used for decades in Europe (Photo 11 in the Detailed Contents) and are becoming more common in the United States.

Another important element of energy-efficient design is *superinsulation.* A house can be so heavily insulated and airtight that heat from direct sunlight, appliances, and human bodies can warm it with little or no need for a backup heating system, even in extremely cold climates. An air-to-air heat exchanger prevents buildup of indoor air pollution. The building cost for such a house is typically 5% more than that for a conventional house of the same size. The extra cost is paid back by energy savings within about 5 years, and the homeowner can save $50,000–100,000 over a 40-year period. Superinsulated houses in Sweden use 90% less energy for heating and cooling than typical American homes of the same size use.

Since the mid-1980s, there has been growing interest in *straw bale houses* (Photos 9 and 10 in the Detailed Contents). The walls of these superinsulated houses are made by stacking compacted bales of low-cost straw (a renewable resource) and then covering the bales on the outside and inside with plaster or adobe. (See the Guest Essay about straw bale and solar energy houses by Nancy Wicks at CengageNOW.)

Green building certification standards now exist in 21 countries, spurred by the World Green Building Council, established in 1999. Since 2001, the U.S. Green Building Council's Leadership in Energy and Environmental Design (LEED) program has accredited more than 25,000 building professionals in energy and environmental design. It has established guidelines, and it awards its much-coveted silver, gold, and platinum standard certifications to buildings meet-

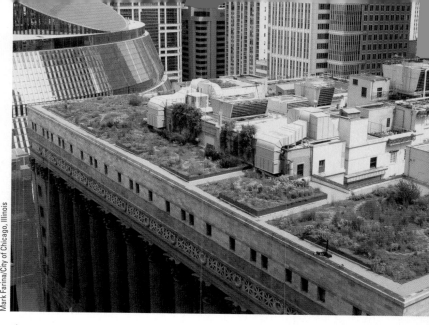

Figure 16-7 A *green* or *living roof* on City Hall in the U.S. city of Chicago, Illinois. It saves energy, improves air quality, reduces storm water runoff and thus pollution of waterways, and provides habitat for birds.

ing certain standards. **GREEN CAREER:** Environmental architect

One platinum standard building is China's Ministry of Science and Technology in Beijing. Its surrounding area is paved with porous bricks made of fly ash left over from burning coal. These bricks allow water to flow through them and to help replenish the city's aquifer. Solar cells made in China provide about 10% of the building's electricity, and it has a solar hot water heating system also made in China. A soil substitute used in its energy-saving roof garden is 75% lighter and holds three to four times more water per cubic meter than dirt can hold. The use of concrete building blocks filled with insulating foam also saves energy.

This is an impressive showcase building, but China lags far behind other countries in energy-efficient building design. Nevertheless, within 20 years, China expects to be the world's leader in this area and to sell its innovative designs and materials in the global marketplace.

We Can Save Energy and Money in Existing Buildings

Most of the world's buildings were not built with energy efficiency as a priority, but many of them can be retrofitted to save energy (**Concept 16-2**). Here are some ways to do so:

- *Insulate and plug leaks.* About one-third of the heated air in typical U.S. homes and buildings escapes through holes, cracks, and closed single pane windows (Figure 16-8, p. 408). The resulting energy loss is roughly equal to the energy in all the oil flowing through the Alaska pipeline every year. During hot weather, these windows and cracks let heat in, increasing the use of air conditioning.

Figure 16-8 A *thermogram,* or infrared photo, showing heat loss (red, white, and orange) around the windows, doors, roofs, and foundations of houses and stores in Plymouth, Michigan. Many homes and buildings in the United States and other countries are so full of leaks that their heat loss in cold weather and heat gain in hot weather are equivalent to what would be lost through a large window-sized hole in a wall of the house. **Question:** How do you think the place where you live would compare to these houses in terms of heat loss?

Adding insulation and plugging leaks are two of the quickest, cheapest, and best ways to save energy and money in any building.

- *Use energy-efficient windows.* Replacing leaky windows with energy-efficient windows can cut expensive heat losses from a house by two-thirds, lessen cooling costs in the summer, and reduce CO_2 emissions. Widely available superinsulating windows do the job of 8–12 panes of glass. They can be expensive, but energy savings usually pay back investment costs within a few years and then save money for their owners.

- *Stop other heating and cooling losses.* Leaky heating and cooling ducts in attics and crawl spaces beneath houses allow 20–30% of a home's heating and cooling energy to escape and draw unwanted moisture and heat into the home. Careful sealing of duct joints can reduce this loss. Some new home designs place the air ducts inside the home's thermal envelope so that escaping hot or cool air is fed back into the living space. Also, using light-colored roofing shingles instead of dark singles—or using living roofs (Photo 11 in the Detailed Contents)—can cut electricity use for air conditioning.

- *Heat houses more efficiently.* In order, the most energy-efficient devices we can use to heat space are superinsulation; a geothermal heat pump that transfers heat stored in the earth to a home (discussed later in this chapter); passive solar heating; a high-efficiency, conventional heat pump (in warm climates only); small cogenerating microturbines; and a high-efficiency (85–98%) natural gas furnace. The most wasteful and expensive way to heat a space is to use electric resistance heating with electricity produced by a coal-fired or nuclear power plant.

- *Heat water more efficiently.* One approach is a roof-mounted solar hot water heater, now being used widely in China and a number of other countries. Another option is a *tankless instant water heater* (about the size of a suitcase) fired by natural gas or LPG. (Using electricity for this purpose is not efficient.) These devices, used widely in many parts of Europe, heat water instantly as it flows through a small burner chamber, providing hot water only when it is needed, and using about 25–30% less energy than traditional water heaters use. They cost 2–4 times more than conventional water heaters, but save money in the long run because they last 3–4 times longer and cost less to operate than conventional water heaters cost. They work. One of the authors (Miller) used them along with passive and active solar hot water heaters in an office and living space for 15 years.

- *Use energy-efficient appliances.* According to the Environmental Protection Agency, if all U.S. households used the most efficient frost-free refrigerator now available, 18 large power plants could close. Microwave ovens can cut electricity use for cooking by 25–50% and convection ovens cut energy use by about 20%. Clothes dryers with moisture sensors cut energy use by 15%, and front-loading washers use 55% less energy and 67% less water than do top-loading models. TV sets, computers, and other appliances in their standby mode use about 5% of U.S. residential electricity. Consumers can reduce this energy waste by plugging their computers, TV sets, and other devices having a standby feature into a power strip (such as Smart Strip) that cuts off power when it detects that the device has been turned off.

- *Use energy-efficient lighting.* A compact fluorescent bulb produces as much light as a regular incandescent bulb, but lasts up to ten times longer and

uses one-fourth as much energy, which pays for its higher price in a few months. According to the U.S. Department of Energy, replacing 30 incandescent bulbs with compact fluorescent bulbs can save consumers more than $1,000 over the life of the bulbs. According to a 2005 estimate by the U.S. National Renewable Energy Laboratory, replacing a ten-watt incandescent lightbulb with an equivalent compact fluorescent bulb saves over its lifetime enough energy to drive a Toyota Prius hybrid car from New York City to San Francisco, California. The world may be on the way to phasing out this outdated, energy-wasting device. Over the next 5–10 years, Australia, Canada, Brazil, China, and a growing list of other countries are phasing out the sale of incandescent bulbs. And the United States is phasing in higher efficiency standards for lighting, which will basically phase out traditional incandescent bulbs by 2014. In 2008, Lester Brown estimated that shifting to more energy-efficient lighting could cut the world's electricity use enough to avoid building more than 700 large coal-fired power plants,

a number equal to about one-third of the world's existing coal-burning plants.

One problem is that the average compact fluorescent bulb contains a small amount of toxic mercury—roughly the amount that would fit on the tip of a ballpoint pen. However the total amount of mercury in all such bulbs is a tiny fraction of the amount of mercury released by coal-fired power plants, a large portion of which is produced to light energy-wasting incandescent bulbs. And the mercury in compact fluorescent bulbs can be recycled, whereas the mercury continuously spewed into the atmosphere by coal-burning power plants can never be retrieved, except by ingesting it through food and water from the environment.

Within the next 2 decades, both incandescent and fluorescent bulbs may be replaced by even more efficient, pea-sized light-emitting diodes (LEDs) and organic LEDs (OLEDs) when their prices come down (see **westinghouselighting.com**).

Figure 16-9 summarizes ways in which you can save energy in the place where you live.

Attic
- Hang reflective foil near roof to reflect heat.
- Use house fan.
- Be sure attic insulation is at least 30 centimeters (12 inches).

Bathroom
- Install water-saving toilets, faucets, and shower heads.
- Repair water leaks promptly.

Kitchen
- Use microwave rather than stove or oven as much as possible.
- Run only full loads in dishwasher and use low- or no-heat drying.
- Clean refrigerator coils regularly.

Outside
Plant deciduous trees to block summer sun and let in winter sunlight.

Other rooms
- Use compact fluorescent lightbulbs or LEDs and avoid using incandescent bulbs wherever possible.
- Turn off lights, computers, TV, and other electronic devices when they are not in use.
- Use high efficiency windows; use insulating window covers and close them at night and on sunny, hot days.
- Set thermostat as low as you can in winter and as high as you can in summer.
- Weather-strip and caulk doors, windows, light fixtures, and wall sockets.
- Keep heating and cooling vents free of obstructions.
- Keep fireplace damper closed when not in use.
- Use fans instead of, or along with, air conditioning.

Basement or utility room
- Use front-loading clothes washer. If possible run only full loads with warm or cold water.
- Hang clothes on racks for drying.
- Run only full loads in clothes dryer and use lower heat setting.
- Set water heater at 140° if dishwasher is used and 120° or lower if no dishwasher is used.
- Use water heater thermal blanket.
- Insulate exposed hot water pipes.
- Regularly clean or replace furnace filters.

Figure 16-9 Individuals Matter: Ways in which you can save energy where you live.

Why Are We Still Wasting So Much Energy?

With such an impressive array of benefits, why is there so little emphasis on improving energy efficiency? One reason is a glut of relatively low-cost fossil fuels. As long as energy remains artificially cheap, primarily because market prices do not include the harmful environmental and health costs of its production and use, people are more likely to waste it and less likely to invest in improving energy efficiency.

Another reason is that there are few large and long-lasting governmental tax breaks, rebates, low-interest, long-term loans, and other economic incentives for consumers and businesses to invest in improving energy efficiency. Also, the U.S. federal government has done a poor job of encouraging fuel efficiency in motor vehicles (Figure 16-5, right).

Some analysts are concerned about what they call a *rebound effect,* in which some people might start using more energy if they save money by improving energy efficiency. For example, a household that saves energy and money by adding more insulation and plugging air leaks might decide that it can raise the thermostat during winter or lower it during the summer. Or people driving an energy efficient car might decide that they can drive more. However, studies reveal that, overall, this rebound effect has been quite small.

HOW WOULD YOU VOTE? ☑

Should the country where you live greatly increase its emphasis on improving energy efficiency? Cast your vote online at **academic.cengage.com/biology/miller**.

We Can Use Renewable Energy in Place of Nonrenewable Energy Sources

One of nature's four **principles of sustainability** (see back cover) is to *rely mostly on solar energy.* We can get renewable solar energy directly from the sun or indirectly from moving water, wind, and biomass, none of which would exist without direct solar energy. Another form of renewable energy is geothermal energy from the earth's interior, which is widely used to provide heat and produce electricity in Iceland (Figure 16-1, **Core Case Study**) and other parts of the world. Studies show that with increased and consistent government backing, renewable energy could provide 20% of the world's electricity by 2025 and 50% by 2050.

Making a major shift toward a variety of locally available renewable energy resources over the next few decades would

- Result in a more decentralized and efficient energy economy that is less vulnerable to supply cutoffs from terrorist attacks and natural disasters such as hurricanes.

- Improve national security for many countries by reducing their need to import oil from the Middle East.

- Reduce trade deficits that grow when a country imports oil.

- Greatly reduce emissions of climate-changing greenhouse gases and other air pollutants.

- Create large numbers of jobs, including high-paying jobs for skilled workers. Already the production, installation, and maintenance of various types of renewable energy systems provide approximately 2.4 million jobs worldwide, according to a 2007 study by the Worldwatch Institute.

- Save consumers money.

If renewable energy is so great, why does it provide only 18% of the world's energy and 7% of the energy used in the United States? One reason is that, since 1950, government tax breaks, subsidies, and funding for research and development of renewable energy resources have been much lower than those for fossil fuels (especially oil) and nuclear power, although subsidies for renewables have increased in recent years. Another reason is that the prices we pay for fossil fuels and nuclear power do not include the environmental and health costs of producing and using them.

If these two economic handicaps—*inequitable subsidies* and *inaccurate pricing*—were eliminated, energy analysts say that many forms of renewable energy would be cheaper than fossil fuels or nuclear energy and would quickly take over the marketplace.

Some governments are increasing their use of renewable energy. The European Union got 10% of its electricity from renewable energy in 2007 and aims to get 34% of its electricity from renewable energy by 2020, mostly by relying more on wind power. In 2006, China got 8% of its total energy and 17% of its electricity from renewable energy. By contrast, the United States got 7% of its total energy and 9% of its electricity from renewable energy in 2006.

In 2005, China set a goal of getting 15% of its total energy and 21% of its electricity from renewable energy sources by 2020. By contrast, the U.S. Energy Information Agency projects that the United States will get only 10% of its total energy from renewable sources by 2030. But several studies show that with a crash program, the United States could get 20% of its total energy and 25% of its electricity from renewable sources by 2020.

Denmark now gets 20% of its electricity from wind and has plans to increase this to 50% by 2030. Brazil gets 40% of its automotive fuel from ethanol made from sugarcane residue, and could phase out its use of gasoline within a decade. And Iceland plans to get all of its energy from renewable resources by 2050 (**Core Case Study**).

Throughout the rest of this chapter, we explore these growing renewable energy options.

16-3 What Are the Advantages and Disadvantages of Solar Energy?

▶ **CONCEPT 16-3** Passive and active solar heating systems can heat water and buildings effectively, and the costs of using direct sunlight to produce high-temperature heat and electricity are coming down.

We Can Heat Buildings and Water with Solar Energy

Buildings and water can be heated by passive and active solar heating systems (Figure 16-10).

A **passive solar heating system** absorbs and stores heat from the sun directly within a well-insulated structure without the need for pumps or fans to distribute the heat (Figure 16-10, left). Energy-efficient windows or attached greenhouses face the sun to collect solar energy directly. Walls and floors of concrete, adobe, brick, stone, or salt-treated timber, and metal or plastic water tanks can store much of the collected solar energy as heat and release it slowly throughout the day and night. A small backup heating system such as a vented natural gas or propane heater may be used, but is not necessary in many climates. (See the Guest Essay by Nancy Wicks at CengageNOW.)

Using passive solar energy is not new. For thousands of years, people have intuitively followed this **principle of sustainability**. They have oriented their dwellings to take advantage of heat and light from the sun and used adobe and thick stonewalls that collect and store heat during the day and gradually release it at night. On a life cycle cost basis, good passive solar and superinsulated design is the cheapest way to heat a home or small building in sunny areas (see Case Study, p. 412).

An **active solar heating system** absorbs energy from the sun by pumping a heat-absorbing fluid (such as water or an antifreeze solution) through special collectors usually mounted on a roof or on special racks to face the sun (Figure 16-10, right). Some of the collected heat can be used directly. The rest can be stored in a large insulated container filled with gravel, water, clay, or a heat-absorbing chemical for release as needed.

Figure 16-11 (p. 412) lists the major advantages and disadvantages of using passive or active solar heating systems for heating buildings (**Concept 16-3**). They can be used to heat new homes in areas with adequate sunlight. (See Figures 11 and 13, p. S66 and p. S67, in Supplement 10 for maps of solar energy availability in the world and in North America.) But solar energy cannot be used to heat existing homes and buildings that are not oriented to receive sunlight or that are blocked from sunlight by other buildings or trees.

Active solar collectors using a fairly simple technology can also provide hot water. In China, some 2,000 companies have built and installed more inexpensive rooftop solar water heaters than have been installed in the rest of the world combined. They are used in homes, apartment buildings (Figure 16-12, p. 412), office buildings, schools, and hotels. A villager or apartment dweller can install such a rooftop system for as little as $200. Once the initial cost is paid, the hot water is essentially free.

PASSIVE

ACTIVE

Figure 16-10 Solutions: passive and active solar heating for a home.

TRADE-OFFS

Passive or Active Solar Heating

Advantages	Disadvantages
Energy is free	Need access to sun 60% of time
Net energy is moderate (active) to high (passive)	Sun can be blocked by trees and other structures
Quick installation	Environmental costs not included in market price
No CO_2 emissions	Need heat storage system
Very low air and water pollution	High cost (active)
Very low land disturbance (built into roof or windows)	Active system needs maintenance and repair
Moderate cost (passive)	Active collectors unattractive

Figure 16-11 Advantages and disadvantages of heating a house with passive or active solar energy (**Concept 16-3**). **Question:** Which single advantage and which single disadvantage do you think are the most important? Why?

Figure 16-12 Rooftop solar hot water heaters on apartment buildings in the Chinese city of Kunming in the Yunnan province. Such heaters are now required on all new buildings and their use is growing rapidly in urban and rural areas of China.

© U.S. National Renewable Energy Laboratory

In 2007, one in ten Chinese households used the sun to heat their water, and by 2030, half of all households in China may be doing so—an excellent example of applying the solar energy **principle of sustainability** (see back cover). Such systems are also widely used in Germany, Japan, Greece, Austria, and Turkey. And in Spain and Israel, all new buildings are required by law to have rooftop systems for heating water and space. Soon millions of households in developing countries such as India and Brazil may turn to this simple and inexpensive technology.

■ CASE STUDY

The Rocky Mountain Institute— A Solar-Powered Office and Home

In 1984, energy analyst Amory Lovins built a large, solar-heated, solar-powered, superinsulated, partially earth-sheltered home and office for the Rocky Mountain Institute in Snowmass, Colorado (USA) (Figure 16-13), an area with extremely cold winters.

This 372-square-meter (4,000-square-foot) office–home has no conventional heating system because of a combination of passive solar energy from the sun, heavy roof insulation, thick stonewalls, energy-efficient windows with the equivalent to 8–12 panes of glass, and a heat-waste recovery system. The structure gets 99% of its hot water, 95% of its daytime lighting, and 90% of its household electricity from the sun. Its heating bill, in an area with very severe winters, is less than $50 a year.

Each year, the home portion of the structure consumes a little more electricity than a single 100-watt incandescent lightbulb would consume if it were lit all year, due mostly to use of energy-efficient lights, refrigerators, and other electrical devices. The institute uses solar cells to generate its electricity, and the excess power not used by the home and office is sold to an electric utility. The savings from these energy-efficiency investments repaid their costs in only 10 months and have been making money ever since.

We Can Cool Buildings Naturally

Direct solar energy actually works against us when we want to keep a building cool, but we can use indirect solar energy (mainly wind) and other natural services to help cool buildings. For example, we can open windows to take advantage of breezes, and use fans to keep the air moving. And a living roof (Figure 16-7 and Photo 11 in the Detailed Contents) can make a huge difference in keeping a building cool.

Many technologies are available to help us make use of natural cooling services. For example, when there is no breeze, superinsulation and high-efficiency

windows help to keep hot air outside. Here are some other ways to keep cool:

- Block the high summer sun with window overhangs or awnings.

- Use a light-colored roof to reflect as much as 80% of the sun's heat (compared to only 8% for a dark-colored roof).

- Suspend reflective insulating foil in an attic to block heat from radiating down into a house.

- Place plastic *earth tubes* underground where the earth is cool year-round. In this inexpensive geothermal cooling system, a tiny fan can pipe cool and partially dehumidified air into a house. One of the authors (Miller) used them for 15 years in his passively heated and cooled office and home at a cost of about $1 per summer. Adding an air purification system is important for people who are allergic to pollen and molds, but this is also necessary with conventional cooling systems.

- Use geothermal heat pumps for cooling (and heating in winter)

We Can Use Sunlight to Produce High-Temperature Heat and Electricity

Solar thermal systems concentrate and transform energy from the sun into high-temperature thermal energy (heat), which can then be used directly or to heat water and produce steam to generate electricity. These systems are used mostly in desert areas with ample sunlight. Figure 16-14 summarizes some advantages and disadvantages of concentrating solar energy to produce high-temperature heat or electricity.

One type uses a *central receiver system,* such as that shown in the top drawing in Figure 16-14 and in Figure 16-15 (p. 414). In very large systems, the central receiver is called a *power tower.* Huge arrays of computer-controlled mirrors called *heliostats* track the sun and focus sunlight on this central heat collection tower. Such a plant is in operation in southern Spain (Figure 16-15), which has the world's largest concentrated solar power research center and is the second largest user of solar energy in Europe, after Germany.

In another type of system, sunlight is collected and focused on oil-filled pipes running through the middle of a large array of curved solar collectors (bottom drawing, Figure 16-14). This concentrated sunlight can generate temperatures high enough to produce steam for running turbines and generating electricity. Such a plant is operating in southern Spain, and several others have been operating in the California desert since the 1980s.

Most analysts do not expect widespread use of such technologies over the next few decades because of their

Figure 16-13 *Sustainable energy:* This facility at the Rocky Mountain Institute in the U.S. state of Colorado is a home and headquarters for a nonprofit center for the study of energy efficiency and sustainable use of energy and other resources. It is also an example of energy-efficient passive solar design.

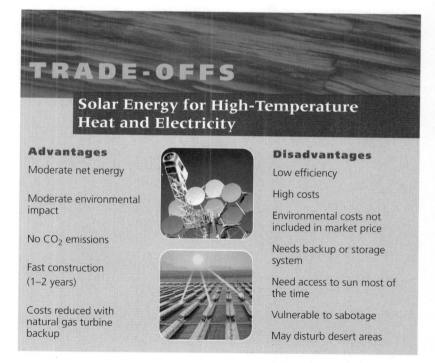

TRADE-OFFS

Solar Energy for High-Temperature Heat and Electricity

Advantages	Disadvantages
Moderate net energy	Low efficiency
Moderate environmental impact	High costs
No CO$_2$ emissions	Environmental costs not included in market price
Fast construction (1–2 years)	Needs backup or storage system
Costs reduced with natural gas turbine backup	Need access to sun most of the time
	Vulnerable to sabotage
	May disturb desert areas

Figure 16-14 Advantages and disadvantages of using solar energy to generate high-temperature heat and electricity (**Concept 16-3**). **Question:** Which single advantage and which single disadvantage do you think are the most important? Why?

Figure 16-15 Solutions: Commercial solar power tower plant operated by Solúcar, an Abengoa company, near the city of Seville in southern Spain. This plant, which began operating in 2007, uses an array of 624 mirrors to reflect and focus sunlight on a receiver atop a 35-story tower. The resulting concentrated heat converts water into steam that spins turbines and generates enough electricity to power 6,000 homes.

Courtesy of Abengoa

high costs, low net energy yields, and the limited number of suitable sites. But they can be useful in especially sunny desert areas (see Figures 11, 13, and 19 in Supplement 10, pp. S66, S67, and S70). The dream of tapping the vast input of solar energy in Africa's Sahara Desert to provide electricity for Europe may soon become a reality. Algeria has begun construction of a series of solar thermal power plants in the Sahara (to be backed up natural gas at night) and plans to send the electricity by cable to Europe. One advantage of such desert solar systems is that their electricity production peaks during the day when air conditioning and other electrical needs also peak.

THINKING ABOUT

Using Solar Energy to Produce High Temperature Heat

↩ CONCEPT LINKS

Use the first and second laws of thermodynamics (**Concepts 2-4A** and **2-4B**, p. 40) and the related concept of net energy (Science Focus, p. 374) to explain why generating high-temperature heat and electricity from direct solar energy is likely to be quite costly without significant government subsidies and tax breaks.

On a smaller scale, inexpensive *solar cookers* can focus and concentrate sunlight to cook food, especially in rural, sunny areas (Figure 16-16). A solar cooker can be built for $2–10 by fitting an insulated box big enough to hold three or four pots with a transparent, removable top. Solar cookers help to reduce deforestation from fuelwood harvesting, and they save time and labor needed to collect firewood (Figure 16-3, p. 135). They also reduce indoor air pollution and premature deaths from smoky fires.

Mark Edwards/Peter Arnold, Inc.

Figure 16-16 Solutions: woman in India uses a solar cooker to prepare a meal for her family.

We Can Use Solar Cells to Produce Electricity

Solar energy can be converted directly into electrical energy by **photovoltaic (PV) cells,** commonly called **solar cells** (Figure 16-17). Most solar cells are thin

Single solar cell

Boron-enriched silicon

Junction

Phosphorus-enriched silicon

Solar-cell roof

Roof options

Panels of solar cells

Solar shingles

Figure 16-17 Solutions: Photovoltaic (PV) or solar cells can provide electricity for a house or other building using solar-cell roof shingles, as shown in this house in Richmond Surrey, England. Solar-cell roof systems that look like a metal roof are also available. In addition, new thin-film solar cells can be applied to windows and outside walls.

wafers of purified silicon with trace amounts of metals that allow them to function as semiconductors to produce electricity. A typical solar cell has a thickness ranging from less than that of a human hair to a sheet of paper. When sunlight strikes these transparent cells, they emit electrons, and many cells wired together in a panel can produce electrical power.

The cells can be connected to existing electrical grid systems or to batteries that store the electrical energy until it is needed. Some countries require power companies to pay homeowners and businesses for any excess electrical energy that they produce and feed back into the grid system.

Solar cells have no moving parts, are safe and quiet, require little maintenance, produce no pollution or greenhouse gases during operation, and last as long as a conventional fossil fuel or nuclear power plant. The semiconductor material used in solar cells can be made into paper-thin rigid or flexible sheets that can be incorporated into traditional-looking roofing materials (Figure 16-17, right) and attached to walls, windows, and clothing.

Easily expandable banks of solar cells can be used in developing countries to provide electricity for many of the nearly 1.6 billion people in rural villages not connected to an electrical grid (Figure 16-18). Such systems provide lighting at a lower cost than do increasingly costly kerosene lamps, and they do so without CO_2 emissions. With financing from the World Bank, India is installing solar-cell systems in 38,000 villages that are located long distances from power grids.

One of the largest solar-cell power plants went on line in 2007 in Portugal. On a sunny day, it generates enough electricity to power 8,000 homes. There are other large systems in southern Spain and in the desert near Tucson, Arizona (USA) (Figure 16-19, p. 416). Other even larger systems are being built in South Korea and in Leipzig, Germany. One U.S. utility plans to install an array of solar cell systems atop the rooftops

Figure 16-18 Solutions: solar cells used to provide electricity for a remote village in Niger, Africa. **Question:** Do you think your government should provide aid so poor countries can obtain such solar cells? Explain.

Figure 16-19 Solutions: This solar-cell power plant in the U.S. state of Arizona near the city of Springerville has been in operation since 2000 and is the world's largest solar-cell power plant. Analysis shows that the plant, which is attached to the area's electrical grid, paid back the energy needed to build it in less than 3 years.

Tucson Electric Power Company

California headquarters that will generate enough electricity to power 1,000 homes and allow Google to save 30% of its current power use.

Electricity from solar cells in China is still in its infancy. But China plans to use solar cells and wind turbines to provide electricity for 29,000 rural villages and is building one of the world's largest solar-cell power stations. The Chinese city of Shanghai has plans to install 100,000 rooftop, solar-cell systems and to rapidly expand such systems to an increasing number of the city's 6 million rooftops. By 2007, China was the world's third largest producer of solar cells, after Germany and Japan, and it plans to be the world's largest producer by 2025. Figure 16-20 lists the advantages and disadvantages of using solar cells (**Concept 16-3**).

The high cost of using solar cells to produce electricity is the key problem. Table 16-1 compares the total costs of producing electricity from various renewable and nonrenewable sources. Note that when their generating costs and their estimated harmful environmental costs are combined, wind, geothermal, and hydropower are the three cheapest ways to produce electricity, and nuclear power, coal, and solar cells are the most expensive. Still, using solar cells is cheaper than building centralized power plants and electric transmission grids in areas of developing countries that currently have little or no electricity (Figure 16-18).

The current high cost of producing electricity from solar cells is expected to drop considerably because of increased mass production and new and more efficient designs. Advances in nanotechnology (Science Focus, p. 362) may lead to mass production of energy-efficient nanosolar cells. The possible result: a dirt-cheap, flex-

of commercial buildings in Southern California that would produce enough electricity to power 162,000 homes.

Such systems are also an option for private companies, which can sell excess electricity to public utilities. Google is building a huge solar cell project for its

Table 16-1

Total Costs of Electricity from Different Sources in 2004 (in U.S. cents per kilowatt-hour)			
Electricity Source	**Generating Costs**	**Environmental Costs**	**Total Costs**
Wind	4.7–6.3	0.1–0.3	4.8–6.6
Geothermal	4.8	1.0 (approximately)	5.8
Hydropower	4.9–8.5	0.3–1.1	5.2–9.6
Natural gas	5.2–6.5	1.1–4.5	6.3–11.0
Biomass	5.5–6.4	1.0–3.4	6.5–9.8
Nuclear*	5.9–12.0	0.2–0.7	6.1–12.7
Coal	4.5–5.4	3.0–17.0	7.5–22.4
Solar cells	12.4–26.0	0.7	13.1–26.7

*Plant only. Costs are much higher if entire nuclear fuel cycle is included.
Source: Data from U.S. Energy Information Administration and a variety of sources compiled by the Worldwatch Institute.

ible, solar-cell material 50 times thinner than today's solar panels. **GREEN CAREER:** Solar cell technology

─ RESEARCH FRONTIER ─

Developing more efficient and affordable solar cells. See **academic.cengage.com/biology/miller**.

The Solar Power Industry Is Expanding Rapidly

Currently, solar cells supply less than 0.2% of the world's electricity, although annual production of solar cells has been growing rapidly (see Figure 12, p. S67, in Supplement 10). Energy analysts say that with increased research and development, plus much greater and more consistent government tax breaks and other subsidies, solar cells could provide 16% of the world's electricity by 2040.

In 2007, Nanosolar, a company in California's Silicon Valley, began mass-producing wafer-thin solar cells printed like newspaper ink on flexible and light aluminum foil. According to Nanosolar, they can be produced at half the cost of producing more rigid conventional solar cells. And further cost reductions from mass production should allow these cells to produce electricity at a price comparable to that of coal-generated electricity. If this happens on a large scale, solar cells could become the world's most widely used way to produce electricity within 2 decades.

According to an International Energy Agency (IEA), installing large-scale solar-cell systems on just 4% of the world's desert land could generate enough electricity to meet the entire world's annual power demand. If the costs come down and these projections are correct, the production, sale, and installation of solar cells could become one of the world's largest and fastest-growing businesses and would represent a major step on the path to sustainability.

Germany, where the weather is often cloudy and rainy, seems an unlikely country to be the world's leading user and producer of solar cells. Nevertheless, more than 300,000 solar-cell systems in Germany supply 55% of the electricity produced by all the solar cells in the world. One reason for this is that the German government has spent $1.9 billion on solar-cell research and development since the 1990s. The government also passed a law that requires utility companies to buy excess electricity produced by solar cells in homes and businesses at almost four times the market prices. As a result, solar panels are sources of income for homes and businesses, some of which have generated solar-cell electricity for 20 years.

The government's investment has paid off. The country's revenues from selling solar cells and wind turbines throughout the world have grown. And

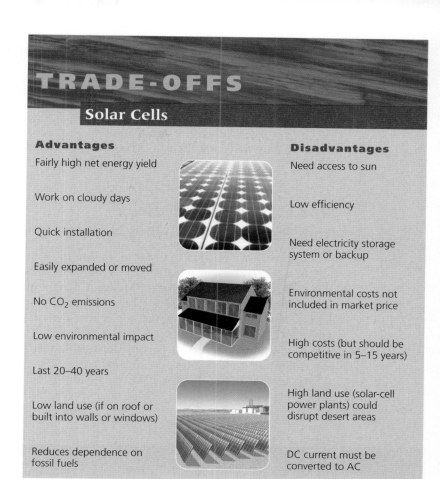

Figure 16-20 Advantages and disadvantages of using solar cells to produce electricity (**Concept 16-3**). **Question:** Which single advantage and which single disadvantage do you think are the most important? Why?

TRADE-OFFS

Solar Cells

Advantages	Disadvantages
Fairly high net energy yield	Need access to sun
Work on cloudy days	Low efficiency
Quick installation	Need electricity storage system or backup
Easily expanded or moved	
No CO_2 emissions	Environmental costs not included in market price
Low environmental impact	High costs (but should be competitive in 5–15 years)
Last 20–40 years	
Low land use (if on roof or built into walls or windows)	High land use (solar-cell power plants) could disrupt desert areas
Reduces dependence on fossil fuels	DC current must be converted to AC

German cities and towns have benefited from more jobs and increased income.

General Electric, the world's largest power generation company, has gotten into the solar-cell business and may help the United States to become a major player in the solar-cell market. The U.S. state of New Mexico will have the world's largest solar-cell power plant, which when completed, will have 65 times the electrical output of the solar-cell plant in Arizona (Figure 16-19)—currently the largest. It will generate enough electricity to power 240,000 homes.

In 2007, Jim Lyons, chief engineer for General Electric, projected that solar cells will be the world's number one source of electricity by the end of this century. If that happens, it will represent a huge global application of the solar energy **principle of sustainability** (see back cover).

─ HOW WOULD YOU VOTE? ─

Should the country where you live greatly increase its dependence on solar cells for producing electricity? Cast your vote online at **academic.cengage.com/biology/miller**.

16-4 What Are the Advantages and Disadvantages of Producing Electricity from the Water Cycle?

▶ **CONCEPT 16-4** Water flowing over dams, tidal flows, and ocean waves can be used to generate electricity, but environmental concerns and limited availability of suitable sites may limit the use of these energy resources.

We Can Produce Electricity from Falling and Flowing Water

Water flowing from higher to lower elevations in rivers and streams can be controlled by dams and reservoirs and used to produce electricity. This form of energy is called *hydropower*. It is an indirect form of solar energy because it is based on the evaporation of water, which is deposited at higher elevations where it can flow to lower elevations in rivers as part of the earth's solar-powered water cycle (Figure 3-17, p. 66).

The most common approach to harnessing hydropower is to build a high dam across a large river to create a reservoir. Some of the water stored in the reservoir is allowed to flow through huge pipes at controlled rates, spinning turbines and producing electricity (Figure 13-12, p. 325, and Figure 13-15, p. 327).

Hydropower is the world's leading renewable energy source used to produce electricity, and it is the third cheapest way to produce electricity when environmental costs are included (Table 16-1). In order, the world's top five producers of hydropower are Canada, China, Brazil, the United States, and Russia. In 2006, hydropower supplied about 16% of the world's electricity, including 99% of Norway's, 75% of New Zealand's, and 21% of China's electricity. It supplied 7% of electricity used in the United States, (but about 50% of that used on the West Coast).

Hydropower is also widely used to produce electricity in Iceland and is an important part of that country's plan to develop the world's first renewable energy economy (**Core Case Study**). Hydroelectric power production in Iceland will increase as global warming slowly melts its glaciers and increases the amount of water flowing down its rivers.

According to the United Nations, only about 13% of the world's technically exploitable potential for hydropower has been developed. Much of this untapped potential is in China (which plans to more than double its hydropower output by 2020), India, South America, Central Africa, and parts of the former Soviet Union.

Some analysts expect the contribution of large-scale hydropower plants to fall slowly over the next several decades as many existing reservoir systems fill with silt and become useless faster than new plants are built. Also, there is growing concern over emissions of methane, a potent greenhouse gas, from the anaerobic decomposition of submerged vegetation in reservoirs.

Another problem is that global warming is projected to greatly increase prolonged drought in many areas of the world throughout this century. This, coupled with the fact that less water will be stored as ice and snow in mountainous areas, will reduce the flow of water in rivers that have been dammed to provide flood control, irrigation water, and electric power (Figure 13-15, p. 327). As the water flow drops in such rivers, electric power production from their hydropower facilities will drop.

Figure 16-21 lists the advantages and disadvantages of using large-scale hydropower plants to produce electricity (**Concept 16-4**).

TRADE-OFFS

Large-Scale Hydropower

Advantages	Disadvantages
Moderate to high net energy	High construction costs
High efficiency (80%)	High environmental impact from flooding land to form a reservoir
Large untapped potential	Environmental costs not included in market price
Low-cost electricity	High CO_2 emissions from rapid biomass decay in shallow tropical reservoirs
Long life span	
No CO_2 emissions during operation in temperate areas	Danger of collapse
Can provide flood control below dam	Uproots people
Provides irrigation water	Decreases fish harvest below dam
Reservoir useful for fishing and recreation	Decreases flow of natural fertilizer (silt) to land below dam

Figure 16-21 Advantages and disadvantages of using large dams and reservoirs to produce electricity (**Concept 16-4**). **Question:** Which single advantage and which single disadvantage do you think are the most important? Why?

The use of *micro-hydropower generators* may become an increasingly important way to produce electricity. These floating turbines—about the size of an overnight suitcase—use the power of flowing water to turn a rotor with blades that feed generators to produce electric current. Such a generator can be placed in any stream or river without altering its course to provide electricity at a very low cost with almost zero environmental impact.

Tides and Waves Can Be Used to Produce Electricity

We can also produce electricity from flowing water by tapping into the energy from *ocean tides* and *waves*. In some coastal bays and estuaries, water levels can rise or fall by 6 meters (20 feet) or more between daily high and low tides. Dams have been built across the mouths of some bays and estuaries to capture the energy in these flows for hydropower. However, due to limited satisfactory locations and the high costs of such projects, only two large tidal energy dams are currently operating, one at La Rance on the northern coast of

France, and the other in Nova Scotia's Bay of Fundy. South Korea, North Korea, China, New Zealand, India, Great Britain, and Russia plan to build tidal flow plants at suitable sites. But globally, sites with large enough daily tidal flows are limited.

In 2006, Verdant Power began installing six underwater turbines to tap the tidal flow of the East River near New York City. The turbines resemble wind turbines as they swivel to face the incoming and outgoing tides. If the project is successful, as many as 300 turbines may be used in the river. Such a system powers a town in Norway. However, such systems are limited.

For decades, scientists and engineers have been trying to produce electricity by tapping wave energy along seacoasts where there are almost continuous waves. Large snake-like chains of floating steel tubes are being installed off the coast of Portugal. Their up and down motion with the wave action generates electricity. In 2008, the system generated enough electricity to power 15,000 homes.

Wave power facilities are being developed in northern California and in some coastal areas in Ireland and near Cornwall in Great Britain. However, most analysts expect tidal and wave power sources to make only a small contribution to world electricity supplies, primarily because there are few suitable sites, the costs are high, and the equipment is vulnerable to corrosion and storm damage (**Concept 16-4**). However, improved technology could greatly increase the production of electricity from waves sometime during this century.

16-5 What Are the Advantages and Disadvantages of Producing Electricity from Wind?

▶ **CONCEPT 16-5** When environmental costs of energy resources are included in market prices, wind energy is the least expensive and least polluting way to produce electricity.

Using Wind to Produce Electricity Is an Important Step toward Sustainability

The difference in solar heating of the earth between the equator and the poles together with the earth's rotation create flows of air called *wind* (Figure 7-3, p. 142). This indirect form of solar energy can be captured by wind turbines and converted into electrical energy (Figure 16-22, p. 420, and the front cover of this book). Because today's wind turbines can be as tall as 30 stories and can have blades as long as a jumbo jet plane, they can tap into the stronger, more reliable, and less turbulent winds found at higher altitudes. This

allows a single modern wind turbine to generate about 20 times more electricity than a turbine installed in the 1980s.

Since 1990, wind power has been the world's second fastest-growing source of energy (see Figure 15, p. S68, in Supplement 10); solar-cell production is growing the fastest (Figure 12, p. S67, in Supplement 10). In Germany, 20,000 land-based wind turbines generate 5% of the country's electricity. Germany's goal is to get 30% of its electricity from wind by 2030, mostly by developing offshore wind farms.

In 2004, Stanford University engineers Cristina L. Archer and Mark Z. Jacobson mapped the global potential for wind energy (see Figure 14, p. S68, in Supplement 10). Their data indicate that capturing only

Wind turbine

Gearbox

Electrical generator

Power cable

Wind farm

Wind farm (offshore)

Figure 16-22 Solutions: A single wind turbine (left) can be used to produce electricity. But increasingly, they are being used in interconnected arrays of ten to hundreds of turbines. These *wind farms* or *wind parks* can be located on land (middle and front cover of this book) or offshore (right). Land lying under these turbines can still be used to grow crops and raise cattle. **Question:** Would you object to having a wind farm located near where you live? Why or why not?

one-fifth of the wind energy at the world's best sites could generate more than seven times the amount of electricity currently used in the world and thus help to phase out energy-wasting coal-burning and nuclear power plants (p. 402) during this century. The key is to develop modern power grids to distribute the electricity to users.

Analysts expect increasing use of offshore wind farms because wind speeds over water are often stronger and steadier than those over land and some countries such as Germany and Great Britain are running out of acceptable land sites for wind turbines. Offshore installation costs are higher and some coastal towns and cities oppose such installations because of their visual pollution. But this problem may be solved by the development of huge turbines that could be anchored on floating platforms (not unlike large floating oil rigs, Figure 15-1, p. 370) far enough from shore to be invisible to coastal residents and to take advantage of stronger and more constant offshore winds.

Figure 16, p. S69, and Figure 19, p. S70, in Supplement 10 show the areas where wind power potential is highest in the United States and in the world. Wind farms operating in 36 states produce almost 1% of the country's electricity, with Texas and California leading the way.

This percentage could well be expanded dramatically. The DOE calls the four Great Plains states of North Dakota, South Dakota, Kansas, and Texas the "Saudi Arabia of wind power." Scientists calculate that wind farms in favorable sites in these four states alone could produce nearly three times the total amount of electricity currently generated from all U.S. power plants. These wind resources are in some of the country's least densely populated areas. In addition, a 2005 study by the U.S. Department of Energy estimated that offshore wind re-

sources within 9–18 kilometers (5–10 nautical miles) of the U.S. coastline could produce enough electricity to meet all of the country's current electricity needs.

Producing Electricity from Wind Energy Is a Rapidly Growing Global Industry

Europe is leading the world into the *age of wind energy*, producing about three-fourths of the world's wind-generated power. European companies—mostly in Denmark, Germany, and Spain—manufacture 80% of the wind turbines sold in the global marketplace. They are aided by strong and consistent government subsidies, tax breaks, and low-cost loans.

In order, the countries with the highest total installed wind power capacity in 2007 were Germany, the United States, Spain, India, and Denmark. Wind power capacity is also increasing dramatically in Canada, France, Portugal, and Australia. China plans to continue its rapid expansion of wind power and to sell the turbines it manufactures at prices lower than those charged by most other countries in the global marketplace. By mid-2008, the amount of electricity produced globally by wind turbines was enough to meet the electricity needs of 150 million people, according to the Earth Policy Institute.

A 2007 National Academy of Sciences study estimated that, with enough government support, the United States could become the world's largest producer and consumer of wind power over the next 20 years. Wind turbines could be mass-produced on assembly lines using some of the idled factory capacity of the U.S. steel and automobile industries. This would create many jobs for laid-off skilled workers and bring economic recovery to such areas.

Wind Energy Is Booming but Still Faces Challenges

Wind is abundant, widely distributed, and cannot run out. A wind farm can be built fairly quickly and expanded as needed, can be controlled by a single laptop computer, and at good sites, can produce electricity at costs that are competitive with those of coal, natural gas, and hydropower (Table 16-1). Within a few years, wind is expected to be the cheapest way to produce electricity. If the environmental costs of various energy resources are included, wind energy is already the cheapest and least polluting way to produce electricity (Table 16-1) (**Concept 16-5**).

Like any energy source, wind power has some drawbacks. Areas with the greatest wind power potential are often sparsely populated, and located far from energy-thirsty cities. Thus, to take advantage of the enormous potential for electricity from wind energy, countries such as the United States will have to invest in a long overdue upgrading and expansion of their outdated electrical grid systems.

Another problem is that winds can die down and thus require a backup source of power, such as natural gas, for generating electricity. Scientists are also working on ways to store wind energy. One way is to store it in the batteries of a nationwide fleet of plug-in hybrid-electric vehicles (Figure 16-6, right). They would be recharged, mostly at night by cheaper off-peak, wind-generated electricity. Electricity produced by wind can also be passed through water and used to produce hydrogen fuel, which could be thought of as "stored" wind power.

In 2007, a group of U.S. utilities announced a plan to use electricity from wind to power compressors that will pump pressurized air deep underground into aquifers in the state of Iowa. Compressed air could also be stored in underground caverns and abandoned natural gas wells. The wind energy stored in the compressed air could then be released slowly to spin turbines and generate electricity when wind power is not available. This process is being used in Germany.

Studies indicate that wind turbines kill as many as 40,000 birds and bats each year in the United States. Most wind turbines involved in these deaths were built 20 years ago from now outdated designs, and some were built in bird migration corridors. This problem is being solved and is far from being the highest cause of death for birds. Each year, according to Defenders of Wildlife, glass windows, buildings, and electrical transmission towers in the United States kill more than 1 billion birds; electric transmission lines, up to 175 million; housecats and feral cats, 100 million; hunters, more than 100 million; and cars and trucks, 50–100 million.

Most studies show that as long as wind farms are not located along bird migration routes, birds fly around them. Wind power developers now make sophisticated studies of bird migration paths in order to avoid them when building wind farms. Newer turbines also reduce this problem by using slower blade rotation speeds and by not providing places for birds to perch or nest. According to the Audubon Society's vice president Betsy Loyless, "If we don't find ways to reduce global warming pollution, far more birds and people will be threatened by climate change than by wind turbines."

Some people in populated areas and in coastal areas oppose wind farms as being unsightly and noisy—a "not in my backyard (NIMBY)" attitude. But in windy parts of the U.S. Midwest, many farmers and ranchers have a "put it in my backyard (PIMBY)" attitude, and some have become wind developers themselves. A single wind turbine on 0.1 hectare (0.25 acre) of land can produce about $300,000 worth of electricity a year. Farmers, without having to put up any money, typically receive $3,000–10,000 a year in royalties for each turbine erected on a small plot of their land. And they can still use that land for growing crops or grazing cattle. Many no longer do this, because they can make much more money leasing their land for wind turbines.

Figure 16-23 lists advantages and disadvantages of using wind to produce electricity. According to energy

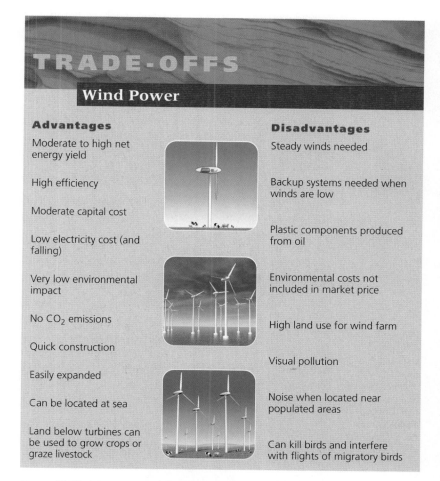

TRADE-OFFS

Wind Power

Advantages	Disadvantages
Moderate to high net energy yield	Steady winds needed
High efficiency	Backup systems needed when winds are low
Moderate capital cost	Plastic components produced from oil
Low electricity cost (and falling)	Environmental costs not included in market price
Very low environmental impact	High land use for wind farm
No CO_2 emissions	
Quick construction	Visual pollution
Easily expanded	
Can be located at sea	Noise when located near populated areas
Land below turbines can be used to grow crops or graze livestock	Can kill birds and interfere with flights of migratory birds

Figure 16-23 Advantages and disadvantages of using wind to produce electricity (**Concepts 16-5**). With sufficient and consistent incentives, wind power could supply more than 10% of the world's electricity and 10–25% of the electricity used in the United States by 2020. **Question:** Which single advantage and which single disadvantage do you think are the most important? Why?

analysts, wind power has more benefits and fewer serious drawbacks than any other energy resource, except for energy efficiency.

16-6 What Are the Advantages and Disadvantages of Biomass as an Energy Source?

▶ **CONCEPT 16-6A** Solid biomass is a renewable resource, but burning it faster than it is replenished produces a net gain in atmospheric greenhouse gases, and creating biomass plantations can degrade soil and biodiversity.

▶ **CONCEPT 16-6B** Liquid biofuels derived from biomass can be used in place of gasoline and diesel fuels, but creating biofuel plantations could degrade soil and biodiversity and increase food prices and greenhouse gas emissions.

We Can Get Energy by Burning Solid Biomass

Biomass consists of plant materials (such as wood and agricultural waste) and animal wastes that can be burned directly as a solid fuel or converted into gaseous or liquid **biofuels.** Biomass is an indirect form of solar energy because it consists of combustible organic (carbon-containing) compounds produced by photosynthesis.

Solid biomass is burned mostly for heating and cooking, but also for industrial processes and for generating electricity. Wood, wood wastes, charcoal (made from wood), animal manure, and other forms of solid biomass used for heating and cooking, supply 10% of the world's energy, 35% of the energy used in developing countries, and 95% of the energy needs in the poorest countries.

In urban areas, wood chips and urban waste can be used to produce both electricity and heat for *district heating systems*. Such systems supply nearly half of the heat for residential and commercial buildings in Sweden. A similar system supplies heat for about 80% of the downtown area of the U.S. city of St. Paul, Minnesota.

But wood is a renewable fuel only as long if it is not harvested faster than it is replenished. The problem is, about 2.7 billion people in 77 developing countries face a *fuelwood crisis* and often are forced to meet their fuel needs by harvesting wood faster than it can be replenished.

One way to produce solid biomass fuel is to plant fast-growing trees (such as cottonwoods, poplars, and sycamores), shrubs, perennial grasses (such as switchgrass and miscanthus), and water hyacinths in *biomass plantations.* But repeated cycles of growing and harvesting these plantations can deplete the soil of key nutrients. And clearing forests and grasslands for such plantations destroys or degrades biodiversity (**Concept 16-6A**). In agricultural areas, *crop residues* (such as sugarcane residues, rice husks, cotton stalks, and coconut shells) and *animal manure* can be collected and burned or converted into gaseous or liquid biofuels.

Figure 16-24 lists the general advantages and disadvantages of burning solid biomass as a fuel (**Con-**

Figure 16-24 General advantages and disadvantages of burning solid biomass as a fuel (**Concept 16-6A**). **Question:** Which single advantage and which single disadvantage do you think are the most important? Why?

TRADE-OFFS

Solid Biomass

Advantages	Disadvantages
Large potential supply in some areas	Nonrenewable if harvested unsustainably
Moderate costs	Moderate to high environmental impact
No net CO_2 increase if harvested, burned, and replanted sustainably	Environmental costs not included in market price
	Increases CO_2 emissions if harvested and burned unsustainably
Plantation can be located on semiarid land not needed for crops	Low photosynthetic efficiency
	Soil erosion, water pollution, and loss of wildlife habitat
Plantation can help restore degraded lands	Plantations could compete with cropland
Can make use of agricultural, timber, and urban wastes	Often burned in inefficient and polluting open fires and stoves

cept 16-6A). One problem is that clearing forests reduces the uptake of CO_2, and burning biomass produces CO_2. However, if the rate of use of biomass does not exceed the rate at which it is replenished by new plant growth (which takes up CO_2), there is no net increase in CO_2 emissions, but often this is a big if.

We Can Convert Plants and Plant Wastes to Liquid Biofuels

Liquid biofuels such as *biodiesel* and *ethanol* (ethyl alcohol), which are produced from plants and plant wastes, can be used in place of petroleum-based diesel fuel and gasoline (see Case Study at right and second Case Study, p. 424). The biggest producers of biofuels—Brazil, the United States, the European Union, and China—plan to double their production by 2020. Brazil already runs 45% of its cars on ethanol, and within a decade, could run all of its vehicles on this biofuel and eliminate its oil imports. U.S. government agencies estimate that biodiesel and ethanol could fuel 25–50% of U.S. motor vehicles by 2030.

Biofuels have some major advantages over gasoline and diesel fuel produced from oil. *First,* while oil resources are concentrated in a small number of countries, biofuel crops can be grown almost anywhere. This means that most countries can improve their energy and economic security by reducing their dependence on imported oil.

Second, if these crops are not used faster than they are replenished by new plant growth, there is no net increase in CO_2 emissions, unless existing grasslands or forests are cleared to plant biofuel crops. *Third,* biofuels are available now, are easy to store and transport, can be distributed through existing fuel networks, and can be used in vehicles at little or no additional cost.

However, in a 2007 U.N. report on bioenergy and in another study by R. Zahn and his colleagues, scientists warned that the benefits of biofuels could be outweighed by resulting problems. These reports noted that large-scale biofuel-crop farming could do the following: decrease biodiversity by clearing more natural forests and grasslands; increase soil degradation, erosion, and nutrient leaching; push small farmers off their land; and raise food prices (**Concept 16-6B**). A number of ecologists warn that crops used to produce biofuels should be grown only on land that is already degraded. They argue against clearing existing forests and grasslands and filling in wetlands that support biodiversity and store enormous amounts of carbon.

A 2007 paper published by Nobel Prize–winning chemist Paul Crutzen warned that intensive farming of crops to make biofuels could also speed up global warming by producing more greenhouse gases than are produced by burning conventional diesel fuel and gasoline. This would happen if nitrogen fertilizers were used to grow such crops. Such fertilizers, when applied to the soil, release large amounts of nitrous oxide (N_2O), a greenhouse gas 300 times more potent per molecule than CO_2.

■ CASE STUDY

Is Biodiesel the Answer?

Biodiesel is produced from vegetable oil extracted from soybeans, rapeseeds, sunflowers, oil palms, and jatropha shrubs, and fats such as used vegetable oils from restaurants. European Union countries (primarily Germany, France, and Italy) produce about 95% of the world's biodiesel, mostly from rapeseeds and sunflower seeds, and these countries hope to get 20% of their diesel fuel from this source by 2020. In Europe, where government taxes on vehicle fuel are high, almost half of all cars run on conventional diesel or biodiesel, mostly because they are as much as 40% more efficient than conventional gasoline engines.

In the United States, biodiesel production is growing rapidly, aided by government subsidies. Biodiesel provides only 1% of U.S. diesel consumption and the Department of Energy estimates that biodiesel at best could supply only 10% of the country's diesel fuel needs. This is because soybean and canola crops grown in the United States for biodiesel production require huge areas of land and have low yields. And using industrialized agriculture to produce these crops results in topsoil loss, fertilizer runoff, and increased emissions of the greenhouse gas nitrous oxide.

Brazil, Malaysia, and Indonesia produce biodiesel from palm oil extracted from large oil palm plantations, and they export much of it to Europe. The oil yield for biodiesel from oil palm is five times that from rapeseeds used in Europe and eight to nine times that from soybeans used in the United States. But increased burning and clearing of tropical forests to plant oil palm plantations in these countries poses a serious threat to their biodiversity. This also increases net emissions of CO_2 by replacing rain forests that store lots of carbon with crops that store much less carbon, according to a 2006 study by scientist Alexander Farrell at the University of California, Berkeley.

Another promising source of biodiesel is the *jatropha shrub,* a plant found in tropical areas of Africa, India, and Brazil. The plentiful oil in its golf-ball size fruit can be burned without being refined, and the plant grows in hot, dry, tropical areas. Jatropha crops are unlikely to threaten rain forests, displace food crops, or require extensive use of nitrogen fertilizers. In its natural state,

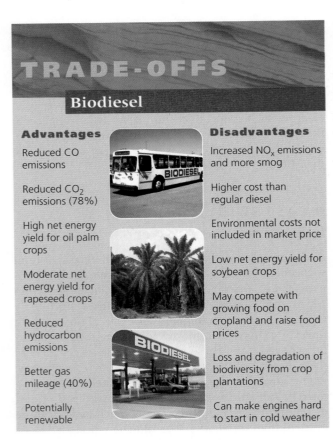

TRADE-OFFS

Biodiesel

Advantages	Disadvantages
Reduced CO emissions	Increased NO_x emissions and more smog
Reduced CO_2 emissions (78%)	Higher cost than regular diesel
High net energy yield for oil palm crops	Environmental costs not included in market price
Moderate net energy yield for rapeseed crops	Low net energy yield for soybean crops
Reduced hydrocarbon emissions	May compete with growing food on cropland and raise food prices
Better gas mileage (40%)	Loss and degradation of biodiversity from crop plantations
Potentially renewable	Can make engines hard to start in cold weather

Figure 16-25 General advantages and disadvantages of using biodiesel as a vehicle fuel, compared to gasoline. **Question:** Which single advantage and which single disadvantage do you think are the most important? Why?

the jatropha shrub needs little water and no fertilizer, although this is not the case if it is grown as a monoculture crop in large plantations. One problem is that the jatropha is an invasive species that could threaten native plants in some areas.

Some scientists are looking for ways to produce biodiesel from various types of oil-rich algae. Algae grow rapidly at any time of the year and can be cultivated in seawater or wastewater ponds. The algae remove CO_2 from the atmosphere and convert it to oil, proteins, and other useful products and require much less land than growing crops requires. The challenge is to cut the very high cost of producing oil by this method.

Figure 16-25 lists the advantages and disadvantages of using biodiesel as a vehicle fuel, compared to gasoline.

■ CASE STUDY

Is Ethanol the Answer?

Ethanol can be made through the fermentation and distillation of sugars in plants such as sugarcane, corn, and switchgrass, and from agricultural, forestry, and municipal wastes. This process involves converting starch in plant material to simple sugars that are fer-

mented by yeast into a slurry that is distilled to remove its ethanol.

Gasoline mixed with 10–23% pure ethanol makes *gasohol,* which can be burned in conventional gasoline engines. Pure ethanol or E85 (a mixture of 85% ethanol and 15% gasoline) can be burned in *flexible-fuel cars,* which have engines designed to run on a variety of fuels.

Brazil, the Saudi Arabia of sugarcane, is the world's second largest ethanol producer after the United States. When burned, ethanol made from *bagasse,* a residue produced when sugarcane is crushed, yields 8 times the amount of energy used to produce it. That compares with a net energy yield of 4.1 for gasoline (Figure 15-A, p. 374). About 45% of Brazil's motor vehicles run on ethanol or ethanol-gasoline mixtures produced from a residue of sugar cane grown on only 1% of the country's arable land.

Since the 1970s, Brazil has saved almost $50 billion in imported oil costs—nearly ten times the government's investment in ethanol production subsidies, which are no longer needed. In addition, ethanol production has created about 1 million rural jobs. Within a decade, Brazil could expand its sugarcane production, eliminate all oil imports, and greatly increase ethanol exports to other countries.

Brazil plans to greatly expand its production of sugarcane to produce ethanol and to grow more soybeans to produce biodiesel and cattle feed. However, this could threaten some of the country's biodiversity. To do this, Brazil plans to clear larger areas of its rapidly disappearing wooded savanna Cerrado region—one of the world's biodiversity hotspots (Figure 10-26, p 241). This area—three times the size of the U.S. state of Texas—is the world's most biodiverse savanna, with 10,000 plant species, nearly half of them found nowhere else.

In the United States, most ethanol is made from corn. U.S. farmers profit from growing corn to produce ethanol because they receive generous government subsidies as part of the nation's energy policy. But studies indicate that using fossil fuel–dependent industrialized agriculture to grow corn and then using more fossil fuel to convert the corn to ethanol provides a net energy yield of only about 1.1–1.5 units of energy per unit of fossil fuel input (Figure 15-A, p. 374).

A growing number of analysts warn that producing ethanol from corn will not significantly reduce the country's oil imports or help to slow global warming. According to environmental economist Stephen Polansky, processing all of the corn grown in the United States into ethanol each year, with high government subsidies, would cover only about 12% (or about 30 days worth) of the country's current demand for gasoline, at an extremely high cost to taxpayers.

A 2008 study by Tim Searchinger at Princeton University and other researchers estimated that clearing and planting grasslands and forests to grow corn for producing ethanol would increase CO_2 emissions by 93% over those from burning conventional gasoline over a

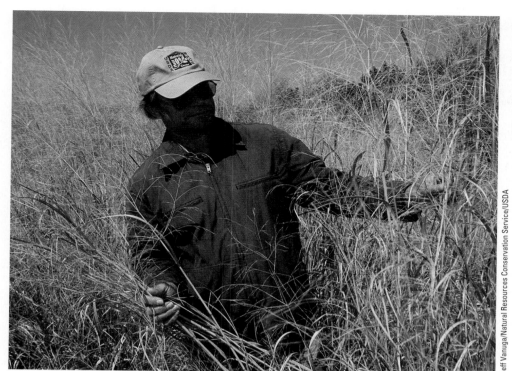

Figure 16-26 Natural capital: the cellulose in this rapidly growing switchgrass in Manhattan, Kansas (USA) can be converted into ethanol. This perennial plant can also help to reduce global warming by removing carbon dioxide from the atmosphere and storing it as organic compounds in the soil.

Jeff Vanuga/Natural Resources Conservation Service/USDA

30-year period. And extensive use of nitrogen fertilizers to grow corn releases large amounts of the greenhouse gas nitrous oxide into the atmosphere. If these estimates are correct, switching from gasoline to biofuels such as ethanol produced from corn will increase global warming and the resulting climate change.

Ethanol production also requires large amounts of water and releases about the same amount of carbon dioxide and much higher levels of nitrous oxide than gasoline production releases, and it offers little if any reduction in air pollution. In addition, fewer than 2% of U.S. gas stations offer ethanol, and most U.S. vehicles do not have flex-fuel engines that can burn ethanol.

Driven by U.S. ethanol production, the price of corn has risen sharply, as have the prices of corn-based foods—including bread, pasta, tortilla flour, poultry, beef, pork, and dairy products. This in turn is increasing the number of hungry and malnourished people who can no longer afford to buy enough food, and this has led to social unrest. In 2007, there were protests in Mexico because of an almost 60% increase in the price of corn meal, followed by pasta protests in Italy. In 2008, environmental expert Lester Brown estimated that filling a 95-liter (25-gallon) tank of an SUV with ethanol would use enough corn grain to feed the average person for a year.

An alternative to corn ethanol is *cellulosic ethanol,* which is produced from inedible cellulose that makes up most of the biomass of plants (see *The Habitable Planet,* Video 10, at **www.learner.org/resources/series209 .html**). In this process, acids or other processes are used to isolate cellulose and lignin from material. Then various enzymes convert the cellulose to a variety of sugars that can be fermented and distilled to produce ethanol. A possible plant that could be used for cellulosic ethanol production is *switchgrass* (Figure 16-26), a tall perennial grass native to North American prairies. It grows faster than corn, is disease resistant and drought tolerant, and can be grown on land unfit for crops without the use of nitrogen fertilizers. According to a 2008 article by U.S. Department of Agriculture scientist Ken Vogel and his colleagues, using switchgrass to produce ethanol yields about 5.4 times more energy than it takes to grow it—a yield much greater than the 1.1–1.5 net energy yield for corn.

While most corn-based ethanol producers burn fossil fuels to provide heat for fermentation, cellulosic ethanol producers can reduce such energy inputs by burning plant wastes. This and other aspects of cellulosic ethanol production and use could reduce net greenhouse gas emissions below current levels.

However, researchers have also found that large areas of land would be needed to produce enough cellulosic ethanol to make a significant dent in gasoline consumption and that clearing and planting large areas of land with switchgrass would likely increase greenhouse gas emissions. Thus scientists differ on how much of a net reduction in greenhouse gas emissions would result from extensive cellulosic ethanol production.

Other sources for producing cellulosic ethanol include crop residues, such as leaves and stalks, and municipal wastes such as sawdust, cardboard, waste paper, and sewage. But cellulose has properties that make it difficult to break down, and affordable chemical processes for converting cellulosic material to ethanol are still being developed (see *The Habitable Planet,* Video 10,

Ethanol Fuel

Advantages	Disadvantages
High octane	Lower driving range
Some reduction in CO_2 emissions (sugarcane bagasse)	Low net energy yield (corn)
	Higher CO_2 emissions (corn)
High net energy yield (bagasse and switchgrass)	Much higher cost
	Environmental costs not included in market price
Reduced CO emissions	May compete with growing food and raise food prices
	Higher NO_x emissions and more smog
Can be sold as E85 or pure ethanol	Corrosive
	Can make engines hard to start in cold weather
Potentially renewable	

Figure 16-27 General advantages and disadvantages of using ethanol as a vehicle fuel, compared to gasoline. (**Concept 16-6B**). **Question:** Which single advantage and which single disadvantage do you think are the most important? Why?

at **www.learner.org/resources/series209.html**) and are probably at least a decade away.

Figure 16-27 lists the advantages and disadvantages of using ethanol as a vehicle fuel, compared to gasoline.

Some scientists are studying the genetic makeup of bacteria living in the guts of termites that use enzymes to break down the cellulose in wood for their food. And California-based Amyris Biotechnologies and several other companies are carrying out research to bypass some of the problems associated with using plants to produce biofuels such as ethanol and biodiesel. These scientists are trying to use genetic engineering to develop bacteria that can turn sugar directly into hydrocarbon fuels like gasoline, jet fuel, and diesel fuel, which, unlike ethanol and biodiesel, can easily be sent through existing oil pipelines and burned in any motor vehicle. Stay tuned to see if we can learn how to turn bacteria into tiny fuel factories.

HOW WOULD YOU VOTE?

Do the advantages of using liquid ethanol as a fuel outweigh its disadvantages? Cast your vote online at **academic.cengage.com/biology/miller**.

RESEARCH FRONTIER

Developing more energy-efficient, cheaper, and more sustainable ways to produce liquid biofuels. See **academic.cengage.com/biology/miller**.

16-7 What Are the Advantages and Disadvantages of Geothermal Energy?

▶ **CONCEPT 16-7** Geothermal energy has great potential for supplying many areas with heat and electricity and generally has a low environmental impact, but locations where it can be exploited economically are limited.

Getting Energy from the Earth's Internal Heat

Geothermal energy is heat stored in soil, underground rocks, and fluids in the earth's mantle (Figure 14-3, p. 346, and Figure 15-2, p. 372). We can tap into this stored energy to heat and cool buildings and to produce electricity. Scientists estimate that using just 1% of the heat stored in the uppermost 5 kilometers (8 miles) of the earth's crust would provide 250 times more energy than that stored in all the earth's oil and natural gas reserves.

A *geothermal heat pump* system (Figure 16-28) can heat and cool a house by exploiting the temperature differences between the earth's surface and underground, almost anywhere in the world at a depth of 3–6 meters (10–20 feet). In winter, a closed loop of buried pipes circulates a fluid (usually water or an antifreeze solution), which extracts heat from the ground and carries it to a heat pump, which transfers the heat to a home's heat distribution system (usually a blower and air ducts). In summer, this system works in reverse, removing heat from a home's interior and storing it in the ground. These systems can also be modified to provide hot water.

According to the EPA, after superinsulation, a well-designed geothermal heat pump system is the most energy-efficient, reliable, environmentally clean, and

cost-effective way to heat or cool a space. It produces no air pollutants and emits no CO_2. For more information, see **www.ghpc.org** and **www.econar.com**. Installation costs are recouped after 3–5 years, and then such systems save their owners money.

We have also learned to tap into deeper, more concentrated *hydrothermal reservoirs* of geothermal energy, as Iceland has done for decades (Figure 16-1, **Core Case Study**). Wells are drilled into these reservoirs to extract their dry steam, wet steam, or hot water, which are used to heat buildings, provide hot water, grow vegetables in greenhouses, raise fish in aquaculture ponds, and spin turbines to produce electricity. Cool water left over can be pumped back into the reservoirs to be reheated.

In France, some 70 district geothermal heating facilities provide both heat and hot water for about 200,000 residences. China, Turkey, and Japan also have a number of district heating systems that use geothermal energy. Currently, about 40 countries (most of them in the developing world) extract enough energy from hydrothermal reservoirs to produce about 1% of the world's electricity, enough to meet the needs of 60 million people and equal to the electrical output of all 104 nuclear power plants in the United States. (See Figure 17, p. S69, in Supplement 10 for a map of the global reserves of usable hydrothermal geothermal energy.)

The United States and the Philippines account for half of the world's electricity produced from geothermal energy; the Philippines gets one-fourth of its electricity from this energy resource. Other countries with a high potential for using geothermal energy to produce electricity are Japan, Indonesia, China, Canada, Mexico, and Russia (Figure 17, p. S69, in Supplement 10).

The United States is the world's largest producer of geothermal electricity from hydrothermal reservoirs. Most of it is produced in California, Nevada, Utah, and Hawaii (see Figure 18, p. S70, in Supplement 10) and meets the electricity needs of about 6 million Americans. The largest operation, called The Geysers, powers about 1 million homes near San Francisco, California. California gets about 6% of its electricity from geothermal energy and 15 new projects are under development.

Using geothermal energy generally has a much lower environmental impact than using fossil fuel energy. At concentrated and accessible hydrothermal sites, electricity can be produced at a low cost compared to other alternatives (Table 16-1). On average, a geothermal power plant releases about one-sixth as much CO_2 as a power plant burning natural gas emits, and one-tenth the amount emitted by a coal-burning power plant. China has a large potential for geothermal power, which could help to reduce its dependence on coal-fired power plants.

Geothermal energy has two main problems. One is that the current cost of tapping large-scale hydrother-

Figure 16-28 Natural capital: a geothermal heat pump system can heat or cool a house almost anywhere. It heats the house in winter by transferring heat from the ground into the house (shown here). In the summer, it cools the house by transferring heat from the house to the ground.

mal reservoirs is too high for all but the most concentrated and accessible sources, although new drilling and extraction technologies may bring these costs down. The other is that some dry- or wet-steam geothermal reservoirs could be depleted if their heat is removed faster than natural processes can renew it. Recirculating the water back into the underground reservoirs for reheating could slow such depletion.

Another potential source of geothermal energy is *hot, dry rock* that lies 5 or more kilometers (3 or more miles) underground almost everywhere. Deep drilling and seismic exploration techniques developed by the oil industry might be used to tap this source. Water could be injected into a deep well and pressurized to create fractures in the intensely hot rock and to form a reservoir of very hot geothermal fluid. Then other wells would be drilled to bring the boiling water and steam to the surface for use in generating electricity.

According to the U.S. Geological Survey, tapping just 2% of this hot, dry rock geothermal energy in the United States could produce more than 2,000 times the country's current annual use of electricity. The limiting factor is cost. Estimates range from 3 to 30 times the cost of electricity produced from suitable hydrothermal sites. More research and improved technology could bring this cost down. Iceland (**Core Case Study**) is drilling deep wells to evaluate this source of the earth's internal heat.

— **RESEARCH FRONTIER** —

Finding better and affordable ways to tap different sources of geothermal energy. See **academic.cengage.com/biology/miller**.

Figure 16-29 (p. 428) lists the advantages and disadvantages of using geothermal energy (**Concept 16-7**).

Some analysts see geothermal energy, coupled with improvements in energy efficiency and electricity produced by solar cells and wind farms, as keys to a more sustainable energy future.

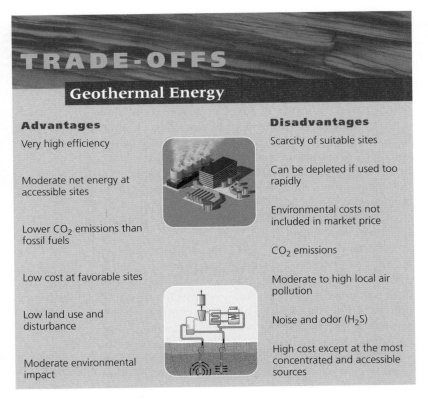

Figure 16-29 Advantages and disadvantages of using geothermal energy for space heating and for producing electricity or high-temperature heat for industrial processes (**Concept 16-7**). **Question:** Which single advantage and which single disadvantage do you think are the most important? Why?

16-8 What Are the Advantages and Disadvantages of Hydrogen as an Energy Source?

▶ **CONCEPT 16-8** Hydrogen fuel holds great promise for powering cars and generating electricity, but to be environmentally beneficial, it would have to be produced without the use of fossil fuels.

Hydrogen Is a Promising Fuel but There Are Challenges

Many scientists and executives of major oil and automobile companies say the fuel of the future is hydrogen gas (H_2)—first envisioned as a fuel in 1874 by science fiction writer Jules Verne in his book, *The Mysterious Island*. Iceland (**Core Case Study**) envisions hydrogen as an important component of the renewable energy economy it hopes to develop by 2050–2060.

Most attention has been focused on fuel cells (Figure 16-30) that combine hydrogen gas (H_2) and oxygen gas (O_2) to produce electricity and emit water vapor into the atmosphere.

Widespread use of hydrogen as a fuel would eliminate most of the air pollution problems we face today. It would also greatly reduce the threats of global warming and climate change, because using it emits no CO_2—as long as the hydrogen is not produced with the use of fossil fuels or other carbon-containing compounds.

Hydrogen also provides more energy per gram than does any other fuel, making it the ideal aviation fuel.

So what is the catch? There are three challenges in turning the vision of widespread use of hydrogen as a fuel into reality. *First*, hydrogen is chemically locked up in water and in organic compounds such as methane and gasoline, so it takes energy and money to produce hydrogen from these compounds. In other words, hydrogen is not an energy resource like coal or oil. It is a fuel produced by using energy, and thus its net energy yield will always be negative. *Second*, fuel cells are the best way to use hydrogen to produce electricity, but current versions of such cells are expensive. However, progress in the development of nanotechnology (Science Focus, p. 362) could lead to improved membranes and catalysts that would make fuel cells more efficient and cheaper.

Third, whether a hydrogen-based energy system produces less air pollution and CO_2 than a fossil fuel system depends on how the hydrogen is produced. We could use electricity from coal-burning and conven-

Figure 16-30 A *fuel cell* takes in hydrogen gas and separates the hydrogen atoms' electrons from their protons. The electrons flow through wires to provide electricity, while the protons pass through a membrane and combine with oxygen gas to form water vapor ($2H_2 + O_2 \longrightarrow 2H_2O$ + energy). Note that this process is the reverse of electrolysis, which is used to produce hydrogen fuel ($2H_2O$ + energy $\longrightarrow 2H_2 + O_2$).

tional nuclear power plants to decompose water into hydrogen and oxygen gas. But this approach does not avoid the harmful environmental effects associated with using these fuels (Figure 15-15, p. 385, and Figure 15-21, p. 391) (**Concept 16-8**). We can also make hydrogen from coal and strip it from organic compounds found in fuels such as gasoline or natural gas. However, according to a 2002 study by a team of scientists, producing hydrogen from coal and organic compounds will add much more CO_2 to the atmosphere per unit of heat generated than does burning these carbon-containing fuels directly.

Most proponents of hydrogen believe that if we are to receive its very low pollution and low CO_2 emission benefits, the energy used to produce H_2 must come from low-polluting, renewable sources that emit little or no CO_2, such as solar cell power plants (Figure 16-19), wind farms (Figure 16-22, middle and right), geothermal energy (Figure 16-1), and micro-hydropower plants. Also, some naturally occurring bacteria and algae can produce hydrogen by biodegrading almost any organic material in a microbiological fuel cell.

Once produced, hydrogen can be stored in a pressurized tank as liquid hydrogen or in solid metal hydride compounds and sodium borohydride, which when heated release hydrogen gas. Scientists are also evaluating ways to store H_2 by absorbing it onto the surfaces of activated charcoal or carbon nanofibers, which release hydrogen gas when heated. Another possibility is to store it inside nano-size glass microspheres that can easily be filled and refilled. More research is needed to convert these possibilities into reality.

Metal hydrides, sodium borohydride, charcoal powders, ammonia borane, carbon nanotubes, and glass microspheres containing hydrogen will not explode or burn if a vehicle's fuel tank or system is ruptured in an accident. Also, use of ultralight car bodies made of composites and energy-efficient aerodynamic design would improve fuel efficiency so that large hydrogen fuel tanks would not be needed.

This makes hydrogen stored in such ways a much safer fuel than gasoline, diesel fuel, natural gas, and concentrated ethanol. Indeed, gasoline is 22 times more explosive than hydrogen, and if hydrogen burns, the heat produced dissipates much more rapidly than it does from gasoline or diesel fires. For 30 years, liquefied hydrogen has been shipped all over the United States with no fires reported.

Most automobile companies have developed prototype hydrogen-powered cars, trucks, and buses (such as those used in Iceland, **Core Case Study**). More than 500 such vehicles are on the road today and the number is expected to increase rapidly. Currently, hydrogen can be produced from natural gas at a cost equivalent to $2.50 per gallon of gasoline—less than what consumers are paying now. Some analysts project that a variety of fuel-cell cars running on affordable hydrogen produced from natural gas could be introduced between 2010 and 2020 and take over the market by 2030.

In 2007, engineering professor Jerry Woodall invented a new way to produce hydrogen on demand by exposing pellets of an aluminum-gallium alloy to water, without producing toxic fumes. If this process is perfected and proves economically feasible, hydrogen could be generated as needed inside a tank about the same size as an average gasoline tank, and thus would not have to be transported or stored. Merely replacing the gasoline fuel injector with a hydrogen injector would allow current internal combustion engines to run on hydrogen. The engine would then be 25% more fuel efficient than it was when burning gasoline, and it would emit no pollutants. Stay tuned while this possibility is evaluated.

Larger stationary fuel cells can provide electricity and heat for commercial and industrial users. A 45-story office building in New York City gets much of its heat from two large fuel-cell stacks. And Japan has built a large fuel cell that produces enough electricity to run a small town.

Canada's Toronto-based Stuart Energy is developing a home fueling unit about the size of a dishwasher that will allow homeowners to use electricity to produce their own hydrogen from tap water. The

TRADE-OFFS

Hydrogen

Advantages	Disadvantages
Can be produced from plentiful water	Not found as H_2 in nature
Low environmental impact	Energy is needed to produce fuel
Renewable if produced from renewable energy resources	Negative net energy
No CO_2 emissions if produced from water	CO_2 emissions if produced from carbon-containing compounds
Good substitute for oil	Environmental costs not included in market price
Competitive price if environmental and social costs are included in cost comparisons	Nonrenewable if generated by fossil fuels or nuclear power
Easier to store than electricity	High costs (that may eventually come down)
Safer than gasoline and natural gas	Will take 25 to 50 years to phase in
Nontoxic	Short driving range for current fuel-cell cars
High efficiency (45–65%) in fuel cells	No fuel distribution system in place
	Excessive H_2 leaks may deplete ozone in the atmosphere

Figure 16-31 Advantages and disadvantages of using hydrogen as a fuel for vehicles and for providing heat and electricity (**Concept 16-8**). **Question:** Which single advantage and which single disadvantage do you think are the most important? Why?

unit could be installed in a garage and used to fuel a hydrogen-powered vehicle overnight, when electricity rates are sometimes lower. In sunny areas, homeowners could install rooftop panels of solar cells to produce and store hydrogen fuel for their cars, especially when solar cells become more affordable or with the help of generous government subsidies for such installations.

Another promising application is in homes, where a fuel-cell stack about the size of a refrigerator could provide heat, hot water, and electricity. The extra cost involved could be included in the purchase price of the home, just as conventional heating and cooling systems are, and spread out over the life of the mortgage. These systems could also be leased to homeowners. Honda has developed a home unit that produces hydrogen from the methane in natural gas. In 2007, about 2,200 Japanese homeowners got their electricity and hot water from such units. Japanese government plans call for using such fuel cell systems for about one-fourth of Japanese households by 2020.

Most current fuel cell units convert natural gas to hydrogen, but this releases CO_2. A better long-term solution would be to produce H_2 using electricity generated by wind farms, large solar-cell farms, and hydroelectric and geothermal power plants, and to distribute it using upgraded natural gas pipelines. **GREEN CAREER:** Hydrogen energy

— RESEARCH FRONTIER —

Developing better and more affordable ways to produce hydrogen from renewable energy resources and ways to store and distribute it. See **academic.cengage.com/biology/miller**.

Figure 16-31 lists the advantages and disadvantages of using hydrogen as an energy resource (**Concept 16-8**).

— HOW WOULD YOU VOTE? **—**

Do the advantages of producing and burning hydrogen as an energy resource outweigh the disadvantages? Cast your vote online at **academic.cengage.com/biology/miller**.

16-9 How Can We Make a Transition to a More Sustainable Energy Future?

▶ **CONCEPT 16-9** We can make a transition to a more sustainable energy future if we greatly improve energy efficiency, use a mix of renewable energy resources, and include environmental costs in the market prices of all energy resources.

Choosing Energy Paths

Energy policies must be developed with the future in mind, because experience shows that it usually takes at least 50 years and huge investments to phase in new energy alternatives. Creating energy policy involves trying to answer the following questions for *each* energy alternative:

- How much of the energy resource is likely to be available in the near future (the next 25 years) and the long term (the next 50 years)?

- What is the estimated net energy yield (Science Focus, p. 374) for the resource?
- How much will it cost to develop, phase in, and use the resource?
- What government research and development subsidies and tax breaks will be used to help develop the resource?
- How will dependence on the resource affect national and global economic and military security?
- How vulnerable is the resource to terrorism?
- How will extracting, transporting, and using the resource affect the environment, human health, and the earth's climate? Can these harmful costs be included in the market price of the resource through mechanisms like taxing and reducing environmentally harmful subsidies?

Our energy future depends primarily on which energy resources the government and private companies decide to promote, coupled with political and economic pressure from citizens and consumers. Energy expert Amory Lovins describes this process as "choosing an energy path."

Fossil fuel and nuclear power interests generally favor a *supply-side* or *hard-path* approach. It is built around scouring a country to find more nonrenewable oil, natural gas, and coal, and building more energy-wasting coal-burning and nuclear power plants.

In contrast, many environmental scientists and economists favor a *demand-side* or *soft-path* approach, described in 1977 by Lovins. (See his Guest Essay at CengageNOW.) This approach emphasizes reducing energy waste and depending more on a mix of renewable resources such as solar, wind, geothermal, and biomass. Proponents see this energy path as more sustainable than relying on nonrenewable fossil energy and nuclear energy resources.

In considering possible energy paths, scientists and energy experts who have evaluated energy alternatives have come to some general conclusions. First, *there will be a gradual shift from large, centralized macropower systems to smaller, decentralized micropower systems* (Figure 16-32) such as wind turbines, fuel cells for cars, household solar panels, rooftop solar water heaters, small natural gas turbines, and stationary fuel cells for houses and commercial buildings.

Figure 16-32 Solutions: *decentralized power system* in which electricity is produced by a large number of dispersed, small-scale *micropower systems.* Some such systems would produce power on site; others would feed the power they produce into an updated electrical distribution system. Over the next few decades, many energy and financial analysts expect a shift to this type of power system often based on locally available renewable energy resources. **Question:** Can you think of any disadvantages of a decentralized power system?

Making the Transition to a More Sustainable Energy Future

Improve Energy Efficiency

Increase fuel-efficiency standards for vehicles, buildings, and appliances

Mandate government purchases of efficient vehicles and other devices

Provide large tax credits or feebates for buying efficient cars, houses, and appliances

Offer large tax credits for investments in energy efficiency

Reward utilities for reducing demand for electricity

Greatly increase energy efficiency research and development

More Renewable Energy

Greatly increase use of renewable energy

Provide large subsidies and tax credits for use of renewable energy

Include environmental costs in prices for all energy resources

Encourage government purchase of renewable energy devices

Greatly increase renewable energy research and development

Reduce Pollution and Health Risk

Cut coal use 50% by 2020

Phase out coal subsidies

Levy taxes on coal and oil use

Phase out nuclear power subsidies, tax breaks, and loan guarantees

Figure 16-33 Suggestions of various energy analysts for helping us to make the transition to a more sustainable energy future (**Concept 16-9**). **Question:** Which five of these solutions do you think are the most important? Why?

Currently, most countries have a centralized and geographically concentrated system of large power plants, refineries, pipelines, and other infrastructure that is vulnerable to disruption from events such as terrorist attacks and natural disasters. For example, in 2005, Hurricane Katrina crippled about 10% of America's oil and gas producing wells (see Figure 6, bottom, p. S62, in Supplement 10) and oil refineries in the Gulf of Mexico for more than a year.

This shift from centralized macropower to dispersed micropower would be similar to the computer industry's shift from large, centralized mainframe computers to increasingly smaller, widely dispersed PCs, laptops, and handheld computers. It would improve national and economic security, because countries would rely on diverse, dispersed, domestic, and renewable energy resources instead of on a small number of large coal and nuclear power plants, imported oil, and oil refineries that are vulnerable to storm damage and sabotage.

The second general conclusion of experts is that *a combination of greatly improved energy efficiency and the use of natural gas and sustainably produced biofuels will best help us to make the transition to a diverse mix of locally available renewable energy resources* (**Concept 16-9**) *over the next several decades.* Natural gas is the most clean-burning fossil fuel, and it can be used easily in the short term during this transition. Instead of depending mostly on nonrenewable fossil fuels produced elsewhere, people will eventually make use of a variety of often locally available renewable energy resources, applying the diversity **principle of sustainability** by not putting all of their "energy eggs" in one basket.

Third, *because of their supplies and artificially low prices, fossil fuels will continue to be used in large quantities.* The challenge is to find ways to reduce the harmful environmental impacts of widespread fossil fuel use, with special emphasis on reducing air pollution and emissions of greenhouse gases and including their harmful environmental costs in their market prices, as less environmentally harmful alternatives are phased in.

Figure 16-33 summarizes these and other strategies for making the transition to a more sustainable energy future over the next 50 years (**Concept 16-9**).

Economics, Politics, and Education Can Help Us to Shift to More Sustainable Energy Resources

To most analysts, economics, politics, and consumer education hold the key to making a shift to a more sustainable energy future. It will require maintaining

consistent and sustained energy policies at the local, state, and national levels so that businesses can make long-range plans. Governments can use three strategies to help stimulate or dampen the short-term and long-term use of particular energy resources.

First, they can *keep the prices of selected energy resources artificially low to encourage use of those resources.* They do this by providing research and development (R&D) subsidies and tax breaks to encourage the development of those resources, and by enacting regulations to favor them. For decades, this approach has been employed to stimulate the development and use of fossil fuels and nuclear power in the United States and in most other developed countries. The U.S. oil industry received almost half of the $600 billion in R&D subsidies provided by taxpayers between 1950 and 2003, according to the U.S. Department of Energy and the Congressional Budget Office.

And in 2007, the Department of Energy allocated $159 million for solar energy R&D. At the same time, it allocated nearly double this amount, $303 million, for nuclear energy R&D and $427 million for coal R&D. For decades, this sort of policy has created an uneven economic playing field that *encourages* energy waste and rapid depletion of nonrenewable energy resources, while it *discourages* improvements in energy efficiency and the development of renewable energy resources.

To many energy analysts, one of the most important steps governments can take to level the economic playing field is to phase out the $250–300 billion in annual subsidies now provided worldwide for fossil fuels and nuclear energy—both mature industries that could be left to stand on their own, economically. These analysts call for greatly increasing subsidies for developing renewable energy and energy-efficiency technologies. If this had been done beginning in 1980, they say, the world probably could have greatly increased its energy sustainability, sharply decreased its dependence on imported oil, avoided two wars in the Middle East, and be well on the way to slowing global warming and projected climate change.

The *second* major strategy that governments can use is to *keep energy prices artificially high for selected resources to discourage their use.* They can do this by eliminating existing tax breaks and other subsidies that favor use of the targeted resource, and by enacting restrictive regulations or taxes on its use. Canada, in 2007 introduced rebates for hybrid vehicles, a tax on gas-guzzlers, and subsidies for development of renewable fuels. China is also taxing gas-guzzlers and raising energy-efficiency requirements for homes and office buildings. Such measures can increase government revenues, encourage improvements in energy efficiency, reduce dependence on imported energy, and decrease use of energy resources that have limited supplies. To make such changes acceptable to the public, analysts suggest that governments can offset energy taxes by reducing income and payroll taxes and providing an energy safety net for low-income users.

Third, governments can *emphasize consumer education.* Even if governments offer generous financial incentives for energy efficiency and renewable energy use, people will not make such investments if they are uninformed—or misinformed—about the availability, advantages, disadvantages, and relative costs of energy resources (Table 16-1). For example, cloudy Germany has more solar water heaters and solar cell panels than sunny France and Spain, mostly because the German government has made the public aware of the environmental benefits of these technologies. It has also provided consumers with substantial economic incentives for using them.

The good news is that we have the technology, creativity, and wealth to make the transition to a more sustainable energy future within your lifetime, as the state of California is proving (see following Case Study). Making this transition depends primarily on *education* and *politics*—on how well individuals understand ecological and environmental problems, and on how they vote and then influence their elected officials. People can also vote with their pocketbooks by refusing to buy inefficient and environmentally harmful products and by letting company executives know about their choices. Figure 16-34 (p. 434) lists some ways in which you can contribute to making this transition.

■ CASE STUDY
California's Efforts to Improve Energy Efficiency

The U.S. state of California has a population of 37 million people and is the world' sixth largest economy. It uses less energy per person than any other U.S. state. While overall per capita energy use in the United States has grown by half since 1974, California's energy use per person has stayed about even. The state has accomplished this through a mix of state regulations and high electricity prices, and at the same time it has kept its economy growing.

Because California promotes the use of cleaner, renewable sources of power, state residents pay close to the highest rates for electricity anywhere in the United States. This helps reduce energy waste and encourages the use of energy-efficient devices. In the long run, because of increased efficiency, it also saves people money. While the average American uses 12,000 kilowatt-hours (kWh) of electricity in a year, the average California resident uses less than 7,000 kWh. That translates to about $800 in savings for the average

WHAT CAN YOU DO?

Shifting to Sustainable Energy Use

- Get an energy audit done for your house or office

- Drive a car that gets at least 15 kilometers per liter (35 miles per gallon)

- Use a carpool to get to work or to school

- Walk, bike, and use mass transit

- Superinsulate your house and plug all air leaks

- Turn off lights, TV sets, computers, and other electronic equipment when they are not in use

- Wash laundry in warm or cold water

- Use passive solar heating

- For cooling, open windows and use ceiling fans or whole-house attic or window fans

- Turn thermostats down in winter and up in summer

- Buy the most energy-efficient home, lights, and appliances available

- Turn down the thermostat on water heaters to 43–49 °C (110–120 °F) and insulate hot water heaters and pipes

Figure 16-34 Individuals matter: ways to reduce your use and waste of energy. **Questions:** Which three of these items do you think are the most important? Why? Which things in this list do you already do or plan to do?

Californian, even with the state's high electricity prices. And while U.S. per capita CO_2 emissions stayed about the same between1974 and 2004, such emissions in California fell by 30%.

Actions taken by the state of California to reduce energy waste include establishing the nation's first and strictest building standards for energy efficiency, setting stringent appliance efficiency standards, and giving rebates for purchases of solar energy equipment. The state has also created a strategy called *decoupling*, in which utility profits were disconnected from the amount of electricity sold, and instead, are now tied into the amount of energy conserved.

> **THINKING ABOUT**
> **California's Strategies for Energy Efficiency**
>
> Do you think that California would have the lowest per capita energy use in the country if the state government had not acted as it did? Explain.

REVISITING Iceland and Sustainability

Iceland's attempt to convert completely to renewable resources (**Core Case Study**) shows that it is possible for a society to run at least partly on renewable energy such as geothermal heat (Figure 16-1) and hydropower. It also reveals that it is not easy to make such a conversion, and that all players—government, industry, financial institutions, and citizens—must make a commitment to shifting to a more sustainable energy future.

For example, the use of hydrogen as fuel is neither an easy nor an inexpensive solution. But Iceland is tackling this obstacle and, within a decade or two, could be using hydrogen as the fuel for its transportation sector.

Other countries and cities can follow the lead of Icelanders and make the transition to a more sustainable energy future by

applying the four **scientific principles of sustainability**. This means:

- Relying much more on direct and indirect forms of solar energy

- Recycling and reusing materials and thus reducing wasteful and excessive consumption of energy and matter

- Mimicking nature's reliance on biodiversity by using a diverse mix of locally and regionally available renewable energy resources

- Reducing use and waste of energy and other resources by slowing population growth.

A transition to renewable energy is inevitable,
not because fossil fuel supplies will run out—
large reserves of oil, coal, and gas remain in the world—
but because the costs and risks of using these supplies will continue
to increase relative to renewable energy.

MOHAMED EL-ASHRY

1. Review the Key Questions and Concepts for this chapter on p. 400. Describe Iceland's attempt to develop a renewable energy economy by 2050.

2. Distinguish between **energy conservation** and **energy efficiency.** Explain why energy efficiency can be thought of as an energy resource. How much of the energy used in the United States is wasted unnecessarily? What are the major advantages of reducing energy waste? List three reasons why this source of energy has been neglected?

3. What is net energy efficiency and why is it important? Describe three ways to save energy and money in **(a)** industry, **(b)** transportation, and **(c)** buildings. What is **cogeneration (combined heat and power or CHP)?** Describe how Dow Chemicals has saved energy and money. Describe the trends in fuel efficiency in the United States since the 1970s. Explain why the price of gasoline is much higher than what consumers pay at the pump. What is a feebate? Distinguish among hybrid, plug-in hybrid, and fuel-cell motor vehicles. Describe five ways to save energy in an existing building.

4. List five advantages of relying more on a variety of renewable sources of energy and describe two factors holding back such a transition.

5. Distinguish between **passive solar heating** and **active solar heating** and discuss the major advantages and disadvantages of such systems. What are three ways to cool houses naturally? Discuss the major advantages and disadvantages of using solar energy to generate high-temperature heat and electricity. What is a **solar cell (photovoltaic** or **PV cell)** and what are the major advantages and disadvantages of using such cells to produce electricity?

6. What are the major advantages and disadvantages of using flowing water to produce electricity in hydropower plants? What is the potential for using tides and waves, to produce electricity?

7. What is a wind turbine? What is a wind farm? What are the major advantages and disadvantages of using wind to produce electricity? What are the major advantages and disadvantages of using wood to provide heat and electricity? What are **biofuels** and what are the major advantages and disadvantages of using **(a)** biodiesel and **(b)** ethanol to power motor vehicles? Evaluate the use of corn, sugar cane, and cellulose plants to produce ethanol.

8. What is **geothermal energy** and what are three sources of such energy? What are the major advantages and disadvantages of using geothermal energy as a source of heat and to produce electricity? What are the major advantages and disadvantages of burning hydrogen gas to provide heat, to produce electricity, and to fuel cars?

9. List three general conclusions of energy experts about possible future energy paths for the world. List five major strategies for making the transition to a more sustainable energy future. Describe three roles that governments play in determining which energy resources we use. Describe what the U.S. state of California has done to improve energy efficiency and rely more on various forms of renewable energy.

10. Describe how Iceland's pursuit of a renewable energy economy (**Core Case Logo**) applies the four **scientific principles of sustainability.**

Note: Key Terms are in **bold** type.

1. Imagine that the country where you live has completed a transition to using only renewable energy, as Iceland plans to do (**Core Case Study**). Do you think that you would have to give up any of the conveniences you now enjoy? If so, what are they? Describe any adjustments you might have to make in your way of living.

2. List five ways in which you unnecessarily waste energy during a typical day, and explain how these actions violate any of the four **scientific principles of sustainability** (see back cover).

3. Congratulations! You have won $500,000 to build a more sustainable house of your choice. With the goal of maximizing energy efficiency, what type of house would you build? How large would it be? Where would you locate it? What types of materials would you use? What types of materials would you *not* use? How would you heat and cool the house? How would you heat water? What types of lighting, stove, refrigerator, washer, and dryer would you use? Which, if any, of these appliances could you do without?

4. A homebuilder installs electric baseboard heat and claims, "It is the cheapest and cleanest way to go." Apply your understanding of the second law of thermodynamics (**Concept 2-4B**, p. 40) and net energy (Figure 16-4), to evaluate this claim.

5. Should buyers of energy-efficient motor vehicles receive large rebates, funded by taxes on gas-guzzlers? Explain.

6. Explain why you agree or disagree with the following proposals made by various energy analysts:

 a. Government subsidies for all energy alternatives should be eliminated so that all energy choices can compete in a true free-market system.

 b. All government tax breaks and other subsidies for conventional fossil fuels (oil, natural gas, and coal), synthetic natural gas and oil, and nuclear power (fission and fusion) should be phased out. They should be replaced with subsidies and tax breaks for improving energy efficiency and developing solar, wind, geothermal, hydrogen, and biomass energy alternatives.

 c. Development of solar, wind, and hydrogen energy should be left to private enterprise and should receive little or no help from the federal government, but nuclear energy and fossil fuels should continue to receive large federal government subsidies.

7. Imagine that you are in charge of the U.S. Department of Energy (or the energy agency in the country where you live). What percentage of your research and development budget will you devote to fossil fuels, nuclear power, renewable energy, and improving energy efficiency? How would you distribute your funds among the various types of renewable energy (wind, solar, hydropower, geothermal)? Explain your thinking.

8. List three ways in which you could apply **Concept 16-9** to making your lifestyle more environmentally sustainable.

9. Congratulations! You are in charge of the world. List the five most important features of your energy policy.

10. List three questions that you would like to have answered as a result of reading this chapter.

Note: See Supplement 13 (p. S78) for a list of Projects related to this chapter.

ECOLOGICAL FOOTPRINT ANALYSIS

Make calculations to fill in the missing data in this table. Show all calculations. (*Note:* 1 liter = 0.265 gallon; 1 kilogram = 2.20 pounds; 1 hectare = 10,000 square meters = 2.47 acres.)

EPA Size Class/ Model	Compact/ Honda Civic Hybrid	Midsize/ Toyota Camry Hybrid	Sports Utility Vehicle (SUV)/ Hummer H3
Combined highway and city fuel efficiency in kilometers per liter (miles per gallon)	17.8 (42.0)	14.4 (34.0)	6.4 (15.0)
Liters (gallons) of gasoline consumed per year, assuming an average mileage of 19,300 kilometers (12,000 miles)			
Kilograms (pounds) of CO_2 produced per year, assuming combustion of gasoline releases 2.3 kilograms per liter (19 pounds per gallon)			
Hectares (acres) of tropical rain forest needed to take up CO_2 produced per year, assuming the annual uptake of an undisturbed forest is 0.5 kilograms of CO_2 per square meter			

1. About how many times as much CO_2 per year is produced by the SUV as is produced by the compact car?

2. About how many times as much CO_2 per year is produced by the SUV as is produced by the midsize car?

3. How many hectares (acres) of tropical rain forest are needed to take up the CO_2 produced annually by 1 million SUVs?

4. How many hectares (acres) of tropical rain forest are needed to take up the CO_2 produced annually by 1 million midsize cars?

5. How many hectares (acres) of tropical rain forest are needed to take up the CO_2 produced annually by 1 million compact cars?

LEARNING ONLINE

Log on to the Student Companion Site for this book at **academic.cengage.com/biology/miller**, and choose Chapter 16 for many study aids and ideas for further reading and research. These include flash cards, practice quizzing, Weblinks, information on Green Careers, and InfoTrac® College Edition articles.

CORE CASE STUDY The Global HIV/AIDS Epidemic

The spread of *acquired immune deficiency syndrome* (AIDS), caused by infection with the *human immunodeficiency virus* (HIV), is a serious global health threat. The virus itself is not deadly, but it cripples the immune system and leaves the body vulnerable to infections such as tuberculosis (TB) and rare forms of cancer such as *Kaposi's sarcoma* (Figure 17-1). The virus is transmitted from one person to another by unsafe sex, sharing of needles by drug users, infected mothers who pass the virus on to their offspring before or during birth, and exposure to infected blood.

Since the HIV virus was identified in 1981, this viral infection has spread exponentially around the globe. According to the World Health Organization (WHO), in 2007, about 33 million people worldwide (1.1 million in the United States) were infected with HIV. Almost two-thirds of them were in African countries located south of the Sahara Desert (sub-Saharan Africa) (see Figure 14, p. S18, in Supplement 3).

In 2007, about 2.5 million people (58,000 to 63,000 in the United States) became infected with HIV—an average of 6,800 new cases per day—half of them between the ages of 15 and 24. Within 7–10 years, at least half of all HIV-infected people will develop AIDS. This long incubation period means that infected people often spread the virus for several years without knowing they are infected.

Currently, there is no vaccine to prevent HIV and no cure for AIDS. If you get AIDS, you will almost certainly die from it. Drugs help some infected people live longer, but 90% of those suffering from AIDS cannot afford to use these drugs.

Between 1981 and 2007, more than 25 million people (584,000 in the United States) died of AIDS-related diseases. Each year AIDS claims about 2.1 million more lives (17,000 in the United States)—a daily average of 5,750 premature deaths.

AIDS has reduced the life expectancy of the 750 million people living in sub-Saharan Africa from 62 to 47 years—40 years in the seven countries most severely affected by AIDS. The premature deaths of teachers, health-care workers, soldiers, and other young productive adults in such countries leads to diminished education and health care, decreased food production and economic development, and disintegrating families.

This means that countries like Botswana are expected to lose half of their adult populations within a decade. Such death rates drastically alter a country's age structure (Figure 17-2). AIDS has also left more than 15 million children orphaned—roughly equal to the number of children under age 5 in the United States. Many of these orphans are forced into child labor or the sex trade.

Figure 17-1 Lesions that are a sign of Kaposi's sarcoma, a rare form of cancer common among AIDS patients.

U.S. National Cancer Institute

Figure 17-2 *Global outlook:* Worldwide, AIDS is the leading cause of death for people of ages 15–49. This loss of productive working adults affects the age structure of a population. In Botswana, more than 24% of this age group is infected with HIV. This figure shows the projected age structure of Botswana's population in 2020 with and without AIDS. (Data from the U.S. Census Bureau) **Question:** How might this affect Botswana's economic development?

In this chapter, we will look at connections between environmental hazards and human health and at what we can do to reduce the deadly global pandemic of AIDS and other environmental health threats from diseases and exposure to harmful chemicals.

17-1 What major health hazards do we face?

CONCEPT 17-1 People face health hazards from biological, chemical, physical, and cultural factors, and from the lifestyle choices they make.

17-2 What types of biological hazards do we face?

CONCEPT 17-2 In terms of death rates, the most serious infectious diseases are flu, AIDS, diarrheal diseases, malaria, and tuberculosis; most of these deaths occur in developing countries.

17-3 What types of chemical hazards do we face?

CONCEPT 17-3 There is growing concern about chemicals that can cause birth defects and cancers and disrupt the human immune, nervous, and endocrine systems.

17-4 How can we evaluate and deal with chemical hazards?

CONCEPT 17-4A Scientists use live laboratory animals, non-animal tests, case reports of poisonings, and epidemiological studies to estimate the toxicity of chemicals, but these methods have limitations.

CONCEPT 17-4B Many health scientists call for much greater emphasis on pollution prevention to reduce our exposure to potentially harmful chemicals.

17-5 How do we perceive risks and how can we avoid the worst of them?

CONCEPT 17-5 We can reduce the major risks we face if we become informed, think critically about risks, and make careful choices.

Note: Supplements 2 (p. S4), 3 (p. S10), 6 (p. S39), and 13 (p. S78) can be used with this chapter.

The dose makes the poison.

PARACELSUS, 1540

17-1 What Major Health Hazards Do We Face?

▶ **CONCEPT 17-1** People face health hazards from biological, chemical, physical, and cultural factors, and from the lifestyle choices they make.

Risks Are Usually Expressed as Probabilities

A **risk** is the *probability* of suffering harm from a hazard that can cause injury, disease, death, economic loss, or damage. It is usually expressed in terms of *probability*—a mathematical statement about how likely it is that harm will be suffered from a hazard. Scientists often state a probability in terms such as "The lifetime probability of developing lung cancer from smoking one pack of cigarettes per day is 1 in 250." This means that 1 of every 250 people who smoke a pack of cigarettes every day will likely develop lung cancer over a typical lifetime (usually considered to be 70 years).

It is important to distinguish between *possibility* and *probability*. When we say that it is *possible* that a smoker can get lung cancer, we are saying that this event could happen. *Probability* gives us an estimate of the likelihood of such an event.

Risk assessment is the scientific process of using statistical methods to estimate how much harm a particular hazard can cause to human health or to the environment. Scientists use it to estimate the probability of a risk, to compare it with the probability of other risks, and to establish priorities for avoiding or managing risks. **Risk management** involves deciding whether or how to reduce a particular risk to a certain level and at what cost. Figure 17-3 (p. 440) summarizes how risks are assessed and managed.

A major problem is that most people are not good at understanding and comparing risks. Because of sensational news coverage about the latest scare, many people worry about the highly unlikely possibility of minor risks and ignore the significant probability of harm from major risks.

For example, some Americans worry about getting widely publicized avian flu, which by mid-2008 had killed no one in the United States, but they do not get vaccinated for the common flu, which contributes to the deaths of about 36,000 Americans each year. Thus, educating people and especially news reporters about the meaning of risk assessments and

Figure 17-3 Science: *risk assessment* and *risk management.*
Question: What is an example of how you have applied this process in your daily living?

teaching them how to make risk comparisons is an important priority.

We Face Many Types of Hazards

All of us take risks every day. Examples include choosing to drive or ride in a car instead of using a bicycle or mass transit; eating foods with a high cholesterol or fat content that contribute to heart attacks, which kill far more people each year than any other risk; drinking alcohol; smoking or being in an enclosed space with a smoker; lying out in the sun or going to a tanning parlor, which increases the risk of getting skin cancer and wrinkled skin; and living in a hurricane-prone area. The key questions are, how serious are the risks we face, and do the benefits of certain activities outweigh the risks?

We can suffer harm from five major types of hazards (**Concept 17-1**):

- *Biological hazards* from more than 1,400 pathogens that can infect humans. A **pathogen** is a living organism that can cause disease in another organism. Examples are bacteria, viruses, parasites, protozoa, and fungi.

- *Chemical hazards* from harmful chemicals in air, water, soil, and food.

- *Physical hazards* such as fire, earthquakes, volcanic eruptions, floods, and storms.

- *Cultural hazards* such as unsafe working conditions, unsafe highways, criminal assault, and poverty.

- *Lifestyle choices* such as smoking, eating too much, drinking too much alcohol, and having unsafe sex.

THINKING ABOUT
Hazards

Think of a hazard from each of these categories that you may have faced recently. Which one was the most threatening?

17-2 What Types of Biological Hazards Do We Face?

▶ **CONCEPT 17-2** In terms of death rates, the most serious infectious diseases are flu, AIDS, diarrheal diseases, malaria, and tuberculosis; most of these deaths occur in developing countries.

Some Diseases Can Spread from One Person to Another

A **nontransmissible disease** is caused by something other than a living organism and does not spread from one person to another. Such diseases tend to develop slowly and have multiple causes. Examples include cardiovascular (heart and blood vessel) diseases, most cancers, asthma, diabetes, and malnutrition. As average life expectancy increases, people are more likely to suffer and die from nontransmissible diseases such as cardiovascular diseases and cancers.

An **infectious disease** is caused when a pathogen such as a bacterium, virus, or parasite invades the body and multiplies in its cells and tissues. Examples are flu, HIV (**Core Case Study**), malaria, tuberculosis, and measles. A **transmissible disease** (also called a *contagious* or *communicable disease*) is an infec-

CORE
CASE
STUDY

tious disease that can be transmitted from one person to another. Examples are flu, HIV, tuberculosis, and measles.

A bacterial disease such as tuberculosis spreads as the bacteria multiply. A viral disease such as flu or HIV spreads when a virus takes over a cell's genetic machinery to copy itself. Figure 17-4 shows the percentages of deaths in the world and in the United States in 2005 caused by cardiovascular diseases (mostly heart attacks and strokes), infectious diseases, and cancers.

In 1900, infectious disease was the leading cause of death in the world and in the United States. Since then, and especially since 1950, the incidences of infectious diseases and the death rates from such diseases have been greatly reduced. This has been achieved mostly by a combination of better health care, the use of antibiotics to treat infectious diseases caused by bacteria, and the development of vaccines to prevent the spread of

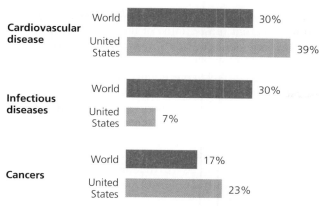

Cardiovascular disease		
World		30%
United States		39%

Infectious diseases		
World		30%
United States		7%

Cancers		
World		17%
United States		23%

Figure 17-4 Major causes of death in the world and in the United States in 2005. **Questions:** Why do you think that a higher percentage of the people in the United States die of cardiovascular disease than in the world? Why do you think that a much higher percentage of people in the world die from infectious diseases than in the United States? (Data from the World Health Organization and the U.S. Centers for Disease Control and Prevention)

some infectious viral diseases. As a result, average life expectancy has increased in most countries, the leading cause of death has shifted to nontransmissible cardiovascular disease, and the percentage of people dying from cancers is increasing in both developed and developing countries.

Traditionally, health organizations have described the state of global and national health by focusing mainly on the leading causes of death such as those shown in Figure 17-4. But such data do not give us a

picture of the nonfatal effects of disease and injury like prolonged illnesses and forms of disability like blindness, paralysis, and impaired learning ability. Health agencies now use *disability-adjusted life years* (DALYs) as a measure of the total disease burden in a population. A DALY is a measure designed to assess the amount of ill health, including premature death and disability due to specific diseases and injuries. However, the use of DALYs to make health care decisions has been criticized because of the shortage of reliable data on disabilities, and because of disagreements about how to evaluate the severity of various disabilities and how to apply them to different age groups.

Infectious Diseases Are Still Major Health Threats

Despite the shift in higher risk levels from transmissible to nontransmissible diseases, infectious diseases remain as serious health threats, especially in developing countries. Figure 17-5 shows major pathways for infectious diseases in humans. Such diseases can then be spread through air, water, food, and body fluids such as feces, urine, blood, and droplets sprayed by sneezing and coughing.

A large-scale outbreak of an infectious disease in an area or country is called an *epidemic* and a global epidemic, such as AIDS (**Core Case Study**) is called a *pandemic.* Figure 17-6 (p. 442) shows the annual ⟶ CORE CASE STUDY

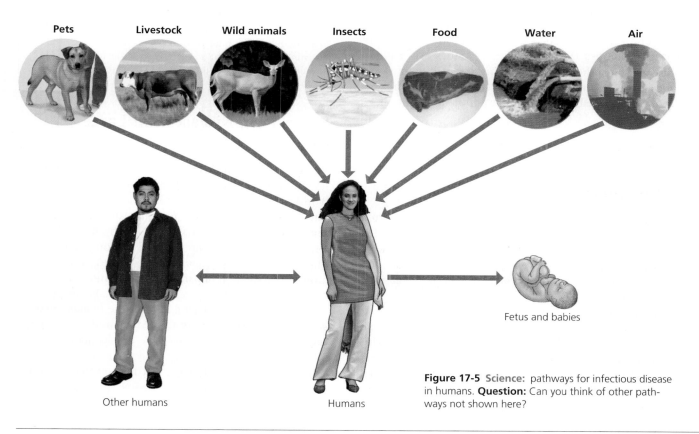

Pets Livestock Wild animals Insects Food Water Air

Other humans Humans Fetus and babies

Figure 17-5 Science: pathways for infectious disease in humans. **Question:** Can you think of other pathways not shown here?

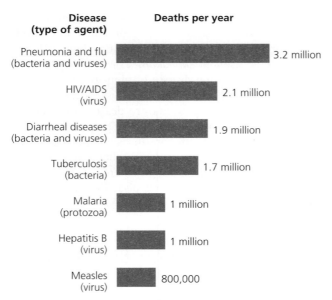

Disease (type of agent)	Deaths per year
Pneumonia and flu (bacteria and viruses)	3.2 million
HIV/AIDS (virus)	2.1 million
Diarrheal diseases (bacteria and viruses)	1.9 million
Tuberculosis (bacteria)	1.7 million
Malaria (protozoa)	1 million
Hepatitis B (virus)	1 million
Measles (virus)	800,000

Figure 17-6 *Global outlook:* The World Health Organization estimates that each year, the world's seven deadliest infectious diseases kill 12.5 million people—most of them poor people in developing countries (**Concept 17-2**). This amounts to an average of about 34,200 mostly preventable deaths every day. **Question:** How many people, on average, die prematurely from these diseases every hour? (Data from the World Health Organization, 2007)

death toll from the world's seven deadliest infectious diseases (**Concept 17-2**). The World Health Organization (WHO) is developing a Global Health Atlas, which contains a global database and maps based on health statistics, such as child mortality and the incidence of major infectious diseases in the world's countries (see **www.who.int/GlobalAtlas/**).

A growing problem is that many disease-carrying bacteria have developed genetic immunity to widely used antibiotics (Science Focus, at right). Also, many disease-transmitting species of insects such as mosquitoes have become immune to widely used pesticides that once helped to control their populations.

■ CASE STUDY
The Growing Global Threat from Tuberculosis

Since 1990, one of the world's most underreported stories has been the rapid spread of tuberculosis (TB). According to the WHO, this highly infectious bacterial disease strikes about 9.2 million people per year and kills 1.7 million—about 84% of them in developing countries.

Many TB-infected people do not appear to be sick, and about half of them do not know they are infected. Left untreated, each person with active TB typically infects 10–15 other people.

Several factors account for the recent increase in TB incidence. One is that there are too few TB screening and control programs, especially in developing coun-

tries, where 95% of the new cases occur. A second problem is that most strains of the TB bacterium have developed genetic resistance to the majority of the effective antibiotics. Also, population growth, urbanization, and air travel have greatly increased person-to-person contacts, and TB has spread, especially in areas where large numbers of poor people crowd together. In addition, AIDS (**Core Case Study**) greatly weakens its victims' immune systems, which allows TB bacteria to multiply in AIDS victims.

Slowing the spread of the disease requires early identification and treatment of people with active TB, especially those with a chronic cough. Treatment with a combination of four inexpensive drugs can cure 90% of individuals with active TB. To be effective, the drugs must be taken every day for 6–8 months. Because the symptoms disappear after a few weeks, many patients think they are cured and stop taking the drugs, allowing the disease to recur in drug-resistant forms and to spread to other people.

In recent years, a deadly and apparently incurable form of tuberculosis, known as multidrug-resistant TB has been increasing, especially in parts of Africa, China, India, and Russia. This form of TB is now found in 45 countries, and each year, there are about 500,000 new cases. Because this disease cannot be treated effectively with antibiotics, victims must be permanently quarantined from the rest of society. They also pose a threat to health workers and to potentially every person on the planet who takes a train or a plane and is exposed to undiagnosed people with this incurable form of TB.

THINKING ABOUT
Dealing with Tuberculosis

If you were a public health official, what would you do to ensure that people infected with TB received and took the treatments necessary to rid themselves of the disease and to lessen their chances of spreading it to others?

Some Viral Diseases Kill Large Numbers of People

Viruses evolve quickly, are not affected by antibiotics, and can kill large numbers of people. What are the world's three most widespread and dangerous viruses? The biggest killer is the *influenza* or *flu* virus (**Concept 17-2**), which is transmitted by the body fluids or airborne emissions of an infected person. Easily transmitted and especially potent flu viruses could spread around the world in a pandemic that could kill millions of people in only a few months (Science Focus, p. 444).

The second biggest viral killer is the *human immunodeficiency virus* (HIV) (**Core Case Study** and **Concept 17-2**). On a global scale, HIV infects about 2.5 million people each year, and the resulting complications from AIDS kill about 2.1 million people annually. AIDS is a serious and growing threat, but fortunately it is not as easily spread as is the common flu.

SCIENCE FOCUS

Genetic Resistance to Antibiotics Is Increasing

We risk falling behind in our efforts to prevent infectious bacterial diseases because of the astounding reproductive rate of bacteria, some of which can produce well over 16 million offspring in 24 hours. This allows bacteria to become genetically resistant to an increasing number of antibiotics through natural selection (**Concept 4-2A**, p. 80, and Figure 4-5, p. 83). In addition, some drug-resistant bacteria can quickly transfer their resistance to nonresistant bacteria by exchanging genetic material.

CONCEPT LINK

Other factors play a key role in fostering such genetic resistance. One is the spread of bacteria (some beneficial and some harmful) around the globe by human travel and international trade. Another is the overuse of pesticides, which increases populations of pesticide-resistant insects and other carriers of bacterial diseases.

Yet another factor is overuse of antibiotics. According to a 2000 study by Richard Wenzel and Michael Edward, at least half of all antibiotics used to treat humans are prescribed unnecessarily. In many countries, antibiotics are available without a prescription, which promotes unnecessary use. Resistance to some antibiotics has increased because of their widespread use in livestock and dairy animals to control disease and to promote growth. Also, the growing use of antibacterial hand soaps and other cleansers is probably promoting genetic resistance.

As a result of these factors acting together, every major disease-causing bacterium now has strains that resist at least one of the roughly 160 antibiotics used to treat bacterial infections such as tuberculosis (Case Study, at left). Each year, genetic resistance to antibiotics plays a role in the deaths of at least 90,000 of the 2 million people who pick up mostly preventable infections while they are in U.S. hospitals, according to the U.S. Centers for Disease Control and Prevention. This serious problem is much worse in hospitals in many other countries.

A bacterium known as methicillin-resistant *staphylococcus aureus,* or MRSA, has become resistant to most common antibiotics. This staph infection first appears on the skin as a red, swollen pimple or boil that may be painful and have pus. Many victims think they have a spider bite that will not heal. MRSA can cause a vicious type of pneumonia, flesh-eating wounds, and a quick death if it gets into the bloodstream.

This staph germ typically thrives on the body (mostly on the skin and in the nose) in health care settings where people have open wounds and tubes. But in recent years, it has been increasingly found in the general population. It can be picked up on playgrounds, in meeting rooms, and in gyms. It can spread through skin contact, tattoo needles, and contact with poorly laundered clothing items or shared items such as towels.

In 2005, MRSA infections contributed to the premature deaths of 18,650 people in the United States (more than the 17,000 who died from AIDS).

Ways to reduce the chances of infection from MRSA include frequent, thorough washing of hands; careful cleansing of even superficial wounds with soap and water; covering all wounds with a clean, dry bandage; and not sharing towels or other linens.

Most antibiotics work by crippling key proteins inside bacteria. But bacteria often develop immunity to such chemicals by modifying their protein receptor sites in ways that prevent the antibiotic molecules from entering their cells. In 2008, University of Pennsylvania scientists developed a compound that mimics molecules they identified in frog skin that essentially stab staph bacteria to death. This has inspired a new approach toward killing harmful bacteria by poking thousands of tiny holes in the bacterium's membranes. Ironically, at a time when chemical knowledge gained by studying amphibians could save millions of human lives, our activities are threatening many of the world's amphibian species. (Case Study, p. 93).

Critical Thinking

What are three things that could be done to slow the rate at which disease-causing organisms develop resistance to antibiotics?

CENGAGENOW™ Examine the HIV virus and how it replicates by using a host cell at CengageNOW™.

Antiviral drugs can slow the progress of AIDS, but they are expensive. With such drugs, an American with AIDS, on average, can expect to live about 24 years at a cost of about $25,200 a year. Such drugs cost too much for extensive use in developing countries with widespread AIDS infections.

According to the WHO, a global strategy to slow the spread of AIDS should have six major priorities. *First,* reduce the number of new infections below the number of deaths. *Second,* concentrate on the groups in a society that are most likely to spread the disease, such as sex workers, intravenous drug users, and soldiers. *Third,* provide free HIV testing and pressure people from high-risk groups to get tested.

Fourth, implement a mass-advertising and education program geared toward adults and schoolchildren to help prevent the disease, and emphasize abstinence, condom use, and male circumcision (which apparently can reduce the transmission of HIV to men from infected female partners by up to 60%). *Fifth,* provide free or low-cost drugs to slow the progress of the disease. *Sixth,* increase funding for research on the development of microbiocides such as a vaginal gel that could help women to protect themselves from getting HIV/AIDS.

HOW WOULD YOU VOTE? ☑

Should developed and developing nations mount an urgent global campaign to reduce the spread of HIV (**Core Case Study**) and to help countries afflicted by the disease? Cast your vote online at **academic.cengage.com/biology/miller**.

CORE CASE STUDY

The third largest viral killer is the *hepatitis B virus* (HBV), which damages the liver and kills about a

A Nightmare Flu Scenario

Common flu viruses kill up to 2% of the people they infect, most of them very young, old, weak, or sick. Most die from secondary infections of bacterial pneumonia. Flu viruses regularly contribute to the deaths of about 1 million people a year—36,000 of them in the United States.

Every now and then, an especially potent flu virus develops that can kill up to 80% of it victims, including healthy young adults. The result: a global flu pandemic that can kill millions of people within a few months and cause economic and social chaos. This happened in 1918 when a virus called Spanish flu spread rapidly around the globe and, within a few months, killed 20–50 million people—including 250,000–500,000 in the United States. Some people woke up healthy and were dead by nightfall.

Many health scientists believe that, sooner or later, a mass infection from a new and very potent flu virus will sweep the world

again. Infected people crisscrossing the world every day in airliners would hasten its spread. Health officials project that within a few months, such a flu pandemic could infect up to one-fourth of the world's population and kill anywhere from 51–360 million people, with the highest death tolls in South Asia, East Asia, Sub-Saharan Africa, and the Middle East. According to the U.S. Centers for Disease Control and Prevention (CDC), a worst-case pandemic could kill as many as 1.9 million and hospitalize 8.5 million Americans.

Pigs, chickens, ducks, and geese are the major reservoirs of flu viruses. As these viruses move from one animal species to another, they can mutate and exchange genetic materials with other flu viruses to create new flu viruses.

In 1997, a new H5N1 avian strain of flu virus genetically related to the 1918 killer strain emerged in Asia. This strain, commonly known as *bird flu* or *avian flu*, first showed

up in chickens that were probably infected by wild bird droppings. The virus then spread to people in Hong Kong, and since then, has spread to chickens and wild birds, including migratory birds that can spread the viruses far and wide. Between 2003 and 2008, it is known to have infected 376 people in at least 61 countries and killed 238, most of them in Vietnam and Indonesia.

Only one genetic mutation hurdle keeps this flu from becoming a pandemic: the current forms of this virus do not have the ability to spread easily from person to person. However, health officials tracking this virus say it is probably only a matter of time before strains with such an ability emerge.

Critical Thinking

What would you do to protect yourself if a global flu pandemic occurred?

million people each year. Like HIV, it is transmitted by unsafe sex, sharing of needles by drug users, infected mothers who pass the virus to their offspring before or during birth, and exposure to infected blood.

In recent years, several other viruses that cause previously unknown diseases have received widespread media coverage. One is the *West Nile virus,* transmitted to humans by the bite of a common mosquito that becomes infected by feeding on birds that carry the virus. Since 1992 when this virus emerged in the United States, it has spread from coast to coast. Between 1999 and 2007, it caused severe illness, including encephalitis and meningitis, in 11,016 people and killed 1,062 people. Fortunately, the chance of being infected and killed by this disease is low (about 1 in 2,500).

A second highly publicized virus is the *severe acute respiratory syndrome (SARS) virus,* which first appeared in humans in China in 2002. SARS, which has flu-like symptoms, can quickly turn into life-threatening pneumonia and is easily spread from person to person. During 6 months in 2003, the disease began spreading beyond China, where it infected at least 8,500 people and caused 812 deaths. Swift local action by the WHO and other health agencies helped to contain the spread of this disease by July 2003. But without careful vigilance, it might break out again.

West Nile and SARS are examples of diseases that move from animals to humans, a problem that is increasing as more people move into undeveloped areas to settle and interact with wildlife species (Science Focus, p. 446). Health officials are concerned about the

spread of West Nile virus, SARS, and other *emerging viral diseases,* and are working hard to control them. But in terms of annual infection rates and deaths, the three most dangerous viruses by far are still flu, HIV, and HBV (**Concept 17-2**). In 2007, for example, flu killed about 36,000 Americans and West Nile virus killed 109.

RESEARCH FRONTIER

Ecological medicine. See **academic.cengage.com/biology/miller**.

You can greatly reduce your chances of getting infectious diseases by practicing good, old-fashioned hygiene. Wash your hands thoroughly and frequently, avoid touching your face, and stay away from people who have flu or other viral diseases. Another growing hazard is infectious diseases caused by parasites, especially malaria (see the following Case Study).

■ CASE STUDY

Malaria—Death by Parasite-Carrying Mosquitoes

About one of every five people in the world—most of them living in poor African countries—is at risk from malaria (Figure 17-7). This should concern anyone traveling to malaria-prone areas because there is no vaccine for this disease.

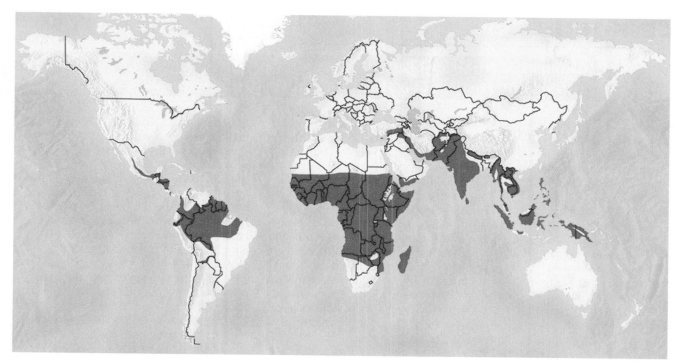

Figure 17-7 *Global outlook:* distribution of malaria. About 40% of the world's population lives in areas in which malaria is prevalent. Malaria kills at least 1 million—and perhaps as many as 2 million—people a year. More than 80% of these victims live in Sub-Saharan Africa—most of them children under the age of 5 (**Concept 17-2**). (Data from the World Health Organization and U.S. Centers for Disease Control and Prevention)

Malaria is caused by a parasite that is spread by the bites of certain mosquito species. It infects and destroys red blood cells, causing intense fever, chills, drenching sweats, anemia, severe abdominal pain, headaches, vomiting, extreme weakness, and greater susceptibility to other diseases. It kills at least 1 million and perhaps as many as 2 million people each year—an average of 2,700–5,400 deaths per day (**Concept 17-2**). About 90% of those dying are children younger than age 5. Many of the children who survive suffer brain damage or impaired learning ability.

Four species of protozoan parasites in the genus *Plasmodium* cause malaria. Most infections occur when an uninfected female of any of about 60 *Anopheles* mosquito species bites a person (usually at night) who is infected with *Plasmodium* parasite, ingests blood that contains the parasite, and later bites an uninfected person (Figure 17-8). *Plasmodium* parasites then move out of the mosquito and into the human's bloodstream and liver where they multiply. Malaria can also be transmitted by blood transfusions and by drug users sharing needles.

The malaria cycle repeats itself until immunity develops, treatment is given, or the victim dies. *Over the course of human history, malarial protozoa probably have killed more people than all the wars ever fought.*

During the 1950s and 1960s, the spread of malaria was sharply curtailed when swamplands and marshes where mosquitoes were breeding were drained or sprayed with insecticides, and drugs were used to kill the parasites in victims' bloodstreams. Since 1970, however, malaria has come roaring back. Most species of the *Anopheles* mosquito have become genetically resistant to most insecticides. Worse, the *Plasmodium* parasites have become genetically resistant to common antimalarial drugs. In addition, clearing and developing tropical forests (Figure 10-5, p. 218, and Figure 10-13, p. 224) leads to the spread of malaria among workers and the settlers who follow. Global warming is also

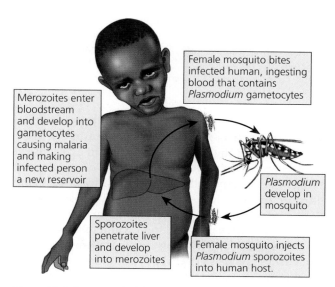

Merozoites enter bloodstream and develop into gametocytes causing malaria and making infected person a new reservoir

Female mosquito bites infected human, ingesting blood that contains *Plasmodium* gametocytes

Plasmodium develop in mosquito

Sporozoites penetrate liver and develop into merozoites

Female mosquito injects *Plasmodium* sporozoites into human host.

Figure 17-8 Science: the life cycle of malaria. *Plasmodium* parasites circulate from mosquito to human and back to mosquito.

Ecological Medicine: Tracking the Spread of Infectious Diseases to Humans from Other Animals

Infectious diseases are moving from one animal species to another and from wild and domesticated animal species to humans. Examples of infectious diseases transmitted from animals to humans include avian flu (Science Focus, p. 444), HIV (**Core Case Study**), SARS, West Nile virus, and Lyme disease. A 2008 study by Peter Daszak and other scientists found that about 60% of 325 diseases emerging between 1940 to 2004 were transmitted from animals to humans and the majority of those came from wild animals. And most people do not realize that pets such as rabbits, hamsters and other "pocket pets," and turtles can transfer various infectious diseases to humans.

Ecological medicine is a new interdisciplinary field devoted to tracking down these connections between wildlife and humans. Scientists in this new field are looking for ways to slow or prevent the spread of such diseases, and they have identified several human practices that encourage the spread of diseases from animals to humans.

One is the clearing of forests, which forces wild animals to move to other areas. For example, cutting down tropical rain forests has increased the spread of malaria by increasing the range of *Anopheles* mosquito species, which live in sunlit ponds and carry the parasite that infects humans. There is also concern that global warming will expand tropical conditions under which malaria-carrying species thrive (Figure 17-7).

Where forests have been cleared and fragmented to make way for suburbs in the eastern United States, the chances of many suburbanites becoming infected with debilitating Lyme disease have increased. The bacterium that causes this disease lives in the bodies of deer and white mice and is passed between these two animals and to humans by certain types of ticks.

Lyme disease usually starts with a red rash shaped like a bull's-eye around the tick bite. Antibiotics can cure people of the disease if they are administered within a month after the bite. Left untreated, though, Lyme disease can cause debilitating arthritis, heart disease, and nervous disorders. A vaccine is available, but does not give full protection. The best defense against the disease is to avoid tick bites by using insect repellent and wearing light-color clothing that is tucked in.

Expanding suburbs while fragmenting woodland areas has led to greatly reduced populations of foxes and wildcats, which had kept down populations of the white-footed mouse, which carries the Lyme bacterium. The result: white-footed mouse and tick populations have exploded and more suburbanites and hikers are becoming infected. Fortunately, the Lyme bacterium does not spread from person to person.

Another practice that encourages the spread of diseases takes place in parts of Africa and Asia where local people kill monkeys and other animals for bush meat (Figure 9-22, p. 206). They come in regular contact with primate blood and can be exposed to a simian strain of the HIV virus that causes AIDS and has infected 65 million people. Some monkeys also carry dangerous viruses such as herpes B that can be transferred to humans.

Another important factor is the legal and illegal international trade in wild species. In 2005, some 210 million wild animals— from kangaroos to iguanas, kinkajous, and tropical fish—were legally imported into the United States with limited quarantining and minimal screening for disease. Each year, countless other animals, bush meats, and animal parts are smuggled illegally across U.S. borders. A backpack carried off a flight from Nigeria may contain bush meat that could harbor the lethal Ebola virus. Salted duck eggs from South Korea could carry the deadly bird flu virus. Exotic birds or monkey paws, which are often smuggled internationally, could harbor various diseases that could jump to humans.

Factory meat production and the global trade of livestock animals can also increase the spread of food-borne infectious diseases to humans. For example, a deadly form of *E. coli* bacteria sometimes spreads from livestock animals and animals such as wild boars to people when they eat meat contaminated by animal manure. Salmonella bacteria, found on animal hides and in poorly processed meat can also cause food-borne disease.

Finally, global trade in general contributes to the spread of diseases among species. Infectious organisms can be transferred in shipping crates and containers, agricultural products, and ship ballast water. And global travel, especially ecotourism in wilderness areas, can spread infectious organisms around the world. **GREEN CAREER:** Ecological medicine

Critical Thinking

Do you think that air travelers should be more carefully monitored and possibly more restricted in the face of threats such as the Ebola and bird flu viruses? Explain.

likely to increase cases of malaria as populations of malaria-carrying mosquitoes spread to warmer areas.

Malaria and the AIDS virus interact in a vicious cycle because of the weakened condition of people with these diseases, especially in parts of Africa where both diseases are prevalent. People with HIV (**Core Case Study**) are more vulnerable to malaria, and people with malaria are more vulnerable to HIV.

Researchers are working to develop new antimalarial drugs (such as *artemisinins,* a Chinese remedy derived from the sweet wormwood plant), vaccines, and biological controls for *Anopheles* mosquitoes. But these approaches receive too little funding and have proved more difficult to implement than they were originally thought to be. In 2005, malaria researchers were evaluating the use of two fungi, which are apparently harmless to humans and the environment, to kill malaria-carrying mosquitoes. If effective, they could be sprayed on walls or soaked into mosquito nets.

U.S. researchers are also developing mosquitoes genetically engineered to resist the malaria parasite and outbreed conventional malaria-carrying mosquitoes. And biochemist David Schooley and his colleagues are looking for a way to kill mosquitoes through dehydra-

Figure 17-9 A boy in Brazil's Amazon sleeps under an insecticide-treated mosquito net to reduce the risk of getting bitten by malaria-carrying mosquitoes. Such nets cost about $5 each and can be donated through groups such as **www.MalariaNoMore.org**.

tion, before they have a chance to bite and infect a human, by exposing the mosquitoes to a hormone that dramatically increases their urination.

Another approach is to provide poor people in malarial regions with free or inexpensive, long-lasting, insecticide-treated bed nets (Figure 17-9) and window screens. Also, zinc and vitamin A supplements could be used to boost resistance to malaria in children. And we can greatly reduce the incidence of malaria by spraying the insides of homes with low concentrations of the pesticide DDT twice a year at a cost of about $10. Under an international treaty enacted in 2002, DDT and five similar pesticides are being phased out in developing countries. However, in 2006 the WHO supported the use of DDT for malaria control.

Columbia University scientist Jeffrey Sachs estimates that spending $2–3 billion a year on preventing and treating malaria might save more than a million lives a year. Sachs notes, "This is probably the best bargain on the planet."

RESEARCH FRONTIER

Finding new drugs and other treatments for malaria and finding ways to hinder the mosquitoes that transmit the malaria parasite. See **academic.cengage.com/biology/miller**.

CENGAGENOW™ Watch through a microscope what happens when a mosquito infects a human with malaria at CengageNOW.

We Can Reduce the Incidence of Infectious Diseases

Good news. According to the WHO, the global death rate from infectious diseases decreased by about two-thirds between 1970 and 2000 and is projected to continue dropping. Also, between 1971 and 2000, the percentage of children in developing countries immunized with vaccines to prevent tetanus, measles, diphtheria, typhoid fever, and polio increased from 10% to 84%—saving about 10 million lives each year. It costs about $30 to get a basic package of vaccines to a child—an affordable way to save a child's life for roughly the price of a single night out at the movies in a developed country.

Figure 17-10 lists measures promoted by health scientists and public health officials to help prevent or reduce the incidence of infectious diseases—especially in developing countries. An important breakthrough has

SOLUTIONS
Infectious Diseases

- Increase research on tropical diseases and vaccines

- Reduce poverty

- Decrease malnutrition

- Improve drinking water quality

- Reduce unnecessary use of antibiotics

- Educate people to take all of an antibiotic prescription

- Reduce antibiotic use to promote livestock growth

- Require careful hand washing by all medical personnel

- Immunize children against major viral diseases

- Provide oral rehydration for diarrhea victims

- Conduct global campaign to reduce HIV/AIDS

Figure 17-10 Solutions: ways to prevent or reduce the incidence of infectious diseases, especially in developing countries. **Question:** Which three of these approaches do you think are the most important?

been the development of simple *oral rehydration therapy* to help prevent death from dehydration for victims of severe diarrhea, which causes about one-fourth of all deaths of children younger than age 5 (**Concept 17-2**). It involves administering a simple solution of boiled water, salt, and sugar or rice, at a cost of only a few cents per person. It has been the major factor in reducing the annual number of deaths from diarrhea from 4.6 million in 1980 to 1.9 million in 2006. Few investments have saved so many lives at such a low cost. In 2006, the WHO estimated that implementing the solutions in Figure 17-10 could save the lives of as many as 4 million children younger than age 5 each year.

Recall that more than a third of the world's people—2.6 billion—do not have decent bathroom facilities, and more than a billion get their water for drinking, washing, and cooking from sources polluted by animal and human feces. A key to reducing sickness and premature death from infectious disease is to focus on providing people with simple latrines and access to safe drinking water. The U.N. estimates that this could be done for about $20 billion a year—about what rich countries with almost universal access to clean water spend each year on bottled water.

Bad news. The WHO estimates that only 10% of global medical research and development money goes toward preventing infectious diseases in developing countries, even though more people worldwide suffer and die from these diseases than from all other diseases combined. Fortunately, the problem is getting more attention. In recent years, philanthropists including Bill and Melinda Gates and Warren E. Buffet have donated billions of dollars to improve global health, with primary emphasis on preventing infectious diseases in developing countries. **GREEN CAREER:** Infectious disease prevention

17-3 What Types of Chemical Hazards Do We Face?

▶ **CONCEPT 17-3** There is growing concern about chemicals that can cause birth defects and cancers and disrupt the human immune, nervous, and endocrine systems.

Some Chemicals Can Cause Cancers, Mutations, and Birth Defects,

A **toxic chemical** is one that can cause temporary or permanent harm or death to humans and animals. In 2004, the U.S. Environmental Protection Agency (EPA) listed arsenic, lead, mercury, vinyl chloride (used to make PVC plastics), and polychlorinated biphenyls (PCBs, Case Study, at right) as the five most toxic substances in terms of threats to human and environmental health. Note that three of them are metallic chemical elements.

There are three major types of potentially toxic agents. **Carcinogens** are chemicals, types of radiation, or certain viruses that can cause or promote *cancer*—a disease in which malignant cells multiply uncontrollably and create tumors that can damage the body and often lead to premature death.

Examples of carcinogens are arsenic, benzene, chloroform, formaldehyde, gamma radiation, nickel, PCBs, radon, certain chemicals in tobacco smoke, UV radiation, X-rays, and vinyl chloride. (For a longer list, see **ntp.niehs.nih.gov/**.)

Typically, 10–40 years may elapse between the initial exposure to a carcinogen and the appearance of detectable symptoms. Partly because of this time lag, many healthy teenagers and young adults have trouble believing that their smoking, drinking, eating, and other lifestyle habits today could lead to some form of cancer before they reach age 50.

The second major type of toxic agent, **mutagens,** includes chemicals or forms of radiation that cause mutations, or changes, in the DNA molecules found in cells, or that increase the frequency of such changes. Most mutations cause no harm but some can lead to cancers and other disorders. For example, nitrous acid (HNO_2), formed by the digestion of nitrite (NO_2^-) preservatives in foods, can cause mutations linked to stomach cancer in people who consume large amounts of processed foods and wine with such preservatives. Harmful mutations occurring in reproductive cells can be passed on to offspring and to future generations. (For a list of mutagens see **www.evol.nw.ru/~spirov/hazard/mutagen_lst.html**.)

Third, **teratogens** are chemicals that cause harm or birth defects to a fetus or embryo. Ethyl alcohol is a teratogen. Drinking during pregnancy can lead to offspring with low birth weight and a number of physical, developmental, behavioral, and mental problems. Other teratogens are angel dust, benzene, cadmium, formaldehyde, lead, mercury, mescaline, PCBs, phthalates, thalidomide, vinyl chloride, and hundreds of other organic compounds. (For a list, see **ptcl.chem.ox.ac.uk/MSDS/teratogens.html**.) Between 2001 and 2006, birth defects in Chinese infants soared by nearly 40%. Officials link this to the country's growing

pollution, especially from coal-burning power plants and industries.

PCBs Are Everywhere—A Legacy from the Past

PCBs are a class of more than 200 chlorine-containing organic compounds that are very stable and nonflammable. They exist as oily liquids or solids that can enter the air as a vapor. Between 1929 and 1977, PCBs were widely used as lubricants, hydraulic fluids, and electrical insulators in transformers and capacitors. They also became ingredients in a variety of products including paints, fire retardants in fabrics, preservatives, adhesives, and pesticides.

The U.S. Congress banned the domestic production of PCBs in 1977 after research showed that they could cause liver and other cancers in test animals and, according to the EPA, probably can cause cancers in humans. A 1996 study related exposure of fetuses to PCBs in the womb to learning disabilities in children.

Production of PCBs has also been banned in most other countries. But the potential health threats from these chemicals will be with us for a long time. For decades, PCBs entered the air, water, and soil during their manufacture, use, and disposal, as well as from accidental spills and leaks. Because PCBs breakdown very slowly in the environment, they can travel long distances in the air and be deposited far away from where they were released. Because they are fat soluble, PCBs can also be biologically magnified (Figure 9-19, p. 202) in food chains and webs.

As a result, PCBs are now found everywhere—in soil, air, lakes, rivers, fish, birds, your body, and the bodies of polar bears and penguins in the Arctic. PCBs are present even in the milk of nursing mothers, although most scientists say the health benefits of mother's milk

outweigh the risk from exposure to trace amounts of PCBs in breast milk.

According to the EPA, about 70% of all the PCBs made in the United States are still in the environment. Figure 17-11 shows potential pathways on which persistent toxic chemicals such as PCBs move through the living and nonliving environment.

THINKING ABOUT
PCBs

We are stuck with long-term exposure to trace amounts of PCBs with mostly unknown long-term health effects, but what environmental lesson can we learn from the widespread use of these persistent chemicals?

Some Chemicals May Affect Our Immune, Nervous, and Endocrine Systems

Since the 1970s, a growing body of research on wildlife and laboratory animals, along with some studies of humans, suggest that long-term exposure to some chemicals in the environment can disrupt the body's immune, nervous, and endocrine systems (**Concept 17-3**).

The *immune system* consists of specialized cells and tissues that protect the body against disease and harmful substances by forming antibodies that render invading agents harmless. Some chemicals such as arsenic, methylmercury, and dioxins can weaken the human immune system and leave the body vulnerable to attacks by allergens and infectious bacteria, viruses, and protozoa.

Some natural and synthetic chemicals in the environment, called *neurotoxins*, can harm the human *nervous system* (brain, spinal cord, and peripheral nerves). Effects can include behavioral changes, learning disabilities, retardation, attention deficit disorder, paralysis, and death. Examples of neurotoxins are PCBs, methyl mercury (Science Focus, pp. 450–451), arsenic, lead, and certain pesticides. For example, a 2007 study by scientists Beate Ritz and Caroline Tanner found that farm workers exposed to the widely used weed killer paraquat had two to three times the normal risk of suffering from Parkinson's disease, a degenerative brain disease that eventually paralyzes victims.

Figure 17-11 Potential pathways on which toxic chemicals move through the living and nonliving environment.

Mercury's Toxic Effects

Mercury (Hg) is a teratogen and a potent neurotoxin that interferes with the nervous system and brain function. This toxic metal is released into the air from rocks, soil, and volcanoes and by vaporization from the ocean. Such natural sources account for about one-third of the mercury reaching the atmosphere each year.

According to the EPA, the remaining two-thirds come from human sources—mostly coal-burning power plants, waste incinerators, and chemical manufacturing plants. Other lesser sources gold and silver mines, metal ore smelters, and facilities that inciner-

ate or crush products containing mercury, such as batteries, electronic switches, and relays.

Mercury is persistent and, because it is an element, cannot be degraded. Therefore this global pollutant accumulates in soil, water, and the bodies of people and other animals that feed high on food chains. This includes polar bears, toothed whales, and seals living in the Arctic, which is a global mercury hotspot. In 2007, scientists from the EPA and Oregon State University surveyed 2,707 fish randomly collected from 626 rivers in 12 U.S. states. They found mercury in every fish and every

river, but generally at levels considered safe for people to eat occasionally.

In the atmosphere, some elemental mercury is converted to more toxic inorganic and organic mercury compounds that can be deposited in aquatic environments. In acidic aquatic systems, bacteria can convert inorganic mercury compounds to highly toxic *methylmercury* (CH_3Hg^+), which can be biologically magnified in food chains and webs (Figure 17-A). As a result, high levels of methylmercury are often found in the tissues of predatory fishes such as large albacore (white) tuna, sharks, swordfish, king mackerel, tilefish,

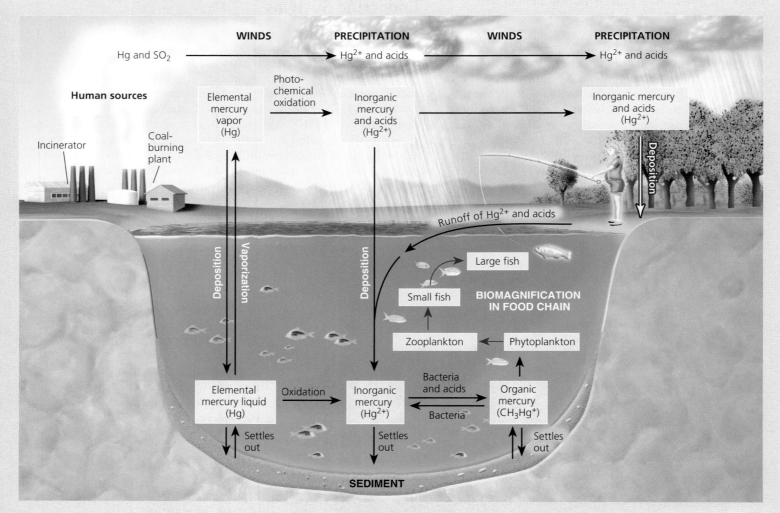

Figure 17-A Science: cycling of mercury in aquatic environments, in which mercury is converted from one form to another. The most toxic form to humans is methylmercury (CH_3Hg^+), which can be biologically magnified in aquatic food chains and webs. Some mercury is also released back into the atmosphere as mercury vapor.

SOLUTIONS

Mercury Pollution

Figure 17-B Ways to prevent or control inputs of mercury into the environment from human sources—mostly coal-burning power plants and incinerators. **Question:** Which four of these solutions do you think are the most important?

Prevention

Phase out waste incineration

Remove mercury from coal before it is burned

Switch from coal to natural gas and renewable energy resources such as wind, solar cells, and hydrogen

Convert coal to liquid or gaseous fuel

Phase out use of mercury in batteries, TVs, compact fluorescent lightbulbs, and all other products unless they are recycled

Control

Sharply reduce mercury emissions from coal-burning plants and incinerators

Tax each unit of mercury emitted by coal-burning plants and incinerators

Require labels on all products containing mercury

Collect and recycle mercury-containing electric switches, relays, and dry-cell batteries

walleye, and marlin, which feed at high trophic levels in food chains and webs.

In 2007, two studies identified five mercury hotspots in New England, New York, and Nova Scotia, apparently related to mercury emissions by coal-burning power plants. Another 2007 study found unusually high levels of mercury in 178 wild songbirds in New York State, which do not live in aquatic settings and never eat fish.

Humans are exposed to mercury in two ways. *First,* they may inhale vaporized elemental mercury (Hg) or particulates of inorganic mercury salts, such as HgS and $HgCl_2$. *Second,* they may eat fish contaminated with highly toxic methylmercury.

The greatest risk from exposure to low levels of methylmercury is brain damage in fetuses and young children. Scientists estimate that mercury exposure for a fetus in the womb can cause neurological problems, which can eventually lead to a lowered IQ and poor school performance. Roughly 30,000 to 60,000 of the children born each year in the

United States are likely to have such problems because of mercury exposure. In addition methylmercury can harm the heart, kidneys, and immune system in adults.

Since 2004, the U.S. Food and Drug Administration (FDA) and the EPA have advised nursing mothers, pregnant women, and women who may become pregnant not to eat shark, swordfish, king mackerel, or tilefish and to limit their consumption of albacore tuna to no more than 170 grams (6 ounces) per week. The EPA also warned that one-fourth of the nation's rivers, one-third of its lakes (including all of the Great Lakes), and three-fourths of its coastal waters are contaminated with mercury and other pollutants.

In 2007, the EPA estimated that one of every six women of childbearing age in the United States has enough mercury in her blood to harm a developing fetus. This means that 630,000 of the 4 million babies born in the United States each year could suffer neurological damage from mercury exposure

before birth. Figure 17-B lists ways to prevent or control human exposure to mercury.

In its 2003 report on global mercury pollution, the U.N. Environment Programme recommended phasing out coal-burning power plants and waste incinerators throughout the world as rapidly as possible. In 2007, 16 U.S. states filed a lawsuit in federal court against the EPA for not requiring sharp reductions in mercury emissions from all U.S. coal-burning power plants.

Other recommended goals are to reduce or eliminate use of mercury in batteries and paints and in factories that produce chlorine by no later than 2020. Substitute materials and processes are available for all of these uses.

Critical Thinking

Should we phase out all coal burning and waste incineration as rapidly as possible as a way to sharply reduce mercury pollution? Explain. How might your lifestyle change if this were done?

The *endocrine system* is a complex network of glands that release tiny amounts of *hormones* into the bloodstreams of humans and other vertebrate animals. Low levels of these chemical messengers turn on and turn off bodily systems that control sexual reproduction, growth, development, learning, and behavior. Each type of hormone has a unique molecular shape that allows it to attach to certain cells, using a part of the cell called a *receptor,* and to transmit its chemical message (Figure 17-12, left). In this "lock-and-key" relationship, the receptor is the lock and the hormone is the key.

Certain chemicals, called *hormonally active agents* (HAAs), can mimic hormones. Exposure to low levels of HAAs can impair reproductive systems and sexual development and cause physical and behavioral disorders. Examples of HAAs include aluminum, atrazine and several other herbicides, DDT, mercury, PCBs (Case Study, p. 449), phthalates, and bisphenol A (Science Focus, at right). Natural biological evolution has not equipped us to deal with these synthetic hormone imposters.

Some hormone mimics are chemically similar to estrogens (female sex hormones). They can disrupt the endocrine system by attaching to estrogen receptor molecules (Figure 17-12, center). Others, called *hormone blockers,* disrupt the endocrine system by preventing natural hormones such as androgens (male sex hormones) from attaching to their receptors (Figure 17-12, right).

Estrogen mimics and hormone blockers are sometimes called *gender benders* because of their possible effects on sexual development and reproduction. In males, excessive levels of female hormones can cause feminization, smaller penises, lower sperm counts, and the presence of both male and female sex organs (hermaphroditism). And there is growing concern about still another group of HAAs—pollutants that can act as *thyroid disrupters*—which cause growth, weight, brain, and behavioral disorders.

The harmful effects of certain *phthalates* (pronounced "thall-eights") are also of concern. These chemicals are used to soften polyvinyl chloride (PVC) plastic found in a variety of products such as soft vinyl toys, teething rings, and blood storage bags, IV bags, and medical tubes used in hospitals. And they are used as solvents in many consumer products, including perfumes, cosmetics, body lotions, hair sprays, deodorants, nail polishes and shampoos. A 2008 study by Sheela Sathyanarayana and her colleagues tested urine from the diapers of 163 infants aged 2 months to 28 months. They found that 81% of the urine had measurable amounts of seven or more phthalates contained in baby lotions, baby powders, and baby shampoos.

Exposure of laboratory animals to high doses of various phthalates has caused birth defects, liver cancer, kidney and liver damage, premature breast development, immune suppression, and abnormal sexual development in these animals. The European Union and at least 14 other countries have banned phthalates. But scientists in the United States are divided on its risks to human health and reproductive systems. U.S. toy makers say that phthalates, which have been used for more than 20 years in baby items, pose no threats. In addition, they warn that substitutes could make plastic toys more brittle and subject to breaking, which could lead to harm to children.

Some scientists hypothesize that certain problems may be related to increased levels of hormone disruptors in our bodies. Such problems include sharp drops in male sperm counts and male sperm mobility occurring in 20 countries on six continents, rising rates of testicular cancer and genital birth defects in men, and increased breast cancer rates in women.

Normal Hormone Process

Hormone Mimic

Hormone Blocker

Figure 17-12 Hormones are molecules that act as messengers in the endocrine system to regulate various bodily processes, including reproduction, growth, and development. Each type of hormone has a unique molecular shape that allows it to attach to specially shaped receptors on the surfaces of, or inside, cells and to transmit its chemical message (left). Molecules of certain pesticides and other synthetic chemicals have shapes similar to those of natural hormones and can disrupt the endocrine systems in humans and other animals (center and right). These molecules are called *hormonally active agents* (HAAs). Because of the difficulty of determining the harmful effects of long-term exposure to low levels of HAAs, there is uncertainty about their effects on human health.

A widely used estrogen mimic is bisphenol A (BPA). It is a chemical building block in certain plastics (especially polycarbonates) used in a variety of products including water and baby bottles, sports drink and juice bottles, microwave dishes, food storage containers, food and beverage can liners (including cans of infant formula), and dental fillings.

Studies show that this chemical can leach out of many of these products into food and water, especially when exposed to heat or acidic and basic liquids and foods. Research indicates that 95% of Americans and most of the citizens in other industrialized countries have trace levels of bisphenol A in their bodies.

Ninety-four studies by independent laboratories have found a number of adverse effects on test animals from exposure to very low levels of bisphenol A. These effects include brain damage, prostrate disease, breast cancer, early puberty, reduced sperm count, impaired immune function, type 2 diabetes, hyperactivity, increased aggressiveness, impaired learning, increased addiction to drugs such as amphetamines, decreased sex drive in males, and obesity in unborn test animals exposed to bisphenol A. On the other hand, 12 studies funded by the chemical industry found no evidence or weak evidence for adverse effects from low-level exposure to bisphenol A on test animals. In recent years, two groups of scientists convened by the U.S. National Institutes of Health have reached opposite conclusions about whether exposure to trace amounts of BPA poses a threat to human health.

This controversy may take decades to resolve. Meanwhile, some scientists call for banning or phasing out many uses of bisphenol A (especially in baby bottles and sipping cups used by more vulnerable infants and in the plastic lining of baby formula cans) as a preventive or precautionary measure and encouraging scientists to look for less potentially harmful substitutes. Others say we need more research before taking such measures that would cause huge economic losses for the chemical industry.

Critical Thinking

Should we ban or phase out the use of BPA? Defend your choice. What beneficial or harmful effects might such a ban have on your life or on the lives of future generations?

Other scientists disagree and point out that there have not been enough scientific studies and that we lack statistical evidence to link these medical problems to HAA levels in humans. They call for doing more research before banning or severely restricting HAAs, which could cause huge economic losses for companies that make them. Such research will take decades. Meanwhile, some scientists believe that as a precaution, we should sharply reduce our use of potential hormone disrupters.

RESEARCH FRONTIER

Evaluating the health effects of HAAs. See **academic.cengage.com/biology/miller**.

THINKING ABOUT
Hormone Disrupters

Should we ban or severely restrict the use of potential hormone disrupters? What beneficial or harmful effects might this have in your life?

17-4 How Can We Evaluate and Deal with Chemical Hazards?

▶ **CONCEPT 17-4A** Scientists use live laboratory animals, non-animal tests, case reports of poisonings, and epidemiological studies to estimate the toxicity of chemicals, but these methods have limitations.

▶ **CONCEPT 17-4B** Many health scientists call for much greater emphasis on pollution prevention to reduce our exposure to potentially harmful chemicals.

Many Factors Determine the Harmful Health Effects of a Chemical

We are exposed to small amounts of potentially harmful chemicals every day in the air we breathe, the water we drink, and the food we eat. **Toxicology** is the study of the harmful effects of chemicals on humans and other organisms. In effect, it is a study of poisons.

Toxicity is a measure of how harmful a substance is—its ability to cause injury, illness, or death to a living organism. A basic principle of toxicology is that *any synthetic or natural chemical can be harmful if ingested in a large enough quantity*. But the critical question is this: *At*

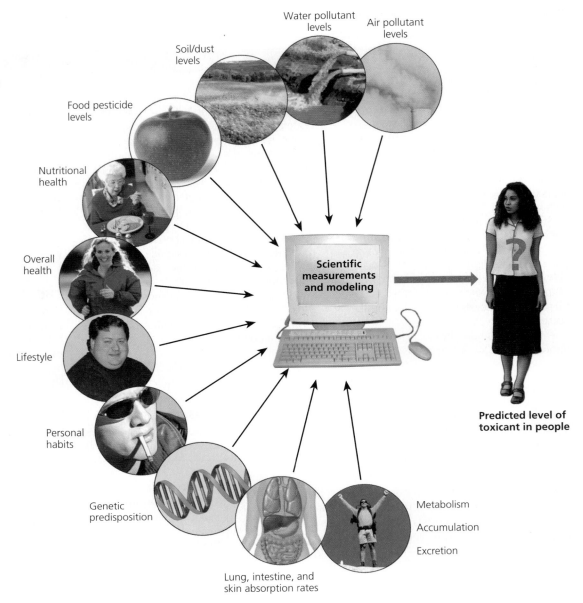

Figure 17-13 Science: Estimating human exposure to chemicals and measuring their effects are very difficult because of the many and often poorly understood variables involved. **Question:** What three of these factors might make you more vulnerable to the harmful effects of chemicals?

Water pollutant levels

Air pollutant levels

Soil/dust levels

Food pesticide levels

Nutritional health

Overall health

Lifestyle

Personal habits

Genetic predisposition

Scientific measurements and modeling

Lung, intestine, and skin absorption rates

Metabolism

Accumulation

Excretion

Predicted level of toxicant in people

what level of exposure to a particular toxic chemical will the chemical cause harm? This is the meaning of the chapter-opening quote by the Swiss scientist Paracelsus about the dose making the poison.

This is a difficult question to answer because of the many variables involved in estimating the effects of human exposure to chemicals (Figure 17-13). A key factor is the **dose,** the amount of a harmful chemical that a person has ingested, inhaled, or absorbed through the skin.

The effects of a particular chemical can also depend upon the age of the person exposed to it. For example, toxic chemicals usually have a greater effect on children than on adults (Case Study, at right). Toxicity also depends on *genetic makeup,* which determines an individual's sensitivity to a particular toxin. Some individuals are sensitive to a number of toxins—a condition known

as *multiple chemical sensitivity* (MCS). Another factor is how well the body's detoxification systems (such as the liver, lungs, and kidneys) work.

Several other factors can affect the level of harm caused by a chemical. One is its *solubility.* Water-soluble toxins (which are often inorganic compounds) can move throughout the environment and get into water supplies and the aqueous solutions that surround the cells in our bodies. Oil- or fat-soluble toxins (which are usually organic compounds) can penetrate the membranes surrounding cells, because the membranes allow similar oil-soluble chemicals to pass through them. Thus, oil- or fat-soluble toxins can accumulate in body tissues and cells.

Another factor is a substance's *persistence,* or resistance to breakdown. Many chemicals, such as DDT and PCBs (Case Study, p. 449), have been widely used

because they are not easily broken down in the environment. This means that more people and wildlife are likely to come in contact with them and that they are more likely to remain in the body and have long-lasting harmful health effects.

Another factor is *biological magnification,* in which the concentrations of some potential toxins in the environment increase as they pass through the successive trophic levels of food chains and webs. Organisms at low trophic levels might ingest only small amounts of a toxin, but each animal on the next trophic level up that eats many of those organisms will take in increasingly larger amounts of that toxin (Figure 9-19, p. 202). Examples of chemicals that can be biomagnified include long-lived, fat-soluble, organic compounds, such as DDT and PCBs.

The damage to health resulting from exposure to a chemical is called the **response.** An *acute effect* is an immediate or rapid harmful reaction to an exposure—ranging from dizziness and nausea to death. A *chronic effect* is a permanent or long-lasting consequence (kidney or liver damage, for example) of exposure to a single dose or to repeated lower doses of a harmful substance.

■ CASE STUDY
Protecting Children from Toxic Chemicals

Everyone on the planet is exposed to an array of toxic chemicals whose long-term effects are largely unknown. In 2005, the Environmental Working Group analyzed umbilical cord blood from 10 randomly selected newborns in U.S. hospitals. Of the 287 chemicals detected, 180 cause cancers in humans or animals, 217 damage the brain and nervous systems in test animals, and 208 cause birth defects or abnormal development in test animals. Scientists do not know what harm, if any, might be caused by the very low concentrations of these chemicals found in infants' blood.

But infants and young children are more susceptible to the effects of toxic substances than are adults for three major reasons. *First,* children generally breathe more air, drink more water, and eat more food per unit of body weight than do adults. *Second,* they are exposed to toxins in dust or soil when they put their fingers, toys, or other object in their mouths (as they frequently do). *Third,* children usually have less well-developed immune systems and body detoxification processes than adults have.

In 2003, the U.S. EPA proposed that in determining any risk, regulators should assume children face a risk 10 times higher than that faced by adults. Some health scientists contend that these guidelines are too weak. They suggest that, to be on the safe side, we should assume that this risk for children is 100 times that of adults.

Scientists Use Live Laboratory Animals and Nonanimal Tests to Estimate Toxicity

The most widely used method for determining toxicity is to expose a population of live laboratory animals to measured doses of a specific substance under controlled conditions. Laboratory-bred mice and rats are widely used because, as mammals, their systems function somewhat like humans, they are small, and can reproduce rapidly under controlled laboratory conditions.

Animal tests take 2–5 years, involve hundreds to thousands of test animals, and cost as much as $2 million per substance tested. Such tests can be painful to the test animals and can kill or harm them. Animal welfare groups want to limit or ban the use of test animals or ensure that they are treated in the most humane manner possible.

Scientists estimate the toxicity of a chemical by determining the effects of various doses of the chemical on test organisms and then by plotting the results in a **dose-response curve** (Figure 17-14). One approach is to determine the lethal dose—the amount needed to kill an animal. A chemical's *median lethal dose (LD50)* is the dose that can kill 50% of the animals (usually rats and mice) in a test population within an 18-day period.

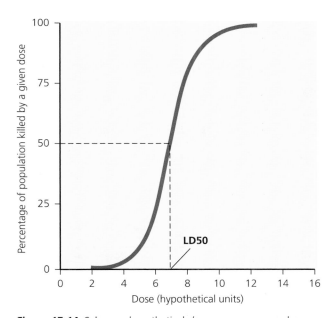

Figure 17-14 Science: hypothetical *dose-response curve* showing determination of the LD50, the dosage of a specific chemical that kills 50% of the animals in a test group. Toxicologists use this method to compare the toxicities of different chemicals.

Table 17-1

Toxicity Ratings and Average Lethal Doses for Humans

Toxicity Rating	LD50 (milligrams per kilogram of body weight)*	Average Lethal Dose**	Examples
Supertoxic	Less than 5	Less than 7 drops	Nerve gases, botulism toxin, mushroom toxin, dioxin (TCDD)
Extremely toxic	5–50	7 drops to 1 teaspoon	Potassium cyanide, heroin, atropine, parathion, nicotine
Very Toxic	50–500	1 teaspoon to 1 ounce	Mercury salts, morphine, codeine
Moderately toxic	500–5,000	1 ounce to 1 pint	Lead salts, DDT, sodium hydroxide, sodium fluoride, sulfuric acid, caffeine, carbon tetrachloride
Slightly toxic	5,000–15,000	1 pint to 1 quart	Ethyl alcohol, Lysol, soaps
Essentially nontoxic	15,000 or greater	More than 1 quart	Water, glycerin, table sugar

*Dosage that kills 50% of individuals exposed.
**Amounts of substances in liquid form at room temperature that are lethal when given to a 70-kilogram (150-pound) human.

Chemicals vary widely in their toxicity (Table 17-1). Some poisons can cause serious harm or death after a single exposure at very low dosages. Others cause such harm only at dosages so huge that it is nearly impossible to get enough into the body to cause injury or death. Most chemicals fall between these two extremes.

There are two general types of dose-response curves (Figure 17-15). With the *nonthreshold dose-response model* (Figure 17-15, left), any dosage of a toxic chemical causes harm that increases with the dosage. With the *threshold dose-response model* (Figure 17-15, right), a threshold dosage must be reached before any detectable harmful effects occur, presumably because the body can repair the damage caused by low dosages of some substances.

Establishing which model applies at low dosages is extremely difficult and controversial. To be on the safe side, scientists often assume that even a small dose will have an effect, and they choose the nonthreshold dose-response model.

In testing, scientists often use fairly high dosages to reduce the number of test animals needed, obtain results quickly, and lower costs. Otherwise, manufacturers would need to run tests on millions of laboratory animals for many years, which means they could not afford to test most chemicals. For the same reasons, scientists usually use mathematical models to *extrapolate,* or estimate the effects of low-dose exposures based on the measured results of high-dose exposures. Then they extrapolate these results from test organisms to humans to estimate LD50 values for acute toxicity (Table 17-1).

Some scientists challenge the validity of extrapolating data from test animals to humans because human physiology and metabolism differ from those of the test animals. Other scientists say that such tests and models work fairly well (especially for revealing cancer risks) when the correct experimental animal is chosen or when a chemical is toxic to several different test-animal species.

More humane methods for carrying out toxicity tests are available and are increasingly being used to replace live animal testing. They include computer simulations and testing with tissue cultures, chicken egg membranes, and individual animal cells, instead of with whole, live animals. High-speed robot testing devices can now measure the biological activity of more than one million compounds a day to help determine their possible toxic effects.

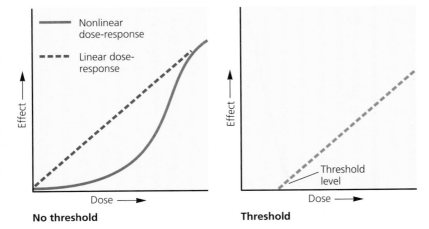

Figure 17-15 Science: two types of *dose-response curves.* The linear and nonlinear curves in the left graph apply if even the smallest dosage of a chemical has a harmful effect that increases with the dosage. The curve on the right applies if a harmful effect occurs only when the dosage exceeds a certain *threshold level.* Which model is better for a specific harmful agent is often uncertain and controversial because of the difficulty in estimating the responses to very low dosages.

In 2008, U.S. government labs began making much greater use of such nonanimal testing methods because they are faster and cheaper than animals tests and because of growing public concern over animal testing. For example, government toxicity testing labs can run between 10 and 100 tests a year using live rodents such as rats and mice but can run more than 10,000 tests a day using specialized cells and computerized testing measurements.

RESEARCH FRONTIER

Computer modeling and other alternatives to animal testing. See **academic.cengage.com/biology/miller**.

There are several problems involved with using laboratory experiments to estimate toxicities (**Concept 17-4A**). In real life, each of us is exposed to a variety of chemicals, some of which can interact in ways that decrease or enhance their short- and long-term individual effects. Toxicologists already have great difficulty in estimating the toxicity of a single substance. And adding the problem of evaluating *mixtures of potentially toxic substances*, separating out which ones are the culprits, and determining how they can interact with one another is overwhelming from a scientific and economic standpoint. For example, just studying the interactions of three of the 500 most widely used industrial chemicals would take 20.7 million experiments—a physical and financial impossibility.

THINKING ABOUT
Animal Testing

Should laboratory-bred mice, rats, and other animals be used to determine toxicity and other effects of chemicals? Explain.

There Are Other Ways to Estimate the Harmful Effects of Chemicals

Scientists use several other methods to get information about the harmful effects of chemicals on human health. For example, *case reports*, usually made by physicians, provide information about people suffering adverse health effects or death after exposure to a chemical. Such information often involves accidental or deliberate poisonings, drug overdoses, homicides, or suicide attempts.

Most case reports are not reliable sources for estimating toxicity because the actual dosage and the exposed person's health status are often unknown. But such reports can provide clues about environmental hazards and suggest the need for laboratory investigations.

Another source of information is *epidemiological studies*, which compare the health of people exposed to a particular chemical (the *experimental group*) with

the health of a similar group of people not exposed to the agent (the *control group*). The goal is to determine whether the statistical association between exposure to a toxic chemical and a health problem is strong, moderate, weak, or undetectable.

Four factors can limit the usefulness of epidemiological studies. *First*, in many cases, too few people have been exposed to high enough levels of a toxic agent to detect statistically significant differences. *Second*, they usually take a long time. *Third*, conclusively linking an observed effect with exposure to a particular chemical is difficult because people are exposed to many different toxic agents throughout their lives and can vary in their sensitivity to such chemicals. And *fourth*, we cannot use epidemiological studies to evaluate hazards from new technologies or chemicals to which people have not yet been exposed.

Are Trace Levels of Toxic Chemicals Harmful?

Almost everyone is now exposed to potentially harmful chemicals (Figure 17-16, p. 458) that have built up to trace levels in their blood and other parts of their bodies. Trace amounts of birth control pills, blood pressure medicines, antidepressants, painkillers, and a host of other chemicals with largely unknown effects on human health are being released into waterways from sewage treatment plants or are leaching into groundwater from home septic sewage systems.

Should we be concerned about trace amounts of various synthetic chemicals in air, water, food, and our bodies? The honest answer is that, in most cases, we do not know, because there is too little data and because of the difficulty of determining the effects of low levels of these chemicals (**Concept 17-4A**).

Some scientists view trace amounts of such chemicals with alarm, especially because of their potential long-term effects on the human immune, nervous, and endocrine systems. Others view the risks from trace levels as minor. They point out that average life expectancy has been increasing in most countries, especially developed ones, for decades. And many industry officials contend that the concentrations of such chemicals are so low that they are harmless.

Chemists are also able to detect increasingly smaller amounts of potentially toxic chemicals in air, water, and food. This is good news, but it can give the false impression that dangers from toxic chemicals are increasing. In reality, we may simply be uncovering levels of chemicals that have been around for a long time.

RESEARCH FRONTIER

Learning more about the long-term effects of trace amounts of potentially harmful chemicals on human health. See **academic.cengage.com/biology/miller**.

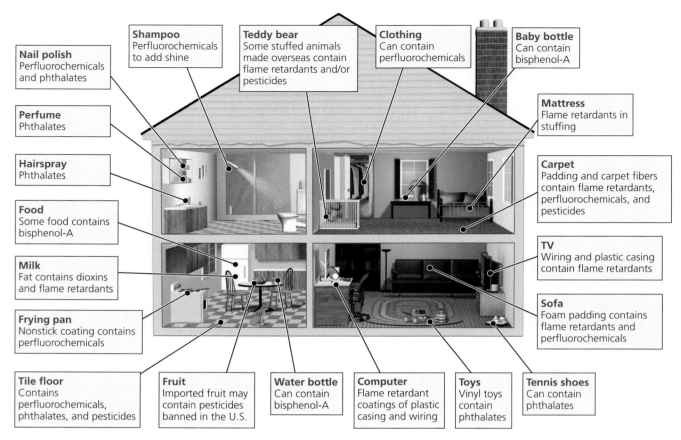

Nail polish
Perfluorochemicals and phthalates

Perfume
Phthalates

Hairspray
Phthalates

Food
Some food contains bisphenol-A

Milk
Fat contains dioxins and flame retardants

Frying pan
Nonstick coating contains perfluorochemicals

Tile floor
Contains perfluorochemicals, phthalates, and pesticides

Shampoo
Perfluorochemicals to add shine

Teddy bear
Some stuffed animals made overseas contain flame retardants and/or pesticides

Clothing
Can contain perfluorochemicals

Baby bottle
Can contain bisphenol-A

Mattress
Flame retardants in stuffing

Carpet
Padding and carpet fibers contain flame retardants, perfluorochemicals, and pesticides

TV
Wiring and plastic casing contain flame retardants

Sofa
Foam padding contains flame retardants and perfluorochemicals

Fruit
Imported fruit may contain pesticides banned in the U.S.

Water bottle
Can contain bisphenol-A

Computer
Flame retardant coatings of plastic casing and wiring

Toys
Vinyl toys contain phthalates

Tennis shoes
Can contain phthalates

Figure 17-16 Some potentially harmful chemicals found in most homes. Most people have traces of these chemicals in their blood and body tissues. We do not know the long-term effects of exposure to low levels of such chemicals. (Data from U.S. Environmental Protection Agency, Centers for Disease Control and Prevention, and New York State Department of Health) **Question:** Does the fact that we do not know much about long-term harmful effects of these chemicals make you more likely or less likely to minimize your exposure to them? Why?

Some people have the mistaken idea that all natural chemicals are safe and all synthetic chemicals are harmful. In fact, many synthetic chemicals, including many of the medicines we take, are quite safe if used as intended; and many natural chemicals, such as mercury and lead, are deadly.

Why Do We Know So Little about the Harmful Effects of Chemicals?

As we have seen, all methods for estimating toxicity levels and risks have serious limitations (**Concept 17-4A**). But they are all we have. To take this uncertainty into account and to minimize harm, scientists and regulators typically set allowed levels of exposure to toxic substances and ionizing radiation at 1/100 or even 1/1,000 of the estimated harmful levels.

According to risk assessment expert Joseph V. Rodricks, "Toxicologists know a great deal about a few chemicals, a little about many, and next to nothing about most." The U.S. National Academy of Sciences estimates that only 10% of 100,000 registered synthetic chemicals in commercial use have been thoroughly screened for toxicity, and only 2% have been adequately tested to determine whether they are carcinogens, teratogens, or mutagens. Hardly any of the chemicals in commercial use have been screened for possible damage to the human nervous, endocrine, and immune systems. Because of insufficient data and the high costs of regulation, federal and state governments do not regulate about 99.5% of the commercially used chemicals in the United States.

How Far Should We Go in Using Pollution Prevention and the Precautionary Principle?

So where does this leave us? We do not know a lot about the potentially toxic chemicals around us and inside of us, and estimating their effects is very difficult, time-consuming, and expensive. Is there a way to deal with this problem?

Some scientists and health officials, especially those in European Union countries, are pushing for much greater emphasis on *pollution prevention* (**Concept 1-4**, p. 16). They say we should not release into the environment chemicals that we know or suspect can cause significant harm. This means looking for harmless or less harmful substitutes for toxic and hazardous chemicals (Individuals Matter, below). Another option is to recycle them within production processes to keep them from reaching the environment, as the U.S. companies 3M and DuPont have been doing (**Concept 17-4B**).

Pollution prevention is a strategy for implementing the *precautionary principle* (p. 210). According to this principle, when there is reasonable but incomplete scientific evidence of significant or irreversible harm to humans or to the environment from a proposed or existing chemical or technology, we should take action to prevent or reduce the risk instead of waiting for more conclusive evidence (**Concept 9-4C**, p. 206). Note that the precautionary principle is a general guideline that does not specify what should trigger action, or what specific action should be taken.

There is controversy over how far we should go in implementing pollution prevention based on the precautionary principle. If we applied it strictly, those proposing to introduce a new chemical or technology would bear the burden of establishing its safety. This would require two major changes in the way we evaluate risks. *First,* new chemicals and technologies would be assumed to be harmful until scientific studies could show otherwise. *Second,* existing chemicals and technologies that appear to have a strong chance of causing significant harm would be removed from the market until their safety could be established. For example, after decades of research revealed the harmful effects of lead, especially in children, lead-based paints and leaded gasoline were phased out in most developed countries.

Some movement is being made in this direction, especially in the European Union. In 2000, negotiators agreed to a global treaty that would ban or phase out use of 12 of the most notorious *persistent organic pollutants* (POPs), also called the *dirty dozen* (See **www.pops.int/**). The list includes DDT and eight other persistent pesticides, PCBs, dioxins, and furans. Animal studies have shown that the harmful effects of various POPS include tumors and cancers, birth defects, compromised immune systems, feminization of males and masculinization of females, abnormally functioning thyroid glands, and reproductive failure. There is also concern that some of these chemicals may play a role in malformed penises in boys, increased testicular cancers, and a 50% decline in sperm counts and sperm quality in men in a number of countries.

Although such evidence is not clear and is controversial, phasing these chemicals out could be a reasonable precaution. New chemicals will be added to the list when the harm they could potentially cause is seen as outweighing their usefulness. This treaty went into effect in 2004 but has not been ratified and implemented by the United States.

INDIVIDUALS MATTER

Ray Turner and His Refrigerator

Life as we know it could not exist on land or in the upper layers of the oceans and other bodies of water without the thin layer of ozone (O_3) found in the lower stratosphere (Figure 3-8, p. 56). In other words, a basic rule of sustainability relating to pollution prevention is: *Do not mess with the ozone layer.*

However, for decades we violated this principle of pollution prevention by releasing large amounts of chemicals such as chlorofluorocarbons (CFCs) into the troposphere. These chemicals have drifted into the stratosphere where they react with and destroy some of the ozone that protects life from harmful UV radiation.

In 1974, scientists alerted the world to this threat. After further research and lengthy debate, in 1992, most of the world's nations signed a landmark international agreement to phase out the use of CFCs and other ozone-destroying chemicals. The discovery of these chemicals led scientists to use the principle of pollution prevention and search for less harmful alternatives.

Ray Turner, a manager at Hughes Aircraft in the U.S. state of California, was concerned about this. His company was using CFCs as cleaning agents to remove films of oxidation from the electronic circuit boards they manufactured. Turner's concern for the environment led him to search for a cheap and simple substitute for these chemicals. He found it in his refrigerator.

Turner decided to put drops of some common kitchen substances on a corroded penny to see whether any of them would remove the film of oxidation. Then he used his soldering gun to see whether solder would stick to the surface of the penny, indicating that the film had been cleaned off.

First he tried vinegar. No luck. Then he tried some ground-up lemon peel. Another failure. Next he tried a drop of lemon juice and watched as the solder took hold. The rest, as they say, is history.

Today, Hughes Aircraft uses inexpensive, CFC-free, citrus-based solvents to clean circuit boards. This new cleaning technique has reduced circuit board defects by about 75% at the company. And Turner got a hefty bonus. Now, other companies clean computer boards and chips using acidic chemicals extracted from cantaloupes, peaches, and plums. Maybe you can find a solution to an environmental problem in your refrigerator.

In 2006, the European Union enacted new regulations that require the registration of 30,000 untested and unregulated potentially harmful chemicals. The most hazardous substances will no longer be approved for use if safer alternatives exist. And when there is no alternative, producers must present a research plan aimed at finding one. Many environmental scientists applaud this use of pollution prevention to implement the precautionary principle, but some say that the regulation does not go far enough and has too many loopholes.

Manufacturers and businesses contend that widespread application of this approach would make it too expensive and almost impossible to introduce any new chemical or technology. They argue that we can never have a risk-free society.

Proponents of increased reliance on pollution prevention agree that we can go too far, but argue we have an ethical responsibility to reduce known or potentially serious threats to human health and to our life support system (**Concept 17-4B**). They also point out that using the precautionary principle focuses the efforts and creativity of scientists, engineers, and businesses on finding solutions to pollution problems based on prevention rather than on cleanup. This in turn reduces health risks for employees and society, frees businesses from having to deal with pollution regulations, reduces the threat of lawsuits from harmed parties, increases profits, in some cases, from sales of safer products and innovative technologies, and improves the public image of businesses operating in this manner.

— **HOW WOULD YOU VOTE?**
Should we rely more on pollution prevention to implement the precautionary principle as a way to reduce the potential risks from chemicals and technologies? Cast your vote online at **academic.cengage.com/biology/miller**.

17-5 How Do We Perceive Risks and How Can We Avoid the Worst of Them?

▶ **CONCEPT 17-5** We can reduce the major risks we face if we become informed, think critically about risks, and make careful choices.

The Greatest Health Risks Come from Poverty, Gender, and Lifestyle Choices

Risk analysis involves identifying hazards and evaluating their associated risks (*risk assessment;* Figure 17-3, left), ranking risks (comparative risk analysis), determining options and making decisions about reducing or eliminating risks (*risk management;* Figure 17-3, right), and informing decision makers and the public about risks (*risk communication*).

Statistical probabilities based on past experience, animal testing, and other tests are used to estimate risks from older technologies and chemicals. To evaluate new technologies and products, risk evaluators use more uncertain statistical probabilities, based on models rather than actual experience and testing.

Figure 17-17 lists the results of a *comparative risk analysis* and summarizes the greatest ecological and health risks identified by a panel of scientists acting as advisers to the EPA.

The greatest risks many people face today are rarely dramatic enough to make the daily news. In terms of the number of premature deaths per year (Figure 17-18) and reduced life span (Figure 17-19, p. 462), *the greatest risk by far is poverty.* The high death toll ultimately resulting from poverty is caused by malnutrition, increased susceptibility to normally nonfatal infectious diseases, and often-fatal infectious diseases transmitted by unsafe drinking water.

After poverty and gender, the greatest risks of premature death mostly result from lifestyle choices that people make (Figures 17-18 and 17-19, p. 462) (**Concept 17-1**). The best ways to reduce one's risk of premature death and serious health problems are to avoid smoking and exposure to smoke (Case Study, p. 462), lose excess weight, reduce consumption of foods containing cholesterol and saturated fats, eat a variety of fruits and vegetables, exercise regularly, drink little or no alcohol (no more than two drinks in a single day), avoid excess sunlight (which ages skin and can cause skin cancer), and practice safe sex (**Concept 17-5**). A 2005 study by Majjid Ezzati with participation by 100 scientists around the world estimated that one-third of the 7 million annual cancer deaths could be prevented if individuals followed these guidelines.

COMPARATIVE RISK ANALYSIS

Most Serious Ecological and Health Problems

High-Risk Health Problems
 Indoor air pollution
 Outdoor air pollution
 Worker exposure to industrial or farm chemicals
 Pollutants in drinking water
 Pesticide residues on food
 Toxic chemicals in consumer products

High-Risk Ecological Problems
 Global climate change
 Stratospheric ozone depletion
 Wildlife habitat alteration and destruction
 Species extinction and loss of biodiversity

Medium-Risk Ecological Problems
 Acid deposition
 Pesticides
 Airborne toxic chemicals
 Toxic chemicals, nutrients, and sediment in surface waters

Low-Risk Ecological Problems
 Oil spills
 Groundwater pollution
 Radioactive isotopes
 Acid runoff to surface waters
 Thermal pollution

Figure 17-17 Science: *comparative risk analysis* of the most serious ecological and health problems, according to scientists acting as advisers to the EPA. Risks under each category are not listed in rank order. **Question:** Which two risks in each of the two high-risk problem lists do you think are the most serious? Why? (**Concept 17-5**) (Data from Science Advisory Board, *Reducing Risks,* Washington, D.C.: Environmental Protection Agency, 1990)

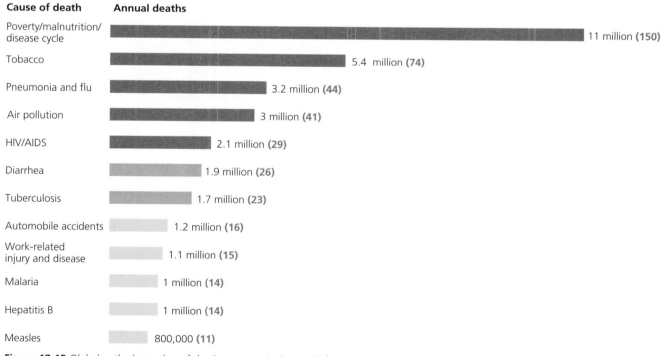

Cause of death	Annual deaths
Poverty/malnutrition/disease cycle	11 million (150)
Tobacco	5.4 million (74)
Pneumonia and flu	3.2 million (44)
Air pollution	3 million (41)
HIV/AIDS	2.1 million (29)
Diarrhea	1.9 million (26)
Tuberculosis	1.7 million (23)
Automobile accidents	1.2 million (16)
Work-related injury and disease	1.1 million (15)
Malaria	1 million (14)
Hepatitis B	1 million (14)
Measles	800,000 (11)

Figure 17-18 *Global outlook:* number of deaths per year in the world from various causes (**Concept 17-5**). Numbers in parentheses give these deaths in terms of the number of fully loaded 200-passenger jet airplanes crashing *every day of the year* with no survivors. Because of sensational media coverage, most people are misinformed about the largest annual causes of death. **Question:** Which three of these items are most likely to shorten your life span? (Data from World Health Organization)

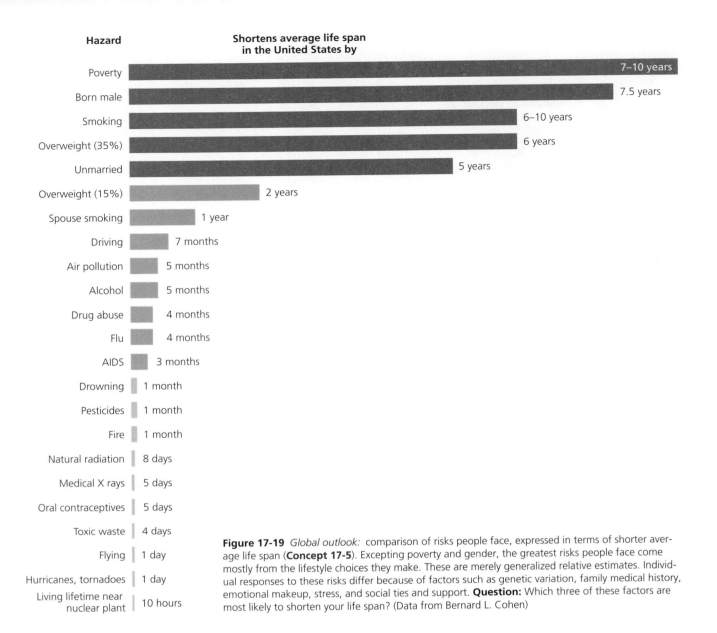

Hazard	Shortens average life span in the United States by
Poverty	7–10 years
Born male	7.5 years
Smoking	6–10 years
Overweight (35%)	6 years
Unmarried	5 years
Overweight (15%)	2 years
Spouse smoking	1 year
Driving	7 months
Air pollution	5 months
Alcohol	5 months
Drug abuse	4 months
Flu	4 months
AIDS	3 months
Drowning	1 month
Pesticides	1 month
Fire	1 month
Natural radiation	8 days
Medical X rays	5 days
Oral contraceptives	5 days
Toxic waste	4 days
Flying	1 day
Hurricanes, tornadoes	1 day
Living lifetime near nuclear plant	10 hours

Figure 17-19 *Global outlook:* comparison of risks people face, expressed in terms of shorter average life span (**Concept 17-5**). Excepting poverty and gender, the greatest risks people face come mostly from the lifestyle choices they make. These are merely generalized relative estimates. Individual responses to these risks differ because of factors such as genetic variation, family medical history, emotional makeup, stress, and social ties and support. **Question:** Which three of these factors are most likely to shorten your life span? (Data from Bernard L. Cohen)

■ CASE STUDY

Death from Smoking

What is roughly the diameter of a 30-caliber bullet, can be bought almost anywhere, is highly addictive, and kills an average of about 14,800 people every day, or about one every 6 seconds? It is a cigarette. *Cigarette smoking is the world's most preventable major cause of suffering and premature death among adults.*

In 2007, the WHO estimated that tobacco use helped to kill 100 million people prematurely during the 20th century and could kill 1 billion people during this century unless governments act now to dramatically reduce smoking. The WHO estimates that each year, tobacco contributes to the premature deaths of at least 5.4 million people (about half from developed and half from developing countries) from 25 illnesses including heart disease, lung cancer, other cancers, bron-

chitis, emphysema, and stroke. By 2030, the annual death toll from smoking-related diseases is projected to reach more than 8 million—an average of 21,900 preventable deaths per day. About 80% of these deaths are expected to occur in developing countries, especially China, with 30% of the world's smokers, and India with 11% (compared to 4.5% in the United States).

According to the CDC, smoking kills about 442,000 Americans per year prematurely—an average of 1,210 deaths per day, or nearly one every minute (Figure 17-20). This death toll is roughly equivalent to six fully loaded 200-passenger jet planes crashing *every day* with no survivors! Yet, this ongoing major human tragedy in the United States and throughout the world rarely makes the news.

The overwhelming consensus in the scientific community is that the nicotine inhaled in tobacco smoke is highly addictive. Only one in ten people who try to

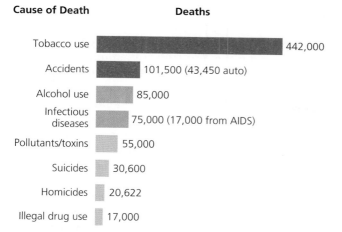

Cause of Death	Deaths
Tobacco use	442,000
Accidents	101,500 (43,450 auto)
Alcohol use	85,000
Infectious diseases	75,000 (17,000 from AIDS)
Pollutants/toxins	55,000
Suicides	30,600
Homicides	20,622
Illegal drug use	17,000

Figure 17-20 Annual deaths in the United States from tobacco use and other causes in 2004. Smoking is by far the nation's leading cause of preventable death, causing more premature deaths each year than all the other categories in this figure combined. (Data from U.S. National Center for Health Statistics, the Centers for Disease Control and Prevention, and the U.S. Surgeon General)

quit smoking succeeds. Smokers suffer about the same relapse rate as do recovering alcoholics and those addicted to heroin or crack cocaine. A British government study showed that adolescents who smoke more than one cigarette have an 85% chance of becoming smokers.

Passive smoking, or breathing secondhand smoke, also poses health hazards for children and adults. Children who grow up with smokers are more likely to develop allergies and asthma. Among adults, nonsmoking spouses of smokers have a 30% higher risk of both heart attack and lung cancer than do spouses of nonsmokers. In 2006, the CDC estimated that each year, secondhand smoke causes an estimated 3,000 lung cancer deaths and 46,000 deaths from heart disease in the United States.

A 50-year study published in 2004 by Richard Doll and Richard Peto found that cigarette smokers die, on average, 10 years earlier than nonsmokers, but that kicking the habit—even at 50 years old—can cut a person's risk in half. If people quit smoking by the age of 30, they can avoid nearly all the risk of dying prematurely, but again, the longer one smokes, the harder it is to quit.

Many health experts urge that a $3–5 federal tax be added to the price of a pack of cigarettes in the United States (and in other countries). Then users of tobacco products—not the rest of society—would pay a much greater share of the $158 billion per year (an average of $301,000 per minute) in health, economic, and social costs associated with smoking in the United States.

Analysts also call for classifying and regulating the use of nicotine as an addictive and dangerous drug under the U.S. Food and Drug Administration, eliminating all federal subsidies and tax breaks to tobacco farmers and tobacco companies, and using cigarette tax revenues to finance an aggressive antitobacco adver-

tising and education program. After a rally one day in 2005, activists left 1,210 pairs of empty shoes in front of the U.S. capital to remind lawmakers that *each day* tobacco kills that many people in the United States. So far, Congress has not enacted such reforms.

Some other countries are enacting smoking bans. In 2004, Ireland, Norway, and Scotland enacted bans on smoking in all indoor workplaces, bars, and restaurants. And in 2004, India banned smoking in public places, as well as tobacco advertising in the mass media and tobacco sales to minors. Studies had shown that smoking was killing 2,200 people a day in India. In 2006, France banned smoking in public facilities, and Great Britain did so in 2007.

There is some encouraging news: the average number of cigarettes smoked per person in the United States declined by 56% between 1976 and 2006 and dropped globally by 16% between 1988 and 2004. This decline in smoking can be attributed to media coverage of the harmful health effects of smoking, sharp increases in taxes on cigarette sales, mandatory health warnings on cigarette packs, and the banning of smoking in workplaces, bars, restaurants, and public buildings in some countries. This is an encouraging trend, but there is still a long way to go in dealing with this serious global health problem.

┌───┐
HOW WOULD YOU VOTE?

Do you favor classifying and regulating nicotine as an addictive and dangerous drug? Cast your vote online at **academic .cengage.com/biology/miller**.
└───┘

Estimating Risks from Technologies Is Not Easy

The more complex a technological system and the more people needed to design and run it, the more difficult it is to estimate the risks of using the system. The system's overall *reliability,* or the probability (expressed as a percentage) that a person or device will complete a task without failing, is the product of two factors:

System reliability (%) = Technological reliability × Human reliability

With careful design, quality control, maintenance, and monitoring, a highly complex system such as a nuclear power plant or space shuttle can achieve a high degree of technological reliability. But human reliability usually is much lower than technological reliability and is almost impossible to predict: To err is human.

Suppose the technological reliability of a nuclear power plant is 95% (0.95) and human reliability is 75% (0.75). Then the overall system reliability is 71% (0.95 × 0.75 = 71%). Even if we could make the technology 100% reliable (1.0), the overall system reliability would still be only 75% (1.0 × 0.75 = 75%). The crucial dependence of even the most carefully designed systems on unpredictable human reliability helps to

explain tragedies that were deemed almost impossible, such as the Chernobyl nuclear power plant explosion (Case Study, p. 390), the Three Mile Island nuclear power accident (Case Study, p. 390), and the *Challenger* and *Columbia* space shuttle accidents.

One way to make a system more foolproof is to move more of the potentially fallible elements from the human side to the technological side. However, chance events such as a lightning strike can knock out an automatic control system, and no machine or computer program can completely replace human judgment. Also, the parts in any automated control system are manufactured, assembled, tested, certified, and maintained by fallible human beings. In addition, computer software programs used to monitor and control complex systems can be flawed because of human error or can be deliberately caused to malfunction.

Most People Do Not Know How to Evaluate Risks

Most of us are not good at assessing the relative risks from the hazards that surround us. Many people deny or shrug off the high-risk chances of death (or injury) from voluntary activities they enjoy, such as *motorcycling* (1 death in 50 participants), *smoking* (1 in 250 by age 70 for a pack-a-day smoker), *hang gliding* (1 in 1,250), and *driving* (1 in 3,300 without a seatbelt and 1 in 6,070 with a seatbelt). Indeed, the most dangerous thing that most people in many countries do each day is to drive or ride in a car.

Yet some of these same people may be terrified about the possibility of being killed by a *gun* (1 in 28,000 in the United States), *flu* (1 in 130,000), *nuclear power plant accident* (1 in 200,000), *West Nile virus* (1 in 1 million), *lightning* (1 in 3 million), *commercial airplane crash* (1 in 9 million), *snakebite* (1 in 36 million), or *shark attack* (1 in 281 million).

Five factors can cause people to see a technology or a product as being more or less risky than experts judge it to be. First is *fear*. Research going back 3 decades shows that fear causes people to overestimate risks and to worry more about unusual risks than they do about common everyday risks. Studies show that people tend to overestimate the numbers of deaths caused by tornadoes, floods, fires, homicides, cancer, and terrorist attacks, and to underestimate numbers of deaths from flu, diabetes, asthma, stroke, and automobile accidents. Many people also fear a new, unknown product or technology more than they do an older, more familiar one. For example, some people fear genetically modified food and trust food produced by traditional plant-breeding techniques. Most people have a greater fear of nuclear power plants than of more familiar and highly polluting coal-fired power plants (Figure 15-22, p. 392).

A second factor is the *degree of control* we have. Most of us have a greater fear of things over which we do not have personal control. For example, some individuals feel safer driving their own cars over long distances and through bad traffic than they do traveling the same distance on a plane. But look at the numbers. The risk of dying in a car accident in the United States while using a seatbelt is 1 in 6,070 whereas the risk of dying in a commercial airliner crash is 1 in 9 million.

Third is *whether a risk is catastrophic,* not chronic. We usually are more frightened by news of a single catastrophic accident such as a plane crash than we are of a cause of death such as smoking, which has an even larger death toll spread out over time.

Fourth, some people suffer from *optimism bias,* the belief that risks that apply to other people do not apply to them. Some people get upset when they see someone driving erratically while talking on a cell phone. But they may believe that their own talking on the cell phone does not impair their driving ability.

A fifth factor is that many of the risky things we do are highly pleasurable and give *instant gratification,* while the potential harm from such activities comes later. Examples are smoking cigarettes, eating lots of ice cream, and getting a tan.

There is also concern about the *unfair distribution of risks* in some cases. For example, citizens are often outraged when government officials decide to put a hazardous waste landfill or incinerator in or near their neighborhood. Even when the decision is based on careful risk analysis, it is usually seen as a political, not a scientific, decision. Residents will not be satisfied by an estimate putting the lifetime risk of cancer death from the facility at no greater than, say, 1 in 100,000. Instead, they point out that living near the facility means that they will have a much higher risk of dying from cancer than will people living farther away.

Several Principles Can Help Us to Evaluate and Reduce Risk

Here are some guidelines for evaluating and reducing risk (**Concept 17-5**):

- *Compare risks.* Is there a risk of getting cancer from eating a charcoal-broiled steak once or twice a week for a lifetime? Yes, because in theory, anything can harm you. The question is whether this danger is great enough for you to worry about. In evaluating a risk, the question is not "Is it safe?" but rather *"How risky is it compared to other risks?"*

- *Determine how much risk you are willing to accept.* For most people, a 1 in 100,000 chance of dying or suffering serious harm from exposure to an environmental hazard is a threshold for changing their behavior. However, in establishing standards and reducing risk, the U.S. EPA generally assumes that a 1 in 1 million chance of dying from an environmental hazard is acceptable. People

exposed to such risks believe that this standard is too high.

- *Determine the actual risk involved.* The news media usually exaggerate the daily risks we face in order to capture our interest and sell newspapers and magazines or gain television viewers. As a result, most people believe that the world is much more risky than it really is.

- *Concentrate on evaluating and carefully making important lifestyle choices,* and you will have a much greater chance of living a longer, healthier, happier, and

less fearful life. When you worry about a risk, the most important question to ask is, "Do I have any control over this?" There is no point worrying about risks over which you have no control. But you do have control over risks from heart attack, stroke, and many forms of cancer, for example, because you can decide whether to smoke, what to eat, and how much alcohol to drink. Other factors under your control are whether you practice safe sex, how much exercise you get, how safely you drive, and how often you expose yourself to ultraviolet rays from the sun or in tanning booths.

REVISITING AIDS and Sustainability

In this chapter, we have seen that on a global basis, the greatest threat to human health is the tragic poverty-malnutrition-disease cycle, followed by the threats from smoking, pneumonia and flu, air pollution, and HIV/AIDS (**Core Case Study**). These five global threats prematurely kill about 24.7 million people a year (Figure 17-18)—an average of nearly 67,700 a day or 2,820 an hour—half of them children younger than age 5.

These major global health risks are largely preventable if governments, under pressure from concerned citizens, choose to make them global priorities. We can use the four **scientific principles of sustainability** (see back cover) to help us reduce these and other major risks to human health. One way to do this involves shifting from nonrenewable fossil fuels to renewable

energy, thereby reducing pollution and the threats from global warming. Another solution is to cut down on waste of energy and matter resources by reusing and recycling them, thereby helping to provide enough resources for most people to avoid poverty. We can also emphasize the use of diverse strategies for solving environmental and health problems, and especially for reducing poverty and controlling population growth.

Is this idealistic? Sure. But if creative and caring people throughout human history had not acted to improve the world by doing things that others said were impossible or too idealistic, we would have accomplished very little on this marvelous planet. Each of us can make a difference.

The burden of proof imposed on individuals, companies, and institutions should be to show that pollution prevention options have been thoroughly examined, evaluated, and used before lesser options are chosen.

JOEL HIRSCHORN

REVIEW

1. Review the Key Questions and Concepts for this chapter on p. 439. Describe how an HIV/AIDs epidemic (**Core Case Study**) in a country can affect the age structure of its population.

2. Distinguish among **risk, risk assessment,** and **risk management.** Distinguish between possibility and probability. What is a **pathogen?** Give an example of a risk from each of the following: biological hazards, chemical hazards, physical hazards, cultural hazards, and lifestyle choices.

3. Distinguish among a **nontransmissible disease, infectious disease,** and **transmissible disease** and give an

example of each. In terms of death rates, what are the world's four most serious infectious diseases? Distinguish between an epidemic and a pandemic of an infectious disease. Describe the causes and possible solutions for the increasing genetic resistance in microbes to commonly used antibiotics.

4. Describe the global threat from tuberculosis. Describe the threat from flu and the effects of a global flu pandemic. Describe the health threats from the global HIV/AIDS pandemic and list six ways to reduce this threat. Describe the threats from the hepatitis B, West Nile, and SARS viruses. Describe the threat from malaria for 40% of the world's people and how we can reduce this threat.

5. Give three examples of problems being studied in the new field of ecological medicine. What is Lyme disease, and how can individuals reduce their chances of getting it? List five major ways to reduce the global threat from infectious diseases.

6. What is a **toxic chemical?** Discuss the threat from PCBs. Distinguish among **mutagens, teratogens,** and **carcinogens,** and give an example of each. Describe the toxic legacy from PCBs. Describe the human immune, nervous, and endocrine systems and give an example of a chemical that can threaten each of these systems. Describe the toxic effects of the various forms of mercury and ways to reduce these threats. What are hormonally active agents, what risks do they pose, and how can we reduce these risks? Describe the potential threats from bisphenol A.

7. Define **toxicology, toxicity, dose,** and **response.** Give three reasons why children are more vulnerable to harm from toxic chemicals. Describe how the toxicity of a substance can be estimated by using laboratory animals, and discuss the limitations of this approach. What is a **dose-response curve?** Describe how toxicities are estimated

by case reports and epidemiological studies and discuss the limitations of these approaches. Why do we know so little about the harmful effects of chemicals? Discuss the use of pollution prevention and the precautionary principle in dealing with health threats from chemicals.

8. What is **risk analysis?** In terms of premature deaths, what are the three greatest threats that humans face? Describe the health threats from smoking and what can be done to reduce these threats.

9. How can we reduce the threats from the use of various technologies? What five factors can cause people to misjudge risks? List five principles that can help us evaluate and reduce risk.

10. Discuss how lessening the threats of HIV/AIDS and other major infectious diseases (**Core Case Study**) can be achieved by applying the four **scientific principles of sustainability.**

Note: Key Terms are in **bold** type.

CRITICAL THINKING

1. List three ways in which you could apply **Concept 17-5** to make your lifestyle more environmentally sustainable while reducing the major risks you face.

2. How can changes in the age structure of a human population increase the spread of infectious diseases? How can the spread of infectious diseases such as HIV/AIDS affect the age structure of human populations (**Core Case Study** and Figure 17-2)?

3. What are three actions you would take to reduce the global threats to human health and life from **(a)** HIV/AIDS (**Core Case Study**), **(b)** tuberculosis, and **(c)** malaria?

4. Evaluate the following statements:
 a. We should not get worked up about exposure to toxic chemicals because almost any chemical, at a large enough dosage, can cause some harm.
 b. We should not worry much about exposure to toxic chemicals because, through genetic adaptation, we can develop immunity to such chemicals.
 c. We should not worry much about exposure to toxic chemicals because we can use genetic engineering to reduce our susceptibility to the effects of toxic chemicals.

5. Workers in a number of industries are exposed to higher levels of various toxic substances than are the general public. Should workplace levels allowed for such chemicals be reduced? What economic effects might this have?

6. Explain why you agree or disagree with the proposals for reducing the death toll and other harmful effects of smoking listed in the Case Study on p. 462. Do you believe there should be a ban on smoking indoors in all public places? Explain.

7. What are the three major risks you face from **(a)** your lifestyle, **(b)** where you live, and **(c)** what you do for a living? Which of these risks are voluntary and which are involuntary? List the three most important things you can do to reduce these risks. Which of these things do you plan to do?

8. Would you support legislation requiring the use of pollution prevention based on the precautionary principle in deciding what to do about risks from chemicals in the country where you live? Explain.

9. Congratulations! You are in charge of the world. List the three most important features of your program to reduce the risks from exposure to **(a)** infectious disease organisms and **(b)** toxic and hazardous chemicals.

10. List two questions that you would like to have answered as a result of reading this chapter.

Note: See Supplement 13 (p. S78) for a list of Projects related to this chapter.

The graph below shows the effects of AIDS on life expectancy at birth in Botswana from 1950 through 2000 and projects the effects to 2050. Answer the questions below.

Source: Data from United Nations and U.S. Census Bureau

1. **a.** By what percentage did life expectancy in Botswana increase between 1950 and 1995?

 b. By what percentage is life expectancy in Botswana projected to decrease between 1995 and 2015?

2. **a.** By what percentage is life expectancy in Botswana projected to increase between 2015 and 2050?

 b. By what percentage is life expectancy in Botswana projected to decrease between 1995 and 2050?

LEARNING ONLINE

Log on to the Student Companion Site for this book at **academic.cengage.com/biology/miller**, and choose Chapter 17 for many study aids and ideas for further reading and research. These include flash cards, practice quizzing, Weblinks, information on Green Careers, and InfoTrac® College Edition articles.

South Asia's Massive Brown Cloud

Air pollution is no longer viewed as mostly a localized urban problem. Satellite images, along with a 2002 study by the U.N. Environment Programme (UNEP), revealed a massive brown cloud of pollution—called the *Asian Brown Cloud*—stretching nearly continuously across much of India, Bangladesh, the industrial heart of China, and the open sea east of this area (Figure 18-1).

The cloud is about 3 kilometers (2 miles) thick, hangs at an elevation of 2–5 kilometers (1–3 miles), and covers an area about the size of the continental United States. About a third of it is dust, smoke, and ash resulting from drought and the clearing and burning of forests for planting crops. The rest is made up of acidic compounds, soot, toxic metals such as mercury and lead, hundreds of organic compounds, and fly ash produced by the burning of coal, diesel, and other fossil fuels in industries, motor vehicles, and homes.

Beneath the cloud, photosynthesis has been reduced in China by about 7–10% and in India by about 7% because less sunlight reaches the ground, according to a 2008 report by atmospheric scientist Veerabhadran Ramanathan, who heads the UNEP's Atmospheric Brown Cloud Project. Acids in the haze fall to the surface and damage crops, trees, and aquatic life in lakes.

Instead of blue skies, many of the people living under this cloud see brown or gray polluted skies much of the year (Figure 18-2). UNEP scientists estimate that the pollution in the cloud contributes to at least 700,000 premature deaths every year.

Originally, scientists thought that the brown cloud deflected sunlight and tended to cool the atmosphere above it. But a study of black carbon particles in the cloud, done in 2006 and 2008 by a team of climate scientists led by Ramanathan, found that these particles helped to warm the atmosphere above the cloud by about as much as CO_2 and other greenhouse gases do.

The Asian Brown Cloud is also bad news for other parts of the world because it doesn't stay put. In 2006, a U.S. satellite tracked the spread of a dense cloud of pollutants from northern China to Seoul, South Korea, and then across the Pacific Ocean to the United States. The U.S. Environmental Protection Agency (EPA) estimates that on certain days, nearly 25% of the particulate matter, 77% of the black carbon, and 33% of the toxic mercury in the skies above Los Angeles, California, can be traced to coal-fired power plants, smelters, diesel trucks, and dust storms caused by drought and deforestation in China.

Satellite measurements show that it takes about 2 weeks for long-lived air pollutants to circle the world. Thus, long-lived air pollutants from China, India, the United States, or anywhere can affect the entire world. Air pollution connects us all.

The history of air pollution control in Europe and the United States shows that pollution such as that in the Asian cloud can be cleared up fairly quickly by setting strict pollution control standards for coal-burning industries and utilities and by shifting from coal to cleaner-burning natural gas in industries and homes. China is beginning to take such steps but has a long way to go. India's capital city of Delhi, under orders from India's Supreme Court, has also made progress in reducing air pollution.

Figure 18-1
The *Asian Brown Cloud*. A gigantic cloud of dust, smoke, soot, and other pollutants stretches over much of Asia, visible here over eastern China.

The SeaWiFS Project, NASA/Goddard Space Flight Center, and Orbimage

ullstein-Hiss/Peter Arnold, Inc.

Figure 18-2 Air pollution in Shanghai, China, in 2004.

18-1 What is the nature of the atmosphere?

CONCEPT 18-1 The atmosphere is structured in layers, including the troposphere, which supports life, and the stratosphere, which contains the protective ozone layer.

18-2 What are the major outdoor air pollution problems?

CONCEPT 18-2 Pollutants mix in the air to form *industrial smog*, mostly the result of burning coal, and *photochemical smog*, caused by motor vehicle, industrial, and power plant emissions.

18-3 What is acid deposition and why is it a problem?

CONCEPT 18-3 Acid deposition is caused mainly by coal-burning power plant and motor vehicle emissions, and in some regions, threatens human health, aquatic life and ecosystems, forests, and human-built structures.

18-4 What are the major indoor air pollution problems?

CONCEPT 18-4 The most threatening indoor air pollutants are smoke and soot from wood and coal cooking fires (a hazard found mostly in developing countries) and chemicals used in building materials and products.

18-5 What are the health effects of air pollution?

CONCEPT 18-5 Air pollution can contribute to asthma, chronic bronchitis, emphysema, lung cancer, heart attack, and stroke.

18-6 How should we deal with air pollution?

CONCEPT 18-6 Legal, economic, and technological tools can help to clean up air pollution, but much greater emphasis should be focused on preventing air pollution.

Note: Supplements 2 (p. S4), 6 (p. S39) and 13 (p. S78) can be used with this chapter.

*I thought I saw a blue jay this morning.
But the smog was so bad that it turned out to be
a cardinal holding its breath.*

MICHAEL J. COHEN

18-1 What Is the Nature of the Atmosphere?

▶ **CONCEPT 18-1** The atmosphere is structured in layers, including the troposphere, which supports life, and the stratosphere, which contains the protective ozone layer.

The Atmosphere Consists of Several Layers

We live at the bottom of a thin envelope of gases surrounding the earth, called the *atmosphere*. It is divided into several spherical layers (Figure 18-3, p. 470), each characterized by abrupt changes in temperature caused by differences in the absorption of incoming solar energy.

Two other factors that vary throughout the atmosphere are density and atmospheric pressure, both influenced by gravity, which pulls the gas molecules in the atmosphere toward the earth's surface. Because of this, the air we breathe at sea level has a higher **density**—more molecules per liter—than the air we would inhale on top of the world's highest mountain. **Atmospheric pressure** is the force, or mass, per unit area of a column of air. This force is caused by the bombardment of a surface such as your skin by air molecules.

Atmospheric pressure decreases with altitude because there are fewer gas molecules at higher altitudes.

Air Movements in the Troposphere Play a Key Role in the Earth's Weather and Climate

About 75–80% of the earth's air mass is found in the **troposphere,** the atmospheric layer closest to the earth's surface. This layer extends only about 17 kilometers (11 miles) above sea level at the equator and 8 kilometers (5 miles) over the poles. If the earth were the size of an apple, this lower layer containing the air we breathe would be no thicker than the apple's skin.

Take a deep breath. About 99% of the air you inhaled consists of two gases: nitrogen (78%) and oxygen

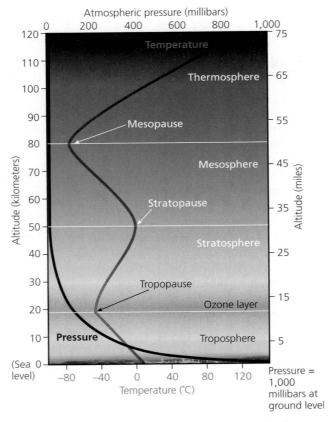

Atmospheric pressure (millibars)

Figure 18-3 Natural capital: The earth's atmosphere is a dynamic system that includes four layers. The average temperature of the atmosphere varies with altitude (red line). Most UV radiation from the sun is absorbed by ozone (O_3), found primarily in the stratosphere in the *ozone layer* 17–26 kilometers (10–16 miles) above sea level. **Question:** Why do you think the temperature falls and rises twice, going from lower to higher altitudes?

other gases including methane (CH_4), ozone (O_3), and nitrous oxide (N_2O).

The troposphere is a dynamic system involved in the chemical cycling of the earth's vital nutrients (**Concept 3-5**, p. 65). Its rising and falling air currents and winds are largely responsible for the planet's short-term *weather* and long-term *climate* (**Concept 7-1**, p. 141).

CONCEPT LINKS

The Stratosphere Is Our Global Sunscreen

The atmosphere's second layer is the **stratosphere,** which extends from about 17 to about 48 kilometers (from 11 to 30 miles) above the earth's surface (Figure 18-3). Although the stratosphere contains less matter than the troposphere, its composition is similar, with two notable exceptions: its volume of water vapor is about 1/1,000 that of the troposphere, and its concentration of ozone (O_3) is much higher.

Much of the atmosphere's small amount of ozone (O_3) is concentrated in a portion of the stratosphere called the **ozone layer,** found roughly 17–30 kilometers (11–19 miles) above sea level. Stratospheric ozone is produced when some of the oxygen molecules there interact with ultraviolet (UV) radiation emitted by the sun ($3 O_2 + UV \rightleftharpoons 2 O_3$). This "global sunscreen" of ozone in the stratosphere keeps about 95% of the sun's harmful UV radiation (Figure 3-8, p. 56) from reaching the earth's surface.

The UV filter of ozone in the lower stratosphere allows us and other forms of life to exist on land and helps to protect us from sunburn, skin and eye cancer, cataracts, and damage to our immune systems. It also prevents much of the oxygen in the troposphere from being converted to photochemical ozone, a harmful air pollutant when found near the ground.

(21%). The remainder consists of water vapor (varying from 0.01% at the frigid poles to 4% in the humid tropics), 0.93% argon (Ar), 0.038% carbon dioxide (CO_2), and trace amounts of dust and soot particles and

18-2 What Are the Major Outdoor Air Pollution Problems?

▶ **CONCEPT 18-2** Pollutants mix in the air to form *industrial smog,* mostly the result of burning coal, and *photochemical smog,* caused by motor vehicle, industrial, and power plant emissions.

Air Pollution Comes from Natural and Human Sources

Air pollution is the presence of chemicals in the atmosphere in concentrations high enough to harm organisms, ecosystems, or human-made materials. The effects of air pollution range from annoying to lethal.

Air pollutants come from natural and human sources. Natural sources include dust blown by wind (Figure 7-1, p. 140), pollutants from wildfires and volcanic eruptions, and volatile organic chemicals released by some plants. Most natural air pollutants are spread out over the globe or removed by chemical cycles, precipitation, and gravity. However, chemicals emitted

from volcanic eruptions and some natural forest fires can temporarily reach harmful levels in areas where they occur.

Human inputs of outdoor air pollutants occur mostly in industrialized and urban areas where people, cars, and factories are concentrated. Most of these pollutants are generated by the burning of fossil fuels in power and industrial plants (*stationary sources,* Figure 1-11, p. 17) and in motor vehicles (*mobile sources*). However, human inputs began long before the industrial era, and have steadily increased. Air pollution is a very old problem (see Case Study that follows).

■ CASE STUDY
Air Pollution in the Past: The Bad Old Days

Modern civilization did not invent air pollution. It probably began when humans discovered fire and used it to burn wood in poorly ventilated caves for warmth and cooking, as they inhaled unhealthy smoke and soot.

During the Middle Ages, a haze of wood smoke hung over densely packed urban areas. The Industrial Revolution, starting in the late 1700s, brought even worse air pollution as coal was burned to power factories and heat homes. As a result, there were great increases in respiratory diseases such as asthma, bronchitis, and allergies. Many people died from these ailments, especially children and elderly people.

By the 1850s, dense mixtures of coal smoke and fog as "thick as pea soup" engulfed London in a sunless gloom during winter months. In 1880, a prolonged coal fog killed an estimated 2,200 people. In 1905, a physician used the word *smog* to describe the deadly mixture of smoke and fog that afflicted London. Another episode in 1911 killed more than 1,100 Londoners.

In December 1952, an even worse yellow fog lasted for 5 days and killed 4,000–12,000 Londoners. Visibility was so low that people walking outside during the day could not see their feet. So many people died that undertakers ran out of coffins.

This tragedy prompted the British Parliament to pass the Clean Air Act of 1956. Before the beneficial effects of the law could be realized, additional air pollution disasters in 1956, 1957, and 1962 killed 2,500 more people. Because of strong air pollution laws, London's air today is much cleaner, and "pea soup" fogs are a thing of the past.

The Industrial Revolution, powered by coal, brought air pollution to the United States. Large industrial cities such as Pittsburgh, Pennsylvania, and St. Louis, Missouri, were known for their smoky air. By the 1940s, the air over some cities was so polluted that people had to turn their automobile headlights on during the day.

The first documented air pollution disaster in the United States occurred in October of 1948, in the small industrial town of Donora in Pennsylvania's Monongahela River Valley south of Pittsburgh. Pollutants from the area's coal-burning factories, steel mill, zinc smelter, and sulfuric acid plant became trapped in a dense fog that stagnated over the valley for 5 days. This killer fog resulted from a combination of mountainous terrain surrounding the valley and weather conditions that trapped and concentrated deadly pollutants. About 6,000 of the town's 14,000 inhabitants became sick, and 20 of them died.

In 1963, high concentrations of air pollutants in New York City killed about 300 people and injured thousands. Incidents like these finally resulted in city, state, and federal air pollution control programs in the United States, with the U.S. state of California leading the way. As a result, air quality has dramatically improved throughout the country.

However, many major urban areas in developing countries, such as China (Figure 18-2), India (Figure 15-13, p. 384), and parts of Eastern Europe that depend on burning coal in industries and in some homes face air pollution levels similar to those in London, England, and in American industrial cities in the 1950s.

┌─ **THINKING ABOUT** ─────────────
Outdoor Air Pollution—Past and Present

Explain why you agree or disagree with the statement: "Air pollution in the United States should no longer be a major concern because of the significant progress made in reducing outdoor air pollution since 1970."
└──────────────────────────────────

Some Pollutants in the Atmosphere Combine to Form Other Pollutants

Scientists classify outdoor air pollutants into two categories. **Primary pollutants** are harmful chemicals emitted directly into the air from natural processes and human activities (Figure 18-4, center, p. 472). While in the atmosphere, some primary pollutants react with one another and with the basic components of air to form new harmful chemicals, called **secondary pollutants** (Figure 18-4, right).

With their high concentrations of cars and factories, urban areas normally have higher outdoor air pollution levels than rural areas have. But prevailing winds can spread long-lived primary and secondary air pollutants from urban and industrial areas to the countryside and to other urban areas.

Good news. Over the past 30 years, the quality of outdoor air in most developed countries has improved greatly. This occurred mostly because grassroots pressure from citizens caused governments to pass and enforce air pollution control laws.

Bad news. According to the World Health Organization (WHO), more than 1.1 billion people (one of every six people on the earth) live in urban areas where outdoor air is unhealthy to breathe. Most of them live in densely populated cities in developing countries where air pollution control laws do not exist or are poorly enforced. However, the biggest pollution threat to poor

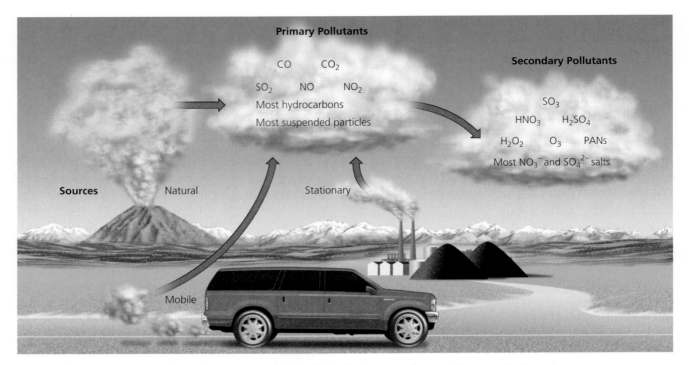

Primary Pollutants

CO CO_2

SO_2 NO NO_2

Most hydrocarbons

Most suspended particles

Secondary Pollutants

SO_3

HNO_3 H_2SO_4

H_2O_2 O_3 PANs

Most NO_3^- and SO_4^{2-} salts

Sources Natural Stationary

Mobile

Figure 18-4 *Sources and types of air pollutants.* Human inputs of air pollutants come from *mobile sources* (such as cars) and *stationary sources* (such as industrial and power plants). Some *primary air pollutants* react with one another and with other chemicals in the air to form *secondary air pollutants.*

people is *indoor air pollution* caused by their burning of wood, charcoal, coal, or dung in open fires or poorly designed stoves to heat their dwellings and cook their food. Also, they often must work in poorly ventilated and highly polluted spaces (Figure 18-5).

Air pollution was once a regional problem limited mostly to cities. Now it is a global problem, largely due to the sheer volume of pollutants produced. Pollutants entering the atmosphere in India and China now find their way across the Pacific where they affect the west coast of North America (**Core Case Study**). Even in arctic regions with virtually no population, air pollutants from northern Eurasia flow north and collect to form arctic haze. There is no place on the planet that has not been affected by air pollution.

What Are the Major Outdoor Air Pollutants?

Carbon oxides. *Carbon monoxide* (CO) is a colorless, odorless, and highly toxic gas that forms during the incomplete combustion of carbon-containing materials (see Table 18-1). Major sources are motor vehicle exhaust, burning of forests and grasslands, tobacco smoke, and open fires and inefficient stoves used for cooking.

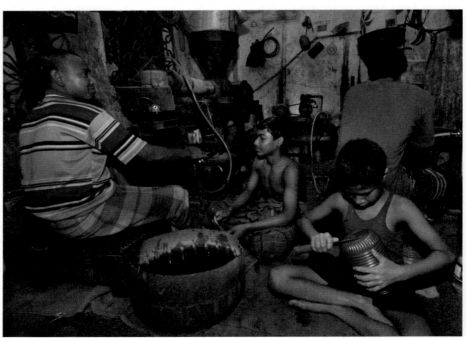

Figure 18-5 *Indoor air pollution.* These children are working in a plastic factory in an urban slum in Dhaka, Bangladesh. They have very long working days in this highly polluted room with little ventilation.

CO reacts with hemoglobin in red blood cells and reduces the ability of blood to transport oxygen to body cells and tissues. Chronic exposure can trigger heart attacks and aggravate lung diseases such as asthma and emphysema. At high levels, CO can cause headache, nausea, drowsiness, mental impairment, collapse, coma, and death. CO detectors, similar to smoke detectors, can warn people of dangerously high indoor levels of CO.

Carbon dioxide (CO_2) is a colorless, odorless gas. About 93% of the CO_2 in the atmosphere is the result of the natural carbon cycle (Figure 3-18, p. 68). The rest comes from human activities, mostly burning fossil fuels and clearing CO_2-absorbing forests and grasslands. Such emissions have been rising since the industrial revolution and especially since 1950 (Figure 20, p. S71, in Supplement 10). There is considerable scientific evidence that increasing levels of CO_2 from human activities are contributing to global warming and climate change (Science Focus, p. 33). We discuss this problem in detail in Chapter 19.

Nitrogen oxides and nitric acid. *Nitric oxide* (NO) is a colorless gas that forms when nitrogen and oxygen gas in air react at the high-combustion temperatures in automobile engines and coal-burning plants (Table 18-1). Lightning and certain bacteria in soil and water also produce NO as part of the nitrogen cycle (Figure 3-19, p. 69).

In the air, NO reacts with oxygen to form *nitrogen dioxide* (NO_2), a reddish-brown gas. Collectively, NO and NO_2 are called *nitrogen oxides* (NO_x). Some of the NO_2 reacts with water vapor in the air to form *nitric acid* (HNO_3) and nitrate salts (NO_3^-)—components of harmful *acid deposition*, which we discuss later in this chapter. Both NO and NO_2 play a role in the formation of *photochemical smog*—a mix of chemicals formed under the influence of sunlight in cities with heavy traffic (discussed further below). *Nitrous oxide* (N_2O), a greenhouse gas, is emitted from fertilizers and animal wastes and is produced by burning fossil fuels.

Nitrogen oxides can irritate the eyes, nose, and throat; aggravate lung ailments such as asthma and bronchitis; and increase susceptibility to respiratory infections by impairing the immune system. They can also suppress plant growth and reduce visibility when they are converted to nitric acid and nitrate salts.

Sulfur dioxide and sulfuric acid. *Sulfur dioxide* (SO_2) is a colorless gas with an irritating odor. About one-third of the SO_2 in the atmosphere comes from natural sources as part of the sulfur cycle (Figure 3-22, p. 72). The other two-thirds (and as much as 90% in some urban areas) come from human sources, mostly combustion of sulfur-containing coal in electric power and industrial plants (Table 18-1) and oil refining and smelting of sulfide ores. Sulfur dioxide emissions feeding the Asian Brown Cloud (**Core Case Study**) have

CORE
CASE
STUDY

Table 18-1

Chemical Reactions That Form Major Outdoor Air Pollutants

Pollutant	Chemical Reaction
Carbon monoxide (CO)	$2C + O_2 \rightarrow 2CO$
Carbon dioxide (CO_2)	$C + O_2 \rightarrow CO_2$
Nitric oxide (NO)	$N_2 + O_2 \rightarrow 2\,NO$
Nitrogen dioxide (NO_2)	$2NO + O_2 \rightarrow 2NO_2$
Sulfur dioxide (SO_2)	$S + O_2 \rightarrow SO_2$

increased by over a third in the past decade, according to a 2007 U.S. National Academy of Sciences report.

In the atmosphere, SO_2 can be converted to *aerosols*, which are microscopic suspended droplets of *sulfuric acid* (H_2SO_4) and suspended particles of sulfate (SO_4^{2-}) salts that return to the earth as a component of acid deposition. Sulfur dioxide, sulfuric acid droplets, and sulfate particles reduce visibility and aggravate breathing problems. SO_2 and H_2SO_4 can damage crops, trees, soils, and aquatic life in lakes. They also corrode metals and damage paint, paper, leather, and stone on buildings and statues (Figure 18-6).

Figure 18-6 Statue in Newport, Rhode Island (USA), corroded by acid deposition and other forms of air pollution.

Particulates. *Suspended particulate matter* (SPM) consists of a variety of solid particles and liquid droplets small and light enough to remain suspended in the air for long periods. About 62% of the SPM in outdoor air comes from natural sources such as dust, wild fires, and sea salt. The remaining 38% comes from human sources such as coal-burning power and industrial plants, motor vehicles, plowed fields, road construction, unpaved roads, and tobacco smoke.

Scientists are increasingly concerned about particulates. More than 2,000 studies published since 1990 link SPM with adverse health effects, according to the American Lung Association. The most harmful forms of SPM are *fine particles* (PM-10, with an average diameter of less than 10 micrometers) and *ultrafine particles* (PM-2.5, with an average diameter of less than 2.5 micrometers, about one-fortieth the diameter of a human hair). These particles can irritate the nose and throat, damage the lungs, aggravate asthma and bronchitis, and shorten life. Toxic particulates, such as lead (Case Study, at right), cadmium, and polychlorinated biphenyls (PCBs, Case Study, p. 449), can cause mutations, reproductive problems, and cancer. Particulates also reduce visibility, corrode metals, and discolor clothes and paints. In the United States, particulate air pollution—mostly from fine and ultrafine particles—is responsible for 60,000–70,000 premature deaths a year, according to the U.S. Environmental Protection Agency (EPA) and the Harvard School of Public Health.

Ozone. *Ozone* (O_3), a colorless and highly reactive gas, is a major component of photochemical smog. It can cause coughing and breathing problems, aggravate lung and heart diseases, reduce resistance to colds and pneumonia, and irritate the eyes, nose, and throat. It also damages plants, rubber in tires, fabrics, and paints.

Ozone in the troposphere near ground level is often referred to as "bad" ozone, whereas we think of ozone in the stratosphere as "good" ozone that protects us from harmful UV radiation. Both are the same chemical. Much evidence indicates that some human activities are *decreasing* the amount of beneficial ozone in the stratosphere and *increasing* the amount of harmful ozone in the troposphere near ground level—especially in some urban areas. We examine the issue of stratospheric ozone thinning in the next chapter.

Volatile organic compounds (VOCs). Organic compounds that exist as gases in the atmosphere are called *volatile organic compounds* (VOCs). Most are hydrocarbons, such as *isoprene* (C_3H_8) and *terpenes* such as $C_{10}H_{15}$ emitted by the leaves of many plants, and *methane* (CH_4), a greenhouse gas. About a third of global methane emissions come from natural sources, mostly plants, wetlands, and termites. The rest comes from human sources, primarily rice paddies, landfills, oil and natural gas wells, and cows (mostly from their belching).

Other VOCs, including benzene, vinyl chloride, and trichloroethylene (TCE), are used as industrial solvents, dry-cleaning fluids, and components of gasoline, plastics, drugs, synthetic rubber, and other products. Benzene (C_6H_6) is found in motor vehicle and power plant emissions and tobacco smoke. Long-term exposure to benzene can cause leukemia, numerous blood disorders, and immune system damage. Short-term exposure to high levels can cause dizziness, unconsciousness, and death. Scientists use a variety of methods to detect the presence and concentrations of air pollutants (Science Focus, at right).

■ CASE STUDY
Lead Is a Highly Toxic Pollutant

Because it is a chemical element, lead (Pb) does not break down in the environment. This potent neurotoxin can harm the nervous system, especially in young children. Each year, 12,000–16,000 American children under age 9 are treated for acute lead poisoning, and about 200 die. About 30% of the survivors suffer from palsy, partial paralysis, blindness, and mental retardation.

Children under age 6 and unborn fetuses even with low blood levels of lead are especially vulnerable to nervous system impairment, lowered IQ (by an average of 7.4 points), shortened attention span, hyperactivity, hearing damage, and various behavior disorders. A 1993 study by the U.S. National Academy of Sciences and numerous other studies indicate there is no safe level of lead in children's blood.

Good news. Between 1976 and 2000, the percentage of U.S. children ages 1–5 with blood lead levels above the safety standard dropped from 85% to 2.2%; at least 9 million childhood lead poisonings were prevented. The primary reason for this drop was that government regulations banned leaded gasoline in 1976 (with a complete phase-out by 1986) and lead-based paints in 1970 (but illegal use continued until about 1978). This is an excellent example of the effectiveness of pollution prevention.

But the U.S. Centers for Disease Control and Prevention estimates that at least 310,000 U.S. children still have unsafe blood levels of lead caused by exposure to a number of sources. Major sources are particles of peeling lead-based paint found in about 38 million houses built in the United States before 1960 and lead-contaminated dust in deteriorating buildings.

In 2007, major U.S. toy companies had to recall various toys made in China that contained lead paint. Also in 2007, tests revealed that almost two-thirds of the red, long-lasting lipsticks manufactured in the United States contained surprisingly high levels of lead, although none of them listed lead as an ingredient.* Lead can also

*Here is a way to test a lipstick for lead. Put some on your hand and scratch it with a gold ring. If the lipstick color changes to black, it contains lead.

SCIENCE FOCUS

Detecting Air Pollutants

The presence of pollutants in the air can be detected with the use of chemical instruments and satellites. The scientists who discovered the components and effects of the Asian Brown Cloud (Figure 18-1, **Core Case Study**) used small unmanned aircraft and miniaturized instruments to study the cloud. The aircraft flew in stacked formations and measured chemical concentrations and temperatures at different altitudes.

CORE CASE STUDY

Aerodyne Research in the U.S. city of Boston, Massachusetts, has developed a mobile laboratory that uses sophisticated instruments to make instant measurements of primary and secondary air pollutants from motor vehicles, factories, and other sources. It also records how their concentrations change throughout a day or under different weather conditions (see *The Habitable Planet*, Video 11, at **www.learner.org/resources/series209.html**). In addition to providing real-time measurements of key air pollutants, the mobile van can measure the effectiveness of various air pollution control devices used in cars, trucks, and buses.

Scientists are also using nanotechnology (Science Focus, p. 362) to try to develop inexpensive nanodetectors for various air pollutants. Another way to detect air pollutants is through biological indicators, including lichens (Figure 18-A).

A lichen consists of a fungus and an alga living together, usually in a mutually beneficial (mutualistic) relationship. These hardy pioneer species are good biological indicators of air pollution because they

Milton Rand/Tom Stack & Associates

Gerald & Buff Corsi/Visuals Unlimited

Figure 18-A **Natural capital:** Old man's beard (*Usnea trichodea*) lichen growing on a branch of a larch tree in Gifford Pinchot National Park, Washington (USA) (left) and red and yellow crustose lichens growing on slate rock in the foothills of the Sierra Nevada near Merced, California (USA) (right). The vulnerability of various lichen species to specific air pollutants can help researchers detect levels of these pollutants and track down their sources.

continually absorb air as a source of nourishment. A highly polluted area around an industrial plant may have only gray-green crusty lichens or none at all. An area with moderate air pollution may have orange crusty lichens on outdoor walls. Walls and trees in areas with fairly clean air can support leafy lichens.

Some lichen species are sensitive to specific air-polluting chemicals. Old man's beard (*Usnea trichodea*, Figure 18-A, left) and yellow *Evernia* lichens, for example, sicken or die in the presence of excess sulfur dioxide, even if the pollutant originates far away. For example, scien-

tists discovered sulfur dioxide pollution on Isle Royale, Michigan (USA) in Lake Superior, an island where no car or smokestack has ever intruded. They used *Evernia* lichens to point the finger northward to coal-burning facilities at Thunder Bay, Ontario, Canada.

Critical Thinking

Look for lichens on rocks, trees, and buildings in the area where you live or go to school. What do they indicate about air pollution levels?

leach from water pipes and faucets that contain it. And in 2008, Brian Schwartz and other researchers at Johns Hopkins University reported that high lifetime exposure to lead appears to decrease mental functions, like verbal and visual memory and language ability, in older people.

Health scientists have proposed a number of ways to help protect children from lead poisoning, as listed in Figure 18-7 (p. 476). Although the threat from lead has been greatly reduced in the United States, this is not the case in many developing countries. About 80% of the gasoline sold in the world today is unleaded, but about 100 countries still use leaded gasoline. The WHO estimates that 130 million–200 million children around the world are at risk from lead poisoning, and

15 million–18 million children in developing countries have permanent brain damage because of lead poisoning—mostly because of the use of leaded gasoline in their countries.

Environmental and health scientists call for global bans on leaded gasoline and lead-based paints. Some good news is that China recently phased out leaded gasoline in less than 3 years.

THINKING ABOUT
Reductions in Lead Poisoning

Why do you think the decline in lead poisoning in the United States since 1976 is such an excellent example of the power of pollution prevention?

Today, urban industrial smog is rarely a problem in most developed countries where coal and heavy oil are burned only in large boilers with reasonably good pollution control or with tall smokestacks that transfer the pollutants to downwind rural areas. However, industrial smog remains a problem in industrialized urban areas of China, India (Figure 15-13, p. 384), Ukraine, and some eastern European countries, where large quantities of coal are still burned in houses, power plants, and factories with inadequate pollution controls. This is contributing to the gigantic Asian pollution cloud (**Core Case Study**).

Because of its heavy reliance on coal, China has some of the world's highest levels of industrial smog and 16 of the world's 20 most polluted cities. In 2007, China's State Environmental Protection Administration (SEPA) reported that 62% of China's cities were plagued by industrial smog and that each year, air pollution prematurely kills 358,000 Chinese—an average of 981 deaths per day. However, a 2007 study by the World Bank puts the annual death toll from air pollution in China at about 750,000 people a year, about 500,000 of them from outdoor air pollution and 250,000 from indoor air pollution (mostly from burning coal for heating and cooking).

Sunlight Plus Cars Equals Photochemical Smog

A *photochemical reaction* is any chemical reaction activated by light. **Photochemical smog** is a mixture of

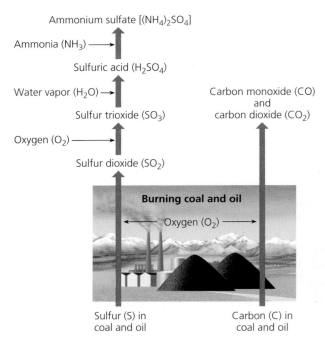

Figure 18-8 How pollutants are formed from burning of coal and oil. The result is industrial smog (**Concept 18-2**).

Figure 18-7 Ways to help protect children from lead poisoning. **Question:** Which two of these solutions do you think are the most important? Why?

Burning Coal Produces Industrial Smog

Fifty years ago, cities such as London, England, and the U.S. cities of Chicago, Illinois, and Pittsburgh, Pennsylvania, burned large amounts of coal in power plants, factories, and homes. People in such cities, especially during winter, were exposed to **industrial smog** consisting mostly of sulfur dioxide, suspended droplets of sulfuric acid, and a variety of suspended solid particles.

The chemistry of industrial smog is fairly simple. When burned, most of the carbon in coal and oil is converted to carbon monoxide (CO) and carbon dioxide (CO_2). Unburned carbon in coal also ends up in the atmosphere as suspended particulate matter (soot).

When coal and oil are burned (Figure 18-8), the sulfur compounds they contain react with oxygen to produce sulfur dioxide (SO_2) gas, some of which reacts with water vapor and is converted to tiny suspended droplets of sulfuric acid (H_2SO_4). Some of these droplets react with ammonia (NH_3) in the atmosphere to form solid particles of ammonium sulfate [$(NH_4)_2SO_4$]. The suspended particles of such salts and soot give the result-

primary and secondary pollutants formed under the influence of UV radiation from the sun. In greatly simplified terms,

$$VOCs + NO_x + heat + sunlight \longrightarrow \begin{array}{l} \text{ground level ozone } (O_3) \\ + \text{ other photochemical} \\ \quad \text{oxidants} \\ + \text{ aldehydes} \\ + \text{ other secondary} \\ \quad \text{pollutants} \end{array}$$

The formation of photochemical smog (Figure 18-9) begins when exhaust from morning commuter vehicles releases large amounts of NO and VOCs into the air over a city. The NO is converted to reddish-brown NO_2, explaining why photochemical smog is sometimes called *brown-air smog*. When exposed to ultraviolet radiation from the sun, some of the NO_2 reacts in complex ways with VOCs released by certain trees (such as some oak species, sweet gums, and poplars), motor vehicles, and businesses (such as bakeries and dry cleaners).

The resulting photochemical smog is a mixture of ozone, nitric acid, aldehydes, peroxyacyl nitrates (PANs), and other secondary pollutants. Collectively, NO_2, O_3, and PANs in this chemical brew are called *photochemical oxidants* because these damaging chemicals can react with and oxidize certain compounds in the atmosphere or inside your lungs.

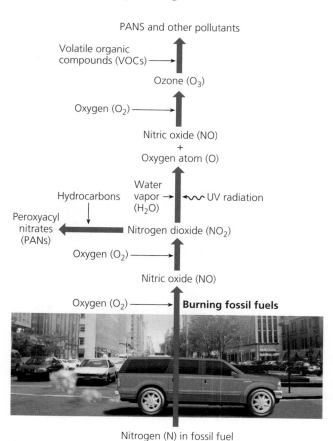

Figure 18-9 A greatly simplified model of how pollutants that make up photochemical smog are formed.

Figure 18-10 *Global outlook:* photochemical smog in Santiago, Chile. **Question:** How serious is photochemical smog where you live?

Hotter days lead to higher levels of ozone and other components of smog. As traffic increases on a sunny day, photochemical smog (dominated by ozone) usually builds up to peak levels by late morning, irritating people's eyes and respiratory tracts.

All modern cities have some photochemical smog, but it is much more common in cities with sunny, warm, and dry climates and lots of motor vehicles. Examples are Los Angeles, Denver, and Salt Lake City in the United States; Sydney, Australia; São Paulo, Brazil; Buenos Aires, Argentina; Bangkok, Thailand; Jakarta, Indonesia; Mexico City, Mexico (Photo 12 in the Detailed Contents); and Santiago, Chile (Figure 18-10). According to a 1999 study, if 400 million people in China are driving conventional gasoline-powered cars by 2050 (a worst-case scenario projected by some), the resulting photochemical smog could regularly cover the entire western Pacific, extending to North America.

CENGAGENOW˜ See how photochemical smog forms and how it affects us at CengageNOW™.

Several Factors Can Decrease or Increase Outdoor Air Pollution

Five natural factors help to *reduce* outdoor air pollution. First, particles heavier than air settle out as a result of gravitational attraction to the earth. Second, *rain and snow* help cleanse the air of pollutants. Third, *salty sea*

spray from the oceans wash out much of the particulates and other water-soluble pollutants from air that flows from land over the oceans. Fourth, *winds* sweep pollutants away, diluting them by mixing them with cleaner air, and bringing in fresh air. Fifth, some pollutants are removed by *chemical reactions*. For example, SO_2 can react with O_2 in the atmosphere to form SO_3, which reacts with water vapor to form droplets of H_2SO_4 that fall out of the atmosphere as acid precipitation.

Six other factors can *increase* outdoor air pollution. First, *urban buildings* can slow wind speed and reduce dilution and removal of pollutants. Second, *hills and mountains* can reduce the flow of air in valleys below them (Figure 18-10) and allow pollutant levels to build up at ground level. Third, *high temperatures* promote the chemical reactions leading to photochemical smog formation, which means that global warming could increase photochemical smog in many of the world's cities. Fourth, *emissions of volatile organic compounds (VOCs) from certain trees and plants* such as some oak species, sweet gums, poplars, and kudzu (Figure 9-15, p. 200) in heavily wooded urban areas can play a large role in the formation of photochemical smog.

A fifth factor—the so-called *grasshopper effect*—occurs when volatile air pollutants are transported by evaporation and winds from tropical and temperate areas through the atmosphere to the earth's polar areas, where they are deposited. This happens mostly during winter. It explains why, for decades, pilots have reported seeing dense layers of reddish-brown haze over the Arctic. It also explains why polar bears, sharks, and other top carnivores and native peoples in remote arctic areas have high levels of toxic pollutants in their bodies.

Sixth, *temperature inversions* can cause pollutants to build to high levels. During daylight, the sun warms the air near the earth's surface. Normally, this warm air and most of the pollutants it contains rise to mix and disperse the pollutants with the cooler air above it. Under certain atmospheric conditions, however, a layer of warm air can temporarily lie atop a layer of cooler air nearer the ground. This is called a **temperature inversion.** Because the cooler air is denser than the warmer air above it, the air near the surface does not rise and mix with the air above. This allows pollutants to build up in the stagnant layer of cool air near the ground.

Two types of areas are especially susceptible to prolonged temperature inversions. The first is a town or city located in a valley surrounded by mountains where the weather turns cloudy and cold during part of the year (Figure 18-11, left). In such a case, the surrounding mountains and the clouds block much of the winter sunlight that causes air to heat and rise, and the mountains block the wind. As long as these stagnant conditions persist, pollutants in the valley below will build up to harmful and even lethal concentrations. The tragic pollution event in Donora, Pennsylvania (Case Study, p. 471), was partly the result of such a temperature inversion.

The other type of area vulnerable to temperature inversions is a city with several million motor vehicles in an area with a sunny climate, light winds, mountains on three sides, and an ocean on the other side (Figure 18-11, right). Here, the conditions are ideal for photochemical smog worsened by frequent thermal inversions, and the surrounding mountains prevent the polluted surface air from being blown away by sea breezes. This describes the U.S. state of California's heavily populated Los Angeles basin, which has prolonged temperature inversions, mostly during summer and fall.

CENGAGENOW™ Learn more about thermal inversions and what they can mean for people in some cities at CengageNOW.

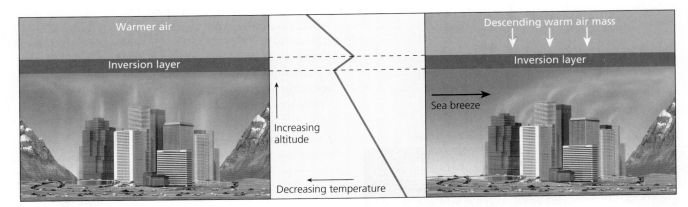

Figure 18-11 A *temperature inversion,* in which a warm air layer sits atop a cooler air layer, can take place in either of two sets of topography and weather conditions. Air pollutants can build to harmful levels during an inversion, which can occur during cold, cloudy weather in a valley surrounded by mountains (left). Frequent and prolonged temperature inversions can also occur in an area with a sunny climate, light winds, mountains on three sides, and the ocean on the other (right). A layer of descending warm air from a high-pressure system prevents ocean-cooled air near the ground from ascending enough to disperse and dilute pollutants. Because of their topography, Los Angeles, California (USA), and Mexico City, Mexico, have frequent temperature inversions, many of them prolonged, during the summer and fall.

▶ **CONCEPT 18-3** Acid deposition is caused mainly by coal-burning power plant and motor vehicle emissions, and in some regions, threatens human health, aquatic life and ecosystems, forests, and human-built structures.

Acid Deposition Is a Serious Regional Air Pollution Problem

Most coal-burning power plants, ore smelters, and other industrial plants in developed countries use tall smokestacks to emit sulfur dioxide, suspended particles, and nitrogen oxides high into the atmosphere where wind can mix, dilute, and disperse them.

These tall smokestacks reduce *local* air pollution, but can increase *regional* air pollution downwind. The primary pollutants (sulfur dioxide and nitrogen oxides) emitted high into the troposphere may be transported as far as 1,000 kilometers (600 miles) by prevailing winds. During their trip, they form secondary pollutants such as droplets of sulfuric acid (H_2SO_4), nitric acid vapor (HNO_3), and particles of acid-forming sulfate (SO_4^{2-}) and nitrate (NO_3^-) salts.

These acidic substances remain in the atmosphere for 2–14 days, depending mostly on prevailing winds, precipitation, and other weather patterns. During this period, they descend to the earth's surface in two forms: *wet deposition* consisting of acidic rain, snow, fog, and cloud vapor with a pH less than 5.6* and *dry deposition* consisting of acidic particles. (See Figure 5, p. S41, in Supplement 6.) The resulting mixture is called **acid deposition** (Figure 18-12)—sometimes termed *acid rain*. Most dry deposition occurs within 2–3 days fairly near the emission sources, whereas most wet deposition takes place within 4–14 days in more distant downwind areas.

Acid deposition has been occurring since the Industrial Revolution. In 1872, British chemist Robert A. Smith coined the term *acid rain* after observing that rain was eating away stone in the walls of buildings in major industrial areas. Acid deposition occurs when human activities disrupt the natural nitrogen and sulfur cycles (Figures 3-19, p. 69, and 3-22, p. 72) by adding excessive amounts of nitrogen oxides and sulfur dioxide to the atmosphere.

Acid deposition is a *regional* air pollution problem (**Concept 18-3**) in areas that lie downwind from coal-burning facilities and in urban areas with large

* Unpolluted rain is acidic with a pH of about 5.6 because of the reaction of CO_2 and water to form carbonic acid (H_2CO_3).

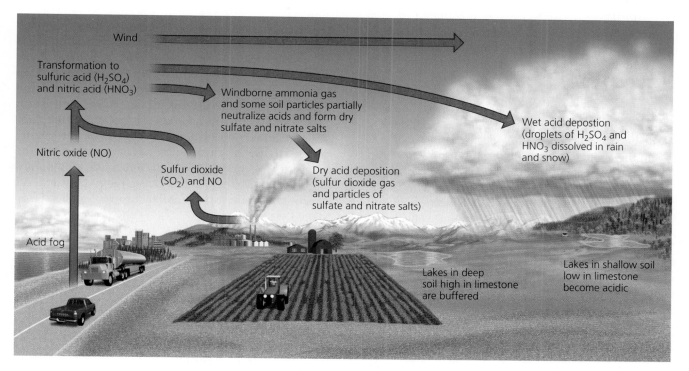

CENGAGENOW **Active Figure 18-12 Natural capital degradation:** *acid deposition*, which consists of rain, snow, dust, or gas with a pH lower than 5.6, is commonly called acid rain. Soils and lakes vary in their ability to neutralize excess acidity. *See an animation based on this figure at* CengageNOW. **Question:** What are three ways in which your daily activities contribute to acid deposition?

	Potential problem areas because of sensitive soils
	Potential problem areas because of air pollution: emissions leading to acid deposition
	Current problem areas (including lakes and rivers)

Figure 18-13 Regions where acid deposition is now a problem and regions with the potential to develop this problem (**Concept 18-3**). Such regions have large inputs of air pollution (mostly from power plants, industrial plants, and ore smelters) or are sensitive areas with soils and bedrock that cannot neutralize (buffer) inputs of acidic compounds. **Question:** Do you live in or near an area that is affected by acid deposition or any area that is likely to be affected by acid deposition in the future? (Data from World Resources Institute and U.S. Environmental Protection Agency)

numbers of motor vehicles. Such areas include the eastern United States and other parts of the world shown in Figure 18-13. In some areas, soils contain *basic* compounds (Figure 5, p. S41, in Supplement 6) such as calcium carbonate ($CaCO_3$) or limestone that can react with and neutralize, or *buffer,* some inputs of acids. The areas most sensitive to acid deposition are those with thin, acidic soils that provide no such natural buffering (Figure 18-13, green and most red areas) and those where the buffering capacity of soils has been depleted by decades of acid deposition.

In the United States, older, coal-burning power and industrial plants without adequate pollution controls in the Midwest emit the largest quantities of sulfur dioxide and other pollutants that cause acid deposition. Because of these emissions, and those of motor vehicles and other urban sources, typical precipitation in the eastern United States is at least 10 times more acidic than natural precipitation is. (Figure 14, p. S9, in Supplement 2 is a map of pH values in the continental United States in 2005.) Some mountaintop forests in the eastern United States and east of Los Angeles, California, are bathed in fog and dews as acidic as lemon juice—with about 1,000 times the acidity of normal precipitation.

Many acid-producing chemicals generated in one country are exported to other countries by prevailing winds. For example, acidic emissions from the United Kingdom and Germany blow into Switzerland, Austria, Norway, and neighboring countries. Some SO_2 and other emissions from coal-burning power and industrial plants in the United States end up in southeastern Canada.

The worst acid deposition occurs in Asia, especially in China, which gets 70% of its total energy and 80% of its electricity from burning coal. According to its government, China is the world's top emitter of SO_2. The resulting acid precipitation is damaging crops and threatening food security in China, Japan, and North and South Korea. In addition, air pollution that contributes to acid deposition is produced by the greatly increased use of cheap diesel generators to provide electricity for rural villages and to run irrigation pumps in China, India, and other developing countries. All of this contributes to the Asian Brown Cloud (**Core Case Study**).

CORE CASE STUDY

CENGAGENOW™ Learn more about the sources of acid deposition, how it forms, and what it can do to lakes and soils at CengageNOW.

Acid Deposition Has a Number of Harmful Effects

Acid deposition causes harm in several ways. It contributes to human respiratory diseases, and damages statues (Figure 18-6), national monuments, buildings,

metals, and car finishes. Also, acidic particles in the air can decrease visibility.

One of the most alarming and often unseen effects of acid deposition is that it can leach toxic metals (such as lead and mercury) from soils and rocks into lakes used as sources of drinking water. These toxic metals can accumulate in the tissues of fish eaten by people, other mammals, and birds. Currently, 45 U.S. states have issued statements warning people (especially pregnant women) not to eat fish caught from some of their waters because of mercury contamination (Figure 17-A, p. 450).

THINKING ABOUT
Acid Deposition and Mercury

Do you live in or near an area where government officials have warned people not to eat fish caught from waters contaminated with mercury? If so, what do you think are the specific sources of the mercury pollution?

Acid deposition harms aquatic ecosystems. Most fish cannot survive in water with a pH less than 4.5. Acid deposition can also release aluminum ions (Al^{3+}), which are attached to minerals in nearby soil, into lakes. These ions asphyxiate many kinds of fish by stimulating excessive mucus formation, which clogs their gills.

Because of excess acidity, several thousand lakes in Norway and Sweden contain no fish, and many more

lakes there have lost most of their acid-neutralizing capacity. In Ontario, Canada, at least 1,200 acidified lakes contain few if any fish, and fish populations in thousands of other lakes are declining because of increased acidity. In the United States, several hundred lakes (most in the Northeast) are threatened in this way. And scientists are just beginning to study the effects of the Asian Brown Cloud on oceans (**Core Case Study**). Aerosols in the cloud get pulled into thunderstorms that dump acid rain into the Indian and Pacific Oceans, possibly harming marine life and ecosystems.

Acid deposition (often along with other air pollutants such as ozone) can harm crops, especially when the soil pH is below 5.1. It reduces plant productivity and the ability of soils to buffer or neutralize acidic inputs. An estimated 30% of China's cropland suffers from excess acidity.

Acid deposition can affect forests by leaching essential plant nutrients, such as calcium and magnesium, from soils and releasing ions of aluminum, lead, cadmium, and mercury, which are toxic to the trees (Figure 18-14). This rarely kills trees directly, but it can weaken them and leave them vulnerable to stresses such as severe cold, diseases, insect attacks, and drought.

Mountaintop forests are the terrestrial areas hardest hit by acid deposition (see photo insert in Figure 18-14). These areas tend to have thin soils without

CENGAGENOW™ **Active Figure 18-14**
Natural capital degradation: Air pollution is one of several interacting stresses that can damage, weaken, or kill trees and pollute surface and groundwater. The photo insert shows air pollution damage to trees at a high elevation in Mount Mitchell State Park, North Carolina (USA). *See an animation based on this figure at* CengageNOW.

Carolina Biological/Visuals Unlimited

SCIENCE FOCUS

Revisiting Hubbard Brook to Study Effects of Acid Deposition

In the Core Case Study for Chapter 2 (p. 28), we discussed controlled experiments by scientists in the Hubbard Brook Experimental Forest in the White Mountains of the northeastern U.S. state of New Hampshire—an area where forests are suffering the effects of acid deposition.

Ecologist Gene Likens and his colleagues considered the effects of acid deposition on the experimental forest. They noticed that after SO_2 and particulate levels in the atmosphere declined as a result of pollution controls mandated by the Clean Air Acts, damaged trees in the forest did not recover as expected. One hypothesis they considered to explain this was that nutrients required for tree health and growth—particularly calcium and magnesium—had been stripped from the soil by acid precipitation. It takes decades for soil to rebuild these nutrients, which would explain the lag in recovery of the trees.

The scientists first turned to data they had collected since the 1950s and found that, since that time, most of the calcium ions (Ca^{2+}) in the soils had been leached into streams, which carried them away. They hypothesized that acid rain and snow had stripped the soils of these nutrients, and that, by the 1990s, this had essentially stopped the growth of trees and other vegetation.

In 1998, Likens and his team began a new experiment to further test this hypothesis. For 2 years, they made measurements in control and experimental forests (Figure 2-1, p. 28). Then they used a helicopter to drop quantities of a calcium salt into the experimental forest to add calcium to the soil. Subsequent examination of the soil showed that soil nitrate concentrations increased significantly, along with soil pH (showing lower soil acidity) in the experimental forest. The researchers projected that changes to plant growth patterns would not be recorded for

several years. But in 2003, they found that during the winter, red spruce trees in the untreated forest lost about three times more of their needles than did the red spruce in the calcium-treated forest.

These studies support the hypothesis that trees do not suffer from direct contact with acid precipitation, but rather from insufficient nutrients in depleted soil (Figure 18-14). The researchers also concluded that, in order for damaged forests to recover, their soil nutrients must be restored, and that it could take decades for natural processes to repair the damage to these forests from acid precipitation.

Critical Thinking

Explain why it will take decades for natural processes to repair forests damaged by acid precipitation? (*Hint:* think about nutrient cycling, see Section 3-5, p. 65.)

much buffering capacity. And trees on mountaintops (especially conifers such as red spruce and balsam fir) are bathed almost continuously in highly acidic fog and clouds. However, uncontrolled emissions of sulfur dioxide and other pollutants can devastate the vegetation in an area.

CENGAGENOW™ Examine how acid deposition can harm a pine forest and what it means to surrounding land and waters at CengageNOW.

Most of the world's forests and lakes are not being destroyed or seriously harmed by acid deposition. Rather, this regional problem is harming forests and lakes that lie downwind from large car-dominated cities and from coal-burning facilities without adequate pollution controls (**Concept 18-3**). Also, acid deposition has not reduced overall tree growth in the vast majority of forests in the United States and Canada, partly because of significant reductions in SO_2 and NO_x emissions from coal-fired power and industrial plants under 1990 amendments to the U.S. Clean Air Act.

However, acid deposition has accelerated the leaching of plant nutrients from soils in some areas, which has hindered tree growth, as researchers found in the on-going Hubbard Brook studies (see Science Focus, above, and the Chapter 2 Core Case Study, p. 28). Scientists estimate that an additional 80% reduction in SO_2 emissions from coal-burning power and industrial

plants in the midwestern United States will be needed before northeastern lakes, forests, and streams can recover from past and projected effects of acid deposition.

RESEARCH FRONTIER

Learning more about the extent and effects of acid deposition throughout the world. See **academic.cengage.com/biology/miller**.

We Know How to Reduce Acid Deposition

Figure 18-15 summarizes ways to reduce acid deposition. According to most scientists studying the problem, the best solutions are *prevention approaches* that reduce or eliminate emissions of sulfur dioxide, nitrogen oxides, and particulates.

Controlling acid deposition is politically difficult. One problem is that the people and ecosystems it affects often are quite distant from those who cause the problem. Also, countries with large supplies of coal (such as China, India, Russia, and the United States) have a strong incentive to use it as a major energy resource. Owners of coal-burning power plants resist taking measures such as adding the latest pollution control equipment, using low-sulfur coal, or removing sulfur from coal before burning it. They argue that these precautions would increase the cost of electricity for consumers.

Environmental scientists counter that affordable and much cleaner resources are available to produce electricity, including wind, hydropower, and natural gas. They also point out that the largely hidden health and environmental costs of burning coal are up to five times its market price (Table 16-1, p. 416). Including these costs in the market prices of coal would reduce coal use, spur the use of cleaner ways to generate electricity, help to prevent acid deposition, and reduce CO_2 emissions.

As for technological fixes, large amounts of limestone or lime are used to neutralize some acidified lakes and surrounding soil—the only cleanup approach now being used. However, this expensive and temporary remedy usually must be repeated annually. Also, it can kill some types of plankton and aquatic plants and can harm wetland plants that need acidic water. And it is difficult to know how much lime to put where (in the water or at selected locations on the ground). In 2002, researchers in England found that adding a small amount of phosphate fertilizer can neutralize excess acidity in a lake. The effectiveness of this approach is being evaluated.

Air pollution laws in the United States have reduced the acidity of rainfall in parts of the northeast, mid-Atlantic, and midwest regions, but there is still a long way to go in reducing emissions from older coal-burning power and industrial plants. Some plants have lowered SO_2 emissions by switching from high-sulfur to low-sulfur coals. However, this has increased CO_2 emissions that contribute to global warming, because low-sulfur coal has a lower heat value, which means that more coal must be burned to generate a given amount of electricity. Low-sulfur coal also has higher levels of toxic mercury and other trace metals, so burning it emits more of these hazardous chemicals into the atmosphere.

SOLUTIONS

Acid Deposition

Prevention

Reduce coal use

Burn low-sulfur coal

Increase natural gas use

Increase use of renewable energy resources

Remove SO_2 particulates and NO_x from smokestack gases

Remove NO_x from motor vehicular exhaust

Tax emissions of SO_2

Reduce air pollution by improving energy efficiency

Cleanup

Add lime to neutralize acidified lakes

Add phosphate fertilizer to neutralize acidified lakes

Figure 18-15 Methods for reducing acid deposition and its damage. **Question:** Which two of these solutions do you think are the most important? Why?

> **THINKING ABOUT**
> **Low-Sulfur Coal**
>
> Do you think that the advantages of burning low-sulfur coal outweigh the disadvantages? Explain. Are there better options?

18-4 What Are the Major Indoor Air Pollution Problems?

> ▶ **CONCEPT 18-4** The most threatening indoor air pollutants are smoke and soot from wood and coal cooking fires (a hazard found mostly in developing countries) and chemicals used in building materials and products.

Indoor Air Pollution Is a Serious Problem

In developing countries, the indoor burning of wood, charcoal, dung, crop residues, coal, and other cooking and heating fuels in open fires or in unvented or poorly vented stoves exposes people to dangerous levels of particulate air pollution. And workers, including children (Figure 18-5), can be exposed to high levels of indoor air pollution where there are few if any laws or regulations for controlling such pollution. According to

the WHO and the World Bank, *indoor air pollution is, for poor people, the world's most serious air pollution problem.* This is a glaring example of the relationship between poverty, environmental quality, and human health.

Indoor air pollution is also a serious problem in developed countries. If you live in such a country and are reading this book indoors, you may be inhaling more air pollutants than you would if you were outside. Figure 18-16 (p. 484) shows some typical sources of indoor air pollution in a modern home.

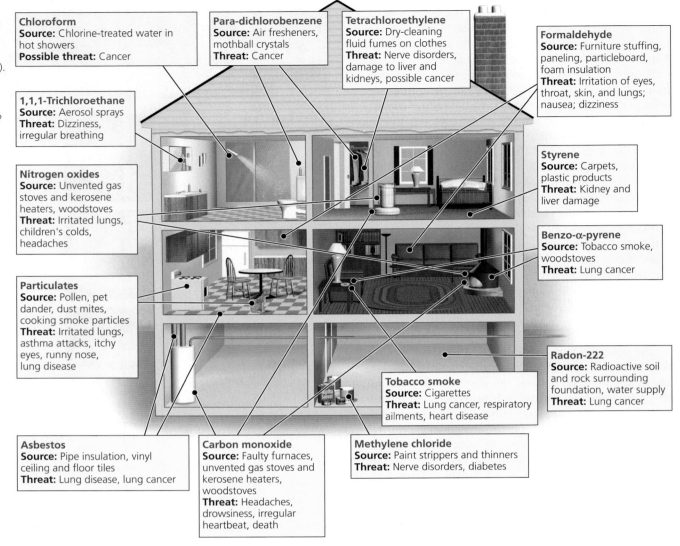

Figure 18-16
Some important indoor air pollutants (**Concept 18-4**). **Question:** Which of these pollutants are you exposed to? (Data from U.S. Environmental Protection Agency)

Chloroform
Source: Chlorine-treated water in hot showers
Possible threat: Cancer

Para-dichlorobenzene
Source: Air fresheners, mothball crystals
Threat: Cancer

Tetrachloroethylene
Source: Dry-cleaning fluid fumes on clothes
Threat: Nerve disorders, damage to liver and kidneys, possible cancer

Formaldehyde
Source: Furniture stuffing, paneling, particleboard, foam insulation
Threat: Irritation of eyes, throat, skin, and lungs; nausea; dizziness

1,1,1-Trichloroethane
Source: Aerosol sprays
Threat: Dizziness, irregular breathing

Nitrogen oxides
Source: Unvented gas stoves and kerosene heaters, woodstoves
Threat: Irritated lungs, children's colds, headaches

Styrene
Source: Carpets, plastic products
Threat: Kidney and liver damage

Benzo-α-pyrene
Source: Tobacco smoke, woodstoves
Threat: Lung cancer

Particulates
Source: Pollen, pet dander, dust mites, cooking smoke particles
Threat: Irritated lungs, asthma attacks, itchy eyes, runny nose, lung disease

Radon-222
Source: Radioactive soil and rock surrounding foundation, water supply
Threat: Lung cancer

Tobacco smoke
Source: Cigarettes
Threat: Lung cancer, respiratory ailments, heart disease

Asbestos
Source: Pipe insulation, vinyl ceiling and floor tiles
Threat: Lung disease, lung cancer

Carbon monoxide
Source: Faulty furnaces, unvented gas stoves and kerosene heaters, woodstoves
Threat: Headaches, drowsiness, irregular heartbeat, death

Methylene chloride
Source: Paint strippers and thinners
Threat: Nerve disorders, diabetes

Indoor air pollution usually poses a much greater threat to human health than does outdoor air pollution. EPA studies have revealed some alarming facts about indoor air pollution in the United States and in other developed countries. *First,* levels of 11 common pollutants generally are two to five times higher inside U.S. homes and commercial buildings than they are outdoors, and as much as 100 times higher in some cases. *Second,* pollution levels inside cars in traffic-clogged urban areas can be up to 18 times higher than outside levels. *Third,* the health risks from exposure to such chemicals are magnified because most people in developed countries spend 70–98% of their time indoors or inside vehicles.

Since 1990, the EPA has placed indoor air pollution at the top of the list of 18 sources of cancer risk. It causes as many as 6,000 premature cancer deaths per year in the United States. At greatest risk are smokers, children younger than age 5, the elderly, the sick, pregnant women, people with respiratory or heart problems, and factory workers.

Pesticide residues and lead particles brought in on shoes can collect in carpets, which is why some people remove their shoes after entering a house. According to the EPA, three of every four U.S. homes use pesticides indoors at least once a year. And many chemicals containing potentially harmful organic solvents are found in paints and various sprays.

Living organisms and their excrements can also pollute indoor air. Evidence indicates that exposure to allergens such as *dust mites* (Figure 18-17) and *cockroach droppings* found in some homes plays an important role in the almost threefold increase between 1972 and 2002 in the number of people suffering from asthma in the United States.

Another living source of indoor air pollution is toxic *airborne spores of molds* (fungal growths) and *mildew* that can cause headaches and allergic reactions and aggravate asthma and other respiratory diseases. Some evidence suggests that spores from molds and mildew growing underneath houses and on inside walls are the single greatest cause of allergic reactions to indoor air.

Danish and U.S. EPA studies have linked various air pollutants found in buildings to a number of health effects, a phenomenon known as the *sick-building syndrome* (SBS). Such effects include dizziness, headaches,

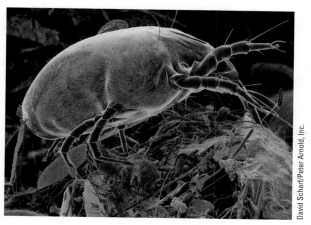

Figure 18-17 Science: magnified view of a household dust mite in a dust ball. This minute eight-legged relative of spiders feeds on dead human skin and household dust and lives in materials such as bedding and furniture fabrics. Dead dust mites and their excrement can cause asthma attacks and allergic reactions in some people. **Questions:** Were you aware that you are exposed to large numbers of dust mites? How could you reduce such exposure?

coughing, sneezing, shortness of breath, nausea, burning eyes, sore throats, chronic fatigue, irritability, skin dryness and irritation, respiratory infections, flu-like symptoms, and depression. EPA and Labor Department studies indicate that almost one in five commercial buildings in the United States is considered "sick," exposing employees to these health risks. **GREEN CAREER:** Indoor air pollution specialist

According to the EPA and public health officials, the four most dangerous indoor air pollutants in developed countries are *tobacco smoke* (Case Study, p. 462); *formaldehyde* found in many building materials and household products; *radioactive radon-222 gas,* which can seep into houses from underground rock deposits (Case Study, right); and *very small particles.*

The chemical that causes most people in developed countries difficulty is *formaldehyde* (CH_2O, a colorless, extremely irritating chemical). According to the EPA and the American Lung Association, 20–40 million Americans suffer from chronic breathing problems, dizziness, rash, headaches, sore throat, sinus and eye irritation, skin irritation, wheezing, and nausea caused by daily exposure to low levels of formaldehyde emitted from common household materials. In 2008, Marc Weisskopf at the Harvard School of Public Health, reported on a 15-year study of more than 1,100 people exposed to formaldehyde. Those people had a 34% higher than normal risk of getting Lou Gehrig's disease, which progressively kills nerve cells that control muscle movement.

Formaldehyde is found in building materials (such as plywood, particleboard, paneling, and high-gloss wood used in floors and cabinets), furniture, drapes, upholstery, adhesives in carpeting and wallpaper, urethane-formaldehyde foam insulation, fingernail hardener, and wrinkle-free coating on permanent-press clothing. The EPA estimates that 1 of every 5,000 people who live in manufactured homes for more than 10 years will develop cancer from formaldehyde exposure. In 2008, the U.S. Centers for Disease Control and Prevention (CDC) measured unhealthy indoor air levels of formaldehyde in many of the trailers and mobile homes provided by the government to house thousands of people displaced from their homes in 2005 by Hurricane Katrina.

RESEARCH FRONTIER

Learning more about indoor air pollutants and how to prevent them. See **academic.cengage.com/biology/miller**.

■ CASE STUDY
Radioactive Radon Gas

Radon-222 is a colorless, odorless, radioactive gas that is produced by the natural radioactive decay of uranium-238, small amounts of which are contained in most rocks and soils. But this isotope is much more concentrated in underground deposits of minerals such as uranium, phosphate, granite, and shale.

When radon gas from such deposits seeps upward through the soil and is released outdoors, it disperses quickly in the air and decays to harmless levels. However, in buildings above such deposits, radon gas can enter through cracks in foundations and walls, openings around sump pumps and drains, and hollow concrete blocks (Figure 18-18). Once inside it can build up to high levels, especially in unventilated lower levels of homes and buildings.

Radon-222 gas quickly decays into solid particles of other radioactive elements such as polonium-210, which if inhaled, expose lung tissue to large amounts of ionizing radiation from alpha particles. This exposure can damage lung tissue and lead to lung cancer over the course of a 70-year lifetime. Your chances of

Outlet vents for furnaces and dryers

Figure 18-18 Science: sources and paths of entry for indoor radon-222 gas. **Question:** Have you tested the indoor air where you live for radon-222? (Data from U.S. Environmental Protection Agency)

getting lung cancer from radon depend mostly on how much radon is in your home, how much time you spend in your home, and whether you are a smoker or have ever smoked. About 90% of radon-related lung cancers occur among current or former smokers.

Ideally, radon levels should be monitored continuously in the main living areas (not basements or crawl spaces) for 2 months to a year. But less than 10% of U.S. households have followed the EPA's recommendation to conduct radon tests (most lasting only 2–7 days and costing $20–100 per home).

For information about radon testing, visit the EPA website at **www.epa.gov/iaq/radon**. According to the EPA, radon control could add $350–500 to the cost of a new home, and correcting a radon problem in an existing house could run $800–2,500. Remedies include sealing cracks in the foundation and walls, increasing ventilation by cracking a window or installing vents, and using a fan to create cross ventilation.

> **THINKING ABOUT**
> **Preventing Indoor Air Pollution**
>
> What are some steps you could take to prevent indoor air pollution where you live, especially regarding the four most dangerous indoor air pollutants listed above.

18-5 What Are the Health Effects of Air Pollution?

▶ **CONCEPT 18-5** Air pollution can contribute to asthma, chronic bronchitis, emphysema, lung cancer, heart attack, and stroke.

Your Body's Natural Defenses against Air Pollution Can Be Overwhelmed

Your respiratory system (Figure 18-19) has a number of mechanisms that help to protect you from air pollution. Hairs in your nose filter out large particles. Sticky mucus in the lining of your upper respiratory tract captures smaller (but not the smallest) particles and dissolves some gaseous pollutants. Sneezing and coughing expel contaminated air and mucus when pollutants irritate your respiratory system.

In addition, hundreds of thousands of tiny mucus-coated hair-like structures called *cilia* line your upper

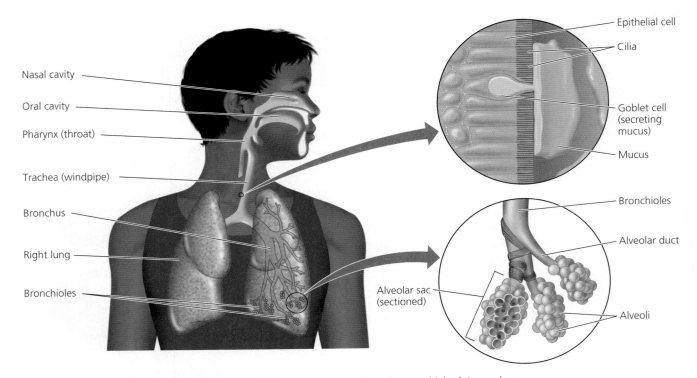

Figure 18-19 Major components of the human respiratory system. **Question:** Can you think of times when pollution might have affected your respiratory system?

Figure 18-20 Normal human lungs (left) and the lungs of a person who died of emphysema (right). Prolonged smoking and exposure to air pollutants can cause emphysema in anyone, but about 2% of emphysema cases result from a defective gene that reduces the elasticity of the air sacs in the lungs. Anyone with this hereditary condition, for which testing is available, should not smoke and should not live or work in a highly polluted area.

respiratory tract. They continually wave back and forth and transport mucus and the pollutants they trap to your throat where they are swallowed or expelled.

Prolonged or acute exposure to air pollutants, including tobacco smoke, can overload or break down these natural defenses. Fine and ultrafine particulates get lodged deep in the lungs, contributing to lung cancer, asthma attack, heart attack, and stroke. For example, in the United States, about 17 million people, 6% of the population, suffer from asthma, and about 14 of them die each day from asthma attacks. A French study found that asthma attacks increased by about 30% on smoggy days.

Years of smoking or breathing polluted air can lead to other lung ailments such as chronic bronchitis. Another such disease is *emphysema,* in which irreversible damage to air sacs or alveoli (Figure 18-19, bottom right) leads to abnormal dilation of air spaces, loss of lung elasticity, and acute shortness of breath (Figure 18-20).

Air Pollution Is a Big Killer

According to the WHO, at least 3 million people worldwide (most of them in Asia) die prematurely each year from the effects of air pollution—an average of 8,200 deaths per day. About 2.2 million of these deaths (73%) result from indoor air pollution, typically from heart attacks, respiratory diseases, and lung cancer related to daily breathing of polluted air.

In the United States, the EPA estimates that annual deaths related to indoor and outdoor air pollution range from 150,000 to 350,000 people—equivalent to 2–5 fully loaded, 200-passenger airliners crashing *each day* with no survivors. Millions more suffer from asthma attacks and other respiratory disorders, and they lose work time. Studies indicate that 60,000–70,000 of these

deaths are related to inhalation of very small, fine, and ultrafine particulates from coal-burning power plants (Figure 18-21).

According to recent EPA studies, each year, more than 125,000 Americans (96% of them in urban areas) get cancer from breathing soot-laden diesel fumes from buses and trucks. Other sources of these fumes include tractors, bulldozers and other construction equipment, and trains. A large diesel truck emits as much particulate matter as 150 cars, and particulate emissions from a diesel train engine equal those from 1,500 cars.

The EPA has proposed stricter emission standards for diesel-powered vehicles. The agency estimates that these standards should reduce diesel-fuel emissions by more than 90% and prevent as many as 12,000 prema-

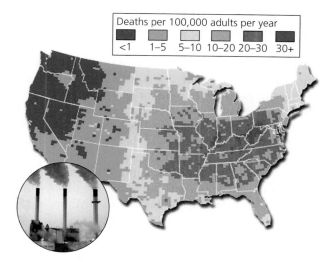

Figure 18-21 Premature deaths from air pollution in the United States, mostly from very small, fine, and ultrafine particles added to the atmosphere by coal-burning power plants. **Question:** What is the risk where you live or go to school? (Data from U.S. Environmental Protection Agency)

ture deaths. Manufacturers of diesel-powered engines and vehicles dispute the EPA findings and hope to relax or delay the standards.

There is also a connection between international trade and air pollution. Currently, about 75% of the world's international trade is transported in large rectangular aluminum containers carried by increasingly larger cargo ships. As they come into harbors, the massive diesel motors propelling these ships pollute the air. Long lines of diesel-powered trucks idling their engines while waiting to be loaded with the arriving containers emit additional air pollution.

For example, the current average daily emissions of PM-10 particulates at California's Port of Los Angeles equal the daily particulate emissions from 500,000 U.S. cars. In 2005, California's Air Resources Board estimated that diesel fuel emissions from the state's booming cargo shipping industry cause as many as 750 premature deaths per year. In a 2007 study, a team of scientists led by James Corbett and James Winebrake estimated that emissions from ocean-going ships are responsible for about 60,000 premature deaths per year worldwide.

18-6 How Should We Deal with Air Pollution?

▶ **CONCEPT 18-6** Legal, economic, and technological tools can help to clean up air pollution, but much greater emphasis should be focused on preventing air pollution.

Laws and Regulations Can Reduce Outdoor Air Pollution

The United States provides an excellent example of how a regulatory approach can reduce pollution (**Concept 18-6**). The U.S. Congress passed the Clean Air Acts in 1970, 1977, and 1990. With these laws, the federal government established air pollution regulations that are enforced by states and major cities, and that focus on key pollutants.

Congress directed the EPA to establish *national ambient air quality standards* (*NAAQS*) for six outdoor *criteria pollutants*—carbon monoxide, nitrogen oxides, sulfur dioxide, suspended particulate matter (less than PM-10), ozone, and lead. One limit, called a *primary standard,* is set to protect human health. Another limit, called a *secondary standard,* is intended to prevent environmental and property damage. Each standard specifies the maximum allowable level, averaged over a specific period, for a certain pollutant in outdoor (ambient) air.

The EPA has also established national emission standards for more than 188 *hazardous air pollutants* (*HAPs*) that may cause serious health and ecological effects. Most of these chemicals are chlorinated hydrocarbons, volatile organic compounds, or compounds of toxic metals. An important public source of information about hazardous air pollutants is the annual *Toxic Release Inventory* (*TRI*). The TRI law requires 21,500 refineries, power plants, hard rock mines, chemical manufacturers, and factories to report their releases and waste management methods for 667 toxic chemicals.

The TRI, which is now available on the Internet, lists this information by community, which allows individuals and local groups to evaluate potential threats to their health. Since the first TRI report was released in 1988, reported emissions of toxic chemicals have dropped sharply.

In 2005, the EPA proposed that industries be required to report their HAP emissions every other year and that the thresholds for reporting these chemicals be raised tenfold. This means that more than 922 communities would no longer have numerical data on local polluters. Environmental and health scientists, emergency responders, and many citizens strongly oppose this weakening of the TRI, which has saved many lives and stimulated industries to reduce their emissions of such chemicals.

The good news is that, according to a 2007 EPA report (see **www.epa.gov/air/airtrends/sixpoll.html**), the combined emissions of the six major air pollutants decreased 49% between 1980 and 2006, even with significant increases during the same period in gross domestic product (up 121%), vehicle miles traveled (up 101%), population (up 32%), and energy consumption (up 29%). The decreases in emissions during this period were 97% for lead (Pb), 52% for volatile organic compounds (VOCs), 50% for carbon monoxide (CO), 47% for sulfur dioxide (SO$_2$), 33% for nitrogen oxides (NO$_x$), 28% for suspended particulate matter (PM10), and 21% for ground-level ozone.

According to the EPA, in 2006, about one of every three Americans lived in an area where the air was unhealthy to breathe *during part of the year.* This occurred mostly because levels of ozone and particulate matter exceeded the national air quality standards at various times.

An encouraging development in 2006 and 2007 was the arrival of low-sulfur diesel fuel, mandated by a regulation written in the 1990s. With it came a push for a new generation of diesel engines that will eventu-

ally cut particulate pollution by up to 95%. The new rules added about 5 cents per gallon to diesel prices but are projected to result in annual health care cost savings of $150 billion.

Elsewhere in the world, especially in developing countries, pollution levels are much worse. But some countries, such as Taiwan, Singapore, Malaysia, South Korea, Thailand, and parts of China, are making progress in improving outdoor urban air quality.

While the U.S. regulatory model has been successful in reducing outdoor air pollution, environmental scientists point to areas where it could stand improvement (see the following Case Study).

■ CASE STUDY

U.S. Air Pollution Laws Can Be Improved

The reduction of outdoor air pollution in the United States since 1970 has been a remarkable success story that can be attributed to two factors. *First,* U.S. citizens insisted that laws be passed and enforced to improve air quality. *Second,* the country was affluent enough to afford such controls and improvements.

Environmental scientists applaud the success of U.S. air pollution control laws but point to the following deficiencies:

- *The United States continues to rely mostly on pollution cleanup rather than prevention.* The power of prevention is clear (**Concept 18-6**). In the United States, the air pollutant with the largest drop in its atmospheric level was lead (97% between 1980 and 2006 and 99% since 1970), which was largely banned in gasoline (Case Study, p. 474). This has prevented a generation of children from suffering lead poisoning and is viewed as one the country's greatest environmental success stories. Despite such success, in 2007, environmental scientists were shocked that the EPA was considering removing lead from the list of regulated pollutants.

- *For decades, the U.S. Congress failed to increase fuel-efficiency standards for cars, sport utility vehicles (SUVs), and light trucks* (Figure 16-5, left, p. 404) and is far behind European Union countries, Japan, and China (Figure 16-5, right). According to environmental scientists, increased fuel efficiency would reduce air pollution from motor vehicles more quickly and effectively than any other measure. It would also reduce CO_2 emissions, reduce dependence on imported oil, save energy, and save consumers enormous amounts of money. Many scientists and economists believe that a *feebate program* (p. 405) is the best way to encourage the use of fuel-efficient motor vehicles.

- *Regulation of emissions from motorcycles and two-cycle gasoline engines remains inadequate.* A 2005 Swiss study indicated that motorcycles collectively emit 16 times more hydrocarbons and three times more carbon monoxide than do cars. And two-cycle engines used in lawn mowers, leaf blowers, chain saws, personal watercraft, outboard motors, and snowmobiles emit high levels of pollutants (although less-polluting versions are becoming available). According to the California Air Resources Board, a 1-hour ride on a typical personal watercraft creates more air pollution than the average U.S. car does in a year. And for each unit of fuel burned, a 2006-model, gasoline-powered lawn mower emits 93 times the amount of smog-forming pollutants that a 2006 car emits, on average. In 2007, the EPA proposed rules to cut such emissions from lawn mowers and boat motors by 35% to 70%. Manufacturers of two-cycle engines regularly push for extending legal deadlines and for weakening these standards.

- *There is little or no regulation of air pollution from oceangoing ships in American ports.* According to the Earth Justice Legal Defense Fund, a single cargo ship emits more air pollution than 2,000 diesel trucks or 350,000 cars. Ships burn the dirtiest grades of diesel fuel and threaten the health of millions of dockworkers and other people living in port cities. The EPA has proposed regulations to cut such diesel emissions by 90% between 2008 and 2015.

- *Airports are exempt from many air pollution regulations.* Major U.S. airports in cities such as New York; Los Angeles, California; and Chicago, Illinois, are among the top polluters in these metropolitan areas. A study of the health records of people living downwind from the airport in Seattle, Washington, found that they had significantly higher rates of respiratory diseases, infant mortality, and pregnancy complications than people not exposed to pollution from the airport.

- *The laws do not regulate emissions of the greenhouse gas CO_2,* which can alter climate and cause numerous harmful ecological, health, and economic effects. This may change, because in 2007, the U.S. Supreme Court ruled that the EPA can use the Clean Air Act to regulate CO_2.

- *Ultrafine particulates are not regulated.* These extremely small particles can penetrate the lung's defenses and are a major contributor to premature deaths from air pollution.

- *Urban ozone levels are still too high in many areas.* Because O_3 is a secondary pollutant, the only way to reduce its levels is to decrease the

emissions of NO_x and VOCs from point and non-point sources—something that is politically difficult to accomplish.

- *The laws have failed to deal seriously with indoor air pollution,* even though it is by far the most serious air pollution problem in terms of poorer health, premature death, and economic losses due to lost work time and increased health costs.

- *There is a need for better enforcement of the Clean Air Acts.* Under the Clean Air Acts, state and local officials have primary responsibility for implementing federal clean air standards, with federal funding. However, a 2006 study by the Center for American Progress found that, since 1993, enforcement has become lax because of a sharp drop in federal grants to state and local air quality agencies and relaxed federal inspection standards. According to a 2002 government study, more rigorous enforcement would save about 6,000 lives and prevent 140,000 asthma attacks each year in the United States.

Executives of companies that would be affected by implementing stronger air pollution regulations claim that correcting these deficiencies would cost too much and would harm economic growth. But history has shown that most industry estimates of the costs of implementing U.S. air pollution control standards have been many times higher than the actual costs. In addition, implementing such standards has boosted economic growth and created jobs by stimulating companies to develop new technologies for reducing air pollution emissions—many of which can be sold in the global marketplace. Without intense pressure from citizens, it is unlikely that the U.S. Congress will strengthen the Clean Air Acts.

HOW WOULD YOU VOTE?

Should the 1990 U.S. Clean Air Act be strengthened? Cast your vote online at **academic.cengage.com/biology/ miller**.

We Can Use the Marketplace to Reduce Outdoor Air Pollution

Allowing producers of air pollutants to buy and sell government air pollution allotments in the marketplace can help reduce emissions (**Concept 18-6**). To help reduce SO_2 emissions, the Clean Air Act of 1990 authorizes an *emissions trading,* or *cap-and-trade, program,* which enables the 110 most polluting power plants in 21 states (primarily in the midwestern and eastern United States, Figure 14, p. S9, in Supplement 2) to

buy and sell SO_2 pollution rights. Each of these coal-burning plants is annually given a number of pollution credits, which allow it to emit a certain amount of SO_2. A utility that emits less than its allotment of SO_2 has a surplus of pollution credits. It can use these credits to avoid reductions in SO_2 emissions at another of its plants, keep them for future plant expansions, or sell them to other utilities, private citizens, or environmental groups.

Proponents argue that this approach is cheaper and more efficient than having the government dictate how to control air pollution. Critics of this plan contend that it allows utilities with older, dirtier power plants to buy their way out of their environmental responsibilities and continue polluting. This approach also can encourage cheating, because it is based largely on self-reporting of emissions.

The ultimate success of any emissions trading approach depends on how low the initial cap is set and then on the lowering of the cap every few years, which should promote continuing innovation in air pollution prevention and control. Without these two provisions, emissions trading programs mostly move air pollutants from one area to another without achieving any overall reduction in air quality.

Good news. Between 1990 and 2006, the emissions trading system helped to reduce SO_2 emissions from electric power plants in the United States by 53% at a cost of less than one-tenth the cost projected by industry. By 2010, the EPA estimates that this approach will annually generate health and environmental benefits that are 60 times higher than the annual cost of the program.

Emissions trading is also being tried for NO_x and perhaps will be used in the future for other air pollutants. However, environmental and health scientists strongly oppose the cap-and-trade program set to begin in 2010 for controlling emissions of mercury by coal-burning power plants and industries. They note that mercury is highly toxic and does not break down in the environment, and that coal-burning plants choosing to buy permits instead of sharply reducing their mercury emissions will create toxic hotspots with unacceptably high levels of mercury.

The jury is still out for emissions trading programs in general. In 2002, the EPA reported results from the country's oldest and largest emissions trading program, in effect since 1993 in southern California. According to the EPA, this cap-and-trade model fell far short of projected emissions reductions. The same study also found accounting abuses. This highlights the need for more careful monitoring of all cap-and-trade programs.

HOW WOULD YOU VOTE?

Should emissions trading be used to help control emissions of all major air pollutants? Cast your vote online at **academic.cengage.com/biology/miller**.

Prevention	Dispersion or Cleanup
Burn low-sulfur coal	Disperse emissions above thermal inversion layer with tall smokestacks
Remove sulfur from coal	
Convert coal to a liquid or gaseous fuel	Remove pollutants after combustion
Shift to less polluting energy sources	Tax each unit of pollution produced

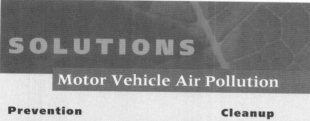

Figure 18-22 Methods for reducing emissions of sulfur oxides, nitrogen oxides, and particulate matter from stationary sources such as coal-burning electric power plants and industrial plants (**Concept 18-6**). **Question:** Which two of these solutions do you think are the most important? Why?

There Are Many Ways to Reduce Outdoor Air Pollution

Figure 18-22 summarizes ways to reduce emissions of sulfur oxides, nitrogen oxides, and particulate matter from stationary sources such as electric power plants and industrial plants that burn coal.

Between 1980 and 2006, emissions of SO_2 from U.S. electric power plants were decreased by 66%, emissions of NO_x by 41%, and particulate (PM10) emissions by 28%, mostly through use of output cleanup methods (Figure 18-22, right). However, approximately 20,000 older coal-burning plants, industrial plants, and oil refineries in the United States have not been required to meet the air pollution standards required for new facilities under the Clean Air Acts. Officials of states subject to pollution from such plants have been trying to get Congress to correct this shortcoming since 1970. But they have not been successful because of strong lobbying efforts by U.S. coal and electric power industries.

HOW WOULD YOU VOTE? ☑

Should older coal-burning power and industrial plants in the country where you live have to meet the same air pollution standards that new facilities have to meet? Cast your vote online at **academic.cengage.com/biology/ miller**.

Figure 18-23 lists ways to reduce emissions from motor vehicles, the primary factor in the formation of photochemical smog.

Unfortunately, the already poor air quality in urban areas of many developing countries is worsening as the number of motor vehicles in these countries rises. Many of these vehicles are 10 or more years old, have no pollution control devices, and burn leaded gasoline.

Prevention	Cleanup
Use mass transit	Require emission control devices
Walk or bike	
Use less polluting fuels	
Improve fuel efficiency	Inspect car exhaust systems twice a year
Get older, polluting cars off the road	
Give large tax write-offs or rebates for buying low-polluting, energy efficient vehicles	Set strict emission standards

Figure 18-23 Methods for reducing emissions from motor vehicles (**Concept 18-6**). To find out what and how much your car emits, go to **www .cleancarsforkids.org**. **Question:** Which two of these solutions do you think are the most important? Why?

Prevention	Cleanup or Dilution
Clean ceiling tiles and line AC ducts to prevent release of mineral fibers	Use adjustable fresh air vents for work spaces

Ban smoking or limit it to well-ventilated areas	Increase intake of outside air

Set stricter formaldehyde emissions standards for carpet, furniture, and building materials	Change air more frequently

Prevent radon infiltration	Circulate a building's air through rooftop greenhouses
Use office machines in well-ventilated areas	Use efficient venting systems for wood-burning stoves

Use less polluting substitutes for harmful cleaning agents, paints, and other products	Use exhaust hoods for stoves and appliances burning natural gas

Figure 18-24 Ways to prevent and reduce indoor air pollution (**Concept 18-6**). **Question:** Which two of these solutions do you think are the most important? Why?

The good news is that because of the Clean Air Acts, a new car today in the United States emits 75% less pollution than did pre-1970 cars. Over the next 10–20 years, technology will bring more gains through improved engine and emission systems, conventional hybrid-electric vehicles and plug-in hybrids (Figure 16-6, p. 405), and vehicles powered by hydrogen fuel cells (Figure 16-30, p. 429).

Reducing Indoor Air Pollution Should Be a Priority

Little effort has been devoted to reducing indoor air pollution, even though it poses a much greater threat to human health than does outdoor air pollution. Air pollution experts suggest several ways to prevent or reduce indoor air pollution, as shown in Figure 18-24.

In developing countries, indoor air pollution from open fires and leaky and inefficient stoves that burn wood, charcoal, or coal could be reduced. People could use inexpensive clay or metal stoves that burn biofuels more efficiently, while venting their exhaust to the outside. They could use stoves that run on biogas (mostly methane) produced in biodigesters from animal and other organic wastes, or they could use stoves that use solar energy to cook food (Figure 16-16, p. 414). This would also reduce deforestation by cutting demand for fuelwood and charcoal.

In developed countries, where VOCs present the greatest indoor air pollution threats, houseplants may provide some relief. Scientists have found that some common houseplants—such as the Boston fern, peace lily, rubber plant, and bamboo palm—absorb VOCs and use them as nutrients.

Figure 18-25 lists some ways in which you can reduce your exposure to indoor air pollution.

- Test for radon and formaldehyde inside your home and take corrective measures as needed
- Do not buy furniture and other products containing formaldehyde
- Remove your shoes before entering your house to reduce inputs of dust, lead, and pesticides
- Test your house or workplace for asbestos fiber levels, and check for any crumbling asbestos materials if it was built before 1980
- Do not store gasoline, solvents, or other volatile hazardous chemicals inside a home or attached garage
- If you smoke, do it outside or in a closed room vented to the outside
- Make sure that wood-burning stoves, fireplaces, and kerosene- and gas-burning heaters are properly installed, vented, and maintained
- Install carbon monoxide detectors in all sleeping areas

Figure 18-25 Individuals matter: ways to reduce your exposure to indoor air pollution. **Question:** Which three of these actions do you think are the most important? Why?

SOLUTIONS

Air Pollution

Outdoor	Indoor
Improve energy efficiency to reduce fossil fuel use	Reduce poverty
Rely more on lower-polluting natural gas	Distribute cheap and efficient cookstoves or solar cookers to poor families in developing countries
Rely more on renewable energy (especially solar cells, wind, and solar-produced hydrogen)	Reduce or ban indoor smoking
Transfer energy efficiency, renewable energy, and pollution prevention technologies to developing countries	Develop simple and cheap tests for indoor pollutants such as particulates, radon, and formaldehyde

Figure 18-26 Ways to prevent outdoor and indoor air pollution over the next 30–40 years (**Concept 18-6**). **Question:** Which two of these solutions do you think are the most important? Why?

We Need to Put More Emphasis on Pollution Prevention

Since 1970, most of the world's developed countries have enacted laws and regulations that have significantly reduced outdoor air pollution. Most of these laws emphasize controlling outdoor air pollution by using *output approaches*. To environmental and health scientists, the next step is to shift to *preventing* air pollution (**Concept 1-4**, p. 16). With this approach, the question is not *"What can we do about the air pollutants we produce?"* but rather *"How can we avoid producing these pollutants in the first place?"*

CONCEPT LINK

Figure 18-26 shows ways to prevent outdoor and indoor air pollution over the next 30–40 years (**Concept 18-6**). Like the shift to *controlling outdoor air pollution* between 1970 and 2007, this new shift to *preventing outdoor and indoor air pollution* will not take place unless individual citizens and groups put political pressure on elected officials and economic pressure on companies through their purchases.

REVISITING

The Asian Brown Cloud and Sustainability

CORE CASE STUDY

The Asian Brown Cloud (**Core Case Study**) is a striking example of how bad air pollution can get. It results from large and dense populations of people relying mostly on fossil fuels (especially coal) for their energy, and wasting too many resources, rather than recycling and reusing products. It is an example of what can happen when people violate all four **scientific principles of sustainability** (see back cover) on a massive scale.

However, we can use these four principles to help reduce air pollution. We can reduce inputs of conventional air pollutants into the atmosphere by relying more on direct and indirect forms of solar energy than on fossils fuels; reducing the waste of matter and energy resources and recycling and reusing matter resources; mimicking biodiversity by using a diversity of nonpolluting or low-polluting renewable energy resources; and reducing human population growth and wasteful resource consumption.

A key strategy for solving air pollution problems is to focus on preventing or sharply reducing outdoor and indoor air pollution at global, regional, national, local, and individual levels. Each of us has an important role to play in protecting the atmosphere that sustains life and supports our economies.

The atmosphere is the key symbol of global interdependence.
If we can't solve some of our problems in the face of threats to this global commons,
then I can't be very optimistic about the future of the world.

MARGARET MEAD

REVIEW

1. Review the Key Questions and Concepts for this chapter on p. 469. Describe the nature and harmful effects of the massive Asian Brown Cloud (**Core Case Study**).
CORE CASE STUDY

2. Define **density, atmospheric pressure, troposphere, stratosphere,** and **ozone layer.** Describe how the troposphere and stratosphere differ.

3. What is **air pollution?** Summarize the history of air pollution. Distinguish between **primary pollutants** and **secondary pollutants** and give an example of each. List the major outdoor air pollutants and their harmful effects.

4. Describe the effects of lead as a pollutant and how we can reduce our exposure to this chemical. Describe a chemical method and a biological method for detecting air pollutants.

5. Distinguish between **industrial smog** and **photochemical smog** in terms of their chemical composition and formation. List and briefly describe five natural factors that help to reduce outdoor air pollution and six natural factors that help to worsen it. What is a **temperature inversion** and how can it affect air pollution levels?

6. What is **acid deposition** and how does it form? Briefly describe its major environmental impacts on vegetation, lakes, human-build structures, and human health. List three major ways to reduce acid deposition.

7. What are the top four indoor air pollutants in the United States? What is the major indoor air pollutant in many developing countries? Describe indoor air pollution by radon-222 and what can be done about it.

8. Briefly describe the human body's defenses against air pollution, how they can be overwhelmed, and illnesses that can result. About how many people die prematurely from air pollution each year?

9. Describe air pollution laws in the United States. Summarize the accomplishments of such laws and discuss how they can be improved. List the advantages and disadvantages of using an emissions trading program. Summarize the major ways to reduce emissions from power plants and motor vehicles. What are four ways to reduce indoor air pollution?

10. Discuss the relationship between the Asian Brown Cloud (**Core Case Study**) and the ways in which people have violated the four **scientific principles of sustainability**.

Note: Key Terms are in **bold** type.

CRITICAL THINKING

1. The Asian Brown Cloud (**Core Case Study**) exists mainly in southern and eastern Asia, and some people blame Asians for its existence. Others argue that its causes, as well as its effects, are global in nature. What is your position on this? Explain your reasoning.

2. Investigate rocks, trees, and buildings in the area where you live for the presence of various lichens. Use their presence, color, and type to identify high- and low-pollution areas, and explain your reasoning.

3. China relies on coal for 70% of its commercial energy usage, partly because the country has abundant supplies of this resource. Yet China's coal burning has caused innumerable and growing problems for China and neighboring countries, and now, because of the Asian Brown Cloud (**Core Case Study**), for the Pacific Ocean and west coast of North America. Do you think China is justified in developing this resource to the maximum, as other countries including the United States have done with their coal resources? Explain. What are China's alternatives?

4. Photochemical smog is largely the result of motor vehicle emissions. Considering your use, now and in the future, of motor vehicles, what are three ways in which you could reduce your contribution to photochemical smog?

5. Should all tall smokestacks be banned in an effort to promote greater emphasis on preventing air pollution and acid deposition? Explain.

6. Explain how sulfur impurities in coal can increase the acidity of rainwater and deplete soil nutrients.

7. List three important ways in which your life would be different today if grassroots actions by U.S. citizens between the 1970s and 1990s had not led to the Clean Air Acts of 1970, 1977, and 1990, despite strong political opposition by the affected industries. List three important ways in which your life in the future might be different if grassroots actions now do not lead to strengthening of the U.S. Clean Air Act or a similar act in the country where you live. Do you think the U.S. Clean Air Act should be strengthened? Explain.

8. List three ways in which you could apply **Concept 18-6** to making your lifestyle more environmentally sustainable.

9. Congratulations! You are in charge of the world. List your three most important strategies for dealing with the problems of **(a)** outdoor air pollution and **(b)** indoor air pollution.

10. List two questions that you would like to have answered as a result of reading this chapter.

Note: See Supplement 13 (p. S78) for a list of Projects related to this chapter.

Coal often contains sulfur (S) as an impurity that is released as gaseous SO_2 during combustion. SO_2 is one of six primary air pollutants monitored by the U.S. Environmental Protection Agency. The U.S. Clean Air Act limits sulfur emissions from large coal-fired boilers to 0.54 kilograms (1.2 pounds) of sulfur per million Btus of heat generated. (*Note:* 1 metric ton = 1,000 kilograms = 2,200 pounds = 1.1 ton; 1 kilogram = 2.20 pounds)

1. Given that coal used by electric power plants has a heating value of 27.5 million Btus per metric ton (25 million Btus per ton), determine the number of kilograms (and pounds) of coal needed to produce 1 million Btus of heat.

2. Assuming that all of the sulfur in the coal is released to the atmosphere during combustion, what is the maximum percentage of sulfur the coal can contain and still allow the utility to meet the standards of the U.S. Clean Air Act?

3. About 10,000 Btus of heat input are required for an electric utility to produce 1 kilowatt-hour (kwh) of electrical energy. How many metric tons (and how many tons) of coal must be supplied each hour to provide the input heat requirements for a 1,000-megawatt (1 million-kilowatt) power plant?

Log on to the Student Companion Site for this book at **academic.cengage.com/biology/miller**, and choose Chapter 18 for many study aids and ideas for further reading and research. These include flash cards, practice quizzing, Weblinks, information on Green Careers, and InfoTrac® College Edition articles.

Climate Change and Ozone Depletion

Studying a Volcano to Understand Climate Change

In June of 1991, after 600 years of slumber, Mount Pinatubo in the Philippines exploded (Figure 19-1). A huge amount of volcanic material blasted out of the mountain, sending a cloud of air pollutants and ash to a height of 35 kilometers (22 miles). Avalanches of hot gases and ash roared down the sides of the mountain, killing hundreds of people and filling valleys with volcanic deposits. It was the second-largest volcanic eruption of the 20th century. (The largest took place in Alaska in 1912.)

Figure 19-1 An enormous cloud of air pollutants and ash rises above Mount Pinatubo in the Philippines on June 12, 1991. The volcano exploded in a catastrophic eruption, killing hundreds. Sulfur dioxide and other gases emitted into the atmosphere by the eruption circled the globe, polluted the air, reduced sunlight reaching the earth's surface, and cooled the atmosphere for 15 months.

Rick Hoblitt/U.S. Geological Service

Besides killing many people, the eruption of Mount Pinatubo destroyed homes and farmland, and caused hundreds of millions of dollars in damage. At the same time, it enabled scientists to test whether they understood the global climate well enough to estimate how the eruption would affect temperatures on the earth.

By the late 1980s, most of the world's climate scientists had become concerned that human actions, especially fossil fuel use, were enhancing the world's natural greenhouse effect and contributing to a rise in the average temperature of the atmosphere, which in turn would begin changing the earth's climate. Some stated publicly that climate change from such global warming was very likely to occur and could have disastrous ecological and economic effects. Their concerns were based in part on results from computer models of the global climate. But were these models reliable?

Although their complex global climate models mimicked past and present climates well, Mount Pinatubo provided them with an opportunity to perform a more rigorous test of such models. Soon after the volcano erupted, James Hansen, a leading U.S. National Aeronautics and Space Administration (NASA) scientist, estimated that the Pinatubo explosion would probably cool the average temperature of the earth by 0.5 C° (1 F°) over a 19-month period. The earth would then begin to warm, Hansen said, and by 1995 would return to the temperatures observed before the explosion. His projections turned out to be correct.

To make his forecasts, Hansen added the estimated amount of sulfur dioxide released by the volcano's eruption to a global climate model and then used the model to forecast how the earth's temperature would change. His model passed the test with flying colors. Its success helped to convince most scientists and policy makers that climate model projections—including those relating to the impact of human actions—should be taken seriously.

Hansen's model and 18 other climate models indicate that global temperatures are likely to rise several degrees during this century—mostly because of human actions—and to affect global and regional climates and economies and human ways of life. To many scientists and a growing number of business executives, global climate change represents the biggest challenge that humanity faces during this century. The primary question is, "What should we do about it?"

19-1 How might the earth's temperature and climate change in the future?

CONCEPT 19-1 The overwhelming scientific consensus is that the earth's atmosphere is warming rapidly, mostly because of human activities, and that this will lead to significant climate change during this century.

19-2 What are some possible effects of a warmer atmosphere?

CONCEPT 19-2 The projected rapid change in the atmosphere's temperature during this century is very likely to increase drought and flooding, shift areas where food can be grown, raise sea levels, result in intense heat waves, and cause the premature extinction of many species.

19-3 What can we do to slow climate change?

CONCEPT 19-3A To slow the rate of global warming and climate change, we can increase energy efficiency, sharply reduce greenhouse gas emissions, rely more on renewable energy resources, and slow population growth.

CONCEPT 19-3B Governments can subsidize energy efficiency and renewable energy use, tax greenhouse gas emissions, set up cap-and-trade emissions reduction systems, and help to slow population growth.

19-4 How have we depleted ozone in the stratosphere and what can we do about it?

CONCEPT 19-4A Widespread use of certain chemicals has reduced ozone levels in the stratosphere, which allows for more harmful ultraviolet radiation to reach the earth's surface.

CONCEPT 19-4B To reverse ozone depletion, we must stop producing ozone-depleting chemicals and adhere to the international treaties that ban such chemicals.

Note: Supplements 2 (p. S4), 3 (p. S10), 4 (p. S20), 6 (p. S39), 8 (p. S47), 10 (p. S59), and 13 (p. S78) can be used with this chapter.

Civilization has evolved during a period of remarkable climate stability, but this era is drawing to a close. We are entering a new era, a period of rapid and often unpredictable climate change. The new climate norm is change.

LESTER R. BROWN

19-1 How Might the Earth's Temperature and Climate Change in the Future?

▶ **CONCEPT 19-1** The overwhelming scientific consensus is that the earth's atmosphere is warming rapidly, mostly because of human activities, and that this will lead to significant climate change during this century.

Global Warming and Global Cooling Are Not New

Changes in the earth's climate are neither new nor unusual. Over the past 4.5 billion years, the planet's climate has been altered by volcanic emissions (**Core Case Study**), changes in solar input, continents moving slowly atop shifting tectonic plates (Figure 4-6, p. 85, and **Concept 4-3**, p. 84), impacts by large meteors, and other factors.

Over the past 900,000 years, the atmosphere has experienced prolonged periods of *global cooling* and *global warming* (Figure 19-2, top left, p. 498). These alternating cycles of freezing and thawing are known as *glacial and interglacial* (between ice ages) *periods.*

For roughly 10,000 years, we have had the good fortune to live in an interglacial period characterized by a fairly stable climate and a fairly steady average global surface temperature (Figure 19-2, bottom left). These conditions allowed agriculture, and then cities to flourish, as the human population grew.

For the past 1,000 years, the average temperature of the atmosphere has remained fairly stable but began rising during the last century (Figure 19-2, bottom right) when people began clearing more forests and burning more fossil fuels. Figure 19-2, top right, shows that most of the recent increase in temperature has taken place since 1975.

Past temperature changes such as those depicted in Figure 19-2 are estimated by analysis of radioisotopes in

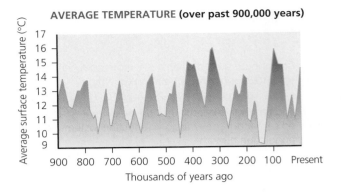

AVERAGE TEMPERATURE (over past 900,000 years)

Thousands of years ago

AVERAGE TEMPERATURE (over past 130 years)

Year

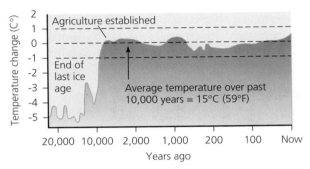

TEMPERATURE CHANGE (over past 22,000 years)

Years ago

TEMPERATURE CHANGE (over past 1,000 years)

Year

Figure 19-2 Science: estimated changes in the average global temperature of the atmosphere near the earth's surface over different periods of time. The graphs in this figure are rough estimates of global average temperatures and temperature changes based on limited evidence, but they do indicate general trends. **Question:** Assuming these estimates are correct, what are two conclusions you can draw from these diagrams? (Data from Goddard Institute for Space Studies, Intergovernmental Panel on Climate Change, National Academy of Sciences, National Aeronautics and Space Agency, National Center for Atmospheric Research, and National Oceanic and Atmospheric Administration)

Figure 19-3 Science: *Ice cores* are extracted by drilling deep holes in ancient glaciers at various sites such as this one in Antarctica (the South Pole). Scientists analyze tiny air bubbles, layers of soot, and other materials trapped in different segments of such ice cores to uncover information about past composition of the lower atmosphere, temperature trends such as those shown in Figure 19-2, greenhouse gas concentrations, solar activity, snowfall, and forest fire frequency.

rocks and fossils; plankton and radioisotopes in ocean sediments; tiny bubbles of ancient air found in ice cores from glaciers (Figure 19-3); temperature measurements taken at different depths from boreholes drilled deep into the earth's surface; pollen from the bottoms of lakes and bogs; tree rings; historical records; insects, pollen, and minerals in different layers of bat dung deposited in caves over thousands of years; and temperature measurements taken regularly since 1861. Such measurements have limitations, but they show general changes in temperature, which in turn can affect the earth's climate. See *The Habitable Planet,* Video 12, at **www.learner.org/resources/series209.html** for a discussion of how scientists are analyzing ice cores from mountain glaciers to understand past climate change.

Our Climate, Lives, and Economies Depend on the Natural Greenhouse Effect

Along with solar energy, a natural process called the *greenhouse effect* warms the earth's lower atmosphere and surface (Figure 3-8, p. 56). It occurs when some of the solar energy absorbed by the earth radiates into the atmosphere as infrared radiation (heat). About 1% of the earth's lower atmosphere is composed of greenhouse

gases, primarily water vapor (H_2O), carbon dioxide (CO_2), methane (CH_4), and nitrous oxide (N_2O). Heat radiated into the atmosphere by the earth causes molecules of these greenhouse gases to vibrate and release infrared radiation with an even longer wavelength into the lower atmosphere. As this radiation interacts with molecules in the air, it increases their kinetic energy and warms the lower atmosphere and the earth's surface, which in turn affects the earth's climate.

Swedish chemist Svante Arrhenius first recognized the natural greenhouse effect in 1896. Since then, numerous laboratory experiments and measurements of temperatures at different altitudes have confirmed this effect—now one of the most widely accepted theories in the atmospheric sciences.

Life on the earth and the world's economies are totally dependent on the natural greenhouse effect—one of the planet's most important forms of natural capital. Without this natural greenhouse effect, the world would be a cold, uninhabitable place.

CENGAGENOW™ See how greenhouse gases trap heat in the lower atmosphere and raise the earth's temperature at CengageNOW™.

Human Activities Emit Large Quantities of Greenhouse Gases

Since the beginning of the Industrial Revolution about 275 years ago, human actions have led to significant increases in the concentration of earth-warming, and thus climate-changing, CO_2, CH_4, and N_2O in the lower atmosphere. These increases came mainly from agriculture, deforestation, and burning of fossil fuels. Measurements of CO_2 and CH_4 in bubbles at various depths in ancient glacial ice (Figure 19-3) indicate that changes in the levels of CO_2 and CH_4 in the lower atmosphere correlate fairly closely with changes in the average global temperature near the earth's surface during the past 400,000 years, and with changes in the global sea level (Figure 19-4, p. 500).

Carbon dioxide levels are increasing exponentially at a rapid rate. For example, the atmospheric concentration of carbon dioxide emitted largely by burning fossil fuels has risen from a level of 280 parts per million at the start of the Industrial Revolution about 275 years ago to 384 parts per million in 2007 (see Figure 20, p. S71, in Supplement 10) and is increasing by about 2 parts per million each year.

According to a 2007 study by scientists Christopher Field and Gregg Marland, if CO_2 emissions continue to increase at the current exponential rate of about 3.3% a year, levels in the atmosphere will rise to 560 ppm by 2050 and soar to 1,390 ppm by 2100, bringing about significant changes in the earth's climate and causing major ecological and economic disruption in the latter half of this century. Scientific studies and models indicate that we should try to prevent CO_2 levels from exceeding 450 ppm—an estimated threshold, or irreversible *tipping point*, that could set into motion large-scale climate changes for hundreds to thousands of years. And according to climate expert James Hansen (**Core Case Study**), ideally we need to reduce the atmospheric concentration of CO_2 to 350 parts per million—which is below current levels—to help stabilize the earth's climate.

In 2007, the largest CO_2 emitting countries were, in order, the United States, China, the European Union (with 27 countries), Indonesia, Russia, Japan, and India. The United States has been responsible for 25% of the world's cumulative CO_2 emissions, compared to China's 5% contribution. A 2007 study by Tao Wang and Jim Watson estimated that about one-fourth of China's rapidly rising CO_2 emissions are a result of its export trade to Europe and the United States. Without this demand for goods from industrialized nations, China's economy would not have developed so rapidly and its CO_2 emissions would not have risen so sharply.

In 2008, U.S. economists Maximillian Auffhammer and Richard Carson found that China's growth in emissions of CO_2 and other air pollutants was much greater than previously estimated. They found that some of China's more affluent provinces were building cleaner and more efficient coal-burning power plants. However, other provinces with fewer financial resources have been building polluting and less efficient coal-burning power plants by replicating inefficient 1950s Soviet technology. The problem is that once they are built, China and the rest of the world are stuck with these polluting plants throughout their 40- to 75-year lifetimes.

It is also important to compare the per capita emissions of CO_2 emitted by various countries, as shown in Figure 21, p. S71, in Supplement 10. Although China's total CO_2 emissions are high and growing rapidly, its per capita emissions are low. The United States emits about five times more CO_2 per person than China.

Ice core analysis also reveals that about 60% of methane emissions during the last 275 years are the result of human activities such as extracting fossil fuels, creating landfills, raising cattle and sheep (which belch methane), and constructing reservoirs behind large dams in tropical countries. Some good news is that methane emissions have leveled off since 1990, but they are expected to rise in the future as the atmosphere warms and melts some of the permafrost in the arctic tundra, which will release more methane into the atmosphere.

Nitrous oxide levels have risen about 20% during the last 275 years, mostly as a result of increased use of nitrogen fertilizers. A given volume of N_2O traps 3–10 times more heat than the same amount of CO_2 traps and accounts for about 9% of greenhouse gas emissions from human activities. Fertilizers and other sources of nitrogen have greatly increased nitrogen inputs into the environment (Figure 3-20, p. 70).

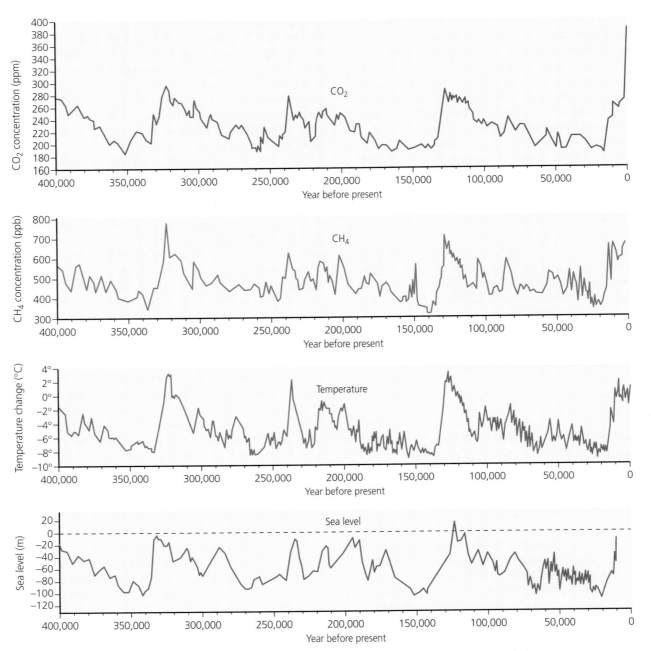

Figure 19-4 **Science:** atmospheric levels of carbon dioxide (CO_2) and methane (CH_4), changes in average global temperature of the atmosphere near the earth's surface, and changes in average sea level over the past 400,000 years. These data were obtained by analysis of ice cores removed at Russia's Vostok Research Station in Antarctica. More recent ice core analyses from Antarctica in 2007 indicate that current levels of CO_2 in the atmosphere are higher than at any time during the past 800,000 years. Carbon dioxide remains in the atmosphere for about 120 years compared to 15 years for methane. However, each molecule of methane can warm the atmosphere 23 times more than a molecule of carbon dioxide. **Question:** What are two conclusions that you can draw from these data? (Data from Intergovernmental Panel on Climate Change, National Center for Atmospheric Research, and F. Vimeux, et al. 2002. *Earth and Planetary Science Letters,* vol. 203: 829–843)

Scientific Consensus: The Atmosphere Is Warming Mostly Because of Human Activities

In 1988, the United Nations and the World Meteorological Organization established the Intergovernmental Panel on Climate Change (IPCC) to document past climate changes and project future changes. The IPCC network includes more than 2,500 climate experts from more than 130 countries. Its 2007 report was based on more than 29,000 sets of data, much of it collected since 2002. In this report, the IPCC listed a number of findings indicating that it is *very likely* (a 90–99% probability) that the lower atmosphere is getting warmer (Science Focus, p. 502) (**Concept 19-1**) and that human activities are responsible for most of the recent temperature

increase and will be responsible for most of the larger increase projected for this century.

In 2007, former U.S. Vice President Al Gore shared the Nobel Peace Prize with the IPCC for alerting the world to the reality and dangers of global warming and its effects on the world's climate. In his acceptance speech he said, " . . . the Earth has a fever. And the fever is rising. . . . We are what is wrong, and we must make it right."

┌─ RESEARCH FRONTIER ─────────────────┐
Computer modeling of climate change. See **academic
.cengage.com/biology/miller**.

Here is some of the evidence that supports the major conclusions of the 2007 IPCC report:

- Between 1906 and 2005, the average global surface temperature has risen by about 0.74 C° (1.3 F°). Most of this increase has taken place since 1980 (Figure 19-2, top right).

- Annual greenhouse gas emissions from human activities rose 70% between 1970 and 2005 and average CO_2 concentrations are higher than they have been in 650,000 years (and 800,000 years, according to a 2007 study).

- Over the past 50 years, arctic temperatures have risen almost twice as fast as average temperatures in the rest of the world have risen.

- In some parts of the world, glaciers (Figure 19-5) and floating sea ice (Figure 19-6) are melting and shrinking at increasing rates, rainfall patterns are changing, and extreme and prolonged drought is increasing.

- During the last century, the world's average sea level rose by 10–20 centimeters (4–8 inches), mostly because of runoff from melting land-based ice and the expansion of ocean water as its temperature increased.

Figure 19-5 Melting of Alaska's Muir Glacier in the popular Glacier Bay National Park and Preserve between 1948 and 2004. Mountain glaciers are now melting almost everywhere in the world. **Question:** How might melting glaciers in Alaska and other parts of the world affect your lifestyle?

Sept. 1979

Sept. 2007

Figure 19-6 *The big melt.* Each summer, some of the floating sea ice in the Arctic Sea melts. But in recent years, rising atmospheric and ocean temperatures have caused more and more ice to melt. Satellite data show a large drop in the average cover of summer arctic sea ice between 1979 and 2007. In 2007 alone, the sea ice shrank by an area equal to that of six Californias or two Alaskas, much more than in any year since 1979 when scientists began taking satellite measurements. Such summer ice may be gone by 2030, and perhaps by as early as 2013, according to an estimate made by NASA climate scientist Jay Zwally in 2007. **Question:** Do you think that the increased melting of floating arctic sea ice is part of a positive or negative feedback loop (p. 45)? Explain? (Data U.S. Goddard Space Flight Center, NASA, National Snow and Ice Data Center)

What Is the Scientific Consensus about Future Temperature Change?

To project the effects of increasing levels of greenhouse gases on average global temperatures, scientists develop complex *mathematical models,* which simulate interactions among the earth's sunlight, clouds, landmasses, oceans, ocean currents, concentrations of greenhouse gases and pollutants, and positive and negative feedback loops (Figures 2-11 and 2-12, p. 45) within the climate system. Then they run these continually improving models on supercomputers and compare the results to known past climate changes, from which they project future changes in the atmosphere's average temperature. Figure 19-A gives a greatly simplified summary of some of the interactions in the global climate system.

Such models provide scenarios or projections of what is *very likely* (90–99% level of confidence) or *likely* (66–89% level of confidence) to happen to the average temperature of the lower atmosphere. How well the results correspond to the real world depends on the validity of the assumptions and

Figure 19-A Science: simplified model of some major processes that interact to determine the average temperature and greenhouse gas content of the lower atmosphere and thus the earth's climate. Red arrows show processes that warm the atmosphere and blue arrows show those that cool the atmosphere. **Question:** Why do you think a decrease in snow and ice cover would increase global warming?

variables built into the models and on the accuracy of the data used.

In 1990, 1995, 2001, and 2007, the IPCC published reports on how global temperatures have changed in the past (Figure 19-2) and made forecasts of how they are likely to change during this century and how such changes can affect the earth's climate. According to the 2007 report, based on analysis of past climate data and use of 19 climate models, it is very likely (a 90–99% probability) that human activities, especially the burning of fossil fuels, have been the main cause of the observed atmospheric warming during the past 50 years.

The 2007 report and recent runs of 19 different climate models suggest that it is *very likely* that the earth's mean surface temperature will increase by 2–4.5 C° (3.6–8.1 F°) between 2005 and 2100 (Figure 19-B), with about 3 C° (5.4 F°) the most likely rise, unless the world halts deforestation and makes drastic cuts in greenhouse gas emissions from fossil fuel burning power plants, factories, and cars. The lower temperature in this range is likely only if global greenhouse gas emissions fall 50–85% by 2050.

There is an overwhelming consensus among the world's climate scientists that global warming is occurring at a rapid rate,

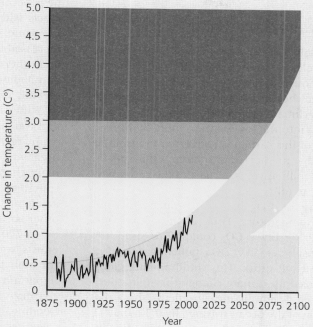

Figure 19-B Science: comparison of measured changes in the average temperature of the atmosphere at the earth's surface between 1860 and 2007 and the projected range of temperature increase during the rest of this century (**Concept 19-1**). (Data from U.S. National Academy of Sciences, National Center for Atmospheric Research, Intergovernmental Panel on Climate Change, and Hadley Center for Climate Prediction and Research)

that human activities are the major factor in this temperature increase since 1950, and that human activities will play an even greater role in the warming projected to take place during this century and thus in the resulting changes to the earth's climate (**Concept 19-1**).

Critical Thinking

If projected temperature increases shown in Figure 19-B take place, what are some major ways in which this will affect your lifestyle and that of any children or grandchildren you might have?

Data reveal that the melting of ice in the Arctic (Figure 19-6) has accelerated since the 2007 IPCC report. According to climate scientist Allan Robock, "the climate is changing even faster than the current models said. In fact, arctic sea ice is melting much faster than any models predicted, and sea level is rising faster than the IPCC previously predicted."

— HOW WOULD YOU VOTE?

Do you think that we will experience significant global warming and climate change during this century? Cast your vote online at **academic.cengage.com/biology/miller**.

Scientists have identified several natural and human-influenced factors that might *amplify* (give positive feedback to) or *dampen* (give negative feedback to) the projected changes in the average temperature of the atmosphere shown in Figure 19-B.

Recall that a positive feedback loop occurs when a small change leads to an even larger change of the same type and the spiral of change keeps growing (Figure 2-11, p. 45). The earth's complex climate system

has both positive (harmful or amplifying) and negative (corrective) feedback loops (Figure 19-A), but positive feedback loops are more common than negative feedback loops. Let us look more closely at a few factors that could affect the atmospheric temperature projected in Figure 19-B.

Is a Hotter Sun the Culprit?

The energy output of the sun affects the earth's temperature, and this output has varied over millions of years. However, in 2007, Swiss, British, and American climate researchers (Claus Froelich, Mike Lockwood, and Ben Santer, respectively) concluded that the rapid rise in global mean temperatures since 1980 (Figure 19-2, top right, and Figure 19-B) could not be the result of increased solar output.

According to satellite and weather balloon measurements, since 1975, the lower atmosphere, or troposphere, has warmed while the stratosphere has cooled. This is not what a warmer sun would do. According to researcher Ben Santer, "If you increase output from

the sun, you increase the amount of energy that arrives at the top of the Earth's atmosphere and get heating throughout the atmosphere." Instead, the atmosphere is now heating from the bottom up, which indicates that inputs at the earth's surface (from human activities) are the main cause.

Can the Oceans Save Us?

The world's oceans have absorbed about half of all the carbon dioxide released to the atmosphere since the beginning of the Industrial Revolution. Today, the oceans help to moderate the earth's average surface temperature and thus its climate by removing about 25–30% of the CO_2 pumped into the lower atmosphere by human activities. Some of this carbon is converted to insoluble carbonate salts that are buried in bottom sediments for several hundred million years, as part of the carbon cycle (Figure 3-18, p. 68).

The solubility of CO_2 in ocean water decreases with increasing temperature. Thus, as the oceans heat up, some of their dissolved CO_2 could be released into the lower atmosphere—like CO_2 bubbling out of a warm carbonated soft drink. This could amplify global warming and speed up climate change through a positive feedback loop. Scientific measurements show that the upper portion of the ocean warmed by 0.32–0.67 C° (0.6–1.2 F°) during the last century—an astounding increase considering the huge volume of water involved.

According to a 2007 report of a study carried out for 10 years by a team of researchers, led by Corinne Le Quere, and a 2008 report by European Union scientists, some of the world's oceans now appear to be absorbing less CO_2 from the atmosphere as they have started to warm. The 2007 study found that the rate of CO_2 absorption in the Southern Ocean around Antarctica—the world's largest sink for removing CO_2—has been dropping in every decade since 1981 and had dropped by half between the mid-1990s and 2005. This could help to accelerate the rise of CO_2 levels in the atmosphere, causing it to warm more rapidly in another positive feedback loop.

The 2008 study confirmed the findings about a decrease in CO_2 uptake by Antarctica's Southern Ocean. It also found a decrease in CO_2 uptake in the North Atlantic Ocean. But it will take at least another 20 years of data to establish the drop in CO_2 uptake as a trend.

In 2008, oceanographer Jeffrey Polovin and his colleagues reported that between 1998 and 2007, the net primary productivity (NPP) of the least productive open ocean areas (Figure 3-16, p. 64) in the Atlantic and Pacific oceans increased by about 15%. This represented an increase in phytoplankton, and the researchers hypothesized that it was the result of warmer ocean water temperatures. It could increase the absorption of CO_2 from the atmosphere. But the authors cautioned that the study covered just nine years, too short a time to determine any relationship between the NPP increase and long-term climate change caused by human activities. And it is possible that the change may be due to a shorter, natural cycle.

In 2005, the U.K. Royal Society reported that higher levels of CO_2 in the ocean have increased the acidity of the ocean surface by 30% since preindustrial times and could reach dangerous levels before 2050. This happens because much of the CO_2 absorbed by the ocean reacts with water to produce carbonic acid (H_2CO_3)—the same weak acid found in carbonated drinks. The increased acidity could decrease the ability of corals and other organisms to make calcium carbonate shells and bodies, and it could dissolve these shells. You can see this effect by dropping a piece of chalk (made of calcium carbonate) in a glass of vinegar (a weak acid) and watching it rapidly dissolve.

Increasing acidity also reduces the ability of the oceans to help regulate atmospheric warming and climate change by removing CO_2 from the lower atmosphere and storing it in bottom sediments. It thus can accelerate global warming and climate change in another positive feedback loop.

In 2007, a report by Ken Calderia and 16 other scientists warned that warmer and more acidic oceans threaten to destroy much of the world's ecologically important coral reef ecosystems before the end of this century. This would expose more people to coastal flooding, increase coastal erosion, reduce yields and revenues for reef-based fisheries and tourism, and greatly decrease aquatic biodiversity.

In 2006, satellite tracking by NASA found that as ocean water temperatures increased between 1999 and 2004, there was a sharp drop in populations of phytoplankton, which serve as the base of ocean food webs and, through photosynthesis, remove CO_2 from the atmosphere. In addition to disrupting food webs, reduced phytoplankton populations would remove less CO_2 from the atmosphere, leading to amplified global warming and more rapid climate change in yet another positive feedback loop. However, a preliminary study in 2008 by a team of British scientists led by M. Debora Iglesias-Rodríguez found that the populations of coccolithophore algae increased as the acidity of ocean water increased. More research is needed to understand the effects of acidity on populations of various phytoplankton species.

Bottom line: Temperature, acidity, the ability to absorb CO_2 from the atmosphere, and other properties of the oceans are changing as a result of human activities, and if these trends continue, they are likely to intensify and accelerate global warming and climate change.

There Is Uncertainty about the Effects of Cloud Cover on Global Warming

A major unknown in global climate models is the effect that changes in the global distribution of clouds might have on the temperature of the atmosphere. Warmer

temperatures increase evaporation of surface water and create more clouds. Depending on their content and reflectivity, these additional clouds could have two effects. An increase in thick and continuous light-colored clouds at low altitudes could *decrease* surface warming by reflecting more sunlight back into space. But an increase in thin and discontinuous cirrus clouds at high altitudes could warm the lower atmosphere.

In addition, infrared satellite images indicate that the wispy condensation trails (contrails) left behind by jet planes might have a greater impact on atmospheric temperatures than scientists once thought. Although air travel is responsible for less than 2% of global greenhouse gas emissions, NASA scientists found that jet contrails expand and turn into large cirrus clouds that tend to release heat into the upper troposphere. If these preliminary results are confirmed, emissions from jet planes could be responsible for as much as half of the warming of the lower atmosphere in the northern hemisphere. Air travel is increasing rapidly and there is no technological fix for this problem unless hydrogen is phased in as a fuel for planes. Much more research is needed to evaluate the effects of clouds on global warming and climate change.

Outdoor Air Pollution Can Temporarily Slow Global Warming

Aerosols (suspended microscopic droplets and solid particles) of various air pollutants are released or formed in the troposphere by volcanic eruptions (**Core Case Study**) and human activities (Figure 18-8, p. 476). They can either warm or cool

the air and hinder or enhance cloud formation depending on factors such as their size and reflectivity.

Most aerosols, such as light-colored sulfate particles produced by fossil fuel combustion, tend to reflect incoming sunlight and cool the lower atmosphere. Sulfate particles also cool the lower atmosphere by serving as condensation nuclei that form cooling clouds. Scientists estimate that sulfate particles played a roll in slowing global warming between 1880 and 1970. However, a 2008 study by atmospheric scientist V. Ramanathan and his colleagues found that the black carbon particulate matter emitted into the air by diesel exhaust, burning forests (Figure 10-16, p. 226) and grasslands, and cooking with solid fuels (such as coal, wood, charcoal, and cow dung) has a warming effect on the atmosphere four times greater than was estimated earlier.

Climate scientists do not expect aerosol and soot pollutants to counteract or enhance projected global warming and the resulting climate change very much in the next 50 years for two reasons. *First*, aerosols and soot fall back to the earth or are washed out of the lower atmosphere within weeks or months, whereas CO_2 remains in the lower atmosphere for about 120 years. *Second*, aerosol and soot inputs into the lower atmosphere are being reduced because of their harmful impacts on plants and human health—especially in developed countries.

According to the IPCC, the fall in sulfate concentrations in most developed countries since 1970 has played a role in the warming of the atmosphere, especially since 1990. This trend will allow further increased global warming as sulphate concentrations continue to drop because of improved air pollution regulations.

19-2 What Are Some Possible Effects of a Warmer Atmosphere?

▶ **CONCEPT 19-2** The projected rapid change in the atmosphere's temperature during this century is very likely to increase drought and flooding, shift areas where food can be grown, raise sea levels, result in intense heat waves, and cause the premature extinction of many species.

Enhanced Global Warming Could Have Severe Consequences: Some Worst–Case Scenarios

So what is the big deal? Why should we worry about the projected rise of only a few degrees in the earth's average surface temperature? We often have that much change between May and July, or between yesterday and today. The key distinction is that we are not considering normal swings in *local weather* (see pp. S47–S48

in Supplement 8), but a projected, very rapid, *global change in climate*—weather measurements averaged over decades, centuries, and millennia (**Concept 19-2**).

Climate scientists are concerned not only about how much the temperature changes, but also about how rapidly it occurs. Most past changes in the temperature of the lower atmosphere took place over thousands to hundreds of thousands of years (Figure 19-2, top left). *The key problem we face is a projected rapid increase in the average temperature of the lower atmosphere during this century* (Figure 19-B).

Climate models indicate that we must deal with a rapidly changing climate that will determine where food can be grown and how much food can be grown; which areas will suffer from increased drought and which will experience increased flooding; and where people and many forms of wildlife will live. And we must deal with this within this century—warp speed, in terms of the earth's overall climate history (Figure 19-2).

A 2003 U.S. National Academy of Sciences report laid out a nightmarish *worst-case scenario* in which human activities, alone or in combination with natural factors, trigger new and abrupt climate and ecological changes. At that point, the global climate system would reach an irreversible *tipping point* after which it would be too late to reverse catastrophic changes for tens of thousands of years. The report describes ecosystems suddenly collapsing, low-lying cities being flooded, forests being consumed in vast wildfires, and grasslands drying out from prolonged drought and turning into dust bowls. It speculates on rivers that now supply drinking and irrigation water drying up as mountain glaciers that feed them melt. And it describes premature extinction of up to half of the world's species, prolonged heat waves and droughts, increased flooding, more destructive storms, and tropical infectious diseases spreading rapidly beyond their current ranges.

Climate change already threatens peace and economic and military security, as changing patterns of rainfall are increasing competition for water and food resources, especially in Africa. The much greater changes in climate projected for this century could cause migrations of tens of millions of people, and lead to extensive economic and social disruption.

These possibilities were supported by a 2003 analysis carried out by Peter Schwartz and Doug Randall for the U.S. Department of Defense. They concluded that global warming and the resulting climate change "must be viewed as a serious threat to global stability and should be elevated to a U.S. national security concern." In 2004, the United Kingdom's chief science adviser, David A. King, wrote, "In my view, climate change is the most severe problem we are facing today—more serious even than the threat of terrorism."

In 2007, the International Institute for Strategic Studies, a military security think-tank, warned that if the emission of greenhouse gases is allowed to continue, "the effects will be catastrophic—on the level of nuclear war." In 2007, International Alert, a London-based conflict resolution group, identified 102 countries with a high risk of becoming failed states during this century. (See Figure 17, p. S19, Supplement 3 for a list of current failed states.) These failures would stem partly from violent conflict and political instability resulting from shrinking supplies of key resources such as food, water, and land, due largely to climate change.

Figure 19-7 summarizes some of the projected effects of global warming and the resulting changes in global climate. These projected changes will affect parts of the world differently, as summarized in the maps in Figures 22 and 23 on pp. S72–S73 in Supplement 10.

According to the IPCC, a 2 C° (3.6 F°) warming appears to be inevitable because we have waited too long to prevent it, ignoring warnings from a number of prestigious committees of climate scientists for more than 25 years. Such a temperature increase is probably manageable. But as temperatures increase beyond this level, the projected harmful effects and costs of the resulting irreversible climate change will escalate rapidly (Figure 19-7, middle and right). Climate scientists warn that a 4 C° (7.2 F°) warming will threaten human civilization as we know it and much of the earth's biodiversity. If we do little or nothing, the projected warming could be 5 C° (9 F°) by the end of this century. Scientists warn that the resulting irreversible changes in global climate could lead to a 33–50% decrease in global food production, a significant drop in the human population, and widespread loss of biodiversity.

It is very important to distinguish between short-term daily and annual changes in the *weather* in an area and long-term changes in an area's *climate,* based mostly on temperature and precipitation. Climate is described by an area's weather factors averaged over several decades to thousands of years. It is clear that the average temperature of the lower atmosphere has increased during the last 35 years (Figure 19-B). But this does not necessarily mean that the area where you live is getting hotter each year. Some years the weather will be warmer, and some years it will be cooler.

Thus, a warmer-than-average year cannot necessarily be attributed to climate change from global warming, and a cooler-than-average year is not necessarily a sign of global cooling or a sign that global warming is not taking place. The latter is a common misconception that people get when they do not distinguish between weather and climate. Nevertheless, average atmospheric temperatures are clearly on the rise.

The good news is that we can avoid or sharply reduce the projected harmful effects of global warming and the resulting climate change if the world takes strong global emergency action, beginning now, to sharply decrease greenhouse gas emissions and to stabilize greenhouse gas concentrations in the 450–550 ppm range, and thus to slow the rate of climate change. Climate scientists urge immediate emergency action, because if we exceed certain tipping points, we will set into motion *irreversible* climate change that will last for hundreds to thousands of years.

In 2007, economist and climate change expert Nicholas Stern prepared a report for the British government that formed the basis for much of the information in Figure 19-7. In 2008, Stern said that he had underestimated the threat from global warming. According to Stern,

Emissions are growing much faster than we'd thought, the absorptive capacity of the planet is less than we thought,

2°C (3.6°F) Warming with 450 ppm CO$_2$ (now unavoidable effects)	3°C (5.4°F) Warming with 550 ppm CO$_2$ (potentially avoidable effects)	4°C (7.2°F) Warming with 650 ppm CO$_2$ (potentially avoidable effects)
■ Forest fires worsen	■ Forest fires get much worse	■ Forest fires and drought increase sharply
■ Prolonged droughts intensify	■ Prolonged droughts get much worse	■ Water shortages affect almost all people
■ Deserts spread	■ Deserts spread more	■ Crop yields fall sharply in all regions and cease in some regions
■ Major heat waves more common	■ Major heat waves and deaths from heat increase	■ Tropical diseases spread even faster and further
■ Fewer winter deaths in higher latitudes	■ Irrigation and hydropower decline	■ Water wars, environmental refugees, terrorism, and economic collapse increase sharply
■ Conflicts over water supplies increase	■ 1.4 billion people suffer water shortages	■ Methane emissions from melting permafrost accelerate
■ Modest increases in crop production in temperate regions	■ Water wars, environmental refugees, and terrorism increase	■ Ecosystems such as coral reefs, tropical forests, alpine and Arctic tundra, polar seas, coastal wetlands, and high-elevation mountaintops begin collapsing
■ Crop yields fall by 5–10% in tropical Africa	■ Malaria and several other tropical diseases spread faster and further	■ Glaciers and ice sheets melt faster
■ Coral reefs affected by bleaching	■ Crop pests multiply and spread	■ Sea levels rise faster and flood many low-lying cities and agricultural areas
■ Many glaciers melt faster and threaten water supplies for up to 100 million people	■ Crop yields fall sharply in many areas, especially Africa	■ At least half of plant and animal species face premature extinction
■ Sea levels rise enough to flood low-lying coastal areas such as Bangladesh	■ Coral reefs severely threatened	
■ More people exposed to malaria	■ Amazon rainforest may begin collapsing	
■ High risk of extinction for Arctic species such as the polar bear	■ Up to half of Arctic tundra melts	
	■ Sea levels continue to rise	
	■ 20–30% of plant and animal species face premature extinction	

Figure 19-7 Some projected effects of global warming and the resulting changes in global climate, based on the extent of warming and the total atmospheric concentrations of greenhouse gases in parts per million. According to the IPCC, a warming of 2 C° (3.6 F°) over 2005 levels is unavoidable, and an increase of at least 3 C° (5.4 F°) is likely sometime during this century (Figure 19-B). (Data from 2007 Intergovernmental Panel on Climate Change Report and Nicolas Stern, *The Economics of Climate Change: The Stern Report,* Cambridge University Press, 2006)

the risks of greenhouse gases are potentially bigger than more cautious estimates, and the speed of climate change seems to be faster.

Let us look more closely at some of the current signs of, and projected effects, of global warming on the earth's climate.

THINKING ABOUT
Weather and Climate

In 2008, some radio talk show hosts said that a colder winter in parts of the United States was evidence that all the talk about global warming and climate change was wrong. How would you respond to such reasoning?

Severe Drought Is Increasing: The Browning of the Earth

Recall that *drought* occurs when evaporation from increased temperatures greatly exceeds precipitation for a prolonged period. According to a 2005 study by Aiguo

Dai and his colleagues, between 1979 and 2002, the area of the earth's land (excluding Antarctica) experiencing severe drought increased from about 15% to 30%—a total area about the size of Asia.

Prolonged drought over several decades is caused by a combination of natural changes and cycles in the earth's climate system and human activities such as widespread deforestation and increased greenhouse gas emissions. According to the 2007 IPCC report, these human influences are very likely to increase throughout this century.

As this browning of the land increases, in the affected areas, there will be less moisture in the soil; stream flows and available surface water will decline; net primary productivity will fall; growth of trees and other plants will slow, which will reduce CO$_2$ removal from the atmosphere and intensify global warming; forest and grassland fires will increase, which will add CO$_2$ to the atmosphere; water tables will fall with more evaporation and irrigation; some lakes and seas will shrink or disappear; more rivers will fail to reach the sea; 1–3 billion people will face a severe shortage of water; biodiversity will decrease; and the area of dry climate biomes, such as savannas, chaparral, and

SCIENCE FOCUS

Melting Ice in Greenland

Greenland, the world's largest island, has a population of about 60,000 people. Glaciers as deep as 3.2 kilometers (2 miles) cover about 80% of this mountainous island, which is roughly one-fourth the size of the continental United States.

Greenland's ice is a result of the last ice age and survives only because of its huge mass. Its glaciers contain about 10% of the world's freshwater—enough water to raise the global sea level by as much as 7 meters (23 feet) if the glaciers all melt. This would flood many coastal cities and much of the earth's farmland. The large moving ice mass in a glacier scrapes along very slowly, but it can pick up speed when meltwater flowing downward through its crevices lubricates its bottom, which sits on bedrock. As the thickness of the glacier decreases, its grip on the land weakens, further accelerating its movement toward the sea.

Recent satellite measurements show that Greenland's net loss of ice more than doubled between 1996 and 2007 and is not being replaced by increased snowfall (Figure 19-C). According to glacial ice expert Konrad Steffen, the record amount of ice melted from Greenland's ice sheet during

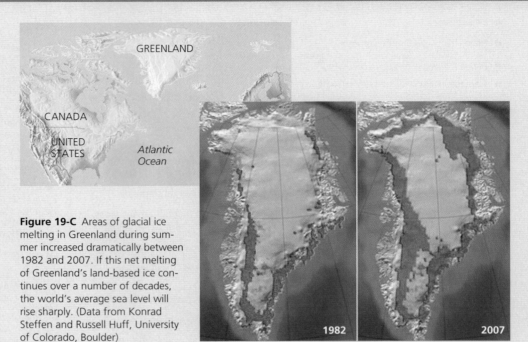

Figure 19-C Areas of glacial ice melting in Greenland during summer increased dramatically between 1982 and 2007. If this net melting of Greenland's land-based ice continues over a number of decades, the world's average sea level will rise sharply. (Data from Konrad Steffen and Russell Huff, University of Colorado, Boulder)

the summer of 2007 was equivalent to two times all of the ice in the Alps of south-central Europe. If this trend continues throughout much of this century, more of Greenland's land-based ice will melt, helping to raise the world's average sea level.

Critical Thinking

List three ways in which the rapid melting of ice in Greenland over the next few decades could affect your lifestyle or that of any children or grandchildren you might have.

deserts, will increase. In other words, some of the effects of prolonged drought over several decades create conditions that, through positive feedback, accelerate global warming and climate change and lead to even more drought.

Ice and Snow Are Melting

Climate models predict that global warming will be the most severe in the world's polar regions, the Arctic and Antarctica. Light colored ice and snow in the polar regions help to cool the earth by reflecting incoming solar energy. The melting of such ice and snow exposes much darker land and sea, which absorb more solar energy. This will likely cause polar regions to warm faster than lower latitudes, which will further accelerate global warming and the resulting climate change, which in turn will melt more sea ice, which will raise atmospheric temperatures more, and faster, in a runaway positive feedback loop.

The world is losing ice in mountain glaciers and in the vast polar ice sheets much faster than scientists thought possible only a few years ago. Over the past

30 years, snow cover in the Arctic has declined by about 10%, and mountain glaciers are melting and retreating (Figure 19-5) in many parts of the world. In 2006, the U.S. National Oceanic and Atmospheric Administration issued a *State of the Arctic* report in which researchers predicted arctic summers without floating sea ice (Figure 19-6) by 2040, and perhaps as early as 2013 according to some recent studies.

So why should we care if arctic sea ice is melting? The answer is that this sea ice plays an important role in the earth's climate by affecting the average amount of precipitation that falls in certain areas to the south over several decades or centuries. For example, a loss of arctic sea ice could reduce long-term average rainfall and snowfall in the already arid American West (Figure 13-5, p. 318), thereby affecting food production by reducing the availability of irrigation water. A loss of arctic sea ice could also increase long-term average precipitation and flooding in western and southern Europe.

Some good news is that because sea ice floats, it does not contribute to a rising sea level when it melts. The Arctic's contribution to a rising sea level will come

from land-based ice that melts and runs into the sea faster than new ice forms. This is especially true of Greenland (Science Focus, at left).

Mountain glaciers are affected by two climatic factors: average snowfall, which adds to their mass during the winter, and average warm temperatures, which spur their melting during the summer. These high-elevation reservoirs play a vital role in the water cycle (Figure 3-17, p. 66) by storing water as ice during cold wet seasons and releasing it slowly as meltwater during warmer dry seasons. Such glaciers are a major source of water for large rivers such as the Ganges, which provides water for 407 million people in India and Bangladesh, and the Yangtze and Yellow Rivers in China. In 2004, Yao Tandong, a leading Chinese glaciologist, predicted that by 2060, two-thirds of China's glaciers will be gone, a development that "will eventually lead to an ecological catastrophe."

During the last 25 years, many of the world's mountain glaciers have been melting and shrinking at accelerating rates. For example, climate models predict that by 2070, Glacier National Park in the United States will have no glaciers for the first time in at least 7,000 years.

In 2007, scientists projected that at their current rate of melting, most glaciers will disappear from Europe's Alps somewhere between 2037 and 2059. Other scientists estimate that 80% of the mountain glaciers in South America will be gone by 2025. As these mountain glaciers disappear, millions of people in countries such as Bolivia, Peru, and Ecuador, who rely on meltwater from the glaciers for irrigation and hydropower, could face severe water, power, and food shortages.

Since 2006, veteran glaciologists and ice watchers have been amazed and alarmed at how fast ice in the polar regions and on the world's mountaintops is melting. According to a 2008 survey of more than 100 leading climate scientists, the two irreversible climate tipping points most likely to be exceeded during this century are the disappearance of floating ice in the Arctic Ocean during the Arctic summers and the accelerating loss of ice from the Greenland ice sheet (Science Focus, at left) and from many of the world's mountaintop glaciers.

Sea Levels Are Rising

According to the 2007 IPCC report, the world's average sea level is *very likely* (with 90–99% certainty) to rise 18–59 centimeters (0.6–1.9 feet) during this century, and to keep rising for centuries. About two-thirds of the increase will result from expansion of water as it warms, and the other third from the melting of land-based ice, especially from the melting of thousands of small mountaintop glaciers.

Newer models developed in 2008 suggest that sea levels could rise as much as 1–2 meters (3.3–6.6 feet) sometime between 2050 and 2100, if glaciers in Greenland (Figure 19-C) reach an irreversible tipping

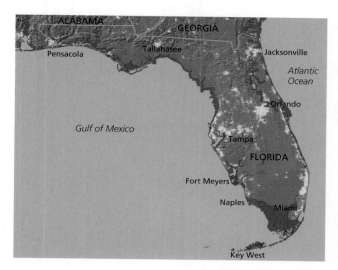

Figure 19-8 Areas of the U.S. state of Florida that will be flooded (red) if the average sea level rises by 1 meter (3.2 feet). (Data from Jonathan Overpeck and Jeremy Weiss based on U.S. Geological Service Data)

point and continue melting at their current or higher rates as the atmosphere warms. The total area of land threatened by rising sea levels increases significantly when we factor in storm surges of 6 meters (20-feet) or higher, which can accompany tropical cyclones and tsunamis. Figure 19-8 shows areas of the U.S. state of Florida that would be flooded with an average sea level rise of 1 meter (3.2 feet).

According to the IPCC, the projected rise in sea levels during this century (excluding the additional effects of storm surges) could cause the following essentially irreversible effects:

- Degradation or destruction of at least one third of the world's coastal estuaries, wetlands, and coral reefs.

- Disruption of many of the world's coastal fisheries.

- Flooding of low-lying barrier islands and erosion and retreat of gently sloping coastlines (especially on the U.S. Eastern and Gulf Coasts). U.S. states that would loose the most land to flooding are Louisiana (Figure 8-19, p. 178), Florida (Figure 19-8), North Carolina, Texas, and South Carolina.

- Flooding of agricultural lowlands and deltas in coastal areas where much of the world's rice is grown.

- Contamination of freshwater coastal aquifers with saltwater and brackish water and decreased supplies of groundwater currently used for irrigation, drinking, and cooling power plants in such areas.

- Submergence of low-lying islands in the Pacific Ocean, the Caribbean Sea, and the Indian Ocean (Figure 19-9, p. 510), which are home to 1 of every 20 of the world's people.

Figure 19-9 For a low-lying island nation like the Maldives in the Indian Ocean, even a small rise in sea level could spell disaster for most of its 295,000 people. About 80% of the 1,192 small islands that make up this country lie less than 1 meter (3.2 feet) above sea level. Rising sea levels and higher storm surges during this century could flood most of these islands and their coral reefs.

- Flooding of coastal areas, including some of the world's largest cities, and displacement of at least 100 million people, especially in China, India, Bangladesh, Vietnam, Indonesia, Japan, Egypt, the United States, Thailand, and the Philippines. A 2007 study by the Organization for Economic Cooperation and Development (OECD) estimated that, by 2070, coastal flooding from a sea level rise of 0.5 meter (1.6 feet) would affect 150 million people and cause property and other damages of $35 trillion (roughly equal to the current global world product). The United States would suffer the highest estimated monetary loss—over

Figure 19-10 Projected decline in arctic tundra (Figure 7-12, bottom, p. 151) in portions of eastern Russia between 2004 and 2100 as a result of global warming. The melting of permafrost in such tundra soils could release the greenhouse gases CH_4 and CO_2 and accelerate global warming, which would melt more tundra. This loss of arctic tundra could reduce grazing lands for caribou and breeding areas for a number of tundra-dwelling bird species. Shrubs and small trees and eventually boreal, or northern coniferous, forests (Figure 7-15, bottom, p. 154) would gradually replace the lost tundra. (Data from Intergovernmental Panel on Climate Change and 2004 Arctic Climate Impact Assessment)

$3.5 trillion. The study projected that the following cities, in order, as most likely to suffer serious damages from such flooding: Calcutta (India), Mumbai (India), Dhaka (Bangladesh), Guangzhou (China), Ho Chi Minh City (Vietnam), Shanghai (China), Bangkok (Thailand), Rangoon (Myanmar), Miami (USA), and Hai Phong (Vietnam). See *The Habitable Planet*, Video 5, at **www.learner.org/resources/ series209.html** for a discussion of the effects of projected rising sea levels on densely populated, low-elevation coastal areas in Vietnam, and on New York City in the United States.

Such changes are unlikely to take place soon. But climate scientists warn that they are very likely to occur during your lifetime, unless the world takes immediate emergency action to slow or prevent such irreversible changes. Many scientists argue that our most urgent priority is *to do all we can to avoid any and all irreversible climate or ecological tipping points*. Once we reach such a point, there is no going back. It is like going over a cliff.

Permafrost Is Likely to Melt: Another Dangerous Scenario

The amount of carbon locked up as methane in permafrost soils is 50–60 times the amount emitted as carbon dioxide from burning fossil fuels each year. If the permafrost in soil and lake bottoms in parts of the rapidly warming Arctic melts, significant amounts of methane (CH_4) and carbon dioxide (CO_2) will be released into the atmosphere, and this will accelerate global warming and the resulting climate change.

According to the 2004 Arctic Climate Impact Assessment, 10–20% of the Arctic's current permafrost might thaw during this century, decreasing the total area of arctic tundra (Figure 19-10). The resulting increase in emissions of CH_4 and CO_2 would cause more warming, which would in turn melt more permafrost and cause still more warming and climate change in yet another positive feedback loop.

Ocean Currents Are Changing but the Threat Is Unknown

Shallow and deep ocean currents are connected and move like a gigantic conveyor belt, transferring CO_2 and warm and cool water between the surface and the depths and between the tropics and the poles (Figure 7-5, p. 143).

Scientists are concerned that melting of land-based glaciers (especially in Greenland) and increased rain in the North Atlantic, both due to global warming, could add enough freshwater to the ocean in the arctic area to slow or disrupt this conveyor belt. Reaching this irreversible tipping point would drastically alter the climates of northern Europe, northeastern North

America, and probably Japan (**Concept 19-2**). Most climate scientists do not see this as a threat in the near future, based on projected temperature increases (Figure 19-B). But we still have much to learn about this pattern of ocean circulation, and ice in land-based glaciers is melting faster than most scientists expected.

Extreme Weather Will Increase in Some Areas

According to the IPCC, global warming will increase the incidence of extreme weather such as heat waves and droughts in some areas, which could kill large numbers of people, reduce crop production, and expand deserts. At the same time, because a warmer atmosphere can hold more moisture, other areas will experience increased flooding (especially flash floods) from heavy and prolonged precipitation. (See Figures 22 and 23, pp. S72 and S73, in Supplement 10 for maps of how different parts of the world may be affected by projected global warming.)

There is controversy over the question of whether global warming will increase the frequency and intensity of tropical storms and hurricanes (Figure 9, p. S51, Supplement 8). In 2008, climatologists Mark Saunders and Adam Lea analyzed data collected since 1950 and found that for every increase of about 0.8 C° (1 F°) in the water temperature of the Atlantic Ocean, the overall number of hurricanes and tropical storms increased by about a third. The number of intense hurricanes, with winds over 177 kilometers per hour (110 miles per hour), increased by 45%.

An example of such increasing hurricane intensity was Hurricane Katrina, which occurred in 2005, a year when Atlantic water temperatures were especially warm. With an 8.5-meter-(28-foot-) high storm surge, Katrina caused massive damage and flooding in New Orleans, Louisiana (USA) (Figure 8-18, p. 177) and the surrounding area and killed more than 1,500 people. The 2005 hurricane season was the most active on record.

Satellite imaging revealed that wind and long-term exposure to water from hurricanes Katrina and Rita in 2005 killed or severely damaged more trees in Mississippi and Louisiana than any recorded forestry disaster in U.S. history. This contributed to global warming, according to a 2007 study by Jeffrey Q. Chambers and his colleagues. They found that the estimated loss of over 320 million big trees sharply reduced the amount of CO_2 removed from the atmosphere. In addition, the researchers estimated that as the dead and damaged trees decayed, they emitted CO_2 equal to the total amount that all forest trees in the United States absorb in a year.

A 2005 statistical analysis by MIT climatologist Kerry Emmanuel and six other peer-reviewed studies published in 2006 also indicated that global warming, on average, could increase the size and strength of Atlantic storms and their storm surges by warming the ocean's surface water. However, some researchers blame the ups and downs in ferocity of tropical Atlantic hurricanes on natural climate cycles that can change ocean circulation patterns. And in 2007, NASA climate researchers hypothesized that warmer water in the Atlantic could increase vertical wind shear—the difference in wind direction or speed at different altitudes—which can help to inhibit hurricane formation. More research is needed to resolve the scientific controversy over the effects of global warming on hurricane frequency and intensity.

Global Warming Is a Major Threat to Biodiversity

According to the 2007 IPCC report, changes in climate resulting from global warming are now affecting physical and biological systems on every continent and are altering ecosystem services (Figure 1-3, p. 8) in some areas. A warmer climate could expand ranges and populations of some plant and animal species that can adapt to warmer climates, including certain weeds; insect pests, such as cockroaches, fire ants, and ticks; and some disease-carrying organisms.

According to the 2007 IPCC study, approximately 30% of the land-based plant and animal species assessed so far could disappear if the average global temperature change exceeds 1.5–2.5 C° (2.7–4.5 F°). This percentage could grow to 70% if the temperature change exceeds 3.5 C° (6.3 F°) (**Concept 19-2**). The hardest hit will be plant and animal species in colder climates, such as the polar bear in the Arctic and penguins in Antarctica; species at higher elevations; plant and animal species with limited ranges, such as some amphibians (Figure 4-9, p. 87); and those with limited tolerance for temperature change.

A 2007 study by the Convention on Migratory Species warned that global warming could also disrupt the biological clocks of birds, whales, and other migratory species. This could put many of them in the wrong places at the wrong times and make them more vulnerable to food shortages, heat waves, droughts, or cold snaps that would accompany climate change in various parts of the world.

The ecosystems most likely to suffer disruption and species loss from climate change are coral reefs (Figure 19-11, p. 512), polar seas, coastal wetlands, high-elevation mountaintops, and alpine and arctic tundra (Figure 19-10). Some types of forests unable to migrate fast enough to keep up with climate shifts will decline, and others, such as oak–pine and oak–hickory forests in the United States, may expand northward. Mostly because of drier conditions, forest fires may increase in some areas such as the southeastern and western United States. This would severely degrade some forest ecosystems, add more CO_2 to the atmosphere, reduce total CO_2 uptake by plants, and accelerate global

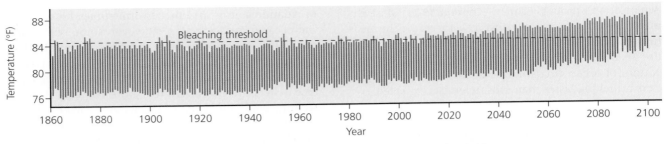

Figure 19-11 Changes in average temperatures of ocean water, relative to the coral bleaching threshold, past and projected, 1860–2100. As temperatures consistently remain above the bleaching threshold, global losses of coral reefs due to coral bleaching (Figure 8-1, right, p. 162) are projected to increase dramatically. Other threats to reefs are increasing ocean acidity and the spread of infectious diseases as ocean temperatures rise due to projected global warming. However, a 2007 study by researchers at the Australian Institute of Marine Science suggested that some reefs store several types of algae, including heat-resistant types that can cope with warmer water. Nevertheless, this would not help to prevent disintegration of some coral reefs caused by the increasing acidity of ocean water. (Data from Intergovernmental Panel on Climate Change)

warming and climate change through still another positive feedback loop.

A warmer climate can also greatly increase populations of insects and fungi that damage trees. In the Canadian province of British Columbia, for example, warmer winters have led to surges in mountain pine beetle populations that have infected huge areas of lodgepole pine forests, which are now dying (Figure 19-12). Pine beetles have also damaged about 60% of the lodgepole pines in the U.S. state of Colorado, which has been experiencing warmer winters. In Yellowstone Park in the United States, global warming has increased beetle infestations of white bark pine trees that grow at high altitudes. This threatens the park's grizzly bears, which feed on white bark pine seeds.

Figure 19-12 With warmer winters, exploding populations of mountain pine beetles have munched their way through large areas of lodgepole pine forest (orange colored trees) in the Canadian province of British Columbia. Foresters are trying to reduce the threat by planting a mix of trees less susceptible to the pest—an example of applying the biodiversity **principle of sustainability**.

Shifts in regional climate would also threaten many existing parks, wildlife reserves, wilderness areas, and wetlands—wiping out more biodiversity. In other words, slowing global warming and the resulting climate change would help to sustain the earth's biodiversity, which in turn supports us and our economies (Science Focus, p. 218).

Climate Change Will Shift Areas Where Crops Can Be Grown

Farming, probably more than any other human activity, depends on a stable climate. Thus, farmers will face dramatic changes due to shifting climates and a faster hydrologic cycle, if global warming continues as projected (Figure 19-B).

Agricultural productivity may increase in some areas and decrease in others. According to the 2007 IPCC report, crop productivity is projected to increase slightly at middle to high latitudes if global temperatures rise by 1–3 C° (1.8–5.4 F°), but productivity would likely decrease at higher temperatures (Figure 22, p. S72, in Supplement 10). Models project that moderately warmer temperatures and increased precipitation at northern latitudes may lead to a northward shift of some agricultural production to parts of midwestern Canada, Russia, and Ukraine. But overall food production could decrease because of unsuitable soils in these northern regions. There could be a 10–15% drop in rainfall in the United States and several other parts of the world. But as long as the temperature does not rise by more than 3 C° (5.4 F°), scientists hope that new genetically modified varieties of key food crops could tolerate this drier climate.

Climate change models predict a decline in agricultural productivity in tropical and subtropical regions, especially in Southeast Asia and Central America, where many of the world's poorest people live. In addition, flooding of river deltas due to rising sea levels could reduce crop and fish production in these produc-

tive agricultural lands and nearby coastal aquaculture ponds. Food production could also decrease in farm regions dependent on rivers fed by snowmelt and glacier melt; arid and semiarid areas where prolonged drought will increase; and humid areas in southeastern Asia that are vulnerable to changes in monsoon patterns, which could bring more devastating storms and heavier flooding.

According to the IPCC, for a time, food will be plentiful because of the longer growing season in northern regions. But by 2050, the IPCC warns that some 200–600 million of the world's poorest and most vulnerable people could face starvation and malnutrition from the effects of climate change.

Climate Change Will Threaten the Health of Many People

According to the IPCC and a 2006 study by U.S National Center for Atmospheric Research, heat waves in some areas will be hotter, more frequent and longer. This will increase the number of deaths and illnesses, especially among older people, those with poor health, and the urban poor who cannot afford air conditioning. During the summer of 2003, a major heat wave killed about 52,000 people in Europe (an estimate based on a detailed analysis in 2006 by the Earth Policy Institute)—almost two-thirds of them in Italy and France.

On the other hand, in a warmer world, fewer people will die from cold weather. However, a 2007 study by Mercedes Medin-Ramon and his colleagues suggests that increased numbers of heat-related deaths will be greater than the projected drop in cold-related deaths in a warmer world.

A warmer, CO_2-rich world will be a great place for rapidly multiplying insects, microbes, toxic molds, and fungi that make us sick, and for plants that produce allergenic pollens. Longer and more intense pollen seasons will mean more itchy eyes, runny noses, and asthma attacks. Insect pests and weeds will likely multiply, spread, and reduce crop yields.

In a warmer world, microbes that cause tropical infectious diseases such as dengue fever, yellow fever, and malaria (Figure 17-7, p. 445) are likely to expand their ranges and their prevalence, if mosquitoes that carry them spread to temperate and higher elevation areas that are getting warmer. And while more frequent prolonged droughts would sharply reduce populations of mosquitoes, populations of their predators, such as dragonflies and damselflies would also decline. In addition, hunger and malnutrition will increase in areas where agricultural production drops.

Higher atmospheric temperatures will also increase some forms of air pollution. The greatest effect will be to speed up the rate of the chemical reactions that produce ozone and other harmful chemicals in photochemical smog in urban areas (Figure 18-10, p. 477).

Increasing illness, hunger, flooding, and drought will likely lead to forced migrations of tens of millions of people. Environmental scientist Norman Myers says that climate change during this century could produce at least 150 million, and perhaps 250 million, environmental refugees. The higher estimate would be equal to about four-fifths of the current U.S. population. (See Myers's Guest Essay on this topic on the website for this chapter.)

A 2005 WHO study estimates that each year, climate change already contributes to the premature deaths of more than 150,000 people—an average of 410 people a day—and that this number could double by 2030. Most of these deaths are the result of increases in malaria, diarrhea, malnutrition, and floods that can be traced to climate change. In addition, the WHO estimates that climate change causes 5 million sicknesses each year. By the end of this century, the annual death toll from climate change could be in the millions.

19-3 What Can We Do to Slow Climate Change?

▶ **CONCEPT 19-3A** To slow the rate of global warming and climate change, we can increase energy efficiency, sharply reduce greenhouse gas emissions, rely more on renewable energy resources, and slow population growth.

▶ **CONCEPT 19-3B** Governments can subsidize energy efficiency and renewable energy use, tax greenhouse gas emissions, set up cap-and-trade emissions reduction systems, and help to slow population growth.

Dealing with Climate Change Is Difficult

It is becoming increasingly clear that addressing climate change could be one of the most urgent scientific, political, economic, and ethical issues that humanity faces.

But the following characteristics of this complex problem make it difficult to tackle:

- *The problem is global.* Dealing with this threat will require unprecedented and prolonged international cooperation.

- *The effects will last a long time.* Carbon dioxide molecules emitted by burning coal for the past 120 years are still in the atmosphere and those we emit today will also be around for about 120 years. Once set into motion, irreversible climate change due to excessive emissions of greenhouse gases will last hundreds to thousands of years.

- *The problem is a long-term political issue.* Voters and elected officials generally respond well to short-term problems but have difficulty acknowledging and coping with long-term threats. Most of the people who will suffer the most serious harm from projected climate change during the latter half of this century, caused by our failure to act, have yet to be born.

- *The harmful and beneficial impacts of climate change are not spread evenly.* There will be winners and losers in the event of moderate climate change. Higher latitude nations such as Canada, Russia, Scandinavia, Greenland, and New Zealand, could have higher crop yields [with temperature increases no greater than 3 C° (5.4 F°)], fewer deaths in winter, lower heating bills, and more tourism. The catch: We will not know who will benefit and who will suffer until it is too late to avoid harmful effects. And at some temperature threshold (Figure 19-7), essentially everyone will be harmed directly or indirectly.

- *Many proposed actions that might reduce the threat of climate change, such as phasing out use of fossil fuels, are controversial because they would disrupt economies and lifestyles.* But waiting too long to slow climate change would also disrupt economies and lifestyles, probably to an even greater extent, according to a 2008 study by the OECD.

RESEARCH FRONTIER

Predicting the effects of climate change caused mostly by global warming in different parts of the world. See **academic .cengage.com/biology/miller**.

What Are Our Options?

There are two basic approaches to dealing with the projected harmful effects of global climate change. One is to drastically reduce greenhouse gas emissions to slow down the rate of warming in time to prevent major climate changes. The other is to recognize that some warming is unavoidable and to devise strategies to reduce its harmful effects. Most analysts believe we need a mix of both approaches.

In 2006, NASA climate scientist, James Hansen (**Core Case Study**), warned that we probably have no longer than a decade to mount a massive global effort to prevent an irreversible change in the earth's climate that will fundamentally alter the planet, cause ecological and economic havoc, and threaten civilization as we know it. He and other prominent climate scientists urge policymakers and business leaders to mount an unprecedented crash program to cut global carbon dioxide emissions by 50–85% by 2050 in an effort to slow down or avoid major climate change and its harmful effects (Figure 19-7).

The question is, will enough individuals exert sufficient bottom up political pressure and consumer pressure (through their purchases) to reach a *political tipping point?* At that point, such pressure could force elected officials and business leaders to implement well-known solutions on an urgent basis to avoid reaching various irreversible *climate change tipping points.*

HOW WOULD YOU VOTE?

Should we take serious action now to help slow climate change resulting from human activities? Cast your vote online at **academic.cengage.com/biology/miller**.

Avoiding Catastrophe: We Can Reduce the Threat of Climate Change

The good news is that we know a number of ways to slow the rate and degree of global warming and the resulting climate change caused by our activities, as summarized in Figure 19-13.

These solutions come down to three major *input* or *prevention* strategies: *improve energy efficiency to reduce fossil fuel use; shift from nonrenewable carbon-based fossil fuels to a mix of carbon-free renewable energy resources; and stop cutting down tropical forests* (**Concept 19-3A**). A fourth strategy—which is an *output strategy*—is to keep burning fossils fuels but to *capture and store as much CO_2 as possible in soil, in vegetation, underground, and in the deep ocean and to hope that it will never leak out.* If a substantial amount of stored CO_2 somehow leaked into the environment, it would rapidly increase global warming and accelerate global climate change.

The effectiveness of these strategies would be enhanced by *reducing population,* which would decrease the number of fossil fuel consumers and CO_2 emitters. It would also help to *reduce poverty,* which would decrease the need of the poor to clear more land for crops and fuelwood. The three input strategies and the population control strategy follow the four **scientific principles of sustainability** (see back cover).

U.S. scientists Robert Socolow and Stephen Pacala at Princeton University have outlined a plan for holding 2057 CO_2 levels to those in 2007 in order to help us avoid harmful effects such as those shown in the middle section of Figure 19-7. They have identified 15 different strategies, which they call *climate stabilization*

SOLUTIONS

Global Warming

Prevention		Cleanup

Prevention

Cut fossil fuel use (especially coal)

Shift from coal to natural gas

Improve energy efficiency

Shift to renewable energy resources

Transfer energy efficiency and renewable energy technologies to developing countries

Reduce deforestation

Use more sustainable agriculture and forestry

Limit urban sprawl

Reduce poverty

Slow population growth

Cleanup

Remove CO_2 from smokestack and vehicle emissions

Store (sequester) CO_2 by planting trees

Sequester CO_2 deep underground (with no leaks allowed)

Sequester CO_2 in soil by using no-till cultivation and taking cropland out of production

Sequester CO_2 in the deep ocean (with no leaks allowed)

Repair leaky natural gas pipelines and facilities

Use animal feeds that reduce CH_4 emissions from cows (belching)

Figure 19-13 Methods for slowing atmospheric warming and the resulting climate change during this century (**Concept 19-3A**). **Question:** Which five of these solutions do you think are the most important? Why?

wedges (Figure 19-14). Phasing in each wedge would reduce CO_2 emissions by roughly the same amount during the coming 50-year period. They estimate that getting CO_2 emissions to 2007 levels by 2057, and holding them there would require implementing any 8 of the 15 wedges during the next 5 decades or phasing in amounts of all 15 wedges sufficient to be the equivalent of implementing 8 wedges.

Socolow and Pacala have turned their proposals into a role-playing wedges game that is being adapted and used in some schools. A 2007 study by the American Solar Energy Association showed how implementing just the energy efficiency and renewable energy wedge strategies alone could lead to a 60–80% reduction in greenhouse gas emissions by 2050.

─── **HOW WOULD YOU VOTE?** ☑ ───

Should we drastically reduce the use of fossil fuels over the next 50 years? Cast your vote online at **academic.cengage .com/biology/miller**.

Environmental expert Lester R. Brown, believes that the CO_2 reductions proposed by Socolow and Pacala are not enough. In his 2008 book, *Plan B 3.0: Mobilizing to Save Civilization* (Norton), he outlines a global emergency plan to cut CO_2 emissions by 80% by 2020. Brown argues that such a course is necessary in order to keep climate change from spiraling out of control and threatening human civilization as we know it.

We have looked at several ways to reduce the release of greenhouse gases to the atmosphere in order to slow climate change, involving energy efficiency, renewable energy, nuclear power, synfuels from coal, stopping or sharply reducing tropical deforestation,

Energy Efficiency and Conservation

Increase the average fuel economy of the 2 billion cars projected to exist by 2057 to 26 kpl (60 mpg).

Cut the average distance traveled by all cars in 2057 to no more than 8,000 kilometers (5,000 miles) per year.

Cut electricity use in homes, offices, and stores by 25%.

Increase efficiency of 1,600 large coal-fired power plants from 40% to 60%.

Renewable Energy

Increase solar power 700-fold to displace coal-fired plants.

Increase wind power 25-fold to displace coal-fired plants.

Increase wind power 50-fold to make hydrogen for fuel-cell cars.

Lower Carbon Emitting Fuels

Increase ethanol biofuel production 50-fold by growing crops on one-sixth of the world's cropland.

Replace 1,400 coal-fired plants with natural-gas-fired plants.

Displace coal-fired plants by tripling electricity production from nuclear power plants.

Forestry and Agriculture

Stop all deforestation.

Use conservation tillage on all cropland.

Carbon Capture and Storage (CCS)

Install Carbon Capture and Storage (CCS) at 800 large coal-fired plants

Install CCS at enough large coal-fired plants to produce hydrogen for 1 million fuel-cell cars.

Install CSS at enough coal-to-syngas plants to produce 30 million barrels of synfuels per day.

Figure 19-14 Fifteen ways to cut CO_2 emissions. Each strategy, called a *climate stabilization wedge*, cuts CO_2 emissions by roughly the same amount. Princeton University scientists, Robert Socolow and Stephen Pacala, who developed this strategy, estimate that to keep global CO_2 emissions from doubling as projected between 2007 and 2057, the world would have to phase in eight of these wedges or the equivalent of that by 2057. Wedges based on other technologies or strategies could also be used. **Questions:** Which of these wedges are input strategies that prevent or reduce CO_2 emissions and which are output strategies that try to deal with CO_2 after it is produced? Which eight of these wedges would you choose to implement over the next 50 years? Explain. (Data from Robert H. Socolow and Stephen W. Pacala, 2006, 2007)

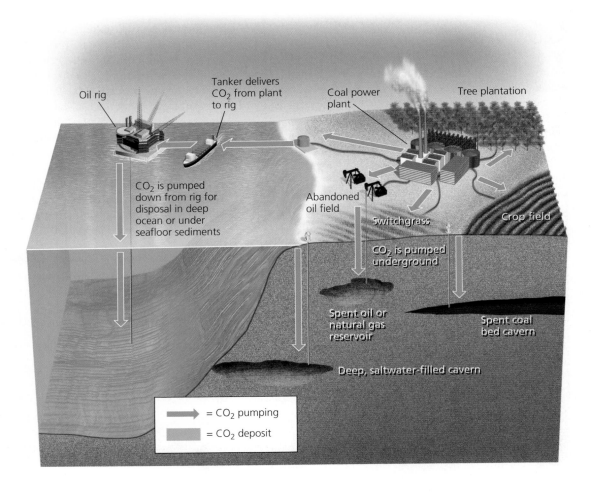

and using no-till soil cultivation. Let us now look more closely at some of the output solutions shown in Figure 19-15—strategies for removing some of the CO_2 from the atmosphere or from smokestacks and storing (sequestering) it in other parts of the environment.

One way to increase the uptake of CO_2 is by implementing a massive global tree-planting program, especially on degraded land in the tropics, on an emergency basis. The 2007 global campaign to plant a billion trees, inspired by Nobel laureate Wangari Maathai (Individuals Matter, p. 230), is a start. China and Africa each have a program to plant a Great Green Wall of Trees to stop the spread of desertification.

However, according to estimates by Lester R. Brown and the Swedish energy firm Vattenfall, to effectively slow global warming and the resulting climate change, the world must plant at least 4 billion trees, assuming that half of them will survive. This will cost about $200 billion or $20 billion a year for a decade. Tree planting would have to continue to a lesser degree, indefinitely, because trees decrease their CO_2 uptake as they mature and release their stored CO_2 back into the atmosphere when they die and decompose or are burned.

A second output approach is to plant large areas of degraded land with fast-growing perennial plants such as switchgrass (Figure 16-26, p. 425), which remove CO_2 from the air and store it in the soil and can be harvested to produce biofuels such as ethanol.

This approach would not involve clearing forests. (See *The Habitable Planet*, Video 10, at **www.learner.org/ resources/series209.html** for a discussion of how scientists are using an outdoor grassland as a laboratory to measure the carbon uptake of plants.)

In 2007, biologist Renton Righelato and climate scientist Dominick Spracklen analyzed a number of strategies for reducing or avoiding CO_2 emissions. They concluded that the amount of carbon that could be sequestered by restoring forests is greater than the amount of carbon in CO_2 emissions that would be avoided by the use of biofuels, such as ethanol and biodiesel (pp. 423–426). The Case Study that follows explores another output approach.

■ **CASE STUDY**

Is Capturing and Storing CO_2 the Answer?

A third output approach is **carbon capture and storage (CCS).** It involves *removing CO_2 from the smokestacks of coal-burning power and industrial plants and then storing it somewhere.* CO_2 gas could be pumped deep underground into coal beds and abandoned oil and gas fields. (See *The Habitable Planet*, Video 10, at **www.learner .org/resources/series209.html** for a discussion of how scientists are evaluating this form of CCS.) Or the

gas could be liquefied and injected into thick sediments under the sea floor (Figure 19-15).

Analysts point to several problems with this approach. One is that power plants using CCS are much more expensive to build and operate than conventional coal-burning plants and thus would sharply raise the price of electricity for consumers. Without strict government regulation of CO_2 emissions, carbon taxes to bring coal prices in line with environmental costs, or generous subsidies and tax breaks, coal-burning utilities and industries have no incentive to build such plants. According to the U.S. Department of Energy, the current costs of CCS systems will have to be reduced by a factor of ten before these systems will be widely used.

A second problem is that CCS is an unproven technology that would remove only part (perhaps 25–35%) of the CO_2 from smokestack emissions. No plants using CCS exist, and building and testing them could take 20–30 years and huge amounts of money with no guaranteed successes. A third problem is that this process requires large inputs of energy, which could increase CO_2 emissions and cancel out some of the gains made from collecting and storing some of the CO_2.

A fourth problem is that CCS promotes the continued use of coal, which should probably be phased out. Coal companies talk about a future based on greatly increased use of *clean coal technologies,* such as coal-to-synfuels. But even with successful CCS and cleaner coal technologies, *coal is by far the world's dirtiest fuel* to dig up and burn (Figures 15-15, right, p. 385, and 15-16, right, p. 386). And if coal's harmful environmental costs were included in its price, burning coal would be a very costly way to produce electricity compared to most other alternatives (Table 16-1, p. 416).

It is not surprising that coal companies are pushing for a shift to CCS coal-fired plants to be funded with the help of generous taxpayer subsidies and tax breaks. Indeed, without CCS, the conventional coal industry probably will not survive in the long term. And because converting coal to synfuels produces twice as much CO_2 per volume of fuel as burning gasoline, CCS also helps to make the coal-to-synfuels industry more feasible. This helps to assure a future for the coal industry, which will ensure continued increasing CO_2 emissions.

A fifth problem is that providing huge government subsidies and taxbreaks for developing and testing CCS technology would divert or reduce the huge subsidies and taxbreaks needed for the rapid development of solar, wind, geothermal, and other forms of renewable energy that reduce rather than attempt to deal with CO_2 emissions.

A sixth very serious potential problem with CCS is that essentially *no leaks are allowed.* In effect, the stored CO_2 would have to remain sealed from the atmosphere forever. Any large-scale leaks due to earthquakes, other geological events, or wars, as well as any number of smaller continuous leaks from storage sites around the world, could dramatically increase global warming and the resulting climate change in a very short time.

According to a 2007 estimate by environmental scientist Peter Montague, if 25% of the carbon in the world's estimated remaining fossil fuels were sequestered, any leakage greater than 0.16% of the total amount stored per year could eventually result in runaway global warming and climate change. And if 75% of the world's estimated remaining carbon in fossil fuels were sequestered, it would take a leakage of only 0.05% of the amount stored per year to lead to the same result. Montague contends that we cannot bury several trillion tons of CO_2 in the ground or under the sea with complete confidence that leaks totaling 0.05% of the total amount stored per year will not occur at any time in the future.

According to the precautionary principle (**Concept 9-4C**, p. 206), we should not rely on a technology that commits us to an essentially irreversible threat. Reliance on nuclear power commits human societies to fail-safe storage of dangerous radioactive wastes for up to 240,000 years. But Montague points out that relying on CCS to store much of the CO_2 we produce commits human societies to fail-safe storage of the CO_2 *forever.*

CONCEPT LINK

To coal companies, CCS is the wave of the future that will help to keep them in business. To scientists like Peter Montague, CCS is an extremely risky output solution to a serious problem that can be dealt with by using a variety of cheaper, quicker, and safer input approaches (Figure 19-13, left). To these scientists, when we face a problem such as CO_2 coming out of a smokestack or exhaust pipe, the most important question to ask is not what do we do with it, but how do we avoid producing the CO_2 in the first place?

THINKING ABOUT
Carbon Capture and Storage

Are you in favor of relying on carbon capture and storage to deal with some of the carbon dioxide that we produce by burning coal? Explain.

Should We Use Geo-Engineering Schemes to Help Slow Climate Change?

Carbon capture and storage (CCS) is one proposed *geo-engineering* scheme for helping us to slow global warming and the resulting climate change. Most scientists oppose using such large-scale solutions, because the long-term effects of such projects on the earth's energy flow, chemical cycling processes, and vital biodiversity are unknown.

However, in recent years, some scientists have become discouraged by the glacially slow response of governments to what they see as the global emergency of climate change with its projected serious harmful effects. Some of these scientists are suggesting that we at least look at the possible implications and costs of using

large-scale geo-engineering schemes as a last resort, if humanity fails to deal with the world's climate change emergency soon enough.

For example, some scientists have suggested using balloons, large jet planes, or giant cannons to inject sulfate particles into the stratosphere where they might reflect some of the incoming sunlight into space and thus cool the troposphere. It is thought that the effect would be similar to the cooling effect that lasted about 15 months after the 1991 volcanic eruption of Mt. Pinatubo (**Core Case Study**, Figure 19-1). Huge amounts of SO_2 would have to be injected into the stratosphere about every 2 years.

CORE CASE STUDY

Other scientists reject this idea as being too risky because of our limited knowledge about possible unknown effects. In addition, such a scheme could accelerate ozone depletion by boosting levels of ozone-destroying chlorine in the stratosphere; it could also increase acid rain in the troposphere (Figure 18-12, p. 479). This short-term technological fix would also allow CO_2 levels in the lower atmosphere to continue rising, which would increase the acidity of the oceans, thereby decreasing their ability to absorb CO_2 and disrupting ocean life (p. 504). This could then accelerate global warming and climate change.

Some scientists would deal with this problem by building a global network of thousands of chemical plants that would remove hydrochloric acid from seawater to reduce ocean acidity. But this also could have unpredictable and possibly harmful ecological effects.

THINKING ABOUT
Tinkering with the Stratosphere

Would you support the proposal to inject large quantities of sulfate particles into the stratosphere every 2 years to help cool the troposphere? Explain.

Scientist James Lovelock has suggested that we anchor huge vertical pipes in the sea as part of a system that would allow wave motion to pump nutrient-rich water up from the deep ocean to fertilize algae on the ocean surface. He contends that the resulting algal blooms would remove CO_2 from the atmosphere and emit dimethyl sulfide (Figure 3-22, p. 72), which would contribute to the formation of low clouds that would reflect sunlight.

Another scheme is to tow 8,000 ice-making barges to the Arctic each year to re-ice the Arctic Sea. And another is to wrap large areas of glaciers with insulating blankets to slow down their melting and to help preserve ski resort businesses.

The major problem with these techno fixes is that if they ever fail while we continue adding CO_2 to the atmosphere, the rebound effects could be calamitous. Geo-engineering schemes all depend on complex machinery running constantly and flawlessly, and essentially forever, mostly to pump something from one place to another in the environment. Once the machines break down, natural processes would overwhelm such a system, and atmospheric temperatures would soar at a rapid rate and accelerate climate change.

Critics of large-scale geo-engineering schemes argue for slowing climate change by using prevention approaches, such as improving energy efficiency, replacing fossil fuels with already available renewable energy resources, and drastically reducing tropical deforestation. They say this makes more sense than gambling on large-scale, costly changes to the global environment that could have unknown and potentially long-lasting harmful effects.

How Much Will It Cost to Slow Climate Change?

Estimates of the financial costs of damages from projected global warming and the resulting climate change vary widely, based on different economic assumptions and on differing rates and degrees of temperature change. In 2007, the IPCC estimated that by the year 2030, the global economic cost of stabilizing greenhouse gas emissions to keep the global temperature from rising more than 2 C° (3.6 F°) (Figure 19-7, left) would cost only about 0.12% of the projected 2030 gross world product.

Other economists dispute these estimates and project that preventing average atmospheric temperatures from increasing more than 2 C° (3.6 F°) would reduce the gross world product by as much as 3% in 2030. However, there is considerable agreement that the short-term costs of slowing climate change will be much lower than the long-term costs of climate change caused largely by human activities. A number of economic studies indicate that implementing the strategies listed in Figures 19-13 and 19-14 would boost global and national economies, provide millions of much-needed jobs, and cost much less than struggling with the harmful effects that these problems would ultimately cost.

Governments Can Help Reduce the Threat of Climate Change

Governments can use four major methods to promote the solutions listed in Figures 19-13 and 19-14 (**Concept 19-3B**).

One is to *strictly regulate carbon dioxide and methane as pollutants*. Second, governments could phase in *carbon taxes* on each unit of CO_2 or CH_4 emitted by fossil fuel use, or they could levy *energy taxes* on each unit of fossil fuel that is burned. Decreasing taxes on income, wages, and profits to offset such taxes could help make such a

strategy more politically acceptable. In other words, *tax pollution, not payrolls and profits.* Some European countries are phasing in such a tax shift.

A related approach is to place a cap on total human-generated CO_2 and CH_4 emissions in a country or region, issue permits to emit these pollutants, and then let polluters trade their permits in the marketplace. This *cap-and-trade approach* has a political advantage over taxes, but it would be difficult to manage because there are so many emitters of greenhouse gases, including industries, power plants, motor vehicles, buildings, and homes. And according to a 2008 study by Goldman Sachs, one of the world's largest investment banks, a cap-and-trade strategy is an important way to cut CO_2 emissions, but by itself would not be enough to achieve the desired drop in such emissions.

Environmental economists argue that, regardless of whether governments use taxes or a cap-and-trade system, the most important goal is to get all emitters to pay the full environmental and social costs of their carbon emissions. The resulting higher costs for fossil fuels would spur innovation in finding ways to reduce carbon emissions, improve energy efficiency, and phase in noncarbon renewable energy alternatives.

A third strategy is to *level the economic playing field* by greatly increasing government subsidies to businesses and individuals to encourage their use of energy-efficiency technologies, carbon-free renewable energy sources, and more sustainable agriculture. This would also include phasing out or sharply reducing subsidies and tax breaks that encourage use of fossil fuels and nuclear power, unsustainable agriculture, and clearing of forests. In other words, we could shift from environmentally-degrading to environmentally-sustaining subsidies and tax breaks.

A fourth strategy would focus on *technology transfer.* Governments of developed countries could help to fund the transfer of the latest green technologies to developing countries so that they could bypass older, energy-wasting and polluting technologies. Helping poorer countries to deal with the harmful effects of climate change would make sense, because these are the countries that will suffer the most from these effects, which have been caused mostly by developed countries. Increasing the current tax on each international currency transaction by a quarter of a penny could finance this technology transfer, which would then generate wealth for developing countries and help to stimulate a more environmentally sustainable global economy.

Governments Can Enter into International Climate Negotiations: The Kyoto Protocol

In December 1997, more than 2,200 delegates from 161 nations met in Kyoto, Japan, to negotiate a treaty to slow climate change. The first phase of the resulting *Kyoto Protocol* went into effect in February 2005 with 174 of the world's 194 countries (but not the United States) ratifying the agreement by mid-2008. It requires 36 participating developed countries to cut their emissions of CO_2, CH_4, and N_2O to an average of at least 5.2% below their 1990 levels by 2012. Developing countries were excluded from having to reduce greenhouse gas emissions in this first phase, because such reductions would curb their economic growth. In 2005, countries began negotiating a second phase that is supposed to go into effect after 2012.

The protocol also allows trading of greenhouse gas emissions among participating countries. For example, a country or business that reduces its CO_2 emissions or plants trees receives a certain number of credits. It can use these credits to avoid having to reduce its emissions in other areas, or it can bank them for future use or sell them to other countries or businesses.

In 2005, the European Union instituted such a cap-and-trade system for carbon emissions. However, in 2007, critics pointed out that the system was not working well because the caps were set too high and thus have been encouraging greenhouse gas emissions. Environmental economists warn that the success of any cap-and-trade emissions system depends on setting caps low enough to increase the value of the tradable allowances and periodically reducing the caps to encourage further innovation in reducing emissions. They also advise that such a system by itself will not achieve the desired reductions in greenhouse gas emissions.

Some analysts praise the Kyoto agreement as a small but important step in attempting to slow projected global warming. They hope that rapidly developing nations such as China, Brazil, India, and Indonesia will agree to reduce their greenhouse gases in the second phase of the protocol. Others see the agreement as a weak and slow response to an urgent global problem.

In 2001, President George W. Bush withdrew the United States from participation in the Kyoto Protocol, arguing that it would harm the U.S. economy. He also objected to the agreement because it did not require emissions reductions by rapidly developing countries such as China, India, Brazil, and Indonesia, which were producing large and increasing emissions of greenhouse gases. Most analysts, and 59% of Americans responding to a 2007 poll, believe that the United States, which has the world's highest overall and per capita CO_2 emissions, should use its influence to improve the treaty rather than to weaken and abandon it.

HOW WOULD YOU VOTE? ☑

Should the United States participate in the Kyoto Protocol and try to strengthen it? Cast your vote online at **academic .cengage.com/biology/miller**.

We Can Move Beyond the Kyoto Protocol

In 2004, environmental law experts Richard B. Stewart and Jonathan B. Wiener proposed that countries work together to develop a new strategy for slowing climate change. They concluded that the Kyoto Protocol would have little effect on future global warming without support and action by the United States and by China, India, Brazil, and other developing countries, which will soon be emitting more than half of the world's greenhouse gases.

Stewart and Wiener urge the development of a new climate treaty among the United States, China, India, Russia, Australia, Japan, South Korea, Brazil, Indonesia, the European Union, and other major greenhouse gas emitters. The treaty would create a cap-and-trade emissions program that includes developing countries omitted from the trading plan under the first phase of the Kyoto Protocol. In addition, it would set achievable 10-year goals for reducing emissions over the next 40 years. It would include evaluation of global and national strategies for adapting to the harmful ecological and economic effects of climate change caused by a warmer atmosphere. Stewart and Wiener call for the United States to lead in developing such a treaty and to take strong steps to drastically cut its own greenhouse gas emissions.

Some Governments Are Leading the Way

Some nations are leading the way. Costa Rica aims to be the first country to become *carbon neutral* by cutting its net carbon emissions to zero by 2030. The country generates 78% of its electricity with renewable hydroelectric power and another 18% from renewable wind and geothermal energy. In 2007, Norway announced that it aims to become carbon neutral by 2050.

Some analysts are urging rapidly developing countries such as China and India to shift toward more sustainable, low-carbon economic development by leapfrogging over the traditional forms of economic development that have led to the global warming problem. If China and India continue their massive coal burning and other high-carbon activities, the world will not be able to avoid the projected harmful effects of climate change set in motion by global warming (Figure 19-7, middle and right). And such effects will be especially severe for China and India. Both countries can benefit economically by becoming leaders in designing, manufacturing, and selling cleaner power systems, appliances, cars, and homes. Such green technologies will make up much of the world's global industry during this century.

Growing weary of waiting for the federal government to get serious about slowing projected climate change, some U.S. states have taken action. By 2007, 21 states and the District of Columbia had required utilities to increase their reliance on renewable energy resources, especially solar and wind power. California has led the way (Case Study, below).

Since 1990, local governments in more than 650 cities around the world (including 453 U.S. cities) have established programs to reduce their greenhouse gas emissions. The first major U.S. city to do this was Portland, Oregon. Between 1993 and 2005, the city cut its greenhouse gas emissions back to 1990 levels, while national levels rose by 16%. The city promotes energy-efficient buildings and the use of electricity from wind and solar sources. It has also built bicycle trails and has greatly expanded its mass transit system. Far from hurting Portland's economy, these strategies have produced an economic boom and have saved the city $2 million a year on its energy bills.

Other U.S. cities striving to be more sustainable and to reduce their greenhouse gas emissions are Seattle, Washington; San Francisco, California; New York City; Chattanooga, Tennessee (Case Study, p. 21); Boulder, Colorado; Chicago, Illinois; Minneapolis, Minnesota; and Salt Lake City, Utah.

■ CASE STUDY
Reducing Greenhouse Gas Emissions in California

The U.S. state of California, where one of every eight Americans lives, is the world's sixth largest economy and twelfth largest producer of greenhouse gases. Many of the most important advances in reducing air pollution and improving energy efficiency started in California, spread across the United States, and eventually were enacted into federal legislation and regulations. Until these federal regulations went into effect, many businesses lost money trying to comply with a patchwork of different standards imposed by various states.

California has set increasingly higher standards for the energy efficiency of appliances and buildings and has created incentives for utilities to encourage consumers to use less power (Case Study, p. 433). This has helped to slow CO_2 emissions and saved California from having to build 24 large power plants.

In 2006, California passed a law to cut its greenhouse gas emissions to 25% below its 1990 levels by 2020 and to 80% below its 1990 levels by 2050. Its approach is to set fuel efficiency and carbon emission standards and let the free market find the best ways to meet the standards. However, in 2008, the EPA refused California's request to lead the way in setting tougher CO_2 emission standards than those enacted by other states, even though it had approved such requests in the past for setting tougher standards for other air pollutants. California and 17 other states are suing the federal government to overturn this decision and lobbying

Congress to amend the Clean Air Act so that it allows California and other states to set tougher CO_2 emission standards.

THINKING ABOUT
What States and Localities Can Do

What are three steps that you think the state or locality where you live should take to help slow climate change from global warming caused mostly by human actions?

Some Companies and Schools Are Reducing Their Carbon Footprints

A growing number of major global companies, including Alcoa, DuPont, IBM, Toyota, General Electric, Johnson & Johnson, Volvo, and British Petroleum (BP), have established targets for reducing their greenhouse gas emissions to 10–65% below 1990 levels by 2010. Between 1990 and 2006, DuPont slashed its energy usage and cut its greenhouse emissions by 72%. In the process, it saved $3 billion while increasing its business by 30%.

Wal-Mart has become the world's largest seller of energy-saving, more climate-friendly, compact fluorescent lightbulbs. It saves $12 million a year using energy-efficient LED bulbs in its refrigeration units. The company, which has the country's second largest corporate truck fleet, saved $22 million in 2006 by installing auxiliary power units that allow drivers to operate their trucks' electrical systems without idling their motors. In 2007, the company ordered its truck manufacturers to double the gas mileage of its entire truck fleet by 2015. In 2007, Wal-Mart also announced a partnership with the Carbon Disclosure Project to measure the energy used to create and transport the products it sells (many of them from China) and the resulting CO_2 emissions. It plans to use the data to find suppliers that are more energy efficient and that produce lower CO_2 emissions.

These and many other major companies see an enormous profit opportunity in developing or using energy-efficient and cleaner-energy technologies, such as fuel-efficient cars, wind turbines, and solar cells. They understand that there is gold in going green. A 2006 study found that companies lagging behind in these efforts are putting their stockholders at risk of losses and lawsuits for failure to take advantage of the rapidly growing international marketplace for green technologies.

Some colleges and universities are also taking action. Students and faculty at Oberlin College in Ohio (USA) have asked their board of trustees to reduce the college's CO_2 emissions to zero by 2020 by buying or producing renewable energy. In the U.S. state of

Pennsylvania, 25 colleges have joined to purchase wind power and other forms of mostly carbon-free, renewable energy. In 2005, the president of Yale University committed the school to cutting its considerable greenhouse gas emissions by 44% by 2020.

In February 2008, the United States had the largest teach-in in its history, with more than 1,500 colleges focusing on educating their students and local citizens about climate change and sustainability. If you are interested in finding out more about how to help your school cut its CO_2 emissions, see the 2008 National Wildlife Federation report on this topic, which can be found at **www.nwf.org/campusecology/Business Case/index.cfm**.

To calculate your *carbon footprint*, the amount of carbon dioxide generated by your lifestyle, you can go to sites like the following:

nature.org/climatecalculator

epa.gov/climatechange/emissions/ind_ calculator.html

carbonfootprint.com

climatecrisis.net/takeaction/ carboncalculator

gocarbonzero.org

Most of these websites and others suggest ways for you to reduce, or *offset*, some of your carbon dioxide emissions by helping to fund or participate in projects such as forest restoration and renewable energy development. Other sites are

nativeenergy.com

my-climate.com

conservationfund.org/gozero

climatecare.org

carbon-clear.com

epa.gov/climatechange/wycd/ actionsteps.html

But buyers need to do their homework, because some carbon offset programs are not real and do not actually produce emissions cuts. For evaluations of carbon-offset providers, go to **cleanair-coolplanet .org** and **tufts.edu/tie/tci/carbonoffsets**. The non-profit Consumers Union also has a website (**Greener Choices.org/calculators.cfm**) to help consumers evaluate conflicting claims of travel agencies over the greenest way to travel.

Critics of such carbon-offset schemes say that most of them primarily offer a way for consumers to ease their guilt. Some critics argue that such programs actually encourage people to continue producing greenhouse gases, instead of making carbon-cutting lifestyle changes, such as driving less, using public transit, and using less electricity.

Figure 19-16 (p. 522) lists some ways in which you can cut your CO_2 emissions. Taking each of these steps

Figure 19-16 Individuals matter: ways to reduce your annual emissions of CO_2.
Question: Which of these steps, if any, do you take now or plan to take?

the larger policy changes needed to deal with climate change and other urgent environmental problems.

We Can Prepare for the Harmful Effects of Climate Change

According to the latest global climate models, the world needs to make a 50–85% cut in emissions of greenhouse gases by 2050 to stabilize concentrations of these gases in the atmosphere and prevent the planet from heating up more than 2C° (3.6F°) (Figures 19-B). This will be necessary in order to prevent rapid climate changes and the resulting projected harmful effects (Figure 19-7).

However, because of the difficulty of making such large reductions, many analysts believe that, while we work to slash emissions, we should also begin to prepare for the projected harmful effects of essentially irreversible climate change. Figure 19-17 shows some ways to implement this strategy.

Some analysts and religious leaders call for the world's richer nations to increase technological and monetary aid to poorer regions at risk from climate change in order to help them deal with the changes. Emphasis could be on developing genetically engineered crops that could thrive in a warmer world and constructing flood defenses in low-lying coastal areas of countries such as India, Indonesia, and Bangladesh, which may experience more severe flooding due to global warming.

makes a small contribution to reducing greenhouse gas emissions. But when millions of people take such steps, the cumulative result is quite large. In addition, political scientists estimate that it would take only 5–10% of the public exerting political pressure to bring about

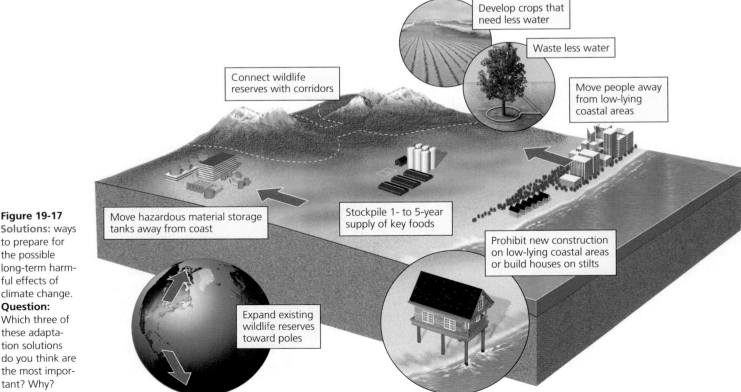

Figure 19-17 Solutions: ways to prepare for the possible long-term harmful effects of climate change. **Question:** Which three of these adaptation solutions do you think are the most important? Why?

Develop crops that need less water

Waste less water

Connect wildlife reserves with corridors

Move people away from low-lying coastal areas

Move hazardous material storage tanks away from coast

Stockpile 1- to 5-year supply of key foods

Prohibit new construction on low-lying coastal areas or build houses on stilts

Expand existing wildlife reserves toward poles

Relief organizations, including the International Red Cross and Oxfam are turning their attention to projects such as expanding mangrove forests as buffers against storm surges, building shelters on high ground, and planting trees on slopes to help prevent landslides. Sea wall design and construction will be a major growth industry. And low-lying countries such as Bangladesh are trying to figure out what to do with millions of environmental refuges who would be displaced by rising sea levels. Some cities plan to establish cooling centers to shelter residents during increasingly intense heat waves.

Some U.S. cities, including New York City and Seattle, Washington, have developed adaptation plans, as have some states, including California, Alaska, Maryland, Washington, and Oregon. Alaska has plans to relocate coastal villages at risk from rising sea levels and storm surges. California is beefing up its forest fire-fighting capabilities and is proposing desalination plants to help relieve projected water shortages, which will worsen as mountain glaciers melt. And some coastal communities require that new houses and other new buildings be built high enough off of the ground to survive projected higher storm surges; others are prohibiting new construction in especially vulnerable areas.

Some people fear that emphasizing these adaptation approaches will distract us from the more urgent need to reduce greenhouse gas emissions. However, to some analysts, projected climate change is already such a serious threat that we have no alternative but to implement both prevention and adaptation strategies, and we have no time to lose.

19-4 How Have We Depleted Ozone in the Stratosphere and What Can We Do about It?

▶ **CONCEPT 19-4A** Widespread use of certain chemicals has reduced ozone levels in the stratosphere, which allows for more harmful ultraviolet radiation to reach the earth's surface.

▶ **CONCEPT 19-4B** To reverse ozone depletion, we must stop producing ozone-depleting chemicals and adhere to the international treaties that ban such chemicals.

Our Use of Certain Chemicals Threatens the Ozone Layer

A layer of ozone in the lower stratosphere keeps about 95% of the sun's harmful ultraviolet (UV-A and UV-B) radiation from reaching the earth's surface (Figure 18-3, p. 470). But measurements taken by meteorologists using weather balloons, aircraft, and satellites show considerable seasonal depletion (thinning) of ozone concentrations in the stratosphere above Antarctica and the Arctic. Similar measurements reveal a lower overall thinning everywhere except over the tropics. Figure 19-18 shows the average global concentrations of ozone in the stratosphere between 1979 and 2005.

In 1984, researchers analyzing satellite data discovered unexpectedly that each year, 40–50% of the ozone in the upper stratosphere over Antarctica (100% in some places) disappears during October and November. This observed loss of ozone has been called an *ozone hole*. A more accurate term is *ozone thinning* because the ozone depletion varies with altitude and location. Figure 19-19 (p. 524) shows a colorized satellite image of ozone thinning over Antarctica in 2007.

When the seasonal thinning ends each year, huge masses of ozone-depleted air above Antarctica flow northward, and these masses linger for a few weeks over parts of Australia, New Zealand, South America, and South Africa. This raises biologically damaging UV-B levels in these areas by 3–10% and in some years as much as 20%.

In 1988, scientists discovered that similar but usually less severe ozone thinning occurs over the Arctic from February to June, resulting in a typical ozone loss of 11–38% (compared to a typical 50% loss above Antarctica). When this mass of air above the Arctic breaks up each spring, large masses of ozone-depleted air flow south to linger over parts of Europe, North America, and Asia.

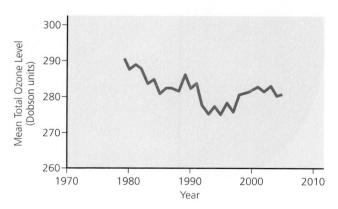

Figure 19-18 Global average total ozone values in the stratosphere at a certain latitude, 1979–2005 (Data from National Oceanic and Atmospheric Administration)

Based on these measurements and mathematical and chemical models, the overwhelming consensus of researchers in this field is that ozone depletion in the stratosphere poses a serious threat to humans, other animals, and some primary producers (mostly plants) that use sunlight to support the earth's food webs (**Concept 19-4A**).

This problem began when Thomas Midgley, Jr., a General Motors chemist, discovered the first chlorofluorocarbon (CFC) in 1930. Chemists soon developed similar compounds to create a family of highly useful CFCs, known by their trade name as Freons. These chemically unreactive, odorless, nonflammable, nontoxic, and noncorrosive compounds seemed to be dream chemicals. Inexpensive to manufacture, they became popular as coolants in air conditioners and refrigerators, propellants in aerosol spray cans, cleaners for electronic parts such as computer chips, fumigants for granaries and ship cargo holds, and gases used to fill tiny bubbles in plastic foam used for insulation and packaging.

It turned out that CFCs were too good to be true. Starting in 1974 with the work of chemists Sherwood Rowland and Mario Molina (Science Focus, at right), scientists demonstrated that CFCs are persistent chemicals that destroy protective ozone in the stratosphere. Measurements and models indicate that 75–85% of the observed ozone losses in the stratosphere since 1976 resulted from people releasing CFCs and other *ozone-depleting chemicals* into the atmosphere, beginning in the 1950s.

CFCs are not the only ozone-depleting chemicals. Others are *halons* and *hydrobromoflurocarbons* (*HBFCs*) (used in fire extinguishers), *methyl bromide* (a widely used fumigant), *hydrogen chloride* (emitted into the

NATURAL CAPITAL DEGRADATION

Effects of Ozone Depletion

Human Health

- Worse sunburns
- More eye cataracts
- More skin cancers
- Immune system suppression

Food and Forests

- Reduced yields for some crops
- Reduced seafood supplies from reduced phytoplankton
- Decreased forest productivity for UV-sensitive tree species

Wildlife

- Increased eye cataracts in some species
- Decreased populations of aquatic species sensitive to UV radiation
- Reduced populations of surface phytoplankton
- Disrupted aquatic food webs from reduced phytoplankton

Air Pollution and Materials

- Increased acid deposition
- Increased photochemical smog
- Degradation of outdoor paints and plastics

Global Warming

- While in troposphere, CFCs act as greenhouse gases

Figure 19-20 Science: expected effects of decreased levels of ozone in the stratosphere (**Concept 19-4A**). **Question:** Which three of these effects do you think are the most threatening? Why?

stratosphere by space shuttles), and cleaning solvents such as *carbon tetrachloride, methyl chloroform, n-propyl bromide,* and *hexachlorobutadiene.*

Models indicate that the Arctic is unlikely to develop the large-scale ozone thinning found over the Antarctic. They also project that ozone depletion over the Antarctic and Arctic will be at its worst between 2010 and 2019.

Why Should We Worry about Ozone Depletion?

Why should we care about ozone loss? Figure 19-20 lists some of the expected effects of decreased levels of ozone in the stratosphere. One effect is that more bio-

Figure 19-19 Natural capital degradation: massive ozone thinning over Antarctica in 2007. The deep blue in this colorized satellite image shows a large area where the concentration of ozone has decreased by 50% or more. (NASA)

Total ozone (Dobson units)

110 220 330 440 550

SCIENCE FOCUS

Sherwood Rowland and Mario Molina—A Scientific Story of Courage and Persistence

In 1974, calculations by chemists Sherwood Rowland and Mario Molina at the University of California–Irvine indicated that chlorofluorocarbons (CFCs) were lowering the average concentration of ozone in the stratosphere. They shocked both the scientific community and the $28-billion-per-year CFC industry by calling for an immediate ban of CFCs in spray cans, for which substitutes were available.

The research of these two scientists led them to four major conclusions. *First,* once injected into the atmosphere, these persistent CFCs remain there. *Second,* over 11–20 years, these compounds rise into the stratosphere through convection, random drift, and the turbulent mixing of air in the lower atmosphere.

Third, once they reach the stratosphere, the CFC molecules break down under the influence of high-energy UV radiation. This releases highly reactive chlorine atoms (Cl), as well as atoms of fluorine (F) and bromine (Br), all of which accelerate the breakdown of ozone (O_3) into O_2 and O in a cyclic chain of chemical reactions, one of which is shown in Figure 19-D. As a consequence, ozone is destroyed faster than it forms in some parts of the stratosphere.

Fourth, each CFC molecule can last in the stratosphere for 65–385 years, depending on its type. During that time, each chlorine atom released during the breakdown of CFC can convert hundreds of O_3 molecules to O_2.

The CFC industry (led by DuPont), a powerful, well-funded adversary with a lot of profits and jobs at stake, attacked Rowland's and Molina's calculations and conclusions. The two researchers held their ground, expanded their research, and explained their results to other scientists, elected officials, and the media. After 14 years of delaying tactics, DuPont officials acknowledged in 1988 that CFCs were depleting the ozone layer and they agreed to stop producing them.

Summary of Reactions
$$CFCl_3 + UV \longrightarrow Cl + CFCl_2$$
$$Cl + O_3 \longrightarrow ClO + O_2$$
$$ClO + O \longrightarrow Cl + O_2$$
Repeated many times

Ultraviolet light hits a chlorofluorocarbon (CFC) molecule, such as $CFCl_3$, breaking off a chlorine atom and leaving $CFCl_2$.

Once free, the chlorine atom is off to attack another ozone molecule and begin the cycle again.

The chlorine atom attacks an ozone (O_3) molecule, pulling an oxygen atom off it and leaving an oxygen molecule (O_2).

A free oxygen atom pulls the oxygen atom off the chlorine monoxide molecule to form O_2.

Ozone

The chlorine atom and the oxygen atom join to form a chlorine monoxide molecule (ClO).

Figure 19-D Simplified summary of how chlorofluorocarbons (CFCs) and other chlorine-containing compounds can destroy ozone in the stratosphere faster than it is formed. Note that chlorine atoms are continuously regenerated as they react with ozone. Thus, they act as *catalysts*—chemicals that speed up chemical reactions without being used up by the reactions. Bromine atoms released from bromine-containing compounds that reach the stratosphere also destroy ozone by a similar mechanism.

In 1995, Rowland and Molina received the Nobel Prize in chemistry for their work on CFCs. In awarding the prize, the Royal Swedish Academy of Sciences said that the two scientists contributed to "our salvation from a global environmental problem that could have catastrophic consequences."

Critical Thinking

How does the problem of ozone depletion in the stratosphere differ from the problem of global warming?

logically damaging UV-A and UV-B radiation will reach the earth's surface (**Concept 19-4A**). This will give people more eye cataracts, worse sunburns, and more skin cancers (Science Focus, p. 526).

The most serious threat from ozone depletion is that the resulting increase in UV radiation could impair or destroy phytoplankton, especially in Antarctic waters (Figure 3-14, p. 63). These tiny marine plants play a key role in removing CO_2 from the atmosphere. They also form the base of ocean food webs. Destroying them would eliminate the vital ecological services they provide. Global warming could make this problem worse by slowing down the upwelling of nutrients that support these populations of phytoplankton. This in turn would accelerate global warming and lead to another runaway positive feedback loop.

Skin Cancer

Research indicates that exposure to the UV-B ionizing radiation in sunlight is the primary cause of *squamous cell* (Figure 19-E, left) and *basal cell* (Figure 19-E, center) *skin cancers*. Together, these two types account for 95% of all skin cancers. Typically, a 15- to 40-year lag separates excessive exposure to UV-B and the development of these cancers.

Caucasian children and adolescents who experience only a single severe sunburn double their chances of getting these two types of cancers. Some 90–95% of these types of skin cancer can be cured if detected early enough, although their removal may leave disfiguring scars. These cancers kill 1–2% of their victims, which, in the United States, amounts to about 2,300 deaths each year.

A third type of skin cancer, *malignant melanoma* (Figure 19-E, right), occurs in pigmented areas such as moles. Within a few months, this type of cancer can spread to other organs. Melanoma kills about one-fourth of its victims (most of them younger than age 40) within 5 years, despite surgery, chemotherapy, and radiation treatments. Each year, it kills about 100,000 people (including 7,700 Americans), mostly Caucasians, and the number of cases and deaths is rising in many countries.

A 2003 study found that women who visited tanning parlors once a month or

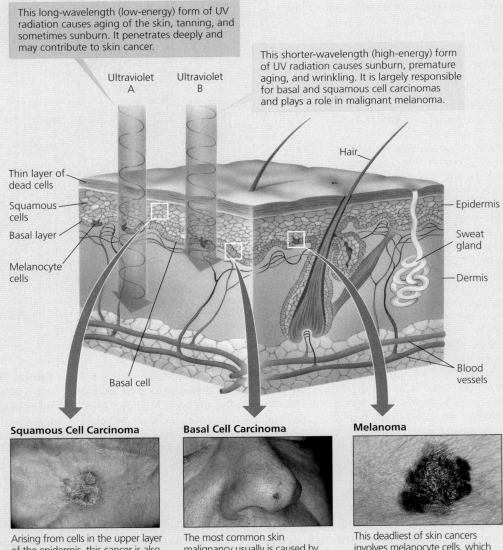

This long-wavelength (low-energy) form of UV radiation causes aging of the skin, tanning, and sometimes sunburn. It penetrates deeply and may contribute to skin cancer.

This shorter-wavelength (high-energy) form of UV radiation causes sunburn, premature aging, and wrinkling. It is largely responsible for basal and squamous cell carcinomas and plays a role in malignant melanoma.

Ultraviolet A

Ultraviolet B

Hair

Thin layer of dead cells

Squamous cells

Basal layer

Melanocyte cells

Epidermis

Sweat gland

Dermis

Basal cell

Blood vessels

Squamous Cell Carcinoma

Basal Cell Carcinoma

Melanoma

Figure 19-E Structure of the human skin and the relationships between ultraviolet (UV-A and UV-B) radiation and the three types of skin cancer. (Data and photos from the Skin Cancer Foundation)

Arising from cells in the upper layer of the epidermis, this cancer is also caused by exposure to sunlight or tanning lamps. It is usually curable if treated early. It grows faster than basal cell carcinoma and can spread to other parts of the body (metastasize).

The most common skin malignancy usually is caused by excessive exposure to sunlight or tanning lamps. It develops slowly, rarely metastasizes and is nearly 100% curable if diagnosed early and treated properly.

This deadliest of skin cancers involves melanocyte cells, which produce pigment. It can develop from a mole or on blemished skin, grows quickly, and can spread to other parts of the body (metastasize).

more increased their chances of developing malignant melanoma by 55%. The risk was highest for young adults. A 2004 study by Dartmouth College scientists found that people using tanning beds were also more likely to develop basal cell and squamous cell skin cancers.

People (especially Caucasians) who experience three or more blistering sunburns before age 20 are five times more likely to develop malignant melanoma than are those who have never had severe sunburns. About 10% of all people who get malignant melanoma have an inherited gene that makes them especially susceptible to the disease.

Figure 19-F lists ways for you to protect yourself from harmful UV radiation.

Critical Thinking

Which three of the measures listed in Figure 19-F do you think are the most important? Why? Which ones do you take?

WHAT CAN YOU DO?

Reducing Exposure to UV Radiation

- Stay out of the sun, especially between 10 A.M. and 3 P.M.
- Do not use tanning parlors or sunlamps.
- When in the sun, wear protective clothing and sunglasses that protect against UV-A and UV-B radiation.
- Be aware that overcast skies do not protect you.
- Do not expose yourself to the sun if you are taking antibiotics or birth control pills.
- When in the sun, use a sunscreen with a protection factor of at least 15.
- Examine your skin and scalp at least once a month for moles or warts that change in size, shape, or color and sores that keep oozing, bleeding, and crusting over. If you observe any of these signs, consult a doctor immediately.

Figure 19-F Individuals matter: ways to reduce your exposure to harmful UV radiation as suggested by dermatologists (skin experts).

We Can Reverse Stratospheric Ozone Depletion

According to researchers in this field, we should immediately stop producing all ozone-depleting chemicals (**Concept 19-4B**). However, models indicate that even with immediate and sustained action, it will take about 60 years for the earth's ozone layer to recover the levels of ozone it had in 1980, and it could take about 100 years for recovery to pre-1950 levels. Scientists have also discovered an important connection between warming of the troposphere and repair of the ozone layer. Warming of the troposphere makes the stratosphere cooler, which slows down the rate of its ozone repair.

In 1987, representatives of 36 nations met in Montreal, Canada, and developed the *Montreal Protocol.* This treaty's goal was to cut emissions of CFCs (but not other ozone-depleting chemicals) by about 35% between 1989 and 2000. After hearing more bad news about seasonal ozone thinning above Antarctica in 1989, representatives of 93 countries met in London in 1990 and then in Copenhagen, Denmark, in 1992. They adopted the *Copenhagen Protocol,* an amendment that accelerated the phase-out of key ozone-depleting chemicals. These landmark international agreements, now signed by 191 countries, are important examples of global cooperation in response to a serious global environmental problem. If nations continue to follow these agreements, ozone levels should return to 1980 levels by 2068 (18 years later than originally projected) and to 1950 levels by 2108 (**Concept 19-4B**).

The ozone protocols set an important precedent by using *prevention* to solve a serious environmental problem (**Concept 1-4**, p. 16). Nations and companies agreed to work together to solve this global problem for three reasons. *First,* there was convincing and dramatic scientific evidence of a serious problem. *Second,* CFCs were produced by a small number of international companies. *Third,* the certainty that CFC sales would decline over a period of years unleashed the economic and creative resources of the private sector to find even more profitable substitute chemicals.

Good news. Substitutes are available for most uses of CFCs, and others are being developed (see Individuals Matter, p. 459). However, the most widely used substitutes, such as HCFC-22, cause some ozone depletion. In 2007, delegates from 191 nations met in Montreal, Canada, and agreed to phase out the production of HCFCs by 2020 in developed countries and by 2030 in developing nations—10 years earlier than they had agreed to under the 1992 Copenhagen Protocol. On its trip to the stratosphere, a typical HCFC molecule helps warms the troposphere by as much as 10,000 times more than does a molecule of CO_2. Thus, phasing out HCFCs to help protect the stratosphere will also help us to reduce global warming in the troposphere.

Volcanic Eruptions, Climate Change, Ozone Depletion and Sustainability

In this chapter, we have seen that humans play a major role in changing the earth's climate by emitting huge quantities of chemicals that warm the troposphere and deplete ozone in the stratosphere. Occasional large volcanic eruptions also emit CO_2 and other pollutants into the lower atmosphere (**Core Case Study**). But about three-fourths of current emissions of CO_2 come from human activities, especially the burning of fossil fuels. Thus, energy policy (Figure 16-33, p. 432) and climate policy (Figures 19-13, 19-14, and 19-17) are closely connected.

The four **scientific principles of sustainability** (see back cover) can be applied to help reduce the harmful effects of global climate change and stratospheric ozone depletion. We can reduce inputs of greenhouse gases and ozone-depleting chemicals into the atmosphere by relying more on direct and indirect forms of solar energy than on fossils fuels; reducing the waste of matter and energy resources and recycling and reusing matter resources; mimicking biodiversity by using a variety of carbon-free renewable energy resources determined partly by local and regional availability; and reducing human population growth and wasteful resource consumption. We can also find substitutes for ozone-depleting chemicals and emphasize pollution prevention.

According to the scientific consensus, we need to implement known solutions to the problems of climate change and ozone depletion, and we must do this globally and on an emergency basis. Each of us has an important role to play in protecting the atmosphere—an irreplaceable resource that sustains all life on earth.

> *The atmosphere is the key symbol of global interdependence.*
> *If we can't solve some of our problems*
> *in the face of threats to this global commons,*
> *then I can't be very optimistic*
> *about the future of the world.*
>
> MARGARET MEAD

REVIEW

1. Review the Key Questions and Concepts for this chapter on p. 497. Describe how a major volcanic eruption (**Core Case Study**, Figure 19-1) allowed scientists to test the validity of climate models.

2. Describe global warming and cooling over the past 900,000 years and during the last century. How do scientists get information about past temperatures and climates? What is the greenhouse effect and why is it so important to life on the earth? What is the scientific consensus about global temperature change during the last half of the 20th century and about projected temperature changes during this century?

3. How can positive feedback loops affect future temperature changes and thus global climate? Give two examples of such loops. Describe the role played by oceans in the regulation of atmospheric temperatures. What are three factors that could decrease its effect in moderating temperature increases?

4. Describe how each of the following might affect global warming and its resulting effects on global climate: **(a)** cloud cover and **(b)** air pollution. Briefly describe the projections of scientists on how global warming is likely to affect: drought; ice cover; flooding; sea levels; permafrost; ocean currents; extreme weather; biodiversity; crop yields; and human health during this century.

5. What are five reasons for the fact that it is difficult to deal with the problem of climate change due to global warming caused mostly by human activities? What are four major strategies for slowing projected climate change? What is **carbon capture and storage (CCS)?** Describe six problems associated with capturing and storing carbon dioxide emissions.

6. List four things that governments could do to help slow projected climate change. What are the pros and cons of the Kyoto Protocol? What have the U.S. state of California and the U.S. city of Portland, Oregon, done to help reduce their greenhouse gas emissions?

7. Give two examples of what some major corporations and some schools have done to reduce their carbon footprints. List five ways in which you can reduce your carbon footprint. List five ways in which we can prepare for the possible long-term harmful effects of climate change.

8. Describe how human activities have depleted ozone in the stratosphere, and list five harmful effects of such depletion. Describe how scientists Sherwood Roland and Mario Molina helped to awaken the world to this threat. Describe the relationships between higher UV levels and three types of skin cancer. What has the world done to help reduce the threat from ozone depletion in the stratosphere?

9. Describe how the four **scientific principles of sustainability** can be applied to deal with the problems of climate change (**Core Case Study**) and ozone depletion.

Note: Key Terms are in **bold** type.

CRITICAL THINKING

1. What might happen to the earth's climate if several large-scale eruptions like the one on Mount Pinatubo in 1991 (**Core Case Study**) took place at the same time?

2. A top U.S. presidential economic adviser once gave a speech in Williamsburg, Virginia (USA), to representatives of governments from a number of countries. He told his audience not to worry about global warming because the average global temperature increases predicted by scientists were much less than the temperature increase he had experienced that day in traveling from Washington, D.C., to nearby Williamsburg. What was the flaw in his reasoning? Write an argument that you could use to counter his claim.

3. How might the earth's climate change if the land area of the planet were larger than the planet's ocean area?

4. List three ways in which you could apply **Concept 19-3A** to making your lifestyle more environmentally sustainable.

5. Explain why you agree or disagree with each of the proposals listed in Figure 19-13 for slowing projected climate change caused by atmospheric warming.

6. What changes might occur in **(a)** the global hydrologic cycle (Figure 3-17, p. 66) and **(b)** the global carbon cycle (Figure 3-18, p. 68) if the atmosphere experiences significant warming? Explain.

7. One way to slow the rate of CO_2 emissions is to reduce the clearing of forests—especially in tropical developing countries where intense deforestation is taking place. Should the United States and other developed countries pay poorer countries to stop cutting their forests? Explain.

8. What are three consumption patterns or other aspects of your lifestyle that directly add greenhouse gases to the atmosphere? Which, if any, of these habits would you be willing to give up to help slow projected climate change?

9. Congratulations! You are in charge of the world. List your three most important strategies for dealing with the problems of **(a)** global climate change due to atmospheric warming caused mostly by human activities and **(b)** depletion of ozone in the stratosphere.

10. List two questions that you would like to have answered as a result of reading this chapter.

Note: See Supplement 13 (p. S78) for a list of Projects related to this chapter.

ECOLOGICAL FOOTPRINT ANALYSIS

Largely because of the intense use of fossil fuels, per capita CO_2 emissions for the United States are nearly five times the world average. According to a recent report from the International Energy Agency, the average American is responsible for adding 19.6 metric tons (21.6 tons) of CO_2 per year to the atmosphere, compared with a world average of 4.23 metric tons (4.65 tons). The table on p. 530 is designed to help you understand the sources of your personal inputs of CO_2 into the atmosphere and how you can reduce your inputs.

Some typical data are provided in the "Typical Quantity per Year" column of the table. However, the calculations will be more accurate if you can substitute for these typical values information based on your own personal lifestyle, which you can enter in the blank "Personal Quantity" column. For example,

you could add up your monthly utility bills for a year and divide the total by the number of persons in your household to determine your utility use, and you could analyze your driving habits to determine how much fuel you use in automobile transportation.

After completing the table, you can compare your emissions against the per capita U.S. average. Your answer should be considerably less—roughly half the per capita value, because this computation only accounts for direct emissions. For instance, CO_2 resulting from driving a car is included, but the CO_2 emitted in manufacturing or disposing of the car is not.

Finally, you can check your result against the greenhouse gas calculator provided on the web by the EPA at **epa.gov/climatechange/emissions/ind_calculator.html**.

	Units per Year	Personal Quantity per Year	Typical Quantity per Year	Multiplier	Emissions per Year (lbs. CO_2)
Residential Utilities					
Electricity	kwh		4,500	1.5	
Heating oil	gallons		37	22	
Natural gas	hundreds of cubic feet (ccf)		400	12	
Propane	gallons		8	13	
Coal	tons		–	4,200	
Transportation					
Automobiles	gallons		600	19	
Air travel	miles		2,000	0.6	
Bus, urban	miles		12	0.07	
Bus, intercity	miles		0	0.2	
Rail or subway	miles		28	0.6	
Taxi or limousine	miles		2	1	
Other motor fuel	gallons		9	22	
Household Waste					
Trash	pounds		780	0.75	
Recycled Items	pounds		337	−2	
				Total (pounds)	
				Total (tons)	
				Total (kilograms)	
				Total (metric tons)	

Source: This CO_2 calculator was developed by Thomas B. Cobb, Bowling Green State University, Bowling Green, Ohio (USA).

1. Calculate your carbon footprint. To calculate your emissions, first complete the blank "Personal Quantity" column using your personal information (for a more accurate outcome) or using data from a typical utility bill and from other personally known information. If your information is not available, use the data listed in the "Typical Quantity" column. Then, for each activity, calculate your annual consumption (using the units specified in the "Units per Year" column), and multiply your annual consumption by the associated number in the "Multiplier" column to obtain an estimate of the pounds of CO_2 resulting from that activity. Finally, add the numbers in the "Emissions" column to find your carbon footprint, and express the final CO_2 result in both pounds and tons (1 ton = 2,000 lbs) and in kilograms and metric tons (1 kilogram = 2.2 pounds and 1 metric ton = 1.1 tons).

2. Compare your emissions with the per capita U.S. average of 19.6 metric tons (21.6 tons) of CO_2 per person per year and with those of your classmates.

3. Consider what actions you might take to reduce your carbon footprint by 20%.

LEARNING ONLINE

Log on to the Student Companion Site for this book at **academic.cengage.com/biology/miller**, and choose Chapter 19 for many study aids and ideas for further reading and research. These include flash cards, practice quizzing, Weblinks, information on Green Careers, and InfoTrac® College Edition articles.

Water Pollution

Lake Washington

The U.S. city of Seattle, Washington, was founded on the western shore of Puget Sound (Figure 20-1). As it grew in the early 20th Century, it expanded eastward toward Lake Washington, which became a popular recreation site for its citizens.

The lake also became attractive to growing suburban municipal governments as a place to dispose of wastewater. By the mid-1950s, Seattle's suburbs surrounded the lake, and ten sewage treatment plants were operating near its shores, dumping huge amounts of treated wastewater into the lake every day. In 1955, researchers working with the late Dr. W. T. Edmondson of the University of Washington discovered the presence in the lake of a species of cyanobacteria, commonly called blue-green algae.

Masses of these algae quickly grew and darkened the lake's waters. Dead algae accumulated on the lakeshores where they rotted and fouled the air. The water became cloudy and populations of desirable fish declined. Edmondson, who spent decades studying the biochemistry of Lake Washington, hypothesized that the chief nutrient feeding the algae was phosphorus coming from the area's sewage treatment plants.

Edmondson and his colleagues wrote a technical paper for use by sewage treatment plant managers but did not get much attention from these officials. So they wrote letters and articles to educate the general public about the nature of Lake Washington's pollution and its likely sources.

The researchers' public education efforts paid off. Within 3 years, citizen pressure on elected officials led to the development of a scheme to divert nutrient-rich effluents from Seattle's sewage treatment plants into the nearby Puget Sound (Figure 20-1), where tides would mix and dilute them with ocean water. This diversion was completed by 1968, and by 1976, the blue-green algae were virtually gone. The clarity of the lake water improved dramatically, fish populations in the lake recovered, and recreationists returned to Lake Washington (Figure 20-2).

However, with the increasing popularity of the Seattle area, the human population has steadily grown. This is posing new challenges to the lake and to the sound, as we discuss throughout this chapter.

Figure 20-2 With the city of Seattle in the background, this kayaker enjoys the waters of Lake Washington. Although the once badly polluted lake has recovered, population pressures now threaten it again.

Lee Foster/Bruce Coleman USA

Lake Washington

Puget Sound

Seattle

NASA/Goddard Space Flight Center Scientific Visualization Studio

Figure 20-1 *The Seattle, Washington (USA) area.* Seattle was founded on the shore of Puget Sound and quickly expanded eastward toward Lake Washington.

20-1 What are the causes and effects of water pollution?

CONCEPT 20-1A Water pollution causes illness and death in humans and other species and disrupts ecosystems.

CONCEPT 20-1B The chief sources of water pollution are agricultural activities, industrial facilities, and mining, but growth in population and resource use makes it increasingly worse.

20-2 What are the major water pollution problems in streams and lakes?

CONCEPT 20-2A While streams are extensively polluted worldwide by human activities, they can cleanse themselves of many pollutants if we do not overload them or reduce their flows.

CONCEPT 20-2B Addition of excessive nutrients to lakes from human activities can disrupt lake ecosystems, and prevention of such pollution is more effective and less costly than cleaning it up.

20-3 What are the major pollution problems affecting groundwater and other drinking water sources?

CONCEPT 20-3A Chemicals used in agriculture, industry, transportation, and homes can spill and leak into groundwater and make it undrinkable.

CONCEPT 20-3B There are simple ways and complex ways to purify drinking water, but protecting it through pollution prevention is the least expensive and most effective strategy.

20-4 What are the major water pollution problems affecting oceans?

CONCEPT 20-4A The great majority of ocean pollution originates on land and includes oil and other toxic chemicals and solid wastes, which threaten aquatic species and other wildlife and disrupt marine ecosystems.

CONCEPT 20-4B The key to protecting the oceans is to reduce the flow of pollutants from land and air and from streams emptying into these waters.

20-5 How can we best deal with water pollution?

CONCEPT 20-5 Reducing water pollution requires preventing it, working with nature to treat sewage, cutting resource use and waste, reducing poverty, and slowing population growth.

Note: Supplements 2 (p. S4), 6 (p. S39), and 13 (p. S78) can be used with this chapter.

Today everybody is downwind or downstream from somebody else.

WILLIAM RUCKELSHAUS

20-1 What Are the Causes and Effects of Water Pollution?

▶ **CONCEPT 20-1A** Water pollution causes illness and death in humans and other species and disrupts ecosystems.

▶ **CONCEPT 20-1B** The chief sources of water pollution are agricultural activities, industrial facilities, and mining, but growth in population and resource use makes it increasingly worse.

Water Pollution Comes from Point and Nonpoint Sources

Water pollution is any chemical, biological, or physical change in water quality that harms living organisms or makes water unsuitable for desired uses.

Water pollution can come from single (point) sources, or from larger and dispersed (nonpoint) sources. **Point sources** discharge pollutants at specific locations through drain pipes (Figure 20-3), ditches, or sewer lines into bodies of surface water. Examples include factories, sewage treatment plants (which remove some, but not all, pollutants), underground mines, and oil tankers.

Because point sources are located at specific places, they are fairly easy to identify, monitor, and regulate. Most developed countries have laws that help to control point-source discharges of harmful chemicals into aquatic systems. In most developing countries, there is little control of such discharges.

Nonpoint sources are broad, and diffuse areas, rather than points, from which pollutants enter bodies of surface water or air. Examples include runoff of chemicals and sediments from cropland (Figure 20-4), livestock feedlots, logged forests, urban streets, parking lots, lawns, and golf courses. We have made little progress in controlling water pollution from nonpoint sources because of the difficulty and expense of iden-

Figure 20-3 Point source of polluted water in Gargas, France.

tifying and controlling discharges from so many diffuse sources.

Agricultural activities are by far the leading cause of water pollution. Sediment eroded from agricultural lands (Figure 20-4) is the largest source. Other major agricultural pollutants include fertilizers and pesticides, bacteria from livestock and food processing wastes, and excess salt from soils of irrigated cropland. *Industrial facilities,* which emit a variety of harmful inorganic and organic chemicals, are a second major source of water pollution. *Mining* is the third biggest source. Surface mining disturbs the land (Figures 14-18, p. 357, and 14-19, p. 358), creating major erosion of sediments and runoff of toxic chemicals (Chapter 14 Core Case Study, p. 344).

A 2007 study by Purdue University (Indiana, USA) researchers found that parking lots are a major source of nonpoint pollution for rivers and lakes because of grease, toxic metals, and sediments that collect on their impervious surfaces. Because parking lots also disrupt the hydrologic cycle by preventing rain from soaking into the ground, they can worsen local flooding and erosion.

Another relatively new form of water pollution is caused by the widespread use of human-made materials such as plastics that make up millions of products, all of which eventually end up in the environment. The polymers that make up the plastics break down very slowly and, in the process, pollute many waterways where they have been discarded improperly. Plastics products can also harm various forms of wildlife (Figures 11-5, p. 254, and 11-10, p. 260).

Climate change from global warming will also contribute to water pollution in some areas. In a warmer world, some regions will get more precipitation and other areas will get less. Intense downpours will flush more harmful chemicals, plant nutrients, and microorganisms into waterways. Prolonged drought will reduce river flows that dilute wastes.

Major Water Pollutants Have Harmful Effects

Table 20-1 (p. 534) lists the major types of water pollutants along with examples of each and their harmful effects and sources (**Concept 20-1A**).

One of the major water pollution problems people face is exposure to infectious disease organisms (pathogens) mostly through contaminated drinking water. Scientists have identified more than 500 types of disease-causing bacteria, viruses, and parasites that can be transferred into water from the wastes of humans and animals. Table 20-2 (p. 534) lists some common diseases that can be transmitted to humans through drinking water contaminated with infectious agents (**Concept 20-1A**). Various methods are used to measure water quality (Science Focus, p. 535).

The World Health Organization (WHO) estimates that 3.2 million people—most of them children younger than age 5—die prematurely every year from infectious diseases that they get by drinking contaminated water or by not having enough clean water for adequate hygiene. This amounts to an average of almost 8,700 premature deaths a day. The WHO also estimates that about 1.2 billion people—one of every six in the world—have no access to clean drinking water. Each year, diarrhea alone kills about 1.9 million people—about 90% of them children under age 5—in developing countries. This means that diarrhea, caused mostly by exposure to polluted water, on average, kills a young child every 18 seconds.

Figure 20-4 Nonpoint sediment from unprotected farmland flows into streams and sometimes changes their courses or dams them up. As measured by weight, it is the largest source of water pollution. **Question:** What do you think the owner of this farm could have done to prevent such sediment pollution?

Table 20-1

Major Water Pollutants and Their Sources

Type and Effects	Examples	Major sources
Infectious agents (pathogens) *Cause diseases*	Bacteria, viruses, protozoa, parasites	Human and animal wastes
Oxygen-demanding wastes *Deplete dissolved oxygen needed by aquatic species*	Biodegradable animal wastes and plant debris	Sewage, animal feedlots, food processing facilities, pulp mills
Plant nutrients *Cause excessive growth of algae and other species*	Nitrates (NO_3^-) and phosphates (PO_4^{3-})	Sewage, animal wastes, inorganic fertilizers
Organic chemicals *Add toxins to aquatic systems*	Oil, gasoline, plastics, pesticides, cleaning solvents	Industry, farms, households
Inorganic chemicals *Add toxins to aquatic systems*	Acids, bases, salts, metal compounds	Industry, households, surface runoff
Sediments *Disrupt photosynthesis, food webs, other processes*	Soil, silt	Land erosion
Heavy metals *Cause cancer, disrupt immune and endocrine systems*	Lead, mercury, arsenic	Unlined landfills, household chemicals, mining refuse, industrial discharges
Thermal *Make some species vulnerable to disease*	Heat	Electric power and industrial plants

Table 20-2

Common Diseases Transmitted to Humans through Contaminated Drinking Water

Type of Organism	Disease	Effects
Bacteria	Typhoid fever	Diarrhea, severe vomiting, enlarged spleen, inflamed intestine; often fatal if untreated
	Cholera	Diarrhea, severe vomiting, dehydration; often fatal if untreated
	Bacterial dysentery	Diarrhea, bleeding; rarely fatal except in infants without proper treatment
	Enteritis	Severe stomach pain, nausea, vomiting; rarely fatal
Viruses	Infectious hepatitis (Type B)	Fever, severe headache, loss of appetite, abdominal pain, jaundice, enlarged liver; rarely fatal but may cause permanent liver damage
	Poliomyelitis	Fever, diarrhea, backache, sore throat, aches in limbs; can infect spinal chord and cause paralysis and muscle weakness
Parasitic protozoa	Amoebic dysentery	Severe diarrhea, headache, abdominal pain, chills, fever; if not treated can cause liver abscess, bowel perforation, and death
	Giardiasis	Diarrhea, abdominal cramps, flatulence, belching, fatigue
	Cryptosporidum	Severe diarrhea, cramps for up to 3 weeks, and possible death for people with weakened immune systems
Parasitic worms	Schistosomiasis	Abdominal pain, skin rash, anemia, chronic fatigue, and chronic general ill health
	Ancylostomiasis	Severe anemia and possible symptoms of bronchial infection

SCIENCE FOCUS

Testing Water for Pollutants

Scientists use a variety of methods to measure water quality. For example, Dr. Edmondson and his students tested samples of water from Lake Washington (**Core Case Study**) for the presence of various infectious agents such as certain strains of coliform bacteria *Escherichia coli,* or *E. coli,* which live in the colons and intestines of humans and other animals and thus are present in their fecal wastes. Although most strains of coliform bacteria do not cause disease, their presence indicates that water has been exposed to human or animal wastes that are likely to contain disease-causing agents.

To be considered safe for drinking, a 100-milliliter (about 1/2 cup) sample of water should contain no colonies of coliform bacteria. To be considered safe for swimming, such a water sample should contain no more than 200 colonies of coliform bacteria. By contrast, a similar sample of raw sewage may contain several million coliform bacterial colonies.

Another indicator of water quality is its *level of dissolved oxygen* (DO). Excessive inputs of oxygen-demanding wastes can deplete DO levels in water. Figure 20-A shows the relationship between dissolved oxygen content and water quality.

Scientists can use *chemical analysis* to determine the presence and concentrations of

specific organic chemicals in polluted water. They can also monitor water pollution by using living organisms as *indicator species.* For example, they remove aquatic plants such as cattails from areas contaminated with fuels, solvents, and other organic chemicals, and analyze them to determine the exact pollutants contained within them. Scientists also determine water quality by analyzing bottom-dwelling species such as mussels, which feed by filtering water through their bodies.

Genetic engineers are working to develop bacteria and yeasts (single-celled fungi) that glow in the presence of specific pollutants such as toxic heavy metals in the ocean, toxins in the air, and carcinogens in food.

Scientists measure the amount of sediment in polluted water by evaporating the water in a sample and weighing the resulting sediment. They also use instruments such as colorimeters and turbidimeters to measure the color and *turbidity,* or cloudiness, of water samples containing sediment.

Critical Thinking

Runoff of fertilizer into a lake such as Lake Washington (**Core Case Study**) from farm fields, lawns, and sewage treatment plants can overload the water with nitrogen and phosphorus plant nutrients that can cause algae population explosions. How could this process lower the dissolved oxygen level of the water and lead to fish kills?

Figure 20-A Water quality as measured by dissolved oxygen (DO) content in parts per million (ppm) at 20 °C (68 °F). Only a few fish species can survive in water with less than 4 ppm of dissolved oxygen at this temperature. Some warmer water species have evolved ways to tolerate low DO levels better than cold water species can. **Question:** Would you expect the dissolved oxygen content of polluted water to increase or decrease if the water is heated? Explain.

20-2 What Are the Major Water Pollution Problems in Streams and Lakes?

> **CONCEPT 20-2A** While streams are extensively polluted worldwide by human activities, they can cleanse themselves of many pollutants if we do not overload them or reduce their flows.

> **CONCEPT 20-2B** Addition of excessive nutrients to lakes from human activities can disrupt lake ecosystems, and prevention of such pollution is more effective and less costly than cleaning it up.

Streams Can Cleanse Themselves If We Do Not Overload Them

Flowing rivers and streams can recover rapidly from moderate levels of degradable, oxygen-demanding wastes through a combination of dilution and biodegradation of such wastes by bacteria. But this natural recovery process does not work when streams become overloaded with such pollutants or when drought, damming, or water diversions reduce their flows (**Concept 20-2A**). Also, while this process can remove biodegradable wastes, it does not eliminate slowly degradable and nondegradable pollutants.

In a flowing stream, the breakdown of biodegradable wastes by bacteria depletes dissolved oxygen and creates an *oxygen sag curve* (Figure 20-5, p. 536). This

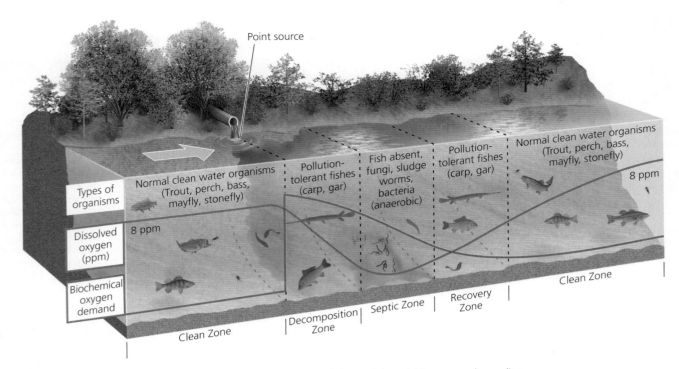

Point source

Types of organisms | Normal clean water organisms (Trout, perch, bass, mayfly, stonefly) | Pollution-tolerant fishes (carp, gar) | Fish absent, fungi, sludge worms, bacteria (anaerobic) | Pollution-tolerant fishes (carp, gar) | Normal clean water organisms (Trout, perch, bass, mayfly, stonefly)

Dissolved oxygen (ppm) 8 ppm 8 ppm

Biochemical oxygen demand

Clean Zone | Decomposition Zone | Septic Zone | Recovery Zone | Clean Zone

CENGAGENOW **Active Figure 20-5 Natural capital:** dilution and decay of degradable, oxygen-demanding wastes (or heated water) in a stream, showing the oxygen sag curve (blue) and the curve of oxygen demand (red). Depending on flow rates and the amount of biodegradable pollutants, streams recover from oxygen-demanding wastes and from injection of heated water if they are given enough time and are not overloaded (**Concept 20-2A**). *See an animation based on this figure at* CengageNOW™. **Question:** What would be the effect of putting another biodegradable waste discharge pipe to the right of the one in this picture?

reduces or eliminates populations of organisms with high oxygen requirements until the stream is cleansed of oxygen-demanding wastes, at which place or time, such populations can recover. Similar oxygen sag curves can be plotted when heated water from industrial and power plants is discharged into streams, because heating water decreases its levels of dissolved oxygen.

CENGAGENOW Learn more about how pollution affects the water in a stream and the creatures living there at CengageNOW™.

Stream Pollution in Developed Countries

Water pollution control laws enacted in the 1970s have greatly increased the number and quality of waste-water treatment plants in the United States and most other developed countries. Such laws also require industries to reduce or eliminate their point-source discharges of harmful chemicals into surface waters. This has enabled the United States to hold the line against increased pollution by disease-causing agents and oxygen-demanding wastes in most of its streams. This is an impressive accomplishment given the country's increased economic activity, resource consumption, and population growth since passage of these laws.

One success story is the cleanup of Ohio's Cuyahoga River. It was so polluted with flammable chemicals that it caught fire several times in the 1950s and 1960s as it flowed through the U.S. city of Cleveland. A highly publicized photo of this burning river in 1969 prompted elected officials to enact laws that limited the discharge of industrial wastes into the river and into local sewage systems and provided funds to upgrade sewage treatment facilities. Today, the river is cleaner, no longer flammable, and is widely used by boaters and anglers. This accomplishment illustrates the power of bottom-up pressure by citizens, who prodded elected officials to change a severely polluted river into an economically and ecologically valuable public resource.

Another spectacular cleanup occurred in Great Britain. In the 1950s, the Thames River was little more than a flowing, smelly sewer. Now, after 50 years of effort and large inputs of money from British taxpayers and private industry, the Thames has made a remarkable recovery. Commercial fishing is thriving and the number of fish species has increased 20-fold since 1960. In addition, many species of waterfowl and wading birds have returned to their former feeding grounds.

Large fish kills and drinking water contamination still occasionally occur in parts of developed countries. One cause of such problems is accidental or deliberate releases of toxic inorganic and organic chemicals by industries or mines (Chapter 14 Core Case Study, p. 344).

See *The Habitable Planet*, Video 6, at **www.learner.org/resources/series209.html** for discussion of how scientists measure water pollution that includes toxic heavy metals from mining wastes and abandoned underground mines. Another cause of such pollution is malfunctioning sewage treatment plants. A third cause is nonpoint runoff of pesticides and excess plant nutrients from cropland and animal feedlots.

Still, streams can recover if given the chance (**Concept 20-2A**). Stream restoration can be done on a large scale, as in the cases of the Cuyahoga and the Thames, or on a smaller scale, as in the case of Hamm Creek (Individuals Matter, below). In either case, anyone and everyone can play a role.

Global Outlook: Stream Pollution in Developing Countries

In most developing countries, stream pollution from discharges of untreated sewage and industrial wastes is a serious and growing problem. According to a 2003 report by the World Commission on Water in the 21st Century, half of the world's 500 rivers are heavily polluted, and most of them run through developing countries. Most of these countries cannot afford to build waste treatment plants and do not have, or do not enforce, laws for controlling water pollution.

According the Global Water Policy Project, most cities in developing countries discharge 80–90% of their untreated sewage directly into rivers, streams (Figure 20-6), and lakes whose waters are then used for drinking water, bathing, and washing clothes.

Industrial wastes and sewage pollute more than two-thirds of India's water resources (Case Study, p. 538) and 54 of the 78 rivers and streams monitored in China (Figure 20-7, p. 538). One-third of China's rivers are judged unfit for agricultural use, and even for

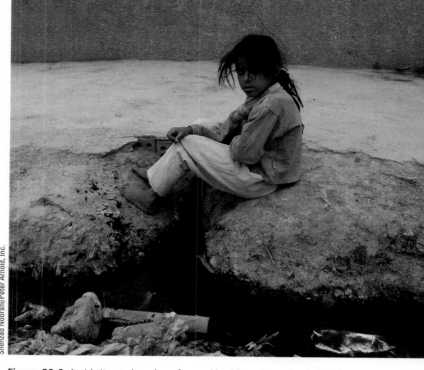

Shehzad Noorani/Peter Arnold, Inc.

Figure 20-6 A girl sits on the edge of a road beside a stream loaded with raw sewage, near her home in Baghdad, Iraq.

industrial uses. According to a 2007 report by Chinese officials, more than half of China's 1.3 billion people, including those in 278 cities, live without any form of sewage treatment. And 300 million Chinese—an amount almost equal to the entire U.S. population—do not have access to drinkable water. In Latin America and Africa, most streams passing through urban or industrial areas suffer from severe pollution. Garbage is also dumped into rivers in some places (Figure 20-8, p. 538).

INDIVIDUALS MATTER

The Man Who Planted Trees to Restore a Stream

In 1980, heart problems forced John Beal, an engineer with the Boeing Company, to take some time off. To improve his health, he began taking daily walks. His strolls took him by Hamm Creek, a smallstream that flows from the southwest hills of Seattle, Washington (USA), into the Duwamish River, which empties into Puget Sound (Figure 20-1, **Core Case Study**). He remembered when the stream was a spawning ground for salmon and when evergreen trees lined its banks. By 1980, the polluted

stream had no fish and the trees were gone.

Beal decided to restore Hamm Creek. He persuaded companies to stop polluting the creek, and he hauled out many truckloads of garbage. Then he began a 15-year project of planting thousands of trees along the stream's banks. He also restored natural waterfalls and ponds and salmon spawning beds.

At first he worked alone, but word spread and other people joined him. TV news reports and newspaper articles about

the restoration project brought more volunteers.

The creek's water now runs clear, its vegetation has been restored, and salmon have returned to spawn. His reward is the personal satisfaction he feels about having made a difference for Hamm Creek and his community. His dedication to making the world a better place is an outstanding example of *stewardship* based on the idea that *all sustainability is local.*

Figure 20-7 Natural capital degradation: highly polluted river in China. Water in many of central China's rivers is greenish-black from uncontrolled pollution by thousands of factories. Water in some rivers is too toxic to touch, much less drink. The cleanup of some modernizing Chinese cities such as Beijing and Shanghai is forcing polluting refineries and factories to move to rural areas where two-thirds of China's population resides. Liver and stomach cancer, linked in some cases to water pollution, are among the leading causes of death in the countryside. Farmers too poor to buy bottled water must often drink polluted well water.

Figure 20-8 Trash truck disposing of garbage into a river in Peru.

■ **CASE STUDY**

India's Ganges River: Religion, Poverty, Population Growth, and Health

To India's Hindu people, the Ganges is a holy river. Each day, large numbers of Hindus bathe, drink from, or take a dip in the river for religious reasons (Figure 20-9).

But the Ganges is highly polluted. About 350 million people—almost one-third of the country's population—live in the Ganges River basin. Very little of the sewage produced by these people and by the industries and 29 large cities in the basin is treated.

This situation is complicated by the Hindu belief in cremating the dead to free the soul and in throwing the ashes into the holy Ganges to increase the chances of the soul getting into heaven. Traditionally, wood fires are used to burn most bodies. This creates air pollution and helps deplete India's forests.

It also causes water pollution. Because many people cannot afford enough wood for cremation, many unburned or partially burnt bodies are dumped into the river where they mingle with large numbers of livestock corpses. Decomposition of these bodies depletes dissolved oxygen and adds disease-carrying bacteria and viruses to the water. This problem is expected to get worse as India's population grows; about 18 million people are added to the population each year—about a third of them to the Ganges River basin.

To help clean up the river, the Indian government plans to build waste treatment plants in the basin's 29 large cities and construct along the banks of the river 32 electric crematoriums, which can burn bodies more efficiently and at a cost lower than that of wood-fired cremation. The government also introduced 25,000 snapping turtles to devour corpses.

But most of the sewage treatment plants are not yet completed or do not work very well, and only a few of the crematoriums are in operation. There is also concern that many Hindus will not abandon the traditional ritual of wood-fired cremation or will not be able to afford any type of cremation.

Another religious custom involves throwing vividly painted small statues into the river. These create another source of pollution, because paints and coatings on these objects often contain toxic metals, such as lead and mercury, and various potentially harmful organic compounds.

Global warming is almost certain to make this situation worse. About 70% of the water flowing into the Ganges comes from the country's Gangotri Glacier, which is now melting at an accelerating rate. If this continues, within decades, the Ganges will become a seasonal river that flows only during the rainy season. The resulting loss of water poses a severe threat to the more than 400 million people living within the Ganges basin in India and Bangladesh.

THINKING ABOUT
Ancient Rituals and Water Pollution

What arguments would you use to convince someone to use a newer, cleaner technology for observing an ancient religious ritual in order to help protect the Ganges River?

Low Water Flow and Too Little Mixing Makes Lakes Vulnerable to Water Pollution

Lakes and reservoirs are generally less effective at diluting pollutants than streams are, for two reasons. *First,* lakes and reservoirs often contain stratified layers (Figure 8-15, p. 175) that undergo little vertical mixing. *Second,* they have little or no flow. The flushing and changing of water in lakes and large artificial reservoirs can take from 1 to 100 years, compared to several days or weeks for streams.

As a result, lakes and reservoirs are more vulnerable than streams are to contamination by runoff or discharge of plant nutrients, oil, pesticides, and nondegradable toxic substances, such as lead, mercury, and selenium. These contaminants can kill bottom life and fish and birds that feed on contaminated aquatic organisms. Many toxic chemicals and acids also enter lakes and reservoirs from the atmosphere (Figure 18-12, p. 479).

As they pass through food webs in lakes, the concentrations of some harmful chemicals are biologically magnified. Examples include DDT (Figure 9-19, p. 202), PCBs (Case Study, p. 449), some radioactive isotopes, and some mercury compounds (Figure 17-A, p. 450).

Cultural Eutrophication Is Too Much of a Good Thing

Eutrophication is the name given to the natural nutrient enrichment of a shallow lake, estuary, or slow-moving stream, mostly from runoff of plant nutrients such as nitrates and phosphates from surrounding land. In the case of Lake Washington (**Core Case Study**), the major nutrient was phosphorous in treated wastewater that was dumped into the lake from all sides.

An *oligotrophic lake* is low in nutrients and its water is clear (Figure 8-16, left, p. 175). Over time, some lakes become more eutrophic (Figure 8-16, right) as nutrients are added from natural and human sources in the surrounding watersheds.

Near urban or agricultural areas, human activities can greatly accelerate the input of plant nutrients to a lake—a process called **cultural eutrophication** involving mostly nitrate- and phosphate-containing effluents from various sources. These sources include runoff from farmland, animal feedlots, urban areas, chemically fertilized suburban yards, and mining sites,

Figure 20-9 *India's Ganges River.* Each day more than 1 million Hindus in India bathe, drink from, or carry out religious ceremonies in the highly polluted Ganges River, which flows from the southern slopes of the Himalayas through India and into the Bay of Bengal. This photo shows people engaging in a Hindu Puja ritual ceremony in the river.

and discharges of treated and untreated municipal sewage. Some nitrogen also reaches lakes by deposition from the atmosphere.

During hot weather or drought, this nutrient overload produces dense growths or "blooms" of organisms such as algae and cyanobacteria (Figure 8-16, right, p. 175) and thick growths of water hyacinth (Figure 11-4, p. 252), duckweed, and other aquatic plants. These dense colonies of plant life can reduce lake productivity and fish growth by decreasing the input of solar energy needed for photosynthesis by phytoplankton that support fish (**Concept 20-2B**).

When the algae die, they are decomposed by swelling populations of aerobic bacteria, which deplete dissolved oxygen in the surface layer of water near the shore and in the bottom layer. This can kill fish and other aerobic aquatic animals. This is what was happening to Lake Washington (Figure 20-1) before scientists and citizens worked together to clean it up (**Core Case Study**). If excess nutrients continue to flow into a lake, anaerobic bacteria take over and produce gaseous products such as smelly, highly toxic hydrogen sulfide and flammable methane.

According to the U.S. Environmental Protection Agency (EPA), about one-third of the 100,000 medium to large lakes and 85% of the large lakes near major U.S. population centers have some degree of cultural eutrophication. And the International Water Association estimates that more than half of the lakes in China suffer from cultural eutrophication.

There are several ways to *prevent* or *reduce* cultural eutrophication. We can use advanced (but expensive) waste treatment to remove nitrates and phosphates before wastewater enters lakes. In Lake Washington, this approach, plus diversion of lake water to nearby Puget Sound (Figure 20-1, **Core Case Study**), were used to reduce eutrophication. We can also use a preventive approach by banning or limiting the use of phosphates in household detergents and other cleaning agents and by employing soil conservation and land-use control to reduce nutrient runoff (**Concept 20-2B**).

There are several ways to *clean up* lakes suffering from cultural eutrophication. We can mechanically remove excess weeds, control undesirable plant growth with herbicides and algicides, and pump air through lakes and reservoirs to prevent oxygen depletion, all of which are expensive and energy-intensive methods.

As usual, pollution prevention is more effective and usually cheaper in the long run than cleanup. The good news is that a lake usually can recover from cultural eutrophication, if excessive inputs of plant nutrients are stopped.

Revisiting Lake Washington and Puget Sound

We can learn two lessons from the story of Lake Washington (**Core Case Study**), both related to themes we explore throughout this book. *First,* severe water pollution can be reversed in a fairly short time, if pollutant inputs are sharply reduced. *Second,* citizen action combined with scientific research works.

However, recall that the wastewater treatment plant effluents that had been flowing into Lake Washington were diverted to the Puget Sound (Figure 20-1). Today, mostly because of continued population and economic growth in the Seattle area, the sound is becoming overloaded with these effluents. There is growing concern also about overflows of increased urban runoff and raw sewage during storms and about large inputs of toxic materials into the sound.

Despite the ecological and political success story of Lake Washington, the relentless growth of population, resource use, and urbanization are again overwhelming the lake, as well as the sound. This brings a third lesson to the Lake Washington story: Even good solutions to environmental problems cannot work indefinitely if we keep overwhelming the natural systems involved. Ultimately, **scientific principles of sustainability** (see back cover) require reducing population growth and resource use.

In 2007, the Washington state government appointed the Puget Sound Partnership to determine the condition of the sound and to develop strategies for meeting the state's goal of having a healthy Puget Sound by 2020.

■ **CASE STUDY**
Pollution in the Great Lakes

The five interconnected Great Lakes of North America (Figure 20-10) contain at least 95% of the fresh surface water in the United States and one-fifth of the world's fresh surface water. At least 38 million people in the United States and Canada obtain their drinking water from these lakes.

Despite their enormous size, these lakes are vulnerable to pollution from point and nonpoint sources. One reason is that less than 1% of the water entering these lakes flows out to the St. Lawrence River each year, meaning that pollutants can take as long as 100 years to be flushed out to sea.

By the 1960s, many areas of the Great Lakes were suffering from severe cultural eutrophication, huge fish kills, and contamination from bacteria and a variety of toxic industrial wastes. The impact on Lake Erie was particularly intense because it is the shallowest of the Great Lakes and has the highest concentrations of people and industrial activity along its shores.

Since 1972, Canada and the United States have joined forces and spent more than $20 billion on a Great Lakes pollution control program. This program has decreased algal blooms, increased dissolved oxygen levels and sport and commercial fishing catches in Lake Erie, and allowed most swimming beaches to reopen. These improvements occurred mainly because of new or upgraded sewage treatment plants, better treatment of industrial wastes, and bans on use of detergents, household cleaners, and water conditioners that contain phosphates—all instituted mostly because of bottom-up citizen pressure.

Despite this important progress, many problems remain. In 2006, Canadian scientists reported that cities around the lakes were releasing, on average, the equivalent of more than 100 Olympic swimming pools of raw sewage into the lakes each day. Dozens of municipal sewage systems combine storm water with wastewater and allow emergency overflows into the lakes. These systems overflow far too easily and too often, according to the scientists. Cities that use systems that separate sewage from storm water, which include Green Bay, Wisconsin, and Duluth, Minnesota, contribute the least to this problem.

Increasing nonpoint runoff of pesticides and fertilizers resulting from urban sprawl, fueled by population growth, now surpasses industrial pollution as the greatest threat to the lakes. Sediments in 26 toxic hotspots remain heavily polluted. And *biological pollution* in the form of invasions by zebra mussels and more than 180 other alien species, threaten some native species and cause billions of dollars in damages (Case Study, p. 269).

Air quality over the Great Lakes has generally improved, according to the 2007 State of the Great Lakes

Figure 20-10 *The Great Lakes of North America.* Two Canadian provinces to the north and eight U.S. states to the south surround these five lakes, which make up the world's largest freshwater system. Dozens of growing cities lie on their shores, and water pollution is a growing problem. (Provided by the SeaWiFS project, NASA/Goddard Space Flight Center, and ORBIMAGE)

report, but about half of the toxic compounds entering the lakes still come from atmospheric deposition of pesticides, mercury from coal-burning plants, and other toxic chemicals from as far away as Mexico and Russia. A survey by Wisconsin biologists found that one fish in four taken from the Great Lakes is unsafe for human consumption. Despite ongoing pollution problems, EPA funding for cleanup of the Great Lakes has dropped 80% since 1992.

The 2007 State of the Great Lakes report also notes:

- New pollutants found in the lakes include toxic polybrominated diphenyl ethers used in flame retardants and various pharmaceutical and personal care products.

- Continuing wetland loss and degradation is shrinking the habitats of birds, amphibians, and some fishes.

- Populations of native species near the base of the food web, such as some plankton, are declining in some of the lakes.

- Native populations of carnivorous fish species, such as lake trout, are declining in most of the lakes.

Some environmental and health scientists call for taking a prevention approach and banning the use of toxic chlorine compounds, such as bleach used in the pulp and paper industry, which is prominent around the Great Lakes. They would also ban new incinerators, which can release toxic chemicals into the atmosphere, and they would stop the discharge into the lakes of 70 toxic chemicals that threaten human health and wildlife. So far, officials in the industries involved have successfully opposed such bans.

In 2007, a group of economists estimated that improving the health of the Great Lakes, by rebuilding antiquated sewer systems, restoring degraded wetlands, dealing with the invasive species, and cleaning up contaminated lake and tributary sediment, would cost $26 billion. They also argued that the estimated resulting gain of $50 billion in real estate values would justify the investment.

THINKING ABOUT
Pollution in the Great Lakes

What are three steps you would take to sharply reduce pollution in the Great Lakes?

20-3 What Are the Major Pollution Problems Affecting Groundwater and Other Drinking Water Sources?

▶ **CONCEPT 20-3A** Chemicals used in agriculture, industry, transportation, and homes can spill and leak into groundwater and make it undrinkable.

▶ **CONCEPT 20-3B** There are simple ways and complex ways to purify drinking water, but protecting it through pollution prevention is the least expensive and most effective strategy.

Groundwater Cannot Cleanse Itself Very Well

Drinking water for about half of the U.S. population and 95% of those who live in rural areas comes from groundwater. According to many scientists, groundwater pollution is a serious threat to human health.

Common pollutants such as fertilizers, pesticides, gasoline, and organic solvents can seep into groundwater from numerous sources (Figure 20-11). People who dump or spill gasoline, oil, and paint thinners and other organic solvents onto the ground also contaminate groundwater (**Concept 20-3A**).

Once a pollutant from a leaking underground storage tank or other source contaminates groundwater, it fills the aquifer's porous layers of sand, gravel, or bedrock like water saturates a sponge. This makes removal of the contaminant difficult and costly. The slowly flowing groundwater disperses the pollutant in a widening *plume* of contaminated water. If the plume reaches a well used to extract groundwater, the toxic pollutants can get into drinking water and into water used to irrigate crops (Figure 20-12).

When groundwater becomes contaminated, it cannot cleanse itself of *degradable wastes* as quickly as flowing surface water does. Groundwater flows so slowly—

Figure 20-11 Natural capital degradation: principal sources of groundwater contamination in the United States (**Concept 20-3A**). Another source is salt-water intrusion from excessive groundwater withdrawal in coastal areas. (Figure is not drawn to scale.) **Question:** What are three sources shown in this picture that might be affecting groundwater in your area?

Figure 20-12 Natural capital degradation: groundwater contamination from a leaking gasoline tank. As the contaminated water spreads from its source in a widening plume, it can be extracted by wells used to provide water for drinking and irrigation.

usually less than 0.3 meter (1 foot) per day—that contaminants are not diluted and dispersed effectively. In addition, groundwater usually has much lower concentrations of dissolved oxygen (which helps to decompose many contaminants) and smaller populations of decomposing bacteria. Also, the usually cold temperatures of groundwater slow down chemical reactions that decompose wastes.

Thus, it can take decades to thousands of years for contaminated groundwater to cleanse itself of *slowly degradable wastes* (such as DDT). On a human time scale, *nondegradable wastes* (such as toxic lead and arsenic) remain in the water permanently.

Groundwater Pollution Is a Serious Hidden Threat

On a global scale, we do not know much about groundwater pollution because few countries go to the great expense of locating, tracking, and testing aquifers. But the results of scientific studies in scattered parts of the world are alarming.

China has limited water resources for its huge population. Groundwater is crucial because it provides about 70% of the country's drinking water. In 2006, the Chinese government reported that aquifers in about nine of every ten Chinese cities are polluted or overexploited, and could take hundreds of years to recover.

In the United States, an EPA survey of 26,000 industrial waste ponds and lagoons found that one-third of them had no liners to prevent toxic liquid wastes from seeping into aquifers. One-third of these sites are within 1.6 kilometers (1 mile) of a drinking water well. In addition, almost two-thirds of America's liquid hazardous wastes are injected into deep disposal wells underground, some of which leak water into aquifers used as sources of drinking water.

By 2006, the EPA had completed the cleanup of about 350,000 of the more than 460,000 underground tanks in the United States that were leaking gasoline, diesel fuel, home heating oil, or toxic solvents into groundwater. During this century, scientists expect many of the millions of such tanks, which have been

installed around the world, to corrode, leak, contaminate groundwater, and become a major global health problem. Determining the extent of a leak from a single underground tank can cost $25,000–250,000, and cleanup costs range from $10,000 to more than $250,000. If the chemical reaches an aquifer, effective cleanup is often not possible or is too costly. *Bottom line:* Wastes that we think we have thrown away or stored safely can escape and come back to haunt us.

Another problem in the United States is groundwater pollution by MTBE (methyl tertiary butyl ether)—a gasoline additive used since 1979. MTBE is a suspected carcinogen. By the time this was discovered in the 1990s, about 250,000 leaking gasoline tanks had contaminated aquifers in many parts of the country. Use of MTBE is being phased out, but plumes of contaminated groundwater will move through aquifers for decades. Oil companies may face thousands of lawsuits from health problems related to this chemical.

Groundwater used as a source of drinking water can also be contaminated with *nitrate ions* (NO_3^-), especially in agricultural areas where nitrates in fertilizer are often leached into groundwater. Nitrite ions (NO_2^-) in the stomach, colon, and bladder can convert some of the nitrate ions in drinking water to organic compounds that have been shown to cause cancer in more than 40 test animal species. The conversion of nitrates in tap water to nitrites in infants under 6 months old can cause a potentially fatal condition known as "blue baby syndrome," in which a baby's blood cannot carry sufficient oxygen to body cells.

■ CASE STUDY
A Natural Threat from Arsenic in Groundwater

Toxic *arsenic* contaminates drinking water when a well is drilled into aquifers where soils and rock are naturally rich in arsenic. Some rivers used for drinking water also are contaminated naturally, having originated in springs that have high levels of arsenic. Human activities such as mining and ore processing can also release arsenic into drinking water supplies. The accepted standard for safe levels of arsenic is 10 parts per billion (ppb). But according to a 2007 study by the WHO, more than 140 million people in 70 countries are drinking water with arsenic concentrations of 5–100 times that level. Levels are especially high in parts of Bangladesh and China and India's state of West Bengal.

The WHO estimates that long-term exposure to nondegradable arsenic in drinking water is likely to cause hundreds of thousands of premature deaths from cancer of the skin, bladder, and lung. In 2007, researchers reported that bladder and lung cancer death rates in a part of northern Chile were 3 to 14 times higher than normal. People there had been exposed to high levels of arsenic in their drinking water for decades.

There is also concern over arsenic levels in drinking water in parts of the United States. According to the EPA, some 13 million people in several thousand communities, mostly in the western half of the country, have arsenic levels of 3–10 ppb in their drinking water. Scientists from the WHO and other organizations warn that even the 10 ppb standard is not safe. At that level, say EPA scientists, the lifelong cancer risk from ingesting arsenic is 30 times as high as that for any carcinogen regulated by the EPA. Many scientists call for lowering the standard to 3–5 ppb, but it would be costly to comply with such a lower standard.

In 2006, researchers from Rice University in Houston, Texas (USA), reported that transforming a common mineral similar to rust into a powder of tiny nanocrystals (see p. S45 in Supplement 6) and using it to purify drinking water could greatly reduce the threat of arsenic in the water at a cost of a few cents a day for families. Stay tuned while this process is evaluated.

THINKING ABOUT
Preventing Arsenic Pollution

Would you be willing to pay more for drinking water if the higher price was necessary to guarantee arsenic concentrations lower than 5 ppb? How much more (expressed as a percent over what is paid now) would you pay? Explain.

Pollution Prevention Is the Only Effective Way to Protect Groundwater

Treating a contaminated aquifer involves eliminating the source of pollution and drilling monitoring wells to determine how far, in what direction, and how fast the contaminated plume is moving. Then a computer model is used to project future dispersion of the contaminant in the aquifer. The final step is to develop and implement a strategy to clean up the contamination. Pumping polluted groundwater to the surface, cleaning it up, and returning it to the aquifer is very expensive.

Figure 20-13 lists ways to prevent and clean up groundwater contamination (**Concept 20-3B**). Because of the difficulty and expense of cleaning up a contaminated aquifer, *preventing contamination is the least expensive and most effective way to protect groundwater resources* (Figure 20-13, left, and **Concept 20-3B**).

There Are Many Ways to Purify Drinking Water

Most developed countries have laws establishing drinking water standards, but most developing countries do not have such laws or, if they do have them, they do not enforce them. There are many simple ways and many complex ways to purify drinking water (**Concept 20-3B**).

SOLUTIONS

Groundwater Pollution

Prevention

Find substitutes for toxic chemicals

Keep toxic chemicals out of the environment

Install monitoring wells near landfills and underground tanks

Require leak detectors on underground tanks

Ban hazardous waste disposal in landfills and injection wells

Store harmful liquids in aboveground tanks with leak detection and collection systems

Cleanup

Pump to surface, clean, and return to aquifer (very expensive)

Inject microorganisms to clean up contamination (less expensive but still costly)

Pump nanoparticles of inorganic compounds to remove pollutants (still being developed)

In developed countries, wherever people depend on surface water, it is usually stored in a reservoir for several days. This improves clarity and taste by increasing dissolved oxygen content and allowing suspended matter to settle. The water is then pumped to a purification plant and treated to meet government drinking water standards. In areas with very pure groundwater or surface water sources, little treatment is necessary. Some cities have found that protecting watersheds that supply their drinking water is a lot cheaper than building water purification plants (see Case Study, p. 546).

Japan and several other countries are beginning to develop plants that process sewer water into drinking water. El Paso, Texas (USA), gets 40% of its drinking water from recycling and purifying wastewater. In 2007, Orange County, California, completed the world's largest plant devoted to making sewer water as pure as distilled water. If it receives approval by state health officials, it will be used to supply drinking water and to recharge aquifers. We have the technology to convert sewer water into pure drinking water and to help reduce water shortages. However, using such toilet-to-tap systems is expensive. And it faces opposition from citizens and health officials who are unaware of the advances in this technology.

Simpler measures can be used to purify drinking water. In tropical countries that lack centralized water treatment systems, the WHO urges people to purify drinking water by exposing a clear plastic bottle filled with contaminated water to intense sunlight. The sun's heat and UV rays can kill infectious microbes in as little

as 3 hours. Painting one side of the bottle black can improve heat absorption in this simple solar disinfection method, which applies one of the four **scientific principles of sustainability** (see back cover). Where this measure has been used, incidence of dangerous childhood diarrhea has decreased by 30–40%.

Researchers are also developing nanofilters to clean contaminated water. The goal is to develop a low-cost water filter that can be cleaned and reused. Vermont-based Seldon Technologies, for example, is developing a hand-held, carbon, nanotube filter that can quickly purify water from any source—a river, mud puddle, or sample of groundwater.

Vestergaard Frandsen, a Danish company, has developed the LifeStraw, a similar inexpensive, portable water filter that eliminates many viruses and parasites from water drawn into it (Figure 20-14). It has been particularly useful in Africa, where aid agencies are distributing it.

Figure 20-14 The *LifeStraw*, designed by Torben Vestergaard Frandsen, is a personal water purification device that gives many poor people access to safe drinking water. **Question:** Do you think the development of such devices should make prevention of water pollution less of a priority? Explain.

■ CASE STUDY

Protecting Watersheds Instead of Building Water Purification Plants

Several major U.S. cities have avoided building expensive water treatment facilities by investing in protection of the forests and wetlands in the watersheds that provide their water (**Concept 20-3B**). Examples are New York City, N.Y; Boston, Massachusetts; Seattle, Washington; and Portland, Oregon.

New York City's drinking water is known for its purity. The city gets 90% of the water for its 9 million residents from reservoirs in New York State's Catskill Mountains. Forests cover more than three-fourths of this watershed. Underground tunnels transport the water to the city.

To continue providing quality drinking water for its citizens, the city faced spending $6 billion to build water purification facilities. Instead, the city decided to negotiate an agreement with towns, farmers, the state, and other interests in the Catskills watershed. The city would pay this diverse group of governments and private citizens $1.5 billion over 10 years for protecting and restoring the forests, wetlands, and streams in the watershed.

After many years of negotiations, this historic agreement was signed in 1997. The $1.5 billion to be spent on watershed protection will save New York City the $6 billion cost of building water purification facilities plus $300 million a year in filtration costs. This is an excellent example of working with nature to provide a sustainable supply of clean drinking water.

--- THINKING ABOUT ---
Protecting the Sources of Drinking Water

Where does the community in which you live get its drinking water? Could it save money and help to protect biodiversity by finding ways to protect its watershed or the aquifers that supply this water?

Using Laws to Protect Drinking Water Quality

About 54 countries, most of them in North America and Europe, have standards for safe drinking water. The U.S. Safe Drinking Water Act of 1974 requires the EPA to establish national drinking water standards, called *maximum contaminant levels*, for any pollutants that may have adverse effects on human health. But such laws do not exist or are not enforced in most developing countries.

Despite passage of the Clean Water Act in 1972, the United Nations estimates that 5.6 million Americans drink water that does not meet EPA safety standards for one or more contaminants. And according to the EPA, one in five Americans drinks water supplied by a water treatment plant that has violated one or more safety standards during part of a year.

Health scientists call for strengthening the U.S. Safe Drinking Water Act in several ways. One way is to combine many of the drinking water treatment systems that serve fewer than 3,300 people with nearby larger systems to make it cheaper for small systems to meet federal standards. Another is to strengthen and enforce public notification requirements about violations of drinking water standards. Scientists also call for banning all toxic lead in new plumbing pipes, faucets, and fixtures. Current law allows for fixtures with up to 10% lead content to be sold as lead-free. According to the Natural Resources Defense Council (NRDC), such improvements would cost each U.S. household an average of about $30 a year.

However, water-polluting industries are pressuring elected officials to weaken the Safe Drinking Water Act. One proposal is to eliminate national tests of drinking water and public notification requirements about violations of drinking water standards. Another such proposal is to allow states to give waivers to drinking water providers, allowing them a permanent right to violate the standard for a given contaminant if they claim they cannot afford to comply. Another suggestion is to eliminate the requirement that water systems use affordable, feasible technology to remove cancer-causing contaminants. Finally, there are suggestions to greatly reduce the EPA's already low budget for enforcing the U.S. Safe Drinking Water Act.

--- HOW WOULD YOU VOTE? ---
Should the U.S. Safe Drinking Water Act be strengthened? Cast your vote online at **academic.cengage.com/biology/miller**.

Is Bottled Water the Answer?

Despite some problems, experts say the United States has some of the world's cleanest drinking water. Municipal water systems in the United States are required to test their water regularly for a number of pollutants and to make the results available to citizens. Yet about half of all Americans worry about getting sick from tap water contaminants, and many drink bottled water or install expensive water purification systems. Some other countries rely on bottled water wherever their tap water is too polluted to drink.

Between 1976 and 2006, average bottled water consumption per person in the United States increased from 7.5 liters (2 gallons) to 113 liters (30 gallons) a year. Studies reveal that in the United States, bottled water costs 240 to 100,000 times more than tap water. Yet studies also indicate that about one-fourth of it is ordinary tap water in a bottle, and that bacteria or fungi contaminate about 40% of bottled water. And the government testing standards for bottled water in the United States are not as high as those for tap water.

Each year, consumers worldwide spend about $100 billion on bottled water ($1–2 billion in the United States). Compare this to the cost of providing access to safe drinking water for the 1.1 billion people who now lack it, which the U.N. estimates to be $11.9 billion.

Use of bottled water also causes environmental problems, according to a 2007 study by the Worldwatch Institute. Each year, the number of plastic water bottles thrown away, if lined up end-to-end, could circle the earth's equator eight times. Toxic gases and liquids are released during the manufacture of plastic water bottles, and greenhouse gases and other air pollutants are emitted by the fossil fuels burned to make them and to deliver bottled water to suppliers. For example, a bottle of Fiji water used in the United States travels on average about 8,900 kilometers (5,500 miles). According to the Pacific Institute, the oil used to produce the plastic for the nearly 30 billion water bottles used in the United States each year would fuel 100,000 cars for a year. If we include the energy used to pump, process, transport, and refrigerate the water, it would be enough to run 3 million cars a year. And withdrawing water for bottling is helping to deplete some underground aquifers.

Because of these harmful environmental impacts and the high cost of bottled water, there is a growing *back-to-the-tap* movement based on boycotting bottled water. From San Francisco to New York to Paris, city governments, high-class restaurants, schools, religious groups, and many consumers are refusing to buy bottled water as this trend picks up steam.

Health officials suggest that, before drinking expensive bottled water or buying costly home water purifiers, consumers have their water tested by local health departments or private labs (but not by companies trying to sell water purification equipment). The goals for such testing are to identify what contaminants (if any) must be removed and to determine the type of purification needed to remove such contaminants. Independent experts contend that unless tests show otherwise, for most urban and suburban people served by large municipal drinking water systems, home water treatment systems are not worth the expense and maintenance hassles.

Buyers should check out companies selling water purification equipment and be wary of claims that the EPA has approved a treatment device. Although it does *register* such devices, the EPA neither tests nor approves them.

HOW WOULD YOU VOTE?

Should we establish standards for bottled water that are as strict as those for water from public tap water systems? Cast your vote online at **academic.cengage.com/biology/ miller**.

20-4 What Are the Major Water Pollution Problems Affecting Oceans?

▶ **CONCEPT 20-4A** The great majority of ocean pollution originates on land and includes oil and other toxic chemicals and solid wastes, which threaten aquatic species and other wildlife and disrupt marine ecosystems.

▶ **CONCEPT 20-4B** The key to protecting the oceans is to reduce the flow of pollutants from land and air and from streams emptying into these waters.

Ocean Pollution Is a Growing and Poorly Understood Problem

Coastal areas—especially wetlands, estuaries, coral reefs, and mangrove swamps—bear the brunt of our enormous inputs of pollutants and wastes into the ocean (**Concept 20-4A**) (Figure 20-15, p. 548). This is not surprising, because about 40% of the world's population (53% in the United States) lives on or near the coast. See *The Habitable Planet*, Video 5, at **www.learner.org/ resources/series209.html** to learn how scientists are studying the effects of population growth and development on nitrogen pollution of coastal aquatic systems in Cape Cod, Massachusetts (USA). Of the world's 15 largest metropolitan areas (each with 10 million people or more), 14 are near coastal waters. Coastal populations are expected to double by 2050.

According to a 2006 *State of the Marine Environment* study by the U.N. Environment Programme (UNEP), an estimated 80% of marine pollution originates on land (**Concept 20-4A**), and this percentage could rise significantly by 2050 if coastal populations double as projected. The report says that 80–90% of the municipal sewage from most coastal developing countries and from some coastal developed countries is dumped into oceans untreated. This dumping often overwhelms the ability of some coastal waters to biodegrade such wastes.

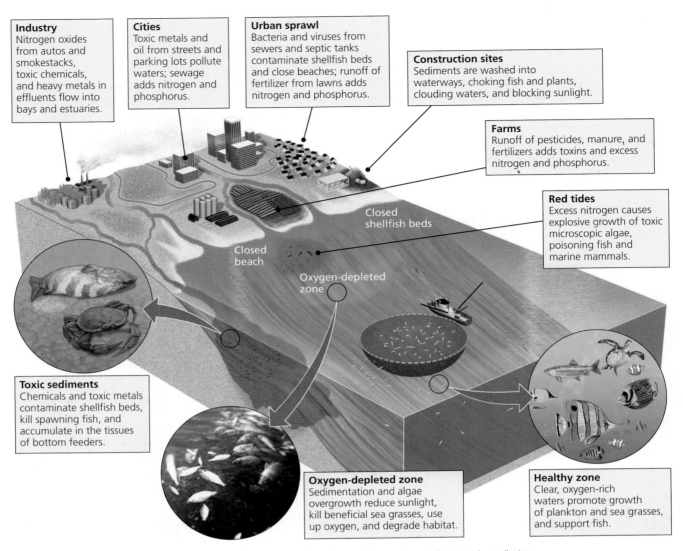

Industry
Nitrogen oxides from autos and smokestacks, toxic chemicals, and heavy metals in effluents flow into bays and estuaries.

Cities
Toxic metals and oil from streets and parking lots pollute waters; sewage adds nitrogen and phosphorus.

Urban sprawl
Bacteria and viruses from sewers and septic tanks contaminate shellfish beds and close beaches; runoff of fertilizer from lawns adds nitrogen and phosphorus.

Construction sites
Sediments are washed into waterways, choking fish and plants, clouding waters, and blocking sunlight.

Farms
Runoff of pesticides, manure, and fertilizers adds toxins and excess nitrogen and phosphorus.

Red tides
Excess nitrogen causes explosive growth of toxic microscopic algae, poisoning fish and marine mammals.

Closed shellfish beds

Closed beach

Oxygen-depleted zone

Toxic sediments
Chemicals and toxic metals contaminate shellfish beds, kill spawning fish, and accumulate in the tissues of bottom feeders.

Oxygen-depleted zone
Sedimentation and algae overgrowth reduce sunlight, kill beneficial sea grasses, use up oxygen, and degrade habitat.

Healthy zone
Clear, oxygen-rich waters promote growth of plankton and sea grasses, and support fish.

Figure 20-15 Natural capital degradation: residential areas, factories, and farms all contribute to the pollution of coastal waters and bays. According to the U.N. Environment Programme, coastal water pollution costs the world $16 billion annually—more than $30,000 a minute—due to ill health and premature death. **Question:** What are three changes you could make in your lifestyle that might help to prevent this pollution?

This problem is much worse than what we experienced in Lake Washington (**Core Case Study**), and its effects are equally apparent. Lake Washington had the potential to recover and had a larger body of water nearby that could dilute and help degrade such wastes. But the coastline of China, for example, is so choked with algae growing on the nutrients provided by sewage, that some scientists believe large areas of China's coastal waters can no longer sustain marine ecosystems.

In deeper waters, the oceans can dilute, disperse, and degrade large amounts of raw sewage and other types of degradable pollutants. Some scientists suggest that it is safer to dump sewage sludge and most other harmful wastes into the deep ocean than to bury them on land or burn them in incinerators. Other scientists disagree, pointing out that we know less about the deep ocean than we do about the moon. They add that dumping harmful wastes into the ocean would delay urgently needed pollution prevention measures and promote further degradation of this vital part of the earth's life-support system.

Scientists also point to a little known problem of cruise ship pollution. A cruise liner can carry as many as 2,000 passengers and 1,000 crew members and generate as much waste as a small city produces. Much of this waste, including perchloroethylene from dry-cleaning and benzene from paint and solvents, is highly toxic. Cruise ships also generate huge amounts of plastic garbage and waste oil. For decades, cruise ships and other ocean vessels, which often sail through fragile ecosystems such as Florida's coral reefs, have been dumping their wastes at sea.

In U.S. waters, such dumping is illegal, but some ships continue dumping secretively, usually at night. Ship owners can save large amounts of money by such illegal dumping. However, since 2002, a few of these companies have been caught in the act of illegal dump-

ing and fined millions of dollars. And some vacationers are refusing to go on cruise ships that do not have sophisticated systems for dealing with the wastes they produce.

THINKING ABOUT
Ocean Pollution

Should we dump sewage sludge and other harmful pollutants into the deep ocean? Explain.

Recent studies of some U.S. coastal waters have found vast colonies of viruses thriving in raw sewage and in effluents from sewage treatment plants (which do not remove viruses) and leaking septic tanks. According to one study, one-fourth of the people using coastal beaches in the United States develop ear infections, sore throats, eye irritations, respiratory disease, or gastrointestinal disease.

In its 2005 report on the environmental health of coastal areas in the 48 continental U.S. states, the EPA classified four of five estuaries as threatened or impaired. It found one in four coastal sites unsuitable for swimming, and one in five of the sites had fish contaminated with unsafe levels of mercury and various other pollutants.

Runoffs of sewage and agricultural wastes into coastal waters introduce large quantities of nitrate (NO_3^-) and phosphate (PO_4^{3-}) plant nutrients, which can cause explosive growths of harmful algae. These *harmful algal blooms* are called red, brown, or green toxic tides (Figure 20-16). They release waterborne and airborne toxins that damage fisheries, kill some fish-eating birds, reduce tourism, and poison seafood. Each year, harmful algal blooms lead to the poisoning of about 60,000 Americans who eat shellfish contaminated by the algae.

Each year, because of harmful algal blooms, at least 200 *oxygen-depleted zones* form in coastal waters around the world, according to a 2006 report by the UNEP They occur mostly in temperate coastal waters and in landlocked seas such as the Baltic and Black Seas. About 43 of these zones occur in U.S. waters (Science Focus, p. 550).

These zones are incorrectly called *dead zones*. Because of low oxygen levels (hypoxia), they contain few oxygen-consuming fish and bottom-dwelling organisms, but they abound with decomposing bacteria. The low oxygen levels are caused by the rapid growth of algae in nutrient-rich waters, which are decomposed by colonies of oxygen-consuming bacteria. Evidence indicates that oxygen-depleted zones result mostly from excessive inputs of nitrates and phosphates from runoff of fertilizers and animal wastes, and also deposition of nitrogen compounds from the atmosphere.

Ocean Oil Pollution Is a Serious Problem

Crude petroleum (oil as it comes out of the ground) and *refined petroleum* (fuel oil, gasoline, and other processed petroleum products, Figure 15-4, p. 375) reach the ocean from a number of sources and become highly disruptive pollutants (**Concept 20-4A**).

In 1989, the *Exxon Valdez* oil tanker went off course, hit rocks, and released 40.8 million liters (10.8 million gallons) of oil into Alaska's Prince William Sound—an accident that ended up costing Exxon Mobil billions of dollars in cleanup costs, fines, and damages to 34,000 fishers and other Alaskans. The oil killed large numbers of seabirds, fish, and sea otters. The cleanup cost around $2.5 billion and the estimated total cost, including claims settlements and fines, was $4 billion.

In 2002, the oil tanker *Prestige* sank off the coast of Spain and, during the next 2 years, leaked about twice as much oil as the *Exxon Valdez* had lost. This and other spills also killed and harmed large numbers of seabirds (Figure 15-7, p. 379).

Tanker accidents and blowouts at offshore drilling rigs (when oil escapes under high pressure from a borehole in the ocean floor) get most of the publicity because of their high visibility. But *studies show that the largest source of ocean oil pollution is urban and industrial runoff from land,* much of it from leaks in pipelines and oil-handling facilities.

At least 37%—and perhaps even half—of the oil reaching the oceans is waste oil, dumped, spilled, or leaked onto the land or into sewers by cities, industries, and people changing their own motor oil. Some good news: according to a 2006 UNEP study, since the mid-1980s the amount of oil entering the marine

Kevin Schafer/Peter Arnold, Inc.

Figure 20-16 *A brown tide.* This harmful algal bloom off the cost of Mexico contains organisms that give the water a reddish brown tint. They make the water unappealing to tourists and can be toxic to fish, wildlife, people, and their pets. **Question:** What are two ways in which this sort of pollution could be prevented?

SCIENCE FOCUS

Oxygen Depletion in the Northern Gulf of Mexico

The world's third largest oxygen-depleted zone (after those in the Baltic Sea and the northwestern Black Sea) forms every spring and summer in a narrow stretch of the northern Gulf of Mexico off the mouth of the Mississippi River (Figure 20-B). The low oxygen levels suffocate fish, crabs, and shrimp that cannot move to less polluted areas.

The Mississippi River basin drains all or parts of 31 U.S. states and two Canadian provinces. Its watershed contains almost two-thirds of the continental U.S. land area and more than half of all U.S. croplands; it is one of the world's most productive agricultural regions.

According to a 2005 study of sediment cores by geologist Lisa Osterman, seasonal oxygen-depleted zones in the northern Gulf of Mexico existed as long ago as the 1800s. But since 1950, when fertilizer use began increasing sharply, the sizes of the zones and levels of oxygen-depletion have been increasing on average, even though it shrank in some years. In many years, it covers an area larger than the U.S. state of Connecticut.

In 2007, scientists projected that the 15% increase in the size of the heavily fertilized U.S. corn crop, resulting from increased demand for corn to make ethanol fuel for cars (Figure 16-27, p. 426), would cause the zone to grow in size. They later found that the zone was larger that summer than it had been in the previous 22 years. Thus, despite a commitment in 2001 by state and federal governments and Native American tribes to reduce the size of the zone by 75% by 2015, it is still growing.

Because of the size and agricultural importance of the Mississippi River basin, there are no easy solutions to the severe cultural eutrophication of this overfertilized coastal zone. Preventive measures include applying less fertilizer, injecting fertilizer below the soil surface, using controlled-release fertilizers that have water-insoluble coatings, planting strips of forests and grasslands along waterways to soak up excess nitrogen, and restoring and creating wetlands between crop fields and streams emptying into the Mississippi River.

Other measures involve improving flood control, to prevent the release of nitrogen from floodplains during major floods, and upgrading sewage treatment to reduce discharges of nitrates into waterways. In addition, deposition of nitrogen compounds from the atmosphere could be reduced by requiring lower emissions of nitrogen oxides from motor vehicles and phasing in forms of renewable energy to replace the burning of fossil fuels.

Some scientists who have studied this problem fear that it could reach a tipping point where many of the organisms living in this part of the gulf simply can no longer move far enough away to avoid the oxygen-depletion or to enable their populations to recover.

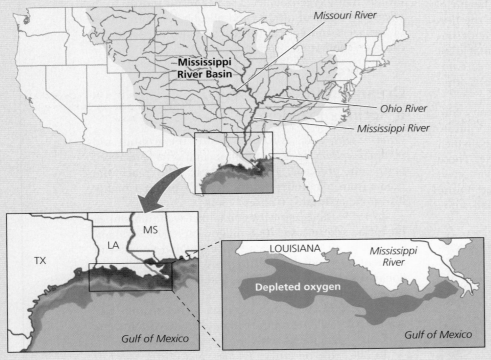

Figure 20-B **Natural capital degradation:** a large zone of oxygen-depleted water (containing less than 2 ppm dissolved oxygen) forms each year during the spring and summer in the Gulf of Mexico as a result of oxygen-depleting algal blooms. Evidence indicates that it is created mostly by huge inputs of nitrate (NO_3^-) plant nutrients from farms, cities, factories, and sewage treatment plants in the vast Mississippi River basin. The drawing (bottom left) based on a satellite image, shows the inputs of such nutrients into the Gulf of Mexico during the summer of 2006. In the image, reds and greens represent high concentrations of phytoplankton and river sediment. This problem was worsened by losses of wetlands, which would have filtered out some of these plant nutrients. **Question:** Can you think of a product you used today that was directly connected to this sort of pollution? (NASA)

Critical Thinking

How do you think each of the preventive measures described above would help to prevent pollution in the Gulf of Mexico? Can you think of other possible preventive solutions?

environment from oil tanker accidents has decreased 75% and oil discharges from industry and cities have dropped by nearly 90%.

Volatile organic hydrocarbons in oil immediately kill many aquatic organisms, especially in their vulnerable larval forms. Other chemicals in oil form tar-like globs that float on the surface and coat the feathers of birds (especially diving birds) and the fur of marine mammals. This oil coating destroys their natural heat insulation and buoyancy, causing many of them to drown or die of exposure from loss of body heat.

Heavy oil components that sink to the ocean floor or wash into estuaries can smother bottom-dwelling organisms such as crabs, oysters, mussels, and clams,

or make them unfit for human consumption. Some oil spills have killed coral reefs. See *The Habitable Planet,* Video 9, at **www.learner.org/resources/series209 .html** for discussion of how scientists measure the effects of oil spills on coral reefs and fish populations.

Research shows that populations of many forms of marine life recover from exposure to large amounts of *crude oil* within about 3 years. But recovery from exposure to *refined oil,* especially in estuaries and salt marshes, can take 10–20 years. Oil slicks that wash onto beaches can have a serious economic impact on coastal residents, who lose income normally gained from fishing and tourist activities. In 2006, some 17 years after the *Exxon Valdez* spill, researchers found patches of oil remaining on some parts of the shoreline of Prince William Sound. These areas continue to be hazardous for sea otters, shore birds, and other wildlife.

If they are not too large, oil spills can be partially cleaned up by mechanical means including floating booms, skimmer boats, and absorbent devices such as large pillows filled with feathers or hair. Chemical, fire, and natural methods, such as using bacteria to speed up oil decomposition, are also used.

But scientists estimate that current cleanup methods can recover no more than 15% of the oil from a major spill. Thus, *preventing* oil pollution is the most effective and, in the long run, the least costly approach (**Concept 20-4B**). One of the best ways to prevent tanker spills is to use only oil tankers with double hulls. After the *Exxon Valdez* accident, oil companies promised that they would do so. But 19 years later, in 2008, about half of the world's 10,000 oil tankers still had the older and more vulnerable single hulls.

Figure 20-17 lists ways to prevent and reduce pollution of coastal waters. The key to protecting the oceans is to reduce the flow of pollution from land and air and from streams emptying into these waters

SOLUTIONS

Coastal Water Pollution

Prevention

Reduce input of toxic pollutants

Separate sewage and storm lines

Ban dumping of wastes and sewage by ships in coastal waters

Ban ocean dumping of sludge and hazardous dredged material

Regulate coastal development, oil drilling, and oil shipping

Require double hulls for oil tankers

Cleanup

Improve oil-spill cleanup capabilities

Use nanoparticles on sewage and oil spills to dissolve the oil or sewage (still under development)

Require secondary treatment of coastal sewage

Use wetlands, solar-aquatic, or other methods to treat sewage

Figure 20-17 Methods for preventing and cleaning up excessive pollution of coastal waters (**Concept 20-4B**). **Question:** Which two of these solutions do you think are the most important? Why?

(**Concept 20-4B**). Thus, ocean pollution control must be linked with land-use and air pollution control policies, which in turn are linked to energy policies (Figure 16-33, p. 432) and climate policies (Figures 19-13 and 19-14, p. 515).

20-5 How Can We Best Deal with Water Pollution?

▶ **CONCEPT 20-5** Reducing water pollution requires preventing it, working with nature to treat sewage, cutting resource use and waste, reducing poverty, and slowing population growth.

We Need to Reduce Surface Water Pollution from Nonpoint Sources

There are a number of ways to reduce nonpoint-source water pollution, most of which comes from agriculture. Farmers can reduce soil erosion by keeping cropland covered with vegetation. They can also reduce the amount of fertilizer that runs off into surface waters and leaches into aquifers by using slow-release fertilizer, using no fertilizer on steeply sloped land,

and planting buffer zones of vegetation between cultivated fields and nearby surface water. See *The Habitable Planet,* Video 7, at **www.learner.org/resources/ series209.html** to learn how scientists have reduced excessive nitrogen runoff from fertilizers to decrease their harmful impacts on aquatic systems.

Organic farming techniques also offer ways to prevent water pollution. For example, organic farmers use manure for fertilizer, in which nitrogen is contained within organic matter that clings to the soil.

Industrialized agriculture applies fertilizer as granules to cropland, which can more easily wash into streams.

Applying pesticides only when needed and relying more on integrated pest management (p. 300) can reduce pesticide runoff. Farmers can control runoff and infiltration of manure from animal feedlots by planting buffers and locating feedlots and animal waste sites away from steeply sloped land, surface water, and flood zones.

Tougher pollution control regulations for U.S. livestock operations are spurring scientists to come up with better ways to deal with animal waste. They are exploring techniques for converting it to natural gas, recycling undigested nutrients in manure back into animal feed, and extracting valuable chemicals from manure to make plastics or even cosmetics. For example, Smithfield Foods, a large pork producer, plans to build a facility in the state of Utah to convert the wastes from 500,000 hogs into renewable biodiesel fuel for vehicles.

> ── **HOW WOULD YOU VOTE?** ☑ ──
>
> Should we greatly increase efforts to reduce water pollution from nonpoint sources even though this could be quite costly? Cast your vote online at **academic.cengage.com/biology/miller**.

Laws Can Help Reduce Water Pollution from Point Sources

The Federal Water Pollution Control Act of 1972 (renamed the Clean Water Act when it was amended in 1977) and the 1987 Water Quality Act form the basis of U.S. efforts to control pollution of the country's surface waters. (See Case Study at right.) The Clean Water Act sets standards for allowed levels of key water pollutants and requires polluters to get permits limiting how much of various pollutants they can discharge into aquatic systems.

The EPA is experimenting with a *discharge trading policy,* which uses market forces to reduce water pollution (as has been done with sulfur dioxide for air pollution control, p. 490) in the United States. Under this program, a permit holder can pollute at higher levels than allowed in its permit if it buys credits from permit holders who are polluting below their allowed levels.

Environmental scientists warn that the effectiveness of such a system depends on how low the cap on total pollution levels in any given area is set, along with how regularly the cap is lowered. They also warn that discharge trading could allow pollutants to build up to dangerous levels in areas where credits are bought. They call for careful scrutiny of the cap levels and for gradual lowering of the caps to encourage prevention of water pollution and development of better pollution control technology. Neither adequate scrutiny of the cap levels nor gradual lowering of caps is a part of the current EPA discharge trading system.

■ **CASE STUDY**

The U.S. Experience with Reducing Point-Source Pollution

According to the EPA, the Clean Water Act of 1972 led to numerous improvements in U.S. water quality. Between 1992 and 2002 (the latest figures available):

- The number of Americans served by community water systems that met federal health standards increased from 79% to 94%.

- The percentage of U.S. stream lengths found to be fishable and swimmable increased from 36% to 60% of those tested.

- The amount of topsoil lost through agricultural runoff was cut by about 1.1 billion metric tons (1 billion tons) annually.

- The proportion of the U.S. population served by sewage treatment plants increased from 32% to 74%.

- Annual wetland losses decreased by 80%.

These are impressive achievements given the increases in the U.S. population and per capita consumption of water and other resources since 1972. But there is more work to be done. In 2006, the EPA, found that 45% of the country's lakes and 40% of the streams surveyed were still too polluted for swimming or fishing, and that runoff of animal wastes from hog, poultry, and cattle feedlots and meat processing facilities pollutes seven of every ten U.S. rivers.

Even where sewage treatment plants are in place, treated wastewater can serve as a nutrient leading to algal blooms as it did in Lake Washington (**Core** CORE CASE STUDY **Case Study**). Population growth and increasing levels of resource use and waste can overwhelm these sewage treatment systems.

Fish caught in more than 1,400 different waterways and more than a fourth of the nation's lakes are unsafe to eat because of high levels of pesticides, mercury, and other toxic substances. Also, a 2003 study by the EPA found that at least half of the country's 6,600 largest industrial facilities and municipal wastewater treatment plants have illegally discharged toxic or biological wastes into waterways for years without government enforcement actions or fines. And according to a 2007 study by the U.S. Public Interest Research Group, more than half of the country's industrial and wastewater facilities exceeded the limits of their Clean Water Act pollution permits one or more times in 2005, with the average facility discharging close to four times its legal limit of water pollutants.

Finally, the U.S. government reported in 2007 that tens of thousands of gasoline storage tanks are leaking (Figure 20-12), possibly affecting groundwater, and cleanup of most of these leaks has yet to begin. In 43 states, these numbers are expected to increase by thousands before 2020. The estimated cost of cleanup is

$12 billion, but the government in 2007 was planning to spend about 3% of that amount on the problem.

Some environmental scientists call for strengthening the Clean Water Act. Suggested improvements include shifting the emphasis to water pollution prevention instead of focusing mostly on end-of-pipe removal of specific pollutants. One EPA inspector general's report called for increased funding and authority to control nonpoint sources of pollution, greatly increased monitoring for compliance with the law, much larger mandatory fines for violators, and stronger programs to prevent and control toxic water pollution.

Other suggestions include providing more funding and authority for integrated watershed and airshed planning to protect groundwater and surface water from contamination, and regulating irrigation water quality, for which there is no federal regulation. Another suggestion is to expand the rights of citizens to bring lawsuits to ensure that water pollution laws are enforced. Studies have shown that many violators of federal water pollution standards receive no fines or only small ones. The National Academy of Sciences has called for prohibiting the destruction of wetlands, instituting higher standards for wetland restoration (Figure 11-13, p. 267), and requiring that before any natural wetland is destroyed, a new one be created.

Many people oppose these proposals, contending that the Clean Water Act's regulations and government wetlands regulations are already too restrictive and costly. Farmers and developers see these laws as limitations on their rights as property owners to fill in wetlands. They also believe they should be compensated for any property value losses due to federal wetland protection.

In 2006, the EPA weakened its application of the Clean Water Act by ruling that pesticides can be applied over or near bodies of water without a permit, if the application is needed to control aquatic weeds, mosquitoes or other pests. Critics fear that this ruling will result in more toxic pollutants winding up in streams and lakes. Funding cuts have also weakened the act.

Also among critics of the Clean Water Act are some state and local officials who want more discretion in testing for and meeting water quality standards. They argue that in many communities, it is unnecessary and too expensive to test for all the water pollutants as required by federal law. In this and other areas, federal laws require states and localities to spend a great deal of money to meet federal water pollution standards without reimbursing them for most of the costs from federal tax revenues. Many local officials object to these so-called "unfunded mandates" from Congress. Many small cities cannot afford such costs.

┌─ **HOW WOULD YOU VOTE?** ✓ ─────
│ Should the U.S. Clean Water Act be strengthened? Cast
│ your vote online at **academic.cengage.com/biology/**
│ **miller**.
└──────────────────────────────────

Sewage Treatment Reduces Water Pollution

In rural and suburban areas with suitable soils, sewage from each house usually is discharged into a **septic tank** with a large drainage field (Figure 20-18). In this system, household sewage and wastewater is pumped into a settling tank, where grease and oil rise to the top and solids fall to the bottom and are decomposed by bacteria. The resulting partially treated wastewater is discharged in a large drainage (absorption) field through small holes in perforated pipes embedded in porous gravel or crushed stone just below the soil's surface. As these wastes drain from the pipes and percolate downward, the soil filters out some potential pollutants and soil bacteria decompose biodegradable materials.

Every few years, when the settling tank becomes full, it must be pumped out into a tank truck and taken to a municipal sewage treatment plant for proper disposal. About one-fourth of all homes in the United States are served by septic tanks.

Manhole cover (for cleanout)

Gas

Scum

Wastewater

Sludge

Septic tank

Distribution box

Drain field (gravel or crushed stone)

Vent pipe

Perforated pipe

Figure 20-18
Solutions: *septic tank system* used for disposal of domestic sewage and wastewater in rural and suburban areas.

Figure 20-19 Solutions: primary and secondary sewage treatment.
Question: What do you think should be done with the sludge produced by sewage treatment plants?

Labels in figure:
Primary
Secondary
Bar screen
Grit chamber
Settling tank
Aeration tank
Settling tank
Chlorine disinfection tank
Raw sewage from sewers
Sludge
Activated sludge
Air pump
Sludge digester
To river, lake, or ocean
(kills bacteria)
Sludge drying bed
Disposed of in landfill or ocean or applied to cropland, pasture, or rangeland

If these systems are not installed correctly or maintained properly, they can cause sewage to backup into homes or to pollute nearby groundwater and surface water. Chlorine bleaches, drain cleaners, and antibacterial soaps should not be used in these systems, because they can kill the bacteria that decompose the wastes. Kitchen sink garbage disposals should not be used either, because they can overload septic systems.

In urban areas in the United States and most developed countries, most waterborne wastes from homes, businesses, and storm runoff flow through a network of sewer pipes to *wastewater* or *sewage treatment plants*. Raw sewage reaching a treatment plant typically undergoes one or two levels of wastewater treatment. The first is **primary sewage treatment**—a *physical* process that uses screens and a grit tank to remove large floating objects and to allow solids such as sand and rock to settle out. Then the waste stream flows into a primary settling tank where suspended solids settle out as sludge (Figure 20-19, left). By itself, primary treatment removes about 60% of the suspended solids and 30–40% of the oxygen-demanding organic wastes from sewage. It removes no pathogens, phosphates, nitrates, salts, radioisotopes, or pesticides.

The second level is **secondary sewage treatment**—a *biological* process in which aerobic bacteria remove as much as 90% of dissolved and biodegradable, oxygen-demanding organic wastes (Figure 20-19, right). A combination of primary and secondary treatment removes 95–97% of the suspended solids and oxygen-demanding organic wastes, 70% of most toxic metal compounds and nonpersistent synthetic organic chemicals, 70% of the phosphorus, and 50% of the nitrogen. But this process removes only a tiny fraction of

long-lived radioactive isotopes and persistent organic substances such as some pesticides, and it does not kill pathogens.

A third level of cleanup, *advanced* or *tertiary sewage treatment*, uses a series of specialized chemical and physical processes to remove specific pollutants left in the water after primary and secondary treatment. Its most common form makes use of special filters to remove phosphates and nitrates from wastewater before it is discharged into surface waters to help reduce nutrient overload from nitrates and phosphates. Because of its high costs it is not widely used.

Before discharge, water from sewage treatment plants usually undergoes *bleaching* to remove water coloration and *disinfection* to kill disease-carrying bacteria and some (but not all) viruses. The usual method for accomplishing this is *chlorination*. But chlorine can react with organic materials in water to form small amounts of chlorinated hydrocarbons. Some of these chemicals cause cancers in test animals, can increase the risk of miscarriages, and may damage the human nervous, immune, and endocrine systems. Use of other disinfectants, such as ozone and ultraviolet light, is increasing, but they cost more and their effects do not last as long as those of chlorination.

Officials in Peru decided to stop chlorinating the country's drinking water because of concern over the increased risk of cancer from organic compounds that form during chlorination. However, the country resumed chlorinating its drinking water after a 1991 cholera outbreak infected more than 300,000 people and caused at least 3,500 deaths.

U.S. federal law requires primary and secondary treatment for all municipal sewage treatment plants,

but exemptions from secondary treatment are possible when the cost of installing such treatment poses an excessive financial burden. And according to the EPA, at least two-thirds of the country's sewage treatment plants have sometimes violated water pollution regulations. Also, 500 cities have failed to meet federal standards for sewage treatment plants, and 34 East Coast cities simply screen out large floating objects from their sewage before discharging it into coastal waters.

Some cities have a separate network of pipes for carrying runoff of storm water from streets and parking lots. But 1,200 U.S. cities have combined the sewer lines for these two systems because it is cheaper. Heavy rains or too many users hooked up to such combined systems can cause them to overflow and discharge untreated sewage directly into surface waters such as the Great Lakes (Case Study, p. 540). According to the EPA, at least 40,000 such overflows occur each year in the United States.

The EPA estimates that each year, 7.1 million people get sick from swimming in waters contaminated by sewage overflows and storm-water runoff. These numbers may increase. In 2005, the EPA authorized sewage treatment plants to bypass secondary treatment, blend the partially treated sewage with fully treated wastewater, and dump the mixture into waterways anytime it rains or snows. Before 2005, they could do this only under the most extreme emergency circumstances, such as during hurricanes and tropical storms.

Research by health and environmental scientists indicates that the health risks from swimming in waters containing blended sewage wastes will be 100 times greater than they would be if the wastewater were fully treated. They say that this new policy may save sewage treatment plants money, but it will cause more illnesses, close more beaches, kill more fish, destroy more shellfish beds, and hurt the fishing and tourism industries.

THINKING ABOUT
Sewage Treatment

Should the EPA rule allowing U.S. sewage treatment plants to blend partially treated sewage with fully treated sewage and to dump it into waterways anytime it rains or snows be overturned? Explain.

We Can Improve Conventional Sewage Treatment

Environmental scientist Peter Montague calls for redesigning the conventional sewage treatment system shown in Figure 20-19. The idea is to prevent toxic and hazardous chemicals from reaching sewage treatment plants and thus from getting into sludge and water discharged from such plants (**Concept 1-4**, p. 16).

CONCEPT LINK

Montague suggests several ways to do this. One is to require industries and businesses to remove toxic and

hazardous wastes from water sent to municipal sewage treatment plants. Another is to encourage industries to reduce or eliminate use and waste of toxic chemicals.

HOW WOULD YOU VOTE?

Should we ban the discharge of toxic chemicals into pipes leading to sewage treatment plants? Cast your vote online at **academic.cengage.com/biology/miller**.

Another suggestion is to require or encourage more households, apartment buildings, and offices to eliminate sewage outputs by switching to waterless, odorless *composting toilet systems*, to be installed, maintained, and managed by professionals. These systems, pioneered several decades ago in Sweden, convert nutrient-rich human fecal matter into a soil-like humus that can be used as a fertilizer supplement. About once a year, vendors collect the humus and sell it as a soil conditioner. This process returns plant nutrients in human waste to the soil and thus mimics the natural chemical cycling **principle of sustainability**. It also reduces the need for energy-intensive commercial inorganic fertilizers.

Such systems would be cheaper to install and maintain than current sewage systems are, because they do not require vast systems of underground pipes connected to centralized sewage treatment plants. They also save large amounts of water, reduce water bills, and decrease the amount of energy used to pump and purify water. The EPA lists several brands of dry composting toilets approved for use in the United States. One of the authors of this book (Miller) used a composting toilet for over a decade with no problems in his office-home facility. This more environmentally sustainable replacement for the conventional toilet is now being used in more than a dozen countries, including China, India, Mexico, Syria, and South Africa.

Many communities are using unconventional, but highly effective, *wetland-based sewage treatment systems*, which work with nature (Science Focus, p. 556). This approach will become increasingly important, according to sustainability expert Lester Brown. In what Brown calls the conventional "flush and forget" water-based systems, we take nutrients originating in the soil and processed through our food, and we eventually dump them into streams, lakes, and oceans. These systems are expensive and water-intensive, and they spread disease and disrupt nutrient cycling.

Wetland-based sewage treatment systems would be ideal in some developing countries where water is short and disease is rampant. They would also be useful in areas where valuable wetlands have been destroyed, such as on the U.S. Gulf Coast. In these areas, such wetland systems could replace devastated coastal marshlands, which were economically and ecologically vital natural resources.

Oysters, mussels, clams, scallops, and other shellfish consume algae and help to filter out and reduce

Treating Sewage by Working with Nature

Some communities and individuals are seeking better ways to purify sewage by working with nature (**Concept 20-5**). Biologist John Todd has developed an ecological approach to treating sewage, which he calls *living machines* (Figure 20-C).

This purification process begins when sewage flows into a passive solar greenhouse or outdoor site containing rows of large open tanks populated by an increasingly complex series of organisms. In the first set of tanks, algae and microorganisms decompose organic wastes, with sunlight speeding up the process. Water hyacinths, cattails, bulrushes, and other aquatic plants growing in the tanks take up the resulting nutrients. After flowing though several of these natural purification tanks, the water passes through an artificial marsh of sand, gravel, and bulrushes, which filters out algae and remaining organic waste. Some of the plants also absorb (sequester) toxic metals such as lead and mercury and secrete natural antibiotic compounds that kill pathogens.

Next, the water flows into aquarium tanks, where snails and zooplankton consume microorganisms and are in turn consumed by crayfish, tilapia, and other fish that can be eaten or sold as bait. After 10 days, the clear water flows into a second artificial marsh for final filtering and cleansing. The water can be made pure enough to drink by using ultraviolet light or by passing the water through an ozone generator, usually immersed out of sight in an attractive pond or wetland habitat. Operating costs are about the same as those of a conventional sewage treatment plant.

Ocean Arks International

More than 800 cities and towns around the world and 150 in the United States (including West Palm Beach, Florida, and Phoenix, Arizona) use natural or artificially created wetlands to treat sewage as a lower-cost alternative to expensive waste treatment plants. For example, Arcata, California—a coastal town of 16,000 people—created some 65 hectares (160 acres) of wetlands between the town and the adjacent Humboldt Bay. The marshes and ponds, developed on land that was once a dump, act as a natural waste treatment plant. The project cost was

Figure 20-C Solutions: ecological wastewater purification by a *living machine*. At the Providence, Rhode Island, Solar Sewage Treatment Plant, biologist John Todd demonstrates an ecological process he invented for purifying wastewater by using the sun and a series of tanks containing living organisms. Todd and others are conducting research to perfect such solar-aquatic sewage treatment systems based on working with nature.

less than half the estimated price of a conventional treatment plant.

This system returns purified water to Humboldt Bay, and the sludge that is removed is processed for use as fertilizer. The marshes and ponds also serve as an Audubon Society bird sanctuary, which provides habitats for thousands of otters, seabirds, and marine animals. The town even celebrates its natural sewage treatment system with an annual "Flush with Pride" festival.

This approach and the living machine system developed by John Todd apply three of the four **scientific principles of sustainability**: using solar energy, using natural processes to remove and recycle nutrients and other chemicals, and relying on a diversity of organisms and natural processes.

Critical Thinking

Can you think of any disadvantages of using such a nature-based system instead of a conventional sewage treatment plant? Do you think any such disadvantages outweigh the advantages? Why or why not?

excessive inputs of nitrogen and phosphorus nutrients that cause oxygen-depleting algal blooms in coastal waters. For this reason, some coastal communities use shellfish farms to help remove excessive plant nutrients from coastal waterways. Studies show that using this approach is less expensive than using sewage treatment plants for this purpose.

There Are Sustainable Ways to Reduce and Prevent Water Pollution

It is encouraging that, since 1970, most developed countries have enacted laws and regulations that have significantly reduced point-source water pollution. These improvements were largely the result of

bottom-up political pressure on elected officials by individuals and groups.

Conversely, little has been done to reduce water pollution in most developing countries. However, within a decade, China plans to provide all of its cities with small sewage treatment plants that will cleanse wastewater enough to be recycled back into the urban water supply systems. If China is successful in this ambitious plan, it could become a world leader in developing systems that tackle both pollution and water scarcity by recycling water, in keeping with the nutrient cycling **principle of sustainability**. Still, technological solutions can be temporary if they become overwhelmed by increased population growth and resource use, as the **Core Case Study** on Lake Washington illustrates.

Figure 20-20 Methods for preventing and reducing water pollution (**Concept 20-5**). **Question:** Which two of these solutions do you think are the most important? Why?

To environmental and health scientists, the next step is to increase efforts to reduce and prevent water pollution in developed and developing countries. They would begin by asking the question: *How can we avoid producing water pollutants in the first place?* (**Concept 20-5**).

Figure 20-20 lists ways to achieve this goal over the next several decades.

This shift to pollution prevention will not take place unless citizens put political pressure on elected officials. Also, developing countries will need financial and technical aid from developed countries. Finally, the daily choices of each and every individual will help to determine whether the shift to pollution prevention can be achieved. Figure 20-21 lists some actions you can take to help reduce water pollution.

Figure 20-21 Individuals matter: ways to help reduce water pollution. **Question:** Which three of these actions do you think are the most important? Why?

REVISITING

Lake Washington and Sustainability

The story of Lake Washington (**Core Case Study**) is an example of people loving a natural resource and abusing it at the same time. For many years, Seattle residents assumed the lake could easily absorb their treated sewage and remain a good place for swimming and boating. Eventually, with the help of scientific inquiry and investigation, they learned that the numbers of people using it could overwhelm this natural system. A technical solution was found for dealing with sewage treatment effluent, but now continually increasing pressures due to a growing population, are again overwhelming the natural systems of Lake Washington and Puget Sound.

This story is instructive for dealing with pollution problems in other developed countries and in rapidly growing developing countries. Pollution control for the world's water supplies is within

our reach. But even more hopeful is the possibility of shifting our emphasis from cleaning up water pollution to reducing and preventing it.

The four **scientific principles of sustainability** (back cover) can guide us in reducing and preventing pollution. We can use solar energy to purify the water we use. Recycling more water will help us to reduce water waste, and natural nutrient cycles can be used to treat our waste in wetland-based sewage treatment systems (Science Focus, p. 556). Preserving biodiversity by avoiding disruption of aquatic systems and their bordering terrestrial systems, which in turn help to reduce pollution, is a key factor in maintaining water supplies and water quality. And controlling human population growth and levels of resource use and waste is fundamental to maintaining water quality.

It is a hard truth to swallow,
but nature does not care if we live or die.
We cannot survive without the oceans,
for example, but they can do just fine without us.

ROGER ROSENBLATT

1. Review the Key Questions and Concepts for this chapter on p. 532. Describe the cleanup of Lake Washington near Seattle (**Core Case Study**) and list the three lessons learned from this process. CORE CASE STUDY

2. What is **water pollution?** Distinguish between **point sources** and **nonpoint sources** of water pollution and give an example of each. List nine major types of water pollutants and give an example of each. List three diseases transmitted to humans by polluted water. Describe chemical and biological methods that scientists use to measure water quality.

3. Describe how streams can cleanse themselves and how these cleansing processes can be overwhelmed. Describe the state of stream pollution in developed and developing countries. Describe the pollution problems of the Ganges River, which runs through part of India.

4. Give two reasons why lakes cannot cleanse themselves as readily as streams can. Distinguish between **eutrophication** and **cultural eutrophication.** List three ways to prevent or reduce cultural eutrophication. Describe pollution of the Great Lakes and the progress made in reducing this pollution.

5. Explain why groundwater cannot cleanse itself very well. What are the major sources of groundwater contamination in the United States? Describe the threat from arsenic in groundwater. List ways to prevent or clean up groundwater contamination.

6. Describe U.S. laws for protecting drinking water quality. Describe the environmental problems caused by the widespread use of bottled water.

7. How are coastal waters and deeper ocean waters polluted? What causes harmful algal blooms and what are their harmful effects? Describe oxygen depletion in the northern Gulf of Mexico. How serious is oil pollution of the oceans, what are its effects, and what can be done to reduce such pollution?

8. List two ways to reduce water pollution from (**a**) nonpoint sources and (**b**) point sources. Describe the U.S. experience with reducing point-source water pollution. What is a **septic tank** and how does it work? Describe how **primary sewage treatment** and **secondary sewage treatment** are used to help purify water.

9. How would Peter Montague improve conventional sewage treatment? What is a composting toilet system? Describe how wetlands can be used to treat sewage. Describe the use of living machines to treat sewage. List six ways to prevent and reduce water pollution. List five steps you can take to reduce water pollution.

10. Describe connections between the cleanup of Lake Washington (**Core Case Study**) and the four **scientific principles of sustainability**. CORE CASE STUDY SUSTAINABILITY

Note: Key Terms are in **bold** type.

CRITICAL THINKING

1. What were two important roles played by the scientists who studied Lake Washington as discussed in the **Core Case Study** that opens this chapter? Explain how the story might have been different if the scientists had not fulfilled each of these roles. CORE CASE STUDY

2. Lake Washington and Puget Sound now face new problems similar to those of the past, as suggested in the **Core Case Study**. Describe the nature of those problems and suggest possible solutions. CORE CASE STUDY

3. A large number of dead fish are found floating in a lake. How would you determine whether they died from cultural eutrophication or from exposure to toxic chemicals?

4. If you were a regulator charged with drawing up plans for controlling water pollution, briefly describe one idea for controlling water pollution from each of the following sources: (**a**) an effluent pipe from a factory going into a stream, (**b**) a parking lot at a shopping mall bordered by a stream, (**c**) a farmer's field on a slope next to a stream.

5. What role does population growth play in (**a**) groundwater pollution problems and (**b**) coastal water pollution problems?

6. When you flush your toilet, where does the wastewater go? Trace the actual flow of this water in your community from your toilet through sewers to a wastewater treatment plant and from there to the environment. Try to visit a local sewage treatment plant to see what it does with your wastewater. Compare the processes it uses with those shown in Figure 20-19. What happens to the sludge produced by this plant? What improvements, if any, would you suggest for this plant?

7. In your community,
 a. What are the principal nonpoint sources of contamination of surface water and groundwater?
 b. What is the source of drinking water?
 c. How is drinking water treated?
 d. How many times during each of the past 5 years have levels of tested contaminants violated federal standards? Were violations reported to the public?
 e. What problems related to drinking water, if any, have arisen in your community? What actions, if any, has your local government taken to solve such problems?
 f. Is groundwater contamination a problem? If so, where, and what has been done about the problem?

g. Is there a vulnerable aquifer or critical recharge zone that needs protection to ensure the quality of ground-water? Is your local government aware of this? What action (if any) has it taken?

8. List three ways in which you could apply **Concept 20-5** to make your lifestyle more environmentally sustainable.

9. Congratulations! You are in charge of the world. What are three actions you would take to (a) sharply reduce point-source water pollution in developed countries, (b) sharply reduce nonpoint-source water pollution throughout the world, (c) sharply reduce groundwater pollution throughout the world, and (d) provide safe drinking water for the poor and for other people in developing countries?

10. List two questions that you would like to have answered as a result of reading this chapter.

Note: See Supplement 13 (p. S78) for a list of Projects related to this chapter.

DATA ANALYSIS

You have inherited the family cabin on the southerly coast of Maine overlooking the Atlantic Ocean where your source of drinking water is a well.

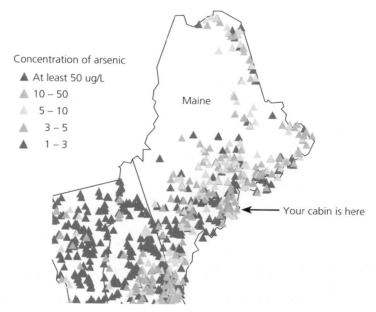

Concentration of arsenic
▲ At least 50 ug/L
▲ 10 – 50
▲ 5 – 10
▲ 3 – 5
▲ 1 – 3

Maine

Your cabin is here

1. You learn that your drinking water could be contaminated with unsafe levels of arsenic. Referring to the above map, what range of arsenic levels can you expect to find in your well water?

2. The EPA requires all public water utilities to have no more than 10 micrograms of arsenic per liter of water (10 ug/L) in drinking water. Although domestic wells are not regulated, you must comply if you decide to sell your family cabin. Does your well comply? Explain.

3. To reduce your exposure to arsenic, you could install a reverse osmosis treatment system. It is 95% effective and can produce 11 liters (3.0 gallons) of water a day. Assuming your water contains 50 micrograms of arsenic per liter, what would the arsenic concentration be after the reverse osmosis treatment? Does this comply with the EPA-accepted safe level?

LEARNING ONLINE

Log on to the Student Companion Site for this book at **academic.cengage.com/biology/miller**, and choose Chapter 20 for many study aids and ideas for further reading and research. These include flash cards, practice quizzing, Weblinks, information on Green Careers, and InfoTrac® College Edition articles.

CORE CASE STUDY E-waste—an Exploding Problem

Electronic waste, or *e-waste*, consists of discarded television sets, cell phones, computers, e-toys, and other electronic devices (Figure 21-1). It is the fastest-growing solid waste problem in the United States and in the world. Each year, Americans discard an estimated 155 million cell phones, 48 million personal computers, and many more millions of television sets, iPods, Blackberries, and other electronic products.

Most e-waste ends up in landfills and incinerators. It includes high-quality plastics and valuable metals such as aluminum, cop-

per, nickel, platinum, silver, and gold. E-waste is also a source of toxic and hazardous pollutants, including polyvinylchloride (PVC), brominated flame retardants, lead, and mercury, which can contaminate air, surface water, groundwater, and soil and cause serious health problems and even early death for e-waste workers.

According to a 2005 report by the U.N.-sponsored Basel Action Network, about 70% of the world's e-waste is shipped to China, while most of the rest goes to India and poor African nations where labor is cheap and environmental regulations are weak. Workers there—many of them children—dismantle such products to recover valuable metals and reusable parts. As they do this, they are exposed to toxic metals and other harmful chemicals. The remaining scrap is dumped in waterways and fields or burned in open fires, exposing many people to toxic dioxins.

Transferring hazardous waste from developed to developing countries is banned by the International Basel Convention. Even so, much e-waste is not classified as hazardous waste or is illegally smuggled to countries such as China. The United States can export this waste legally because it is one of only three countries that have not ratified the Basel Convention (the other two are Afghanistan and Haiti).

The European Union (EU) has led the way in dealing with e-waste. Its *cradle-to-grave* approach requires manufacturers to take back electronic products at the ends of their useful lives for repair, remanufacture, or recycling, and e-waste is banned from landfills and incinerators. Japan is also adopting cradle-to-grave standards for electronic devices and appliances.

The United States produces roughly half of the world's e-waste and recycles only about 10–15% of it; but that is beginning to change. Massachusetts and five other states have banned the disposal of computers and TV sets in landfills and incinerators. Some electronics manufacturers, including Apple, Intel, Hewlett-Packard, Dell, Sharp, Panasonic, and Sony, have free recycling programs for consumers, which are described on their websites. Some manufacturers will arrange for pickups or pay shipping costs. A growing consumer awareness of the problem has spawned highly profitable *e-cycling* businesses. In addition, nonprofit groups, such as Free Geek in Portland, Oregon, are motivating many people to donate, recycle, and reuse old electronic devices.

But recycling and reuse probably will not keep up with the explosive growth of e-waste. According to Jim Puckett, coordinator of the Basel Action Network, the only real long-term solution is a *prevention* approach that gets toxic materials out of electrical and electronic products through green design. Electronic waste is just one of many types of solid and hazardous waste discussed in this chapter.

© Fotopic/Miles Simmons/PhototakeUSA

Figure 21-1 *Rapidly growing electronic waste* (e-waste) from discarded computers and other electronic devices represents a waste of resources and pollutes the air, water, and land with harmful compounds. **Question:** Have you disposed of an electronic device lately? If so, how did you dispose of it?

21-1 What are solid waste and hazardous waste, and why are they problems?

CONCEPT 21-1 Solid waste represents pollution and unnecessary waste of resources, and hazardous waste contributes to pollution, natural capital degradation, health problems, and premature deaths.

21-2 How should we deal with solid waste?

CONCEPT 21-2 A sustainable approach to solid waste is first to reduce it, then to reuse or recycle it, and finally to safely dispose of what is left.

21-3 Why is reusing and recycling materials so important?

CONCEPT 21-3 Reusing items decreases the use of matter and energy resources and reduces pollution and natural capital degradation; recycling does so to a lesser degree.

21-4 What are the advantages and disadvantages of burning or burying solid waste?

CONCEPT 21-4 Technologies for burning and burying solid wastes are well developed, but burning contributes to pollution and greenhouse gas emissions, and buried wastes eventually contribute to pollution and land degradation.

21-5 How should we deal with hazardous waste?

CONCEPT 21-5 A sustainable approach to hazardous waste is first to produce less of it, then to reuse or recycle it, then to convert it to less hazardous materials, and finally, to safely store what is left.

21-6 How can we make the transition to a more sustainable low–waste society?

CONCEPT 21-6 Shifting to a low-waste society requires individuals and businesses to reduce resource use and to reuse and recycle wastes at local, national, and global levels.

Note: Supplements 2 (p. S4), 3 (p. S10), 6 (p. S39) and 13 (p. S78) can be used with this chapter.

Solid wastes are only raw materials we're too stupid to use.

ARTHUR C. CLARKE

21-1 What Are Solid Waste and Hazardous Waste, and Why Are They Problems?

▶ **CONCEPT 21-1** Solid waste represents pollution and unnecessary waste of resources, and hazardous waste contributes to pollution, natural capital degradation, health problems, and premature deaths.

We Throw Away Huge Amounts of Useful Things and Hazardous Materials

In nature, there is essentially no waste because the wastes of one organism become nutrients for others (Figure 3-14, p. 63, and **Concept 3-3B**, p. 57). This recycling of nutrients is the basis for one of the four scientific principles of sustainability (see back cover).

Humans, on the other hand, produce huge amounts of wastes that go unused and pollute the environment. Because of the law of conservation of matter (**Concept 2-3**, p. 39) and the nature of human lifestyles, we will always pro- duce some waste, but the amount can be drastically reduced.

One major category of waste is **solid waste**—any unwanted or discarded material we produce that is not a liquid or a gas. Solid waste can be divided into two types. One type is **industrial solid waste** pro- duced by mines, agriculture, and industries that supply people with goods and services. The other is **munici- pal solid waste (MSW),** often called *garbage* or *trash,* which consists of the combined solid waste produced by homes and workplaces. Examples include paper and cardboard, food wastes, cans, bottles, yard wastes, furniture, plastics, metals, glass, wood, and e-waste (**Core Case Study**). In developed countries, most MSW is buried in landfills or burned in

What Harmful Chemicals Are in Your Home?

Cleaning
- Disinfectants
- Drain, toilet, and window cleaners
- Spot removers
- Septic tank cleaners

Paint Products
- Paints, stains, varnishes, and lacquers
- Paint thinners, solvents, and strippers
- Wood preservatives
- Artist paints and inks

General
- Dry-cell batteries (mercury and cadmium)
- Glues and cements

Gardening
- Pesticides
- Weed killers
- Ant and rodent killers
- Flea powders

Automotive
- Gasoline
- Used motor oil
- Antifreeze
- Battery acid
- Brake and transmission fluid

Figure 21-2 Harmful chemicals found in many homes. The U.S. Congress has exempted disposal of these materials from government regulation. **Question:** Which of these chemicals are in your home?

incinerators. In many developing countries, much of it ends up in open dumps, where poor people eke out a living finding items they can sell for reuse or recycling (Figure 1-6, p. 11).

Another major category of waste is **hazardous,** or **toxic, waste,** which threatens human health or the environment because it is poisonous, dangerously chemically reactive, corrosive, or flammable. Examples include industrial solvents, hospital medical waste, car batteries (containing lead and acids), household pesticide products, dry-cell batteries (containing mercury and cadmium), and incinerator ash. Figure 21-2 lists some of the harmful chemicals found in many homes. The two largest classes of hazardous wastes are *organic compounds* (such as various solvents, pesticides, PCBs, and dioxins) and nondegradable *toxic heavy metals* (such as lead, mercury, and arsenic).

Another form of extremely hazardous waste is highly radioactive waste produced by nuclear power plants (p. 392) and nuclear weapons facilities. Such wastes must be stored safely for 10,000 to 240,000 years depending on what radioactive isotopes are present. After 60 years of research, there is still considerable scientific disagreement over how to store such dangerous wastes and political controversy over where to store them (Case Study, p. 393).

According to the U.N. Environment Programme (UNEP), developed countries produce 80–90% of the world's hazardous wastes. The United States produces more of such wastes than any other country; chemical and mining industries and the military are the top three producers. As China continues to industrialize with inadequate pollution controls, it may take over the number one spot. Ways to deal with hazardous wastes are discussed later in this chapter in Section 21-5 (p. 577).

There are two reasons for sharply reducing the amount of solid and hazardous wastes we produce. One reason is that at least three-fourths of these materials represent an unnecessary waste of the earth's resources. Studies show that we could reduce resource use and reuse and recycle up to 90% of the MSW we produce, using existing technology and waste prevention, reuse, and recycling systems.

Instead we collect, mix, crush, and bury many of these potentially valuable resources in holes or landfills all over the planet. Once these materials are mixed, it is usually too expensive to recover them. Mixing trash also disperses hazardous materials with the rest of the trash and prohibits separating them out for safe disposal or recycling. We also burn hazardous wastes in incinerators, which pollutes the air and leaves a toxic ash that we usually have to bury. Another approach, sometimes misleadingly called pollution prevention or waste reduction, is for wealthy countries to move polluting industries and hazardous wastes to poor countries where environmental laws are weak or nonexistent.

A second reason for sharply reducing our output of solid waste is that, in producing the products we use

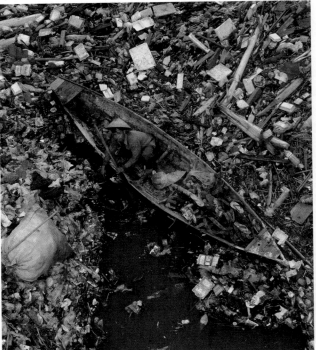

Figure 21-3 Natural capital degradation: solid wastes polluting a river in Jakarta, Indonesia, a city of more than 11 million people. The man in the boat is looking for items to salvage or sell.

Lineair Fotochief/Peter Arnold, Inc.

and often discard, we create huge amounts of air pollution, greenhouse gases, water pollution (Figure 21-3), and land degradation (**Concept 21-1**).

Solid Waste in the United States

The United States leads the world in producing solid waste. With only 4.6% of the world's population, the U.S. produces about one-third of the world's solid waste. About 98.5% of all solid waste produced in the United States is industrial solid waste from mining (76%), agriculture (13%), and industry (9.5%).

Suppose you buy a desktop computer. You may not know that manufacturing it required 700 or more different materials obtained from mines, oil wells, and chemical factories all over the world. You may also be unaware that for every 0.5 kilogram (1 pound) of electronics it contains, approximately 3,600 kilograms (8,000 pounds) of solid and liquid wastes were created. The manufacture of a single semiconductor computer chip generates about 630 times its weight in solid and hazardous wastes.

Extracting these resources and using them to make your computer also required large amounts of energy that was produced mostly by burning fossil fuels, which emitted CO_2 and numerous other pollutants into the atmosphere. If you consider all the computers and other products made every year, you can see how industrial solid wastes and hazardous wastes mount up.

The remaining 1.5% of the solid waste produced in the United States is municipal solid waste (MSW), the largest categories of which are paper and cardboard (37%), yard waste (12%), food waste (11%), plastics (11%), and metals (8%). Figure 15, p. S18, in Supplement 3 shows the total and per capita production of MSW in the United States between 1960 and 2005. Each year, the United States generates enough MSW to fill a bumper-to-bumper convoy of garbage trucks long enough to encircle the globe almost eight times!

Consider some of the solid wastes that consumers throw away in the high-waste economy of the United States:

- Enough tires each year to encircle the planet almost three times (Figure 21-4)

- An amount of disposable diapers each year that, if linked end to end, would reach to the moon and back seven times

- Enough carpet each year to cover the U.S. state of Delaware

- About 2.5 million nonreturnable plastic bottles *every hour*

- About 274 million plastic shopping bags per day, an average of about 3,200 every second

Figure 21-4 Hundreds of millions of discarded tires have accumulated in this massive tire dump in Midway, Colorado (USA). Lehigh Technologies has developed a way to freeze the material in a scrap tire with liquid nitrogen. This converts the rubber to a brittle form that can be pulverized into a fine powder, which can be used in a variety of products such as paints, sealants, and coating. A prevention approach would be to double the average lifetime of tires in order to reduce the number thrown away each year.

- Enough office paper each year to build a wall 3.5 meters (11 feet) high across the country from New York City to San Francisco, California

- Some 186 billion pieces of junk mail (an average of 660 pieces per American) each year, about 45% of which are thrown into the trash unopened*

- Around 132,000 personal computers and 425,000 cell phones each day (**Core Case Study**).

The United States also leads the world in trash production (by weight) per person, and Canada takes second place. Each day the average American produces more than 2.0 kilograms (4.5 pounds) of MSW; three-fourths of it is dumped into landfills or incinerated. That is about twice the amount of solid waste produced per person in other industrial countries such as Japan and Germany, and 5–10 times that amount produced in most developing countries.

Some encouraging news is that since 1990, the average annual production of MSW by weight per American has leveled off (Figure 15, p. S18, Supplement 3), mostly because of increased recycling (Figure 16, p. S19, in Supplement 3) and the use of lighter products. Historical analysis reveals some surprises about U.S. waste production (Case Study, p. 564).

*You can register to stop receiving direct marketing mail from most companies for five years at **www.the-dma.org/consumers/offmailinglist.html**.

Trash Production and Recycling in New York City: Past, Present, and Future

You might guess that trash production in New York City has been rising steadily. You would be wrong. In 2002, Columbia University adjunct professor Daniel C. Walsh discovered some surprising facts when analyzing detailed records about what residents of New York City threw away between 1900 and 2000.

He found that the per person output, by weight, of trash dumped by New Yorkers was higher between 1920 and 1940 than it is today—mostly because of the coal ash produced by people burning coal for heat and cooking. The city's highest trash output per person was in 1940, when the rate was more than two times today's output.

During 1962 and 1963, the trash output per New Yorker was at its lowest level during the 20th century, as household coal burning had been phased out and paper was the largest component of trash. However, coal ash has not gone away. Now this waste is laced with toxic metals, as it is produced in large quantities by electric power plants. Instead of being classified as MSW, coal ash is now a form of toxic waste that must be disposed of safely.

Between 1964 and 1974, the city's trash output per person rose to slightly above today's levels as returnable, refillable bottles were phased out and the use of throwaway items increased. Since 1975, the weight of trash per New Yorker has remained about the same because materials thrown away are lighter and because of an increase in recycling.

In 1999, New York City passed a mandatory recycling law, but it was not the city's first experience with such recycling. Between 1896 and 1914, the city had a mandatory recycling program that required curbside separation of trash. But this recycling effort faded and died before World War I.

Professor Walsh also found that trash output per person rose in good economic times when people could buy more and fell in bad times as people reduced their spending.

Despite its efforts to limit MSW, New York City was one of the first U.S. cities to run out of landfill space. Until 2001, most of the city's garbage was buried in its Fresh Kills landfill on Staten Island, the world's largest public landfill. At its busiest in 2001, this landfill, a monument to a throwaway society, was deeper than the city's Statue of Liberty is tall. However, after filling up in 2001, it was closed. Now it is being transformed into recreational facilities, restored wetlands, and a large public parkland.

Since 2001, the city has been hauling its massive amounts of garbage to landfill sites in New Jersey, Pennsylvania, and Virginia. Each day, some 600 energy-inefficient and polluting tractor trailers, which if lined up would form a convoy nearly 14 kilometers (9 miles) long, haul trash out of New York City to landfills as far as 480 kilometers (300 miles) away. Similarly, in 2002, Canada's largest city Toronto closed its last landfill and, since then, has been shipping all of its garbage to Wayne County, Michigan (USA), for burial.

As oil prices rise and concerns over CO_2 emissions increase, it may become too expensive at some point for New York (and for other cities) to haul garbage long distances to burial sites. Then what?

THINKING ABOUT
Analyzing Trash

What are two lessons that we can learn from this analysis of data on New York City's trash?

21-2 How Should We Deal with Solid Waste?

▶ **CONCEPT 21-2** A sustainable approach to solid waste is first to reduce it, then to reuse or recycle it, and finally to safely dispose of what is left.

We Can Burn or Bury Solid Waste or Produce Less of It

We can deal with the solid wastes we create in two ways. One is **waste management**, in which we attempt to reduce the environmental impact of MSW without seriously trying to reduce the amount of waste produced. This output approach begins with the question: "What do we do with solid waste?" It typically involves mixing wastes together and then transferring them from one part of the environment to another, usually by burying them, burning them, or shipping them to another location. This is the most common approach to dealing with e-waste (**Core Case Study**).

The second approach is **waste reduction**, in which much less waste and pollution are produced, and the wastes that are produced are viewed as potential resources that can be reused, recycled, or composted (**Concept 21-2**). It begins with the question: "How can we avoid producing so much solid waste?" With

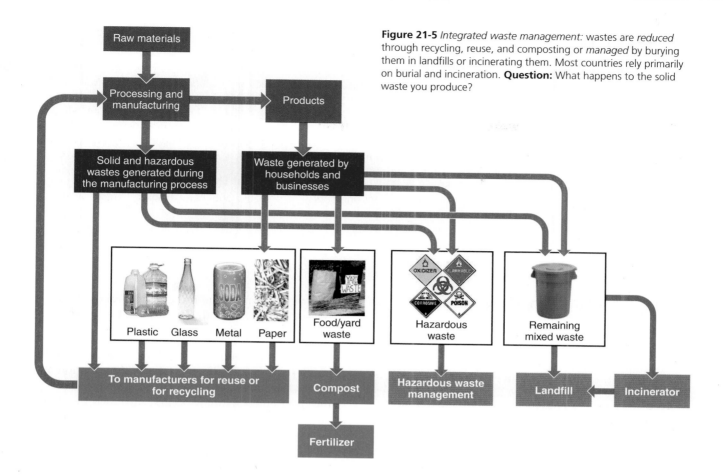

Figure 21-5 *Integrated waste management:* wastes are *reduced* through recycling, reuse, and composting or *managed* by burying them in landfills or incinerating them. Most countries rely primarily on burial and incineration. **Question:** What happens to the solid waste you produce?

this prevention approach (**Concept 1-4**, p. 16), we could think of trash cans and garbage trucks as *resource containers* that are on their way to recycling or composting facilities.

There is no single solution to the solid waste problem. Most analysts call for using **integrated waste management**—a variety of strategies for both waste reduction and waste management (Figure 21-5). Scientists call for much greater emphasis on waste reduction

(Figure 21-6). But this is not done in the United States (or in most industrialized countries) where 54% of the MSW is buried in landfills (Science Focus, p. 566), 25% is recycled, 14% is incinerated, and 7% is composted.

Some scientists and economists have estimated that 75–90% of the solid waste we produce could be eliminated by a combination of the strategies shown in Figure 21-5. Let us look more closely at these options in the order of the priorities suggested by these scientists.

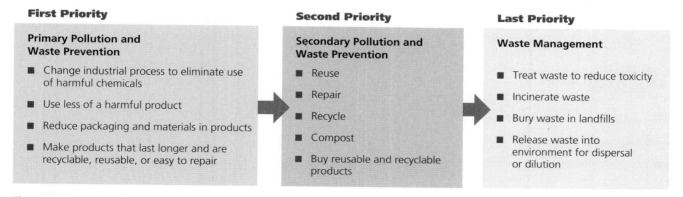

Figure 21-6 *Integrated waste management:* priorities suggested by the U.S. National Academy of Sciences for dealing with solid waste. To date, these waste reduction priorities have not been followed in the United States or in most other countries. Instead, most efforts are devoted to waste management (bury it or burn it). **Question:** Why do you think most countries do not follow these priorities, even though they are based on reliable science? (Data from U.S. Environmental Protection Agency and U.S. National Academy of Sciences)

Garbology

How do we know the composition of trash in landfills? Much of this knowledge comes from research by *garbologists* such as William Rathje, who was a pioneer in his field in the 1970s at the University of Arizona. These scientists work in the fashion of archaeologists, training their students to sort, weigh, and itemize people's trash, and to bore holes in garbage dumps and analyze what they find.

Many people think of landfills as huge compost piles where biodegradable wastes are decomposed within a few months. But garbologists looking at the contents of landfills found 50-year-old newspapers that were still readable and hot dogs and pork chops buried for decades that still looked edible. In landfills (as opposed to open dumps, Figure 1-6, p. 11), trash can resist decomposition for perhaps centuries because it is tightly packed and protected from sunlight, water, and air.

Critical Thinking

Should landfills be exposed to more air and water to hasten decomposition of their wastes? Explain.

We Can Cut Solid Wastes by Reducing, Reusing, and Recycling

Waste reduction (**Concept 21-2**) is based on three Rs:

- **Reduce:** consume less and live a simpler lifestyle.

- **Reuse:** rely more on items that can be used repeatedly instead of on throwaway items. Buy necessary items secondhand or borrow or rent them. Take a refillable coffee cup to class or to the coffee shop and use it instead of using throwaway cups.

- **Recycle:** separate and recycle paper, glass, cans, plastics, metal, and other items, and buy products made from recycled materials.

From an environmental standpoint, the first two Rs are preferred because they are input or prevention approaches that tackle the problem of waste production at the front end—before it occurs. Any such input approach also saves matter and energy resources, reduces pollution (including greenhouse gas emissions), helps protect biodiversity, and saves money. Recycling is important, but it is an output approach based on dealing with wastes after they have been produced.

Figure 21-7 lists some ways in which you can use the 3Rs to reduce your output of solid waste.

Here are seven strategies with which industries and communities can reduce resource use, waste, and pollution.

First, *redesign manufacturing processes and products to use less material and energy.* The weight of a typical car has been reduced by about one-fourth since the 1960s through use of lighter steel and lightweight plastics and composite materials. Plastic milk jugs contain less plastic and weigh 40% less than they did in the 1970s, and soft drink cans contain one-third less aluminum. Disposable diapers contain 50% less paper pulp because of improved absorbent-gel technology. Dry-cell batteries contain much less toxic mercury than they did in the 1980s, and the small amount of mercury in today's compact fluorescent lightbulbs is being reduced. Downloading music from the Internet reduces the use of materials and energy in packaging records and CDs and in shipping them long distances.

Second, *redesign manufacturing processes to produce less waste and pollution.* In the *ecoindustrial revolution* (Case Study, p. 366), manufacturing processes are being redesigned to mimic how nature reduces and recycles wastes (**Concept 1-6**, p. 23). This includes designing industrial ecosystems in which the wastes of some businesses are used as raw materials by

WHAT CAN YOU DO?

Solid Waste

- Follow the three Rs of resource use: Reduce, Reuse, and Recycle

- Ask yourself whether you really need a particular item, and refuse packaging where possible

- Rent, borrow, or barter goods and services when you can, buy secondhand, and donate or sell unused items

- Buy things that are reusable, recyclable, or compostable, and be sure to reuse, recycle, and compost them

- Avoid disposables, and do not use throwaway paper and plastic plates, cups, and eating utensils, and other disposable items when reusable or refillable versions are available

- Use e-mail or text-messaging in place of conventional paper mail

- Read newspapers and magazines online

- Buy products in bulk or concentrated form whenever possible

Figure 21-7 Individuals matter: ways to save resources by reducing your output of solid waste and pollution. **Questions:** Which three of these actions do you think are the most important? Why? Which of these things do you do?

CONCEPT LINK

other businesses (Figure 14-25, p. 367) in what is in effect, an industrial resource web that mimics a natural food web (Figure 3-14, p. 63). In addition, to reducing waste production and pollution, these changes also save energy and reduce resource use.

Ways to reduce the waste outputs of industrial processes include recycling most toxic organic solvents within factories or replacing them with water-based or citrus-based solvents (Individuals Matter, p. 459) and using hydrogen peroxide instead of toxic chlorine to bleach paper and other materials. In addition, toxic chemicals used in dry cleaning can be replaced. One method uses a nontoxic silicone solvent in conventional dry-cleaning machines and another involves submersing clothes in liquid carbon dioxide. Check your local phone directory to locate dry cleaners that use these alternative methods. A promising new approach is to develop nanotechnology coatings (Case Study, p. 362) that would eliminate the need for dry cleaning, although these materials would have to be carefully tested to avoid any harmful consequences of using them.

Third, *develop products that are easy to repair, reuse, remanufacture, compost, or recycle.* For example, a Xerox photocopier made of reusable or recyclable parts that allow for easy remanufacturing could eventually save the company $1 billion in manufacturing costs. And, the automobile industry, led by European automakers, now designs motor vehicles with bar coded parts so that the vehicles can be more easily disassembled and recycled. At a plant in Corinth, Mississippi (USA), the heavy equipment manufacturer Caterpillar disassembles some 17 truckloads of diesel engines a day, repairs worn parts, and reassembles the engines, which are then as good as new. Sales of the reassembled engines are about $1 billion a year and are growing at 15% a year.

┌─ RESEARCH FRONTIER ─────────────────────────

Inventing less wasteful and less polluting manufacturing processes and products. See **academic.cengage.com/biology/ miller**.

Fourth, *eliminate or reduce unnecessary packaging.* Use the following hierarchy for packaging: no packaging, minimal packaging, reusable packaging, and recyclable packaging. Canada has set a goal of using the first three of these priorities to cut excess packaging in half. The European Union has instituted a requirement for recycling 55–80% of all packaging waste. In 2007, Wal-Mart asked its suppliers to cut down on the amount of packaging used in products sold in its stores and introduced a scorecard to rate its vendors. The company estimates that it could save $3.4 billion by reducing the packaging of its products by 5% by 2013.

Fifth, *use fee-per-bag waste collection systems* that charge consumers for the amount of waste they throw away but provide free pickup of recyclable and reusable items.

Sixth, *establish cradle-to-grave responsibility laws* that require companies to take back various consumer products such as electronic equipment (**Core Case Study**), appliances, and motor vehicles, as Japan and many European countries do.

CORE CASE STUDY

Seventh, *restructure urban transportation systems* to rely more on mass transit and bicycles than on cars. An urban bus can carry about as many people as 60 cars. And using a bicycle represents a small fraction of the resource use and solid waste involved with manufacturing and using a motor vehicle.

21-3 Why Is Reusing and Recycling Materials So Important?

▶ **CONCEPT 21-3** Reusing items decreases the use of matter and energy resources and reduces pollution and natural capital degradation; recycling does so to a lesser degree.

Reuse Is an Important Way to Reduce Solid Waste and Pollution and to Save Money

In today's modern societies, we have increasingly substituted throwaway items for reusable ones, which has resulted in growing masses of waste. For example, if

the 1 billion throwaway paper coffee cups used by one famous chain of donut shops each year were lined up end-to-end, they would encircle the earth two times. Rewarding those who bring their own refillable coffee mugs would help to reduce this waste.

Reuse involves cleaning and using materials over and over and thus increasing the typical life span of a product (see following Case Study). This form of waste

reduction decreases the use of matter and energy resources, cuts pollution, creates local jobs, and saves money (**Concept 21-3**).

Reuse is alive and well in most developing countries, but it has a downside. The poor who scavenge in open dumps (Figure 1-6, p. 11) for food scraps and items they can reuse or sell are often exposed to toxins and infectious diseases.

Traditional forms of reuse include salvaging automobile parts from older cars in junkyards and recovering materials from old houses and buildings. By 2015, the European Union will require that 95% of any discarded car must be reused or recycled. Other reuse strategies involve yard sales, flea markets, secondhand stores, traditional and online auctions, classified newspaper ads, and online sites such as e-Bay and Craigslist. An international website at **www.freecycle.org** links people who want to give away household belongings free to people in their area who want or need them.

Technology allows reuse of many items such as batteries. The latest rechargeable batteries come fully charged, can hold a charge for up to 2 years when they are not used, can be recharged in as few as 15 minutes, and greatly reduce toxic waste when used in place of discarded conventional batteries. They cost more than conventional batteries but the extra cost is recovered quickly, because replacement batteries do not have to be purchased as often.

■ CASE STUDY
Use of Refillable Containers

Two examples of reusable items are refillable glass beverage bottles and refillable soft drink bottles made of polyethylene terephthalate (PET) plastic. Typically, such bottles make 15 round-trips before they become too damaged for reuse and then are recycled. Reusing these containers saves energy (Figure 21-8) reduces CO_2 emissions, air pollution, water pollution, and solid wastes, and stimulates local economies by creating local jobs related to their collection and refilling. Moreover, studies by Coca-Cola and PepsiCo of Canada show that their soft drinks in 0.5-liter (21-ounce) bottles, on average, cost consumers one-third less in refillable bottles than in throwaway bottles.

But big companies make more money by producing and shipping beverages and food in throwaway containers at centralized facilities. This shift has put many small local bottling companies, breweries, and canneries out of business and hurt local economies. However, some U.S. dairies now deliver milk to their customers in refillable glass or plastic acrylic bottles that can be reused 50 to 100 times. And a few U.S. breweries have switched back to refillable bottles.

Parts of Canada and 11 U.S. states have bottle laws that place a deposit fee on all beverage containers. Retailers must accept the used containers and pass them

Figure 21-8 Energy consumption involved with using different types of 350-milliliter (12-fluid-ounce) beverage containers. (Data from Argonne National Laboratory)

on for recycling or reuse. Large beverage industries have used their political and financial clout to keep most U.S. states from passing bottle laws, arguing that they lead to a loss of jobs and higher beverage costs for consumers. But experience in Canada and U.S. states with bottle bills shows that more jobs are gained than lost, costs to consumers have not risen, resources are saved, and roadside litter decreases.

Some analysts call for a national bottle bill in the United States, while others would ban all beverage containers that cannot be reused, as Denmark, Finland, and Canada's Prince Edward Island have done. Ecuador levies a refundable beverage container deposit fee that amounts to 50% of the cost of the drink. In Finland, 95% of the soft drink, beer, wine, and spirits containers are refillable.

HOW WOULD YOU VOTE? ☑

Do you support banning all beverage containers that cannot be reused as Denmark has done? Cast your vote online at **academic.cengage.com/biology/miller**.

Reusable cloth bags can be used instead of throwaway paper or plastic bags to carry groceries and other items. Both paper and plastic bags are environmentally harmful, and the question of which is more damaging has no clear-cut answer.

According to Vincent Cobb, founder of **reusablebags .com**, an estimated 500 billion to 1 trillion plastic bags are used and usually discarded each year throughout the world. Producing them requires large amounts of oil because most are made from ethylene, a petroleum byproduct. And each discarded bag can take from 400 to 1,000 years to break down. Less than 1% of the estimated 100 billion plastic bags used each year in the United States are recycled.

In a number of African countries, the landscape is littered with plastic bags. In addition to being an eyesore and a waste of resources the bags block drains and sewage systems and can kill wildlife and livestock that eat them. They also kill plants and spread malaria by holding mini-pools of warm water where mosquitoes can breed.

There is a growing backlash against plastic shopping bags. To encourage people to use reusable bags, the governments of Ireland, Taiwan, and the Netherlands tax plastic shopping bags. In Ireland, a tax of about 25¢ per bag has cut plastic bag litter by 90% as people have switched to reusable bags. Bangladesh, Bhutan, Rwanda, South Africa, parts of India, Taiwan, Kenya, South Africa, Uganda, China, Australia, France, Italy, and the U.S. city of San Francisco have banned the use of all or most types of plastic shopping bags. But the plastics industry and some businesses have mounted a successful campaign against such bans in most of the United States.

┌─── **HOW WOULD YOU VOTE?** ☑ ───
│ Should consumers have to pay for plastic or paper bags at
│ grocery and other stores? Cast your vote online at **academic**
│ **.cengage.com/biology/miller.**
└

There are many other ways to reuse various items (Figure 21-9). For example, an increasing number of coffeehouses and university food services offer discounts to customers who bring their own refillable mugs. Some environmentally conscious coffee consumers are boycotting coffeehouses that do not provide discounts for customers bringing their own mugs.

There Are Two Types of Recycling

Recycling involves reprocessing discarded solid materials into new, useful products. In addition to saving resources and reducing solid waste and pollution, recycling also reduces unsightly and environmentally harmful litter.

Households and workplaces produce five major types of materials that can be recycled: paper products, glass, aluminum, steel, and some plastics. Such materials can be reprocessed in two ways. In **primary** or **closed-loop recycling,** these materials are recycled into new products of the same type—turning used aluminum cans into new aluminum cans, for example. In **secondary recycling,** waste materials are converted into different products. For example, used tires can be shredded and turned into rubberized road surfacing, newspapers can be reprocessed into cellulose insulation, and plastics can be reprocessed into various items. Engineer Henry Liu has developed a process for making bricks from recycled fly ash produced by coal-burning

WHAT CAN YOU DO?
Reuse

- Buy beverages in refillable glass containers instead of cans or throwaway bottles
- Use reusable plastic or metal lunchboxes
- Carry sandwiches and store food in the refrigerator in reusable containers instead of wrapping them in aluminum foil or plastic wrap
- Use rechargeable batteries and recycle them when their useful life is over
- Carry groceries and other items in a reusable basket, a canvas or string bag, or a small cart
- Use reusable sponges and washable cloth napkins, dish towels, and handkerchiefs instead of throwaway paper ones
- Buy used furniture, computers, cars, and other items instead of buying new
- Give away or sell items you no longer use

Figure 21-9 Individuals matter: ways to reuse some of the items you buy. **Question:** Which three of these actions do you think are the most important? Why?

power plants. The process saves energy, reduces air pollution, and costs at least 20% less than the cost of making conventional bricks.

Scientists distinguish between two types of wastes that can be recycled: *preconsumer* or *internal waste* generated in a manufacturing process and *postconsumer* or *external waste* generated by consumer use of products. Preconsumer waste makes up more than three-fourths of the total.

Just about anything is recyclable, but there are two key questions. *First,* are the items separated for recycling actually recycled? Sometimes they are mixed with other wastes and sent to landfills or incinerated. *Second,* will businesses and individuals complete the recycling loop by buying products that are made from recycled materials?

Switzerland and Japan recycle about half of their MSW. The United States recycles about 25% of its MSW—up from 6.4% in 1960. This increase has gotten a boost by almost 9,000 curbside pickup recycling programs, which serve about half of the U.S. population.

In 2007, the United States recycled about 60% of its steel, 56% of its aluminum cans, 56% of its paper and cardboard, 36% of its tires, 22% of its glass, and 5% of its plastics. Experts say that with education and proper incentives, the United States could recycle 60–70% of these and many other forms of solid waste, in keeping with one of the four **scientific principles of sustainability** (see back cover).

We Can Mix or Separate Household Solid Wastes for Recycling

One way to recycle is to send mixed urban wastes to centralized *materials-recovery facilities* (MRFs or "murfs"). There, machines or workers separate the mixed waste to recover valuable materials for sale to manufacturers as raw materials. The remaining paper, plastics, and other combustible wastes are recycled or burned to produce steam or electricity to run the recovery plant or to sell to nearby industries or homes.

Such plants are expensive to build, operate, and maintain. If not operated properly, they can emit CO_2 and toxic air pollutants, and they produce a toxic ash that must be disposed of safely, usually in landfills. Because MRFs require a steady diet of garbage to make them financially successful, their owners have a vested interest in increasing the throughput of matter and energy resources to produce more trash—the reverse of what prominent scientists believe we should be doing (Figure 21-6).

To many experts, it makes more environmental and economic sense for households and businesses to separate their trash into recyclable categories such as glass, paper, metals, certain types of plastics, and compostable materials. This *source separation* approach produces much less air and water pollution and costs less to implement than MRFs cost. It also saves more energy, provides more jobs per unit of material, and yields cleaner and usually more valuable recyclables. In addition, sorting material educates people about the need for recycling.

To promote separation of wastes for recycling, more than 4,000 communities in the United States use a *pay-as-you-throw* or *fee-per-bag* waste collection system. It charges households and businesses for the amount of mixed waste picked up, but does not charge for pickup of materials separated for recycling or reuse. In the U.S. city of Ft. Worth, Texas, the proportion of households recycling went from 21% to 85% when such a system was implemented. And the city went from losing $600,000 in its recycling program to making $1 million a year because of increased sales of recycled materials to industries.

Using a similar program, the U.S. city of San José, California reuses or recycles 62% of its MSW and is becoming a model for cities everywhere. It is now focusing on reducing the large flow of waste from demolition and construction sites by separating such waste materials into recyclable piles of scrap metal, wood, and cement. The salvaged scrap metal goes to recycling plants, wood is converted into wood chips or mulch for fueling power plants, and concrete is recycled to build road banks.

HOW WOULD YOU VOTE? ☑

Should households and businesses be charged for the amount of mixed waste picked up but not for pickup of materials separated for recycling? Cast your vote online at **academic .cengage.com/biology/miller**.

We Can Copy Nature and Recycle Biodegradable Solid Wastes

Composting is a form of recycling that mimics nature's recycling of nutrients—one of the four **scientific principles of sustainability**. It involves allowing decomposer bacteria to recycle yard trimmings, food scraps, and other biodegradable organic wastes. The resulting organic material can be added to soil to supply plant nutrients, slow soil erosion, retain water, and improve crop yields. Homeowners can compost such wastes in simple backyard containers, in composting piles that must be turned over occasionally, or in small composting drums (Figure 21-10) that can be rotated to mix the wastes and to speed up the decomposition process. Over 6 million homes in North America compost their organic wastes and the number is increasing. For details on composting, see the website for this chapter.

The United States has about 3,300 municipal composting programs that recycle about 37% of country's yard wastes. This is likely to rise as the number of states (now 20) that ban yard wastes from sanitary landfills increases. The resulting compost can be used as organic soil fertilizer, topsoil, or landfill cover. It can also be used to help restore eroded soil on hillsides and along highways, and on strip-mined land, overgrazed areas, and eroded cropland.

Figure 21-10 Backyard composter drum in which bacteria convert kitchen waste into rich compost. When the compost is ready, the device can be wheeled out to the garden or flowerbeds.

Courtesy of Green Culture

To be successful, a large-scale composting program must be located carefully and odors must be controlled, because people do not want to live near a giant compost pile or plant. Some cities in Canada and many European Union countries compost more than 85% of their biodegradable wastes in centralized community facilities. Sometimes composting takes place in huge indoor buildings. In the Canadian city of Edmonton, Alberta, an indoor composting facility the size of eight football fields composts 50% of the city's organic solid waste. Composting programs must also exclude toxic materials that can contaminate the compost and make it unsafe for use as fertilizer.

■ CASE STUDY
Recycling Paper

About 55% of the world's industrial tree harvest is used to make paper. The pulp and paper industry is the world's fifth largest energy user and uses more water to produce a metric ton of its product than any other industry. In both Canada and the United States, it is the third-largest industrial energy user and polluter, and paper is the dominant material in the MSW of both countries. The Internet was supposed to reduce paper use by giving users access to vast amounts of electronic information. However, because most users print out their search results there has been no drop in per capita paper consumption in the United States (and in most other countries).

Paper (especially newspaper and cardboard) is easy to recycle. Recycling newspaper involves removing its ink, glue, and coating and then reconverting it to pulp, which is pressed into new paper. Making recycled paper uses 64% less energy and produces 35% less water pollution and 74% less air pollution than does making paper from wood, and, of course, no trees are cut down.

In 2007, the United States recycled about 56% of its wastepaper (up from 25% in 1989). At least 10 other countries recycle 50–97% of their wastepaper and paperboard, and the global recycling rate is 43%. Paper recycling leaders are Denmark (97%), South Korea (77%), and Germany (72%).

Despite a 56% recycling rate, the amount of paper thrown away each year in the United States is more than all of the paper used in China. Also, about 95% of books and magazines produced in the United States are printed on virgin paper. In producing this textbook we strive to use paper with a high percentage of recycled fibers. However, recycled paper of the quality required is often hard to get and costs more than conventional paper, which adds to the price of this book. Being green often involves trade-offs.

One problem associated with making paper is the chlorine (Cl_2) and chlorine compounds (such as chlorine dioxide, ClO_2), used to bleach about 40% of the world's pulp for making paper. These compounds are corrosive to processing equipment, hazardous for workers, hard to recover and reuse, and harmful when released into the environment. A growing number of paper mills (mostly in the European Union) are replacing chlorine-based bleaching chemicals with chemicals such as hydrogen peroxide (H_2O_2) or oxygen (O_2).

■ CASE STUDY
Recycling Plastics

Plastics consist of various types of large polymers, or *resins*—organic molecules made by chemically linking monomer molecules produced mostly from oil and natural gas (Figure 15-4, p. 375). About 46 different types of plastics are used in consumer products, and some products contain several kinds of plastic.

Many plastic containers and other items are thrown away and end up as litter on roadsides, beaches (Figure 21-11), and oceans and other bodies of water. Each year they threaten millions of seabirds, marine mammals (Figure 11-5, p. 254), and sea turtles, which can mistake a floating plastic sandwich bag for a jellyfish or get caught in discarded plastic nets (Figure 11-10, right, p. 260). About 80% of the plastics in the ocean are

Figure 21-11 Discarded solid waste litters beaches, poses a threat to beach users, and washes into the ocean and threatens marine animals.

Mike Biddle's Contribution to Recycling Plastics

Only 4% of the huge amount of plastics discarded each year in the United States is recycled. This is beginning to change.

In 1994, Mike Biddle, a former PhD engineer with Dow Chemical, and Trip Allen founded MBA Polymers, Inc. Their goal was to develop a commercial process for recycling high-value plastics from complex streams of goods such as computers, electronics, appliances, and automobiles. They succeeded by designing a 21-step automated process that separates plastics from nonplastic items in mixed waste streams, and then separates plastics from each other by type and grade and converts them to pellets that can be used to make new products.

The pellets are cheaper than virgin plastics, because the company's process uses 90% less energy than that needed to make a new plastic, and because the raw material is cheap or free junk. The environment also wins because greenhouse gas emissions are much lower than those generated in the making of virgin plastics. Also, recycling waste plastics reduces the need to incinerate them or bury them in landfills.

MBA Polymers is considered a world leader in plastics recycling. It operates a large state-of-the-art research and recycling plant in Richmond, California, and recently opened the world's two most advanced plastics recycling plants in China and Austria. It has won many awards, including the 2002 Thomas Alva Edison Award for Innovation, and was selected by *Inc.* magazine as one of "America's Most Innovative Companies."

Those who grew up with Mike Biddle are not surprised that he played a major role in developing an innovative and money-making process that many scientists and engineers thought impossible. As a kid growing up in Kentucky, he hated waste. He says he drove his parents crazy by following them around the house turning off lights in rooms that were not being used. Maybe you can be an environmental entrepreneur by using your brainpower to develop an environmentally beneficial and financially profitable process or business.

blown or washed in from beaches, rivers, storm drains, and other sources, and the rest gets dumped into the ocean from vessels and fishing boats.

Plastics discarded on beaches or dumped into the ocean from ships can disintegrate into particles the size of sand grains that resemble the prey of a variety of organisms. These particles can fill the stomachs of birds and other sea creatures and cause dehydration, malnutrition, and eventually starvation. Because tiny plastic particles can accumulate as they move through food webs, some level of plastic is found in most of the seafood people eat.

Currently, only about 4% by weight of all plastic wastes in the United States is recycled. As American comedian Lily Tomlin observes, "We buy a wastebasket and take it home in a plastic bag. Then we take the wastebasket out of the bag, and put the bag in the wastebasket." The percentage of plastic waste that is recycled is low for three reasons. *First,* many plastics are hard to isolate from other wastes because the many different resins used to make them are often difficult to identify, and some plastics are composites of different resins. For example, a plastic ketchup bottle might have as many as six different layers of plastics bonded together. Most plastics also contain stabilizers and other chemicals that must be removed before recycling.

Second, recovering individual plastic resins does not yield much material because only small amounts of any given resin are used in each product.

Third, the inflation-adjusted price of oil used to produce petrochemicals for making plastic resins is low enough to make the cost of virgin plastic resins much lower than that of recycled resins. An exception is PET (polyethylene terephthalate), used mostly in plastic drink bottles. However, PET collected for recycling must not have other plastics mixed with it; a single PVC (polyvinyl chloride) bottle in a truckload of PET can render it useless for recycling. Despite its economic value, only about 20% of the PET used in plastic containers in the United States is recycled. However, in 2007, Coca-Cola announced a goal of reusing or recycling 100% of the PET bottles it sells in the United States.

Progress is being made in the recycling of plastics (Individuals Matter, above) and in the development of more degradable bioplastics (Science Focus, at right).

Recycling Has Advantages and Disadvantages

Figure 21-12 lists the advantages and disadvantages of recycling (**Concept 21-3**). Whether recycling makes economic sense depends on how you look at its economic and environmental benefits and costs.

Critics say recycling does not make sense if it costs more to recycle materials than to send them to a landfill or incinerator. They concede that recycling may make economic sense for valuable and easy-to-recycle materials such as aluminum, paper, and steel, but probably not for cheap or plentiful resources such as glass made from silica. They also argue that recycling should pay for itself.

Proponents of recycling point out that conventional garbage disposal systems are funded by charges to households and businesses. So why should recycling be held to a different standard and forced to compete on an uneven playing field? Proponents also point to studies showing that the net economic, health, and envi-

ronmental benefits of recycling (Figure 21-12, left) far outweigh the costs. They argue that the U.S. recycling industry employs about 1.1 million people and that its annual revenues are much larger than those of the waste management industry.

Cities that make money by recycling and have higher recycling rates tend to use a *single-pickup system* for both recyclable and nonrecyclable materials, instead of a more expensive dual-pickup system. Successful systems also tend to use a pay-as-you-throw approach. San Francisco, California (USA), uses such a system to recycle almost half of its MSW.

We Can Encourage Reuse and Recycling

Three factors hinder reuse and recycling. *First,* we have a faulty and misleading accounting system in which the market price of a product does not include the harmful environmental and health costs associated with the product during its life cycle.

Second, there is an uneven economic playing field, because in most countries, resource-extracting industries receive more government tax breaks and subsidies than recycling and reuse industries get.

Third, the demand and thus the price paid for recycled materials fluctuates, mostly because buying goods made with recycled materials is not a priority for most governments, businesses, and individuals.

How can we encourage reuse and recycling? Proponents say that leveling the economic playing field is the

TRADE-OFFS
Recycling

Advantages

Reduces air and water pollution

Saves energy

Reduces mineral demand

Reduces greenhouse gas emissions

Reduces solid waste production and disposal

Helps protect biodiversity

Can save landfill space

Important part of economy

Disadvantages

Can cost more than burying in areas with ample landfill space

May lose money for items such as glass and some plastics

Reduces profits for landfill and incinerator owners

Source separation is inconvenient for some people

Figure 21-12 Advantages and disadvantages of recycling solid waste (**Concept 21-3**). **Question:** Which single advantage and which single disadvantage do you think are the most important? Why?

best way to start. Governments can *increase* subsidies and tax breaks for reusing and recycling materials (the carrot) and *decrease* subsidies and tax breaks for making items from virgin resources (the stick).

SCIENCE FOCUS

Bioplastics

Most of today's plastics are made from organic polymers produced from petroleum-based chemicals (petrochemicals). This may change as scientists develop more *bioplastics* or plastics made from biologically based chemicals.

The search for biologically based plastics dates from 1913 when a French scientist and a British scientist each filed for patents on a soy-based plastic. At that time, there was intense competition between the petrochemical and agricultural industries to dominate the market for plastics made from organic polymers.

Henry Ford, who developed the first motorcar, supported research on the development of a bioplastic made from soybeans. A 1914 photograph showed him using an ax to strike the body of a car made from soy bioplastic to demonstrate its strength.

But as oil became widely available, petrochemical plastics took over the market. Now with climate change and other environmental problems associated with the use of oil, chemists are stepping up efforts to make biodegradable and more environmentally sustainable plastics from a variety of green polymers. Such bioplastics can be made from corn, soy, sugarcane, switchgrass (Figure 16-26, p. 425), chicken feathers, some components of garbage, and CO_2 extracted from coal-burning power plant emissions.

The key to making such biopolymers is to find chemicals called *catalysts,* which accelerate reactions that form polymers from biologically based chemicals without having to use high temperatures. With proper design and mass production, bioplastics could be

lighter, stronger, and cheaper, and could use less energy and produce less pollution per unit of weight than conventional petroleum-based plastics do.

Instead of being sent to landfills, packaging made from bioplastics could be composted to produce a soil conditioner, thus mimicking nature by implementing the nutrient recycling **principle of sustainability**.

Toyota is investing $38 billion in a process that makes plastics from plants. By 2020, it expects to control two-thirds of the world's supply of such bioplastics.

Critical Thinking

What might be some disadvantages of more rapidly degradable bioplastics? Do you think they outweigh the advantages?

Other strategies are to greatly increase use of the fee-per-bag waste collection system and to encourage or require government purchases of recycled products to help increase demand and lower prices. Governments can also pass laws requiring companies to take back and recycle or reuse packaging and electronic waste discarded by consumers (**Core Case Study**), as is done in Japan and some European Union countries.

CORE
CASE
STUDY

— HOW WOULD YOU VOTE? ☑ ————

Should governments pass laws requiring manufacturers to take back and reuse or recycle all packaging waste, appliances, electronic equipment (**Core Case Study**), and motor vehicles at the end of their useful lives? Cast your vote online at **academic.cengage.com/biology/ miller**.

CORE
CASE
STUDY

Citizens can pressure governments to require labels on all products listing recycled content and the types and amounts of any hazardous materials they contain. This would help consumers to be better informed about the environmental consequences of buying certain products.

One reason for the popularity of recycling is that it helps to soothe people's consciences in a throwaway society. Many people think that recycling their newspapers and aluminum cans is all they need do to meet their environmental responsibilities. Recycling is important, but reducing resource consumption and reusing resources are more effective ways to reduce the flow and waste of resources (**Concept 21-3**).

21-4 What Are the Advantages and Disadvantages of Burning or Burying Solid Waste?

▶ **CONCEPT 21-4** Technologies for burning and burying solid wastes are well developed, but burning contributes to pollution and greenhouse gas emissions, and buried wastes eventually contribute to pollution and land degradation.

Burning Solid Waste Has Advantages and Disadvantages

Globally, MSW is burned in more than 600 large *waste-to-energy incinerators* (98 in the United States), which burn MSW to boil water to make steam for heating water or space, or for producing electricity. Great Britain burns about 90% of its MSW in incinerators, compared to 16% in the United States and 8% in Canada. Trace the flow of materials through this process, as diagrammed in Figure 21-13.

In addition to producing energy, incinerators reduce the volume of solid waste by 90%. However, without expensive air pollution control devices and careful monitoring, incinerators pollute the air with particulates, carbon monoxide, toxic metals such as mercury, and other toxic materials. They also add CO_2 to the atmosphere, although they emit about 38% less CO_2 per unit of energy than coal-burning power plants emit. Incinerators produce large quantities of toxic bottom ash and fly ash (removed by air pollution control devices), which must be disposed of safely, ideally in specially licensed hazardous waste landfills.

To be economically feasible, incinerators must be fed huge volumes of trash every day. This encourages trash production and discourages reuse, recycling, and

waste reduction. Since 1985, more than 280 new incinerator projects have been delayed or canceled in the United States because of high costs, concern over air pollution, and intense citizen opposition.

Figure 21-14 lists the advantages and disadvantages of using incinerators to burn solid waste.

— HOW WOULD YOU VOTE? ————

Do the advantages of incinerating solid waste outweigh the disadvantages? Cast your vote online at **academic.cengage .com/biology/miller**.

Burying Solid Waste Has Advantages and Disadvantages

About 54% by weight of the MSW in the United States is buried in sanitary landfills, compared to 80% in Canada, 15% in Japan, and 12% in Switzerland.

There are two types of landfills. **Open dumps** are essentially fields or holes in the ground where garbage is deposited and sometimes burned. They are rare in developed countries, but are widely used near major cities in many developing countries (Figure 1-6, p. 11).

In newer landfills, called **sanitary landfills** (Figure 21-15, p. 576), solid wastes are spread out in thin

Figure 21-13 Solutions: a waste-to-energy incinerator with pollution controls. It burns mixed solid wastes and recovers some of the energy to produce steam used for heating or producing electricity. **Question:** Would you invest in such a project? Why or why not?

Ash for treatment, disposal in landfill, or use as landfill cover

TRADE-OFFS

Incineration

Advantages	Disadvantages
Reduces trash volume	Expensive to build
Less need for landfills	Costs more than short-distance hauling to landfills
Low water pollution	Difficult to site because of citizen opposition
Concentrates hazardous substances into ash for burial	Some air pollution and CO_2 emissions
Sale of energy reduces cost	Older or poorly managed facilities can release large amounts of air pollution
Modern controls reduce air pollution	Output approach that encourages waste production
Some facilities recover and sell metals	Can compete with recycling for burnable materials such as newspaper

layers, compacted, and covered daily with a fresh layer of clay or plastic foam, which helps to keep the material dry and reduces leakage of contaminated water (leachate) from the landfill. This covering also lessens the risk of fire, decreases odor, and reduces accessibility to vermin.

Figure 21-16 (p. 576) lists the advantages and disadvantages of using sanitary landfills to dispose of solid waste. According to the EPA, all landfills eventually leak, passing both the effects of contamination and cleanup costs on to future generations. In the United States, some people were shocked to learn that radioactive materials from nuclear weapons facilities run by the Department of Energy were being dumped into regular landfills with little tracking of their dispersal, despite intense public opposition.

— **HOW WOULD YOU VOTE?** ✓ —

Do the advantages of burying solid waste in sanitary landfills outweigh the disadvantages? Cast your vote online at **academic.cengage.com/biology/miller**.

Figure 21-14 Advantages and disadvantages of incinerating solid waste (**Concept 21-4**). These trade-offs also apply to the incineration of hazardous waste. **Question:** Which single advantage and which single disadvantage do you think are the most important? Why?

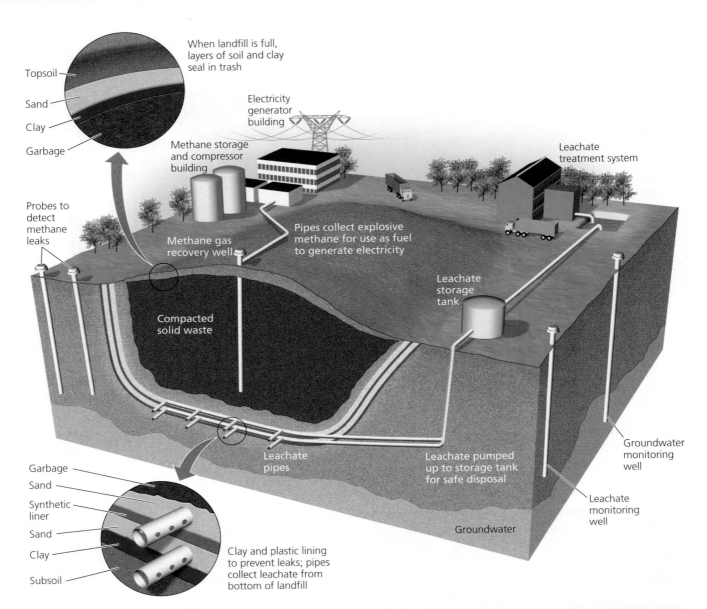

Topsoil

Sand

Clay

Garbage

When landfill is full, layers of soil and clay seal in trash

Electricity generator building

Methane storage and compressor building

Leachate treatment system

Probes to detect methane leaks

Methane gas recovery well

Pipes collect explosive methane for use as fuel to generate electricity

Leachate storage tank

Compacted solid waste

Garbage

Sand

Synthetic liner

Sand

Clay

Subsoil

Clay and plastic lining to prevent leaks; pipes collect leachate from bottom of landfill

Leachate pipes

Leachate pumped up to storage tank for safe disposal

Groundwater monitoring well

Leachate monitoring well

Groundwater

Figure 21-15 Solutions: state-of-the-art *sanitary landfill,* which is designed to eliminate or minimize environmental problems that plague older landfills. Since 1997, only modern sanitary landfills are allowed in the United States. As a result, many small, older local landfills have been closed and replaced with larger regional landfills. **Question:** How do you think sanitary landfills could develop leaks of toxic liquids?

Figure 21-16 Advantages and disadvantages of using sanitary landfills to dispose of solid waste (**Concept 21-4**). **Question:** Which single advantage and which single disadvantage do you think are the most important? Why?

TRADE-OFFS

Sanitary Landfills

Advantages	Disadvantages
No open burning	Noise and traffic
Little odor	Dust
Low groundwater pollution if sited properly	Air pollution from toxic gases and trucks
Can be built quickly	Releases greenhouse gases (methane and CO_2) unless they are collected
Low operating costs	
Can handle large amounts of waste	Slow decomposition of wastes
Filled land can be used for other purposes	Output approach that encourages waste production
No shortage of landfill space in many areas	Eventually leaks and can contaminate groundwater

21-5 How Should We Deal with Hazardous Waste?

▶ **CONCEPT 21-5** A sustainable approach to hazardous waste is first to produce less of it, then to reuse or recycle it, then to convert it to less hazardous materials, and finally, to safely store what is left.

We Can Use Integrated Management of Hazardous Waste

Figure 21-17 shows an integrated management approach suggested by the U.S. National Academy of Sciences that establishes three levels of priorities for dealing with hazardous waste: produce less of it; convert as much of it as possible to less hazardous substances; and put the rest in long-term, safe storage (**Concept 21-5**). Denmark follows these priorities, but most countries do not.

As with solid waste, the top priority should be pollution prevention and waste reduction. With this approach, industries try to find substitutes for toxic or hazardous materials, reuse or recycle them within industrial processes, or use them as raw materials for making other products (Figure 14-25, p. 367). (See Case Study, p. 366, and the Guest Essays on this subject by Lois Gibbs and Peter Montague at CengageNOW™.)

At least one-third of industrial hazardous wastes produced in Europe are exchanged through clearinghouses where they are sold as raw materials for use by other industries. The producers of these wastes do not have to pay for their disposal, and recipients get low-cost raw materials. About 10% of the hazardous waste in the United States is exchanged through such clearinghouses, a figure that could be raised significantly.

Integrated waste management could be used more for dealing with postconsumer hazardous waste, but such waste is not well managed, globally. For example, most e-waste recycling efforts (**Core Case Study**) create further hazards, especially for workers in some developing countries (see following Case Study).

■ CASE STUDY
Recycling E–Waste

In e-waste recycling operations in some countries, workers—many of them children—are often exposed to toxic chemicals as they dismantle e-wastes to extract valuable metals or parts that can be reused or recycled (**Core Case Study**).

According to the United Nations, more than 70% of the world's e-waste ends up in China. A center for such waste is the small port city of Guiyu, where the air reeks of burning plastic and acid fumes. There, more than 5,500 small-scale e-waste businesses employ over 30,000 people (sometimes including children) who work at very low wages in dangerous conditions to extract valuable metals like gold and copper from millions of discarded computers, television sets, and cell phones.

These workers usually wear no masks or gloves, often work in rooms with no ventilation, and are usually exposed to a cocktail of toxic chemicals. They carry out dangerous activities, such as smashing TV picture tubes by hand to recover glass and electronic parts—a method that releases large amounts of toxic lead dust. They also burn computer wires to expose copper, melt circuit boards in metal pots to extract lead and other metals, and douse the boards with strong acid to extract gold.

After the valuable metals are removed, leftover parts are burned or dumped in rivers or on the land. Atmospheric levels of deadly dioxin in Guiyu are up to 86 times as high as WHO safety standards, and an estimated 82% of the area's children younger than age 6 suffer from lead poisoning.

Produce Less Hazardous Waste	Convert to Less Hazardous or Nonhazardous Substances	Put in Perpetual Storage
■ Change industrial processes to reduce or eliminate hazardous waste production	■ Natural decomposition	■ Landfill
■ Recycle and reuse hazardous waste	■ Incineration	■ Underground injection wells
	■ Thermal treatment	■ Surface impoundments
	■ Chemical, physical, and biological treatment	■ Underground salt formations
	■ Dilution in air or water	

Figure 21-17 *Integrated hazardous waste management:* priorities suggested by the U.S. National Academy of Sciences for dealing with hazardous waste (**Concept 21-5**). To date, these priorities have not been followed in the United States and in most other countries. **Question:** Why do you think that most countries do not follow these priorities? (Data from U.S. National Academy of Sciences)

Partly in response to this problem, Hewlett Packard and Dell Inc., which together sell more than half of the personal computers in the United States, have reduced the use of toxic components and will take back their products for recycling when consumers discard them. Still, in 2007, only 10–15% of the e-waste in the United States was recycled, and up to 80% of that was shipped overseas to dismantling shops such as the one in Guiyu.

Sarah Westervelt of the Basal Action Network (BAN) warns, in fact, that "most of those businesses calling themselves recyclers are little more than international waste distributors. They take your old equipment for free, or pocket your recycling fee, and then simply load it into a sea-going container and ship it to China, India, or Nigeria." In 2008, BAN implemented its *e-Stewards Initiative*, designed to help people find responsible recyclers who work according to strict environmental and health standards.

You can find your closest e-Steward recycler at **ban .org/pledge1.html**. Also see **phones4charity.org** to find out how to recycle or donate your used cell phone. And see **recycle.org** to find out how to donate your used computer for reuse or recycling.

THINKING ABOUT
E-Waste Recycling

Would you pay more for a computer if you knew the higher price would keep the computer out of recycling shops such as those described above? If so, what percentage increase in the price would you accept?

We Can Detoxify Hazardous Wastes

The first step in dealing with hazardous wastes is to collect them. In Denmark, all hazardous and toxic waste from industries and households is delivered to 21 transfer stations throughout the country. From there it is taken to a large treatment facility, where three-fourths of the waste is detoxified by physical, chemical, and biological methods. The rest is buried in a carefully designed and monitored landfill.

Physical methods for detoxifying hazardous wastes include using charcoal or resins to filter out harmful solids and distilling liquid mixtures to separate out harmful chemicals. Especially deadly wastes can be encapsulated in glass, cement, or ceramics and then put in secure storage sites.

Chemical methods are used to convert hazardous chemicals to harmless or less harmful chemicals through chemical reactions. Currently, for example, chemists are testing the use of *cyclodextrin* to remove toxic materials such as solvents and pesticides from contaminated soil and groundwater. After this molecular-sponge-type material moves through the soil or groundwater, picking up various toxic chemicals, it

is pumped out of the ground, stripped of its contaminants, and reused.

Another approach is the use of *nanomagnets*, magnetic nanoparticles coated with certain compounds that can remove various pollutants from water. For example, nanomagnets coated with *chitosan*, a chemical derived from the exoskeletons of shrimps and crabs, can be used to remove oil and other organic pollutants from contaminated water. Magnetic fields are then used to remove the pollutant coated magnetic nanoparticles. The pollutants are then separated out and disposed of or recycled. And the magnetic nanoparticles can be reused.

Some scientists and engineers consider *biological methods* for treatment of hazardous waste to be the wave of the future. One such approach is *bioremediation*, in which bacteria and enzymes help to destroy toxic or hazardous substances or convert them to harmless compounds. (See the Guest Essay by John Pichtel on this topic on the website for this chapter.)

In bioremediation, a contaminated site is inoculated with an army of microorganisms that break down specific hazardous chemicals, such as organic solvents, PCBs, pesticides, and oil, and leave behind harmless substances such as water and water-soluble chloride salts. So far, more than 1,000 different types of bacteria and fungi have been used to detoxify various types of hazardous waste. Bioremediation takes a little longer to work than most physical and chemical methods, but it costs much less.

Another approach is *phytoremediation*, which involves using natural or genetically engineered plants to absorb, filter, and remove contaminants from polluted soil and water, as shown in Figure 21-18. Various plants have been identified as "pollution sponges," which can help to clean up soil and water contaminated with chemicals such as pesticides, organic solvents, and radioactive or toxic metals. Figure 21-19 lists advantages and disadvantages of phytoremediation.

HOW WOULD YOU VOTE?

Do the advantages of using phytoremediation to detoxify hazardous waste outweigh the disadvantages? Cast your vote online at **academic.cengage.com/biology/miller**.

Hazardous wastes can be incinerated to break them down and convert them to harmless or less harmful chemicals such as carbon dioxide and water. This has the same mixture of advantages and disadvantages as burning solid wastes (Figure 21-14). But incinerating hazardous waste can release air pollutants such as toxic dioxins, and it produces a highly toxic ash that must be safely and permanently stored in a landfill or vault especially designed for hazardous waste.

RESEARCH FRONTIER

Improving current methods and finding new ways to detoxify wastes. See **academic.cengage.com/biology/miller**.

Radioactive contaminants	Organic contaminants	Inorganic metal contaminants

Sunflower

Poplar tree

Willow tree

Indian mustard

Brake fern

Polluted groundwater in

Decontaminated water out

Landfill

Oil spill

Polluted leachate

Soil

Groundwater

Soil

Groundwater

Rhizofiltration

Roots of plants such as sunflowers with dangling roots on ponds or in greenhouses can absorb pollutants such as radioactive strontium-90 and cesium-137 and various organic chemicals.

Phytostabilization

Plants such as willow trees and poplars can absorb chemicals and keep them from reaching groundwater or nearby surface water.

Phytodegredation

Plants such as poplars can absorb toxic organic chemicals and break them down into less harmful compoinds which they store or release slowly into the air.

Phytoextraction

Roots of plants such as Indian mustard and brake ferns can absorb toxic metals such as lead, arsenic, and others and store them in their leaves. Plants can then be recycled or harvested and incinerated.

Figure 21-18 Solutions: *phytoremediation.* Various types of plants can be used as pollution sponges to clean up soil and water and radioactive substances (left), organic compounds (center), and toxic metals (right). (Data from American Society of Plant Physiologists, U.S. Environmental Protection Agency, and Edenspace)

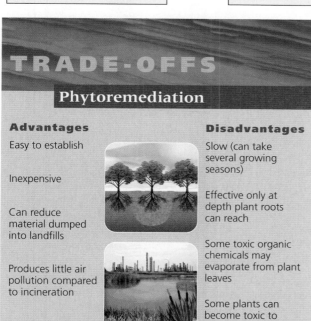

TRADE-OFFS

Phytoremediation

Advantages

Easy to establish

Inexpensive

Can reduce material dumped into landfills

Produces little air pollution compared to incineration

Low energy use

Disadvantages

Slow (can take several growing seasons)

Effective only at depth plant roots can reach

Some toxic organic chemicals may evaporate from plant leaves

Some plants can become toxic to animals

Figure 21-19 Advantages and disadvantages of using *phytoremediation* to remove or detoxify hazardous waste. **Question:** Which single advantage and which single disadvantage do you think are the most important? Why?

TRADE-OFFS

Plasma Arc

Advantages	Disadvantages
Small	High cost
	Produces CO_2 and CO
Mobile. Easy to move to different sites	Can release particulates and chlorine gas
Produces no toxic ash	Can vaporize and release toxic metals and radioactive elements

Figure 21-20 Advantages and disadvantages of using a *plasma arc torch* to detoxify hazardous wastes. **Question:** Which single advantage and which single disadvantage do you think are the most important? Why?

We can also detoxify hazardous wastes by using a *plasma arc torch,* somewhat similar to a welding torch, to incinerate them at very high temperatures. Passing an electrical current through a gas to generate an electric arc and very high temperatures creates *plasma*—an ionized gas made up of electrically conductive ions and electrons.

This process decomposes liquid or solid hazardous organic waste into ions and atoms that can be converted into simple molecules of a synthetic gas (syngas) consisting mostly of hydrogen (H_2) and carbon monoxide (CO). The syngas can then be used to make fuels such as hydrogen, natural gas (CH_4), or ethanol. The high temperatures can also convert hazardous inorganic matter into a molten glassy material that can be used to encapsulate toxic metals to keep them from leaching into groundwater. Figure 21-20 lists the advantages and disadvantages of using this process.

— **HOW WOULD YOU VOTE?** —

Do the advantages of using a plasma arc torch to detoxify hazardous waste outweigh the disadvantages? Cast your vote online at **academic.cengage.com/biology/miller**.

We Can Store Some Forms of Hazardous Waste

Ideally, burial on land or long-term storage of hazardous and toxic wastes should be used only as the third resort after the first two priorities have been exhausted (Figure 21-17 and **Concept 21-5**). But currently, burial on land is the most widely used method in the United States and most countries, largely because it is the least costly method for waste producers.

The most common form of burial is *deep-well disposal*, in which liquid hazardous wastes are pumped through a pipe into dry, porous rock formations far beneath aquifers, many of which are tapped for drinking and irrigation water. Theoretically, these liquids soak into the porous rock and are isolated from overlying groundwater by essentially impermeable layers of clay and rock.

However, there are a limited number of such sites and limited space within them. Sometimes the wastes leak into groundwater from the well shaft or migrate into groundwater in unexpected ways. In the United States, roughly 64% of liquid hazardous wastes are injected into deep disposal wells. Many scientists believe that current regulations for deep-well disposal in the United States are inadequate and should be improved. Figure 21-21 lists the advantages and disadvantages of deep-well disposal of liquid hazardous wastes.

— **HOW WOULD YOU VOTE?** —

Do the advantages of deep-well disposal of hazardous waste outweigh the disadvantages? Cast your vote online at **academic.cengage.com/biology/miller**.

Surface impoundments are ponds, pits, or lagoons into which liners are placed and liquid hazardous wastes are stored (Figure 21-22). As the water evaporates, the waste settles and becomes more concentrated. But inadequate seals can allow such wastes to percolate into the groundwater, and volatile harmful chemicals can

TRADE-OFFS

Deep-Well Disposal

Advantages	Disadvantages
Safe method if sites are chosen carefully	Leaks or spills at surface
	Leaks from corrosion of well casing
Wastes can often be retrieved if problems develop	Existing fractures or earthquakes can allow wastes to escape into groundwater
Easy to do	Output approach that encourages waste production
Low cost	

Figure 21-21 Advantages and disadvantages of injecting liquid hazardous wastes into deep underground wells. **Question:** Which single advantage and which single disadvantage do you think are the most important? Why?

Figure 21-22 *Surface impoundment* in Niagara Falls, New York (USA). Such sites can pollute the air and nearby groundwater and surface water.

Ken Sherman/Bruce Coleman USA, Inc.

TRADE-OFFS

Surface Impoundments

Advantages	Disadvantages
Low construction costs	Groundwater contamination from leaking liners (or no lining)
Low operating costs	Air pollution from volatile organic compounds
Can be built quickly	Overflow from flooding
Wastes can often be retrieved if necessary	Disruption and leakage from earthquakes
Can store wastes indefinitely with secure double liners	Output approach that encourages waste production

Figure 21-23 Advantages and disadvantages of storing liquid hazardous wastes in surface impoundments. **Question:** Which single advantage and which single disadvantage do you think are the most important? Why?

evaporate into the air. Also, powerful storms can cause these impoundments to overflow. Figure 21-23 lists the advantages and disadvantages of this method.

EPA studies found that 70% of these storage basins in the United States have no liners, and up to 90%

of them may threaten groundwater. According to the EPA, eventually all liners probably will leak and could contaminate groundwater.

HOW WOULD YOU VOTE?

Do the advantages of storing hazardous wastes in surface impoundments outweigh the disadvantages? Cast your vote online at **academic.cengage.com/biology/miller**.

There are some highly toxic materials (such as mercury, Science Focus, p. 450) that we cannot destroy, detoxify, or safely bury. The best way to deal with such materials is to prevent or reduce their use and to put waste from such materials in special containers. The containers are then stored aboveground in specially designed buildings or underground in salt mines or bedrock caverns, where they can be inspected on a regular basis and retrieved if necessary. Carefully designed aboveground storage buildings are a good option in areas where the water table is close to the surface and in areas that lie above aquifers used for drinking water. Such storage structures should be built to withstand storms and to prevent the release of toxic gases, and leaks should be carefully monitored.

Sometimes, liquid and solid hazardous wastes are put into drums or other containers and buried in carefully designed and monitored *secure hazardous waste landfills* (Figure 21-24, p. 582). This is the least used method because of the expense involved.

Some developed countries are careless with their hazardous wastes. In the United Kingdom, most such wastes are mixed with household garbage and stored

Bulk waste | Gas vent | Topsoil | Plastic cover
Earth
Sand | Impervious clay cap | Clay cap
Impervious clay

Water table

Earth

Groundwater

Double leachate collection system | Plastic double liner | Reactive wastes in drums | Leak detection system | Groundwater monitoring well

Figure 21-24 Solutions: secure hazardous waste landfill.

in hundreds of conventional landfills throughout the country. Most developing countries do little to regulate and control what happens to the hazardous wastes they produce.

Figure 21-25 lists some ways in which you can reduce your output of hazardous waste—the first step in dealing with it.

WHAT CAN YOU DO?

Hazardous Waste

■ Avoid using pesticides and other hazardous chemicals, or use them in the smallest amounts possible

■ Use less harmful substances instead of commercial chemicals for most household cleaners. For example, use vinegar to polish metals, clean surfaces, and remove stains and mildew; baking soda to clean household utensils and to deodorize and remove stains; and borax to remove stains and mildew.

■ Do not dispose of pesticides, paints, solvents, oil, antifreeze, or other hazardous chemicals by flushing them down the toilet, pouring them down the drain, burying them, throwing them into the garbage, or dumping them down storm drains. Instead, use hazardous waste disposal services available in many cities.

Figure 21-25 Individuals matter: ways to reduce your output of hazardous waste (**Concept 21-5**). **Questions:** When was the last time you created hazardous waste? Could you have avoided it, and if so, how?

■ **CASE STUDY**

Hazardous Waste Regulation in the United States

About 5% of all hazardous waste produced in the United States is regulated under the Resource Conservation and Recovery Act (RCRA, pronounced "RICK-ra"), passed in 1976 and amended in 1984. The EPA sets standards for management of several types of hazardous waste and issues permits to companies allowing them to produce and dispose of a certain amount of wastes in acceptable ways. Permit holders must use a *cradle-to-grave* system to keep track of waste they transfer from a point of generation (cradle) to an approved disposal facility (grave), and they must submit proof of this disposal to the EPA.

RCRA is a good start, but this and other laws regulate only about 5% of the hazardous and toxic wastes, including e-waste, produced in the United States. In most other countries, especially developing countries, even less of this waste is regulated.

THINKING ABOUT
Hazardous Waste

CORE CASE STUDY

Why is it that 95% of the hazardous waste, including the growing mounds of e-waste (**Core Case Study**) produced in the United States, is not regulated? Do you favor regulating such wastes? What are the economic consequences of doing this? How would this change the way waste producers deal with the hazardous wastes they produce?

In 1980, the U.S. Congress passed the *Comprehensive Environmental Response, Compensation, and Liability Act,* commonly known as the *CERCLA* or *Superfund* program. Its goals are to identify sites where hazardous wastes have contaminated the environment (Figure 21-26) and to clean them up on a priority basis. The worst sites, which represent an immediate and severe threat to human health, are put on a *National Priorities List* and scheduled for total cleanup using the most cost-effective method.

In 2008, there were about 1,240 sites on this list. The Waste Management Research Institute estimates that at least 10,000 sites should be on the priority list and that cleanup of these sites would cost about $1.7 trillion, not including legal fees. This shows the economic and environmental value of emphasizing waste reduction and pollution prevention.

Since 1980, the EPA has dealt with 1,579 Superfund sites. So far 321 have been cleaned up and removed from the Superfund list. The degree of cleanup at the remaining sites varies.

In 1984, Congress amended the Superfund Act to give citizens the right to know what toxic chemicals are being stored or released in their communities. This required 23,800 large manufacturing facilities to report their annual releases of any of nearly 650 toxic chemicals into the environment. If you live in the United

States, you can find out what toxic chemicals are being stored and released in your neighborhood by going to the EPA's *Toxic Release Inventory* website at **www.epa .gov/tri/**.

The Superfund law, designed to have polluters pay for cleaning up abandoned hazardous waste sites, has virtually made illegal dump sites (Figure 21-26) relics of the past. It has forced waste producers, fearful of future liability claims, to reduce their production of such waste and to recycle or reuse much more of it. However, facing pressure from polluters, the U.S. Congress refused to renew the tax on oil and chemical companies that had financed the Superfund after it expired in 1995. The Superfund is now broke, and taxpayers, not polluters, are footing the bill for future cleanups when the responsible parties cannot be found. As a result, the pace of cleanup has slowed.

— HOW WOULD YOU VOTE? ☑ —

Should the U.S. Congress reinstate the polluter-pays principle by using taxes from chemical, oil, mining, and smelting companies to reestablish a fund for cleaning up existing and new Superfund sites? Register your vote online at **academic .cengage.com/biology/miller**.

The U.S. Congress and several state legislatures have also passed laws that encourage the cleanup of *brownfields*—abandoned industrial and commercial sites such as factories, junkyards, older landfills, and gas stations. In most cases, they are contaminated with hazardous wastes. Brownfields can be cleaned up and reborn as parks, nature reserves, athletic fields, ecoindus-

Figure 21-26 Leaking barrels of toxic waste at a Superfund site in the United States. This site has since been cleaned up.

trial parks (Figure 14-25, p. 367), and neighborhoods. By 2008, more than 42,000 former brownfield sites had been redeveloped in the United States.

In addition to laws that are focused on cleaning up hazardous waste, other laws have been passed to try to prevent creation of such wastes. One of the most successful was the 1976 law requiring that use of leaded gasoline be phased out in the United States (Case Study, p. 474).

21-6 How Can We Make the Transition to a More Sustainable Low-Waste Society?

▶ **CONCEPT 21-6** Shifting to a low-waste society requires individuals and businesses to reduce resource use and to reuse and recycle wastes at local, national, and global levels.

Grassroots Action Has Led to Better Solid and Hazardous Waste Management

In the United States, individuals have organized to prevent the construction of hundreds of incinerators, landfills, treatment plants for hazardous and radioactive wastes, and polluting chemical plants in or near their communities. Health risks from incinerators and landfills, when averaged for the entire country, are quite low, but the risks for people living near such facilities are much higher.

Manufacturers and waste industry officials point out that something must be done with the toxic and hazardous wastes produced to provide people with certain goods and services. They contend that even if local citizens adopt a "not in my back yard" (NIMBY) approach, the waste will always end up in someone's back yard.

Many citizens do not accept this argument. To them, the best way to deal with most toxic and hazardous waste is to produce much less of it, as suggested by the U.S. National Academy of Sciences (Figure 21-17). For such materials, they believe that the goal should be "not in anyone's back yard" (NIABY) or "not on planet Earth" (NOPE), which calls for drastically reducing

Providing Environmental Justice for Everyone Is an Important Goal

Environmental justice is an ideal whereby every person is entitled to protection from environmental hazards regardless of race, gender, age, national origin, income, social class, or any political factor. (See the Guest Essay on this subject by Robert Bullard on the website for this chapter.)

Studies have shown that a disproportionate share of polluting factories, hazardous waste dumps, incinerators, and landfills in the United States are located in communities populated mostly by African Americans, Asian Americans, Latinos, and Native Americans and the working poor. Studies have also shown that, in general, toxic waste sites in white communities have been cleaned up faster and more completely than such sites in African American and Latino communities have.

Such environmental discrimination in the United States and in other parts of the world has led to a growing grassroots movement known as the *environmental justice movement*. Members of this group have pressured governments, businesses, and environmental groups to become aware of environmental injustice and to act to prevent it. They have made some progress toward their goals, but there is a long way to go.

THINKING ABOUT
Environmental Injustice

Have you or anyone in your family ever been a victim of environmental injustice? If so, describe what happened. What would you do to help prevent environmental injustice?

Countries Have Developed International Treaties to Reduce Hazardous Waste

Environmental justice also applies at the international level. For decades, some developed countries had been shipping hazardous wastes to developing countries. In 1989, the UNEP developed an international treaty known as the Basel Convention. It banned developed countries that participate in the treaty from shipping hazardous waste (including e-waste) to or through other countries without their permission. In 1995, the treaty was amended to outlaw all transfers of hazardous wastes from industrial countries to developing countries. By 2008, this agreement had been ratified by 152 countries, but not by the United States, Afghanistan, and Haiti.

This ban will help, but it will not wipe out the very profitable illegal waste trade. Smugglers evade the laws by using an array of tactics, including bribes, false permits, and mislabeling of hazardous wastes as materials to be recycled.

In 2000, delegates from 122 countries completed a global treaty to control 12 *persistent organic pollutants* (POPs). These widely used toxic chemicals are persistent, insoluble in water, and soluble in fat. This means that they can be concentrated in the fatty tissues of humans and other organisms feeding at high trophic levels in food webs to levels hundreds of thousands of times higher than levels in the general environment (Figure 9-19, p. 202). Because they are persistent, POPs can also be transported long distances by wind and water.

The original list of 12 chemicals, called the *dirty dozen*, includes DDT and 8 other chlorine-containing persistent pesticides, PCBs (Case Study, p. 449), dioxins, and furans. Using blood tests, medical researchers at New York City's Mount Sinai School of Medicine found that nearly every person on earth has detectable levels of POPs. The long-term health effects of this involuntary global chemical experiment are largely unknown.

The treaty seeks to ban or phase out use of these chemicals and to detoxify or isolate stockpiles of them. It allows 25 countries to continue using DDT to combat malaria until safer alternatives are available. The United States has not ratified this treaty.

Environmental scientists consider the POPs treaty to be an important milestone in international environmental law and pollution prevention because it uses the precautionary principle to manage and reduce the risks from toxic chemicals. The list of persistant organic pollutants is expected to grow.

In 2000, the Swedish Parliament enacted a law that, by 2020, will ban all chemicals that are persistent and can bioaccumulate in living tissue. This law also requires industries to perform risk assessments on the chemicals they use and to show that these chemicals are safe to use, as opposed to requiring the government to show that they are dangerous. In other words, chemicals are assumed to be guilty until proven innocent—the reverse of the current policy in the United States and in most countries. There is strong opposition to this approach in the United States, especially from most industries producing potentially dangerous chemicals. The European Union is also considering legislation that puts the burden of proof on manufacturers to show that about 30,000 industrial chemicals and substances are safe.

We Can Make the Transition to Low-Waste Societies

According to physicist Albert Einstein, "A clever person solves a problem, a wise person avoids it." Some are taking these words seriously. The governments of Norway, Austria, and the Netherlands have committed to reduc-

ing their resource waste by 75%. In a pilot study, residents of the U.S. city of East Hampton, New York, cut their solid waste production by 85%.

To prevent pollution and reduce waste, many environmental scientists urge us to understand and follow several key principles:

- Everything is connected.

- There is no *away*, as in *to throw away*, for the wastes we produce.

- Dilution is not always the solution to pollution.

- Polluters and producers should pay for the wastes they produce.

- Different categories of hazardous waste and recyclable waste should not be mixed.

- We should mimic nature by reusing, recycling, or composting at least 75% of the solid wastes we produce.

- The best and cheapest ways to deal with solid and hazardous wastes are waste reduction and pollution prevention.

CENGAGENOW™ Learn more about how shifting to a low-waste (low-throughput) economy would be the best long-term solution to environmental and resource problems at CengageNOW.

REVISITING

E-Waste and Sustainability

The growing problem of e-waste (**Core Case Study**) and other topics discussed in this chapter represent the problems of maintaining a high-waste society. The challenge is to make the transition from a high-waste, throwaway mode to a low-waste, reducing-reusing-recycling economy.

Such a transition will require applying the four **scientific principles of sustainability**. Shifting from reliance on fossil fuels and nuclear power (which produces long-lived, hazardous, radioactive wastes) to greater use of renewable solar energy,

wind, and flowing water will reduce our outputs of solid and hazardous waste, as will reusing and recycling materials by mimicking nature's chemical cycling processes. Integrated waste management (Figures 21-5 and 21-17), using a diversity of approaches with emphasis on waste reduction and pollution prevention, is another useful way to mimic nature. Reducing the human population and the resources used per person would also decrease the demand for materials that eventually become solid and hazardous wastes.

The key to addressing the challenge of toxics use and wastes
rests on a fairly straightforward principle:
harness the innovation and technical ingenuity
that has characterized the chemicals industry from its beginning
and channel these qualities in a new direction
that seeks to detoxify our economy.

ANNE PLATT MCGINN

REVIEW

1. Review the Key Questions and Concepts for this chapter on p. 561. Describe the problems associated with electronic waste (e-waste) (**Core Case Study**).

2. Distinguish among **solid waste, industrial solid waste, municipal solid waste (MSW),** and **hazardous (toxic) waste** and give an example of each. Give two reasons for sharply reducing the amount of solid and hazardous waste we produce. Describe the production of solid waste in the United States and what happens to such waste.

3. Distinguish among **waste management, waste reduction,** and **integrated waste management.** Describe the priorities that prominent scientists believe we should use for dealing with solid waste. What is garbology? Distinguish among **reducing, reusing,** and **recycling** as strate-

gies for waste reduction. Describe six ways in which industries and communities can reduce resource use, waste, and pollution.

4. Explain why reusing and recycling materials are so important and give two examples of each. Describe the importance of using refillable containers and list five other ways to reuse various items. Distinguish between **primary (closed-loop)** and **secondary recycling** and give an example of each. Describe two approaches to recycling household solid wastes and evaluate each approach. What is a materials recovery facility? What is composting?

5. Describe the recycling of paper and the problems involved. Describe the recycling of plastics and the problems involved. Describe progress in recycling plastics. What are bioplastics? What are the major advantages and

disadvantages of recycling? What are three factors that discourage recycling? Describe three ways to encourage recycling and reuse.

6. What are the major advantages and disadvantages of using incinerators to burn solid and hazardous waste? Distinguish between **open dumps** and **sanitary landfills.** What are the major advantages and disadvantages of burying solid waste in sanitary landfills?

7. What are the priorities that scientists from the National Academy of Sciences believe we should use in dealing with hazardous waste? What is phytoremediation and what are the major advantages and disadvantages of using it to remove or detoxify hazardous wastes? What are the major advantages and disadvantages of using a plasma torch to detoxify hazardous wastes?

8. What are the major advantages and disadvantages of disposing of liquid hazardous wastes in **(a)** deep underground wells and **(b)** surface impoundments? What is

a secure hazardous waste landfill? Describe the regulation of hazardous waste in the United States under the Resource Conservation and Recovery Act and the Comprehensive Environmental Response, Compensation, and Liability (or Superfund) Act. What is a brownfield?

9. How has grassroots action improved solid and hazardous waste management in the United States? What is **environmental justice** and how well has it been applied in locating and cleaning up hazardous waste sites in the United States? Describe regulation of hazardous wastes at the global level through the Basel Convention and the treaty to control persistent organic pollutants.

10. Describe connections between dealing with the growing problem if e-waste (**Core Case Study**) and the four **scientific principles of sustainability.**

Note: Key Terms are in **bold** type.

CRITICAL THINKING

1. Do you think that manufacturers of computers and television sets and other forms of e-waste (**Core Case Study**) should be required to take them back at the ends of their useful lives for repair, remanufacture, or recycling? Explain. Would you be willing to pay more for these products to cover the costs of such a takeback program? If so, what percent more per purchase would you be willing to pay?

2. Find three items you regularly use once and then throw away. Are there other reusable items that you could use in place of these disposable items? Compare the cost of using the disposable option for a year versus the cost of using the alternatives.

3. Use the second law of thermodynamics (p. 43) to explain why a *properly designed* source-separation recycling program takes less energy and produces less pollution than a centralized program that collects mixed waste over a large area and hauls it to a centralized facility where workers or machinery separate the wastes for recycling.

4. Changing World Technologies has built a pilot plant to test a process it has developed for converting a mixture of computers, old tires, turkey bones and feathers, and other wastes into oil by mimicking and speeding up natural processes for converting biomass into oil. If this recycling process turns out to be technologically and economically feasible, explain why it could increase waste production.

5. Would you oppose having a hazardous waste landfill, waste treatment plant, deep-injection well, or incinerator in your community? For each of these facilities, explain your answer. If you oppose these disposal facilities, how do you believe the hazardous waste generated in your community should be managed?

6. How does your school dispose of its solid and hazardous waste? Does it have a recycling program? How well does it work? Does it have a hazardous waste collection system? If so, what does it do with these wastes? List three ways to improve your school's waste reduction and management system.

7. Give your reasons for agreeing or disagreeing with each of the following proposals for dealing with hazardous waste:
 a. Reduce the production of hazardous waste and encourage recycling and reuse of hazardous materials by charging producers a tax or fee for each unit of waste generated.
 b. Ban all land disposal and incineration of hazardous waste to protect air, water, and soil from contamination and to encourage reuse, recycling, and treatment of wastes to make them less hazardous.
 c. Provide low-interest loans, tax breaks, and other financial incentives to encourage industries that produce hazardous waste to reduce, reuse, recycle, treat, and decompose such waste.

8. List three ways in which you could apply **Concept 21-6** to making your lifestyle more environmentally sustainable.

9. Congratulations! You are in charge of the world. List the three most important components of your strategy for dealing with **(a)** solid waste and **(b)** hazardous waste.

10. List two questions you would like to have answered as a result of reading this chapter.

Note: See Supplement 13 (p. S78) for a list of Projects related to this chapter.

This pie chart diagram from an EPA report shows the typical composition of U.S. municipal solid waste (MSW) in 2006.

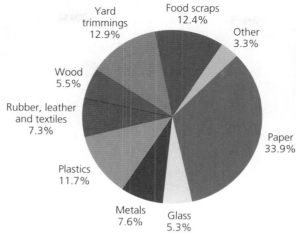

Yard trimmings 12.9%

Food scraps 12.4%

Other 3.3%

Wood 5.5%

Rubber, leather and textiles 7.3%

Plastics 11.7%

Metals 7.6%

Glass 5.3%

Paper 33.9%

Source: U.S. Environmental Protection Agency, 2008, **www.epa.gov/ epaoswer/non-hw/muncpl/pubs/msw06.pdf**

1. The average daily municipal solid waste production per person in the United States in 2006 was 2.10 kilograms (4.60 pounds). Use the data in the figure above to understand what makes up the annual MSW ecological footprint for a typical American. For each category in the pie chart, calculate the total weight in kilograms (and pounds) generated in 2006 by the average American. (*Note:* 1 kilogram = 2.20 pounds)

The Ecocity Concept in Curitiba, Brazil

Urban planners envision large portions of the growing human population living in more environmentally sustainable cities, called *ecocities* or *green cities*. This is not just a futuristic dream. One such ecocity is Curitiba ("koor-i-TEE-ba"), a city of 3.2 million people known as the "ecological capital" of Brazil.

Planners in this city decided in 1969 to focus on an inexpensive and efficient mass transit rather than on the car. Curitiba now has the world's best bus system, in which clean and modern buses transport about 72% of the population every day throughout the city along express lanes dedicated to buses (Figure 22-1). Only high-rise apartment buildings are allowed near major bus routes, and each building must devote its bottom two floors to stores—a practice that reduces the need for residents to travel.

Cars are banned from 49 blocks in the center of the downtown area, which has a network of pedestrian walkways connected to bus stations, parks, and bicycle paths running throughout most of the city. Consequently, Curitiba uses less energy per person and has less emissions of greenhouse gases and other air pollutants and traffic congestion than do most comparable cities.

The city transformed flood-prone areas along its rivers into a series of interconnected parks. Volunteers have planted more than 1.5 million trees throughout the city, none of which can be cut down without a permit. And two trees must be planted for each one that is cut down.

Curitiba recycles roughly 70% of its paper and 60% of its metal, glass, and plastic, which is collected from households three times a week. Recovered materials are sold mostly to the city's more than 500 major industries, which must meet strict pollution standards. Most of these businesses are located in an industrial park outside the city limits. A major bus line runs to the industrial park, but many of the workers live nearby and can walk or bike to work.

The poor receive free medical and dental, child care, and job training, and 40 feeding centers are available for street children. Poor people living in squatter settlements that garbage trucks cannot reach can exchange filled garbage bags for surplus food, bus tokens, and school supplies. Also, the city has a *build-it-your-self* program that gives a poor family a plot of land, building materials, two trees, and an hour's consultation with an architect.

This internationally acclaimed model of urban planning and sustainability is the brainchild of architect and former college professor Jaime Lerner, who has served as the city's mayor three times since 1969.

Curitiba now faces new challenges, as do all cities, mostly due to increasing population pressure. Throughout this chapter, we will examine such challenges and possible solutions for Curitiba and other urban areas.

John Maier, Jr./Peter Arnold, Inc.

Route
━ Express ━ Interdistrict ━ Direct ━ Feeder ━ Workers

Figure 22-1 Solutions: bus rapid transit (BRT) system in Curitiba, Brazil. This system moves large numbers of passengers around rapidly because each of the five major spokes has two express lanes used only by buses. Double- and triple-length bus sections are hooked together as needed to carry up to 300 passengers. Boarding is speeded up by the use of extra-wide doors and boarding platforms under glass tubes where passengers can pay before getting on the bus (top).

22-1 What are the major population trends in urban areas?

CONCEPT 22-1 Urbanization continues to increase steadily and the numbers and sizes of urban areas are growing rapidly, especially in developing countries.

22-2 What are the major urban resource and environmental problems?

CONCEPT 22-2 Most cities are unsustainable because of high levels of resource use, waste, pollution, and poverty.

22-3 How does transportation affect urban environmental impacts?

CONCEPT 22-3 A combination of plentiful land, inexpensive fuel, and expanding networks of highways in some countries has resulted in dispersed cities whose residents depend on motor vehicles for most transportation.

22-4 How important is urban land use planning?

CONCEPT 22-4 Urban land use planning can help to reduce uncontrolled sprawl and to slow the resulting degradation of air, water, land, biodiversity, and other natural resources.

22-5 How can cities become more sustainable and livable?

CONCEPT 22-5 An *ecocity* allows people to: choose walking, biking, or mass transit for most transportation needs; recycle or reuse most of their wastes; grow much of their food; and protect biodiversity by preserving surrounding land.

Note: Supplements 2 (p. S4), 3 (p. S10) and 13 (p. S78) can be used with this chapter.

The city is not an ecological monstrosity.
It is rather the place where both the problems and the opportunities
of modern technological civilization are most potent and visible.

PETER SELF

22-1 What Are the Major Population Trends in Urban Areas?

▶ **CONCEPT 22-1** Urbanization continues to increase steadily and the numbers and sizes of urban areas are growing rapidly, especially in developing countries.

Half of the World's People Live in Urban Areas

Urbanization is the creation and growth of *urban areas,* or cities and their surrounding developed land. It is measured as the percentage of the people in a country or in the world living in urban areas. **Urban growth** is the *rate* of increase of urban populations. The world's first cities emerged about 6,000 years ago. Since then, the world has become increasingly urbanized. Today 79% of Americans and almost 50% of the world's people live in urban areas (Figure 12, p. S16, Supplement 3).

Urban areas grow in two ways—by *natural increase* (more births than deaths) and by *immigration,* mostly from rural areas. Rural people are *pulled* to urban areas in search of jobs, food, housing, educational opportunities, better health care, entertainment, and freedom from religious, racial, and political conflicts. Some are also *pushed* from rural to urban areas by factors such as

poverty, lack of land for growing food, declining agricultural jobs, famine, and war.

People are also pushed and pulled to cities by government policies that favor urban over rural areas. For example, developing countries tend to spend most of their budgets on economic development and job creation in urban areas and especially in capital cities where their leaders live. Some governments establish lower food prices in urban areas, which rewards city dwellers, helps to keep leaders in power, and attracts the rural poor.

Four major trends in urban population dynamics do seem clear, and they are important for understanding the problems and challenges of urban growth (**Concept 22-1**). First, *the proportion of the global population living in urban areas is increasing.* Between 1850 and 2008, the percentage of people living in urban areas increased from 2% to almost 50% and could reach 60% by 2030 (Figure 12, p. S16, Supplement 3). About 88% of this

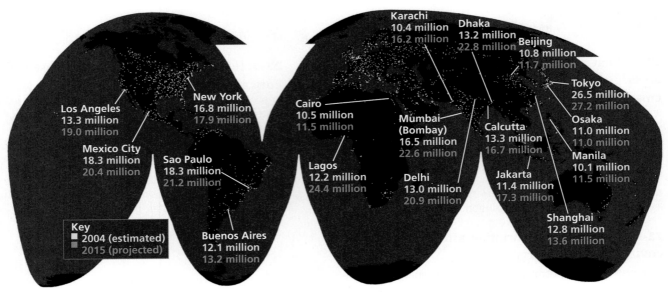

Figure 22-2 *Global outlook:* major urban areas throughout the world revealed in satellite images of the earth at night, showing city lights. Currently, the 50% of the world's people living in urban areas occupy about 2% of the earth's land area. Note that most of the urban areas are found along the coasts of continents, which explains why most of Africa and much of the interior of South America, Asia, and Australia are dark at night. This figure also shows the populations of the world's 18 *megacities* (each with 10 million or more people) in 2008 and their projected populations in 2015. All but three are located in developing countries. **Question:** In order, what were the world's five most populous cities in 2008, and what will likely be the five most populous cities in 2015? (Data from National Geophysics Data Center, National Oceanic and Atmospheric Administration, and United Nations)

growth will occur in already overcrowded and stressed cities in developing countries, where 44% of the people live in urban areas.

Second, *urban areas are expanding rapidly in number and size.* Each week 1 million people are added to the world's urban areas. Between 2008 and 2015, the number of urban areas with a million or more people is projected to increase from 400 to 564. And there are 18 *megacities* or *megalopolises*—cities with 10 million or more people—up from 8 in 1985. Fifteen of them are in developing countries (Figure 22-2). Such megacities will soon be eclipsed by *hypercities* with more than 20 million people each. So far, Tokyo, Japan, with 26.5 million people, is the only city in this category. But according to U.N. projections, Mumbai (formerly Bombay) in India, Lagos in Nigeria, Dakar in Bangladesh, and São Paulo in Brazil will become hypercities by 2015.

Third, *urban growth is much slower in developed countries than in developing countries.* Still, developed countries, now with 75% urbanization, are projected to reach 81% urbanization by 2030.

Fourth, *poverty is becoming increasingly urbanized, mostly in developing countries* (Figure 12, p. S16, Supplement 3). The United Nations estimates that at least 1 billion people in developing countries live in crowded and unsanitary slums and shantytowns within most cities or on their outskirts; within 30 years this number may double.

Figure 22-3 Typical daily traffic jam of people, carts, and other vehicles in an older section of Delhi, India—a city with 13 million people.

Mark Edwards/Peter Arnold, Inc.

THINKING ABOUT
Urban Trends

If you could reverse one of the four urban trends discussed here, which one would it be? Which of these trends has Curitiba, Brazil, (**Core Case Study**) reversed?

CORE CASE STUDY

If you visit a poor area of a typical city in a developing country, your senses may be overwhelmed by a chaotic but vibrant crush of people, vehicles of all types, congestion, noise, traffic jams (Figure 22-3), and smells, including smoke from burning trash and wood and coal cooking fires and raw sewage. Many people sleep on the streets or live in crowded, unsanitary, rickety, and unsafe slums and shantytowns with little or no access to safe drinking water or modern sanitation facilities.

■ CASE STUDY

Urbanization in the United States

Between 1800 and 2008, the percentage of the U.S. population living in urban areas increased from 5% to 79%. This population shift has occurred in four phases.

First, *people migrated from rural areas to large central cities.* Currently, three-fourths of Americans live in cities with at least 50,000 people, and nearly half live in urban areas with 1 million or more residents (Figure 22-4).

Second, *many people migrated from large central cities to suburbs and smaller cities.* Currently, about half of urban Americans live in the suburbs, nearly a third in central cities, and the rest in rural housing developments beyond suburbs.

Third, *many people migrated from the North and East to the South and West.* Since 1980, about 80% of the U.S. population increase has occurred in the South and West. Between 2008 and 2043, demographers project that the fastest growing U.S. states will continue to be Nevada, Arizona, and Florida, although increased drought and heat waves due to climate change may alter this trend. According to a 2006 study by the Center for Environment and Population, the South and West also contain many of the country's ecological hotspots (Figure 10-27, p. 241) where biodiversity is being threatened.

Fourth, since the 1970s, and especially since 1990, *some people have fled both cities and suburbs and migrated to developed rural areas.* The result is rapid growth of *exurbs*—housing developments scattered over vast areas that lie beyond suburbs and have no socio-economic centers.

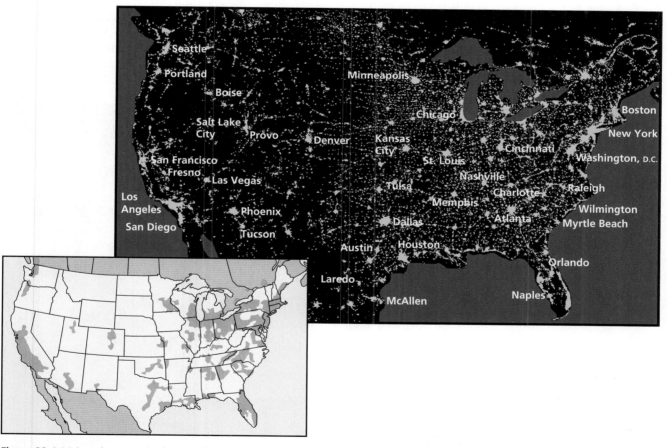

Figure 22-4 Major urban areas in the United States revealed in satellite images of the earth at night showing city lights (top). Almost eight of every ten Americans live in urban areas, which occupy a small but growing fraction of the country's land area. Areas with names in white are the fastest-growing metropolitan areas. Nearly half (48%) of all Americans live in cities of 1 million or more people, which are projected to merge into huge urban areas shown as shaded areas in the bottom map. **Questions:** Why are most of the largest urban areas located near water? What effect might climate change due to global warming have on these cities? (Data from National Geophysical Data Center/National Oceanic and Atmospheric Administration, U.S. Census Bureau)

Since 1920, many of the worst urban environmental problems in the United States have been reduced significantly. Most people have better working and housing conditions, and air and water quality have improved. Better sanitation, public water supplies, and medical care have slashed death rates and incidences of sickness from malnutrition and infectious diseases (Figure 6-6, p. 128). Concentrating most of the population in urban areas also has helped protect the country's biodiversity by reducing the destruction and degradation of wildlife habitat.

However, a number of U.S. cities—especially those that are older—have deteriorating services and aging *infrastructures* (streets, bridges, schools, housing, water supply pipes, dams, and sewers). Indeed, according to a 2007 study by the U.S. American Society of Civil Engineers, the United States has fallen $1.5 trillion behind in maintaining its vital public infrastructure. This maintenance backlog amounts to an average of $5,000 for each U.S. citizen. Maintenance is not glamorous, but without it, urban areas and much of the country can suffer from physical and economic collapse as the costs of long-neglected repairs rise sharply.

At a time when there are urgent needs to upgrade rapidly declining infrastructure, many U.S. cities face budget crunches and decreasing public services, as businesses and people move to the suburbs and exurbs and city revenues from property taxes decline. And there is rising poverty in the centers of many older cities, where unemployment rates are typically 50% or higher.

Urban Sprawl Gobbles Up the Countryside

In the United States and some other countries, **urban sprawl**—the growth of low-density development on the edges of cities and towns—is eliminating surrounding agricultural and wild lands (Figure 22-5). It results in a far-flung hodgepodge of housing developments, shopping malls, parking lots, and office complexes that are loosely connected by multilane highways and freeways.

Six major factors promoted urban sprawl in the United States. *First,* ample land was available for most cities to spread outward. *Second,* federal government loan guarantees for new single-family homes for World War II veterans stimulated the development of suburbs starting around 1950. *Third,* low-cost gasoline and federal and state funding of highways encouraged automobile use and the development of outlying tracts of land. *Fourth,* tax laws encouraged home ownership. *Fifth,* most state and local zoning laws favored large residential lots and separation of residential and commercial areas. And *sixth,* most urban areas consist of multiple political jurisdictions, which rarely work together to develop an overall plan for managing urban growth.

UNEP/GRID-Sioux Falls

UNEP/GRID-Sioux Falls

1973

2000

Figure 22-5 *Urban sprawl* in and around the U.S. city of Las Vegas, Nevada, between 1973 and 2000—a process that has continued. Between 1970 and 2008, the population of water-short Clark Country, which includes Las Vegas, more than quadrupled from 463,000 to around 2 million, making it one of the nation's fastest-growing urban areas. And the growth is expected to continue. **Question:** What might be a limiting factor on population growth in the Las Vegas area?

NATURAL CAPITAL DEGRADATION

Urban Sprawl

Land and Biodiversity	Water	Energy, Air, and Climate	Economic Effects
Loss of cropland	Increased use of surface water and groundwater	Increased energy use and waste	Decline of downtown business districts
Loss of forests and grasslands	Increased runoff and flooding	Increased air pollution	Increased unemployment in central city
Loss of wetlands	Increased surface water and groundwater pollution	Increased greenhouse gas emissions	Loss of tax base in central city
Loss and fragmentation of wildlife habitats	Decreased natural sewage treatment	Enhanced global warming	

In a nutshell, urban sprawl is the product of affordable land, automobiles, cheap gasoline, and poor urban planning. It has caused or contributed to a number of environmental problems. Because of nonexistent or inadequate mass transportation in most such areas, sprawl forces people to drive everywhere and, in the process, to emit greenhouse gases and other forms of air pollution. Sprawl has decreased energy efficiency, increased traffic congestion, and destroyed prime cropland, forests, and wetlands. It has also led to the economic death of many central cities as people and businesses moved out of these areas. Figure 22-6 summarizes some of the undesirable consequences of urban sprawl.

On the other hand, many people prefer living in suburbs and exurbs. Compared to central cities, these areas provide lower density living and access to larger lot sizes and single-family homes. Often they also have newer public schools and lower crime rates.

THINKING ABOUT
Urban Sprawl

Do you think the advantages of sprawl outweigh its disadvantages? What might happen to suburban and exurban areas as the prices of oil and gasoline continue to rise sharply? How might this affect your lifestyle?

As they grow outward, separate urban areas sometimes merge to form a *megalopolis*. For example, the remaining open spaces between the U.S. cities of Boston, Massachusetts, and Washington, D.C., are rapidly urbanizing and coalescing. The result is an almost 800-kilometer-long (500-mile-long) urban area that contains about 35 million people and is sometimes called *Bowash* (Figure 22-7 and Figure 22-4).

Megalopolises are developing all over the world and include the areas between Amsterdam and Paris in Europe, Japan's Tokyo–Yokohama–Osaka–Kobe corridor known as Tokohama (with nearly 50 million people), and the Brazilian industrial triangle made up of São Paulo, Rio de Janeiro, and Belo Horizonte.

CENGAGENOW™ Examine how the San Francisco Bay area in the U.S. state of California grew in population between 1900 and 1990 at CengageNow™.

Bowash (Boston to Washington)

Figure 22-7 *U.S. megalopolis of Bowash,* consisting of urban sprawl and coalescence between Boston, Massachusetts, and Washington, D.C. **Question:** What are two ways in which this development might be harming ecosystems along the Atlantic Coast?

22-2 What Are the Major Urban Resource and Environmental Problems?

▶ **CONCEPT 22-2** Most cities are unsustainable because of high levels of resource use, waste, pollution, and poverty.

Urbanization Has Advantages

Urbanization has many benefits. From an *economic standpoint,* cities are centers of economic development, innovation, education, technological advances, and jobs. They serve as centers of industry, commerce, and transportation.

Urban residents in many parts of the world tend to live longer than do rural residents and have lower infant mortality rates and fertility rates. They also have better access to medical care, family planning, education, and social services than do their rural counterparts. However, the health benefits of urban living are usually greater for the rich than for the poor.

Urban areas also have some environmental advantages. Recycling is more economically feasible because concentrations of recyclable materials and funding for recycling programs tend to be higher in urban areas. Concentrating people in cities helps to preserve biodiversity by reducing the stress on wildlife habitats. And central cities can save energy if residents rely more on energy-efficient mass transportation, walking, and bicycling, as they do in Curitiba, Brazil (**Core Case Study**).

CORE CASE STUDY

Urbanization Has Disadvantages

Half of the world's 6.8 billion people live in urban areas and each year, about 63 million people—an average of nearly 7,200 per hour—are added to such areas. Such intense population pressure makes most of the world's cities more environmentally unsustainable every year (**Concept 22-2**).

Even in more sustainable cities such as Curitiba, Brazil (**Core Case Study**), this pressure is taking its toll. Curitiba's once clear streams are often overloaded with pollutants. The bus system is nearing capacity, and car ownership is on the rise. While the recycling rate is still among the highest in the world, the region's only landfill will soon be full. Architect Jorge Wilheim, who helped to create Curitiba's master plan, points out that the city's population is now more than five times as large as it was in 1965 when the plan was drafted. With a metropolitan area population of 3.2 million, Wilheim believes it is time to adjust the plan.

CORE CASE STUDY

We will touch on aspects of a possible new vision for Curitiba and similar solutions for other cities throughout this chapter. Here we take a closer look at some problems faced by cities.

They Have Huge Ecological Footprints. Although urban populations occupy only about 2% of the earth's land area, they consume 75% of its resources and produce 75% of all carbon dioxide emissions from human activities, according to the Worldwatch Institute. Because of this high resource input of food, water, and materials and the resulting high waste output (Figure 22-8), *most of the world's cities have huge ecological footprints and are not self-sustaining systems* (**Concept 22-2**).

By concentrating people in a small area urbanization helps to preserve some of the earth's biodiversity. But at the same time large areas of land that must be disturbed and degraded to provide urban dwellers with food, water, energy, minerals, and other resources. This decreases and degrades the earth's overall biodiversity (**Concept 4-4B**, p. 86).

Thus, urban areas have huge ecological footprints (**Concept 1-3**, p. 12) that extend far beyond their boundaries. If you live in a city, you can calculate its ecological footprint by going to the website **www.redefiningprogress.org/**. Also, see the Guest Essay on this topic by Michael Cain at CengageNOW.

CONCEPT LINKS

Recent studies also reveal that most urban dwellers live in an artificial environment that isolates them from forests, grasslands, streams, and other natural areas that contain the world's biodiversity. As a result, many of them tend to have little passion for protecting biodiversity and other forms of natural capital (Figure 1-3, p. 8) that support their lives and the cities where they live.

They Lack Vegetation. In urban areas, most trees, shrubs, or other plants are destroyed to make way for buildings, roads, parking lots, and housing developments. So most cities do not benefit from vegetation that would absorb air pollutants, give off oxygen, cool the air through transpiration, provide shade, muffle noise, provide wildlife habitats, and give aesthetic pleasure. As one observer remarked, "Most cities are places where they cut down most of the trees and then name the streets after them."

They Have Water Problems. As cities grow and water demands increase, expensive reservoirs and canals must be built and deeper wells must be drilled. This can

Inputs		Outputs
Energy		Solid wastes
Food		Waste heat
Water		Air pollutants
Raw materials		Water pollutants
Manufactured goods		Greenhouse gases
Money		Manufactured goods
Information		Noise
		Wealth
		Ideas

Figure 22-8 Natural capital degradation: urban areas rarely are sustainable systems (**Concept 22-2**). The typical city depends on large nonurban areas for huge inputs of matter and energy resources and for large outputs of waste matter and heat and thus have large ecological footprints that extend far beyond their boundaries. According to an analysis by Mathis Wackernagel and William Rees, an area 58 times as large as that of London, England, is needed to supply its residents with resources. They estimate that meeting the needs of all the world's people at the same rate of resource use as that of London would take at least three more planet Earths. **Question:** How would you apply the four **scientific principles of sustainability** (see back cover) to lessen some of these impacts?

deprive rural and wild areas of surface water and deplete groundwater.

Flooding also tends to be greater in some cities, because they are built on floodplains near rivers or along low-lying coastal areas subject to natural flooding. And covering land with buildings, asphalt, and concrete causes precipitation to run off quickly and overload storm drains. In addition, urban development has often destroyed or degraded wetlands that have served as natural sponges to help absorb excess water. Many of the world's largest cities face another threat because they are located in coastal areas (Figure 22-2) that could be partially flooded some time in this century as sea levels rise due to projected climate change.

Other cities in arid areas that depend on withdrawing water from rivers and the reservoirs behind dams may face severe water shortages as global warming reduces the mountaintop glaciers and snow that melt each year to provide river flows and fill reservoirs. As a result, in some hotspot arid areas in the western United States (Figure 13-5, p. 318) fed by the Colorado River (Figure 13-14, p. 327) and California's water distribution system (Figure 13-17, p. 330), irrigated agriculture may have to be abandoned. And urban populations in some of these areas may drop sharply as people are forced to move elsewhere because of a lack of water. Greater re-

liance on groundwater is not an option in many areas where aquifers are being depleted faster than they are being replenished.

They Concentrate Pollution and Health Problems. Because of their high population densities and high resource consumption, cities produce most of the world's air pollution, water pollution, and solid and hazardous wastes. Pollutant levels are generally higher because pollution is produced in a smaller area and cannot be dispersed and diluted as readily as pollution produced in rural areas.

The concentration of motor vehicles and industry in urban centers, where about three-fourths of the world's CO_2 from human-related sources are emitted, causes disruption of local and regional portions of the carbon cycle (Figure 3-18, p. 68). This urban concentration also disrupts the nitrogen cycle (Figure 3-19, p. 69) because of emissions of large quantities of nitrogen oxides, which play a key role in the formation of photochemical smog, and nitric acid and nitrates, which are major components of acid deposition in urban areas and beyond. And nitrogen nutrients in urban runoff and discharges from urban sewage treatment plants can disrupt the nitrogen cycle in nearby lakes and other bodies of water and cause excessive eutrophication.

In addition, high population densities in urban areas can increase the spread of infectious diseases, especially where there are no adequate drinking water and sewage systems.

They Have Excessive Noise. Most urban dwellers are subjected to **noise pollution:** any unwanted, disturbing, or harmful sound that impairs or interferes with hearing, causes stress, hampers concentration and work efficiency, or causes accidents. Noise levels are measured in decibel-A (dbA) sound pressure units that vary with different human activities (Figure 22-9).

Sound pressure becomes damaging at about 85 dbA and painful at around 120 dbA. At 180 dbA it can kill. Prolonged exposure to sound levels above 85 dbA can cause permanent hearing damage. About one of every eight children and teens in the United States has some permanent hearing loss, mostly from listening to music at loud levels.

They Have a Different Climate and Experience Light Pollution. Cities generally are warmer, rainier, foggier, and cloudier than suburbs and nearby rural areas. The enormous amount of heat generated by cars, factories, furnaces, lights, air conditioners, and heat-absorbing dark roofs and streets in cities creates an *urban heat island* that is surrounded by cooler suburban and rural areas. As cities grow and merge (Figure 22-7), their heat islands merge, which can reduce the natural dilution and cleansing of polluted air.

In addition to changing local and regional climate, the urban heat island effect in hot climates puts heat stress on humans and many other organisms, can increase the formation of photochemical smog, and greatly increases dependence on air conditioning for cooling. This in turn increases energy consumption, greenhouse gas emissions, and other forms of air pollution in a positive feedback loop.

Also, the artificial light created by cities (Figures 22-2 and 22-4) hinders astronomers from conducting their research and makes it difficult for casual observers to enjoy the night sky. *Light pollution* also affects some plant and animal species. For example, endangered sea turtles lay their eggs on beaches at night and require darkness. And each year, large numbers of migrating birds, lured off course by the lights of high-rise buildings, fatally collide with the buildings.

— **THINKING ABOUT**
Disadvantages of Urbanization

Which two of the disadvantages of living in urban areas discussed here do you think are the most serious? Explain.

Life Is a Desperate Struggle for the Urban Poor in Developing Countries

Poverty is a way of life for many urban dwellers in developing countries. About 1 billion people live in cities in developing countries under crowded and unsanitary conditions (Figure 22-10), and according to a 2006 U.N. study, that number could reach 1.4 billion by 2020. In Sub-Saharan Africa, more than seven of every ten urban residents live in such conditions, according to the U.N.

Some of these people live in *slums*—areas dominated by tenements and rooming houses where several people might live in a single room. The world's largest slum, Dharavi, in Mumbai (Bombay), India, is home to 2.8 million people. Others live in *squatter settlements* and *shantytowns* on the outskirts of these cities. They build shacks from corrugated metal, plastic sheets, scrap wood, and other scavenged building materials, or they live in rusted shipping containers and junked cars. Still others live or sleep on the streets (Photo 2 in the Detailed Contents). When it rains, the usually unpaved alleys can become clogged with dead rats, garbage, and sewage.

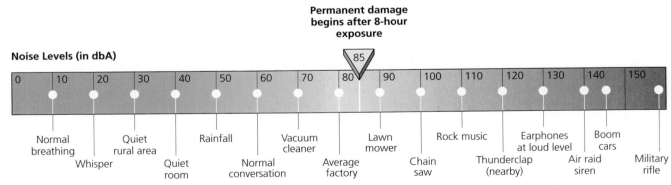

Figure 22-9 *Noise levels* (in decibel-A sound pressure units) of some common sounds. You are being exposed to a sound level high enough to cause permanent hearing damage if you need to raise your voice to be heard above the racket, if a noise causes your ears to ring, or if nearby speech seems muffled. Prolonged exposure to lower noise levels and occasional loud sounds may not damage your hearing, but it can greatly increase internal stress.
Question: How many times per day are your ears subjected to noise levels of 85 or more dbA?

Figure 22-10 *Global outlook:* extreme poverty forces hundreds of millions of people to live in slums and shantytowns such as this one in Rio de Janeiro, Brazil, where adequate water supplies, sewage disposal, and other services do not exist.

United Nations

Some shantytowns are illegal subdivisions where landowners rent land to the poor without city approval. Others are illegal *squatter settlements* where people take over unoccupied land without the owners' permission, simply because it is their only option for survival.

Poor people living in shantytowns and squatter settlements usually lack clean water supplies, sewers, electricity, and roads, and are subject to severe air and water pollution and hazardous wastes from nearby factories. Many of these settlements are in locations especially prone to landslides, flooding, or earthquakes. Many city governments regularly bulldoze squatter shacks and send police to drive illegal settlers out. The people usually move back in or develop another shantytown somewhere else.

There are ways in which governments can address these problems. For example, they can slow the migration from rural to urban areas by improving education, health care, and family planning in the countryside and encouraging investment in small towns.

Governments can also designate land for squatter settlements and supply them with clean water, as Curitiba, Brazil, has done (**Core Case Study**). In addition to providing access to safe drink-

CORE CASE STUDY

ing water, they can place composting toilets, which require no water, at locations throughout such settlements. Guaranteeing regular bus service enables workers who live in such settlements to travel to and from their workplaces. In Brazil and Peru, governments legally recognize existing slums (favelas) and grant legal titles to the land. This is based on understanding that the poor will usually improve their living conditions and develop schools, day care centers, and other social improvements once they know they can stay there.

--- **HOW WOULD YOU VOTE?** ---

Should squatters around cities of developing countries be given title to land they live on? Cast your vote online at **academic.cengage.com/biology/miller**.

■ CASE STUDY
Mexico City

Mexico City—the world's second most populous city—is an urban area in crisis. About 19 million people—nearly equal to the entire population of Australia or the U.S. state of New York—live there, and each year at least 400,000 new residents arrive.

Mexico City suffers from severe air pollution, close to 50% unemployment, deafening noise, overcrowding, traffic congestion, inadequate public transportation, and a soaring crime rate. More than one-third of its residents live in slums called *barrios* or in squatter settlements that lack running water and electricity.

At least 3 million people have no sewer facilities. As a consequence, huge amounts of human waste are deposited in gutters, vacant lots, and open sewers every day, attracting armies of rats and swarms of flies. When the winds pick up dried excrement, a *fecal snow* blankets parts of the city. This bacteria-laden fallout leads to widespread salmonella and hepatitis infections, especially among children.

Mexico City has one of the world's worst air pollution problems because of a combination of factors: too many cars, polluting factories, a sunny climate and thus more smog, and topographical bad luck (Photo 12 in the Detailed Contents). The city sits in a high-elevation, bowl-shaped valley surrounded on three sides by mountains—conditions that trap air pollutants at ground level (Figure 18-11, left, p. 478). Breathing its air is said to be roughly equivalent to smoking three packs of cigarettes per day, and respiratory diseases are rampant. See *The Habitable Planet,* Video 11, at **www.learner .org/resources/series209.html** to learn how scientists are measuring air pollution levels in Mexico City.

The city's air and water pollution cause an estimated 100,000 premature deaths per year. Writer Carlos Fuentes has nicknamed it "Makesicko City."

Large-scale water withdrawals from the city's aquifer have caused parts of the city to subside by 9 meters

(30 feet) during the last century. Some areas are now subsiding as much as 30 centimeters (1 foot) a year. The city's growing population increasingly relies on pumping in water from as far away as 150 kilometers (93 miles) and then using large amounts of energy to pump the water 1,000 meters (3,300 feet) uphill to reach the city.

Progress has been made in solving some of Mexico City's problems. The percentage of days each year in which air pollution standards are violated has fallen from 50% to 20%. The city government has banned cars in its central zone, required catalytic converter air pollu-

tion control devices on all cars made after 1991, phased out use of leaded gasoline, and replaced old buses, taxis, and delivery vehicles with cleaner ones. The city also bought land for use as green space and planted more than 25 million trees to help absorb pollutants.

> **THINKING ABOUT**
> **Mexico City and Curitiba**
>
> What are two sustainability strategies used in Curitiba, Brazil, (**Core Case Study**) that might be helpful in Mexico City?
>
> CORE CASE STUDY

22-3 How Does Transportation Affect Urban Environmental Impacts?

▶ **CONCEPT 22-3** A combination of plentiful land, inexpensive fuel, and expanding networks of highways in some countries has resulted in dispersed cities whose residents depend on motor vehicles for most transportation.

Cities Can Grow Outward or Upward

If a city cannot spread outward, it must grow vertically—upward and downward (below ground)—so that it occupies a small land area with a high population density. Most people living in *compact cities,* such as Hong Kong, China, and Tokyo, Japan, get around by walking, biking, or using mass transit such as rail or buses. Many new high-rise apartment buildings in these Asian cities contain everything from grocery stores to fitness centers. People living in the apartments hardly have to leave the buildings to meet their needs.

In other parts of the world, a combination of plentiful land, cheap gasoline, and networks of highways have produced *dispersed cities* whose residents depend on motor vehicles for most travel (**Concept 22-3**). This dependency is a growing problem for members of the increasing elderly population who live in urban and suburban areas with no sidewalks and who can no longer drive. Many of them essentially become prisoners in their homes.

Such car-centered cities are found in the United States, Canada, Australia, and other countries where ample land often is available for outward expansion. The resulting urban sprawl (Figure 22-5) can have a number of undesirable effects (Figure 22-6).

The United States is a prime example of a car-centered nation. With 4.6% of the world's people, the country has almost one-third of the world's 900 million passenger cars and commercial vehicles. In its dispersed urban areas, passenger vehicles are used for 98% of all transportation, and about three of every four residents drive alone to work every day. Largely because of urban sprawl, all Americans combined drive about the same distance each year as the total distance driven by all other drivers in the world, and in the process, use about 43% of the world's gasoline.

Motor Vehicles Have Advantages and Disadvantages

Motor vehicles provide mobility and offer a convenient and comfortable way to get from one place to another. They also are symbols of power, sex appeal, social status, and success for many people. And much of the world's economy is built on producing motor vehicles and supplying fuel, roads, services, and repairs for them.

Despite their important benefits, motor vehicles have many harmful effects on people and the environment. Globally, automobile accidents kill approximately 1.2 million people a year—an average of nearly 3,300 deaths per day—and injure another 15 million people. They also kill about 50 million wild animals and family pets every year.

In the United States, motor vehicle accidents kill more than 40,000 people per year and injure another 5 million, at least 300,000 of them severely. *Car accidents have killed more Americans than have all wars in the country's history.*

Motor vehicles are the world's largest source of outdoor air pollution, which causes 30,000–60,000 prema-

ture deaths per year in the United States, according to the Environmental Protection Agency. They are also the fastest-growing source of climate-changing carbon dioxide emissions. In addition, they account for two-thirds of the oil used in the form of gasoline in the United States and one-third of the world's oil consumption.

Motor vehicles have helped to create urban sprawl and the car commuter culture. At least a third of urban land worldwide and half in the United States is devoted to roads, parking lots, gasoline stations, and other automobile-related uses. This prompted urban expert Lewis Mumford to suggest that the U.S. national flower should be the concrete cloverleaf.

Another problem is congestion. As the number of cars increases in an urban area, at some point they contribute to immobility not mobility. If current trends continue, U.S. motorists will spend an average of 2 years of their lives in traffic jams, as streets and freeways, begin to look like parking lots. Traffic congestion in some cities in developing countries is much worse. Commuter distances increase as cities sprawl out. Building more roads may not be the answer. Many analysts agree with economist Robert Samuelson that "cars expand to fill available concrete."

Reducing Automobile Use Is Not Easy, but It Can Be Done

Some environmental scientists and economists suggest that one way to reduce the harmful effects of automobile use is to make drivers pay directly for most environmental and health costs of their automobile use—a *user-pays* approach, based on honest environmental accounting.

One way to phase in such *full-cost pricing* would be to charge a tax on gasoline to cover the estimated harmful costs of driving. According to a study by the International Center for Technology Assessment, such a tax would amount to about $3.18 per liter ($12 per gallon) of gasoline in the United States (p. 404). Gradually phasing in such a tax would spur the use of more energy-efficient motor vehicles and mass transit and decrease dependence on imported oil, thereby increasing economic and military security. It would also reduce pollution and environmental degradation and help to slow climate change.

┌─ **RESEARCH FRONTIER** ──────────────────
Determining the full costs of using gasoline and how to include such costs in is market price. See **academic.cengage .com/biology/miller**.
└──

Proponents of this approach urge governments to use gasoline tax revenues to help finance mass transit systems, bike lanes, and sidewalks as alternatives to cars. They also urge reduction of taxes on income,

wages, and wealth to offset the increased taxes on gasoline. Such a *tax shift* would make higher gasoline taxes more politically acceptable.

Europe, Japan, Singapore, and some rapidly developing Chinese cities have developed rapid mass transit systems within and between urban areas and networks of sidewalks and bike lanes in urban and suburban areas. Analysts warn that countries such as the United States, which lags in developing such systems, will face great economic difficulties in the face of a projected decline in oil production and rapidly rising oil and gasoline prices. The end result could be mass abandonment of sprawling car-dependent suburbs and shopping centers as they become environmentally and economically unsustainable.

Heavily taxing gasoline is difficult in the United States, for three reasons. *First*, it faces strong opposition from the public, which is largely unaware of the huge hidden costs they are paying for gasoline, and from powerful transportation-related industries such as oil and tire companies, road builders, car makers, and real estate developers. *Second*, fast, efficient, reliable, and affordable mass transit options and bike lanes are not widely available in most of the United States. And *third*, the dispersed nature of most U.S. urban areas makes people dependent on cars. These factors make it politically difficult to raise gasoline taxes. But U.S. taxpayers might accept sharp increases in gasoline taxes if a tax shift were employed, as mentioned above.

Another way to reduce automobile use and urban congestion is to raise parking fees and charge tolls on roads, tunnels, and bridges leading into cities—especially during peak traffic times. Densely populated Singapore is rarely congested, because it auctions the rights to buy a car, and car owners must pay a high tax to use any of the roads leading into the city center.

London, England, which has a similar road congestion fee of $49 per day on gas-guzzling cars (with no fee for fuel-efficient vehicles), has reduced the number of cars entering the city during the day. Stockholm, Sweden, and Milan, Italy, have also imposed stiff fees for motor vehicles entering their central cities. And in an effort to reduce fuel use, CO_2 emissions, and local air pollution, the mayors of the U.S. cities of New York City and San Francisco, California, are considering such vehicle entry fees, and have announced that all taxis in their cities must be hybrid vehicles by 2012.

Paris, France, one of Europe's most congested cities, has upgraded its mass transit system and created express lanes for buses and bicycles on main thoroughfares while reducing the lanes for cars. The city also established a program with almost 21,000 bikes available for rental at 1,450 rental stations throughout the city at a cost of just over $1 a day.

In Germany, Austria, Italy, Switzerland, and the Netherlands more than 300 cities have *car-sharing* networks. Members reserve a car in advance or call the network and are directed to the closest car. They are billed monthly for the time they use a car and the

distance they travel. In Berlin, Germany, car sharing has cut car ownership by 75%. According to the Worldwatch Institute, car-sharing in Europe has reduced the average driver's carbon dioxide emissions by 40–50%. Car-share companies have sprouted up in the United States since 2000 in cities such as Portland, Oregon, and San Diego and Los Angeles in California, among others.

Some Cities Are Promoting Alternatives to Car Ownership

Mayors and urban planners in many parts of the world are beginning to rethink the role of the car in urban transportation systems and are providing a mix of other options, as they have in Curitiba, Brazil (**Core Case Study**). Leaders in developing countries are recognizing that most of their people could never afford to own automobiles, and they are questioning the development of expensive car-oriented systems that mostly benefit the affluent minority.

CORE CASE STUDY

There are several alternatives to motor vehicles, each with its own advantages and disadvantages. One widely used alternative is the *bicycle* (Figure 22-11). Bicycles reduce congestion, promote physical fitness, emit no CO_2 or other air pollutants, cost little to buy and operate, and reduce the need for parking space. Each year, the number of bicycles produced globally is about 2.5 times the number of cars produced. Also growing is

Figure 22-12 Advantages and disadvantages of *mass transit rail systems in urban areas.* **Question:** Which single advantage and which single disadvantage do you think are the most important?

the use of electric bicycles with lightweight electric motors (bottom drawing in Figure 22-11).

Bicycles are widely used for urban trips in many countries. Bicycling and walking account for about a third of all urban trips in the Netherlands and in Copenhagen, Denmark. Japan and the Netherlands strive to integrate travel by bicycle and by commuter rail by providing secure bicycle parking at rail stations.

By contrast, bicycles account for only about 1% of urban trips in the United States. But one of five Americans say they would bicycle to work if safe bike lanes were available and if their employers provided secure bike storage and showers at work.

THINKING ABOUT
Bicycles

Do you or would you use a conventional or an electric bicycle to go to and from work? Explain.

Heavy-rail systems (subways, elevated railways, and metro trains) and *light-rail* systems (streetcars, trolley cars, and tramways) have their advantages and disadvantages (Figure 22-12). At one time, all major U.S. cities had effective light-rail systems, but they were dismantled to promote car and bus use (Case Study, p. 602). However, such systems are being built again in several U.S. cities. An outstanding example is the light-rail system in Minneapolis, Minnesota, which opened in 2005 and surprised planners by drawing 50% more riders than they had expected.

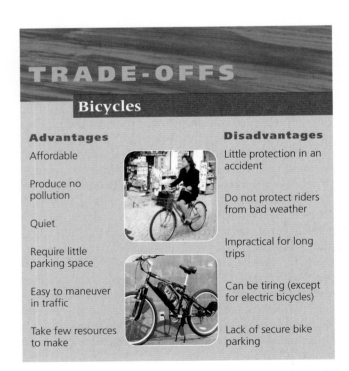

Figure 22-11 Advantages and disadvantages of *bicycles.* The key to increased bicycle use is the creation of bicycle-friendly systems, including bike lanes. **Question:** Which single advantage and which single disadvantage do you think are the most important?

The rail system in Hong Kong is one of the world's most successful, primarily because the city is densely populated, which means that half the population can walk to a subway station within 5 minutes. And owning a car is extremely expensive in this crowded city, which helps to make the rail system attractive to almost everyone.

Buses are the most widely used form of mass transit within urban areas worldwide, mainly because they have more advantages than disadvantages (Figure 22-13). Curitiba, Brazil, has one of the world's best bus rapid transit systems, which carries about 2.3 million passengers a day (**Core Case Study**, Figure 22-1). But the system is now feeling the strain of population pressures, and car use is increasing in Curitiba. Transportation officials there are considering building a light-rail system to help carry the rapidly growing passenger load.

Similar bus rapid transit systems have been developed in Bogotá, Colombia, and are being planned for Mexico City; São Paulo, Brazil; Seoul, South Korea; Beijing and 20 other cities in China; Ottawa and Toronto in Canada; and the U.S. cities of Los Angeles, California; Minneapolis, Minnesota; and Las Vegas, Nevada.

A *rapid-rail system between urban areas* is another option. In western Europe and Japan, high-speed bullet trains travel between cities at up to 306 kilometers (190 miles) per hour. Such trains began operating in Japan in 1964 and now and carry almost a million passengers a day. Since 1964, there has not been a single casualty, and late arrivals average only 6 seconds, explaining why some analysts consider this system to be one of the new wonders of the world.

Figure 22-14 lists the major advantages and disadvantages of such rapid rail systems. In 2004, Shanghai, China, began operating the world's first commercial high-speed magnetic levitation train between its airport and downtown. The train, suspended in air slightly above the track and propelled forward by strong repulsive and attractive magnetic forces, travels much faster than bullet trains do.

HOW WOULD YOU VOTE?

Should half the U.S. gasoline tax be used to develop mass transit, bike lanes, and other alternatives to the car? Cast your vote online at **academic.cengage.com/biology/miller**.

In the United States, a high-speed bullet train network could replace airplanes, buses, and private cars for most medium-distance travel between major American cities (Figure 22-15, p. 602). Critics say such a system would cost too much in government subsidies. But this argument ignores the fact that motor vehicle transportation receives government (taxpayer) subsidies of $300–600 billion per year in the United States. The Urban Land Institute estimates that building a U.S. sys-

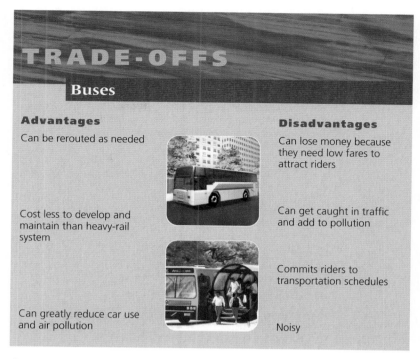

Figure 22-13 Advantages and disadvantages of *bus rapid transit (BRT) and conventional bus systems in urban areas.* **Question:** Which single advantage and which single disadvantage do you think are the most important?

tem comparable to the best European systems would cost $250 billion per year over 2 decades.

HOW WOULD YOU VOTE?

Should the United States (or the country where you live) develop rapid rail systems between some of its urban areas, even if it were quite costly? Cast your vote online at **academic.cengage.com/biology/miller**.

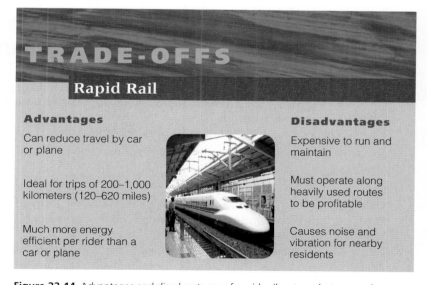

Figure 22-14 Advantages and disadvantages of *rapid-rail systems between urban areas.* **Question:** Which single advantage and which single disadvantage do you think are the most important?

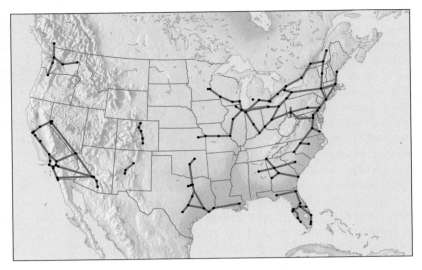

Figure 22-15 Solutions: potential routes for high-speed bullet trains in the United States and parts of Canada. Such a system would allow rapid, comfortable, safe, and affordable travel between major cities in a region. It would greatly reduce dependence on cars, buses, and airplanes for trips between these urban areas, and it would reduce greenhouse gas emissions and other forms of air pollution. **Question:** Why do you think such a system has not been developed in the United States? (Data from High Speed Rail Association)

■ CASE STUDY
Destroying a Great Mass Transit System in the United States

In 1917, all major U.S. cities had efficient electric trolley or streetcar (light-rail) systems. Many people think of Los Angeles, California, as the original car-dominated city. But in the early 20th century, Los Angeles had the largest electric-rail mass transit system in the United States.

That changed when General Motors, Firestone Tire, Standard Oil of California, Phillips Petroleum, and Mack Truck (which also made buses) formed a holding company called National City Lines. By 1950, the holding company had purchased privately owned streetcar systems in 83 major cities. It then dismantled these systems to increase sales of cars and buses.

The courts found the companies guilty of conspiracy to eliminate the country's light-rail system, but the damage had already been done. The executives responsible were each fined $1, and each company paid a fine of $5,000—less than the profit returned by replacing a single streetcar with a bus.

During this same period, National City Lines worked to convert electric-powered locomotives, which were pulling commuter trains between cities and suburbs, to much more expensive and less reliable diesel-powered locomotives. The resulting increased costs contributed significantly to the sharp decline of the nation's railroad system.

> **THINKING ABOUT**
> **Light-Rail Systems**
>
> If National City Lines had not dismantled many light-rail systems, do you think those systems would still be operating today? Explain.

22-4 How Important Is Urban Land Use Planning?

▶ **CONCEPT 22-4** Urban land-use planning can help to reduce uncontrolled sprawl and slow the resulting degradation of air, water, land, biodiversity, and other natural resources.

Conventional Land–Use Planning

Most urban and some rural areas use some form of **land-use planning** to determine the best present and future use of each parcel of land.

Much land-use planning encourages future population growth and economic development, regardless of the environmental and social consequences. Typically, this leads to uncontrolled or poorly controlled urban growth and sprawl.

A major reason for this often destructive process in the United States and in some other countries is that 90% of the revenue local governments use to provide public services, such as schools, police and fire protection, and water and sewer systems, comes from *property taxes* levied on all buildings and property according to their economic value. As a result, local governments often raise money by promoting urban growth because they usually cannot raise property tax rates enough to meet expanding needs.

Once a land-use plan is developed, governments control the uses of various parcels of land by legal and economic methods. The most widely used approach is **zoning,** in which various parcels of land are designated for certain uses. Zoning can be used to control growth and protect areas from certain types of development. For example, cities such as Portland, Oregon (USA), and Curitiba, Brazil, (**Core Case Study**) have used zoning to encourage high-density development along major mass transit corridors to reduce automobile use and air pollution.

Despite its usefulness, zoning has several drawbacks. One problem is that some developers often influence or modify zoning decisions in ways that threaten or destroy wetlands, prime cropland, forested areas, and open space. Another problem is that zoning often favors high-priced housing, factories, hotels, and other businesses over protecting environmentally sensitive areas and providing low-cost housing, because higher-priced developments provide more property tax revenues that most local governments depend on.

THINKING ABOUT
Property Taxes

What are some alternatives for local governments to depending on property taxes for revenue? Which, if any, of these alternatives would you support? Explain.

In addition, overly strict zoning can discourage innovative approaches to solving urban problems. For example, the pattern in the United States and in some other countries has been to prohibit businesses in residential areas, which increases suburban sprawl. Some urban planners want to return to *mixed-use zoning* to help reduce sprawl. For example, in the 1970s, Portland, Oregon, managed to cut driving and gasoline consumption by promoting the establishment of neighborhood grocery stores.

Smart Growth Works

Smart growth is one way to encourage more environmentally sustainable development that reduces dependence on cars, controls and directs sprawl, and cuts wasteful resource use. It recognizes that urban growth will occur. At the same time, it uses zoning laws and other tools to channel growth into areas where it will cause less harm. Smart growth can discourage sprawl, reduce traffic, protect ecologically sensitive and important lands and waterways, and develop neighborhoods that are more enjoyable places to live. Figure 22-16 lists popular smart growth tools that are used to prevent and control urban growth and sprawl.

The U.S. city of Portland, Oregon, has used some of these strategies to control sprawl and reduce dependence on automobiles. Since 1975, Portland's population has grown by about 50%, but its urban area has

SOLUTIONS

Smart Growth Tools

Limits and Regulations

Limit building permits

Urban growth boundaries

Greenbelts around cities

Public review of new development

Zoning

Encourage mixed use of housing and small businesses

Concentrate development along mass transportation routes

Promote high-density cluster housing developments

Planning

Ecological land-use planning

Environmental impact analysis

Integrated regional planning

State and national planning

Protection

Preserve existing open space

Buy new open space

Buy development rights that prohibit certain types of development on land parcels

Taxes

Tax land, not buildings

Tax land on value of actual use (such as forest and agriculture) instead of on highest value as developed land

Tax Breaks

For owners agreeing not to allow certain types of development (conservation easements)

For cleaning up and developing abandoned urban sites (brownfields)

Revitalization and New Growth

Revitalize existing towns and cities

Build well-planned new towns and villages within cities

Figure 22-16 *Smart growth* or *new urbanism tools* that are used to control urban growth and sprawl. **Questions:** Which five of these tools do you think are the most important ways to prevent or control urban sprawl? Which, if any, of these tools are used in your community?

increased by only 2%. Yet abundant green space and natural beauty is just 20 minutes from downtown. The city has built an efficient light-rail and bus system that carries 45% of all commuters to downtown jobs—a much higher percentage than in most U.S. cities. By reducing traffic, Portland was able to convert a former expressway and huge parking lot into a waterfront park.

The city encourages clustered, mixed-use neighborhood development, with stores, light industries, professional offices, high-density housing, and access to mass transit, which allows most people to meet their daily needs without a car. In addition to its excellent light-rail and bus lines, Portland has further reduced car use by developing an extensive network of bike lanes and walkways. Employers are encouraged to give their employees bus passes instead of providing parking spaces.

Since 1975, the city's air pollution has decreased by 86%. Portland is frequently cited by analysts as one

of the most livable cities in the United States, and in 2008, one such analyst, Elizabeth Svoboda, ranked Portland as the greenest city in the United States. In order, the next nine greenest U.S. cities were San Francisco, California; Boston, Massachusetts; Oakland, California; Eugene, Oregon; Cambridge, Massachusetts; Berkeley, California; Chicago, Illinois; Austin, Texas; and Minneapolis, Minnesota.

Curitiba, Brazil, (**Core Case Study**) has also used a variety of smart growth strategies to control sprawl and reduce dependence on the car. Several studies have shown that, compared to conventional economic growth, most forms of smart growth provide more jobs and spur more economic renewal. See *The Habitable Planet*, Video 5, at **www.learner.org/resources/series209.html** to learn how smart growth was used to control population growth and urban sprawl in Cape Cod, Massachusetts (USA).

 CORE CASE STUDY

THINKING ABOUT
Portland and Curitiba

CORE CASE STUDY

How is the urban strategy of Portland, Oregon, similar to that of Curitiba, Brazil (**Core Case Study**)? How do they differ?

China has taken the strongest stand of any country against urban sprawl. The government has designated 80% of the country's arable land as *fundamental land*. Building on such land requires approval from local and provincial governments and from the State Council. This is somewhat like having to get congressional approval for a new subdivision in the United States. However, China's approach does go to extremes; developers violating the rules can face the death penalty.

Most European countries have been successful in discouraging urban sprawl and encouraging compact cities. They have controlled development at the national level and imposed high gasoline taxes to discourage car use and to encourage people to live closer to workplaces and shops. High taxes on heating fuel encourage people to live in apartments and small houses. Governments have used most of the resulting gasoline and heating fuel tax revenues to develop efficient mass transit systems within and between cities.

Preserving and Using Open Space

One way to preserve open space outside a city is to draw an urban growth line around each community and to allow no urban development outside those boundaries. This *urban growth boundary* approach, is used in the U.S. states of Oregon, Washington, and Tennessee. However, the Seattle region of Washington state found that using growth boundaries to increase housing densities inside the boundaries had the unintended effect of encouraging low-density housing and sprawl in rural and natural areas just beyond the growth boundaries.

A more traditional way is to preserve significant blocks of open space in the form of municipal parks. Examples of large urban parks in the United States are Central Park in New York City (Figure 22-17); Golden Gate Park in San Francisco, California; and Grant Park in Chicago, Illinois.

In 1883, officials in the U.S. city of Minneapolis, Minnesota, vowed to create "the finest and most beautiful system of public parks and boulevards of any city in America." In the eyes of many, this goal has been achieved. Today, the city has 170 parks strategically located so that most homes in Minneapolis are within six blocks of a green space. Similarly, Jacksonville, Florida (USA), has a high ratio of parkland to total city area.

Some cities provide open space and control urban growth by surrounding a large city with a *greenbelt*—an open area reserved for recreation, sustainable forestry, or other nondestructive uses. Satellite towns are often built outside these greenbelts. Ideally, the outlying towns are self-contained, not sprawling, and are linked to the central city by a public transport system that does minimal damage to the greenbelt. Many cities in western Europe and the Canadian cities of Toronto and Vancouver have used this approach. Seattle, Washington, could have minimized urban sprawl just beyond its urban grow boundary by establishing a greenbelt outside the boundary. Greenbelt areas can provide vital ecological services such as absorption of CO_2 and other air pollutants, which can make urban air more breathable and help to cut a city's contribution to climate change.

Figure 22-17 Central Park, New York City, USA.

age fotostock/SuperStock

▶ **CONCEPT 22-5** An *ecocity* allows people to: choose walking, biking, or mass transit for most transportation needs; recycle or reuse most of their wastes; grow much of their food; and protect biodiversity by preserving surrounding land.

New Urbanism Is Growing

Since World War II, the typical approach to suburban housing development in the United States has been to bulldoze a tract of woods or farmland and build rows of houses on standard-size lots (Figure 22-18, middle). Many of these developments and their streets, with names like Oak Lane, Cedar Drive, Pheasant Run, and Fox Fields, are named after the trees and wildlife they displaced.

In recent years, builders have increasingly used a pattern, known as *cluster development,* in which high-density housing units are concentrated on one portion of a parcel, with the rest of the land (often 30–50%) used for commonly shared open space (Figure 22-18, bottom). When this is done properly, residents live with more open and recreational space, aesthetically pleasing surroundings, and lower heating and cooling costs because some walls are shared. And developers can cut their costs for site preparation, roads, utilities, and other forms of infrastructure.

Some communities are going further and using the principles of *new urbanism,* which is a modern form of what could be called *old villageism,* to develop entire villages and promote mixed-use neighborhoods within existing cities. These principles include

- *walkability,* with most stores and recreational activities located within a 10-minute walk of homes and apartments

- *mixed-use and diversity,* which provides a mix of pedestrian-friendly shops, offices, apartments, and homes to encourage people of different ages, classes, cultures, and races to move in

- *quality urban design* emphasizing beauty, aesthetics, and architectural diversity

- *environmental sustainability* based on development with minimal environmental impact

- *smart transportation* with well-designed train and bus systems connecting neighborhoods, towns, and cities.

Examples of new urban villages in the United States are Mayfaire in Wilmington, North Carolina; Phillips Place in Charlotte, North Carolina; Mizner Place in Boca Raton, Florida; Middleton Hills near Madison, and Hawksnest near Waukesha, Wisconsin; Yardley, Pennsylvania; Kentlands in Gaithersberg, Maryland; Valencia, California (near Los Angeles); and Stapleton, Colorado (built on an old airport site in Denver).

Undeveloped land
Creek
Marsh

Typical housing development

Cluster housing development
Cluster
Creek
Cluster
Pond

Figure 22-18 *Conventional and cluster housing developments* as they might appear if constructed on the same land area. With cluster development, houses, town houses, condominiums, and two- to six-story apartments are built on part of the tract. The rest, typically 30–50% of the area, is left as open space, parks, and paths for biking and walking.

── **RESEARCH FRONTIER** ──

Evaluating, improving, and implementing more environmentally sustainable new urbanism. See **academic.cengage .com/biology/miller.**

THINKING ABOUT
New Urbanism

Is there an example of new urbanism near where you live? Would you like to live in such an area? Explain.

The Ecocity Concept: Cities for People Not Cars

According to a growing number of urban planners, the primary problem is not urban growth; rather, it is our failure to make cities more sustainable and livable. These planners call for us to make new and existing urban areas more self-reliant, sustainable, and enjoyable places to live through good ecological design. (See the Guest Essay on this topic by David Orr on the website for this chapter.)

New urbanism is a step in the right direction, but we can go further by creating more environmentally sustainable cities, called *ecocities* or *green cities,* which emphasize the following goals built around the four **scientific principles of sustainability** (See back cover):

- Build and redesign cities for people not cars

- Use solar and other locally available, renewable energy resources and design buildings to be heated and cooled as much as possible by nature

- Use solar-powered living machines (Figure 20-C, p. 556) and wetlands to treat sewage

- Depend largely on recycled water that is purified and used again and again

- Use energy and matter resources efficiently

- Prevent pollution and reduce waste

- Recycle, reuse, and compost at least 60% of all municipal solid waste

- Protect and support biodiversity by preserving surrounding land and protecting and restoring natural systems and wetlands

- Promote urban gardens and farmers markets

- Use zoning and other tools to keep the human population at environmentally sustainable levels

An ecocity is a people-oriented city, not a car-oriented city. Its residents are able to walk, bike, or use low-polluting mass transit for most of their travel. Its buildings, vehicles, and appliances meet high energy-efficiency standards. Trees and plants adapted to the local climate and soils are planted throughout the city to provide shade, beauty, and wildlife habitats, and to reduce air pollution, noise, and soil erosion. Small organic gardens and a variety of plants adapted to local climate conditions often replace monoculture grass lawns.

In an ecocity, abandoned lots, industrial sites, and polluted creeks and rivers are cleaned up and restored. Nearby forests, grasslands, wetlands, and farms are preserved. Parks are easily available to everyone. Indeed, environmental leader Lester R. Brown says that "the ratio of parks to parking lots may be the single best indicator of the livability of a city—an indication of whether the city is designed for people or for cars."

Much of an ecocity's food comes from nearby organic farms, solar greenhouses, community gardens, and small gardens on rooftops, in yards, and in window boxes. In 2005 the U.N. Food and Agriculture Organization (FAO) estimated that such urban food production now supplies food for about one-fifth of the world's 3.4 billion urban residents. The FAO projects that with organized programs and financial incentives, much more of the world's food could be grown in or near cities. Increased food transportation costs because of rising oil prices could cause a shift from globalized to more localized food production in urban areas (Science Focus, at right).

Finally, an ecocity provides plenty of educational opportunities for its residents to inform them of environmental problems and solutions, as does the city of Curitiba, Brazil (**Core Case Study**). This includes not only school curricula, but also botanical gardens, humane and sustainable zoos and aquaria, and outdoor learning centers.

This may sound like a futuristic dream, but ecocities are real, as you saw in this chapter's opening core case study. Another example of an emerging ecocity is Bogotá, Colombia. Its transformation began in 1998 when Enrique Peñalosa was elected mayor. His major goal was to improve life for the 70% of the city's 8 million people who did not own cars. Within a short time under his leadership and that of his successor, Antanas Mockus, the city built an efficient and widely used, based rapid transit system; established hundreds of kilometers of bicycle lanes and pedestrian-only streets throughout the city; renovated or created 1,200 parks; planted more than 100,000 trees; and encouraged citizens to become involved in improving their local neighborhoods. As a result, between 1998 and 2006, rush-hour traffic decreased by 40%, traffic related deaths dropped by 20%, crime decreased sharply because of greatly increased enforcement, and its citizens developed a strong sense of civic pride.

Other examples of cities that have attempted to become more environmentally sustainable and to become more livable include Waitakere City, New Zealand; Stockholm, Sweden; Helsinki, Finland; Neerlands, the Netherlands; Leichester, England; and the U.S. cities of Portland, Oregon (p. 603); Davis, California; Olympia, Washington; and Chattanooga, Tennessee (Case Study, p. 21). In 2008, the oil-producing United Arab Emirates, with the world' largest per capita ecological footprint,

Urban Indoor Farming

Here are three approaches being researched for growing more food indoors in urban areas. One that could be implemented quickly is the use of rooftop greenhouses to provide city dwellers with most of their fruits and vegetables. Such greenhouses could use as little as 10% of the water and 5% of the area used by conventional farms.

In 2007, Sun Works, a company in New York City that designs energy-efficient urban greenhouses, built a demonstration greenhouse on a barge floating on the Hudson River. It used solar power and recycled water to grow vegetables and fruits. In 2008, the company installed a greenhouse on top of a city school, which serves as a science teaching area and supplies produce for the school cafeteria. Ted Caplow, executive director of Sun Works, projects that the rooftops of New York City could provide roughly twice the greenhouse area needed to supply the city with fruits and vegetables.

A more far-reaching idea is to use hydroponic gardens to provide the water and nutrients to grow fruits and vegetables on glass walls of office buildings. The crops would be grown in small containers in a space between two glass walls that run up the sides of buildings. Such walls would provide light and passive solar heating during winter and sunlight for growing the crops. In hot weather, one or both glass walls could have automatic shades to keep the interior cool and the plants from being overheated. Vertical conveyor belts could lower the plants for harvesting and replanting on a year-round basis.

Some scientists envision skyscraper farms in 30-story buildings, each occupying a single city block. Each vertical farm could provide vegetables, fruits, chickens, eggs, and fish (grown in aquaculture tanks) for 50,000 people. Hydroponic crops would be grown on the upper floors. Chickens (and the eggs they produce) and fish that feed on plant waste would be grown on lower floors.

Nitrogen and other plant nutrients could be extracted from the animal wastes, and perhaps from city sewage treatment plants, and recycled to the plants. Electricity for heat and lighting could be provided by geothermal, solar, wind, or other forms of renewable energy or by fuel cells powered by hydrogen produced from such forms of renewable energy. Such a system would mimic nature by applying the solar energy and recycling **principles of sustainability**.

Critics point out that the costs of these indoor urban farming systems could be high. But as oil and natural gas prices rise, as water in many crop-growing areas becomes scarce, and as more environmental costs are included in the prices of goods and services, food prices will rise sharply and make urban farming more profitable and affordable. Currently, relatively low food prices are based on cheap water, cheap fuels for transportation, and cheap energy for producing nitrogen fertilizer—a situation that is beginning to change.

Critical Thinking

What are three ecological advantages and three ecological disadvantages of greatly increasing urban farming? Do you believe that the advantages outweigh the disadvantages? Explain.

announced plans to build a green town in the desert by 2015 for up to 50,000 residents, with the goals of having zero-carbon emissions and zero wastes. And China has plans for a half-dozen or so ecocities on the drawing boards (Case Study, below).

■ CASE STUDY

China's Vision for an Ecocity

In 2008, China began constructing an ecocity, called Dongtan. It is located within a 30-minute drive from Shanghai, a city with 17 million people and 300,000 more people arriving each year.

Dongtan will be built at the eastern tip of Chongming Island at the mouth of the Yangtze River where it flows into the East China Sea. This undeveloped area consists of tidal flats and coastal marshes that serve as feeding areas for millions of migratory birds.

The Shanghai government, which owns the site, aims to develop Dongtan in an environmentally responsible way, such that it will exist in harmony with its environment. It wants Dongtan to become the world's first "carbon neutral" city, in which all carbon emissions will be offset by its carbon absorption. The first phase, to be completed by 2010, will provide housing for 10,000 residents. By 2020, the goal is a have a community of 80,000 people with a full array of shops, entertainment, and cultural amenities, and businesses that will provide 50,000 jobs. Eventually the city would be expanded to house half a million people.

The goal is to cut the city's ecological footprint to half that of comparable conventional cities. Buildings will have to meet high standards for construction, insulation, lighting, and appliances. All of the city's energy is to come from renewable sources. Electricity used for power, heating, and cooling will come from rooftop solar-cell panels, wind turbines, and a biomass cogeneration plant, fueled by rice husks and vegetable peelings collected throughout the Yangtze delta.

One piping system will provide drinking water and another will supply treated wastewater to flush toilets and to irrigate crops. Vegetables, fruits, chickens, eggs, and fish will be produced in high-rise vertical farm buildings (Science Focus, above). The initial goal is to put no more than 10% of the city's trash in landfills, and the long-term goal is to make Dongtan the world's first zero-solid-waste city.

The city plans to reduce dependence on private autos by erecting apartment buildings in clusters so that

all residents are within a 10-minute walk of a shopping center and a public transportation system, which will consist of an electric light-rail system and pollution-free, fuel-cell buses running on hydrogen. All vehicles in Dongtan will be required to have zero tailpipe emissions. Thus, residents choosing to drive in Dongtan will have to use electric- or hydrogen-powered, fuel-cell vehicles. Conventional vehicles running on fossil fuel will have to be parked outside the city limits.

Some of these goals may not be met. But Dongtan and other ecocities on the drawing boards in China will be urban laboratories in which people will experiment with new and more sustainable approaches to urban planning and development. During this century, these plans will be modified as needed and used in other parts of China and in other countries, as the world continues to urbanize.

The Ecovillage Movement Is Growing

Another innovative approach is the growing global *ecovillage movement* in which small groups of people have come together to design and live in more sustainable villages and *eco-hoods* in rural, suburban, and urban areas. By 2008, there were more than 375 ecovillages, over half of them in Europe and North America. Examples are found near Ithaca, New York, and Asheville, North Carolina, in the United States; Munksård near Copenhagen, Denmark; Porto Alegre in Brazil; Mbam in Senegal, Africa; Sieben Linden in Germany; and Findhorn in Scotland.

Ecovillagers use a diversity of methods to live more sustainably and to decrease their ecological footprints. Strategies include generating electricity from solar cells, small wind turbines, and small hydropower systems; collecting rainwater; using composting toilets; using solar cookers (Figure 16-16, p. 414) and rooftop solar collectors to provide hot water; cooperating in the development of organic farming plots; and using passive solar design, energy-efficient houses, and living roofs. In Findhorn, Scotland, residents have reduced their average ecological footprint per person by about 40%, and other cities claim similar reductions.

One example in the United States is Los Angeles Ecovillage, started in 1993 as a project to show how people in the middle of a sprawling and unsustainable city such as Los Angeles, California, can establish small pockets of sustainability. It consists of two small apartment buildings with about 55 residents and a common courtyard. Working together, its members use solar panels to get much of their hot water. They compost their organic wastes and use the compost on a large courtyard garden that provides vegetables and fruits and serves as a community commons area for relaxation and social interaction. The ecovillage is within walking distance of subway and bus stops, and members who live without a car pay a lower rent. The group has also established a bicycle repair shop.

— RESEARCH FRONTIER —

Evaluating and improving the design and expansion of more ecologically sustainable cities, villages, and neighborhoods. See **academic.cengage.com/biology/miller**.

REVISITING **Brazil's Curitiba and Sustainability**

CORE CASE STUDY SUSTAINABILITY

This chapter explored issues related to human population growth and its distribution. As elsewhere in this text, we focused on *sustainability* in discussing the issues. For example, we know that, each week, well over a million people are added to urban areas, just in developing countries. For the human population to live more sustainably, cities will have to accommodate this sort of growth without seriously depleting or degrading the natural capital available to them locally and throughout the world.

In this chapter, we have seen that despite the environmental problems plaguing cities, there are a growing number of inspiring examples of cities and villages around the world that are striving to become environmentally sustainable. While cities can magnify environmental problems, they can also gather people with the energy and imagination needed to solve some of those problems.

Curitiba, Brazil (**Core Case Study**), is one city in a developing country that has made great strides toward becoming environmentally sustainable. Its 3.2 million people have opportunities for living well and improving their lives, and its ecological footprint is considerably smaller than that of most cities its size. Curitiba's story gives us hope for developing more sustainable cities. The

most difficult challenge is to convert cities with little or no hope to cities full of hope, as inspiring and innovative leaders have done in Curitiba and Bogotá, Colombia.

For the long run, some experts say, we ought to ask what is the optimal level of human population that the planet or a city or village can support *sustainably?* That is, what are the population numbers at which all people can live comfortably and freely without jeopardizing the earth's ability to provide the same comforts and freedoms for future generations? Curitiba has to grapple with this problem, because its population has grown to the point where it threatens to overwhelm many of the sustainability solutions its founders and citizens implemented.

In the first six chapters of this book, you have learned how ecosystems and species have sustained themselves throughout more than 3 billion years of the earth's history by use of four **scientific principles of sustainability**—relying on solar energy, nutrient recycling, biodiversity, and population control. In this chapter, you may have gained a sense of the need for humans to apply these sustainability principles to their economies, to urban design, and to their lifestyles.

*A sustainable world will be powered by the sun;
constructed from materials that circulate repeatedly;
made mobile by trains, buses, and bicycles;
populated at sustainable levels; and
centered around just, equitable, and tight-knot communities.*

GARY GARDNER

REVIEW

1. Review the Key Questions and Concepts for this chapter on p. 589. Describe how Curitiba, Brazil, has attempted to become a more sustainable city (**Core Case Study**). CORE CASE STUDY

2. Distinguish between **urbanization** and **urban growth.** Describe two factors that increase the population of a city.

3. List four trends in global urban growth. Describe four phases of urban growth in the United States.

4. What is **urban sprawl?** List six factors that have promoted urban sprawl in the United States. List five undesirable effects of urban sprawl.

5. What are four advantages of urbanization? What are six disadvantages of urbanization? What is **noise pollution** and how can it be reduced? Explain why most cities and urban areas are not sustainable.

6. What are squatter settlements and shantytowns? Describe some of the problems faced by the poor who live in urban areas. How can governments help to reduce these problems? Describe the urban problems of Mexico City, Mexico.

7. Distinguish between compact and dispersed cities and give an example of each. What are the major advantages and disadvantages of motor vehicles? List four ways to reduce dependence on motor vehicles. Describe the major advan-

tages and disadvantages of relying more on **(a)** bicycles, **(b)** mass transit rail systems, and **(c)** bus rapid transit systems within urban areas, and **(d)** rapid-rail systems between urban areas. Describe the destruction of the early mass transit system in the United States.

8. What is **land-use planning?** What is **zoning** and what are its limitations? What is **smart growth?** List five tools used to promote smart growth. Describe strategies used by the U.S. city of Portland, Oregon, to help control urban sprawl and reduce dependence on automobiles.

9. What are the five guiding principles of new urbanization? What is cluster development? List ten goals of ecocity and ecovillage design. Describe Shanghai, China's plan to build an ecocity called Dongtan. Describe three strategies used within ecovillages to make their neighborhoods more sustainable.

10. Explain how people in Curitiba, Brazil (**Core Case Study**) have applied each of the four **scientific principles of sustainability** to make their city more sustainable.

Note: Key Terms are in **bold** type.

CRITICAL THINKING

1. Curitiba, Brazil (**Core Case Study**), has made significant progress in becoming a more environmentally sustainable and desirable place to live. If you live in an urban area, what steps, if any, has your community taken along these lines? CORE CASE STUDY

2. In Curitiba, Brazil (**Core Case Study**), recycling rates have fallen in recent years, even though they are still among the highest in the world. Former Mayor Jaime Lerner attributes this decline to failure on the part of the city's leadership to encourage recycling. List five ways in which you think Curitiba, or any city, could encourage residents to recycle. CORE CASE STUDY

3. Do you think that urban sprawl is a problem and something that should be controlled? Develop three arguments to support your answer. Compare your arguments with those of your classmates.

4. Write a brief essay that includes at least three reasons why you **(a)** enjoy living in a large city, **(b)** would like to live in a large city, or **(c)** do not wish to live in a large city. Be specific in your reasoning. Compare your essay with those of your classmates.

5. One issue debated at a U.N. conference was the question of whether housing is a universal *right* (a position

supported by most developing countries) or just a *need* (supported by the United States and several other developed countries). What is your position on this issue? Defend your choice.

6. If you own a car or hope to own one, what conditions, if any, would encourage you to rely less on the automobile and to travel to school or work by bicycle, on foot, by mass transit, or by carpool?

7. Do you believe the United States or the country where you live should develop a comprehensive and integrated mass transit system over the next 20 years, including an efficient rapid-rail network for travel within and between its major cities? How would you pay for such a system?

8. Consider the goals listed on p. 606 as part of the ecocity concept. How close to reaching each of these goals is the city where you live or the city nearest to where you live? Pick what you think are the five most important of these

goals and describe a way in which that city could meet each of those five goals.

9. Congratulations! You are in charge of the world. List the three most important components of your strategy for dealing with urban growth and sustainability in **(a)** developed countries and **(b)** developing countries.

10. List two questions that you would like to have answered as a result of reading this chapter.

Note: See Supplement 13 (p. S78) for a list of Projects related to this chapter.

ECOLOGICAL FOOTPRINT ANALYSIS

Under normal driving conditions, an internal combustion engine produces approximately 200 grams (0.2 kilograms, or 0.44 pounds) of CO_2 per kilometer driven. In one large city, there are 7 million cars, and each car is driven an average of 20,000 kilometers (12,400 miles) per year.

1. Calculate the carbon footprint per car, that is, the number of metric tons (and tons) of CO_2 produced by a typical car in one year. Calculate the total carbon footprint for all cars in the city, that is, the number of metric tons (and tons) of CO_2 produced by all the cars in one year. (*Note:* 1 metric ton = 1,000 kilograms = 1.1 tons)

2. If 20% of the people were to use a carpool or to take mass transit in the city instead of driving their cars, how many metric tons (and how many tons) of these CO_2 emissions would be eliminated?

3. By what percentage and by what average distance driven would use of cars in the city have to be cut in order to reduce by half the carbon footprint calculated in question 1?

LEARNING ONLINE

Log on to the Student Companion Site for this book at **academic.cengage.com/biology/miller**, and choose Chapter 22 for many study aids and ideas for further reading and research. These include flash cards, practice quizzing, Weblinks, information on Green Careers, and InfoTrac® College Edition articles.

Economics, Environment, and Sustainability

A New Economic and Environmental Vision

Economic growth is concerned with making an economy bigger. *Economic development* is devoted to making an economy better by improving human well-being—meeting fundamental human needs for items such food, shelter, physical and economic security, and good health. For more than 100 years, there has been a debate over whether there are limits to economic growth. Now the debate is shifting to questions about what kinds of economic growth and development we should encourage.

Neoclassical economists such as Robert Samuelson and the late Milton Friedman view natural resources as important but not indispensable because of our ability to find substitutes. They also contend that continuing economic growth is necessary for providing profits and jobs and is essentially unlimited.

Ecological economists such as Herman Daly and Robert Costanza disagree. They point out that there are no substitutes for many vital natural resources, such as air, water, fertile soil, and biodiversity or for nature's free ecological services such as climate control, air and water purification, pest control, and nutrient recycling. They also believe that conventional economic growth eventually will become unsustainable because it can deplete or degrade various forms of natural capital (Figure 1-3, p. 8) on which economic systems depend. They argue that such growth will also exceed the capacity of the environment to handle the pollutants and wastes we produce. They call for us to redesign our economic and political systems to encourage

environmentally beneficial and sustainable forms of economic development and to discourage environmentally harmful forms of economic growth.

Taking the middle in this debate are *environmental economists*. They generally agree with ecological economists that some forms of economic growth are not sustainable and should be discouraged. They believe we can encourage more environmentally sustainable economic development by reforming current economic systems rather than having to redesign them.

Ecological economists and many environmental economists call for making a shift from our current economy based on unlimited economic growth to a more *environmentally sustainable economy,* or *eco-economy,* over the next several decades. Figure 23-1 shows some of the goods and services these economists would encourage in an eco-economy.

Because the environment and the economy are intimately linked, improving environmental quality and sustainability is good for the economy. Thus, *the sustainability revolution is also an economic revolution.* According to visionary economic and environmental leaders, making a shift to more environmentally sustainable economies is the biggest investment opportunity in this century. It will promote cleaner energy, cleaner production, and more sustainable agriculture. It will create new jobs and provide profits by spurring investment in efforts to sustain the natural capital that supports life and all economies.

Forest conservation

Production of energy-efficient fuel-cell cars

Underground CO_2 storage using abandoned oil wells

High-speed trains

No-till cultivation

Deep-sea CO_2 storage

Solar-cell fields

Bicycling

Communities of passive solar homes

Wind farms

Water conservation

Cluster housing development

Recycling plant

Landfill

Recycling, reuse, and composting

Figure 23-1 Solutions: some components of more environmentally sustainable economic development favored by ecological and environmental economists. The goal is to get economic systems to put more emphasis on conserving and sustaining the air, water, soil, biodiversity, and other natural resources that sustain all life and all economies. Such a shift toward more efficient resource use, cleaner energy, cleaner production, ecocities (Chapter 22 Core Case Study, p. 588), and natural capital preservation could stimulate economies, create jobs, and be profitable. **Question:** What are three new types of jobs that could be generated by such an economy?

23-1 How are economic systems related to the biosphere?

CONCEPT 23-1A Ecological economists and most sustainability experts regard human economic systems as subsystems of the biosphere and subject to its processes and limiting factors.

CONCEPT 23-1B Governments intervene in market economies to help correct or prevent market failures such as the failure to protect natural capital, which supports life and all economies.

23-2 How can we put values on natural capital, pollution control, and resource use?

CONCEPT 23-2A Economists have developed several ways to estimate the present and future values of a resource or ecological service and optimum levels of pollution control and resource use.

CONCEPT 23-2B Comparing the likely costs and benefits of an environmental action is useful but involves many uncertainties.

23-3 How can we use economic tools to deal with environmental problems?

CONCEPT 23-3A Using resources more sustainably will require including the harmful environmental and health costs of resource use in the market prices of goods and services (*full-cost pricing*).

CONCEPT 23-3B Governments can help to improve and sustain environmental quality by subsidizing environmentally beneficial activities and taxing pollution and waste instead of wages and profits.

23-4 How can reducing poverty help us to deal with environmental problems?

CONCEPT 23-4 Reducing poverty can help us to reduce population growth, resource use, and environmental degradation.

23-5 How can we make the transition to more environmentally sustainable economies?

CONCEPT 23-5 We can use the four *principles of sustainability* and various economic and environmental strategies to develop more environmentally sustainable economies.

Note: Supplements 2, (p. S4), 3 (p. S10), 5 (p. S31), 11(p. S74) and 13 (p. S78) can be used with this chapter.

*When it is asked how much it will cost to protect the environment,
one more question should be asked:
How much will it cost our civilization if we do not?*

GAYLORD NELSON

23-1 How Are Economic Systems Related to the Biosphere?

▶ **CONCEPT 23-1A** Ecological economists and most sustainability experts regard human economic systems as subsystems of the biosphere and subject to its processes and limiting factors.

▶ **CONCEPT 23-1B** Governments intervene in market economies to help correct or prevent market failures such as the failure to protect natural capital, which supports life and all economies.

Economic Systems Are Supported by Three Types of Resources

An **economic system** is a social institution through which goods and services are produced, distributed, and consumed to satisfy people's needs and wants, ideally in the most efficient possible way.

Three types of capital, or resources, are used to produce goods and services (Figure 23-2). **Natural capital (Concept 1-1A**, p. 6, and Figure 1-3, p. 8), includes resources and services produced by the earth's natural processes, which support all economies and all life. (See the Guest Essay on natural capital by Paul Hawken at CengageNOW™.)

Human capital, or **human resources,** includes people's physical and mental talents, which provide labor, innovation, culture, and organization. **Manufactured capital,** or **manufactured resources,** are items such as machinery, equipment, and factories made from natural resources with the help of human resources.

Natural Resources **Manufactured Resources** **Human Resources** **Goods and Services**

Figure 23-2 Three types of resources are used to produce goods and services in an economic system.

Market Economic Systems Depend on Interactions between Buyers and Sellers

In a market-based economic system, buyers and sellers interact competitively in markets to make economic decisions about which goods and services are produced, distributed, and consumed. In a pure version of such a system, all economic decisions are governed solely by the competitive interactions of *demand* (the amount of a good or service that people want), *supply* (the amount of a good or service that is available), and *price* (the market cost of a good or service).

In Figure 23-3, the point at which the supply and demand curves intersect is called the *market price equilibrium point,* where the supplier's price matches what buyers are willing to pay for some quantity of a good or service. Changes in supply and demand can shift one or both curves back and forth, and thus change the equilibrium price. For example, when supply is increased (shifting the blue curve to the right) and demand remains the same, the equilibrium price will go down. Similarly, when demand is increased (shifting the red curve to the right) and supply remains the same, the equilibrium price will increase. Try mentally moving the curves in Figure 23-3 in different ways that represent changes in supply and demand, and notice how the equilibrium price changes.

Most free-market economists believe that an economic system based on private ownership of all resources and operating without government interference is the best system for solving environmental problems for two reasons. One argument is that the true values of resources are determined best through unregulated market interactions based only on supply and demand. The other is that when resources are owned privately, their owners are more likely to cherish them and to want to protect them.

A truly free-market economic system operating without government interference requires that buyers and sellers not coerce or mislead one another and that the prices of all goods and services include all of their direct and indirect costs (full-cost pricing). But this rarely happens in today's capitalist market systems. To increase their profits, most businesses strive to drive their competitors out of business and to exert as much control as possible over the prices of the goods and services they provide. Companies lobby for government subsidies, tax breaks, or regulations that give their products a market advantage over their competitors' products. Such advantages create an uneven economic playing field that can undermine true free-market competition.

Some companies also withhold information from consumers about health or environmental dangers posed by their products, unless the government requires them to provide such information. Some companies also lobby against having to include harmful environmental or health costs in the prices of goods or services they provide. Thus, buyers often do not get accurate information about the harmful aspects of goods and services they buy.

Figure 23-3 Supply, demand, and market equilibrium for a good such as oil in a market economic system. If all factors except price, supply, and demand are held fixed, market equilibrium occurs at the point at which the demand and supply curves intersect. This is the price at which sellers are willing to sell and buyers are willing to pay for the good or service provided. **Question:** How would a decrease in the available supply of oil shift the market equilibrium point on this diagram?

Governments Intervene to Help Correct Market Failures

Markets often work well in guiding the efficient production and distribution of *private goods.* But experience shows that they cannot be relied upon to provide adequate levels of *public services,* such as national security and environmental protection. Markets also cannot prevent the degradation of open-access, renewable resources, such as clean air and the open ocean, which are not bought and sold in the marketplace, because they are owned by no one and available for use by everyone at little or no charge. Therefore, governments intervene in market systems to provide various public services, to protect open-access resources that cannot be privately owned, and to help correct other *market failures* (**Concept 23-1B**). For example, without government intervention, a single seller or buyer (monopoly) or a single group of sellers or buyers (oligopoly) might come to dominate the market and thus control supply, demand, and price for a good or service.

Markets can also fail when they do not value the beneficial effects provided by the earth's natural capital and the harmful health or environmental effects of some types of economic activity. Scientists and ecological and environmental economists have attempted to place economic values on various forms of the earth's natural capital (Science Focus, p. 218), but these values are seldom included in market prices. Governments can try to prevent or correct these and other market failures through laws, regulations, taxes, subsidies and tax breaks, and a number of other tools.

Other reasons for government intervention into the marketplace are to:

- Provide public services such as national security and education

- Provide an economic safety net for people who, because of health, age, and other factors, cannot meet their own basic needs

- Protect people from fraud, trespass, theft, and bodily harm

- Establish and enforce civil rights and property rights

- Protect the health and safety of workers and consumers

- Prevent or reduce pollution, degradation, and depletion of natural capital

- Manage public land resources such as national forests, parks, and wildlife reserves

THINKING ABOUT
The Government's Role in an Economy

In a pure free-market economy the government plays no role in an economy, and some economists see this as the best possible economic system. Explain why you agree or disagree with their position.

Economists Disagree over the Importance of Natural Capital and whether Economic Growth Is Sustainable

Neoclassical economists view the earth's natural capital as a subset or part of a human economic system (Figure 23-2) and assume that the potential for economic growth is essentially unlimited (**Core Case Study**).

Most of today's advanced industrialized countries have **high-throughput economies,** which attempt to boost economic growth by increasing the flow of matter and energy resources extracted from the environment through their economic systems to produce goods and services (Figure 23-4). These resources flow through economies and end up in planetary *sinks* (air, water, soil, and organisms), where some pollutants and wastes can accumulate to harmful levels.

Ecological economists and some *environmental economists* disagree with the neoclassical economic model and its assumptions. They view economic systems as subsystems of the biosphere that depend heavily on the earth's irreplaceable natural resources (**Concept 23-1A**) (Figure 23-5).

According to ecological economist Herman Daly, the neoclassical model of an economy "ignores the origin of natural resources flowing into the system and the fate of wastes flowing out of the system. It is as if a biologist had a model of an animal that contained a circulatory system but had no digestive system that tied it firmly to the environment at both ends." (See Herman Daly's Guest Essay on this topic at CengageNOW.)

The models of ecological economists are built on three major assumptions:

- Resources are limited and should not be wasted, and there are no substitutes for most types of natural capital (Figure 1-3, p. 8).

CENGAGENOW™ **Active Figure 23-4** The *high-throughput economies* of most developed countries rely on continually increasing the rates of energy and matter flow to increase economic growth. This practice produces valuable goods and services, but it also converts high-quality matter and energy resources into waste, pollution, and low-quality heat, and in the process, can deplete or degrade various forms of natural capital that support all life and economies. *See an animation based on this figure at* CengageNOW. **Question:** What are three things that you regularly do that add to this throughput of matter and energy?

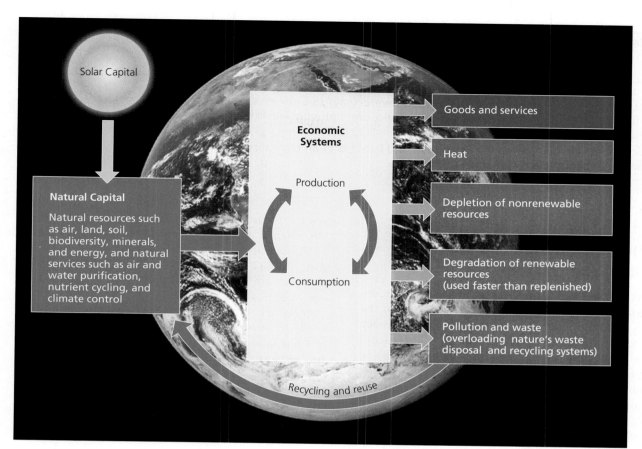

Solar Capital

Natural Capital

Natural resources such as air, land, soil, biodiversity, minerals, and energy, and natural services such as air and water purification, nutrient cycling, and climate control

Economic Systems

Production

Consumption

Goods and services

Heat

Depletion of nonrenewable resources

Degradation of renewable resources (used faster than replenished)

Pollution and waste (overloading nature's waste disposal and recycling systems)

Recycling and reuse

CENGAGENOW
Active Figure 23-5
Ecological economists see all economies as human subsystems of the biosphere that depend on natural resources and services provided by the sun and earth. *See an animation based on this figure at* CengageNOW. **Question:** Do you agree or disagree with this model? Explain.

- We should encourage environmentally beneficial and sustainable forms of economic development and discourage environmentally harmful and unsustainable forms of economic growth.

- The harmful environmental and health effects of producing economic goods and services should be included in their market prices (*full-cost pricing*), so that consumers can have accurate information about the harmful effects of the goods and services they buy.

THINKING ABOUT
Economic Growth

CORE CASE STUDY

Do you think that the economy of the country where you live is sustainable or unsustainable? (See the **Core Case Study** that opens this chapter.) Explain.

CENGAGENOW Learn more about how ecological economists view market-based systems, and contrast their views with those of conventional economists at CengageNOW.

Ecological and environmental economists have suggested several strategies for implementing these principles and making the transition to more sustainable eco-economies over the next several decades.

- Use and publicize indicators that monitor economic and environmental health.

- Use *eco-labeling* and *certification* to identify products produced by environmentally sound methods and thus to help consumers make informed choices.

- Phase out environmentally harmful government subsidies and tax breaks while increasing subsidies and tax breaks for environmentally beneficial goods and service (*subsidy shifting*).

- Decrease taxes on wages, income, and wealth while increasing taxes on pollution, resource waste, and environmentally harmful goods and services (*tax shifting*).

- Use laws and regulations to help prevent pollution and resource depletion and degradation.

- Use tradable permits or rights to pollute or to use resources, in order to limit overall pollution and resource use (*cap-and-trade,* p. 490).

- Sell services instead of things.

- Reduce poverty, one of the basic causes of environmental degradation, pollution, poor health, and premature death (Figure 17-18, p. 461).

The rest of this chapter looks at these proposed solutions in more detail.

Practicing Natural Capitalism: Taking into Account the Monetary Value of Natural Capital

Environmental and ecological economists have developed various tools for estimating the values of the earth's natural capital (Figure 1-3, p. 8). One approach involves estimating the monetary worth of the earth's natural ecological services and the biological income they provide (Science Focus, p. 218). For example, using this approach a team of ecologists and economists estimated that the ecological services provided by the earth's forests (Figure 10-4, left, p. 217) are worth at least $4.7 trillion a year. This is hundreds of times greater than the estimated value of their economic services (Figure 10-4, right).

However, the estimated economic values of the ecological services provided by forests, oceans, and other ecosystems are not included in the market prices of goods, such as timber and fish, that we get from them. Ecological economists point out that until this economic imbalance is corrected we will continue to use forests and many of nature's other irreplaceable forms of natural capital unsustainably.

However, according to neoclassical economists (**Core Case Study**), a product or service has no economic value until it is sold in the marketplace, and thus because they are not sold in the marketplace, ecological services have no economic value. Ecological economists view this as a misleading circular argument designed to exclude serious evaluation of ecological services in the marketplace.

Environmental economists and ecologists have also developed ways to estimate *nonuse values* of natural resources and ecological services not represented in market transactions (**Concept 23-2A**). One is an *existence value*—a monetary value placed on a resource such as an old-growth forest or endangered species just because it exists, even though we may never see it or use it. Another is *aesthetic value*—a monetary value placed on a forest, species, or part of nature because of its beauty. A third type, called a *bequest* or *option value,* is based on the willingness of people to pay to protect some forms of natural capital for use by future generations.

There are different ways to estimate these types of monetary values. One approach is to estimate a *mitiga-tion cost* of how much it would take to offset any environmental damage. For example, we could estimate how much it would cost to protect a forest from cutting, to move an endangered species to a new habitat, or to restore a statue damaged by air pollution (Figure 18-6, p. 473).

Another method involves estimating *willingness to pay* by surveying public opinion to determine how much people would pay to maintain natural resources and services. The survey might ask, for example, how much people would pay to keep a particular species from becoming extinct, a particular forest from being cut down, or a specific river or beach from being polluted. This approach is controversial because people could inflate their estimates and not indicate the prices they would actually pay to preserve nonuse values.

> ┌─ **RESEARCH FRONTIER**
>
> Developing better ways to estimate the economic values of the earth's natural capital. See **academic.cengage.com/ biology/miller**.

Estimating the Future Value of a Resource Is Controversial

The **discount rate** is an estimate of a resource's future economic value compared to its present value (**Concept 23-2A**). It is based on the idea that having something today may be worth more than it will be in the future. The size of the discount rate (usually given as a percentage) is a primary factor affecting how a resource such as a forest or fishery is used or managed.

At a zero discount rate, for example, a stand of redwood trees worth $1 million today will still be worth $1 million 50 years from now. However, the U.S. Office of Management and Budget, the World Bank, and most businesses typically use a 10% annual discount rate to evaluate how resources should be used. At this rate, the stand of redwood trees will be worth only $10,000 in 50 years. With this discount rate, it makes sense from an economic standpoint for an owner to cut these trees down as quickly as possible and invest the money in something else.

This economic analysis does not take into account the immense economic value of the ecological services

provided by forests, including flood protection, air and water purification, and several other services (Figure 10-4, left, p. 217). This one-sided and incomplete economic evaluation loads the dice against sustaining these important ecological services.

The process of choosing a discount rate is controversial. Proponents cite several reasons for using a high (5–10%) discount rate. One is that inflation may reduce the value of their future earnings on a resource. Another is that innovation or changes in consumer preferences could make a product or resource obsolete. For example, composites made from plastics to look like redwood may reduce the future use and market value of these trees. In addition, owners of resources such as forests argue that without a high discount rate, they can make more money by investing their capital in some other venture.

Critics point out that high discount rates encourage rapid exploitation of resources for immediate payoffs, thus making sustainable use of most renewable natural resources virtually impossible. These critics believe that a 0% or even a negative discount rate should be used to protect unique, scarce, and irreplaceable resources. They also point out that moderate discount rates of 1–3% would make it profitable to use nonrenewable and renewable resources more sustainably (more slowly). In addition, suppose that an acceptable substitute for a resource such as redwood does not become available. Then these resources and the ecological services they provide could become priceless. This is another major disagreement between neoclassical economists and environmental and ecological economists (**Core Case Study**).

Economic return is not always the determining factor in how resources are used or managed. In some cases, farmers and owners of forests, wetlands, and other resources weigh their *ethical concerns* in determining how they use and manage such resources. Their respect for the land and nature, or what they believe to be their responsibility to future generations, can override their desire for short-term profit at the expense of long-term resource and environmental sustainability.

THINKING ABOUT
Discount Rates

If you owned a forested area, would you want the discount rate for resources such as trees from the forest to be high, moderate, or zero? Explain.

We Can Estimate Optimum Levels of Pollution Control and Resource Use

An important concept in environmental economics is that of *optimum levels* for pollution control and resource use (**Concept 23-2A**). In the early days of a new coal mining operation, for example, the cost of removing

Figure 23-6 *Optimum resource use.* The cost of mining coal (blue line) from a particular mine rises with each additional unit removed. Mining a certain amount of coal is profitable, but at some point the marginal cost of further mining exceeds the monetary benefits (red line) unless some factor such as scarcity raises the value of the coal remaining in a mine (in which case, the red curve shifts to the right). Where the two curves meet is theoretically the *optimum level of resource use.* **Question:** How would the location of the optimum level of resource use shift if the value (and therefore the price) of the coal increased?

coal is low enough to make it easy for developers to recover their investments by selling their product. However, after most of the readily accessible coal has been removed from a mine, taking what is left can become too costly. In this case, the cost of removal goes up with each additional unit of coal taken—what economists call its *marginal cost.* Figure 23-6 shows this in terms of supply, demand, and equilibrium. The point at which removing more coal is not worth the cost is where the demand curve crosses the supply curve.

You might think that the best solution for pollution is total cleanup. In fact, there are optimum levels for various kinds of pollution. The main reason is that the cost of pollution control goes up for each additional unit of a pollutant removed from the environment, primarily because it takes increasing amounts of energy to remove increasingly lower concentrations of a pollutant from the air, water, or soil. Figure 23-7 (p. 618) shows various possible optimum levels of pollution control. In this case, cleanup costs are shown on the blue curve. This is the supply curve, because the service supplied is removal of pollutants. Note that it slopes up more sharply (costing more) as we get closer to removing 100% of the pollutant in question.

The red line in Figure 23-7 represents the *demand* for cleanup by users of water or air. When the air they breathe or the water they drink are polluted, consumers are initially up in arms, but as the pollutant is removed, their concern decreases and their demand for cleanup approaches zero. At some point, the cost of removing additional units of a pollutant becomes higher than what people are willing to pay, as their demand for cleanup lessens. That point is the equilibrium point, or the *optimum level* for pollution cleanup.

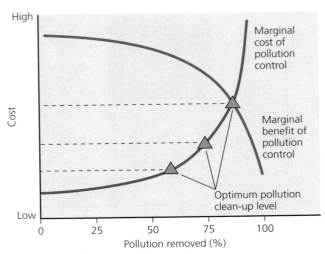

Figure 23-7 *Optimum pollution control.* The cost of cleaning up pollution (blue line) rises with each additional unit removed. Cleanup of a certain amount of pollution is affordable, but at some point the cost of additional pollution control is greater than the benefit of the pollution control to society. Where the blue curve intersects with the red curve is a level of optimum pollution control for the pollutant represented. The other two green triangles show where the curves would intersect if the red curve shifted to the left (if some factor lessened the value of pollution control). **Question:** How does the fact that the cost of pollution control goes up for each additional unit of a pollutant cleaned up relate to the second law of thermodynamics (p. 43)?

The shapes and locations of these curves vary with different types of pollutants. Another factor determining the shape and placement of the demand curve is the degree to which people value a particular resource. If no one cares whether the water in a lake is clear, the optimum level of cleanup will be close to zero. But if they demand a clean lake, the optimum level will rise. In other words, pollution control (or resource use) that is optimal for some people will be high or low for others.

Cost–Benefit Analysis Is a Useful but Crude Tool

Another widely used tool for making economic decisions about how to control pollution and manage resources is **cost–benefit analysis.** This is done by comparing estimated costs and benefits for actions such as implementing a pollution control regulation, building a dam on a river, or preserving an area of forest. It involves trying to estimate the optimum level of pollution cleanup (Figure 23-7) or resource use (Figure 23-6).

Making a cost-benefit analysis involves determining who or what might be affected by a particular regulation or project, projecting potential outcomes, evaluating alternative actions, and establishing who benefits and who is harmed. Then an attempt is made to assign monetary costs and benefits to each of the factors and components involved.

Direct costs involving land, labor, materials, and pollution-control technologies are often fairly easy to estimate. But estimates of indirect costs of clean air and water that are not traded in the marketplace are difficult to make and are controversial. We can put estimated price tags on human life, good health, clean air and water, and natural capital, such as an endangered species, a forest, and a wetland. However, the monetary values assigned to these estimates vary widely depending on individual assumptions, value judgments, and the discount factors used.

Because of these drawbacks, a cost-benefit analysis can lead to wide ranges of benefits and costs with a lot of room for error, and this is another source of controversy between neoclassical economists and environmental and ecological economists (**Core Case Study**). For example, one U.S. industry-sponsored, cost-benefit analysis estimated that compliance with a standard to protect U.S. workers from vinyl chloride would cost $65–90 billion. In the end, meeting the standard cost the industry less than $1 billion. A study by the Washington-based Economic Policy Institute found that the estimated costs made by industries for complying with proposed environmental regulations in the United States are almost always more (and often much more) than the actual costs of implementing the regulations. In such cases, inflated cost-benefit analyses are used as an economic and political ploy by industries to prevent, weaken, or delay compliance with pollution standards.

If conducted fairly and accurately, cost-benefit analysis is a useful tool for helping to make economic decisions, but it always includes uncertainties (**Concept 23-2B**). To minimize possible abuses and errors, environmental economists advocate using the following guidelines for a cost-benefit analysis:

- Use uniform standards.

- Clearly state all assumptions used.

- Include estimates of the ecological services provided by the resources involved.

- Rate the reliability of data used.

- Estimate short- and long-term benefits and costs for all affected population groups.

- Compare the costs and benefits of alternative courses of action.

- Summarize the range of estimated costs and benefits.

Because of the uncertainties and multiple variables involved in cost-benefit analysis, environmental and ecological economists argue that no major decision affecting the environment should be made on the basis of any such analysis by itself.

23-3 How Can We Use Economic Tools to Deal with Environmental Problems?

▶ **CONCEPT 23-3A** Using resources more sustainably will require including the harmful environmental and health costs of resource use in the market prices of goods and services (*full-cost pricing*).

▶ **CONCEPT 23-3B** Governments can help to improve and sustain environmental quality by subsidizing environmentally beneficial activities and taxing pollution and waste instead of wages and profits.

Most Things Cost a Lot More Than You Think

The *market price*, or *direct price*, that you pay for something does not include most of the *indirect*, or *external*, *costs* of harm to the environment and human health associated with its production and use. For this reason, such costs are also called *hidden costs*. For example, if you buy a car, the direct price you pay includes the costs of raw materials, labor, and shipping, as well as a markup to allow the car company and its dealers to make a profit. In using the car, you pay additional direct costs for gasoline, maintenance, repair, and insurance.

However, you do not pay the harmful external costs, some of which affect people other than you and the seller. For example, to extract and process raw materials to make a car, manufacturers use nonrenewable energy and mineral resources, produce solid and hazardous wastes, disturb land, pollute the air and water, and release greenhouse gases into the atmosphere. These hidden external costs can have short- and long-term harmful effects on other people, on future generations, and on the earth's life-support systems.

Because these harmful external costs are not included in the market price of a car, most people do not connect them with car ownership. Still, the car buyer and other people in a society pay these hidden costs sooner or later, in the forms of poorer health, higher costs of health care and insurance, higher taxes for pollution control, traffic congestion, and land used for highways and parking.

In general, neoclassical economists do not believe that these costs should be included in the prices of goods and service. Environmental and ecological economists (**Core Case Study**) and economists who believe in a truly free market economic system believe that they should be included as part of full-cost pricing (**Concept 23-3A**). They contend that excluding the harmful environmental costs from the market prices of goods and services has three major effects: it hides these harmful costs from consumers; hinders the development of more environmentally beneficial, or green, goods and services; and promotes pollution, resource waste, and environmental degradation.

Many economists and environmental experts cite this failure in today's market-based economic systems as one of the major causes of the environmental problems we face (Figure 1-12, p. 18). For example, in his 2006 analysis of the economics of climate change, Nicholas Stern, former chief economist of the World Bank, described the failure to include the harmful costs of climate change in the prices of fossil fuels as "a market failure on the greatest scale the world has ever seen."

THINKING ABOUT
Hidden Costs

Were you aware of the external costs involved in producing, buying, and owning a car (or any other item)? Do you believe that such hidden costs should be included in the prices of goods and services? Explain.

Using Environmental Economic Indicators Can Help Reduce Our Environmental Impact

Gross domestic product (GDP) and *per capita GDP* indicators (p. 10) provide a standardized and useful method for measuring and comparing the economic outputs of nations. The GDP is deliberately designed to measure the annual economic value of all goods and services produced within a country without distinguishing between goods and services that are environmentally or socially beneficial and those that are harmful.

Environmental and ecological economists and environmental scientists call for the development and widespread use of new indicators to help monitor environmental quality and human well-being. One such indicator is the *genuine progress indicator* (GPI). Redefining Progress, a nonprofit organization that develops economic and policy tools to promote environmental sustainability, introduced the genuine progress indicator in 1995. (This group also developed the concept of ecological footprints, Figure 1-10, p. 15.)

Within the genuine progress indicator, the estimated value of beneficial transactions that meet basic needs, but in which no money changes hands, are

added to the GDP. Examples include unpaid volunteer work, health care for family members, child care, and housework. Then the estimated harmful environmental costs (such as pollution and resource depletion and degradation) and social costs (such as crime) are subtracted from the GDP.

$$\begin{array}{c}\text{Genuine}\\\text{progress}\\\text{indicator}\end{array} = \text{GDP} + \begin{array}{c}\text{benefits not}\\\text{included in}\\\text{market transactions}\end{array} - \begin{array}{c}\text{harmful}\\\text{environmental}\\\text{and social costs}\end{array}$$

Figure 23-8 compares the per capita GDP and GPI for the United States between 1950 and 2004 (the latest data available). While the per capita GDP rose sharply over this period, the per capita GPI stayed nearly flat and even declined slightly and fluctuated between 1975 and 2004.

Another green economic indicator is the *happy planet index* (HPI), first published by Friends of the Earth and the New Economics Foundation in 2006. Calculating the values of this indicator involves multiplying a country's self-reported life satisfaction index (determined through surveys) by its average life expectancy and then dividing this number by it ecological footprint. In the first HPI assessment, the Central American region had the highest overall score because of its high life satisfaction scores, relatively long life expectancy, and a fairly low ecological footprint compared to other regions of the world.

Bhutan is a South Asian nation in the Himalaya Mountains and is located between China and India. Since 1972, its government has been using the concept of *gross national happiness* (GNH) as a measure of its efforts toward sustainable economic development. Its GNH is based on an evaluation of the country's conservation of its natural environment, preservation of cultural values, fairness in access to its wealth (equity), and good governance.

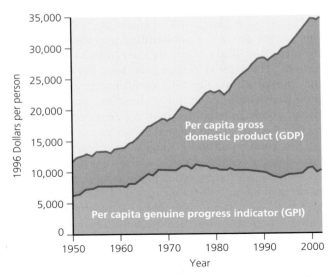

Figure 23-8 *Monitoring environmental progress:* Comparison of the per capita gross domestic product (GDP) and the per capita genuine progress indicator (GPI) in the United States between 1950 and 2004. **Questions:** Would you favor making widespread use of this or similar green economic indicators? Why hasn't this been done? (Data from Redefining Progress, 2006)

The genuine progress indicator, happy planet index, gross national happiness indicator, and other environmental indicators under development are far from perfect. But without such indicators, we will not know much about what is happening to people, the environment, and the planet's natural capital, nor will we have a way to evaluate which policies work best in improving environmental quality and life satisfaction. According to ecological and environmental economists, without these indicators, we are steering national and global economies through treacherous economic and environmental waters at ever increasing speeds without a good navigation system.

— RESEARCH FRONTIER —
Developing and refining environmental and social quality-of-life indicators. See **academic.cengage.com/biology/ miller**.

We Can Include Harmful Environmental Costs in the Prices of Goods and Services

Environmental and ecological economists argue that an *environmentally honest market system* would include the harmful environmental and health costs of goods and services in their market prices to reflect as close as possible their *full costs*—internal costs plus external costs (**Concept 23-3A**). Again, this is a fundamental source of controversy between neoclassical economists and environmental and ecological economists (**Core Case Study**). CORE CASE STUDY

According to environmental and ecological economists, full-cost pricing would reduce resource waste, pollution, and environmental degradation and improve human health by encouraging producers to invent more resource-efficient and less-polluting methods of production. It would also allow consumers to make more informed choices—one of the goals of a truly free market economic system. Jobs and profits would be lost in environmentally harmful businesses as consumers more often chose green products, but jobs and profits would be created in environmentally beneficial businesses. Such shifts in profits and in types of businesses and jobs is the way market-based capitalism works.

Full-cost pricing seems to make a lot of sense. So why is it not used more widely? *First,* many producers of harmful and wasteful products would have to charge more, and some would go out of business. Naturally, they oppose such pricing. *Second,* it is difficult to estimate many environmental and health costs. *Third,* because these harmful costs are not included in the market prices of goods and services, most consumers do not connect them with the things they buy. Correcting this requires consumer education.

—— RESEARCH FRONTIER ——

Refining methods for estimating and implementing full-cost pricing. See **academic.cengage.com/biology/miller**.

—— HOW WOULD YOU VOTE? ——

Should full-cost pricing be used in setting market prices for goods and services? Cast your vote online at **academic.cengage.com/biology/miller**.

Phasing in such a system would require government action prodded by bottom-up citizen pressure. Few if any companies will volunteer to reduce their short-term profits to become more environmentally responsible unless it eventually becomes more profitable for them. Governments can use several strategies to encourage or force producers to work toward full-cost pricing: certifying and labeling environmentally beneficial goods and services; phasing out environmentally harmful subsidies; levying taxes on environmentally harmful goods and services; passing laws to regulate pollution and resource depletion; and using tradable permits for reducing pollution and resource use. Let us take a closer look at each of these strategies.

Environmentally Informed Consumers Can Vote with Their Wallets

Product eco-labeling and *certification programs* encourage companies to develop green products and services and help consumers to select more environmentally beneficial products and services. Eco-labeling programs have been developed in Europe, Japan, Canada, and the United States. The U.S. *Green Seal* labeling program has certified more than 300 products as environmentally friendly.

Eco-labels are also being used to identify fish caught by sustainable methods (certified by the Marine Stewardship Council) and to certify timber produced and harvested by sustainable methods (evaluated by organizations such as the Forest Stewardship Council; see Science Focus, p. 228). You can buy coffee that was produced under sustainable conditions, certified by the Rainforest Alliance and the Organic Trade Association.

We Can Reward Environmentally Sustainable Businesses

One way to encourage a shift to full-cost pricing is to *phase out* environmentally harmful subsidies and tax breaks, which according to environmental scientist Norman Myers cost the world's governments (taxpay-

ers) at least $2 trillion a year—an average $3.8 million a minute! (See Myer's Guest Essay on perverse subsidies at CengageNOW.) Such subsidies distort the economic playing field and create a huge economic incentive for unsustainable resource waste, depletion, and degradation. Examples include depletion subsidies and tax breaks for extracting minerals and oil, cutting timber on public lands, and irrigating with low-cost water.

A typical American taxpayer unknowingly pays at least $2,500 a year to provide these and other subsidies and then pays another $1,000 per year for environmental degradation, pollution cleanup, and higher health and insurance costs related to the impacts from harmful environmental subsidies. According to a 1997 study by the Earth Council, "There is something unbelievable about the world spending hundreds of billions of dollars annually to subsidize its own destruction."

On paper, phasing out such subsidies may seem like a great idea. In reality, the economically and politically powerful interests receiving them want to keep—and if possible increase—these mostly hidden economic benefits provided to them by taxpayers. In addition to lobbying to keep or increase their subsidies, such businesses often lobby against subsidies and tax breaks for more environmentally beneficial competitors. Environmentally destructive subsidies and tax breaks will continue until enough individuals work together to counteract such lobbying by electing and then influencing elected officials who support phasing out environmentally harmful subsidies.

Some countries have begun reducing environmentally harmful subsidies. Japan, France, and Belgium have phased out all coal subsidies. Germany cut its coal subsidies in half between 1989 and 2002 and plans to phase them out by 2018. China has cut coal subsidies by about 73% and has imposed a tax on high-sulfur coals. New Zealand has eliminated virtually all of its agricultural subsidies.

At the same time, governments could phase in environmentally beneficial subsidies and tax breaks for pollution prevention (**Concept 1-4**, p. 16), ecocities (pp. 588 and 606), sustainable forestry (Figure 10-19, p. 231), sustainable agriculture (Figure 12-31, p. 307), sustainable water use (Figure 13-23, p. 337), energy conservation and renewable energy use (Figure 16-33, p. 432), and actions to slow climate change (Figure 19-13, p. 515) (**Concept 23-3B**).

Making such subsidy shifts over 2 decades would encourage the rise of new environmentally beneficial businesses. It would also give current environmentally harmful businesses enough time to transform themselves into profitable businesses that benefit the environment and their stockholders. This would be capitalism at its best, and some business leaders have increased their companies' profits by applying sustainability principles (Individuals Matter, p. 622).

INDIVIDUALS MATTER

Ray Anderson

Ray Anderson (Figure 23-A) is CEO of Interface, a company based in Atlanta, Georgia (USA). The company is the world's largest commercial manufacturer of carpet tiles, with 26 factories in six countries, customers in 110 countries, and more than $1 billion in annual sales.

Anderson changed the way he viewed the world—and his business—after reading Paul Hawken's book *The Ecology of Commerce*. In 1994, he announced plans to develop the nation's first totally sustainable green corporation by applying some of the four **principles of sustainability** to the carpet business.

Since then, he has implemented hundreds of projects with goals of producing zero waste, greatly reducing energy use, reducing fossil fuel use, relying on solar energy, and copying nature. Between 1996 and 2006, his company cut water usage by 73%, reduced solid waste by 63%, cut greenhouse gas emissions by 56%, and lowered energy use by 28%. These efforts have eliminated $316 million in waste. Interface has also invented a variety of recyclable and compostable carpet fabrics that have diverted enough carpet from landfills to carpet New York City's Empire State Building 45 times. One of Interface's factories in California runs on solar cells and produced the world's first solar-made carpet. And one of its

Figure 23-A Ray Anderson.

factories near Atlanta, Georgia, gets 20% of its energy by burning recovered natural gas (methane) piped in from a municipal landfill.

Ray Anderson sent his carpet design team into the forest and told them to learn how nature would design floor covering. What the team discovered is nature's biodiversity sustainability principle. They observed that there were no regular repeating patterns on the forest floor. Instead of order, there was diversity and chaos.

They came back and created a line of carpet tiles, no two of which had the same design. Within 18 months after it

was introduced, this new product line was the company's top-selling design. Buyers like it because it seems to bring nature indoors. Similar product lines based on the diversity principle now make up 52% of the company's sales.

With this product line, the company has also applied nature's no-waste sustainability principle. The tiles are made from recycled fibers, and because they do not have to look alike and can be taken out of a box and laid down randomly, there is no need for large orders of one style, some of which are usually wasted. And the tile factory runs partly on solar energy, thus applying another of nature's sustainability principles.

To achieve the goal of zero waste, Interface plans to stop selling carpet and to lease it. The company will install, clean, and inspect the carpet on a monthly basis, repair or replace worn carpet tiles overnight, and recycle worn-out tiles into new carpeting.

Anderson is one of a growing number of business leaders committed to finding more economically and ecologically sustainable, yet profitable, ways to do business. Sales have increased to $1 billion, making Interface the world's largest seller of carpet tiles, and profits have tripled.

Anderson has created a new consulting group as part of Interface to help other businesses start on the path to sustainability. One of his clients is Wal-Mart.

THINKING ABOUT
Subsidies

Do you favor phasing out environmentally harmful government subsidies and tax breaks and phasing in environmentally beneficial ones? Explain. What are three things you could you do to help bring this about? How might doing this affect your lifestyle?

We Can Tax Pollution and Wastes instead of Wages and Profits

Another way to discourage pollution and resource waste is to use *green taxes*, or *ecotaxes*, to help include many of the harmful environmental and health costs of production and consumption in market prices (**Concept 23-3B**). Taxes can be levied on a per-unit basis on the amount of pollution and hazardous waste produced, and on the use of fossil fuels, nitrogen fertilizer,

timber, minerals, and other resources. Figure 23-9 lists advantages and disadvantages of using green taxes.

HOW WOULD YOU VOTE?

Do the advantages of green taxes outweigh their disadvantages? Cast your vote online at **academic.cengage.com/biology/miller**.

To many analysts, the tax system in most countries is backward. It *discourages* what we want more of—jobs, income, and profit-driven innovation—and *encourages* what we want less of—pollution, resource waste, and environmental degradation. A more environmentally sustainable economic and political system would *lower* taxes on labor, income, and wealth and *raise* taxes on environmentally harmful activities that produce pollution, wastes, and environmental degradation.

With such a *tax shift,* a tax on coal, for example, would cover the health costs of breathing polluted air

TRADE-OFFS

Environmental Taxes and Fees

Advantages

Help bring about full-cost pricing

Encourage businesses to develop environmentally beneficial technologies and goods to save money

Easily administered by existing tax agencies

Fairly easy to detect cheaters

Disadvantages

Low-income groups are penalized unless safety nets are provided

Hard to determine optimal level for taxes and fees

Governments may use money as general revenue instead of improving environmental quality and reducing taxes on income, payroll, and profits

Figure 23-9 Advantages and disadvantages of using green taxes to help reduce pollution and resource waste (**Concept 23-3B**). **Question:** Which single advantage and which single disadvantage do you think are the most important? Why?

and costs of damage from acid deposition (p. 479). Then taxes on wages and wealth could be reduced by an equal amount. Some 2,500 economists, including eight Nobel Prize winners in economics, have endorsed this tax shifting concept.

Proponents point out three requirements for successful implementation of green taxes. *First,* green taxes would have to be phased in over 10–20 years to allow businesses to plan for the future. *Second,* income, payroll, or other taxes would have to be reduced or replaced so that there is no net increase in taxes. *Third,* the poor and middle class would need a safety net to reduce the regressive nature of any new taxes on essentials such as fuel and food.

In Europe and the United States, polls indicate that, once such tax shifting is explained to voters, 70% of them support the idea. The U.S. Congress has not enacted green taxes, mostly because politically powerful auto, oil, coal, and other industries claim that such taxes will reduce their competitiveness and harm the economy. In addition, most voters have been conditioned to oppose any new taxes and have not been educated about the economic and environmental benefits of tax-shift proposals.

Meanwhile, 15 European countries have green taxes on activities such as waste disposal, air and water pollution, CO_2 emissions, energy consumption, and vehicles entering congested cities. In Europe, green taxes account for more than 7% of total tax revenues.

Germany's green tax on fossil fuels, introduced in 1999, has reduced pollution and greenhouse gas emis-

sions, created up to 250,000 new jobs, lowered taxes on wages, and greatly increased dependence on renewable energy resources such as wind farms and solar cells. Between 2001 and 2006, Sweden shifted about $2 billion from income taxes to taxes on environmentally harmful activities. Some countries are taxing automobile ownership to help reduce the car's environmental impact, to promote the shift to more energy-efficient vehicles and mass transit, and to help reduce congestion in cities. In Denmark, the tax on a new car is greater than the price of the car.

Several European countries have found that green taxes can increase economic competitiveness by stimulating companies to come up with environmentally beneficial technologies such as improved wind turbines and solar cells. These companies can then sell their products in the global marketplace, as well as in their own countries.

HOW WOULD YOU VOTE?

Do you favor shifting taxes from wages and profits to pollution and waste? Cast your vote online at **academic.cengage.com/biology/miller**.

Environmental Laws and Regulations Can Discourage or Encourage Innovation

Regulation is a form of government intervention in the marketplace that is widely used to help control or prevent pollution and reduce resource waste and environmental degradation. It involves enacting and enforcing laws that set pollution standards, regulate harmful activities such as the release of toxic chemicals into the environment, and protect certain irreplaceable or slowly replenished resources such as forests from unsustainable use.

So far, most environmental regulation in the United States and many other developed countries has involved passing laws that are enforced through a

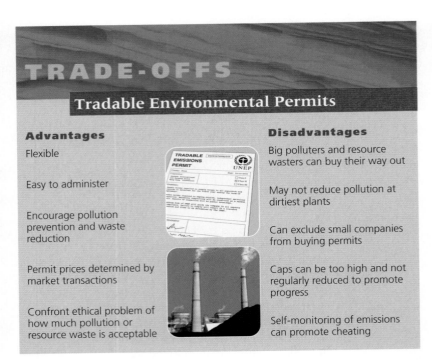

TRADE-OFFS

Tradable Environmental Permits

Advantages	Disadvantages
Flexible	Big polluters and resource wasters can buy their way out
Easy to administer	May not reduce pollution at dirtiest plants
Encourage pollution prevention and waste reduction	Can exclude small companies from buying permits
Permit prices determined by market transactions	Caps can be too high and not regularly reduced to promote progress
Confront ethical problem of how much pollution or resource waste is acceptable	Self-monitoring of emissions can promote cheating

Figure 23-10 Advantages and disadvantages of using tradable pollution and resource-use permits to reduce pollution and resource waste. **Question:** Which two advantages and which two disadvantages do you think are the most important? Why?

command and control approach. Critics say that this approach can unnecessarily increase costs and discourage innovation. They argue that many of these regulations concentrate on cleanup instead of prevention, set compliance deadlines that are too short to allow companies to find innovative solutions, or require use of specific technologies where less costly but equally effective alternatives are available.

A different approach, favored by many economists and environmental and business leaders, is to use *incentive-based regulations*. Rather than requiring all companies to follow the same fixed procedures, this approach uses the economic forces of the marketplace to encourage businesses to be innovative in reducing pollution and resource waste. Experience in several European nations shows that *innovation-friendly regulation* sets goals, frees industries to meet them in any way that works, and allows enough time for innovation. This can motivate companies to develop green products and industrial processes that create jobs, increase company profits, and make the companies more competitive in national and international markets.

For years, many companies mostly resisted environmental regulation and developed an adversarial relationship with government regulators. But in recent years, a growing number of companies have realized the economic and competitive advantages of making environmental improvements. They now recognize that their shareholder value depends in part on having a good environmental record. In other words, environmental stewardship and sustainability can be good for business (Individuals Matter, p. 622). They also realize

that if they do not produce a waste or pollutant, they do not have to spend money on managing it and filling out government reports, and they cannot be sued for any harm that their outputs may cause. At the same time, many consumers have begun buying green products, which further reinforces this trend.

We Can Use the Marketplace to Reduce Pollution and Resource Waste

In one incentive-based regulation system, the government decides on acceptable levels of total pollution or resource use, sets limits, or caps, to maintain these levels, and gives companies a certain number of tradable pollution permits or resource-use permits governed by the caps.

With this *cap-and-trade* approach, a permit holder not using its entire allocation can save credits for future expansion, use them in other parts of its operation, or sell them to other companies. In the United States, this approach has been used to reduce the emissions of sulfur dioxide (p. 490) and several other air pollutants. Tradable rights can also be established among countries to help preserve biodiversity and reduce emissions of greenhouse gases and other regional and global pollutants.

Figure 23-10 lists the advantages and disadvantages of using tradable pollution and resource-use permits. The effectiveness of such programs depends on how high or low the initial cap is set and on the rate at which the cap is reduced to encourage further innovation.

HOW WOULD YOU VOTE? ☑

Do the advantages of using tradable permits to reduce pollution and resource waste outweigh the disadvantages? Cast your vote online at **academic.cengage.com/biology/ miller**.

We Can Reduce Pollution and Resource Waste by Selling Services instead of Things

In the mid-1980s, German chemist Michael Braungart and Swiss industry analyst Walter Stahel independently proposed a new economic model that would provide profits while greatly reducing resource use and waste for a number of goods. Their idea for more sustainable economies focuses on shifting from the current *material-flow economy* (Figure 23-4) to a *service-flow economy*. Instead of buying most goods outright, customers *eco-lease*, or rent the *services* that such goods provide. In a service-flow economy, a manufacturer makes more money if its product uses the minimum amount of materials, lasts as long as possible, produces as little pollution (including

greenhouse gases) as possible in its production and use, and is easy to maintain, repair, reuse, or recycle.

Such an economic shift based on eco-leasing is under way in some businesses. Since 1992, Xerox has been leasing most of its copy machines as part of its mission to provide *document services* instead of selling photocopiers. When a customer's service contract expires, Xerox takes the machine back for reuse or remanufacture. It has a goal of sending no material to landfills or incinerators. To save money, machines are designed to have few parts, be energy efficient, and emit as little noise, heat, ozone, and chemical waste as possible. Canon in Japan and Fiat in Italy are taking similar measures.

In Europe, Carrier has begun shifting from selling heating and air conditioning equipment to providing customers with the indoor temperatures they want. It makes higher profits by installing energy-efficient heat-

ing and air conditioning equipment that lasts as long as possible and is easily rebuilt or recycled. Carrier also makes money helping clients to save energy by adding insulation, eliminating heat losses, and boosting energy efficiency in offices and homes.

Ray Anderson, CEO of a large carpet and tile company, plans to lease rather than sell carpet (Individuals Matter, p. 622). This will reduce his company's resource use and harmful environmental impacts and will likely increase profits for stockholders.

THINKING ABOUT
Selling Services

Do you favor a shift from selling goods to selling the services that the goods provide? List three services that you would be willing to lease.

23-4 How Can Reducing Poverty Help Us Deal with Environmental Problems?

▶ **CONCEPT 23-4** Reducing poverty can help us to reduce population growth, resource use, and environmental degradation.

The Gap between the Rich and the Poor Is Getting Wider

Poverty is defined as the inability to meet basic economic needs. According to the World Bank and the United Nations, 1.1 billion people—almost four times the entire U.S. population—struggle to survive on an income equivalent to less than $1 a day. And about half of the world's people try to live on a daily income of less than $2 (Figure 23-11).

Poverty has numerous harmful health and environmental effects (Figure 1-13, p. 18, and Figure 17-18, p. 461) and has been identified as one of the five major causes of the environmental problems we face (Figure 1-12, p. 18). Reducing poverty benefits individuals, economies, and the environment and helps to slow population growth (**Concept 23-4**).

Most neoclassical economists believe that a growing economy can help the poor by creating more jobs and providing greater tax revenues, which can be used to help the poor to help themselves. Economists call this the *trickle-down effect*. However, since 1960, most of the benefits of global economic growth, as measured by income, have been *flooding up* to the rich, rather than trickling down to workers at the bottom fifth of the economic ladder. Since 1980, this *wealth gap* between the rich and the poor has grown. In 2006, the world's three richest people had a combined wealth greater than that of the world's 48 poorest countries.

South African President Thabo Mbeki told delegates at the 2003 Johannesburg World Summit on Sustainable Development, "A global human society based on poverty for many and prosperity for a few,

Figure 23-11 Poor family members struggling to live in Mumbai (formerly known as Bombay), India, on the edge of a thriving urban setting.

characterized by islands of wealth, surrounded by a sea of poverty, is unsustainable."

We Can Reduce Poverty

Some nations have climbed out of abject poverty in a short time. For example, South Korea and Singapore made the journey from being poor developing nations to achieving world-class industrial status in only 2 decades by focusing on education, hard work, and discipline, all of which attracted investment capital. China and India are on a similar path. Between 1981 and 2004, the estimated number of people living in absolute poverty in China dropped from 648 million to 200 million.

Analysts point out that reducing poverty will require that the governments of most developing countries make policy changes. For one, they will have to emphasize education to develop a trained workforce that can participate in an increasingly global economy.

Currently, developing countries pay an average of about $571,000 a minute in interest on their debt to developed countries. Analysts have suggested that developed countries and international lending agencies forgive all of this debt for the poorest, most heavily indebted countries, on the condition that the money saved on debt interest be devoted to meeting basic human needs. In 2005, the heads of the eight major industrial countries agreed to cancel the debts of 18 of the poorest debt-ridden countries (14 of them in Africa and 4 in Latin America).

Governments, businesses, international lending agencies, and wealthy individuals in developed countries could also:

- Mount a massive global effort to combat malnutrition and the infectious diseases that kill millions of people prematurely (Figure 17-6, p. 442, and Figure 17-10, p. 447).

- Provide universal primary school education and eradicate illiteracy for nearly 800 million of the world's adults. According to Nobel Prize–winning economist Amartya Sen, "Illiteracy and innumeracy are a greater threat to humanity than terrorism." These problems can foster terrorism and strife within countries by creating large numbers of unemployed individuals who have little hope.

- Stabilize population growth in developing countries as soon as possible, mostly by investing in family planning, reducing poverty, and elevating the social and economic status of women.

- Sharply reduce the total and per capita ecological footprints of developed countries and rapidly developing countries such as China and India (Figure 1-10, p. 15); these growing footprints pose a threat to the world's environmental and economic security.

- Make large investments in small-scale infrastructure such as solar-cell power facilities in villages (Figure 16-18, p. 415) and sustainable agriculture projects; such investments could allow developing nations to leapfrog over the conventional development path to more eco-efficient economies.

- Encourage lending agencies to make small loans to poor people who want to increase their income (see the following Case Study).

Some hopeful news is that in 2007, the World Bank reported that all regions of the developing world, with the notable exception of sub-Saharan Africa, were on track to cut their numbers of people living in poverty in half by 2015.

■ CASE STUDY
Making Microloans to the Poor

Most of the world's poor want to earn more, become more self-reliant, and have a better life. But they have no credit record and lack the assets needed for collateral to secure a loan to buy seeds and fertilizer for farming or tools and materials for a small business.

For almost 3 decades, an innovation, called *microlending* or *microfinance*, has helped large numbers of people to deal with this problem. For example, since economist Muhammad Yunus started it in 1983, the Grameen (Village) Bank in Bangladesh has provided microloans ranging from $50 to $500 to more than 6.6 million Bangladeshi villagers. About 97% of the loans are used by women to plant crops, buy cows, or start their own small businesses as seamstresses, weavers, bookbinders, vendors, and phone service providers. This has empowered women in a male-dominated society and helped to reduce poverty.

To stimulate repayment and provide support, the Grameen Bank organizes microborrowers into five-member "solidarity" groups. If one member of the group misses a weekly payment or defaults on the loan, the other members of the group must make the payments.

The Grameen Bank's microlending has been successful and profitable. The repayment rate on these microloans is an astounding 95% compared to 45–50% for loans made by traditional banks in Bangladesh. About half of Grameen's borrowers move above the poverty line within 5 years, and domestic violence, divorce, and birth rates are lower among most borrowers. These microloans have helped more than 5 million borrowers to work their way out of poverty by empowering them and giving them a sense of dignity and hope.

Grameen Bank microloans are also being used to develop daycare centers, health care clinics, reforestation projects, drinking water supply projects, liter-

acy programs, group insurance programs, and small-scale solar and wind power systems in rural villages. Grameen's model has inspired the development of microcredit projects that have helped some 50 million people in more than 58 countries, and the numbers are growing rapidly. In 2006, Citibank and TIAA-CREF (a giant academic pension investment fund) each committed $100 million to microloans.

In 2006, Muhammad Yunus and his Grameen Bank jointly won the Nobel Peace Prize for their pioneering use of microcredit to assist millions of people in lifting themselves out of poverty.

> **THINKING ABOUT**
> **Microloans**
>
> Why do you think microlending works so well? What might be some of its limitations?

We Can Achieve the World's Millennium Development Goals

In 2000, the nations of the world set goals—called Millennium Development Goals—for sharply reducing hunger and poverty, improving health care, achieving universal primary education, empowering women, and moving toward environmental sustainability by 2015. Developed countries agreed to devote 0.7% of their annual national income toward achieving the goals.

However, the average amount donated in most years has been 0.25% of national income. The United States—the world's richest country—gives only 0.16% of its national income to help poor countries. Meeting the minimum development goals by increasing aid to 0.7% is an ethical decision that requires developed countries and individuals to reevaluate their priorities (Figure 23-12).

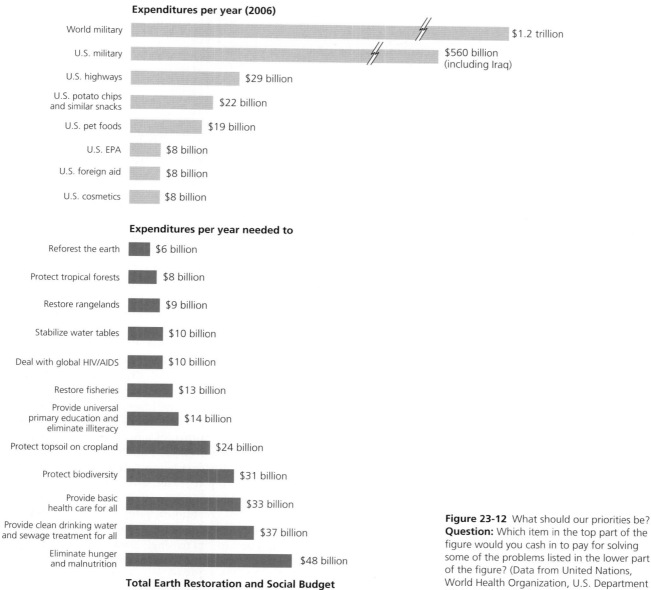

Expenditures per year (2006)

World military	$1.2 trillion
U.S. military	$560 billion (including Iraq)
U.S. highways	$29 billion
U.S. potato chips and similar snacks	$22 billion
U.S. pet foods	$19 billion
U.S. EPA	$8 billion
U.S. foreign aid	$8 billion
U.S. cosmetics	$8 billion

Expenditures per year needed to

Reforest the earth	$6 billion
Protect tropical forests	$8 billion
Restore rangelands	$9 billion
Stabilize water tables	$10 billion
Deal with global HIV/AIDS	$10 billion
Restore fisheries	$13 billion
Provide universal primary education and eliminate illiteracy	$14 billion
Protect topsoil on cropland	$24 billion
Protect biodiversity	$31 billion
Provide basic health care for all	$33 billion
Provide clean drinking water and sewage treatment for all	$37 billion
Eliminate hunger and malnutrition	$48 billion

Total Earth Restoration and Social Budget

$245 billion

Figure 23-12 What should our priorities be? **Question:** Which item in the top part of the figure would you cash in to pay for solving some of the problems listed in the lower part of the figure? (Data from United Nations, World Health Organization, U.S. Department of Commerce, and U.S. Office of Management and Budget, World Bank, Earth Policy Institute)

The estimated $245 billion annual budget that would be needed for providing global health and environmental security is less than half of the current annual U.S. military budget and about one-fifth of the world's annual military budget. National and global military and economic security depend on environmental security. Thus, to proponents of this plan, spending $245 billion a year on environmental security is not only a bargain, it is essential for sustaining civilization as we know it.

As environmental expert Lester R. Brown puts it, "It is decision time. Like earlier civilizations that got into environmental trouble, we can decide to stay with business as usual and watch our modern economy decline and eventually collapse, or we can consciously move onto a new path, one that will sustain economic progress." (See Supplement 5, p. S31, for stories of past civilizations referred to by Brown.)

THINKING ABOUT
The Millennium Development Goals

Do you think the country where you live should devote 0.7% of its annual national income toward achieving the Millennium Development goals? Explain.

23-5 How Can We Make the Transition to More Environmentally Sustainable Economies?

▶ **CONCEPT 23-5** We can use the four principles of sustainability and various economic and environmental strategies to develop more environmentally sustainable economies.

We Can Shift from High–Throughput to Matter Recycling and Reuse Economies

What happens if more people continue to use and waste more energy and matter resources at an increasing rate in today's high-throughput economies (Figure 23-4)? The law of conservation of matter (p. 40) and the two laws of thermodynamics (pp. 42–43) tell us that eventually, this resource consumption will exceed the capacity of the environment to provide sufficient renewable resources, to dilute and degrade waste matter, and to absorb waste heat.

A temporary solution to this problem is to convert our linear high-throughput economies into circular **matter recycling and reuse economies,** which mimic nature by recycling and reusing most matter outputs instead of dumping them into the environment. This involves applying the recycling **principle of sustainability.**

Although changing to a matter recycling and reuse economy would buy some time, it would not allow ever more people to use ever more resources indefinitely, even if all matter resources were somehow perfectly recycled and reused. The two laws of thermodynamics tell us that recycling and reusing matter resources will always require using high-quality energy (which cannot be recycled) and will always add waste heat and various pollutants to the environment (Figure 14-15, p. 356). In the long run, this is not sustainable.

We Can Put More Emphasis on Localizing Economic Production

For several decades, the push has been to globalize economic production, often at the expense of local jobs and economies. What will happen if declining oil production leads to sharper increases in oil prices in a world where most countries are addicted to oil? We may no longer be able to afford shipping food, bottled water, and other goods thousands of kilometers from producers to consumers who might no longer be able to afford such goods because of their much higher prices.

This could lead to a shift back to more localized production of food and other resources and to greater reliance on a diversity of locally available renewable energy resources such as wind, solar energy, hydropower (including microhydropower), and geothermal energy. This would have a number of benefits including: a more stable source of jobs and income from the sale of locally produced food and other goods and services; more profits circulating within each community instead of flowing to external centers of wealth; reduced use of transportation fuel and the accompanying emissions of greenhouse gases and other pollutants; and reduced dependence on a global economic system that could become increasingly unstable and unsustainable, largely because of rising energy costs. This shift would increase local environmental and economic sustainability. At the same time, businesses that depend on selling goods and services in the global economy would

likely decline, as capitalistic economic systems respond to changing economic conditions.

We Can Use Lessons from Nature to Shift to More Sustainable Economies

In this chapter, we have seen a sharp contrast between the beliefs of neoclassical economists and ecological economists (**Core Case Study**). In 2001, environmental scientist Donella Meadows contrasted these two views as follows:

- *The first commandment of economics is: Grow. Grow forever. . . . The first commandment of the Earth is: Enough. Just so much and no more.*

- *Economics says: Compete. . . . The Earth says: Compete, yes, but keep your competition in bounds. Don't annihilate. Take only what you need. Leave your competitor enough to live. Wherever possible, don't compete, cooperate.*

- *Economics says: Use it up fast. Don't bother with repair; the sooner something wears out, the sooner you'll buy another. This makes the gross national product go round . . . Throw things out when you get tired of them. . . . Get the oil out of the ground and burn it now. . . . The Earth says: What's the hurry? . . . When something wears out, don't discard it, turn it into food for something else.*

- *Economics discounts the future [p. 616]. . . . Take your profits from a resource such as a forest now. The earth says: Nonsense. . . . Give to the future. . . . Never take more in your generation than you give back to the next.*

- *The economic rule is: Do whatever makes sense in monetary terms. The Earth says: Money measures nothing more than the relative power of some humans over other humans, and that power is puny compared with the power of the climate, the oceans, the uncounted multitudes of one-celled organisms that created the atmosphere, that recycle the waste, and that have lasted for 3 billion years. The fact that the economy, which has lasted about 200 years, puts zero values on these things means that the economy knows nothing about value—or lasting.*

The three scientific laws governing matter and energy changes and the four **scientific principle of sustainability** (see back cover) suggest that the best long-term solution to our environmental and resource problems is to shift from an economy based on high and increasing matter and energy flow (throughput, Figure 23-4) to a more sustainable **low-throughput (low-waste) economy,** as summarized in Figure 23-13. In other words, we can live more sustainably by using and wasting less matter and energy, reusing, recycling, or composting most matter resources, and controlling human population growth.

CENGAGENOW™ Compare how energy is used in high- and low-throughput economies at CengageNow.

We Can Make Money and Create Jobs by Shifting to an Eco-Economy

Figure 23-14 (p. 630) summarizes suggestions by Paul Hawken, Lester R. Brown, and other environmental and business leaders for using the economic tools discussed in this chapter to make the transition to more environmentally sustainable economies over the next several decades. Such strategies would apply the four **scientific principles of sustainability** (see back cover). Hawken has a simple

CENGAGENOW™ **Active Figure 23-13 Solutions**: lessons from nature. A *low-throughput economy*, based on energy flow and matter recycling, works with nature to reduce the throughput and unnecessary waste of matter and energy resources (items shown in green). This is done by **(1)** reusing and recycling most nonrenewable matter resources, **(2)** using renewable resources no faster than they are replenished, **(3)** reducing resource waste by using matter and energy resources efficiently, **(4)** reducing unnecessary and environmentally harmful forms of consumption, **(5)** emphasizing pollution prevention and waste reduction, and **(6)** controlling population growth to reduce the number of matter and energy consumers. *See an animation based on this figure at* CengageNOW. **Question:** What are three ways in which your community could operate more like a low-throughput economy?

Economics

Reward (subsidize) environmentally sustainable economic development

Penalize (tax and do not subsidize) environmentally harmful economic growth

Shift taxes from wages and profits to pollution and waste

Use full-cost pricing

Sell more services instead of more things

Do not deplete or degrade natural capital

Live off income from natural capital

Reduce poverty

Use environmental indicators to measure progress

Certify sustainable practices and products

Use eco-labels on products

Resource Use and Pollution

Cut resource use and waste by reducing, reusing, and recycling

Improve energy efficiency

Rely more on renewable solar and geothermal energy

Shift from a nonrenewable carbon-based (fossil fuel) economy to a non-carbon renewable energy economy

Ecology and Population

Mimic nature

Preserve biodiversity

Repair ecological damage

Stabilize human population

Environmentally Sustainable Economy (Eco-Economy)

Figure 23-14 Solutions: principles for shifting to more environmentally sustainable economies, or eco-economies, during this century. **Question:** Which five of these solutions do you think are the most important?

golden rule for such an economy: *"Leave the world better than you found it, take no more than you need, try not to harm life or the environment, and make amends if you do."*

In making the shift to eco-economies, some industries and businesses will disappear or remake themselves, and new ones will appear—a normal process in a dynamic capitalist economy. Improving environmental quality and striving for environmental sustainability is now a major growth industry that is creating profits and large numbers of new jobs.

An increasing number of economists and business leaders say that making the shift to an eco-economy could be the greatest investment opportunity of this century. According to Worldwatch Institute estimates, sales in global environmental industries average about $1.3 million per minute—on a par with the global car industry. Green businesses employ more than 11 million people—2 million in the United States. And these numbers are growing rapidly as companies are scrambling to provide more environmentally friendly goods and services.

In 2004, Goldman Sachs, one of the world's largest global investment banks, found that in the previous 2 years, the stock values of companies considered to be sustainability leaders outperformed the general stock market by 25%. And they beat their more conventional peer companies in the same market sector by 75%. In other words, investing in sustainability can be profitable. Figure 23-15 might give you some ideas for a green career choice in the rapidly emerging eco-economy. Also, see the companion website for this book for more information about these career options.

Environmentally Sustainable Businesses and Careers

Aquaculture

Biodiversity protection

Biofuels

Climate change research

Conservation biology

Eco-industrial design

Ecotourism management

Energy efficient product design

Environmental chemistry

Environmental (green) design

Environmental economics

Environmental education

Environmental engineering

Environmental health

Environmental law

Environmental nanotechnology

Fuel cell technology

Geographic information systems (GIS)

Geothermal geologist

Hydrogen energy

Marine science

Pollution prevention

Reconciliation ecology

Selling services in place of products

Solar cell technology

Sustainable agriculture

Sustainable forestry

Waste reduction

Watershed hydrologist

Water conservation

Wind energy

Figure 23-15 Green careers: some key environmentally sustainable, or eco-friendly, businesses and careers. These businesses are expected to flourish during this century, while environmentally harmful, or *sunset*, businesses are expected to decline. See the website for this book for more information on some of these environmental careers.

Making the shift to more sustainable economies will require governments and industries to greatly increase their spending on research and development—especially in environmental, energy-efficiency, and renewable-energy areas—and on science and engineering education. Also, analysts urge major corporations to play a much greater role in shifting to an eco-economy by becoming more socially and ecologically responsible, for both ethical and economic reasons. In 2006, Erik Assadourian, a researcher at the Worldwatch Institute, warned that "either corporations become more sustainable and responsible or the quality of life on Earth—and corporations' bottom lines—will inevitably decline."

In 2004, General Electric (GE), the world's ninth largest corporation, launched its "ecoimagination plan," which commits the company to greatly increasing its investments in green technology research. By 2010,

General Electric hopes to generate $20 billion a year in revenues from its new ecoimagination products.

Educational institutions around the world can also play a key role in making the transition to more sustainable economies serviced by more responsible corporations. They can do this by giving all students a basic environmental education and by developing business schools that integrate sustainable business planning and management into their curricula.

In the United States, there are two recently established business schools—Bainbridge Graduate Institute (Bainbridge Island, Washington) and Presidio World College—dedicated to teaching sustainable business. These and the business schools at Stanford and Harvard Universities emphasize business and environmental sustainability and social responsibility. Graduates from these business schools talk about the *triple bottom line:* people, planet, and profit.

REVISITING Economic Growth and Sustainability

The **core case study** that opens this chapter discusses the long-standing debate over economic growth as a primary economic goal, versus sustainable economic development, which discourages environmentally harmful forms of economic growth and encourages environmentally sustainable forms of economic development.

Shifting to eco-economies by applying the four **scientific principles of sustainability** and using the strategies listed in Figure 23-14 over the next several decades will result in a new reduce-reuse-recycle economy (Figure 23-13) powered by a diversity of solar and other renewable energy resources. It will also have a diverse transportation system that relies more on rail, buses, and bicycles and less on cars. And slowing population growth will

reduce the consumption of environmentally harmful goods and services.

According to environmental and ecological economists, a key to accomplishing this goal is to create an honest marketplace based on full-cost pricing that helps consumers to make informed choices. Making this shift will require bold leadership by business leaders and elected officials and bottom-up political pressure from informed and concerned citizens. People will also vote with their wallets by refusing to buy environmentally harmful goods and services and by supporting environmentally beneficial businesses. Forward-looking investors, corporate executives, and political leaders are recognizing that sustainability is good for the environment and the economy.

> *The challenge for our generation is to build a new economy,*
> *one that is powered largely by renewable sources of energy,*
> *that has a highly diversified transport system,*
> *and that reuses and recycles everything.*
> *And to do it with unprecedented speed.*
>
> LESTER R. BROWN

REVIEW

1. Review the Key Questions and Concepts for this chapter on p. 612. Describe the disagreements among some economists over economic growth versus economic development (**Core Case Study**). Explain why the sustainability revolution is also an economic revolution.

2. What is an **economic system?** Distinguish among **natural capital, human capital (human resources),** and **manufactured capital (manufactured resources).** De-

scribe the interactions among demand, supply, and market prices in a market economic system. Explain why and how governments intervene in market economic systems. What is a **high-throughput economy?**

3. Compare how neoclassical economists and ecological and environmental economists view economic systems. List eight strategies that ecological and economic economists would use to make the transition to more sustainable eco-economies.

4. Describe ways in which economists can estimate the economic values of natural goods and services. Why are such values not included in the market prices of goods and services? Define **discount rate** and discuss the controversy over how to assign such rates.

5. Describe how economists can estimate the optimal levels for pollution control and resource use. Define **cost-benefit analysis** and discuss its advantages and limitations.

6. Why do products and services cost more than most people think? What is the *genuine progress indicator* and how does it differ from the gross domestic product economic indicator? What is *full-cost pricing* and what are some benefits of using it to determine the market values of goods and services? Give three reasons why it is not widely used. What are the advantages of providing consumers with eco-labels on the goods and services they buy?

7. Describe the benefits of shifting from environmentally unsustainable to more environmentally sustainable government subsidies and tax breaks. Describe the proposal to tax pollution and wastes instead of wages and profits. What are the major advantages and disadvantages of using green taxes? What are three requirements for implementing green taxes? Distinguish between command-and-control and incentive-based government regulations, and describe the advantages of the second approach. What is the cap-and-trade approach to implementing environmental regulation, and what are the major advantages and disadvantages of this approach?

8. What are some environmental benefits of selling services instead of goods? Give two examples of this approach. Describe Ray Anderson's progress in developing a more environmentally sustainable carpet business.

9. What is **poverty** and how is it related to population growth and environmental degradation? List three ways in which governments can help to reduce poverty. What are the advantages of making microloans to the poor?

10. What is a **matter recycling and reuse economy?** What is a **low-throughput (low-waste) economy?** List six ways to shift to more environmentally sustainable economies. Name five new businesses and careers that would be important in such eco-economies. Describe connections between more environmentally sustainable economies (**Core Case Study**) and the four **scientific principles of sustainability.**

Note: Key Terms are in **bold** type.

CRITICAL THINKING

1. Should we attempt to maximize economic growth (**Core Case Study**) by producing and consuming more and more economic goods and services? Explain. What are the alternatives?

2. According to one definition, *sustainable economic development* involves meeting the needs of the present human generation without compromising the ability of future generations to meet their needs. What do you believe are the needs referred to in this definition? Compare this definition with the characteristics of a low-throughput economy described in Figure 23-13.

3. Is environmental regulation bad for the economy? Explain. Describe harmful and beneficial forms of environmental regulation. Imagine that you work for a regulatory agency, and outline an innovation-friendly regulation for a business of your choice (such as a coal mining company, a paper mill, or a factory).

4. Do you believe we have an ethical responsibility to leave the earth's natural systems in as good a shape as they are now or better? Explain. How is meeting this responsibility related to the discount factor (p. 616)?

5. Suppose that over the next 20 years, the environmental and health costs of goods and services are internalized until their market prices more closely reflect their total costs. What harmful effects and what beneficial effects might such full-cost pricing have on your lifestyle?

6. Explain why you agree or disagree with the proposals that various analysts have made for sharply reducing poverty, as discussed on pages 625–626. Which two of these proposals do you believe are the most important?

7. Are you for or against shifting to a *service-flow economy* based on eco-leasing many services in the country and community where you live? Explain. Do you believe it will be possible to shift to such an economy over the next 2–3 decades? Explain. What are the three most important strategies for doing this?

8. Explain why you agree or disagree with each of the major principles for shifting to a more environmentally sustainable economy listed in Figure 23-14. Which three of these principles do you believe are the most important?

9. Congratulations! You are in charge of the world. List your five most important strategies for shifting to more environmentally sustainable economies over the next 50 years.

10. List two questions that you would like to have answered as a result of reading this chapter.

Note: See Supplement 13 (p. S78) for a list of Projects related to this chapter.

The figure below compares per capita gross domestic product (GDP) and per capita general progress indicator (GPI) in the United States. On the basis of the data in the figure, answer the questions that follow.

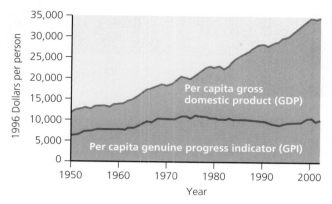

1. What were the monetary differences between per capita GDP and per capita GPI in 1950 and in 2004?

2. By what percentages did per capita GDP increase between 1950 and 1980 and between 1980 and 2004? By what percentages did the per capita GPI increase between 1950 and 1980 and between 1980 and 2004?

3. What conclusion can you draw from your answers to questions 1 and 2?

Log on to the Student Companion Site for this book at **academic.cengage.com/biology/miller**, and choose Chapter 23 for many study aids and ideas for further reading and research. These include flash cards, practice quizzing, Weblinks, information on Green Careers, and InfoTrac® College Edition articles.

CORE CASE STUDY Rescuing a River

In the 1960s, Marion Stoddart (Figure 24-1) moved to Groton, Massachusetts, on the Nashua River, then considered one of the nation's filthiest rivers. For decades, industries and towns along the river had used it as a dump. Dead fish bobbed on its waves, and at times the water was red, green, or blue from pigments discharged by industries.

Instead of assuming that nothing could be done, Stoddart committed herself to restoring the Nashua and establishing public parklands along its banks. She started by creating a careful cleanup plan and approaching state officials with her ideas. They laughed, but she was not discouraged and began practicing the most time-honored skill of politics: one-on-one persuasion. She identified power brokers in the riverside communities and began to educate them, win them over, and gain their cooperation in cleaning up the river.

Stoddart then began working at higher levels of government. She got the state to ban open dumping in the river. When federal matching funds promised for building a treatment plant failed to materialize, she gathered 13,000 signatures on a petition that was sent to President Richard Nixon. The funds arrived soon after.

Stoddart's next success was getting a federal grant to beautify the river. She hired unemployed young people to clear away mounds of debris. When the cleanup was completed, she persuaded communities along the river to create a riverside park and woodlands along both banks.

Now, more than 4 decades later, the Nashua River is still clean. Several new water treatment plants have been built, and a citizens' group founded by Stoddart keeps watch on water quality. The river supports many kinds of fish and other wildlife, and its waters are used for canoeing and recreation.

This success story is testimony to what a committed individual can do to bring about change from the bottom up by getting people to work together. For her efforts, the U.N. Environment Programme named Stoddart an outstanding worldwide worker for the environment.

Politics is concerned with who has power over the distribution of resources—who gets what, and when and how they get it. Many people think of politics mostly as a process of big government, but what happens in their own communities is what directly affects and motivates most people, as illustrated by the story of Marion Stoddart's efforts. In this chapter, we look more closely at how politics can help or hinder us in making the transition to a more sustainable world.

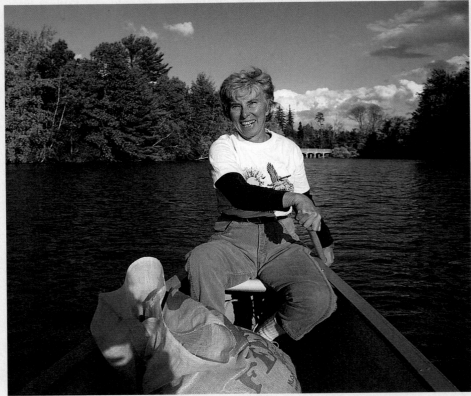

Figure 24-1 Individuals matter: Marion Stoddart canoeing on the Nashua River near Groton, Massachusetts. She spent more than 2 decades spearheading successful efforts to have this river cleaned up.

© Seth Resnick

24-1 What is the role of government in making the transition to more sustainable societies?

CONCEPT 24-1 A government can help to protect environmental and public interests and encourage more environmentally sustainable economic development through its policies.

24-2 How is environmental policy made?

CONCEPT 24-2A Policy making involves law making, program funding, rule writing, and enforcement of rules by agencies—a complex process that is affected at each stage by political processes.

CONCEPT 24-2B Individuals can work with each other to become part of political processes that influence how environmental policies are made and whether or not they succeed. (*Individuals matter.*)

24-3 What is the role of environmental law in dealing with environmental problems?

CONCEPT 24-3 Environmental laws and regulations can be used to control pollution, set safety standards, encourage resource conservation, and protect species and ecosystems.

24-4 What are the major roles of environmental groups?

CONCEPT 24-4 Grassroots groups are growing and combining efforts with large environmental organizations in a global sustainability movement.

24-5 How can we improve global environmental security?

CONCEPT 24-5 Environmental security is necessary for economic security and is at least as important as military security.

24-6 How can we implement more sustainable and just environmental policies?

CONCEPT 24-6 Making the transition to more sustainable societies will require that nations implement green plans and that people and nations cooperate, agree on principles, and make the political commitment to achieve this transition.

Note: Supplements 2 (p. S4), 4 (p. S20), 5 (p. S31), 10 (p. S59), 11 (p. S74), and 13 (p. S78) can be used with this chapter.

I dream things that never were;
and I say "Why not?"

GEORGE BERNARD SHAW

24-1 What Is the Role of Government in Making the Transition to More Sustainable Societies?

▶ **CONCEPT 24-1** A government can help to protect environmental and public interests and encourage more environmentally sustainable economic development through its policies.

Government Can Serve Environmental and Other Public Interests

Businesses and industries thrive on change and innovations that lead to new technologies, products, and opportunities for profits. This process, often referred to as *free enterprise,* can lead to higher living standards for many people, but it can also create harmful impacts on other people and on the environment.

Government on the other hand, especially democratic government working in the interest of the public, can act as a brake on business enterprises that might result in harm to people or the environment. Achieving the right balance between free enterprise and gov-

ernment regulation is not easy. Too much government intervention can strangle business enterprise and innovation. Too little can lead to environmental degradation and social injustices and even to a takeover of the government by business and global trade interests (Supplement 11, p. S74). Analysts point out that in today's global economy, some multinational corporations, which often have budgets larger than the budgets of many countries, have greatly increased their economic and political power over national, state, and local governments and ordinary citizens.

However, businesses can also serve environmental and public interests. Green businesses create products and services that help to sustain or improve environmental quality while improving people's lives. They are

one of the world's fastest growing business sectors and are a growing source of new jobs (Figure 23-15, p. 630).

Many argue that government is the best mechanism for dealing with some of the broader economic and political issues, some of which we have discussed in this book. These include:

- Full-cost pricing (p. 620): governments can create subsidies and levy taxes to require inclusion of harmful environmental and health costs in the market prices of some goods and services.

- Market failures (p. 614): governments can use taxes and subsidies to level the playing field wherever the marketplace is not operating freely due to unfair advantages held by some players.

- The tragedy of the commons: government is the only power that can preserve common or open access renewable resources (pp. 12–13), such as clean air, groundwater, and ozone in the stratosphere. This is true even in the case of a cap-and-trade market approach to solving a problem such as air pollution (p. 490), because government is necessary to administer such a program.

The exact role played by a government is determined by its **policies**—the set of laws and regulations it enforces and the programs it funds (**Concept 24-1**). **Politics** is the process by which individuals and groups try to influence or control the policies and actions of governments at local, state, national, and international levels. One important application of this process is the development of **environmental policy**—environmental laws and regulations that are developed, implemented, and enforced and the environmental programs that are funded by one or more government agencies.

According to social scientists, the development of public policy in democracies often goes through a *policy life cycle* consisting of four stages:

- *Recognition:* identify a problem
- *Formulation:* identify specific causes of the problem and develop a solution such as a law or program to deal with it
- *Implementation:* put the solution into effect
- *Control:* monitor progress and make adaptations as needed

The policy process is usually cyclical, because in the fourth stage, a policy is evaluated according to feedback from the market or environment. Seldom is everyone happy with a policy, and usually, adjustments have to be made or a policy might even have to be abandoned, and the process starts again. Ideally, policies are revised and fine-tuned until they succeed in serving all or most of the affected parties in a reasonably balanced way.

Figure 24-2 illustrates this cycle and shows the general positions of some major environmental problems in the policy life cycle in the United States and most other developed countries.

Democracy Does Not Always Allow for Quick Solutions

Democracy is government by the people through elected officials and representatives. In a *constitutional democracy,* a constitution provides the basis of government authority and in most cases, limits government

Figure 24-2
The policy life cycle, including positions of some major environmental problems within the cycle in most developed countries.

Recognition
Identify the problem

- Nonpoint-source water pollution
- Indoor air pollution
- Electronic waste
- Mining wastes
- Groundwater contamination
- Environmentally harmful subsidies
- Exclusion of environmentally harmful costs from market prices

Formulation
Look for solutions

- Global warming
- Urban sprawl
- Nuclear wastes
- Biodiversity losses
- Toxic wastes
- Aquifer depletion

Implementation
Implement solutions

- Acid deposition
- Ozone depletion
- Municipal solid waste
- Endangered species
- Pest damage
- Soil erosion

Control
Monitor and adjust

- Outdoor air pollution
- Sewage treatment problems
- Drinking water pollution
- Point-source water pollution
- Municipal solid waste
- Some infectious diseases

power by mandating free elections and guaranteeing free speech.

Political institutions in most constitutional democracies are designed to allow for gradual change so as to ensure economic and political stability. In the United States, for example, rapid and destabilizing change is curbed by a system of checks and balances that distributes power among three branches of government—*legislative, executive,* and *judicial*—and among federal, state, and local governments.

In passing laws, developing budgets, and formulating regulations, elected and appointed government officials must deal with pressure from many competing *special-interest groups.* Each group advocates passing laws, providing subsidies or tax breaks, or establishing regulations favorable to its cause, and weakening or repealing laws, subsidies, tax breaks, and regulations unfavorable to its position. Some special-interest groups such as corporations are *profit-making organizations.* Others are *nongovernmental organizations (NGOs),* most of which are nonprofit, such as labor unions and environmental organizations.

The design for stability and gradual change in democracies is highly desirable. But several features of democratic governments hinder their ability to deal with environmental problems. For example, problems such as climate change and biodiversity loss are complex and difficult to understand. Such problems also have long-lasting effects, are interrelated, and require integrated, long-term solutions that emphasize prevention. But because elections are held every few years, most politicians seeking reelection tend to focus on short-term, isolated issues rather than on complex, time-consuming, and long-term problems.

One of our greatest challenges is to place more emphasis on long-term thinking and policies and to educate political leaders and the public about the need for such thinking and actions. Another problem is that many political leaders, with hundreds of issues to deal with, have too little understanding of how the earth's natural systems work and how those systems support all life, economies, and societies. Again, educating politicians and voters about these matters is an urgent priority.

Certain Principles Can Guide Us in Making Environmental Policy

Because most people do not have the time to master the details of every environmental problem and proposed solution, analysts suggest that legislators and individuals evaluating existing or proposed environmental policies should be guided by several principles. It is hoped that such guidelines will help decision-makers to create policies that will satisfy most people while minimizing environmental harm:

- *The humility principle:* Our understanding of nature and how our actions affect nature is quite limited.

- *The reversibility principle:* Avoid taking actions that cannot be reversed if a decision turns out to be wrong. One example of an essentially irreversible policy is the use of nuclear power to produce electricity, which commits civilization to having to safely store the deadly radioactive wastes it produces for 10,000 to 240,000 years. An example of such a policy now being considered is the plan for capturing and storing carbon dioxide underground or under the ocean to help slow climate change (Figure 19-15, p. 516), which would commit civilization to insuring that these deposits will never leak out (p. 517).

- *The precautionary principle:* When substantial evidence indicates that an activity threatens human health or the environment, take precautionary measures to prevent or reduce such harm, even if some of the cause-and-effect relationships are not fully established scientifically (**Concept 9-4C**, p. 206).

- *The net energy principle:* Avoid the widespread use of energy alternatives or technologies with low net energy yields (Science Focus, p. 374). Such energy-inefficient alternatives go against the first and second laws of thermodynamics (pp. 42–43)—which we can never violate—and thus require massive government subsidies and tax breaks to compete in the open marketplace. Such subsidies often waste limited government funds that could be spent on encouraging more energy efficient alternatives.

- *The prevention principle:* Whenever possible, make decisions that help to prevent a problem from occurring or becoming worse.

- *The polluter-pays principle:* Develop regulations and use economic tools such as green taxes to ensure that polluters bear the costs of the pollutants and the wastes they produce. For example, if this principle were applied in policy, users of coal-burning power plants and factories and drivers of motor vehicles would pay the full bill for the harmful health and environmental effects of burning of fossil fuels.

- *The public access and participation principle:* Citizens should have open access to environmental data and the right to participate in developing, critiquing, and modifying environmental policies. Environmental education should receive a high priority.

- *The human rights principle:* All people have a right to live in an environment that does not harm their health and well-being.

- *The environmental justice principle:* Establish environmental policy so that no group of people bears an unfair share of the burden created by pollution, environmental degradation, or the execution of environmental laws. (See the Guest Essay on this subject by Robert D. Bullard at CengageNOW™.)

Implementing such principles will require that policy makers throughout the world, especially in developed countries:

- Become more environmentally literate and insist that they receive the latest scientific information about the environmental problems we face, along with a diagnosis of their basic causes and an evaluation of possible solutions to such problems.

- Support scientific research to understand and monitor changes in environmental conditions throughout the world, their rate of change, and the roles of human activities in such changes.

- Support scientific research to identify early warning signs of ecological crises, and have the ethical and political will to act rapidly to try to prevent us from reaching irreversible critical tipping points.

- Support technological innovation in preventing and dealing with environmental problems, encourage incentive-based environmental regulations (p. 623), and provide funding to transfer innovative environmental technologies to developing countries throughout the world.

- Enact economic measures such as subsidy shifts, taxes on pollution and waste with offsetting reductions in taxes on wages and profits, full-cost pricing, and other measures to stimulate the shift to eco-economies during this century (Figure 23-14, p. 630).

24-2 How Is Environmental Policy Made?

▶ **CONCEPT 24-2A** Policy making involves law making, program funding, rule writing, and enforcement of rules by agencies—a complex process that is affected at each stage by political processes.

▶ **CONCEPT 24-2B** Individuals can work with each other to become part of political processes that influence how environmental policies are made and whether or not they succeed. (*Individuals matter.*)

How Democratic Government Works: The U.S. Model

The U.S. federal government consists of three separate but interconnected branches: legislative, executive, and judicial. The *legislative branch,* called the Congress, consists of the House of Representatives and the Senate, which jointly have two main duties. One is to approve and oversee government policy-making by passing laws to establish government programs and funding for such programs. The other is to oversee the creation, functioning, and funding of agencies in the executive branch concerned with carrying out government policies.

The *executive branch* consists of the president, and a staff that together direct the agencies authorized by Congress to carry out government policies. The president also proposes annual budgets, legislation, and appointees for executive positions, which must be approved by the Senate, and tries to persuade Congress and the public to support executive policy proposals.

The *judicial branch* consists of the Supreme Court and lower federal courts. These courts, along with state and local courts, enforce and interpret different laws passed by legislative bodies at the local, state, and federal levels. Decisions made by the various courts make up a body of law known as *case law.* Previous court rulings are used as legal guidelines, or *precedents,* to help make new legal decisions and rulings.

The major function of the federal government in the United States (and in other democratic countries) is to develop and implement *policy* for dealing with various issues. Policy is typically composed of *laws* passed by the legislative branch, *regulations* instituted by the executive branch to put laws into effect, and *funding* approved by Congress and the president to implement and enforce the laws and regulations (**Concept 24-2A**).

Converting a bill introduced in the U.S. Congress into a law is a complex process. An important part of this process is **lobbying,** in which individuals or groups use public pressure, personal contacts, and political action to persuade legislators to vote or act in their favor. Figure 24-3 presents a greatly simplified overview of how individuals and organizations interact with the three branches of government in the United States.

Most environmental bills are evaluated by as many as ten committees in the U.S. Congress. Effective proposals often are weakened by this fragmentation and by lobbying from groups opposing the law. Nonetheless, since the 1970s, a number of important environmental laws have been passed in the United States, as discussed throughout this text. Figure 24-4 (p. 640)

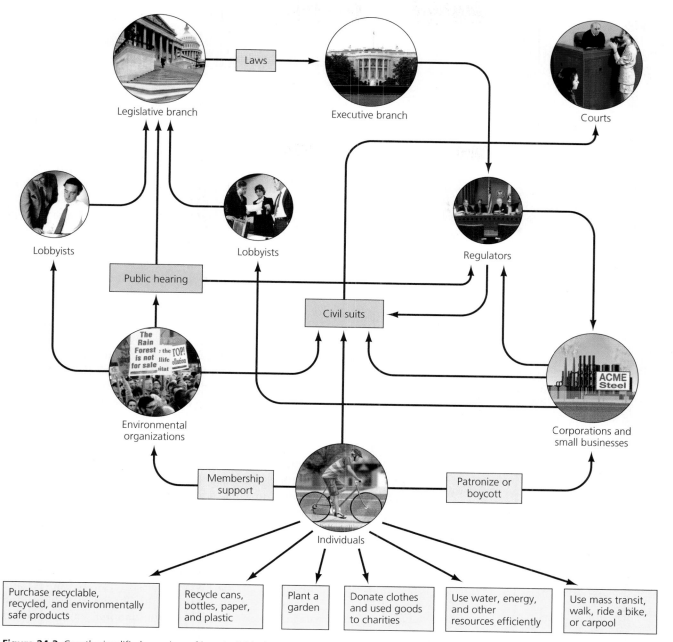

Laws

Legislative branch → Executive branch → Courts

Lobbyists

Lobbyists

Regulators

Public hearing

Civil suits

Environmental organizations

Corporations and small businesses
ACME Steel

Membership support

Patronize or boycott

Individuals

Purchase recyclable, recycled, and environmentally safe products

Recycle cans, bottles, paper, and plastic

Plant a garden

Donate clothes and used goods to charities

Use water, energy, and other resources efficiently

Use mass transit, walk, ride a bike, or carpool

Figure 24-3 Greatly simplified overview of how individuals, corporations, and environmental organizations interact with each other and with the legislative, executive, and judicial branches of the U.S. government. The bottom of this diagram also shows some ways in which individuals can bring about environmental change through their own lifestyles. (See the website for this book for details on contacting elected representatives.)

lists some of the major environmental laws passed in the United States since 1969.

Developing Environmental Policy Is a Controversial Process

Passing a law is not enough to make policy (**Concept 24-2A**). The next step involves trying to get enough funds appropriated by Congress to implement and enforce the new law. Indeed, developing and adopting a budget is the most important and controversial activity of the executive and legislative branches.

Once a law has been passed and funded, the appropriate government department or agency must draw up regulations or rules for implementing it. An affected group may take the agency to court for failing to implement and enforce the regulations effectively or for enforcing them too rigidly.

Politics plays an important role in the policies and staffing of environmental regulatory agencies—depending on what political party is in power and the prevailing environmental attitudes. Businesses facing environmental regulations often put political pressure on regulatory agencies and executives to have people from the regulated industries or groups appointed to

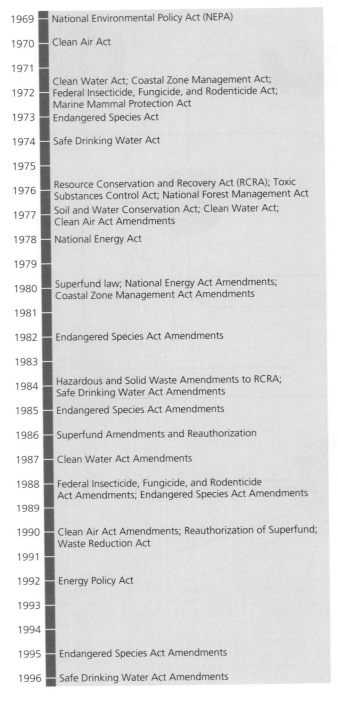

Figure 24-4
Some major environmental laws and their amended versions enacted in the United States since 1969. Supplement 5 (p. S31) contains information on these laws, as does the website for this chapter.
Question: Why do you think so few new U.S. environmental laws have been enacted since 1980?

1969	National Environmental Policy Act (NEPA)
1970	Clean Air Act
1971	
1972	Clean Water Act; Coastal Zone Management Act; Federal Insecticide, Fungicide, and Rodenticide Act; Marine Mammal Protection Act
1973	Endangered Species Act
1974	Safe Drinking Water Act
1975	
1976	Resource Conservation and Recovery Act (RCRA); Toxic Substances Control Act; National Forest Management Act
1977	Soil and Water Conservation Act; Clean Water Act; Clean Air Act Amendments
1978	National Energy Act
1979	
1980	Superfund law; National Energy Act Amendments; Coastal Zone Management Act Amendments
1981	
1982	Endangered Species Act Amendments
1983	
1984	Hazardous and Solid Waste Amendments to RCRA; Safe Drinking Water Act Amendments
1985	Endangered Species Act Amendments
1986	Superfund Amendments and Reauthorization
1987	Clean Water Act Amendments
1988	Federal Insecticide, Fungicide, and Rodenticide Act Amendments; Endangered Species Act Amendments
1989	
1990	Clean Air Act Amendments; Reauthorization of Superfund; Waste Reduction Act
1991	
1992	Energy Policy Act
1993	
1994	
1995	Endangered Species Act Amendments
1996	Safe Drinking Water Act Amendments

high positions in such agencies. In other words, the regulated try to take over regulatory agencies and become the regulators—described by some as "putting foxes in charge of the henhouse."

In addition, people in regulatory agencies work closely with officials in the industries they are regulating, often developing friendships with them. And some industries and other regulated groups offer high-paying jobs to regulatory agency employees in an attempt to influence their regulatory decisions. This has created what some critics have labeled a "revolving door" relationship between some agencies and the industries they regulate.

■ **CASE STUDY**

Managing Public Lands in the United States—Politics in Action

No nation has set aside as much of its land for public use, resource extraction, enjoyment, and wildlife habitat as has the United States. The federal government manages roughly 35% of the country's land, which belongs to every American. About three-fourths of this federal public land is in Alaska and another fifth is in the western states (Figure 24-5).

Some federal public lands are used for a variety of purposes. For example, the *National Forest System* consists of 155 national forests and 22 national grasslands. These lands, managed by the U.S. Forest Service (USFS), are used for logging, mining, livestock grazing, farming, oil and gas extraction, recreation, and conservation of watershed, soil, and wildlife resources.

The Bureau of Land Management (BLM) manages *National Resource Lands.* These lands are used primarily for mining, oil and gas extraction, and livestock grazing.

The U.S. Fish and Wildlife Service (USFWS) manages 549 *National Wildlife Refuges.* Most refuges protect habitats and breeding areas for waterfowl and big game to provide a harvestable supply of such animals for hunters. A few protect endangered species, such as the brown pelican (Photo 1 in the Detailed Contents) from extinction. Permitted activities in most refuges include hunting, trapping, fishing, oil and gas development, mining, logging, grazing, some military activities, and farming.

On other public lands, uses are more restricted. The *National Park System,* managed by the National Park Service (NPS), includes 58 major parks and 331 national recreation areas, monuments, memorials, battlefields, historic sites, parkways, trails, rivers, seashores, and lakeshores. Only camping, hiking, sport fishing, and boating can take place in the national parks, whereas sport hunting, mining, and oil and gas drilling are allowed in national recreation areas.

The most restricted public lands are 702 roadless areas that make up the *National Wilderness Preservation System.* These areas lie within the other public lands and are managed by the agencies in charge of those lands. Most of these areas are open only for recreational activities such as hiking, sport fishing, camping, and nonmotorized boating.

Many federal public lands contain valuable oil, natural gas, coal, geothermal, timber, and mineral resources (see Figures 6, p. S62, and 18, p. S70, in Supplement 10). The use and management of the resources on these lands has been debated since the 1800s.

Most conservation biologists, environmental economists, and many free-market economists believe that four principles should govern use of public lands:

• They should be used primarily for protecting biodiversity, wildlife habitats, and ecosystems.

Figure 24-5 Natural capital: national forests, national parks, and wildlife refuges managed by the U.S. federal government. U.S. citizens jointly own these and other public lands. **Question:** Do you think U.S. citizens should jointly own more or less of the nation's land? Why or why not? (Data from U.S. Geological Survey)

Legend:
- National parks and preserves
- National forests
- (and Xs) National wildlife refuges

- No one should receive government subsidies or tax breaks for using or extracting resources on public lands.

- The American people deserve fair compensation for the use of their property.

- All users or extractors of resources on public lands should be fully responsible for any environmental damage they cause.

There is strong and effective opposition to these ideas. Developers, resource extractors, many economists, and many citizens tend to view public lands in terms of their usefulness in providing minerals, timber, and other resources and in increasing short-term eco-nomic growth. They have succeeded in blocking implementation of these four principles. For example, in recent years, analysis of budgets and funding reveals that the government has given an average of $1 billion a year in subsidies and tax breaks to privately owned interests that use U.S. public lands for mining, fossil fuel extraction, logging, and grazing.

Some developers and resource extractors have sought to go further. Here are some of the proposals that such interests have made to get the U.S. Congress to open up more federal lands for development:

- Sell public lands or their resources to corporations or individuals, usually at less than market value, or turn over their management to state and local

governments. Attempts to do this have been going on since the 1930s (see Supplement 5, p. S31–S39).

- Slash federal funding for administration of regulations over public lands.
- Cut old-growth forests in the national forests and replace them with tree plantations.
- Open national parks, national wildlife refuges, and wilderness areas to oil drilling, mining, off-road vehicles, and commercial development.
- Eliminate or take power away from the National Park Service and launch a 20-year construction program of new concessions and theme parks run by private firms in the national parks.

Since 2002, the U.S. Congress has expanded the extraction of mineral, timber, and fossil fuel resources on U.S. public lands and weakened environmental laws and regulations protecting such lands from abuse and exploitation. In 2006, the USFS issued a new rule eliminating formal environmental impact statements from the process of writing management plans for national forests. This cuts out the public's right to participate in national forest management planning. The timber industry pushed for this change. Conservation biologists and other environmental scientists argue for putting more emphasis on use of scientific principles in the management of forests and other public lands (Science Focus, at right) to help ensure that they are used in more sustainable ways.

This discussion has focused on the debate over the use of public lands in the United States. But the same

issues apply to the use of government or publicly owned lands in other countries.

Individuals Can Influence Environmental Policy

A major theme of this book is that *individuals matter*. History shows that significant change usually comes from the *bottom up* when individuals join with each other to bring about change. Without grassroots political action by millions of individual citizens and organized citizen groups (Figure 24-6), the air you breathe and the water you drink today would be much more polluted, and much more of the earth's biodiversity would have disappeared (**Concept 24-2B**).

With the growth of the Internet and digital technology, individuals have become more empowered. For example, in a highly unusual chain of events in 2007, Chinese citizens using mobile phone text messaging organized to oppose construction of a chemical plant that would threaten the safety of 1.5 million people in a port city. By building opposition from the ground up—circulating nearly a million phone text messages—they persuaded the Chinese government to freeze the construction project and consider less hazardous alternatives.

Figure 24-7 lists ways in which you can influence and change local, state, and national government policies in constitutional democracies. Many people recycle, buy eco-friendly products, and do other important things to help the environment. But when people work together, starting at the local level, just as Marion Stoddard did (**Core Case Study**), they can influence environmental policy. Her story and other cases have demonstrated the validity of Aldo Leopold's (Individuals Matter, p. 22) statement: "All ethics rest upon a single premise: that the individual is a member of a community of interdependent parts."

At a fundamental level, all politics is local. What we do to improve environmental quality in our own neighborhoods, schools, and work places has national and global implications, much like the ripples spreading outward from a pebble dropped in a pond. This is the meaning of the slogan, "Think globally; act locally."

Environmental Leaders Can Make a Big Difference

Not only can we participate, but each of us can also provide environmental leadership in several different ways. First, we can *lead by example,* using our own

Figure 24-6 *Global outlook:* Children in Turin, Italy, wear gas masks as part of an organized protest against high levels of air pollution. **Question:** What environmental issue, if any, would lead you to participate in a demonstration related to the issue? Explain.

Angelo Doto-UNEP/Peter Arnold, Inc.

SCIENCE FOCUS

Logging in U.S. National Forests Is Controversial

According to federal law, U.S. national forests are supposed to be managed according to two scientifically based principles: the *principle of sustainable yield*, which holds that trees in national forests should not be harvested faster than they are replenished; and the *principle of multiple use*, which calls for a variety of uses on the same forest land at the same time.

However, for decades, there has been controversy over the principle of multiple use. Timber companies want timber cutting to be the primary use, and they push for the ability to cut as much timber in national forests as possible at low prices and with taxpayer subsidies. Biodiversity experts and many environmental scientists argue that

national forests should be managed primarily to provide recreation and to sustain biodiversity, water resources, and other ecological services.

By law, the U.S. Forest Service (USFS) must sell timber for no less than the cost of reforesting the land. However, the USFS timber-cutting program loses money because the revenues from timber sales do not cover the expenses of road building, timber sale preparation, administration, and other overhead costs provided by the nation's citizens, who jointly own these forests. More than 644,000 kilometers (400,000 miles) of roads have been cut through the national forests at taxpayer's expense, primarily to facilitate logging.

Because of such government subsidies, a 2005 study by the Earth Island Institute found that timber sales from U.S. federal lands have lost money for their taxpayer owners in 97 of the last 100 years. Not even the value of the timber is being realized. And according to a 2000 study by the accounting firm ECONorthwest, recreation, hunting, and fishing in national forests add ten times more money to the national economy and provide seven times more jobs than does extraction of timber and other resources.

Critical Thinking

Should national forests be managed primarily according to scientific principles or according to market forces? Explain.

lifestyles and values to show others that change is possible and can be beneficial. As Mahatma Gandhi observed: "An ounce of practice is worth more than tons of preaching." While many of us do not end up working in an environmental career, we can all strive to be greener in whatever work we do. For example, wherever we work or attend school, we can cut consumption of disposable products, reuse and recycle, walk, bike, or take mass transit to work, and encourage others to do the same.

Second, we can *work within existing economic and political systems to bring about environmental improvement* by campaigning and voting for informed and eco-friendly candidates and by communicating with elected officials. We can also send a message to companies that we feel are harming the environment through their products or policies by *voting with our wallets*—not buying their products or services and letting them know why. Another way to work within the system is to choose one of the many rapidly growing green careers highlighted throughout this book (see Figure 23-15, p. 630) and described in the book's companion website.

Third, we can *run for some sort of local office.* Look in the mirror. Maybe you are one who can make a difference as an office holder.

Fourth, we can *propose and work for better solutions to environmental problems*. Leadership is more than being against something. It also involves coming up with solutions to problems and persuading people to work together to achieve them. If we care enough, each of us can make a difference, as did Marion Stoddart (**Core Case Study**) and Wangari Maathai (Individuals Matter, p. 230).

> CORE CASE STUDY

Here are two pieces of hopeful news. First, research by social scientists indicates that social change requires

active support by only 5–10% of the population, which is often enough to lead to a political tipping point. Second, experience has shown that reaching such a critical mass can bring about social change much faster than most people think.

THINKING ABOUT

Environmental Leadership

> CORE CASE STUDY

What types of environmental leadership did Marion Stoddart (**Core Case Study**) practice? What type of environmental leadership interests you?

WHAT CAN YOU DO?

Influencing Environmental Policy

- Become informed on issues

- Make your views known at public hearings

- Make your views known to elected representatives, and understand their positions on environmental issues

- Contribute money and time to candidates for office

- Vote

- Run for office (especially at local level)

- Form or join nongovernment organizations (NGOs) seeking change

- Support reform of election campaign financing that reduces undue influence by corporations and wealthy individuals

Figure 24-7 Individuals matter: ways in which you can influence environmental policy (**Concept 24-2B**). **Questions:** Which three of these actions do you think are the most important? Which ones, if any, do you take?

24-3 What Is the Role of Environmental Law in Dealing with Environmental Problems?

▶ **CONCEPT 24-3** Environmental laws and regulations can be used to control pollution, set safety standards, encourage resource conservation, and protect species and ecosystems.

Environmental Law Forms the Basis for Environmental Policy

Environmental law is a body of statements defining what is acceptable environmental behavior for individuals and groups, according to the larger community, and attempting to balance competing social and private interests. This section of the chapter deals primarily with the U.S. legal system as a model that reveals the advantages and disadvantages of using a legal and regulatory approach to dealing with environmental problems.

The body of law includes statutory laws, administrative laws, and common laws. **Statutory laws** are those developed and passed by legislative bodies such as federal and state governments. **Administrative laws** consist of administrative rules and regulations, executive orders, and enforcement decisions related to the implementation and interpretation of statutory laws. **Common law** is a body of unwritten rules and principles derived from thousands of past legal decisions along with commonly accepted practices, or *norms,* within a society. Most of it consists of *case law,* a body of legal opinions derived from past court decisions. The body of environmental law is continuously evolving, as almost every major environmental regulation is challenged in court.

Most environmental lawsuits are **civil suits** brought to settle disputes or damages between one party and another. Many common law cases are settled using the legal concept of *nuisance,* which occurs when people use their property in a way that causes annoyance or injury to others. For example, a homeowner may bring a nuisance suit against a nearby factory because of the noise it generates. In such a suit, the **plaintiff,** the party bringing the charge, seeks to collect damages from the **defendant,** the party being charged, for injuries to health or for economic loss. The plaintiff may also seek an *injunction,* by which the defendant would be required to stop whatever action is causing the harm.

Any individual or clearly identified group may bring a civil suit. A *class action suit* is a civil suit filed by a group, often a public interest or environmental group, on behalf of a larger number of citizens who allege similar damages but who need not be listed and represented individually.

In the case of the nuisance suit example above, using the principles of common law, the court may side with the plaintiff if it finds that the loss of sleep, health problems, or other damage from the noise is greater than the cost of eliminating or reducing the noise. Short of closing the factory, often the court tries to find a reasonable or balanced solution to the problem. For example, it may order the factory to reduce noise to certain levels or to eliminate it during certain periods, such as at night.

Another concept used in common law cases is *negligence,* in which a party causes damage by knowingly acting in an unlawful or unreasonable manner. For example, a company may be found negligent if it fails to handle hazardous waste in a way required by a statutory law (Figure 24-8). A court may also find a company negligent if it fails to do something a reasonable person would do, such as testing waste for certain harmful chemicals before dumping it into a sewer, landfill, or river. Generally, negligence is harder to prove than nuisance.

Environmental Lawsuits Are Difficult to Win

Several factors limit the effectiveness of environmental lawsuits. *First,* plaintiffs bringing the suit must establish that they have the legal right or *legal standing* to do so in a particular court. To have such a right, plaintiffs must show that they have personally suffered health problems or financial losses allegedly caused by the defendant's actions. *Second,* bringing any lawsuit is expensive—too much so for most individuals.

Third, public interest law firms cannot recover attorneys' fees unless Congress has specifically authorized it in the laws that those firms are seeking to have enforced. By contrast, corporations can reduce their taxes by deducting their legal expenses—in effect getting a government subsidy to pay for part of their legal fees. In other words, the legal playing field is uneven and, in financial terms, is stacked against individuals and groups of private citizens filing environmental lawsuits.

Fourth, to stop a nuisance or to collect damages resulting from a nuisance or an act of negligence, plaintiffs must establish that they have been harmed in some significant way and that the defendant caused the harm. Doing this can be difficult and costly. Suppose a company (the defendant) is charged with causing cancer in individuals (the plaintiffs) who get their

Figure 24-8 An open ditch containing acid runoff from a closed coal mine in the U.S. state of West Virginia. Often, it is difficult or impossible to hold polluters responsible in a court of law.

drinking water from a river into which the company dumped hazardous chemicals. If other industries and cities also dump waste into that river, establishing that the defendant is the culprit is very difficult and requires expensive investigation, scientific research, and expert testimony. In addition, it is hard to establish that a particular chemical caused the plaintiffs' cancers.

Fifth, most states have *statues of limitations*, laws that limit how long a plaintiff can take to sue after a particular event occurs. These statutes often make it essentially impossible for victims of cancer, which may take 10–20 years to develop, to file or win a negligence suit.

Sixth, the court, or series of courts if the case is appealed, may take years to reach a decision. During that time a defendant may continue the allegedly damaging action unless the court issues a temporary injunction against it until the case is decided.

Another problem is that corporations and developers sometimes file *strategic lawsuits against public participation* (SLAPPs) against citizens who publicly criticize a business for some activity, such as polluting or filling in a wetland. For example, in Texas when a woman publicly called a nearby landfill "a dump" the landfill owners sued her husband for $5 million for failing to "control his wife."

SLAPPs range from $100,000 to $100 million, but average $9 million per suit. Judges throw out about 90% of the SLAPPs that go to court. But individuals and groups hit with SLAPPs must hire lawyers, and typically spend 1–3 years defending themselves.

Most SLAPPs are not meant to be won, but are intended to intimidate individuals and activist groups. The plaintiff company may offer to drop the lawsuit if the defendants agree to stop their protest and never

discuss the case or oppose the plaintiff again. In the face of rising legal costs, defendants often are forced to accept such terms.

Some citizen activists have fought back with countersuits and have been awarded damages. For example, a Missouri woman who was sued for criticizing a medical waste incinerator won an $86.5 million judgment against the incinerator's owner. However, even after paying such awards, corporations and developers generally save money by filing such SLAPP suits, because they can count legal and liability insurance costs as business expenses and write them off on their taxes.

**THINKING ABOUT
SLAPPS**

What are two things that you think the court system should do to reduce the threat of SLAPPs to well-meaning citizens?

Despite the difficulties of winning environmental lawsuits, some committed and persistent individuals have successfully used the court system to improve environmental protection (Individuals Matter, p. 646).

Analysts have suggested three major reforms to help level the legal playing field for citizens suffering environmental damage. *First*, pressure Congress to pass a law allowing juries and judges to award citizens their attorney fees, to be paid by the defendants, in successful lawsuits.

Second, establish rules and procedures for identifying frivolous SLAPP suits so that cases without factual or legal merit could be dismissed within a few weeks rather than years.

Third, raise the fines for violators of environmental laws and punish more violators with jail sentences. Polls indicate that 84% of Americans consider damaging the environment to be a serious crime. Under the current U.S. political climate, it is unlikely that these reforms will be implemented without strong bottom-up political pressure from concerned citizens.

Arbitration and Mediation Are Alternatives to Battling in Court

Because of the complexities and difficulties of filing suit, an increasing number of environmental lawsuits are being settled out of court. One alternative is **arbitration,** a formal effort, somewhat similar to a trial, to resolve a dispute. It is overseen by a trained and certified arbitrator who takes a more active role than a judge takes and is not constrained by legal precedent. There are clearly defined rules of evidence and cross-examination of witnesses, but the key goal of the arbitrator is to resolve the dispute. The end result of the process is a legally binding decision.

Arbitration can save the time, money, anguish, and uncertainty of a jury trial. Although it does create a legally binding decision for the parties involved,

INDIVIDUALS MATTER

Diane Wilson

Diane Wilson (Figure 24-A) was a fourth-generation shrimp boat captain and a working-class mother of five, when, in 1989, she had to stop shrimping in Lavaca Bay near Seadrift, Texas, along the U.S. Gulf Coast because the number of shrimp there had dropped significantly.

One day, another shrimper brought her a newspaper clipping reporting that impoverished Calhoun County, where Seadrift is located, was one of the most polluted counties in the United States. The article went on to say that Lavaca Bay had a large underwater mercury superfund site.

Despite living in Seadrift all of her life, Wilson had never heard anything about this toxic pollution that could be threatening the health of the town's residents and the shrimp and fish harvests on which her business depended. She was outraged and set up a public meeting to discuss pollution of the bay caused by local chemical plants. Local officials and business leaders were furious about Wilson's complaints because the chemical plants were the county's largest employer. Her character was called into question, neighbors

Figure 24-A Diane Wilson, shrimp boat captain and author of *An Unreasonable Woman*.

Photo courtesy of Kate McConnico

shunned her, thugs threatened her, and she even received death threats.

Wilson and environmental lawyer Jim Blackburn filed a lawsuit charging Formosa Plastics, a multinational Taiwanese chemical company, with dumping toxic chemicals into the bay. With only a high school education,

Wilson taught herself to file legal briefs and analyze mountains of highly technical EPA scientific documents. After years of legal proceedings, protests, and bad publicity, Formosa Plastics and Dow Chemical agreed to stop emitting pollutants into the bay.

Wilson is now an activist for environmental and social justice. She has practiced non-violent protest and civil disobedience in the tradition of Henry David Thoreau, Mahatma Gandhi, and Martin Luther King, and she has gone on hunger strikes to protest releases of toxic chemicals by companies in her community. As a result, she has been arrested and put in jail 13 times.

Wilson wrote a book about her struggle entitled, *An Unreasonable Woman: A True Story of Shrimpers, Politicos, Polluters and the Fight for Seadrift, Texas* (Chelsea Green, 2005). She has received numerous environmental awards and acclaim as an important new writer with an inspiring message. Wilson, who likes to call herself "nobody particular," urges individuals to help change the world by being unreasonable and standing up for what they believe.

arbitration does not establish a legally binding precedent for deciding future cases. There is also less opportunity for appeal and less publicity from the process, which is often a satisfying result for the companies or agencies that have been sued. Whereas a court case usually results in one party winning and the other losing, the outcome of arbitration is usually some sort of compromise.

Another approach for settling a dispute is **mediation,** in which the parties involved are encouraged to sit down and talk under the guidance of a professional mediator. The goal is to see if they can resolve the dispute in a way that is acceptable to both parties. Mediation is much less costly and time consuming and may provide a more satisfactory resolution of a dispute than a court case could provide. But a settlement drawn up by mediation is not legally binding unless the terms of the agreement make it so. Thus, months of mediation can result in an agreement that polluters may ignore.

> **THINKING ABOUT**
> **Environmental Lawsuits**
>
> If you were bringing about a controversial environmental lawsuit, would you prefer a court trial, arbitration, or mediation? Explain.

Major Types of Environmental Laws in the United States

Concerned citizens have persuaded Congress to enact a number of important federal environmental and resource protection laws (Figure 24-4 and Supplement 5, p. S35). One type of such legislation *sets standards for pollution levels* (as in the Clean Air Acts, p. 488, and the Federal Water Pollution Control Act, p. 552). Another *screens new substances for safety* (as in the Toxic Substances Control Act, p. 582). A third type of legislation *encourages resource conservation* (the Resource Conservation and Recovery Act and the National Energy Act, p. 582). A fourth type *sets aside or protects certain species, resources, and ecosystems* (the Endangered Species Act, p. 207, and the Wilderness Act, p. 238) (**Concept 24-3**).

A fifth approach in legislation *requires evaluation of the environmental impact of an activity proposed by a federal agency,* as in the National Environmental Policy Act, or NEPA, passed in 1970. Under NEPA, an *environmental impact statement* (EIS) must be developed for every major federal project likely to have an important effect on environmental quality. The EIS must describe why the proposed project is needed, its short-term and long-term beneficial and harmful environmental impacts, ways to lessen harmful impacts, and an evaluation of

alternatives. The documents must be published and are open to public comment. NEPA does not prohibit environmentally harmful government projects. But it requires federal agencies, with wide-ranging influence over the environment and management of more than one-third of the country's land, to take environmental consequences into account in making decisions. It also exposes proposed projects and their likely harmful effects to public scrutiny. Opponents have targeted NEPA as a law to weaken or repeal.

At least 36 U.S. states have passed similar laws that apply to potentially environmentally disruptive actions by state agencies. A number of other countries, including Canada, Sweden, France, New Zealand, and Australia, have also passed laws similar to NEPA. In 2006, the Vermont Law School formed a partnership with China's Sun Yat-sen University to help China strengthen its environmental laws and policies.

THINKING ABOUT
Environmental Impact Statements

Are you in favor of having government agencies develop an environmental impact statement for any major government project that is likely to have an important effect on environmental quality? Explain.

U.S. Environmental Laws and Regulations Have Been under Attack

U.S. environmental laws have been highly effective, especially in controlling pollution. However, since 1980, a well-organized and well-funded movement has mounted a strong campaign to weaken or repeal existing environmental laws and regulations and to change the ways in which public lands (Figure 24-5) are used.

Three major groups are strongly opposed to various environmental laws and regulations: some corporate leaders and other powerful people who see them as threats to their profits, wealth, and power; citizens who see them as threats to their private property rights and jobs; and state and local government officials who resent having to implement federal laws and regulations with little or no federal funding (unfunded mandates) or who disagree with certain federal regulations.

One problem working against strong regulations is that the focus of environmental issues has shifted from easy-to-see dirty smokestacks and filthy rivers to more complex, controversial, long-term, and often invisible environmental problems such as climate change, biodiversity loss, and groundwater pollution. Explaining such complex issues to the public and mobilizing support for often controversial, long-range solutions to such problems is difficult. (See the Guest Essay on environmental reporting by Andrew C. Revkin at CengageNOW.)

Another problem is that some environmentalists have mostly brought bad news about the state of the environment. History shows that bearers of bad news are not well received, and opponents of the environmental movement have used this to undermine environmental concerns. History also shows that people are moved to bring about change mostly by an inspiring and hopeful vision of what the world could be like. So far, environmental groups have not worked together to develop broad, compelling, and positive visions of a more environmentally sustainable future for humans and other species.

Since 2000, efforts to weaken environmental laws and regulations have escalated. Most major U.S. federal environmental laws and regulatory agencies have been weakened by a combination of executive orders and congressional actions. According to the Natural Resources Defense Council (NRDC), many regulatory agencies have been staffed largely with officials who favor weakening them, decreasing their funding, ignoring sound scientific consensus, and stifling dissent. Environmental leaders warn that the entire environmental legal and regulatory structure, built with bipartisan consensus between 1965 and 1980 (see Supplement 5, p. S35), is being systematically undermined.

Most U.S. citizens are unaware that the increasingly weakened foundation of the nation's environmental laws and regulations could crumble. Independent polls show that more than 80% of the U.S. public strongly support environmental laws and regulations and do not want them weakened. And Gallup polling data show that the percentage of Americans who say they worry about the environment "a great deal" or "a fair amount" increased from 62% to 77% between 2004 and 2006. However, polls also show that less than 10% of the U.S. public consider the environment to be one of the nation's most pressing problems. As a result, environmental concerns often do not get transferred to the ballot box or the pocketbook.

THINKING ABOUT
Environmental Law and Politics

Do you think any environmental laws in the United States or in the country where you live should be weakened or strengthened? Explain.

We need to prevent further weakening of the U.S. environmental laws and regulations, repair the damages done, improve existing laws and regulations, and develop new laws based on sustainability principles. This will require a combination of science-based environmental education, education about the current state of environmental laws, and the same kind of organized bottom-up political pressure from concerned citizens that led to passage of important environmental laws in the first place.

Also, as in the cases of Chattanooga, Tennessee (Case Study, p. 21) and the Nashua River (**Core Case Study**), people on all sides of important environmental issues can build *social capital* by listening

to one another's concerns, trying to find areas of agreement, and working together to find solutions and to improve environmental laws and regulations. An important political component of the transition to a more environmentally sustainable society is for U.S. citizens to elect ecologically literate and concerned leaders. Citizens must also insist that leaders work across party lines to end the political deadlock on most environmental issues, which has characterized the U.S. Congress and the executive branch since 1980.

Beginning in the late 1960s, elected officials in the United States—under intense political pressure from its citizens—led the world into the first phase of the environmental revolution. Members of both political parties in the Congress worked together with the executive branch to pass and implement major environmental laws (Figure 24-4), mostly during the 1970s, which became known as the *environmental decade* (Figure 7, p. S36, in Supplement 5).

Since 1980, the U.S. federal government has largely abandoned its leadership of the environmental revolution it helped to create and has weakened many of the environmental laws it passed. However, because the United States is the world's leading economic power, environmental leaders call for the country to resume its national and global environmental leadership role, with no time to lose, in making the transition to a more sustainable society. This will not happen unless enough U.S. citizens, including a growing group of environmentally aware business leaders, insist that members of Congress and the executive branch work together to plan for a more sustainable future.

24-4 What Are the Major Roles of Environmental Groups?

▶ **CONCEPT 24-4** Grassroots groups are growing and combining efforts with large environmental organizations in a global sustainability movement.

Citizen Environmental Groups Play Important Roles

The spearheads of the global conservation, environmental, and environmental justice movements are the tens of thousands of nonprofit nongovernmental organizations (NGOs) working at the international, national, state, and local levels. The growing influence of these organizations is one of the most important trends influencing environmental decisions and policies (**Concept 24-4**).

NGOs range from grassroots groups with just a few members to organizations like the World Wildlife Fund (WWF), a 5-million-member global conservation organization, which operates in 100 countries. Other international groups with large memberships include Greenpeace, The Nature Conservancy, Conservation International, and the Grameen Bank (Case Study, p. 626).

Using e-mail, text messages, and the Internet, some environmental NGOs have organized themselves into an array of influential international networks. Examples include the Pesticide Action, Climate Action, International Rivers, and Women's Environment and Development Networks. They collaborate across national borders and monitor the environmental activities of governments, corporations, and international agencies such as the World Bank and the World Trade Organization (WTO) (Supplement 11, p. S74). They also help to expose corruption and violations of national and international environmental agreements, such as the Convention on International Trade in Endangered Species (CITES), which prohibits international trade of endangered species (p. 206).

In recent years, *global public policy networks* have formed in response to rapidly changing conditions in a globalizing world. These innovative groups focus on a particular environmental problem by bringing together governments, the private sector, international organizations, and NGOs. Since the 1990s, more than 50 such networks have emerged. Examples include the *International Forum on Forests*, which develops proposals for more sustainable forest management; the *Global Water Partnership*, working toward integrated water resources management; the *Renewable Energy Policy Network*, which develops policies to spur development of renewable energy; and the *Millennium Ecosystem Assessment*, designed to evaluate and distribute information on changes in ecosystems at the global, regional, and national levels.

In the United States, more than 8 million citizens belong to more than 30,000 NGOs that deal with environmental issues. They range from small *grassroots* groups to large heavily funded *mainline* groups, the latter usually staffed by expert lawyers, scientists, economists, lobbyists, and fund raisers. The largest of these groups are the World Wildlife Fund, Sierra Club (whose membership increased by about a third to

800,000 between 2002 and 2006), National Wildlife Federation, Audubon Society, Greenpeace, Friends of the Earth, Natural Resources Defense Council (Case Study, below), Wilderness Society, Ducks Unlimited, Izaak Walton League, Conservation International, Environmental Defense, and The Nature Conservancy.

The largest groups have become powerful and important forces within the U.S. political system. They have helped to persuade Congress to pass and strengthen environmental laws (Figure 24-4 and Supplement 5, p. S35), and they fight attempts to weaken or repeal such laws.

Some industries and environmental groups are working together to find solutions to environmental problems. For example, Environmental Defense worked with McDonald's to redesign its packaging system to eliminate its plastic hamburger containers, with General Motors to help remove high-pollution cars from the road, and with FedEx to lower the air pollution emissions of its planes and trucks. The World Resources Institute is collaborating with leading businesses to build a market among corporations for electrical power produced from renewable energy resources. And the Rainforest Alliance has worked with Chiquita Banana to certify the health, labor, and environmental practices on its farms.

Some environmental groups have shifted resources from demonstrating and litigating to publicizing research on innovative solutions to environmental problems. For example, to promote the use of chlorine-free paper, Greenpeace Germany printed a magazine using such paper and encouraged readers to demand that magazine publishers switch to chlorine-free paper. Shortly thereafter, several major magazines made that shift.

■ CASE STUDY
The Natural Resources Defense Council

One of the stated purposes of the Natural Resources Defense Council (NRDC) is

to establish sustainability and good stewardship of the Earth as central ethical imperatives of human society. . . . We work to foster the fundamental right of all people to have a voice in decisions that affect their environment. We seek to break down the pattern of disproportionate environmental burdens borne by people of color and others who face social or economic inequities. Ultimately, NRDC strives to help create a new way of life for humankind, one that can be sustained indefinitely without fouling or depleting the resources that support all life on Earth.

To those ends, NRDC goes to court to stop environmentally harmful practices. It also informs and organizes millions of environmental activists through its website (**www.nrdc.org**), magazines, and newsletters, to take actions to protect the environment—globally, regionally, and locally. For example, its BioGems Network, accessible through the website, regularly informs subscribers about environmental threats all over the world, and helps people to take action by donating money, signing petitions, and writing letters to corporate and government officials and newspaper editors.

In 2005, with NRDC's help, U.S. citizens organized massive opposition to a proposed government policy that would have allowed sewer operators to routinely dump barely treated sewage into the nation's lakes, rivers, and streams. Because of well-informed, very vocal opposition to this proposal, the U.S. House of Representatives voted overwhelmingly to block the Environmental Protection Agency from finalizing this so-called "blending" proposal.

In another case, in 2001, NRDC helped forge an agreement among Canadian timber companies, environmentalists, native peoples, and the government of British Columbia (Canada) to protect a vast area of the Great Bear Rainforest from destructive logging. This followed years of pressure from NRDC activists on logging companies, their U.S. corporate customers, and provincial officials to protect the habitat of the rare all-white spirit bear as well as the habitats of eagles, grizzly bears, and wild salmon.

Grassroots Environmental Groups Bring about Change from the Bottom Up

The base of the environmental movement in the United States and throughout the world consists of thousands of grassroots citizens' groups organized to improve environmental quality, often at the local level (Core Case Study). According to political analyst Konrad von Moltke, "There isn't a government in the world that would have done anything for the environment if it weren't for the citizen groups." Taken together, a loosely connected worldwide network of grassroots NGOs working for bottom-up political, social, economic, and environmental change can be viewed as an emerging citizen-based *global sustainability movement* (**Concept 24-4**).

These groups have worked with individuals and communities to oppose harmful projects such as landfills, waste incinerators, and nuclear waste dumps, as

Butterfly in a Redwood Tree

"Butterfly" is the nickname given to Julia Hill. This young woman spent 2 years of her life on a small platform near the top of a giant redwood tree in California to protest the clear-cutting of a forest of these ancient trees, some of them more than 1,000 years old. She and other protesters were illegally occupying these trees as a form of *nonviolent civil disobedience*, similar to that used decades ago by Mahatma Gandhi in his efforts to end the British occupation of India and by Martin Luther King in the U.S. civil rights movement.

Butterfly had never participated in an environmental protest or act of civil disobedience. She went to the site to express her belief that it was wrong to cut down these ancient giants for short-term economic gain,

even if you own them. She planned to stay for only a few days.

But after seeing the destruction and climbing one of these magnificent trees, she ended up staying in the tree for 2 years to publicize what was happening and to try to save the surrounding trees. She became a symbol of the protest and, during her stay, used a cell phone to communicate with members of the mass media throughout the world to help develop public support for saving the trees.

Can you imagine spending 2 years of your life in a tree on a platform not much bigger than a king-sized bed, hovering 55 meters (180 feet) above the ground, and enduring high winds, intense rainstorms, snow, and ice? All around her was noise from trucks, chainsaws, and helicopters.

Although Butterfly lost her courageous battle to save the surrounding forest, she persuaded Pacific Lumber MAXXAM to save her tree (called Luna) and a 60-meter (200-foot) buffer zone around it. Not too long after she descended from her perch, someone used a chainsaw to seriously damage the tree. Cables and steel plates are now used to preserve it.

But maybe Butterfly did not lose, after all. The book she wrote about her stand, along with her subsequent travels to campuses all over the world, have inspired a number of young people to stand up for protecting biodiversity and for other environmental causes.

Butterfly led others by following in the tradition of Gandhi, who said, "My life is my message." Would you spend a day or a week of your life protesting something that you believed to be wrong?

well as to fight against the clear-cutting of forests and pollution from factories and power plants. They have also taken action against environmental injustice. (See the Guest Essay on this topic by Robert D. Bullard at CengageNOW.)

Grassroots groups have organized *conservation land trusts*, where owners of a piece of property agree to protect it from development or other harmful environmental activities, often in return for tax break on the value of the land. And they have generated other efforts to save wetlands, forests, farmland, and ranchland from development. They have also helped to restore degraded forests, wetlands, and rivers (**Core Case Study**).

CORE CASE STUDY

Some grassroots environmental groups use the nonviolent and nondestructive tactics of protest marches, tree sitting (see Individuals Matter, above), and other devices for generating publicity to help educate and sway members of the public to oppose various environmentally harmful activities. Such tactics often work because they produce bad publicity for environmentally harmful practices and businesses. For example, after 2 years of pressure and protests, Home Depot agreed to sell only wood products made from certified sustainably grown timber. Within a few months, Lowes and eight other major building supply chains in the United States developed similar practices.

In another case, Citigroup (the largest bank holding company in the United States), after being targeted for 3 years by a coalition of investors, stockholders, activists, and NGOs, led by the Rainforest Action Network, changed its policy and reduced its lending to unsus-

tainable industries. After seeing the bad publicity that Citigroup got, Bank of America and JPMorgan Chase (two of the other largest lenders in the United States) agreed to become more eco-friendly banks.

Much more controversial are militant environmental groups that use violent means, such as breaking into labs to free animals used to test drugs and destroying property such as bulldozers and SUVs. Most environmentalists strongly oppose such tactics.

HOW WOULD YOU VOTE?

Do you support the use of nonviolent and nondestructive civil disobedience tactics by environmental groups and individuals? Cast your vote online at **academic.cengage.com/ biology/miller**.

Students and Educational Institutions Can Play Important Environmental Roles

Since the 1980s, there has been a boom in environmental awareness on college campuses and in public and private schools across the United States. Most student environmental groups work with members of their school's faculty and administration to bring about environmental improvements in their schools and local communities.

Many of these groups make *environmental audits* of their campuses or schools. They gather data on practices affecting the environment and use it to propose changes

that will make their campuses or schools more environmentally sustainable while usually saving money in the process. Such audits have resulted in numerous improvements (Science Focus, below). In 2008, the National Wildlife Federation published a useful guide for students wishing to perform such audits. See **www.nwf .org/campusecology/BusinessCase/index.cfm**.

More than 3,800 U.S. universities and colleges make up a $315 billion sector of the U.S. economy. Because their job is to train future leaders and citizens, they bear a special responsibility to promote environmental literacy and encourage environmental stewardship and sustainability, beginning with their own campuses and curricula. Any school without a serious sustainability program is now viewed as outdated.

No campus in the United States is even close to achieving environmental sustainability. However, partly because of student pressure, many campuses are making progress. Since June 2007, 458 university presidents, including some from the most influential schools in the United States, have signed an agreement to make their schools carbon neutral by resolving to add no net emissions of CO_2 to the atmosphere. Arizona State University has opened its Global Institute for Sustainability, which began admitting undergraduates in 2008. Berea College in Kentucky has a sustainable residential and learning

SCIENCE FOCUS

Greening American Campuses

Environmental audits by students at American colleges and universities have focused on increasing recycling, getting university food services to buy more food from local organic farms, reducing reliance on fossil fuels, increasing reliance on renewable energy, improving energy efficiency, and reducing dependence on cars. They also focus on reducing water waste, maintaining campus grounds more sustainably, building environmentally sustainable new buildings, retrofitting existing buildings to make them more sustainable, and implementing concepts of environmental sustainability throughout the curriculum.

Here are a few of a growing number of examples. Morris A. Pierce, a graduate student at the University of Rochester in New York State, developed an energy management plan that was adopted by that school's board of trustees and is projected to save the university $60 million over 20 years. In similar efforts, students have helped to convince almost 80% of universities and colleges in the United States to develop recycling programs.

In 2007, *Sierra Magazine* rated Oberlin College in Ohio as the nation's greenest college. Its students helped to design a more sustainable environmental studies building powered by solar panels, which produce 30% more electricity than the building uses. A living machine (Figure 20-C, p. 556) in the building's lobby purifies all of its wastewater, and half of the school's electricity comes from green sources. The school has a car-sharing program, and student activity fees subsidize public transportation.

At Northland College in Wisconsin, students helped to design a "green" residence

Figure 24-C The *Environmental Living and Learning Center* is an eco-friendly residence hall and meeting space at Northland College in Ashland, Wisconsin. The building, which houses more than 150 students, features a wind turbine and three large solar panels for generating some of the electricity used by the residents. Other green features include passive solar heating, furniture and carpet made from recycled plastic, recycled Mylar window shades that reduce heat loss in winter and prevent excessive solar gain in summer, and composting toilets in some of the bathrooms. Northland students had a major roll in designing the building, which is used by a variety of organizations for meetings and educational programs.

Courtesy of Northland College

hall (Figure 24-C) that features a wind turbine, panels of solar cells, furniture made from recycled materials, and waterless (composting) toilets. A wind turbine at Minnesota's Carleton College generates 40% of that school's electricity. New York University gets all of the electricity it uses from windpower. And Middlebury College in Vermont has built a biomass power plant that runs on wood chips to help to reduce its greenhouse gas emissions.

Many college cafeterias are buying food locally, much of it organic. Yale University gets 40% of its food service items from local farms, including its student-run organic farm. At California's Santa Clara University, local farms provide 80% of the produce served in its dining halls.

In June 2007, a group of 11 students at Dartmouth College in New Hampshire drove across the country in a big green retired school bus fueled by waste cooking oil from fast-food restaurants. Stopping at parks and music festivals, the bus became a mobile science fair for renewable biomass energy (see **thebiggreenbus.org** for more information).

Critical Thinking

What major steps is your school taking to increase its own environmental sustainability and to educate its students about environmental sustainability?

complex with 50 apartments, known as Ecovillage. And most U.S. business schools now require students to take at least one course with an emphasis on sustainability, ethics, and corporate and social responsibility.

Such environmental improvements and research studies are also underway at universities in at least 42 other countries.

24-5 How Can We Improve Global Environmental Security?

▶ **CONCEPT 24-5** Environmental security is necessary for economic security and is at least as important as military security.

Environmental Security Is as Important as Military and Economic Security

Countries are legitimately concerned with *military security* and *economic security*. However, ecologists and many economists point out that all economies are supported by the earth's natural capital (Figure 1-3, p. 8, and Figure 23-5, p. 615). And, according to environmental expert Norman Myers,

If a nation's environmental foundations are degraded or depleted, its economy may well decline, its social fabric deteriorate, and its political structure become destabilized as growing numbers of people seek to sustain themselves from declining resource stocks. Thus, national security is no longer about fighting forces and weaponry alone. It relates increasingly to watersheds, croplands, forests, genetic resources, climate, and other factors that, taken together, are as crucial to a nation's security as are military factors.

[See his Guest Essay on this subject at CengageNOW (**Concept 24-5**).]

Research by Thomas Homer-Dixon, director of Canada's Trudeau Center for Peace and Conflict Studies, has revealed a strong correlation between growing scarcities of resources, such as cropland, water, forests, and fish, and the spread of civil violence and dysfunction in government that can cause countries to become *failed states* (Figure 17, p. S19, Supplement 3), which can no longer ensure the personal security of their people. Failed states suffer from a breakdown of law and order with gangs ruling the streets, violent internal conflict, and a deterioration of roads, electrical power, food and water supplies, sewage systems, and other vital forms of infrastructure.

Such states often degenerate into civil war with segments of the population shifting their allegiance from a central government to tribal chieftains, warlords, or religious leaders, each claiming a piece of what once was a country. Many failing states also become training grounds for terrorists (as in Afghanistan and Iraq), weapon traders (as in Nigeria and Somalia), and drug producers (as in Afghanistan and Myanmar). They often generate large numbers of refugees displaced from their land and homes and often fleeing for their lives.

Economist Jeffrey D. Sachs argues that the deadly violence in Darfur, Sudan (Africa), which headed the 2007 list of failing states, is rooted in an ecological crisis brought about by drought and rapid population growth. Darfur lies in an arid, drought-prone region where two competing populations of impoverished people—one raising livestock and the other farming land—are growing exponentially. Their agricultural activities have strained the limited water supply, and when precipitation became scarce in the 1980s, violent competition began, as communities fought to survive by raiding each other's food and water supplies.

Some analysts call for all countries to make environmental security a major focus of diplomacy and government policy at all levels. This could be implemented by a council of advisers made up of highly qualified experts in environmental, economic, and military security, who could integrate all three security concerns in making major decisions, including funding priorities (Figure 23-12, p. 627). This involves taking into consideration the enormous harmful environmental impacts of wars (Science Focus, at right).

SCIENCE FOCUS

Environmental Impacts of War

Regardless of their cause, military conflicts always lead to human suffering and environmental degradation. In the last 60 years, there has been a general increase in the amount of environmental damage due to wars, as weapons have become more powerful and destructive.

For example, when a large bomb explodes it releases toxic chemicals, annihilates surrounding flora and fauna, and destroys layers of soil that can take thousands of years to regenerate. When bombs destroy oil tankers, oil production facilities, or chemical plants, they release huge quantities of toxic chemicals into the air, water, and soil, threatening humans, plants, birds, and other wildlife species. Oil facilities set afire release enormous amounts of CO_2 and other pollutants into the atmosphere.

Another problem is contamination of military sites. In the United States, over 20,000 sites on more than 1,700 military facilities are contaminated with toxic wastes. Hundreds of sites used to produce nuclear weapons are contaminated with long-lived and dangerous radioactive wastes.

The harmful environmental, economic, and human impacts of war include loss of biodiversity, air and water pollution, higher levels of CO_2 contributing to climate change, soil and groundwater contamination, contaminated fish and wildlife, economic losses, loss of jobs, and large numbers of displaced refugees.

In addition, according to the International Committee of the Red Cross, unexploded land mines kill or maim 1,000–2,000 people every month. And an estimated 100 million such mines are scattered throughout the world, as are radioactive depleted uranium shells.

Since 1998, the U.N. Environment Programme has been sending teams of scientists and other workers to affected areas after conflicts, such as those in the Balkans, Afghanistan, and Iraq, to assess damages to the natural and human environments and to develop plans for dealing with these harmful impacts. The teams identify environmental hotspots in need of immediate cleanup to avoid further threats to human health. These efforts led the international community for the first time to include environmental cleanup as part of humanitarian aid provided after a war.

Critical Thinking

What are three measures the international community might take to deal with the harmful environmental effects of war?

We Can Develop Stronger International Environmental Policies

A number of international environmental organizations help to shape and set global environmental policy. Perhaps the most influential is the United Nations, which houses a large family of organizations including the U.N. Environment Programme (UNEP), the World Health Organization (WHO), the U.N. Development Programme (UNDP), and the U.N. Food and Agriculture Organization (FAO).

Other organizations that make or influence environmental decisions are the World Bank, the Global Environment Facility (GEF), and the World Conservation Union (also known as the IUCN). The website for this chapter has a list of major international environmental organizations. Despite their often limited funding, these and other organizations have played important roles in

- Expanding understanding of environmental issues
- Gathering and evaluating environmental data
- Developing and monitoring international environmental treaties
- Providing grants and loans for sustainable economic development and reducing poverty
- Helping more than 100 nations to develop environmental laws and institutions

One such NGO, the NRDC (Case Study, p. 649) has focused on helping China in its efforts to become more sustainable. In 2007, NRDC opened an office in Beijing, from which it acts as a leading advisor to Chinese officials, helping the country to launch an energy-efficiency campaign, invest in clean-burning vehicles, and implement a carbon-capture program. NRDC states that one of its top priorities is to help see that China's staggering economic growth does not become an environmental disaster for China and for the planet.

Since the 1972 U.N. Conference on the Human Environment in Stockholm, Sweden, progress has been made in addressing environmental issues at the global level. The conference also created the UNEP to help develop the global environmental agenda. However, it is a small and underfunded agency.

In 1992, governments of more than 178 nations and hundreds of NGOs met at the U.N. Conference on Environment and Development (UNCED) in Rio de Janeiro, Brazil. The major policy outcome of this conference was Agenda 21, a global agenda for sustainable development in the 21st century, which was adopted by 178 governments, not including the United States. This ambitious agenda established goals for addressing the world's social, economic, and environmental problems. The conference also established the Commission on Sustainable Development to follow up on the Agenda 21 goals and to monitor and report on the agreements at the local, regional, and international levels. In 1997, an Earth Summit was held to review the somewhat limited progress toward implementing the Agenda 21 goals.

Figure 24-9 (p. 654) lists some of the good and bad news about international efforts to deal with global

TRADE-OFFS

Global Efforts to Solve Environmental Problems

Good News	Bad News
Environmental protection agencies in 115 nations	Most international environmental treaties lack criteria for evaluating their effectiveness
Over 500 international environmental treaties and agreements	
UN Environment Programme (UNEP) created in 1972 to negotiate and monitor international environmental treaties	1992 Rio Earth Summit led to nonbinding agreements with inadequate funding
1992 Rio Earth Summit adopted key principles for dealing with global environmental problems	By 2008 there was little improvement in the major environmental problems discussed at the 1992 Rio summit
2002 Johannesburg Earth Summit attempted to implement 1992 Rio summit policies and goals and reduce poverty	2002 Johannesburg Earth Summit failed to deal with global environmental problems such as climate change, biodiversity loss, and poverty

Figure 24-9 Good and bad news about international efforts to deal with global environmental problems. **Question:** What single piece of good news and what single piece of bad news do you think are the most important?

environmental problems such as poverty, climate change, biodiversity loss, and ocean pollution.

The primary focus of the international community on environmental problems has been the development of various international environmental laws and nonbinding policy declarations called *conventions*. There are more than 500 international environmental treaties and agreements—known as *multilateral environmental agreements* (MEAs).

To date, the Montreal and Copenhagen Protocols for protecting the ozone layer (p. 527) are the most successful examples of how the global community can work together to deal with a serious global environmental challenge. Figure 24-10 lists some major problems with MEAs and solutions to these problems.

In 2004 and in 2008, environmental leader Gus Speth argued that global environmental problems are getting worse and that international efforts to solve them are inadequate. This analysis was confirmed by

the 2006 Global Governance Initiative Report, which gave the world's governments and businesses a score of 2 out of 10 in dealing with environmental issues, especially in addressing declining biodiversity and climate change and providing clean water and sanitation. Speth and other environmental leaders propose the creation of a World Environmental Organization, on the order of the World Health Organization and the World Trade Organization, to deal with global environmental challenges.

The Yale Center for Environmental Law and Policy and the Columbia University Center for International Earth Science Information Network have developed the Environmental Performance Index (EPI), which evaluates 149 countries using 25 environmental indicators. (See Figure 8, p. S30, in Supplement 4 for a map of EPIs by country.) In 2008, the top ten scoring countries, in order, were Switzerland, Norway, Sweden, Finland, Costa Rica, Austria, New Zealand, Latvia, Colombia, and France. Other rankings were Iceland (11) (Chapter 16 Core Case Study, p. 399), Canada (12), Germany (13), United Kingdom (14), Japan (21), Denmark (26), Russia (28), Brazil (34), United States (39), Mexico (47), China (104), and India (120). The five lowest ranked countries were Mali, Mauritania, Sierra Leone, Angola, and Niger.

Corporations Can Play a Key Role in Moving toward Environmental Sustainability

In our increasingly globalized economy, it has become more apparent that governments and corporations must work together to achieve goals for increased environmental sustainability. Governments can set environmental standards and goals through legislation and regulations, but corporations generally have highly efficient ways of accomplishing such goals. Making a transition to more environmentally sustainable societies and economies will require huge amounts of investment capital and research and development. Most of this money is likely to come from profitable corporations, including banks and other lending agencies. Thus corporations must play a vital role in achieving a more sustainable future.

The good news is that an increasing number of corporate chief executive officers (CEOs) and investors are aware that there is much money to be made from developing and selling green products and services during this century. This is guided by the concept of *eco-efficiency*, which is about finding ways to create more economic value with less environmental impact by reducing the amount of energy, water, chemicals and other raw materials used per unit of output. In addition to reducing harmful health and environmental impacts, improving eco-efficiency can save money for businesses and help them to meet their responsibilities to stockholders and investors.

International Environmental Treaties

Problems

Take a long time to develop and are weakened by requiring full consensus

Poorly monitored and enforced

Lack of funding for monitoring and enforcement

Treaties are not integrated with one another

Solutions

Do not require full consensus among regulating parties

Establish procedures for monitoring and enforcement

Increase funding for monitoring and enforcement

Harmonize or integrate existing agreements

Figure 24-10 *Global outlook:* major problems with global environmental treaties and agreements and solutions to these problems. **Question:** Which problem and which solution do you think are the most important?

Stephan Schmidheiny, a Swiss billionaire, organized a group made up of the CEOs of 48 of the world's largest corporations in the early 1990s. It eventually became the World Business Council for Sustainable Development (WBCSD), which is dedicated to promoting sustainable development built around improving eco-efficiency. Today, this council is a highly influential coalition of more than 170 international companies, involving some 700 global business leaders.

24-6 How Can We Implement More Sustainable and Just Environmental Policies?

▶ **CONCEPT 24-6** Making the transition to more sustainable societies will require that nations implement green plans and that people and nations cooperate, agree on principles, and make the political commitment to achieve this transition.

Green Planning Can Be a National Priority

As stated at the beginning of this chapter, governments have the power to play strong roles in making a transition to a more sustainable future. Many argue that they have the responsibility to do so, through national policies (**Concept 24-6**). In countries such as New Zealand, Sweden, and the Netherlands, the national governments are doing just that in a process called *green planning*—the creation of long-term environmental management strategies with the ultimate goal of achieving greater environmental and economic sustainability and a high quality of life for a country's citizens.

The longest running national green plan is that of the Netherlands—a country of about 16 million people. In 1989, the Netherlands began implementing a National Environmental Policy Plan, or Green Plan, as a result of widespread public alarm over declining environmental quality. The goal was to slash production of many pollutants by 70–90% and achieve the world's first environmentally sustainable economy, ideally within 2 or 3 decades.

The government began by identifying eight major areas for improvement: climate change, acid deposi-

tion, eutrophication, toxic chemicals, waste disposal, groundwater depletion, unsustainable use of renewable and nonrenewable resources, and local nuisances (mostly noise and odor pollution). Then the government formed target groups, with members drawn from industry, government, and citizens' groups for each of the eight areas, and asked each group to develop a voluntary agreement on establishing targets and timetables for drastically reducing pollution. Each group was free to pursue whatever policies or technologies it wanted. But if a group could not agree, the government would impose its own targets and timetables and stiff taxes and other penalties for industries not meeting certain pollution reduction goals.

The government identified four general themes for each group to focus on. *First,* implement life-cycle management, which makes companies responsible for the remains of their products after users are finished using them—a goal that has promoted the design of products that can be reused or recycled. *Second,* improve energy efficiency with help from government grants of $385 million per year. *Third,* invent more environmentally sustainable technologies or improve existing technologies with support from a government program for that purpose. *Fourth,* improve public awareness through a massive government-sponsored public education program.

In 2001, the Dutch government published the fourth version of the country's national environmental policy plan. More than 250,000 Dutch businesses have been participating in this plan and have made legally binding performance agreements with the government. Many of the country's leading industrialists like the Green Plan for several reasons. One is that they can make investments in pollution prevention and pollution control with less financial risk, because they have a high degree of certainty about long-term environmental policy. Another is that they are free to deal with the problems in ways that make the most sense for their businesses.

In addition, industrial leaders have learned that creating more efficient and environmentally sound products and processes can often reduce costs and increase profits, because they can sell such innovations at home and abroad. Since the Green Plan's inception in 1989, the Netherlands has achieved 70% of the plan's initial goals.

The Netherlands plan is the first attempt by any country to foster a national debate on the issue of environmental sustainability and to encourage innovative solutions to environmental problems. Is the plan working? Since 1989, more than a dozen government–industry environmental covenants have been signed. Most of the target groups have met or exceeded their goals on schedule. Some of the accomplishments include phasing out ozone-depleting substances (p. 525), cutting industrial waste disposal by 60%, increasing the waste recycling rate to 70%, and reducing sulfur dioxide emissions from power plants by 70%.

The government and private sector have conducted a huge amount of environmental research. This has led to an increase in organic agriculture, greater reliance on bicycles in some cities, and more ecologically sound new housing developments. Significant progress has been made, but there is still a long way to go as evidenced by the fact that the 2008 Environmental Performance Index ranked Netherlands number 55 out of 149 countries.

THINKING ABOUT
Green Planning

Do you know of a green plan in the city, state, or country where you live? Is it meeting its goals?

We Can Shift to More Environmentally Sustainable Societies

A study of nature reveals that all parts of the biosphere are ecologically interdependent. Similarly, all parts of the human *culturesphere* are economically and politically interdependent. This interdependence requires citizens, business leaders, and elected officials to cooperate in trying to find and implement innovative short-term and long-term solutions to local, national, and global environmental, economic, and social problems (**Concept 24-6**).

Several guidelines have been suggested for fostering cooperation instead of confrontation as we struggle to make such a transition. *First,* emphasize *preventing or minimizing* environmental problems instead of letting them build to crisis levels. *Second,* use well-designed and carefully monitored *marketplace solutions* (Chapter 23) to help prevent or reduce the impacts of most environmental problems. *Third,* cooperate and innovate to find *win–win solutions* or *tradeoffs* to environmental problems and injustices. And *fourth,* be *honest and objective.* People on both sides of thorny environmental issues should take a vow not to exaggerate or distort their positions in attempts to play win–lose or winner-take-all games.

The story of Marion Stoddart (**Core Case Study**) illustrates how these guidelines for cooperation can be applied. Stoddart did not confront businesses and government as enemies. Instead, with positive persistence, she enlisted them as partners. She worked hard with them to build social capital (p. 20), and together they arrived at a win–win solution.

Making the transition to sustainability during your lifetime will require governments and their citizens to rethink their priorities. Figure 23-12 (p. 627) shows that it would take about $245 billion a year for the world to meet basic social and health goals and to provide environmental security. This amounts to about one-eighth of what countries spend each year on environmentally harmful subsidies. Converting that portion of those subsidies to environmentally beneficial subsidies and other forms of aid would go a long way toward solving major social and environmental problems.

Also, this $245 billion per year amounts to a fraction of what is spent on wars. Dwight D. Eisenhower, a World War II general and later president of the United States, offered the following advice in helping us to set our priorities for solving such problems:

> *Every gun that is made, every warship launched, every rocket fired, signifies in the final sense a theft from those who hunger and are not fed, those who are cold and are not clothed . . . I hate war as only a soldier who has lived it can . . . as one who has seen its brutality, its futility, its stupidity.*

The world has the knowledge, technologies, and financial resources to eradicate poverty and malnutrition, eliminate illiteracy, sharply reduce infectious diseases, stabilize human populations, and restore and protect the earth's natural capital. Making the shift to more equitable and environmentally sustainable global and national societies is primarily an economic, political, and ethical decision process. It involves shifting to more sustainable forestry (Figure 10-17, p. 227), agri-

culture (Figure 12-31, p. 307), water resources (Figure 13-23, p. 337), energy resources (Figure 16-33, p. 432), and economies (Figure 23-14, p. 630). It also means slowing climate change (Figure 19-13, p. 515), and educating the public and elected officials about the urgent need to make such a shift. Analysts call for the United States, the wealthiest society that has ever existed, to take the lead in making this transition.

Some say that the call for making this shift is idealistic and unrealistic. Others say that it is unrealistic and dangerous to keep assuming that our present course is sustainable, and they warn that we have precious little time to change (see this chapter's opening quote).

Regardless, a shift is underway. Some thoughtful business and political leaders are realizing that "business as usual" is no longer a viable option (Individu-

als Matter, p. 622). The governments of some countries have begun shifting taxes from income to pollution and carbon emissions. In the United States, various states and cities, choosing not to wait for the federal government, have enacted their own laws and programs to improve energy efficiency, use more renewable energy, and reduce pollution and greenhouse gas emissions.

Making the transition, on an urgent basis, to more sustainable societies is a way to restore hope for the world's people and to avoid undermining the earth's natural capital and the futures of our children and grandchildren. It will depend on the choices made and the actions taken by everyone, from business and political leaders to ordinary citizens, as they work together to sustain the human civilization during this century.

REVISITING The Nashua River and Sustainability

The work of Marion Stoddart and her colleagues in the Nashua River watershed (**Core Case Study**) is a model that could be applied on a global scale to bring about environmentally beneficial change. Environmental scientists increasingly believe that an inclusive, cooperative process, such as that started by Stoddart, is the best way to link science and scientific findings to decisions made when creating environmental policy. Science has been sidelined, at least in much recent U.S. environmental policy-making. But scientists can play a role in the *formation* of policy if they are included in the policy processes from the start, just as they were in Chattanooga, Tennessee (Case Study, p. 21) and in the case of the Nashua River cleanup.

Policy-making can be guided by the four **scientific principles of sustainability** (see back cover). Because environmental and social problems are interrelated, their solutions also are interrelated. For example, relying on solar energy and other renewable energy sources reduces oil and coal use, but it also reduces air

pollution and greenhouse gas emissions, thus helping to slow climate change, protect biodiversity, and reduce the need for military action to protect oil supplies. Likewise, reforestation benefits biodiversity, but also increases aquifer recharge, reduces soil erosion, and slows climate change by removing carbon dioxide from the atmosphere. And efforts to eradicate poverty help us to stabilize populations, sharply reduce hunger and malnutrition, and decrease environmental degradation.

Once we initiate these environmentally positive trends, they will reinforce one another and speed up the transition to more sustainable societies. Beneficial change can occur much more rapidly than we think.

In working to make the earth a better and more sustainable place to live, we should be guided by George Bernard Shaw's reminder that "People who say it cannot be done should not interrupt those who are doing it." Marion Stoddart would agree.

As the wagon driver said when they came to a long hard hill,
"Them that's going on with us get out and push.
Them that ain't get out of the way."

THOMAS FULGHUM

REVIEW

1. Review the Key Questions and Concepts for this chapter on p. 635. Describe what Marion Stoddart did to help clean up the Nashua River in the U.S. state of Massachusetts (**Core Case Study**).

2. What key roles can governments play in improving environmental quality? What is a government **policy?** What is **politics?** What is **environmental policy?** What are the four stages of a policy life cycle in

democracies? What is a **democracy?** Describe two features of democratic governments that hinder their ability to deal with environmental problems. Explain nine principles that decision makers can use in making environmental policy and five strategies for implementing these principles.

3. What are the three branches of government in the United States and what major role does each play? What

is **lobbying?** What are three major environmental laws? Explain why developing environmental policy is a difficult and controversial process. What are four major types of public lands in the United States? Describe the controversy over managing these lands. Describe the controversy over logging in U.S. National Forests.

4. Describe four ways in which individuals in democracies can help to develop or change environmental policy. What does it mean to say that we should *think globally and act locally?* Give an example of such an action. What are four ways to provide environmental leadership?

5. What is **environmental law?** Distinguish among a **statutory law,** an **administrative law,** and a **common law.** What is a **civil suit?** What are the **plaintiff** and the **defendant** in a lawsuit? List six reasons why it is difficult to win an environmental lawsuit. What is a SLAPP? Describe the role that Diane Wilson played in helping to reduce the dumping of toxic chemicals into the bay near where she lived. Distinguish between **arbitration** and **mediation** in dealing with a lawsuit. List three general types of U.S. environmental laws. What is an environmental impact statement?

6. Explain how and why U.S. environmental laws have been under attack since 1980. How effective have the attacks been? Describe Julia "Butterfly" Hill's efforts to save giant redwood trees in California.

7. Describe the roles of grassroots and mainstream environmental organizations and give an example of each type of organization. Describe the role and effectiveness of the Natural Resources Defense Council (NRDC). Give two examples of how students have played strong roles in improving environmental quality.

8. Explain the importance of environmental security, relative to economic and military security. Describe some of the harmful environmental impacts of war. List two pieces of good news and two pieces of bad news about international efforts to deal with global environmental problems. List three problems with global environmental treaties and agreements. Describe roles that corporations can play in helping to achieve environmental sustainability, and give an example of such an effort.

9. Describe efforts by the Netherlands to develop and implement a national environmental plan, or green plan. What are four guidelines for shifting to more environmentally sustainable societies?

10. Describe connections between Marion Stoddart's efforts to restore and protect a river (**Core Case Study**) and the four **scientific principles of sustainability.**

Note: Key Terms are in **bold** type.

CRITICAL THINKING

1. If you could, would you work to improve some aspect of the environment in your local community as Marion Stoddart did (**Core Case Study**)? Would you do anything differently from the way she did it?

2. Pick an environmental problem that affects the area where you live and decide where in the policy cycle (Figure 24-2) the problem could best be placed. Apply the cycle to this problem and describe how the problem has progressed (or will likely progress) through each stage. If your problem has not progressed to the control stage, describe how you think the problem would best be dealt with in that stage.

3. Explain why you agree or disagree with each of the **(a)** nine principles listed on p. 637, which are recommended by some analysts for use in making environmental policy decisions, and **(b)** five suggested strategies for implementing these principles (p. 638).

4. What are two ways in which the scientific process described in Chapter 2 (Figure 2-2, p. 30) parallels the policy life cycle (Figure 24-2)? What are two ways in which they differ?

5. Explain why you agree or disagree with **(a)** each of the four principles that biologists and some economists have suggested for deciding how to use public lands in the United States (pp. 640–641), and **(b)** each of the five suggestions made by developers and resource extractors for managing and using U.S. public lands (pp. 641–642).

6. Do you think that corporations and government bodies are ever justified in filing SLAPP lawsuits (p. 645)? Give three reasons for your answer. Do you think that potential defendants of SLAPP suits should be protected in any way from such suits? Explain.

7. Government agencies can help to keep an economy going or to boost certain types of economic development by, for example, building or expanding a major highway through an undeveloped area. Proponents of such development have argued that requiring environmental impact statements (p. 646) for these projects interferes with efforts to help an economy. Do you agree? Is this a problem? Why or why not?

8. List three ways in which you could apply **Concept 24-6** to making your lifestyle more environmentally sustainable.

9. You are in charge of the country where you live. List the five most important components of your environmental policy.

10. List two questions that you would like to have answered as a result of reading this chapter.

Note: See Supplement 13 (p. S78) for a list of Projects related to this chapter.

DATA ANALYSIS

Choose an environmental issue that you have studied in the course. It could be anything you are passionate about, such as global warming or nuclear power. Conduct a poll of students, faculty, staff, and local residents by asking them the questions below as they relate to your particular environmental issue. Poll as many people as you can in order to get a large sample.

Create categories. For example, note whether each respondent is male or female. By creating such categories, you are placing each person into a *respondent pool.* You can add other questions about age, political leaning, and other factors to refine your pools.

Poll Questions

Question 1 On a scale of 1 to 10, how knowledgeable are you about environmental issue X?

Question 2 On a scale of 1 to 10, how aware are you of ways in which you, as an individual, impact environmental issue X?

Question 3 On a scale of 1 to 10, how important is it for you to learn more about environmental issue X?

Question 4 On a scale of 1 to 10, how sure are you that an individual can have a positive influence on environmental issue X?

Question 5 On a scale of 1 to 10, how sure are you that the government is providing the appropriate level of leadership with regard to environmental issue X?"

1. Collect your data, and analyze your findings to measure any differences among the respondent pools.

2. List any major conclusions you would draw from the data.

3. Publicize your findings on your school's website or local newspaper.

LEARNING ONLINE

Log on to the Student Companion Site for this book at **academic.cengage.com/biology/miller**, and choose Chapter 24 for many study aids and ideas for further reading and research. These include flash cards, practice quizzing, Weblinks, information on Green Careers, and InfoTrac® College Edition articles.

Environmental Worldviews, Ethics, and Sustainability

Biosphere 2—A Lesson in Humility

In 1991, eight scientists (four men and four women) were sealed inside Biosphere 2, a $200 million glass and steel enclosure designed to be a self-sustaining life-support system (Figure 25-1) that would increase our understanding of Biosphere 1: the earth's life-support system.

The sealed system of interconnected domes was built in the desert near Tucson, Arizona (USA). It contained artificial ecosystems including a tropical rain forest, savanna, desert, lakes, streams, freshwater and saltwater wetlands, and a mini-ocean with a coral reef.

Biosphere 2 was designed to mimic the earth's natural chemical recycling systems. Water evaporated from its ocean and other aquatic systems, then condensed to provide rainfall over the tropical rain forest. The precipitation trickled through soil filters into the marshes and the ocean before evaporating again. Human and animal excrement and other wastes were treated and recycled to help support plant growth.

The facility was stocked with more than 4,000 species of plants and animals, including small primates, chickens, cats, and insects, selected to help maintain life-support functions. Sunlight and external natural gas–powered generators provided energy. The Biospherians were to be isolated for 2 years and to raise their own food using intensive organic agriculture. They were to breathe air recirculated by plants and to drink water cleansed by natural recycling processes.

From the beginning, many unexpected problems cropped up and the life-support system began unraveling. Large amounts of oxygen disappeared when soil organisms converted it to carbon dioxide and the Biospherians suffered from headaches, shortness of breath, and the threat of CO_2 and nitrous oxide poisoning. Addi-

tional oxygen had to be pumped in from the outside to keep the Biospherians from suffocating.

Tropical birds died after the first freeze. An ant species got into the enclosure, proliferated, and killed off most of the system's original insect species. As a result, the facility was overrun with cockroaches, katydids, and ants. In total, 19 of the Biosphere's 25 small animal species became extinct. Before the 2-year period was up, all plant-pollinating insects became extinct, thereby dooming to extinction most of the plant species.

Despite many problems, the facility's waste and wastewater were recycled. With much hard work, the Biospherians were also able to produce 80% of their food supply, despite rampant weed growths, spurred by higher CO_2 levels, that crowded out food crops. However, they suffered from persistent hunger and weight loss.

In the end, an expenditure of $200 million failed to maintain a life-support system for eight people for 2 years. Scientists Joel Cohen and David Tilman, who evaluated the project, concluded, "No one yet knows how to engineer systems that provide humans with life-supporting services that natural ecosystems provide for free." Since 2007, the University of Arizona has been leasing the facility for biological research and to provide environmental education for school teachers.

Figure 25-1 Biosphere 2, constructed near Tucson, Arizona, was designed to be a self-sustaining life-support system for eight people sealed into the facility in 1991. The experiment failed because of a breakdown in its nutrient cycling systems.

PRNewsFoto/Huron Valley Travel

25-1 What are some major environmental worldviews?

CONCEPT 25-1 Major environmental worldviews differ on which is more important—human needs and wants, or the overall health of ecosystems and the biosphere.

25-2 What is the role of education in living more sustainably?

CONCEPT 25-2 The first step to living more sustainably is to become environmentally literate, partly by learning from nature.

25-3 How can we live more sustainably?

CONCEPT 25-3A We can live more sustainably by using certain guidelines to convert environmental literacy and concerns into action.

CONCEPT 25-3B We can live more sustainably by living more simply and lightly on the earth and by becoming informed and active environmental citizens.

Note: Supplements 2 (p. S4), 12 (p. S76) and 13 (p. 78) can be used with this chapter.

*The main ingredients of an environmental ethic
are caring about the planet and all of its inhabitants,
allowing unselfishness to control the immediate self-interest that harms others,
and living each day so as to leave the lightest possible footprints on the planet.*

ROBERT CAHN

25-1 What Are Some Major Environmental Worldviews?

▶ **CONCEPT 25-1** Major environmental worldviews differ on which is more important—human needs and wants, or the overall health of ecosystems and the biosphere.

What Is an Environmental Worldview?

People disagree on how serious various environmental problems are and on what we should do about them. These conflicts arise mostly out of differing **environmental worldviews**—how people think the world works and what they believe their role in the world should be. Part of that perceived role is determined by a person's **environmental ethics**—what one believes about what is right and what is wrong in our behavior toward the environment.

Worldviews, including one's religious beliefs, are built from the answers to fundamental questions: Who am I? Why am I here? What should I do with my life? These views provide key principles or values that give us a sense of meaning and purpose and help us to sort out and evaluate the daily flood of information and misinformation that bombards us. People with widely differing environmental worldviews can take the same data, be logically consistent, and arrive at quite differ-

ent conclusions, because they start with different assumptions and values.

Figure 25-2 (p. 662) summarizes the major beliefs of each of three major environmental worldviews.

Some environmental worldviews are *human-centered* (anthropocentric), focusing on the needs and wants of people; others are *life-centered* (biocentric), focusing on individual species, the entire biosphere, or some level in between, as shown in Figure 25-3 (p. 662). As you move from the center to the outside rings in the figure, the worldviews become less human-centered and more life- or earth-centered. Also, some worldviews are based on environmental knowledge and understanding, and some are guided more by one's moral or spiritual beliefs.

A brief overview of various philosophies, the values we attach to different forms of life, and cultural and religious perspectives can help us understand the nature and roots of different environmental worldviews. (Supplement 12, p. S76–S77, discusses the environmental aspects of various philosophies and religions.)

Figure 25-2
Comparison of three major environmental worldviews (**Concept 25-1**).
Questions: Which of these descriptions most closely fits your worldview? Which of them most closely fits the worldviews of your parents?

Environmental Worldviews

Planetary Management

We are apart from the rest of nature and can manage nature to meet our increasing needs and wants.

Because of our ingenuity and technology, we will not run out of resources.

The potential for economic growth is essentially unlimited.

Our success depends on how well we manage the earth's life-support systems mostly for our benefit.

Stewardship

We have an ethical responsibility to be caring managers, or stewards, of the earth.

We will probably not run out of resources, but they should not be wasted.

We should encourage environmentally beneficial forms of economic growth and discourage environmentally harmful forms.

Our success depends on how well we manage the earth's life-support systems for our benefit and for the rest of nature.

Environmental Wisdom

We are a part of and totally dependent on nature, and nature exists for all species.

Resources are limited and should not be wasted.

We should encourage earth-sustaining forms of economic growth and discourage earth-degrading forms.

Our success depends on learning how nature sustains itself and integrating such lessons from nature into the ways we think and act.

Figure 25-3
Environmental worldviews lie on a continuum running from more self- and human-centered (center) to life-, biosphere- or earth-centered (outer rings).

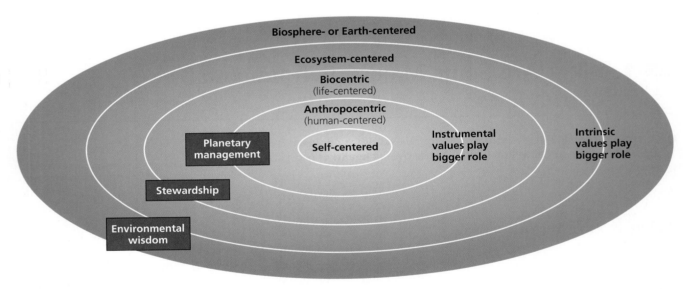

Biosphere- or Earth-centered

Ecosystem-centered

Biocentric
(life-centered)

Anthropocentric
(human-centered)

Planetary management

Self-centered

Stewardship

Environmental wisdom

Instrumental values play bigger role

Intrinsic values play bigger role

Most People Have Human–Centered Environmental Worldviews

One human-centered worldview held by many people is the **planetary management worldview.** Figure 25-2 (left) summarizes the major beliefs of this worldview.

According to this view, we are the planet's most important and dominant species, and we can and should manage the earth mostly for our own benefit. Other species and parts of nature are seen as having only *instrumental value* based on how useful they are to us (Figure 25-3). According to this view of nature, human well-being depends on the degree of control that we have over natural processes. This view holds that, as

the world's most important and intelligent species, we can bypass the biological limitations imposed by nature on other species and redesign the planet and its life-support systems to support us and our ever-growing economies.

Here are three variations of the planetary management environmental worldview:

- The *no-problem school*. We can solve any environmental, population, or resource problem with more economic growth and development, better management, and better technology.

- The *free-market school*. The best way to manage the planet for human benefit is through a free-market

global economy with minimal government interference and regulations. All public property resources should be converted to private property resources and the global marketplace, governed only by free-market competition, should decide essentially everything.

- The *spaceship-earth school*. The earth is like a spaceship: a complex machine that we can understand, dominate, change, and manage, in order to provide a good life for everyone without overloading natural systems. This view was developed after people saw photographs taken from outer space showing the earth as a finite planet, or an island in space (Figure 1-1, p. 5).

Another largely human-centered environmental worldview is the **stewardship worldview.** It assumes that we have an ethical responsibility to be caring and responsible managers, or *stewards*, of the earth. Figure 25-2 (center) summarizes the major beliefs of this worldview.

According to the stewardship view, as we use the earth's natural capital, we are borrowing from the earth and from future generations. We have an ethical responsibility to pay this debt by leaving the earth in at least as good a condition as what we now enjoy. In applying stewardship, parents act in their children's long-term interests to provide a better future for them. When thinking about our responsibility toward future generations, some analysts suggest we consider the wisdom of the 18th century Iroquois Confederation of Native Americans: *In our every deliberation, we must consider the impact of our decisions on the next seven generations.*

Can We Manage the Earth?

Some people believe any human-centered worldview will eventually fail because it wrongly assumes we now have or can gain enough knowledge to become effective managers or stewards of the earth. To biologist and environmental philosopher René Dubos (1901–1982), "The belief that we can manage the earth and improve on nature is probably the ultimate expression of human conceit, but it has deep roots in the past and is almost universal." According to environmental leader Gus Speth, "This view of the world—that nature belongs to us rather than we to nature—is powerful and pervasive—and it has led to much mischief."

According to some critics of human-centered worldviews, the unregulated global free-market approach will not work because it is based on increased degradation and depletion of the earth's natural capital, and it focuses on short-term economic benefits with little regard for the long-term harmful environmental, health, and social consequences.

The image of the earth as an island or spaceship has played an important role in raising global environmental awareness. But critics argue that thinking of the earth as a spaceship that we can manage is an oversimplified, arrogant, misleading way to view an incredibly complex and ever-changing planet. This criticism was supported by the failure of the Biosphere 2 experiment (**Core Case Study**).

Critics of human-centered worldviews point out that we do not even know how many species live on the earth, much less what their roles are and how they interact with one another and with their nonliving environment. We have only an inkling of what goes on in a handful of soil, a meadow, or any other part of the earth. As biologist David Ehrenfeld puts it, "In no important instance have we been able to demonstrate comprehensive successful management of the world, nor do we understand it well enough to manage it even in theory." As a newcomer species, we have much to learn about how the earth works, as illustrated by the failure of the Biosphere 2 project.

Some People Have Life-Centered and Earth-Centered Environmental Worldviews

Critics of human-centered environmental worldviews argue that they should be expanded to recognize the *inherent* or *intrinsic value* of all forms of life, regardless of their potential or actual use to humans. Many of the world's religions have such an inherent respect for life (see Supplement 12, p. S76).

However, people disagree about how far we should extend our ethical concerns for various forms of life (Figure 25-4, p. 664). Most people with a life-centered worldview believe we have an ethical responsibility to avoid causing the premature extinction of species through our activities, for two reasons. *First*, each species is a unique storehouse of genetic information that should be respected and protected simply because it exists (*intrinsic value*). *Second*, each species has potential economic value (*instrumental value*). The latter is also an important part of the more human-centered stewardship environmental worldview.

Some people think we should go beyond focusing mostly on species. They believe we have an ethical responsibility to prevent degradation of the earth's ecosystems, biodiversity, and the biosphere. This *earth-centered* environmental worldview (Figure 25-3) is devoted to preserving the earth's biodiversity and the functioning of its life-support systems for all forms of life, now and in the future. This worldview sees the earth as a single community whose members are bound together through a web of life—a complex and adaptable network of interdependent relationships.

People with earth-centered worldviews believe that humans are not in charge of the world and that human economies are subsystems of the earth's life-support systems (Figure 23-5, p. 615). They understand that the earth's natural capital keeps us and other species

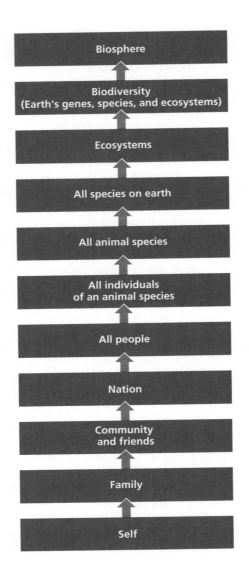

Figure 25-4 Levels of ethical concern. People disagree about how far we should extend our ethical concerns on this scale. **Question:** How far up this scale would you extend your own ethical concerns?

Biosphere

Biodiversity
(Earth's genes, species, and ecosystems)

Ecosystems

All species on earth

All animal species

All individuals
of an animal species

All people

Nation

Community
and friends

Family

Self

alive and supports our economies. They argue that preserving the earth's natural capital requires mimicking nature by applying the four **principles of sustainability** (see back cover) to human economies and lifestyles.

One earth-centered worldview is called the **environmental wisdom worldview.** Figure 25-2 (right) summarizes its major beliefs. According to this view, we are part of—not apart from—the community of life and the ecological processes that sustain all life. Therefore, we should work with the earth to promote environmental sustainability (Figure 25-5) instead of trying to conquer and manage it mostly for our own benefit. In many respects, it is the opposite of the planetary management worldview (Figure 25-2, left).

This worldview is based on the belief that we should strive to care for all species and all humanity. It also suggests that the earth does not need us managing it in order to go on, whereas we need the earth in order to survive. *It is futile to try to save the earth, because it does not need saving.* What we need to save is the existence of our own species and cultures—which have been around for less than an eyeblink of the earth's 4.5-billion-year history—and other species that may become extinct because of our activities.

Another earth-centered environmental worldview is the **deep ecology worldview.** It consists of eight premises developed in 1972 by Norwegian philosopher Arne Naess, in conjunction with philosopher George Sessions and sociologist Bill Devall.

- Each nonhuman form of life on the earth has inherent value, independent of its value to humans.

- The fundamental interdependence of diverse life forms contributes to the flourishing of human and nonhuman life on earth.

Figure 25-5 The earth flag is a symbol of commitment to promoting environmental sustainability by working with the earth at the individual, local, national, and international levels. Chief Seattle (1786–1866), leader of the Suquamish and Duwamish Native American tribes in what is now the U.S. state of Washington, summarized this ethical belief: "The earth does not belong to us. We belong to the earth." **Question:** Explain why you agree or disagree with Chief Seattle's view.

Courtesy of Earth Flag Co.

- Humans have no right to reduce this interdependence and diversity except to satisfy vital needs.

- Present human interference with the nonhuman world is excessive, and the situation is worsening rapidly.

- Because of the damage caused by this interference, it would be better for humans, and much better for other species, if there were a substantial decrease in the human population.

- Basic economic and technological policies must therefore be changed.

- The predominant ideology must change such that measurements of the *quality of life* focus on the overall health of the environment and all living things, rather than on the material wealth of human individuals and societies.

- Those who subscribe to these points have an ethical obligation to try to implement the necessary changes.

Naess also described some lifestyle guidelines compatible with the basic beliefs of deep ecology. They include appreciating all forms of life, consuming less, emphasizing satisfaction of vital needs rather than wants, working to improve the standard of living for the world's poor, taking steps to eliminate injustice toward fellow humans and other species, and acting nonviolently.

Deep ecology is not an eco-religion, nor is it antireligious or antihuman, as some of its critics have claimed. It is a set of beliefs that would have us think more deeply about the inherent value of all life on the earth and about our obligations toward all life. Global and national polls reveal a shift toward the stewardship, environmental wisdom, and deep ecology worldviews.

THINKING ABOUT

Environmental Worldviews and Biosphere 2 CORE CASE STUDY

What environmental worldview is the most compatible with the failure of Biosphere 2 (**Core Case Study**)?

HOW WOULD YOU VOTE? ☑

Which one of the following comes closest to your environmental worldview: planetary management, stewardship, environmental wisdom, or deep ecology? Cast your vote online at **academic.cengage.com/biology/miller**.

25-2 What Is the Role of Education in Living More Sustainably?

▶ **CONCEPT 25-2** The first step to living more sustainably is to become environmentally literate, partly by learning from nature.

We Can Become More Environmentally Literate

There is widespread evidence and agreement that we are a species in the process of degrading our own life-support system and that, during this century, this behavior threatens human civilization and the existence of up to half of the world's species. Part of the problem stems from ignorance about how the earth works, what we are doing to its life-sustaining systems, and how we can change our behavior toward the earth. This begins by understanding three important ideas:

- *Natural capital matters* because it supports the earth's life and our economies.

- *Our threats to natural capital are immense and growing.* Our rapidly expanding ecological footprints (Figure 1-10, p. 15) currently are about 25% larger than the earth's ecological capacity to support us indefinitely and could be 50% larger by 2050.

- *Ecological and climate change tipping points are irreversible and should never be crossed.* Once a tipping point (p. 46) is reached, there is no going back, and neither money nor technology will save us from harmful consequences that can last for hundreds to thousands of years.

Learning how to live more sustainably (**Concept 1-6**, p. 23, and **Concept 2-5B**, p. 44) requires a foundation of environmental education aimed at producing environmentally literate citizens. CONCEPT LINKS

Here are some key goals for each person seeking *environmental literacy:*

- Develop respect or reverence for all life.

- Understand as much as we can about how the earth works and sustains itself (Figure 1-17, p. 23), and use such knowledge to guide our lives, communities, and societies.

- See the big picture by looking for connections within the biosphere and between our activities and the biosphere.

- Understand the relationships between the economy and the earth's natural life-support systems (Figure 23-5, p. 615) and the role of economics in

making the transition to more sustainable economies (Figure 23-14, p. 630) and societies.

- Use critical thinking skills (p. 2) to become seekers of environmental wisdom instead of overfilled vessels of environmental information.

- Understand and explain our environmental worldviews and continue this as a lifelong process.

- Learn how to evaluate the beneficial and harmful environmental consequences that result from our lifestyles and professional choices, today and in the future.

- Foster a desire to make the world a better place for all and act on this desire.

Specifically, an environmentally literate person should have a basic comprehension of the items listed in Figure 25-6.

According to environmental educator Mitchell Thomashow, four basic questions lie at the heart of environmental literacy. *First,* where do the things I consume come from? *Second,* what do I know about the place where I live? *Third,* how am I connected to the earth and to other living things? *Fourth,* what is my purpose and responsibility as a human being? How we answer these questions determines our *ecological identity.*

THINKING ABOUT
Environmental Literacy and Your Ecological Identity
How would you answer the four questions listed above?

We Can Learn from the Earth

Formal environmental education is important, but is it enough? Many analysts say no. They call for us to appreciate not just the economic value of nature, but also its ecological, aesthetic, and spiritual values. To these analysts, the problem is not just a lack of environmental literacy but also a lack of intimacy with nature and an incomplete understanding of how it sustains us.

We face a dangerous paradox. At a time when humans have more power than ever before to disrupt nature, most people know little about, and have little direct contact with, nature. A growing chorus of analysts urge us to escape the cultural and technological body armor we use to insulate ourselves from nature. They call for us to experience and learn directly from nature as a way to understand why we have an ethical obligation to protect the earth's biodiversity and other components of its natural capital in order to sustain our own cultures and economies (**Concept 25-2**).

Some would suggest that we kindle a sense of awe, wonder, mystery, excitement, and humility by standing under the stars, sitting in a forest, or taking in the majesty and power of the sea. We might pick up a handful of soil and try to sense the teeming microscopic life within it. We might look at a tree, mountain, rock, or bee or listen to the sound of a bird and try to sense how they are a part of us and we are a part of them as interdependent participants in the earth's life-sustaining processes.

Such direct experiences with nature reveal parts of the complex web of life that cannot be built with technology, created in a chemical lab, bought with money, or reproduced with genetic engineering. Understanding and directly experiencing the precious gifts we receive from nature can help us to foster an ethical commitment to live more sustainably.

Many psychologists believe that, consciously or unconsciously, we spend much of our lives searching for roots—something to anchor us in a bewildering and frightening sea of change. As philosopher Simone Weil observed, "To be rooted is perhaps the most important and least recognized need of the human soul."

Earth-focused philosophers say that to be rooted, each of us needs to find a *sense of place*—near a stream, a mountain, a neighborhood lot—any piece of the earth with which we can feel as one, a place to know, experience emotionally, and love. According to biologist

Major Components of Environmental Literacy

- Concepts such as environmental sustainability, natural capital, exponential growth, carrying capacity, risk, and risk analysis

- Four scientific principles of sustainablility

- Environmental history (to help us to keep from repeating past mistakes)

- The two laws of thermodynamics and the law of conservation of matter

- Basic principles of ecology, such as food webs, nutrient cycling, biodiversity, ecological succession, and population dynamics

- Human population dynamics

- Ways to sustain biodiversity

- Sustainable agriculture

- Sustainable forestry

- Soil conservation

- Sustainable water use

- Nonrenewable mineral resources

- Nonrenewable and renewable energy resources

- Climate change and ozone depletion

- Pollution prevention and waste reduction

- Sustainable cities

- Environmentally sustainable economic and political systems

- Environmental worldviews and ethics

Figure 25-6 Major components of environmental literacy (**Concept 25-2**). **Question:** After taking this course do you feel that you have a basic understanding of each of these items?

Stephen Jay Gould, "We will not fight to save what we do not love." When we become part of a place, it becomes a part of us.

In other words, find out what you really care about and live a life that shows it. As Mahatma Gandhi, a political and spiritual leader of India put it: "We must become the change we want to see in the world."

This might lead us to recognize that the healing of the earth and the healing of the human spirit are one and the same. We might discover and tap into what Aldo Leopold (Individuals Matter, p. 22) called "the green fire that burns in our hearts" and use this as a force for respecting and working with the earth and with one another.

25-3 How Can We Live More Sustainably?

▶ **CONCEPT 25-3A** We can live more sustainably by using certain guidelines to convert environmental literacy and concerns into action.

▶ **CONCEPT 25-3B** We can live more sustainably by living more simply and lightly on the earth and by becoming informed and active environmental citizens.

Some Guidelines for Living More Sustainably: A Vision Based on Hope

Living more sustainably must be based on respect for the natural processes that keep us alive and support our economies. An overall goal is to protect the earth's precious biodiversity and provide a high quality of life for all life forms. Sustainability is not only about sustaining resources for us; it is about sustaining the entire web of life, because all past, present, and future forms of life are connected.

Figure 25-7 summarizes guidelines, which have been discussed throughout this book, for achieving more sustainable and compassionate societies. They represent ways to convert environmental literacy and concerns into environmentally responsible actions (**Concept 25-3**).

Let us look at some ways in which individuals can put such guidelines into action in their lifestyles.

We Can Live More Simply and Lightly on the Earth

People living in poverty have to use most of their meager income to meet their basic survival needs. This is not the case for affluent people, who are more often concerned about meeting an expanding list of wants beyond their basic needs.

Many analysts urge people who have a habit of consuming excessively to *learn how to live more simply and sustainably.* Seeking happiness through the pursuit of material things is considered folly by almost every major religion and philosophy. Yet, modern advertising persistently encourages people to buy more and more things to fill a growing list of wants as a way to achieve happiness. As American humorist and writer Mark

Twain put it: "Civilization is the limitless multiplication of unnecessary necessities."

According to research by psychologists, deep down what a growing number of people really want is more community, not more stuff. They want greater and more fulfilling interactions with family, friends, and neighbors and a greater opportunity to express their creativity and to have more fun.

SOLUTIONS

Some Guidelines for Living More Sustainably

- Learn about, respect, and mimic how nature sustains itself
- Do not degrade or deplete the earth's natural capital
- Take no more from nature than what nature can replenish
- Do not waste matter and energy resources
- Protect biodiversity
- Avoid climate-changing activities
- Help maintain the earth's capacity for self-repair
- Repair ecological damage that we have caused
- Leave the world in as good a condition as we found or better
- Cultivate a passion for sustaining all life and let this passion energize your actions

Figure 25-7 Some guidelines for achieving more environmentally sustainable and compassionate societies. **Question:** Which three guidelines do you think are the most important? Why?

Living more sustainably is about living well, but within certain limits of consumption. Some affluent people in developed countries are adopting a lifestyle of *voluntary simplicity*. Some are *downshifters* who have decided to reduce their incomes and alter their lifestyles so they can spend more time with family and friends or pursue community or personal interests.

People voluntarily adopting simpler lifestyles are doing and enjoying more with less by using products and services that have a smaller environmental impact (**Concept 25-3B**). Voluntary simplicity applies Mahatma Gandhi's *principle of enoughness:*

> *The earth provides enough to satisfy every person's need but not every person's greed. . . . When we take more than we need, we are simply taking from each other, borrowing from the future, or destroying the environment and other species.*

Most of the world's major religions (see Supplement 12, p. S76) have similar teachings: "You cannot be the slave both of God and money." (Christianity: Matthew, 6:24); "Eat and drink, but waste not by excess." (Islam: Q'uran 7.31); "One should abstain from acquisitiveness." (Hinduism: Acarangasutra 2.119); "He who knows he has enough is rich." (Taoism: Tao Te Ching, Chapter 33); "Give me neither poverty nor riches." (Judaism: Proverbs, 30.8); "Whosoever in this world overcomes his selfish cravings, his sorrows fall away from him." (Buddhism: Dhammapada, 336)

How Much Is Enough?

Living more lightly starts with asking the question: "How much is enough? This is not an easy question to answer, because people in affluent societies are conditioned to want more and more material possessions. And as a result of a lifetime of exposure to commercial advertising, they often think of such wants as needs. For example, in increasingly affluent societies, certain goods that were once considered luxuries—cars, television sets, cell phones, personal computers, and iPods—now are considered necessities by a rapidly growing number of people.

So what are our basic needs? Obviously, each of us needs food, clean air, clean water, shelter, and clothing to keep us alive and in good health. According to various psychologists and other social scientists, beyond these basic needs, each of us also has other qualitative needs, including:

- A secure and meaningful livelihood to provide our basic material needs

- Good physical and mental health

- The opportunity to learn and give expression to our intellectual, mechanical, and artistic talents

- A nurturing family and friends and a peaceful, just, and secure community

- The freedom to make personal choices

- A sense of belonging to and caring for a particular place and community

- A clean and healthy environment that is vibrant with biological and cultural diversity

- An assurance that our children and grandchildren will be able to meet these same basic and qualitative needs

How much of the stuff we are all urged to buy really helps us to meet these needs? This is a difficult but important question. Indeed, psychologists point out that many people buy things in the hope or belief that their purchases will make up for their lack of one or more of the needs listed above.

THINKING ABOUT
Your Basic and Qualitative Needs

How do your needs compare with those listed above? Which needs would you drop or add to this list? Is your list of needs compatible with your environmental worldview?

We Can Each Choose Ways to Live More Lightly

Throughout this text, you have encountered lists of steps you can take to reduce the *size* and *impact* of your ecological footprints on the earth. It would be difficult for most of us to do all or even most of these things, so which ones are the most important?

The human activities that have the greatest harmful impacts on the environment are *agriculture, transportation, home energy use, water use,* and our *overall resource consumption and waste.* Based on this fact, Figure 25-8 lists the *sustainability dozen*—key ways in which some people are choosing to walk more lightly on the earth.

Living more sustainably is not easy. We will not make this transition by relying only on technological fixes such as recycling, buying carbon offsets to reduce our carbon footprints, changing from incandescent to compact fluorescent lightbulbs, and driving energy-efficient hybrid cars. These are important things to do and can help us feel less guilty. But they still involve considerable consumption of matter and energy resources. Some of them are essentially output approaches that are concerned with what to do with the wastes we produce. More effective long-term solutions will start with the question: How do we greatly reduce the amounts of matter and energy resources we use and thus our ecological footprints?

A growing number of individuals are finding self-fullfillment and satisfaction by participating in local efforts to preserve natural areas and restore degraded ones, often in their own neighborhoods. Others are finding such rewards by improving the energy effi-

Insulate your house and plug air leaks

Use renewable energy, especially wind and direct solar

Reduce meat consumption

Use energy-efficient heating and cooling systems, lights, and appliances

Buy locally grown food

Reduce, reuse, and recycle

Buy or grow organic food

Use water-saving appliances and irrigation methods

Don't use pesticides on your garden or lawn

EARTH

Reduce car use

Drive an energy-efficient vehicle

Walk, bike, carpool, or take mass transit whenever possible

Figure 25-8 *The sustainability dozen*—12 ways in which people can live more lightly on the earth (**Concept 25-3B**). **Question:** Which of these things do you already do? Which, if any, do you hope to do?

ciency of their homes, planting organic gardens, and converting monoculture grass lawns to biologically diverse arrays of plants adapted to local climate conditions. They are improving the quality of their lives and walking more lightly on the earth by learning that, for them, less is more.

We Can Become Better Environmental Citizens

In the end, it comes down to what you and I do to make the earth a better place to live for current and future generations, for other species, and for the ecosystems that support us. This means becoming and staying environmentally informed, evaluating and reducing the environmentally harmful aspects of our own lifestyles, and becoming politically involved. We can vote with both our ballots and our pocketbooks to influence governments and businesses to help steer our societies onto a more sustainable path.

We all make some direct or indirect contributions to the environmental problems we face. However, because we do not want to feel guilty about it, we try not to think about it too much—taking a path that can lead to denial, indifference, and inaction. Some suggest that in order to avoid this path, we move beyond blame, guilt, fear, and apathy by recognizing and avoiding some common mental traps. These traps include:

• *gloom-and-doom pessimism* (it is hopeless)

• *blind technological optimism* (science and technofixes will save us)

• *paralysis by analysis* (searching for the perfect worldview, philosophy, scientific information, or solutions before doing anything)

- *faith in simple, easy answers* (which some political leaders tend to offer)

Avoiding these traps can help us to hold on to and be inspired by empowering feelings of realistic hope, rather than to be immobilized by feelings of fear and despair.

THINKING ABOUT
Mental Traps

Which of these traps, if any, have you fallen into? Were you aware you had been ensnared by any of them? How do you think you could free yourself from these traps?

According to primatologist and conservationist Jane Goodall (Figure 9-A, p. 205), "The greatest danger to our future is apathy. . . . Can we overcome apathy? Yes, but only if we have hope. . . . Technology alone is not enough. We must engage with our hearts also." Technological innovations that promote environmental sustainability are necessary and should be encouraged. But by themselves, such innovations will not solve the environmental challenges we face. These innovations must also be accompanied by organizational, political, economic, and cultural changes that promote environmental stainability.

It is also important to recognize that there is no single correct or best solution to each of the environmental problems we face. Indeed, one of nature's four **principles of sustainability** holds that preserving diversity—in this case, being flexible and adaptable in trying a variety of cultural and technological solutions to our problems—is the best way to adapt to the earth's largely unpredictable, ever-changing conditions.

Finally, we should have fun and take time to enjoy life. Laugh every day and enjoy and celebrate nature, beauty, connectedness, friendship, and love. This can empower us to become good earth citizens who practice *good earthkeeping*. As Mahatma Gandhi reminded us, "Power based on love is a thousand times more effective and permanent than power derived from fear."

The Earth Charter Is a Blueprint for Building More Sustainable Societies

After the 1992 Earth Summit in Rio de Janeiro, Brazil, a U.N. Council was established to develop an *Earth Charter* listing the basic principles for environmental protection and environmentally sustainable development (see **www.earthcharter.org**). After many drafts, in March 2000, the Earth Charter was finalized. More than 100,000 people in 51 countries and 25 global leaders in environmental science, business, politics, religion, and education took part in creating this charter.

The charter calls for us to respect and care for life and biodiversity and to build more sustainable, just, democratic, and peaceful societies for present and future generations. Its preamble reflects many of the ideas found in this book:

In an increasingly interdependent world, it is imperative that we, the citizens of the Earth, declare our responsibility to one another, the greater community of life, and future generations. The Earth Community stands at a defining moment. . . . Fundamental changes in our attitudes, values, and ways of living are necessary. A shared vision of basic values is urgently needed to provide an ethical foundation for the emerging world community. . . . As never before in human history, common destiny beckons us to redefine our priorities and seek a new beginning. . . . Every individual, family, organization, business enterprise and government has a critical role to play. . . . We can, if we will, take advantage of the creative possibilities before us and inaugurate an era of fresh hope."

The Earth Charter uses 49 principles as part of an ethical and moral framework to guide the conduct of people and nations toward each other and toward the earth. Here are its four guiding principles:

- Respect earth and life in all its diversity.
- Care for life with understanding, love, and compassion.
- Build societies that are free, just, participatory, sustainable, and peaceful.
- Secure earth's bounty and beauty for present and future generations.

Religion Can Play an Important Role in the Sustainability Revolution

It is encouraging that a growing number of religious leaders and members of all faiths are calling for governments and individuals to focus on the environmental problems we face. They point out that these problems bring a sense of urgency and shared purpose matched by few other issues, and that the teachings of the world's great religions remind us of our moral and spiritual obligations to help sustain the earth for current and future generations (see Supplement 12, p. S76). They see the environment as a great unifier of religious people throughout the world—a force that can help them to strengthen their faith and to learn about the importance of stewardship (Individuals Matter, at right).

According to Gary Gardner, Director of Research at the Worldwatch Institute, "Without ethics, without a deep spiritual respect for life and for the planet that

Sister Diane Cundiff and Her Sister Teachers

In his book *Inspiring Progress: Religions' Contributions to Sustainable Development* (Norton, 2006), Gary T. Gardner described environmental lessons taught at the Colegio Santa Maria, a private religious school in São Paulo, Brazil. It is run by the Sisters of the Holy Cross, and environmental ideas and concerns are integrated throughout the entire curriculum.

According to Sister Diane Cundiff, the school's principal, environmental issues are closely linked to the faith and justice values on which the school is based. The school's teachers are motivated by their religious beliefs and the ethical guidelines of the Earth Charter (p. 670), which connect taking care of people to taking care of the earth. Their curriculum is built around respecting and caring for the community of life, preserving environmental sustainability, and promoting social and economic justice, democracy, nonviolence, and peace.

Here are three examples of the school's environmental instruction. *First,* as part of their science education, students test water quality in a channel running through the campus that carries sewage water coming from a slum 20 kilometers (12 miles) away. To help the people there to stay healthy, students go to the slum and teach them to test the water before drinking it.

Second, students separate the schools garbage into paper, metal, and glass that can be recycled, and they pull out items such as toys and clothing that can be reused.

Third, during a few weeks of their summer vacation, students work with poor people where they live. They teach them how to recycle and reuse materials, how to build a compost heap for recycling food wastes, and how to plant and tend a garden that can help them to reduce their food costs.

With such teachings, the Sisters of Colegio Santa Maria are putting their faith to work for a more sustainable future.

supports us, progress over the next 50 years will likely be little more than progressive decline."

According to this view, religion is a largely untapped but powerful force for helping to bring about a sustainability revolution. Part of this power lies in the sheer numbers of religious followers. It is estimated that 50–75% of the world's population are active members of one of the world's more than 10,000 religions. And about 150 of these religions each have more than a million followers. This includes the world's 2.1 billion Christians, 1.4 billion Muslims, 900 million Hindus, 400 million Buddhists, and 17 million Jews. The number of religious adherents is so large, that mobilizing just a fraction—say 10% of them—to the cause of building more just and environmentally sustainable societies would create a powerful force for education and change.

Despite their significant philosophical differences, most of the world's religions share a common core of moral values such as compassion, love, kindness, and justice. Most of the world's major religions also have a deep reverence and respect for nature and all life as summarized on pp. S76–S77 in Supplement 12.

In recent years, a number of Islamic, Jewish, and Christian groups have worked to protect nature. The World Council of Churches has had an active group working on climate change for over a decade. Since 2004, a movement called *creation care* has been rapidly gaining momentum. In 2006, two dozen scientists and American evangelical leaders met to draft a joint statement calling for immediate action on climate change. In 2007, a group of 85 American evangelical Christian leaders kicked off a campaign to mobilize religious conservatives to work on slowing climate change as a way to practice stewardship of the earth, which is called for in their religious teachings.

We Can Bring About a Sustainability Revolution during Your Lifetime

The industrial revolution took place during the past 275 years. Now in this century, environmental leaders say it is time for an *environmental* or *sustainability revolution.* Figure 1-18 on p. 24 lists some of the major shifts involved in bringing about such a revolution. It would have several interrelated components:

- *Biodiversity protection*—dedication to protecting and sustaining the genes, species, natural systems, and chemical and biological processes that make up the earth's biodiversity.

- A *commitment to eco-efficiency,* based on sharply reducing our waste of matter and energy resources and reducing the amount of matter resources needed to produce the products we use.

- An *energy transformation* devoted to decreasing our dependence on carbon-based, nonrenewable fossil fuels and increasing our reliance on renewable energy from the sun, wind, flowing water, biomass, and geothermal sources, which could also be used to produce clean-burning hydrogen fuel (p. 429). Emphasis should be on supporting renewable energy alternatives with high net energy yields (Figure 15-A, p. 374).

- *Pollution prevention* based on a commitment to reducing pollution and environmental degradation by applying the precautionary principle (p. 210).

- An *emphasis on sufficiency,* dedicated to meeting the basic needs of all people on the planet while

affluent societies learn to live more sustainably by living with less.

- *Demographic equilibrium* based on bringing the size and growth rate of the human population into balance with the earth's ability to support humans and other species sustainably.

- *Economic and political transformations* through which we use economic and political systems to reward environmentally beneficial behavior and to discourage environmentally harmful behavior.

We know what to do. Currently, we have the knowledge and the natural and financial wealth to make a transition to more sustainable lifestyles and societies. But waiting to deal with the environmental problems we face could drain these forms of wealth. As the

Canadian writer Ronald Wright put it in his 2004 book *A Short History of Progress,* "The future of everything we have accomplished since our intelligence evolved will depend on the wisdom of our actions over the next few years. . . . The world has grown too small to forgive us any big mistakes."

Here is some hopeful news. We can change. According to social science research, we need to convince only 5–10% of the people in the world or in a country to bring about a major social change. We are already approaching this critical mass in terms of political awareness of major environmental issues. History also shows that we can change faster than we might think, once we have the courage to leave behind ideas and practices that no longer work. Making this challenging cultural transition will require a combination of environmental wisdom, courage, and an ethical commitment.

REVISITING Biosphere 2 and Sustainability

In Biosphere 2 (**Core Case Study**), scientists tried to create a microcosm of the earth that would help us understand how to live more sustainably. What they learned was that nature is so complex that predicting and controlling what will happen in the environment is essentially impossible.

As we explore different paths to sustainability we must first understand that our lives and societies depend on *natural capital* and that one of the biggest threats to our ways of life is our active role in *natural capital degradation.* With that understanding, we begin the search for *solutions* to difficult environmental problems. Competing interests working together to find the solutions must make *trade-offs,* because this is the essence of the political process.

This requires understanding that *individuals matter.* Virtually all of the environmental progress made during the last few decades occurred because individuals banded together to insist that we can do better. This means also that the journey begins in your own community and with your own lifestyle, because in the final analysis, *all sustainability is local.* This is the meaning of the motto, "Think globally; act locally."

In working to make the earth a better and more sustainable place to live, we should be guided by George Bernard Shaw's reminder that "indifference is the essence of inhumanity" and by historian Arnold Toynbee's observation: "If you make the world ever so little better, you will have done splendidly, and your life will have been worthwhile." Each of us has to decide whether we want to be part of the problem or part of the solution to the environmental challenges we face. What and incredible and exciting time to be alive!

When there is no dream, the people perish.

PROVERBS 29:18

REVIEW

1. Review the Key Questions and Concepts for this chapter on p. 661. Describe the Biosphere 2 project (**Core Case Study**) and the major lessons learned from this project.

2. What is an **environmental worldview?** What are **environmental ethics?** Distinguish among the following environmental worldviews: **planetary management, stewardship, environmental wisdom,** and **deep ecology**

3. List three issues involved in deciding which species to protect from premature extinction as a result of our activities.

Discuss the controversy over whether we can effectively manage the earth.

4. List eight goals for a person seeking environmental literacy. List four questions that lie at the heart of environmental literacy.

5. Describe three ways in which we can learn from the earth. What does it mean to find a *sense of place?*

6. What is sustainability, if not all about sustaining resources for human use? List six guidelines for achieving more sustainable and compassionate societies.

7. Describe the relationship between owning things and being happy. What is *voluntary simplicity?* What are three examples of basic needs? What are five examples of more qualitative needs? List a dozen steps that are among the important steps that can be taken to make the transition to more sustainable societies.

8. Describe three traps that lead to denial, indifference, and inaction concerning the environmental problems we face. What are the four basic principles of the Earth Charter?

9. What role might religion play in making the transition to more sustainable societies? List seven components of the environmental or sustainability revolution.

10. Describe connections between Biosphere 2 (**Core Case Study**), the transition to more environmentally sustainable societies, and the four **scientific principles of sustainability**.

Note: Key Terms are in **bold** type.

CRITICAL THINKING

1. Some analysts argue that the problems with Biosphere 2 (**Core Case Study**) resulted mostly from inadequate design and that a better team of scientists and engineers could make it work. Explain why you agree or disagree with this view.

2. Do you believe that we have an ethical responsibility to leave the earth's natural systems in as good a condition as they are now or better? Explain. List three aspects of your lifestyle that hinder implementing this ideal and three aspects that promote this ideal.

3. This chapter summarized several different environmental worldviews. Go through these worldviews and find the beliefs you agree with, and then describe your own environmental worldview. Which of your beliefs were added or modified as a result of taking this course? Compare your answer with those of your classmates.

4. Explain why you agree or disagree with the following ideas: **(a)** everyone has the right to have as many children as they want; **(b)** all people have a right to use as many resources as they want; **(c)** individuals should have the right to do whatever they want with land they own, regardless of whether such actions harm the environment, their neighbors, or the local community; **(d)** other species exist to be used by humans; **(e)** all forms of life have an intrinsic value and therefore have a right to exist. Are your answers consistent with the beliefs making up your environmental worldview, which you described in question 3?

5. Theologian Thomas Berry calls the industrial–consumer society built on the human-centered, planetary management environmental worldview the "supreme pathology of all history." He says, "We can break the mountains apart; we can drain the rivers and flood the valleys. We can turn the most luxuriant forests into throwaway paper products. We can tear apart the great grass cover of the western plains, and pour toxic chemicals into the soil and pesticides onto the fields, until the soil is dead and blows away in the wind. We can pollute the air with acids, the rivers with sewage, the seas with oil. We can invent computers capable of processing 10 million calculations per second. And why? To increase the volume and speed with which we move natural resources through the consumer economy to the junk pile or the waste heap. . . . If, in these activities, the topography of the planet is damaged, if the environment is made inhospitable for a multitude of living species, then so be it. We are, supposedly, creating a technological wonderworld. . . . But our supposed progress . . . is bringing us to a wasteworld instead of a wonderworld." Explain why you agree or disagree with this assessment.

6. Some analysts believe that trying to gain environmental wisdom by experiencing the earth and forming an emotional bond with its life forms and processes is unscientific, mystical nonsense based on a romanticized view of nature. They believe that gaining a better scientific understanding of how the earth works and inventing or improving technologies are the best ways to achieve sustainability. Do you agree or disagree? Explain.

7. In 1988, Patriach Batholomew, a high-ranking member of the Eastern Orthodox Church caused a stir by proclaiming that human actions against the environment are a new class of sins. He wrote, "For humans to cause species to become extinct and to destroy the biological diversity of God's creation, to degrade the integrity of the earth by causing changes in its climate, by stripping the earth of its natural forests or destroying its wetlands, for humans to contaminate the Earth's waters, its land, its air, and its life with poisonous substances—these are sins." Explain why you agree or disagree with this statement. What effect, if any, might this statement have on the way you view the world and on the choices you make?

8. List the five most important aspects of living on earth that you would want your grandchildren to experience.

9. If you could use television or the Internet to speak to everyone in the world today, what are five pieces of environmental wisdom that you would give in your speech? What beliefs from your environmental worldview and ecological identity influenced your selection of these five items?

10. List two questions that you would like to have answered as a result of reading this chapter.

Note: See Supplement 13 (p. S78) for a list of Projects related to this chapter.

ECOLOGICAL FOOTPRINT ANALYSIS

Working with classmates, conduct an ecological footprint analysis of your campus. Work with a partner, or in small groups, to research and investigate an area of your school, such as material resources–recycling/composting; water use; food service practices; energy use; building management and energy conservation; transportation (both on- and off-campus trips); grounds maintenance; and institutional environmental aware- ness and education. Depending on your school and its location, you may be able to add more areas to your investigation. You may decide to study the campus as a whole, or you may decide to break down the campus into smaller research areas, such as dorms, library, administration buildings, classrooms and class- room buildings, grounds, or other areas.

1. After deciding on your group's research area, conduct your analysis. As part of your analysis, develop a list of questions that will help to determine the ecological im- pact related to your chosen topic. Each question item in the audit could have a range of responses on a scale of 1 (poor) to 10 (excellent).

2. Analyze your results, and share them with the class to determine what can be done to shrink the ecological foot- print of your school.

3. Arrange a meeting with school officials to share your ac- tion plan with them.

LEARNING ONLINE

Log on to the Student Companion Site for this book at **academic.cengage.com/biology/miller**, and choose Chapter 25 for many study aids and ideas for further read- ing and research. These include flash cards, practice quiz- zing, Weblinks, information on Green Careers, and InfoTrac® College Edition articles.

Supplements

LENGTH

Metric

1 kilometer (km) = 1,000 meters (m)
1 meter (m) = 100 centimeters (cm)
1 meter (m) = 1,000 millimeters (mm)
1 centimeter (cm) = 0.01 meter (m)
1 millimeter (mm) = 0.001 meter (m)

English

1 foot (ft) = 12 inches (in)
1 yard (yd) = 3 feet (ft)
1 mile (mi) = 5,280 feet (ft)
1 nautical mile = 1.15 miles

Metric–English

1 kilometer (km) = 0.621 mile (mi)
1 meter (m) = 39.4 inches (in)
1 inch (in) = 2.54 centimeters (cm)
1 foot (ft) = 0.305 meter (m)
1 yard (yd) = 0.914 meter (m)
1 nautical mile = 1.85 kilometers (km)

AREA

Metric

1 square kilometer (km^2) = 1,000,000 square meters (m^2)
1 square meter (m^2) = 1,000,000 square millimeters (mm^2)
1 hectare (ha) = 10,000 square meters (m^2)
1 hectare (ha) = 0.01 square kilometer (km^2)

English

1 square foot (ft^2) = 144 square inches (in^2)
1 square yard (yd^2) = 9 square feet (ft^2)
1 square mile (mi^2) = 27,880,000 square feet (ft^2)
1 acre (ac) = 43,560 square feet (ft^2)

Metric–English

1 hectare (ha) = 2.471 acres (ac)
1 square kilometer (km^2) = 0.386 square mile (mi^2)
1 square meter (m^2) = 1.196 square yards (yd^2)
1 square meter (m^2) = 10.76 square feet (ft^2)
1 square centimeter (cm^2) = 0.155 square inch (in^2)

VOLUME

Metric

1 cubic kilometer (km^3) = 1,000,000,000 cubic meters (m^3)
1 cubic meter (m^3) = 1,000,000 cubic centimeters (cm^3)
1 liter (L) = 1,000 milliliters (mL) = 1,000 cubic centimeters (cm^3)
1 milliliter (mL) = 0.001 liter (L)
1 milliliter (mL) = 1 cubic centimeter (cm^3)

English

1 gallon (gal) = 4 quarts (qt)
1 quart (qt) = 2 pints (pt)

Metric–English

1 liter (L) = 0.265 gallon (gal)
1 liter (L) = 1.06 quarts (qt)
1 liter (L) = 0.0353 cubic foot (ft^3)
1 cubic meter (m^3) = 35.3 cubic feet (ft^3)
1 cubic meter (m^3) = 1.30 cubic yards (yd^3)
1 cubic kilometer (km^3) = 0.24 cubic mile (mi^3)
1 barrel (bbl) = 159 liters (L)
1 barrel (bbl) = 42 U.S. gallons (gal)

MASS

Metric

1 kilogram (kg) = 1,000 grams (g)
1 gram (g) = 1,000 milligrams (μg)
1 gram (g) = 1,000,000 micrograms ((g)
1 milligram (mg) = 0.001 gram (g)
1 microgram (μg) = 0.000001 gram (g)
1 metric ton (mt) = 1,000 kilograms (kg)

English

1 ton (t) = 2,000 pounds (lb)
1 pound (lb) = 16 ounces (oz)

Metric–English

1 metric ton (mt) = 2,200 pounds (lb) = 1.1 tons (t)
1 kilogram (kg) = 2.20 pounds (lb)
1 pound (lb) = 454 grams (g)
1 gram (g) = 0.035 ounce (oz)

ENERGY AND POWER

Metric

1 kilojoule (kJ) = 1,000 joules (J)
1 kilocalorie (kcal) = 1,000 calories (cal)
1 calorie (cal) = 4.184 joules (J)

Metric–English

1 kilojoule (kJ) = 0.949 British thermal unit (Btu)
1 kilojoule (kJ) = 0.000278 kilowatt-hour (kW-h)
1 kilocalorie (kcal) = 3.97 British thermal units (Btu)
1 kilocalorie (kcal) = 0.00116 kilowatt-hour (kW-h)
1 kilowatt-hour (kW-h) = 860 kilocalories (kcal)
1 kilowatt-hour (kW-h) = 3,400 British thermal units (Btu)
1 quad (Q) = 1,050,000,000,000,000 kilojoules (kJ)
1 quad (Q) = 293,000,000,000 kilowatt-hours (kW-h)

Temperature Conversions

Fahrenheit (°F) to Celsius °C):
 °C (°F − 32.0) ÷ 1.80
Celsius (°C) to Fahrenheit (°F):
 °F = (°C × 1.80) + 32.0

Uncertainty, Accuracy, and Precision in Scientific Measurements

How do we know whether a scientific measurement is correct? All scientific observations and measurements have some degree of uncertainty because people and measuring devices are not perfect.

However, scientists take great pains to reduce the errors in observations and measurements by using standard procedures and testing (calibrating) measuring devices. They also repeat their measurements several times, and then find the average value of these measurements.

It is important to distinguish between accuracy and precision when determining the uncertainty involved in a measurement. *Accuracy* is how well a measurement conforms to the accepted correct value for the measured quantity, based on many careful measurements made over a long time. *Precision* is a measure of reproducibility, or how closely a series of measurements of the same quantity agree with one another.

Good accuracy and good precision

Poor accuracy and poor precision

Poor accuracy and good precision

Figure 1 The distinction between accuracy and precision. In scientific measurements, a measuring device that has not been calibrated to determine its accuracy may give precise or reproducible results that are not accurate.

A dartboard analogy (Figure 1) shows the difference between precision and accuracy. Accuracy depends on how close the darts are to the bull's-eye. Precision depends on how close the darts are to each other. Note that good precision is necessary for accuracy but does not guarantee it. Three closely spaced darts may be far from the bull's-eye.

Reading Graphs and Maps (All Chapters)

Graphs and Maps are Important Visual Tools

A *graph* is a tool for conveying information that can be summarized numerically by illustrating the information in a visual format. This information, called *data,* is collected in experiments, surveys, historical studies, and other information-gathering activities. Graphing can be a powerful tool for summarizing and conveying complex information, especially in the ever-expanding fields of environmental science.

In this textbook and accompanying web-based Active Graphing exercises, we use three major types of graphs: line graphs, bar graphs, and pie graphs. Here you will explore each of these types of graphs and learn how to read them. In the web-based Active Graphing exercises, you can try your hand at creating some graphs.

Another important visual tool that can serve the same purpose of communicating complex information is a *map.* Maps can be used to summarize data that vary over small or large areas—from ecosystems to the biosphere, and from backyards to continents. We discuss some aspects of reading maps relating to environmental science at the end of this supplement.

Line Graphs

Line graphs usually represent data that fall in some sort of sequence, such as a series of measurements over time or distance. In most such cases, units of time or distance lie on the horizontal *x-axis.* The possible measurements of some quantity or variable, such as temperature that changes over time or distance, usually lie on the vertical *y-axis.* In Figure 1, the x-axis shows the years between 1950 and 2010, and the y-axis displays the possible values for the annual amounts of oil consumed worldwide in millions of tons, ranging from 0 to 4,000 million (or 4 billion) tons. Usually, the y-axis appears on the left end of the x-axis, although y-axes can appear on the right end, in the middle, or on both ends of the x-axis.

The line on a line graph, sometimes referred to as the *curve,* represents the measurements taken at certain time or distance intervals. In Figure 1, the line represents changes in oil consumption between 1950 and 2007. To find the oil consumption for any year, find that year on the x-axis (a point called the *abscissa*) and run a vertical line from the axis to the consumption line. At the point where your line intersects the consumption line, run a horizontal line to the y-axis. The value at that point on the y-axis, called the *ordinate,* is the amount you are seeking. You can

go through the same process in reverse to find a year in which oil consumption was at a certain point.

Questions

1. What was the total amount of oil consumed in the world in 1990?
2. In about what year between 1950 and 2000 did oil consumption first start declining?
3. About how much oil was consumed in 2007? Roughly how many times more oil was consumed in 2007 than in 1970? How many times more oil was consumed in 2007 than in 1950?

Line graphs have several important uses. One of the most common applications is to compare two or more variables. For example, while Figure 1 shows worldwide oil consumption over a certain time period, Figure 2 enables

us to compare consumption of coal and natural gas during the same period.

Questions

1. In what year was the gap between coal use and natural gas use the widest? In what year was it narrowest? Describe the trend in coal use since 1995.
2. Compare Figures 1 and 2. Among trends in oil, coal, and natural gas use, which one grew the most sharply between 1950 and 1980?

Likewise, we can compare many variables on the same scales. Figure 3 depicts U.S. energy consumption for six different energy resources between 1980 and 2007 and projections to 2030. It uses scales to measure time on the x-axis and energy consumption in quadrillions of British thermal units (a standard unit of energy required to produce a certain amount of heat) on the y-axis.

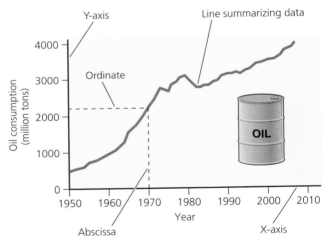

Figure 1 World oil consumption, 1950–2007. (Data from U.S. Energy Information Administration, British Petroleum, International Energy Agency, and United Nations)

Figure 2 World coal and natural gas consumption, 1950–2007. (Data from U.S. Energy Information Administration, British Petroleum, International Energy Agency, and United Nations)

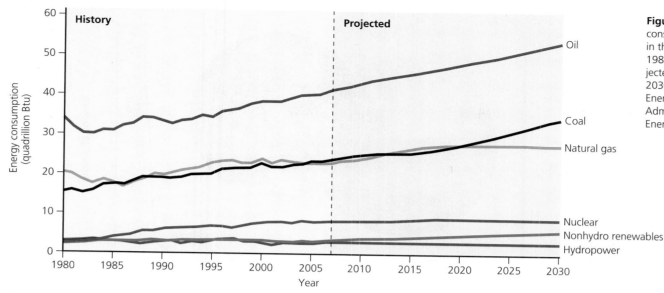

Figure 3 Energy consumption by fuel in the United States, 1980–2007, with projected consumption to 2030. (Data from U.S. Energy Information Administration/Annual Energy Outlook 2007)

Questions

1. Which of the energy sources has seen the least growth in consumption through 2007? Which two sources are projected to have the sharpest growth after 2020? Coal and hydropower are both used mostly to generate electricity. The U.S. consumed about how many times as much coal as hydropower in 2007?

2. Compare Figures 2 and 3. Has the U.S. increased its use of coal and natural gas more sharply or less sharply than the world as a whole?

Figure 4 compares two variables—monthly temperature and precipitation (rain and snowfall) during a typical year in a temperate deciduous forest. However, in this case, the variables are measured on two different scales, so there are two y-axes. The y-axis on the left end of the graph shows a Centigrade temperature scale, while the y-axis on the right shows the range of precipitation measurements in millimeters. The x-axis displays the first letters of each of the 12 month names.

Questions

1. In what month does most precipitation fall? What is the driest month of the year? What is the hottest month?

2. If the temperature curve were almost flat, running throughout the year at about its highest point of about 30 °C, how do you think this forest would be different from what it is? (See Figure 7-15, p. 154.) If the annual precipitation suddenly dropped and remained under 25 centimeters all year, what do you think would eventually happen to this forest?

Line graphs can also show dramatic differences between two types of the same phenomenon, such as growth. *Linear growth* is growth by a given amount in each interval. *Exponential growth* is growth by a fixed percentage of a growing amount in each interval. For example, if you save $1,000 per year under your mattress for 70 years, you will end up with $70,000. But if you invest $1,000 per year and earn 10% interest on your total amount invested every year, and if you keep it all invested for 70 years, you

will end up with $1,024,000. Figure 5 shows the difference between these two types of growth.

Questions

1. How do you think the upper curve would differ if the exponential growth rate were 1% instead of 10%? How would it differ if the rate were 50%? Explain why any exponential growth curve eventually becomes nearly vertical.

Temperate deciduous forest

Figure 4 Climate graph showing typical variations in annual temperature (red) and precipitation (blue) in a temperate deciduous forest.

2. Which one of the curves in Figure 2 most closely resembles the upper curve in Figure 5? What, if anything, might stop that curve from becoming nearly vertical?

It is important to know that line graphs can give different impressions of data, depending on how they are designed. Changing the measurement ranges on either of the x- or y-axes can change the shape of the curve, and two curves

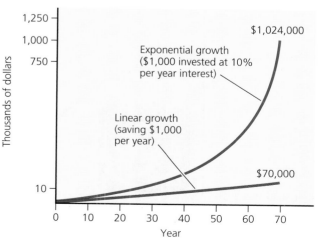

Figure 5 Linear and exponential growth. If resource use, economic growth, or money invested grows exponentially for 70 years at 10%, it will increase 1,024-fold.

representing the same data can look quite differ-ent. For example, in Figure 6, the upper graph shows the growth of the human population from 8000 B.C. to the present, while the lower graph shows human population growth between 1950 and 2010. The latter would be only a small segment on the right end of the curve in the upper graph.

Questions

1. What would be your overall impression of human population growth if you saw only the upper graph? What would be your overall impression if you saw only the lower graph?

2. On the upper graph, mark what you would estimate to be the left and right ends of the segment of the curve that fall between 1950 and 2007. Why do you think the slope (steepness) of this segment varies so much from the slope of the curve in the lower graph? Describe the differences between the two graphs that might explain this difference in slopes.

3. You can see from these graphs that by ad-justing a graph's time span and the height of its y-axis, you can change the slope of the curve. This can give the reader a different first impression of the data. Does this make the changed graph inaccurate or somehow wrong? Explain. What does this tell you about what you need to look for when read-ing a graph?

It is also important to consider what aspect of a data set is being displayed on a graph. The cre-ator of a graph can take two different aspects of one data set and create two very different look-ing graphs that would give two different impres-sions of the same phenomenon. For example, we must be careful when talking about any type of growth to distinguish the question of whether something is growing from the question of how fast it is growing. While a quantity can keep growing continuously, its rate of growth can go up and down.

One of many important examples of growth used in this book is human population growth. Look again at Figure 6. The two graphs in this figure give you the impression that human population growth has been continuous and uninterrupted, for the most part. However con-sider Figure 7, which plots the rate of growth of the human population since 1950. Note that all of the numbers on the y-axis, even the smallest ones, represent growth. The lower end of the scale represents slower growth, the higher end, faster growth.

Questions

1. If this graph were presented to you as a picture of human population growth, what would be your first impression? Do you think that reaching a growth rate of 0.5% would relieve those who are con-cerned about overpopulation? Why or why not?

2. As the curve in Figure 7 proceeds to the right and downward, what do you think will hap-pen to the curve in the lower graph in Fig-ure 6? Explain.

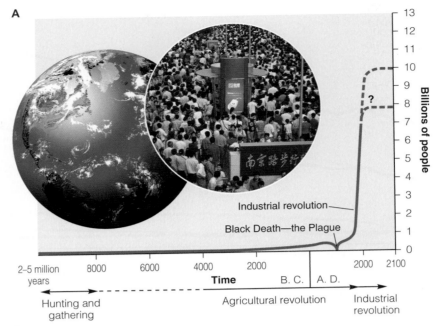

Industrial revolution

Black Death—the Plague

2–5 million years

8000 6000 4000 2000 **Time** B. C. | A. D. 2000 2100

Billions of people

Hunting and gathering

Agricultural revolution

Industrial revolution

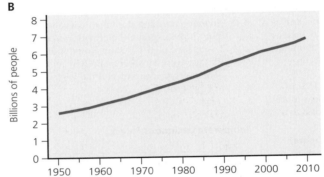

Billions of people

1950 1960 1970 1980 1990 2000 2010

Figure 6 Two graphs show-ing human population growth. The upper graph spans the time between 8000 B.C. and 2100 A.D. The lower graph runs from 1950 A.D. to 2010.

Finally, but no less important, a common scientific use of the line graph is to show experi-mental results. Usually, such graphs represent *variables,* which are factors or values that can change. Experimenters measure changes to a *dependent variable*—a variable that changes in re-sponse to changes in another variable called the *independent variable.* The latter may be manipu-lated by experimenters in order to cause changes in the dependent variable. For example, in the Core Case Study of Chapter 2 (p. 28), we report

on the Hubbard Brook experiment in which sci-entists measured changes over time in the pres-ence of soil nutrients in a forest (the dependent variable) in response to removal of trees from the forest (the independent variable).

On a line graph, the range of values for the independent variable is usually placed on the x-axis, while range of values for the independent variable usually appears on the y-axis (for example, see Figure 17-14, p. 455). However, another way to represent changes

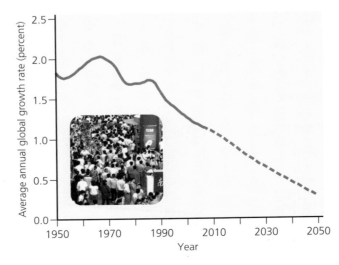

Average annual global growth rate (percent)

1950 1970 1990 2010 2030 2050
Year

Figure 7 Annual growth rate in world population, 1950–2007, with projections to 2050. (Data from U.N. Population Division and U.S. Census Bureau)

to the independent and dependent variables is to show them on two separate curves. This is useful when an experiment takes place over a long period of time, as did the Hubbard Brook experiment. In Figure 8, the years in which this experiment was conducted appear on the x-axis. The range of values for presence of a soil nutrient called nitrate appears on the y-axis. And two curves were plotted: one showing the values of the dependent variable in the uncut forest (the *control* site), and the other showing the values of the dependent variable in the clear-cut forest (the *experimental* site).

Questions

1. Approximately what was the maximum amount of nitrate lost from the undisturbed (control) forest? Approximately what was the maximum amount of nitrate lost from the clear-cut (experimental) forest? At the point of maximum nitrate loss from the experimental forest, about how many times more nitrate was lost there than in the control forest?

2. In what year do you think the experimental forest was cut? How long did it take, once nutrient loss started there, for the losses to reach their maximum? How long did it take for the forest to regain its pre-experimental level of nutrients?

Bar Graphs

The *bar graph* is used to compare measurements for one or more variables across categories. Unlike the line graph, a bar graph typically does not involve a sequence of measurements over time or distance. The measurements compared on a bar graph usually represent data collected at some point in time or during a well-defined period. For instance, we can compare the

Figure 8 Loss of nitrate ions from a deforested watershed (upper curve), mostly due to precipitation washing away the nutrients, compared with loss of nitrate ions from an undisturbed forest (lower curve). (Data from F. H. Bormann and Gene Likens)

net primary productivity (or NPP, a measure of chemical energy produced by plants in an ecosystem) for different ecosystems, as represented in Figure 9.

In most bar graphs, the categories to be compared are laid out on the x-axis, while the range of measurements for the variable under consideration lies along the y-axis. In our example in Figure 9, the categories (ecosystems) are on the y-axis, and the variable range (NPP) lies on the x-axis. In either case, reading the graph is straightforward. Simply run a line perpendicular to the bar you are reading from the top of that bar (or the right or left end, if it lies horizontally) to the variable value axis. In Figure 9, you can see that the NPP for continental shelf, for example, is close to 1,600 kcal/m²/yr.

Questions

1. About how many times greater is the NPP in a tropical rain forest than the NPP in a savannah?

2. What is the most productive of aquatic ecosystems shown here? What is the least productive?

An important application of the bar graph used in this book is the *age structure diagram* (Figure 10, p. S8), which describes a population by showing the numbers of males and females in certain age groups (see pp. 130–132). Environmental scientists are concerned about human population growth, and one of the key factors determining a particular population's growth rate is the relative numbers of people in various age categories.

In particular, the number of women of childbearing age and younger gives an important clue to whether the population might grow rapidly. If most of the women fall into that category, the population will grow much more quickly than it would if most of the women are beyond their child-bearing years. Likewise, a population with most of its women beyond childbearing age will likely shrink in future years.

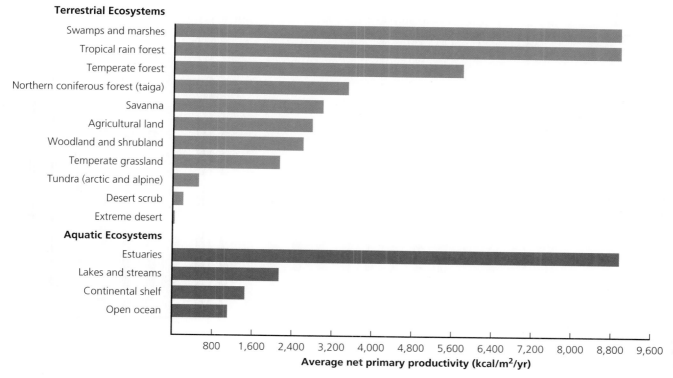

Figure 9
Estimated annual average net primary productivity (NPP) in major life zones and ecosystems, expressed as kilocalories of energy produced per square meter per year (kcal/m²/yr). (Data from R. H. Whittaker, *Communities and Ecosystems*, 2nd ed., New York: Macmillan, 1975)

Note that in Figure 10, the bars are placed horizontally and run both left and right from the center. This allows comparison of two parts of the population across age groups. In this case, values for males lie to the left and values for females lie to the right of the y-axis. The figure contains two separate graphs, one for developing countries and the other for developed countries.

Questions

1. What are the three largest age groups in developing countries? What are the three largest age groups in developed countries? Which group of countries will likely grow more rapidly in coming years?
2. If all girls under the age of 15 had only one child during their lifetimes, how do you think these structures would change over time?

Another interesting application of the bar graph is called a *stacked bar graph*. In such a graph, each bar is actually a combination of several smaller bars representing multiple data groups, stacked up to make one bar divided by different colors or shades to depict the subgroups. Figure 11 is a good example of this, combining population data for six different regions of the world. Each bar contains six different sets of data, all collected in, or projected for, a particular year. This bar graph is a powerful illustration of comparative growth rates of the human populations in the various regions, and it effectively shows the cumulative effects of growth.

Questions

1. Which region will likely grow the fastest between 2010 and 2050? Which region's population will likely stay about the same size?
2. About what proportion of the world's population lived in Asia, Oceania, Africa, and Europe (the eastern hemisphere) and what proportion lived in North America, Latin America, and the Caribbean (the western hemisphere) in 2005? How will these projected proportions change, if at all, by 2050?

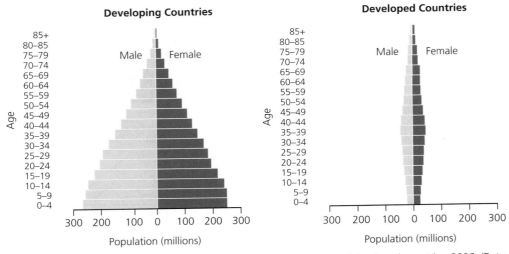

Figure 10 Population structure by age and sex in developing countries and developed countries, 2006. (Data from United Nations Population Division and Population Reference Bureau)

Pie Graphs

Like bar graphs, *pie graphs* illustrate numerical values for two or more categories. But in addition to that, they can also show each category's proportion of the total of all measurements. Usually, the categories are ordered on the graph from largest to smallest, for ease of comparison, although this is not always the case. Also, as with bar graphs, pie graphs are generally snapshots of a set of data at a point in time or during a defined time period. Unlike line graphs, one pie graph cannot show changes over time.

For example, Figure 12 shows how much each major energy source contributes to the world's total amount of energy used in 2006. This graph includes the numerical data used to construct it—the percentages of the total taken up by each part of the pie. But pie graphs can be used without the numerical data included, and such percentages can be estimated roughly. The pie graph thereby provides a generalized picture of the composition of a set of data.

Figure 15-3 (p. 373) shows this and other data in more detail, and illustrates how pie graphs can be used to compare different groups of categories and different data sets. Also, see the Active Graphing exercise for Chapter 15 on the website for this book.

Questions

1. Do you think that the fact that use of natural gas, coal, and oil have all grown over the years (see Figures 1 and 2, p. S4) means that other categories on this graph have shrunk in that time? Explain.
2. Use the projected data in Figure 3 (p. S5) to estimate how the relative sizes of the pie slices for each energy resource might change by 2030.

Reading Maps

Maps can be used for considerably more than showing where places are relative to one another. They can also show comparisons among

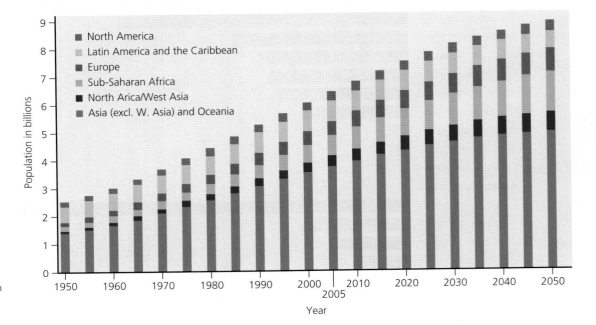

Figure 11 Projected world population growth, by region. (Data from United Nations)

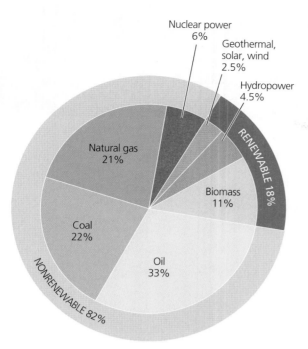

Nuclear power
6%

Geothermal,
solar, wind
2.5%

Hydropower
4.5%

Figure 12 World energy use by source in 2006. (Data from U.S. Department of Energy, British Petroleum, Worldwatch Institute, and Inernational Energy Institute)

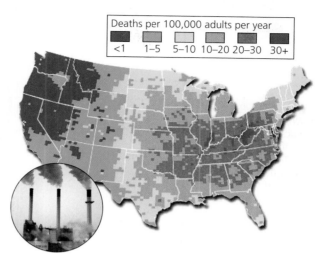

Deaths per 100,000 adults per year

| <1 | 1–5 | 5–10 | 10–20 | 20–30 | 30+ |

Figure 13 Premature deaths from air pollution in the United States, mostly from very small particles added to the atmosphere by coal-burning power plants. (Data from U.S. Environmental Protection Agency)

different areas with regard to any number of factors or conditions. This is the basis for the field of geographical information systems (GIS, see Figure 3-23, p. 73). Using powerful GIS tools, scientists can create detailed maps that show where natural resources are concentrated, for example, within a given region or in the world.

In environmental science, such maps can be very helpful. For example, maps can be used to compare how people or different areas are affected by environmental problems such as air pollution and acid deposition (a form of air pollution). Figure 13 is a map of the United States showing the relative numbers of premature deaths due to

air pollution in the various regions of the country. Figure 14 compares various regions of the country in terms of levels of acidity in precipitation. Study these figures and their captions.

Questions

1. Generally, what part of the country has the lowest level of premature deaths due to air pollution? What part of the country has the highest level? What is the level in the area where you live or go to school?
2. Generally, what part of the country has the highest levels of acidity in its precipitation? What area has the lowest? Do you see any

similarities between the maps in Figures 13 and 14? If so, what are they?

The Active Graphing exercises available for various chapters on the website for this textbook will help you to apply this information. Register and log on to CengageNOW™ using the access code card in the front of your book. Choose a chapter with an Active Graphing exercise, click on the exercise, and begin learning more about graphing. There is also a data analysis exercise at the end of each chapter in this book. Some of these exercises involve analysis of various types of graphs and maps.

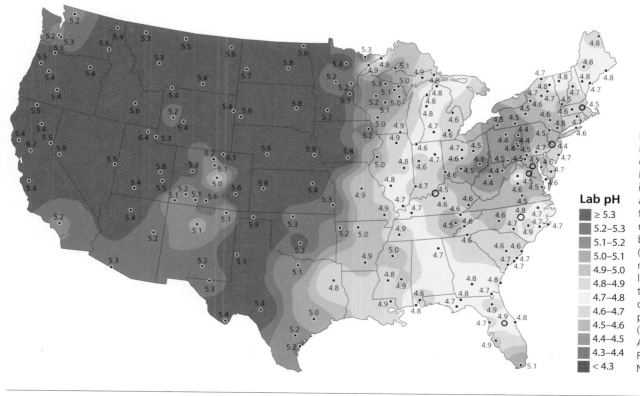

Figure 14 Measurements of the pH of precipitation at various sites in the lower 48 states in 2005 as a result of acid deposition, mostly from a combination of motor vehicles and coal burning power plants (red dots). The pH is a measure of acidity; the lower the pH, the higher the acidity. For more details see Figure 5, p. S41, in Supplement 6. (Data from National Atmospheric Deposition Program/National Trends Network, 2006)

Lab pH

	≥ 5.3
	5.2–5.3
	5.1–5.2
	5.0–5.1
	4.9–5.0
	4.8–4.9
	4.7–4.8
	4.6–4.7
	4.5–4.6
	4.4–4.5
	4.3–4.4
	< 4.3

Economic, Population, Hunger, Health, and Waste Production Data and Maps

Figure 1 Countries of the world.

Map Analysis:

1. What is the largest country (in area) in **(a)** North America, **(b)** Central America, **(c)** South America, **(d)** Europe, and **(e)** Asia?
2. What countries surround **(a)** China, **(b)** Mexico, **(c)** Germany, and **(d)** Sudan?

Figure 2 Gross world product, 1950–2006. (Data from International Monetary Fund and the World Bank)

Data and Graph Analysis

1. Roughly how many times bigger than the gross world product of 1985 was the gross world product in 2006?
2. If the current trend continues, about what do you think the gross world product will be in 2010, in trillions of dollars?

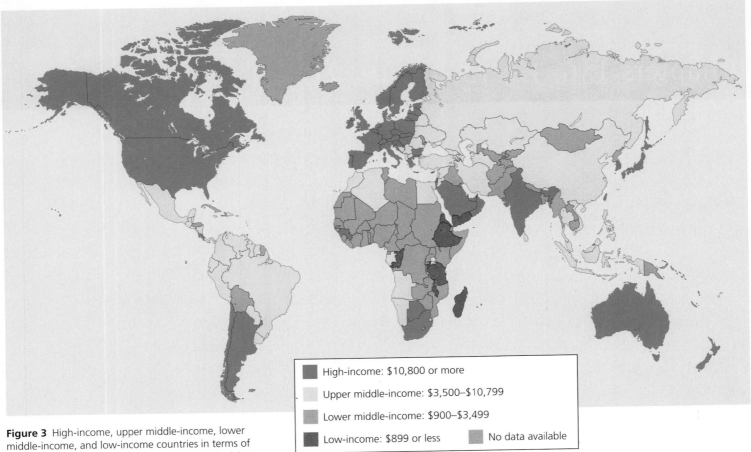

	High-income: $10,800 or more
	Upper middle-income: $3,500–$10,799
	Lower middle-income: $900–$3,499
	Low-income: $899 or less No data available

Figure 3 High-income, upper middle-income, lower middle-income, and low-income countries in terms of gross national income (GNI) PPP per capita (U.S. dollars) in 2006. (Data from World Bank and International Monetary Fund)

Data and Map Analysis

1. In how many countries is the per capita average income $899 or less? Look at Figure 1 and find the names of three of these countries.
2. In how many instances does a lower-middle- or low-income country share a border with a high-income country? Look at Figure 1 and find the names of the countries involved in three of these instances.

Figure 4 Population timeline, 10,000 B.C.–2008.

Data Analysis

1. About how many years did it take the human population to reach 1 billion? How long after that did it take to reach 2 billion?
2. In about what year was the population half of what it is projected to be in 2011?

Year	Event	Human population
50,000 B.C.	Hunter-gatherer societies	1.2 million
10,000 B.C.	End of last Ice Age	4 million
8,000 B.C.	Agricultural Revolution	5 million
2,000 B.C.	Contraceptives in use in Egypt	
500 B.C.		100 million
1,000 A.D.		250 million
1347–1351	Black Death (Plague); 75 million people die	
1500		450 million
1750	Industrial Revolution begins in Europe	791 million
1800	Industrial Revolution begins in the United States	
1804		1 billion
1845–1849	Irish potato famine: 1 million people die	
1927		2 billion
1943	Penicillin used against infection	
1952	Contraceptive pill introduced	
1957	Great famine in China; 20 million die	
1961		3 billion
1974		4 billion
1984		5 billion
1987		6 billion
2011	Projected human population:	7 billion
2024	Projected human population:	8 billion
2042	Projected human population:	9 billion

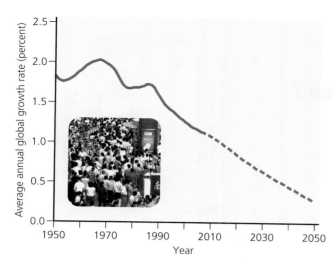

Figure 5 Annual growth rate in world population, 1950–2007 with projections to 2050. (Data from U.N. Population Division and U.S. Census Bureau

Data and Graph Analysis

1. In about what year since 1950 did the growth rate first start increasing? In about what year did it first start decreasing?
2. In about what year will the growth rate reach half of what it was at its peak since 1950, according to projections?

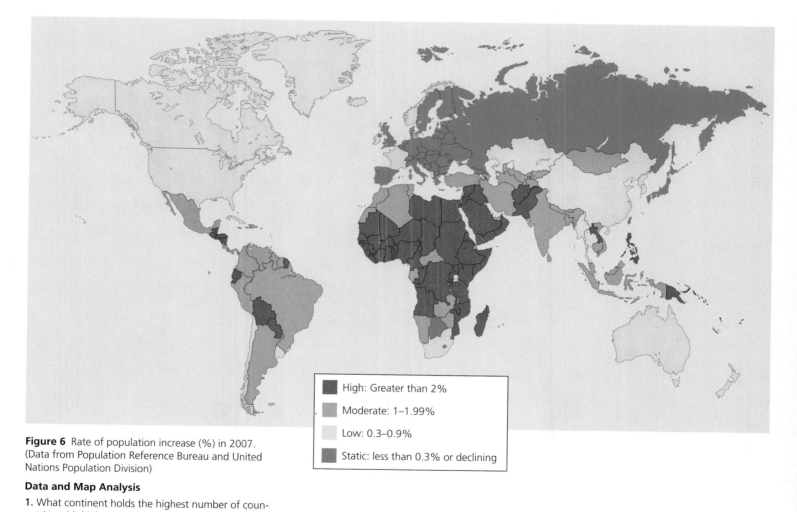

■	High: Greater than 2%
■	Moderate: 1–1.99%
■	Low: 0.3–0.9%
■	Static: less than 0.3% or declining

Figure 6 Rate of population increase (%) in 2007. (Data from Population Reference Bureau and United Nations Population Division)

Data and Map Analysis

1. What continent holds the highest number of countries with high rates of population increase? What continent has the highest number of countries with static rates? (See Figure 1 on pp. S10–S11 for country and continent names.)
2. For each category on this map, name the two countries that you think are largest in terms of total area?

Figure 7 Total fertility rate for the world, developed regions, and less developed regions, 1950–2007, with projection to 2050 (based on medium population projections). (Data from U.N. Population Division)

Data and Graph Analysis

1. What are two conclusions you can draw from comparing these curves?
2. In 2010, about how many more children will be born to each woman in the developing regions than will be born to each woman in the developed regions?

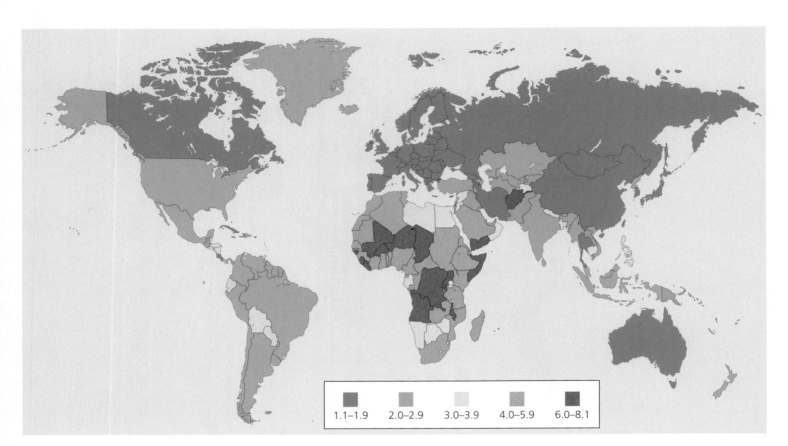

Figure 8 Total fertility rate (TFR), or average number of children born to the world's women throughout their lifetimes, as measured in 2007. (Data from Population Reference Bureau and United Nations Population Division)

Data and Map Analysis

1. Which country in the highest TFR category borders two countries in the lowest TFR category? What are those two countries? (See Figure 1 on pp. S10–S11.)
2. Describe two geographic patterns that you see on this map.

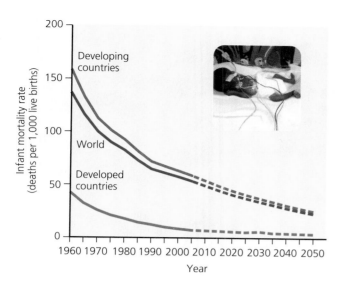

Figure 9 Infant mortality rate for the world, developed regions, and less developed regions, 1950–2007, with projection to 2050 (based on medium population projections). (Data from United Nations Population Division)

Data and Graph Analysis

1. What are two conclusions you can draw from comparing these curves?
2. When the world infant mortality was 100 deaths per 1,000 live births, what were the approximate infant mortality rates for developed and developing regions?

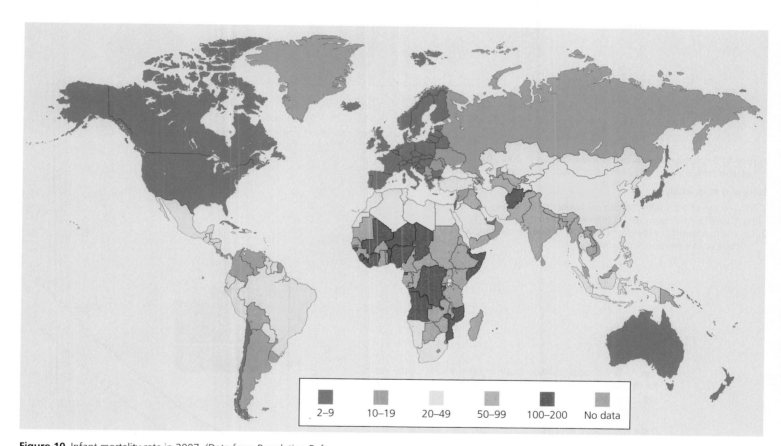

Figure 10 Infant mortality rate in 2007. (Data from Population Reference Bureau and United Nations Population Division)

Data and Map Analysis

1. Describe a geographic pattern that you can see, related to infant mortality rates as reflected on this map.
2. Describe any similarities that you see in geographic patterns between this map and the one in Figure 8.

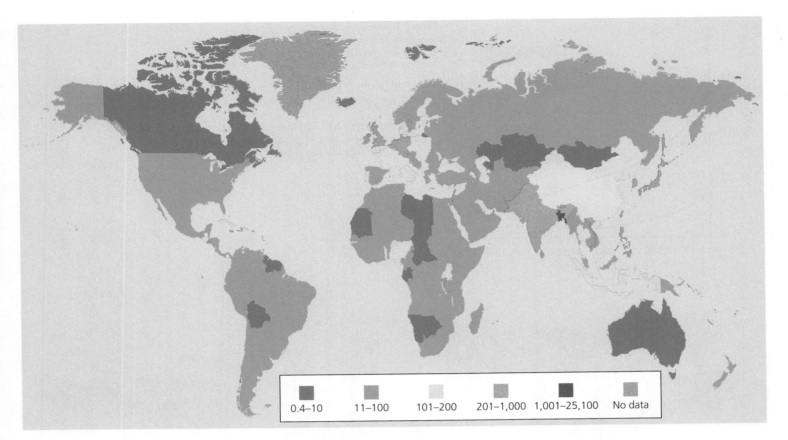

| 0.4–10 | 11–100 | 101–200 | 201–1,000 | 1,001–25,100 | No data |

Figure 11 Population density per square kilometer in 2007. (Data from Population Reference Bureau and United Nations Population Division)

Data and Map Analysis

1. What is the country with the densest population? (See Figure 1 on pp. S10–S11 for country and continent names.)
2. List the continents in order from the most densely populated, overall, to the least densely populated, overall.

Figure 12 Urban population totals and projections for the world, for developing countries, and for developed countries, 1950–2007 with projections to 2030 (Data from United Nations Population Division)

Data and Graph Analysis

1. In about what year did the urban population in developing countries surpass the urban population in developed countries? In about what year was the former twice that of the latter?
2. About how many people will be living in urban areas in developing countries in 2030? In 2030, about how many people will be living in a developing country urban area for every person living in a developed country urban area?

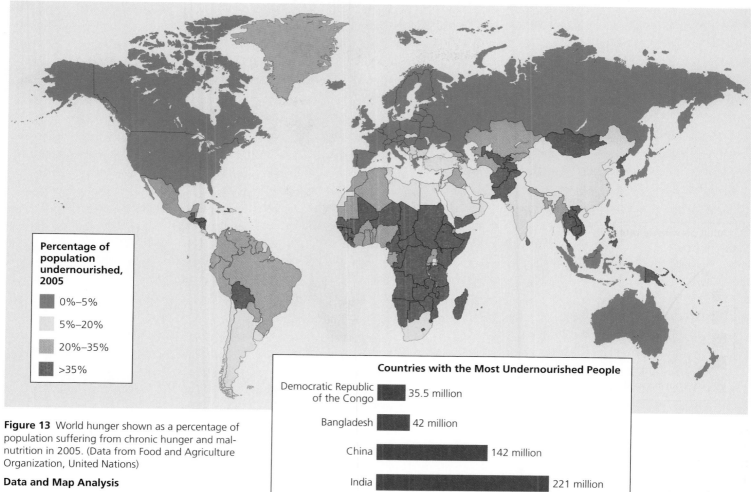

Figure 13 World hunger shown as a percentage of population suffering from chronic hunger and malnutrition in 2005. (Data from Food and Agriculture Organization, United Nations)

Countries with the Most Undernourished People

Democratic Republic of the Congo — 35.5 million
Bangladesh — 42 million
China — 142 million
India — 221 million

Percentage of population undernourished, 2005

- 0%–5%
- 5%–20%
- 20%–35%
- >35%

Data and Map Analysis

1. List the continents in order from the highest percentage of undernourished people to the lowest such percentage. (See Figure 1 on pp. S10–S11 for country and continent names.)

2. On which continent is the largest block of countries that suffer the highest levels of undernourishment? List five of these countries.

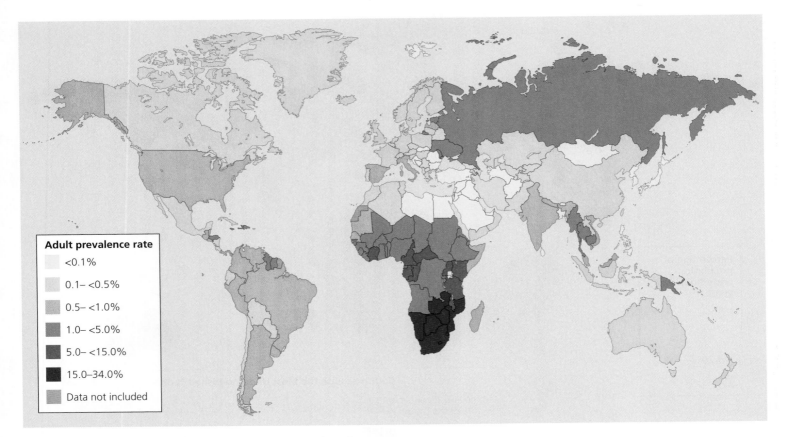

Adult prevalence rate

- [] <0.1%
- [] 0.1– <0.5%
- [] 0.5– <1.0%
- [] 1.0– <5.0%
- [] 5.0– <15.0%
- [] 15.0–34.0%
- [] Data not included

Figure 14 Percentage of adults infected with HIV in 2005. (Data from the World Health Organization)

Data and Map Analysis

1. How many countries have an adult HIV prevalence rate of 5% or higher? List ten of these countries. (See Figure 1 on pp. S10–S11 for country and continent names.)
2. Find an instance where a country with a high rate of HIV prevalence borders a country or countries with a low rate. List the countries involved.

Figure 15 Total and per capita production of municipal solid waste in the United States, 1960–2005. (Data from the U.S. Environmental Protection Agency)

Data and Graph Analysis

1. How much more municipal solid waste was generated in 2005 than in 1960, in millions of tons?
2. In what year did per capita solid waste generation reach a level four times as high as it was in 1960?

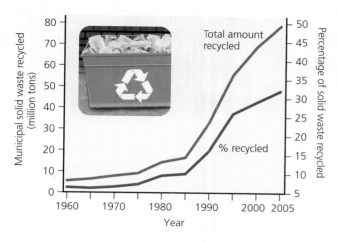

Figure 16 Total amount and percent of municipal solid waste recycled in the United States, 1960–2005. (Data from the U.S. Environmental Protection Agency)

Data and Graph Analysis

1. After 1980, how long did it take for the United States to triple the total amount of materials recycled in 1980?
2. In what 10-year period was the sharpest increase in the percentage of solid waste recycled in the United States?

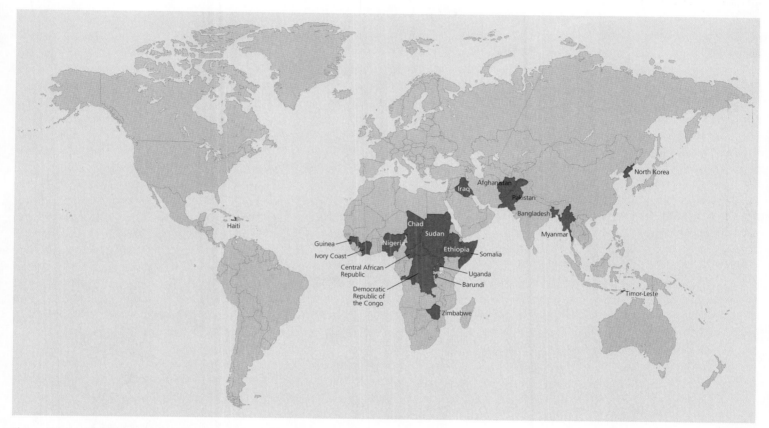

Figure 17 *Top 20 failing states in 2006.* According to a 2007 study by International Alert, 56 countries could become failing or failed states during this century as a result of conflict and instability caused by projected climate change. (Data from the U.S. Central Intelligence Agency, Carnegie Endowment for International Peace, and Fund for Peace)

Biodiversity, Ecological Footprints, and Environmental Performance Maps

Figure 1 Composite satellite view of the earth showing its major terrestrial and aquatic features.

Data and Map Analysis

1. On what continent does desert make up the largest percentage of the continent's total land area?

2. Which two continents contain large areas of polar ice?

NASA Goddard Space Flight Center Image by Reto Stöcki (land surface, shallow water, clouds). Enhancements by Robert Simmon (ocean color, compositing, 3D globes, animation)

Figure 2 Global map of plant biodiversity. (Used by permission from Kier, et al. 2005. "Global Patterns of Plant Diversity and Floristic Knowledge." *Journal of Biogeography,* Vol. 32, Issue 6, pp. 921–1106, and Blackwell Publishing)

Data and Map Analysis

1. What continent holds the largest continuous area of land that hosts more than 5,000 species per eco-region? On what continent is the second largest area of such land? (See Figure 1, pp. S10–S11, in Supplement 3 for the names of countries and continents.)
2. Of the six categories represented by six different colors on this map, which category seems to occupy the most land area in the world (not counting Antarctica, the large land mass on the bottom of the map, and Greenland)?

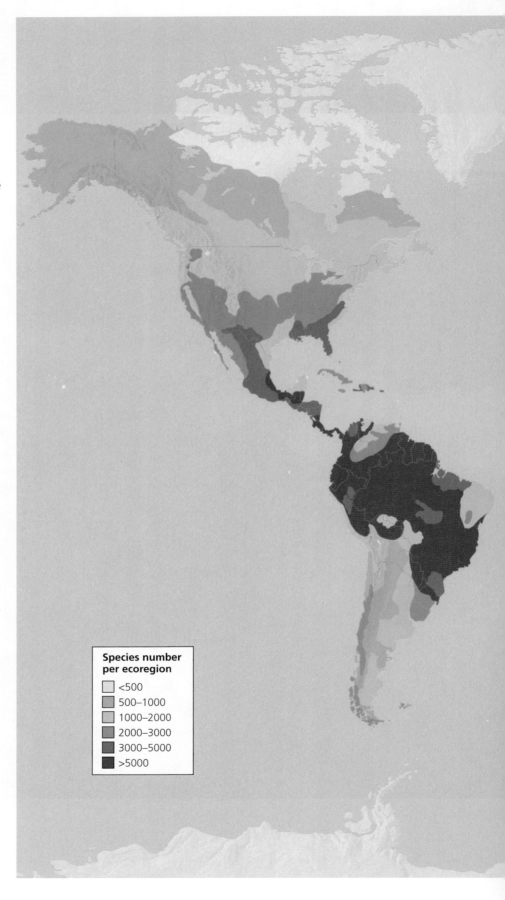

Species number per ecoregion

- <500
- 500–1000
- 1000–2000
- 2000–3000
- 3000–5000
- >5000

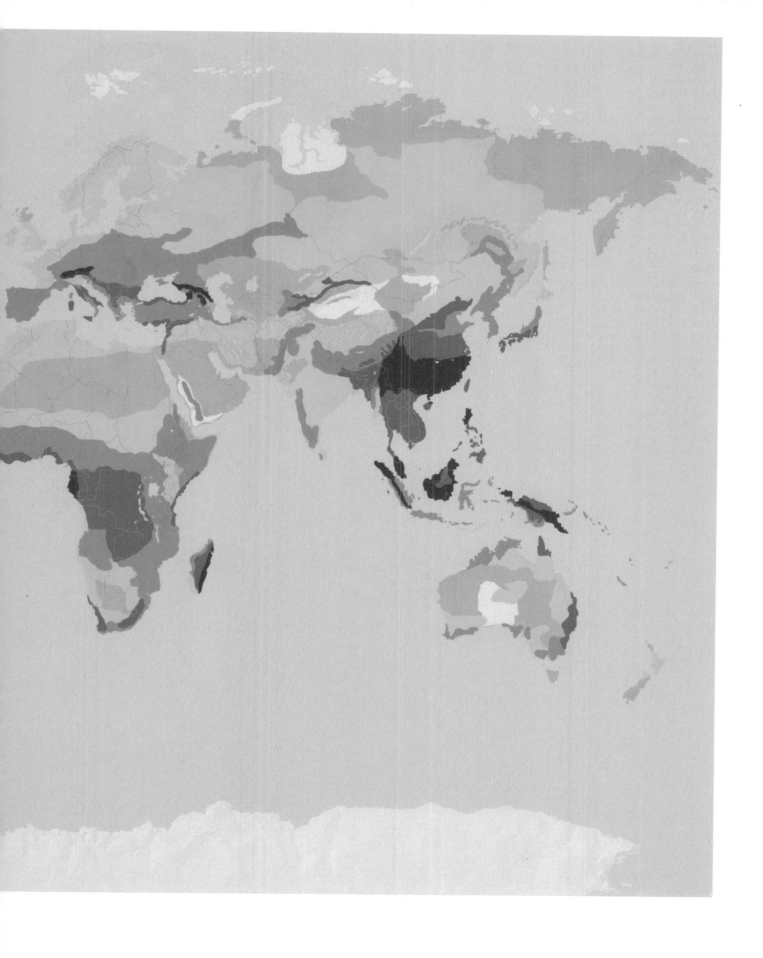

Figure 3 **Natural capital degradation:** the human footprint on the earth's land surface—in effect the sum of all ecological footprints (Figure 1-10, p. 15) of the human population. Colors represent the percentage of each area influenced by human activities. Excluding Antarctica and Greenland, human activities have directly affected to some degree about 83% of the earth's land surface and 98% of the area where it is possible to grow rice, wheat, or corn. (Data from Wildlife Conservation Society and the Center for International Earth Science Information Network at Columbia University)

Data and Map Analysis

1. What is the approximate human footprint value for the area in which you live? List three other countries in the world that have about the same human footprint value as that of the area where you live.
2. Compare this map with that of Figure 2 and list three countries in which the species number per eco-region is 2000 or more and the human footprint value is higher than 40 in parts of each country. (See Figure 1 in Supplement 3, pp. S10–S11, for names of countries and continents.)

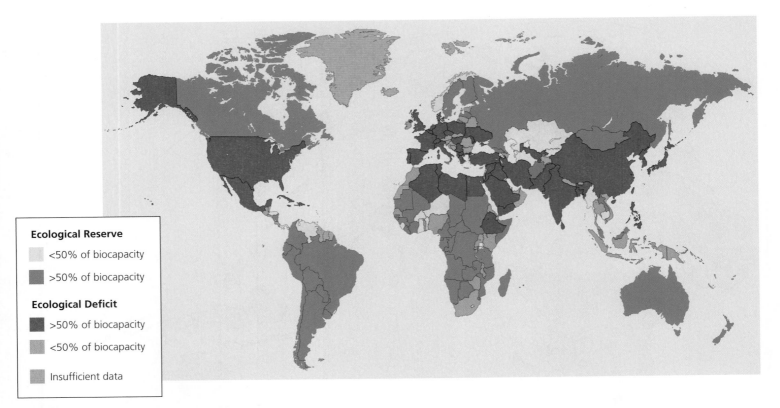

Ecological Reserve

- <50% of biocapacity
- >50% of biocapacity

Ecological Deficit

- >50% of biocapacity
- <50% of biocapacity
- Insufficient data

Figure 4 *Ecological Debtors and Creditors.* The ecological footprints of some countries exceed their biocapacity, while others still have ecological reserves. (Data from Global Footprint Network)

Data and Map Analysis

1. List five countries, including the three largest, in which the ecological deficit is greater than 50% of biocapacity. (See Figure 1, pp. S10–S11, in Supplement 3 for names of countries and continents.)
2. On which two continents does land with ecological reserves of more than 50% of biocapacity occupy the largest percentage of total land area? Look at Figure 3 and for each of these two continents, list the highest human footprint value that you see on the map.

Figure 5 Natural capital: biomes of North America.

Data and Map Analysis

1. What type of biome occupies the most coastal area?
2. Which biome is the rarest in North America?

North American Biomes

- Tundra
- Mountain zone
- Mountain forest
- Taiga
- Temperate deciduous forest
- Tropical forest
- Temperate grassland
- Chaparral
- Desert
- Semidesert

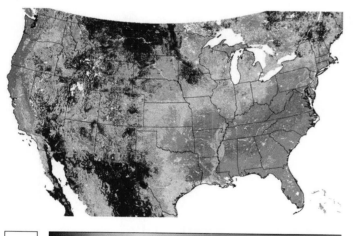

0 0.1 4.5 8.6 12.6 16.7 20.8 24.9 28.9 32.9

Gross primary productivity
(grams of carbon per square meter)

Figure 6 Gross primary productivity across the continental United States, based on remote satellite data. The differences roughly correlate with variations in moisture and soil types. (NASA's Earth Observatory)

Data and Map Analysis

1. Comparing the five northwestern-most states with the five southeastern-most states, which of these regions has the greater variety of levels of gross primary productivity? Which of the regions has the highest levels, overall?
2. Compare this map with that of Figure 5. Which biome in the United States is associated with the highest level of gross primary productivity?

Figure 7 Natural capital degradation: the human ecological footprint in North America. Colors represent the percentage of each area influenced by human activities. This is an expanded portion of Figure 2, which shows the human footprint on the earth's entire land surface. (Data from Wildlife Conservation Society and the Center for International Earth Science Information Network at Columbia University)

Data and Map Analysis

1. Considering the northwest, southwest, middle third, northeast, and southeast regions of the United States, which region of the country has the greatest concentration of areas with the highest human footprint values?
2. Comparing this map with that of Figure 5, which biome in North America is located in areas with the highest overall human footprint values?
3. If you live in North America, what is the approximate value of the human ecological footprint **(a)** where you live and **(b)** where you go to school?

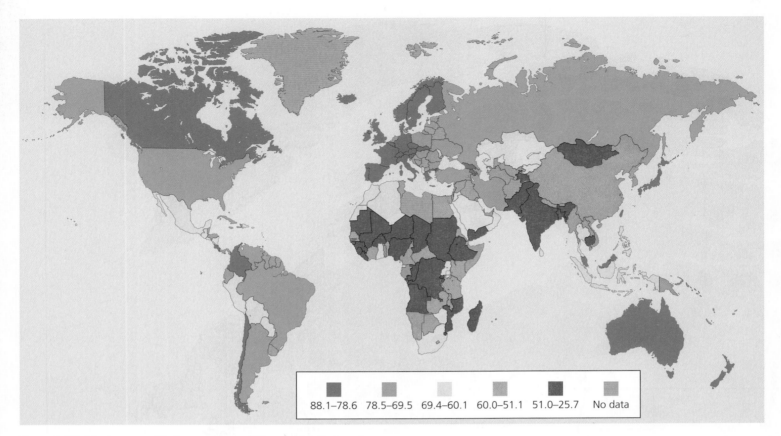

Figure 8 2006 Environmental Performance Index (EPI) uses 16 indicators of environmental health, air quality, water resources, biodiversity and habitat quality, renewable natural resources, and sustainable energy resources to evaluate countries in terms of their ecosystem health and environmental stresses on human health. This map shows the scores by quintiles. (Data from Yale Center for Environmental Law and Policy and the Center for International Earth Science Information Network)

Data and Map Analysis

1. List six countries that lie within the highest EPI range. List six countries that lie within the lowest EPI range. (See Figure 1, pp. S10–S11, in Supplement 3 for country names.)
2. Name a country in the 78.5–69.5 quintile that is completely surrounded by countries in lower quintiles?

Environmental History
(Chapters 1, 2, 5, 7, 8, 10, 12, 13, 23, 24)

S5-1 A Look at Some Past Civilizations

The Norse Greenland Civilization Destroyed Its Resource Base

Greenland is a vast and mostly ice-covered island about three times the size of the U.S. state of Texas. During the 10th century, Viking explorers settled a small, flat portion of this island that was covered with vegetation and located near the water.

In his 2005 book, *Collapse: How Societies Choose to Fail or Succeed,* biogeographer Jared Diamond describes how this 450-year-old Norse settlement in Greenland collapsed in the 1400s from a combination of colder weather in the 1300s and abuse of its soil resources.

Diamond suggests the Norse made three major errors. First, they cut most of the trees and shrubs to clear fields, make lumber, and gather firewood. Without that vegetation, cold winds dried and eroded the already thin soil. The second error was overgrazing, which meant the depletion of remaining vegetation and trampling of the fragile soil.

Finally, when wood used for lumber was depleted, the Norse removed chunks of their turf and used it to build thick walls in their houses to keep out cold winds. Because they removed the turf faster than it could be regenerated, there was less land for grazing so livestock numbers fell. As a result, their food supply and civilization collapsed. Archeological evidence suggests the last residents starved or froze to death.

After about 500 years, nature healed the ecological wounds that the Norse had caused, and Greenland's meadows recovered. In the 20th century, Danes who settled in Greenland reintroduced livestock. Today, more than 56,000 people make their living there by mining, fishing, growing crops, and grazing livestock.

But there is evidence that Greenland's green areas—about 1% of its total land area—are again being overused and strained to their limits. Now Greenlanders have the scientific knowledge to avoid the tragedy of the commons by reducing livestock numbers to a sustainable level, cutting trees no faster than they can be replenished, and practicing soil conservation.

Because of global warming, some of the ice that covers most of Greenland is melting (Figure 19-C, p. 508). This may increase agricultural activity and extraction of mineral and energy resources. Time will tell what beneficial and harmful environmental effects such changes will bring.

The Sumerian Civilization Collapsed Because of Unsustainable Farming

By around 4000 B.C., a highly advanced urban and literate Sumerian civilization had begun emerging on the flood plains of the lower reaches of the Tigris and Euphrates Rivers in parts of what is now Iraq. This civilization developed science and mathematics and built a well-engineered crop irrigation system, which used dams to divert water from the Euphrates River through a network of gravity-fed canals.

The irrigated cropland produced a food surplus and allowed Sumerians to develop the world's first cities and written language (the cunneiform script). But the Sumerians also learned the painful lesson that long-term irrigation can lead to salt buildup in soils and sharp declines in food production.

Poor underground drainage slowly raised the water table to the surface, and evaporation of the water left behind salts that sharply reduced crop productivity—a form of environmental degradation we now call *soil salinization* (Figure 12-14, p. 289). As wheat yields declined, the Sumerians slowed the salinization by shifting to more salt-tolerant barley. But as salt concentrations continued to increase, barley yields declined and food production was undermined.

Around 2000 B.C., this once-great civilization disappeared as a result of such environmental degradation, along with economic decline and invasion by Semitic peoples.

Iceland Has Had Environmental Struggles and Triumphs

Iceland is a Northern European island country slightly smaller than the U.S. state of Kentucky. This volcanic island is located in the North Atlantic Ocean just south of the Arctic Circle between Greenland and Norway, Ireland, and Scotland. Glaciers cover about 10% of the country, and it is subject to earthquakes and volcanic activity.

Immigrants from Scandinavia, Ireland, and Scotland began settling the country during the late 9th and 10th centuries A.D. Since these settlements began, most of the country's trees and other vegetation have been destroyed and about half of its original soils have eroded into the sea. As a result, Iceland suffers more ecological degradation than any other European country.

The early settlers saw what appeared to be a country with deep and fertile soils, dense forests, and highland grasslands similar to those in their native countries. They did not realize that the soils built up by ash from volcanic eruptions were replenished very slowly and were highly susceptible to water and wind erosion when protective vegetation was removed for growing crops and grazing livestock. Within a few decades, the settlers degraded much of this natural capital that had taken thousands of years to build up.

When the settlers realized what was happening, they took corrective action to save the remaining trees, and they stopped raising ecologically destructive pigs and goats. The farmers joined together to slow soil erosion and preserve their grasslands. They estimated how many sheep the communal highland grasslands could sustain and divided the allotted quotas among themselves.

Icelanders also learned how to fish, how to tap into an abundance of hot springs and heated rock formations for geothermal power, and how to use hydroelectric power from the many rivers. Renewable hydropower and geothermal energy provide about 95% of the country's electricity, and geothermal energy is used to heat 80% of its buildings and to grow most of its fruits and vegetables in greenhouses.

In terms of per capita income, Iceland is one of the world's ten richest countries, and in 2006 it had the world's fifth highest Environmental Sustainability Index. By 2050, Iceland plans to become the world's first country to run its entire economy on renewable hydropower, geothermal energy, and wind, and to use these resources to produce hydrogen for running all of its motor vehicles and ships (see Chapter 16 Core Case Study, p. 399).

> **THINKING ABOUT**
> **Past Civilizations**
>
> What are two ecological lessons that we could learn from these three stories?

S5-2 An Overview of U.S. Environmental History

There Have Been Four Major Eras of U.S. Environmental History

The environmental history of the United States can be divided into four eras. During the *tribal era,* 5–10 million tribal people (now called Native Americans) occupied North America for at least 10,000 years before European settlers began arriving in the early 1600s. These hunter–gatherers generally had sustainable, low-impact ways of life because of their low numbers and low resource use per person.

Next was the *frontier era* (1607–1890) when European colonists began settling North America. Faced with a continent offering seemingly inexhaustible resources, the early colonists developed a **frontier environmental worldview.** They saw a wilderness to be conquered and managed for human use.

Next came the *early conservation era* (1832–1870), which overlapped the end of the frontier era. During this period some people became alarmed at the scope of resource depletion and degradation in the United States. They argued that part of the unspoiled wilderness on public lands should be protected as a legacy to future generations. Most of these warnings and ideas were not taken seriously.

This period was followed by an era—lasting from 1870 to the present—featuring an increased role of the federal government and private citizens in resource conservation, public health, and environmental protection.

The Frontier Era (1607–1890)

During the frontier era, European settlers spread across the land, cleared forests for cropland and settlements, and displaced the Native Americans who generally had lived on the land sustainably for thousands of years.

The U.S. government accelerated this settling of the continent and use of its resources by transferring vast areas of public land to private interests. Between 1850 and 1890, more than half of the country's public land was given away or sold cheaply by the government to railroad, timber, and mining companies, land developers, states, schools, universities, and homesteaders to encourage settlement. This era came to an end when the government declared the frontier officially closed in 1890.

Early Conservationists (1832–1870)

Between 1832 and 1870, some people became alarmed at the scope of resource depletion and degradation in the United States. They urged the government to preserve part of the unspoiled wilderness on public lands owned jointly by all people (but managed by the government) and to protect it as a legacy to future generations.

Two of these early conservationists were Henry David Thoreau (1817–1862) and George Perkins Marsh (1801–1882). Thoreau (Figure 1) was alarmed at the loss of numerous wild species from his native eastern Massachusetts. To gain a better understanding of nature, he built a cabin in the woods on Walden Pond near Concord, Massachusetts, lived there alone for 2 years, and wrote *Life in the Woods,* an environmental classic.*

In 1864, George Perkins Marsh, a scientist and member of Congress from Vermont, published *Man and Nature,* which helped legislators and citizens see the need for resource conservation. Marsh questioned the idea that the country's resources were inexhaustible. He also used scientific studies and case studies to show how the rise and fall of past civilizations were linked to the use and misuse of their soils, water supplies, and other resources. Some of his resource conservation principles are still used today.

What Happened between 1870 and 1930?

Between 1870 and 1930, a number of actions increased the role of the federal government and private citizens in resource conservation and public health (Figure 2). The *Forest Reserve Act of 1891* was a turning point in establishing the responsibility of the federal government for protecting public lands from resource exploitation.

In 1892, nature preservationist and activist John Muir (1838–1914)(Figure 3) founded the Sierra Club. He became the leader of the *preservationist movement,* which called for protecting large areas of wilderness on public lands from human exploitation, except for low-impact recreational activities such as hiking and camping. This idea was not enacted into law until 1964. Muir also proposed and lobbied for creation of a national park system on public lands.

Mostly because of political opposition, effective protection of forests and wildlife did not begin until Theodore Roosevelt (Figure 4, p. S34), an ardent conservationist, became president. His term of office, 1901–1909, has been called the country's *Golden Age of Conservation.*

While in office he persuaded Congress to give the president power to designate public land as federal wildlife refuges. In 1903, Roosevelt established the first federal refuge at Pelican Island off the east coast of Florida for preservation of the endangered brown pelican (Photo 1 in the Detailed Contents), and he added 35 more reserves by 1904. He also more than tripled the size of the national forest reserves.

In 1905, Congress created the U.S. Forest Service to manage and protect the forest reserves. Roosevelt appointed Gifford Pinchot (1865–1946) as its first chief. Pinchot pioneered scientific management of forest resources on public lands. In 1906, Congress passed the *Antiquities Act,* which allows the president to protect areas of scientific or historical interest on federal lands as national monuments. Roosevelt used this act to protect the Grand Canyon and other areas that would later become national parks.

Congress became upset with Roosevelt in 1907, because by then he had added vast tracts to the forest reserves. Congress passed a law banning further executive withdrawals of public forests. On the day before the bill became law, Roosevelt defiantly reserved another large block of land. Most environmental historians view Roosevelt (a Republican) as the country's best environmental president.

Figure 1 Henry David Thoreau (1817–1862) was an American writer and naturalist who kept journals about his excursions into wild areas in parts of the northeastern United States and Canada and at Walden Pond in Massachusetts. He sought self-sufficiency, a simple lifestyle, and a harmonious coexistence with nature.

Early in the 20th century, the U.S. conservation movement split into two factions over how public lands should be used. The *wise-use,* or *conservationist,* school, led by Roosevelt and Pinchot, believed all public lands should be managed wisely and scientifically to provide needed resources. The *preservationist* school, led by Muir wanted wilderness areas on public lands to be left untouched. This controversy over use of public lands continues today.

In 1916, Congress passed the *National Park Service Act.* It declared that parks are to be maintained in a manner that leaves them unimpaired for future generations. The act also established the National Park Service (within the Department of the Interior) to manage the system. Under its first head, Stephen T. Mather (1867–1930), the dominant park policy was to encourage tourist visits by allowing private concessionaires to operate facilities within the parks.

After World War I, the country entered a new era of economic growth and expansion. During the Harding, Coolidge, and Hoover administrations, the federal government promoted increased sales of timber, energy, mineral and other resources found on public lands at low prices to stimulate economic growth.

President Herbert Hoover (a Republican) went even further and proposed that the federal government return all remaining federal lands to the states or sell them to private interests for economic development. But the Great Depression (1929–1941) made owning such lands unattractive to state governments and private investors. The depression was bad news for the country. But some say that without it we might have little if any of the public lands that make up about one-third of the country's land today (Figure 24-5, p. 641).

1891–97 Timber cutting banned on large tracts of public land.

1920–27 Public health boards established in most cities.

1880
Killer fog in London kills 700 people.

1893
Few remaining American bison given refuge in Yellowstone National Park.

1918 Migratory Bird Act restricts hunting of migratory birds.

1872
Yellowstone National Park established. American Forestry Association organized by private citizens to protect forests. American Public Health Association formed.

1916 National Park Service Act creates National Park System and National Park Service.

1892 Sierra Club founded by John Muir to promote increased preservation of public land. Killer fog in London kills 1,000 people.

1915 Ecologists form Ecological Society of America.

1912 Public Health Service Act authorizes government investigation of water pollution.

1870
First official wildlife refuge established at Lake Merritt, California.

1911 Weeks Act allows Forest Service to purchase land at headwaters of navigable streams as part of National Forest System.

1891
Forest Reserve Act authorized the president to set aside forest reserves.

1908 Swedish chemist Svante Arrhenius argues that increased emissions from burning fossil fuels will lead to global warming.

1906 Antiquities Act allows president to set aside areas on federal lands as national monuments. Pure Food and Drug Act enacted.

1890
Government declares the country's frontier closed. Yosemite National Park established.

1905 U.S. Forest Service created. Audubon Society founded by private citizens to preserve country's bird species.

1904 Child lead poisoning linked to lead-based paints.

1903 First National Wildlife Refuge at Pelican Island, Florida, established by President Theodore Roosevelt.

1902 Reclamation Act promotes irrigation and water development projects in arid West.

1900 Lacey Act bans interstate shipment of birds killed in violation of state laws.

1870–1930

Figure 2 Examples of the increased role of the federal government in resource conservation and public health and the establishment of key private environmental groups, 1870–1930.
Question: Which two of these events do you think were the most important?

Figure 3 John Muir (1838–1914) was a geologist, explorer, and naturalist. He spent 6 years studying, writing journals, and making sketches in the wilderness of California's Yosemite Valley and then went on to explore wilderness areas in Utah, Nevada, the Northwest, and Alaska. He was largely responsible for establishing Yosemite National Park in 1890. He also founded the Sierra Club and spent 22 years lobbying actively for conservation laws.

Figure 4 Theodore (Teddy) Roosevelt (1858–1919) was a writer, explorer, naturalist, avid birdwatcher, and twenty-sixth president of the United States. He was the first national political figure to bring the issues of conservation to the attention of the American public. According to many historians, he has contributed more than any other president to natural resource conservation in the United States.

What Happened between 1930 and 1960?

A second wave of national resource conservation and improvements in public health began in the early 1930s (Figure 5) as President Franklin D. Roosevelt (1882–1945) strove to bring the country out of the Great Depression. He persuaded Congress to enact federal government programs to provide jobs and to help restore the country's degraded environment.

During this period, the government purchased large tracts of land from cash-poor landowners, and established the *Civilian Conservation Corps* (CCC) in 1933. It put 2 million unemployed people to work planting trees and developing and maintaining parks and recreation areas. The CCC also restored silted waterways and built levees and dams for flood control.

The government built and operated many large dams in the Tennessee Valley and in the arid western states, including Hoover Dam on the Colorado River (Figure 13-14, p. 327). The goals were to provide jobs, flood control, cheap irrigation water, and cheap electricity for industry.

Congress enacted the Soil Conservation Act in 1935. It established the *Soil Erosion Service* as part of the Department of Agriculture to correct the enormous erosion problems that had ruined many farms in the Great Plains states during the depression, as discussed on pp. 303–305. Its name was later changed to the *Soil Conservation Service,* now called the *Natural Resources Conservation Service.* Many environmental historians praise Roosevelt (a Democrat) for his efforts to get the country out of a major economic depression and to help restore environmentally degraded areas.

Federal resource conservation and public health policy during the 1940s and 1950s changed little, mostly because of preoccupation with World War II (1941–1945) and economic recovery after the war.

Between 1930 and 1960, improvements in public health included establishment of public health boards and agencies at the municipal, state, and federal levels; increased public education about health issues; introduction of vaccination programs; and a sharp reduction in the incidence of waterborne infectious diseases, mostly because of improved sanitation and garbage collection.

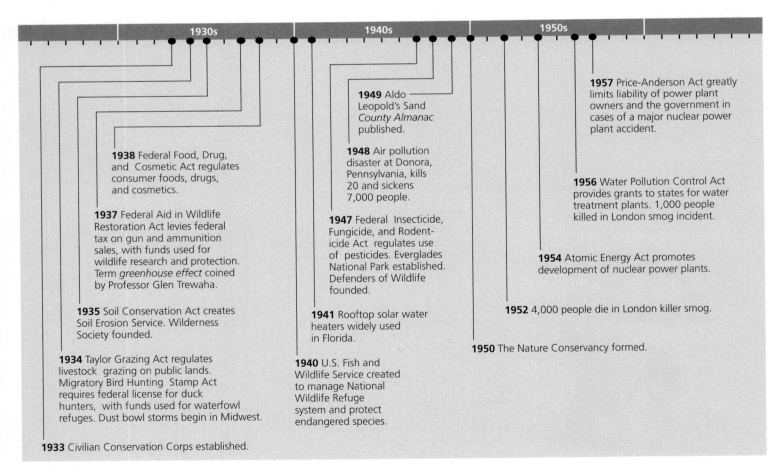

1930s **1940s** **1950s**

1957 Price-Anderson Act greatly limits liability of power plant owners and the government in cases of a major nuclear power plant accident.

1949 Aldo Leopold's Sand *County Almanac* published.

1948 Air pollution disaster at Donora, Pennsylvania, kills 20 and sickens 7,000 people.

1938 Federal Food, Drug, and Cosmetic Act regulates consumer foods, drugs, and cosmetics.

1956 Water Pollution Control Act provides grants to states for water treatment plants. 1,000 people killed in London smog incident.

1937 Federal Aid in Wildlife Restoration Act levies federal tax on gun and ammunition sales, with funds used for wildlife research and protection. Term *greenhouse effect* coined by Professor Glen Trewaha.

1947 Federal Insecticide, Fungicide, and Rodenticide Act regulates use of pesticides. Everglades National Park established. Defenders of Wildlife founded.

1954 Atomic Energy Act promotes development of nuclear power plants.

1935 Soil Conservation Act creates Soil Erosion Service. Wilderness Society founded.

1941 Rooftop solar water heaters widely used in Florida.

1952 4,000 people die in London killer smog.

1934 Taylor Grazing Act regulates livestock grazing on public lands. Migratory Bird Hunting Stamp Act requires federal license for duck hunters, with funds used for waterfowl refuges. Dust bowl storms begin in Midwest.

1940 U.S. Fish and Wildlife Service created to manage National Wildlife Refuge system and protect endangered species.

1950 The Nature Conservancy formed.

1933 Civilian Conservation Corps established.

1930–1960

Figure 5 Some important conservation and environmental events, 1930–1960. **Question:** Which two of these events do you think were the most important?

1965–69
Severe pollution of Lake Erie kills fish and closes beaches.

1960s

1961 World Wildlife Fund founded.

1962 Rachel Carson publishes *Silent Spring* to alert the public about harmful effects of pesticides. 750 people die in London smog incident.

1963 300 deaths and thousands of illnesses in New York City from air pollution. Clean Air Act begins regulation of air pollution with stricter amendments in 1965, 1970, and 1990.

1964 Wilderness Act establishes National Wilderness System.

1965 Land and Water Conservation Act authorizes federal funds for local, state, and federal purchase of open space and parkland.

1967 Environmental Defense Fund formed.

1968 Biologist Paul Ehrlich publishes *The Population Bomb*. Biologist Garrett Hardin publishes *Tragedy of the Commons* article. UN Biosphere Conference to discuss global environmental problems.

1969 Oil-polluted Cuyahoga River, flowing through Cleveland, Ohio, catches fire. Leaks from offshore oil well off coast of Santa Barbara, California, kill wildlife and pollute beaches. Environmental Policy Act requires federal agencies to evaluate environmental impact of their actions. *Apollo* mission photo of the earth from space leads to spaceship-earth environmental worldview.

1960s

Figure 6 Some important environmental events during the 1960s. **Question:** Which two of these events do you think were the most important?

What Happened during the 1960s?

A number of milestones in American environmental history occurred during the 1960s (Figure 6). In 1962, biologist Rachel Carson (1907–1964) published *Silent Spring,* which documented the pollution of air, water, and wildlife from use of pesticides such as DDT (see Individuals Matter, p. 295). This influential book helped to broaden the concept of resource conservation to include preservation of the *quality* of the air, water, soil, and wildlife.

Many environmental historians mark Carson's wake-up call as the beginning of the modern **environmental movement** in the United States. It flourished when a growing number of citizens organized to demand that political leaders enact laws and develop policies to curtail pollution, clean up polluted environments, and protect unspoiled areas from environmental degradation.

In 1964, Congress passed the *Wilderness Act,* inspired by the vision of John Muir more than 80 years earlier. It authorized the government to protect undeveloped tracts of public land as part of the National Wilderness System, unless Congress later decides they are needed for the national good. Land in this system is to be used only for nondestructive forms of recreation such as hiking and camping.

Between 1965 and 1970, the emerging science of *ecology* received widespread media attention. At the same time, the popular writings of biologists such as Paul Ehrlich, Barry Commoner, and Garrett Hardin awakened people to the interlocking relationships among population growth, resource use, and pollution.

During that period, a number of events increased public awareness of pollution (Figure 6).

The public also became aware that pollution and loss of habitat were endangering well-known wildlife species such as the North American bald eagle, grizzly bear, whooping crane, and peregrine falcon.

During the 1968 U.S. Apollo 8 mission to the moon, astronauts photographed the earth for the first time from lunar orbit. This allowed people to see the earth as a tiny blue and white planet in the black void of space (Figure 1-1, p. 5), and it led to the development of the *spaceship-earth environmental worldview.* It reminded us that we live on a planetary spaceship that we should not harm because it is the only home we have.

What Happened during the 1970s? The Environmental Decade

During the 1970s, media attention, public concern about environmental problems, scientific research, and action to address environmental concerns grew rapidly. This period is sometimes called the *environmental decade,* or the *first decade of the environment* (Figure 7).

The first annual *Earth Day* was held on April 20, 1970. During this event, proposed by Senator Gaylord Nelson (1916–2005), some 20 million people in more than 2,000 communities took to the streets to heighten awareness and to demand improvements in environmental quality.

Republican President Richard Nixon (1913–1994) responded to the rapidly growing environmental movement. He established the *Environmental Protection Agency* (EPA) in 1970 and supported passage of the *Endangered Species Act of 1973.* This greatly strengthened the role of the federal government in protecting endangered species and their habitats.

In 1978, the *Federal Land Policy and Management Act* gave the *Bureau of Land Management* (BLM) its first real authority to manage the public land under its control, 85% of which is in 12 western states. This law angered a number of western interests whose use of these public lands was restricted for the first time.

In response, a coalition of ranchers, miners, loggers, developers, farmers, some elected officials, and others launched a political campaign known as the *sagebrush rebellion.* It had two major goals. *First,* sharply reduce government regulation of the use of public lands. *Second,* remove most public lands in the western United States from federal ownership and management and turn them over to the states. Then the plan was to persuade state legislatures to sell or lease the resource-rich lands at low prices to ranching, mining, timber, land development, and other private interests. This represented a return to President Hoover's plan to get rid of all public land, which had been thwarted by the Great Depression.

Jimmy Carter (a Democrat), president between 1977 and 1981, was very responsive to environmental concerns. He persuaded Congress to create the Department of Energy in order to develop a long-range energy strategy to reduce the country's heavy dependence on imported oil. He appointed respected environmental leaders to key positions in environmental and resource agencies and consulted with environmental interests on environmental and resource policy matters.

In 1980, Carter helped to create a *Superfund* as part of the *Comprehensive Environment Response, Compensation, and Liability Act* (pp. 582–583) to clean up abandoned hazardous waste sites, including the Love Canal housing development in

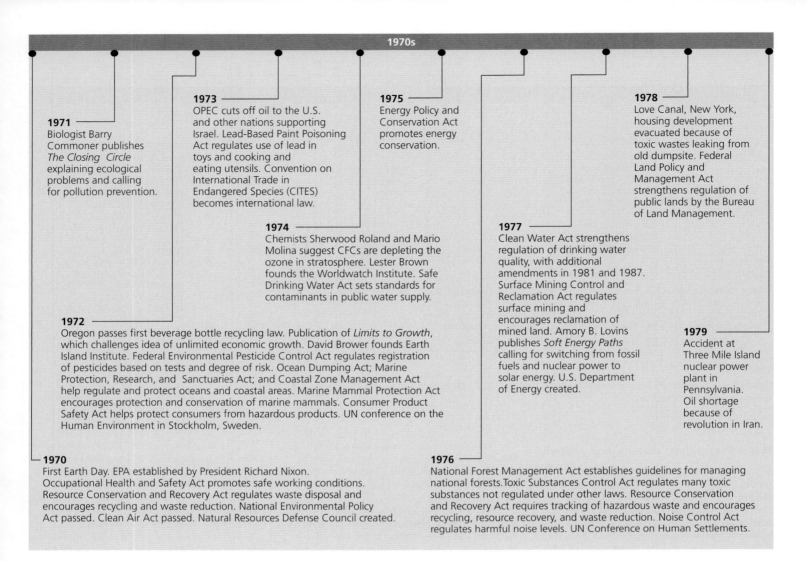

1971
Biologist Barry Commoner publishes *The Closing Circle* explaining ecological problems and calling for pollution prevention.

1973
OPEC cuts off oil to the U.S. and other nations supporting Israel. Lead-Based Paint Poisoning Act regulates use of lead in toys and cooking and eating utensils. Convention on International Trade in Endangered Species (CITES) becomes international law.

1975
Energy Policy and Conservation Act promotes energy conservation.

1978
Love Canal, New York, housing development evacuated because of toxic wastes leaking from old dumpsite. Federal Land Policy and Management Act strengthens regulation of public lands by the Bureau of Land Management.

1974
Chemists Sherwood Roland and Mario Molina suggest CFCs are depleting the ozone in stratosphere. Lester Brown founds the Worldwatch Institute. Safe Drinking Water Act sets standards for contaminants in public water supply.

1977
Clean Water Act strengthens regulation of drinking water quality, with additional amendments in 1981 and 1987. Surface Mining Control and Reclamation Act regulates surface mining and encourages reclamation of mined land. Amory B. Lovins publishes *Soft Energy Paths* calling for switching from fossil fuels and nuclear power to solar energy. U.S. Department of Energy created.

1972
Oregon passes first beverage bottle recycling law. Publication of *Limits to Growth*, which challenges idea of unlimited economic growth. David Brower founds Earth Island Institute. Federal Environmental Pesticide Control Act regulates registration of pesticides based on tests and degree of risk. Ocean Dumping Act; Marine Protection, Research, and Sanctuaries Act; and Coastal Zone Management Act help regulate and protect oceans and coastal areas. Marine Mammal Protection Act encourages protection and conservation of marine mammals. Consumer Product Safety Act helps protect consumers from hazardous products. UN conference on the Human Environment in Stockholm, Sweden.

1979
Accident at Three Mile Island nuclear power plant in Pennsylvania. Oil shortage because of revolution in Iran.

1970
First Earth Day. EPA established by President Richard Nixon. Occupational Health and Safety Act promotes safe working conditions. Resource Conservation and Recovery Act regulates waste disposal and encourages recycling and waste reduction. National Environmental Policy Act passed. Clean Air Act passed. Natural Resources Defense Council created.

1976
National Forest Management Act establishes guidelines for managing national forests. Toxic Substances Control Act regulates many toxic substances not regulated under other laws. Resource Conservation and Recovery Act requires tracking of hazardous waste and encourages recycling, resource recovery, and waste reduction. Noise Control Act regulates harmful noise levels. UN Conference on Human Settlements.

1970s

Figure 7 Some important environmental events during the 1970s, sometimes called the *environmental decade*. **Question:** Which two of these events do you think were the most important?

Niagara Falls, New York, which was abandoned when hazardous wastes began leaking into yards, school grounds, and basements.

Carter also used the *Antiquities Act of 1906* to triple the amount of land in the National Wilderness System and double the area in the National Park System (primarily by adding vast tracts in Alaska). He used the Antiquities Act to protect more public land, in all 50 states, than any president before him had done.

What Happened during the 1980s? Environmental Backlash

Figure 8 summarizes some key environmental events during the 1980s that shaped U.S. environmental policy. During this decade, farmers and ranchers and leaders of the oil, coal, automobile, mining, and timber industries strongly opposed many of the environmental laws and regulations developed in the 1960s and 1970s. They organized and funded multiple efforts to defeat environmental laws and regulations—efforts that persist today.

In 1981, Ronald Reagan (a Republican, 1911–2004), a self-declared *sagebrush rebel* and advocate of less federal control, became president. During his 8 years in office, he angered environmentalists by appointing to key federal positions people who opposed most existing environmental and public land-use laws and policies.

Reagan greatly increased private energy and mineral development and timber cutting on public lands. He also drastically cut federal funding for research on energy conservation and renewable energy resources and eliminated tax incentives for residential solar energy and energy conservation enacted during the Carter administration. In addition, he lowered automobile gas mileage standards and relaxed federal air and water quality pollution standards.

Although Reagan was immensely popular, many people strongly opposed his environmental and resource policies. This resulted in strong opposition in Congress, public outrage, and legal challenges by environmental and conservation organizations, whose memberships soared during this period.

In 1988, an industry-backed, coalition called the *wise-use movement* was formed. Its major goals were to weaken or repeal most of the country's environmental laws and regulations and destroy the effectiveness of the environmental movement in the United States. Politically powerful coal, oil, mining, automobile, timber, and ranching interests helped back this movement.

Upon his election in 1989, George H. W. Bush (a Republican) promised to be "the environmental president." But he received criticism from environmentalists for not providing leadership on such key environmental issues as population growth, global warming, and loss of biodiversity. He also continued support of exploitation of valuable resources on public lands at giveaway prices. In addition, he allowed some environmental laws to be undercut by the politi-

1980–89
Rise of a strong anti-environmental movement.

1980s

1989 *Exxon Valdez* oil tanker accident in Alaska's Prince William Sound.

1985 Scientists discover annual seasonal thinning of the ozone layer above Antarctica.

1986 Explosion of Chernobyl nuclear power plant in Ukraine. Times Beach, Missouri, evacuated and bought by EPA because of dioxin contamination.

1987 Montreal Protocol to halve emissions of ozone-depleting CFCs signed by 24 countries. International Basel Convention controls movement of hazardous wastes from one country to another.

1984 Toxic fumes leaking from pesticide plant in Bhopal, India, kill at least 6,000 people and injure 50,000–60,000. Lester R. Brown publishes first annual *State of the World* report.

1988 Industry-backed wise-use movement established to weaken and destroy U.S. environmental movement. Biologist E. O. Wilson publishes *Biodiversity*, detailing how human activities are affecting the earth's diversity of species.

1983 U.S. EPA and National Academy of Sciences publish reports finding that buildup of carbon dioxide and other greenhouse gases will lead to global warming.

1980 Superfund law passed to clean up abandoned toxic waste dumps. Alaska National Interest Lands Conservation Act protects 42 million hectares (104 million acres) of land in Alaska.

1980s

Figure 8 Some important environmental events during the 1980s. **Question:** Which two of these events do you think were the most important?

cal influence of industry, mining, ranching, and real estate development interests. They argued that environmental laws had gone too far and were hindering economic growth.

What Happened from 1990 to 2008?

Between 1990 and 2008, opposition to environmental laws and regulations gained strength. This occurred because of continuing political and economic support from corporate backers, who argued that environmental laws were hindering economic growth, and because federal elections gave Republicans (many of whom were generally unsympathetic to environmental concerns) a majority in Congress.

Consequently, during this period, leaders and supporters of the environmental movement have had to spend much of their time and funds fighting efforts to discredit the movement and weaken or eliminate most environmental laws passed during the 1960s and 1970s. They also have had to counter claims by anti-environmental groups that problems such as global warming and ozone depletion are hoaxes or are not very serious and that environmental laws and regulations have hindered economic growth.

In 1993, Bill Clinton (a Democrat) became president and promised to provide national and global environmental leadership. During his 8 years in office, he appointed respected environmental leaders to key positions in environmental and resource agencies and consulted with

environmental interests about environmental policy, as Carter had done.

He also vetoed most of the anti-environmental bills (or other bills passed with anti-environmental riders attached) passed by a Republican-dominated Congress between 1995 and 2000. He announced regulations requiring sport utility vehicles (SUVs) to meet the same air pollution emission standards as cars. Clinton also used executive orders to make forest health the primary priority in managing national forests and to declare many roadless areas in national forests off limits to the building of roads and to logging.

In addition, he used the Antiquities Act of 1906 to protect various parcels of public land in the West from development and resource exploitation by declaring them national monuments. He protected more public land as national monuments in the lower 48 states than any other president, including Teddy Roosevelt and Jimmy Carter. However, environmental leaders criticized Clinton for failing to push hard enough on key environmental issues such as global warming energy policy and global and national biodiversity protection.

During the 1990s, many small and mostly local grassroots environmental organizations sprang up to deal with environmental threats in their local communities. Interest in environmental issues increased on many college campuses and environmental studies programs at colleges and universities expanded. In addition, awareness of important, complex environmental issues, such as sustainability, population growth,

biodiversity protection, and threats from global warming, increased.

In 2001, George W. Bush (a Republican) became president. Like Reagan in the 1980s, he appointed to key federal positions people who opposed or wanted to weaken many existing environmental and public land-use laws and policies because they were alleged to threaten economic growth. Also like Reagan, he did not consult with environmental groups and leaders in developing environmental policies, and he greatly increased private energy and mineral development and timber cutting on public lands.

Bush also opposed increasing automobile gas mileage standards as a way to save energy and reduce dependence on oil imports, and he supported relaxation of various federal air and water quality standards. Like Reagan, he developed an energy policy that emphasized use of fossil fuels and nuclear power with much less support during his first term for reducing energy waste and relying more on renewable energy resources.

In addition, he withdrew the United States from participation in the international Kyoto treaty, designed to help reduce carbon dioxide emissions that can promote global warming and lead to long-lasting climate change. He also repealed or tried to weaken most of the pro-environmental measures established by Bill Clinton. On the other hand, in 2006, he created the world's second largest marine reserve in waters around some of the Hawaiian Islands.

According to leaders of a dozen major environmental organizations Bush, backed by a

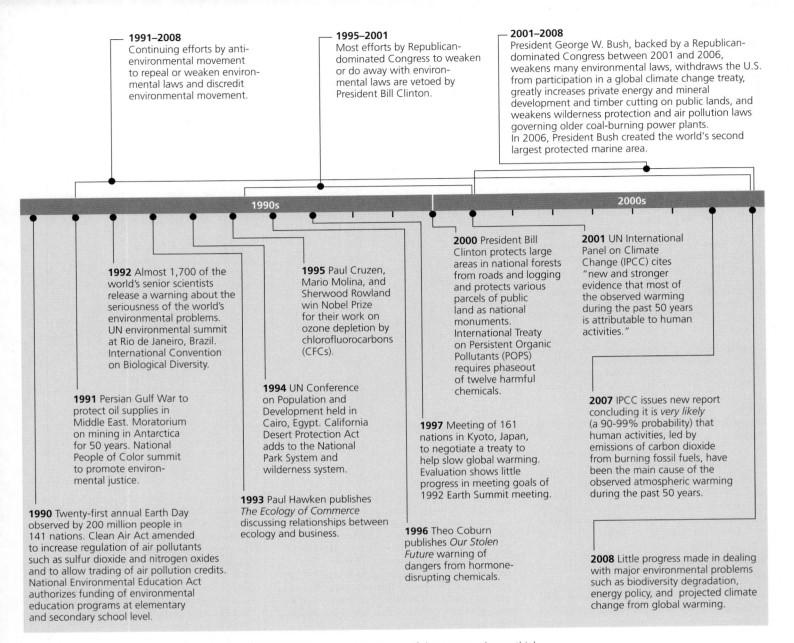

1991–2008
Continuing efforts by anti-environmental movement to repeal or weaken environmental laws and discredit environmental movement.

1995–2001
Most efforts by Republican-dominated Congress to weaken or do away with environmental laws are vetoed by President Bill Clinton.

2001–2008
President George W. Bush, backed by a Republican-dominated Congress between 2001 and 2006, weakens many environmental laws, withdraws the U.S. from participation in a global climate change treaty, greatly increases private energy and mineral development and timber cutting on public lands, and weakens wilderness protection and air pollution laws governing older coal-burning power plants.
In 2006, President Bush created the world's second largest protected marine area.

1990s

2000s

1992 Almost 1,700 of the world's senior scientists release a warning about the seriousness of the world's environmental problems. UN environmental summit at Rio de Janeiro, Brazil. International Convention on Biological Diversity.

1995 Paul Cruzen, Mario Molina, and Sherwood Rowland win Nobel Prize for their work on ozone depletion by chlorofluorocarbons (CFCs).

2000 President Bill Clinton protects large areas in national forests from roads and logging and protects various parcels of public land as national monuments. International Treaty on Persistent Organic Pollutants (POPS) requires phaseout of twelve harmful chemicals.

2001 UN International Panel on Climate Change (IPCC) cites "new and stronger evidence that most of the observed warming during the past 50 years is attributable to human activities."

1991 Persian Gulf War to protect oil supplies in Middle East. Moratorium on mining in Antarctica for 50 years. National People of Color summit to promote environmental justice.

1994 UN Conference on Population and Development held in Cairo, Egypt. California Desert Protection Act adds to the National Park System and wilderness system.

1997 Meeting of 161 nations in Kyoto, Japan, to negotiate a treaty to help slow global warming. Evaluation shows little progress in meeting goals of 1992 Earth Summit meeting.

2007 IPCC issues new report concluding it is *very likely* (a 90-99% probability) that human activities, led by emissions of carbon dioxide from burning fossil fuels, have been the main cause of the observed atmospheric warming during the past 50 years.

1993 Paul Hawken publishes *The Ecology of Commerce* discussing relationships between ecology and business.

1990 Twenty-first annual Earth Day observed by 200 million people in 141 nations. Clean Air Act amended to increase regulation of air pollutants such as sulfur dioxide and nitrogen oxides and to allow trading of air pollution credits. National Environmental Education Act authorizes funding of environmental education programs at elementary and secondary school level.

1996 Theo Coburn publishes *Our Stolen Future* warning of dangers from hormone-disrupting chemicals.

2008 Little progress made in dealing with major environmental problems such as biodiversity degradation, energy policy, and projected climate change from global warming.

Figure 9 Some important environmental events, 1990–2008 **Question:** Which two of these events do you think were the most important?

Republican-dominated Congress during his first term, compiled the worst environmental record of any president in the history of the country during his two terms in office.

A few moderate Republican members of Congress have urged their party to return to its environmental roots, which were put down during Teddy Roosevelt's presidency, and to shed its anti-environmental approach to legislation. Most Democrats agree and assert that the environmental problems we face are much too serious to be held hostage by political squabbling. They call for cooperation, not confronta-

tion. These Democrats and Republicans urge elected officials, regardless of party, to enter into a new pact in which the United States becomes the world leader in making this the *environmental century*. This would help to sustain the country's rich heritage of natural capital and provide economic development, jobs, and profits in rapidly growing businesses such as solar and wind energy, energy-efficient vehicles and buildings, ecological restoration, and pollution prevention.

In 2007, the Intergovernmental Panel on Climate Change (IPCC), which includes more

than 2,500 of the world's climate experts, issued its third report on global climate change. According to this overwhelming consensus among the world's climate scientists, global warming is occurring and it is very likely (a 90–99% probability) that human activities, led by the burning of fossil fuels, have been the main cause of the observed atmospheric warming during the past 50 years. By 2008, the U.S. government had made little progress in facing up to the threat of climate change from global warming.

Some Basic Chemistry (Chapters 1–5, 14, 16–21)

Chemists Use the Periodic Table to Classify Elements on the Basis of Their Chemical Properties

Chemists have developed a way to classify the elements according to their chemical behavior, in what is called the *periodic table of elements* (Figure 1). Each horizontal row in the table is called a *period*. Each vertical column lists elements with similar chemical properties and is called a *group*.

The partial periodic table in Figure 1 shows how the elements can be classified as *metals*, *nonmetals*, and *metalloids*. Most of the elements found to the left and at the bottom of the table are *metals*, which usually conduct electricity and heat, and are shiny. Examples are sodium (Na), calcium (Ca), aluminum (Al), iron (Fe), lead (Pb), silver (Ag), and mercury (Hg).

Atoms of metals tend to lose one or more of their electrons to form positively charged ions such as Na^+, Ca^{2+}, and Al^{3+}. For example, an atom of the metallic element sodium (Na, with

atomic number 11) with 11 positively charged protons and 11 negatively charged electrons can lose one of its electrons. It then becomes a sodium ion with a positive charge of 1 (Na^+) because it now has 11 positive charges (protons) but only 10 negative charges (electrons).

Nonmetals, found in the upper right of the table, do not conduct electricity very well. Examples are hydrogen (H), carbon (C), nitrogen (N), oxygen (O), phosphorus (P), sulfur (S), chlorine (Cl), and fluorine (F).

Atoms of some nonmetals such as chlorine, oxygen, and sulfur tend to gain one or more electrons lost by metallic atoms to form negatively charged ions such as O^{2-}, S^{2-}, and Cl^-. For example, an atom of the nonmetallic element chlorine (Cl, with atomic number 17) can gain an electron and become a chlorine ion. The ion has a negative charge of 1 (Cl^-) because it has 17 positively charged protons and 18 negatively charged electrons. Atoms of nonmetals can also combine with one another to form molecules in which they share one or more pairs of their electrons. Hydrogen, a nonmetal,

is placed by itself above the center of the table because it does not fit very well into any of the groups

The elements arranged in a diagonal staircase pattern between the metals and nonmetals have a mixture of metallic and nonmetallic properties and are called *metalloids*.

Figure 1 also identifies the elements required as *nutrients* (black squares) for all or some forms of life and elements that are moderately or highly toxic (red squares) to all or most forms of life. Six nonmetallic elements—carbon (C), oxygen (O), hydrogen (H), nitrogen (N), sulfur (S), and phosphorus (P)—make up about 99% of the atoms of all living things.

THINKING ABOUT
The Periodic Table

Use the periodic table to identify by name and symbol two elements that should have chemical properties similar to those of **(a)** Ca, **(b)** potassium, **(c)** S, **(d)** lead.

Figure 1 Abbreviated periodic table of elements. Elements in the same vertical column, called a *group*, have similar chemical properties. To simplify matters at this introductory level, only 72 of the 118 known elements are shown.

Figure 2 A solid crystal of an ionic compound such as sodium chloride consists of a three–dimensional array of oppositely charged ions held together by *ionic bonds* resulting from the strong forces of attraction between opposite electrical charges. They are formed when an electron is transferred from a metallic atom such as sodium (Na) to a nonmetallic element such as chlorine (Cl).

Na⁺ Cl⁻

Sodium ion in solution

Water molecules

Sodium chloride (NaCl) salt

Chloride ion in solution

Figure 3 How a salt dissolves in water.

Ionic and Covalent Bonds Hold Compounds Together

Sodium chloride (NaCl) consists of a three-dimensional network of oppositely charged *ions* (Na⁺ and Cl⁻) held together by the forces of attraction between opposite charges (Fig-

ure 2). The strong forces of attraction between such oppositely charged ions are called *ionic bonds*. Because ionic compounds consist of ions formed from atoms of metallic (positive ions) and nonmetallic (negative ions) elements (Figure 1), they can be described as *metal–nonmetal compounds*.

Sodium chloride and many other ionic compounds tend to dissolve in water and break apart into their individual ions (Figure 3).

NaCl Na⁺ + Cl⁻
sodium chloride sodium ion + chloride ion
(in water)

Water, a *covalent compound*, consists of molecules made up of uncharged atoms of hydrogen (H) and oxygen (O). Each water molecule consists of two hydrogen atoms chemically bonded to an oxygen atom, yielding H_2O molecules. The bonds between the atoms in such molecules are called *covalent bonds* and form when the atoms in the molecule share one or more pairs of their electrons. Because they are formed from atoms of nonmetallic elements (Figure 1), covalent compounds can be described as *nonmetal–nonmetal compounds*. Figure 4 shows the chemical formulas and shapes of the molecules that are the building blocks for several common *covalent compounds*.

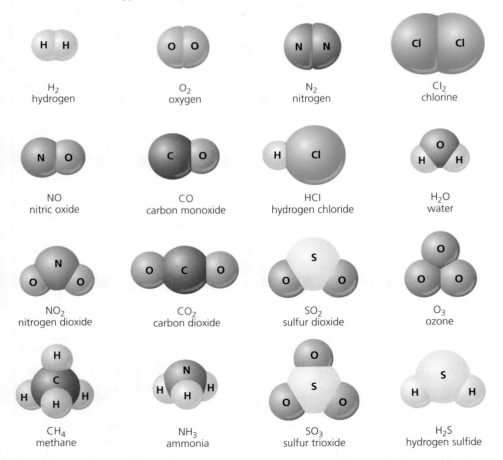

H_2 hydrogen	O_2 oxygen	N_2 nitrogen	Cl_2 chlorine
NO nitric oxide	CO carbon monoxide	HCl hydrogen chloride	H_2O water
NO_2 nitrogen dioxide	CO_2 carbon dioxide	SO_2 sulfur dioxide	O_3 ozone
CH_4 methane	NH_3 ammonia	SO_3 sulfur trioxide	H_2S hydrogen sulfide

Figure 4 Chemical formulas and shapes for some *covalent compounds* formed when atoms of one or more nonmetallic elements combine with one another and share one or more pairs of their electrons. The bonds between the atoms in such molecules are called *covalent bonds*.

What Makes Solutions Acidic? Hydrogen Ions and pH

The *concentration*, or number of hydrogen ions (H⁺) in a specified volume of a solution (typically a liter), is a measure of its acidity. Pure water (not tap water or rainwater) has an equal number of hydrogen (H⁺) and hydroxide (OH⁻) ions. It is called a **neutral solution**. An **acidic solution** has more hydrogen ions than hydroxide ions per liter. A **basic solution** has more hydroxide ions than hydrogen ions per liter.

Scientists use **pH** as a measure of the acidity of a solution based on its concentration of hydrogen ions (H⁺). By definition, a neutral solution has a pH of 7, an acidic solution has a pH of less than 7, and a basic solution has a pH greater than 7.

Each single unit change in pH represents a 10-fold increase or decrease in the concentration of hydrogen ions per liter. For example, an acidic solution with a pH of 3 is 10 times more acidic than a solution with a pH of 4. Figure 5

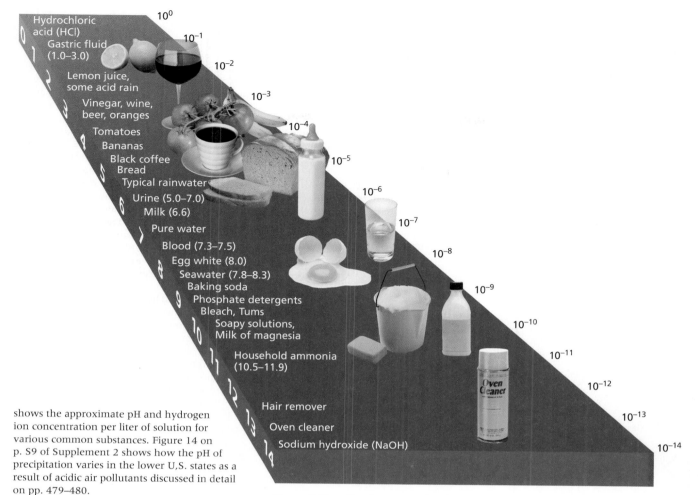

shows the approximate pH and hydrogen ion concentration per liter of solution for various common substances. Figure 14 on p. S9 of Supplement 2 shows how the pH of precipitation varies in the lower U.S. states as a result of acidic air pollutants discussed in detail on pp. 479–480.

Figure 5 The *pH scale,* representing the concentration of hydrogen ions (H$^+$) in one liter of solution is shown on the righthand side. On the left side are the approximate pH values for solutions of some common substances. A solution with a pH less than 7 is *acidic,* one with a pH of 7 is *neutral,* and one with a pH greater than 7 is *basic.* A change of 1 on the pH scale means a tenfold increase or decrease in H$^+$ concentration. (Modified from Cecie Starr, *Biology: Today and Tomorrow,* Pacific Grove, Calif.: Brooks/Cole, © 2005)

THINKING ABOUT
pH

A solution has a pH of 2. How many times more acidic is this solution than one with a pH of 6?

There Are Weak Forces of Attraction between Some Molecules

Ionic and covalent bonds form between the ions or atoms *within* a compound. There are also weaker forces of attraction *between* the molecules of covalent compounds (such as water) resulting from an unequal sharing of electrons by two atoms.

For example, an oxygen atom has a much greater attraction for electrons than does a hydrogen atom. Thus, in a water molecule the electrons shared between the oxygen atom and its two hydrogen atoms are pulled closer to the oxygen atom, but not actually transferred to the oxygen atom. As a result, the oxygen atom in a water molecule has a slightly negative partial charge and its two hydrogen atoms have a slightly positive partial charge (Figure 6).

The slightly positive hydrogen atoms in one water molecule are then attracted to the slightly negative oxygen atoms in another water molecule. These forces of attraction *between* water molecules are called *hydrogen bonds* (Figure 6).

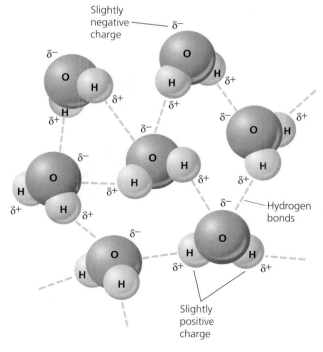

Figure 6 *Hydrogen bond:* slightly unequal sharing of electrons in the water molecule creates a molecule with a slightly negatively charged end and a slightly positively charged end. Because of this electrical polarity, the hydrogen atoms of one water molecule are attracted to the oxygen atoms in other water molecules. These fairly weak forces of attraction *between* molecules (represented by the dashed lines) are called *hydrogen bonds.*

They account for many of water's unique properties (Science Focus, p. 67). Hydrogen bonds also form between other covalent molecules or portions of such molecules containing hydrogen and nonmetallic atoms with a strong ability to attract electrons.

Four Types of Large Organic Compounds Are the Molecular Building Blocks of Life

Larger and more complex organic compounds, called *polymers*, consist of a number of basic structural or molecular units (*monomers*) linked by chemical bonds, somewhat like rail cars linked in a freight train. Four types of macro-molecules—complex carbohydrates, proteins, nucleic acids, and lipids—are molecular building blocks of life.

Complex carbohydrates consist of two or more monomers of *simple sugars* (such as glucose, Figure 7) linked together. One example is the starches that plants use to store energy and also to provide energy for animals that feed on plants. Another is cellulose, the earth's most abundant organic compound, which is found in the cell walls of bark, leaves, stems, and roots.

Proteins, are large polymer molecules formed by linking together long chains of monomers called *amino acids* (Figure 8). Living organisms use about 20 different amino acid molecules to build a variety of proteins, which play different roles. Some help to store energy. Some are components of the *immune system* that protects the body against diseases and harmful substances by forming antibodies that make invading agents harmless. Others are *hormones* that are used as chemical messengers in the bloodstreams of animals to turn various bodily functions on or off. In animals, proteins are also components of hair, skin, muscle, and tendons. In addition, some proteins act as *enzymes* that catalyze or speed up certain chemical reactions.

Nucleic acids are large polymer molecules made by linking hundreds to thousands of four types of monomers called *nucleotides*. Two nucleic acids—DNA (**d**eoxyribo**n**ucleic **a**cid) and RNA (**rib**o**n**ucleic **a**cid)—participate in the building of proteins and carry hereditary information used to pass traits from parent to offspring. Each nucleotide consists of a *phosphate group*, a *sugar molecule* containing five carbon atoms (deoxy-ribose in DNA molecules and ribose in RNA molecules), and one of four different *nucleotide bases* (represented by A, G, C, and T, the first letter in each of their names, or A, G, C, and U in RNA; see Figure 9). In the cells of living organisms, these nucleotide units combine in different numbers and sequences to form *nucleic acids* such as various types of RNA and DNA (Figure 10).

Hydrogen bonds formed between parts of the four nucleotides in DNA hold two DNA strands together like a spiral staircase, forming a double helix (Figure 10). DNA molecules can unwind and replicate themselves.

The total weight of the DNA needed to reproduce all of the world's people is only about 50 milligrams—the weight of a small match. If the DNA coiled in your body were unwound, it would stretch about 960 million kilometers (600 million miles)—more than six times the distance between the sun and the earth.

The different molecules of DNA that make up the millions of species found on the earth are like a vast and diverse genetic library. Each species is a unique book in that library. The *genome* of a species is made up of the entire sequence of DNA "letters" or base pairs that combine to "spell out" the chromosomes in typical members of each species. In 2002, scientists were able to map out the genome for the human species by

General structure ➡ **Chain of glucose units** ➡ **Starch**

Glucose ($C_6H_{12}O_6$)

Figure 7 Straight-chain and ring structural formulas of glucose, a simple sugar that can be used to build long chains of complex carbohydrates such as starch and cellulose.

General structure of amino acid ➡ **Chain of amino acids** ➡ **Protein**

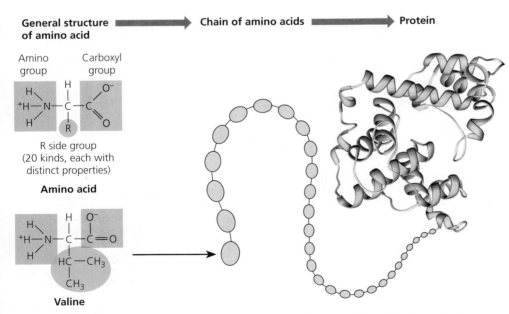

Figure 8 General structural formula of amino acids and a specific structural formula of one of the 20 different amino acid molecules that can be linked together in chains to form proteins that fold up into more complex shapes.

Figure 9 Generalized structures of the nucleotide molecules linked in various numbers and sequences to form large nucleic acid molecules such as various types of DNA (deoxyribonucleic acid) and RNA (ribonucleic acid). In DNA, the 5-carbon sugar in each nucleotide is deoxyribose; in RNA it is ribose. The four basic nucleotides used to make various forms of DNA molecules differ in the types of nucleotide bases they contain—guanine (G), cytosine (C), adenine (A), and thymine (T). (Uracil, labeled U, occurs instead of thymine in RNA.)

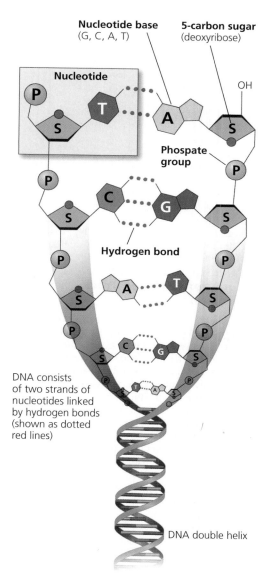

Nucleotide base
(G, C, A, T)

5-carbon sugar
(deoxyribose)

Nucleotide

Phospate group

Hydrogen bond

DNA consists of two strands of nucleotides linked by hydrogen bonds (shown as dotted red lines)

DNA double helix

Figure 10 Portion of the double helix of a DNA molecule. The double helix is composed of two spiral (helical) strands of nucleotides. Each nucleotide contains a unit of phosphate (P), deoxyribose (S), and one of four nucleotide bases: guanine (G), cytosine (C), adenine (A), and thymine (T). The two strands are held together by hydrogen bonds formed between various pairs of the nucleotide bases. Guanine (G) bonds with cytosine (C), and adenine (A) with thymine (T).

analyzing the 3.1 billion base sequences in human DNA.

Lipids, a fourth building block of life, are a chemically diverse group of large organic compounds that do not dissolve in water. Examples are *fats and oils* for storing energy (Figure 11), *waxes* for structure, and *steroids* for producing hormones.

Figure 12 shows the relative sizes of simple and complex molecules, cells, and multicelled organisms.

Fatty acid
(lipid)

Fat molecule
(triglyceride)

Fatty tissue
(adipose cells)

Figure 11 Structural formula of fatty acid that is one form of lipid (left). Fatty acids are converted into more complex fat molecules that are stored in adipose cells (right).

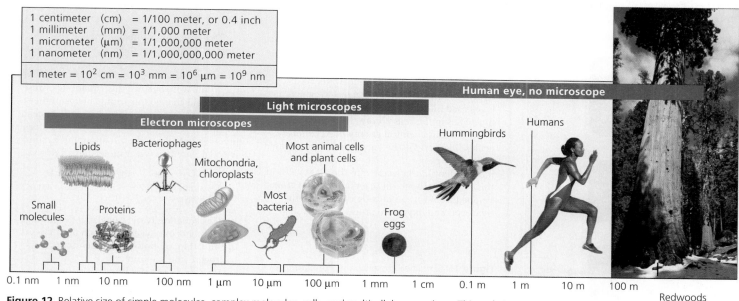

1 centimeter (cm)	= 1/100 meter, or 0.4 inch
1 millimeter (mm)	= 1/1,000 meter
1 micrometer (μm)	= 1/1,000,000 meter
1 nanometer (nm)	= 1/1,000,000,000 meter

1 meter = 10^2 cm = 10^3 mm = 10^6 μm = 10^9 nm

Human eye, no microscope

Light microscopes

Electron microscopes

Lipids

Bacteriophages

Mitochondria, chloroplasts

Most animal cells and plant cells

Hummingbirds

Humans

Small molecules

Proteins

Most bacteria

Frog eggs

0.1 nm 1 nm 10 nm 100 nm 1 μm 10 μm 100 μm 1 mm 1 cm 0.1 m 1 m 10 m 100 m

Redwoods

Figure 12 Relative size of simple molecules, complex molecules, cells, and multicellular organisms. This scale is exponential, not linear. Each unit of measure is 10 times larger than the unit preceding it. (Used by permission from Cecie Starr and Ralph Taggart, *Biology*, 11th ed, Belmont, Calif.: Thomson Brooks/Cole, © 2006)

Figure 13 Energy storage and release in cells.

ATP synthesis:
Energy is stored in ATP

ADP Phosphate

Energy

ATP

ATP breakdown:
Energy stored in ATP is released

ATP

ADP Phosphate

Energy

Sun

Chloroplast in leaf cell

Chlorophyll

H_2O → **Light-dependent reaction** → O_2

Energy storage and release (ATP/ADP)

CO_2 → **Light-independent reaction** → Glucose

$$6CO_2 + 6H_2O \xrightarrow{\text{Sunlight}} C_6H_{12}O_6 + 6O_2$$

Certain Molecules Store and Release Energy in Cells

Chemical reactions occurring in photosynthesis (pp. 58–59) release energy that is absorbed by adenosine diphosphate (ADP) molecules and stored as chemical energy in adenosine triphosphate (ATP) molecules (Figure 13, left). When cellular processes require energy, ATP molecules release it to form ADP molecules (Figure 13, right).

A Closer Look at Photosynthesis

In photosynthesis, sunlight powers a complex series of chemical reactions that combine water taken up by plant roots and carbon dioxide from the air to produce sugars such as glucose. This process converts solar energy into chemical energy in sugars for use by plant cells, with the solar energy captured, stored, and released as chemical energy in ATP and ADP molecules (Figure 13). Figure 14 is a greatly simplified summary of the photosynthesis process.

Photosynthesis takes place within tiny enclosed structures called *chloroplasts* found within plant cells. Chlorophyll, a special compound in chloroplasts, absorbs incoming visible light mostly in the violet and red wavelengths. The green light that is not absorbed is reflected back, which is why photosynthetic plants look green. The absorbed wavelengths of solar energy initiate a sequence of chemical reactions with other molecules in what are called *light-dependent reactions.*

This series of reactions splits water into hydrogen ions (H^+) and oxygen (O_2) which is released into the atmosphere. Small ADP molecules in the cells absorb the energy released and store it as chemical energy in ATP molecules (Figure 13). The chemical energy released by the ATP molecules drives a series of *light-independent (dark) reactions* in the plant cells. In this second sequence of reactions, carbon atoms stripped from carbon dioxide combine with hydrogen and oxygen to produce sugars such

Figure 14 Simplified overview of *photosynthesis.* In this process, chlorophyll molecules in the chloroplasts of plant cells absorb solar energy. This initiates a complex series of chemical reactions in which carbon dioxide and water are converted to sugars such as glucose and oxygen.

as glucose ($C_6H_{12}O_6$, Figure 7, p. S42) that plant cells can use as a source of energy and carbon.

Chemists Balance Chemical Equations to Keep Track of Atoms

Chemists use a shorthand system to represent chemical reactions. These chemical equations are also used as an accounting system to verify that no atoms are created or destroyed in a chemical reaction as required by the law of conservation of matter (p. 39 and **Concept 2-3**, p. 40). As a consequence, each side of a chemical equation must have the same number of atoms or ions of each element involved. Ensuring that this condition is met leads to what chemists call a *balanced chemical equation.* The equation for the burning of carbon ($C + O_2 \longrightarrow CO_2$) is balanced because one atom of carbon and two atoms of oxygen are on both sides of the equation.

Consider the following chemical reaction: When electricity passes through water (H_2O), the latter can be broken down into hydrogen (H_2) and oxygen (O_2), as represented by the following equation:

$$H_2O \longrightarrow H_2 + O_2$$
2 H atoms 2 H atoms 2 O atoms
1 O atom

This equation is unbalanced because one atom of oxygen is on the left side of the equation but two atoms are on the right side.

We cannot change the subscripts of any of the formulas to balance this equation because that would change the arrangements of the atoms, leading to different substances. Instead, we must use different numbers of the molecules involved to balance the equation. For example, we could use two water molecules:

$$2 H_2O \longrightarrow H_2 + O_2$$
4 H atoms 2 H atoms 2 O atoms
2 O atoms

This equation is still unbalanced. Although the numbers of oxygen atoms on both sides of the equation are now equal, the numbers of hydrogen atoms are not.

We can correct this problem by having the reaction produce two hydrogen molecules:

$$2 H_2O \longrightarrow 2 H_2 + O_2$$
4 H atoms 4 H atoms 2 O atoms
2 O atoms

Now the equation is balanced, and the law of conservation of matter has been observed. For every two molecules of water through which we pass electricity, two hydrogen molecules and one oxygen molecule are produced.

| THINKING ABOUT
| **Chemical Equations**

Try to balance the chemical equation for the reaction of nitrogen gas (N_2) with hydrogen gas (H_2) to form ammonia gas (NH_3).

Scientists Are Learning How to Build Materials from the Bottom Up: Nanotechnology

Nanotechnology (Science Focus, p. 362) uses atoms and molecules to build materials from the bottom up using atoms of the elements in the periodic table as its raw materials. A *nanometer* (nm) is one-billionth of a meter—equal to the length of about 10 hydrogen atoms lined up side by side. A DNA molecule (Figure 10) is about 2.5 nanometers wide. A human hair has a width of 50,000 to 100,000 nanometers.

For objects smaller than about 100 nanometers, the properties of materials change dramatically. At this *nanoscale* level, materials can exhibit new properties, such as extraordinary strength or increased chemical activity, that they do not exhibit at the much larger *macroscale* level that we are all familiar with.

For example, scientists have learned how to make tiny tubes of carbon atoms linked together in hexagons. Experiments have shown that these carbon nanotubes are the strongest material ever made—60 times stronger than high-grade steel. Such nanotubes have been linked together to form a rope so thin that it is invisible, but strong enough to suspend a pickup truck.

At the macroscale, zinc oxide (ZnO) can be rubbed on the skin as a white paste to protect against the sun's harmful UV rays; at the nanoscale it becomes transparent and is being used as invisible coatings to protect the skin and fabrics from UV damage. Because silver (Ag) can kill harmful bacteria, silver nanocrystals are being incorporated into bandages for wounds.

Researchers hope to incorporate nanoparticles of hydroxyapatite, with the same chemical structure as tooth enamel, into toothpaste to put coatings on teeth that prevent bacteria from penetrating. Nanotech coatings now being used on cotton fabrics form an impenetrable barrier that causes liquids to bead and roll off. Such stain-resistant fabrics used to make clothing, rugs, and furniture upholstery could eliminate the need to use harmful chemicals for removing stains.

Self-cleaning window glass coated with a layer of nanoscale titanium dioxide (TiO_2) particles is now available. As the particles interact with UV rays from the sun, dirt on the surface of the glass loosens and washes off when it rains. Similar products can be used for self-cleaning sinks and toilet bowls.

Scientists are working on ways to replace the silicon in computer chips with carbon-based nanomaterials that greatly increase the processing power of computers. Biological engineers are working on nanoscale devices that could deliver drugs. Such devices could penetrate cancer cells and deliver nanomolecules that could kill the cancer cells from the inside. Researchers also hope to develop nanoscale crystals that could change color when they detect tiny amounts (measured in parts per trillion) of harmful substances such chemical and biological warfare agents and food pathogens. For example, a color change in food packaging could alert a consumer when a food is contaminated or has begun to spoil. The list of possibilities could go on.

By 2008, more than 1,000 products containing nanoscale particles were commercially available and thousands more were in the pipeline. Examples are found in cosmetics, sunscreens, fabrics, pesticides, and food additives.

So far, these products are unregulated and unlabeled. This concerns many health and environmental scientists because the tiny size of nanoparticles can allow them to penetrate the natural defenses of the body against invasions by foreign and potentially harmful chemicals and pathogens. Nanoparticles of a chemical tend to be much more chemically reactive than macroparticles of the chemical, largely because the tiny nanoparticles have relatively large surface areas for their small mass. This means that a chemical that is harmless at the macroscale may be hazardous at the nanoscale when they are inhaled, ingested, or absorbed through the skin.

We know little about such effects and risks at a time when the use of untested and unregulated nanoparticles is increasing exponentially. A few toxicological studies are sending up red flags:

- In 2004, Eva Olberdorster, an environmental toxicologist at Southern Methodist University, found that fish swimming in water loaded with a certain type of carbon nanomolecule called buckyballs experienced brain damage within 48 hours.
- In 2005, NASA researchers found that injecting commercially available carbon nanotubes into rats caused significant lung damage.
- A 2005 study by researchers at the U.S. National Institute of Occupational Safety and Health found substantial damage to the heart and aortic arteries of mice exposed to carbon nanotubes.
- In 2005, researchers at New York's University of Rochester found increased blood clotting in rabbits inhaling carbon buckyballs.

In 2004, the British Royal Society and Royal Academy of Engineering recommended that we avoid the environmental release of nanoparticles and nanotubes as much as possible until more is known about their potential harmful impacts. They recommended as a precautionary measure that factories and research laboratories treat manufactured nanoparticles and nanotubes as if they were hazardous to their workers and to the general public. **GREEN CAREER:** Nanotechnology

| THINKING ABOUT
| **Nanotechnology**

Do you think that the benefits of nanotechnology outweigh its potentially harmful effects? Explain. What are three things you would do to reduce its potentially harmful effects?

| RESEARCH FRONTIER

Learning more about nanotechnology and how to reduce its potentially harmful effects. See **academic.cengage.com/biology/miller**.

Classifying and Naming Species (Chapters 3, 4, 8)

All organisms on the earth today are descendants of single-cell organisms that lived almost 4 billion years ago. As a result of biological evolution through natural selection, life has evolved into six major groups of species, called *kingdoms: eubacteria, archaebacteria, protists, fungi, plants,* and *animals* (Figure 4-3, p. 81).

Eubacteria are prokaryotes with single cells that lack a nucleus and other internal compartments (Figure 3-2b, p. 52) found in the cells of species from other kingdoms. Examples include various cyanobacteria and bacteria such as *staphylococcus* and *streptococcus.*

Archaebacteria are single-celled bacteria that are closer to eukaryotic cells (Figure 3-2a, p. 52) than to eubacteria. Examples include methanogens, which live in oxygen-free sediments of lakes and swamps and in animal guts; halophiles, which live in extremely salty water; and thermophiles, which live in hot springs, hydro-thermal vents, and acidic soil. These organisms live in extreme environments.

The remaining four kingdoms—protists, fungi, plants, and animals (Figure 4-3, p. 81) are eukarotes with one or more cells that have a nucleus and complex internal compartments (Figure 3-2a, p. 52). *Protists* are mostly single-celled eukaryotic organisms, such as diatoms, dinoflagellates, amoebas, golden brown and yellow-green algae, and protozoans. Some protists cause human diseases such as malaria (pp. 444–447) and sleeping sickness.

Fungi are mostly many-celled, sometimes microscopic, eukaryotic organisms such as mushrooms, molds, mildews, and yeasts. Many fungi are decomposers (Figure 3-11, p. 60). Other fungi kill various plants and animals and cause huge losses of crops and valuable trees.

Plants are mostly many-celled eukaryotic organisms such as red, brown, and green algae and mosses, ferns, and flowering plants (whose flowers produce seeds that perpetuate the species). Some plants such as corn and marigolds are *annuals,* meaning that they complete their life cycles in one growing season. Others are *perennials,* which can live for more than 2 years, such as roses, grapes, elms, and magnolias.

Animals are also many-celled, eukaryotic organisms. Most have no backbones and hence are called *invertebrates.* Invertebrates include sponges, jellyfish, worms, arthropods (e.g., insects, shrimp, and spiders), mollusks (e.g., snails, clams, and octopuses), and echinoderms (e.g., sea urchins and sea stars). *Vertebrates* (animals with backbones and a brain protected by skull bones) include fishes (e.g., sharks and tuna), amphibians (e.g., frogs and salamanders), reptiles (e.g., crocodiles and snakes), birds (e.g., eagles and robins), and mammals (e.g., bats, elephants, whales, and humans).

Within each kingdom, biologists have created subcategories based on anatomical, physiological, and behavioral characteristics. Kingdoms are divided into *phyla,* which are divided into subgroups called *classes.* Classes are subdivided into *orders,* which are further divided into *families.* Families consist of *genera* (singular, *genus*), and each genus contains one or more *species.* Note that the word *species* is both singular and plural. Figure 1 shows this detailed taxonomic classification for the current human species.

Most people call a species by its common name, such as robin or grizzly bear. Biologists use scientific names (derived from Latin) consisting of two parts (printed in italics, or underlined) to describe a species. The first word is the capitalized name (or abbreviation) for the genus to which the organism belongs. It is followed by a lowercase name that distinguishes the species from other members of the same genus. For example, the scientific name of the robin is *Turdus migratorius* (Latin for "migratory thrush") and the grizzly bear goes by the scientific name *Ursus horribilis* (Latin for "horrible bear").

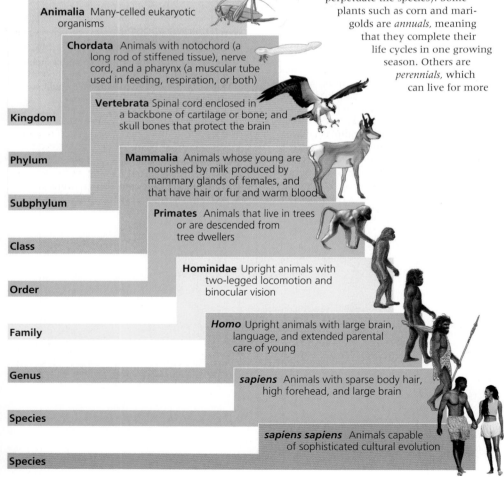

Animalia Many-celled eukaryotic organisms

Chordata Animals with notochord (a long rod of stiffened tissue), nerve cord, and a pharynx (a muscular tube used in feeding, respiration, or both)

Vertebrata Spinal cord enclosed in a backbone of cartilage or bone; and skull bones that protect the brain

Mammalia Animals whose young are nourished by milk produced by mammary glands of females, and that have hair or fur and warm blood

Primates Animals that live in trees or are descended from tree dwellers

Hominidae Upright animals with two-legged locomotion and binocular vision

Homo Upright animals with large brain, language, and extended parental care of young

sapiens Animals with sparse body hair, high forehead, and large brain

sapiens sapiens Animals capable of sophisticated cultural evolution

Kingdom

Phylum

Subphylum

Class

Order

Family

Genus

Species

Species

Figure 1 Taxonomic classification of the latest human species, *Homo sapiens sapiens.*

Weather Basics: El Niño, Tornadoes, and Tropical Cyclones (Chapters 4, 7, 11, 19)

Weather Is Affected by Moving Masses of Warm and Cold Air

Weather is the set of short-term atmospheric conditions—typically those occurring over hours or days—for a particular area. Examples of atmospheric conditions include temperature, pressure, moisture content, precipitation, sunshine, cloud cover, and wind direction and speed.

Meteorologists use equipment mounted on weather balloons, aircraft, ships, and satellites, as well as radar and stationary sensors, to obtain data on weather variables. They then feed these data into computer models to draw weather maps. Other computer models project the weather for a period of several days by calculating the probabilities that air masses, winds, and other factors will change in certain ways.

Much of the weather we experience results from interactions between the leading edges of moving masses of warm or cold air. Weather changes as one air mass replaces or meets another. The most dramatic changes in weather occur along a **front,** the boundary between two air masses with different temperatures and densities.

A **warm front** is the boundary between an advancing warm air mass and the cooler one it is replacing (Figure 1, left). Because warm air is less dense (weighs less per unit of volume) than cool air, an advancing warm front rises up over a mass of cool air. As the warm front rises, its moisture begins condensing into droplets, forming layers of clouds at different altitudes. Gradually the clouds thicken, descend to a lower altitude, and often release their moisture as rainfall. A moist warm front can bring days of cloudy skies and drizzle.

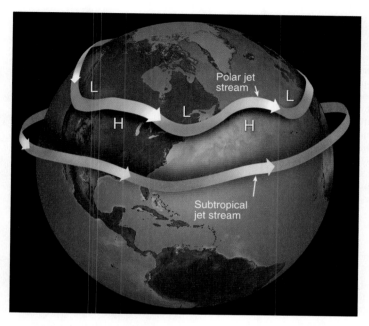

A **cold front** (Figure 1, right) is the leading edge of an advancing mass of cold air. Because cold air is denser than warm air, an advancing cold front stays close to the ground and wedges underneath less dense warmer air. An approaching cold front produces rapidly moving, towering clouds called *thunderheads.*

As a cold front passes through, we may experience high surface winds and thunderstorms. After it leaves the area, we usually have cooler temperatures and a clear sky.

Near the top of the troposphere, hurricane-force winds circle the earth. These powerful winds, called *jet streams,* follow rising and falling paths that have a strong influence on weather patterns (Figure 2).

Figure 2 A *jet stream* is a rapidly flowing air current that moves west to east in a wavy pattern. This figure shows a polar jet stream and a subtropical jet stream in winter. In reality, jet streams are discontinuous and their positions vary from day to day. (Used by permission from C. Donald Ahrens, *Meteorology Today,* 8th ed. Belmont, Calif.: Brooks/Cole, 2006)

Weather Is Affected by Changes in Atmospheric Pressure

Changes in atmospheric pressure also affect weather. *Atmospheric pressure* results from molecules of gases (mostly nitrogen and oxygen) in the atmosphere zipping around at very high speeds and hitting and bouncing off everything they encounter.

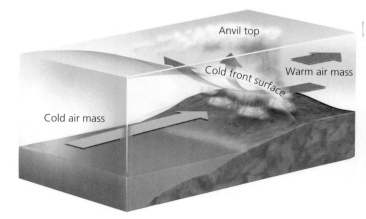

Figure 1 *Weather fronts:* a *warm front* (left) arises when an advancing mass of warm air meets and rises up over a mass of denser cool air. A *cold front* (right) forms when a moving mass of cold air wedges beneath a mass of less dense warm air.

Atmospheric pressure is greater near the earth's surface because the molecules in the atmosphere are squeezed together under the weight of the air above them. An air mass with high pressure, called a **high,** contains cool, dense air that descends toward the earth's surface and becomes warmer. Fair weather follows as long as this high-pressure air mass remains over the area.

In contrast, a low-pressure air mass, called a **low,** produces cloudy and sometimes stormy weather. Because of its low pressure and low density, the center of a low rises, and its warm air expands and cools. When the temperature drops below a certain level where condensation takes place, called the *dew point,* moisture in the air condenses and forms clouds.

If the droplets in the clouds coalesce into larger drops or snowflakes heavy enough to fall from the sky, then precipitation occurs. The condensation of water vapor into water drops usually requires that the air contain suspended tiny particles of material such as dust, smoke, sea salts, or volcanic ash. These so-called *condensation nuclei* provide surfaces on which the droplets of water can form and coalesce.

Every Few Years Major Wind Shifts in the Pacific Ocean Affect Global Weather Patterns

An **upwelling,** or upward movement of ocean water, can mix the water, bringing cool and nutrient-rich water from the bottom of the ocean to the surface where it supports large populations of phytoplankton, zooplankton, fish, and fish-eating seabirds.

Figure 7-2 (p. 142) shows the oceans' major upwelling zones. Upwellings far from shore occur when surface currents move apart and draw water up from deeper layers. Strong upwellings are also found along the steep western coasts of some continents when winds blowing along the coasts push surface water away from the land and draw water up from the ocean bottom (Figure 3).

Every few years in the Pacific Ocean, normal shore upwellings (Figure 4, left) are affected by changes in weather patterns called the *El Niño–Southern Oscillation,* or *ENSO* (Figure 4, right). In an ENSO, often called simply *El Niño,* prevailing tropical trade winds blowing east to west weaken or reverse direction. This allows

Figure 3 *A shore upwelling* occurs when deep, cool, nutrient-rich waters are drawn up to replace surface water that has been moved away from a steep coast by wind flowing along the coast toward the equator.

Figure 4 Normal trade winds blowing east to west cause shore upwellings of cold, nutrient-rich bottom water in the tropical Pacific Ocean near the coast of Peru (left). A zone of gradual temperature change called the *thermocline* separates the warm and cold water. Every few years a shift in trade winds known as the *El Niño–Southern Oscillation* (*ENSO*) disrupts this pattern. Trade winds blowing from east to west weaken or reverse direction, which depresses the coastal upwellings and warms the surface waters off South America (right). When an ENSO lasts 12 months or longer, it severely disrupts populations of plankton, fish, and seabirds in upwelling areas and can alter weather conditions over much of the globe (Figure 5).

Drought

Unusually high rainfall

Unusually warm periods

El Niño

Figure 5 Typical global weather effects of an El Niño–Southern Oscillation. During the 1996–1998 ENSO, huge waves battered the coast in the U.S. state of California and torrential rains caused widespread flooding and mudslides. In Peru, floods and mudslides killed hundreds of people, left about 250,000 people homeless, and ruined harvests. Drought in Brazil, Indonesia, and Australia led to massive wildfires in tinder-dry forests. India and parts of Africa also experienced severe drought. A catastrophic ice storm hit Canada and the northeastern United States, but the southeastern United States had fewer hurricanes. (Data from United Nations Food and Agriculture Organization)

Data and Map Analysis

1. How might an ENSO affect the weather where you live or go to school?
2. Why do you think the area to the west of El Niño suffers drought?

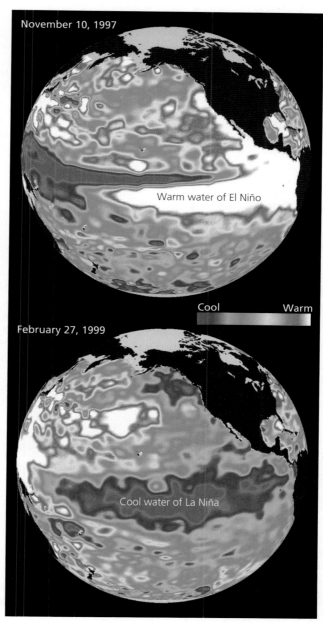

November 10, 1997

Warm water of El Niño

Cool Warm

February 27, 1999

Cool water of La Niña

Figure 6 Locations of flowing masses of warm and cold water in the Pacific Ocean during an *El Niño* (top) and a *La Niña* (bottom). (Data from Jet Propulsion Lab, NASA)

the warmer waters of the western Pacific to move toward the coast of South America, which suppresses the normal upwellings of cold, nutrient-rich water (Figure 4, right). The decrease in nutrients reduces primary productivity and causes a sharp decline in the populations of some fish species.

A strong ENSO can alter the weather of at least two-thirds of the globe (Figure 5)—especially in lands along the Pacific and Indian Oceans. Scientists do not know for sure the causes of an ENSO, but they know how to detect its formation and track its progress.

La Niña, the reverse of El Niño, cools some coastal surface waters, and brings back upwellings. Typically, La Niña means more Atlantic Ocean hurricanes, colder winters in Canada and the northeastern United States, and warmer and drier winters in the southeastern and southwestern United States. It also usually leads to wetter winters in the Pacific Northwest, torrential rains in Southeast Asia, lower wheat yields in Argentina, and more wildfires in Florida. Figure 6 uses satellite data to show changes in the locations of masses of warm and cold water in the Pacific Ocean during an *El Niño* (top) and a *La Niña* (bottom).

Tornadoes and Tropical Cyclones Are Violent Weather Extremes

Sometimes we experience *weather extremes*. Two examples are violent storms called *tornadoes* (which form over land) and *tropical cyclones* (which form over warm ocean waters and sometimes pass over coastal land).

Tornadoes or *twisters* are swirling funnel-shaped clouds that form over land. They can destroy houses, cause other serious damage, and kill people in areas where they touch down on the earth's surface. The United States is the world's most tornado-prone country, followed by Australia.

Tornadoes in the plains of the midwestern United States usually occur when a large, dry, cold-air front moving southward from Canada runs into a large mass of humid air moving northward from the Gulf of Mexico. Most tornadoes occur in the spring and summer when fronts of cold air from the north penetrate deeply into the midwestern plains.

As the large warm-air mass moves rapidly over the more dense cold-air mass, it rises swiftly and forms strong vertical convection currents that suck air upward, as shown in Figure 7. Scientists hypothesize that the rising vortex of air starts spinning because the air near the ground in the funnel is moving more slowly than the air above. This difference causes the air ahead of the advancing front to roll or spin in a vertically rising air mass or vortex.

Figure 8 shows the areas of greatest risk from tornadoes in the continental United States.

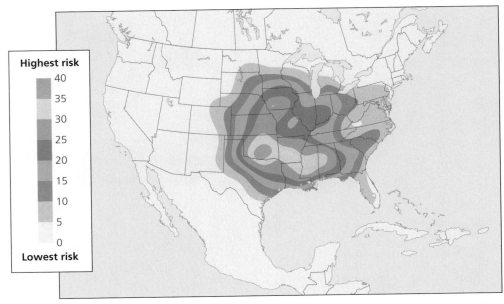

Highest risk

40
35
30
25
20
15
10
5
0

Lowest risk

Figure 8 States with *very high* and *high* tornado risk in the continental United States. (Data from NOAA)

Data and Map Analysis

1. How many states have areas with a risk factor of 25 or higher? How many have areas with a risk factor of 20 or higher?
2. What is the level of risk where you live? If you live in a far western state, does this mean you are guaranteed never to see a tornado in your area?

Tropical cyclones are spawned by the formation of low-pressure cells of air over warm tropical seas. Figure 9 shows the formation and structure of a tropical cyclone. *Hurricanes* are tropical cyclones that form in the Atlantic Ocean; those forming in the Pacific Ocean usually are called *typhoons*. Tropical cyclones take a long time to form and gain strength. As a result, meteorologists can track their paths and wind speeds and warn people in areas likely to be hit by these violent storms.

For a tropical cyclone to form, the temperature of ocean water has to be at least 27 °C (80 °F) to a depth of 46 meters (150 feet). A tropical cyclone forms when areas of low pressure over the warm ocean draw in air from surrounding higher-pressure areas. The earth's rotation makes these winds spiral counterclockwise in the northern hemisphere and clockwise in the southern hemisphere (Figure 7-3, p. 142). Moist air warmed by the heat of the ocean rises in a vortex through the center of the storm until it becomes a tropical cyclone (Figure 9).

The intensities of tropical cyclones are rated in different categories based on their sustained wind speeds: *Category 1:* 119–153 kilometers per hour (74–95 miles per hour); *Category 2:* 154–177 kilometers per hour (96–110 miles per hour); *Category 3:* 178–209 kilometers per hour (111–130 miles per hour); *Category 4:* 210–249 kilometers per hour (131–155 miles per hour); and *Category 5:* greater than 249 kilometers per hour (155 miles per hour). The longer a tropical cyclone stays over warm waters, the stronger it gets. Significant hurricane-force winds can extend 64–161 kilometers (40–100 miles) from the center, or eye, of a tropical cyclone.

Figure 10 shows the change in the average surface temperature of the global ocean between 1871 and 2000. Note the rise in this temperature since 1980. These higher temperatures, especially in tropical waters, may explain why

Descending cool air

Severe thunderstorm

Rising warm air

Severe thunderstorms can trigger a number of smaller tornadoes

Tornado forms when cool downdraft and warm updraft of air meet and interact

Rising updraft of air

Warm moist air drawn in

Figure 7 Formation of a *tornado* or *twister*. Although twisters can form at any time of the year, the most active tornado season in the United States is usually March through August. Meteorologists cannot tell us with great accuracy when and where most tornadoes will form.

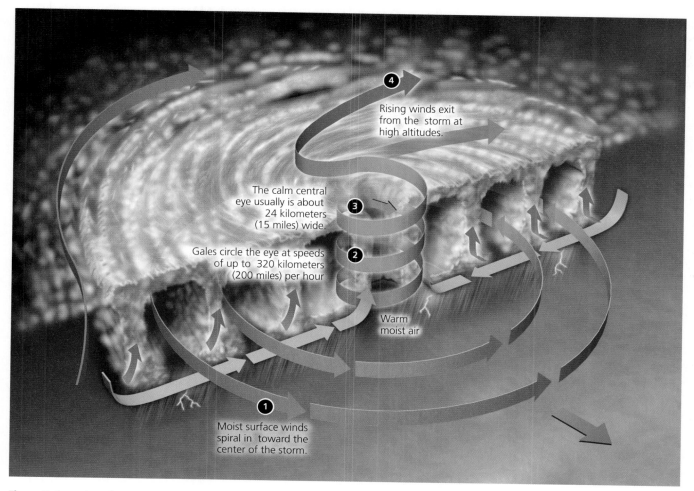

Figure 9 Formation of a *tropical cyclone.* Those forming in the Atlantic Ocean usually are called *hurricanes;* those forming in the Pacific Ocean usually are called *typhoons.*

the average intensity of tropical cyclones has increased since 1990. With the number of people living along the world's coasts increasing, the danger to lives and property has risen dramatically. The greatest risk from hurricanes in the continental United States is along the gulf and eastern coasts, as shown in Figure 11 (p. S52).

Hurricanes and typhoons kill and injure people and damage property (Figure 8-18, p. 177) and agricultural production. Sometimes, however, the long-term ecological and economic

benefits of a tropical cyclone exceed its short-term harmful effects.

For example, in parts of the U.S. state of Texas along the Gulf of Mexico, coastal bays and marshes normally are closed off from freshwater and saltwater inflows. In August 1999, Hurricane Brett struck this coastal area. According to marine biologists, it flushed out excess nutrients from land runoff and swept dead sea grasses and rotting vegetation from the coastal bays and marshes. It also carved out 12 channels through

the barrier islands along the coast, allowing huge quantities of seawater to flood the bays and marshes.

This flushing of the bays and marshes reduced brown tides consisting of explosive growths of algae that had fed on excess nutrients. It also increased growth of sea grasses, which serve as nurseries for shrimp, crabs, and fish and provide food for millions of ducks wintering in Texas bays. Production of commercially important species of shellfish and fish also increased.

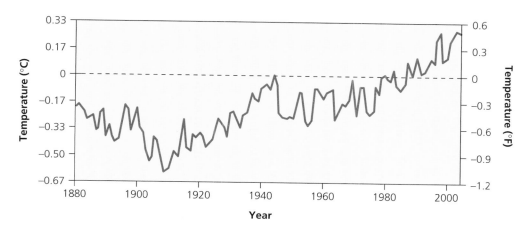

Figure 10 Change in global ocean temperature from its average baseline temperature from 1880 to 2000. (Data from National Oceanic and Atmospheric Administration)

Data and Graph Analysis

1. In 1900, the global ocean temperature dropped from its baseline temperature by how many degrees? (Give answer in both Centigrade and Fahrenheit.)

2. Since about what year after 1980 have global ocean temperatures consistently increased (with all changes being positive)? What has been the highest temperature increase since then, and in about what year did it happen?

Figure 11 The number of hurricanes expected to occur during a 100-year period in the continental United States (based on historical data from U.S. Geological Survey)

Data and Map Analysis

1. What is the degree of risk where you live or go to school?
2. How many states include an area with some risk of hurricanes?

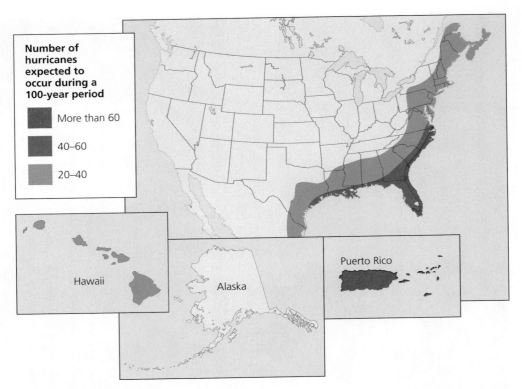

Number of hurricanes expected to occur during a 100-year period

More than 60

40–60

20–40

Hawaii

Alaska

Puerto Rico

Components and Interactions in Major Biomes (Chapters 8, 9, 10, 15)

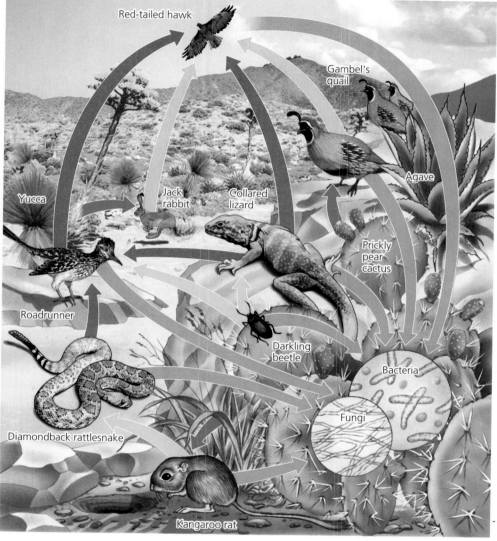

Figure 1 Some components and interactions in a *temperate desert ecosystem*. When these organisms die, decomposers break down their organic matter into minerals that plants use. Colored arrows indicate transfers of matter and energy among producers, primary consumers (herbivores), secondary or higher-level consumers (carnivores), and decomposers. Organisms are not drawn to scale. **Question:** What species might undergo population growth and what species might suffer a population decline if the diamondback rattlesnake were eliminated from this ecosystem?

 Producer to primary consumer

 Primary to secondary consumer

 Secondary to higher-level consumer

All producers and consumers to decomposers

CENGAGENOW™ **Active Figure 2** Some components and interactions in a *temperate tall-grass prairie ecosystem* in North America. When these organisms die, decomposers break down their organic matter into minerals that plants can use. Colored arrows indicate transfers of matter and energy among producers, primary consumers (herbivores), secondary or higher-level consumers (carnivores), and decomposers. Organisms are not drawn to scale. *See an animation based on this figure at* CengageNOW™. **Question:** What species might increase and what species might decrease in population size if the threatened prairie dog were eliminated from this ecosystem?

Golden eagle

Pronghorn antelope

Coyote

Grasshopper sparrow

Grasshopper

Blue stem grass

Prairie dog

Bacteria

Fungi

Prairie coneflower

Producer to primary consumer

Primary to secondary consumer

Secondary to higher-level consumer

All producers and consumers to decomposers

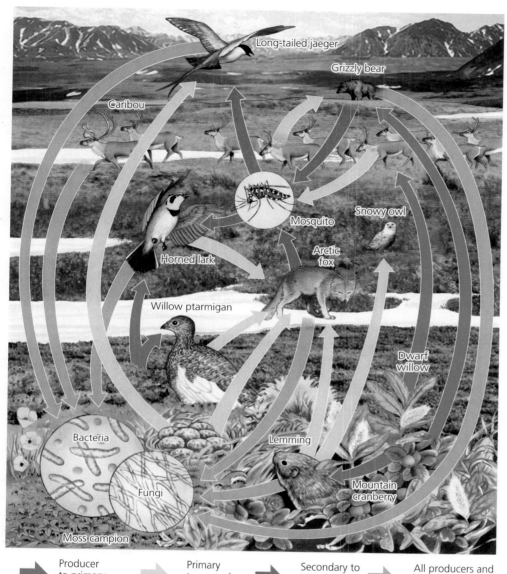

Figure 3 Some components and interactions in an *arctic tundra (cold grassland) ecosystem*. When these organisms die, decomposers break down their organic matter into minerals that plants use. Colored arrows indicate transfers of matter and energy among producers, primary consumers (herbivores), secondary or higher-level consumers (carnivores), and decomposers. Organisms are not drawn to scale. **Question:** What species might increase and what species might decrease in population size if the arctic fox were eliminated from this ecosystem?

Long-tailed jaeger

Grizzly bear

Caribou

Mosquito

Snowy owl

Horned lark

Arctic fox

Willow ptarmigan

Dwarf willow

Bacteria

Lemming

Fungi

Mountain cranberry

Moss campion

→ Producer to primary consumer

→ Primary to secondary consumer

→ Secondary to higher-level consumer

→ All producers and consumers to decomposers

Figure 4 Some components and interactions in a *temperate deciduous forest ecosystem.* When these organisms die, decomposers break down their organic matter into minerals that plants use. Colored arrows indicate transfers of matter and energy among producers, primary consumers (herbivores), secondary or higher-level consumers (carnivores), and decomposers. Organisms are not drawn to scale. **Question:** What species might increase and what species might decrease in population size if the broad-winged hawk were eliminated from this ecosystem?

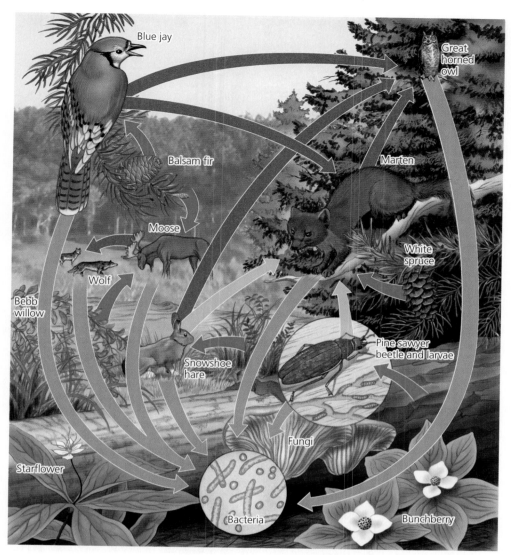

Figure 5 Some components and interactions in an *evergreen coniferous* (*boreal* or *taiga*) *forest ecosystem*. When these organisms die, decomposers break down their organic matter into minerals that plants use. Colored arrows indicate transfers of matter and energy among producers, primary consumers (herbivores), secondary or higher-level consumers (carnivores), and decomposers. Organisms are not drawn to scale. **Question:** What species might increase and what species might decrease in population size if the great horned owl were eliminated from this ecosystem?

Blue jay

Balsam fir

Moose

Wolf

Bebb willow

Snowshoe hare

Starflower

Great horned owl

Marten

White spruce

Pine sawyer beetle and larvae

Fungi

Bacteria

Bunchberry

➡ Producer to primary consumer

➡ Primary to secondary consumer

➡ Secondary to higher-level consumer

➡ All producers and consumers to decomposers

Figure 6 Components and interactions in a *coral reef ecosystem*. When these organisms die, decomposers break down their organic matter into minerals used by plants. Colored arrows indicate transfers of matter and energy among producers, primary consumers (herbivores), secondary or higher-level consumers (carnivores), and decomposers. Organisms are not drawn to scale. See the photo of a coral reef in Figure 8-1, left, p. 162. **Question:** How would the species in this ecosystem be affected if phytoplankton populations suffered a sharp drop?

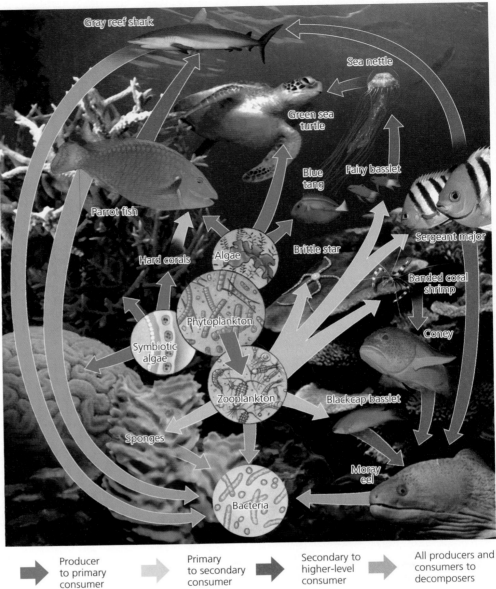

Gray reef shark

Sea nettle

Green sea turtle

Blue tang

Fairy basslet

Parrot fish

Sergeant major

Hard corals

Algae

Brittle star

Banded coral shrimp

Phytoplankton

Coney

Symbiotic algae

Zooplankton

Blackcap basslet

Sponges

Moray eel

Bacteria

Producer to primary consumer

Primary to secondary consumer

Secondary to higher-level consumer

All producers and consumers to decomposers

Energy and Climate Data and Maps (Chapters 7, 15, 16, 19, 24)

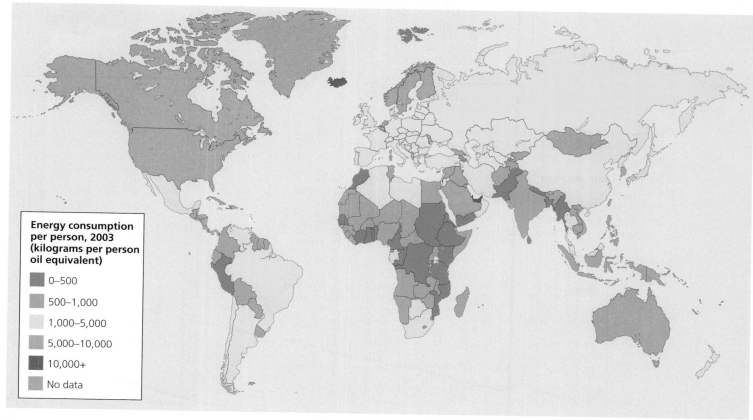

Energy consumption per person, 2003 (kilograms per person oil equivalent)

- 0–500
- 500–1,000
- 1,000–5,000
- 5,000–10,000
- 10,000+
- No data

Figure 1 Energy use per person in 2003 (latest data available) in kilograms per person oil equivalent. **Question:** What are two conclusions you can draw from these data? (Data from International Energy Agency, 2006)

Data and Map Analysis

1. How many countries have energy consumption levels of 5,000 or more kilograms per person? How many have levels of 500 or less?
2. Find an example where a country with consumption levels of 5,000 or more kilograms per person borders a country or countries with levels of 1,000 or less, and list the countries. (See Figure 1, pp. S10–S11, in Supplement 3 for country names.)

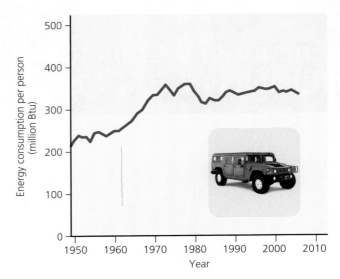

Figure 2 Total (left) and per capita (right) energy consumption in the United States, 1999–2007. (Data from U.S. Energy Information Administration/ Annual Energy Review 2008)

Data and Graph Analysis

1. In what year or years did the total U. S. energy consumption reach 80 quadrillion Btu?
2. In what year did energy consumption per person reach its highest level shown on this graph, and about what was that level of consumption?

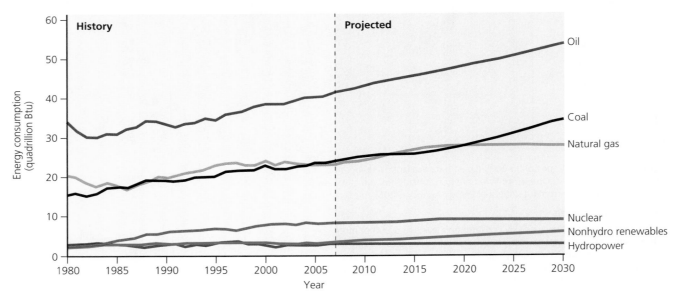

Figure 3 Energy consumption by fuel in the United States, 1980–2007, with projections to 2030. (Data from U.S. Energy Information Administration/ Annual Energy Outlook 2008)

Data and Graph Analysis

1. Which energy source grew the most between 1990 and 2007? How much did it grow (in quadrillion Btu)?
2. In what years after 2007 was it projected that coal use would be higher than the use of natural gas?

Figure 4 World oil consumption, 1950–2007. (Data from U.S. Energy Information Administration, British Petroleum, International Energy Agency, and United Nations)

Data and Graph Analysis

1. In 2000, the world consumed about how many times as much oil as it consumed in 1950?
2. In what years since 1950 was oil consumption declining?

Figure 5 World coal and natural gas consumption, 1950–2007. (Data from U.S. Energy Information Administration, British Petroleum, International Energy Agency, and United Nations)

Data and Graph Analysis

1. Which energy source has grown more steadily—coal or natural gas? In what years has coal use grown the most sharply?
2. In what year did coal use reach a level twice as high as it was in 1960?

Figure 6 Locations of the major known deposits of oil, natural gas, and coal in North America, and offshore areas where more crude oil and natural gas might be found. Geologists do not expect to find very much new oil and natural gas in North America. Offshore drilling for oil accounts for about one-fourth of U.S. oil production. Nine of every ten barrels of this oil comes from the Gulf of Mexico, where there are 4,000 oil drilling platforms and 53,000 kilometers (33,000 miles) of underwater pipeline (see insert). (Data from Council on Environmental Quality and U.S. Geological Survey)

Data and Map Analysis

1. If you live in North America, where are the major oil, coal, and natural gas deposits closest to where you live?
2. Which country has the longest total coastal area with high potential for oil and natural gas deposits?

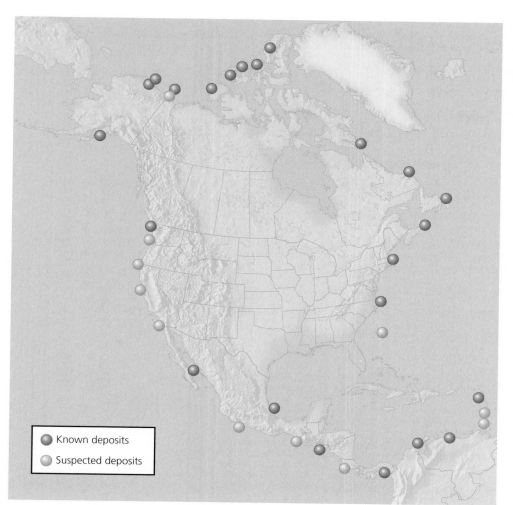

Figure 7 Known and suspected deposits of methane hydrates in North American, Central American, and some South American coastal waters. Developing the technology to tap into only 1% of these deposits could double the nation's supply of natural gas. (Data from U.S. Geological Survey)

Data and Map Analysis

1. How many known deposits lie in U.S. and Mexican coastal waters?
2. Most suspected deposits lie along which coast of North America?

Year	Event
1857	First commercial oil well drilled near Titusville, Pennsylvania.
1905	Oil supplies 10% of U.S. energy.
1925	The United States produces 71% of the world's oil.
1930	Because of an oil glut, oil sells for 10¢ per barrel.
1953	U.S. oil companies account for about half the world's oil production, and the United States is the world's leading oil exporter.
1955	The United States has 20% of the world's estimated oil reserves.
1960	OPEC is formed so that developing countries, with most of the world's known oil and projected oil reserves, can get a higher price for their oil.
1973	The United States uses 30% of the world's oil, imports 36% of this oil, and has only 5% of the world's proven oil reserves.
1973–1974	OPEC reduces oil imports to the West and bans oil exports to the United States because of its support for Israel in the 18-day Yom Kippur War with Egypt and Syria. World oil prices rise sharply and lead to double-digit inflation in the United States and many other countries and a global economic recession.
1975	Production of estimated U.S. oil reserves peaks.
1979	Iran's Islamic Revolution shuts down most of Iran's oil production and reduces world oil production.
1981	The Iran-Iraq war pushes global oil prices to an historic high.
1983	Facing an oil glut, OPEC cuts its oil prices.
1985	U.S. domestic oil production begins to decline and is not expected to increase enough to affect the global price of oil or to reduce U.S. dependence on oil imports.
August 1990–June 1991	The United States and its allies fight the Persian Gulf War to oust Iraqi invaders of Kuwait and to protect Western access to Saudi Arabian and Kuwaiti oil supplies.
2004–???	The United States and a small number of allies fight a second Persian Gulf War to oust Saddam Hussein from power in Iraq and to protect Western access to Saudi Arabian, Kuwaiti, and Iraqi oil.
2008	OPEC has 67% of the world oil reserves and produces 40% of the world's oil. The United States has only 2.9% of oil reserves, uses 26% of the world's oil production, and imports 60% of its oil. Oil prices surpass $140 per barrel for the first time.
2020	The United States could be importing at least 70% of the oil it uses, as consumption continues to exceed production.
2010–2030	Production of oil from the world's estimated oil reserves is expected to peak as half of the world's oil reserves are used up. Oil prices are expected to increase gradually as the demand for oil increasingly exceeds the supply—unless the world decreases its demand by wasting less energy and shifting to other sources of energy.
2010–2048	Domestic U.S. oil reserves are projected to be 80% depleted.
2042–2083	A gradual decline in dependence on oil is expected.

Figure 8 Brief history of the Age of Oil.

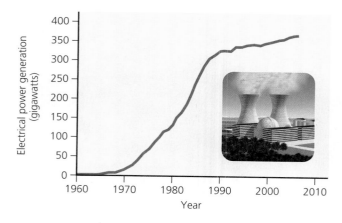

Figure 9 Worldwide electrical generating capacity of nuclear power plants, 1960–2006. (Data from International Energy Agency and Worldwatch Institute)

Data and Graph Analysis

1. After 1980, how long did it take to double the generating capacity that existed in 1980?
2. Considering the decades of the 1970s, 1980s, and 1990s, which decade saw the sharpest growth in generating capacity? During which decade did this growth level off?

Population per square kilometer

- 1,158–25,839
- 116–1,157
- 62–115
- 31–61
- 3–30
- 1–2
- <1

▲ Commercial power reactors

Figure 10 Population density and location of the 104 commercial nuclear power reactors in the United States and proposed Yucca Mountain nuclear waste storage site in Nevada. (Data from U.S. Nuclear Regulatory Commission)

Data and Map Analysis

1. If you live in the Untied States, do you live or go to school within about 97 kilometers (60 miles) of a commercial nuclear power reactor?
2. Which state has the largest number of commercial nuclear power reactors?

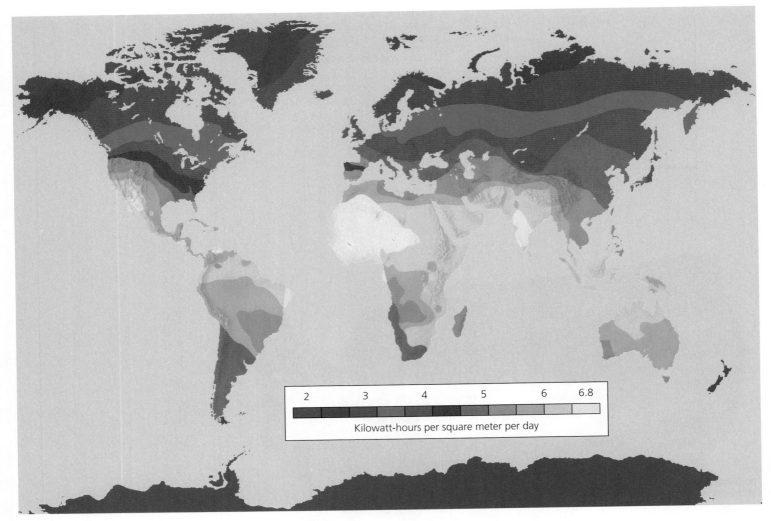

Figure 11 Global availability of direct solar energy. Areas with more than 3.5 kilowatt-hours per square meter per day (see scale) are good candidates for passive and active solar heating systems and use of solar cells to produce electricity. The United Nations is mapping the potential wind and solar energy resources of 13 developing countries in Africa, Asia, and South and Central America. (Data from U.S. Department of Energy)

Data and Map Analysis

1. What is the potential for making greater use of solar energy to provide heat and produce electricity (with solar cells) where you live or go to school?
2. List the continents in order of overall availability of direct solar energy, from those with the highest to those with the lowest. (See Figure 1 of Supplement 3, pp. S10–S11, for continent names.)

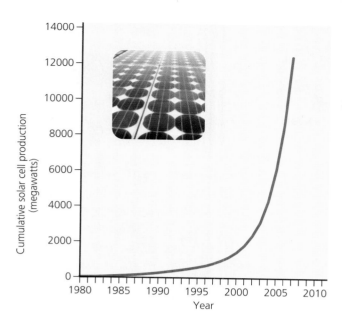

Figure 12 Global cumulative production of electricity by solar (photovoltaic) cells, 1980–2006. (Data from PV News, Worldwatch Institute).

Data and Graph Analysis

1. About how many times as much solar cell production had occurred by 2005 as had occurred by 2000?
2. How long did it take the world to go from zero to 6,000 megawatts in its cumulative production of electricity by solar cells?

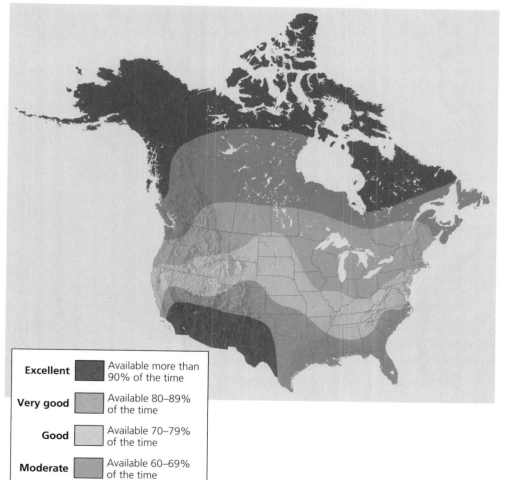

Excellent		Available more than 90% of the time
Very good		Available 80–89% of the time
Good		Available 70–79% of the time
Moderate		Available 60–69% of the time
Fair		Available 50–59% of the time
Poor		Available less than 50% of the time

Figure 13 Availability of direct solar energy in the continental United States and Canada. If prices come down as expected, large banks of solar cells in desert areas in the southwestern United States could produce enough energy to meet all U.S. electricity needs. Electricity produced by such solar-cell power plants would be distributed through an upgraded and expanded version of the country's electric power grid. (Data from the U.S. Department of Energy and the National Wildlife Federation)

Data and Map Analysis

1. If you live in the United States, what is the potential for making increased use of solar energy to provide heat and electricity (with solar cells) where you live or go to school?
2. How many states have areas with excellent, very good, or good availability of direct solar energy?

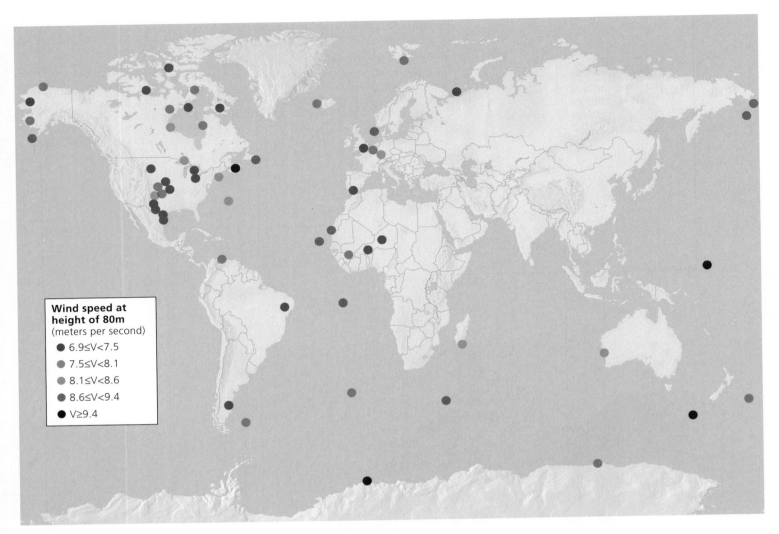

Figure 14 Global potential of winds (an indirect form of solar energy) at a height of 80 meters (260 feet) above the earth's surface based on analysis of over 8,000 wind speed measurements around the world in 2000. This map shows locations of Category 3 through 7 wind speeds, with Category 3 (wind speeds of 6.9–7.5 meters per second or 15–17 miles per hour) being the minimum speed required for economically useful wind generation. The U.S. Department of Energy, National Renewable Energy Laboratory website at **www.nrel.gov/wind/resource_assessment.html** shows that areas with great wind potential are found in northern Europe along the North Sea, the southern tip of the South American continent, the island of Tasmania in Australia, the Great Lakes region, and the northeastern and northwestern coasts of North America. The United Nations is mapping the potential wind (and solar) energy resources of 13 developing countries in Africa, Asia, and South and Central America. (Data from Cristina L. Archer and Mark Z. Jacobson, Department of Civil and Environmental Engineering, Stanford University, Stanford, California, USA)

Legend (Wind speed at height of 80m, meters per second):
- $6.9 \leq V < 7.5$
- $7.5 \leq V < 8.1$
- $8.1 \leq V < 8.6$
- $8.6 \leq V < 9.4$
- $V \geq 9.4$

Data and Map Analysis

1. What are five large countries with little or no wind energy potential? (See Figure 1, pp. S10–S11, in Supplement 3 for country names.)
2. Name a country on each continent (excluding Antarctica) with category 6 or 7 wind energy potential.

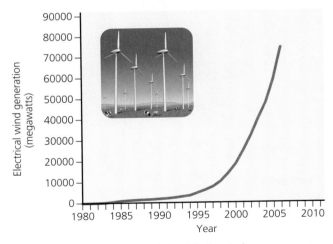

Figure 15 Global generation of electricity by wind energy, 1980–2006. (Data from Global Wind Energy Council, European Wind Energy Association, American Wind Energy Association, and Worldwatch Institute)

Data and Graph Analysis

1. How long did it take for the world to go from zero to 40,000 megawatts in generation of electricity by wind energy?
2. In 2006, the world generated about how many times as much electricity by wind power as it generated in 1995?

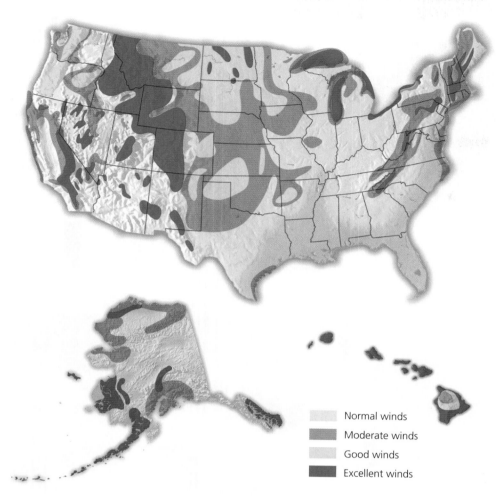

Figure 16 Potential supply of land-based wind energy (an indirect form of solar energy) in the United States. Locate the areas with the highest potential for wind power. Other excellent sites are found offshore along coasts in parts of the United States. Electricity produced by wind farms at such sites can be distributed through an updated and expanded version of the country's electric power grid. For more detailed maps by state see the U.S. Department of Energy, National Renewable Energy Laboratory website at **www.nrel.gov/wind/resource_assessment.html**.

Data and Map Analysis

1. If you live in the United States, what is the general wind energy potential where you live or go to school?
2. How many states have areas with good or excellent potential for wind energy?

Normal winds
Moderate winds
Good winds
Excellent winds

Figure 17 Known global reserves of moderate- to high-temperature geothermal energy. (Data from Canadian Geothermal Resources Council)

Data and Map Analysis

1. Between North and South America, which continent appears to have the greatest total potential for geothermal energy?
2. What country in Asia has the greatest known reserves of geothermal energy? (See Figure 1, pp. S10–S11, in Supplement 3 for country and continent names.)

Figure 18 Potential geothermal energy resources in the continental United States. (Data from U.S. Department of Energy)

Data and Map Analysis

1. If you live in the United States, what is the potential for using geothermal energy to provide heat or to produce electricity where you live or go to school?
2. How many states have areas with very good or excellent potential for using geothermal energy?

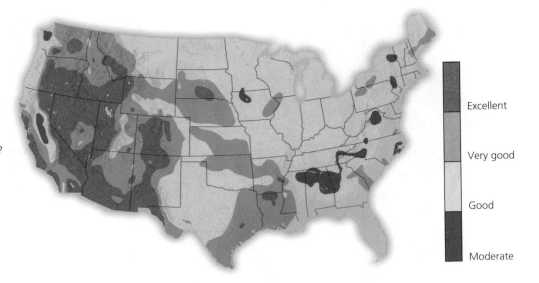

Excellent

Very good

Good

Moderate

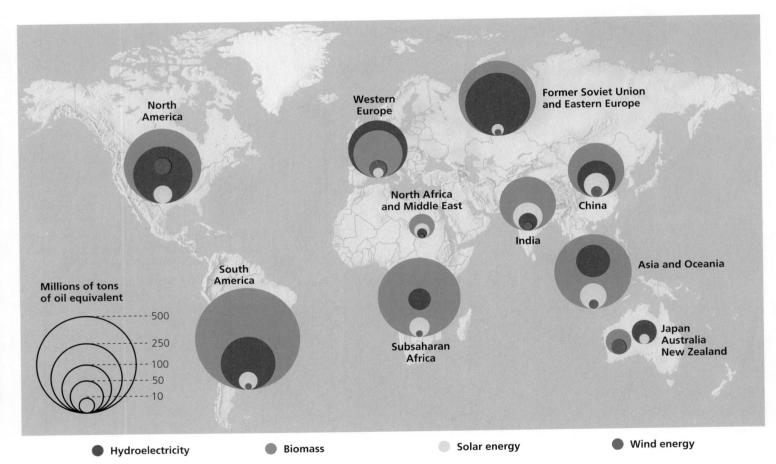

● **Hydroelectricity** ● **Biomass** ● **Solar energy** ● **Wind energy**

Figure 19 Global potential for renewable solar, wind, hydroelectric, and biomass energy. The circle sizes represent the energy potentially available from these resources in millions of metric tons of oil. (Data from Worldwatch Institute)

Data and Map Analysis

1. Which two renewable energy sources have the most potential in the country or region where you live?
2. Which two continents appear to have the greatest potential for wind energy? Which two continents appear to have the greatest potential for solar energy? (See Figure 1, pp. S10–S11, in Supplement 3 for continent names.)

Figure 20 Atmospheric concentration of carbon dioxide (CO_2), 1950–2007. (Data from Scripps Institute of Oceanography and U.S. Energy Information Agency)

Data and Graph Analysis

1. How much did atmospheric CO_2 concentrations grow between 1960 and 2007 (in parts per million)?
2. Assuming that atmospheric CO_2 concentrations continue growing as is reflected on this graph, estimate the year in which such concentrations will reach 400 parts per million.

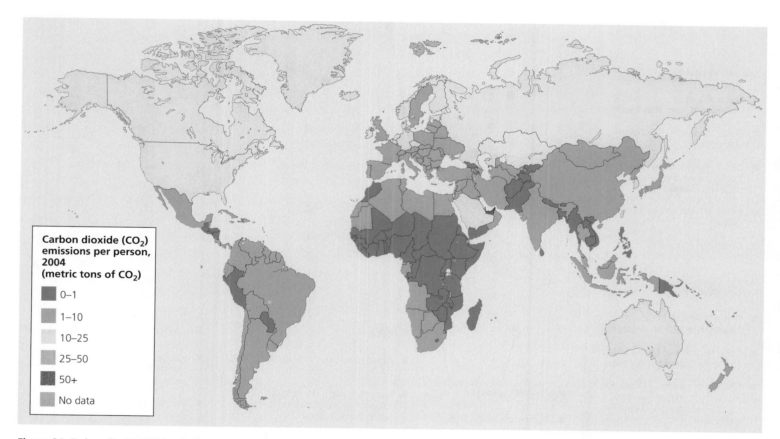

Carbon dioxide (CO_2) emissions per person, 2004 (metric tons of CO_2)

- 0–1
- 1–10
- 10–25
- 25–50
- 50+
- No data

Figure 21 Carbon dioxide (CO_2) emissions per person, 2004 (Data from World Resources Institute and International Energy Agency)

Data and Map Analysis

1. What two countries have the highest CO_2 emissions per person? (See Figure 1, pp. S10–S11, in Supplement 3 for country and continent names.)
2. Which continent has the largest area with CO_2 emissions of 0–1 metric tons?

North America

- In the western mountains, decreased snowpack, earlier snowmelt, more winter flooding, and reduced summer water flows
- Increased forest growth and increased forest fires
- Rapid melting of Alaska glaciers
- Increased intensity, duration and number of heat waves in many cities
- Sea-level rises, tidal surges, and flooding along Gulf and Atlantic coasts
- More intense Atlantic and Gulf hurricanes

Europe

- Increased risk of inland flash floods and coastal flooding
- Disappearance of alpine glaciers
- In the south, more health-threatening heat waves and wildfires, reduced forest area, reduced water availability and hydropower potential, reduced crop yields, and reduced summer tourism
- In the central and eastern areas, more heat waves and peatland fires and reduced summer rainfall and forest productivity
- In the north, negative impacts eventually outweigh such initial benefits as reduced heating demand, less severe winters, increased forest area, and increased crop yields and forest growth

Asia

- Increased flooding, rock avalanches and water resource disruptions from melting Himalayan glaciers
- Ongoing risk of hunger in several developing regions because of crop productivity declines combined with rapid population growth and urbanization
- Coastal flooding from rising sea levels

Small Islands

- Severe flooding and loss of low-lying islands from seal-level rise
- Reduced water resources in many places
- Beach erosion, coral bleaching, loss of mangroves, and declining fisheries
- Increased invasion by nonnative species
- Decreased tourism

Latin America

- Replacement of tropical forest by savannah in eastern Amazonia
- Desertification of farmland from increased drought in dry areas
- Melting of glaciers in tropical Andes with reduced water supplies and hydropower generation
- Rice yields may fall but soybean yields may increase

Polar Regions

- Thinning and shrinking of glaciers and ice sheets
- Decreased Arctic summer sea ice
- Decreased permafrost
- Increased agriculture and forest cover in Siberia
- Arctic tundra replaced by forest

Africa

- Decreased water availability by 2020 for 75 million–250 million people
- Loss of arable land, reduced growing seasons and reduced crop yields in some areas
- Decreased fish stocks in large lakes
- Coastal flooding from rising sea levels
- Flooding, drought, spread of disease

Australia and New Zealand

- Intensified water shortages in southern and eastern Australia and parts of New Zealand by 2030
- Increased loss of biodiversity by 2020
- Increased bush fires
- Increased storm severity and frequency in several places
- Coastal flooding from sea-level rise

Figure 22 Projected regional impacts of global warming during this century. (Data from Intergovernmental Panel on Climate Change, 4th Assessment, 2007)

Data and Map Analysis

1. In the area where you were born, what do you think will be the two most serious projected impacts of global warming?
2. Find three impacts listed on this map that you think may be happening already. Describe these effects, and tell exactly where you think they are occurring.

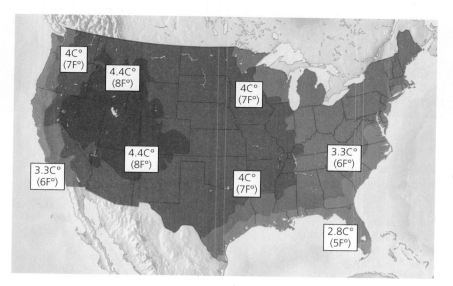

Figure 23 Projected June–July–August temperature changes in the continental United States, 2091–2100, relative to 1971–2000 averages. (Data from Martin Hoerling and Jon Eischeid, Earth System Research Laboratory National Oceanic and Atmospheric Administration, 2007)

Data and Map Analysis

1. If you live in one of these lower 48 states, what will be the projected temperature change between 2091 and 2100 where you live or go to school?
2. If these temperature projections are correct, what might be the most desirable and least desirable states to live in between 2091 and 2100?

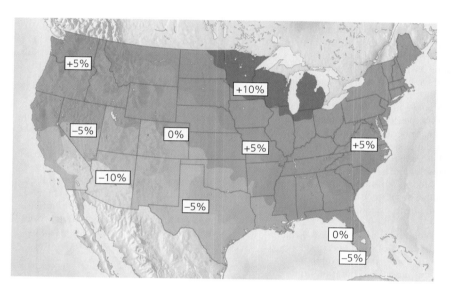

Figure 24 Projected annual precipitation changes in the continental United States, 2091–2100, relative to 1971–2000 averages. (Data from Martin Hoerling and Jon Eischeid, Earth System Research Laboratory National Oceanic and Atmospheric Administration, 2007)

Data and Map Analysis

1. If you live in one of these lower 48 states, what will be the projected change between 2091 and 2100 in annual precipitation where you live or go to school?
2. How many states will have areas with declines of 5% or more in annual precipitation, according to these projections?
3. If these projections are correct, what might be the most desirable and least desirable states to live in between 2091 and 2100. Taking into account these projections based on precipitation and the projection in Figure 23 based on temperature, what might be the most desirable and least desirable states to live in between 2091 and 2100.

Global Trade and the Environment (Chapters 23, 24)

There Are Some Environmental Concerns about Free Trade

Expanding international trade is seen as a way to stimulate economies around the world, help distribute wealth, and decrease poverty. International free trade allows companies providing goods and services to take advantage of the cheapest labor and resources anywhere in the world. In theory, this should also lower the market prices of goods and services for consumers.

On April 15, 1994, representatives of 120 nations signed the Uruguay Round of the General Agreement on Tariffs and Trade (GATT). This is a revised version of the 1948 GATT convention, which attempted to lower tariff barriers to world trade between member nations. The new GATT established a World Trade Organization (WTO) and gave it the status of a major international organization similar in stature and power to the United Nations and the World Bank. The WTO's role is to enforce the new GATT rules of world trade and settle disputes about these rules between participating nations.

Currently, the WTO has 151 member countries. Representatives of the Quad Countries— the United States, Canada, Japan, and the European Union—determine most WTO policy.

Most WTO officials are trade experts and corporate lawyers, representing primarily the interests of transnational corporations. The WTO has no elected representatives and is not subject to freedom of information laws or public review of its proceedings and decisions. Any member country can charge any other WTO member country with violating one of the WTO's complex international trade rules.

When this occurs, several important conditions apply. First, the case is decided in closed session by a tribunal of three anonymous WTO judges, usually corporate lawyers with no particular expertise in the issues being decided. There are no conflict-of-interest restraints on tribunal members, and no information about their possible conflicts of interest is available to the public.

Second, all documents, transcripts, and details of the proceedings are kept secret and only the results are announced.

Third, only official government representatives of the countries involved can submit documents or appear before the tribunal.

Fourth, decisions are binding worldwide and can be appealed only to another tribunal of judges within the WTO. A final panel ruling can be appealed to the entire WTO, but the ruling can be overturned only by agreement of all WTO members. This is virtually impossible because the winning country is unlikely to vote to overturn a ruling in its favor.

Any country (or part of a country) that violates a ruling of WTO panels has four choices: amend its laws to comply with WTO rules, pay annual compensation to the winning country, pay high tariffs imposed on the disputed goods by the WTO, or find itself shunned and locked out of global commerce.

Proponents of this arrangement argue that the significant transfer of power from nations to the WTO is necessary and beneficial for several reasons. First, globalization of trade is inevitable and we have to take part in guiding it. Second, reducing global trade barriers will benefit developing countries, whose products often are at a competitive disadvantage in the global marketplace because of trade barriers erected by developed countries. Third, reducing global trade barriers can stimulate economic growth in all countries by allowing consumers to buy more things at lower prices. Fourth, globalization of trade will raise the environmental and health standards of developing countries.

Opponents contend that current WTO rules will have several harmful consequences. First, they will increase the economic and political power of transnational corporations and decrease the power of small businesses, citizens, and democratically elected governments. Second, they will eliminate many jobs and will lower wages in developed countries, and eventually in developing countries, as transnational companies move their operations throughout the world in search of cheap labor and natural resources and lower environmental standards. Third, current WTO rules will weaken environmental and health and safety standards in developed countries.

According to critics, some WTO rules and omissions of important environmental and health principles can limit the ability of national, state, and local governments to protect the environment, the health of citizens, and worker health and safety. Here are some examples:

- Governments cannot set standards for how imported products are produced or harvested. This means, for example, that government purchasing policies cannot discriminate against materials produced by child labor or slave labor. Also, they cannot require that items be manufactured from recycled materials or that fish-harvesting methods be protective of dolphins or turtles.

- Current WTO rules do not recognize the rights of countries to take action to protect the atmosphere, the oceans, and other parts of the global commons. There is concern that some provisions of international treaties to protect biodiversity and the ozone layer and to reduce the threats of global warming might be ruled illegal under WTO rules.

- All national, state, or local environmental, health, and safety laws and regulations must be based on globally accepted scientific evidence and risk analysis showing there is a worldwide scientific consensus on the danger. Otherwise, they are considered to be trade barriers that exceed WTO international standards. Because of the inherent scientific uncertainties in determining health risks and carrying out risk analysis, this requirement is almost impossible to meet. In other words, the burden of proof falls on those trying to prevent pollution rather than on polluters. This means that chemicals, products, and technologies cannot be banned on the basis of the pollution prevention and precautionary principles—two of the foundation stones of modern environmental protection.

- National laws covering packaging, recycling, and eco-labeling of items involved in international markets are illegal barriers to trade. Enforcing this rule can effectively cancel the third mainstay of modern environmental protection: consumers' right to know about the safety and content of products through labeling.

Environmental scientists and environmental groups contend that the four WTO rules described above, if they are allowed to stand, will force us back to an earlier and less effective era of using an end-of-pipe pollution cleanup strategy, based on uncertain and easily manipulated risk assessment, as the primary way to deal with pollution.

On a more positive note, nations may be able to use the WTO to reduce environmentally harmful subsidies that distort the economic playing field. But this would also prevent using subsidies to reward companies producing environmentally beneficial goods and services—an important part of proposed strategies for developing eco-economies (Figure 23-14, p. 630).

We Can Improve Global Trade Agreements

Critics call for changing trade agreements to make WTO proceedings more open and to establish global standards for protecting the environment, consumers, and workers. They would correct what they believe are serious weaknesses in GATT, turning it into GAST: the General Agreement for Sustainable Trade.

They offer several suggestions for doing this:

- Set minimum environmental, consumer protection, and worker health and safety standards for all participating countries.

- Open all discussions and findings of any GATT panel or other WTO body to global public scrutiny and inputs from experts on the issues involved.

- Incorporate the precautionary and pollution prevention principles into WTO rules.

- Protect the rights of consumers to know about the health and environmental impacts of imported products they purchase by allowing eco-labeling programs.

- Recognize the right of countries to use trade measures to protect the global commons.

- Allow countries to require that imported or exported items be manufactured totally or partially from recycled materials.

- Allow national, state, or local governments to restrict imports of products from countries shown by international investigation to have used child or slave labor or to have violated universally recognized human rights.

- Allow international environmental agreements and treaties to prevail when they conflict with WTO rules or the rules of any other trade agreement.

- Set up and fund a World Environmental Organization (WEO) as a counterbalance to the power and influence of the WTO over global and national environmental policy.

Unless citizens and nongovernmental organizationals exert intense pressure on legislators in the Quad Countries governing the WTO, critics warn that such safeguards will not be incorporated into GATT and other international trade agreements.

HOW WOULD YOU VOTE?

Do you favor changing trade agreements to make WTO proceedings more open and to establish global standards for protecting the environment, consumers, and workers? Cast your vote online at **academic.cengage.com/ biology/miller.**

There Are Different Philosophies Concerning Right and Wrong Behavior

Several major moral philosophical systems deal with the concepts of what constitutes right and wrong environmental (or other) behavior.

One is *universalism,* developed by philosophers such as Plato (427–347 B.C.) and Immanuel Kant (1724–1804). According to this view, there are basic principles of ethics, or rules of right and wrong, that are universal and unchanging. Some believe that God or other sources of wisdom have provided ethical guidelines. Others believe that these rules can be discovered through reason, experience, and knowledge.

Another philosophy is *utilitarianism,* developed by Jeremy Bentham (1748–1832) and John Stuart Mill (1806–1873). It says that an action is right if it produces the greatest satisfaction or pleasure for the greatest number of people. Critics contend that this assumes that we can somehow quantitatively measure happiness and compare it among different people.

A related moral philosophy is *consequentialism,* which proposes that whether an act is morally right or wrong depends on its consequences. In effect, we determine correct moral conduct *solely* by analyzing the beneficial and harmful consequences of our actions. Thus, an action is morally right if its consequences as a whole are more favorable than unfavorable.

Another philosophy of right and wrong is *relativism,* promoted by ancient Greek teachers called Sophists who disagreed with Plato's universalist ethics. It asserts that moral values of right and wrong are relative to cultures, eras, or situations and that there are no absolute principles of right and wrong.

According to the philosophy of *rationalism,* we can develop principles of right and wrong by using logic to analyze ideas and arguments. This philosophy based on the power of reason was developed by philosophers such as René Descartes (1596–1650), Benedict De Spinoza (1632–1677), and Gottfried W. Leibniz (1646–1716).

According to *nihilism,* often associated with Friedrich Nietzsche (1844–1900), the concepts of values and moral beliefs are useless because nothing can be known or communicated. There is no purpose or meaning to life except the struggle to survive in which "might is right."

Because this is not a philosophy textbook, these brief summaries of several moral philosophies are superficial and incomplete. There are many complex arguments involved in evaluating these and other moral philosophies, which you can explore by studying philosophy.

THINKING ABOUT
Philosophies of Right and Wrong

Based on the limited discussion provided here, which one of these philosophies has the greatest appeal for you? Why? You might want to use the library or the Internet to explore these philosophies in greater detail.

Many Cultures and Religions Have a Deep Respect for Life and Nature

Several of the world's cultures and religions have a reverence for life and nature. Examples of such nature-based cultural beliefs or religions are *Shamanism, Shintoism, Buddhism, Taoism,* and *Hinduism,* which date back for thousands of years.

Shamanism holds that all living creatures have souls or sacred spirits that should be revered and respected. Before killing an animal or cutting a tree, one should honor and placate the spirit or deity residing in these forms of life. Shamanism provides the basis for many different forms of religion practiced among indigenous tribal societies of North and South America, Africa, Asia, and Australia.

Shintoism, the earliest Japanese religion, holds that all trees, animals, and other forms of life and all natural objects such as mountains, seas, rivers, and rocks have divine spirits or deities that one should respect and revere to secure their continued favor. Its most prominent values are a sense of gratitude and respect for life and a deep appreciation of the beauty and power of nature. Shintos put this belief into practice by establishing sacred shrines such as groves of trees or parts of a river that are to be protected from human degradation and destruction. Currently, the only green spaces in crowded and heavily industrialized urban Japan are groves of trees that surround Shinto shrines.

Buddhism, a later Japanese religion, holds that all human beings have the potential to attain the wisdom that can bring an end to suffering, primarily by dismissing their attachments to material things. Most Buddhists have a respect and compassion for life and the earth, and they believe that we should not unnecessarily kill trees, animals, and other forms of life, including spiders, rodents, and forms of life many people regard as pests. To Buddhists, the desire to own land and things is a form of greed that promotes violent conflict, environmental destruction and degradation, and chaos. They see all life as integrated into a web of mutual interdependence where humans can develop an ecological ethic based on restraint, simplicity, compassion, loving kindness, calmness, patience, generosity, nonviolence, and wisdom.

Taoism views all systems as coexisting in an interdependent network, which adapts and changes through dynamic interactions between two opposite forces, Yin and Yang. Taoists believe that through personal and social change we can convert the destructive elements of life into constructive elements and thus achieve harmony and peace through nonaggressive and peaceful coexistence. Taoists call for us to respect and revere the earth and all life. They do not believe in trying to control or manage everything; instead they believe in preventing trouble before it arises, focusing on simplicity, and encouraging those who have too much to give to those who have too little.

Islam, the world's second largest religion after Christianity, is based on the teachings of the prophet Muhammad in the 7th century as laid down in the Qur'an. It charges its believers to submit themselves to the will of Allah (God). Allah has commanded Muslims to respect and revere the environment and nature's many bounties given to them by the Almighty. Islam puts humans on a higher plane of life than other species and puts the earth in the hands of Muslims. However, the Qur'an calls for Muslims to "walk on the Earth in humility" and recognize that they have a profound responsibility to be good stewards of the earth. To Muslims, the color green is the most blessed of all colors and some Muslin leaders call for a "green *jihad*" to deal with the global environmental crisis.

Hinduism, the world's third largest religion, consists of several thousand different religious groups that have developed in India during the past 3,000 years. Hinduism calls for respecting nature by living in harmony with all life, instead of trying to change or conquer it. Hindus believe that a universal God (Brahman) is present in all nature (including rivers, mountains, lakes, and minerals), in all creatures, and in all human beings regardless of their faith or lack of it. Because God is in all things and all things are in God, all of nature is sacred and should be respected and celebrated with gratitude based a tradition of

simple living and nonviolence to nature and one another.

Hinduism has great respect for the tree, which helps to explain India's long history of forest protection. It also includes respect for rivers, such as India's Ganges, as holy bodies of water (Figure 20-9, p. 539). According to some scholars, Hinduism places more emphasis on environmental ethics than any other religion.

Christianity, the world's largest religion, is based on a belief in a single, supreme God as revealed by the teachings of the Bible. In general, Christians believe that God gave humans dominion over the world as evidenced by the biblical command to "Be fruitful and multiply, and replenish the earth, and subdue it; and have dominion over the fish of the sea, and over the fowl of the air, and over every living thing that moveth upon the earth." (Genesis 1:28). Christian scholars disagree on how to interpret this biblical passage. Most agree that humans are more important than other species, but many argue that subduing the earth does not mean destroying or severely degrading it. Later in Genesis 2:15, God installs Adam in the Garden of Eden "to cultivate and to preserve it." According to this interpretation, nature is a sacred gift from God.

Biblical scholars also point out that Genesis 1:28 must be taken in context with other Bible passages that call for humans to love and nurture creation rather than to dominate and exploit it. Many Biblical passages, especially in the New Testament, call for Christians to practice *stewardship* by caring for and responsibly managing the earth with humility and reverence for places and species. According to many Christian leaders, Christians have a moral responsibility to sustain life and to make the earth a better and more socially just place for humans and other forms of life, out of respect for the earth as God's creation.

Judaism includes belief in a single and unique God as revealed by the teachings of the *written Torah* (the first 5 books of the Bible) and the *oral Torah* (teachings now contained in the *Talmud* and other writings). Like Christianity, Judaism believes that humans are superior to other forms of life, were created in God's image, and have dominion over the earth as described in Genesis 1:28. However, scholars point to various writings in the Torah and other Jewish texts that call for believers in Judaism to respect and care for the earth with humility and to live lightly on the earth. The earth belongs to God and we are given permission to enjoy it, but we must not waste or wantonly destroy the earth's abundance. Judaist teachings also call for us to preserve the earth for our descendants.

Of course, the real test of any religion's effectiveness is whether people following the religion put its beliefs into practice. For example, Buddhism and Shintoism are widespread religions in Japan. This has led the country to revere and protect its own forests. At the same time, however, Japan is one of the world's largest importers of timber harvested from tropical areas where deforestation is a threat.

Because this is not a religious history textbook, these brief summaries of various major religions and their relationships to nature are superficial and incomplete. They do not reflect the many different sects and beliefs within each of these religions. Nor do they cover more geographically limited, but no less important, religious views such as those of the many native American tribes and indigenous cultures in other parts of the world. However, this overview might help you understand some of the roots of various environmental worldviews.

THINKING ABOUT
Religion and Environmental Problems

What religion do you believe has the greatest potential for dealing with the environmental problems we face? What role, if any, does religion play in your environmental worldview and in how you deal with environmental problems?

CHAPTER 3 Ecosystems: What Are They and How Do They Work?

1. Visit a nearby terrestrial ecosystem or aquatic life zone and try to identify major producers, primary and secondary consumers, detritus feeders, and decomposers.

2. Write a brief scenario describing the major consequences for us and other species if each of the following biogeochemical cycles were to stop functioning: **(a)** water; **(b)** carbon; **(c)** nitrogen; **(d)** phosphorus; and **(e)** sulfur.

CHAPTER 4 Biodiversity and Evolution

1. Visit a nearby terrestrial or aquatic ecosystem and try to identify a native, nonnative, indicator, keystone, and foundation species.

2. Visit a nearby terrestrial or aquatic ecosystem, try to identify a generalist species and a specialist species, and try to estimate the area's species richness and species evenness.

3. Use the library or the Internet to learn about the emerging field of synthetic biology, which combines biology, genetics, and engineering. How might synthetic biology get around some of the problems raised by genetic engineering? What problems does it pose?

CHAPTER 5 Biodiversity, Species Interactions, and Population Control

1. Use the library or Internet to find and describe two species not discussed in this textbook that are engaged in a **(a)** commensalistic interaction, **(b)** mutualistic interaction, and **(c)** parasite–host relationship.

2. Visit a nearby natural area and try to identify examples of **(a)** mutualism and **(b)** resource partitioning.

3. Do some research to identify the parasites likely to be found in your body.

4. Choose one wild plant species and one wild animal species and use the library or the Internet to help you analyze the factors that are likely to limit the population of each species.

5. Visit a nearby land area such as a partially cleared or burned forest or grassland or an abandoned crop field and record signs of secondary ecological succession. Was the disturbance that led to this succession natural or caused by humans? Study the area carefully to see whether you can find patches that are at different stages of succession because of various disturbances.

CHAPTER 6 The Human Population and Its Impact

1. Assume your entire class (or each of a number of groups from your class) is charged with coming up with a plan for cutting the world's population growth rate by half within the next 20 years. Develop a detailed plan that would achieve this goal, including any differences between policies in developing countries and those in developed countries. Justify each part of your plan. Try to anticipate what problems you might face in implementing the plan, and devise strategies for dealing with these problems.

2. Prepare an age structure diagram for your community. Use the diagram to project future population growth and economic and social problems.

CHAPTER 7 Climate and Terrestrial Biodiversity

1. How has the climate changed in the area where you live during the past 50 years? Investigate the beneficial and harmful effects of these changes. How have these changes benefited or harmed you personally?

2. How have human activities over the past 50 years affected the characteristic vegetation and animal life normally found where you live?

CHAPTER 8 Aquatic Biodiversity

1. Develop three guidelines for preserving the earth's aquatic biodiversity based on the four **scientific principles of sustainability** (see back cover).

2. Visit a nearby lake or reservoir. Would you classify it as oligotrophic, mesotrophic, eutrophic, or hypereutrophic? What are the primary factors contributing to its nutrient enrichment? Which of these factors are related to human activities? Try to determine the specific activities in your area that may be affecting this body of water.

3. Developers want to drain a large area of inland wetland in your community and build a large housing development. List **(a)** the main arguments the developers would use to support this project and **(b)** the main arguments ecologists would use in opposing it. If you were an elected city official, would you vote for or against this project? Can you come up with a compromise plan?

CHAPTER 9 Sustaining Biodiversity: The Species Approach

1. Make a record of your own consumption of all products for a single day. Relate your level and types of consumption to the decline of wildlife species and the increased destruction, degradation, and fragmentation of wildlife habitats in **(a)** the country where you live and **(b)** tropical forests. Compare your results with those of your classmates.

2. Identify examples of habitat destruction or degradation in your community that have had harmful effects on the populations of various wild plant and animal species. Develop a management plan for rehabilitating these habitats and species.

3. Choose a particular animal or plant species that interests you and use the library or the Internet to find out (a) its numbers and distribution, (b) whether it is threatened with extinction, (c) the major future threats to its survival, (d) actions that are being taken to help sustain this species, and (e) a type of reconciliation ecology that might be useful in sustaining this species.

CHAPTER 10 Sustaining Terrestrial Biodiversity: The Ecosystem Approach

1. If possible, try to visit (a) a diverse old-growth forest, (b) an area that has been recently clear-cut, and (c) an area that was clear-cut 5–10 years ago. Compare the biodiversity, soil erosion, and signs of rapid water runoff in each of these areas.

2. For many decades, New Zealand has had a policy of meeting all its demand for wood and wood products by growing timber on intensively managed tree plantations. Use the library or Internet to evaluate the effectiveness of this approach and its major advantages and disadvantages.

3. Try to find an area near where you live that includes a degraded ecosystem. Assume you are the conservation biologist in charge of restoring the ecosystem, and write a five-step plan for completing the project. Would your plan involve use of reconciliation ecology? Explain.

4. Use the library or the Internet to find one example of a successful ecological restoration project not discussed in this chapter and an example of one that failed. For each example, describe the strategy used, the ecological principles involved, and why the project succeeded or failed.

CHAPTER 11 Sustaining Aquatic Biodiversity

1. Survey the condition of a nearby wetland, coastal area, river, or stream and research its history. Has its condition improved or deteriorated during the last 10 years? What local, state, or national efforts are being used to protect this aquatic system? Develop a plan for protecting it.

2. Pick a major seafood species and describe its life cycle, including its reproductive cycle. Find out if the species has been overfished, and if so, where in the world it has been depleted, to what degree, and by what methods. Describe your findings. Write a brief script for telling a friend or relative why you would or would not recommend choosing that species from a seafood restaurant menu.

3. Work with your classmates to develop an experiment in aquatic reconciliation ecology for your campus or local community.

CHAPTER 12 Food, Soil, and Pest Management

1. Use public health records and other local government records to estimate how many people in your community suffer from undernutrition or malnutrition. Has this problem increased or decreased since 1980? What are the basic causes of this problem, and what is being done to alleviate it? Share the results of your study with local officials, and present a plan for reducing hunger in your community.

2. Make a list of all the food you eat in one day and read all labels or look up the amount of calories, fat, protein, and carbohydrates in each food. Then determine how many calories you took in that day and the percentages that came from fat, protein, and carbohydrates. Rate your diet as healthy, borderline healthy, or unhealthy. Compare your results with those of your classmates.

3. If possible, visit both a conventional industrialized farm and an organic or low-input farm. Compare (a) soil erosion and other forms of land degradation, (b) use and costs of energy, (c) use and costs of pesticides and inorganic fertilizer, (d) use and costs of natural pest control and organic fertilizer, (e) yields per hectare for the same crops, and (f) overall profit per hectare for the same crops.

4. If you live in an urban area, survey vacant lots that could be used for food production. Pick a lot and write a plan for how you would use it for growing food, including what you would grow and how you would do it.

5. Conduct a survey of soil erosion and soil conservation in and around your community on cropland, construction sites, mining sites, grazing land, and deforested land. Use these data to develop a plan for reducing soil erosion in your community.

6. Weigh the food purchased in your home and the food thrown out for a week. Develop a plan for cutting the resulting food waste in half. Make a similar study for your school cafeteria and reporting the results and your recommendations to school officials.

7. Use the library or the Internet to learn about the four types of vegetarians and the advantages and disadvantages of a vegetarian diet in terms of your health and the environment.

8. Interview a sample of people and/or food providers in your community in order to make an estimate of how much of the food consumed in your community is grown locally (within a 160-kilometer (100-mile) radius). Develop a plan for increasing the amount of food grown in the area where you live.

CHAPTER 13 Water Resources

1. In your community,

 (a) What are the major sources of the water supply?

 (b) How is water use divided among agricultural, industrial, power plant cooling, and public uses?

 (c) Who are the biggest consumers of water?

 (d) What has happened to water prices (adjusted for inflation) during the past 20 years? Are they too low to encourage water conservation and reuse?

 (e) What water supply problems are projected?

2. Who owns and manages the water supply and distribution system in your community? If it is privately owned or managed, is the company a domestic or foreign corporation? Does this make any difference to you? Explain.

3. The direct use of water per person in the United States is 570 liters (150 gallons) per day. Many of the world's poor have to live on only 4 liters (1 gallon) of water per day. To get an idea of what this means try to survive on this daily amount of water for 2 days. Describe your experience and how it differs from your normal water usage.

4. Develop a water conservation plan for your school and submit it to school officials.

5. Consult with local officials to identify any floodplain areas in your community. Develop a map showing these areas and the types of activities (such as housing, manufacturing, roads, and recreational use) found on these lands. Evaluate management of such floodplains in your community and come up with suggestions for improvement.

CHAPTER 14 Geology and Nonrenewable Minerals

1. Write a brief scenario describing the series of consequences that would arise for us and other life forms if the rock cycle were to stop functioning.

2. What nonrenewable mineral resources, if any, are extracted in your area? What mining methods are used, and what have been their harmful environmental impacts? How has mining benefited the local economy?

3. Do some research to determine what each of the following is made of and how much of those materials are required to make each item: **(a)** a class ring, **(b)** a wide-screen plasma television screen, and **(c)** a large pickup truck.

4. Use the library or Internet to find out what major advances have recently been made in the nanotechnology industry and what attempts, if any, have been made to control its potentially harmful effects.

CHAPTER 15 Nonrenewable Energy

1. How is the electricity in your community produced? How has the inflation-adjusted cost of that electricity changed since 1970?

2. Write a two-page scenario of what your life might be like without oil (**Core Case Study**, p. 399). Compare it with scenarios developed by other members of your class.

 CORE CASE STUDY

3. Use the library or the Internet to find information about the accident at the Three Mile Island (TMI) nuclear power plant near Harrisburg, Pennsylvania (USA), in 1979. According to the nuclear power industry, the TMI accident showed that its safety systems work because the accident caused no known deaths. Other analysts argue that the accident was a wake-up call about the potential dangers of nuclear power plants and that it led to tighter and better safety regulations. Use the information you find to choose one of these positions, and defend your choice.

CHAPTER 16 Energy Efficiency and Renewable Energy

1. Make a study of energy use in your school and use the findings to develop an energy-efficiency improvement program. Present your plan to school officials.

2. Learn how easy it is to produce hydrogen gas from water using a battery, some wire for two electrodes, and a dish of water. Hook a wire to each of the poles of the battery, immerse the electrodes in the water, and observe bubbles of hydrogen gas being produced at the negative electrode and bubbles of oxygen at the positive electrode.

3. Study the maps in Figures 11, 13, 14, 16, and 19 in Supplement 10 to determine what renewable-energy resources could be used in your local community or region. Which, if any, of these renewable forms of natural capital are being used in your community? How would you increase your own use of these resources and their use in your community?

4. Use the library or the Internet to compare the energy policies of the United States, Germany, and China. Write a brief description of each, and compare key points to determine which of the three you think is the most sustainable.

5. Visit a local appliance store and compare energy-efficiency labels for several products, including, but not limited to, refrigerators, clothes washers, driers, and dehumidifiers. Compile a table listing the most and least efficient brands for each of these appliances.

CHAPTER 17 Environmental Hazards and Human Health

1. Use the library or the Internet to find recent articles describing the increasing genetic resistance in disease-causing bacteria to commonly used antibiotics. Evaluate the evidence and claims in these articles.

2. Pick an emerging viral disease not discussed in this chapter and use the library or Internet to find out **(a)** how it spreads, **(b)** its effects, **(c)** strategies for controlling its spread, and **(d)** possible treatments.

3. Pick a potentially harmful chemical not discussed in detail in this chapter and use the library or Internet to evaluate **(a)** its potential harm, **(b)** the scientific evidence for such claims, and **(c)** solutions for dealing with this threat.

CHAPTER 18 Air Pollution

1. Identify climate and topographic factors in your local community that **(a)** intensify air pollution and **(b)** help reduce air pollution.

2. Evaluate your exposure to some or all of the indoor air pollutants in Figure 18-16 (p. 484) in your school, workplace, and home. Come up with a plan for reducing your exposure to these pollutants.

3. Have buildings at your school been tested for radon? If so, what were the results? What has been done about areas with unacceptable levels? If this testing has not been done, talk with school officials about having it done.

CHAPTER 19 Climate Change and Ozone Depletion

1. As a class, conduct a poll of students at your school to determine **(a)** whether they understand the difference between global warming of the troposphere and ozone depletion in the stratosphere, and **(b)** whether they believe global warming from an enhanced greenhouse effect is a very serious problem, a moderately serious problem, or of little concern. Tally the results to see whether there are differences related to each poll participant's year in school, political leaning (liberal, conservative, independent), age, and sex.

2. As a class, conduct a poll of students at your school to determine whether they believe stratospheric ozone depletion is a very serious problem, a moderately serious problem, or of little concern. Tally the results to see whether there are differences related to each poll participant's year in school, political leaning (liberal, conservative, independent), age, and sex.

3. Use the library or the Internet to determine how the current government policy on global warming in the country where you live compares with the policy suggestions made by various analysts and listed in Figures 19-13 (p. 515), 19-15 (p. 516), and 19-17 (p. 522).

4. If you drive a car, calculate how much CO_2 it emits per day by taking the number of miles you drive, multiplying it by 20, and dividing the result by the number of miles per gallon your car gets. If you use the metric system, multiply the kilometers driven by the number of liters of gasoline it takes to drive your car 100 kilometers and divide the result by 42 to get your daily CO_2 emissions in kilograms. In either case, add another 20% to include the CO_2 emitted in manufacturing the gasoline you used. Compare your results with those of other members of your class.

5. Write a one- to two-page scenario of what your life could be like by 2060 if nations, companies, and individuals do not take steps to reduce global warming and the resulting climate change caused mostly by human activities. Compare and critique scenarios written by other members of your class.

CHAPTER 20 Water Pollution

1. Are the storm drains and sanitary sewers in your community combined or separate? Are there plans to reduce pollution from runoff of storm water? If not, evaluate the costs and benefits of developing separate storm drains and sanitary sewers, and present your findings to local officials.

2. Use library research, the Internet, and user interviews to evaluate the relative effectiveness and costs of home water purification devices. Determine the types of pollutants each device removes and the effectiveness of this process.

3. Find out the price of tap water where you live. Then go to a grocery or other store and record prices per liter (or other volume unit) on all the available types of bottled water. Use these data to compare the prices of various brands of bottled water with the price of tap water.

CHAPTER 21 Solid and Hazardous Waste

1. Collect all of the trash (excluding food waste) that you generate in a typical week. Measure its total weight and volume. Sort it into major categories such as paper, plastic, metal, and glass. Then weigh each category and calculate the percentage by weight of the trash in each category. What percentage by weight of this waste consists of materials that could be recycled? What percentage consists of materials for which you could have used a reusable substitute, such as a coffee mug instead of a disposable cup? What percentage by weight of the items could you have done without? Tally and compare your results with those of your classmates.

2. What percentage of the municipal solid waste (MSW) in your community is (a) placed in a landfill, (b) incinerated, (c) composted, and (d) recycled? What technology is used in local landfills and incinerators? What leakage and pollution problems have local landfills or incinerators had? Does your community have a recycling program? Is it voluntary or mandatory? Does it have curbside collection? Drop-off centers? A hazardous waste collection system? Devise a plan for improving the MSW system and reducing hazardous waste in your community and submit it to local officials.

3. Make a survey of the hazardous materials (Figure 21-2, p. 562) found in your house or apartment, or in your family home if you live in a dorm. Which of these materials are actually used? Contact city officials to find out how you can dispose of hazardous chemicals you do not need.

4. What hazardous wastes are produced at your school? What happens to these wastes?

5. If you live in the United States, go to the EPA Superfund website at **www.epa.gov/superfund/sites/npl/npl.htm**. Click on your state to find out how many hazardous sites it has on the National Priority List. Find any sites close to where you live or go to school. Click on each site near you to learn about its history, what types of pollutants it contains, the sources of these pollutants, how it is being cleaned up, and progress toward this goal.

6. If you live in the United States, go to the EPA's Toxic Release Inventory website at **www.epa.gov/tri/** to find out what toxic chemicals are being stored and released in your neighborhood.

7. If possible, visit a local recycling center or materials-recovery facility to find out how it works, where the materials separated out for recycling go, how the prices of these separated materials have fluctuated in the last 3 years, and the major problems faced by the facility.

CHAPTER 22 Sustainable Cities

1. Consult local officials to determine how land use is decided in your community. What roles do citizens play in this process?

2. For a class or group project, borrow one or more decibel meters from your school's physics or engineering department or from a local electronics repair shop. Make a survey of sound pressure levels at various times of day and at several locations. Plot the results on a map. Also, measure sound levels in a room with a sound system and from the earphones of an iPod or other listening device at several different volume settings. If possible, measure sound levels at an indoor concert, a club, and inside and outside a boom car at various distances from the speakers. Correlate your findings with those in Figure 22-9, p. 596.

3. Pick a city (the one in which you live or the one nearest you) that has experienced sprawl. Do some research to determine (a) the increased monetary costs to taxpayers of providing city services to the areas of sprawl and (b) the increase in tax revenues resulting from the growth of the city. Compare the two numbers. List three other costs and three other benefits that cannot be measured monetarily that resulted from the city's growth. (For example, consider health costs of increased air pollution versus the health benefits to those living outside of the city center.)

4. As a class project, use the following criteria to rate the community where you live or go to school on a green index from 0 to 10. Rate the community on each of the following questions and average the results to get an overall score. Are existing trees protected and new ones planted throughout the city? Do you have parks to enjoy? Can you swim in any nearby lakes and rivers? What is the quality of your water and air? Is there an effective noise pollution reduction program? Does your city have a recycling program, a composting program, and a hazardous waste collection program, with the goal of reducing the current solid waste output by at least 60%? Is there an effective mass transit system? Are there bicycle paths? Are all buildings required to meet high energy-efficiency standards? How much of the energy is obtained from locally available renewable resources? Are environmental regulations for existing industry tough enough and enforced well enough to protect citizens? Do local officials look carefully at an industry's environmental record and plans before encouraging it to locate in your city or county? Are city officials actively planning to improve the quality of life for all its citizens? If so, what is the plan? (Add any other questions that you think should be included.)

CHAPTER 23 Economics, Environment, and Sustainability

1. List all the goods you use, and then identify those that meet your basic needs and those that satisfy your wants. Identify any economic wants that (a) you would be willing to give up and (b) you believe you should give up but are unwilling to

give up. Relate the results of this analysis to your personal impact on the environment. Compare your results with those of your classmates.

2. Pick one of the suggestions listed under "Eco-Friendly Businesses" in Figure 23-15 (p. 630) and develop a business plan for a company that would provide the service you selected. For example, assume you'll go into the business of ecological restoration and describe: (a) your service, (b) your customers, (c) your mission statement, and (d) your strategy for promoting the business.

3. Pick an environmental regulation in the state or country where you live (such as a water pollution law or regulation) and examine how it affects businesses and other organizations. Determine whether it is an innovation-friendly regulation and explain why or why not. If not, how could it be made more innovation-friendly?

4. Interview officials at a company in your town or region to get their views on an environmental regulation affecting their industry. Describe the effects of that regulation on the company and the company's approach toward dealing with it.

CHAPTER 24 Politics, Environment, and Sustainability

1. Polls have identified five categories of U.S. citizens in terms of their concern over environmental quality: (a) those involved in a wide range of environmental activities, (b) those who do not want to get involved but are willing to pay more for a cleaner environment, (c) those who are not involved because they disagree with many environmental laws and regulations, (d) those who are concerned but do not believe individual action will make much difference, and (e) those who strongly oppose the environmental movement. To which group do you belong? Compare your answer to those of other members in your class, and determine the percentage for each category of the total number of people in your class. As a class, conduct a similar poll on your campus.

2. Rate the last three presidents of the United States (or leaders of the country where you live) on a scale of 1–10 in terms of their performance as environmental leaders. Explain your rankings, giving three reasons for each. See Supplement 5, pp. S35–S38, for information on the environmental policies of various U.S. presidents.

3. Obtain a topographic map of the region where you live and use it to identify federally, locally, and state-owned lands designated as parks, rangeland, forests, and wilderness areas. Identify the government agency or agencies responsible for managing each of these areas, and try to evaluate how well these agencies are preserving the natural resources on this public land on your behalf.

4. Pick a particular environmental law in the United States or in the country where you live. Use the library or the Internet to evaluate the law's major strengths and weaknesses. Decide whether the law should be weakened, strengthened, or abolished, and explain why. List the three most important ways you believe the law should be improved.

5. What student environmental groups, if any, are active at your school? How many people actively participate in them?

Pick one of these groups and find out what they have done that is environmentally beneficial. What actions taken by this group, if any, do you disagree with? Why?

6. Use the library or the Internet to do some research on a large national or global environmental group. Describe three projects undertaken by the group in recent years. Assume you are in charge of the group and, for each of these projects, explain why you would or would not have chosen to undertake the project. For any that you would not have chosen, explain where you would instead have spent the group's resources.

CHAPTER 25 Environmental Worldviews, Ethics, and Sustainability

1. Using major societal, economic, and environmental trends, possible new trends, and your imagination, construct three different scenarios of what the world might be like in 2060. Identify the scenario you favor, and outline a strategy for achieving this alternative future. Compare your scenarios and strategies with those of your classmates.

2. Interview three people (friends, relatives or acquaintances) to determine for each which of the following corresponds most closely to their environmental worldviews: a planetary management worldview, a stewardship worldview, or an environmental wisdom worldview (Figure 25-2, p. 662). Construct four or five questions to use in determining this. Ask each person the same questions and record the answers with equal care. Compare your results with those of your classmates.

3. Does your school's curriculum provide *all* graduates with the basic elements of environmental literacy (Figure 25-6, p. 666)? Over the past 20 years, what important roles have its graduates played in making the world a better and more sustainable place to live? Using such information, rate your school on a 1–10 scale in terms of its contributions to environmental awareness and sustainability. Develop a detailed plan illustrating how your school could become better at achieving such goals, and present this information to school officials, alumni, parents, and financial backers.

4. Try to answer the following fundamental ecological questions about the corner of the world you inhabit: Where does your water come from? Where does the energy you use come from? What kinds of soils are under your feet? What types of wildlife are your neighbors? Where does your food come from? Where does your waste go? Does any of this information surprise you? If so, explain.

5. Write an essay in which you identify key environmental experiences that have influenced your life and thus helped you to form your current ecological identity. Examples may include (a) fond childhood memories of special places where you connected with the earth in some way; (b) places you knew and cherished that have been polluted, developed, or destroyed; (c) key events that forced you to think about environmental values or worldviews; (d) people or educational experiences that influenced your understanding of and concern about environmental problems and challenges; and (e) direct experience and contemplation of wild places. Share your experiences with other members of your class.

Key Concepts (by Chapter)

CHAPTER 1 Environmental Problems, Their Causes, and Sustainability

Concept 1-1A Our lives and economies depend on energy from the sun (*solar capital*) and on natural resources and natural services (*natural capital*) provided by the earth.

Concept 1-1B Living sustainably means living off the earth's natural income without depleting or degrading the natural capital that supplies it.

Concept 1-2 Societies can become more environmentally sustainable through economic development dedicated to improving the quality of life for everyone without degrading the earth's life support systems.

Concept 1-3 As our ecological footprints grow, we are depleting and degrading more of the earth's natural capital.

Concept 1-4 Preventing pollution is more effective and less costly than cleaning up pollution.

Concept 1-5A Major causes of environmental problems are population growth, wasteful and unsustainable resource use, poverty, exclusion of environmental costs of resource use from the market prices of goods and services, and attempts to manage nature with insufficient knowledge.

Concept 1-5B People with different environmental worldviews often disagree about the seriousness of environmental problems and what we should do about them.

Concept 1-6 Nature has sustained itself for billions of years by using solar energy, biodiversity, population control, and nutrient cycling—lessons from nature that we can apply to our lifestyles and economies.

CHAPTER 2 Science, Matter, Energy, and Systems

Concept 2-1 Scientists collect data and develop theories, models, and laws about how nature works.

Concept 2-2 Matter consists of elements and compounds, which are in turn made up of atoms, ions, or molecules.

Concept 2-3 When matter undergoes a physical or chemical change, no atoms are created or destroyed (the law of conservation of matter).

Concept 2-4A When energy is converted from one form to another in a physical or chemical change, no energy is created or destroyed (first law of thermodynamics).

Concept 2-4B Whenever energy is changed from one form to another, we end up with lower-quality or less usable energy than we started with (second law of thermodynamics).

Concept 2-5A Systems have inputs, flows, and outputs of matter and energy, and their behavior can be affected by feedback.

Concept 2-5B Life, human systems, and the earth's lifesupport systems must conform to the law of conservation of matter and the two laws of thermodynamics.

CHAPTER 3 Ecosystems: What Are They and How Do They Work?

Concept 3-1 Ecology is the study of how organisms interact with one another and with their physical environment of matter and energy.

Concept 3-2 Life is sustained by the flow of energy from the sun through the biosphere, the cycling of nutrients within the biosphere, and gravity.

Concept 3-3A Ecosystems contain living (biotic) and nonliving (abiotic) components.

Concept 3-3B Some organisms produce the nutrients they need, others get their nutrients by consuming other organisms, and some recycle nutrients back to producers by decomposing the wastes and remains of organisms.

Concept 3-4A Energy flows through ecosystems in food chains and webs.

Concept 3-4B As energy flows through ecosystems in food chains and webs, the amount of chemical energy available to organisms at each succeeding feeding level decreases.

Concept 3-5 Matter, in the form of nutrients, cycles within and among ecosystems and the biosphere, and human activities are altering these chemical cycles.

Concept 3-6 Scientists use field research, laboratory research, and mathematical and other models to learn about ecosystems.

CHAPTER 4 Biodiversity and Evolution

Concept 4-1 The biodiversity found in genes, species, ecosystems, and ecosystem processes is vital to sustaining life on earth.

Concept 4-2A The scientific theory of evolution explains how life on earth changes over time through changes in the genes of populations.

Concept 4-2B Populations evolve when genes mutate and give some individuals genetic traits that enhance their abilities to survive and to produce offspring with these traits (natural selection).

Concept 4-3 Tectonic plate movements, volcanic eruptions, earthquakes, and climate change have shifted wildlife habitats, wiped out large numbers of species, and created opportunities for the evolution of new species.

Concept 4-4A As environmental conditions change, the balance between formation of new species and extinction of existing species determines the earth's biodiversity.

Concept 4-4B Human activities can decrease biodiversity by causing the premature extinction of species and by destroying or degrading habitats needed for the development of new species.

Concept 4-5 Species diversity is a major component of biodiversity and tends to increase the sustainability of ecosystems.

Concept 4-6A Each species plays a specific ecological role called its niche. Any given species may play one or more of five important roles—native, nonnative, indicator, keystone, or foundation-roles—in a particular ecosystem.

CHAPTER 5 Biodiversity, Species Interactions, and Population Control

Concept 5-1 Five types of species interactions—competition, predation, parasitism, mutualism, and commensalism—affect the resource use and population sizes of the species in an ecosystem.

Concept 5-2 Some species develop adaptations that allow them to reduce or avoid competition with other species for resources.

Concept 5-3 No population can continue to grow indefinitely because of limitations on resources and because of competition among species for those resources.

Concept 5-4 The structure and species composition of communities and ecosystems change in response to changing environmental conditions through a process called ecological succession.

CHAPTER 6 The Human Population and Its Impact

Concept 6-1 We do not know how long we can continue increasing the earth's carrying capacity for humans without seriously degrading the life-support system for humans and many other species.

Concept 6-2A Population size increases because of births and immigration and decreases through deaths and emigration.

Concept 6-2B The average number of children born to women in a population (*total fertility rate*) is the key factor that determines population size.

Concept 6-3 The numbers of males and females in young, middle, and older age groups determine how fast a population grows or declines.

Concept 6-4 Experience indicates that the most effective ways to slow human population growth are to encourage family planning, to reduce poverty, and to elevate the status of women.

CHAPTER 7 Climate and Terrestrial Biodiversity

Concept 7-1 An area's climate is determined mostly by solar radiation, the earth's rotation, global patterns of air and water movement, gases in the atmosphere, and the earth's surface features.

Concept 7-2 Differences in average annual precipitation and temperature lead to the formation of tropical, temperate, and cold deserts, grasslands, and forests, and largely determine their locations.

Concept 7-3 In many areas, human activities are impairing ecological and economic services provided by the earth's deserts, grasslands, forests, and mountains.

CHAPTER 8 Aquatic Biodiversity

Concept 8-1A Saltwater and freshwater aquatic life zones cover almost three-fourths of the earth's surface with oceans dominating the planet.

Concept 8-1B The key factors determining biodiversity in aquatic systems are temperature, dissolved oxygen content, availability of food, and availability of light and nutrients necessary for photosynthesis.

Concept 8-2 Saltwater ecosystems are irreplaceable reservoirs of biodiversity and provide major ecological and economic services.

Concept 8-3 Human activities threaten aquatic biodiversity and disrupt ecological and economic services provided by saltwater systems.

Concept 8-4 Freshwater ecosystems provide major ecological and economic services and are irreplaceable reservoirs of biodiversity.

Concept 8-5 Human activities threaten biodiversity and disrupt ecological and economic services provided by freshwater lakes, rivers, and wetlands.

CHAPTER 9 Sustaining Biodiversity: The Species Approach

Concept 9-1A We are degrading and destroying biodiversity in many parts of the world, and these threats are increasing.

Concept 9-1B Species are becoming extinct 100 to 1,000 times faster than they were before modern humans arrived on the earth (the *background rate*), and by the end of this century, the extinction rate is expected to be 10,000 times the background rate.

Concept 9-2 We should prevent the premature extinction of wild species because of the economic and ecological services they provide and because they have a right to exist regardless of their usefulness to us.

Concept 9-3 The greatest threats to any species are (in order) loss or degradation of its habitat, harmful invasive species, human population growth, pollution, climate change, and overexploitation.

Concept 9-4A We can use existing environmental laws and treaties and work to enact new laws designed to prevent premature species extinction and protect overall biodiversity.

Concept 9-4B We can help to prevent premature species extinction by creating and maintaining wildlife refuges, gene banks, botanical gardens, zoos, and aquariums.

Concept 9-4C According to the *precautionary principle,* we should take measures to prevent or reduce harm to the environment and to human health, even if some of the cause-and-effect relationships have not been fully established, scientifically.

CHAPTER 10 Sustaining Terrestrial Biodiversity: The Ecosystem Approach

Concept 10-1A Forest ecosystems provide ecological services far greater in value than the value of raw materials obtained from forests.

Concept 10-1B Unsustainable cutting and burning of forests, along with diseases and insects, made worse by global warming, are the chief threats to forest ecosystems.

Concept 10-1C Tropical deforestation is a potentially catastrophic problem because of the vital ecological services at risk, the high rate of tropical deforestation, and its growing contribution to global warming.

Concept 10-2 We can sustain forests by emphasizing the economic value of their ecological services, protecting old-growth forests, harvesting trees no faster than they are replenished, and using sustainable substitute resources.

Concept 10-3 We can sustain the productivity of grasslands by controlling the number and distribution of grazing livestock and by restoring degraded grasslands.

Concept 10-4 Sustaining biodiversity will require protecting much more of the earth's remaining undisturbed land area as parks and nature reserves.

Concept 10-5A We can help to sustain biodiversity by identifying severely threatened areas and protecting those with high plant diversity (biodiversity hotspots) and those where ecosystem services are being impaired.

Concept 10-5B Sustaining biodiversity will require a global effort to rehabilitate and restore damaged ecosystems.

Concept 10-5C Humans dominate most of the earth's land, and preserving biodiversity will require sharing as much of it as possible with other species.

CHAPTER 11 Sustaining Aquatic Biodiversity

Concept 11-1 Aquatic species are threatened by habitat loss, invasive species, pollution, climate change, and overexploitation, all made worse by the growth of the human population.

Concept 11-2 We can help to sustain marine biodiversity by using laws and economic incentives to protect species, setting aside marine reserves to protect ecosystems, and using community-based integrated coastal management.

Concept 11-3 Sustaining marine fisheries will require improved monitoring of fish populations, cooperative fisheries management among communities and nations, reduction of fishing subsidies, and careful consumer choices in seafood markets.

Concept 11-4 To maintain the ecological and economic services of wetlands, we must maximize preservation of remaining wetlands and restoration of degraded and destroyed wetlands.

Concept 11-5 Freshwater ecosystems are strongly affected by human activities on adjacent lands, and protecting these ecosystems must include protection of their watersheds.

Concept 11-6 Sustaining the world's biodiversity and ecosystem services will require mapping terrestrial and aquatic biodiversity, maximizing protection of undeveloped terrestrial and aquatic areas, and carrying out ecological restoration projects worldwide.

CHAPTER 12 Food, Soil, and Pest Management

Concept 12-1A Many of the poor suffer health problems from chronic lack of food and poor nutrition, while many people in developed countries have health problems from eating too much food.

Concept 12-1B The greatest obstacles to providing enough food for everyone are poverty, political upheaval, corruption, war, and the harmful environmental effects of food production.

Concept 12-2A We have sharply increased crop production using a mix of industrialized and traditional agriculture.

Concept 12-2B We have used industrialized and traditional methods to greatly increase supplies of meat, fish, and shellfish.

Concept 12-3 Food production in the future may be limited by its serious environmental impacts, including soil erosion and degradation, desertification, water and air pollution, greenhouse gas emissions, and degradation and destruction of biodiversity.

Concept 12-4 We can sharply cut pesticide use without decreasing crop yields by using a mix of cultivation techniques, biological pest controls, and small amounts of selected chemical pesticides as a last resort (integrated pest management).

Concept 12-5 We can improve food security by creating programs to reduce poverty and chronic malnutrition, relying more on locally grown food, and cutting food waste.

Concept 12-6A Sustainable food production will require reducing topsoil erosion, sharply reducing overgrazing and overfishing, irrigating more efficiently, using integrated pest management, promoting agrobiodiversity, and providing government subsidies for more sustainable farming, fishing, and aquaculture.

Concept 12-6B Producing enough food to feed the rapidly growing human population will require growing crops in a mix of monocultures and polycultures and decreasing the enormous environmental impacts of industrialized food production.

CHAPTER 13 Water Resources

Concept 13-1A We are using available freshwater unsustainably by wasting it, polluting it, and charging too little for this irreplaceable natural resource.

Concept 13-1B One of every six people does not have sufficient access to clean water, and this situation will almost certainly get worse.

Concept 13-2 Groundwater that is used to supply cities and grow food is being pumped from aquifers in some areas faster than it is renewed by precipitation.

Concept 13-3 Building dam and reservoir systems has greatly increased water supplies in some areas, but it has disrupted ecosystems and displaced people.

Concept 13-4 Transferring water from one place to another has greatly increased water supplies in some areas, but it has also disrupted ecosystems.

Concept 13-5 We can convert salty ocean water to freshwater, but the cost is high, and the resulting salty brine must be disposed of without harming aquatic or terrestrial ecosystems.

Concept 13-6 We can use water more sustainably by cutting water waste, raising water prices, slowing population growth, and protecting aquifers, forests, and other ecosystems that store and release water.

Concept 13-7 We can lessen the threat of flooding by protecting more wetlands and natural vegetation in watersheds and by not building in areas subject to frequent flooding.

CHAPTER 14 Geology and Nonrenewable Minerals

Concept 14-1A Gigantic plates in the earth's crust move very slowly atop the planet's mantle, and wind and water move matter from place to place across the earth's surface.

Concept 14-1B Natural geological hazards such as earthquakes, tsunamis, volcanoes, and landslides can cause considerable damage.

Concept 14-2 The three major types of rocks found in the earth's crust—sedimentary, igneous, and metamorphic—are recycled very slowly by the processes of erosion, melting, and metamorphism.

Concept 14-3A Some naturally occurring materials in the earth's crust can be extracted and made into useful products in processes that provide economic benefits and jobs.

Concept 14-3B Extracting and using mineral resources can disturb the land, erode soils, produce large amounts of solid waste, and pollute the air, water, and soil.

Concept 14-4A All nonrenewable mineral resources exist in finite amounts, and as we get closer to depleting any mineral resource, the environmental impacts of extracting it generally become more harmful.

Concept 14-4B An increase in the price of a scarce mineral resource can lead to increased supplies and more efficient use of the mineral, but there are limits to this effect.

Concept 14-5 We can try to find substitutes for scarce resources, reduce resource waste, and recycle and reuse minerals.

CHAPTER 15 Nonrenewable Energy

Concept 15-1A About three-quarters of the world's commercial energy comes from nonrenewable fossil fuels and the rest comes from nonrenewable nuclear fuel and renewable sources.

Concept 15-1B Net energy is the amount of high-quality usable energy available from a resource after the amount of energy needed to make it available is subtracted.

Concept 15-2A Conventional oil is currently abundant, has a high net energy yield, and is relatively inexpensive, but using it causes air and water pollution and releases greenhouse gases to the atmosphere.

Concept 15-2B Heavy oils from oil sand and oil shale exist in potentially large supplies but have low net energy yields and higher environmental impacts than conventional oil has.

Concept 15-3 Conventional natural gas is more plentiful than oil, has a high net energy yield and a fairly low cost, and has the lowest environmental impact of all fossil fuels.

Concept 15-4A Conventional coal is very plentiful and has a high net energy yield and low cost, but it has a very high environmental impact.

Concept 15-4B Gaseous and liquid fuels produced from coal could be plentiful, but they have lower net energy yields and higher environmental impacts than conventional coal has.

Concept 15-5 Nuclear power has a low environmental impact and a very low accident risk, but high costs, a low net energy yield, long-lived radioactive wastes, vulnerability to sabotage, and the potential for spreading nuclear weapons technology have limited its use.

CHAPTER 16 Energy Efficiency and Renewable Energy

Concept 16-1 We could save as much as 43% of all the energy we use by improving energy efficiency.

Concept 16-2 We have a variety of technologies for sharply increasing the energy efficiency of industrial operations, motor vehicles, and buildings.

Concept 16-3 Passive and active solar heating systems can heat water and buildings effectively, and the costs of using direct sunlight to produce high-temperature heat and electricity are coming down.

Concept 16-4 Water flowing over dams, tidal flows, and ocean waves can be used to generate electricity, but environmental concerns and limited availability of suitable sites may limit the use of these energy resources.

Concept 16-5 When environmental costs of energy resources are included in market prices, wind energy is the least expensive and least polluting way to produce electricity.

Concept 16-6A Solid biomass is a renewable resource, but burning it faster than it is replenished produces a net gain in atmospheric greenhouse gases, and creating biomass plantations can degrade soil and biodiversity.

Concept 16-6B Liquid biofuels derived from biomass can be used in place of gasoline and diesel fuels, but creating biofuel plantations could degrade soil and biodiversity and increase food prices and greenhouse gas emissions.

Concept 16-7 Geothermal energy has great potential for supplying many areas with heat and electricity and generally has a low environmental impact, but locations where it can be exploited economically are limited.

Concept 16-8 Hydrogen fuel holds great promise for powering cars and generating electricity, but to be environmentally beneficial, it would have to be produced without the use of fossil fuels.

Concept 16-9 We can make a transition to a more sustainable energy future if we greatly improve energy efficiency, use a mix of renewable energy resources, and include environmental costs in the market prices of all energy resources.

CHAPTER 17 Environmental Hazards and Human Health

Concept 17-1 People face health hazards from biological, chemical, physical, and cultural factors, and from the lifestyle choices they make.

Concept 17-2 In terms of death rates, the most serious infectious diseases are flu, AIDS, diarrheal diseases, malaria, and tuberculosis; most of these deaths occur in developing countries.

Concept 17-3 There is growing concern about chemicals that can cause birth defects and cancers and disrupt the human immune, nervous, and endocrine systems.

Concept 17-4A Scientists use live laboratory animals, nonanimal tests, case reports of poisonings, and epidemiological studies to estimate the toxicity of chemicals, but these methods have limitations.

Concept 17-4B Many health scientists call for much greater emphasis on pollution prevention to reduce our exposure to potentially harmful chemicals.

Concept 17-5 We can reduce the major risks we face if we become informed, think critically about risks, and make careful choices.

CHAPTER 18 Air Pollution

Concept 18-1 The atmosphere is structured in layers, including the troposphere, which supports life, and the stratosphere, which contains the protective ozone layer.

Concept 18-2 Pollutants mix in the air to form *industrial smog*, mostly the result of burning coal, and *photochemical smog*, caused by motor vehicle, industrial, and power plant emissions.

Concept 18-3 Acid deposition is caused mainly by coalburning power plant and motor vehicle emissions, and in some regions, threatens human health, aquatic life and ecosystems, forests, and human-built structures.

Concept 18-4 The most threatening indoor air pollutants are smoke and soot from wood and coal cooking fires (a hazard found mostly in developing countries) and chemicals used in building materials and products.

Concept 18-5 Air pollution can contribute to asthma, chronic bronchitis, emphysema, lung cancer, heart attack, and stroke.

Concept 18-6 Legal, economic, and technological tools can help to clean up air pollution, but much greater emphasis should be focused on preventing air pollution.

CHAPTER 19 Climate Change and Ozone Depletion

Concept 19-1 The overwhelming scientific consensus is that the earth's atmosphere is warming rapidly, mostly because of human activities, and that this will lead to significant climate change during this century.

Concept 19-2 The projected rapid change in the atmosphere's temperature during this century is very likely to increase drought and flooding, shift areas where food can be grown, raise sea levels, result in intense heat waves, and cause the premature extinction of many species.

Concept 19-3A To slow the rate of global warming and climate change, we can increase energy efficiency, sharply reduce greenhouse gas emissions, rely more on renewable energy resources, and slow population growth.

Concept 19-3B Governments can subsidize energy efficiency and renewable energy use, tax greenhouse gas emissions, set up cap-and-trade emissions reduction systems, and help to slow population growth.

Concept 19-4A Widespread use of certain chemicals has reduced ozone levels in the stratosphere, which allows for more harmful ultraviolet radiation to reach the earth's surface.

Concept 19-4B To reverse ozone depletion, we must stop producing ozone-depleting chemicals and adhere to the international treaties that ban such chemicals.

CHAPTER 20 Water Pollution

Concept 20-1A Water pollution causes illness and death in humans and other species and disrupts ecosystems.

Concept 20-1B The chief sources of water pollution are agricultural activities, industrial facilities, and mining, but growth in population and resource use makes it increasingly worse.

Concept 20-2A While streams are extensively polluted worldwide by human activities, they can cleanse themselves of many pollutants if we do not overload them or reduce their flows.

Concept 20-2B Addition of excessive nutrients to lakes from human activities can disrupt lake ecosystems, and prevention of such pollution is more effective and less costly than cleaning it up.

Concept 20-3A Chemicals used in agriculture, industry, transportation, and homes can spill and leak into groundwater and make it undrinkable.

Concept 20-3B There are simple ways and complex ways to purify drinking water, but protecting it through pollution prevention is the least expensive and most effective strategy.

Concept 20-4A The great majority of ocean pollution originates on land and includes oil and other toxic chemicals and solid wastes, which threaten aquatic species and other wildlife and disrupt marine ecosystems.

Concept 20-4B The key to protecting the oceans is to reduce the flow of pollutants from land and air and from streams emptying into these waters.

Concept 20-5 Reducing water pollution requires preventing it, working with nature to treat sewage, cutting resource use and waste, reducing poverty, and slowing population growth.

CHAPTER 21 Solid and Hazardous Waste

Concept 21-1 Solid waste represents pollution and unnecessary waste of resources, and hazardous waste contributes to pollution, natural capital degradation, health problems, and premature deaths.

Concept 21-2 A sustainable approach to solid waste is first to reduce it, then to reuse or recycle it, and finally to safely dispose of what is left.

Concept 21-3 Reusing items decreases the use of matter and energy resources and reduces pollution and natural capital degradation; recycling does so to a lesser degree.

Concept 21-4 Technologies for burning and burying solid wastes are well developed, but burning contributes to pollution and greenhouse gas emissions, and buried wastes eventually contribute to pollution and land degradation.

Concept 21-5 A sustainable approach to hazardous waste is first to produce less of it, then to reuse or recycle it, then to convert it to less hazardous materials, and finally, to safely store what is left.

Concept 21-6 Shifting to a low-waste society requires individuals and businesses to reduce resource use and to reuse and recycle wastes at local, national, and global levels.

CHAPTER 22 Sustainable Cities

Concept 22-1 Urbanization continues to increase steadily and the numbers and sizes of urban areas are growing rapidly, especially in developing countries.

Concept 22-2 Most cities are unsustainable because of high levels of resource use, waste, pollution, and poverty.

Concept 22-3 A combination of plentiful land, inexpensive fuel, and expanding networks of highways in some countries has resulted in dispersed cities whose residents depend on motor vehicles for most transportation.

Concept 22-4 Urban land use planning can help to reduce uncontrolled sprawl and to slow the resulting degradation of air, water, land, biodiversity, and other natural resources.

Concept 22-5 An *ecocity* allows people to: choose walking, biking, or mass transit for most transportation needs; recycle or reuse most of their wastes; grow much of their food; and protect biodiversity by preserving surrounding land.

CHAPTER 23 Economics, Environment, and Sustainability

Concept 23-1A Ecological economists and most sustainability experts regard human economic systems as subsystems of the biosphere and subject to its processes and limiting factors.

Concept 23-1B Governments intervene in market economies to help correct or prevent market failures such as the failure to protect natural capital, which supports life and all economies.

Concept 23-2A Economists have developed several ways to estimate the present and future values of a resource or ecological service and optimum levels of pollution control and resource use.

Concept 23-2B Comparing the likely costs and benefits of an environmental action is useful but involves many uncertainties.

Concept 23-3A Using resources more sustainably will require including the harmful environmental and health costs of resource use in the market prices of goods and services (*full-cost pricing*).

Concept 23-3B Governments can help to improve and sustain environmental quality by subsidizing environmentally beneficial activities and taxing pollution and waste instead of wages and profits.

Concept 23-4 Reducing poverty can help us to reduce population growth, resource use, and environmental degradation.

Concept 23-5 We can use the four *principles of sustainability* and various economic and environmental strategies to develop more environmentally sustainable economies.

CHAPTER 24 Politics, Environment, and Sustainability

Concept 24-1 Government can help to protect environmental and public interests and encourage more environmentally sustainable economic development through its policies.

Concept 24-2A Policy making involves law making, program funding, rule writing, and enforcement of rules by agencies—a complex process that is affected at each stage by political processes.

Concept 24-2B Individuals can work with each other to become part of political processes that influence how environmental policies are made and whether or not they succeed. (*Individuals matter.*)

Concept 24-3 Environmental laws and regulations can be used to control pollution, set safety standards, encourage resource conservation, and protect species and ecosystems.

Concept 24-4 Grassroots groups are growing and combining efforts with large environmental organizations in a global sustainability movement.

Concept 24-5 Environmental security is necessary for economic security and is at least as important as military security.

Concept 24-6 Making the transition to more sustainable societies will require that nations implement green plans and that people and nations cooperate, agree on principles, and make the political commitment to achieve this transition.

CHAPTER 25 Environmental Worldviews, Ethics, and Sustainability

Concept 25-1 Major environmental worldviews differ on which is more important—human needs and wants, or the overall health of ecosystems and the biosphere.

Concept 25-2 The first step to living more sustainably is to become environmentally literate, partly by learning from nature.

Concept 25-3A We can live more sustainably by using certain guidelines to convert environmental literacy and concerns into action.

Concept 25-3B We can live more sustainably by living more simply and lightly on the earth and by becoming informed and active environmental citizens.

Glossary

abiotic Nonliving. Compare *biotic*.

acid See *acid solution*.

acid deposition The falling of acids and acid-forming compounds from the atmosphere to the earth's surface. Acid deposition is commonly known as *acid rain*, a term that refers to the wet deposition of droplets of acids and acid-forming compounds.

acid rain See *acid deposition*.

acid solution Any water solution that has more hydrogen ions (H^+) than hydroxide ions (OH^-); any water solution with a pH less than 7. Compare *basic solution, neutral solution*.

active solar heating system System that uses solar collectors to capture energy from the sun and store it as heat for space heating and water heating. Liquid or air pumped through the collectors transfers the captured heat to a storage system such as an insulated water tank or rock bed. Pumps or fans then distribute the stored heat or hot water throughout a dwelling as needed. Compare *passive solar heating system*.

adaptation Any genetically controlled structural, physiological, or behavioral characteristic that helps an organism survive and reproduce under a given set of environmental conditions. It usually results from a beneficial mutation. See *biological evolution, differential reproduction, mutation, natural selection*.

adaptive radiation Process in which numerous new species evolve to fill vacant and new ecological niches in changed environments, usually after a mass extinction. Typically, this process takes millions of years.

adaptive trait See *adaptation*.

administrative laws Administrative rules and regulations, executive orders, and enforcement decisions related to the implementation and interpretation of statutory laws.

advanced sewage treatment Third level of sewage cleanup, which uses a series of specialized chemical and physical processes to remove specific pollutants left in the water after primary and secondary treatment. See *primary sewage treatment* and *secondary sewage treatment*.

aerobic respiration Complex process that occurs in the cells of most living organisms, in which nutrient organic molecules such as glucose ($C_6H_{12}O_6$) combine with oxygen (O_2) to produce carbon dioxide (CO_2), water (H_2O), and energy. Compare *photosynthesis*.

age structure Percentage of the population (or number of people of each sex) at each age level in a population.

agricultural revolution Gradual shift from small, mobile hunting and gathering bands to settled agricultural communities in which people survived by breeding and raising wild animals and cultivating wild plants near where they lived. It began 10,000–12,000 years ago. Compare *environmental revolution, hunter–gatherers, industrial–medical revolution, information and globalization revolution*.

agroforestry Planting trees and crops together.

air pollution One or more chemicals in high enough concentrations in the air to harm humans, other animals, vegetation, or materials. Excess heat is also considered a form of air pollution. Such chemicals or physical conditions are called air pollutants. See *primary pollutant, secondary pollutant*.

albedo Ability of a surface to reflect light.

alien species See *nonnative species*.

alley cropping Planting of crops in strips with rows of trees or shrubs on each side.

alpha particle Positively charged matter, consisting of two neutrons and two protons, which is emitted as radioactivity from the nuclei of some radioisotopes. See also *beta particle, gamma rays*.

altitude Height above sea level. Compare *latitude*.

anaerobic respiration Form of cellular respiration in which some decomposers get the energy they need through the breakdown of glucose (or other nutrients) in the absence of oxygen. Compare *aerobic respiration*.

ancient forest See *old-growth forest*.

animal manure Dung and urine of animals used as a form of organic fertilizer. Compare *green manure*.

annual Plant that grows, sets seed, and dies in one growing season. Compare *perennial*.

anthropocentric Human-centered.

applied ecology See *reconciliation ecology*.

aquaculture Growing and harvesting of fish and shellfish for human use in freshwater ponds, irrigation ditches, and lakes, or in cages or fenced-in areas of coastal lagoons and estuaries or in the open ocean. See *fish farming, fish ranching*.

aquatic Pertaining to water. Compare *terrestrial*.

aquatic life zone Marine and freshwater portions of the biosphere. Examples include freshwater life zones (such as lakes and streams) and ocean or marine life zones (such as estuaries, coastlines, coral reefs, and the open ocean).

aquifer Porous, water-saturated layers of sand, gravel, or bedrock that can yield an economically significant amount of water.

arable land Land that can be cultivated to grow crops.

arbitration Formal effort, somewhat similar to a trial, to resolve a dispute; overseen by a trained and certified arbitrator who takes a more active role than a judge and is not constrained by legal precedent. Compare *mediation*.

area strip mining Type of surface mining used where the terrain is flat. An earthmover strips away the overburden, and a power shovel digs a cut to remove the mineral deposit. The trench is then filled with overburden, and a new cut is made parallel to the previous one. The process is repeated over the entire site. Compare *mountaintop removal, open-pit mining, subsurface mining*.

arid Dry. A desert or other area with an arid climate has little precipitation.

artificial selection Process by which humans select one or more desirable genetic traits in the population of a plant or animal species and then use *selective breeding* to produce populations containing many individuals with the desired traits. Compare *genetic engineering, natural selection*.

asexual reproduction Reproduction in which a mother cell divides to produce two identical daughter cells that are clones of the mother cell. This type of reproduction is common in single-celled organisms. Compare *sexual reproduction*.

atmosphere Whole mass of air surrounding the earth. See *stratosphere, troposphere*. Compare *biosphere, geosphere, hydrosphere*.

atmospheric pressure Force or mass per unit area of air, caused by the bombardment of a surface by the molecules in air.

atom Minute unit made of subatomic particles that is the basic building block of all chemical elements and thus all matter; the smallest unit of an element that can exist and still have the unique characteristics of that element. Compare *ion, molecule*.

atomic number Number of protons in the nucleus of an atom. Compare *mass number*.

atomic theory Idea that all elements are made up of atoms; the most widely accepted scientific theory in chemistry.

autotroph See *producer*.

background extinction Normal extinction of various species as a result of changes in local environmental conditions. Compare *mass extinction*.

bacteria Prokaryotic, one-celled organisms. Some transmit diseases. Most act as decomposers and get the nutrients they need by breaking down complex organic compounds in the tissues of living or dead organisms into simpler inorganic nutrient compounds.

barrier islands Long, thin, low offshore islands of sediment that generally run parallel to the shore along some coasts.

basic solution Water solution with more hydroxide ions (OH⁻) than hydrogen ions (H⁺); water solution with a pH greater than 7. Compare *acid solution, neutral solution*.

benthos Bottom-dwelling organisms. Compare *decomposer, nekton, plankton*.

beta particle Swiftly moving electron emitted by the nucleus of a radioactive isotope. See also *alpha particle, gamma ray*.

bioaccumulation An increase in the concentration of a chemical in specific organs or tissues at a level higher than would normally be expected. Compare *biomagnification*.

biocentric Life-centered. Compare *anthropocentric*.

biodegradable Capable of being broken down by decomposers.

biodegradable pollutant Material that can be broken down into simpler substances (elements and compounds) by bacteria or other decomposers. Paper and most organic wastes such as animal manure are biodegradable but can take decades to biodegrade in modern landfills. Compare *nondegradable pollutant*.

biodiversity Variety of different species (*species diversity*), genetic variability among individuals within each species (*genetic diversity*), variety of ecosystems (*ecological diversity*), and functions such as energy flow and matter cycling needed for the survival of species and biological communities (*functional diversity*).

biodiversity hot spots Areas especially rich in plant species that are found nowhere else and are in great danger of extinction. These areas suffer serious ecological disruption, mostly because of rapid human population growth and the resulting pressure on natural resources.

biofuel Gas (such as methane) or liquid fuel (such as ethyl alcohol) made from plant material (biomass).

biogeochemical cycle Natural processes that recycle nutrients in various chemical forms from the nonliving environment to living organisms and then back to the nonliving environment. Examples include the carbon, oxygen, nitrogen, phosphorus, sulfur, and hydrologic cycles.

biological amplification See *biomagnification*.

biological community See *community*.

biological diversity See *biodiversity*.

biological evolution Change in the genetic makeup of a population of a species in successive generations. If continued long enough, it can lead to the formation of a new species. Note that populations, not individuals, evolve. See also *adaptation, differential reproduction, natural selection, theory of evolution*.

biological pest control Control of pest populations by natural predators, parasites, or disease-causing bacteria and viruses (pathogens).

biomagnification Increase in concentration of DDT, PCBs, and other slowly degradable, fat-soluble chemicals in organisms at successively higher trophic levels of a food chain or web. Compare *bioaccumulation*.

biomass Organic matter produced by plants and other photosynthetic producers; total dry weight of all living organisms that can be supported at each trophic level in a food chain or web; dry weight of all organic matter in plants and animals in an ecosystem; plant materials and animal wastes used as fuel.

biome Terrestrial regions inhabited by certain types of life, especially vegetation. Examples include various types of deserts, grasslands, and forests.

biosphere Zone of the earth where life is found. It consists of parts of the atmosphere (the troposphere), hydrosphere (mostly surface water and groundwater), and lithosphere (mostly soil and surface rocks and sediments on the bottoms of oceans and other bodies of water) where life is found. Compare *atmosphere, geosphere, hydrosphere*.

biotic Living organisms. Compare *abiotic*.

biotic pollution The effect of invasive species that can reduce or wipe out populations of many native species and trigger ecological disruptions.

biotic potential Maximum rate at which the population of a given species can increase when there are no limits on its rate of growth. See *environmental resistance*.

birth rate See *crude birth rate*.

bitumen Gooey, black, high-sulfur, heavy oil extracted from oil sand and then upgraded to synthetic fuel oil. See *oil sand*.

breeder nuclear fission reactor Nuclear fission reactor that produces more nuclear fuel than it consumes by converting nonfissionable uranium-238 into fissionable plutonium-239.

broadleaf deciduous plants Plants such as oak and maple trees that survive drought and cold by shedding their leaves and becoming dormant. Compare *broadleaf evergreen plants, coniferous evergreen plants*.

broadleaf evergreen plants Plants that keep most of their broad leaves year-round. An example is the trees found in the canopies of tropical rain forests. Compare *broadleaf deciduous plants, coniferous evergreen plants*.

buffer Substance that can react with hydrogen ions in a solution and thus hold the acidity or pH of a solution fairly constant. See *pH*.

calorie Unit of energy; amount of energy needed to raise the temperature of 1 gram of water by 1 C° (unit on Celsius temperature scale). See also *kilocalorie*.

cancer Group of more than 120 different diseases, one for each type of cell in the human body. Each type of cancer produces a tumor in which cells multiply uncontrollably and invade surrounding tissue.

carbon capture and storage (CCS) Process of removing carbon dioxide gas from coal-burning power and industrial plants and storing it somewhere (usually underground or under the seabed) so that it is not released into the atmosphere, essentially forever.

carbon cycle Cyclic movement of carbon in different chemical forms from the environment to organisms and then back to the environment.

carcinogen Chemicals, ionizing radiation, and viruses that cause or promote the development of cancer. See *cancer*. Compare *mutagen, teratogen*.

carnivore Animal that feeds on other animals. Compare *herbivore, omnivore*.

carrying capacity (K) Maximum population of a particular species that a given habitat can support over a given period. Compare *cultural carrying capacity*.

cell Smallest living unit of an organism. Each cell is encased in an outer membrane or wall and contains genetic material (DNA) and other parts to perform its life function. Organisms such as bacteria consist of only one cell, but most organisms contain many cells.

cell theory The idea that all living things are composed of cells; the most widely accepted scientific theory in biology.

CFCs See *chlorofluorocarbons*.

chain reaction Multiple nuclear fissions, taking place within a certain mass of a fissionable isotope, which release an enormous amount of energy in a short time.

chemical One of the millions of different elements and compounds found naturally and synthesized by humans. See *compound, element*.

chemical change Interaction between chemicals in which the chemical composition of the elements or compounds involved changes. Compare *nuclear change, physical change*.

chemical formula Shorthand way to show the number of atoms (or ions) in the basic structural unit of a compound. Examples include H_2O, NaCl, and $C_6H_{12}O_6$.

chemical reaction See *chemical change*.

chemosynthesis Process in which certain organisms (mostly specialized bacteria) extract inorganic compounds from their environment and convert them into organic nutrient compounds without the presence of sunlight. Compare *photosynthesis*.

chlorinated hydrocarbon Organic compound made up of atoms of carbon, hydrogen, and chlorine. Examples include DDT and PCBs.

chlorofluorocarbons (CFCs) Organic compounds made up of atoms of carbon, chlorine, and fluorine. An example is Freon-12 (CCl_2F_2), which is used as a refrigerant in refrigerators and air conditioners and in making plastics such as Styrofoam. Gaseous CFCs can deplete the ozone layer when they slowly rise into the stratosphere and their chlorine atoms react with ozone molecules. Their use is being phased out.

chromosome A grouping of genes and associated proteins in plant and animal cells that carry certain types of genetic information. See *genes*.

chronic undernutrition Condition suffered by people who cannot grow or buy enough food to meet their basic energy needs. Most chronically undernourished children live in developing countries and are likely to suffer from mental retardation and stunted growth and to die from infectious diseases. Compare *malnutrition, overnutrition*.

civil suit Court case brought to settle disputes or damages between one party and another.

clear-cutting Method of timber harvesting in which all trees in a forested area are removed in a single cutting. Compare *selective cutting, strip cutting*.

climate Physical properties of the troposphere of an area based on analysis of its weather records over a long period (at least 30 years). The two main factors determining an area's climate are its average *temperature,* with its seasonal variations, and the average amount and distribution of *precipitation*. Compare *weather*.

climax community See *mature community*.

closed-loop recycling See *primary recycling*.

coal Solid, combustible mixture of organic compounds with 30–98% carbon by weight, mixed with various amounts of water and small amounts of sulfur and nitrogen compounds. It forms in several stages as the remains of plants are subjected to heat and pressure over millions of years.

coal gasification Conversion of solid coal to synthetic natural gas (SNG).

coal liquefaction Conversion of solid coal to a liquid hydrocarbon fuel such as synthetic gasoline or methanol.

coastal wetland Land along a coastline, extending inland from an estuary that is covered with salt water all or part of the year. Examples include marshes, bays, lagoons, tidal flats, and mangrove swamps. Compare *inland wetland*.

coastal zone Warm, nutrient-rich, shallow part of the ocean that extends from the high-tide mark on land to the edge of a shelf-like extension of continental land masses known as the continental shelf. Compare *open sea*.

coevolution Evolution in which two or more species interact and exert selective pressures on each other that can lead each species to undergo adaptations. See *evolution, natural selection*.

cogeneration Production of two useful forms of energy, such as high-temperature heat or steam and electricity, from the same fuel source.

cold front Leading edge of an advancing mass of cold air. Compare *warm front*.

combined heat and power (CHP) production See *cogeneration*.

commensalism An interaction between organisms of different species in which one type of organism benefits and the other type is neither helped nor harmed to any great degree. Compare *mutualism*.

commercial extinction Depletion of the population of a wild species used as a resource to a level at which it is no longer profitable to harvest the species.

commercial forest See *tree plantation*.

commercial inorganic fertilizer Commercially prepared mixture of inorganic plant nutrients such as nitrates, phosphates, and potassium applied to the soil to restore fertility and increase crop yields. Compare *organic fertilizer*.

common law A body of unwritten rules and principles derived from court decisions along with commonly accepted practices, or norms, within a society. Compare *statutory law*.

common-property resource Resource that is owned jointly by a large group of individuals. One example is the roughly one-third of the land in the United States that is owned jointly by all U.S. citizens and held and managed for them by the government. Another example is an area of land that belongs to a whole village and that can be used by anyone for grazing cows or sheep. Compare *open access renewable resource* and *private property resource*. See *tragedy of the commons*. Compare *open access renewable resource, private property resource*.

community Populations of all species living and interacting in an area at a particular time.

competition Two or more individual organisms of a single species (*intraspecific competition*) or two or more individuals of different species (*interspecific competition*) attempting to use the same scarce resources in the same ecosystem.

compost Partially decomposed organic plant and animal matter used as a soil conditioner or fertilizer.

compound Combination of atoms, or oppositely charged ions, of two or more elements held together by attractive forces called chemical bonds. Examples are NaCl, CO2, and $C_6H_{12}O_6$. Compare *element*.

concentration Amount of a chemical in a particular volume or weight of air, water, soil, or other medium.

condensation nuclei Tiny particles on which droplets of water vapor can collect.

coniferous evergreen plants Cone-bearing plants (such as spruces, pines, and firs) that keep some of their narrow, pointed leaves (needles) all year. Compare *broadleaf deciduous plants, broadleaf evergreen plants*.

coniferous trees Cone-bearing trees, mostly evergreens, that have needle-shaped or scale-like leaves. They produce wood known commercially as softwood. Compare *deciduous plants*.

consensus science See *reliable science*.

conservation Sensible and careful use of natural resources by humans. People with this view are called *conservationists*.

conservation biology Multidisciplinary science created to deal with the crisis of maintaining the genes, species, communities, and ecosystems that make up earth's biological diversity. Its goals are to investigate human impacts on biodiversity and to develop practical approaches to preserving biodiversity.

conservationist Person concerned with using natural areas and wildlife in ways that sustain them for current and future generations of humans and other forms of life.

conservation-tillage farming Crop cultivation in which the soil is disturbed little (minimum-tillage farming) or not at all (no-till farming) in an effort to reduce soil erosion, lower labor costs, and save energy. Compare *conventional-tillage farming*.

constancy Ability of a living system, such as a population, to maintain a certain size. Compare *inertia, resilience*.

consumer Organism that cannot synthesize the organic nutrients it needs and gets its organic nutrients by feeding on the tissues of producers or of other consumers; generally divided into *primary consumers* (herbivores), *secondary consumers* (carnivores), *tertiary (higher-level) consumers, omnivores*, and *detritivores* (decomposers and detritus feeders). In economics, one who uses economic goods. Compare *producer*.

contour farming Plowing and planting across the changing slope of land, rather than in straight lines, to help retain water and reduce soil erosion.

contour strip mining Form of surface mining used on hilly or mountainous terrain. A power shovel cuts a series of terraces into the side of a hill. An earthmover removes the overburden, and a power shovel extracts the coal. The overburden from each new terrace is dumped onto the one below. Compare *area strip mining, mountaintop removal, open-pit mining, subsurface mining*.

controlled burning Deliberately set, carefully controlled surface fires that reduce flammable litter and decrease the chances of damaging crown fires. See *ground fire, surface fire*.

conventional-tillage farming Crop cultivation method in which a planting surface is made by plowing land, breaking up the exposed soil, and then smoothing the surface. Compare *conservation-tillage farming.*

convergent plate boundary Area where the earth's lithospheric plates are pushed together. See subduction zone. *Compare divergent plate boundary, transform fault.*

coral reef Formation produced by massive colonies containing billions of tiny coral animals, called polyps, that secrete a stony substance (calcium carbonate) around themselves for protection. When the corals die, their empty outer skeletons form layers and cause the reef to grow. Coral reefs are found in the coastal zones of warm tropical and subtropical oceans.

core Inner zone of the earth. It consists of a solid inner core and a liquid outer core. Compare *crust, mantle.*

corrective feedback loop See *negative feedback loop.*

cost–benefit analysis A comparison of estimated costs and benefits of actions such as implementing a pollution control regulation, building a dam on a river, or preserving an area of forest.

critical mass Amount of fissionable nuclei needed to sustain a nuclear fission chain reaction.

crop rotation Planting a field, or an area of a field, with different crops from year to year to reduce soil nutrient depletion. A plant such as corn, tobacco, or cotton, which removes large amounts of nitrogen from the soil, is planted one year. The next year a legume such as soybeans, which adds nitrogen to the soil, is planted.

crown fire Extremely hot forest fire that burns ground vegetation and treetops. Compare *controlled burning, ground fire, surface fire.*

crude birth rate Annual number of live births per 1,000 people in the population of a geographic area at the midpoint of a given year. Compare *crude death rate.*

crude death rate Annual number of deaths per 1,000 people in the population of a geographic area at the midpoint of a given year. Compare *crude birth rate.*

crude oil Gooey liquid consisting mostly of hydrocarbon compounds and small amounts of compounds containing oxygen, sulfur, and nitrogen. Extracted from underground accumulations, it is sent to oil refineries, where it is converted to heating oil, diesel fuel, gasoline, tar, and other materials.

crust Solid outer zone of the earth. It consists of oceanic crust and continental crust. Compare *core, mantle.*

cultural carrying capacity The limit on population growth that would allow most people in an area or the world to live in reasonable comfort and freedom without impairing the ability of the planet to sustain future generations. Compare *carrying capacity.*

cultural eutrophication Overnourishment of aquatic ecosystems with plant nutrients (mostly nitrates and phosphates) because of human activities such as agriculture, urbanization, and discharges from industrial plants and sewage treatment plants. See *eutrophication.*

culture Whole of a society's knowledge, beliefs, technology, and practices.

currents Mass movements of surface water produced by prevailing winds blowing over the oceans.

dam A structure built across a river to control the river's flow or to create a reservoir. See *reservoir.*

data Factual information collected by scientists.

DDT Dichlorodiphenyltrichloroethane, a chlorinated hydrocarbon that has been widely used as an insecticide but is now banned in some countries.

death rate See *crude death rate.*

debt-for-nature swap Agreement in which a certain amount of foreign debt is canceled in exchange for local currency investments that will improve natural resource management or protect certain areas in the debtor country from environmentally harmful development.

deciduous plants Trees, such as oaks and maples, and other plants that survive during dry seasons or cold seasons by shedding their leaves. Compare *coniferous trees, succulent plants.*

decomposer Organism that digests parts of dead organisms and cast-off fragments and wastes of living organisms by breaking down the complex organic molecules in those materials into simpler inorganic compounds and then absorbing the soluble nutrients. Producers return most of these chemicals to the soil and water for reuse. Decomposers consist of various bacteria and fungi. Compare *consumer, detritivore, producer.*

deductive reasoning Use of logic to arrive at a specific conclusion based on a generalization or premise. Compare *inductive reasoning.*

deep ecology worldview Worldview holding that each form of life has inherent value, that the fundamental interdependence and diversity of life forms helps all life to thrive, that humans have no right to reduce this interdependence and diversity except to satisfy vital needs, and that present human interference with the nonhuman world is excessive, and the situation is worsening rapidly. Compare *environmental wisdom worldview, frontier worldview, planetary management worldview, stewardship worldview.*

defendant The party in a court case being charged with creating a harm. See *plaintiff* and *civil suit.*

deforestation Removal of trees from a forested area.

degree of urbanization Percentage of the population in the world, or in a country, living in urban areas. Compare *urban growth.*

democracy Government by the people through their elected officials and appointed representatives. In a *constitutional democracy,* a constitution provides the basis of government authority and puts restraints on government power through free elections and freely expressed public opinion.

demographic transition Hypothesis that countries, as they become industrialized, have declines in death rates followed by declines in birth rates.

density Mass per unit volume.

depletion time The time it takes to use a certain fraction (usually 80%) of the known or estimated supply of a nonrenewable resource at an assumed rate of use. Finding and extracting the remaining 20% usually costs more than it is worth.

desalination Purification of salt water or brackish (slightly salty) water by removal of dissolved salts.

desert Biome in which evaporation exceeds precipitation and the average amount of precipitation is less than 25 centimeters (10 inches) per year. Such areas have little vegetation or have widely spaced, mostly low vegetation. Compare *forest, grassland.*

desertification Conversion of rangeland, rain-fed cropland, or irrigated cropland to desert-like land, with a drop in agricultural productivity of 10% or more. It usually is caused by a combination of overgrazing, soil erosion, prolonged drought, and climate change.

detritivore Consumer organism that feeds on detritus, parts of dead organisms, and cast-off fragments and wastes of living organisms. Examples include earthworms, termites, and crabs. Compare *decomposer.*

detritus Parts of dead organisms and cast-off fragments and wastes of living organisms.

detritus feeder See *detritivore.*

deuterium (D; hydrogen-2) Isotope of the element hydrogen, with a nucleus containing one proton and one neutron and a mass number of 2.

developed country Country that is highly industrialized and has a high per capita GDP. Compare *developing country.*

developing country Country that has low to moderate industrialization and low to moderate per capita GDP. Most are located in Africa, Asia, and Latin America. Compare *developed country.*

dieback Sharp reduction in the population of a species when its numbers exceed the carrying capacity of its habitat. See *carrying capacity.*

differential reproduction Phenomenon in which individuals with adaptive genetic traits produce more living offspring than do individuals without such traits. See *natural selection.*

dioxins Family of 75 chlorinated hydrocarbon compounds formed as unwanted by-products in chemical reactions involving chlorine and hydrocarbons, usually at high temperatures.

discount rate An estimate of a resource's future economic value compared to its present value; based on the idea that having something today may be worth more than it will be in the future.

dissolved oxygen (DO) content Amount of oxygen gas (O_2) dissolved in a given volume of water at a particular temperature and pressure, often expressed as a concentration in parts of oxygen per million parts of water.

disturbance An event that disrupts an ecosystem or community. Examples of *natural disturbances* include fires, hurricanes, tornadoes, droughts, and floods. Examples of *human-caused disturbances* include deforestation, overgrazing, and plowing.

divergent plate boundary Area where the earth's lithospheric plates move apart in opposite directions. Compare *convergent plate boundary, transform fault.*

DNA (deoxyribonucleic acid) Large molecules in the cells of organisms that carry genetic information in living organisms.

domesticated species Wild species tamed or genetically altered by crossbreeding for use by humans for food (cattle, sheep, and food crops), pets (dogs and cats), or enjoyment (animals in zoos and plants in botanical gardens). Compare *wild species.*

dose Amount of a potentially harmful substance an individual ingests, inhales, or absorbs through the skin. Compare *response.* See *dose-response curve, median lethal dose.*

dose-response curve Plot of data showing the effects of various doses of a toxic agent on a group of test organisms. See *dose, median lethal dose, response.*

doubling time Time it takes (usually in years) for the quantity of something growing exponentially to double. It can be calculated by dividing the annual percentage growth rate into 70.

drainage basin See *watershed.*

drift-net fishing Catching fish in huge nets that drift in the water.

drought Condition in which an area does not get enough water because of lower-than-normal precipitation or higher-than-normal temperatures that increase evaporation.

earthquake Shaking of the ground resulting from the fracturing and displacement of rock, which produces a fault, or from subsequent movement along the fault.

ecological diversity The variety of forests, deserts, grasslands, oceans, streams, lakes, and other biological communities interacting with one another and with their nonliving environment. See *biodiversity.* Compare *functional diversity, genetic diversity, species diversity.*

ecological efficiency Percentage of energy transferred from one trophic level to another in a food chain or web.

ecological footprint Amount of biologically productive land and water needed to supply a population with the renewable resources it uses and to absorb or dispose of the wastes from such resource use. It is a measure of the average environmental impact of populations in different countries and areas. See *per capita ecological footprint.*

ecological niche Total way of life or role of a species in an ecosystem. It includes all physical, chemical, and biological conditions that a species needs to live and reproduce in an ecosystem. See *fundamental niche, realized niche.*

ecological restoration Deliberate alteration of a degraded habitat or ecosystem to restore as much of its ecological structure and function as possible.

ecological succession Process in which communities of plant and animal species in a particular area are replaced over time by a series of different and often more complex communities. See *primary succession, secondary succession.*

ecologist Biological scientist who studies relationships between living organisms and their environment.

ecology Biological science that studies the relationships between living organisms and their environment; study of the structure and functions of nature.

economic depletion Exhaustion of 80% of the estimated supply of a nonrenewable resource. Finding, extracting, and processing the remaining 20% usually costs more than it is worth. May also apply to the depletion of a renewable resource, such as a fish or tree species.

economic development Improvement of human living standards by economic growth. Compare *economic growth, environmentally sustainable economic development.*

economic growth Increase in the capacity to provide people with goods and services; an increase in gross domestic product (GDP). Compare *economic development, environmentally sustainable economic development.* See *gross domestic product.*

economic resources Natural resources, capital goods, and labor used in an economy to produce material goods and services. See *natural resources.*

economic system Method that a group of people uses to choose which goods and services to produce, how to produce them, how much to produce, and how to distribute them to people.

economy System of production, distribution, and consumption of economic goods.

ecosphere See *biosphere.*

ecosystem One or more communities of different species interacting with one another and with the chemical and physical factors making up their nonliving environment.

ecosystem services Natural services or natural capital that support life on the earth and are essential to the quality of human life and the functioning of the world's economies. Examples are the chemical cycles, natural pest control, and natural purification of air and water. See *natural resources.*

electromagnetic radiation Forms of kinetic energy traveling as electromagnetic waves. Examples include radio waves, TV waves, microwaves, infrared radiation, visible light, ultraviolet radiation, X rays, and gamma rays. Compare *ionizing radiation, nonionizing radiation.*

electron (e) Tiny particle moving around outside the nucleus of an atom. Each electron has one unit of negative charge and almost no mass. Compare *neutron, proton.*

element Chemical, such as hydrogen (H), iron (Fe), sodium (Na), carbon (C), nitrogen (N), or oxygen (O), whose distinctly different atoms serve as the basic building blocks of all matter. Two or more elements combine to form the compounds that make up most of the world's matter. Compare *compound.*

elevation Distance above sea level.

endangered species Wild species with so few individual survivors that the species could soon become extinct in all or most of its natural range. Compare *threatened species.*

endemic species Species that is found in only one area. Such species are especially vulnerable to extinction.

energy Capacity to do work by performing mechanical, physical, chemical, or electrical tasks or to cause a heat transfer between two objects at different temperatures.

energy conservation Reducing or eliminating the unnecessary waste of energy.

energy efficiency Percentage of the total energy input that does useful work and is not converted into low-quality, generally useless heat in an energy conversion system or process. See *energy quality, net energy.* Compare *material efficiency.*

energy productivity See *energy efficiency.*

energy quality Ability of a form of energy to do useful work. High-temperature heat and the chemical energy in fossil fuels and nuclear fuels are concentrated high-quality energy. Low-quality energy such as low-temperature heat is dispersed or diluted and cannot do much useful work. See *high-quality energy, low-quality energy.*

enhanced greenhouse effect See *global warming, greenhouse effect.*

environment All external conditions, factors, matter, and energy, living and nonliving, that affect any living organism or other specified system.

environmental degradation Depletion or destruction of a potentially renewable resource such as soil, grassland, forest, or wildlife that is used faster than it is naturally replenished. If such use continues, the resource becomes nonrenewable (on a human time scale) or nonexistent (extinct). See also *sustainable yield.*

environmental ethics Human beliefs about what is right or wrong with how we treat the environment.

environmentalism Social movement dedicated to protecting the earth's life support systems for us and other species.

environmentalist Person who is concerned about the impacts of human activities on the environment.

environmental justice Fair treatment and meaningful involvement of all people regardless of race, color, sex, national origin, or income with respect to the development, implementation, and enforcement of environmental laws, regulations, and policies.

environmentally sustainable economic development Development that meets the basic needs of the current generations of humans and other species without preventing future generations of humans and other species from meeting their basic needs. It is the economic component of an *environmentally sustainable society*. Compare *economic development, economic growth*.

environmentally sustainable society Society that meets the current and future needs of its people for basic resources in a just and equitable manner without compromising the ability of future generations of humans and other species from meeting their basic needs.

environmental law Body of statements defining what is acceptable environmental behavior for individuals and groups, according to the larger community, and attempting to balance competing social and private interests.

environmental movement Citizens organized to demand that political leaders enact laws and develop policies to curtail pollution, clean up polluted environments, and protect unspoiled areas from environmental degradation.

environmental policy Laws, rules, and regulations related to an environmental problem that are developed, implemented, and enforced by a particular government body or agency.

environmental resistance All of the limiting factors that act together to limit the growth of a population. See *biotic potential, limiting factor*.

environmental revolution Cultural change that includes halting population growth and altering lifestyles, political and economic systems, and the way we treat the environment with the goal of living more sustainably. It requires working with the rest of nature by learning more about how nature sustains itself.

environmental science Interdisciplinary study that uses information and ideas from the physical sciences (such as biology, chemistry, and geology) with those from the social sciences and humanities (such as economics, politics, and ethics) to learn how nature works, how we interact with the environment, and how we can to help deal with environmental problems.

environmental scientist Scientist who uses information from the physical sciences and social sciences to understand how the earth works, learn how humans interact with the earth, and develop solutions to environmental problems. See *environmental science*.

environmental wisdom worldview Worldview holding that humans are part of and totally dependent on nature and that nature exists for all species, not just for us. Our success depends on learning how the earth sustains itself and integrating such environmental wisdom into the ways we think and act. Compare *deep ecology worldview, frontier worldview, planetary management worldview, stewardship worldview*.

environmental worldview Set of assumptions and beliefs about how people think the world works, what they think their role in the world should be, and what they believe is right and wrong environmental behavior (environmental ethics). See *deep ecology worldview, environmental wisdom worlview, frontier worldview, planetary management worldview, stewardship worldview*.

EPA U.S. Environmental Protection Agency; responsible for managing federal efforts to control air and water pollution, radiation and pesticide hazards, environmental research, hazardous waste, and solid waste disposal.

epidemiology Study of the patterns of disease or other harmful effects from toxic exposure within defined groups of people to find out why some people get sick and some do not.

epiphyte Plant that uses its roots to attach itself to branches high in trees, especially in tropical forests.

erosion Process or group of processes by which loose or consolidated earth materials are dissolved, loosened, or worn away and removed from one place and deposited in another. See *weathering*.

estuary Partially enclosed coastal area at the mouth of a river where its fresh water, carrying fertile silt and runoff from the land, mixes with salty seawater.

eukaryotic cell Cell that is surrounded by a membrane and has a distinct nucleus. Compare *prokaryotic cell*.

euphotic zone Upper layer of a body of water through which sunlight can penetrate and support photosynthesis.

eutrophication Physical, chemical, and biological changes that take place after a lake, estuary, or slow-flowing stream receives inputs of plant nutrients—mostly nitrates and phosphates—from natural erosion and runoff from the surrounding land basin. See *cultural eutrophication*.

eutrophic lake Lake with a large or excessive supply of plant nutrients, mostly nitrates and phosphates. Compare *mesotrophic lake, oligotrophic lake*.

evaporation Conversion of a liquid into a gas.

evergreen plants Plants that keep some of their leaves or needles throughout the year.

Examples include cone-bearing trees (conifers) such as firs, spruces, pines, redwoods, and sequoias. Compare *deciduous plants, succulent plants*.

evolution See *biological evolution*.

exhaustible resource See *nonrenewable resource*.

exotic species See *nonnative species*.

experiment Procedure a scientist uses to study some phenomenon under known conditions. Scientists conduct some experiments in the laboratory and others in nature. The resulting scientific data or facts must be verified or confirmed by repeated observations and measurements, ideally by several different investigators.

exponential growth Growth in which some quantity, such as population size or economic output, increases at a constant rate per unit of time. An example is the growth sequence 2, 4, 8, 16, 32, 64, and so on, which increases by 100% at each interval. When the increase in quantity over time is plotted, this type of growth yields a curve shaped like the letter J. Compare *linear growth*.

external benefit Beneficial social effect of producing and using an economic good that is not included in the market price of the good. Compare *external cost, full cost*.

external cost Harmful environmental, economic, or social effect of producing and using an economic good that is not included in the market price of the good. Compare *external benefit, full cost, internal cost*.

extinction Complete disappearance of a species from the earth. It happens when a species cannot adapt and successfully reproduce under new environmental conditions or when a species evolves into one or more new species. Compare *speciation*. See also *endangered species, mass extinction, threatened species*.

extinction rate Percentage or number of species that go extinct within a certain time such as a year.

family planning Providing information, clinical services, and contraceptives to help people choose the number and spacing of children they want to have.

famine Widespread malnutrition and starvation in a particular area because of a shortage of food, usually caused by drought, war, flood, earthquake, or other catastrophic events that disrupt food production and distribution.

feedback Any process that increases (positive feedback) or decreases (negative feedback) a change to a system.

feedback loop Occurs when an output of matter, energy, or information is fed back into the system as an input and leads to changes in that system. See *positive feedback loop* and *negative feedback loop*.

feedlot Confined outdoor or indoor space used to raise hundreds to thousands of domesticated livestock.

fermentation See *anaerobic respiration*.

fertility rate Number of children born to an average woman in a population during her life-time. Compare *replacement-level fertility.*

fertilizer Substance that adds inorganic or organic plant nutrients to soil and improves its ability to grow crops, trees, or other vegetation. See *commercial inorganic fertilizer, organic fertilizer.*

first law of thermodynamics In any physical or chemical change, no detectable amount of energy is created or destroyed, but energy can be changed from one form to another; you cannot get more energy out of something than you put in; in terms of energy quantity, you cannot get something for nothing. This law does not apply to nuclear changes, in which energy can be produced from small amounts of matter. See *second law of thermodynamics.*

fishery Concentration of particular aquatic species suitable for commercial harvesting in a given ocean area or inland body of water.

fish farming Form of aquaculture in which fish are cultivated in a controlled pond or other environment and harvested when they reach the desired size. See also *fish ranching.*

fishprint Area of ocean needed to sustain the consumption of an average person, a nation, or the world. Compare *ecological footprint.*

fish ranching Form of aquaculture in which members of a fish species such as salmon are held in captivity for the first few years of their lives, released, and then harvested as adults when they return from the ocean to their freshwater birthplace to spawn. See also *fish farming.*

fissionable isotope Isotope that can split apart when hit by a neutron at the right speed and thus undergo nuclear fission. Examples include uranium-235 and plutonium-239.

floodplain Flat valley floor next to a stream channel. For legal purposes, the term often applies to any low area that has the potential for flooding, including certain coastal areas.

flows See *throughputs.*

flyway Generally fixed route along which waterfowl migrate from one area to another at certain seasons of the year.

food chain Series of organisms in which each eats or decomposes the preceding one. Compare *food web.*

food insecurity Condition under which people live with chronic hunger and malnutrition that threatens their ability to lead healthy and productive lives. Compare *food security.*

food security Condition under which every person in a given area has daily access to enough nutritious food to have an active and healthy life. Compare *food insecurity.*

food web Complex network of many interconnected food chains and feeding relationships. Compare *food chain.*

forest Biome with enough average annual precipitation to support the growth of tree species and smaller forms of vegetation. Compare *desert, grassland.*

fossil fuel Products of partial or complete decomposition of plants and animals; occurs as crude oil, coal, natural gas, or heavy oils as a result of exposure to heat and pressure in he earth's crust over millions of years. See *coal, crude oil, natural gas.*

fossils Skeletons, bones, shells, body parts, leaves, seeds, or impressions of such items that provide recognizable evidence of organisms that lived long ago.

foundation species Species that plays a major role in shaping a community by creating and enhancing a habitat that benefits other species. Compare *indicator species, keystone species, native species, nonnative species.*

free-access resource See *open access renewable resource.*

freons See *chlorofluorocarbons.*

freshwater life zones Aquatic systems where water with a dissolved salt concentration of less than 1% by volume accumulates on or flows through the surfaces of terrestrial biomes. Examples include *standing* (lentic) bodies of fresh water such as lakes, ponds, and inland wetlands and *flowing* (lotic) systems such as streams and rivers. Compare *biome.*

front The boundary between two air masses with different temperatures and densities. See *cold front, warm front.*

frontier science See *tentative science.*

frontier worldview View held by European colonists settling North America in the 1600s that the continent had vast resources and was a wilderness to be conquered by settlers clearing and planting land.

full cost Cost of a good when its internal costs and its estimated short- and long-term external costs are included in its market price. Compare *external cost, internal cost.*

functional diversity Biological and chemical processes or functions such as energy flow and matter cycling needed for the survival of species and biological communities. See *biodiversity, ecological diversity, genetic diversity, species diversity.*

fundamental niche Full potential range of the physical, chemical, and biological factors a species can use if it does not face any competition from other species. See *ecological niche.* Compare *realized niche.*

fungicide Chemical that kills fungi.

game species Type of wild animal that people hunt or fish for, for sport and recreation and sometimes for food.

gamma ray Form of ionizing electromagnetic radiation with a high energy content emitted by some radioisotopes. It readily penetrates body tissues. See also *alpha particle, beta particle.*

GDP See *gross domestic product.*

gene mutation See *mutation.*

gene pool Sum total of all genes found in the individuals of the population of a particular species.

generalist species Species with a broad ecological niche. They can live in many different places, eat a variety of foods, and tolerate a wide range of environmental conditions. Examples include flies, cockroaches, mice, rats, and humans. Compare *specialist species.*

genes Coded units of information about specific traits that are passed from parents to offspring during reproduction. They consist of segments of DNA molecules found in chromosomes.

gene splicing See *genetic engineering.*

genetic adaptation Changes in the genetic makeup of organisms of a species that allow the species to reproduce and gain a competitive advantage under changed environmental conditions. See *differential reproduction, evolution, mutation, natural selection.*

genetically modified organism (GMO) Organism whose genetic makeup has been altered by genetic engineering.

genetic diversity Variability in the genetic makeup among individuals within a single species. See *biodiversity.* Compare *ecological diversity, functional diversity, species diversity.*

genetic engineering Insertion of an alien gene into an organism to give it a beneficial genetic trait. Compare *artificial selection, natural selection.*

geographic isolation Separation of populations of a species for long times into different areas.

geology Study of the earth's dynamic history. Geologists study and analyze rocks and the features and processes of the earth's interior and surface.

geosphere Earth's intensely hot core, thick mantle composed mostly of rock, and thin outer crust that contains most of the earth's rock, soil, and sediment. Compare *atmosphere, biosphere, hydrosphere.*

geothermal energy Heat transferred from the earth's underground concentrations of dry steam (steam with no water droplets), wet steam (a mixture of steam and water droplets), or hot water trapped in fractured or porous rock.

global climate change Broad term referring to changes in any aspects of the earth's climate, including temperature, precipitation, and storm activity. Compare *weather.*

globalization Broad process of global social, economic, and environmental change that leads to an increasingly integrated world.

global warming Warming of the earth's lower atmosphere (troposphere) because of increases in the concentrations of one or more greenhouse gases. It can result in climate change that can last for decades to thousands of years. See *greenhouse effect, greenhouse gases, natural greenhouse effect.*

GMO See *genetically modified organism.*

GPP See *gross primary productivity.*

grassland Biome found in regions where enough annual average precipitation to support the growth of grass and small plants but not enough to support large stands of trees. Compare *desert, forest*.

greenhouse effect Natural effect that releases heat in the atmosphere (troposphere) near the earth's surface. Water vapor, carbon dioxide, ozone, and other gases in the lower atmosphere (troposphere) absorb some of the infrared radiation (heat) radiated by the earth's surface. Their molecules vibrate and transform the absorbed energy into longer-wavelength infrared radiation (heat) in the troposphere. If the atmospheric concentrations of these greenhouse gases increase and other natural processes do not remove them, the average temperature of the lower atmosphere will increase gradually. Compare *global warming*. See also *natural greenhouse effect*.

greenhouse gases Gases in the earth's lower atmosphere (troposphere) that cause the greenhouse effect. Examples include carbon dioxide, chlorofluorocarbons, ozone, methane, water vapor, and nitrous oxide.

green manure Freshly cut or still-growing green vegetation that is plowed into the soil to increase the organic matter and humus available to support crop growth. Compare *animal manure*.

green revolution Popular term for the introduction of scientifically bred or selected varieties of grain (rice, wheat, maize) that, with adequate inputs of fertilizer and water, can greatly increase crop yields.

gross domestic product (GDP) Annual market value of all goods and services produced by all firms and organizations, foreign and domestic, operating within a country. See *per capita GDP*.

gross primary productivity (GPP) Rate at which an ecosystem's producers capture and store a given amount of chemical energy as biomass in a given length of time. Compare *net primary productivity*.

ground fire Fire that burns decayed leaves or peat deep below the ground surface. Compare *crown fire, surface fire*.

groundwater Water that sinks into the soil and is stored in slowly flowing and slowly renewed underground reservoirs called aquifers; underground water in the zone of saturation, below the water table. Compare *runoff, surface water*.

habitat Place or type of place where an organism or population of organisms lives. Compare *ecological niche*.

habitat fragmentation Breakup of a habitat into smaller pieces, usually as a result of human activities.

half-life Time needed for one-half of the nuclei in a radioisotope to emit their radiation. Each radioisotope has a characteristic half-life, which may range from a few millionths of a second to several billion years. See *radioisotope*.

hazard Something that can cause injury, disease, economic loss, or environmental damage. See also *risk*.

hazardous chemical Chemical that can cause harm because it is flammable or explosive, can irritate or damage the skin or lungs (such as strong acidic or alkaline substances), or can cause allergic reactions of the immune system (allergens). See also *toxic chemical*.

hazardous waste Any solid, liquid, or containerized gas that can catch fire easily, is corrosive to skin tissue or metals, is unstable and can explode or release toxic fumes, or has harmful concentrations of one or more toxic materials that can leach out. See also *toxic waste*.

heat Total kinetic energy of all randomly moving atoms, ions, or molecules within a given substance, excluding the overall motion of the whole object. Heat always flows spontaneously from a warmer sample of matter to a colder sample of matter. This is one way to state the second law of thermodynamics. Compare *temperature*.

herbicide Chemical that kills a plant or inhibits its growth.

herbivore Plant-eating organism. Examples include deer, sheep, grasshoppers, and zooplankton. Compare *carnivore, omnivore*.

heterotroph See *consumer*.

high Air mass with a high pressure. Compare *low*.

high-grade ore Ore containing a large amount of the desired mineral. Compare *low-grade ore*.

high-input agriculture See *industrialized agriculture*.

high-quality energy Energy that is concentrated and has great ability to perform useful work. Examples include high-temperature heat and the energy in electricity, coal, oil, gasoline, sunlight, and nuclei of uranium-235. Compare *low-quality energy*.

high-quality matter Matter that is concentrated and contains a high concentration of a useful resource. Compare *low-quality matter*.

high-throughput economy Economic system in most advanced industrialized countries, in which ever-increasing economic growth is sustained by maximizing the rate at which matter and energy resources are used, with little emphasis on pollution prevention, recycling, reuse, reduction of unnecessary waste, and other forms of resource conservation. Compare *low-throughput economy, matter-recycling economy*.

high-waste economy See *high-throughput economy*.

HIPPCO Acronym used by conservation biologists for the six most important secondary causes of premature extinction: **H**abitat destruction, degradation, and fragmentation; **I**nvasive (nonnative) species; **P**opulation growth (too many people consuming too many resources); **P**ollution; **C**limate change; and **O**verexploitation.

host Plant or animal on which a parasite feeds.

human capital People's physical and mental talents that provide labor, innovation, culture, and organization. Compare *manufactured capital, natural capital*.

human resources See *human capital*.

humus Slightly soluble residue of undigested or partially decomposed organic material in topsoil. This material helps retain water and water-soluble nutrients, which can be taken up by plant roots.

hunger See *chronic undernutrition*.

hunter–gatherers People who get their food by gathering edible wild plants and other materials and by hunting wild animals and fish.

hydrocarbon Organic compound made of hydrogen and carbon atoms. The simplest hydrocarbon is methane (CH_4), the major component of natural gas.

hydroelectric power plant Structure in which the energy of falling or flowing water spins a turbine generator to produce electricity.

hydrologic cycle Biogeochemical cycle that collects, purifies, and distributes the earth's fixed supply of water from the environment to living organisms and then back to the environment.

hydropower Electrical energy produced by falling or flowing water. See *hydroelectric power plant*.

hydrosphere Earth's *liquid water* (oceans, lakes, other bodies of surface water, and underground water), *frozen water* (polar ice caps, floating ice caps, and ice in soil, known as permafrost), and *water vapor* in the atmosphere. See also *hydrologic cycle*. Compare *atmosphere, biosphere, geosphere*.

hypereutrophic Result of excessive inputs of nutrients in a lake. See *cultural eutrophication*.

igneous rock Rock formed when molten rock material (magma) wells up from the earth's interior, cools, and solidifies into rock masses. Compare *metamorphic rock, sedimentary rock*. See *rock cycle*.

immature community Community at an early stage of ecological succession. It usually has a low number of species and ecological niches and cannot capture and use energy and cycle critical nutrients as efficiently as more complex, mature communities. Compare *mature community*.

immigrant species See *nonnative species*.

immigration Migration of people into a country or area to take up permanent residence.

indicator species Species that serve as early warnings that a community or ecosystem is being degraded. Compare *foundation species, keystone species, native species, nonnative species*.

inductive reasoning Using specific observations and measurements to arrive at a general conclusion or hypothesis. Compare *deductive reasoning*.

industrialized agriculture Production of large quantities of crops and livestock for domestic and foreign sale; involves use of large inputs of energy from fossil fuels (especially oil and natural gas), water, fertilizer, and pesticides. Compare *subsistence farming*.

industrial–medical revolution Use of new sources of energy from fossil fuels and later from nuclear fuels, and use of new technologies, to grow food and manufacture products. Compare *agricultural revolution, environmental revolution, hunter–gatherers, information and globalization revolution*.

industrial smog Type of air pollution consisting mostly of a mixture of sulfur dioxide, suspended droplets of sulfuric acid formed from some of the sulfur dioxide, and suspended solid particles. Compare *photochemical smog*.

industrial solid waste Solid waste produced by mines, factories, refineries, food growers, and businesses that supply people with goods and services. Compare *municipal solid waste*.

inertia Ability of a living system, such as a grassland or a forest, to survive moderate disturbances. Compare *constancy, resilience*.

infant mortality rate Number of babies out of every 1,000 born each year who die before their first birthday.

infectious disease Disease caused when a pathogen such as a bacterium, virus, or parasite invades the body and multiplies in its cells and tissues. Examples are flu, HIV, malaria, tuberculosis, and measles. See *transmissible disease*. Compare *nontransmissible disease*.

infiltration Downward movement of water through soil.

information and globalization revolution Use of new technologies such as the telephone, radio, television, computers, the Internet, automated databases, and remote sensing satellites to enable people to have increasingly rapid access to much more information on a global scale. Compare *agricultural revolution, environmental revolution, hunter–gatherers, industrial–medical revolution*.

inherent value See *intrinsic value*.

inland wetland Land away from the coast, such as a swamp, marsh, or bog, that is covered all or part of the time with fresh water. Compare *coastal wetland*.

inorganic compounds All compounds not classified as organic compounds. See *organic compounds*.

inorganic fertilizer See *commercial inorganic fertilizer*.

input Matter, energy, or information entering a system. Compare *output, throughput*.

input pollution control See *pollution prevention*.

insecticide Chemical that kills insects.

instrumental value Value of an organism, species, ecosystem, or the earth's biodiversity based on its usefulness to humans. Compare *intrinsic value*.

integrated pest management (IPM) Combined use of biological, chemical, and cultivation methods in proper sequence and timing to keep the size of a pest population below the size that causes economically unacceptable loss of a crop or livestock animal.

integrated waste management Variety of strategies for both waste reduction and waste management designed to deal with the solid wastes we produce.

intercropping Growing two or more different crops at the same time on a plot. For example, a carbohydrate-rich grain that depletes soil nitrogen and a protein-rich legume that adds nitrogen to the soil may be intercropped. Compare *monoculture, polyculture*.

internal cost Direct cost paid by the producer and the buyer of an economic good. Compare *external benefit, external cost, full cost*.

interspecific competition Attempts by members of two or more species to use the same limited resources in an ecosystem. See *competition, intraspecific competition*.

intertidal zone The area of shoreline between low and high tides.

intraspecific competition Attempts by two or more organisms of a single species to use the same limited resources in an ecosystem. See *competition, interspecific competition*.

intrinsic rate of increase (r) Rate at which a population could grow if it had unlimited resources. Compare *environmental resistance*.

intrinsic value Value of an organism, species, ecosystem, or the earth's biodiversity based on its existence, regardless of whether it has any usefulness to humans. Compare *instrumental value*.

invasive species See *nonnative species*.

inversion See *temperature inversion*.

invertebrates Animals that have no backbones. Compare *vertebrates*.

ion Atom or group of atoms with one or more positive (+) or negative (−) electrical charges. Examples are Na^+ and Cl^-. Compare *atom, molecule*.

ionizing radiation Fast-moving alpha or beta particles or high-energy radiation (gamma rays) emitted by radioisotopes. They have enough energy to dislodge one or more electrons from atoms they hit, thereby forming charged ions in tissue that can react with and damage living tissue. Compare *nonionizing radiation*.

isotopes Two or more forms of a chemical element that have the same number of protons but different mass numbers because they have different numbers of neutrons in their nuclei.

J-shaped curve Curve with a shape similar to that of the letter J; can represent prolonged exponential growth. See *exponential growth*.

junk science See *unreliable science*.

kerogen Solid, waxy mixture of hydrocarbons found in oil shale rock. Heating the rock to high temperatures causes the kerogen to vaporize. The vapor is condensed, purified, and then sent to a refinery to produce gasoline, heating oil, and other products. See also *oil shale, shale oil*.

keystone species Species that play roles affecting many other organisms in an ecosystem. Compare *foundation species, indicator species, native species, nonnative species*.

kilocalorie (kcal) Unit of energy equal to 1,000 calories. See *calorie*.

kilowatt (kW) Unit of electrical power equal to 1,000 watts. See *watt*.

kinetic energy Energy that matter has because of its mass and speed, or velocity. Compare *potential energy*.

K-selected species Species that produce a few, often fairly large offspring but invest a great deal of time and energy to ensure that most of those offspring reach reproductive age. Compare *r-selected species*.

K-strategists See *K-selected species*.

lake Large natural body of standing fresh water formed when water from precipitation, land runoff, or groundwater flow fills a depression in the earth created by glaciation, earth movement, volcanic activity, or a giant meteorite. See *eutrophic lake, mesotrophic lake, oligotrophic lake*.

land degradation Decrease in the ability of land to support crops, livestock, or wild species in the future as a result of natural or human-induced processes.

landfill See *sanitary landfill*.

land-use planning Planning to determine the best present and future uses of each parcel of land.

latitude Distance from the equator. Compare *altitude*.

law of conservation of energy See *first law of thermodynamics*.

law of conservation of matter In any physical or chemical change, matter is neither created nor destroyed but merely changed from one form to another; in physical and chemical changes, existing atoms are rearranged into different spatial patterns (physical changes) or different combinations (chemical changes).

law of nature See *scientific law*.

law of tolerance Existence, abundance, and distribution of a species in an ecosystem are determined by whether the levels of one or more physical or chemical factors fall within the range tolerated by the species. See *threshold effect*.

LD50 See *median lethal dose*.

LDC See *developing country*.

leaching Process in which various chemicals in upper layers of soil are dissolved and carried to lower layers and, in some cases, to groundwater.

less developed country (LDC) See *developing country*.

life-cycle cost Initial cost plus lifetime operating costs of an economic good. Compare *full cost*.

life expectancy Average number of years a newborn infant can be expected to live.

limiting factor Single factor that limits the growth, abundance, or distribution of the population of a species in an ecosystem. See *limiting factor principle*.

limiting factor principle Too much or too little of any abiotic factor can limit or prevent growth of a population of a species in an ecosystem, even if all other factors are at or near the optimal range of tolerance for the species.

linear growth Growth in which a quantity increases by some fixed amount during each unit of time. An example is growth that increases by 2 units in the sequence 2, 4, 6, 8, 10, and so on. Compare *exponential growth*.

liquefied natural gas (LNG) Natural gas converted to liquid form by cooling it to a very low temperature.

liquefied petroleum gas (LPG) Mixture of liquefied propane (C_3H_8) and butane (C_4H_{10}) gas removed from natural gas and used as a fuel.

lithosphere Outer shell of the earth, composed of the crust and the rigid, outermost part of the mantle outside the asthenosphere; material found in the earth's plates. See *crust, geosphere, mantle*.

loams Soils containing a mixture of clay, sand, silt, and humus. Good for growing most crops.

lobbying Process in which individuals or groups use public pressure, personal contacts, and political action to persuade legislators to vote or act in their favor.

logistic growth Pattern in which exponential population growth occurs when the population is small, and population growth decreases steadily with time as the population approaches the carrying capacity. See *S-shaped curve*.

low Air mass with a low pressure. Compare *high*.

low-grade ore Ore containing a small amount of a desired mineral. Compare *high-grade ore*.

low-input agriculture See *sustainable agriculture*.

low-quality energy Energy that is dispersed and has little ability to do useful work. An example is low-temperature heat. Compare *high-quality energy*.

low-quality matter Matter that is dilute or dispersed or contains a low concentration of a useful resource. Compare *high-quality matter*.

low-throughput economy Economy based on working with nature by recycling and reusing discarded matter, preventing pollution, conserving matter and energy resources by reducing unnecessary waste and use, not degrading renewable resources, building things that are easy to recycle, reuse, and repair, not allowing population size to exceed the carrying capacity of the environment, and preserving biodiversity. Compare *high-throughput economy, matter-recycling economy*.

low-waste economy See *low-throughput economy*.

LPG See *liquefied petroleum gas*.

magma Molten rock below the earth's surface.

malnutrition Faulty nutrition, caused by a diet that does not supply an individual with enough protein, essential fats, vitamins, minerals, and other nutrients needed for good health. Compare *overnutrition, undernutrition*.

mangrove swamps Swamps found on the coastlines in warm tropical climates. They are dominated by mangrove trees, any of about 55 species of trees and shrubs that can live partly submerged in the salty environment of coastal swamps.

mantle Zone of the earth's interior between its core and its crust. Compare *core, crust*. See *geosphere, lithosphere*.

manufactured capital See *manufactured resources*.

manufactured resources Manufactured items made from natural resources and used to produce and distribute economic goods and services bought by consumers. They include tools, machinery, equipment, factory buildings, and transportation and distribution facilities. Compare *human resources, natural resources*.

manure See *animal manure, green manure*.

mass Amount of material in an object.

mass extinction Catastrophic, widespread, often global event in which major groups of species are wiped out over a short time compared with normal (background) extinctions. Compare *background extinction*.

mass number Sum of the number of neutrons (n) and the number of protons (p) in the nucleus of an atom. It gives the approximate mass of that atom. Compare *atomic number*.

mass transit Buses, trains, trolleys, and other forms of transportation that carry large numbers of people.

mass wasting Process that occurs when rock and soil masses become detached from underlying material and move downhill under the influence of gravity.

material efficiency Total amount of material needed to produce each unit of goods or services. Also called *resource productivity*. Compare *energy efficiency*.

matter Anything that has mass (the amount of material in an object) and takes up space. On the earth, where gravity is present, we weigh an object to determine its mass.

matter quality Measure of how useful a matter resource is, based on its availability and concentration. See *high-quality matter, low-quality matter*.

matter-recycling-and-reuse economy Economy that emphasizes recycling the maximum amount of all resources that can be recycled and reused. The goal is to allow economic growth to continue without depleting matter resources and without producing excessive pollution and environmental degradation. Compare *high-throughput economy, low-throughput economy*.

mature community Fairly stable, self-sustaining community in an advanced stage of ecological succession; usually has a diverse array of species and ecological niches; captures and uses energy and cycles critical chemicals more efficiently than simpler, immature communities. Compare *immature community*.

maximum sustainable yield See *sustainable yield*.

MDC See *developed country*.

median lethal dose (LD50) Amount of a toxic material per unit of body weight of test animals that kills half the test population in a certain time.

mediation Approach for settling a dispute in which the parties involved are encouraged to sit down and talk under the guidance of a professional mediator, with the goal of resolving the dispute in a way that is acceptable to both parties. Compare *arbitration*.

megacity City with 10 million or more people.

meltdown Melting of the highly radioactive core of a nuclear reactor.

mesotrophic lake Lake with a moderate supply of plant nutrients. Compare *eutrophic lake, oligotrophic lake*.

metabolism Ability of a living cell or organism to capture and transform matter and energy from its environment to supply its needs for survival, growth, and reproduction.

metamorphic rock Rock produced when a preexisting rock is subjected to high temperatures (which may cause it to melt partially), high pressures, chemically active fluids, or a combination of these agents. Compare *igneous rock, sedimentary rock*. See *rock cycle*.

metastasis Spread of malignant (cancerous) cells from a tumor to other parts of the body.

metropolitan area See *urban area*.

microorganisms Organisms such as bacteria that are so small that it takes a microscope to see them.

micropower systems Systems of small-scale decentralized units that generate 1–10,000 kilowatts of electricity. Examples include microturbines, fuel cells, wind turbines, and household solar-cell panels and solar-cell roofs.

migration Movement of people into and out of specific geographic areas. Compare *emigration* and *immigration*.

mineral Any naturally occurring inorganic substance found in the earth's crust as a crystalline solid. See *mineral resource*.

mineral resource Concentration of naturally occurring solid, liquid, or gaseous material in or on the earth's crust in a form and amount such that extracting and converting it into useful materials or items is currently or potentially profitable. Mineral resources are classified as *metallic* (such as iron and tin ores) or *nonmetallic* (such as fossil fuels, sand, and salt).

minimum-tillage farming See *conservation-tillage farming*.

mixture Combination of one or more elements and compounds.

model Approximate representation or simulation of a system being studied.

molecule Combination of two or more atoms of the same chemical element (such as O_2) or different chemical elements (such as H_2O) held together by chemical bonds. Compare *atom, ion*.

monoculture Cultivation of a single crop, usually on a large area of land. Compare *polyculture*.

more developed country (MDC) See *developed country*.

mountaintop removal Type of surface mining that uses explosives, massive shovels, and large machines called draglines to remove the top of a mountain to expose seams of coal underneath a mountain. Compare *area strip mining, contour strip mining*.

MSW See *municipal solid waste*.

multiple use Use of an ecosystem such as a forest for a variety of purposes such as timber harvesting, wildlife habitat, watershed protection, and recreation. Compare *sustainable yield*.

municipal solid waste (MSW) Solid materials discarded by homes and businesses in or near urban areas. See *solid waste*. Compare *industrial solid waste*.

mutagen Chemical or form of radiation that causes inheritable changes (mutations) in the DNA molecules in genes. See *carcinogen, mutation, teratogen*.

mutation Random change in DNA molecules making up genes that can alter anatomy, physiology, or behavior in offspring. See *mutagen*.

mutualism Type of species interaction in which both participating species generally benefit. Compare *commensalism*.

native species Species that normally live and thrive in a particular ecosystem. Compare *foundation species, indicator species, keystone species, nonnative species*.

natural capital Natural resources and natural services that keep us and other species alive and support our economies. See *natural resources, natural services*.

natural gas Underground deposits of gases consisting of 50–90% by weight methane gas (CH_4) and small amounts of heavier gaseous hydrocarbon compounds such as propane (C_3H_8) and butane (C_4H_{10}).

natural greenhouse effect Heat buildup in the troposphere caused by the presence of certain gases, called greenhouse gases. Without this effect, the earth would be nearly as cold as Mars, and life as we know it could not exist. See *global warming*.

natural income Renewable resources such as plants, animals, and soil provided by natural capital.

natural law See *scientific law*.

natural radioactive decay Nuclear change in which unstable nuclei of atoms spontaneously shoot out particles (usually alpha or beta particles) or energy (gamma rays) at a fixed rate.

natural rate of extinction See *background extinction*.

natural recharge Natural replenishment of an aquifer by precipitation, which percolates downward through soil and rock. See *recharge area*.

natural resources Materials such as air, water, and soil and energy in nature that are essential or useful to humans. See *natural capital*.

natural selection Process by which a particular beneficial gene (or set of genes) is reproduced in succeeding generations more than other genes. The result of natural selection is a population that contains a greater proportion of organisms better adapted to certain environmental conditions. See *adaptation, biological evolution, differential reproduction, mutation*.

natural services Processes of nature, such as purification of air and water and pest control, which support life and human economies. See *natural capital*.

negative feedback loop Feedback loop that causes a system to change in the opposite direction from which is it moving. Compare *positive feedback loop*.

nekton Strongly swimming organisms found in aquatic systems. Compare *benthos, plankton*.

net energy Total amount of useful energy available from an energy resource or energy system over its lifetime, minus the amount of energy *used* (the first energy law), *automatically wasted* (the second energy law), and *unnecessarily wasted* in finding, processing, concentrating, and transporting it to users.

net primary productivity (NPP) Rate at which all the plants in an ecosystem produce net useful chemical energy; equal to the difference between the rate at which the plants in an ecosystem produce useful chemical energy (gross primary productivity) and the rate at which they use some of that energy through cellular respiration. Compare *gross primary productivity*.

neurotoxins Chemicals that can harm the human *nervous system* (brain, spinal cord, peripheral nerves).

neutral solution Water solution containing an equal number of hydrogen ions (H^+) and hydroxide ions (OH^-); water solution with a pH of 7. Compare *acid solution, basic solution*.

neutron (n) Elementary particle in the nuclei of all atoms (except hydrogen-1). It has a relative mass of 1 and no electric charge. Compare *electron, proton*.

niche See *ecological niche*.

nitric oxide (NO) Colorless gas that forms when nitrogen and oxygen gas in air react at the high-combustion temperatures in automobile engines and coal-burning plants. Lightning and certain bacteria in soil and water also produce NO as part of the nitrogen cycle.

nitrogen cycle Cyclic movement of nitrogen in different chemical forms from the environment to organisms and then back to the environment.

nitrogen dioxide (NO_2) Reddish-brown gas formed when nitrogen oxide reacts with oxygen in the air.

nitrogen fixation Conversion of atmospheric nitrogen gas into forms useful to plants by lightning, bacteria, and cyanobacteria; it is part of the nitrogen cycle.

nitrogen oxides (NO_x) See *nitric oxide* and *nitrogen dioxide*.

noise pollution Any unwanted, disturbing, or harmful sound that impairs or interferes with hearing, causes stress, hampers concentration and work efficiency, or causes accidents.

nondegradable pollutant Material that is not broken down by natural processes. Examples include the toxic elements lead and mercury. Compare *biodegradable pollutant*.

nonionizing radiation Forms of radiant energy such as radio waves, microwaves, infrared light, and ordinary light that do not have enough energy to cause ionization of atoms in living tissue. Compare *ionizing radiation*.

nonnative species Species that migrate into an ecosystem or are deliberately or accidentally introduced into an ecosystem by humans. Compare *native species*.

nonpoint sources Broad and diffuse areas, rather than points, from which pollutants enter bodies of surface water or air. Examples include runoff of chemicals and sediments from cropland, livestock feedlots, logged forests, urban streets, parking lots, lawns, and golf courses. Compare *point source*.

nonrenewable resource Resource that exists in a fixed amount (stock) in the earth's crust and has the potential for renewal by geological, physical, and chemical processes taking place over hundreds of millions to billions of years. Examples include copper, aluminum, coal, and oil. We classify these resources as exhaustible because we are extracting and using them at a much faster rate than they are formed. Compare *renewable resource*.

nontransmissible disease Disease that is not caused by living organisms and does not spread from one person to another. Examples include most cancers, diabetes, cardiovascular disease, and malnutrition. Compare *transmissible disease*.

no-till farming See *conservation-tillage farming*.

NPP See *net primary productivity*.

nuclear change Process in which nuclei of certain isotopes spontaneously change, or are forced to change, into one or more different isotopes. The three principal types of nuclear change are natural radioactivity, nuclear fission, and nuclear fusion. Compare *chemical change, physical change*.

nuclear energy Energy released when atomic nuclei undergo a nuclear reaction such as the spontaneous emission of radioactivity, nuclear fission, or nuclear fusion.

nuclear fission Nuclear change in which the nuclei of certain isotopes with large mass numbers (such as uranium-235 and plutonium-239) are split apart into lighter nuclei when struck by a neutron. This process releases more neutrons and a large amount of energy. Compare *nuclear fusion*.

nuclear fusion Nuclear change in which two nuclei of isotopes of elements with a low mass number (such as hydrogen-2 and hydrogen-3) are forced together at extremely high temperatures until they fuse to form a heavier nucleus (such as helium-4). This process releases a large amount of energy. Compare *nuclear fission*.

nucleus Extremely tiny center of an atom, making up most of the atom's mass. It contains one or more positively charged protons and one or more neutrons with no electrical charge (except for a hydrogen-1 atom, which has one proton and no neutrons in its nucleus).

nutrient Any chemical element or compound an organism must take in to live, grow, or reproduce.

nutrient cycle See *biogeochemical cycle*.

nutrient cycling The circulation of chemicals necessary for life, from the environment (mostly from soil and water) through organisms and back to the environment

oil See *crude oil*.

oil sand Deposit of a mixture of clay, sand, water, and varying amounts of a tar-like heavy oil known as bitumen. Bitumen can be extracted from oil sand by heating. It is then purified and upgraded to synthetic crude oil. See *bitumen*.

oil shale Fine-grained rock containing various amounts of kerogen, a solid, waxy mixture of hydrocarbon compounds. Heating the rock to high temperatures converts the kerogen into a vapor that can be condensed to form a slow-flowing heavy oil called shale oil. See *kerogen, shale oil*.

old-growth forest Virgin and old, second-growth forests containing trees that are often hundreds—sometimes thousands—of years old. Examples include forests of Douglas fir, western hemlock, giant sequoia, and coastal redwoods in the western United States. Compare *second-growth forest, tree plantation*.

oligotrophic lake Lake with a low supply of plant nutrients. Compare *eutrophic lake, mesotrophic lake*.

omnivore Animal that can use both plants and other animals as food sources. Examples include pigs, rats, cockroaches, and humans. Compare *carnivore, herbivore*.

open access renewable resource Renewable resource owned by no one and available for use by anyone at little or no charge. Examples include clean air, underground water supplies, the open ocean and its fish, and the ozone layer. Compare *common property resource, private property resource*.

open dump Fields or holes in the ground where garbage is deposited and sometimes covered with soil. They are rare in developed countries, but are widely used in many developing countries, especially to handle wastes from megacities. Compare *sanitary landfill*.

open-pit mining Removing minerals such as gravel, sand, and metal ores by digging them out of the earth's surface and leaving an open pit behind. Compare *area strip mining, contour strip mining, mountaintop removal, subsurface mining*.

open sea Part of an ocean that lies beyond the continental shelf. Compare *coastal zone*.

ore Part of a metal-yielding material that can be economically extracted from a mineral; typically containing two parts: the ore mineral, which contains the desired metal, and waste mineral material (gangue). See *high-grade ore, low-grade ore*.

organic agriculture Growing crops with limited or no use of synthetic pesticides and synthetic fertilizers; raising livestock without use of synthetic growth regulators and feed additives; using organic fertilizer (manure, legumes, compost) and natural pest control (bugs that eat harmful bugs, plants that repel bugs, and environmental controls such as crop rotation). See *sustainable agriculture*.

organic compounds Compounds containing carbon atoms combined with each other and with atoms of one or more other elements such as hydrogen, oxygen, nitrogen, sulfur, phosphorus, chlorine, and fluorine. All other compounds are called *inorganic compounds*.

organic farming See *organic agriculture* and *sustainable agriculture*.

organic fertilizer Organic material such as animal manure, green manure, and compost, applied to cropland as a source of plant nutrients. Compare *commercial inorganic fertilizer*.

organism Any form of life.

output Matter, energy, or information leaving a system. Compare *input, throughput*.

output pollution control See *pollution cleanup*.

overburden Layer of soil and rock overlying a mineral deposit. Surface mining removes this layer.

overfishing Harvesting so many fish of a species, especially immature individuals, that not enough breeding stock is left to replenish the species and it becomes unprofitable to harvest them.

overgrazing Destruction of vegetation when too many grazing animals feed too long and exceed the carrying capacity of a rangeland or pasture area.

overnutrition Diet so high in calories, saturated (animal) fats, salt, sugar, and processed foods and so low in vegetables and fruits that the consumer runs a high risk of developing diabetes, hypertension, heart disease, and other health hazards. Compare *malnutrition, undernutrition*.

oxygen-demanding wastes Organic materials that are usually biodegraded by aerobic (oxygen-consuming) bacteria if there is enough dissolved oxygen in the water.

ozone (O_3) Colorless and highly reactive gas and a major component of photochemical smog. Also found in the ozone layer in the stratosphere. See *photochemical smog*.

ozone depletion Decrease in concentration of ozone (O_3) in the stratosphere. See *ozone layer*.

ozone layer Layer of gaseous ozone (O_3) in the stratosphere that protects life on earth by filtering out most harmful ultraviolet radiation from the sun.

PANs Peroxyacyl nitrates; group of chemicals found in photochemical smog.

paradigm shift Shift in thinking that occurs when the majority of scientists in a field or related fields agree that a new explanation or theory is better than the old one

parasite Consumer organism that lives on or in, and feeds on, a living plant or animal, known as the host, over an extended period. The parasite draws nourishment from and gradually weakens its host; it may or may not kill the host. See *parasitism*.

parasitism Interaction between species in which one organism, called the parasite, preys on another organism, called the host, by living on or in the host. See *host, parasite*.

particulates Also known as suspended particulate matter (SPM); variety of solid particles and liquid droplets small and light enough to remain suspended in the air for long periods. About 62% of the SPM in outdoor air comes from natural sources such as dust, wild fires, and sea salt. The remaining 38% comes from human sources such as coal-burning electric power and industrial plants, motor vehicles, plowed fields, road construction, unpaved roads, and tobacco smoke.

parts per billion (ppb) Number of parts of a chemical found in 1 billion parts of a particular gas, liquid, or solid.

parts per million (ppm) Number of parts of a chemical found in 1 million parts of a particular gas, liquid, or solid.

parts per trillion (ppt) Number of parts of a chemical found in 1 trillion parts of a particular gas, liquid, or solid.

passive solar heating system System that captures sunlight directly within a structure and

converts it into low-temperature heat for space heating or for heating water for domestic use without the use of mechanical devices. Compare *active solar heating system.*

pasture Managed grassland or enclosed meadow that usually is planted with domesticated grasses or other forage to be grazed by livestock. Compare *feedlot.*

pathogen Living organism that can cause disease in another organism. Examples include bacteria, viruses, and parasites.

PCBs See *polychlorinated biphenyls.*

peer review Process of scientists reporting details of the methods and models they used, the results of their experiments, and the reasoning behind their hypotheses for other scientists working in the same field (their peers) to examine and criticize.

per capita ecological footprint Amount of biologically productive land and water needed to supply each person or population with the renewable resources they use and to absorb or dispose of the wastes from such resource use. It measures the average environmental impact of individuals or populations in different countries and areas. Compare *ecological footprint.*

per capita GDP Annual gross domestic product (GDP) of a country divided by its total population at midyear. It gives the average slice of the economic pie per person. Used to be called per capita gross national product (GNP). See *gross domestic product.*

per capita GDP PPP (Purchasing Power Parity) Measure of the amount of goods and services that a country's average citizen could buy in the United States.

percolation Passage of a liquid through the spaces of a porous material such as soil.

perennial Plant that can live for more than 2 years. Compare *annual.*

permafrost Perennially frozen layer of the soil that forms when the water there freezes. It is found in arctic tundra.

perpetual resource Essentially inexhaustible resource on a human time scale because it is renewed continuously. Solar energy is an example. Compare *nonrenewable resource, renewable resource.*

persistence (1) the ability of a living system, such as a grassland or a forest, to survive moderate disturbances (2) the tendency for a pollutant to stay in the air, water, soil, or body. Compare *constancy, resilience.*

pest Unwanted organism that directly or indirectly interferes with human activities.

pesticide Any chemical designed to kill or inhibit the growth of an organism that people consider undesirable. See *fungicide, herbicide, insecticide.*

petrochemicals Chemicals obtained by refining (distilling) crude oil. They are used as raw materials in manufacturing most industrial chemicals, fertilizers, pesticides, plastics, syn-

thetic fibers, paints, medicines, and many other products.

petroleum See *crude oil.*

pH Numeric value that indicates the relative acidity or alkalinity of a substance on a scale of 0 to 14, with the neutral point at 7. Acid solutions have pH values lower than 7; basic or alkaline solutions have pH values greater than 7.

phosphorus cycle Cyclic movement of phosphorus in different chemical forms from the environment to organisms and then back to the environment.

photochemical smog Complex mixture of air pollutants produced in the lower atmosphere by the reaction of hydrocarbons and nitrogen oxides under the influence of sunlight. Especially harmful components include ozone, peroxyacyl nitrates (PANs), and various aldehydes. Compare *industrial smog.*

photosynthesis Complex process that takes place in cells of green plants. Radiant energy from the sun is used to combine carbon dioxide (CO_2) and water (H_2O) to produce oxygen (O_2), carbohydrates (such as glucose, $C_6H_{12}O_6$), and other nutrient molecules. Compare *aerobic respiration, chemosynthesis.*

photovoltaic (PV) cell Device that converts radiant (solar) energy directly into electrical energy. Also called a solar cell.

physical change Process that alters one or more physical properties of an element or a compound without changing its chemical composition. Examples include changing the size and shape of a sample of matter (crushing ice and cutting aluminum foil) and changing a sample of matter from one physical state to another (boiling and freezing water). Compare *chemical change, nuclear change.*

phytoplankton Small, drifting plants, mostly algae and bacteria, found in aquatic ecosystems. Compare *plankton, zooplankton.*

pioneer community First integrated set of plants, animals, and decomposers found in an area undergoing primary ecological succession. See *immature community, mature community.*

pioneer species First hardy species—often microbes, mosses, and lichens—that begin colonizing a site as the first stage of ecological succession. See *ecological succession, pioneer community.*

plaintiff Party in a court case bringing charges or seeking to collect damages for injuries to health or for economic loss; may also seek an injunction, by which the party being charged would be required to stop whatever action is causing harm. See *defendant* and *civil suit.*

planetary management worldview Worldview holding that humans are separate from nature, that nature exists mainly to meet our needs and increasing wants, and that we can use our ingenuity and technology to manage the earth's life-support systems, mostly for our benefit. It assumes that economic growth is unlimited. Compare *deep ecology worldview, environmental wisdom worldview, stewardship worldview.*

plankton Small plant organisms (phytoplankton) and animal organisms (zooplankton) that float in aquatic ecosystems.

plantation agriculture Growing specialized crops such as bananas, coffee, and cacao in tropical developing countries, primarily for sale to developed countries.

plates See *tectonic plates.*

plate tectonics Theory of geophysical processes that explains the movements of lithospheric plates and the processes that occur at their boundaries. See *lithosphere, tectonic plates.*

point source Single identifiable source that discharges pollutants into the environment. Examples include the smokestack of a power plant or an industrial plant, drainpipe of a meatpacking plant, chimney of a house, or exhaust pipe of an automobile. Compare *nonpoint source.*

poison Chemical that adversely affects the health of a living human or animal by causing injury, illness, or death.

policies Set of laws and regulations that a government enforces and the programs it funds.

politics Process through which individuals and groups try to influence or control government policies and actions that affect the local, state, national, and international communities.

pollutant Particular chemical or form of energy that can adversely affect the health, survival, or activities of humans or other living organisms. See *pollution.*

pollution Undesirable change in the physical, chemical, or biological characteristics of air, water, soil, or food that can adversely affect the health, survival, or activities of humans or other living organisms.

pollution cleanup Device or process that removes or reduces the level of a pollutant after it has been produced or has entered the environment. Examples include automobile emission control devices and sewage treatment plants. Compare *pollution prevention.*

pollution prevention Device, process, or strategy used to prevent a potential pollutant from forming or entering the environment or to sharply reduce the amount entering the environment. Compare *pollution cleanup.*

polychlorinated biphenyls (PCBs) Group of 209 toxic, oily, synthetic chlorinated hydrocarbon compounds that can be biologically amplified in food chains and webs.

polyculture Complex form of intercropping in which a large number of different plants maturing at different times are planted together. See also *intercropping.* Compare *monoculture.*

population Group of individual organisms of the same species living in a particular area.

population change Increase or decrease in the size of a population. It is equal to (Births + Immigration) − (Deaths + Emigration).

population density Number of organisms in a particular population found in a specified area or volume.

population dispersion General pattern in which the members of a population are arranged throughout its habitat.

population distribution Variation of population density over a particular geographic area or volume. For example, a country has a high population density in its urban areas and a much lower population density in rural areas.

population dynamics Major abiotic and biotic factors that tend to increase or decrease the population size and affect the age and sex composition of a species.

population size Number of individuals making up a population's gene pool.

positive feedback loop Feedback loop that causes a system to change further in the same direction. Compare *negative feedback loop*.

potential energy Energy stored in an object because of its position or the position of its parts. Compare *kinetic energy*.

poverty Inability to meet basic needs for food, clothing, and shelter.

ppb See *parts per billion*.

ppm See *parts per million*.

ppt See *parts per trillion*.

prairie See *grassland*.

precautionary principle When there is significant scientific uncertainty about potentially serious harm from chemicals or technologies, decision makers should act to prevent harm to humans and the environment. See *pollution prevention*.

precipitation Water in the form of rain, sleet, hail, and snow that falls from the atmosphere onto land and bodies of water.

predation Interaction in which an organism of one species (the predator) captures and feeds on parts or all of an organism of another species (the prey).

predator Organism that captures and feeds on parts or all of an organism of another species (the prey).

predator–prey relationship Relationship that has evolved between two organisms, in which one organism has become the prey for the other, the latter called the predator. See *predator, prey*.

prey Organism that is captured and serves as a source of food for an organism of another species (the predator).

primary consumer Organism that feeds on all or part of plants (herbivore) or on other producers. Compare *detritivore, omnivore, secondary consumer*.

primary pollutant Chemical that has been added directly to the air by natural events or human activities and occurs in a harmful concentration. Compare *secondary pollutant*.

primary productivity See *gross primary productivity, net primary productivity*.

primary recycling Process in which materials are recycled into new products of the same type—turning used aluminum cans into new aluminum cans, for example.

primary sewage treatment Mechanical sewage treatment in which large solids are filtered out by screens and suspended solids settle out as sludge in a sedimentation tank. Compare *advanced sewage treatment, secondary sewage treatment*.

primary succession Ecological succession in a bare area that has never been occupied by a community of organisms. See *ecological succession*. Compare *secondary succession*.

principles of sustainability Principles by which nature has sustained itself for billions of years by relying on solar energy, biodiversity, population regulation, and nutrient recycling.

private property resource Land, mineral, or other resource owned by individuals or by a firm. Compare *common property resource, open access renewable resource*.

probability Mathematical statement about how likely it is that something will happen.

producer Organism that uses solar energy (green plants) or chemical energy (some bacteria) to manufacture the organic compounds it needs as nutrients from simple inorganic compounds obtained from its environment. Compare *consumer, decomposer*.

prokaryotic cell Cell containing no distinct nucleus or organelles. Compare *eukaryotic cell*.

proton (p) Positively charged particle in the nuclei of all atoms. Each proton has a relative mass of 1 and a single positive charge. Compare *electron, neutron*.

pyramid of energy flow Diagram representing the flow of energy through each trophic level in a food chain or food web. With each energy transfer, only a small part (typically 10%) of the usable energy entering one trophic level is transferred to the organisms at the next trophic level.

radiation Fast-moving particles (particulate radiation) or waves of energy (electromagnetic radiation). See *alpha particle, beta particle, gamma ray*.

radioactive decay Change of a radioisotope to a different isotope by the emission of radioactivity.

radioactive isotope See *radioisotope*.

radioactive waste Waste products of nuclear power plants, research, medicine, weapon production, or other processes involving nuclear reactions. See *radioactivity*.

radioactivity Nuclear change in which unstable nuclei of atoms spontaneously shoot out "chunks" of mass, energy, or both at a fixed rate. The three principal types of radioactivity are gamma rays and fast-moving alpha particles and beta particles.

radioisotope Isotope of an atom that spontaneously emits one or more types of radioactivity (alpha particles, beta particles, gamma rays).

rain shadow effect Low precipitation on the leeward side of a mountain when prevailing winds flow up and over a high mountain or range of high mountains, creating semiarid and arid conditions on the leeward side of a high mountain range.

rangeland Land that supplies forage or vegetation (grasses, grass-like plants, and shrubs) for grazing and browsing animals and is not intensively managed. Compare *feedlot, pasture*.

range of tolerance Range of chemical and physical conditions that must be maintained for populations of a particular species to stay alive and grow, develop, and function normally. See *law of tolerance*.

rare species Species that has naturally small numbers of individuals (often because of limited geographic ranges or low population densities) or that has been locally depleted by human activities.

realized niche Parts of the fundamental niche of a species that are actually used by that species. See *ecological niche, fundamental niche*.

recharge area Any area of land allowing water to percolate down through it and into an aquifer. See *aquifer, natural recharge*.

reconciliation ecology Science of inventing, establishing, and maintaining habitats to conserve species diversity in places where people live, work, or play.

recycling Collecting and reprocessing a resource so that it can be made into new products. An example is collecting aluminum cans, melting them down, and using the aluminum to make new cans or other aluminum products. See *primary recycling, secondary recycling*. Compare *reuse*.

reforestation Renewal of trees and other types of vegetation on land where trees have been removed; can be done naturally by seeds from nearby trees or artificially by planting seeds or seedlings.

reliable runoff Surface runoff of water that generally can be counted on as a stable source of water from year to year. See *runoff*.

reliable science Concepts and ideas that are widely accepted by experts in a particular field of the natural or social sciences. Compare *tentative science, unreliable science*.

renewable resource Resource that can be replenished rapidly (hours to several decades) through natural processes as long as it is not used up faster than it is replaced. Examples include trees in forests, grasses in grasslands, wild animals, fresh surface water in lakes and streams, most groundwater, fresh air, and fertile soil. If such a resource is used faster than it is replenished, it can be depleted and converted into a nonrenewable resource. Compare *nonrenewable resource* and *perpetual resource*. See also *environmental degradation*.

replacement-level fertility Average number of children a couple must bear to replace themselves. The average for a country or the world

usually is slightly higher than two children per couple (2.1 in the United States and 2.5 in some developing countries) mostly because some children die before reaching their reproductive years. See also *total fertility rate*.

reproduction Production of offspring by one or more parents.

reproductive isolation Long-term geographic separation of members of a particular sexually reproducing species.

reproductive potential See *biotic potential*.

reserves Resources that have been identified and from which a usable mineral can be extracted profitably at present prices with current mining or extraction technology.

reservoir Artificial lake created when a stream is dammed. See *dam*.

resilience Ability of a living system to be restored through secondary succession after a moderate disturbance.

resource Anything obtained from the environment to meet human needs and wants. It can also be applied to other species.

resource partitioning Process of dividing up resources in an ecosystem so that species with similar needs (overlapping ecological niches) use the same scarce resources at different times, in different ways, or in different places. See *ecological niche, fundamental niche, realized niche*.

resource productivity See *material efficiency*.

respiration See *aerobic respiration*.

response Amount of health damage caused by exposure to a certain dose of a harmful substance or form of radiation. See *dose, dose-response curve, median lethal dose*.

restoration ecology Research and scientific study devoted to restoring, repairing, and reconstructing damaged ecosystems.

reuse Using a product over and over again in the same form. An example is collecting, washing, and refilling glass beverage bottles. Compare *recycling*.

riparian zones Thin strips and patches of vegetation that surround streams. They are very important habitats and resources for wildlife.

risk Probability that something undesirable will result from deliberate or accidental exposure to a hazard. See *risk analysis, risk assessment, risk management*.

risk analysis Identifying hazards, evaluating the nature and severity of risks associated with the hazards (*risk assessment*), ranking risks (*comparative risk analysis*), using this and other information to determine options and make decisions about reducing or eliminating risks (*risk management*), and communicating information about risks to decision makers and the public (*risk communication*).

risk assessment Process of gathering data and making assumptions to estimate short- and long-term harmful effects on human health or the environment from exposure to hazards associated with the use of a particular product or technology.

risk communication Communicating information about risks to decision makers and the public. See *risk, risk analysis*.

risk management Use of risk assessment and other information to determine options and make decisions about reducing or eliminating risks. See *risk, risk analysis, risk communication*.

rock Any solid material that makes up a large, natural, continuous part of the earth's crust. See *mineral*.

rock cycle Largest and slowest of the earth's cycles, consisting of geologic, physical, and chemical processes that form and modify rocks and soil in the earth's crust over millions of years.

r-selected species Species that reproduce early in their life span and produce large numbers of usually small and short-lived offspring in a short period. Compare *K-selected species*.

r-strategists See *r-selected species*.

rule of 70 Doubling time (in years) = 70/(percentage growth rate). See *doubling time, exponential growth*.

runoff Freshwater from precipitation and melting ice that flows on the earth's surface into nearby streams, lakes, wetlands, and reservoirs. See *reliable runoff, surface runoff, surface water*. Compare *groundwater*.

salinity Amount of various salts dissolved in a given volume of water.

salinization Accumulation of salts in soil that can eventually make the soil unable to support plant growth.

saltwater intrusion Movement of saltwater or brackish (slightly salty) water into freshwater aquifers in coastal and inland areas as groundwater is withdrawn faster than it is recharged by precipitation.

sanitary landfill Waste disposal site on land in which waste is spread in thin layers, compacted, and covered with a fresh layer of clay or plastic foam each day. Compare *open dump*.

scavenger Organism that feeds on dead organisms that were killed by other organisms or died naturally. Examples include vultures, flies, and crows. Compare *detritivore*.

science Attempts to discover order in nature and use that knowledge to make predictions about what is likely to happen in nature. See *reliable science, scientific data, scientific hypothesis, scientific law, scientific methods, scientific model, scientific theory, tentative science, unreliable science*.

scientific data Facts obtained by making observations and measurements. Compare *scientific hypothesis, scientific law, scientific methods, scientific model, scientific theory*.

scientific hypothesis An educated guess that attempts to explain a scientific law or certain scientific observations. Compare *scientific data, scientific law, scientific methods, scientific model, scientific theory*.

scientific law Description of what scientists find happening in nature repeatedly in the same way, without known exception. See *first law of thermodynamics, law of conservation of matter, second law of thermodynamics*. Compare *scientific data, scientific hypothesis, scientific methods, scientific model, scientific theory*.

scientific methods The ways scientists gather data and formulate and test scientific hypotheses, models, theories, and laws. See *scientific data, scientific hypothesis, scientific law, scientific model, scientific theory*.

scientific model A simulation of complex processes and systems. Many are mathematical models that are run and tested using computers.

scientific theory A well-tested and widely accepted scientific hypothesis. Compare *scientific data, scientific hypothesis, scientific law, scientific methods, scientific model*.

secondary consumer Organism that feeds only on primary consumers. Compare *detritivore, omnivore, primary consumer*.

secondary pollutant Harmful chemical formed in the atmosphere when a primary air pollutant reacts with normal air components or other air pollutants. Compare *primary pollutant*.

secondary recycling A process in which waste materials are converted into different products; for example, used tires can be shredded and turned into rubberized road surfacing. Compare *primary recycling*.

secondary sewage treatment Second step in most waste treatment systems in which aerobic bacteria decompose as much as 90% of degradable, oxygen-demanding organic wastes in wastewater. It usually involves bringing sewage and bacteria together in trickling filters or in the activated sludge process. Compare *advanced sewage treatment, primary sewage treatment*.

secondary succession Ecological succession in an area in which natural vegetation has been removed or destroyed but the soil or bottom sediment has not been destroyed. See *ecological succession*. Compare *primary succession*.

second-growth forest Stands of trees resulting from secondary ecological succession. Compare *old-growth forest, tree farm*.

second law of energy See *second law of thermodynamics*.

second law of thermodynamics In any conversion of heat energy to useful work, some of the initial energy input is always degraded to lower-quality, more dispersed, less useful energy—usually low-temperature heat that flows into the environment; you cannot break even in terms of energy quality. See *first law of thermodynamics*.

sedimentary rock Rock that forms from the accumulated products of erosion and in some cases from the compacted shells, skeletons, and other remains of dead organisms. Compare *igneous rock, metamorphic rock*. See *rock cycle*.

selective cutting Cutting of intermediate-aged, mature, or diseased trees in an uneven-aged forest stand, either singly or in small groups. This encourages the growth of younger trees and maintains an uneven-aged stand. Compare *clear-cutting, strip cutting*.

septic tank Underground tank for treating wastewater from a home in rural and suburban areas. Bacteria in the tank decompose organic wastes, and the sludge settles to the bottom of the tank. The effluent flows out of the tank into the ground through a field of drainpipes.

sexual reproduction Reproduction in organisms that produce offspring by combining sex cells or *gametes* (such as ovum and sperm) from both parents. It produces offspring that have combinations of traits from their parents. Compare *asexual reproduction*.

shale oil Slow-flowing, dark brown, heavy oil obtained when kerogen in oil shale is vaporized at high temperatures and then condensed. Shale oil can be refined to yield gasoline, heating oil, and other petroleum products. See *kerogen, oil shale*.

shelterbelt See *windbreak*.

shifting cultivation Clearing a plot of ground in a forest, especially in tropical areas, and planting crops on it for a few years (typically 2–5 years) until the soil is depleted of nutrients or the plot has been invaded by a dense growth of vegetation from the surrounding forest. Then a new plot is cleared and the process is repeated. The abandoned plot cannot successfully grow crops for 10–30 years. See also *slash-and-burn cultivation*.

slash-and-burn cultivation Cutting down trees and other vegetation in a patch of forest, leaving the cut vegetation on the ground to dry, and then burning it. The ashes that are left add nutrients to the nutrient-poor soils found in most tropical forest areas. Crops are planted between tree stumps. Plots must be abandoned after a few years (typically 2–5 years) because of loss of soil fertility or invasion of vegetation from the surrounding forest. See also *shifting cultivation*.

sludge Gooey mixture of toxic chemicals, infectious agents, and settled solids removed from wastewater at a sewage treatment plant.

smart growth Form of urban planning which recognizes that urban growth will occur but uses zoning laws and other tools to prevent sprawl, direct growth to certain areas, protect ecologically sensitive and important lands and waterways, and develop urban areas that are more environmentally sustainable and more enjoyable places to live.

smelting Process in which a desired metal is separated from the other elements in an ore mineral.

smog Originally a combination of smoke and fog but now used to describe other mixtures of pollutants in the atmosphere. See *industrial smog, photochemical smog*.

social capital Result of getting people with different views and values to talk and listen to one another, find common ground based on understanding and trust, and work together to solve environmental and other problems.

soil Complex mixture of inorganic minerals (clay, silt, pebbles, and sand), decaying organic matter, water, air, and living organisms.

soil conservation Methods used to reduce soil erosion, prevent depletion of soil nutrients, and restore nutrients previously lost by erosion, leaching, and excessive crop harvesting.

soil erosion Movement of soil components, especially topsoil, from one place to another, usually by wind, flowing water, or both. This natural process can be greatly accelerated by human activities that remove vegetation from soil. Compare *soil conservation*.

soil horizons Horizontal zones, or layers, that make up a particular mature soil. Each horizon has a distinct texture and composition that vary with different types of soils. See *soil profile*.

soil profile Cross-sectional view of the horizons in a soil. See *soil horizon*.

solar capital Solar energy that warms the planet and supports photosynthesis, the process that plants use to provide food for themselves and for us and other animals. This direct input of solar energy also produces indirect forms of renewable solar energy such as wind and flowing water. Compare *natural capital*.

solar cell See *photovoltaic cell*.

solar collector Device for collecting radiant energy from the sun and converting it into heat. See *active solar heating system, passive solar heating system*.

solar energy Direct radiant energy from the sun and a number of indirect forms of energy produced by the direct input of such radiant energy. Principal indirect forms of solar energy include wind, falling and flowing water (hydropower), and biomass (solar energy converted into chemical energy stored in the chemical bonds of organic compounds in trees and other plants)—none of which would exist without direct solar energy.

solid waste Any unwanted or discarded material that is not a liquid or a gas. See *industrial solid waste, municipal solid waste*.

sound science See *reliable science*.

spaceship-earth worldview View of the earth as a spaceship: a machine that we can understand, control, and change at will by using advanced technology. See *planetary management worldview*. Compare *environmental wisdom worldview, stewardship worldview*.

specialist species Species with a narrow ecological niche. They may be able to live in only one type of habitat, tolerate only a narrow range of climatic and other environmental conditions, or use only one type or a few types of food. Compare *generalist species*.

speciation Formation of two species from one species because of divergent natural selection in response to changes in environmental conditions; usually takes thousands of years. Compare *extinction*.

species Group of similar organisms, and for sexually reproducing organisms, they are a set of individuals that can mate and produce fertile offspring. Every organism is a member of a certain species.

species diversity Number of different species (species richness) combined with the relative abundance of individuals within each of those species (species evenness) in a given area. See *biodiversity, species evenness, species richness*. Compare *ecological diversity, genetic diversity*.

species equilibrium model See *theory of island biogeography*.

species evenness Relative abundance of individuals within each of the species in a community. See *species diversity*. Compare *species richness*.

species richness Number of different species contained in a community. See *species diversity*. Compare *species evenness*.

spoils Unwanted rock and other waste materials produced when a material is removed from the earth's surface or subsurface by mining, dredging, quarrying, or excavation.

S-shaped curve Leveling off of an exponential, J-shaped curve when a rapidly growing population exceeds, reaches, or exceeds the carrying capacity of its environment and ceases to grow.

statistics Mathematical tools used to collect, organize, and interpret numerical data.

statutory laws Laws developed and passed by legislative bodies such as federal and state governments. Compare *common law*.

stewardship worldview Worldview holding that we can manage the earth for our benefit but that we have an ethical responsibility to be caring and responsible managers, or *stewards*, of the earth. It calls for encouraging environmentally beneficial forms of economic growth and discouraging environmentally harmful forms. Compare *deep ecology worldview, environmental wisdom worldview, planetary management worldview*.

stratosphere Second layer of the atmosphere, extending about 17–48 kilometers (11–30 miles) above the earth's surface. It contains small amounts of gaseous ozone (O_3), which filters out about 95% of the incoming harmful ultraviolet radiation emitted by the sun. Compare *troposphere*.

stream Flowing body of surface water. Examples are creeks and rivers.

strip cropping Planting regular crops and close-growing plants, such as hay or nitrogen-fixing legumes, in alternating rows or bands to help reduce depletion of soil nutrients.

strip cutting Variation of clear-cutting in which a strip of trees is clear-cut along the contour of the land, with the corridor being narrow enough to allow natural regeneration within a

few years. After regeneration, another strip is cut above the first, and so on. Compare *clear-cutting, selective cutting.*

strip mining Form of surface mining in which bulldozers, power shovels, or stripping wheels remove large chunks of the earth's surface in strips. See *area strip mining, contour strip mining, surface mining.* Compare *subsurface mining.*

subatomic particles Extremely small particles—electrons, protons, and neutrons—that make up the internal structure of atoms.

subduction zone Area in which oceanic lithosphere is carried downward (subducted) under an island arc or continent at a convergent plate boundary. A trench ordinarily forms at the boundary between the two converging plates. See *convergent plate boundary.*

subsidence Slow or rapid sinking of part of the earth's crust that is not slope-related.

subsistence farming See *traditional subsistence agriculture.*

subsurface mining Extraction of a metal ore or fuel resource such as coal from a deep underground deposit. Compare *surface mining.*

succession See *ecological succession, primary succession, secondary succession.*

succulent plants Plants, such as desert cacti, that survive in dry climates by having no leaves, thus reducing the loss of scarce water. They store water and use sunlight to produce the food they need in the thick, fleshy tissue of their green stems and branches. Compare *deciduous plants, evergreen plants.*

sulfur cycle Cyclic movement of sulfur in various chemical forms from the environment to organisms and then back to the environment.

sulfur dioxide (SO_2) Colorless gas with an irritating odor. About one-third of the SO_2 in the atmosphere comes from natural sources as part of the sulfur cycle. The other two-thirds come from human sources, mostly combustion of sulfur-containing coal in electric power and industrial plants and from oil refining and smelting of sulfide ores.

superinsulated house House that is heavily insulated and extremely airtight. Typically, active or passive solar collectors are used to heat water, and an air-to-air heat exchanger prevents buildup of excessive moisture and indoor air pollutants.

surface fire Forest fire that burns only undergrowth and leaf litter on the forest floor. Compare *crown fire, ground fire.* See *controlled burning.*

surface mining Removing soil, subsoil, and other strata and then extracting a mineral deposit found fairly close to the earth's surface. See *area strip mining, contour strip mining, mountaintop removal, open-pit mining.* Compare *subsurface mining.*

surface runoff Water flowing off the land into bodies of surface water. See *reliable runoff.*

surface water Precipitation that does not infiltrate the ground or return to the atmosphere by evaporation or transpiration. See *runoff.* Compare *groundwater.*

survivorship curve Graph showing the number of survivors in different age groups for a particular species.

suspended particulate matter See *particulates.*

sustainability Ability of earth's various systems, including human cultural systems and economies, to survive and adapt to changing environmental conditions indefinitely.

sustainable agriculture Method of growing crops and raising livestock based on organic fertilizers, soil conservation, water conservation, biological pest control, and minimal use of nonrenewable fossil-fuel energy.

sustainable development See *environmentally sustainable economic development.*

sustainable living Taking no more potentially renewable resources from the natural world than can be replenished naturally and not overloading the capacity of the environment to cleanse and renew itself by natural processes.

sustainable society Society that manages its economy and population size without doing irreparable environmental harm by overloading the planet's ability to absorb environmental insults, replenish its resources, and sustain human and other forms of life over a specified period, usually hundreds to thousands of years. During this period, the society satisfies the needs of its people without depleting natural resources and thereby jeopardizing the prospects of current and future generations of humans and other species.

sustainable yield (sustained yield) Highest rate at which a potentially renewable resource can be used indefinitely without reducing its available supply. See also *environmental degradation.*

synergistic interaction Interaction of two or more factors or processes so that the combined effect is greater than the sum of their separate effects.

synergy See *synergistic interaction.*

synfuels Synthetic gaseous and liquid fuels produced from solid coal or sources other than natural gas or crude oil.

synthetic natural gas (SNG) Gaseous fuel containing mostly methane produced from solid coal.

system Set of components that function and interact in some regular and theoretically predictable manner.

tailings Rock and other waste materials removed as impurities when waste mineral material is separated from the metal in an ore.

tar sand See *oil sand.*

tectonic plates Various-sized areas of the earth's lithosphere that move slowly around with the mantle's flowing asthenosphere. Most earthquakes and volcanoes occur around the boundaries of these plates. See *lithosphere, plate tectonics.*

temperature Measure of the average speed of motion of the atoms, ions, or molecules in a substance or combination of substances at a given moment. Compare *heat.*

temperature inversion Layer of dense, cool air trapped under a layer of less dense, warm air. It prevents upward-flowing air currents from developing. In a prolonged inversion, air pollution in the trapped layer may build up to harmful levels.

tentative science Preliminary scientific data, hypotheses, and models that have not been widely tested and accepted. Compare *reliable science, unreliable science.*

teratogen Chemical, ionizing agent, or virus that causes birth defects. Compare *carcinogen, mutagen.*

terracing Planting crops on a long, steep slope that has been converted into a series of broad, nearly level terraces with short vertical drops from one to another that run along the contour of the land to retain water and reduce soil erosion.

terrestrial Pertaining to land. Compare *aquatic.*

tertiary (higher-level) consumers Animals that feed on animal-eating animals. They feed at high trophic levels in food chains and webs. Examples include hawks, lions, bass, and sharks. Compare *detritivore, primary consumer, secondary consumer.*

tertiary sewage treatment See *advanced sewage treatment.*

theory of evolution Widely accepted scientific idea that all life forms developed from earlier life forms. It is the way most biologists explain how life has changed over the past 3.6–3.8 billion years and why it is so diverse today.

theory of island biogeography Widely accepted scientific theory holding that the number of different species (species richness) found on an island is determined by the interactions of two factors: the rate at which new species immigrate to the island and the rate at which species become *extinct,* or cease to exist, on the island. See *species richness.*

thermal inversion See *temperature inversion.*

third and higher-level consumers Carnivores such as tigers and wolves that feed on the flesh of carnivores.

threatened species Wild species that is still abundant in its natural range but is likely to become endangered because of a decline in numbers. Compare *endangered species.*

threshold effect Harmful or fatal effect of a small change in environmental conditions that exceeds the limit of tolerance of an organism or population of a species. See *law of tolerance.*

throughput Rate of flow of matter, energy, or information through a system. Compare *input, output.*

throwaway society See *high-throughput economy.*

tipping point Threshold level at which an environmental problem causes a fundamental and irreversible shift in the behavior of a system.

tolerance limits Minimum and maximum limits for physical conditions (such as temperature) and concentrations of chemical substances beyond which no members of a particular species can survive. See *law of tolerance.*

total fertility rate (TFR) Estimate of the average number of children who will be born alive to a woman during her lifetime if she passes through all her childbearing years (ages 15–44) conforming to age-specific fertility rates of a given year. More simply, it is an estimate of the average number of children that women in a given population will have during their childbearing years.

toxic chemical See *poison, carcinogen, hazardous chemical, mutagen, teratogen.*

toxicity Measure of the harmfulness of a substance.

toxicology Study of the adverse effects of chemicals on health.

toxic waste Form of hazardous waste that causes death or serious injury (such as burns, respiratory diseases, cancers, or genetic mutations). See *hazardous waste.*

toxin See *poison.*

traditional intensive agriculture Production of enough food for a farm family's survival and perhaps a surplus that can be sold. This type of agriculture uses higher inputs of labor, fertilizer, and water than traditional subsistence agriculture. See *traditional subsistence agriculture.* Compare *industrialized agriculture.*

traditional subsistence agriculture Production of enough crops or livestock for a farm family's survival and, in good years, a surplus to sell or put aside for hard times. Compare *industrialized agriculture, traditional intensive agriculture.*

tragedy of the commons Depletion or degradation of a potentially renewable resource to which people have free and unmanaged access. An example is the depletion of commercially desirable fish species in the open ocean beyond areas controlled by coastal countries. See *common-property resource, open access renewable resource.*

trait Characteristic passed on from parents to offspring during reproduction in an animal or plant.

transform fault Area where the earth's lithospheric plates move in opposite but parallel directions along a fracture (fault) in the lithosphere. Compare *convergent plate boundary, divergent plate boundary.*

transgenic organisms See *genetically modified organisms.*

transmissible disease Disease that is caused by living organisms (such as bacteria, viruses, and parasitic worms) and can spread from one person to another by air, water, food, or body fluids (or in some cases by insects or other organisms). Compare *nontransmissible disease.*

transpiration Process in which water is absorbed by the root systems of plants, moves up through the plants, passes through pores (stomata) in their leaves or other parts, and evaporates into the atmosphere as water vapor.

tree farm See *tree plantation.*

tree plantation Site planted with one or only a few tree species in an even-aged stand. When the stand matures it is usually harvested by clear-cutting and then replanted. These farms normally raise rapidly growing tree species for fuelwood, timber, or pulpwood. Compare *old-growth forest, second-growth forest.*

trophic level All organisms that are the same number of energy transfers away from the original source of energy (for example, sunlight) that enters an ecosystem. For example, all producers belong to the first trophic level, and all herbivores belong to the second trophic level in a food chain or a food web.

troposphere Innermost layer of the atmosphere. It contains about 75% of the mass of earth's air and extends about 17 kilometers (11 miles) above sea level. Compare *stratosphere.*

true cost See *full cost.*

tsunami Series of large waves generated when part of the ocean floor suddenly rises or drops.

turbidity Cloudiness in a volume of water; a measure of clarity of water in lakes, streams, and other bodies of water.

undergrazing Absence of grazing for long periods (at least 5 years), which can reduce the net primary productivity of grassland vegetation and grass cover.

undernutrition Consuming insufficient food to meet one's minimum daily energy needs for a long enough time to cause harmful effects. Compare *malnutrition, overnutrition.*

unreliable science Scientific results or hypotheses presented as reliable science but not having undergone the rigors of the peer review process. Compare *reliable science, tentative science.*

upwelling Movement of nutrient-rich bottom water to the ocean's surface. It can occur far from shore but usually takes place along certain steep coastal areas where the surface layer of ocean water is pushed away from shore and replaced by cold, nutrient-rich bottom water.

urban area Geographic area containing a community with a population of 2,500 or more. The number of people used in this definition may vary, with some countries setting the minimum number of people at 10,000–50,000.

urban growth Rate of growth of an urban population. Compare *degree of urbanization.*

urbanization Creation or growth of urban areas, or cities, and their surrounding developed land. See *degree of urbanization, urban area.*

urban sprawl Growth of low-density development on the edges of cities and towns. See *smart growth.*

utilitarian value See *instrumental value.*

volatile organic compounds (VOCs) Organic compounds that exist as gases in the atmosphere and act as pollutants, some of which are hazardous.

volcano Vent or fissure in the earth's surface through which magma, liquid lava, and gases are released into the environment.

warm front Boundary between an advancing warm air mass and the cooler one it is replacing. Because warm air is less dense than cool air, an advancing warm front rises over a mass of cool air. Compare *cold front.*

waste management Managing wastes to reduce their environmental harm without seriously trying to reduce the amount of waste produced. See *integrated waste management.* Compare *waste reduction.*

waste reduction Reducing the amount of waste produced; wastes that are produced are viewed as potential resources that can be reused, recycled, or composted. See *integrated waste management.* Compare *waste reduction.*

water cycle See *hydrologic cycle.*

waterlogging Saturation of soil with irrigation water or excessive precipitation so that the water table rises close to the surface.

water pollution Any physical or chemical change in surface water or groundwater that can harm living organisms or make water unfit for certain uses.

watershed Land area that delivers water, sediment, and dissolved substances via small streams to a major stream (river).

water table Upper surface of the zone of saturation, in which all available pores in the soil and rock in the earth's crust are filled with water.

watt Unit of power, or rate at which electrical work is done. See *kilowatt.*

weather Short-term changes in the temperature, barometric pressure, humidity, precipitation, sunshine, cloud cover, wind direction and speed, and other conditions in the troposphere at a given place and time. Compare *climate.*

weathering Physical and chemical processes in which solid rock exposed at earth's surface is changed to separate solid particles and dissolved material, which can then be moved to another place as sediment. See *erosion.*

wetland Land that is covered all or part of the time with salt water or fresh water, excluding streams, lakes, and the open ocean. See *coastal wetland, inland wetland.*

wilderness Area where the earth and its ecosystems have not been seriously disturbed by humans and where humans are only temporary visitors.

wildlife All free, undomesticated species. Sometimes the term is used to describe animals only.

wildlife resources Wildlife species that have actual or potential economic value to people.

wild species Species found in the natural environment. Compare *domesticated species*.

windbreak Row of trees or hedges planted to partially block wind flow and reduce soil erosion on cultivated land.

wind farm Cluster of wind turbines in a windy area on land or at sea, built to capture wind energy and convert it into electrical energy.

worldview How people think the world works and what they think their role in the world should be. See *environmental wisdom worldview, planetary management worldview, stewardship worldview*.

zone of aeration Zone in soil that is not saturated with water and that lies above the water table. See *water table, zone of saturation*.

zone of saturation Zone where all available pores in soil and rock in the earth's crust are filled by water. See *water table, zone of aeration*.

zoning Designating parcels of land for particular types of use.

zooplankton Animal plankton; small floating herbivores that feed on plant plankton (phytoplankton). Compare *phytoplankton*.

Index

Peto, Richard, 463
Petrochemical reaction, 476
Petrochemicals, **375**
Petroleum, **375**
pH, **37, S40**–41
Phelps Dodge, 363
Pheromones, 299
Philippines
 geothermal power in, 427
 Mt. Pinatubo, 496
Philosophy and nature, S76–77
Phosphate group, S42
Phosphate salts, 355
Phosphorus cycle, **70,** 71*f*
Photochemical oxidants, 477
Photochemical smog, 473, **476**–77
Photosynthesis, 9, **58**–59, 67, S44
Photovoltaic (PV) cells, **414**–17, 415*f*
Phthalates, 452
Physical change in matter, **39**
Phytoplankton, 58, 61, 164
Phytoremediation, 578, 579*f*
Piaster orchaceus sea stars, 95
Pie graphs, S8, S9*f*
Pierce, David, 327
Pierce, Morris A., 651*b*
Pimental, David, 198, 297, 307
Pimm, Stuart, 186
Pinchot, Gifford, S32
Pioneer (early) successional species, 116
Pioneering species, 115
Plan B 3.0: Mobilizing to Save Civilization
 (Brown), 515
Planetary worldview, **20**
Planetary management worldview, **662**
Plankton, **164**
Plantation agriculture, **279**
Plasma arc torch, 580, 580*f*
Plasmodium parasites, 445
Plastics recycling, 571–72, 572*b*
Plate tectonics, 347–49
Platinum, 355, 361
Plato, S76
Playa Junquillal, Costa Rica, 260
Plug-in hybrid electric vehicle, 405–6, 405*f*
Plume, 542
Plutonium, 392
Poaching, 204, 204*f*
Point sources, **16,** 17*f*, **532,** 552–53
Polar bears and global warming, 203
Policies, **636.** *See also* Environmental policies
 components of, 638
Policy life cycle, 636, 636*f*
Political upheaval
 food insecurity and, 277
Political tipping point, 514
Politics, 634, **636**
 sustainable environment and, 634–57
Pollination, 95
Pollutants, 16
 effects of, 16–17
Polluter pays principle, 637
Pollution, **16**–17
 air, 468–93
 aquatic biodiversity and, 252–53, 254*f*
 cities and, 595–96
 cleanup vs. prevention, 17
 from coal burning, 384–85, 384*f*
 from mining, 358–59
 species extinctions and, 201–2, 202*f*

as threat to kelp forests, 104*b*
 transportation and, 404
 water, 287, 531–57
Pollution cleanup (output pollution control), **17**
 problems with, 17
Pollution control
 optimum levels, 617, 618*f*
 valuing, 616–18
Pollution prevention (input pollution control),
 17, 300, 459, 493
Pollution Prevention Pays (3P) program, 366
Polovin, Jeffrey, 504
Polyaquaculture, 285, 306
Polyculture, **280**–81
Polycultures of perennials, 309*b*
Polymers, 38, S42
Polyps, 162
Poonswad, Pilai, 205
Population(s), **52,** 52*f*, 53*f*
 age structure, 109
 changes in size, 109
 dieback (crash), 111, 112*f*
Population change, **126**
Population change types, 113–14, 114*f*
Population control, 23*f*, 24, 58
Population decline, 132–33, 133*f*
Population density, **113,** S15*f*
Population distribution (dispersion), 108, 109*f*
Population dynamics, **108**
Population growth. *See also* Human population
 age structure and, 130–33
 aquatic biodiversity and, 252–53
 in China, 135–36
 developing countries and, 133–34
 humans and, 114, 122–37, 124*b*
 increase rate, map, S13*f*
 in India, 136–37
 J-curves and S-curves, 109–11, 123
 limiting by abiotic factors, 58
 limiting factors, 108–15, 111*f*
 minimum viable population size, 113
 older people and, 132
 projections of human population, 125*f*, S16*f*
 slowing, 133–37, 307
 slowing by empowering women, 135, 135*f*
 species extinction and, 201
 United States and, 126–27, 127*f*, 128*f*
 urban areas and, 589–93, S16*f*
 world rate, graph, S13*f*
 young people and, 130–31, 131*f*
Population, human, and its impact, 122–37
Population timeline, graph, S12*f*
Pores, 281*b*
Porsche, Ferdinand, 405
Portland, Oregon
 smart growth in, 603–4
Portugal
 solar cell power systems in, 415
 wave power production in, 419
 wind power production in, 420
Possibility, 439
Postconsumer waste, 569
Postel, Sandra, 333, 341
Potential energy, 40, **42**
Potrykus, Ingo, 275
Poverty, **18,** 18*f*, **625**
 in cities, 596–97, 597*f*
 environmental problems and, 625–28
 extreme, 11, 11*f*
 food insecurity and, 277, 301

malnutrition and, 18, 19*f*, 275
 reducing, 626–27
 as risk to people, 460
 water shortages and, 318–19
Power measurements, S2
Power tower, 413
Prairie potholes, 178
Precautionary principle, 210–**11,** 263, 459, 637
Precipitation, 65
Precision in measurements, S3
Preconsumer waste, 569
Predation, **101, 102**
Predator, **102,** 103–4
Predator–prey relationship, **102**
 and natural selection, 103–4
Prescribed fires, 228
Preservationist movement, S32
Preservationist school of conservation, S32
Presidio business school, 631
Prevailing winds, 141, 143
Prevention principle, 637
Prey, **102**
 ways of avoiding predators, 102–3, 103*f*
Price, 613
Prices
 environmental costs in, 620–21
 natural capital and, 19–20
Primary consumers, **59**
Primary ecological succession, **115,** 116–17,
 116*f*
Primary pollutants, **471**
Primary recycling, **569**
Primary sewage treatment, **554**
Primary standard, 488
Principle of enoughness, 668
Principle of multiple use, 643*b*
Principle of sustainable yield, 643*b*
Principles for making environmental policy,
 637–38
Private goods, 614
Private property rights, 12–13
Probability, **34***b*, 35, 439
Producers, in ecosystems, **58**
Profit-making organizations, 637
Profundal zone, 174, 175*f*
Prokaryotic cells, **51,** 52*f*
Propane, 381
Property (resource) rights, 12–13
Property taxes, 602
Proteins, 38, **S42**
Protons (p), **36,** 36*f*
Public access and participation principle, 637
Public Interest Research Group, 552
Public lands, U.S., managing, 640–42, 641*f*
Public services, 614
Puckett, Jim, 560
Puget Sound, 531, 540
Puget Sound Partnership, 540
Purchasing power, 10
Purchasing power parity (PPP), 10
Purdue University, 533
Purification of drinking water, 544–46
Purple loosestrife, 252
Purse-seine fishing, 256–57
Pyramid of energy flow, **62–63,** 63*f*

Qatar
 natural gas reserves in, 381
Quad countries, S74, S75
Quagga mussel, 270